Mathematical Techniques

An Introduction for the Engineering, Physical, and Mathematical Sciences

THIRD EDITION

D. W. Jordan and P. Smith

Department of Mathematics
Keele University

OXFORD

UNIVERSITY PRESS

OXFORD

UNIVERSITY PRESS

Great Clarendon Street, Oxford OX2 6DP

Oxford New York

Athens Auckland Bangkok Bogota Buenos Aires Calcutta Cape Town
Chennai Dar es Salaam Delhi Florence Hong Kong Istanbul Karachi
Kuala Lumpur Madrid Melbourne Mexico City Mumbai Nairobi Paris
São Paulo Singapore Taipei Tokyo Toronto Warsaw
and associated companies in
Berlin Ibadan

Oxford is a trade mark of Oxford University Press

Published in the United States
by Oxford University Press Inc., New York

First edition 1994
Second edition 1997
Third edition 2002

A catalogue record for this book is available from the British Library

Library of Congress Cataloging in Publication Data
(Data available)

ISBN 0 19 924972 5

Typeset by Graphicraft Limited, Hong Kong
Printed in LEGO Print, S.p.A., Trento, Italy

PREFACE TO THE THIRD EDITION

This book is a student text covering the mathematical techniques usually taught in the early stages of science and engineering courses. It also provides the groundwork of 'methods' needed by first and second-year mathematics specialists. The requirements of such students have influenced its content and presentation, helped in many ways by the authors' long and continuous experience of teaching mathematical methods to a variety of joint degree students at Keele University, including many who have a varied background in mathematics.

The main textual differences to be found in the third edition are as follows. Chapter 1, on background techniques, has been extensively revised in the interests of clarity, and has been extended to include sections on permutations, combinations, and the binomial theorem. The Fourier transform now constitutes a new and separate Chapter 27, with a simpler approach through sine and cosine transforms. It now also includes applications to diffraction illustrated largely in terms of directional aerial arrays. Appropriate background material on scalar waves and their phasors is also introduced in Chapters 20 and 21. There are many smaller additions and clarifications, often made in response to suggestions from readers. We have deferred to the majority view on complex number notation and replaced 'j' by 'i' throughout the book. The book has been completely re-set, which has resulted, we feel, in an improved layout of the text and diagrams. However, the sequence of material has not been appreciably altered in this new edition, and the chapter numbering is unaffected up to Chapter 27. Previous users of the text should experience minimum disruption.

The text has been divided into eight parts, each covering a coherent theme. Following Part I, which is an introduction to elementary methods, differentiation, and complex numbers, the calculus sequence continues in Parts III, IV, and V, which cover integration, differential equations, Laplace transforms, Fourier series, Fourier transforms, functions of several variables, multiple and line integrals, and vector fields. The algebra strand in Parts II and VI includes matrices, vectors, linear equations, eigenvalues, sets, Boolean algebra, graph (network) theory, and difference equations. Part VII is an introduction to probability, and descriptive statistics. The final part, Part VIII, consists of a selection of symbolic computing projects related, chapter by chapter, to the main text.

There are over 500 fully-worked examples in the book, a great number of exercises and problems including simple programming applications, over 120 projects which use symbolic computation, and several appendices which include standard results for reference, as well as answers and hints to selected problems. There is also a comprehensive index.

We have organized the book so as to enable students to use it with the minimum of guidance, and as a source of reference. The same features will help teachers to make selections from the book according to the length, emphasis and prerequisites of different courses. From Chapter 2 onwards, every subject treated starts from scratch. For example, it is not assumed that the reader is familiar with calculus, although most science and engineering students are likely to have had some previous encounter with the subject; if so, the text can be used for revision as well as extension of various topics in calculus. Most of the chapters are intentionally short (the average length is about 20 pages), and the sections within the chapters include not more than one or two new ideas. All the principal results are displayed in summary form in shaded 'boxes'. For revision purposes, or in desperate cases, progress could be made by attending only to the boxes.

The reader is encouraged to look out for some kind of geometrical or numerical reality to illustrate symbolic statements. Where possible we make use of graphical justifications; science students in particular should benefit by cultivating geometrical reasoning. Numerical methods for equation-solving, integration, and the solution of differential equations are introduced at the points where they can effectively illustrate the main text, rather than being collected together in a separate chapter on numerical analysis.

Attempts to generate interest by using specialized examples from physics, chemistry, engineering, etc. are liable to misfire with many students, who have enough to do at first in grasping the underlying mathematical processes, and can be confused by layers of scientific vocabulary and unfamiliar notations. We have therefore taken out into separate chapters certain technical applications such as harmonic oscillations, phasors, and circuit analysis, and given these a fuller treatment. Certain other applications are confined to separate sections, so that they can be avoided if they do not suit a particular class. Most of the applications in the main text are drawn from common knowledge, or are as such as can be easily understood.

The last chapter contains a collection of projects which follow the text chapter by chapter and introduce the use of symbolic computation and graphics. A short first-course at Keele University introducing mathematics students to *Mathematica* has proved successful. Symbolic computation is a very useful interactive facility for graph plotting and routine numerical manipulations. To cite just two examples, calculations involving matrices and the sketching of surfaces in three dimensions, which can be time consuming, are easily handled by *Mathematica*. However, computing is no substitute for understanding methods and principles. Almost all the graphs of curves and surfaces in the book have been generated by *Mathematica*. The specific *Mathematica* programs which provide solutions for the projects in Chapter 42 are freely available on the web site of the Mathematics Department of Keele University at:

www.keele.ac.uk/depts/ma

Mostly they incorporate standard *Mathematica* commands, and they are readily adaptable to a variety of alternative inputs for specific functions, matrices, vectors, determinants, Fourier series, Laplace transforms, etc.

Confidence is half the battle in mathematics: it is very encouraging to learn how to do something so as to be able to get it nearly right most times. Continual practice, even at a point of repetitive drill, is the way to achieve this. The beneficial effects of practice can be obtained without using very complicated and difficult exercises, so on the whole we have avoided such problems.

We should like to record our specific thanks to the following: to John Bentin and Robert West for checking earlier versions of the text; to Peter Jones, Keele University, for helpful comments on early drafts of the chapters on probability and statistics; Andrew Looms for setting up and maintaining the web site; to Michael Basler, University of Jena, the translator of the German edition; to Anne Smith for checking a large number of examples and problems; and to colleagues and students from Keele University and many other institutions for comments on previous editions. The development, writing and organization of this text and the associated software over three editions has been a complicated process, and we wish to express our appreciation of the helpfulness of the staff at Oxford University Press during the production of this book.

Keele D.W.J.
February 2002 P.S.

CONTENTS

PART III

Integration and
differential equations

14 Antidifferentiation and area

15 The definite and indefinite integral

16 Applications involving the integral as a sum

17 Systematic techniques for integration

18 Unforced linear differential equations with constant coefficients

19 Forced linear differential equations

20 Harmonic functions and the harmonic oscillator

21 Steady forced oscillations: phasors, impedance, transfer functions

22 Graphical, numerical, and other aspects of first-order equations

PART IV

Transforms and
Fourier series

PART V

Multivariable calculus

32　Double integration

33　Line integrals

34　Vector fields: divergence and curl

PART VI

Discrete mathematics

35　Sets

41   Descriptive statistics

PART VIII

Projects

42   Applications projects using symbolic computing

Answers to selected problems

Appendices

Index

PART I

Elementary methods, differentiation, complex numbers

1

Standard functions and techniques

CONTENTS

This is a long chapter covering a variety of subjects, some of which you will have met before. It is not necessary to work through every section in detail; to a large extent the chapter can be used for reference as required later on. However, you should read it carefully in order to find what is in it, and to pick up terms and notations used regularly in the rest of the book. If you find that a familiar subject is treated in an unfamiliar way, try to understand the fresh approach since the ideas behind it are liable to reappear in later chapters.

1.1 Real numbers, powers, inequalities

The real numbers are the ordinary numbers used in arithmetic and measurement. (We call them 'real' to distinguish them from the so-called 'complex numbers', to be introduced in Chapter 6.) The following terms are used to classify special types of real number:

1. An **integer** is a 'whole number', positive, negative, or zero; integers are the numbers … , $-3, -2, -1, 0, 1, 2, …$.

2. A **rational number** is any number that can be expressed as a fraction having the form p/q, where p and q are *integers*. They consist of all numbers expressible as **finite** or **recurring decimals**. Examples of rational numbers with recurring decimals are $1/3$ which has the decimal representation $0.3333…$ written as $0.\dot{3}$, and $1/7$ which has the recurring decimal form $0.\dot{1}4\dot{2}\,8\dot{5}\dot{7}$ (the dots mark out the decimal repetition pattern). Notice that the integers are rational numbers in this definition.

3. The rest are **irrational numbers**. These are the numbers that cannot be expressed as fractions made up of integers; they are represented by infinite, non-recurring decimals. Although there is an infinite number of rational numbers, there is a sense in which there are infinitely more irrational numbers, so they appear everywhere. For example, the hypotenuse of a right-angled triangle with sides of unit length has length $\sqrt{2}$, and this is known to be an irrational number. The number π is irrational, and so is the number e which we will meet in Section 1.8. Irrational numbers can be approximated as closely as we wish by rational numbers: retain the appropriate number of decimal places, and the resulting approximation is a rational number. For example, $\pi = 3.141$ to 3 decimal places, which is the rational number $3141/1000$.

4. The sign ∞, standing for the word **infinity**, is frequently useful, but it cannot be used in algebra as if it stood for an ordinary number. Claims such as $\infty/\infty = 1$, or $\infty - \infty = 0$, are fallacious. For example, if we take away the infinity of odd integers from the infinity of all integers, an infinity of even integers is left behind.

We have continually to manipulate **powers of numbers**, which take the form a^x. The power x is called the **exponent** or **index** in the expression. It is assumed that you know how to use the rules when x is a positive or negative integer, and can interpret fractional powers by their connection with square roots, cube roots, and so on. For example, $2^{\frac{1}{2}} = \sqrt{2}$, $2^{-\frac{1}{2}} = 1/\sqrt{2}$, and $2^{\frac{3}{2}} = (\sqrt{2})^3 = \sqrt{(2^3)}$. The rules applying to general exponents work in the same way, as follows:

> **Rules for exponents**
>
> a and b are any positive real numbers; x and y are any real numbers, positive, negative, or zero. Then
>
> (a) $a^x a^y = a^{x+y}$.
> (b) $a^0 = 1$.
> (c) $a^{-x} = 1/a^x$.
> (d) $(a^x)^y = a^{xy}$.
> (e) $a^x b^x = (ab)^x$.   (1.1)

The notations $a^{\frac{1}{2}}$ and $\sqrt{a}$ always stand for the *positive square root of a*. If we want the negative square root we must attach a minus sign. Thus the solutions of the equation $x^2 = 2$ are written separately as $\sqrt{2}$ and $-\sqrt{2}$, or as $\pm\sqrt{2}$.

The condition $a > 0$ is necessary if the rules are to apply to *all* exponents; for example, $(-2)^{\frac{1}{2}}$ has no meaning in real-number terms since the square of any real number is always positive. If a is negative, then sometimes a^x is a real number, but only if $x = p/q$ in its lowest terms where p and q are integers and q is an *odd* integer. For example, $(-8)^{\frac{1}{3}} = -2$, because $(-2)^3 = -8$.

The concept of an **identity** needs to be distinguished from that of a mere **equation**. The statement $x^2 + 2x + 1 = 0$ is an *equation*: it is only true conditionally; that is, for particular values of x. On the other hand statements such as

$$x^2 + 2xy + y^2 = (x + y)^2 \quad \text{and} \quad \sin^2 A + \cos^2 A = 1$$

are called *identities*, because they are automatically true for all values of x, y, and A, and to stress the difference we may use the sign $\equiv$ instead of $=$. Anything involving $\leqslant$, $>$, etc., is an **inequality**. An algebraic 'phrase' standing on its own, such as $x^2 + 2x + 1$ or $\sin^2 A + \cos^2 A$, is an **expression**. However, virtually anything with an $=$, $\equiv$, or inequality sign in it is commonly referred to as an equation.

Draw a straight line containing a point O, called the **origin**, and indicate a scale starting at O, as in Fig. 1.1, with positive scale markings above O and negative markings below. Imagine the line to be infinitely long in both directions. This is called a **number line**, and every real number, positive or negative, has a place on it. We shall use x to denote a general number.

The **symbols for inequalities** $<$, $\leqslant$, $>$, $\geqslant$ have the following meanings:

$<$ 'is less than'	$\leqslant$ 'is less than or equal to'
$>$ 'is greater than'	$\geqslant$ 'is greater than or equal to'.

If we are given two numbers, then the one which is higher on the number line is the **greater** one. Therefore $-2 > -3$, $-3 < -2$, $3 > 0$, $-3 < 0$, and so on. Obviously $2 < 3$. But it is also true that $2 \leqslant 3$,

Fig. 1.1 The number line, x axis.

because 2 is certainly *either* less than *or* equal to 3. For similar reasons, all these are true: $1 = 1$, $1 \leqslant 1$, $1 \geqslant 1$.

A single piece, or a segment, of the number line is called an **interval**. The piece of the line between $x = 2$ and $x = 3$ which includes both **end-points** $x = 2$ and 3 can be specified by the expression

the interval $2 \leqslant x \leqslant 3$

which means 'all the values of x between, and including, 2 and 3'. The interval $2 < x < 3$ means all values between 2 and 3, but excluding the end values. **Infinite intervals** can be expressed in two ways: for example, the interval

$$x \geqslant 2 \quad \text{or} \quad 2 \leqslant x < \infty$$

contains all the numbers x which are greater than or equal to 2.

The 'size' of a number is denoted by the symbol

$$|x| = \begin{cases} x & \text{if } x \geqslant 0, \\ -x & \text{if } x < 0, \end{cases} \tag{1.2}$$

which is called the **modulus** or **absolute value** of x. Thus $|3| = 3$, $|-4| = 4$. We can use the modulus notation to define intervals. The inequality $|x| \leqslant 2$ defines the same interval as $-2 \leqslant x \leqslant 2$; $|x - 1| \leqslant 3$ is the same as $-3 \leqslant x - 1 \leqslant 3$ or $-2 \leqslant x \leqslant 4$.

1.2 Coordinates in the plane

The location of a point in a plane can be specified in terms of **right-handed cartesian axes**, as illustrated in Fig. 1.2. These are effectively two number lines, typically labelled x and y, at right angles, meeting at the common **origin** O. Axes are right handed if, when we walk along the x axis in the direction of increasing x, the positive y axis is on our left. If you look at Fig. 1.2 in a mirror you will see **left-handed** axes.

The position of a point is determined by two **coordinates** (x, y). They represent, in order, the *signed 'distances'* of the point from the y and x axes respectively, *as read off from the numbers on the axis scales*. In Fig. 1.2 the point A has coordinates $x = 1.5$, $y = 1$. We shall use the **notation** $A : (1.5, 1)$ for such a point, so as to display the name of the point together with its coordinates. On Fig. 1.2 we also show the point $B : (2.5, -2)$, and a general point $P : (x, y)$. For a point $P : (x, y)$, x is called the **abscissa** and y the **ordinate** of P.

In Fig. 1.3a, the scales are supposed to be equal, so that the distance of $P : (x, y)$ from the origin is OP. By Pythagoras's theorem,

$$OP = \sqrt{(OU^2 + UP^2)},$$

where U is the base of the perpendicular from P on to the x axis (Fig. 1.3a). If we put $OP = r$, then

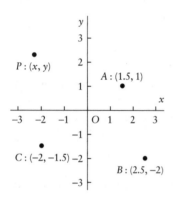

Fig. 1.2 Axes and coordinates.

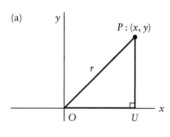

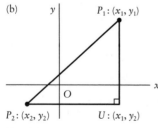

Fig. 1.3

$$r = \sqrt{(x^2 + y^2)}. \tag{1.3a}$$

Note that **distances**, such as OP and r, are always counted as **positive numbers**.

Similarly, for any two points $P_1 : (x_1, y_1)$ and $P_2 : (x_2, y_2)$ in the plane, the distance P_1P_2 between them – see Fig. 1.3b – is given by

$$P_1P_2 = \sqrt{[(x_1 - x_2)^2 + (y_1 - y_2)^2]}. \tag{1.3b}$$

1.3 Graphs

If x and y are connected by an equation, then this relation can be represented by a curve or curves in the (x, y) plane which is known as the **graph** of the equation.

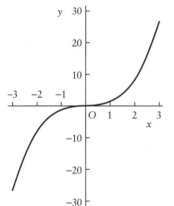

Fig. 1.4 Graph of $y = x^3$.

Example 1.1 *Sketch the graph of $y = x^3$.*

We decide over what interval of values of (say) x we wish to sketch the graph. Let $-3 \leqslant x \leqslant 3$. Construct a table of (x, y) values as shown below:

x	-3	-2	-1	0	1	2	3
y	-27	-8	-1	0	1	8	27

We then plot the points corresponding to this set of coordinates and draw a smooth curve through them as shown in Fig. 1.4. The greater the number of values of x in the interval, the greater is the reliability of the graph. It is assumed that the curve has a smooth or regular behaviour between consecutive plotted points. In Fig. 1.4 the scales are not the same on the two axes. Since y has a much greater spread of values (54) compared with x (6), the vertical scale has been compressed.

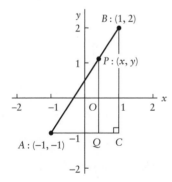

Fig. 1.5

Example 1.2 *Find the equation of the straight line through the points $A : (-1, -1)$ and $B : (1, 2)$.*

The line is shown in Fig. 1.5. Let $P : (x, y)$ be any point on the line. PAQ and BAC are similar triangles, so that

$$\frac{QP}{AQ} = \frac{y + 1}{x + 1} = \frac{CB}{AC} = \frac{3}{2}.$$

Therefore

$$2(y + 1) = 3(x + 1) \qquad \text{or} \qquad y = \tfrac{3}{2}x + \tfrac{1}{2}.$$

This represents the equation of the straight line through the points $(1, 2)$ and $(-1, -1)$.

Any equation of the form

$$y = mx + c$$

has a straight line graph, and any straight line can be expressed in this form unless it is parallel to the y axis, when its equation is

$$x = d.$$

A more general form which includes both of the above cases is

$$ax + by = c$$

where a, b, and c are any constants.

The method of Example 1.2 can be used to find the equation of a line passing through any two given points $A : (x_1, y_1)$ and $B : (x_2, y_2)$ (see Fig. 1.6) provided $x_1 \neq x_2$. Let $P : (x, y)$ be any point on the line. Then

$$\frac{QP}{AQ} = \frac{RB}{AR} \quad \text{or} \quad \frac{y - y_1}{x - x_1} = \frac{y_2 - y_1}{x_2 - x_1}. \tag{1.4}$$

This equation can be rearranged in the form $y = mx + c$, where

$$m = \frac{y_2 - y_1}{x_2 - x_1}, \qquad c = \frac{x_2 y_1 - x_1 y_2}{x_2 - x_1}. \tag{1.5}$$

In the right-angled triangle ABR of Fig. 1.6, the angle α (Greek 'alpha') is given by

$$\tan \alpha = \frac{y_2 - y_1}{x_2 - x_1} \tag{1.6}$$

(provided that the x and y scales are equal; if not, the angle α will not be correct in the figure). The number m or $\tan \alpha$ gives the standard measure of the **slope** or **gradient** of the straight line. The line slopes upwards or downwards from left to right according as $\tan \alpha$ is positive or negative respectively; the slope is zero when α or $\tan \alpha$ is zero; and the larger the size (or modulus – see (1.2)) of $\tan \alpha$, the steeper is the line. Also, according to (1.5) and (1.6), if the equation is in the form $y = mx + c$, then

$$\text{slope} = \tan \alpha = m. \tag{1.7}$$

From (1.7), if we require the line through $A : (x_1, y_1)$ with given slope m, its equation is

$$y - y_1 = m(x - x_1). \tag{1.8}$$

The following result is often needed:

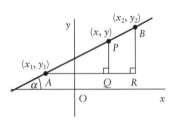

Fig. 1.6

> **Perpendicular straight lines**
> The condition for the straight lines $y = m_1 x + c$ and $y = m_2 x + d$ to be perpendicular is
> $$m_1 m_2 = -1. \tag{1.9}$$

This is proved as follows. First suppose that $m_1 m_2 = -1$; we have to deduce that the lines are perpendicular. **Translate** the two lines (meaning move them without rotation so that their directions are unchanged) to meet at the origin O as in Fig. 1.7. Their *slopes* and *angle of intersection are unaffected by translation*, and the new equations are

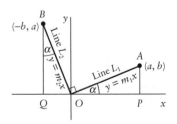

Fig. 1.7

line $L_1 : y = m_1x$,   line $L_2 : y = m_2x$;

with $m_1m_2 = -1$. Let $A : (a, b)$ be any point on L_1. Then $b = m_1a$ so that $m_1 = b/a$. Since $m_1m_2 = -1$, we have $m_2 = -a/b$; therefore the point $B : (-b, a)$ lies on L_2 as shown. We now use the theorem of Pythagoras which states: *if $\widehat{AOB}$ is a right angle then $AB^2 = OA^2 + OB^2$, and conversely if $AB^2 = OA^2 + OB^2$ then $\widehat{AOB}$ is a right angle.*

Using the converse of Pythagoras first,

$$AB^2 - OA^2 - OB^2 = (a + b)^2 + (b - a)^2 - (a^2 + b^2) - (b^2 + a^2) = 0.$$

Hence $AB^2 = OA^2 + OB^2$, which implies that $\widehat{AOB}$ is a right angle. To prove the converse, let A be the point (a, m_1a) on L_1, and B the point (b, m_2b) on L_2. Then, given that $\widehat{AOB}$ is a right angle,

$$AB^2 = OA^2 + OB^2 \quad \text{implies}$$

$$(a - b)^2 + (m_1a - m_2b)^2 = (a^2 + m_1^2a^2) + (b^2 + m_2^2b^2),$$

which, in turn, implies $m_1m_2 = -1$ after simplifying, assuming that neither a nor b is zero.

A **circle** consists of all points which are a constant distance from a given point. In Fig. 1.8, the circle has radius r, and its centre is at (a, b). The point $P : (x, y)$ represents any point on the circle. Equation (1.3b) for the distance between two points gives

$$\sqrt{[(x - a)^2 + (y - b)^2]} = r.$$

Square this expression to get rid of the square root, and we have the **equation of a circle** in its **standard form**:

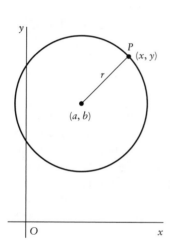

Fig. 1.8

> **Equation of a circle, centre (a, b) and radius r**
> $$(x - a)^2 + (y - b)^2 = r^2. \tag{1.10}$$

Example 1.3   *Find the centre and radius of the circle*

$$4x^2 + 4y^2 - 4x + 8y - 11 = 0. \tag{i}$$

To convert (i) to the form (1.10), rewrite it in the form

$$x^2 - x + y^2 + 2y = \tfrac{11}{4}. \tag{ii}$$

Take the terms involving x and reorganize them:

$$x^2 - x = (x - \tfrac{1}{2})^2 - \tfrac{1}{4}$$

(this process is used in many different contexts, and is called **completing the square**). Treat the terms in y similarly:

$$y^2 + 2y = (y + 1)^2 - 1.$$

Replace the terms in (ii) by the new forms; we get

$$(x - \tfrac{1}{2})^2 - \tfrac{1}{4} + (y + 1)^2 - 1 = \tfrac{11}{4} \quad \text{or} \quad (x - \tfrac{1}{2})^2 + (y + 1)^2 = 4.$$

Therefore the centre is at $(\tfrac{1}{2}, -1)$, and the radius is 2.

(a)

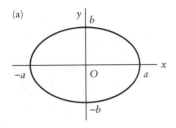

(b)

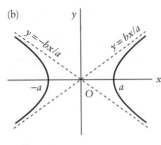

(c)

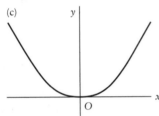

Fig. 1.9 (a) An ellipse
$x^2/a^2 + y^2/b^2 = 1$.
(b) A hyperbola $x^2/a^2 - y^2/b^2 = 1$.
(c) A parabola $y = x^2$.

Fig. 1.10 The function f
represented by a processor.

Notice that (1.10) implies that, if we are given an equation

$$Ax^2 + By^2 + Cx + Dy + E = 0,$$

it can only represent a circle if $A = B$. (The equation might not represent anything, as with $x^2 + y^2 + 1 = 0$, but if it does, it will be a circle.)

Figure 1.9 shows other second-degree curves in simple forms, namely an ellipse, hyperbola, and parabola.

1.4   Functions

The area A of a circle depends on its radius r, and the dependence is expressed in the **formula** $A = \pi r^2$. In general, suppose that the values of a certain **independent variable** x, say, determine the values of a **dependent variable** y in such a way that if a numerical value of x is given, a **single value** of y is determined. Then we say that y is a **function** of x, and write, for example,

$$y = f(x), \qquad y = g(x),$$

and so on, where the letters f, g, etc., can be used to distinguish different forms of dependence which can be thought of pictorially in terms of different graphs. The letters f, g, and so on, standing alone, need not be associated with a *formula* in the usual sense. They can stand for any rule, program, or calculation process which produces a definite *single value* for y when we offer a number x to it. A function can be thought of as an input–output device as in Fig. 1.10.

Functions can be defined **implicitly** by means of formulae. For example,

$$x^2 + y^2 = 1$$

represents a circle, centre the origin and radius 1. But if we solve the equation for y, we obtain $y = \pm\sqrt{(1 - x^2)}$, which is not a single function, but two separate, single-valued, functions

$$y = \sqrt{(1 - x^2)} \quad \text{and} \quad y = -\sqrt{(1 - x^2)},$$

representing the upper and lower semicircles which together make up the circle.

The following result is frequently required. Suppose that c is a positive constant, and we are given a function f, with graph $y = f(x)$. The graph $y = f(x - c)$ is exactly the same as that of $f(x)$, except that it is moved, or **translated a distance c to the right** along the x axis. There is a similar result for $f(x + c)$, the movement being to the left. Therefore

> **Translation of a function along the x axis**
>
> Let c be any positive number, and f any function. Then
>
> $$y = f(x - c) \quad \text{and} \quad y = f(x + c)$$
>
> represent translations of $y = f(x)$ a distance c along the x axis to the right and left respectively.
>
> **(1.11)**

Thus $y = x^2$ and $y = (x + 2)^2$ have the same shape, but the second is a distance 2 to the left of the first.

Sometimes it is helpful to adopt a more formal way of presenting a function. For example, instead of putting simply $f(x) = \sqrt{(1 - x^2)}$ we may say

the function f defined, for $-1 < x < 1$, by $f(x) = \sqrt{(1 - x^2)}$,

or

the function f defined, for $-1 < t < 1$, by $f(t) = \sqrt{(1 - t^2)}$,

which has exactly the same meaning. Any letter may be used as the independent variable to specify the formula or rule that f symbolizes; it is sometimes called a **dummy variable** for this reason. When we call on the function f in the course of a particular problem, we then revert to the symbols that are natural to the problem: we might want $f(r)$ or $f(x^2)$ or $f(x - y)$ or just a single value $f(2\pi)$. In these examples, the symbols r, x^2, $x - y$, and 2π are called **arguments** of the function f. For example, if a function g is defined by

$$g(t) = (1 - t)^2 \quad \text{for all values of } t,$$

then, with the new argument $1 - x$,

$$g(1 - x) = [1 - (1 - x)]^2 = x^2.$$

It is useful to have terms in which symmetry of a graph can be described. For example, the graph of the parabola $y = x^2$ shown in Fig. 1.9c is symmetrical about the y axis; the two halves for $x > 0$ and $x < 0$ are reflections of each other in the y axis. Functions with such graphs are called **even functions**. On the other hand $y = x^3$ (Fig. 1.4) is its own reflection in the origin: the function $f(x) = x^3$ is an example of an **odd function**. The corresponding algebraic properties are defined by

> **Even and odd functions**
>
> (a) $f(x)$ is **even** if $f(-x) = f(x)$
> (b) $f(x)$ is **odd** if $f(-x) = -f(x)$
>
> for all x for which f is specified.
>
> **(1.12)**

For example, in plotting $y = f(x) = x^3$ in Example 1.1, we did not really have to calculate x^3 for negative values of x. All that was necessary was to notice that x^3 is an odd function since $(-x)^3 = -(x^3)$, and this gives the table for negative x by changing the sign of the entries for x positive.

Some functions of practical significance have graphs that are not entirely smooth. For example, we may wish to model a device that is turned on at a given time, being quiescent before that time but active afterwards. A sudden change in the state of the device can be represented by a function which has a **jump** or **discontinuity** in its graph at the critical moment. The basic building block for functions with a jump is the **unit step function** H(t) (also known as the Heaviside function after its inventor, and sometimes denoted by U(t)) which we shall define, using t to represent time, by

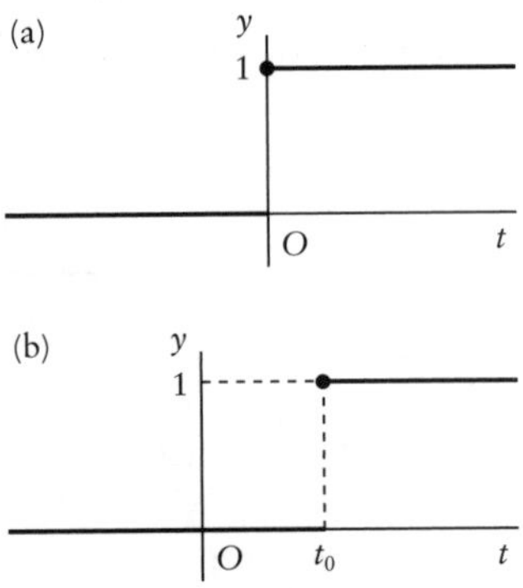

$$H(t) = \begin{cases} 0 & \text{when } t < 0, \\ 1 & \text{when } t \geqslant 0 \end{cases} \tag{1.13}$$

(see Fig. 1.11a).

If switch-on is required at $t = t_0$ then we can use the translation

$$H(t - t_0) = \begin{cases} 0 & \text{when } t < t_0, \\ 1 & \text{when } t \geqslant t_0, \end{cases}$$

shown in Fig. 1.11b: it is the same graph translated to the right a distance t_0 by (1.11).

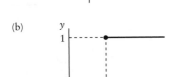

(a)

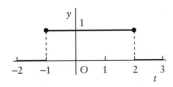

(b)

Fig. 1.11 (a) Graph of $y = H(t)$. (b) Graph of $y = H(t - t_0)$.

Example 1.4 *Sketch the graph of $f(t) = H(2 - t) + H(t + 1) - 1$.*

The function $f(t)$ is a combination of unit functions each of which has a discontinuity where its argument is zero. Thus $f(t)$ has discontinuities at $t = -1$ (from H($t + 1$)) and at $t = 2$ (from H($2 - t$)). Note that

$$H(t + 1) = \begin{cases} 0 & \text{when } t < -1, \\ 1 & \text{when } t \geqslant -1, \end{cases} \qquad H(2 - t) = \begin{cases} 0 & \text{when } t > 2, \\ 1 & \text{when } t \leqslant 2. \end{cases}$$

Hence for

$$\begin{array}{ll} t < -1 & f(t) = 1 + 0 - 1 = 0; \\ -1 \leqslant t \leqslant 2 & f(t) = 1 + 1 - 1 = 1; \\ t > 2 & f(t) = 0 + 1 - 1. \end{array}$$

The graph is shown in Fig. 1.12.

This function would switch a device on at $t = -1$ and switch it off at $t = 2$.

Fig. 1.12

The *odd* function denoted by sgn and defined by

$$\text{sgn } t = H(t) - H(-t) = \begin{cases} -1 & \text{when } t < 0, \\ 0 & \text{when } t = 0, \\ 1 & \text{when } t > 0, \end{cases}$$

is called the **signum function** (the Latin *signum* is used to avoid verbal confusion with the trigonometric sine). Its graph is shown in Fig. 1.13a.

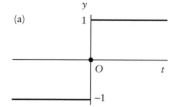
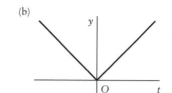
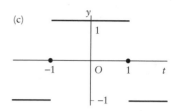

Fig. 1.13 (a) $y = \mathrm{sgn}(t)$. (b) $y = t\,\mathrm{sgn}(t)$. (c) $y = \mathrm{sgn}(1 - t^2)$.

H(t) and sgn t can be used along with other functions to produce a variety of functions having discontinuities in either value or direction at assigned points. Figures 1.13b,c show the *even* functions $y = t\,\mathrm{sgn}(t)$ and $y = \mathrm{sgn}(1 - t^2)$. Note that $\mathrm{sgn}(1 - t^2)$ has discontinuities where $1 - t^2 = 0$; that is, where $t = \pm 1$.

1.5 Radian measure of angles

For everyday purposes angles are measured in degrees, so we are still following the Babylonian practice of dividing the circle into 360 sectors each of which subtends a **degree** (1°). For mathematical purposes, a less arbitrary measure is desirable. The absolute unit is the **radian**, which represents about 57°. The special property which makes the unit valuable is its connection with length.

Figure 1.14 shows a circle of radius R with a **sector** AOB containing an angle θ. The length of the arc $\widehat{AB}$ is obviously *proportional* to R, and it is *proportional* to θ whatever the angular units, so it is *proportional* to the product $R\theta$. One radian is the unit of angle such that $\widehat{AB}$ is *numerically equal* to $R\theta$.

Then if θ is measured in radians

$$\widehat{AB} = R\theta.$$

But when θ covers the whole circle,

$$\widehat{AB} = 2\pi R.$$

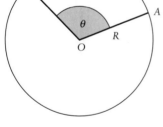

Fig. 1.14

Since the whole circle measures 360 in degrees, we obtain

$$\theta = 360° = 2\pi \text{ radians}, \quad \text{or} \quad 1 \text{ radian} = \tfrac{180}{\pi} \text{ degrees} = 57.295\,78\ldots°.$$

The following summarizes some useful information:

Radians and degrees

(a) α degrees $= \frac{\pi}{180}\alpha$ radians, β radians $= \frac{180}{\pi}\beta$ degrees.

$$360° = 2\pi \text{ rad}, \quad 180° = \pi \text{ rad}, \quad 90° = \tfrac{1}{2}\pi \text{ rad},$$
$$45° = \tfrac{1}{4}\pi \text{ rad}, \quad 60° = \tfrac{1}{3}\pi \text{ rad}, \quad 30° = \tfrac{1}{6}\pi \text{ rad}.$$

(b) On a circle of radius R, the arc-length subtended by θ radians is $R\theta$.

(1.14)

1.6 Trigonometric functions; properties

We assume that you know the meanings of *sine*, *cosine*, and *tangent* for positive acute angles, as in ordinary trigonometry. We shall extend their meaning to angles greater than 90°, and to negative angles. Unless indicated otherwise, angles are given in radians. In Fig. 1.15, X is any point on the *positive x* axis. If we rotate the segment OX about O in the **anticlockwise** direction to arrive at a final position OX', the total angle through which it has turned is counted as a **positive** number. If the rotation is **clockwise**, then the angle is given a **negative** sign. These angles are unlimited in magnitude. We refer to an angle measured from the positive x axis in this way as a **polar angle**. There is an infinite number of polar angles leading to the same direction OX', and differing by multiples of a complete revolution 2π.

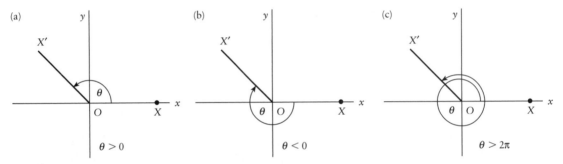

Fig. 1.15 Polar angles θ: (a) $\frac{3}{4}\pi$. (b) $-\frac{5}{4}\pi$. (c) $\frac{11}{4}\pi$. (These all have the same direction OX'.)

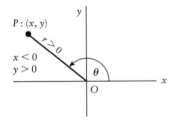

Fig. 1.16 Diagram for cos θ, sin θ, tan θ.

The trigonometric functions sine, cosine, and tangent for *all* angles are defined by the construction in Fig. 1.16, in which $P : (x, y)$ is any point, and θ is treated as a polar angle. The length OP is given by

$$OP = r = \sqrt{(x^2 + y^2)} > 0.$$

Then the *definitions* of the trigonometric functions for arbitrary angles θ are as follows:

> **Trigonometric functions; definitions**
> θ is arbitrary, $r = \sqrt{(x^2 + y^2)} > 0$ (see Fig. 1.16)
>
> $$\cos\theta = \frac{x}{r}, \qquad \sin\theta = \frac{y}{r}, \qquad \tan\theta = \frac{y}{x}.$$
>
> (sec $\theta = 1/\cos\theta$, cosec θ (or csc θ) = $1/\sin\theta$, cot $\theta = 1/\tan\theta$).   **(1.15)**

These definitions are extensions, to all four quadrants of the (x, y) plane, of the familiar geometrical meanings in the first quadrant. The length r is positive, but the coordinates x and y are *signed* quantities

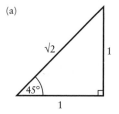

(a)

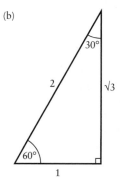

(b)

Fig. 1.17 (a) sin 45° = cos 45° = 1/√2. (b) sin 60° = cos 30° = √3/2. cos 60° = sin 30° = 1/2.

which determine the signs of the trigonometric functions $\sin\theta$, $\cos\theta$, $\tan\theta$ in the four quadrants. The following lists the ones which are positive:

1st quadrant, $x > 0$, $y > 0$:   all are > 0.
2nd quadrant, $x < 0$, $y > 0$: $\sin\theta > 0$.
3rd quadrant, $x < 0$, $y < 0$: $\tan\theta > 0$.
4th quadrant, $x > 0$, $y < 0$: $\cos\theta > 0$.

Example 1.5 *Obtain* (a) $\sin\frac{1}{3}\pi$, (b) $\tan\frac{1}{6}\pi$, (c) $\cos\frac{1}{4}\pi$. (*The angles are in radians.*)

Equation (1.14a) gives the angles in degrees. Use the triangles in Fig. 1.17.

(a) $\sin\frac{1}{3}\pi = \sin 60° = \sqrt{3}/2$.
(b) $\tan\frac{1}{6}\pi = \tan 30° = 1/\sqrt{3}$.
(c) $\cos\frac{1}{4}\pi = \cos 45° = 1/\sqrt{2}$.

Example 1.6 *Obtain* (a) $\cos 2\pi$, (b) $\sin\frac{3}{2}\pi$, (c) $\sin(-\frac{3}{4}\pi)$.

The points P on Fig 1.18 have $OP = r = 1$, and polar angles equal to the angles given in the question. The x, y coordinates of P are easy to find.

(a) P is at $(x, y) = (1, 0)$, so that $\cos 2\pi = x/r = 1$.
(b) P is at $(x, y) = (0, -1)$, so that $\sin\frac{3}{2}\pi = y/r = -1$.
(c) P is at $(x, y) = (-1/\sqrt{2}, -1/\sqrt{2})$, so that $\sin(-\frac{3}{4}\pi) = y/r = -1/\sqrt{2}$.

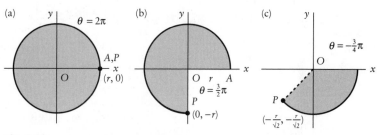

Fig. 1.18

The graphs of $\cos\theta$ and $\sin\theta$ are shown in Fig. 1.19. Observe the following:

1. The curves for $\cos\theta$ and $\sin\theta$ have **identical shape**, but are displaced a distance $\frac{1}{2}\pi$ (radians) from each other. They are related by

$$\sin\theta = \cos(\theta - \tfrac{1}{2}\pi), \qquad \cos\theta = \sin(\theta + \tfrac{1}{2}\pi). \tag{1.16}$$

2. The functions $\cos\theta$ and $\sin\theta$ are said to be **periodic**, with **period** (or **wavelength**) equal to 2π; that is, the curves repeat themselves at intervals of length 2π. This is evident from the definition of a polar angle, because in terms of the polar angle of a point P, an increase or decrease of 2π radians is equivalent to a complete revolution.

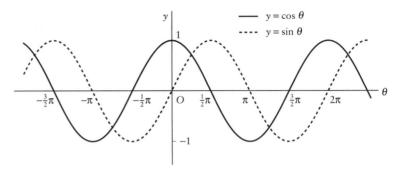

Fig. 1.19 Graphs of $\cos\theta$ and $\sin\theta$.

3. $\cos\theta$ is an **even function** (see eqn (1.12)), so that $\cos(-\theta) = \cos\theta$; $\sin\theta$ is an **odd function**, so that $\sin(-\theta) = -\sin\theta$.

4. The values taken by $\cos\theta$ and $\sin\theta$ oscillate between -1 and $+1$.

The graphs of $\tan\theta$, $\cot\theta$, $\sec\theta$, and $\csc\theta$ are shown in Fig. 1.20. There are many **trigonometric identities** in common use. The following are some of the more important (a more extensive list is given in Appendix B):

(a)

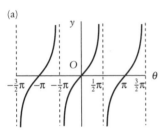

(b)

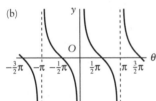

(c)

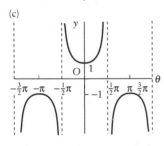

(d)

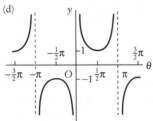

Fig. 1.20 (a) $\tan\theta$. (b) $y = \cot\theta$. (c) $y = \sec\theta$. (d) $y = \csc\theta$.

> **Trigonometric identities**
>
> For all angles A and B:
>
> (a) Sums and differences of angles
> $$\sin(A \pm B) = \sin A \cos B \pm \cos A \sin B,$$
> $$\cos(A \pm B) = \cos A \cos B \mp \sin A \sin B,$$
> $$\tan(A \pm B) = (\tan A \pm \tan B)/(1 \mp \tan A \tan B).$$
>
> (b) Products as sums and differences
> $$\cos A \cos B = \tfrac{1}{2}[\cos(A + B) + \cos(A - B)],$$
> $$\cos A \sin B = \tfrac{1}{2}[\sin(A + B) - \sin(A - B)],$$
> $$\sin A \sin B = \tfrac{1}{2}[-\cos(A + B) + \cos(A - B)].$$
>
> (c) Double angles
> $$\cos^2 A + \sin^2 A = 1,$$
> $$\sin(2A) = 2\sin A \cos A,$$
> $$\cos^2 A = \tfrac{1}{2}(1 + \cos 2A),$$
> $$\sin^2 A = \tfrac{1}{2}(1 - \cos 2A).$$
>
> (d) Cosine rule. In a triangle with side lengths a, b, c and opposite angles A, B, C,
> $$c^2 = a^2 + b^2 - 2ab \cos C.$$
>
> (e) Sine rule. In a triangle with side lengths a, b, c and opposite angles A, B, C,
> $$\frac{\sin A}{a} = \frac{\sin B}{b} = \frac{\sin C}{c}.$$
>
> (Since the identities are the same as for positive acute angles we do not prove them here.)   (1.17)

We have so far encountered $\cos\theta$, $\sin A$, etc., in which θ and A are understood to represent certain angles arising in a geometrical context, but trigonometric functions are used in many applications which have nothing directly to do with angles. For example, expressions such as $\cos\omega t$ will occur, in which t stands for time and ω (Greek 'omega') is another constant. The identities continue to hold, whatever the context.

With this in mind we now obtain an important identity which shows that any function having the form $a\cos u + b\sin u$, where u is the independent variable and a and b are constants, can be put into the form $c\cos(u + \phi)$, in which c and ϕ (Greek 'phi') are constants obtainable in terms of a and b.

Put

$$c = \surd(a^2 + b^2) > 0,$$

and rewrite $a\cos u + b\sin u$ in the form

$$a\cos u + b\sin u = c[(a/c)\cos u + (b/c)\sin u].$$

Let ϕ be an angle such that

$$\cos\phi = a/c \quad \text{and} \quad \sin\phi = -b/c.$$

The angle ϕ can be found by locating the point $Q : (a, -b)$ on cartesian axes, as shown in Fig. 1.21. Any one of the polar angles of Q satisfy the condition above; usually the smallest in absolute magnitude is chosen. (Since $c = \surd(a^2 + b^2) > 0$, the radial and angular **polar coordinates** of Q are c and ϕ respectively. If you do not know about polar coordinates, look forward to the first paragraphs of Section 1.9.)

We then obtain

$$a\cos u + \sin u = c(\cos\phi\cos u - \sin\phi\sin u) = c\cos(u + \phi)$$

from (1.17a) with u and ϕ in place of A and B. We have obtained the identity

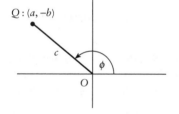

$Q : (a, -b)$

c

ϕ

O

Fig. 1.21 To find ϕ satisfying $a/c = \cos\phi$, $b/c = -\sin\phi$.

Harmonic functions

$$a\cos u + b\sin u = c\cos(u + \phi),$$

where c and ϕ are polar coordinates of the point $(a, -b)$ in cartesian axes.

                                                                    (1.18)

A function having the form $A\cos(ku + \alpha)$, where A, k, and α are *any* constants, is called a **harmonic function** or a **sinusoid** in the variable u. Sine functions are included by virtue of (1.16), since we can change a sine into a cosine by subtracting $\frac{1}{2}\pi$ from the argument of sine.

Consider the harmonic function given by

$$y = c\cos(\omega t + \phi) \quad (c > 0),$$

where the independent variable t represents *time*. The function is used to represent quantities y which have a regular wave-like time

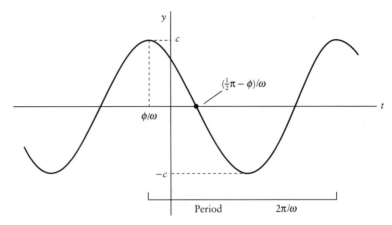

Fig. 1.22 Graph of $y = c\cos(\omega t + \phi)$.

variation (or space variation if the variable is distance x instead of time t). Figure 1.22 displays a function of this type. It is like a cosine function $c\cos t$, translated a distance $-\phi/\omega$ along the t axis, and stretched or compressed horizontally to an extent depending on ω. It is **periodic** with **period** $2\pi/\omega$ because, if we take any value of t and increase it by $2\pi/\omega$, we obtain

$$y = c\cos[\omega(t + 2\pi/\omega) + \phi] = c\cos(\omega t + 2\pi + \phi) = c\cos(\omega t + \phi).$$

Therefore the value of y is repeated across every interval of length $2\pi/\omega$. The number $c > 0$ is called the **amplitude**; y is said to **oscillate** between $\pm c$. The **frequency**, in cycles per unit time, is $\omega/2\pi$, and ω is called the **circular or angular frequency**. The constant ϕ is the **phase angle**.

A **general function** $f(x)$ is said to be **periodic with period** p if $f(x + p) = f(x)$ for every value of x. If p is a period, so obviously are $2p$, $3p$, and so on. The period usually meant when we say that a function has period p is the **smallest positive period**. From Fig. 1.20, $\tan\theta$ and $\cot\theta$ have period π in θ.

1.7 Inverse functions

Let $y = f(x)$, where f is a given function. It is often necessary to find a value of x corresponding to a given value of y, which amounts to solving a certain equation. For example, if f is defined by

$$y = f(x) = x^3 \quad \text{and} \quad y = 8,$$

then the resulting equation, $8 = f(x) = x^3$, is solved uniquely by $x = 8^{\frac{1}{3}} = 2$. In fact there is a unique value of x corresponding to every value of y, positive or negative. These x values each *depend on y*, so we say that x is a *function* of y. Denoting this function by F, then we can write

$$x = F(y) = y^{\frac{1}{3}}.$$

The function F is called the **inverse function** of f.

The following are the fundamental **reciprocal relationships** between F and f:

$$F\{f(x)\} = F(y) = (x^3)^{\frac{1}{3}} = x, \text{ for every value of } x,$$

and

$$f\{F(y)\} = f(x) = (y^{\frac{1}{3}})^3 = y, \text{ for every value of } y.$$

Initially, the values of x and y were connected through the relation $y = f(x)$, but in the final form of these equations:

$$F\{f(x)\} = x, \text{ for every value of } x,$$
$$f\{F(y)\} = y, \text{ for every value of } y.$$

Since x and y are separated and independent, we may use any letters to indicate the variables in place of x and y. In fact, these two equations are *identities* (see Section 1.4). For example, if we were concerned with an application involving an angle θ it might be convenient to write the first one in the form

$$F\{f(\theta)\} = \theta, \text{ for every value of } \theta,$$

or we could substitute x for y in the second equation, without changing their meaning.

Returning to the original problem, if we know the inverse function F we can solve the equation

$$c = f(x)$$

by using the first reciprocal relation $F\{f(x)\} = x$. Taking the inverse F of both sides of the equation, we obtain

$$F(c) = F\{f(x)\} = x,$$

so that the required value of x is $F(c)$.

We shall now give a geometrical description of the operations we have just gone through. Figure 1.23a is the graph of $y = f(x) = x^3$. Choose any number a. To find a^3, locate $x = a$ at A and follow the track ABC. The point C on the y axis represents $y = f(a) = a^3$. Now choose a number b with the aim of obtaining $b^{\frac{1}{3}}$. Read the graph backwards: locate $y = b$ at U and follow the track UVW. Then W represents $b^{\frac{1}{3}}$. Therefore the *same curve* provides values of x^3 and also of its inverse function $F(x) = x^{\frac{1}{3}}$. The two identities given above amount to the obvious fact that if we follow the tracks $ABCBA$ and $CBABC$ respectively in Fig. 1.23a, then we arrive back at the starting point in each case.

In order to obtain cube roots, we might prefer a graph from which the cube root can be read off in the ordinary way: from a horizontal x axis to a vertical y axis. Suppose that we plot the curves $y = f(x) = x^3$, and the inverse in the form $y = F(x) = x^{\frac{1}{3}}$, on the same sheet of paper

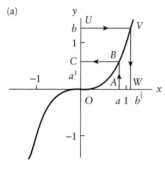

(a)

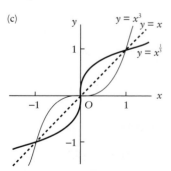

(b)

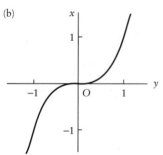

(c)

Fig. 1.23 (a) and (b) The function $f(x) = x^3$. (c) The function $f(x) = x^3$ and its inverse $F(x) = x^{\frac{1}{3}}$: since the scales are equal, $f(x)$ and $F(x)$ reflect each other in the 45° line.

(see Fig. 1.2c). We also arrange that *the x and y scales are equal*. Let $P : (a, b)$, where $b = a^3$, be any point on $y = f(x) = x^3$. The corresponding point on the graph of the inverse function $y = F(x) = x^{\frac{1}{3}}$ is $Q : (b, a)$. Since the x and y scales are equal, Q is the reflection of P in the straight line through the origin inclined at 45° to the x axis. Therefore **the graphs of $y = f(x)$ and of its inverse $y = F(x)$ are reflections of each other in the 45° radial line.** This is shown for $f(x) = x^3$ and its inverse $F(x) = x^{\frac{1}{3}}$ in Fig. 1.23b.

The arguments are basically the same for any function and its inverse. Whatever f may be, we obtain the graph of its inverse function F by plotting $y = f(x)$ with equal scales, and reflecting it in the 45° line. Also, the reciprocal identities

$$F\{f(x)\} \equiv x \quad \text{and} \quad f\{F(x)\} \equiv x \tag{1.19}$$

apply in the general case, though this is usually subject to restrictions on the range of the function F, in order to ensure that a single, unique value is assigned to $F(x)$.

The example $f(x) = x^3$ is particularly straightforward, since the functions $y = x^{\frac{1}{3}}$ and $y = x^3$ are unique inverses of each other for *every* value of x and y. The problem of single-valuedness of F arises if the graph of $f(x)$ 'turns over' at some point, or points. A simple example is $y = f(x) = x^2$ (see Fig. 1.9c) which turns over at $x = 0$, so that the graph falls into two parts, on the left and right sides of the y axis. The inverse function F_1 corresponding to the right-hand branch of $y = x^2$ is $y = F_1(x) = x^{\frac{1}{2}}$, valid only for $x \geqslant 0$ and $y \geqslant 0$. A second inverse function, defined by $y = F_2(x) = -x^{\frac{1}{2}}$ for $x \geqslant 0$ and $y \leqslant 0$, arises from the left-hand branch. The two inverse functions taken together provide, for example, the two expected solutions of the equation $f(x) = x^2 = 2$, namely

$$x = F_1(2) = +2^{\frac{1}{2}} \quad \text{and} \quad x = F_2(2) = -2^{\frac{1}{2}}.$$

The problem arises particularly in Section 1.8 in connection with inverse trigonometric functions.

Figure 1.24 illustrates the general character of positive integer powers x^n and their inverses $x^{1/n}$, the picture being confined to the range $0 \leqslant x \leqslant 1$ for clarity. Notice the symmetry of the inverse pairs about the 45° line $y = x^1 = x$. Graphs corresponding to other powers of x lie between those shown, in a regular way.

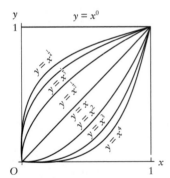

Fig. 1.24 Some positive integer powers x^n and their inverses $x^{\frac{1}{n}}$, showing reflection across the 45° line.

1.8 Inverse trigonometric functions

Consider the inverses of the trigonometric functions given by $x = \sin \theta$, $x = \cos \theta$, and $x = \tan \theta$, in which θ can be thought of as an angle measured in radians. There are two commonly used notations for the inverses. In this book they are denoted by

$$\theta = \arcsin x, \quad \theta = \arccos x, \quad \theta = \arctan x.$$

The *alternative notation* for inverse trigonometric functions is

$$\theta = \sin^{-1}x, \quad \theta = \cos^{-1}x, \quad \theta = \tan^{-1}x.$$

(In this notation the index (−1) does *not* signify a negative power: for example, $\sin^{-1}x$ does not stand for $1/\sin x$. Correspondingly, $1/\sin x$ must be written as $(\sin x)^{-1}$ in index form to distinguish its meaning. It is to avoid this possible source of confusion that we have adopted the more modern notation arcsin etc., which is also consistent with computer notation.)

Firstly, consider the problem of finding the inverse sine function defined by $\theta = \arcsin x$, corresponding to the given function $x = \sin\theta$. The inverse function should answer the question 'what angle θ has its sine equal to x?' Evidently $-1 \leqslant x \leqslant 1$, or there exists no such angle. Moreover, if x does lie in the permitted range, there is an infinite number of such angles. This is illustrated in Fig. 1.25 for the special case where $x = \frac{1}{2}$. The graph $x = \sin\theta$ intersects the line $x = \frac{1}{2}$ at an infinite number of points,

$$\theta = \dots, \ -\tfrac{7}{6}\pi, \ \tfrac{1}{6}\pi, \ \tfrac{5}{6}\pi, \ \dots .$$

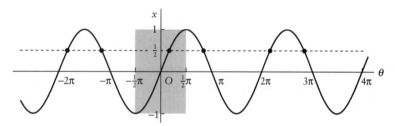

Fig. 1.25 The intersections of the graphs $x = \frac{1}{2}$ and $x = \sin\theta$ give the solutions of the equation $\sin\theta = \frac{1}{2}$.

A computer program or a hand calculator can deliver only a single, definite value of a function such as arcsin x, not an infinite shower of alternatives, but if we are given any single one of these values we can easily use it to construct all of them. Suppose $\theta = \alpha$ is any one solution of the equation $\sin\theta = c$, where c is a constant with $-1 \leqslant c \leqslant 1$. Then *all* the values of θ satisfying the equation $\sin\theta = c$ are obtainable from α by means of the formula

$$\theta = n\pi + (-1)^n\alpha,$$

where n is any integer (see Appendix B(g)).

In order to obtain a single value for $\theta = \arcsin x$ and to exclude the rest, we require an interval on the graph of $x = \sin\theta$ along which every value of x between −1 and 1 occurs once and only once. The standard interval used for this purpose is

$$-\tfrac{1}{2}\pi \leqslant \theta \leqslant \tfrac{1}{2}\pi,$$

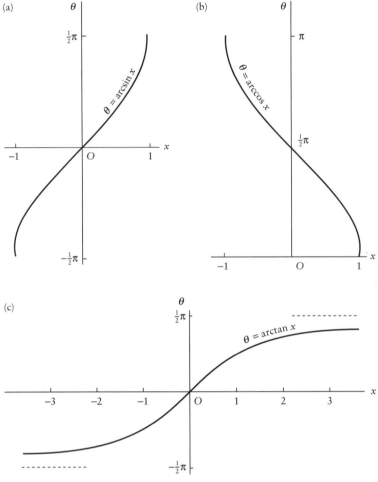

Fig. 1.26 The inverse trigonometric functions. In (a) and (b), the scales of θ and x are the same. (a) $\theta = \arcsin x$, $-\frac{1}{2}\pi \leqslant \theta \leqslant \frac{1}{2}\pi$, $-1 \leqslant x \leqslant 1$. (b) $\theta = \arccos x$, $0 \leqslant \theta \leqslant \pi$, $-1 \leqslant x \leqslant 1$. (c) $\theta - \arctan x$, $-\frac{1}{2}\pi \leqslant \theta \leqslant \frac{1}{2}\pi$, all x.

shown shaded in Fig. 1.25, and in this way the **standard inverse function** $\theta = \arcsin x$ is **restricted** to lie in the range $-\frac{1}{2}\pi \leqslant \theta \leqslant \frac{1}{2}\pi$. Figure 1.26a shows its graph. Given a value of x with $-1 \leqslant x \leqslant 1$, and the equation $\sin\theta = x$, it returns the value of θ which has the smallest absolute magnitude (this is the value that the software Mathematica returns for the command **ArcSin[x]**). If we want other solutions we must derive them from the formula above.

The inverse cosine and tangent functions, arccos and arctan, are approached in a similar way and are displayed in Figs 1.26b,c. They have different standard ranges, and Appendix B(g) contains the formulae for obtaining all other solutions of $x = \cos\theta$ and $x = \tan\theta$. The various inverse functions are connected, as in the following example.

(a)

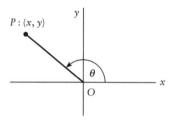

(b)

Fig. 1.27

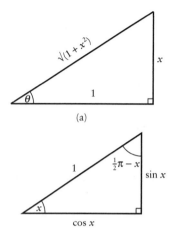

Fig. 1.28 Polar coordinates:
$x = r\cos\theta, y = r\sin\theta.$

Example 1.7 *Simplify* (a) cos(arctan x), (b) arcsin(cos x).

(a) In the right-angled triangle of Fig. 1.27a, $x = \tan\theta$, so that $\theta = \arctan x$.
Therefore

$$\cos(\arctan x) = \cos\theta = 1/\sqrt{(1 + x^2)}.$$

(b) In the right-angled triangle of Fig. 1.27b,

$$\cos x = \sin(\tfrac{1}{2}\pi - x), \text{ so that } \arcsin(\cos x) = \tfrac{1}{2}\pi - x.$$

1.9 Polar coordinates

In Fig. 1.28, P is a general point, with coordinates x, y. The length of
OP (always counted positive) is given by

$$r = OP = \sqrt{(x^2 + y^2)} > 0.$$

The angle θ is any polar angle (see Section 1.7) that locates P; we
choose the simplest one for the illustration, but our statements are
true for any of the valid polar angles (which are given by $\theta + 2\pi n$ for
every integer n).

Definite values for r and θ locate a unique point P just as well
as the cartesian coordinates x, y, so in suitable cases we can use r, θ
as coordinates in place of x, y. They are called **polar coordinates**.
From the definitions of sine and cosine in eqn (1.15), the cartesian
and polar coordinates of P are related in the following way:

> **Cartesian and polar coordinates**
> (a) $x = r\cos\theta, y = r\sin\theta.$
> (b) $r = \sqrt{(x^2 + y^2)} > 0, \cos\theta = x/r, \text{ and } \sin\theta = y/r.$ (1.20)

Polar coordinates are often easier to use than x, y coordinates,
especially for curves that surround the origin. The simplest example
is a circle, centre the origin and radius c, whose **polar equation** is

$$r = c.$$

A **spiral**, such as a track on a compact disk with inner radius a, outer
radius b, and track width h, is described by

$$r = b - \frac{h}{2\pi}\theta,$$

in which θ runs from zero to $2\pi N$, where $N = (b - a)/h$ is the number
of revolutions. Note the enormous size attained by the polar angle as
it follows the rotation through many revolutions.

Example 1.8   (a) *Obtain the polar equation of the central ellipse*

$$\frac{x^2}{a^2} + \frac{y^2}{b^2} = 1$$

(*see* Fig. 1.9a). (b) *Obtain the polar equation for the same ellipse tilted through an angle* α.

(a) Referring to eqn (1.21), the cartesian equation becomes

$$\frac{r^2 \cos^2\theta}{a^2} + \frac{r^2 \sin^2\theta}{b^2} = 1. \tag{i}$$

From the identity (1.17c) we can express $\cos^2\theta$ and $\sin^2\theta$ in terms of $\cos 2\theta$, so that (i) can be written as

$$r^2 \left[\frac{(1 + \cos 2\theta)}{2a^2} + \frac{(1 - \cos 2\theta)}{2b^2} \right] = 1,$$

which simplifies to

$$r = \frac{2ab}{\sqrt{[(a^2 + b^2) - (a^2 - b^2) \cos 2\theta]}}.$$

When θ runs from, say zero to 2π, the complete ellipse is traced out once.

(b) To tilt the ellipse through an angle α, simply replace θ by $\theta - \alpha$ (by analogy with eqn (1.11) for a change of origin of the x axis). Then the equation of the tilted ellipse is

$$r = \frac{2ab}{\sqrt{[(a^2 + b^2) - (a^2 - b^2) \cos 2(\theta - \alpha)]}}. \tag{ii}$$

Equation (1.20b) gives the formulae for converting from x, y coordinates to polar coordinates, and you may wonder why the two simultaneous equations for θ were not written more simply as

$$\tan \theta = \frac{\sin \theta}{\cos \theta} = \frac{y}{x},$$

apparently leading to the explicit solution $\theta = \arctan(y/x)$. To explain why, consider the point P which has the coordinates

$$x = -1, \quad y = 1.$$

The point P is in the *second* quadrant, and the simplest choice for θ (there is an infinite number of valid choices, all differing by multiples of 2π) is

$$\theta = \tfrac{3}{4}\pi.$$

However, if we obtain the value of $\arctan(y/x) = \arctan(-1)$ on a calculator, the value we find is $\theta = -\tfrac{1}{4}\pi$, predicting, wrongly, that P is in the *fourth* quadrant. The reason for the problem is that the *standard definition* of $\arctan(y/x)$ requires that its value always lies in the range $-\tfrac{1}{2}\pi \leqslant \theta \leqslant \tfrac{1}{2}\pi$, and so if $\arctan(y/x) = \theta$ is interpreted geometrically as a polar angle its associated point P will lie in quadrants 1 or 4 (see Section 1.8 and Fig. 1.26c). A different solution of the equation $\tan \theta = y/x$ is therefore needed if P lies in quadrants 2 or 3 (which is the region of negative x). This is provided by the rule

Cartesian–polar rule

Let P be the point (x, y) with polar coordinates r, θ. Then

(i) if $x = r \cos \theta$ is positive, a value of the polar angle of $P : (x, y)$ is $\arctan(y/x)$;
(ii) if $x = r \cos \theta$ is negative, a value for the polar angle is $\arctan(y/x) \pm \pi$.

The result (ii) follows from the fact that $\tan \theta$ has period π; so

$$\tan[\arctan(y/x) \pm \pi] = \tan[\arctan(y/x)] = y/x,$$

as required. As always, we can take the polar angles $\pm 2\pi n$.

Many symbolic computer systems accept x and y arguments to avoid this problem. For example, in Mathematica the polar angle is given correctly by the command **ArcTan**$[x, y]$.

1.10 **Exponential functions; the number** e

Consider the function

$$y = a^x,$$

in which a is any *positive* constant and x can take any value (see Section 1.1). Graphs of y against x for several cases where $a \geqslant 1$ are shown in Fig. 1.29. All graphs pass through the point $x = 0$, $y = 1$.

The number a is called the **base for the exponential function** a^x. Exponential functions having different bases are all closely related. For example, $2^x = 4^{\frac{1}{2}x}$, and $10^x = 2^{3.3219}x$ (to 4 significant figures), and so on. We shall show later (Example 1.9) that all the functions $y = a^x$ can be displayed on the same curve provided that the x-scales are contracted or extended by appropriate factors, so we really only need one, standard, base to describe them all.

A standard base may be chosen according to what is most convenient for later requirements. For example, at one time the base $a = 10$ was adopted in order to simplify the arithmetic involved in large calculations, but nowadays we have better methods. The base $a = 2$ is used in the theory of binary processes such as occur in information theory. The base now in most general use is denoted by the letter e, and the corresponding **standard exponential function** is e^x. It is written alternatively as

$$\exp x \quad \text{or} \quad \exp(x)$$

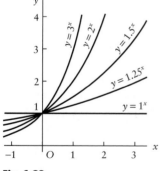

Fig. 1.29

in programming (Mathematica uses **Exp[x]**), on hand calculators, and in text. The numerical value of e is chosen so as to simplfy the algebra associated with exponential functions. We shall identify e by the following requirement:

> e **is the number such that the graph of** $y = e^x$ **cuts the** y **axis at** 45° **(provided that the** x **and** y **scales are equal).**

Reference to Fig. 1.29, in which the scales are equal, indicates that the value of e is between 2 and 3.

To obtain a close estimate for the numerical value of e, see Fig. 1.30. The x and y scales on the graph are equal, and $P : (0, 1)$ is the point where the graph of $y = e^x$ meets the y axis. Q is any nearby point on the graph having coordinates $x = h$, $y = e^h$. PN and NQ are parallel to the axes and meet at N. PT is the tangent line to the curve at P, meeting QN at T. We have said that e must take a value such that the graph cuts the y axis at 45°; by this we mean strictly that the *tangent line PT* must cut the y axis at 45°. This is equivalent to saying that the slope of the tangent line PT (see eqn (1.6)) should equal 1, that is

$$\frac{NT}{PN} = \tan 45° = 1.$$

First suppose that h is quite small, so that Q is close to P. Then $NT \approx NQ$, and

$$\frac{NQ}{NT} \approx 1,$$

the approximation improving as h becomes smaller. We have $NQ = e^h - 1$ and $PN = h$, so that

$$\frac{e^h - 1}{h} \approx 1, \tag{1.21}$$

and, in fact, *the approximation can be made as close as we wish by taking h small enough*. To estimate e, multiply through by h to give $e^h \approx 1 + h$. Finally, raise both sides to the power $1/h$ to isolate the number e:

$$e \approx (1 + h)^{1/h},$$

to any degree of accuracy, provided h is made small enough. The following table, constructed using a calculator or computer, illustrates how the desired value of e gradually emerges as h approaches zero:

h	0.1	0.01	0.001	0.0001	0.000 01
$(1 + h)^{1/h}$ ($\approx$ e)	2.5937...	2.7048...	2.7169...	2.7181...	2.7182...

(To 7 decimal places, the value of e is given by $e = 2.718\ 281\ 8....$)

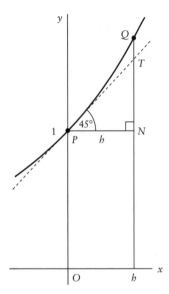

Fig. 1.30 Graph of $y = e^x$, with equal x, y scales. At $P : (0, 1)$ the slope of the tangent line is equal to 1.

The approximation (1.21) can be written briefly by using the notation

$$\frac{e^h - 1}{h} \to 1 \text{ as } h \to 0, \tag{1.22}$$

the sign '$\to$' standing for 'approaches'. A non-geometrical way of describing this requirement is to say that the **rate of increase** or the **growth rate** of the function e^x at $x = 0$ is equal to 1. By using this expression we avoid having to refer continually to graphs having equal x, y scales.

The growth rate of e^x at a **general value of** x can be obtained similarly, by imagining a fixed point P on the graph with coordinates (x, e^x), and a nearby point $Q : (x + h, e^{x+h})$. Then when $h \to 0$,

$$\frac{e^{x+h} - e^x}{h} \to \text{the growth rate of } e^x \text{ at } P.$$

Notice that we cannot shortcut the process of letting $h \to 0$ by putting $h = 0$ directly into the expression: we would obtain 0/0, which is meaningless. To obtain an explicit expression, write $e^{x+h} = e^x e^h$. Then

$$\frac{e^{x+h} - e^x}{h} = \frac{e^x(e^h - 1)}{h} \to e^x \text{ as } h \to 0, \text{ from eqn (1.22)}.$$

Therefore, for any value of x,

the rate of growth of $e^x = e^x$.

This is the property of e^x that makes e the preferred base for the exponential functions, and e is the only base that delivers this property.

The function e^x increases rapidly, as the following table illustrates (to 2 significant figures):

x	-9	-6	-3	0
e^x	1.2×10^{-4}	2.5×10^{-3}	5.0×10^{-2}	1.0

x	3	6	9
e^x	2.0×10^1	4.0×10^2	8.1×10^3

Whenever x is increased by 3, e^x is *multiplied* by a factor of about 20 (because $e^3 \approx 20$).

1.11 The logarithmic function

The **inverse function** corresponding to the exponential function $y = e^x$ or $\exp(x)$ is called the **logarithm** of x (historically, the **natural logarithm**). It is written

$$y = \ln x,$$

or sometimes as $y = \log_e x$, read as 'log with base e of x', or simply as $\log x$ (in Mathematica the notation is **Log**[x]). It fills in the question mark in $e^? = x$, or solves the equation $e^y = x$ for the unknown y. For example, the equation

$$e^y = 3$$

has the unique solution

$$y = \ln 3.$$

A scientific calculator gives $\ln 3 = 1.098\,61$ to 5 decimal places, and it can be confirmed that $e^{1.098\,61} = 3$ (to 5 decimal places).

Since e^x and $\ln x$ are inverses, their graphs are reflections of each other in the 45° radial line (see Section 1.7) provided that the x and y scales are equal. This relationship is shown in Fig. 1.31. Note that if x is negative, $\ln x$ does not have a real value.

The logarithm has the following properties, which are proved below:

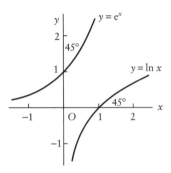

Fig. 1.31 The graph of $\ln x$ obtained from that of e^x.

Properties of the logarithm $\ln x$

a and b are any positive numbers.

(a) Definition of $\ln x$ $(x > 0)$ as the inverse of e^x:

$$e^{\ln x} = x \quad (x > 0), \text{ and } \ln e^x = x \ (\text{any } x)$$

$$(\text{or } \exp(\ln x) = x \ (x > 0), \text{ and } \ln(\exp x) = x \ (\text{any } x).)$$

(b) $\ln 1 = 0$, $\ln e = 1$.

(c) Product rule:

$$\ln ab = \ln a + \ln b.$$

(d) Quotient rules:

$$\ln(a/b) = \ln a - \ln b, \quad \ln(1/b) = -\ln b.$$

(e) Power or exponent rule:

$$\ln a^x = x \ln a \text{ for any } x.$$

(f) $\ln x \to \infty$ as $x \to \infty$; $\ln x \to -\infty$ as $x \to 0 \ (x > 0)$. **(1.23)**

Proof. (a) This is the fundamental property of (1.19) of inverse functions applied to this case.

(b) $e^0 = 1$ (see (1.2b)), so $0 = \ln 1$ (from (1.23a)). Also, $e^1 = e$, so $1 = \ln e$ (from (1.23a)).

(c) From the definition (1.23a), applied three times:

$$e^{\ln ab} = ab = e^{\ln a} e^{\ln b} = e^{\ln a + \ln b} \text{ (from (1.2a)).}$$

By equating powers of e on the two sides, we have

$$\ln ab = \ln a + \ln b.$$

(d) Put $a = (a/b)b$, and take the logarithm of both sides:

$$\ln a = \ln[(a/b)b] = \ln(a/b) + \ln b$$

from the product rule (1.23c). The first result in (1.23d) follows immediately. Put $a = 1$ so that $\ln 1 = 0$ to obtain the second result.

(e) From (1.23a),

$a = e^{\ln a}$, so that $a^x = (e^{\ln a})^x = e^{x \ln a}$.

By the definition (1.23a) with a^x in place of a, $x \ln a$ is the logarithm of a^x.

(f) These follow from (a).

Example 1.9 *Prove that all the graphs $y = a^x$, for $a > 1$, become identical to that of e^x if, for each case, the x axis is scaled by the appropriate factor.*

To fix ideas, think of a as being a given constant such as 2. As in the proof of (1.23e) above, we can write $y = a^x$ in the form

$y = a^x = e^{x \ln a} = e^{kx}$,

say, where $k = \ln a$. The required scale factor is $k > 0$. To make the graph of a^x lie along that of e^x, the x axis must be stretched if $k > 1$ (i.e. if $a > e$), and compressed if $0 < k < 1$ (i.e. if $1 < a < e$).

If k is negative, corresponding to the range $0 < a < 1$, the direction of the x axis has to be reversed as well as being rescaled.

Example 1.10 *Obtain y in terms of x when $\ln(y - 1) = 3 \ln x + 2$.*

Equate the exponential functions of both sides of the equation:
$$e^{\ln(y-1)} = e^{3 \ln x + 2}. \tag{i}$$
From the inverse function property (1.23a),
$$e^{\ln(y-1)} = y - 1. \tag{ii}$$
Also, by using the ordinary rules for exponents (1.2),
$$e^{3 \ln x + 2} = e^{3 \ln x} e^2 = (e^{\ln x})^3 e^2 = x^3 e^2 \tag{iii}$$
from (1.23a). Substitute (ii) and (iii) into (i); we obtain
$$y = 1 + x^3 e^2.$$

1.12 Exponential growth and decay

Here we shall use t (for time) in place of x, and consider the function

$$y = A\, e^{ct}, \tag{1.24}$$

where A and c are constants. This class includes functions such as 2^t, since, by Example 1.8, they can all be expressed in the form (1.24); in this case $2^t = e^{t \ln 2}$.

If $c > 0$, then y is said to have **exponential growth**. To get an idea of what this implies, we shall consider the **doubling period** of y. Choose *any* moment of time, t. At some later time $t + T$, y will have doubled its value, so that

$$A\, e^{c(t+T)} = 2A\, e^{ct} \quad \text{or} \quad A\, e^{ct}\, e^{cT} = 2A\, e^{ct}.$$

After cancelling the factor $A\,e^{ct}$ we have an equation for T:

$$e^{cT} = 2,$$

so that $cT = \ln 2$, from which we obtain the unknown T:

$$T = (1/c)\ln 2.$$

This result is independent of t, so we have

Exponential doubling principle

$y = A\,e^{ct}$ doubles its value in *every* interval of length
$T = (1/c)\ln 2.$

(1.25)

The doubling time T is often quoted as a measure of the rate of growth of populations, investments, etc., which, over a period, behave exponentially as in eqn (1.24). More generally, **the value of $A\,e^{ct}$ is multiplied by a factor N over every interval of length** $(1/c)\ln N$. The successive values form a **geometric progression** (see Section 1.15) with common ratio N.

Example 1.11 *The number N of scientists and engineers in the USA doubled every 10 years between 1900 and 1935, and in 1935 they numbered about 1.5×10^5. This suggests exponential growth $N = A\,e^{ct}$. Find c, and predict the number N for 1990 on the assumption that the trend continued.*

Suppose that we count 1900 as $t = 0$. The doubling period is 10 years; so $N = A\,e^{ct}$, where, by (1.25),

$$c = \tfrac{1}{10}\ln 2 = 0.0693.$$

Thus

$$N = A\,e^{0.0693t}.$$

In 1935, where $t = 35$ (years), $N = 1.5 \times 10^5$, so that

$$1.5 \times 10^5 = A\,e^{0.0693 \times 35},$$

or $A = 13\,265.$

Therefore

$$N = 13\,265\,e^{0.0693t}.$$

In 1990 $t = 90$, from which it follows that

$$N = 6.8 \times 10^6.$$

Exponential growth occurs when a quantity increases at a rate proportional to the amount already accumulated. In the short term, animal populations, epidemics, and some investments have this characteristic.

Exponential decay may also occur. If c is a positive number and

$$y = A\,e^{-ct} = A/e^{ct},$$

then y *halves* itself in *every* interval of length $(1/c)\ln 2$. This occurs in radioactive decay, the period being called the **half-life period** of a radioactive substance. The half-life period provides a convenient, memorable measure of the time it would take for the substance to become less harmful.

1.13 Hyperbolic functions

It is often convenient to represent certain combinations of exponential functions by separate functions. The **hyperbolic cosine** and **hyperbolic sine** functions, denoted by cosh and sinh respectively, are defined by the following formulae.

> **Hyperbolic functions**
> $$\cosh x = \tfrac{1}{2}(e^x + e^{-x}), \qquad \sinh x = \tfrac{1}{2}(e^x - e^{-x}). \tag{1.26}$$

Since

$$\cosh(-x) = \tfrac{1}{2}(e^{-x} + e^{-(-x)}) = \tfrac{1}{2}(e^{-x} + e^x) = \cosh x,$$

it follows that the graph of $\cosh x$ is symmetrical about the y axis, that is, $\cosh x$ is an *even* function. By a similar argument, it can be shown that $\sinh x$ is on *odd* function of x. Graphs of the two functions are shown in Fig. 1.32a.

From the definitions (1.26)

$$\cosh x + \sinh x = e^x, \qquad \cosh x - \sinh x = e^{-x}.$$

The remaining hyperbolic functions are defined in a similar manner to their trigonometric counterparts. Thus

> $$\tanh x = \frac{\sinh x}{\cosh x}, \qquad \coth x = \frac{\cosh x}{\sinh x},$$
> $$\operatorname{sech} x = \frac{1}{\cosh x}, \qquad \operatorname{cosech} x = \frac{1}{\sinh x}. \tag{1.27}$$

Graphs of $\tanh x$, $\coth x$, $\operatorname{sech} x$, and $\operatorname{cosech} x$ are shown in Fig. 1.32.

From the definitions, a number of identities follow which parallel those for trigonometric functions but with important sign differences. Some are derived below.

> (a) $\cosh^2 x + \sinh^2 x = \cosh 2x,$
> (b) $\cosh^2 x - \sinh^2 x = 1.$
> $$\tag{1.28}$$

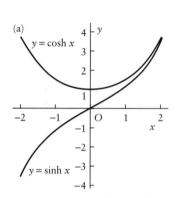

(a)

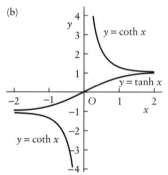

(b)

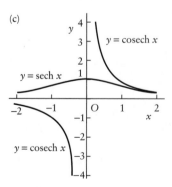

(c)

Fig. 1.32 Graphs of the hyperbolic functions.

For (a):

$$\cosh^2 x + \sinh^2 x = \tfrac{1}{4}(e^{2x} + 2 + e^{-2x}) + \tfrac{1}{4}(e^{2x} - 2 + e^{-2x})$$
$$= \tfrac{1}{2}(e^{2x} + e^{-2x}) = \cosh 2x.$$

For (b):

$$\cosh^2 x - \sinh^2 x = \tfrac{1}{4}(e^{2x} + 2 + e^{-2x}) - \tfrac{1}{4}(e^{2x} - 2 + e^{-2x}) = 1.$$

To obtain the identity

$$\sinh(x_1 + x_2) = \sinh x_1 \cosh x_2 + \cosh x_1 \sinh x_2,$$

start with the right-hand side:

$$\sinh x_1 \cosh x_2 + \cosh x_1 \sinh x_2$$
$$= \tfrac{1}{2}(e^{x_1} - e^{-x_1})\tfrac{1}{2}(e^{x_2} + e^{-x_2}) + \tfrac{1}{2}(e^{x_1} + e^{-x_1})\tfrac{1}{2}(e^{x_2} - e^{-x_2})$$
$$= \tfrac{1}{4}(e^{x_1+x_2} - e^{-x_1+x_2} + e^{x_1-x_2} - e^{-x_1-x_2})$$
$$\quad + \tfrac{1}{4}(e^{x_1+x_2} + e^{-x_1+x_2} - e^{x_1-x_2} - e^{-x_1-x_2})$$
$$= \tfrac{1}{2}(e^{x_1+x_2} - e^{-x_1-x_2}) = \sinh(x_1 + x_2).$$

To sum up similar identities:

$$\sinh(x_1 \pm x_2) = \sinh x_1 \cosh x_2 \pm \cosh x_1 \sinh x_2,$$
$$\cosh(x_1 \pm x_2) = \cosh x_1 \cosh x_2 \pm \sinh x_1 \sinh x_2,$$
$$\tanh(x_1 \pm x_2) = \frac{\tanh x_1 \pm \tanh x_2}{1 \pm \tanh x_1 \tanh x_2}.$$

(1.29)

The inverse hyperbolic functions corresponding to sinh, cosh, and tanh are indicated respectively by the notations

$$\sinh^{-1}, \cosh^{-1}, \text{ and } \tanh^{-1}.$$

The index (−1) is traditional: do not mistake it as standing for a negative power; $\sinh^{-1}x$ does not mean $1/\sinh x$. (Note that the commands **ArcSinh**[x], **ArcCosh**[x], and **ArcTanh**[x] are used for inverse hyperbolic functions in symbolic computation in Mathematica.) The intervals for which they are defined are as follows:

Inverse hyperbolic functions
$y = \sinh^{-1}x$ for all x and y.
$y = \cosh^{-1}x$ for $x \geqslant 1$ and $y \geqslant 0$.
$y = \tanh^{-1}x$ for $-1 < x < 1$ and all y.

(1.30)

These functions are expressible as logarithms. For example, consider $\sinh^{-1}x$. Put

$$y = \sinh^{-1}x,$$

so that

$$x = \sinh y = \tfrac{1}{2}(e^y - e^{-y}).$$

Multiply through by e^y and rearrange to give

$$e^{2y} - 2x\,e^y - 1 = 0 \quad \text{or} \quad (e^y)^2 - 2x(e^y) - 1 = 0.$$

This is a quadratic equation for e^y whose solutions are given by

$$e^y = x \pm \surd(x^2 + 1),$$

where the sign $\surd$ means the positive square root. The negative sign in the right-hand side corresponds to a solution which cannot represent e^y, since e^y is always positive but $x - \surd(x^2 + 1)$ is always negative. Therefore we select the positive sign. By taking the logarithm of both sides and using (1.23a) we obtain

$$y = \sinh^{-1}x = \ln[x + \surd(x^2 + 1)],$$

valid for all x. The inverses of the other hyperbolic functions are obtained similarly, and are shown in the following table:

Inverse hyperbolic functions as logarithms

$y = \sinh^{-1}x = \ln[x + \surd(x^2 + 1)]$ for all x, y.

$y = \cosh^{-1}x = \ln[x + \surd(x^2 - 1)]$ for $x \geqslant 1$, $y \geqslant 0$.

$y = \tanh^{-1}x = \tfrac{1}{2}\ln[(1 + x)/(1 - x)]$ for $-1 < x < 1$, all y. **(1.31)**

1.14 Partial fractions

We shall first reiterate the distinction between an **equation** and an **identity** (see Section 1.4). The word 'equation' has many uses, but for the present we shall think of an *equation* as something like

$$x^2 + 2 = -3x,$$

which is true only for certain particular values of x, namely -1 and -2. On the other hand,

$$x^2 + 3x + 2 \equiv (x + 1)(x + 2)$$

is an *identity*, meaning that it is true automatically, or for all values of x. We shall write $\equiv$ instead of $=$ when we want to draw attention to an identity.

It is easy to test the truth of the following identities by adding up the fractions on the right.

(i) $\dfrac{1}{x^2 - 1} \equiv \dfrac{1}{2}\dfrac{1}{x - 1} - \dfrac{1}{2}\dfrac{1}{x + 1}$,

(ii) $\dfrac{x}{4x^2 - 1} \equiv \dfrac{1}{4}\dfrac{1}{2x - 1} + \dfrac{1}{4}\dfrac{1}{2x + 1}$,

(iii) $\dfrac{3x + 2}{x^2(x + 1)} \equiv \dfrac{1}{x} + \dfrac{2}{x^2} - \dfrac{1}{x + 1}$.

The terms on the right are individually simpler than the functions on the left. This break-up into simpler constituents is useful for many purposes. In this section, we show how to break up a complicated function into simpler terms of the type above.

A **polynomial** in a variable x is an expression such as

$$-2x^2, \quad 3x^3 - x + 16$$

which is the sum of one or more terms of the form ax^n, where n is a positive integer or zero, and a is any number. A **rational function** is a function which takes the form

$$f(x) = P(x)/Q(x),$$

where $P(x)$ and $Q(x)$ are polynomials. For example, $1/x^2(3x - 2)$ and $(2x^3 + 1)/(x - 1)^2$ are rational functions, but $x^{\frac{1}{2}}/(x + 1)$ and $(\cos x)/(x + 1)$ are not rational functions of x. We shall be concerned only with rational functions, and initially we shall suppose that

$$\text{degree of } P(x) < \text{degree of } Q(x),$$

where the **degree** is the highest power occurring in the polynomial. Such functions can be broken up into **partial fractions**, like the examples at the beginning. No proofs will be given here, but the reader should learn the techniques.

It is the *denominator* of $Q(x)$ which determines what the form of the constituent partial fractions will take. Suppose that the denominator is broken up into factors as far as possible. For example,

$$2x^4 + x^3 - 4x^2 + x - 6 = (2x - 3)(x + 2)(x^2 + 1),$$

and it cannot be factorized any further. We shall consider only the cases where the factors are of the type:

$ax + b$ (a **simple factor**), $(cx + d)^n$ (a **repeated factor** of order n), and $px^2 + qx + r$ with $q^2 < 4pr$ (an **irreducible quadratic**).

The rules affecting these are as follows:

**Partial fractions for rational functions $P(x)/Q(x)$
(degree of $P(x)$) < (degree of $Q(x)$)**

Each factor of $Q(x)$ gives rise to a partial fraction (or partial fractions) as below. Capitals denote constants: their values are unique.

(a) *Simple factors*. To each factor $ax + b$ of $Q(x)$, a term $K/(ax + b)$.

(b) *Repeated simple factors*. To each factor $(cx + d)^n$ of $Q(x)$, there are n terms:

$$L_1/(cx + d) + L_2/(cx + d)^2 + \cdots + L_n/(cx + d)^n.$$

(c) *Irreducible quadratic*. To each factor $px^2 + qx + r$ of $Q(x)$, a term $(Mx + N)/(px^2 + qx + r)$.   **(1.32)**

$P(x)$ is involved in these rules only to the extent that it will affect the values of the **coefficients** K etc. The following examples show how to determine the values of the coefficients.

Example 1.12 *Express $x/(x-1)(x+2)$ in partial fractions.*

We can use any convenient letters for the unknown coefficients in the terms. The denominator has two simple factors, $x-1$ and $x+2$, so (1.32a) says that the partial fractions must have the form

$$\frac{x}{(x-1)(x+2)} \equiv \frac{A}{x-1} + \frac{B}{x+2} \tag{i}$$

Multiply through by $(x-1)(x+2)$:

$$x = A(x+2) + B(x-1). \tag{ii}$$

The constants must be chosen so that this becomes an **identity**. An identity has to be true for any x, so if we put *any* value of x into (ii), the result must be correct. Any two substitutions of numbers for x form *two simultaneous equations* for the two unknown constants A and B. For example, if we put $x = -10$ and $x = 100$ we obtain

$$-10 = -8A - 11B, \quad 100 = 102A + 99B.$$

The numbers we chose are inconvenient, but according to (1.32) we get the same A and B whatever values of x we use. Therefore, *choose values that make the equations as simple as possible*:

$$x = -2 \quad \text{gives} \quad -2 = 0 - 3B, \quad \text{so} \quad B = \tfrac{2}{3},$$
$$x = 1 \quad \text{gives} \quad 1 = 3A + 0, \quad \text{so} \quad A = \tfrac{1}{3}.$$

Therefore, from (i),

$$\frac{x}{(x-1)(x+2)} \equiv \frac{\tfrac{1}{3}}{x-1} + \frac{\tfrac{2}{3}}{x+2}.$$

Example 1.13 *Express $(3x-1)/(2x+1)(x-1)^2$ in partial fractions.*

According to (1.32a,b),

$$\frac{3x-1}{(2x+1)(x-1)^2} \equiv \frac{A}{2x+1} + \frac{B}{x-1} + \frac{C}{(x-1)^2}. \tag{i}$$

Multiply by $(2x+1)(x-1)^2$ to give

$$3x - 1 = A(x-1)^2 + B(2x+1)(x-1) + C(2x+1). \tag{ii}$$

We need three values of x to obtain the three equations for A, B, C. Obvious choices are $x = 1$ and $x = -\tfrac{1}{2}$. For the third, choose, say, $x = 0$. From (ii):

$$x = 1 \quad \text{gives} \quad 2 = 0 + 0 + 3C, \quad \text{so} \quad C = \tfrac{2}{3},$$
$$x = -\tfrac{1}{2} \quad \text{gives} \quad -\tfrac{5}{2} = \tfrac{9}{4}A + 0 + 0, \quad \text{so} \quad A = -\tfrac{10}{9},$$
$$x = 0 \quad \text{gives} \quad -1 = A - B + C, \quad \text{so} \quad B = 1 + A + C = \tfrac{5}{9}.$$

Finally,

$$\frac{3x-1}{(2x+1)(x-1)^2} \equiv -\frac{10}{9}\frac{1}{2x+1} + \frac{5}{9}\frac{1}{x-1} + \frac{2}{3}\frac{1}{(x-1)^2}.$$

Example 1.14   *Express $1/x(x^2 + 1)$ in partial fractions.*

Here, $x^2 + 1$ is an irreducible quadratic; so, by (1.32c),

$$\frac{1}{x(x^2 + 1)} \equiv \frac{A}{x} + \frac{Bx + C}{x^2 + 1}.$$

Multiply by $x(x^2 + 1)$:

$$1 = A(x^2 + 1) + (Bx + C)x \qquad \text{(i)}$$

Then $x = 0$ gives $1 = A + 0$, so

$$A = 1.$$

There are no other very easy values of x to choose. Put the value of A just found into (i) and rearrange: we get

$$-x = Bx + C. \qquad \text{(ii)}$$

It is easiest just to notice that (ii) is satisfied for all x only if

$$B = -1 \quad \text{and} \quad C = 0.$$

Therefore

$$\frac{1}{x(x^2 + 1)} \equiv \frac{1}{x} - \frac{x}{x^2 + 1}.$$

If **the degree of the numerator is greater than or equal to the degree of the denominator,** the case is not covered by (1.32), but we can treat it as follows.

Example 1.15   *Put $(x^3 + 1)/x(x - 1)$ into the form of a polynomial plus partial fractions.*

Carry out polynomial division, until the remainder is of lower degree than the divisor:

$$
\begin{array}{r}
x + 1 \\
x^2 - x \,\overline{\big)\, x^3 + 1} \\
\text{subtract} \quad x^3 - x^2 \\
\hline
x^2 + 1 \\
\text{subtract} \quad x^2 - x \\
\hline
\text{remainder} \quad x + 1
\end{array}
$$

Therefore

$$\frac{x^3 + 1}{x(x - 1)} \equiv x + 1 + \frac{x + 1}{x(x - 1)}.$$

The last term is of the right type for partial fractions, and finally

$$\frac{x^3 + 1}{x(x - 1)} \equiv x + 1 - \frac{1}{x} + \frac{2}{x - 1}.$$

1.15 Summation sign: geometric series

The sign Σ (sigma) is a large Greek capital S, standing for 'the sum of ...'. It is used in the following way. Suppose, for example, we are provided with a string of six quantities indexed in order, say

$$u_1, u_2, u_3, \ldots, u_6.$$

This is called a **sequence** consisting of six **terms**. We can denote the **general term** by (say) u_n, where n takes values from 1 to 6. Suppose we want to add them all up. Then

$$u_1 + u_2 + u_3 + u_4 + u_5 + u_6$$

is denoted by

$$\sum_{n=1}^{6} u_n,$$

which is read 'the sum of all the u_n from $n = 1$ through $n = 6$'. Similarly

$$u_2 + u_3 + u_4 + u_5 \quad \text{is written} \quad \sum_{n=2}^{5} u_n.$$

Any letter can be used as the **counting index** instead of n, provided that there is no conflict; so we could also write, for instance,

$$u_3 + u_4 + u_5 + u_6 = \sum_{i=3}^{6} u_i.$$

The letters m, n, r, i, j, k are often used.

We index a sequence according to convenience. The first index does not have to be 1. For example, consider the important sequence

$$1, x, x^2, x^3, \ldots,$$

which is the same as

$$x^0, x^1, x^2, x^3, \ldots.$$

This is called, for historical reasons, a **geometric sequence** or **geometric progression**. Each term in turn is got from its predecessor by multiplying by the **common ratio** x. The natural way to index such a sequence is to start with $n = 0$ instead of $n = 1$. Suppose then we want the sum of the first six terms. It can be expressed as

$$1 + x + x^2 + x^3 + x^4 + x^5 = \sum_{n=0}^{5} x^n,$$

though we could express the sum as

$$\sum_{n=1}^{6} x^{n-1}$$

instead. Such a sum, whether or not it starts with the x^0, or constant, term, is called a **geometric series**.

We will obtain an expression for the sum S of a geometric series having any value of the common ratio x (except $x = 1$) and which runs from the term in x^0 to the term in x^N. Thus $S = \sum_{n=0}^{N} x^n$. Note that it contains $N + 1$ **terms** (i.e. not N terms). Written at length:

$$S = 1 + x + x^2 + \cdots + x^{N-1} + x^N.$$

Then

$$xS = x + x^2 + x^3 + \cdots + x^N + x^{N+1}.$$

Subtract the second line from the first. All the terms cancel except for two; we obtain

$$S(1 - x) = 1 - x^{N+1}, \quad \text{so} \quad S = (1 - x^{N+1})/(1 - x).$$

Sum of a geometric series

$$\sum_{n=0}^{N} x^n = 1 + x + x^2 + \cdots + x^N$$

$$= \frac{1 - x^{N+1}}{1 - x}, \ (x \neq 1).$$

(1.33)

Example 1.16   *Find the following sums.* (a) $\sum_{n=0}^{4} (0.1)^n$, (b) $\sum_{n=0}^{6} \frac{1}{2^n}$, (c) $\sum_{n=0}^{N} e^{nx}$, (d) $\sum_{n=0}^{N} (-1)^n$, (e) $\sum_{n=0}^{5} 2^n$.

(a) $x = 0.1$ and $N = 4$, so

$$S = \frac{1 - (0.1)^5}{1 - (0.1)} = \frac{0.999\,99}{0.9} = 1.1111$$

(as becomes obvious if you write out the terms individually).

(b) $1/2^n = (\tfrac{1}{2})^n$, so $x = \tfrac{1}{2}$, and $N = 6$. By (1.33),

$$S = \frac{1 - (\tfrac{1}{2})^7}{1 - \tfrac{1}{2}} = 2(1 - \tfrac{1}{128}) = \tfrac{127}{64}.$$

(c) $e^{nx} = (e^x)^n$, so the common ratio is e^x, in place of x in (1.33):

$$S = \frac{1 - (e^x)^{N+1}}{1 - e^x} = \frac{1 - e^{(N+1)x}}{1 - e^x}.$$

Example 1.16 *continued*

(d) Here $x = -1$, so

$$S = \frac{1 - (-1)^{N+1}}{1 - (-1)} = \tfrac{1}{2}[1 - (-1)^{N+1}].$$

The sums of $N = 1, 2, 3, 4, \ldots$ terms of the sequence are successively 1, 0, 1, 0, 1, … .

(e) $x = 2$ and $N = 5$. Therefore

$$1 + 2 + 2^2 + 2^3 + 2^4 + 2^5 = \frac{1 - 2^6}{1 - 2} = 63.$$

Example 1.17 *Find an expression for the sum S of*

$$ar^2 + ar^5 + ar^8 + ar^{11} + ar^{14}.$$

This can be written as

$$ar^2(1 + r^3 + r^6 + r^9 + r^{12}).$$

The brackets contain a series of type (1.33), with common ratio r^3. The number of terms $N + 1$ is equal to 5, so that $N = 4$. Then

$$S = ar^2\frac{1 - (r^3)^5}{1 - (r^3)} = ar^2\frac{1 - r^{15}}{1 - r^3}.$$

(In terms of the Σ notation, we have $S = \sum\limits_{n=0}^{4} ar^{2+3n}$. It is perhaps easier to see what to do when the series is written out fully.)

1.16 Infinite geometric series

From (1.33), take the series to N (not $N + 1$) terms, giving

$$S_N = 1 + x + x^2 + \cdots + x^{N-1} = \frac{1}{1 - x} - \frac{x^N}{1 - x}. \tag{1.34}$$

Suppose firstly that the absolute value of x, or $|x|$, is less than 1, meaning $-1 < x < 1$. We shall see what happens when we continue the series to take in more and more terms to obtain (in imagination) an infinite number of terms.

The first term in (1.34), $1/(1 - x)$, is the same for all N. But for $-1 < x < 1$, the second term approaches zero as N increases to infinity:

$$\frac{x^N}{1 - x} \to 0 \text{ as } N \to \infty, \tag{1.35}$$

because $x^N \to 0$ as $N \to \infty$. This can be illustrated numerically by taking any specific value for x in the range $-1 < x < 1$, say $x = 0.1$. The behaviour of x^N as N increases is shown in the following table:

N	1	2	3	4	…
x^N	0.1	0.01	0.001	0.0001	…

The sequence of terms $1, x, x^2, x^3, x^4, x^5, \ldots$ is formed by multiplying each term in turn by x to get the succeeding term, so if $|x| < 1$, the terms become steadily smaller in magnitude, and in fact (though we do not prove it here) can be made as close as we wish to zero if we take a large enough value of N.

Therefore, referring back to (1.34),

$$\text{if } |x| < 1, \text{ then } S_N \to \frac{1}{1-x} \text{ as } N \to \infty.$$

In this way the idea of an infinitely extended geometric series can be given a meaning. Its **sum to infinity**, S_∞, is expressed by

$$S_\infty = \sum_{n=0}^{\infty} x^n = \frac{1}{1-x}. \tag{1.36}$$

On the other hand, if $|x| > 1$, the *magnitude* of the term $x^{N+1}/(1-x)$ in (1.34) will increase to infinity as N increases, so the infinite series cannot be said to have a sum at all. If $x = 1$, the series becomes

$$1 + 1 + 1 + \cdots,$$

which simply continues to grow to infinity as the number of terms increases. The case of $x = -1$ is indeterminate. To summarize:

Geometric series: sum to infinity

The geometric series $1 + x + x^2 + \cdots$ has a sum to infinity S_∞ if, and only if, $-1 < x < 1$; then

$$S_\infty = \sum_{n=0}^{\infty} x^n = \frac{1}{1-x}.$$

(1.37)

The second term on the right in (1.34) is called the **remainder** or **error**: it represents the error incurred by using only the first N terms to approximate to the infinite sum. For the infinite series to be useful, this quantity must approach zero as N approaches infinity.

Example 1.18   *Express the recurring decimals* (a) 0.4444...,
(b) 0.969 696..., *in the form of fractions.*

(a) The decimal can be written as an infinite geometric series:

$$\frac{4}{10} + \frac{4}{100} + \frac{4}{1000} + \cdots = \frac{4}{10}\left(1 + \frac{1}{10} + \frac{1}{100} + \cdots\right) = \frac{4}{10}\frac{1}{1-(1/10)} = \frac{4}{9}$$

from (1.36).
(b) In a similar way

$$\frac{96}{10^2} + \frac{96}{10^4} + \frac{96}{10^6} + \cdots = \frac{96}{10^2}\left(1 + \frac{1}{10^2} + \frac{1}{10^4} + \cdots\right).$$

The series in the brackets has the common ratio $x = 1/10^2$ and by using (1.36) the sum to infinity is $1/[1 - (1/10^2)] = 100/99$. Therefore the decimal is equivalent to 96/99. (Such results can be verified by 'long division'.)

1.17 Permutations and combinations

Suppose that we have four different letters A, B, C, and D, from which we may form **strings** of several letters that we shall call **words** (we shall not require them to be real words) such as BCA, AA, $BBCDA$. The order is important: AB is a different words from BA.

Consider the number of two-letter words that we can form from A, B, C, D if we may *repeat any letter*. There are 16 possible words; listed systematically they are:

AA AB AC AD; BA BB BC BD; CA CB CC CD; DA DB DC DD.

We can obtain the *number of words* without writing them all down by using the following argument. Imagine that we *are* writing them down; then we have four alternatives for the first letter, and following *each* of those choices we have four alternatives for the second letter. Altogether, then,

number of two-letter words $= 4 \times 4 = 4^2 = 16$.

To find the number of five-letter words which can be made up from A, B, C, D, with any number of repetitions allowed within a word, we follow the same reasoning as before, pursuing it through five letters, and obtain

$4 \times 4 \times 4 \times 4 \times 4 = 4^5 = 1024$ different words.

Suppose that repetition of letters within a word is *not* allowed. Or we may imagine n different 'objects', such that we cannot pull duplicates out of the air, as we did with the letters in the discussion above. The objects for selection might be wooden letters or numbers, or a pile of books with different titles, or a group of soldiers, etc. If we select $r \leqslant n$ of the n objects and place them in a certain order, we are said in this case to have a **permutation of r distinct objects, taken from among n distinct objects**. The number of possible permutations is signified by

$$_nP_r.$$

Example 1.19 *How many four-letter words can be made out of the six letters A, B, C, D, E, F, with no repetitions within a word?*

Put another way, how many permutations are there of $n = 6$ distinct objects, taken $r = 4$ at a time, or what is $_6P_4$? There are six choices for the first letter. With each such choice there are only five letters available for the second (no repetition), so there are 6×5 possible choices for the first two letters. There are four letters left to supply the third letter, so there are $6 \times 5 \times 4$ possibilities for the first three letters, and finally

$_6P_4 = 6 \times 5 \times 4 \times 3 = 360$.

Example 1.20 *There are six different books and we must choose one book for each of four children as a present. How many different distributions of books to the children are possible?*

A decision is required as to whether to distinguish by letter the children, or the books, or both. We choose the children, distinguished by W, X, Y, Z, say.

Imagine that we are listing all the possibilities. Child W may receive any one of six books. Whichever one W receives, X will have one of the remaining five; Y will have one of the remaining four; and Z one of the remaining three. The number of entries in our list is therefore

$$_6P_4 = 6 \times 5 \times 4 \times 3 = 360.$$

It may seem at first sight that this treatment favours child W and that child Z is shabbily treated. However, the process describes a systematic way to list *all possible* assignments, only instead of writing them down, we count them. No *choice* of any gift is involved.

Guided by this discussion we can now obtain a formula for $_nP_r$. We firstly need the **factorial notation** $n!$. If $n > 0$ is a positive integer, then the meaning of $n!$ is

$$n! = n(n-1)(n-2) \ldots 2 \cdot 1 \tag{1.38a}$$

(or alternatively $n! = 1 \cdot 2 \cdot 3 \cdot \ldots \cdot n$). We shall need the identity

$$n(n-1)(n-2) \ldots (n-r+1) = \frac{n!}{(n-r)!} \tag{1.38b}$$

for $0 < r < n$, in which *the left-hand side contains r factors*. If $r = n$ in (1.38b), the expression $n!/(n-r)!$ on the right becomes $n!/0!$, and $0!$ has no natural meaning. However, its value is **defined** to equal unity:

$$0! = 1, \tag{1.38c}$$

and then we can use formulae such as (1.38b) without making an exceptional case for $r = n$.

Permutations

The number of possible permutations of r objects, $1 \leqslant r \leqslant n$, taken without repetition from among n distinct objects is given by

$$_nP_r = n(n-1)(n-2) \ldots (n-r+1) = \frac{n!}{(n-r)!}. \tag{1.39}$$

Proof. There are n possibilities for the first place in a permutation. With each of these, the second can contain any of the remaining $(n-1)$ objects, so that there are $n(n-1)$ possibilities for the first two places. The third place can contain any of the remaining $(n-2)$ objects, so there are $n(n-1)(n-2)$ possibilities for the first three places. This continues until we have completed r factors corresponding to the r entries in each permutation; these form the product $n(n-1)(n-2) \dots (n-r+1)$. Then use (1.38b).

Example 1.21 *How many distinct five-letter permutations can be formed by using one A, one B, and three Cs?*

Method (i) (by enumerating the possibilities). Consider firstly the permutations in which A precedes B; the Cs are distributed in all possible ways among the positions marked * in *A*B*. Take the possible distributions of the Cs among the positions case by case:

- The Cs may be adjacent as *CCC*; there are three positions for the group.
- Two adjacent Cs and one separate C; there are three positions for *CC*, and with each there are two positions for the C, giving $3 \times 2 = 6$ possibilities.
- Three separated Cs; there is only one possibility.

Therefore there are $3 + 6 + 1 = 10$ permutations in which A precedes B. Similarly there are a further 10 permutations in which B precedes A, so the total number of permutations is $10 + 10 = 20$.

Method (ii). Make the three Cs distinct in the problem by calling them C_1, C_2, C_3. There are $_5P_5 = 5!$ distinct permutations of $ABC_1C_2C_3$. Imagine that they are all listed. If, in this list, we restore C by putting $C_1 = C_2 = C_3 = C$, we shall see many replications of the same 'word'. For example, consider the entry $C_3C_2BC_1A$. This reduces to $CCBCA$, but so does $C_1C_2BC_3A$, and several others. In fact the number of replications corresponding to *any* distinct word is $_3P_3 = 3! = 6$: this is the number of permutations of the symbols C_1, C_2, C_3, all of which are equivalent. Therefore there are 3! times as many entries in the list of permutations of $ABC_1C_2C_3$ as there are in the list of N (say) distinct words made up from $ABCCC$. Therefore, $3!N = 5!$, and

$$N = 5!/3! = 5 \times 4 = 20.$$

The following example shows Method (ii) in use for a more complicated case.

Example 1.22 *How many distinct permutations exist which use all the 14 letters in the word ASSASSINATIONS?*

It does not matter which string, or anagram, we treat as a source of letters, so start instead from one which displays the repetitions clearly:

$$SSSSSAAAIINNOT. \tag{i}$$

We shall enforce a distinction between the repeated letters of each type by indexing them:

$$S_1S_2S_3S_4S_5A_1A_2A_3I_1I_2N_1N_2OT. \tag{ii}$$

There are $5! \times 3! \times 2! \times 2!$ permutations *within* the indexed groups in (ii) which all correspond to the same word (i). Similarly, there are $5! \times 3! \times 2! \times 2!$ rearrangements of indexed letters corresponding to *any distinct* permutation of the ordinary letters. There are alltogether $_{14}P_{14} = 14!$ permutations of the indexed letters. Therefore, if the number of distinct permutations of the letters in *ASSASSINATIONS* is denoted by n,

↗

Example 1.22 *continued*

number of permutations of the indexed symbols in (ii)
$= 14! = (5! \times 3! \times 2! \times 2!)n$.

Finally

$n = 14!/(5! \times 3! \times 2! \times 2!) = 30\,270\,240$.

Example 1.23   (Circular permutations) *Five people sit round a circular table. In how many distinct orders may they sit?*

The meaning of 'distinct' here is that two arrangements are regarded as being the same if each person has the same person on his or her right (or left would do as well). If the people are named A, B, C, D, E, then rotation of a particular grouping, say *BADEC*, bodily around the table does not count as a new circular permutation: go clockwise (say) from any *fixed* position noting the order of seating; then *BADEC, ADECB, DECBA, ECBAD, CBADE* are to be treated as the same permutation.

The number of ordinary permutations is 5!. Let the number of circular permutations be N_C. To each of the circular permutations there are five ordinary permutations, so that $5N_C = 5!$, and finally

$N_C = 5!/5 = 4! = 24$.

In general, if there are n persons, the number of circular permutations is $(n - 1)!$.

Permutations are **sequences**: if the order of the elements is changed the permutation is counted as a different one. We shall now consider problems in which rearrangements of the same group, collection, or **set** of objects are regarded as equivalent: what defines the set is simply which items it contains, without regard to order. Such a set is called a **combination**. For example, an apple (A), a banana (B), and a carrot (C) in a plastic bag can be regarded as a mere combination of purchases, but the decision to consume them in a certain time order involves consideration of the possible permutations ABC, BAC, and so on.

Suppose there are six distinct objects, A, B, C, D, E, F, and we want to count how many different *combinations* consisting of three elements can be selected. We denote this number by $_6C_3$. A typical combination is any group containing A, B, and C. If we were laying out all the *permutations* our list would include $3! = 6$ entries all arising from the particular combination A, B, and C, namely

$ABC, ACB, BAC, BCA, CBA, CAB$.

We obtain the same factor 3! from every combination of three elements, so in this case there are 3! times as many permutations as there are combinations. Therefore (using (1.39)) the number of different combinations of three objects drawn from six objects is equal to

$$_6C_3 = \frac{_6P_3}{3!} = \frac{6!}{3!(6-3)!} = 20.$$

If we have a set of n distinct objects from which we may select groups of size $r \leqslant n$, the number of combinations is denoted by $_nC_r$. The same argument applies in the general case, and we obtain

---

Combinations

The number of combinations of r distinct objects $(1 \leqslant r \leqslant n)$ selected from n distinct objects is given by

$$_nC_r = \frac{_nP_r}{r!} = \frac{n!}{r!(n-r)!}.$$

(1.40)

Example 1.24 *Find the number of possible combinations made up by selecting some or all of n different objects without repetition of any object.*

To form all combinations of arbitrary size up to n, look at each object in turn, and either take it or reject it, and do this in every possible way. There are two options with regard to the first object; with each of these there are two options for the second object; and so on until we arrive at the nth object. There are 2^n such options in all, but these include rejection of all the objects. The total number of combinations is therefore $2^n - 1$.

Alternatively, we may add the number of combinations consisting of 1 object, 2 objects, ... , n objects, which is

$$_nC_1 + {}_nC_2 + {}_nC_3 + \cdots + {}_nC_n.$$

It will be shown in Example 1.28 that this sum is in fact equal to $2^n - 1$.

Frequently, the permutations/combinations occurring in problems involving *repetition* of objects may, with advantage, be broken up into **types**, which can be considered separately as in Example 1.25.

Example 1.25 *A lucky-dip jar contains seven sweets; there are one each of flavours A, B, C, and D, and three of flavour E. A child reaches into the jar and pulls out four sweets. How many distinct combinations of flavours might the child obtain?*

The combinations may contain three Es, two Es, one E, or no Es:

three Es: one other choice out of four; possible combinations 4.
two Es: two other choices out of four; possible combinations $_4C_2 = 4!/(2!2!) = 6$.
one E: three other choices out of four; possible combinations $_4C_3 = 4!/(3!1!) = 4$.
No Es: four choices out of four; possible combinations 1.

The total number of *distinct* combinations is $4 + 6 + 4 + 1 = 15$.

Example 1.26   *Prove that* (a) $_nC_r = {}_nC_{n-r}$, (b) $_nC_r = {}_{n-1}C_r + {}_{n-1}C_{r-1}$.

(a) From (1.40) $_nC_r = n!/[r!(n-r)!]$ and $_nC_{n-r} = n!/[(n-r)!r!]$.

(b) Starting with the right-hand side

$$_{n-1}C_r + {}_{n-1}C_{r-1} = (n-1)!/[r!(n-r-1)!] + (n-1)!/[(r-1)!(n-r)!]$$
$$= \{(n-1)!/[(r-1)!(n-r-1)!]\} \, [1/r + 1/(n-r)]$$
$$= \{(n-1)!/[(r-1)!(n-r-1)!]\} \, \{n/[r(n-r)]\}$$
$$= n!/[r!(n-r)!] = {}_nC_r.$$

(This result is used in Section 1.18 on the binomial theorem.)

1.18   The binomial theorem

Consider expressions of the form

$$(1+x)^n,$$

where n is a positive integer. For small integers n we can expand $(1+x)^n$ as a polynomial by multiplying it out; for the first few powers we obtain

$$(1+x)^2 = 1 + 2x + x^2,$$
$$(1+x)^3 = 1 + 3x + 3x^2 + x^3,$$
$$(1+x)^4 = 1 + 4x + 6x^2 + 4x^3 + x^4,$$

and so on.

The numbers occurring in the polynomials above (including the first number '1') are called the **binomial coefficients** of the successive powers of x, namely of x^0 (which equals 1), x^1, x^2, x^3, and so on, in order. For example, in the expansion of $(1+x)^4$, the coefficients can be listed as follows:

power of x	0	1	2	3	4
coefficient	1	4	6	4	1

Notice that there are $n+1$ terms in the expansion of $(1+x)^n$, and that the coefficients have a symmetrical pattern, coefficients equidistant from the ends being equal. We shall show later that these properties hold for all positive integer powers n.

The process of repeated multiplication soon becomes arduous. We firstly describe a more efficient method for powers given numerically, and secondly obtain an explicit formula for the coefficients.

Consider, for example, how we obtain the coefficients of $(1+x)^5$ from those of $(1+x)^4$. We have $(1+x)^5 = (1+x)^4(1+x)$. Lay the calculation out like a long multiplication sum:

$$\begin{array}{ll} (1+x)^4: & 1 + \underline{4}x + \underline{6}x^2 + 4x^3 + x^4 \\ \times (1+x): & \underline{1 + x} \\ & 1 + 4x + \underline{6}x^2 + 4x^3 + x^4 \\ & \quad x(1 + \underline{4}x + 6x^2 + 4x^3 + x^4) \\ = (1+x)^5: & 1 + 5x + \underline{10}x^2 + 10x^3 + 5x^4 + x^5. \end{array}$$

Notice how the coefficient of x^2 in $(1+x)^5$ is arrived at: it is the sum of the coefficients of x^2 and x in $(1+x)^4$ (underlined in the sum above). Similarly, the coefficient of x^3 in $(1+x)^5$ is equal to the sum of the coefficients of x^3 and x^2 in $(1+x)^4$, and so on, the only exceptions to the rule being the first and last coefficients, which are both equal to 1. The same rule applies whenever we calculate $(1+x)^{n+1}$ from $(1+x)^n$:

> **Coefficient rule in the binomial theorem**
>
> For $1 < r \leq n-1$, the coefficient of x^r in $(1+x)^n$ = the sum of the coefficients of x^r and x^{r-1} in $(1+x)^{n-1}$. (The first and last coefficients are equal to 1.)

By using this rule we can develop an efficient and rapid method, or **algorithm**, for obtaining the coefficients, called **Pascal's triangle**. It is a triangular array, as shown in (1.41), whose rows consist of the coefficients of $x^0, x^1, x^2, \dots$ in the expansion of $(1+x)^n$ for $n = 1, 2, \dots$. The rows are constructed successively. Each row is obtained from the preceding row by the rules just given: place a '1' at the beginning and end of each new row; then every intermediate entry in that row is equal to the sum of two entries from the previous row, one directly above and one to the left. Two instances are indicated in the table (1.41).

> **Pascal's triangle algorithm**
>
> For the coefficients in the expansion of $(1+x)^n$, n a positive integer:
>
> | power of x | 0 | 1 | 2 | 3 | 4 | 5 | $\dots$ |
> |---|---|---|---|---|---|---|---|
> | $n=1$ | 1 | 1 | | | | | |
> | $n=2$ | 1 | 2 | 1 | | | | |
> | $n=3$ | 1 | $\searrow$ 3 | 3 | 1 | | | |
> | $n=4$ | 1 | 4 | 6 | 4 | 1 | | |
> | $n=5$ | 1 | 5 | 10 | $\searrow$ 10 | 5 | 1 | |
>
> (The construction of two of the coefficients is indicated.)   **(1.41)**

Pascal's triangle works like positional notation in arithmetic. We do not write the number 'three hundred and sixty-five' in the form $5 + (6 \times 10) + (3 \times 10^2)$: the powers of 10 are implicit in the positions of the digits in the sequence 365. Similarly, in Pascal's triangle we temporarily hide the powers x^r, and only have to manipulate coefficients.

The procedure for deriving the nth row from the $(n-1)$th row implies that if the $(n-1)$th row is symmetrical, then the new nth row is also symmetrical. But the row for $n = 2$ is symmetrical; this symmetry is inherited by the row $n = 3$, and so for all subsequent rows.

If we know the coefficients for $(1 + x)^n$ then we can obtain the expansion of $(a + b)^n$ by writing it in the form

$$(a + b)^n = a^n[1 + (b/a)]^n.$$

Taking the case $n = 3$ as an illustration, put $a/b = x$, and use the expansion of $(1 + x)^3$ given above, with coefficients 1, 3, 3, 1:

$$(a + b)^3 = a^3[1 + 3(b/a) + 3(b/a)^2 + (b/a)^3]$$
$$= a^3 + 3a^2b + 3ab^2 + b^3. \tag{1.42}$$

We shall now prove the **binomial theorem**, which provides an explicit formula (rather than an algorithm) for the coefficients in $(a + b)^n$ where n is any positive integer. We shall start with the standard case $(1 + x)^n$.

All the essentials of the general result can be illustrated by the special case $n = 3$. Consider the following expansion, obtained by multiplication:

$$(1 + x_1)(1 + x_2)(1 + x_3)$$
$$= 1 + (x_1 + x_2 + x_3) + (x_1x_2 + x_2x_3 + x_1x_3) + (x_1x_2x_3). \tag{1.43}$$

Each term on the right is really the product of *three* elements (either 1s or xs), one from each of the three brackets on the left. Thus the first term is really $1 \times 1 \times 1$, and x_1x_3 arises from the product $x_1 \times 1 \times x_3$. The terms are then sorted into groups according to the number of x factors. A formula for the *number of terms in each group* can be obtained by using the result (1.40) of Section 1.17. For example, the number of terms having two x factors is equal to the number of ways of choosing two out of the three available x factors. This number is given by

$$_3C_2 = \frac{3!}{2!1!} = 3.$$

Similarly for the other groups: *for $r = 1, 2, 3$ the group containing the products of r x-factors has $_3C_2$ members.* Finally, suppose that all the x elements are made equal by putting

$$x_1 = x_2 = x_3 = x.$$

The bracketed expression in (1.43) containing the products of r of the xs, with $r = 1, 2,$ or 3, collapses into $_3C_rx^r$, so we obtain:

$$(1 + x)^3 = 1 + {}_3C_1x + {}_3C_2x^2 + {}_3C_3x^3.$$

We now prove the general result:

> **Binomial theorem**
>
> If n is a positive integer and a, b are any numbers:
>
> (a) $(1 + x)^n = 1 + {}_nC_1x + {}_nC_2x^2 + \cdots + {}_nC_{n-1}x^{n-1} + {}_nC_nx^n$.
> (b) $(a + b)^n = a^n + {}_nC_1a^{n-1}b + {}_nC_2a^{n-2}b^2 + \cdots + {}_nC_{n-1}ab^{n-1} + {}_nC_nb^n$,
>
> where
>
> $${}_nC_r = \frac{n!}{r!(n-r)!} = \frac{n(n-1)\ldots(n-r+1)}{r!}.$$
>
> The notation
>
> $$\binom{n}{r} = {}_nC_r$$
>
> is also used for binomial coefficients. **(1.44)**

Proof. (a) When expanded

$$(1 + x_1)(1 + x_2)(1 + x_3) \ldots (1 + x_n)$$

$$= 1 + \sum_{r=1}^{n} \text{(all the different } r\text{-fold products of the } xs).$$ **(1.45)**

(The meaning of 'an r-fold product' is one containing exactly r different xs: thus x_px_q, $p \neq q$, is a two-fold product.) For each fixed value of r, the number of r-fold products is equal to the number, ${}_nC_r$, of combinations of r objects selected without repetition from the n different objects x_1 to x_n.

Now put

$$x_1 = x_2 = x_3 = \cdots = x_n = x$$

into (1.45). For each value of r, the sum of the r-fold products collapses into the form

$$(x^r + x^r + x^r + \cdots \text{ to } {}_nC_r \text{ terms}) = {}_nC_rx^r.$$

(b) To obtain the expansion of $(a + b)^n$, we follow the process that led to eqn (1.42), namely

$$(a + b)^n = a^n[1 + (b/a)]^n,$$

and use (1.44a) with $x = b/a$; this becomes

$$(a + b)^n = a^n\left[1 + {}_nC_1\left(\frac{b}{a}\right) + {}_nC_2\left(\frac{b}{a}\right)^2 + \cdots + {}_nC_{n-1}\left(\frac{b}{a}\right)^{n-1} + \left(\frac{b}{a}\right)^n\right].$$

After removing the brackets the sum is as given in (b). An alternative proof of this theorem, using a calculus method, is given at the end of Chapter 4.

Example 1.27   *Expand and simplify the expression* $(x + x^{-1})^6$
$(x \neq 0)$.

It is simplest to add a line to Pascal's triangle (eqn (1.41)) to obtain the coefficients, rather than to use the general form of the binomial theorem (1.44):

power of x:	0	1	2	3	4	5	6	...
$n = 5$	1	5	10	10	5	1		
$n = 6$	1	6	15	20	15	6	1	

Then

$$(x + x^{-1})^6 = x^6(1 + x^{-2})^6$$
$$= x^6[1 + 6(x^{-2}) + 15(x^{-2})^2 + 20(x^{-2})^3 + 15(x^{-2})^4 + 6(x^{-2})^5 + (x^{-2})^6]$$
$$= x^6[1 + 6x^{-2} + 15x^{-4} + 20x^{-6} + 15x^{-8} + 6x^{-10} + x^{-12}]$$
$$= x^6 + 6x^4 + 15x^2 + 20 + 15x^{-2} + 6x^{-4} + x^{-6}.$$

Example 1.28   *Prove that* $_nC_1 + {}_nC_2 + {}_nC_3 + \cdots + {}_nC_n = 2^n - 1$.

Put $x = 1$ into the expansion of $(1 + x)^n$ given in (1.44a). It becomes

$$(1 + 1)^n = 2^n = 1 + {}_nC_1 + {}_nC_2 + {}_nC_3 + \cdots + {}_nC_n,$$

from which the result follows immediately.

Problems

1.1 (Section 1.3). Sketch graphs of the following equations over the intervals stated:
(a) $y = x^4, -1.5 \leqslant x \leqslant 1$;
(b) $y = x(1 - x), -1 \leqslant x \leqslant 2$;
(c) $y = 1 + x + x^2, |x - 1| \leqslant 2$;
(d) $y = |x - 1|, -3 \leqslant x \leqslant 3$;
(e) $y = |x| + |x - 3| + |x + 2|, -3 \leqslant x \leqslant 4$;
(f) $y = ||x| - 1|, -2 \leqslant x \leqslant 2$;
(g) $y = \sqrt{(x^2 + 1)}, |x| \leqslant 2$.

1.2 (Straight lines, Section 1.3). Find the straight lines through the following pairs of points:
(a) $(1, 1), (-1, 5)$;
(b) $(0, 1), (2, 1)$;
(c) $(2, 1), (-1, -1)$.
Sketch the triangle formed by these lines. Find the lengths of each side of the triangle.

1.3 (Straight line, Section 1.3). What are the slopes of the following straight lines, and where do they cut the coordinate axes?
(a) $y = x - 1$;   (b) $3y = x - 2$;
(c) $2x + 5y = 4$.

1.4 (Straight line, Section 1.3). Find the equations of the following straight lines:
(a) passing through $(1, 2)$ inclined at $45°$ to the x axis;
(b) passing through $(-1, -2)$ with slope -2;
(c) with slope 0.5 and x axis intercept $x = 1$;
(d) through $(1, 2)$ parallel to the line $y = 3x - 4$;
(e) through $(-1, 3)$ perpendicular to the line $y = 4x - 1$.

1.5 Show that the following pairs of lines are mutually perpendicular:
(a) $2x + 3y - 2 = 0$ and $-3x + 2y - 3 = 0$;
(b) $y = 2x + 1$ and $y = \frac{1}{2}(3 - x)$;
(c) $y = 2x$ and $y = -\frac{1}{2}x$;   (d) $y = \pm x$.

1.6 (Straight line, Section 1.3). Show that the equation

$$(x + y + 1) + \alpha(2x - 3y - 2) = 0,$$

where α is any constant, represents a straight line through the intersection of $x + y + 1 = 0$ and $2x - 3y - 2 = 0$. Find the line joining this point to the point $(1, 1)$.

1.7 (Circles, Section 1.3). Find the centre and radius for each of the following circles:

(a) $x^2 + y^2 = 9$;
(b) $(x-1)^2 + y^2 = 4$;
(c) $x^2 + y^2 - 2x - 2y - 21 = 0$;
(d) $4x^2 - 4x + 4y^2 + 4y = 9$.

1.8 (Circles, Section 1.3). Find the equation of the circle centred at $(1, -2)$ with radius 3.

1.9 Find the points of intersection of the following circles and lines:
(a) $x^2 + y^2 = 8$ and $x = 2$;
(b) $x^2 + y^2 - 2x + 2y - 4 = 0$ and $y = 2x + 1$;
(c) $x^2 + y^2 = 1$ and $x + y = \sqrt{2}$.

1.10 The following table contains experimental data:

x	1.06	0.84	0.72	0.44	0.23
y	0	0.53	0.71	0.78	1.1

The hypothesis is that the points should lie on a circle centre at the origin in the (x, y) plane. Find the distance of each point from the origin. Calculate the average of these values, and write down the equation of the circle approximation.

1.11 (Functions, Section 1.4). Draw sketches of the following functions in the (x, t) plane over the intervals indicated:
(a) $x = H(t+1) - H(t-1)$ for $-2 \leqslant t \leqslant 2$;
(b) $x = \text{sgn}(1+t) + \text{sgn}(1-t)$ for $-2 \leqslant t \leqslant 2$;
(c) $x = tH(t-1)$ for $0 \leqslant t \leqslant 2$;
(d) $x = (t^2 - 1)[\text{sgn}(t+1) + \text{sgn}(1-t)]$ for $-2 \leqslant t \leqslant 2$.

1.12 (Functions, Section 1.4). Using Heaviside and signum functions, construct a single formula in each case for $f(t)$ where

(a) $f(t) = \begin{cases} 0 & (t < -1), \\ 1 & (-1 < t < 2), \\ 0 & (t > 2); \end{cases}$

(b) $f(t) = \begin{cases} 0 & (t < 0), \\ 2t & (t > 0); \end{cases}$

(c) $f(t) = \begin{cases} 0 & (t < 0), \\ t & (0 < t < 1), \\ 1 & (1 < t < 2), \\ -t + 3 & (2 < t < 3), \\ 0 & (t > 3). \end{cases}$

1.13 (Section 1.5). What are the radian measures of the following angles: (a) 30°, (b) 120°?

1.14 (Trigonometric functions, Section 1.6). Using the methods of Examples 1.5 and 1.6, obtain
(a) $\sin \frac{1}{4}\pi$;   (b) $\sin \frac{1}{2}\pi$;   (c) $\sin \pi$;
(d) $\sin(-\frac{3}{4}\pi)$;   (e) $\cos \frac{1}{6}\pi$;   (f) $\cos \frac{5}{6}\pi$;
(g) $\sin -\frac{1}{3}\pi$;   (h) $\cos -\frac{2}{3}\pi$.

1.15 (Trigonometric functions, Section 1.6). Use (1.18c) to show that
(a) $\cos^4 A = \frac{1}{8}(3 + 4\cos 2A + \cos 4A)$;
(b) $\sin^4 A = \frac{1}{8}(3 - 4\cos 2A + \cos 4A)$.

1.16 (Trigonometric functions, Section 1.6). Use (1.17) to express the following in terms of $\sin x$ and $\cos x$:
(a) $\cos(x + \frac{1}{2}\pi)$;   (b) $\sin(x + \frac{1}{2}\pi)$;   (c) $\sin(x - \frac{1}{2}\pi)$;
(d) $\cos(x \pm \pi)$;   (e) $\sin(x \pm \pi)$.

1.17 (Trigonometric functions, Section 1.6). Use (1.17) to express the following in terms of the cos and sin of $\frac{1}{2}(x + y)$ and $\frac{1}{2}(x - y)$:
(a) $\cos x + \cos y$;   (b) $\sin x - \sin y$;   (c) $\cos x - \cos y$.

1.18 State where the graphs of the following functions cross the x axis:
(a) $\sin x$;   (b) $\cos x$;   (c) $\sin \frac{1}{2}x$;
(d) $\cos 3\pi x$;   (e) $\cos(2x - \frac{1}{2}\pi)$;   (f) $e^{-x} \sin \frac{1}{2}\pi x$.

1.19 State the amplitude, angular frequency, period, and phase of the following harmonic outputs:
(a) $2\cos(0.2t + 3.2)$;   (b) $1.5\sin[0.2(t - 2.4)]$;
(c) $2\cos(0.2t + 0.12) + 2\cos(0.2t - 0.39)$;   (d) $-\cos t$.

1.20 State the inverse $F(x)$ for each of the following functions $f(x)$ for the values of x stated:
(a) $4x^2, x \leqslant 0$;   (b) $2x + 3, -\infty < x < \infty$;
(c) $\sin 2x, 0 \leqslant x \leqslant \frac{1}{4}\pi$;
(d) $2\sin x, 0 \leqslant x \leqslant \frac{1}{2}\pi$;
(e) $\cos x^2, 0 \leqslant x \leqslant \sqrt{\pi}$;
(f) $\sin(\frac{1}{2}\pi \cos x), 0 \leqslant x \leqslant \frac{1}{2}\pi$;
(g) $x^{-\frac{1}{4}}, x > 0$;   (h) $x^2 + x, x > -\frac{1}{2}$.

1.21 Sketch a graph of the function inverse to the function $x^3 - x + 1$. (There is no need to try to solve the equation $x^3 - x + 1 = y$ for x.)

1.22 (Sections 1.10, 1.11). Solve the following equations for x:
(a) $e^{2x} = 3$;   (b) $\ln 3x = 2$;
(c) $\ln x^{-\frac{1}{3}} = 1$;   (d) $3 e^{3x} = 1$;
(e) $e^x + e^{-x} = 2$ (hint: multiply through by e^x first);
(f) $e^{\ln 2x} = 4$;   (g) $\ln e^{2x} + 3\ln e^{5x} = 2$;

(h) $\ln(x+1) + \ln(x-1) = 0$;
(i) $\ln(x+1) + \ln(x-1) = e$;
(j) $2^x = 3$;   (k) $3^{2x} = \frac{1}{3}$;
(l) $\sinh 2x = 4$;   (m) $2\sinh x = 2\cosh x + 3$.

1.23 Express 2^x as a power of e.

1.24 (Section 1.12). Prove that 10^x doubles its value in any interval of length equal to

$$\frac{\ln 2}{\ln 10}.$$

1.25 Sketch regions in the (x, y) plane defined by the following inequalities:
(a) $(x-1)^2 + y^2 \leqslant 9$;
(b) $x \geqslant 0$, $y \geqslant 0$, and $x + y \leqslant 1$;
(c) $\dfrac{x^2}{4} + \dfrac{y^2}{9} \leqslant 1$;
(d) $x^2 + y^2 \leqslant 1$ and $x \geqslant 0$;
(e) $|x| + |y| \leqslant 1$.

1.26 Prove that $\tanh^{-1}x = \frac{1}{2}\ln[(1 + x)/(1 - x)]$ for $-1 < x < 1$.

1.27 Figure 1.33 shows a cross-section of a simple model of a piston and crankshaft. The crankshaft rotates at 4000 rpm (revolutions per minute). If $AB = 2.5$ and $BC = 5$ (in cm), show that the displacement AC is given by

$$AC = 2.5[\sin \omega t + \surd(4 - \cos^2\omega t)]$$

(in cm), where t is measured from $\theta = 0$, and state ω in radians per second.

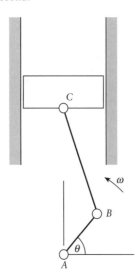

Fig. 1.33

1.28 An oscillation takes the form

$$x = 3 \cos \omega t + 4 \sin \omega t.$$

By finding numbers c and ϕ such that

$$c \cos \phi = 3, \quad c \sin \phi = 4$$

express x as a single cosine term. What are the amplitude and phase of the oscillation?

1.29 The exponential function $f(t) = C\,e^{-\alpha t}$ satisfies the conditions $f(0) = 2$ and $f(1) = 0.5$. Find the constants C and α. What is the value $f(2)$?

1.30 A yacht, which has a draught of 2 metres, is anchored in a tidal estuary, in which the depth of water around the yacht is

$$5 + 4.5 \sin 0.5t$$

(in metres), where the time t is measured in hours. What is the tidal period in hours? Over how many hours in one period can the yacht float free of the estuary?

1.31 Draw sketches of the graphs of the following curves given in polar coordinates, by constructing a table of values of r for equally spaced angles (say 15° intervals):
(a) the cardioid $r = 0.5(1 + \cos \theta)$;
(b) the folium $r = (4 \sin^2\theta - 1) \cos \theta$;
(c) the four-leaved rose $r = \sin 2\theta$;
(d) the Archimedean spiral $r = 0.04\theta$ (extend the interval in θ to $[0, 6\pi]$);
(e) the equiangular spiral $r = 0.1\,e^{0.1\theta}$ (extend the interval in θ to $[0, 6\pi]$).

1.32 Sketch the graphs of the following functions:
(a) sgn $\sin x$;   (b) sgn $\cos 2x$;
(c) $H(x) \sin x$;   (d) $\sin^2x$;
(e) $|\sin x|$;   (f) $\sin|x|$;
(g) $H(x - \pi)\sin x$.

1.33 The coordinates of three vertices of a rectangle are given by

$$(-7, 3), \quad (1, -3), \quad (4, 1).$$

Find the coordinates of the fourth vertex. Determine also the area of the rectangle.

1.34 State which of the following functions are periodic, and, if so, find the (minimum) period:
(a) $\sin 4x$;   (b) $\cos(\pi + t)$;   (c) $\sin t + \cos 2t$;
(d) $\sin(x^2)$;   (e) $e^{-\sin x}$;   (f) $\cos^2x$;
(g) $x \sin x$;   (h) $|\sin x|$;   (i) $1/(4 + \sin^2t)$;
(j) $\sin \frac{1}{4}t$;   (k) $\sin 3t + \cos 9t$;
(l) $\sin(\surd 2 t) + \sin t$ (note: $\surd 2$ is irrational).

1.35 Decide which of the following functions are even, odd, or neither even nor odd:
(a) $x^2 + x^3$;   (b) $x^2 + 2x^4$;
(c) $x + \sin x$;   (d) $\sin x \cos x$;
(e) e^{-x^2};   (f) $\ln(1 + x^2)$;
(g) $e^{\sin x}$.

1.36 (Partial fractions, Section 1.14). Express the following in partial fractions:

(a) $\dfrac{1}{(x - 2)(x + 3)}$;   (b) $\dfrac{x}{(x + 1)(x + 2)}$;

(c) $\dfrac{2x - 1}{x(x - 1)}$;   (d) $\dfrac{1}{x(x + 1)(x + 2)}$;

(e) $\dfrac{1}{x(x^2 - 1)}$;   (f) $\dfrac{1}{x(x + 2)^2}$;

(g) $\dfrac{x^2}{(x + 1)(x + 2)^2}$;   (h) $\dfrac{x - 1}{x^2 - 2x - 3}$;

(i) $\dfrac{1}{(x - 1)^2}$;   (j) $\dfrac{1}{x^3 + x^2}$.

1.37 (Partial fractions, Section 1.14). In the following problems with irreducible factors, express the functions in partial fractions:

(a) $\dfrac{1}{x(x^2 + x + 1)}$;

(b) $\dfrac{x}{(x - 1)(x^2 + 1)}$;

(c) $\dfrac{x}{(x + 1)(x^2 + 2x + 6)}$.

1.38 Express the following in partial fractions:

(a) $\dfrac{1}{x^2(x^2 + 1)}$;

(b) $\dfrac{x^3}{(x + 1)(x + 2)}$;

(c) $\dfrac{x^3}{(x^2 - 9)(x + 1)}$.

1.39 Write down in full all the terms in each of the series:

(a) $\displaystyle\sum_{j=2}^{5} 2^j$;   (b) $\displaystyle\sum_{n=0}^{4} \dfrac{1}{1 + n^2}$;

(c) $\displaystyle\sum_{n=1}^{4} nx^n$.

1.40 (Geometric series, Section 1.15). Using the geometric series formula find the sums of the following series:

(a) $\displaystyle\sum_{n=1}^{7} (0.5)^n$;   (b) $\displaystyle\sum_{n=2}^{6} (\tfrac{1}{3})^n$;   (c) $\displaystyle\sum_{n=0}^{5} e^{-2n}$;

(d) $\displaystyle\sum_{n=1}^{6} n2^n$;   (e) $\displaystyle\sum_{n=1}^{10} (-\tfrac{1}{2})^n$;

(f) $\displaystyle\sum_{n=0}^{6} [2(0.5)^n + 3(0.6)^n]$.

1.41 Find a formula for the sum of
$$x + x^5 + x^9 + \cdots + x^{1+4n} + \cdots + x^{41}.$$

1.42 ABC is a triangle with sides a, b, c opposite to the corresponding angles. Prove the **cosine rule** (see eqn (1.17d)) that
$$c^2 = a^2 + b^2 - 2ab \cos C.$$
(Hint: drop a perpendicular from B on to b. Look for $a \cos C$, and then for an opportunity to use Pythagoras's theorem.)

1.43 Let A, c, t_0, and T be any constants, and put
$$f(t) = A \, e^{ct}.$$
Show that the sequence
$$f(t_0), f(t_0 + T), f(t_0 + 2T), \ldots ,$$
is a geometric progression.

1.44 Express the following recurring decimals as fractions:
(a) $1.111\ldots$;   (b) $0.999\ldots$;
(c) $0.010\ 101\ 0\ldots$;   (d) $0.090\ 909\ 0\ldots$;
(e) $0.666\ldots$;   (f) $2.\dot{7}\dot{2}\ldots$.

1.45 Obtain the sum of the following infinite geometric series:

(a) $\displaystyle\sum_{r=0}^{\infty} (\tfrac{1}{2})^r$;   (b) $\displaystyle\sum_{r=0}^{\infty} (\tfrac{1}{10})^r$;

(c) $\displaystyle\sum_{r=0}^{\infty} e^{-r}$;   (d) $\displaystyle\sum_{r=0}^{\infty} (-1)^r (\tfrac{1}{2})^r$;

(e) $1 - \dfrac{2}{3} + \dfrac{4}{9} - \dfrac{8}{27} + \cdots$.

1.46 Calculate the numerical values of:
(a) $4!$, $6!$, and $7!$;
(b) $12!/11!$;
(c) $10!/7!$;
(d) $12!/(9!3!)$;
(e) $n!/[r!(n - r)!]$ when $n = 10$ and $r = 3$;
(f) $3!/[r!(3 - r)!]$ for each of the cases $r = 0, 1, 2, 3$.

1.47 (a) Simplify (i) $n!/(n-2)!$, (ii) $(n+1)!/(n-1)!$, where n is a positive integer.

(b) Express the following in terms of factorials: (i) $2 \times 4 \times 6 \times \cdots \times (2m)$, (ii) $1 \times 3 \times 5 \times \cdots \times (2m+1)$, where m is a positive integer.

1.48 (a) Calculate the numbers represented by (i) $_5P_4$; (ii) $_9P_3$; (iii) $_6P_3$; (iv) $_7C_3$; (v) $_7C_4$; (vi) $_{10}C_5$; (vii) $_{100}C_{98}$; (viii) $\binom{10}{7}$.

(b) Show that $_nP_n = {_nP_{n-1}}$, and explain why by using an example.

1.49 Given four letters A, B, C, D, obtain:
(a) the number of possible *permutations* of the four letters, without repetitions of letters within a permutation;
(b) the number of three-letter *combinations* of the letters, taken three at a time without repetitions;
(c) the number of distinct four-letter permutations, in which all possible repetitions within a permutation are allowed;
(d) the number of distinct three-letter combinations in which a letter may be repeated up to three times;
(e) the total number of permutations containing from one to four letters without repetition;
(f) the total number of distinct combinations containing from one to four letters, when up to three occurrences of a single letter is allowed.

1.50 Find:
(a) the number of distinct three-letter 'words' obtainable from the letters A, B, C, D, E, in which E may occur 0, 1, or 2 times, but the rest may occur only once;
(b) the possible number of distinct six-letter words in which E occurs exactly twice and the other letters only once.

1.51 (a) How many distinct four-digit numbers may be made up by using the digits 1, 2, 3, 4, 5, no digit being used more than once?
(b) How many of the numbers in (a) are divisible by 5?
(c) How many of the numbers in (a) are divisible by 2?
(d) How many distinct positive numbers are obtainable by using not more than four of the digits taken without repetition from the digits 0, 1, 2, 3, 4?

1.52 There are four women and three men eligible to fill four posts.
(a) What is the total number of distinct combinations of personnel that can be selected?
(b) Split up the combinations according to the number of men/women among them, and obtain the numbers in each such grouping. Check the total against the number in (a).

1.53 Suppose there is a collection of N objects of different **types** A, B, etc., all the separate types being distinct from the others in some way, but objects of a particular type are identical. There are N_A identical objects of type A, N_B identical objects of type B, and so on. Show that:
(a) (A generalization of Example 1.22.) the possible number of *distinct* permutations of the N objects is
$$\frac{N!}{N_A!N_B!N_C!\dots};$$
(b) the combined number of distinct combinations that may be formed out of 1, 2, … , and N of the objects is equal to
$$[(N_A+1)(N_B+1)\dots\,]-1.$$

1.54 (a) Five representatives from each of the countries France, Germany, Italy, and the UK are to be seated along one side of a long table. Each national group should sit together. In how many orders may the individuals be seated?

(b) Suppose that the table is circular with the representatives seated all round it. How many distinct orders are possible then? (Distinct permutations are to be understood as *circular permutations* in the sense of Example 1.27.)

1.55 Three prizes are to be distributed among 10 candidates. How many possible distinct distributions are there in the following cases?
(a) The prizes are all equal with at most one for any person.
(b) The prizes are all unequal, and only one may go to any person.
(c) All the prizes are equal, and any person may receive up to three prizes.
(d) As in (c), but the prizes are all different.

1.56 The field available for a seven-member committee consists of 14 people: 2 accountants, 3 lawyers, 5 doctors, and 4 social workers.
(a) How many committees with at least one member who is *either* an accountant *or* a lawyer may be formed?
(b) How many committees with at least 1 accountant *or* lawyer, at least 1 doctor, and at least 2 social workers may be formed?
(c) How many of the committees in (b) contain *exactly* 1 lawyer?

1.57 (a) Write from memory the binomial expansions of the expressions $(1+x)^n$ and $(a+b)^n$, where n is a positive integer.

(b) Expand the expression $(1 - x)^6$.

(c) Expand and simplfy $(x + x^{-1})^5$ and $(x - x^{-1})^5$ where $x \neq 0$.

1.58 Use the binomial theorem to show that $(1.01)^{10} \approx 1.105$. In a similar way, make an approximation to $(0.99)^8$.

1.59 By giving special values to the constants in the binomial theorem prove that, when n is a positive integer:

(a) $1 + 2\,_nC_1 + 2^2\,_nC_2 + \cdots + 2^n\,_nC_n = 3^n$, and
$1 - \,_nC_1 + \,_nC_2 - \cdots + (-1)^n\,_nC_n = 0$.

(b) $1 + \,_nC_2 + \,_nC_4 + \cdots = 2^{n-1} = \,_nC_1 + \,_nC_3 + \,_nC_5 + \cdots$.

1.60 Let

$$F(n, k) = \,_nC_0 + \,_{n+1}C_1 + \,_{n+2}C_2 + \cdots + \,_{n+k}C_k,$$

where n and k are positive integers. Show that $F(n, k) + \,_{n+k+1}C_{k+1} = F(n, k + 1)$. Check that this identity is satisfied by

$$F(n, k) = \,_{n+k+1}C_k,$$

and that $F(n, 0) = \,_{n+1}C_0$ and $F(n, 1) = \,_{n+2}C_1$ form the original series. Thus since the result is true for $k = 0$ and $k = 1$ it is true for all k by repeated use of the identity. This approach forms a basis of a **proof by induction** which is particularly useful when a result is known intuitively or there is strong circumstantial evidence for a formula.

1.61 Expand $1/(x^2 + 3x + 2)$ in powers of x using partial fractions.

1.62 (a) The value of a single investment of amount A grows by a constant fraction R in each completed year (the annual **growth rate**) from the time it was purchased. Show that its value V_N at the end of the N th completed year is $V_N = A(1 + R)^N$. Calculate V_N over 5, 10, and 15 years when $A = £1000$ and $r = 0.03$ (usually expressed as 3%).

(b) To value an investment when the time t from purchase is not a whole number of complete years the analogous formula $V_t = A(1 + R)^t$ is to be used, where t is measured in years. Show that over *every* period of T years the investment grows by a factor $(1 + R)^T$ (the proposed extension therefore has exactly the property we should hope for).

(c) Obtain the doubling period of the investment when $R = 3\%, 6\%$ and 9%. Obtain the 10-times period when $r = 6\%$.

1.63 Income from an investment is at a rate expressed as R per annum, but it is paid out monthly to the investor at a rate r per month on the current balance. Express r in terms of R. Why is $R > 12r$?

1.64 Money is borrowed from a finance company at an interest rate of r_M per month. What is the equivalent compounded rate per annum? Calculate the annual rate when the monthly rate is 1% and 3%.

1.65 (a) (Geometric series: a model savings scheme.) At the start of every year an amount A is put into a savings scheme. The interest on the current balance at the end of each complete year is reinvested, the (constant) annual rate being R. Show that the value V_N of the fund at the end of year N is given by $V_N = A(1 + R)[(1 + R)^N - 1]/R$.

(b) Calculate V_N after 10 years at 5% interest, the annual subscriptions being £100, and find the percentage gain on the total sum invested.

(c) Find the expression for the fund value if the saver contributes an amount 2A every 2 years, over a period of $2M$ years, where M is an integer. Obtain the value of the fund using the data in (b).

2

Differentiation

CONTENTS

2.1 The slope of a graph

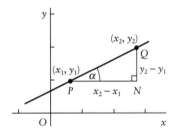

Fig. 2.1

Figure 2.1 shows the graph of a straight line. The x and y coordinates are assumed to have the same scale. Choose any two points $P : (x_1, y_1)$ and $Q : (x_2, y_2)$ which lie on the line. If we measure the angle α *from the positive x direction* then

$$\tan \alpha = \frac{y_2 - y_1}{x_2 - x_1} \tag{2.1}$$

(see (1.6)). The value of $\tan \alpha$ remains the same whether Q is to the right or left of P, since the value of the fraction on the right is unchanged. The angle α itself will differ in the two cases by an amount equal to π (or 180°), but this does not affect the value of $\tan \alpha$. (If we refer to α itself to indicate the steepness of a line, we choose the value that lies between ±90°, but normally we only need $\tan \alpha$.) Notice that if the x and y scales differed, the angle α as depicted would not satisfy (2.1); it would be too great or too small.

The **slope** or **gradient** of a straight line is defined to be the quantity $\tan \alpha$. If the line is horizontal, $\tan \alpha$ is zero. It is positive or negative according as the line slopes upwards or downwards as we go **from left to right**. It increases or decreases as the inclination increases or decreases, becoming ±∞ when $\alpha = \pm 90°$.

Consider now the **slope** or **gradient of a curve** at a point. Figure 2.2a shows a typical curve. By the **slope of the curve** at the point P we mean the **slope of the tangent line to the curve at P**. We can think of

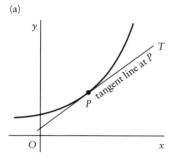

(a)

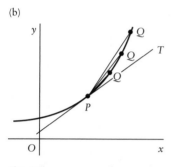

(b)

Fig. 2.2

the tangent line as the line joining two points on the curve which are 'infinitely close together', but it is no use making P and Q coincide, since we simply get $\tan \alpha = 0/0$, which has no definite meaning. It is necessary to carry out an indirect process.

Let P be the fixed point (see Fig. 2.2b). Take any other point Q on the curve and join PQ by a straight line, called the **chord** PQ. If Q is some distance from P, then the slope of PQ will not be close to that of PT, but if we take a succession of points Q closer and closer to P, then the slope of the chord PQ can be made as close as we wish to that of PT. The points Q that we consider are said to **approach** P. The corresponding value of the slope of PQ then **approaches a limit** or a **limiting value**, and this is equal to the slope of the curve at P. We use the sign $\rightarrow$ to signify 'approaches', so we can write:

as $Q \rightarrow P$, slope of $PQ \rightarrow$ slope of the curve at P. (2.2)

We shall be able to obtain the **exact value** of the slope of PT, which is the same as the slope of the curve at P, by carrying out the approach of Q to P in algebraic terms. To do this, we introduce a new symbol

$$\delta x$$

(pronounced 'delta-x'). This is a **single symbol**: the Greek letter δ stands for the words '**the increment in**' or '**the change in**' something, in this case, the increment in the value of x as we move from P to Q. For two points $P : (x_1, y_1)$ and $Q : (x_2, y_2)$, the change in x on moving from P to Q is

$$\delta x = x_2 - x_1. \tag{2.3}$$

(The change in moving from Q to P is given by $\delta x = x_1 - x_2$, which has the opposite sign.) Similarly we use the symbol δy to indicate the change in y as we move from P to Q.

In Fig. 2.3, $P : (x, y)$ is the point on a graph at which we want to find the slope (i.e. we want the slope of the tangent line PT). Take another point Q on the curve, and suppose that it approaches P. The separation between P and Q is indicated by the sides of length δx and δy of the triangle PNQ. Then, by (2.1),

$$\text{slope of } PQ = \frac{\delta y}{\delta x}. \tag{2.4}$$

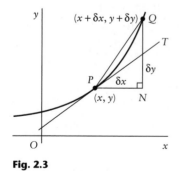

Fig. 2.3

Now let $\delta x \rightarrow 0$ so that $Q \rightarrow P$: the ratio $\delta y/\delta x$ approaches a number which is equal to the value of the slope at P. We first show what happens **numerically** in a particular case.

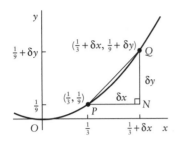

Fig. 2.4

Example 2.1 *Find the slope of the curve $y = x^2$ at the point*
$P : (\frac{1}{3}, \frac{1}{9})$ *on the curve* (Fig. 2.4).

At P, we have $x = \frac{1}{3}$ and $y = \frac{1}{9}$. We shall make a table of values of $\delta y/\delta x$ for
diminishing values of δx; that is, for points Q which are approaching P
(from either side). We first need to express δy in terms of δx:

$$\delta y = (y \text{ at } Q) - (y \text{ at } P) = (\tfrac{1}{3} + \delta x)^2 - (\tfrac{1}{3})^2$$

$$= (\tfrac{1}{9} + \tfrac{2}{3}\delta x + \delta x^2) - \tfrac{1}{9} = \tfrac{2}{3}\delta x + \delta x^2. \tag{2.5}$$

For Q to the left of P:

δx	-0.1	-0.001	-0.0001	...
δy	$-0.05\dot{6}$	$-0.006\,5\dot{6}$	$-0.000\,665\,\dot{6}$	...
$\dfrac{\delta y}{\delta x}$	$0.5\dot{6}$	$0.65\dot{6}$	$0.665\dot{6}$	...

For Q to the right of P:

δx	0.1	0.001	0.0001	...
δy	$0.07\dot{6}$	$0.006\,7\dot{6}$	$0.000\,667\,\dot{6}$	...
$\dfrac{\delta y}{\delta x}$	$0.7\dot{6}$	$0.67\dot{6}$	$0.667\dot{6}$	...

($\dot{6}$ means that the number 6 recurs: e.g. $0.\dot{6} = 0.666\,66...$; see Section 1.1)

First notice that if we put $\delta x = 0$ we obtain $\delta y/\delta x = 0/0$, which gives no
information at all, so we have to look at the *sequence of values* for $\delta y/\delta x$ as
$\delta x \to 0$. Inspection suggests that each term is formed from its predecessor
in a regular way, so we can predict that:

as $\delta x \to 0$, slope of $PQ \to 0.666\,66... = \tfrac{2}{3}$.

The **exact value of the slope** at P is therefore $\tfrac{2}{3}$.

We could have worked out the slope at P in this example without
doing any calculation. From (2.5),

$$\frac{\delta y}{\delta x} = \frac{\tfrac{2}{3}\delta x + \delta x^2}{\delta x} = \tfrac{2}{3} + \delta x$$

provided that $\delta x \neq 0$. Therefore, when $\delta x \to 0$, $\delta y/\delta x \to \tfrac{2}{3}$, as before.

Finally, in exactly the same way, we can find a general formula
giving the slope *at any point* $P : (x, y)$ on the graph of $y = x^2$. The
value of δy corresponding to a value of δx is given by

$$\delta y = (x + \delta x)^2 - x^2 = x^2 + 2x\,\delta x + \delta x^2 - x^2 = 2x\,\delta x + \delta x^2.$$

Therefore

$$\frac{\delta y}{\delta x} = \frac{2x\,\delta x + \delta x^2}{\delta x} = 2x + \delta x.$$

Now let $\delta x \to 0$; we obtain:

the slope of $y = x^2$ at (x, y) is $2x$. \tag{2.6}

This process is a model for treating other functions: for example,
you could now show in the same way that the slope of the graph of
$y = x^3$ at any point is equal to $3x^2$.

2.2 The derivative: notation and definition

We shall need to find the value approached by $\delta y/\delta x$ as $\delta x \to 0$ in many different situations. There is a special notation used to signify the process:

Limit notation

Let $y = f(x)$, so that $\delta y = f(x + \delta x) - f(x)$. Then the value approached by $\delta y/\delta x$ when $\delta x \to 0$ is denoted by

$$\lim_{\delta x \to 0} \frac{\delta y}{\delta x}.$$

(2.7)

Read this as 'The **limit**, or the **limiting value**, of $\delta y/\delta x$ as $\delta x \to 0$'. (The lim sign is used in many other contexts too.)

The result of the process $\lim_{\delta x \to 0} \delta y/\delta x$, where $\delta y = f(x + \delta x) - f(x)$, is called the **derivative of y with respect to x**, or **the derivative of $f(x)$**. The process is called **differentiation**. We worked out earlier that, if $y = f(x) = x^2$, then the derivative is equal to $2x$. The following notations are **standard short ways of indicating a derivative**:

$$\frac{dy}{dx}, \quad \frac{df(x)}{dx}, \quad \text{or} \quad \frac{d}{dx}f(x) \quad \text{signify} \quad \lim_{\delta x \to 0} \frac{\delta y}{\delta x}.$$

The symbol dy/dx is usually pronounced 'dee-y by dee-x'. Notice that the letter used is an ordinary d, not δ.

Derivative, slope, and tangent

(a) Let $y = f(x)$ and $\delta y = f(x + \delta x) - f(x)$. Then the derivative of y with respect to x, signified by

$$\frac{dy}{dx}, \quad \frac{df(x)}{dx}, \quad \frac{d}{dx}f(x), \quad \text{or} \quad \frac{df}{dx},$$

means the result of taking the limit of $\delta y/\delta x$ as $\delta x \to 0$:

$$\frac{dy}{dx} = \lim_{\delta x \to 0} \frac{\delta y}{\delta x}.$$

(b) The slope m of a curve at any point (x_0, y_0), where $y_0 = f(x_0)$, is given by

$$m = \left(\frac{dy}{dx}\right)_{x=x_0},$$

where the derivative is evaluated at $x = x_0$. Therefore the equation of the tangent line at the point is

$$\frac{y - y_0}{x - x_0} = \left(\frac{dy}{dx}\right)_{x=x_0}.$$

(2.8)

Thus our earlier result for $y = x^2$ can be written in several ways:

$$\frac{dy}{dx} \quad \text{or} \quad \frac{d(x^2)}{dx} \quad \text{or} \quad \frac{d}{dx}x^2 = 2x.$$

Strictly speaking, dy/dx should be regarded as a **single shorthand symbol** representing the longer expression $\lim_{\delta x \to 0} \delta y/\delta x$, and not as a ratio which can be taken to pieces. However, its great usefulness is that it often behaves just like an ordinary ratio of nonzero quantities, and we shall later see cases where this property guides us to true results and makes them easy to remember.

It is sometimes useful to think of the symbol

$$\frac{d}{dx}$$

standing alone as meaning: 'differentiate what follows'. It is also called an **operator**, meaning that we operate on one function (x^2 say) to produce another (i.e. $2x$). Sometimes the symbol D is used to stand for the operator d/dx. We would then write

$$Dx^2 = 2x.$$

The **normal** to a curve at a point $x = x_0$ is the straight line which is perpendicular to the tangent at the point. If m is the slope of the tangent then, by (1.9), the slope of the normal is

$$-\frac{1}{m} = -1 \Big/ \left(\frac{dy}{dx}\right)_{x=x_0}.$$

Hence the equation of the normal is

$$y - y_0 = -\frac{1}{m}(x - x_0).$$

2.3 Rates of change

The quantity $\lim_{\delta x \to 0} \delta y/\delta x$ is usually needed to solve problems which have no immediate connection with the slope of graphs: this idea was only introduced to give the reader a picture to hold on to. Moreover, it is not always appropriate to call the variables x and y if other letters arise more naturally.

For example, suppose that a car is moving along a straight road, represented by an x axis, and that at time t its **displacement** from the origin is given by

$$x = f(t).$$

We can deduce its **velocity** from moment to moment from this information.

Choose any moment t, and suppose that, between times t and $t + \delta t$, the car moves from x to $x + \delta x$. Then δx must be given by

$$\delta x = f(t + \delta t) - f(t).$$

The quantities δt and δx could be imagined as being recorded with a stopwatch and distance meter, and the average velocity over the interval δt would be

$$\frac{\text{distance travelled}}{\text{time taken}} = \frac{f(t + \delta t) - f(t)}{\delta t} = \frac{\delta x}{\delta t}.$$

The smaller that δt is, the more nearly will this ratio approximate to the instantaneous velocity v at time t. Therefore, let $\delta t \to 0$; using the notation (2.7), we obtain

$$v = \lim_{\delta t \to 0} \frac{\delta x}{\delta t},$$

or alternatively, by (2.8),

$$v = \frac{dx}{dt}.$$

We can borrow the result (2.6) to complete the calculation in one case. Suppose that

$$x = t^2.$$

Equation (2.6) says in effect that

$$\text{if} \quad y = x^2 \quad \text{then} \quad \frac{dy}{dx} = 2x,$$

and by changing the letters x and y to t and x respectively we obtain:

$$\text{if} \quad x = t^2 \quad \text{then} \quad \frac{dx}{dt} = 2t.$$

Therefore the velocity is

$$v = 2t.$$

Another way of expressing the meaning of velocity is that **velocity is the rate of change of displacement with time**. Similarly, **acceleration a is the rate of change of velocity with time**:

$$a = \frac{dv}{dt}.$$

For the case when $x = t^2$ we have

$$\delta v = 2(t + \delta t) - 2t = 2\,\delta t,$$

so

$$a = \frac{dv}{dt} = \lim_{\delta t \to 0} \frac{\delta v}{\delta t} = 2.$$

The expression 'rate of change' means the same as the term 'growth rate' that we used in Section 1.10. As seen in the next example, the idea of rate of change is quite general and need not involve time.

Example 2.2 *Find the rate of change of the area of a circle with respect to its radius.*

Call the radius r and the area A. The rate of change of A with respect to r is

$$\frac{dA}{dr} \quad \text{or} \quad \lim_{\delta r \to 0} \frac{\delta A}{\delta r}.$$

Since $A = \pi r^2$, we have

$$\delta A = \pi(r + \delta r)^2 - \pi r^2 = \pi(2r\,\delta r + \delta r^2).$$

Therefore

$$\frac{\delta A}{\delta r} = \pi(2r + \delta r).$$

Now let $\delta r \to 0$; we obtain

$$\frac{dA}{dr} = \lim_{\delta r \to 0} \frac{\delta A}{\delta r} = 2\pi r.$$

This result could have been obtained by using our previous result

$$\frac{d(t^2)}{dt} = 2t,$$

with r in place of t, and multiplying it by π. (Notice also that $2\pi r$ is the circumference: the result can be interpreted as meaning that if we increase r by a small amount δr, then the area increase is nearly equal to that of a narrow strip of length $2\pi r$ and breadth δr.)

2.4 Derivative of x^n ($n = 0, 1, 2, 3, \dots$)

The following is our first general result:

(a) if $y = c$, where c is a constant, then

$$\frac{dy}{dx} = 0.$$

(b) If $y = x^n$, where $n = 1, 2, 3, \dots$, then

$$\frac{dy}{dx} = nx^{n-1}.$$

(2.9)

To prove (a): the graph of $y = c$ is a horizontal straight line; therefore its slope is zero, so $dy/dx = (d/dx)c = 0$.

To prove (b) in the most elementary way, we shall use an identity: if n is a positive integer and a, b are any numbers,

$$a^n - b^n = (a - b)(a^{n-1} + a^{n-2}b + a^{n-3}b^2 + \cdots + b^{n-1}).$$

This can be verified by multiplying out the two brackets on the right; everything cancels except for the two terms on the left.

Follow (2.8), with $f(x) = x^n$, so that $\delta y = (x + \delta x)^n - x^n$:

$$\frac{dy}{dx} = \lim_{\delta x \to 0} \frac{\delta y}{\delta x} = \lim_{\delta x \to 0} \frac{1}{\delta x}[(x + \delta x)^n - x^n].$$

Put $a = x + \delta x$ and $b = x$ into the identity, noticing that

$$a - b = (x + \delta x) - x = \delta x.$$

$$\frac{\delta y}{\delta x} = \frac{1}{\delta x}(\delta x)[(x + \delta x)^{n-1} + (x + \delta x)^{n-2}x + \cdots + x^{n-1}]$$

$$= (x + \delta x)^{n-1} + (x + \delta x)^{n-2}x + \cdots + x^{n-1}$$

when $\delta x \neq 0$. Now let $\delta x \to 0$; we obtain

$$\frac{dy}{dx} = \lim_{\delta x \to 0} \frac{\delta y}{\delta x} = x^{n-1} + x^{n-1} + \cdots + x^{n-1}.$$

There are n terms on the right, each equal to x^{n-1}, so finally,

$$\frac{dy}{dx} = nx^{n-1}.$$

(In Section 3.4 we show that (2.9b) is in fact true for *all* values of n.)

Example 2.3   *Obtain* (a) *the general expression for* dy/dx *when* $y = x^3$; (b) *the slope of the curve* $y = x^3$ *at* $P : (2, 8)$; (c) *the angle of inclination of the tangent line to* $y = x^3$ *at the point* P; (d) *the equation of the tangent line through* P; (e) *the velocity* v *and acceleration* a *of a point with coordinate* x, *when* $x = t^3$.

(a) From (2.9) with $n = 3$, we have

$$\frac{dy}{dx} = 3x^{3-1} = 3x^2.$$

(b) The slope of the curve at P is equal to the value of dy/dx at $x = 2$, which is 12.

(c) The slope is equal to $\tan \alpha$, where α is the angle of inclination, so $\alpha = 85.2°$.

(d) Let (x, y) now represent **any point on the tangent line** at $(2, 8)$. The slope of the tangent is equal to 12, so from (2.1).

$$\frac{y - 8}{x - 2} = 12.$$

Therefore the equation of the tangent line is $y = 12x - 16$.

(e) From Section 2.3, $v = dx/dt = (d/dt)t^3 = 3t^2$. Also

$$a = \frac{dv}{dt} = \frac{d}{dt}3t^2 = 3\frac{d}{dt}t^2 = 6t.$$

A little thought about the process of finding $\lim_{\delta t \to 0} \delta v/\delta t$ will persuade the reader that it is right to take the constant 3 from under the differentiation sign in the last line; see also the next section.

2.5 **Derivatives of sums: multiplication by constants**

The following are general rules which become obvious when the definition (2.8) is applied to them.

Linear combinations of functions

(a) If C is a constant, then

$$\frac{d}{dx}[Cf(x)] = C\frac{d}{dx}f(x).$$

(b) $\quad \frac{d}{dx}[f(x) + g(x)] = \frac{d}{dx}f(x) + \frac{d}{dx}g(x).$

(c) If $A, B, C, \ldots$ are constants, then

$$\frac{d}{dx}[Af(x) + Bg(x) + Ch(x) + \cdots]$$

$$= A\frac{d}{dx}f(x) + B\frac{d}{dx}g(x) + C\frac{d}{dx}h(x) + \cdots .$$

(2.10)

The result (2.10c) follows easily by repeated use of (a) and (b). We can use this rule together with (2.9) to obtain the derivatives of polynomials, as in the following example.

Example 2.4 *Obtain* dy/dx *when* (a) $y = 3x^3 - \frac{1}{2}x^2 + 5$;
(b) $y = 3x(x^2 - 2)$.

(a) From (2.10c),

$$\frac{dy}{dx} = \frac{d}{dx}(3x^3 - \tfrac{1}{2}x^2 + 5)$$

$$= 3\frac{d(x^3)}{dx} - \frac{1}{2}\frac{d(x^2)}{dx} + \frac{d(5)}{dx}$$

$$= 3(3x^2) - \tfrac{1}{2}(2x) + 0$$

$$= 9x^2 - x.$$

(b) It is necessary in this case to express y without brackets, that is as a polynomial:

$$\frac{dy}{dx} = \frac{d}{dx}[3x(x^2 - 2)] = \frac{d}{dx}(3x^3 - 6x)$$

$$= 3\frac{d(x^3)}{dx} - 6\frac{dx}{dx} \quad \text{(from 2.10c))}$$

$$= 9x^2 - 6.$$

These derivatives, of course, represent the slopes of the corresponding graphs.

Example 2.5 *A car travels along a straight road with varying velocity v for one hour. At time t hours, its displacement from the starting point O is given by* $x = 60t^2(3 - 2t)$ *kilometres. Find expressions for* (a) *the velocity v;* (b) *the acceleration a.*

(a) The velocity is the rate of change of displacement with time:

$$v = \frac{dx}{dt} = \frac{d}{dt}[60t^2(3 - 2t)]$$

$$= 60\frac{d}{dt}(3t^2 - 2t^3) = 60\left(3\frac{d(t^2)}{dt} - 2\frac{d(t^3)}{dt}\right)$$

$$= 60[3(2t) - 2(3t^2)] = 360(t - t^2) \text{ in km h}^{-1}.$$

(b) Acceleration is the rate of change of velocity with time:

$$a = \frac{dv}{dt}.$$

Therefore

$$a = 360(1 - 2t) \quad \text{(in km h}^{-2}).$$

Example 2.6 *The potential enegy V of a pendulum of length l with a bob of mass m is given approximately by* $V = mgl(\theta - \frac{1}{6}\theta^3)$ *when the angle of inclination* θ *(radians) is small. Find the rate of change of V with respect to* θ*. (This quantity is associated with the moment exerted by gravity.)*

Using the letters suggested by the question, we require

$$\frac{dV}{d\theta} = \frac{d}{d\theta}[mgl(\theta - \frac{1}{6}\theta^3)] = mgl\left(\frac{d\theta}{d\theta} - \frac{1}{6}\frac{d(\theta^3)}{d\theta}\right)$$

$$= mgl(1 - \frac{1}{2}\theta^2).$$

2.6 Three important limits

In order to increase the repertory of functions that we can differentiate, three important limits are needed. You might not need to learn the proofs, but the results (2.11), (2.13), and (2.14) are essential, and you should try to acquire a feeling for what is happening by examining the numerical tables given; or better, by working out tables of your own.

Instead of δx, the letter ε (greek epsilon) will be used to represent the quantity that tends to zero. (The limit is not affected by the letter we use.)

First consider

$$\lim_{\varepsilon \to 0} \frac{e^\varepsilon - 1}{\varepsilon}.$$

If we put $\varepsilon = 0$ we get $0/0$, which is meaningless, but the approach to a limit can be seen in the following table:

ε	0.1	0.01	0.001	...
$\dfrac{e^\varepsilon - 1}{\varepsilon}$	1.0517	1.0050	1.0005	...

(An approach to zero through negative ε values is similar.) It looks as if the limit is equal to 1.

To prove this, recall that in Section 1.10 it was shown that the graph $y = e^x$ intersects the y axis at 45°; that is to say, its slope there is equal to 1. (This is the characteristic property of the base $e = 2.7128...$.) The same thing is true if we plot $y = e^\varepsilon$ against ε, as in Fig. 2.5. Referring to this figure:

$$\frac{e^\varepsilon - 1}{\varepsilon} = \frac{RQ - OP}{PN} = \frac{NQ}{PN},$$

which represents the slope of the chord PQ. When $\varepsilon \to 0$, the slope of the chord PQ approaches the slope of the tangent PT, which is equal to 1. Therefore we have proved that

$$\lim_{\varepsilon \to 0} \frac{e^\varepsilon - 1}{\varepsilon} = 1.$$

(2.11)

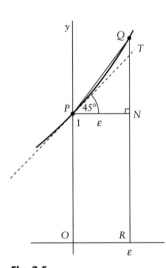

Fig. 2.5

The second limit to be considered is

$$\lim_{\varepsilon \to 0} \frac{\sin \varepsilon}{\varepsilon},$$

ε being measured in radians. The approach to the limit is shown in the following table, which includes negative values of ε:

ε	±0.1	±0.08	±0.06	±0.04	±0.02
$\dfrac{\sin \varepsilon}{\varepsilon}$	±0.998 33	±0.998 93	±0.999 40	±0.999 73	±0.999 93

The limit looks as if it might equal 1.

To prove this, consider Fig. 2.6a. PN is any line segment perpendicular to the base line AB, with Q any point to the left of N, and we allow ε to represent the angle PQN (in radians). The arc PR is a circular arc with centre Q and radius PQ. Then

$$\frac{PN}{PQ} = \sin \varepsilon, \quad \text{so} \quad PN = PQ \sin \varepsilon.$$

Also (radian property, Section 1.5)

$$\text{arc } PR = PQ \times \widehat{PQR} = PQ\,\varepsilon.$$

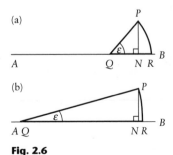

Fig. 2.6

Therefore

$$\frac{\sin \varepsilon}{\varepsilon} = \frac{PN}{\text{arc } PR}. \tag{2.12}$$

Now let Q recede some distance towards the left, as illustrated in Fig. 2.6b. The angle ε decreases, and $\varepsilon \to 0$ as Q recedes to infinity. At the same time the arc PR approaches the straight line PN, tending ultimately to coincide with it. Therefore, when $\varepsilon \to 0$, the length of the arc PR approaches the length of PN; so, from (2.12),

$$\lim_{\varepsilon \to 0} \frac{\sin \varepsilon}{\varepsilon} = 1. \tag{2.13}$$

Finally we consider

$$\lim_{\varepsilon \to 0} \frac{\ln(1 + \varepsilon)}{\varepsilon}.$$

Figure 2.7 shows the graphs of $y = \ln \varepsilon$ and $y = \ln(1 + \varepsilon)$. The graph $y = \ln \varepsilon$ (see Fig. 1.31) passes through the point (1, 0) at 45° to the ε axis. The graph $y = \ln(1 + \varepsilon)$ is the same graph moved over to the left by a distance 1, so it passes through the origin O at 45°: that is to say, it has slope equal to 1 at the origin. Therefore

$$\lim_{\varepsilon \to 0} \frac{\ln(1 + \varepsilon)}{\varepsilon} = 1. \tag{2.14}$$

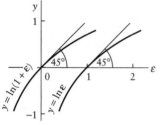

Fig. 2.7

You may be glad to know that there are no more complicated limits to be evaluated.

2.7 Derivatives of e^x, $\sin x$, $\cos x$, $\ln x$

These follow from the definition (2.8) and the limits obtained in the previous section.

First let

$$y = e^x.$$

Then according to the definition (2.8),

$$\frac{dy}{dx} = \lim_{\delta x \to 0} \frac{e^{x + \delta x} - e^x}{\delta x}$$

$$= \lim_{\delta x \to 0} \frac{e^x e^{\delta x} - e^x}{\delta x} = \lim_{\delta x \to 0} e^x \frac{e^{\delta x} - 1}{\delta x}.$$

Now put

$$\delta x = \varepsilon.$$

The previous expression becomes

$$e^x \lim_{\varepsilon \to 0} \frac{e^\varepsilon - 1}{\varepsilon} = e^x,$$

by (2.11). Therefore

$$\frac{d}{dx} e^x = e^x.$$

(2.15)

The rate of increase of e^x is therefore numerically equal to e^x itself. The simplicity of (2.15) is the reason why the number e is a desirable base for exponential functions. The result is not so simple for any other base.

Next, consider

$$y = \sin x.$$

Then

$$\frac{dy}{dx} = \lim_{\delta x \to 0} \frac{\sin(x + \delta x) - \sin x}{\delta x}.$$

From the formula in Appendix B(d)

$$\sin C - \sin D = 2 \sin \tfrac{1}{2}(C - D) \cos \tfrac{1}{2}(C + D)$$

for any C and D. Put $C = x + \delta x$ and $D = x$ into the identity. Then we have

$$\frac{dy}{dx} = \lim_{\delta x \to 0} \frac{2 \sin \tfrac{1}{2}\delta x \cos(x + \tfrac{1}{2}\delta x)}{\delta x}$$

$$= \lim_{\delta x \to 0} \frac{\sin \tfrac{1}{2}\delta x}{\tfrac{1}{2}\delta x} \cos(x + \tfrac{1}{2}\delta x).$$

Putting $\tfrac{1}{2}\delta x = \varepsilon$, we have from (2.13),

$$\frac{dy}{dx} = \lim_{\varepsilon \to 0} \frac{\sin \varepsilon}{\varepsilon} \cos(x + \varepsilon) = \lim_{\varepsilon \to 0} \frac{\sin \varepsilon}{\varepsilon} \lim_{\varepsilon \to 0} \cos(x + \varepsilon)$$

$$= \cos x.$$

Therefore

$$\frac{d}{dx} \sin x = \cos x.$$

(2.16)

By a closely similar argument, it can be shown that

$$\frac{d}{dx}\cos x = -\sin x.$$

(2.17)

(Notice the minus sign which occurs here.)

Finally, suppose that

$$y = \ln x.$$

Then from (2.8),

$$\frac{dy}{dx} = \lim_{\delta x \to 0} \frac{\ln(x + \delta x) - \ln x}{\delta x}$$

$$= \lim_{\delta x \to 0} \frac{1}{\delta x} \ln \frac{x + \delta x}{x} = \lim_{\delta x \to 0} \frac{1}{\delta x} \ln\left(1 + \frac{\delta x}{x}\right).$$

By putting

$$\frac{\delta x}{x} = \varepsilon, \quad \text{or} \quad \delta x = x\varepsilon,$$

the previous equation becomes

$$\frac{dy}{dx} = \lim_{\varepsilon \to 0} \frac{1}{x} \frac{\ln(1 + \varepsilon)}{\varepsilon}.$$

By eqn (2.14) the limit of the part containing ε is 1, so we have

$$\frac{d}{dx}\ln x = \frac{1}{x}.$$

(2.18)

(Remember that x must be positive for $\ln x$ to have a meaning.)

2.8 A basic table of derivatives

We assemble the results (2.15) to (2.18) from Section 2.7, and (2.9) for powers of x, in a short table of derivatives.

Derivatives of the elementary functions

Function	Derivative
$y = f(x)$	dy/dx or $df(x)/dx$
c ($c = $ constant)	zero
x^n ($n = 1, 2, \ldots$)	nx^{n-1}
e^x	e^x
$\sin x$	$\cos x$
$\cos x$	$-\sin x$
$\ln x$ (x positive)	$1/x$ (or x^{-1})

(2.19)

The derivatives of more complicated functions can be obtained from these by using the rules described in the next chapter. A more extensive table is given in Appendix D. Remember rule (2.10) for the addition of functions and multiplication by constants.

Example 2.7   *Obtain the equation of the tangent line at the point* $(\frac{1}{2}\pi, \pi)$ *on the graph of* $y = 2x - 3 \cos x$.

At a general point on the curve,

$$\frac{dy}{dx} = \frac{d}{dx}(2x - 3 \cos x) = 2\frac{dx}{dx} - 3\frac{d}{dx}\cos x \quad \text{(by (2.10))}$$

$$= 2 - 3(-\sin x) = 2 + 3 \sin x$$

(from the table). At $(\frac{1}{2}\pi, \pi)$ this becomes equal to

$$2 + 3 \sin \tfrac{1}{2}\pi = 5,$$

and this is the slope of the tangent line at the point. The equation of the tangent line is therefore

$$\frac{y - \pi}{x - \frac{1}{2}\pi} = 5,$$

or $y = 5x - \frac{3}{2}\pi.$

2.9   Higher-order derivatives

We may differentiate a function, and then differentiate the result. For example:

if $y = x^4,$

then $\dfrac{dy}{dx} = 4x^3,$

which we shall sometimes call the **first derivative** of x^4. By differentiating again, we obtain

$$\frac{d}{dx}\left(\frac{dy}{dx}\right) = 12x^2,$$

which we call the **second derivative** of x^4, and so on.

In general, if $y = f(x)$, we use the notation

$$\frac{d}{dx}\left(\frac{dy}{dx}\right) = \frac{d^2y}{dx^2} \quad \text{or} \quad \frac{d^2f(x)}{dx^2}.$$

(Notice where the indices 2 are placed: the locations are different above and below.) If we differentiate again, we get

$$\frac{d}{dx}\left[\frac{d}{dx}\left(\frac{dy}{dx}\right)\right] = \frac{d}{dx}\left(\frac{d^2y}{dx^2}\right) = \frac{d^3y}{dx^3} \quad \text{or} \quad \frac{d^3f(x)}{dx^3},$$

and so on.

Example 2.8 *Show that* $d^4y/dx^4 = 0$ *when* $y = 2x^3 + 3x^2 - 1$.

Differentiating four times, we have

$$\frac{dy}{dx} = 6x^2 + 6x, \quad \frac{d^2y}{dx^2} = 12x + 6, \quad \frac{d^3y}{dx^3} = 12, \quad \frac{d^4y}{dx^4} = 0.$$

For any polynomial of degree n the $(n+1)$th derivative will be zero.

Example 2.9 *Write down the sequence* $y, dy/dx, d^2y/dx^2, \dots,$
d^7y/dx^7 *when* $y = \sin x$.

The sequence is

$\sin x, \cos x, -\sin x, -\cos x, \sin x, \cos x, -\sin x, -\cos x$

(and it continues in this regular way).

The following example involves the **factorial** $n!$. As we saw in Section 1.17, $n!$ is defined as

$$n! = n(n-1)(n-2) \dots 2 \cdot 1.$$

Remember that 0! is defined to be 1.

Example 2.10 *Let n and r be any integers with $n \geqslant r > 0$. Prove that*

(a) $\dfrac{d^r}{dx^r}(x^n) = \dfrac{n!}{(n-r)!} x^{n-r};$

(b) $\dfrac{d^n}{dx^n}(x^n) = n!.$

(a) From eqn (2.9),

$$\frac{d}{dx}(x^n) = nx^{n-1},$$

and successively

$$\frac{d^2}{dx^2}(x^n) = n(n-1)x^{n-2}, \qquad \frac{d^3}{dx^3}(x^n) = n(n-1)(n-2)x^{n-3},$$

and so on. For $n \geqslant r > 0$, the rth derivative is x^{n-r} with a coefficient of the form $n(n-1)(n-2) \dots$ in which there are r factors. The factors therefore run down from n to $[n - (r-1)]$, so that we have

$$\frac{d^r}{dx^r}(x^n) = n(n-1)(n-2) \dots (n-r+1)x^{n-r}, \tag{i}$$

or

$$\frac{d^r}{dx^r}(x^n) = \frac{n!}{(n-r)!} x^{n-r}. \tag{ii}$$

The step from (i) to (ii) results from the cancellation of terms between the two factorials.

Example 2.10 *continued*

(b) When $r = n$, we obtain from (i)

$$\frac{d^n}{dx^n}(x^n) = n(n-1)(n-2) \dots (n-n+1)x^0 = n(n-1)(n-2) \dots 1 = n!,$$

after putting $x^0 = 1$. (If we try putting $r = n$ directly into (ii) we get $n!/0!$. The conventional value $0! = 1$ does give us $n!$.)

2.10 An interpretation of the second derivative

The second derivative has a simple interpretation. Suppose that $y = f(x)$. From Section 2.9,

$$\frac{d^2y}{dx^2} = \frac{d}{dx}\left(\frac{dy}{dx}\right).$$

dy/dx represents the slope of the graph, so d^2y/dx^2 gives the **rate of change of the slope** with respect to x as we move from left to right on the graph. Where d^2y/dx^2 is positive, the slope is increasing; where it is negative, the slope is decreasing.

If d^2y/dx^2 is consistently positive, then the slope dy/dx steadily increases; it might even increase from negative values (downward slope) through a zero value (tangent horizontal) to positive values (upward slope). If d^2y/dx^2 is consistently negative, then the slope steadily decreases. Figure 2.8 shows two curves upon which, respectively, the second derivative is positive and negative all the way along.

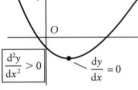

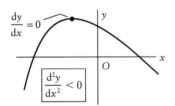

Fig. 2.8

Example 2.11 *Sketch one period 2π of the graph of $\sin x$ and indicate the signs of dy/dx and d^2y/dx^2.*

We have

$$y = \sin x, \qquad \frac{dy}{dx} = \cos x, \qquad \frac{d^2y}{dx^2} = -\sin x.$$

The signs of the derivatives are shown in Fig. 2.9a; dy/dx is zero at the points marked Z. In Fig. 2.9b, dy/dx is sketched to show explicitly how it varies.

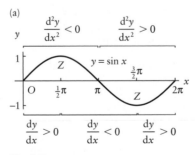

Fig. 2.9

Problems

2.1 (Computational). A point P is given on each of the following curves. Choose a sequence of points Q which lie closer and closer to P on the curve, and make a table giving the slopes of the chords PQ. From this table, estimate the slope of the curve at P. (Consider points on both sides of P.)
(a) $y = x^3$ at $P : (1, 1)$;
(b) $y = x^{\frac{1}{2}}$ at $P : (1, 1)$;
(c) $y = \cos x$ at $P : (\frac{1}{4}\pi, 2^{-\frac{1}{2}})$;
(d) $y = e^x$ at $P : (0, 1)$;
(e) $y = e^{2x}$ at $P : (0, 1)$;
(f) $y = x^3 + x^{\frac{1}{2}}$ at $P : (1, 2)$ (compare (a) and (b));
(g) $y = \ln x$ at $P : (1, 0)$.

2.2 (Sections 2.1, 2.2). Obtain dy/dx in each of the following cases at the given point P. Do this from **first principles**; that is, find δy in terms of δx, simplify $\delta y/\delta x$, and let $\delta x \to 0$ to obtain $\lim_{\delta x \to 0} \delta y/\delta x$, or dy/dx.
(a) $y = 3x$ at $P : (2, 6)$;   (b) $y = 3 - 2x$ at $P : (1, 1)$;
(c) $y = 3x^2$ at $P : (1, 3)$;   (d) $y = x^3$ at $P : (1, 1)$;
(e) $y = 1/x$ at $P : (2, \frac{1}{2})$;   (f) $y = 3x + 2x^2$ at $P : (1, 5)$;
(g) $y = (1 + 2x)^2$ at $P : (-1, 1)$.

2.3 (Sections 2.1, 2.2). Obtain dy/dx from first principles (see Problem 2.2) at a general point $P : (x, y)$ on the given curves.
(a) $y = 3x^2$;   (b) $y = x^3$;
(c) $y = 1/x$;   (d) $y = x + \frac{1}{2}$;
(e) $y = x + 1/x$;   (f) $y = 2x^2 - 3$.

2.4 (Section 2.3). Let x be the displacement of a point moving on a straight line, and let t represent the time elapsed. Form a table by taking the given value of t and calculating the average velocity between t and $t + \delta t$ for diminishing values of δt. Use the table to estimate the velocity at time t.
(a) $x = 3t$ at $t = 1$;   (b) $x = 5t^2$ at $t = 3$;
(c) $x = 2t - 5t^2$ at $t = 1$;   (d) $x = 2t - 5t^2$ at $t = 0.2$.

2.5 Use the formula (2.9) to find dy/dx at the given points in the following cases.
(a) $y = x$ at any point;   (b) $y = x^3$ at $x = 3$;
(c) $y = x^4$ at $x = 2$ and at $x = -2$.

2.6 From (2.9), write down the derivatives, dy/dx or $(d/dx)f(x)$, for the given functions $f(x)$. Use this information to sketch rough graphs of $f(x)$ (notice the sign and the magnitude of the slope of $y = f(x)$).
(a) $y = x$;   (b) $y = x^2$;   (c) $y = x^3$;
(d) $y = x^4$;   (e) $y = x^5$.

2.7 Sketch a velocity–time graph and an acceleration–time graph for a point moving on a straight line with displacement $x = t^3$. Use these to sketch a graph of acceleration against distance. (See Example 2.5.)

2.8 In the following, different letters for the variables are used in place of the usual x and y. Write down the derivatives in the appropriate form. (For example, if $w = r^3$, then $dw/dr = 3r^2$.)
(a) $V = \frac{4}{3}\pi r^3$;   (b) $S = \pi d^2$;
(c) $E = kT^4$ (k is a constant);
(d) $I = V/R$ (R is a constant);
(e) $H = RI^2$ (R is a constant);
(f) $V = RT/P$ (R and P are constant).

2.9 Differentiate the following functions by using (2.10):
(a) $3x^2 - 2x + 1$;   (b) $x^7 - 3x^6 + x + 1$;
(c) $x + C$ (where C is a constant);
(d) $x(x - 1)$;   (e) $x^2(x^2 + 1) - 1$;
(f) $ax^2 + bx + c$ (where a, b, c, are constants);
(g) $(x - 1)^2$.

2.10 Prove that the following pairs of curves intersect in a right angle at the points given. (Hint: find dy/dx at the point for each curve.)
(a) $y = 1 + x - x^2$ and $y = 1 - x + x^2$ at $(1, 1)$;
(b) $y = \frac{1}{2}(1 - x^2)$ and $y = x - 1$ at $(1, 0)$;
(c) $y = 1 - \frac{1}{3}x^3$ and $y = \frac{1}{6} + \frac{1}{2}x^2$ at $(1, \frac{2}{3})$.

2.11 Find the angle between the following curves at their points of intersection. (Hint: the angle of intersection is the angle between the tangents to the curves at the point; then consider (1.7) and the tangent formula of (1.17a) for the difference of angles.)
(a) $y = x^2$ and $y = 1 - x^2$;
(b) $y = \frac{1}{3}x^3$ and $y = x^2 - 2x + \frac{4}{3}$.

2.12 (See Section 2.6.) Find the limits of the following functions when $\varepsilon \to 0$. (Remember: 0/0 has no definite meaning.)

(a) $\dfrac{\varepsilon}{\varepsilon}$;   (b) $\dfrac{\varepsilon}{2\varepsilon}$;   (c) $\dfrac{\varepsilon^2}{\varepsilon}$;   (d) $\dfrac{e^{2\varepsilon} - 1}{2\varepsilon}$;

(e) $\dfrac{e^{2\varepsilon} - 1}{\varepsilon}$;   (f) $\dfrac{\sin 2\varepsilon}{2\varepsilon}$;   (g) $\dfrac{\sin 2\varepsilon}{\varepsilon}$;   (h) $\dfrac{\ln(1 + \varepsilon^2)}{\varepsilon^2}$;

(i) $\dfrac{\sin \varepsilon}{\varepsilon}$ when ε is an angle measured in *degrees*;

(j) $\dfrac{\tan \varepsilon}{\varepsilon}$;   (k) $\dfrac{\sinh \varepsilon}{\varepsilon}$;   (l) $\dfrac{e^{-\varepsilon} - 1}{\varepsilon}$.

2.13 (See Section 2.7.) Obtain $d(\cos x)/dx$ in the same way that (2.16), for $\sin x$, was obtained.

2.14 (See Section 2.7.) (a) Differentiate e^{2x} by following the method leading to (2.15).

(b) Differentiate $\sin 2x$ by following the method leading to (2.16).

(c) Prove that $(d/dx)e^{-x} = -e^{-x}$ by following part-way the method leading to (2.15). (Hint:

$$\lim_{\varepsilon \to 0}[(e^{-\varepsilon} - 1)/(-\varepsilon)] = 1.)$$

Use this result to differentiate $\sinh x$ and $\cosh x$ (see (1.26) for the definitions).

2.15 Differentiate the following functions.
(a) $2 \sin x - 3 \cos x$;
(b) $\ln 3x$ (see Section 1.11 for the properties of the logarithm);
(c) $\ln x^3$ (see Section 1.11);   (d) $\sin x - x$;
(e) $e^x - 1 - x - \frac{1}{2}x^2$.

2.16 Find the equations of the tangent lines in the following cases.
(a) $y = x^3$ at $(1, 1)$;
(b) $y = x^4 - 2x^2 + 1$ at $(2, 9)$;
(c) $y = \cos x$ at $(\frac{1}{2}\pi, 0)$;
(d) $y = \ln x$ at $(e, 1)$;

(e) $y = \dfrac{1}{\sqrt{2}} \sin x + \dfrac{1}{\sqrt{2}} \cos x$ at $(\frac{1}{4}\pi, 1)$;

(f) $y = 3e^x - 4x$ at $(0, 3)$.

2.17 Obtain dy/dx, d^2y/dx^2, d^3y/dx^3 in the following cases.
(a) $y = x^6$;
(b) $y = 3x^2 - 2x + 2$;
(c) $y = x^6 - x^2$;
(d) $y = 2 \sin x - 3 \cos x$;
(e) $e^x - 1 - x - \frac{1}{2}x^2$.

2.18 Show that, if N is a positive whole number, then $(d^N/dx^N)x^N = N!$.

2.19 For the curve $y = x^2(x^2 - 3)$, find the ranges in x for which (a) dy/dx is positive (so that y is increasing); (b) dy/dx is negative (so that y is decreasing); (c) d^2y/dx^2 is positive (so that the slope is increasing); (d) d^2y/dx^2 is negative (so that the slope is decreasing). Deduce the general shape of the curve from these facts. (Hint: if dy/dx changes sign at some point, then dy/dx must be zero at the point. But dy/dx does not *necessarily* change sign where $dy/dx = 0$.)

2.20 Find the equation of the normal to the parabola $y = ax^2$ at any point $x = x_0$.

3 Further techniques for differentiation

The table of elementary derivatives (2.19) is not sufficient to satisfy basic needs; for example, it does not even tell us the derivative of $\sin 2x$. But fortunately we need not start afresh every time we meet a new function. By using the rules of combination given in this chapter it is possible to differentiate functions that are made up using those given in (2.19), no matter how complicated they are.

3.1 The product rule

The derivatives of a product of several functions can be obtained when the derivatives of its individual components are known. Examples of such products are

$$x^2 \, e^x, \quad e^x \sin x, \quad x \, e^x \cos x.$$

Suppose firstly that y takes the form of a product of two functions $u(x)$ and $v(x)$:

$$y(x) = u(x)v(x),$$

where $y(x)$ is written to display the dependence of y on x. We require $\mathrm{d}y/\mathrm{d}x$ in terms of u and v. Fix a value for x, and change it by an amount δx so that

$$x \text{ becomes } x + \delta x.$$

Then u, v, and y all change:

$$u \text{ becomes } u + \delta u, \quad v \text{ becomes } v + \delta v, \quad \text{and } y \text{ becomes } y + \delta y,$$

where u, v, y represent the values at x. Since

$$\delta y = (u + \delta u)(v + \delta v) - uv,$$

we obtain

$$\frac{\delta y}{\delta x} = \frac{(u + \delta u)(v + \delta v) - uv}{\delta x} = \frac{uv + u\,\delta v + v\,\delta u + \delta u\,\delta v - uv}{\delta x}$$

$$= u\frac{\delta v}{\delta x} + v\frac{\delta u}{\delta x} + \delta u\frac{\delta v}{\delta x}.$$

Now let $\delta x \to 0$, so that $\delta y/\delta x$, $\delta u/\delta x$, $\delta v/\delta x$ become dy/dx, du/dy, dv/dx respectively. Also, since $\delta u \to 0$ when $\delta x \to 0$, the final term becomes zero, and we obtain the product rule:

Product rule

If $y(x) = u(x)v(x)$, then

$$\frac{dy}{dx} = \frac{d}{dx}(uv) = u\frac{dv}{dx} + v\frac{du}{dx}.$$

(3.1)

Example 3.1 *Find* dy/dx *when* $y = x^2\,e^x$.

Put $u = x^2$, $v = e^x$, $y = x^2\,e^x = uv$. Then

$$\frac{du}{dx} = 2x, \qquad \frac{dv}{dx} = e^x.$$

Therefore, by (3.1),

$$\frac{dy}{dx} = u\frac{dv}{dx} + v\frac{du}{dx}$$

$$= x^2(e^x) + e^x(2x) = (x^2 + 2x)\,e^x.$$

Example 3.2 *Find* dx/dt *when* $x = e^t \cos t$.

We have to interpret (3.1) in terms of the new symbols. Put

$$u = e^t, \qquad v = \cos t, \qquad x = e^t \cos t = uv.$$

Then (refer if necessary to the table (2.19) with the appropriate changes of letters)

$$\frac{du}{dt} = e^t, \qquad \frac{dv}{dt} = -\sin t.$$

Changing the symbols in (3.1) we have

$$\frac{dx}{dt} = u\frac{dv}{dt} + v\frac{du}{dt}$$

$$= e^t(-\sin t) + (\cos t)\,e^t = e(\cos t - \sin t)$$

$$= e^t(\cos t - \sin t).$$

Example 3.3   *Find* $\mathrm{d}y/\mathrm{d}x$ *when* $y = x\,\mathrm{e}^x \sin x$.

This product has three terms, but we can carry out the differentiation in two stages. Write

$$y = (x\,\mathrm{e}^x)\sin x$$

and put $u = x\,\mathrm{e}^x$ and $v = \sin x$. By (3.1),

$$\frac{\mathrm{d}y}{\mathrm{d}x} = x\,\mathrm{e}^x \frac{\mathrm{d}}{\mathrm{d}x}\sin x + \sin x \frac{\mathrm{d}}{\mathrm{d}x}(x\,\mathrm{e}^x)$$

$$= x\,\mathrm{e}^x \cos x + \sin x \frac{\mathrm{d}}{\mathrm{d}x}(x\,\mathrm{e}^x). \tag{i}$$

To evaluate $(\mathrm{d}/\mathrm{d}x)(x\,\mathrm{e}^x)$, use the product rule again, putting $u = x$ and $v = \mathrm{e}^x$. Then

$$\frac{\mathrm{d}}{\mathrm{d}x}(x\,\mathrm{e}^x) = x\frac{\mathrm{d}}{\mathrm{d}x}\mathrm{e}^x + \mathrm{e}^x\frac{\mathrm{d}}{\mathrm{d}x}x = x\,\mathrm{e}^x + \mathrm{e}^x. \tag{ii}$$

Replace (ii) into (i):

$$\frac{\mathrm{d}y}{\mathrm{d}x} = x\,\mathrm{e}^x \cos x + (\sin x)(x\,\mathrm{e}^x + \mathrm{e}^x)$$

$$= \mathrm{e}^x(x\cos x + x\sin x + \sin x).$$

Another method of dealing with the product of several terms, which is usually more convenient, is given in Section 3.7. You are strongly recommended to write out all the steps completely at first, otherwise mistakes are likely to occur.

3.2 Quotients and reciprocals

Suppose that

$$y(x) = \frac{u(x)}{v(x)}.$$

Proceed as for the product rule: let x change to $x + \delta x$, so that u becomes $u + \delta u$, v becomes $v + \delta v$, and y becomes $y + \delta y$. Then

$$\frac{\delta y}{\delta x} = \left(\frac{u + \delta u}{v + \delta v} - \frac{u}{v}\right)\frac{1}{\delta x} = \frac{uv + v\,\delta u - uv - u\,\delta v}{v(v + \delta v)\,\delta x}$$

$$= \frac{v\,\delta u - u\,\delta v}{v(v + \delta v)\,\delta x} = \frac{1}{v(v + \delta v)}\left(v\frac{\delta u}{\delta x} - u\frac{\delta v}{\delta x}\right).$$

Let $\delta x \to 0$; then $\delta y/\delta x$, $\delta u/\delta x$, $\delta v/\delta x$ become $\mathrm{d}y/\mathrm{d}x$, $\mathrm{d}u/\mathrm{d}x$, $\mathrm{d}v/\mathrm{d}x$, and $\delta v \to 0$. Therefore

$$\frac{\mathrm{d}y}{\mathrm{d}x} = \frac{\mathrm{d}}{\mathrm{d}x}\left(\frac{u}{v}\right) = \frac{1}{v^2}\left(v\frac{\mathrm{d}u}{\mathrm{d}x} - u\frac{\mathrm{d}v}{\mathrm{d}x}\right).$$

It is worth noting the special case of the reciprocal of a function. In that case, $u(x) = 1$, so $du/dx = 0$. Finally we have

Quotient and reciprocal rules

(a) If $y(x) = \dfrac{u(x)}{v(x)}$, then

$$\frac{dy}{dx} = \frac{d}{dx}\left(\frac{u}{v}\right) = \frac{1}{v^2}\left(v\frac{du}{dx} - u\frac{dv}{dx}\right).$$

(b) If $y(x) = \dfrac{1}{v(x)}$ (i.e. if $u(x) = 1$), then

$$\frac{dy}{dx} = \frac{d}{dx}\left(\frac{1}{v}\right) = -\frac{1}{v^2}\frac{dv}{dx}.$$

(3.2)

Example 3.4   *Obtain* dy/dx *when* $y = \tan x$.

Express y in the form

$$y = \tan x = \frac{\sin x}{\cos x}.$$

Put $u = \sin x$, $v = \cos x$, $y = u/v$. Then

$$\frac{du}{dx} = \cos x, \qquad \frac{dv}{dx} = -\sin x.$$

From (3.2),

$$\frac{dy}{dx} = \frac{1}{v^2}\left(v\frac{du}{dx} - u\frac{dv}{dx}\right)$$

$$= \frac{1}{\cos^2 x}[\cos x \cos x - \sin x(-\sin x)]$$

$$= \frac{1}{\cos^2 x}(\cos^2 x + \sin^2 x) = \frac{1}{\cos^2 x} = \sec^2 x.$$

(Remember that $\cos^2 A + \sin^2 A = 1$.)

Example 3.5   *Find* dy/dx *when* $y = (x + 3)/(2x^3 + 1)$.

Put $u = x + 3$, $v = 2x^3 + 1$, $y = u/v$. Then

$$\frac{du}{dx} = 1, \qquad \frac{dv}{dx} = 6x^2.$$

By (3.2),

$$\frac{dy}{dx} = \frac{1}{(2x^3 + 1)^2}[(2x^3 + 1)(1) - (x + 3)(6x^2)]$$

$$= \frac{1 - 18x^2 - 4x^3}{(2x^3 + 1)^2}.$$

Example 3.6 *Obtain* dy/dx *when* (a) $y = 1/x$; (b) $y = 1/x^2$.

(a) Put $v = x$ into the reciprocal rule (3.2b) (or $u = 1$ and $v = x$ into (3.2a)):

$$\frac{dy}{dx} = -\frac{1}{v^2}\frac{dv}{dx} = -\frac{1}{x^2}.$$

(b) Put $v = x^2$ into (3.2b):

$$\frac{dy}{dx} = -\frac{1}{x^4}2x = -\frac{2}{x^3}.$$

If we had put $n = -1$ and -2 respectively into the formula (2.9),

$$\frac{d}{dx}(x^n) = nx^{n-1},$$

proved only for *positive* integer n, the correct result in Example 3.6 is obtained. The formula (2.9) is in fact correct for all values of n, as will be shown in Section 3.4.

3.3 The chain rule

The **chain rule** will be used continually in future chapters. It is also called the **function-of-a-function rule**. Suppose y can be expressed as a function of a variable u, where u is a function of x. We shall express this by the notation

$$y = y(u), \quad \text{where} \quad u = u(x).$$

An example of this is

$$y = \cos(x^3),$$

which we can rewrite in the form

$$y = y(u) = \cos u, \quad \text{where} \quad u = u(x) = x^3.$$

Another example is when $y = \cos^3 x$. Write it in the form $y = (\cos x)^3$, so that

$$y = u^3, \quad \text{where} \quad u = \cos x.$$

The rule for such cases is the following:

> **The chain rule**
> If $y = y(u)$ where $u = u(x)$, then
> $$\frac{dy}{dx} = \frac{dy}{du}\frac{du}{dx}.$$
>
> (3.3)

The form of this result is *easy to remember* if you first write

$$\frac{dy}{dx} = \frac{dy}{\bullet}\frac{\bullet}{dx},$$

then put du in place of the dots. Sometimes it is inconvenient to use u; any letter not already in use can be used in place of u.

To prove (3.3), fix on any value of x. Consider a nearby value $x + \delta x$, and denote the corresponding small changes in u and y by δu and δy. When x becomes $x + \delta x$, then u becomes $u + \delta u$ and y becomes $y + \delta y$. Evidently

$$\frac{\delta y}{\delta x} = \frac{\delta y}{\delta u} \frac{\delta u}{\delta x}$$

since the terms δu cancel. Now let $\delta x \to 0$. Then $\delta u \to 0$, and consequently $\delta y / \delta x$, $\delta y / \delta u$, $\delta u / \delta x$ approach dy/dx, dy/du, du/dx respectively. Thus we obtain

$$\frac{dy}{dx} = \frac{dy}{du} \frac{du}{dx}.$$

The following examples show how to recognize when it is appropriate to use the chain rule. You should lay out every application in the systematic way shown until you are used to it.

Example 3.7   (a) *You are given that* $(d/dx)\, e^x = e^x$ (*see the table,* (2.19)). *Deduce that* $(d/dx)\, e^{ax} = a\, e^{ax}$, *where a is any constant.* (b) *Find the derivative of* e^{-x}. (c) *Use this result to obtain the derivatives of* $\sinh x$ *and* $\cosh x$ (*see* (1.26) *for the definitions of these functions*).

(a) Rewrite $y = e^{ax}$ in the form

$y = e^u$,   where $u = ax$.

To use the chain rule (3.3), we need dy/du and du/dx:

$$\frac{dy}{du} = e^u \quad \text{and} \quad \frac{du}{dx} = a.$$

The chain rule gives

$$\frac{dy}{dx} = \frac{dy}{du} \frac{du}{dx} = e^u\, a = a\, e^{ax},$$

after restoring the variable x.

(b) For e^{-x}, the constant a is -1, so

$$\frac{d}{dx}\, e^{-x} = -e^{-x}.$$

(c) $\sinh x = \frac{1}{2}(e^x - e^{-x})$ and $\cosh x = \frac{1}{2}(e^x + e^{-x})$, so the result (b) gives

$$\frac{d}{dx}(\sinh x) = \frac{d}{dx}\,[\tfrac{1}{2}(e^x - e^{-x})] = \tfrac{1}{2}(e^x + e^{-x}) = \cosh x,$$

$$\frac{d}{dx}(\cosh x) = \frac{d}{dx}\,[\tfrac{1}{2}(e^x + e^{-x})] = \tfrac{1}{2}(e^x - e^{-x}) = \sinh x.$$

Example 3.8   *Find* dy/dx *when* $y = (x^2 + 1)^{10}$.

We could expand $(x^2 + 1)^{10}$ as a polynomial by means of the binomial theorem, but the chain rule is far simpler, Put

$$y = u^{10}, \quad \text{where} \quad u = x^2 + 1.$$

Then

$$\frac{dy}{du} = 10u^9 \quad \text{and} \quad \frac{du}{dx} = 2x.$$

By the chain rule,

$$\frac{dy}{dx} = \frac{dy}{du}\frac{du}{dx} = 10u^9 2x = 20x(x^2 + 1)^9.$$

Example 3.9   *Find* dy/dx *when* (a) $y = \sin(x^3)$; (b) $y = \sin^3 x$.

(a) Put $y = \sin u$, where $u = x^3$. Then

$$\frac{dy}{du} = \cos u \quad \text{and} \quad \frac{du}{dx} = 3x^2,$$

By the chain rule,

$$\frac{dy}{dx} = \frac{dy}{du}\frac{du}{dx} = (\cos u)3x^2 = 3x^2 \cos(x^3).$$

(b) Put $y = u^3$, where $u = \sin x$. Then

$$\frac{dy}{du} = 3u^2 \quad \text{and} \quad \frac{du}{dx} = \cos x.$$

By the chain rule,

$$\frac{dy}{dx} = \frac{dy}{du}\frac{du}{dx} = 3u^2 \cos x = 3 \sin^2 x \cos x.$$

Example 3.10   *Find* dy/dx *when* $y = 1/(x^2 + 1)$.

Put $y = 1/u$, where $u = x^2 + 1$. Then

$$\frac{dy}{du} = -\frac{1}{u^2} \quad \text{and} \quad \frac{du}{dx} = 2x$$

(where the reciprocal rule (3.2b) was used for differentiating $1/u$). By the chain rule,

$$\frac{dy}{dx} = -\frac{1}{u^2} 2x = -\frac{2x}{(x^2 + 1)^2}.$$

Example 3.11   *Find* du/dt *when* $u = a \cos k(x - ct)$ *where* $a, k, c,$ *and* x *are constant,* t *and* u *being the only variables.*

We should not use u for the intermediate variable in the chain rule (3.3), because it is already in use (as the name of the dependent variable). Instead of u, use an uncommitted letter such as w as the intermediate variable, putting

$$u = a \cos w, \quad \text{where} \quad w = kx - kct.$$

Example 3.11 *continued*

The chain rule takes the form

$$\frac{du}{dt} = \frac{du}{dw}\frac{dw}{dt},$$

in which

$$\frac{du}{dw} = -a \sin w \quad \text{and} \quad \frac{dw}{dt} = -kc.$$

Therefore

$$\frac{du}{dt} = (-a \sin w)(-kc) = akc \sin k(x - ct).$$

3.4   Derivative of x^n for any value of n

Consider the derivative of $y = x^n$, where n may have any value, an integer or not, positive or negative. The rule turns out to take the same form as (2.9), in which n was limited to positive integers:

> **Derivative of x^n**
>
> If $y = x^n$, where n may take any value whatever, then
>
> $$\frac{dy}{dx} = nx^{n-1}.$$
>
> (3.4)

To prove (3.4), we use the chain rule (3.3). Note that $x = e^{\ln x}$ (see (1.21)), so that

$$y = x^n = (e^{\ln x})^n = e^{n \ln x}.$$

To use the chain rule, we put this in the form

$$y = e^u, \quad \text{where} \quad u = n \ln x,$$

so that

$$\frac{dy}{du} = e^u \quad \text{and} \quad \frac{du}{dx} = \frac{n}{x}.$$

Then

$$\frac{dy}{dx} = \frac{dy}{du}\frac{du}{dx} = e^u \frac{n}{x} = x^n \frac{n}{x} = nx^{n-1}$$

(where we used $e^u = y = x^n$ again).

Example 3.12   *Find* $\mathrm{d}y/\mathrm{d}x$ *when* (a) $y = x^{\frac{3}{2}}$, (b) $y = 1/x^{\frac{3}{2}}$, (c) $y = 1/\sqrt{x}$, (d) $y = 1/(2x^{\frac{1}{3}} + x)$.

(a) Here $n = \frac{3}{2}$ in (3.4), so $\dfrac{\mathrm{d}}{\mathrm{d}x}(x^{\frac{3}{2}}) = \frac{3}{2}x^{\frac{1}{2}}$.

(b) This may be written $y = x^{-\frac{3}{2}}$, so $n = -\frac{3}{2}$ in (3.4), and

$$\frac{\mathrm{d}}{\mathrm{d}x}(x^{-\frac{3}{2}}) = -\tfrac{3}{2}x^{-\frac{5}{2}}.$$

(c) $y = x^{-\frac{1}{2}}$, so $\dfrac{\mathrm{d}y}{\mathrm{d}x} = -\tfrac{1}{2}x^{-\frac{3}{2}}$.

(d) Write $y = (2x^{\frac{1}{3}} + x)^{-1}$. We can use the chain rule: put $y = u^{-1}$, where $u = 2x^{\frac{1}{3}} + x$, Then

$$\frac{\mathrm{d}y}{\mathrm{d}x} = -u^{-2} \ \text{ (by (3.4))}, \qquad \frac{\mathrm{d}u}{\mathrm{d}x} = \tfrac{2}{3}x^{-\frac{2}{3}} + 1 \ \text{ (by (3.4))}.$$

Therefore, by the chain rule (3.3),

$$\frac{\mathrm{d}y}{\mathrm{d}x} = \frac{\mathrm{d}y}{\mathrm{d}u}\frac{\mathrm{d}u}{\mathrm{d}x} = (-u^{-2})(\tfrac{2}{3}x^{-\frac{2}{3}} + 1) = -\frac{\tfrac{2}{3}x^{-\frac{2}{3}} + 1}{(2x^{\frac{1}{3}} + x)^2}.$$

3.5   Functions of $ax + b$

A frequently occurring application of the chain rule (3.3) is in connection with functions like e^{ax+b}, $\sin(ax+b)$, $(ax+b)^n$, and in general $f(ax + b)$. The spirit of the chain rule is to say: 'If the functions were e^x, $\sin x$, x^n, $f(x)$, then they would be easy. Therefore, try the chain rule with $u = ax + b$.'

Suppose that, in general, we want to differentiate y when

$$y = f(ax + b),$$

and that we know how to differentiate $f(x)$. Write

$$u = ax + b, \qquad y = f(u).$$

Then the chain rule gives

$$\frac{\mathrm{d}y}{\mathrm{d}x} = \frac{\mathrm{d}y}{\mathrm{d}u}\frac{\mathrm{d}u}{\mathrm{d}x} = a\frac{\mathrm{d}f(u)}{\mathrm{d}u},$$

in which the derivative occurring on the right is already known.

The following special cases, in which $b = 0$, should be noticed; they consitute an extension of the table (2.19).

Function	Derivative
e^{ax}	$a\,e^{ax}$
$\sin ax$	$a \cos ax$
$\cos ax$	$-a \sin ax$
$a^x (a > 0)$	$a^x \ln a$
(For a^x see Problem 3.18.)	(3.5)

3.6 An extension of the chain rule

In Section 3.3, the chain rule (3.3) was looked upon as a way of differentiating a function of a function, say $y(u(x))$. Sometimes we need to consider 'a function of a function of a function of ...'. These can always be worked through by repeated applications of (3.3), but it may be less complicated to proceed as in the following example.

Example 3.13   *Obtain* dy/dx *when* $y = e^{\sin(x^2+1)}$.

Instead of using one intermediate variable u, introduce two variables, u and v. Put

$$y = e^v, \qquad v = \sin u, \qquad u = x^2 + 1.$$

Then by exactly the same type of arguments as led to (3.3),

$$\frac{dy}{dx} = \frac{dy}{dv}\frac{dv}{du}\frac{du}{dx}.$$

We have

$$\frac{du}{dx} = 2x, \qquad \frac{dv}{du} = \cos u, \qquad \frac{dy}{dv} = e^v,$$

so

$$\frac{dy}{dx} = (e^v \cos u)2x = 2x\, e^{\sin(x^2+1)} \cos(x^2 + 1).$$

The result can be extended in an obvious way to any number of intermediate variables, but it is seldom that more than two would be needed:

Extended chain rule

Suppose that

$$y = y(v), \quad v = v(u), \quad \text{and} \quad u = u(x).$$

Then

$$\frac{dy}{dx} = \frac{dy}{dv}\frac{dv}{du}\frac{du}{dx}. \tag{3.6}$$

3.7 Logarithmic differentiation

To differentiate a product

$$y = u(x)v(x)w(x)$$

consisting of three terms, the product rule (3.1) can be applied twice, as in Example 3.3. An alternative procedure, which is often simpler, is the following.

Since $y = uvw$,

$$\ln y = \ln(uvw) = \ln u + \ln v + \ln w.$$

By the chain rule (3.3), if $u(x)$ is any function of x,

$$\frac{d}{dx}\ln u = \frac{1}{u}\frac{du}{dx},$$

and we may have y, v, or w in place of u. Therefore

$$\frac{1}{y}\frac{dy}{dx} = \frac{1}{u}\frac{du}{dx} + \frac{1}{v}\frac{dv}{dx} + \frac{1}{w}\frac{dw}{dx}.$$

By multiplying through by $y = uvw$, we obtain

> **Logarithmic differentiation**
>
> If $y = uvw$, then
> $$\frac{dy}{dx} = uvw\left(\frac{1}{u}\frac{du}{dx} + \frac{1}{v}\frac{dv}{dx} + \frac{1}{w}\frac{dw}{dx}\right)$$
> (and so on for any number of terms in the product defining y).
>
> (3.7)

Example 3.14 *Find* dy/dx *when* $y = (x^{\frac{1}{2}}\sin^2 x)/(x^2 + 1)$.

Put $y = uvw$, where

$$u = x^{\frac{1}{2}}, \qquad v = \sin^2 x, \qquad w = (x^2 + 1)^{-1}.$$

Then

$$\ln y = \ln(x^{\frac{1}{2}}) + \ln(\sin^2 x) + \ln(x^2 + 1)^{-1}.$$
$$= \tfrac{1}{2}\ln(x) + 2\ln(\sin x) - \ln(x^2 + 1).$$

Notice that we did not just copy the formula (3.7) rigidly: the logarithm is useful for getting rid of awkward powers and we might have missed this. Differentiate this expression:

$$\frac{1}{y}\frac{dy}{dx} = \frac{1}{2x} + \frac{2}{\sin x}\frac{d}{dx}(\sin x) - \frac{1}{x^2 + 1}\frac{d}{dx}(x^2 + 1)$$

$$= \frac{1}{2x} + \frac{2\cos x}{\sin x} - \frac{2x}{x^2 + 1}.$$

Multiply through by $y = (x^{\frac{1}{2}}\sin^2 x)/(x^2 + 1)$ to give dy/dx:

$$\frac{dy}{dx} = \frac{x^{\frac{1}{2}}\sin^2 x}{x^2 + 1}\left(\frac{1}{2x} + \frac{2\cos x}{\sin x} - \frac{2x}{x^2 + 1}\right).$$

3.8 Implicit differentiation

An equation of the form

$$f(x, y) = c \quad \text{(a constant)}$$

represents a curve or curves; for example, $x^2 + y^2 = 1$ represents a circle, otherwise expressed by $y = y(x) = \pm(1 - x^2)^{\frac{1}{2}}$. This latter relation is **implicit** in the first form, which is called an **implicit equation**.

Suppose that the implicit equation for y is $f(x, y) = c$ and its explicit equation is $y = y(x)$; that is to say, both equations specify the same curve. Then $f(x, y(x)) = c$ *for all values of* x: it is an *identity*. Since the value of $f(x, y(x))$ remains constant, its derivative is zero:

$$\frac{\mathrm{d}}{\mathrm{d}x} f(x, y(x)) = 0,$$

for every relevant value of x.

Notice further that if y is a function of x, then the chain rule (3.3), using y as the intermediate variable instead of u, gives results such as

$$\frac{\mathrm{d}}{\mathrm{d}x} y^2 = 2y \frac{\mathrm{d}y}{\mathrm{d}x} \quad \text{and} \quad \frac{\mathrm{d}}{\mathrm{d}x} \cos y = -\sin y \frac{\mathrm{d}y}{\mathrm{d}x}.$$

This fact can be used in the following way to obtain an expression for $\mathrm{d}y/\mathrm{d}x$, even in cases when we cannot solve the implicit equation to obtain y as a function of x explicitly.

Example 3.15 *Find a general expression for* $\mathrm{d}y/\mathrm{d}x$ *at any point on the curve given by* $f(x, y) = x + y + \sin x + \cos y = 1$.

So long as we stay on the curve, $f(x, y)$ does not change when x changes, so $\mathrm{d}f(x, y)/\mathrm{d}x = 0$. Therefore

$$1 + \frac{\mathrm{d}y}{\mathrm{d}x} + \cos x - \sin y \frac{\mathrm{d}y}{\mathrm{d}x} = 0,$$

so finally

$$\frac{\mathrm{d}y}{\mathrm{d}x} = \frac{\cos x + 1}{\sin y - 1}.$$

Such a result is not quite so simple as its neatness suggests, because we would still find it hard to say what values of y are to be associated with a particular value of x in the new formula: this would in effect involve solving the original equation for y in terms of x.

3.9 Derivatives of inverse functions

The derivatives of functions such as $\ln x$, $\arctan x$, $\arccos x$, and $\arcsin x$, which are the respective inverses of e^x, $\tan x$, $\cos x$, and $\sin x$ can be obtained by a standard procedure. We need the general result

$$\frac{\mathrm{d}y}{\mathrm{d}x} = 1 \bigg/ \frac{\mathrm{d}x}{\mathrm{d}y}$$

$$(3.8)$$

To illustrate what (3.8) means, take as an example the case when $y = \ln x$. As described in Section 1.11, two statements

$$y = \ln x \text{ for } x > 0 \quad \text{and} \quad x = e^y \text{ for all } y$$

are different ways of saying the same thing; the two graphs depicting the relation between x and y are the same graph. For small corresponding increments δx and δy on this graph,

$$\frac{\delta y}{\delta x} = 1 \Big/ \frac{\delta x}{\delta y}.$$

Since δx and δy approach zero together, we obtain $\mathrm{d}y/\mathrm{d}x = 1/(\mathrm{d}x/\mathrm{d}y)$, which is (3.8).

To find $\mathrm{d}y/\mathrm{d}x$ when $y = \ln x$, write equivalently

$$x = e^y.$$

Then

$$\frac{\mathrm{d}x}{\mathrm{d}y} = e^y;$$

so, by (3.8),

$$\frac{\mathrm{d}y}{\mathrm{d}x} = \frac{1}{e^y} = \frac{1}{x}.$$

This, of course, agrees with (2.18), where a more direct method was used.

Example 3.16   *Find* $\mathrm{d}y/\mathrm{d}x$ *when* $y = \arctan x$.

If $y = \arctan x$, then $x = \tan y$. From Example 3.4, interchanging x and y,

$$\frac{\mathrm{d}x}{\mathrm{d}y} = \frac{1}{\cos^2 y};$$

so, by (3.7),

$$\frac{\mathrm{d}y}{\mathrm{d}x} = \cos^2 y.$$

To express $\cos^2 y$ in terms of $\tan y$ (i.e. in terms of x) draw the triangle in Fig. 3.1, in which y is represented by the angle A. This is a right-angled triangle because the sides conform with Pythagoras's theorem. Evidently $\cos^2 y = 1/(1 + \tan^2 y)$, as can be checked by putting $\tan^2 y = (\sin^2 y)/\cos^2 y$. Therefore

$$\frac{\mathrm{d}y}{\mathrm{d}x} = \frac{1}{1 + \tan^2 y} = \frac{1}{1 + x^2}.$$

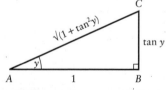

Fig. 3.1

3.10 Derivative as a function of a parameter

If x and y are functions of a **parameter**, or supplementary variable, t, so that

$$x = x(t), \qquad y = y(t),$$

then the point $(x(t), y(t))$ follows a curve as t varies. Suppose that t changes from t to $t + \delta t$; then x changes to $x + \delta x$ and y to $y + \delta y$. Obviously

$$\frac{\delta y}{\delta x} = \frac{\delta y}{\delta t} \Big/ \frac{\delta x}{\delta t},$$

since δt cancels on the right-hand side. Let $\delta t \to 0$; then $\delta x \to 0$, and we have

---

Differentiation in terms of a parameter

If $x = x(t)$ and $y = y(t)$, then

$$\frac{dy}{dx} = \frac{dy}{dt} \Big/ \frac{dx}{dt}.$$

(3.9)

Example 3.17   *A curve is given in polar coordinates by $r = \sin \theta$. Find dy/dx at the point where $\theta = \frac{1}{8}\pi$.*

We can use θ as the parameter in the following way. The universal relation between polar and cartesian coordinates is

$$x = r \cos \theta, \qquad y = r \sin \theta.$$

On the special curve described by $r = \sin \theta$, these equations become

$$x = \sin \theta \cos \theta, \qquad y = \sin^2\theta.$$

Then, from Appendix B(c),

$$\frac{dx}{d\theta} = -\sin^2\theta + \cos^2\theta = \cos 2\theta$$

and

$$\frac{dy}{d\theta} = 2 \sin \theta \cos \theta = \sin 2\theta$$

Therefore

$$\frac{dy}{dx} = \frac{dy}{d\theta} \Big/ \frac{dx}{d\theta} = \tan 2\theta.$$

At the point where $\theta = \frac{1}{8}\pi$,

$$\frac{dy}{dx} = \tan \tfrac{1}{4}\pi = 1.$$

Example 3.18   *The map coordinates of a moving vehicle are given by $x = -t^2$, $y = \frac{1}{3}t^3$, where t is time and $t > 0$. Find the direction the vehicle is facing when $t = 2$.*

From (3.8),

$$\frac{dy}{dx} = \frac{dy}{dt} \Big/ \frac{dx}{dt} = \frac{t^2}{-2t} = -\tfrac{1}{2}t.$$

This equals -1 when $t = 2$. The slope of the curve is negative at this point, so the tangent to the path slopes downwards from left to right as shown in Fig. 3.2. The actual direction in which the vehicle is moving is, however, from right to left. It is facing north west as shown.

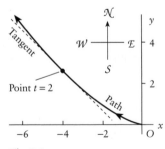

Fig. 3.2

From information such as that given in the previous example, the speed of a moving point can be calculated. Suppose that a point moves so that

$$x = x(t), \qquad y = y(t),$$

where t represents **time**. Figure 3.3 shows the effect of changing t to $t + \delta t$, where δt is small: the point moves from P to Q, a short distance δs say along the curve (δs is called an element of arc-length). Then the **average speed** over this short time is given by

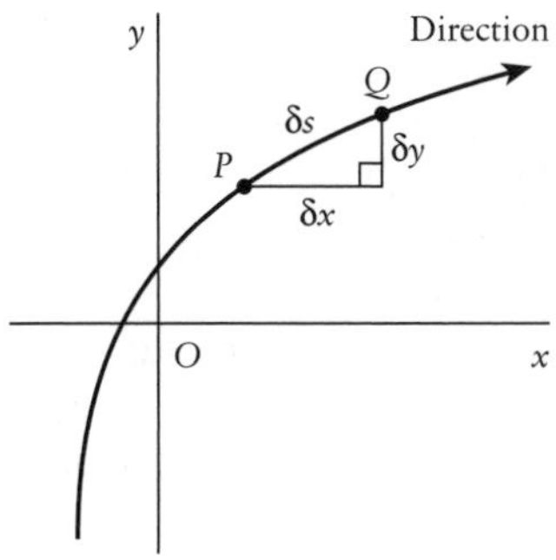

$$\frac{\text{arc-length } PQ}{\delta t} = \frac{\delta s}{\delta t}$$

$$\approx \frac{\text{straight distance } PQ}{\delta t}$$

$$= \frac{(\delta x^2 + \delta y^2)^{\frac{1}{2}}}{\delta t}$$

$$= \left[\left(\frac{\delta x}{\delta t} \right)^2 + \left(\frac{\delta y}{\delta t} \right)^2 \right]^{\frac{1}{2}}.$$

Fig. 3.3

Now let $\delta t \to 0$. Then $\delta x/\delta t$ and $\delta y/\delta t$ become dx/dt and dy/dt, and finally we have the result:

> **Speed of a moving point**
> Let $x = x(t)$ and $y = y(t)$, where t is time.
> The speed of the point is given by
>
> $$\text{speed} = \frac{ds}{dt} = \left[\left(\frac{dx}{dt} \right)^2 + \left(\frac{dy}{dt} \right)^2 \right]^{\frac{1}{2}},$$
>
> where ds stands for an element of arc-length.　　　　(3.10)

Example 3.19 *Find the speed of the vehicle in Example 3.18 when $t = 2$.*

In general

$$\frac{ds}{dt} = \left[\left(\frac{dx}{dt} \right)^2 + \left(\frac{dy}{dt} \right)^2 \right]^{\frac{1}{2}},$$

$$= (4t^2 + t^4)^{\frac{1}{2}}.$$

The speed is therefore $4\sqrt{2}$ when $t = 2$. (Speed is always counted as a non-negative number: when we want to make a distinction as to direction, the word 'velocity' is used.)

Problems

3.1 (Product rule, (3.1)). Obtain $df(x)/dx$ for the following $f(x)$:
(a) $x e^x$;   (b) $x \sin x$;   (c) $x \cos x$;
(d) $e^x \sin x$;   (e) $x \ln x$;   (f) $x^2 \ln x$;
(g) $e^x \ln x$;   (h) $x^2 e^x$;   (i) $\sin x \cos x$;
(j) $x^2 x^3$ (this is the same as x^5: show that the result is the same for both forms).

3.2 (Quotient and reciprocal rule, (3.2)). Obtain $df(x)/dx$ for the following $f(x)$:
(a) $\cot x$;   (b) $x/(x+1)$;   (c) $(\sin x)/x$;
(d) e^x/x;   (e) $(x^2-1)/(x^2+1)$;
(f) $(\tan x)/x^2$;   (g) $(\sin x + \cos x)/(\sin x - \cos x)$;
(h) $\sec x \ (= 1/\cos x)$;   (i) $\operatorname{cosec} x \ (= 1/\sin x)$;
(j) $x/(3x^2-2)$;   (k) $1/x(x^3+1)$;   (l) $1/\ln x$;
(m) x^n where n is a negative whole number $(x^n = 1/x^{-n})$;
(n) $1/(x+1)$;   (o) $e^{-x} (=1/e^x)$;
(p) $1/\tan x$;   (q) $x^{-2} \ln x$.

3.3 Find the first, second, and third derivatives of
(a) $1/(1-x)$;   (b) $x \sin x$;   (c) $x/(x-1)$;
(d) $f(x)g(x)$, where f and g are any functions.

3.4 (Chain rule, (3.3)). Obtain $df(x)/dx$ for the following $f(x)$. (Set out the calculation systematically, as in the examples in Section 3.3.)
(a) $\sin^2 x$;   (b) $\cos^2 x$;   (c) $\sin x^2$;   (d) $\cos x^2$;
(e) $\tan^2 x$;   (f) $\tan x^2$;   (g) $\cos(1/x)$;
(h) e^{-x} (compare Problem 3.2(o));
(i) $(x+1)^5$;   (j) $(x^3+1)^4$;   (k) $\sin 3x$;
(l) $\cos \frac{1}{2}x$;   (m) $\tan \frac{1}{2}x$;   (n) e^{-3x};
(o) $\sin(2x+1)$;   (p) $\cos(3x-2)$;
(q) $\tan(1-2x)$;   (r) $e^{1/x}$;

3.5 (General powers of x, Section 3.4). Differentiate the following.
(a) x^{-2};   (b) x^{-1};   (c) $x^{\frac{1}{3}}$;   (d) $x^{-\frac{1}{3}}$;   (e) $x^{\frac{3}{2}}$;
(f) $\sqrt{x}$;   (g) $\sqrt{(x^3)}$;   (h) $1/x$;   (i) $1/\sqrt{x}$.

3.6 Differentiate the following (the independent variable is not always x, and more than one rule is needed).
(a) $x^{\frac{1}{2}} \sin x$;   (b) $\sin^{\frac{1}{3}}x$;   (c) $(x^2+1)^{-\frac{1}{2}}$;
(d) $\sin^2(3t+1)$;   (e) $e^{-t} \cos t$;   (f) $e^{-t} \sin t$;
(g) $e^{-2t} \cos 3t$;   (h) $e^{-3t} \cos 2t$;   (i) $\sin x \cos^2 x$;
(j) $\sin^2 x \cos x$;   (k) $\left(\dfrac{\sin x}{x}\right)^2$;
(l) $x \sin^3 x$;   (m) $x \cos^3 x$.

3.7 Differentiate $\cos^2 x$ and $\sin^2 x$, (a) by using the identities $\cos^2 A = \frac{1}{2}(1 + \cos 2A$ and $\sin^2 A = \frac{1}{2}(1 - \cos 2A)$, (b) by using the product rule, (c) by using the chain rule.

3.8 Confirm the correctness of the following statements. The letters A, B, C, D, and n stand for any constants.
(a) If $x = A \cos 2t + B \sin 2t$, then $\dfrac{d^2 x}{dt^2} + 4x = 0$
(b) If $x = A \cos nt + B \sin nt$, then $\dfrac{d^2 x}{dt^2} + n^2 x = 0$
(c) If $x = A e^{3t} + B e^{-3t}$, then $\dfrac{d^2 x}{dt^2} - 9x = 0$
(d) If $x = A e^{nt} + B e^{-nt}$, then $\dfrac{d^2 x}{dt^2} - n^2 x = 0$
(e) If $x = A e^{-t} \cos t + B e^{-t} \sin t$, then
$$\frac{d^2 x}{dt^2} + 2\frac{dx}{dt} + 2x = 0$$
(f) If $y = A e^x + B e^{-x} + C \cos x + D \sin x$, then
$$\frac{d^4 y}{dx^4} - y = 0.$$

3.9 (Chain rule (3.3); or, more easily, the extension (3.6)). Differentiate the following functions.
(a) $e^{\cos^2 x}$;   (b) $e^{-\cos x^2}$;   (c) $\ln(\cos x^2)$;   (d) $(e^{x^2}-1)^4$.

3.10 (Logarithmic differentiation, Section 3.7, is easiest.) Differentiate the following.
(a) $x e^x \sin x$;   (b) $t e^t \cos t$;   (c) $x^{\frac{1}{2}} e^{2x} \sin^{\frac{1}{2}} 3x$.

3.11 (Implicit differentiation, Section 3.8). Proceed as in Example 3.15 to obtain expressions for dy/dx in the following.
(a) Show that if $x^2 + y^2 = 4$, then $dy/dx = -x/y$. Check the correctness of the expression by testing it with $y = \pm(4 - x^2)^{\frac{1}{2}}$. Interpret the result geometrically by sketching the circle $x^2 + y^2 = 4$ and considering the meaning of dy/dx in terms of slope.
(b) $x^{\frac{1}{2}} + y^{\frac{1}{2}} = 1$;   (c) $x^3 + xy - y^3 = 0$;
(d) $x \sin y - y \sin x = 1$.

3.12 The same expression for dy/dx in Problem 3.11a is obtained when the radius is changed; for example, if $x^2 + y^2 = 9$, we still get $dy/dx = -x/y$. Is this paradoxical? (Notice that even in the general case of $f(x, y) = c$, a constant, the expression for dy/dx will not depend on c: think of the difference between the form of the expression and the values it takes.)

3.13 Find expressions for dy/dx and then d^2y/dx^2 if $xy^2 - x^2y = 1$.

3.14 Differentiate the following inverse functions, using the method of Section 3.9. The results are quite important, and are included in the table of derivatives, Appendix D.
(a) $\arcsin x$; (b) $\arccos x$;
(c) $\arctan x$; (d) $\sinh^{-1} x$;
(e) $\cosh^{-1} x$; (f) $\tanh^{-1} x$.

3.15 (Parametric differentiation, Section 3.10). The curves in the following are in polar coordinates. Find dy/dx at the point specified.
(a) $r = \sin \frac{1}{2}\theta$ at $\theta = \frac{1}{2}\pi$;
(b) $r = 1 + \sin^2\theta$ at $\theta = \frac{1}{4}\pi$.

3.16 Obtain dy/dx in terms of t, then re-express it in terms of x, when the path of a point is given parametrically by the following.
(a) $x = t^3, y = t^2$; (b) $x = 2\cos t, y = 2\sin t$.

3.17 The path of a point is given parametrically by $x = a\cos t, y = b\sin t$. Show that the point travels around the ellipse

$$\frac{x^2}{a^2} + \frac{y^2}{b^2} = 1.$$

Express dy/dx in terms of t. Suppose that t represents time. Express the speed as a function of t.

3.18 Show that $(d/dx)(a^x) = a^x \ln a$ when $a > 0$. (Hint: write a^x in exponential form with base e.)

4 Applications of differentiation

Reminder. A basic table of derivatives is given in Appendix D at the end of the book.

4.1 Function notation for derivatives

So far, we have used the dy/dx or $(d/dx)f(x)$ notation for derivatives. The usefulness of the dy/dx notation is illustrated by the chain rule (3.3) and by (3.7)–(3.9): it strongly suggests the truth of certain results and makes them easy to remember. However, it is sometimes desirable to use another notation, $f'(x)$, which means exactly the same thing:

$$f'(x) \text{ means the same as } \frac{d}{dx} f(x).$$

By itself the symbol f' stands for the **derivative function**, because it is 'derived' from the original function f. Think of f' in the following way. Choose a 'neutral' letter for the independent variable, u say, which is not being used for anything else at the moment, and specify f in terms of u. For example, suppose that

$$f(u) = u^2 - 3u.$$

Then f' stands for the function specified by

$$f'(u) = \frac{d}{du} f(u) = 2u - 3.$$

Knowing now the form of the function f' (i.e. its formula), we can put anything we like in place of u, so that

$$f'(x) = 2x - 3, \qquad f'(t) = 2t - 3, \qquad f'(5) = 2 \cdot 5 - 3 = 7,$$
$$f'(x^3) = 2x^3 - 3, \qquad f'(x - ct) = 2(x - ct) - 3,$$
$$f'(g(x)) = 2g(x) - 3 \quad \text{(where } g \text{ is any function), and so on.}$$

The following examples show how this notation can be used.

Example 4.1 *The function f is defined by* $f(u) = \sin u$. *Obtain*

(a) $f(x^2)$; (b) $\dfrac{d}{dx} f(x^2)$; (c) $f'(x^2)$.

(a) $f(x^2) = \sin x^2$.

(b) $\dfrac{d}{dx} f(x^2) = \dfrac{d}{dx} \sin x^2 = 2x \cos x^2$

(by using the chain rule (3.3) with $u = x^2$).

(c) The **first thing to do is to obtain the function** f':

$$f'(u) = \frac{d}{du} f(u) = \frac{d}{du} (\sin u) = \cos u.$$

Now put $u = x^2$; then

$$f'(x^2) = \cos x^2.$$

Notice that **the result (c) is different from the result (b):** $f'(x^2)$ **is not the same as** $(d/dx)f(x^2)$. In (b) **we first find** $f(x^2)$ and then differentiate with respect to x; in (c) **we first find** $f'(u)$ and then put $u = x^2$.

Example 4.2 *Express* (a) *the product rule* (3.1); (b) *the quotient rule* (3.2a); (c) *the chain rule* (3.3); *in terms of the 'dash' notation.*

(a) **Product rule**

$$\frac{d}{dx}[u(x)v(x)] = u(x)v'(x) + v(x)u'(x).$$

or simply

$$(uv)' = uv' + vu'.$$

(b) **Quotient rule**

$$\left(\frac{u}{v} \right)' = \frac{1}{v^2}(vu' - uv').$$

(c) **Chain rule**

$$\frac{d}{dx} f(u(x)) = f'(u(x))u'(x).$$

Example 4.3 (a) *Suppose that f is any function. Express* $(d/dx)f(5x - 3)$ *in any terms available.* (b) *Verify the correctness of* (a) *in the special case when* $f(5x - 3) = \sin(5x - 3)$.

(a) Since the particular function f is not specified, the only thing to be done is to express $(d/dx)f(5x - 3)$ in terms of f', which is also unspecified. Then, from the chain rule (c) in Example 4.2, with $u = 5x - 3$,

---

Example 4.3 *continued*

$$\frac{d}{dx} f(5x - 3) = 5f'(5x - 3).$$

It is awkward to express the right-hand side without using the dash notation. One alternative is to write it as

$$\left[\frac{d}{du} f(u) \right]_{u=5x-3}.$$

(b) In this case $f(u) = \sin u$, so

$$f'(u) = \cos u.$$

The result in (a) predicts that

$$\frac{d}{dx} f(5x - 3) = 5 \cos(5x - 3).$$

This is the same as the result obtained by working out

$$(d/dx) \sin(5x - 3)$$

directly by using the chain rule with $u = 5x - 3$.

The dash notation extends to higher derivatives: we put

The dash notation

$$f'(x) = \frac{d}{dx} f(x), \quad f''(x) = \frac{d^2}{dx^2} f(x),$$

$$f'''(x) = \frac{d^3}{dx^3} f(x), \dots .$$

If $y = f(x)$, then the notation $y', y'', y''', \dots$ is also used. **(4.1)**

4.2 Maxima and minima

A prominent feature in the graph of any function

$$y = f(x)$$

is any point at which the graph 'turns over'. For example, in Fig. 2.9, the graph of $y = \sin x$ turns over at $x = \frac{1}{2}\pi$ and $x = \frac{3}{2}\pi$. These are points where **the slope changes sign** from positive to negative or negative to positive. The derivative of $f(x)$ is zero (the tangent is horizontal) at such turnover points: for example, it is easy to verify that, for $y = \sin x$,

$$f'(\tfrac{1}{2}\pi) = 0 \quad \text{and} \quad f'(\tfrac{3}{2}\pi) = 0.$$

Therefore

$$f'(x) = 0$$

can be looked on as an equation whose solutions include all the possible points at which the graph turns over.

However, graphs do not necessarily turn over at points where $f'(x) = 0$. For example, if $y = x^3$, then $f'(x) = 3x^2$. This is zero at $x = 0$, but the graph does not turn over at $x = 0$ (see Fig. 1.3): it flattens instantaneously and then continues upward.

Figure 4.1 sketches two typical cases in which the graph does turn over, at A ($x = a$) in Fig. 4.1a and at B ($x = b$) in Fig. 4.1b. Then $f'(a) = 0$ and $f'(b) = 0$.

If $f'(x) = 0$ at a point $x = c$, that is if

$$f'(c) = 0,$$

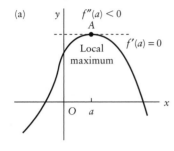

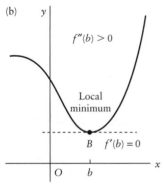

Fig. 4.1

then $x = c$ is called a **stationary point** of $f(x)$. A stationary point such as A in Fig. 4.1a is called a **maximum** of the function $f(x)$. More precisely, the function is said to have a **local maximum** at $x = c$, because the value of $f(x)$ at $x = c$ is greater than its value at any point in the immediate neighbourhood. (There may be local maxima elsewhere that are either greater or smaller than this one.) Similarly a point such as B in Fig. 4.1b is called a **local minimum** of $f(x)$.

To distinguish between types of stationary point algebraically, consider also the second derivative $f''(x)$ at $x = c$. Suppose that

$$f'(c) = 0 \quad \text{and} \quad f''(c) < 0.$$

The second derivative is

$$f''(x) = \frac{d^2y}{dx^2} = \frac{d}{dx}\left(\frac{dy}{dx}\right),$$

and this is **negative** at $x = c$. Therefore the slope, dy/dx, or $f'(x)$, is **decreasing** across $x = c$, and since $f'(x) = 0$ at $x = c$, $f'(x)$ must be positive on the left of c and negative on the right. Thus the graph is of the type shown in Fig. 4.1a, and the point is a **local maximum**.

If $x = c$ is a point where

$$f'(c) = 0 \quad \text{and} \quad f''(c) > 0,$$

then $f'(x)$ is **increasing**, and therefore goes from negative to positive, across $x = c$. The point is therefore a **minimum**, like the point B in Fig. 4.1b.

In the special case when

$$f'(c) = 0 \quad \text{but} \quad f''(c) = 0$$

there might occur a maximum (as with $y = -x^4$ at $x = 0$), or a minimum (as with $y = x^4$ at $x = 0$), or another feature called a **stationary point of inflection** (as with $y = \pm x^3$ at $x = 0$). These cases are illustrated in Fig. 4.2. One way to classify such a point is to examine directly the sign of dy/dx on both sides of the point.

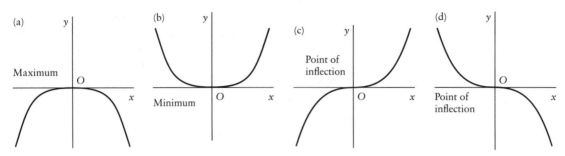

Fig. 4.2 Cases for which $f'(0) = 0$ and $f''(0) = 0$. (a) $y = -x^4$. (b) $y = x^4$. (c) $y = x^3$. (d) $y = -x^3$.

To summarize:

Stationary points of $f(x)$

Let $f'(c) = 0$; that is, $x = c$ is a stationary point of $f(x)$.
Stationary points can be classified by *either* examining
the sign of $f'(x)$ on both sides of $x = c$ *or* looking at the
sign of $f''(c)$.
(a) If $f''(c) < 0$, $f(x)$ has a local maximum at $x = c$.
(b) If $f''(c) > 0$, $f(x)$ has a local minimum at $x = c$.
(c) If $f''(c) = 0$, the stationary point might be a maximum, a
      minimum, or a point of inflection. Examine the sign of $f'(x)$
      on both sides of $x = c$.

(4.2)

Example 4.4   *Classify the stationary points of $f(x) = x^3 - 3x$.*

The stationary points are where $f'(x) = 0$; that is, where

$$3x^2 - 3 = 0, \quad \text{or} \quad x = \pm 1.$$

We need the signs of $f''(\pm 1)$, where

$$f''(x) = 6x.$$

Then

$$f''(1) = 6,$$

which is positive, so there is a minimum at $x = 1$. Also

$$f''(-1) = -6$$

which is negative, so there is a maximum at $x = -1$.
   The values of $f(x)$ at these points are

$$f(1) = -2, \qquad f(-1) = 2;$$

so the graph has the shape shown in Fig. 4.3. Alternatively, we could simply
have checked the signs of $f'(x) = 3x - 3$ on both sides of the stationary
points directly, instead of using the test (4.2).

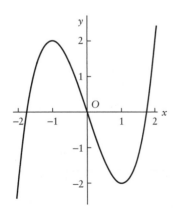

Fig. 4.3

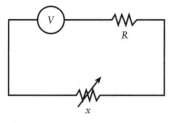

Fig. 4.4

Example 4.5 *In the circuit shown* in Fig. 4.4, *V is a constant voltage and R and x represent two resistances: R is fixed and x is variable. The rate of heat generation y in resistance x is equal to I²x where I is the current. Show that y is a maximum when x = R.*

Current equals voltage divided by total resistance, so

$$I = \frac{V}{R + x}.$$

Therefore the rate of heat generation is

$$y = \frac{V^2 x}{(R + x)^2} = f(x),$$

say. If there is a maximum, it will occur when $f'(x) = 0$. From the quotient rule (3.2)

$$f'(x) = \frac{V^2}{(R + x)^4}[(R + x)^2 - x \cdot 2(R + x)] = V^2\frac{R - x}{(R + x)^3}. \tag{i}$$

This is zero when $x = R$.

To show that $f(x)$ has a maximum when $x = R$ we may work out the sign of $f''(R)$. From (i),

$$f''(x) = \frac{V^2}{(R + x)^6}[(R + x)^3(-1) - (R - x) \cdot 3(R + x)^2]$$

$$= \frac{V^2}{(R + x)^4}(-4R + 2x).$$

Therefore

$$f''(R) = -V^2/8R^3,$$

which is negative, so $x = R$ corresponds to a maximum of y.

However, it is easier to look instead at the expression (i) for $f'(x)$. When $x < R$, we have $f'(x) > 0$, so $f(x)$ is increasing. When $x > R$, we have $f'(x) < 0$, so $f(x)$ is decreasing. This ensures that a maximum has been obtained without the need to differentiate again.

Example 4.6 *x and y are two numbers subject to the restriction that x + y = 1. Find the maximum possible value of xy.*

There are two variables, x and y, but we can reduce the problem to one involving only x by using the fact that $x + y = 1$, so that

$$y = 1 - x. \tag{i}$$

In that case,

$$xy = x(1 - x) = x - x^2 = f(x),$$

say. Now $f(x)$ has a stationary point (a maximum, minimum, or point of inflection) where $f'(x) = 0$, that is to say, where

$$1 - 2x = 0, \quad \text{or} \quad x = \tfrac{1}{2}.$$

By (4.2), this value of x delivers a maximum, because $f''(x) = -2$ (for any value of x) which is negative. From (i), $y = \tfrac{1}{2}$ when $x = \tfrac{1}{2}$, so the maximum value of xy is $\tfrac{1}{4}$.

4.3   Exceptional cases of maxima and minima

The method of finding local maxima and local minima by solving $f'(x) = 0$ reveals only points where the slope of the graph of $y = f(x)$ is horizontal. Sometimes there is a **maximum or minimum at an end-point of an interval**, even if the graph is not horizontal there.

Example 4.7   *Suppose that the values of x to be considered are restricted to lie between 0 and 1 inclusive: that is, $0 \leqslant x \leqslant 1$. Find the points on this interval at which $x - x^2$ takes maximum and minimum values.*

The graph of $y = f(x) = x - x^2$ between $x = 0$ and $x = 1$ is shown is Fig. 4.5. The maximum of $f(x) = x - x^2$ which we found in Example 4.6 at $x = \frac{1}{2}$ can be seen. But, understood in a commonsense way, there are minimum values at $x = 0$ and $x = 1$, the end-points of the restricted interval. These cannot be detected by the method of differentiation. Whether we are interested in them would depend on the demands of any practical problem from which the question originated.

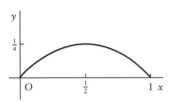

Fig. 4.5 $y = x - x^2, 0 \leqslant x \leqslant 1$.

In problems of the type illustrated in Example 4.6 this situation can arise naturally, as in the following example.

Example 4.8   *Find the maximum and minimum values of $x^2 - y^2$ on the circle $x^2 + y^2 = 1$.*

It is evident that the point (x, y) can only be on the circle if x and y both have values between -1 and 1 inclusive, that is if

$$-1 \leqslant x \leqslant 1 \quad \text{and} \quad -1 \leqslant y \leqslant 1. \tag{i}$$

A restricted interval therefore arises naturally in the problem. On the circle $x^2 + y^2 = 1$, we have

$$y^2 = 1 - x^2, \tag{ii}$$

so

$$x^2 - y^2 = 2x^2 - 1 = f(x), \tag{iii}$$

say. To find the stationary points of $f(x)$ we see that $f'(x) = 4x$, which is zero when $x = 0$. Also $f''(0) = 4 > 0$, so $x = 0$ is a local minimum of $f(x)$, whose value is $f(0) = -1$.

However, we have overlooked something. In Fig. 4.6, we show the graph of $f(x) = 2x^2 - 1$, within the permitted interval $-1 \leqslant x \leqslant 1$. The local minimum at $x = 0$ can be seen, but there are also maxima at the end-points $x = -1$ and $x = 1$, where $f(x)$ takes the values $+1$.

Alternatively, the maxima at $x = \pm 1$ can be found by substituting for x instead of y at the first stage. Put $x^2 = 1 - y^2$, so that

$$x^2 - y^2 = 1 - 2y^2 = g(y),$$

say, and solve $g'(y) = 0$: we then find a local maximum at $y = 0$, where $x = \pm 1$. However, we also lose sight of the minima we found before. The subject is discussed again in Section 28.2.

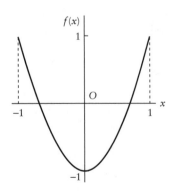

Fig. 4.6 $f(x) = 2x^2 - 1, -1 \leqslant x \leqslant 1$.

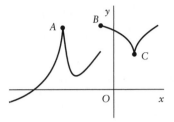

Fig. 4.7

Another possibility is that there may be points at which the graph of $y = f(x)$ does not have a definite tangent. Then $f'(x)$ or dy/dx has no meaning at such points. For example, in Fig. 4.7, there is no tangent at the points A, B, and C. The points A and B could qualify as local maxima, and C as a local minimum, but at A and C the graph suddenly changes direction, and at B there is a jump in the value of $f(x)$. These points cannot be located by solving $f'(x) = 0$, because $f'(x)$ does not exist at A, B, and C.

4.4 Sketching graphs of functions

To **sketch a graph** is to indicate its general shape so as to draw attention to its most important features without being concerned with accurate plotting. To do this it is necessary for you to have a clear idea of the shape of the graphs of the basic functions

$$x^a, \quad e^{ax}, \quad \sin ax, \quad \cos ax, \quad \ln x.$$

Example 4.9   *Sketch the graph $y = 1 - 1/(1 + x)^2$.*

This can be done in stages, as shown in Fig. 4.8. Figure 4.8c is obtained from 4.8b by using the rule (1.11) with $c = 1$; it simply involves sliding the graph $y = -1/x^2$ one unit to the left. To get from 4.8c to 4.8d, we add 1, which moves the graph up the y axis by one unit.

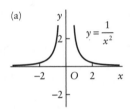

(a)

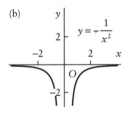

(b)

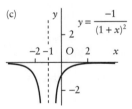

(c)

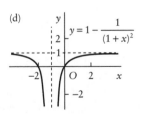

(d)

Fig. 4.8

In Fig. 4.8d of Example 4.9 we see that, as x increases, becoming large and positive, the value of

$$y = 1 - \frac{1}{(1 + x)^2}$$

gets closer and closer to 1. The same is true when x becomes large and negative. This is obviously an important feature of the graph. It can be seen to be true by thinking what happens to y when we put a **large value** of x into the formula for y (think of a *very* large number: $x = 1\,000\,000$, rather than $x = 10$). Then obviously $1/(1 + x)^2$ is very small, so y gets very close to 1, and the larger x becomes, the nearer y is to 1. The same is true when x is large and negative.

We say that, as x increases, the graph **approaches the line** $y = 1$, which in general terms is called an **asymptote** of the graph. When x approaches -1, the graph approaches the vertical line $x = -1$; this is also called an asymptote. The two continuous halves of the graph to the left and right of $x = -1$ are called **branches**.

Suppose that $y = f(x)$ is to be sketched. A general question to be asked is 'What happens to y when x increases towards infinity (or decreases towards minus infinity)?' We normally say 'as x **approaches** $\pm\infty$', and as usual indicate the approach by '$\to$':

$$x \to \pm\infty.$$

For example, $1/x \to 0$ when $x \to -\infty$. Also

$$\frac{x-1}{3x+2} \to \tfrac{1}{3} \quad \text{when} \quad x \to \infty \text{ (or } x \to -\infty).$$

To see this, think of the effect of giving x an immense value. Only the terms x and $3x$ are significant; they are said to **dominate** the expression, so

$$\frac{x-1}{3x+2} \to \frac{x}{3x} = \tfrac{1}{3}.$$

The **limit notation** can be used in this context (see Section 2.2). We can write, for example,

$$\lim_{x\to\infty} \frac{1-2x}{1+x} = -2.$$

The reasoning is the same as in the earlier case: think of a very large value of x.

Very often the function has no definite limit as $x \to \infty$. For example, $\lim_{x\to\infty} \sin x$ *does not exist*; no definite single number is approached, since $\sin x$ simply goes up and down between ± 1 for ever. However, it is quite usual to write, say,

$$\lim_{x\to\infty} x^2 = \infty,$$

even though ∞ is not a number.

Notice the following result:

$$\lim_{x\to\infty} ax^n \, e^{-cx} = 0,$$

where a and n are any constants, and c is a *positive* constant.   **(4.3)**

We shall not prove (4.3); but, to convey the feel of it, a table of values is given for the special case of $x^3 \, e^{-x}$:

x	0	1	2	3	4	8	10
$x^3 \, e^{-x}$	0	0.36	1.08	1.34	1.17	0.18	0.05

Fairly large values are needed before the function settles down to approach zero, because x^3 is increasing, and therefore competes with e^{-x} in the early stage. However, e^{-x} will beat any power of x down to zero eventually. In the following example, we sketch the graph of the function in the table without using the calculated values above.

Example 4.10 *Sketch the graph* $y = x^3 e^{-x}$.

Do it in stages, using any easily obtained facts you can think of.

    1. *Are there any points where it is easy to obtain values?* At $x = 0$, $y = 0$.

    2. *Are there any definite points where $x^3 e^{-x}$ is infinite?* There are no such points.

    3. *Are there any points where the graph crosses the x axis?* Only the point found in (1).

    4. *Are there any maxima/minima?*

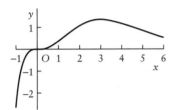

$$\frac{dy}{dx} = 3x^2 e^{-x} - x^3 e^{-x} = x^2(3 - x) e^{-x}.$$

This is zero when $x = 0$ and $x = 3$, so these are stationary points.

    dy/dx is positive when $x < 3$, and negative when $x > 3$, so $x = 3$ is a maximum. Since $e^3 \approx 20$, at this point $y \approx 1\frac{1}{3}$.

    Near the other stationary point $x = 0$, dy/dx is positive on both sides, so $x = 0$ is a point of inflexion.

    5. *Behaviour as $x \to \infty$.* According to (4.3), $y \to 0$ as $x \to \infty$.

    6. *Behaviour as $x \to -\infty$.* As $x \to -\infty$, we have $x^3 \to -\infty$ and $e^{-x} \to \infty$ (think, for example, of $x = -1000$). Therefore, $x^3 e^{-x} \to -\infty$ (very rapidly).

    The sketch is shown in Fig. 4.9.

Fig. 4.9

Example 4.11 *Sketch the graph of* $e^{-\frac{1}{3}x} \sin 2x$ *for* $0 \leq x \leq 2\pi$.

(a)

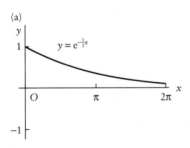

(b)

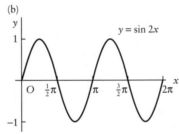

(c)

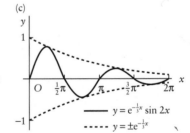

Fig. 4.10

(x is assumed to be in radians.) Split the expression into its two factors, $e^{-\frac{1}{3}x}$ and $\sin 2x$. These are shown in Fig. 4.10a,b. The value of $e^{-\frac{1}{3}x}$ drops to about $\frac{1}{8}$ at $x = 2\pi$. Also, $\sin 2x$ is zero when

$$2x = 0, \pi, 2\pi, \dots, 6\pi,$$

or when

$$x = 0, \tfrac{1}{2}\pi, \pi, \tfrac{3}{2}\pi, 2\pi.$$

    The product of the two is shown in Fig. 4.10c. The graph crosses the x axis (i.e. $y = 0$) where $\sin 2x = 0$, and nowhere else. The height of the peaks and troughs of $e^{-\frac{1}{3}x} \sin 2x$ are estimated by the size of the factor $e^{-\frac{1}{3}x}$, shown as a broken line, which multiplies the maxima and minima of $\sin 2x$. The new maxima and minima do not occur at exactly the same points: it is left to the reader to show that the new maxima and minima occur at values of x which satisfy the equation $\tan 2x = 6$.

(a)

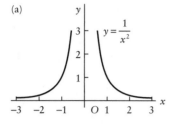

(b)

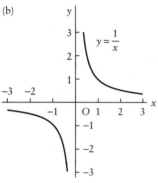

Fig. 4.11

It is useful to be able to distinguish between the behaviour of functions such as those shown in Fig. 4.11a,b. The function $1/x$ of Fig. 4.11b is infinite at $x = 0$, but the sign changes across $x = 0$. The terminology used to describe $y = 1/x$ near $x = 0$ is

$$y \rightarrow -\infty \text{ as } x \rightarrow 0 \text{ from the left;}$$

$$y \rightarrow \infty \ \text{ as } x \rightarrow 0 \text{ from the right.}$$

Example 4.12   *Sketch the graph of* $y = 1/(x-2)(x+1)$.

Look out for the obvious things first. At $x = 0$, we have $y = -\frac{1}{2}$. The function is infinite at $x = 2$ and $x = -1$. It does not cross the x axis anywhere.

Now consider the sign of $1/(x-2)(x+1)$. It is positive when $x < -1$ (try e.g. $x = -3$). It is positive when $x > 2$ (try e.g. $x = 3$). It is negative when $-1 < x < 2$, which is linked with the facts that the graph does not cross the x axis and, as we already know, that y is negative when $x = 0$.

We know now that

$$y \rightarrow \infty \ \text{ as } x \rightarrow -1 \text{ from the left;}$$

$$y \rightarrow -\infty \text{ as } x \rightarrow -1 \text{ from the right;}$$

$$y \rightarrow -\infty \text{ as } x \rightarrow 2 \ \text{ from the left;}$$

$$y \rightarrow \infty \ \text{ as } x \rightarrow 2 \ \text{ from the right;}$$

so Fig. 4.12 is emerging.

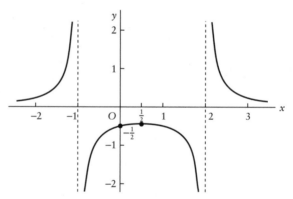

Fig. 4.12 $y = 1/(x-2)\ (x+1)$.

We now locate precisely the obvious maximum between $x = -1$ and 2, and make sure there are no other stationary points. By the reciprocal rule (3.2b), we have

$$\frac{dy}{dx} = \frac{1}{(x-2)^2(x+1)^2} \frac{d}{dx}[(x-2)(x+1)] = \frac{2x-1}{(x-2)^2(x+1)^2}.$$

This is zero at $x = \frac{1}{2}$ and nowhere else. There is no need to use the test (4.1); the point can only be a maximum since there are no other stationary points. The value of y there is $-\frac{4}{9}$.

We now return to asymptotes and show that there can be asymptotes that slope. Consider the function

$$y = \frac{x^2 - 1}{2x + 1},$$

when x is large, positive, or negative. The term $x^2 - 1$ is **dominated** by x^2, meaning that the part -1 is negligible compared with x^2 when x is large. Likewise the **dominant term** in $2x + 1$ is $2x$. It is therefore obvious that

$$y \to \pm\infty \quad \text{when} \quad x \to \pm\infty.$$

However, we can do much better than this, because it can be seen by polynomial division that

$$y = \frac{x^2 - 1}{2x + 1} = \tfrac{1}{2}x - \tfrac{1}{4} - \frac{\tfrac{3}{4}}{2x + 1}.$$

Therefore the graph will approach the straight line

$$y = \tfrac{1}{2}x - \tfrac{1}{4}$$

when x is large. As in the earlier instances we have seen, the line $y = \tfrac{1}{2}x - \tfrac{1}{4}$ is said to be an **asymptote** of the original graph. The notation

$$y \sim \tfrac{1}{2}x - \tfrac{1}{4} \quad \text{when} \quad x \to \pm\infty$$

is sometimes used, meaning that the curve approaches the line $y = \tfrac{1}{2}x - \tfrac{1}{4}$ when x is large. The curve is sketched in Example 4.13.

In the same way, a function may be an **asymptotic to a curve** as $x \to \pm\infty$. For example, if

$$y = \frac{1}{x} - \frac{1}{x^3} \sin x,$$

then

$$y \sim \frac{1}{x} \quad \text{when} \quad x \to \pm\infty.$$

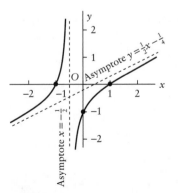

Fig. 4.13

Example 4.13   *Sketch the graph of $y = (x^2 - 1)/(2x + 1)$.*

The curve cuts the x axis ($y = 0$) at $x = \pm 1$. Also $y = -1$ when $x = 0$. The function is infinite at $x = -\tfrac{1}{2}$.

Also, as shown above, the straight line $y = \tfrac{1}{2}x - \tfrac{1}{4}$ is an asymptote for large values of x. This is shown as the broken line in Fig. 4.13.

From the quotient rule (3.2a),

$$\frac{dy}{dx} = \frac{(2x + 1)(2x) - (x^2 - 1)2}{(2x + 1)^2} = 2\frac{x^2 + x + 1}{(2x + 1)^2}.$$

This is never zero, because the equation $x^2 + x + 1 = 0$ has no real solutions. Therefore, there are no stationary points.

4.5 Estimating small changes

Let

$$y = f(x).$$

Suppose that the value of x changes by a small amount δx. Then y will change by a small amount δy. There is a simple approximate relation between δy and δx which is important for practice and theory.

Fix a particular value of x, say $x = a$, the small deviation δx will be made from this value of x. The derivative at $x = a$ is $f'(a)$. According to (2.8), to obtain $f'(a)$ we take a nearby point $x = a + \delta x$ and form the ratio

$$\frac{\delta y}{\delta x} = \frac{f(a + \delta x) - f(a)}{\delta x},$$

and

$$\frac{\delta y}{\delta x} \rightarrow f'(a) \quad \text{as } \delta x \rightarrow 0.$$

If δx is small enough, $\delta y / \delta x$ will become close in value to $f'(a)$:

$$\frac{\delta y}{\delta x} \approx f'(a),$$

so that

$$\delta y \approx f'(a)\, \delta x.$$

This is how to obtain an **approximation to the change δy in y due to a small change from $x = a$ to $x = a + \delta x$**. It is easier to *remember* the result in the form

$$\delta y \approx \frac{\mathrm{d}y}{\mathrm{d}x} \delta x,$$

near a general point x (which again shows the usefulness of the $\mathrm{d}y/\mathrm{d}x$ notation in suggesting true results). We call this the **incremental approximation** for functions of a single variable.

Incremental approximation

For a small increment δx from $x = a$:

(a) $\delta y \approx f'(a)\, \delta x.$
(b) (Mnemonic form)

$$\delta y \approx \frac{\mathrm{d}y}{\mathrm{d}x} \delta x.$$

(4.4)

Example 4.14 *Let $y = x + 1/x$. Estimate the change δy in y when x changes from $x = 2$ to $x = 1.8$. Compare the estimate with the exact value of δy.*

Put

$$y = x + \frac{1}{x} = f(x).$$

Then

$$\frac{dy}{dx} \text{ or } f'(x) = 1 - \frac{1}{x^2},$$

so that $f'(2) = 0.75$. Here $\delta x = 1.8 - 2.0 = -0.2$; so, by (4.4a),

$$\delta y \approx 0.75 \times (-0.2) = -0.15.$$

(The exact value is given by $\delta y = (1.8 + 1/1.8) - (2.0 + 1/2.0) = -0.1444\dots.$)

Example 4.15 *The volume V of a sphere of radius r is given by $V = \frac{4}{3}\pi r^3$. Estimate the change in volume if the radius increases from 2.0 to 2.1 metres.*

We shall use the letters that the question offers, considering δV and δr. Put

$$V = \tfrac{4}{3}\pi r^3 = f(r).$$

Then

$$f'(r) = \frac{dV}{dr} = 4\pi r^2,$$

so, by (4.4b),

$$\delta V \approx 4\pi r^2\, \delta r. \tag{i}$$

(Notice that $4\pi r^2$ is the formula for the *surface area* of a sphere: the change in volume is nearly equal to the surface area times the thickness δr.) Now put $r = 2$:

$$f'(2.0) = 16\pi \quad \text{and} \quad \delta r = 2.1 - 2.0 = 0.1.$$

Then, by (i),

$$\delta V \approx 16\pi \times 0.1 = 5.02.$$

The exact value is $\delta V = 5.282\dots$ (cubic metres), implying an error of 5% in our estimate.

The number $\delta V \approx 5.02$ in Example 4.15 might not seem to qualify as a *small* change. Furthermore, if we express the identical problem in different units, say in centimetres rather than metres, the numbers are even larger; then δr becomes 10 (cm) and δV is about 5×10^6 (cm³). On the other hand, if the units had been kilometres then δr and δV would have looked very small indeed. But nothing at all is changed except the units of measurement. We still get only a 5% error in the estimate. The reason for this is that the ratio

$$\text{Estimated } \delta V / \text{Exact } \delta V = [f'(r)\,\delta r]/[f(r+\delta r) - f(r)]$$

is **dimensionless**; that is to say it is unaffected by the choice of units.

There is no easy way to predict when the method will work well: geometrically speaking, we are content to guess that the graph sticks sufficiently closely to its tangent line at a within the interval $a \pm \delta x$.

Example 4.16 *The cosine rule for a triangle ABC is $c^2 = a^2 + b^2 - 2ab \cos C$. In a triangle for which $a = 3$ and $b = 4$, estimate the change in c when C increases from 60° to 65°.*

Put in the fixed numbers, $a = 3$ and $b = 4$; then

$$c^2 = 25 - 24 \cos C$$

or

$$c = (25 - 24 \cos C)^{\frac{1}{2}} = f(C),$$

say. By the chain rule (3.3) with $u = 25 - 24 \cos C$,

$$f'(C) = \frac{\mathrm{d}c}{\mathrm{d}C} = -(25 - 24 \cos C)^{-\frac{1}{2}}(24 \sin C).$$

The quantity δC must be measured in radians, because radian measure was assumed in obtaining the derivatives of the sin and cos functions. So we put

$$C = 60° = \tfrac{1}{3}\pi \text{ radians}, \qquad \delta C = \tfrac{5}{180}\pi = 0.087 \text{ radians}.$$

We know that $\cos C = \tfrac{1}{2}$ and $\sin C = \tfrac{1}{2}\sqrt{3}$, so

$$f'(\tfrac{1}{3}\pi) = 6\sqrt{3}/\sqrt{13}.$$

Therefore, by the incremental approximation (4.4),

$$\delta c \approx (6\sqrt{3}/\sqrt{13}) \times 0.087 = 0.25.$$

(The exact change is $\delta c = 0.2489\ldots$.)

4.6 Numerical solution of equations: Newton's method

It is often necessary to solve equations for which there is no straightforward method of solution. (In fact, this is true for practically all equations.) Simple examples are the equations

$$x^4 + x^3 - 1 = 0 \quad \text{and} \quad \mathrm{e}^{-x} - x = 0.$$

For such cases, there are many methods for obtaining **numerical solutions**, which are applicable no matter how complicated the equation is. We describe one of them here.

To apply the method, it is necessary first of all to obtain at least a rough idea of the location of the solution we are seeking. There are

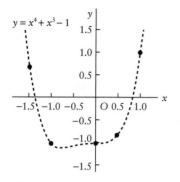

Fig. 4.14 $y = x^4 + x^3 - 1$.

various ways of doing this: for example, we can plot a rough graph. Taking the first example above, the graph

$$y = x^4 + x^3 - 1$$

is sketched in Fig. 4.14 using only five values of x: namely, -1.5, -1, 0, 0.5, and 1. The solutions occur where it crosses the axis; there seems to be one not far from -1.3 and one not far from 0.8.

Suppose now that we have a general equation to solve:

$$f(x) = 0;$$

and that, by drawing its graph, or by some other method, we have established that one of its solutions is not far from the value

$$x = x_0,$$

say. We show how to locate this solution accurately.

Figure 4.15 shows one possibility for the shape of the graph of $y = f(x)$ close to its (unknown) solution $x = c$, say, corresponding to the point C. (If the graph is different from this, the discussion is much the same.) The initial estimate $x = x_0$ corresponds to A_0. The point A_0 could be on the left of the solution C as shown, or on the right; we are not likely to be sure: again the argument is much the same (see Problem 4.14).

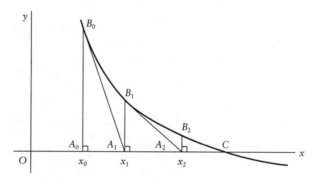

Fig. 4.15

Perform, in imagination, the following steps.

1. Start at the point A_0, where $x = x_0$. Draw the perpendicular A_0B_0, intersecting the curve at B_0. Construct the tangent at B_0, and continue it to intersect the x axis at A_1, where $x = x_1$. Then A_1 is nearer to the solution C than A_0.

2. Repeat the process, starting with the improved estimate A_1. We arrive at A_2, where $x = x_2$, which is a still better approximation.

3. Using the new estimates as starting values as they arise, keep repeating the process to produce a sequence of approximations

$$A_0 \, (x = x_0), \quad A_1 \, (x = x_1), \quad A_2 \, (x = x_2), \quad A_3 \, (x = x_3), \quad \dots$$

and stop when the accuracy attained is satisfactory.

The steps 1–3 can be carried out algebraically:

1. Starting with A_0 $(x = x_0)$, the equation of the tangent line at B_0 is given by

$$\frac{y - f(x_0)}{x - x_0} = f'(x_0).$$

At A_1 $(x = x_1)$, we have $y = 0$, so

$$\frac{-f(x_0)}{x_1 - x_0} = f'(x_0).$$

Therefore, the new approximation A_1 $(x = x_1)$ is given by

$$x_1 = x_0 - \frac{f(x_0)}{f'(x_0)}.$$

2. x_1 takes the place of x_0, and x_2 takes the place of x_1, so

$$x_2 = x_1 - \frac{f(x_1)}{f'(x_1)}.$$

3. Once the nth approximation x_n is available, the $(n + 1)$th value x_{n+1} is given by

$$x_{n+1} = x_n - \frac{f(x_n)}{f'(x_n)}.$$

This process, in which essentially we do exactly the same thing over and over again, using for each step the information obtained from the previous step, is called a **step-by-step process** or an **iterative process**. It is summarized in the following **algorithm** (or recipe), known as **Newton's method**.

Newton's method for the numerical solution of $f(x) = 0$

Find a value $x = x_0$ sufficiently close to the solution required. Then carry out the following step-by-step process until the desired accuracy is obtained:

$$x_{n+1} = x_n - \frac{f(x_n)}{f'(x_n)},$$

for $n = 0, 1, 2, 3, \ldots$, successively.   (4.5)

The following Examples work through the equations with which we opened the Section.

Example 4.17 *The equation $x^4 + x^3 - 1 = 0$ has a solution near $x = 0.8$. Find it to five-decimal accuracy.*

We have

$$x_0 = 0.8, \qquad f(x) = x^4 + x^3 - 1, \qquad f'(x) = 4x^3 + 3x^2.$$

Then, in (4.5),

$$x_{n+1} = x_n - \frac{x_n^4 + x_n^3 - 1}{4x_n^3 + 3x_n^2}.$$

Starting with $x_0 = 0.8$, we obtain the following table:

n	0	1	2	3
x_n	0.800 00	0.819 76	0.819 17	0.819 17
$f(x_n)$	−0.078 40	0.002 47	0.000 00	0.000 00

Evidently we do not have to pursue the sequence any further.

Example 4.18 *The equation $e^{-x} = x$ has a solution near to $x = 0.5$. Find the solution accurately to five decimal places.*

We have

$$x_0 = 0.5, \qquad f(x) = e^{-x} - x, \qquad f'(x) = -e^{-x} - 1.$$

From (4.4),

$$x_{n+1} = x_n - \frac{e^{-x_n} - x_n}{-e^{-x_n} - 1} = \frac{x_n + 1}{e^{x_n} + 1}$$

(the last step for simplicity of calculation). We obtain the following sequence:

x_0	x_1	x_2	x_3
0.500 00	0.566 31	0.567 14	0.567 14

This repetitive process described is easy to program for a computer for individual cases as they arise, and then the complexity of the equation is of no importance. The same program can be adapted to scan a range of x in order to get a provisional idea of where the solutions are to be found. A simple program combined for safety's sake with inspection of the whole sequence of values output would satisfy most requirements.

However, to write a program which will *automatically*, without intervention, find *all* the solutions for *any* function $f(x)$ that might be presented to it is a very different matter. For example, we would have to find means to be absolutely sure that none of the possible tangents would by chance carry us an irrecoverable distance away from the solution we are seeking (see e.g. Problem 4.15) by designing a way of automatically recognizing and rectifying the situation if it occurs.

4.7 The binomial theorem

A proof of the binomial theorem has been given in Section 1.18 (eqn (1.41)), obtained by counting combinations of terms. There follows a simpler proof, obtained by repeated differentiation. Equation (4.6) below is the same as (1.41), with a slight change of expression:

The binomial theorem

If n is a positive integer, then

(a) $(a+b)^n = a^n + c_1 a^{n-1} b + c_2 a^{n-2} b^2 + \cdots + c_n b^n$,

   in which

$$c_r = \frac{n!}{r!(n-r)!} = \frac{1}{r!} n(n-1)(n-2) \cdots (n-r+1),$$

   and $c_r = c_{n-r}$.

(b) Special case

$$(1+x)^n = 1 + c_1 x + c_2 x^2 + \cdots + c_n x^n.$$

   (4.6)

Proof. Consider firstly the case $(1 + x)^n$, where n is any positive integer. It is clear that this can be expressed as a polynomial of degree n. Suppose the coefficients are $c_0, c_1, c_2, \ldots, c_n$. Then

$$(1+x)^n \equiv 1 + c_1 x + c_2 x^2 + \cdots + c_n x^n. \tag{4.7}$$

We have written '$\equiv$' in place of '$=$' to stress that this is an **identity**: that is, it is true for all values of x (see Section 1.1). Differentiate both sides once, twice, $\ldots$, n times:

$$n(1+x)^{n-1} \equiv c_1 + 2c_2 + 3c_3 x^2 + 4c_4 x^3 + \cdots + nc_n x^{n-1},$$

$$n(n-1)(1+x)^{n-2} \equiv 2c_2 + 3 \cdot 2c_3 x + 4 \cdot 3c_4 + \cdots + n(n-1)c_n x^{n-2},$$

$$n(n-1)(n-2)(1+x)^{n-3} \equiv 3 \cdot 2c_3 + 4 \cdot 3 \cdot 2c_4 + \cdots + n(n-1)(n-2)c_n x^{n-3}$$

up to the nth derivative, which is

$$n(n-1)(n-2) \ldots 1 \equiv n(n-1)(n-2) \ldots 1 \cdot c_n.$$

Since these are all identities, they are true when $x = 0$, and for this value they become

$$n = c_1, \quad n(n-1) = 2c_2, \quad n(n-1)(n-2) = 3 \cdot 2c_3, \quad \ldots,$$

and in general, for the rth derivative, with $1 \leqslant r \leqslant n$,

$$n(n-1)(n-2) \ldots (n-r+1) = r!c_r, \tag{4.8}$$

where the term on the left contains r factors. This immediately gives the result (4.6b).

Equation (4.6a) is derived from (4.6b) by writing

$$(a+b)^n = a^n[1 + (b/a)]^n.$$

We now know the coefficients c_r, so we may put $x = b/a$ into (4.7):

$$(a+b)^n = a^n[1 + c_1(b/a) + c_2(b/a)^2 + \cdots + c_n(b/a)^n]$$

$$= a^n + c_1 a^{n-1} b + c_2 a^{n-2} b^2 + \cdots + c_n b^n,$$

as required.

Problems

4.1 (See Section 4.1 on the 'dash' notation.) The function f is defined by $f(u) = u^2$. Obtain the following.

(a) $f'(t)$;   (b) $f'(t^2)$;   (c) $\dfrac{d}{dt} f(t^2)$;

(d) $f'(t^{\frac{1}{2}})$;   (e) $\dfrac{d}{dt} f(t^{\frac{1}{2}})$;   (f) $f''(t^{\frac{1}{2}})$.

4.2 (See Section 4.2.) Find the stationary points of the following functions and classify them as maxima, minima, or points of inflection.

(a) $x^2 - x$;   (b) $x^2 - 2x - 3$;   (c) $x \ln x \ (x > 0)$;
(d) $x e^{-x}$;   (e) $1/(x^2 + 1)$;   (f) $x^2 - 3x + 2$;
(g) $e^x + e^{-x}$;   (h) $x^2 + 4x + 2$;   (i) $x - x^3$;
(j) $x^2(x - 1)$;   (k) $\sin x - \cos x$ (in $0 < x < 2\pi$);
(l) $\sin x \cos x$ $(-\pi < x < \pi)$;   (m) $e^{-x} \sin x$;
(n) $e^{-\frac{1}{3}x} \sin 2x$ (see Example 4.11);   (o) $x - \cos x$;
(p) $2e^x - \frac{1}{2} e^{2x}$;   (q) $x^2 e^{-x}$;   (r) $(\ln x)/x \ (x > 0)$;
(s) $(1 - x)^3$;   (t) $\sin^3 x$;   (u) e^{-x^2};
(v) $e^{x^2 - x}$;   (w) $x + x^{-1}$;   (x) $x^3 e^{-x}$.

4.3 Let $y = f(u(x))$. Use two successive applications of the chain rule in the form of Example 4.2c to show that

$$\frac{d^2 y}{dx^2} = f''(u(x))[u'(x)]^2 + f'(u(x))u''(x).$$

Show that if $f'(u)$ is always greater than zero, or always less than zero, then $f(u(x))$ and $u(x)$ have the same stationary points. Consider, for example, Problem 4.2v in this connection, with $f(u) = e^u$ and $u(x) = x^2 - x$: it becomes rather obvious.

4.4 A rectangular piece of ground is to be marked out, which must have a given area A. Find the dimensions of the plot which requires the minimum length of perimeter fence. (This is a 'restricted' problem, like Example 4.6. Call the sides x and y.)

4.5 A tunnel cross-section is to have the shape of a rectangle surmounted by a semicircular roof. The total cross-sectional area must be A, but the perimeter minimized to save building costs. Find its dimensions.

4.6 A circular-cylindrical oil drum is required to have a given surface area (including its lid and base). Find the proportions of the design which contain the greatest volume.

4.7 Solve Problem 4.6 for the case when the lid is not included in the restriction.

4.8 Sketch the graphs of the following functions.
(a) $1/(x^2 + 1)$ (this is an even function: see (1.12)).
(b) e^{-x^2}.   (c) $x/(x - 1)$.   (d) $x e^{-x}$.
(e) $x^2 e^{-x}$.   (f) $x^3 e^{-x}$.   (g) $e^{2x} - 4 e^x$.
(h) $(\ln x)/x$ for $x > 0$ $((\ln x)/x \to 0$ when $x \to \infty$; this can be proved by putting $x = e^u$ and letting $u \to \infty$).
(i) $[\ln(-x)]/x$ for $x < 0$ (compare (h)).
(j) $x \ln x - x$ for $x > 0$ ($x \ln x \to 0$ when $x \to 0$; this can be seen by writing $x = e^{-u}$ and letting $u \to \infty$).
(k) $\sin 1/x$ (Start by finding where it crosses the axis, using the fact that $\sin u = 0$ when $u = 0, \pm\pi, \pm 2\pi, \dots$.)
(l) $(x^2 - 1)^2$ (This is an even function: see (1.12).)
(m) $x(x^2 - 1)^2$ (This is an odd function: see (1.12).)
(n) $(\sin x)/x$ (You will not be able to find the exact positions of the maxima and minima; be content to indicate the trend. It is an even function: see (1.12). For the value approached at $x = 0$, see (2.13).)

4.9 Sketch the graphs of the following functions.
(a) $1/(x^2 - 1)$ (Hint: write $x^2 - 1 = (x + 1)(x - 1)$, and then follow Example 4.12; alternatively, sketch $y = x^2 - 1$ and imagine taking its reciprocal.)
(b) $x/(x^2 - 1)$.   (c) $1/x(x - 2)$.
(d) $x^3/(1 - x)$ (Hint: see the note on curved asymptotes following Example 4.12.)
(e) $(x + 2)/(x - 1)$ (See the hint in (d).)
(f) $1/(x + 1) + 1/(x + 2)$.

4.10 (See Section 4.5.) Find the approximate value of the change δy in y due to a small change δx in x using the incremental approximation (4.4) in the following cases. Compare the approximate and exact values of δy.
(a) $y = x^3$ when $x = 2$ and $\delta x = 0.1$;
(b) $y = x \sin x$ when $x = \frac{1}{2}\pi$ and $\delta x = -0.2$;
(c) $y = \cos x$ when $x = \frac{1}{4}\pi$ and $\delta x = 0.1$;
(d) $y = (1 + x)/(1 - x)$ when $x = 2$ and $\delta x = -0.2$;
(e) $y = \tan x$ when $x = \frac{1}{4}\pi$ and $\delta x = 0.1$;
(f) $y = 1/(1 - x^2)$ when $x = 0.5$ and $\delta x = \pm 0.1$.

4.11 (a) If the focal length of a lens is f, and a viewed object is at distance u, then the image is at distance v where $v = uf/(u - f)$. Let $f = 0.75$ (m). Find approximately the change in v if u changes from 1.25 to 1.30 (m).

(b) In a Wheatstone bridge circuit, the out-of-balance voltage v is given by

$$v = E(R_1 R_4 - R_2 R_3)/(R_1 + R_2)(R_3 + R_4),$$

where E is the applied voltage and R_1, R_2, R_3, R_4 represent the resistances in the branches. Suppose that $E = 5$, $R_1 = 4$, $R_2 = 2$, $R_3 = 6$, and $R_4 = 3$, so that the circuit is initially balanced. Obtain an approximate expression for δv in terms of a small change δR_1 in R_1.

(c) In a triangle ABC with corresponding sides a, b, c, the formula $a = b \sin A / \sin B$ applies. Show that $\delta a \approx -a \cot B \, \delta B$.

(d) In a triangle ABC with corresponding sides a, b, c, the area A is given by

$$A = [s(s - a)(s - b)(s - c)]^{\frac{1}{2}},$$

where $s = \frac{1}{2}(a + b + c)$. Find an approximate expression for δA in terms of δa. (Hint: use logarithmic differentiation to shorten the working.) Estimate δA when $a = 2$, $b = 4$, $c = 5$, and $\delta c = 0.1$, with a and b remaining constant.

4.12 (Computational). Growth on a deposit by compound interest is given by the formula $C = P(1 + r)^n$, where P is the amount deposited, r is the compound-interest annual growth rate, n is the time of deposit in years or fractions of a year, and C is the accumulated balance. Obtain approximating expressions for δC when (a) r changes by a small amount δr; (b) n changes by a small amount δn. (c) Consider plausible values of P, r, and n, and experiment with the accuracy of the formulae for various values of δr and δn.

4.13 (Newton's method, Section 4.6). It is not too difficult to make these calculations on a hand-held calculator. Find the solutions of the following equations within the broad ranges indicated, which contain exactly one solution.
(a) $x^4 + 2x^2 - x - 1 = 0$ (range $0.5 < x < 1$);
(b) $x^4 + x^{\frac{1}{3}} - 1 = 0$ ($0.5 < x < 0.75$);
(c) $x \ln x = -0.3$ ($0.1 < x < 0.2$);
(d) $e^x = 4x^3$ ($0 < x < 1$);
(e) $\tan x = 2x$ ($0 < x < \frac{1}{2}\pi$);
(f) $(e^x \sin x)/(1 + x) = 2$ ($1.5 < x < 1.9$).

4.14 The equation $f(x) = x \, e^{-x} + 1 = 0$ is known to have exactly one solution (not far from $x = -0.6$). Demonstrate numerically that it is of no use to start off Newton's method for this equation with a value of x greater than 1. Sketch the graph of the function, and make the construction based on Fig. 4.15 to explain why this is so.

4.15 (a) Supposing $y = f(x)$ to have a continuous graph, illustrate graphically that the following principle is true:

If $f(a)$ and $f(b)$ have opposite signs, then there is at least one solution of the equation $f(x) = 0$ in the range $a < x < b$.

(b) (Computational). The equation $e^x - 3x = 0$ has exactly two solutions, and they are in the range $0 < x < 2.5$. Use the principle in (a) to narrow the ranges in which they are known to lie, so as to produce starting values x_0 for Newton's method. One systematic technique is to start with the given end-points $x = 0$ and 2.5, then to halve the interval repeatedly, considering the signs at the ends of the subdivisions.

4.16 (Computational). (a) Suppose that an equation $f(x) = 0$ is known to have exactly one solution in a particular finite interval $a < x < b$. Write a program, using the principle described in Problem 4.15, to obtain a closer starting value x_0 for Newton's method. (Since there is only one solution, any subdivision you find across which the sign of $f(x)$ does not change can be ignored. Arrange for the process to stop when the solution is located within a small preset interval of length E.)

(b) Try this with, say, the equation $x(e^x - 1) = 1$, whose single solution lies between 0 and 1.

(c) By choosing E to be very small, the process can by itself locate the solution to any degree of accuracy if E is small enough. (This is called the **bisection method** for solving equations.) Obtain the number of iterations required to locate the single solution of $f(x) = 0$ in $0 < x < 1$ to two-, four-, and six-decimal accuracy. (The number of iterations is the same for any such equation.)

(d) Solve the equation in (b) by Newton's method to two-, four-, and six-decimal accuracy, starting with $x = 0.5$, and compare the number of iterations required with the number required by the bisection method.

4.17 Consider the neighbouring points $[x_0, f(x_0)]$ and $[x_0 + \delta x_0, f(x_0 + \delta x_0)]$ on the curve $y = f(x)$. Find the normals (see Section 2.2) to the curve at both these points, and the coordinates of their point of intersection. Let $\delta x_0 \to 0$, and show that this point (known as the **centre of curvature** at $[x_0, f(x_0)]$) has the coordinates

$$\left(x_0 + \frac{f'(x_0)[1 + f'(x_0)^2]}{f''(x_0)}, f(x_0) - \frac{[1 + f'(x_0)^2]}{f''(x_0)} \right)$$

provided $f''(x_0) \neq 0$. Show that the distance of the centre of curvature from $[x_0, f(x_0)]$ is

$$R = \frac{[1 + f'(x_0)^2]^{\frac{1}{2}}}{f''(x_0)}.$$

R is known as the **radius of curvature** of the curve (see also Section 10.1).

Find the radius of curvature of the parabola $y = x^2$ at every point on the curve.

4.18 Prove **Leibniz's formula** for the nth derivative of a product of two functions:

$$(fg)^{(n)} = f^{(n)}g + {}_nC_1 f^{(n-1)}g^{(1)} + {}_nC_2 f^{(n-2)}g^{(2)} + \cdots + {}_nC_n fg^{(n)},$$

where ${}_nC_r$ is the rth binomial coefficient, given by $n!/[r!(n - r)!]$. (Hint: try writing out the first three derivatives at full length: notice how the repetitions of terms in $f^{(n-r)}g^{(r)}$ combine to produce the coefficients.)

5 Taylor series and approximations

CONTENTS

5.1   The index notation for derivatives of any order

We shall use yet another standard notation for derivatives in this chapter. Since we shall have to keep track of derivatives of high orders we modify the 'dash' notation of (4.1) as follows to provide a brief form:

> **Index notation for derivatives**
> For the first, second, third, ... derivatives respectively of $f(x)$, write
> $$f'(x) = f^{(1)}(x), \quad f''(x) = f^{(2)}(x), \quad f''' = f^{(3)}(x), \dots.$$
> If $y = f(x)$, the notation $y^{(1)}, y^{(2)}, y^{(3)}, \dots$ is also used.   **(5.1)**

Thus if $f(x) = x^3$, then $f^{(1)}(x) = 3x^2$, $f^{(2)}(x) = 6x$, and $f^{(3)}(x) = 6$. As with the dash notation, we encounter such forms as $f^{(2)}(u) = 6u$, $f^{(2)}(0) = 0$, and $f^{(2)}(x - c) = 6(x - c)$.

5.2   Taylor polynomials

Firstly we shall show how to obtain **approximations to a given $f(x)$ for use when x is a small number.** Suppose, for example, that

$$f(x) = \frac{1}{1 - x}.$$

Since $f(0) = 1$, we can be sure that

$$\frac{1}{1 - x} \approx 1$$

so long as x is small enough. This is shown in Fig. 5.1a. It is, of course,

(a)

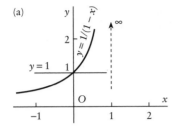

(b)

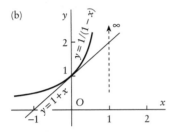

(c)

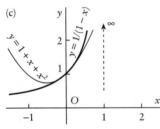

Fig. 5.1

a poor approximation, acceptable only very close to $x = 0$.

A better approximation near $x = 0$ is given by the equation of the tangent line at $x = 0$ (Fig. 5.1b). Since $f^{(1)}(x) = 1/(1 - x)^2$, the slope at $x = 0$ is $f^{(1)}(0) = 1$.

The equation of the tangent line at $x = 0$ is therefore $y = 1 + x$, so

$$\frac{1}{1 - x} \approx 1 + x,$$

when x is small enough.

We need a way to continue improving the approximation, $P(x)$ say, to a further stage and beyond. At present we have reached the tangent approximation $P(x) = 1 + x$, which was chosen so that $P(0) = f(0)$ and $P^{(1)}(0) = f^{(1)}(0)$. To obtain the next approximation choose a $P(x)$ which also **matches the second derivative at $x = 0$**:

$$P(0) = f(0), \qquad P^{(1)}(0) = f^{(1)}(0), \qquad P^{(2)}(0) = f^{(2)}(0).$$

This involves adding a term in x^2, and we can choose its coefficient so that the extra condition is satisfied *without disturbing the two terms we have already found*. Continuing with the example,

$$f^{(2)}(x) = \frac{2}{(1 - x)^3}, \quad \text{so} \quad f^{(2)}(0) = 2.$$

It is easy to check that $P(x) = 1 + x + x^2$ satisfies the three conditions. Therefore

$$\frac{1}{1 - x} \approx 1 + x + x^2$$

is an improved approximation. This represents the parabolic curve shown in Fig. 5.1c.

We can carry out this process for any function $f(x)$, and take it to any level of approximation we wish. Successive approximations will consist of **polynomials** of increasing degree. However, we must not expect too much of it: we cannot go too far from the origin and still expect a good approximation.

To deal with the general case, we need the following simple result.

Derivatives of a polynomial in x at $x = 0$

$$P(x) = a_0 + a_1 x + a_2 x^2 + \cdots + a_N x^N$$

is a polynomial of degree N. Then

$$P(0) = a_0, \quad P^{(1)}(0) = a_1, \quad P^{(2)}(0) = 2! a_2,$$

and in general

$$P^{(n)}(0) = n! a_n$$

for $n = 1, 2, 3, \ldots, N$.

(5.2)

It is easy to verify (5.2) by working out the first few derivatives.

Now suppose that we wish to approximate to a general function $f(x)$ (near $x = 0$) by means of a polynomial

$$P(x) = a_0 + a_1 x + a_2 x^2 + \cdots + a_N x^N.$$

We require that

$$P(0) = f(0), \quad P^{(1)}(0) = f^{(1)}(0), \quad P^{(2)}(0) = f^{(2)}(0), \quad \ldots.$$

According to (5.2), the coefficients are given by

$$a_0 = P(0) = f(0),$$

$$a_1 = \frac{1}{1!} P^{(1)}(0) = \frac{1}{1!} f^{(1)}(0),$$

$$a_2 = \frac{1}{2!} P^{(2)}(0) = \frac{1}{2!} f^{(2)}(0),$$

$$a_3 = \frac{1}{3!} P^{(3)}(0) = \frac{1}{3!} f^{(3)}(0),$$

and so on. By writing the coefficients a_n in terms of the known values $f^{(n)}(0)$ we obtain the **Taylor polynomial approximation**:

Taylor polynomial $P(x)$ of degree N near $x = 0$

Let $P(x)$ be the N th-degree polynomial

$$f(0) + \frac{1}{1!} f^{(1)}(0)x + \frac{1}{2!} f^{(2)}(0)x^2 + \cdots + \frac{1}{N!} f^{(N)}(0)x^N.$$

Then for x sufficiently close to zero,

$$f(x) \approx P(x). \tag{5.3}$$

Example 5.1 *Obtain a fifth-degree polynomial which approximates to e^x for values of x that are not too large.*

Use (5.3), putting $f(x) = e^x$. This case is simple:

$$f(x) = f^{(1)}(x) = f^{(2)}(x) = \cdots = f^{(5)}(x) = e^x,$$

so $f(0) = f^{(1)}(0) = f^{(2)}(0) = \cdots = f^{(5)}(0) = 1$. Therefore

$$e^x \approx 1 + \frac{1}{1!} x + \frac{1}{2!} x^2 + \frac{1}{3!} x^3 + \frac{1}{4!} x^4 + \frac{1}{5!} x^5 = P(x).$$

(If we take higher-degree approximations, the terms continue according to the same rule.) We show e^x and its approximation $P(x)$ in the following table for a few values of x.

↗

Example 5.1 *continued*

x	−4	−3	−2	−1	−0.5
e^x	0.0183	0.0498	0.1353	0.3679	0.6065
$P(x)$	−3.533	−0.6500	0.0667	0.3666	0.6065

x	0	0.5	1	2	3	4
e^x	1	1.6487	2.7183	7.3891	20.086	54.598
$P(x)$	1	1.6487	2.7167	7.2667	18.400	42.867

The approximating polynomial $P(x)$ clings to the true values for a considerable range around the origin.

Example 5.2   (a) *Obtain the Taylor polynomial approximation of any degree N for the function $1/(1-x)$ near $x = 0$. (b) Obtain an expression for the error in the approximation.*

(a) Putting $1/(1-x) = f(x)$, the sequence of derivatives of $f(x)$ is

$$f^{(1)}(x) = \frac{1}{(1-x)^2}, \quad f^{(2)}(x) = \frac{2 \cdot 1}{(1-x)^3}, \quad f^{(3)}(x) = \frac{3 \cdot 2 \cdot 1}{(1-x)^4}$$

and in general

$$f^{(n)}(x) = \frac{n!}{(1-x)^{n+1}}.$$

Therefore, referring to (5.3), the Taylor polynomial of degree N is

$$1 + x + x^2 + x^3 + \cdots + x^N.$$

(b) The error in an estimation using this approximation is equal to

$$P(x) - f(x) = 1 + x + \cdots + x^N - \frac{1}{1-x}$$

$$= \frac{(1-x)(1 + x + x^2 + \cdots + x^N) - 1}{1-x}$$

$$= \frac{-x^{N+1}}{1-x}.$$

You should experiment with this expression using various values of N and x. (i) If x is very small, the error involved is very small even if N is only 2 or 3. (ii) If we take any *fixed* value of x in the range $-1 < x < 1$, the error will approach zero when we take approximations of higher and higher degree (because, when $-1 < x < 1$, x^{N+1} approaches zero as N increases: try this numerically with, say, $x = 0.9$). (iii) The approximation fails altogether if $x > 1$ or $x < -1$. The error will be large, and to increase N will make it still larger because $|x^{N+1}|$ increases when N increases.

5.3 A note on infinite series

In the previous section we did not put any limit on the degree of the approximating polynomial, and there seems to be no reason why we should not let the terms run on for ever: in fact, let the degree N

approach infinity. If we extend the polynomial approximation of Example 5.1 for $f(x) = e^x$, we obtain an example of a so-called **infinite series**:

$$1 + \frac{1}{1!}x + \frac{1}{2!}x^2 + \frac{1}{3!}x^3 + \cdots \quad \text{or} \quad \sum_{n=0}^{\infty} \frac{1}{n!}x^n.$$

It might be that by extending the approximating polynomials in this way, *approximation* will become *equality*, so that the sum of the series will be *equal to* the original function instead of being just an *approximation* to it, but this is only true with reservations.

There are many types of infinite series (see e.g. Chapter 26 on Fourier series). Consider first what is meant by the **sum of an infinite series**. When x is given any particular value, the terms to be added become simply numbers. We cannot in practice add an infinite number of numbers: no matter how many operations we carry out we never reach the end. However, this does not mean that the infinite series does not add up to a definite number, only that we cannot reach it exactly by simply piling on more and more terms.

Consider the simpler infinite series that we get from putting $x = 0.1$ into the Taylor polynomial for $1/(1 - x)$ (see Example 5.2), and letting the degree increase to infinity. It is the geometric series

$$1 + 0.1 + 0.1^2 + 0.1^3 + 0.1^4 + \cdots$$

(see Section 1.16). This is the same as

$$1 + 0.1 + 0.01 + 0.001 + 0.0001 + \cdots .$$

If we record the sum of 1, 2, 3, 4, ... terms successively, we obtain what is called a **sequence of partial sums** ('partial' because we only take a finite number of terms into account). The sequence is

$$1, \quad 1.1, \quad 1.11, \quad 1.111, \quad 1.1111, \quad \dots .$$

The number that is being approached is obviously 1.111 11..., which is equal to 10/9. This number is equal to the value of $1/(1 - x)$ when $x = 0.1$, so in this case the *infinite* series has delivered the value required. Similarly, if we put $x = \frac{1}{2}$, the infinite series is

$$1 + \tfrac{1}{2} + (\tfrac{1}{2})^2 + (\tfrac{1}{2})^3 + (\tfrac{1}{2})^4 + \cdots = 1 + \tfrac{1}{2} + \tfrac{1}{4} + \tfrac{1}{8} + \tfrac{1}{16} + \cdots .$$

For the sum of 1, 2, 3, 4, ... terms, we obtain the sequence of partial sums

$$1, \quad 1\tfrac{1}{2}, \quad 1\tfrac{3}{4}, \quad 1\tfrac{7}{8}, \quad 1\tfrac{15}{16}, \quad \cdots ,$$

which is obviously approaching the value 2, and this is the value of $1/(1 - x)$ when $x = \frac{1}{2}$. Infinite series whose **partial sums approach a definite value** as we take more and more terms are said to **converge** to this value, which is called the **sum of the infinite series**.

However, not all infinite series converge. For example, if we form successively the sum of 1, 2, 3, ... terms of the infinite series

$$1 + 1 + 1 + \cdots,$$

then we obtain

$$1, 2, 3, 4, \ldots,$$

which is obviously going to infinity. The infinite series

$$1 - 1 + 1 - 1 + \cdots$$

has the successive partial sums

$$1, 0, 1, 0, 1, \ldots,$$

which is not going anywhere. Such series are said to **diverge**. You might be surprised to know that the infinite series

$$1 + \tfrac{1}{2} + \tfrac{1}{3} + \tfrac{1}{4} + \cdots$$

diverges: the partial sums go to infinity. (It is worth experimenting with this series: even using a computer you might take a while to convince yourself that it really does diverge.)

5.4   Infinite Taylor expansions

We return to the subject of general Taylor polynomials of the type (5.3) when we extend the polynomial to an infinite number of terms, so that we have an infinite series instead of a polynomial expression. This is called a **Taylor series** or an infinite **Taylor expansion about the origin** $x = 0$ for the function $f(x)$.

The mathematical theory of infinite series, and in particular of Taylor series, cannot be discussed in this book. In the previous section it is indicated that pitfalls might arise when the polynomials are extended into infinite series. Moreover, it seems obvious, for example, that the values of a function and its derivatives *at the origin only* cannot possibly predict values elsewhere if we allow functions to be *completely arbitrary* at other points.

However, ordinary functions do follow the simple pattern illustrated by the case of $f(x) = 1/(1 - x)$ in Example 5.2. Each function has an individual range of values of x, called its **interval of validity**, in which the Taylor series **converges to the exact value of** $f(x)$. Elsewhere, the series must not be used for approximation.

The following table (5.4), displays the infinite Taylor series about the origin for several important functions, together with their ranges of validity. You should confirm the coefficients, as in Examples 5.1 and 5.2.

(i) *Geometric series* (valid for $-1 < x < 1$)

$$\frac{1}{1+x} = 1 - x + x^2 - x^3 + \cdots. \tag{5.4a}$$

(ii) *Exponential series* (valid for all x)

$$e^x = 1 + \frac{1}{1!}x + \frac{1}{2!}x^2 + \cdots. \tag{5.4b}$$

(In particular, $e = 1 + \frac{1}{1!} + \frac{1}{2!} + \cdots.$)

(iii) *Trigonometric series* (valid for all x in radians)

$$\sin x = x - \frac{1}{3!}x^3 + \frac{1}{5!}x^5 - \cdots. \tag{5.4c}$$

$$\cos x = 1 - \frac{1}{2!}x^2 + \frac{1}{4!}x^4 - \cdots. \tag{5.4d}$$

(iv) *Logarithmic series* (valid for $-1 < x \leqslant 1$)

$$\ln(1 + x) = x - \frac{1}{2}x^2 + \frac{1}{3}x^3 - \cdots. \tag{5.4e}$$

(v) *Binomial series* (valid for $-1 < x < 1$ and any α)

$$(1 + x)^\alpha = 1 + \alpha x + \frac{\alpha(\alpha - 1)}{2!}x^2 + \frac{\alpha(\alpha - 1)(\alpha - 2)}{3!}x^3 + \cdots. \tag{5.4f}$$

(By putting $\alpha = -1$, the geometric series (5.4a) is recovered. If $\alpha = N$, a positive integer, the series terminates at the term in x^N, and so the binomial theorem of Sections 1.18 and 4.7 is obtained.)

When a series is used to provide approximations by taking only a finite number of terms, it is necessary to estimate how many terms to take so as to obtain a desired degree of accuracy. It is usually sufficient to observe the size of the terms involved, as in the following example.

Example 5.3 *Find how many terms of the Taylor series for* $\sin x$ *are needed to obtain three-decimal accuracy over the range* $-1 \leqslant x \leqslant 1$ *(in radians).*

The intuitive requirement is that we should stop at the point where we can see that taking further terms is not likely to affect the third decimal place. The magnitude (modulus) of the terms in (5.4c) increases when the magnitude of x increases, so it should be sufficient to provide an approximation good for the largest value, $x = 1$. The magnitudes of successive terms when $x = 1$ are equal to

$$1, \quad 0.1\dot{6}, \quad 0.08\dot{3}, \quad 0.0002, \quad 2 \times 10^{-6}, \quad \cdots,$$

using the recurring decimal notation (Section 1.1). It is therefore enough to retain three terms of the series; that is to say we should retain powers of x up to x^5. To three decimals, then,

$$\sin x \approx x - \frac{1}{3!}x^3 + \frac{1}{5!}x^5 \quad \text{for} \quad -1 \leqslant x \leqslant 1.$$

5.5 Manipulation of Taylor series

We can obtain new Taylor series from the standard ones in (5.4).

Example 5.4 *Find the Taylor expansion about $x = 0$ for the function $(2 - x)^{\frac{1}{2}}$, and state its range of validity.*

Write

$$(2 - x)^{\frac{1}{2}} = 2^{\frac{1}{2}}(1 - \tfrac{1}{2}x)^{\frac{1}{2}} = 2^{\frac{1}{2}}[1 + (-\tfrac{1}{2}x)]^{\frac{1}{2}}.$$

We can use the binomial expansion (5.4f), with $\alpha = \frac{1}{2}$, and with $-\frac{1}{2}x$ in place of x. The expansion will be valid, provided that $-1 < -\frac{1}{2}x < 1$, that is when $-2 < x < 2$. Therefore

$$(2 - x)^{\frac{1}{2}} = 2^{\frac{1}{2}}[1 + (-\tfrac{1}{2}x)]^{\frac{1}{2}}$$

$$= 2^{\frac{1}{2}}\left(1 + \tfrac{1}{2}(-\tfrac{1}{2}x) + \frac{\tfrac{1}{2}(\tfrac{1}{2} - 1)}{2!}(-\tfrac{1}{2}x)^2 + \frac{\tfrac{1}{2}(\tfrac{1}{2} - 1)(\tfrac{1}{2} - 2)}{3!}(-\tfrac{1}{2}x)^3 + \cdots\right)$$

$$= 2^{\frac{1}{2}}(1 - \tfrac{1}{4}x - \tfrac{1}{32}x^2 - \tfrac{1}{128}x^3 + \cdots)$$

when $-2 < x < 2$.

To find the first few terms in the Taylor series for a composite function $f(x)$ such as

$$f(x) = \frac{e^{-x}}{(1 + x)^{\frac{1}{2}}},$$

it is usually best *not* to start from first principles by calculating $f(0)$, $f^{(1)}(0), f^{(2)}(0)$, and so on, which can lead to great complication, but to manipulate standard expansions as in the following examples.

Example 5.5 *Approximate to $(\sin x/x)^2$ by a polynomial of degree 4, and compare the approximate and exact values when $x = 0, \frac{1}{4}, \frac{1}{2}, 1, 2$.*

From (5.4c),

$$\left(\frac{\sin x}{x}\right)^2 = \left(\frac{x - \dfrac{1}{3!}x^3 + \dfrac{1}{5!}x^5 - \cdots}{x}\right)^2$$

$$= \left(1 - \frac{1}{3!}x^2 + \frac{1}{5!}x^4 - \cdots\right)^2$$

$$= 1 - \frac{2}{3!}x^2 + \left(\frac{2}{5!} + \frac{1}{3!^2}\right)x^4 + \cdots,$$

where only terms up to x^4 are retained. Write the approximating polynomial $P(x)$ as

$$P(x) = 1 - 0.3333x^2 + 0.0444x^4$$

Example 5.5 *continued*

to obtain the table

x	0	0.25	0.5	1.0	2.0
$[\sin x/x]^2$	1	0.9793	0.9179	0.6861	0.2067
$P(x)$	1	0.9793	0.9194	0.7111	0.3772

Example 5.6 *Approximate to* $e^{-x}/(1 + x)^{\frac{1}{2}}$ *near* $x = 0$ *by a polynomial of degree 2.*

Write
$$f(x) = e^{-x}(1 + x)^{-\frac{1}{2}}.$$

Use (5.4b) with $-x$ in place of x, and carry it to degree 2:

$$e^{-x} \approx 1 - x + \frac{1}{2!}x^2.$$

Also, by (5.4f) (the binomial theorem) with $\alpha = -\frac{1}{2}$:

$$(1 + x)^{-\frac{1}{2}} \approx 1 + (-\tfrac{1}{2})x + \frac{(-\frac{1}{2})(-\frac{1}{2} - 1)}{2!}x^2.$$

Then by multiplying the two polynomials we obtain
$$f(x) \approx 1 - \tfrac{3}{2}x + \tfrac{11}{8}x^2,$$

when x is small. (Reject powers higher than 2 in the final product – they would not be correct since we neglected such terms in the original approximations.)

Example 5.7 *Obtain the first three nonzero terms of the Taylor expansion for* $1/\cos x$.

There are several ways of doing this problem.
Working from (5.3). You might try this, but it is rather arduous.
Using the power series for cos x *given by* (5.4d). Write

$$\frac{1}{\cos x} = \frac{1}{1 - \dfrac{1}{2!}x^2 + \dfrac{1}{4!}x^4 - \cdots}.$$

The problem is to find the first three terms in the *reciprocal* of the infinite series; we then have a Taylor polynomial. Anticipate that only the even powers of x will occur, as in the expansion of cos x. Then we expect

$$\frac{1}{\cos x} = \frac{1}{1 - \dfrac{1}{2!}x^2 + \dfrac{1}{4!}x^4 - \cdots} = b_0 + b_2 x^2 + b_4 x^4 + \cdots.$$

We have to find b_0, b_2, b_4. To do this, cross-multiply:

$$1 = \left(1 - \frac{1}{2!}x^2 + \frac{1}{4!}x^4 - \cdots\right)(b_0 + b_2 x^2 + b_4 x^4 + \cdots)$$
$$= b_0 + (b_2 - \tfrac{1}{2}b_0)x^2 + (b_4 - \tfrac{1}{2}b_2 + \tfrac{1}{24}b_0)x^4 + \cdots$$

Example 5.7 *continued*

(retaining only powers up to x^4). Match the coefficients of powers of x on both sides, starting with the constant term; we obtain

$$b_0 = 1,$$

and, since the coefficients of x^2 and x^4 on the left are zero,

$$b_2 - \tfrac{1}{2}b_0 = 0 \quad \text{and} \quad b_4 - \tfrac{1}{2}b_2 + \tfrac{1}{24}b_0 = 0.$$

The last two equations can be solved successively to give

$$b_2 = \tfrac{1}{2} \quad \text{and} \quad b_4 = \tfrac{5}{24}.$$

Finally

$$1/\cos x \approx 1 + \tfrac{1}{2}x^2 + \tfrac{5}{24}x^4.$$

Polynomial division. We can evaluate $1/(1 - \tfrac{1}{2}x^2 + \tfrac{1}{24}x^4 - \cdots)$ by long division, setting it out like this, ignoring powers higher than x^4:

$$
\begin{array}{r}
1 + \tfrac{1}{2}x^2 + \tfrac{5}{24}x^4 \\[4pt]
\hline
1 - \tfrac{1}{2}x^2 + \tfrac{1}{24}x^4 \,\big|\, 1 \\
\end{array}
$$

$$
\begin{aligned}
\text{subtract:} \quad & 1 - \tfrac{1}{2}x^2 + \tfrac{1}{24}x^4 \\
\hline
& \tfrac{1}{2}x^2 - \tfrac{1}{24}x^4 \\
\text{subtract:} \quad & \tfrac{1}{2}x^2 - \tfrac{1}{4}x^4 \\
\hline
& \tfrac{5}{24}x^4 \\
& \tfrac{5}{24}x^4
\end{aligned}
$$

5.6 Approximations for large values of x

When x is large, $1/x$ is small. This fact can sometimes be used to obtain approximations valid when x is large, as in the following example.

Example 5.8 *Obtain a three-term approximation to $(1 + 1/x)^{\frac{1}{2}}$ valid when x is **large** enough.*

Translate the binomial series, (5.4f), with $\alpha = \tfrac{1}{2}$, in terms of a neutral variable, say u:

$$(1 + u)^{\frac{1}{2}} = 1 + \tfrac{1}{2}u + \frac{\tfrac{1}{2}(\tfrac{1}{2} - 1)}{2!}u^2 + \cdots \quad \text{when } -1 < u < 1,$$

so $(1 + u)^{\frac{1}{2}} \approx 1 + \tfrac{1}{2}u - \tfrac{1}{8}u^2$ when u is small enough, the approximation improving as u gets smaller. Now put $u = 1/x$; we obtain

$$\left(1 + \frac{1}{x}\right)^{\frac{1}{2}} \approx 1 + \frac{1}{2x} - \frac{1}{8x^2},$$

when x is *large* enough (positively or negatively), the approximation improving as x gets larger.

5.7 Taylor series about other points

The Taylor series about $x = 0$ for

$$f(x) = \frac{1}{1-x} = 1 + x + x^2 + x^3 + \cdots$$

does not work when $x = 2$: we get $1 + 2 + 2^2 + \cdots$, which is infinite. However, we can obtain a *different* Taylor-type series which represents $1/(1-x)$ near $x = 2$ by a process which amounts to changing the origin, as in the following example.

Example 5.9 *Find a Taylor-type series which represents $1/(1-x)$ for values of x near $x = 2$.*

Look for a series of this type:

$$\frac{1}{1-x} = b_0 + b_1(x-2) + b_2(x-2)^2 + \cdots,$$

because we want a series that works when x is close to 2, which is to say when $x - 2$ is small, rather than when x is small as before. Therefore we need a series consisting of powers of $x - 2$. We can bring the element $x - 2$ into view by writing

$$\frac{1}{1-x} = \frac{1}{1-(x-2+2)} = -\frac{1}{1+(x-2)}.$$

Now expand the final term by using (5.4f) (the binomial theorem) with $\alpha = -1$, and $x - 2$ in place of x, obtaining

$$\frac{1}{1-x} = -1 + (x-2) - (x-2)^2 + (x-2)^3 + \cdots,$$

valid if $-1 < x - 2 < 1$, that is if

$$1 < x < 3.$$

Example 5.10 *Obtain a Taylor series about the point $x = \pi$ for the function $\cos x$.*

There exists already the series (5.4d), which is valid at $x = \pi$. However, if we are interested in approximating to $\cos x$ near $x = \pi$, an expansion in powers of $x - \pi$ should be more economical and expressive than one consisting of powers of x. We show two ways of finding the series.

(a) *On the lines of Example 5.9.* Write

$$\cos x = \cos[\pi + (x - \pi)] = \cos \pi \cos(x - \pi) - \sin \pi \sin(x - \pi) = -\cos(x - \pi).$$

We can use (5.4d) to expand this, by putting $x - \pi$ in place of x. We obtain

$$\cos x = -\cos(x - \pi) = -1 + \frac{1}{2!}(x - \pi)^2 - \frac{1}{4!}(x - \pi)^4 + \cdots.$$

This is valid for all values of x. A two-term approximation shows that $\cos x$ has a parabolic shape near $x - \pi = 0$ or $x = \pi$, where $\cos x$ has a local minimum.

↗

Example 5.10 *continued*

(b) *Matching the value and the derivatives at $x = \pi$*. The derivatives of $f(x) = \cos x$ at $x = \pi$ are given by

$$f(\pi) = \cos \pi = -1, \quad f^{(1)}(\pi) = -\sin \pi = 0, \quad f^{(2)}(\pi) = -\cos \pi = 1,$$

and so on. The same relations hold good between the coefficients of a polynomial in powers of $x - \pi$ and the values of its derivatives at $x = \pi$, as was stated in (5.3) for polynomials in x at $x = 0$. We simply put $x - \pi$ in place of x in (5.3). The required Taylor series is

$$f(x) = f(\pi) + \frac{1}{1!} f^{(1)}(\pi)(x - \pi) + \frac{1}{2!} f^{(2)}(\pi)(x - \pi)^2 + \cdots,$$

which is the same as the result obtained in (a).

The general result is the following:

> **Taylor series about a point $x = c$**
>
> $$f(x) = f(c) + \frac{1}{1!} f^{(1)}(c)(x - c) + \frac{1}{2!} f^{(2)}(c)(x - c)^2 + \cdots.$$
>
> (The range of validity depends upon $f(x)$.)
>
> **(5.5)**

5.8 Indeterminate values; l'Hôpital's rule

The relation

$$y = \frac{\sin x}{x}$$

specifies a value for y for all values of x except for $x = 0$. At this point the formula gives $y = 0/0$, which is a meaningless or **indeterminate** expression. The graph of y against x therefore contains a gap at $x = 0$. However, as we approach $x = 0$ from either side, y may approach a single, finite value, $y(0)$ say, that plugs the gap. If such a value exists, it is given by the limiting operation

$$y(0) = \lim_{x \to 0} \frac{\sin x}{x},$$

which can be evaluated in the following way.

Use the Taylor series to write, for $x \neq 0$,

$$\frac{\sin x}{x} = \frac{x - \frac{1}{3!}x^3 + \cdots}{x} = 1 - \frac{1}{3!}x^3 + \cdots,$$

after cancelling the factor x. This new expression has no pecularities. Therefore we put $x = 0$ in the new series, obtaining

$$\lim_{x \to 0} \frac{\sin x}{x} = 1,$$

and this is the missing value $y(0)$.

Example 5.11   *Obtain*

$$\lim_{x \to 0} \frac{\sin^2 3x}{1 - \cos x}.$$

For $x = 0$, we obtain 0/0, which is indeterminate. From (5.4c)

$$\sin(3x) = 3x + \text{higher powers},$$

so that

$$\sin^2(3x) = (3x)^2 + \text{higher powers}. \tag{i}$$

Also

$$1 - \cos x = 1 - \left(1 - \frac{1}{2!}x^2 + \text{higher powers}\right) \tag{ii}$$

$$= \frac{1}{2!}x^2 + \text{higher powers}.$$

Therefore, from (i) and (ii), for $x \neq 0$

$$\frac{\sin^2(3x)}{1 \cos x} = \frac{9x^2 + \text{higher powers}}{\frac{1}{2}x^2 + \text{higher powers}} = \frac{9 + \text{positive powers of } x}{\frac{1}{2} + \text{positive powers of } x}, \tag{iii}$$

after cancelling the common factor x^2. This expression is not problematic; we may put $x = 0$ into it, obtaining

$$\lim_{x \to 0} \frac{\sin^2(3x)}{1 - \cos x} = \frac{9}{\frac{1}{2}} = 18.$$

Observe that only the leading term, or the **dominant terms**, of the Taylor series are needed explicitly in order to obtain the limiting value.

Example 5.12   *The function*

$$\frac{\ln x}{\sin(\pi x)}$$

is indeterminate at $x = 1$. Obtain the limiting value as $x \to 1$.

Here we require the dominant terms of the Taylor series centred on $x = 1$. write

$$\ln x = \ln[1 + (x - 1)] = (x - 1) + \cdots,$$

from eqn (5.4e) in which the variable x is replaced by $(x - 1)$. Also

$$\sin \pi x = \sin \pi[1 + (x - 1)] = \sin \pi \cos \pi(x - 1) + \cos \pi \sin \pi(x - 1)$$

$$= -\sin \pi(x - 1) = -\pi(x - 1) + \cdots.$$

Therefore, for $x \neq 1$,

$$\frac{\ln x}{\sin(\pi x)} = \frac{(x - 1) + \cdots}{-\pi(x - 1) + \cdots} = \frac{1 + \text{positive powers of } (x - 1)}{-\pi + \text{positive powers of } (x - 1)}.$$

The limit is obtained by putting $x = 1$ in the right-hand side, giving

$$\lim_{x \to 1} \frac{\ln x}{\sin(\pi x)} = -\frac{1}{\pi}.$$

Suppose that we require the limit of the ratio $f(x)/g(x)$ as $x \to a$, but $f(a)$ and $g(a)$ are both zero. The dominant terms in the Taylor series can be expressed in terms of derivatives of $f(x)$ and $g(x)$ as in (5.5). This leads to the following formal statement, known as l'Hôpital's rule:

L'Hôpital's rule

Let $f(x)$, $g(x)$ be represented by Taylor series at $x = a$, where $f(a) = g(a) = 0$, and let the first non-vanishing derivatives at a be $f^{(M)}(a)$ and $g^{(N)}(a)$, where $M \geqslant N \geqslant 1$. Then

$$\lim_{x \to a} \frac{f(x)}{g(x)} = \frac{f^{(M)}(a)}{g^{(N)}(a)}.$$

(5.6)

However, for working out particular cases the evaluation of $f^{(M)}$ and $g^{(N)}$ may be very laborious. The methods illustrated earlier are usually more convenient.

Several extensions to this procedure are illustrated in Problems 5.12 to 5.15.

Problems

5.1 Obtain a four-term Taylor polynomial approximation valid near $x = 0$ for each of the following. Estimate the ranges of x over which three-term polynomials will give two-decimal accuracy (you cannot usually tell until you have seen the next-higher term).
(a) $e^{\frac{1}{2}x}$;   (b) $(1 + x)^{\frac{1}{2}}$;   (c) $(1 + x)^{-\frac{1}{3}}$;
(d) $\sin 2x$;   (e) $\cos \frac{1}{2}x$;   (f) $\ln(1 + x)$;
(g) $(1 + x^2)^{\frac{1}{2}}$ (hint: consider $(1 + u)^{\frac{1}{2}}$; then put $u = x^2$);
(h) $\ln(1 + 3x)$ (see the hint in (g)).

5.2 Verify the coefficients of each of the infinite Taylor series shown in (5.4) (taking the ranges of validity for granted); namely
(a) e^x;   (b) $\sin x$;   (c) $\cos x$;
(d) $(1 + x)^\alpha$;   (e) $\ln(1 + x)$.

5.3 For each of the following series give the Taylor polynomial having the lowest degree which you think will safely give four-decimal accuracy over the ranges given.
(a) e^x over $-2 \leqslant x \leqslant 2$;
(b) $\sin x$ over $-2 \leqslant x \leqslant 2$;
(c) $\cos x$ over $-2 \leqslant x \leqslant 2$;
(d) $(1 + x)^{\frac{1}{2}}$ over $-0.5 \leqslant x \leqslant 0.5$;
(e) $\ln(1 + x)$ over $-0.5 \leqslant x \leqslant 0.5$.

5.4 Obtain the first two nonzero terms in the Taylor series at the origin for the following.
(a) $\arcsin x$;   (b) $\arccos x$;   (c) $\arctan x$;
(d) $e^{-x} \sin x$;   (e) $e^{-x} \cos x$.

5.5 (See Section 5.5.) Find three nonzero terms in the Taylor series at $x = 0$ for the following functions and state the ranges of validity.
(a) $1/(1 + 3x)$;   (b) $1/(2 - x)$;   (c) $(3 - x)^{\frac{1}{3}}$;
(d) $(x - 3)^{\frac{1}{3}}$;   (e) $\ln(9 - x)$;   (f) $\cos \frac{1}{2}x$;
(g) $\sin x^{\frac{1}{2}}$ (Consider the series for $\sin u$; then put $u = x^{\frac{1}{2}}$.) This is not strictly a Taylor series – it is spoken of as 'a Taylor series in $x^{\frac{1}{2}}$' – but it still is useful for approximations.
(h) $\cos x^{\frac{1}{2}}$.

5.6 (See Section 5.5.) Find the first three nonzero terms in the Taylor series at $x = 0$ for the following.
(a) $e^{-x}/(1 + x)$;   (b) $(1 - x)^{\frac{1}{2}}e^x$;
(c) $[\ln(1 - x)]^2/x^2$.

5.7 (See Example 5.7.) Find the first three nonzero terms in the Taylor expansions at the origin for the following.

(a) $1/[1 + \ln(1 + x)]$;   (b) $\tan x$;   (c) $1/(1 + e^x)$;

(d) $\tanh x$, or $(e^x - e^{-x})/(e^x + e^{-x})$. (It is less complicated if you firstly reduce this to a more manageable form.)

(e) $x/\sin x$.

5.8 (See Section 5.6.) Find a three-term approximation, valid for *large* enough values of x, in each of the following cases.

(a) $\left(1 - \dfrac{1}{x}\right)^{\frac{1}{2}}$;   (b) $\ln\left(1 + \dfrac{1}{x^{\frac{1}{2}}}\right)$;   (c) $x^{\frac{1}{2}}/(1 + x)^{\frac{1}{2}}$;

(d) $\ln(1 + x + x^2)$;   (e) $1/\sin(x^{-1})$.

5.9 (a) Show that $1/\sin x \approx (1/x) + \frac{1}{6}x$ when x is nonzero but small enough.

(b) Show that, when x is *large* enough, $(1 + x)^{\frac{1}{2}} \approx x^{\frac{1}{2}} + 1/(2x^{\frac{1}{2}})$.

(c) Show that, when x is *large* enough, $(2 + x)^{\frac{1}{2}} - (1 + x)^{\frac{1}{2}} \approx 1/(2x^{\frac{1}{2}})$.

(d) Show that, when x is nonzero but small enough, then $1/(1 - \cos x)^{\frac{1}{2}} \approx (\sqrt{2}/x) + (x\sqrt{2}/24)$.

5.10 (a) (See Section 5.6.) Write

$$\ln x = \ln[1 + (x - 1)],$$

and so obtain the Taylor series for $\ln x$ about $x = 1$. State the range of validity.

(b) Obtain the Taylor series about $x = \frac{1}{2}\pi$ for $\cos x$, and state the range of validity.

(c) Obtain the Taylor series about $x = 1$ for the function $(1 + x)^{\frac{1}{2}}$, and state the range of validity.

5.11 Suppose that $f(x)$ has a stationary point at $x = c$. Write down the form of its Taylor series about $x = c$, taking this into account.

(a) By considering the first three terms, rediscover the conditions on $f''(c)$ which determine the type of stationary point (see (4.2)).

(b) By considering further terms of the Taylor series, extend the criteria to obtain a general rule which covers the case $f''(c) = 0$.

5.12 The following expressions are undetermined at $x = 0$. Obtain the appropriate values there which make up continuous functions.

(a) $(e^x - 1)/x$;   (b) $(1 - \cos x)/x^2$;

(c) $[\ln(1 + x) - x]/\sin x$;   (d) $\sin x/(1 - \cos x)$.

5.13 Obtain

(a) $\lim\limits_{x \to 0} \dfrac{(1 - x)^{12} - 1}{(1 - x)^{10} - 1}$;

(b) $\lim\limits_{x \to 0} \dfrac{\sin x - x}{\sin x - x \cos x}$;

(c) $\lim\limits_{x \to \pi} \dfrac{\cos x + 1}{x - \pi}$;

(d) $\lim\limits_{x \to \frac{1}{2}} \dfrac{\sin x - 1}{\cos 5x}$.

5.14 Show that

$$\lim\limits_{x \to 0} \dfrac{e^x - 1}{e^x - 1 - x}$$

does not exist, but that the function values approach $-\infty$ as x approaches zero from the left, and $+\infty$ as x approaches from the right.

5.15 (Some shortcuts). (a) Obtain $\lim_{x \to 0} \sin^3(3x)/(1 - \cos x)$ by using the results of Examples 5.11 and 5.12.

(b) Obtain $\lim_{x \to 0}[(e^x - 1)/x]^{\frac{1}{2}}$ by using the result of Problem 5.12a.

(c) Find

$$\lim\limits_{x \to 0} \dfrac{\sin x \cdot (2 + \tan x)}{x(3 - \tan^2 x)},$$

assuming that $\lim_{x \to 0}(\sin x)/x = 1$.

5.16 Prove l'Hôpital's rule, eqn (5.6). (Hint: it is easier to see what is happening if you look at the case $M = N$ first.)

5.17 Using (5.4) identify the functions which have the following Taylor series:

(a) $\sum\limits_{n=0}^{\infty} x^n\left[\dfrac{1}{n!} - (-1)^n\right]$;   (b) $\sum\limits_{n=1}^{\infty} \dfrac{x^{n+2}}{n}$;   (c) $\sum\limits_{n=0}^{\infty} \dfrac{x^{2n}}{(2n)!}$;

and indicate the values of x for which the series is valid.

5.18 By identifying the Taylor series find the sums of the following series.

(a) $2 - \dfrac{2^3}{3!} + \dfrac{2^5}{5!} - \dfrac{2^7}{7!} + \cdots$;

(b) $1 + \left(\dfrac{1}{2}\right) + \dfrac{1}{2!}\left(\dfrac{1}{2}\right)^2 + \dfrac{1}{3!}\left(\dfrac{1}{2}\right)^3 + \cdots$;

(c) $1 - \dfrac{1}{4} + \dfrac{1}{16} - \dfrac{1}{64} + \dfrac{1}{256} - \cdots$.

6 Complex numbers

CONTENTS

6.1 Definitions and rules

The quadratic equation

$$x^2 - 2x + 2 = 0$$

can be written in the form $[(x-1)^2 - 1] + 2$, or

$$(x-1)^2 = -2 + 1 = -1$$

using the method known as 'completing the square'. If we solve this equation formally by taking the square root, then

$$x - 1 = \pm\sqrt{-1} \quad \text{or} \quad x = 1 \pm \sqrt{-1}.$$

However, there is no ordinary number whose square is -1. We call $\sqrt{-1}$ an 'imaginary' number, and denote it by the symbol i. It will be treated very like an ordinary number, except that if i^2 appears it may be replaced by -1. Expressions involving i, are called **complex numbers**. The symbol j is common in engineering and electronics, but i is used more widely. The **complex solutions**, or **roots** of the quadratic equation can be expressed as

$$x = 1 + i \quad \text{and} \quad x = 1 - i, \quad \text{or} \quad x = 1 \pm i.$$

The general quadratic equation

$$ax^2 + bx + c = 0 \tag{6.1}$$

can be solved as follows. By completing the square, the left-hand side of this equation can be written as

$$ax^2 + bx + c = a\left(x^2 + \frac{b}{a}x\right) + c = a\left(x + \frac{b}{2a}\right)^2 + c - \frac{b^2}{4a}.$$

The quadratic equation (6.1) becomes

$$\left(x + \frac{b}{2a}\right)^2 = \frac{b^2 - 4ac}{4a^2}.$$

Taking the square root

$$x + \frac{b}{2a} = \pm\frac{\sqrt{(b^2 - 4ac)}}{2a}.$$

Hence the solutions of the quadratic equation (6.1) are

$$x = -\frac{b}{2a} \pm \frac{1}{2a}\sqrt{(b^2 - 4ac)}.$$

The solutions are distinct real numbers if $b^2 > 4ac$, equal and real if $b^2 = 4ac$, and complex numbers if $b^2 < 4ac$. If $b^2 < 4ac$, the solutions can be written in terms of i as

$$x = -\frac{b}{2a} \pm i\frac{1}{2a}\sqrt{(4ac - b^2)},$$

where $\sqrt{(4ac - b^2)}$ is a real number.

A **complex number in standard form** is any number of the form

$$z = x + iy,$$

where x and y are real numbers. In this expression x is the **real part** of z, written as $x = \text{Re } z$, and y is the **imaginary part**, written as $y = \text{Im } z$ (note that it is y, not iy, which is called the imaginary part). If $y = 0$, then z is a **real number**, and if $x = 0$, then z is an **imaginary number**.

We need to put together rules for manipulating complex numbers: the rules are natural consequences of operations of addition, multiplication, etc., on numbers containing i. The only exception to normal algebra is that, *whenever i^2 appears, we can substitute -1.*

Example 6.1 *Express i^2, i^3, i^4, i^5, and i^6 in standard form.*

The standard forms are

$$i^2 = -1, \qquad i^3 = i^2i = -1 \times i = -i.$$

Since $-i$ can be written $0 + (-1)i$, it follows that $\text{Re } i^3 = 0$ and $\text{Im } i^3 = -1$.

$$i^4 = i^2i^2 = (-1)(-1) = 1, \qquad i^5 = ii^4 = i, \qquad i^6 = ii^5 = -i.$$

Example 6.2 *Express the following in standard form, and state the real and imaginary parts in each case:*

(a) $(2 + i) - (3 + 3i)$;   (b) $i(i + 2)$;   (c) $(1 - i)(1 + 2i)$;
(d) $(2 - 3i)(2 + 3i)$.

(a) $(2 + i) - (3 + 3i) = 2 + i - 3 - 3i = -1 - 2i$.
Real part $= -1$; imaginary part $= -2$.

(b) $i(i + 2) = i^2 + 2i = -1 + 2i$.
Real part $= -1$; imaginary part $= 2$.

(c) $(1 - i)(1 + 2i) = 1 + 2i - i - 2i^2 = 1 + 2i - i + 2 = 3 + i$.
Real part $= 3$; imaginary part $= 1$.

(d) $(2 - 3i)(2 + 3i) = 2^2 - (3i)^2 = 4 - 9i^2 = 4 + 9 = 13$.
Real part $= 13$; imaginary part $= 0$.

Let $z_1 = x_1 + iy_1$ and $z_2 = x_2 + iy_2$. In formal terms, the principal rules are as follows.

1. Two complex numbers z_1 and z_2 are said to be **equal** if and only if $x_1 = x_2$ and $y_1 = y_2$: we write $z_1 = z_2$.

2. The **sum** of two complex numbers z_1 and z_2 is given by

$$z_1 + z_2 = (x_1 + iy_1) + (x_2 + iy_2) = (x_1 + x_2) + i(y_1 + y_2),$$

its real part being the sum of the real parts, and its imaginary part the sum of the imaginary parts of z_1 and z_2.

3. Similarly the **difference** $z_1 - z_2$ is

$$z_1 - z_2 = (x_1 + iy_1) - (x_2 + iy_2) = (x_1 - x_2) + i(y_1 - y_2).$$

4. The **product** of z_1 and z_2 is

$$\begin{aligned}
z_1 z_2 &= (x_1 + iy_1)(x_2 + iy_2) \\
&= x_1 x_2 + iy_1 x_2 + x_1 iy_2 + iy_1 iy_2 \\
&= x_1 x_2 + iy_1 x_2 + ix_1 y_2 + i^2 y_1 y_2 \\
&= x_1 x_2 + iy_1 x_2 + ix_1 y_2 - y_1 y_2 \quad (\text{since } i^2 = -1) \\
&= (x_1 x_2 - y_1 y_2) + i(y_1 x_2 + x_1 y_2).
\end{aligned}$$

In order to carry out **division**, a special result is needed. Suppose that $z = x + iy$, where x and y are real. Then the number $x - iy$, where we have changed i to $-i$, is called the **complex conjugate** of z, and will be written $\bar{z}$. The product $z\bar{z}$ is given by

$$z\bar{z} = (x + iy)(x - iy) = x^2 - (iy)^2 = x^2 + y^2,$$

which is a *real positive number*. (This was illustrated in Example 6.2d.)

> **Property of the conjugate**
> Let $z = x + iy$, with x, y real. Then $\bar{z} = x - iy$ and $z\bar{z} = x^2 + y^2$.   (6.2)

5. The **reciprocal** of a complex number in standard form. Let $z = x + iy$, and consider

$$\frac{1}{z} = \frac{1}{x + iy}.$$

This is not in standard form $a + ib$: to reduce it to standard form, multiply it by the factor

$$\frac{x - iy}{x - iy},$$

$x - iy$ being the conjugate of $x + iy$. This factor is equal to 1, and it will not affect the value of $1/z$. Hence

$$\frac{1}{z} = \frac{1}{x + iy} \frac{x - iy}{x - iy} = \frac{x - iy}{x^2 + y^2}$$

$$= \frac{x}{x^2 + y^2} - i\frac{y}{x^2 + y^2} \quad \text{(from (6.2))},$$

which is in standard form. This process also enables us to reduce **quotients** to standard form, as in Example 6.3c below.

Example 6.3 *Reduce to standard form*

(a) $\dfrac{1}{2 + 3i}$; (b) $\dfrac{1}{2 - 3i}$; (c) $\dfrac{1 - i}{1 + i}$; (d) $\dfrac{1}{i}$.

The standard forms are:

(a) $\dfrac{1}{2 + 3i} = \dfrac{1}{2 + 3i} \dfrac{2 - 3i}{2 - 3i} = \dfrac{2 - 3i}{2^2 + 3^2} = \dfrac{2}{13} - \dfrac{3}{13}i$;

(b) $\dfrac{1}{2 - 3i} = \dfrac{1}{2 - 3i} \dfrac{2 + 3i}{2 + 3i} = \dfrac{2 + 3i}{2^2 + 3^2} = \dfrac{2}{13} + \dfrac{3}{13}i$;

(c) $\dfrac{1 - i}{1 + i} = \dfrac{1 - i}{1 + i} \dfrac{1 - i}{1 - i} = \dfrac{1 - 2i + i^2}{1^2 + 1^2} = -i$;

(d) $\dfrac{1}{i} = \dfrac{1}{i} \left(\dfrac{-i}{-i}\right) = -i$.

The general quotient rule is

$$\frac{z_1}{z_2} = \frac{(x_1 + iy_1)(x_2 - iy_2)}{(x_2 + iy_2)(x_2 - iy_2)},$$

$$= \frac{(x_1x_2 + y_1y_2) + i(x_2y_1 - x_1y_2)}{x_2^2 + iy_2x_2 - ix_2y_2 + y_2^2},$$

$$= \frac{(x_1x_2 + y_1y_2) + i(x_2y_1 - x_1y_2)}{x_2^2 + y_2^2}.$$

Example 6.4   *Find the standard form of the complex numbers*
(a) $z_1 + z_2$, (b) $2z_1 - 3z_2$, (c) $z_1 z_2$, (d) z_1^2/z_2, where $z_1 = -1 + 2i$ and
$z_2 = 2 - 3i$.

(a) $z_1 + z_2 = (-1 + 2i) + (2 - 3i) = 1 - i.$

(b) $2z_1 - 3z_2 = 2(-1 + 2i) - 3(2 - 3i)$

$$= -2 + 4i - 6 + 9i = -8 + 13i.$$

(c) $z_1 z_2 = (-1 + 2i)(2 - 3i)$

$$= -2 + 4i + 3i - 6i^2 = -2 + 4i + 3i + 6 = 4 + 7i.$$

(d) First

$$z_1^2 = (-1 + 2i)(-1 + 2i) = 1 - 4i + 4i^2 = -3 - 4i.$$

Then

$$\frac{z_1^2}{z_2} = \frac{-3 - 4i}{2 - 3i} = \frac{(-3 - 4i)}{(2 - 3i)} \frac{(2 + 3i)}{(2 + 3i)}$$

$$= \frac{-6 - 8i - 9i - 12i^2}{4 - 6i + 6i - 9i^2}$$

$$= \frac{6 - 17i}{4 + 9} = \frac{6}{13} - \frac{17}{13}i.$$

The conjugate $\bar{z}$ of z has the following further properties, which
are simply applications of the rule:

to obtain the conjugate, change i to −i wherever it appears.

Properties of the conjugate

(a) $\overline{z_1 + z_2} = \bar{z}_1 + \bar{z}_2.$

(b) $\overline{z_1 z_2} = \bar{z}_1 \bar{z}_2.$

(c) $\overline{\left(\dfrac{z_1}{z_2} \right)} = \dfrac{\bar{z}_1}{\bar{z}_2}.$

(d) $x = \operatorname{Re} z = \dfrac{1}{2}(z + \bar{z}), \quad y = \operatorname{Im} z = \dfrac{1}{2i}(z - \bar{z}).$

(6.3)

Property 6.3(c) was illustrated by Example 6.2a,b. Similarly, for
example,

the conjugate of $\dfrac{(2 + 3i)(4 + i)(3 - 2i)}{(1 - 4i)}$ is $\dfrac{(2 - 3i)(4 - i)(3 + 2i)}{(1 + 4i)}$;

we do not have to work the whole thing out first and then find the
conjugate. These rules become important later.

6.2 The Argand diagram and complex numbers

A complex number $z = x + iy$ can be regarded as a pair of real numbers (x, y) known as an **ordered pair**. The pair of numbers can be interpreted as the cartesian coordinates of a point in the plane in the usual way. The complex number $z = x + iy$ has an abscissa x and an ordinate y. In Fig. 6.1, the x axis is known as the **real axis** and the y axis is the **imaginary axis**. The number $z = x + iy$ is represented by the point $P : (x, y)$. A figure showing complex numbers is known as the **Argand diagram** of the complex numbers.

The length $OP = r = \sqrt{(x^2 + y^2)} \geqslant 0$ is called the **modulus** of z (or simply 'mod z') and written $|z|$. The polar angle θ (see Section 1.6) is called an **argument** of z, and is written arg z. As in Section 1.6, polar angles differing by a multiple of 2π are equivalent.

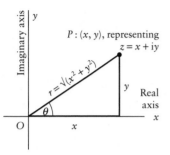

Fig. 6.1 The Argand diagram.

Example 6.5 *Obtain $|z|$ where* (a) $z = 2 + 3i$; (b) $z = 2 - 3i$.

(a) $|z| = |2 + 3i| = (2^2 + 3^2)^{\frac{1}{2}} = \sqrt{13}$.

(b) $|z| = |2 + (-3)i| = [2^2 + (-3)^2]^{\frac{1}{2}} = \sqrt{13}$.

Thus $|2 + 3i| = |2 - 3i|$. (The modulus of a number and the modulus of its conjugate are always equal.)

Example 6.6 *Let $z_1 = 1 - 3i$ and $z_2 = 3 - 2i$. Find*

(a) $|z_1 + z_2|$; (b) $|z_1| + |z_2|$; (c) $|z_1 z_2|$;

(d) $|z_1||z_2|$; (e) $\left|\dfrac{z_1}{z_2}\right|$; (f) $\dfrac{|z_1|}{|z_2|}$.

(a) $z_1 + z_2 = (1 - 3i) + (3 - 2i) = 4 - 5i$

(this *must* be worked out in standard form first). Hence

$\qquad |z_1 + z_2| = |4 - 5i| = [4^2 + (-5)^2]^{\frac{1}{2}} = \sqrt{41}$.

(b) $|z_1| + |z_2| = [1^2 + (-3)^2]^{\frac{1}{2}} + [3^2 + (-2)^2]^{\frac{1}{2}} = \sqrt{10} + \sqrt{13}$.

(c) $|z_1 z_2| = |(1 - 3i)(3 - 2i)| = |-3 - 11i|$

$\qquad\qquad = [(-3)^2 + (-11)^2]^{\frac{1}{2}} = \sqrt{130}$.

(d) $|z_1||z_2| = |1 - 3i||3 - 2i| = \sqrt{10}\sqrt{13} = \sqrt{130}$

$\qquad$ (i.e. the same as (c)).

(e) $\dfrac{z_1}{z_2} = \dfrac{1 - 3i}{3 - 2i}\,\dfrac{3 + 2i}{3 + 2i} = \dfrac{1}{13}(9 - 7i)$.

Therefore

$$\left|\frac{z_1}{z_2}\right| = \frac{1}{13}|9 - 7i| = \frac{1}{13}\sqrt{130} = \frac{\sqrt{10}}{\sqrt{13}}.$$

(f) $\dfrac{|z_1|}{|z_2|} = \dfrac{|1 - 3i|}{|3 - 2i|} = \dfrac{\sqrt{10}}{\sqrt{13}}$ (i.e. the same as (e)).

The following results hold good for the modulus; they are illustrated in Example 6.6 above.

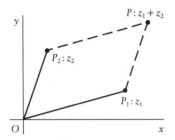

Fig. 6.2 Parallelogram law of addition.

> **Properties of the modulus**
>
> If $z = x + iy$ (x and y real), then $|z| = (x^2 + y^2)^{\frac{1}{2}}$, and
> (a) $|\bar{z}| = |z|$,
> (b) $z\bar{z} = |z|^2$,
> (c) $|z_1 z_2| = |z_1||z_2|$,
> (d) $|z_1/z_2| = |z_1|/|z_2|$,
> (e) the distance between two points z_1 and z_2
> is $|z_1 - z_2| = |z_2 - z_1|$.
>
> $\qquad\qquad\qquad\qquad\qquad\qquad\qquad$ (6.4)

The identity (6.4b) follows directly from (6.2). We shall defer the proof of (c) and (d), but the truth is illustrated in Example 6.6c,d,e,f. The modulus of a sum or difference *cannot* be split in this way: contrast the results of Examples 6.6a and 6.6b.

The sum of two complex numbers can be interpreted by the **parallelogram law of addition** in the Argand diagram, as in Fig. 6.2. Construct a parallelogram on OP_1 and OP_2, where P_1 and P_2 correspond to the complex numbers z_1 and z_2. The corner P of the parallelogram represents the sum $z_1 + z_2$. This follows from the addition rule for complex numbers. If you know anything about vectors, you will recognize that complex numbers add like vectors.

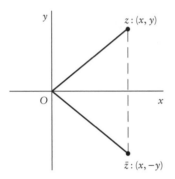

Fig. 6.3 Argand diagram showing z and its conjugate $\bar{z}$.

The conjugate $\bar{z} = x - iy$ is the reflection of z in the x axis, shown in Fig. 6.3.

6.3   Complex numbers in polar coordinates

In Fig. 6.1, r and θ obviously serve as **polar coordinates**, as defined in (1.20). In the context of complex numbers, θ is called the **argument** of z and denoted by arg z. The same point P can be described by the angles $\theta + 2\pi n$, where $n = \pm 1, \pm 2, \dots$. Usually we use the **principal value of the argument**, denoted by Arg z (capital A), which is the smallest numerically; its limits are given by $-\pi < \theta \leq \pi$, that is

$$-\pi < \text{Arg } z \leq \pi.$$

The pair of equations

$$\cos \theta = \frac{x}{r}, \qquad \sin \theta = \frac{y}{r},$$

has exactly one solution for θ within this range.

Example 6.7 *Find the moduli and principal values of the arguments of the following complex numbers:* (a) $z_1 = 2i$; (b) $z_2 = -1 - i$; (c) $z_3 = -2$; (d) $z_4 = \frac{1}{2} + \frac{1}{2}i\sqrt{3}$.

The moduli are given by:
(a) $|z_1| = |2i| = 2$;
(b) $|z_2| = |-1 - i| = \sqrt{(1 + 1)} = \sqrt{2}$;
(c) $|z_3| = |-2| = 2$;
(d) $|z_4| = |\frac{1}{2} + \frac{1}{2}i\sqrt{3}| = \sqrt{(\frac{1}{4} + \frac{3}{4})} = 1$.

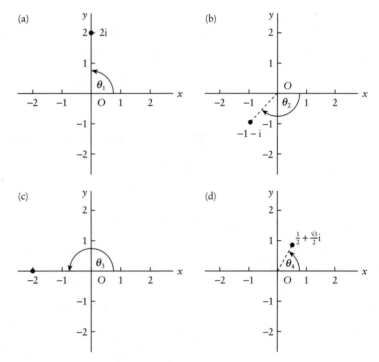

Fig. 6.4

A sketch of the Argand diagram for the complex numbers helps to decide their arguments. Figure 6.4 show their locations. Thus

(a) Arg $z_1 = \theta_1 = \frac{1}{2}\pi$; (b) Arg $z_2 = \theta_2 = -\frac{3}{4}\pi$; (c) Arg $z_3 = \theta_3 = \pi$;
(d) Arg $z_4 = \theta_4 = \frac{1}{3}\pi$.

In Fig. 6.1, the coordinates (x, y) and the polar coordinates (r, θ) are related by

$$x = r \cos \theta, \qquad y = r \sin \theta.$$

Hence the complex number $z = x + iy$ can be written

$$z = r \cos \theta + ir \sin \theta = r(\cos \theta + i \sin \theta), \tag{6.5}$$

which is the **polar form** of the complex number z. Note that $r \geqslant 0$.

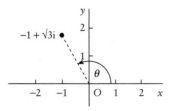

Fig. 6.5

Example 6.8   *Express* $-1 + \sqrt{3}i$ *in polar form.*

Here
$$r = \sqrt{[(-1)^2 + (\sqrt{3})^2]} = \sqrt{(1+3)} = 2$$
and θ is given by (see Fig. 6.5)
$$\cos \theta = -\tfrac{1}{2}, \qquad \sin \theta = \tfrac{1}{2}\sqrt{3}.$$
Hence from Fig. 6.5, $\theta = \tfrac{2}{3}\pi$, and
$$-1 + \sqrt{3}i = 2(\cos \tfrac{2}{3}\pi + i \sin \tfrac{2}{3}\pi).$$

Example 6.9   *Obtain* (a) $|\cos \theta + i \sin \theta|$; (b) $|1/(\cos \theta + i \sin \theta)|$.

(a) $|\cos \theta + i \sin \theta| = (\cos^2\theta + \sin^2\theta)^{\frac{1}{2}} = 1$.
(b) $|1/(\cos \theta + i \sin \theta)| = 1/|\cos \theta + i \sin \theta|$ (by (6.4d))
$$= 1 \quad \text{(from (a))}.$$

6.4   Complex numbers in exponential form

Consider the function
$$f(\theta) = \cos \theta + i \sin \theta,$$

where θ can take any value. Its derivative with respect to θ is

$$\frac{\mathrm{d}f(\theta)}{\mathrm{d}\theta} = -\sin \theta + i \cos \theta = i(\cos \theta + i \sin \theta) = if(\theta).$$

Hence $f(\theta)$ satisfies a relation in which the derivative is 'proportional' to itself notwithstanding that the constant of proportionality is i. As we saw in Chapter 1, a function with this property is the exponential function $k\,\mathrm{e}^{i\theta}$, where k is a constant. We conclude that

$$\cos \theta + i \sin \theta = k\,\mathrm{e}^{i\theta}$$

for some value of k. In particular this must be true for the value $\theta = 0$, from which $k = 1$. Hence, we obtain the important result

$$\mathrm{e}^{i\theta} = \cos \theta + i \sin \theta. \tag{6.6}$$

The conjugate formula is

$$\mathrm{e}^{-i\theta} = \cos \theta - i \sin \theta, \tag{6.7}$$

applying the rule of replacing i by $(-i)$.

From (6.5), any complex number can be written in the **exponential form** ('Euler's formula')

$$z = r(\cos \theta + i \sin \theta) = r\,\mathrm{e}^{i\theta},$$

with its conjugate

$$\bar{z} = r(\cos \theta - i \sin \theta) = r\,\mathrm{e}^{-i\theta}.$$

As an alternative justification of (6.6), we may use the Taylor series for exponential and trigonometric functions in (5.4b,c,d). Formally, by putting $x = i\theta$ into (5.4b), we obtain

$$e^{i\theta} = 1 + \frac{1}{1!}(i\theta) + \frac{1}{2!}(i\theta)^2 + \frac{1}{3!}(i\theta)^3 + \frac{1}{4!}(i\theta)^4 + \cdots$$

$$= 1 + i\frac{1}{1!}\theta - \frac{1}{2!}\theta^2 - i\frac{1}{3!}\theta^3 + \frac{1}{4!}\theta^4 + \cdots$$

$$= \left(1 - \frac{1}{2!}\theta^2 + \frac{1}{4!}\theta^4 - \cdots\right) + i\left(\theta - \frac{1}{3!}\theta^3 + \frac{1}{5!}\theta^5 - \cdots\right)$$

$$= \cos\theta + i\sin\theta$$

as required.

Properties of $e^{i\theta}$ for real θ

(a) $e^{i\theta} = \cos\theta + i\sin\theta$,

(b) $|e^{i\theta}| = 1$,   (c) conjugate of $e^{i\theta}$ is $e^{-i\theta}$.

(6.8)

Exponential form for a complex number

$z = r\,e^{i\theta} = r\cos\theta + ir\sin\theta$,

where $r = |z|$ and θ is any value of arg z.

(6.9)

Example 6.10 *Express the following in standard form:* (a) $2\,e^{\frac{1}{2}\pi i}$; (b) $3\,e^{-\pi i}$; (c) $e^{3\pi i}$; (d) $2\,e^{-\frac{1}{2}\pi i}$; (e) $3\,e^{\frac{1}{4}\pi i}$.

Remember that, in $r\,e^{i\theta}$, the numbers r and θ are polar coordinates. For these simple cases, we can therefore put the points straight on an Argand diagram and read off the coordinates, without needing to work out $\cos\theta$ and $\sin\theta$.

(a) $r = 2, \theta = \frac{1}{2}\pi$ (90°). Hence $2\,e^{\frac{1}{2}\pi i} = 2i$.

(b) $r = 3, \theta = -\pi$ (−180°). Hence $3\,e^{-i\pi} = -3$.

(c) $r = 1, \theta = 3\pi$. Hence $e^{3\pi i} = -1$.

(d) $r = 2, \theta = -\frac{1}{2}\pi$ (−90°). Hence $2\,e^{-\frac{1}{2}\pi i} = -2i$.

(e) $r = 3, \theta = \frac{1}{4}\pi$ (45°). Hence $3\,e^{\frac{1}{4}\pi i} = \frac{3}{2}\sqrt{2} + i\frac{3}{2}\sqrt{2}$.

It follows by treating (6.6) and (6.7) as two simultaneous equations for $\sin\theta$ and $\cos\theta$ that

---

$$\cos\theta = \frac{1}{2}(e^{i\theta} + e^{-i\theta}), \qquad \sin\theta = \frac{1}{2i}(e^{i\theta} - e^{-i\theta}).$$

(6.10)

Equation (6.6) will still be true if we replace he angle θ by $n\theta$, where n is an integer. Hence we obtain De Moivre's theorem:

$$\cos n\theta + i\sin n\theta = e^{ni\theta} = (e^{i\theta})^n = (\cos\theta + i\sin\theta)^n.$$

> **De Moivre's theorem**
>
> If n is any integer, then
>
> $$(\cos \theta + i \sin \theta)^n = \cos n\theta + i \sin n\theta. \tag{6.11}$$

The complex numbers having arguments θ and $\theta + 2n\pi$ are equal for all integer values of n, since 2π is a complete revolution on the Argand diagram. Thus

$$e^{i(\theta + 2n\pi)} = e^{i\pi} \, e^{i2n\pi} = e^{i\theta} \cdot 1 = e^{i\theta}$$

If $z_1 = r_1 e^{i\theta_1}$, $z_2 = r_2 e^{i\theta_2}$, then the product $z_1 z_2 = r_1 r_2 e^{i(\theta_1 + \theta_2)}$: its argument is the sum of the arguments of z_1 and z_2.

Example 6.11 *Express the following complex numbers in exponential form with principal values of the arguments* $(-\pi < \theta \leqslant \pi)$: (a) i; (b) $-5i$; (c) -3; (d) $4 - 4i$; (e) $3 - 4i$.

In each example, we put $r \cos \theta$ equal to the real part and $r \sin \theta$ equal to the imaginary part of the given complex number. In each case we shall find θ by plotting the point on an Argand diagram.

(a) $r \cos \theta = 0$, $r \sin \theta = 1$. Hence $r = 1$ and, in the interval $-\pi < \theta \leqslant \pi$, we obtain $\theta = \frac{1}{2}\pi$. The exponential form is

  $i = e^{\frac{1}{2}i\pi}$.

(b) $r \cos \theta = 0$, $r \sin \theta = -5$. Hence $r = 5$ and $\theta = -\frac{1}{2}\pi$. The exponential form is

  $-5i = 5 \, e^{-\frac{1}{2}i\pi}$.

(c) $r \cos \theta = -3$, $r \sin \theta = 0$. Hence $r = 3$ and $\theta = \pi$. The exponential form is

  $-3 = 3 \, e^{i\pi}$.

(d) $r \cos \theta = 4$, $r \sin \theta = -4$. Hence $r = \sqrt{(16 + 16)} = 4\sqrt{2}$ while $\theta = -\frac{1}{4}\pi$. Hence

  $4 - 4i = 4\sqrt{2} \, e^{-\frac{1}{4}i\pi}$.

(e) $r \cos \theta = 3$, $r \sin \theta = -4$. Hence $r = \sqrt{(9 + 16)} = 5$ while the angle α is the principal value of the argument such that $\cos \alpha = \frac{3}{5}$, $\sin \alpha = -\frac{4}{5}$. The exponential form is

  $3 - 4i = 5 \, e^{i\alpha}$.

where $\alpha = -53.1°$ or -0.927 radians.

Example 6.12 *By expressing* $-1 + i$ *in the form* $r \, e^{i\theta}$, *find* $(-1 + i)^{-8}$ *as a complex number in standard form.*

First $r = |-1 + i| = \sqrt{2}$. From its position on an Argand diagram $\theta = 3 \times 45°$, or $\frac{3}{4}\pi$ in radians. Therefore

  $-1 + i = \sqrt{2} \, e^{\frac{3}{4}\pi i}$.

Then

  $(-1 + i)^{-8} = (\sqrt{2} \, e^{\frac{3}{4}\pi i})^{-8} = (\sqrt{2})^{-8} \, e^{-6\pi i} = \frac{1}{16} \, e^{-6\pi i}$.

Example 6.12 *continued*

On an Argand diagram the polar coordinates are $r = \frac{1}{16}$ and $\theta = -6\pi$ $= -3(2\pi)$. This value of θ, equivalent to three complete revolutions, puts us on the positive real axis again, so that

$$(-1 + i)^{-8} = \tfrac{1}{16}.$$

6.5 The general exponential form

The advantage of the exponential form of a complex number is that it is particularly easy to differentiate, integrate, and to combine with other exponentials, including ordinary ones. We are not tied to the Argand diagram when we manipulate the exponentials, so we shall often use letters other than r and θ.

Example 6.13 *Prove that*

$\cos(A + B) = \cos A \cos B - \sin A \sin B,$

$\sin(A + B) = \sin A \cos B + \cos A \sin B.$

As with an ordinary exponential,

$$e^{i(A+B)} = e^{iA}\, e^{iB}.$$

In terms of the definition (6.8), this becomes

$$\cos(A + B) + i \sin(A + B)$$
$$= (\cos A + i \sin A)(\cos B + i \sin B)$$
$$= (\cos A \cos B - \sin A \sin B) + i(\sin A \cos B + \cos A \sin B).$$

The real and imaginary parts of the two sides of the equation must be respectively equal, and so we have the result immediately.

Example 6.14 *Express $\cos 5\theta$ in terms of power of $\cos \theta$.*

Since

$$e^{5i\theta} = (e^{i\theta})^5,$$

it follows that

$$\cos 5\theta + i \sin 5\theta = (\cos \theta + i \sin \theta)^5.$$

Expand the right-hand side by the binomial theorem. Thus

$$\cos 5\theta + i \sin 5\theta = \cos^5\theta + 5 \cos^4\theta \cdot i \sin \theta + 10 \cos^3\theta(i \sin \theta)^2$$
$$+ 10 \cos^2\theta(i \sin \theta)^3 + 5 \cos \theta(i \sin \theta)^4 + (i \sin \theta)^5$$
$$= \cos^5\theta + 5i \cos^4\theta \sin \theta - 10 \cos^3\theta \sin^2\theta$$
$$- 10i \cos^2\theta \sin^3\theta + 5 \cos \theta \sin^4\theta + i \sin^5\theta.$$

Equate real parts on both sides of the equation:

$$\cos 5\theta = \cos^5\theta - 10 \cos^3\theta \sin^2\theta + 5 \cos \theta \sin^4\theta.$$

Finally replace $\sin^2\theta$ by $1 - \cos^2\theta$ and simplify:

$$\cos 5\theta = \cos^5\theta - 10 \cos^3\theta(1 - \cos^2\theta) + 5 \cos \theta(1 - \cos^2\theta)^2,$$
$$= 16 \cos^5\theta - 20 \cos^3\theta + 5 \cos \theta.$$

(Incidentally, $\sin 5\theta$ can also be found in terms of powers of $\sin \theta$.)

Example 6.15 *Prove the result in (6.4d), that*

$$\left|\frac{z_1}{z_2}\right| = \frac{|z_1|}{|z_2|}.$$

Put $z_1 = r_1\,e^{i\theta_1}$ and $z_2 = r_2\,e^{i\theta_2}$. Then

$$\frac{z_1}{z_2} = \frac{r_1\,e^{i\theta_1}}{r_2\,e^{i\theta_2}} = \frac{r_1}{r_2}\,e^{i(\theta_1-\theta_2)}.$$

Therefore

$$\left|\frac{z_1}{z_2}\right| = \frac{r_1}{r_2} = \frac{|z_1|}{|z_2|},$$

as required.

Consider the number z where

$$z = c\,e^{p+iq},$$

and p, r, and $c > 0$ are real numbers. We have

$$z = r\,e^{p+iq} = r\,e^{p}\,e^{iq} = r\,e^{p}(\cos q + i \sin q)$$

(q is assumed to be in radians). Therefore we have

> **The form** $z = c\,e^{p+iq}$ (p, q, c **real, with** $c > 0$)
>
> (a) $|z| = c\,e^{p}$,   (b) $\arg z = q + 2n\pi$,
> (c) $\operatorname{Re} z = c\,e^{p} \cos q$, $\operatorname{Im} z = c\,e^{p} \sin q$.   **(6.12)**

In science and engineering, complex exponentials of the type in (6.12) are often used to describe oscillations of various kinds. Instead of $c\,e^{p+iq}$, the kind of symbols that occur may look like

$$A\,e^{(-k+i\omega)t}, \quad \text{or} \quad c\,e^{(\alpha+i\beta)t},$$

in which t represents time. Recast $c\,e^{(\alpha+i\beta)t}$ by writing it as $c\,e^{\alpha t+i\beta t}$, and we have

> **The form** $c\,e^{(\alpha+i\beta)t}$ (α, β, c, t **real**)
>
> (a) $c\,e^{\alpha t} \cos \beta t = \operatorname{Re} c\,e^{(\alpha+i\beta)t}$,
> (b) $c\,e^{\alpha t} \sin \beta t = \operatorname{Im} c\,e^{(\alpha+i\beta)t}$.   **(6.13)**

Example 6.16 *The damped vibration of a piece of machinery is described by $x = 0.01\,e^{-0.02t} \cos 15t$. Write this in the form $x = \operatorname{Re}(c\,e^{(\alpha+i\beta)t})$.*

We have

$$x = 0.01\,e^{-0.02t} \cos 15t = 0.01\,e^{-0.02t}\operatorname{Re} e^{i15t}$$
$$= \operatorname{Re}(0.01\,e^{-0.02t}\,e^{i15t}) = \operatorname{Re}(0.01\,e^{(-0.02+15i)t}).$$

Example 6.17   *The current $i(t)$ in a branch of a circuit is given by*

$$i(t) = c\,e^{-kt}\sin(\omega t + \phi).$$

Write this in the form of (a) the imaginary part of a complex function; (b) the real part of a complex function.

(a) $i(t) = \text{Im}(c\,e^{-kt}\,e^{i(\omega t + \phi)}) = \text{Im}(c\,e^{(-k+i\omega)t+i\phi})$.

(b) Note that, if $z = x + iy$, then

$$y = \text{Im}\,z = \text{Re}(-iz).$$

Therefore

$$i(t) = \text{Re}(-i\,e^{(-k+i\omega)t+i\phi}) = \text{Re}(c\,e^{-\frac{1}{2}\pi i}\,e^{(-k+i\omega)t+i\phi})$$
$$= \text{Re}(c\,e^{(-k+i\omega)t+i(\phi-\frac{1}{2}\pi)}).$$

6.6 Hyperbolic functions

The hyperbolic functions cosh and sinh are related to the trigonometric functions cos and sin. The hyperbolic functions were defined in Section 1.13 by

$$\cosh x = \tfrac{1}{2}(e^x + e^{-x}), \qquad \sinh x = \tfrac{1}{2}(e^x - e^{-x}).$$

It follows that

$$\cosh ix = \tfrac{1}{2}(e^{ix} + e^{-ix}) = \cos x,$$
$$\sinh ix = \tfrac{1}{2}(e^{ix} - e^{-ix}) = i\sin x,$$

by (6.10). Similarly

$$\cos ix = \tfrac{1}{2}(e^{i^2x} + e^{-i^2x}) = \tfrac{1}{2}(e^{-x} + e^{x}) = \cosh x,$$

$$\sin ix = \frac{1}{2i}(e^{i^2x} - e^{-i^2x}) = -\tfrac{1}{2}i(e^{-x} - e^{x}) = i\sinh x.$$

Example 6.18   *Solve the equation* $\cosh z = -1$.

Since, for real z, we have $\cosh z \geqslant 1$, we expect the equation to have complex roots. In exponential form,

$$\tfrac{1}{2}(e^z + e^{-z}) = -1,$$

or

$$e^{2z} + 2e^z + 1 = 0,$$

or

$$(e^z + 1)^2 = 0.$$

Hence

$$e^z = -1 = e^{(2n+1)\pi i} \quad (n = 0, \pm 1, \pm 2, \dots).$$

Here we have considered **all the representations** of the number -1, in the form $e^{(2n+1)\pi i}$. It is important when finding *all* the roots of an equation to include **all possible arguments,** not just the principal one. By matching the exponents, the solutions are

$$z = (2n + 1)\pi i \quad (n = 0, \pm 1, \pm 2, \dots).$$

6.7 Miscellaneous applications

The polar form of complex numbers can be used to solve polynomial equations as in the following example.

Example 6.19   *Find all solutions of the equation $z^5 = 4 - 4i$.*

We first express $4 - 4i$ in polar form $\rho\, e^{i\alpha}$. Thus

$$\rho \cos \alpha = 4, \qquad \rho \sin \alpha = -4,$$

from which it follows that $\rho = \sqrt{32}$ and $\alpha = -\frac{1}{4}\pi + 2n\pi$, using an Argand diagram. All the polar representations of $4 - 4i$ are given by

$$4 - 4i = 4\sqrt{2}\, e^{-\frac{1}{4}i\pi + 2n\pi i} = 2^{\frac{5}{2}}\, e^{i\rho(-\frac{1}{4} + 2n)} \quad (n = 0, \pm 1, \pm 2, \dots).$$

Let $z = r\, e^{i\theta}$. Then

$$r^5\, e^{5i\theta} = 2^{\frac{5}{2}}\, e^{i\pi(-\frac{1}{4} + 2n)},$$

so that

$$r^5 = 2^{\frac{5}{2}} \quad \text{and} \quad 5\theta = \pi(-\tfrac{1}{4} + 2n),$$

with $n = 0, \pm 1, \pm 2, \dots$. Therefore

$$r = \sqrt{2}, \qquad \theta = \tfrac{1}{20}(-1 + 8n)\pi.$$

Five successive values of n give distinct solutions; other values of n merely duplicate existing solutions. In full, the solutions (the five fifth roots of $4 - 4i$) are

$$\sqrt{2}\, e^{-\frac{1}{20}i\pi}, \quad \sqrt{2}\, e^{\frac{7}{20}i\pi}, \quad \sqrt{2}\, e^{\frac{3}{4}i\pi}, \quad \sqrt{2}\, e^{\frac{23}{20}i\pi}, \quad \sqrt{2}\, e^{\frac{31}{20}i\pi}.$$

The exponential form (6.8a) can be used to express $\cos n\theta$ and $\sin n\theta$ in terms of powers of $\cos \theta$ and $\sin \theta$ respectively, and, conversely, to express $\cos^n\theta$ and $\sin^n\theta$ in terms of cosines and sines of multiple angles.

Example 6.20   *Expand $\cos^6\theta$ in terms of multiple angles.*

Let $z = \cos \theta + i \sin \theta$. By De Moivre's theorem, with n an integer,

$$z^n = \cos n\theta + i \sin n\theta, \qquad \frac{1}{z^n} = \cos n\theta - i \sin n\theta.$$

By adding these two results, it follows that

$$\cos n\theta = \frac{1}{2}\left(z^n + \frac{1}{z^n} \right). \tag{6.14}$$

Hence

$$(2 \cos \theta)^6 = \left(z + \frac{1}{z} \right)^6$$

$$= z^6 + 6z^4 + 15z^2 + 20 + \frac{15}{z^2} + \frac{6}{z^4} + \frac{1}{z^6}$$

$$= \left(z^6 + \frac{1}{z^6} \right) + 6\left(z^4 + \frac{1}{z^4} \right) + 15\left(z^2 + \frac{1}{z^2} \right) + 20$$

$$= 2 \cos 6\theta + 12 \cos 4\theta + 30 \cos 2\theta + 20,$$

by repeated use of (6.14). Finally

$$\cos^6\theta = \tfrac{1}{32} \cos 6\theta + \tfrac{3}{16} \cos 4\theta + \tfrac{15}{32} \cos 2\theta + \tfrac{5}{16}.$$

We can also use the polar form to sum certain series as in the following example.

Example 6.21 *Find the sum of the series*

$$f(\theta) = 1 + \cos\theta + \frac{\cos 2\theta}{2!} + \frac{\cos 3\theta}{3!} + \cdots .$$

In summation notation, the series can be written

$$f(\theta) = \sum_{n=0}^{x} \frac{\cos n\theta}{n!}.$$

Since $\cos n\theta = \mathrm{Re}\, e^{ni\theta}$, consider the series

$$S(\theta) = 1 + e^{i\theta} + \frac{e^{2i\theta}}{2!} + \frac{e^{3i\theta}}{3!} + \cdots ,$$

the real part of whose sum is the required sum $f(\theta)$. Thus

$$S(\theta) = 1 + e^{i\theta} + \frac{(e^{i\theta})^2}{2!} + \frac{(e^{i\theta})^3}{3!} + \cdots$$

$$= \exp e^{i\theta} = e^{\cos\theta + i\sin\theta}$$

$$= e^{\cos\theta}[\cos(\sin\theta) + i\sin(\sin\theta)],$$

using the formula (5.4b) for the power series of the exponential function. Thus

$$f(\theta) = \mathrm{Re}\, S(\theta) = e^{\cos\theta}\cos(\sin\theta).$$

Problems

6.1 (Section 6.1). Find the solutions of the following quadratic equations:
(a) $x^2 + 2x + 5 = 0$; (b) $x^2 - 6x + 10 = 0$;
(c) $x^2 + 2ix + 3 = 0$.

6.2 (Section 6.1). Find all the complex solutions of $x^4 + 3x^2 - 4 = 0$.

6.3 (Section 6.1). Express the following complex numbers in standard form:
(a) $(1 - i) + (3 + 4i)$; (b) $2(3 - i) + 3(-1 - i)$;
(c) $3(-1 + i) - 4(2 - 3i)$; (d) $3(1 + i)(2 - i)$;
(e) $\dfrac{2 + i}{3 - i}$; (f) $\dfrac{(2 + i)(7 + 5i)}{3 - i}$; (g) $(-1 + 2i)^2$;
(h) $(-1 + 2i)^2 + \dfrac{1}{(-1 + 2i)^2}$; (i) $(1 + i)^5$.

6.4 (Section 6.1). Find the boundary curve in the (p, q) plane which separates the (p, q) values giving real roots from those for the complex roots in the quadratic equation

$$x^2 + px + q = 0,$$

where p and q are real parameters. Of the real solutions, where in the (p, q) plane do the solutions which are both negative lie?

6.5 (Section 6.1). Let $z_1 = 3 - i$ and $z_2 = 1 + 2i$. Find, in standard form, the complex numbers
(a) $z_1 + z_2$; (b) $z_1 z_2$; (c) z_1/z_2; (d) z_1/z_2^2.

6.6 (Section 6.1). Let $z_1 = 2 + 3i$ and $z_2 = -2 + i$. Find the following complex conjugates:
(a) $\overline{z_1 + z_2}$; (b) $\overline{z_1 z_2}$; (c) $\overline{z_1/z_2}$; (d) $\overline{\bar{z}_1/\bar{z}_2}$.

6.7 Let $z = 1 + i$. Find the following complex numbers in standard form and plot their corresponding points in the Argand diagram:
(a) $\bar{z}$; (b) z^2; (c) $\bar{z}^2$; (d) $1/\bar{z}$; (e) $z/\bar{z}$.

6.8 (Section 6.5). Three complex numbers are given by $z_1 = 2\,e^{1+i}$, $z_2 = 3\,e^{-i}$, and $z_3 = \frac{1}{2}\,e^{-1+2i}$. Express the following complex numbers in standard form:
(a) $z_1 + z_2 + z_3$; (b) $z_1 z_2 + \bar{z}_3$; (c) $z_1 z_2 z_3$;
(d) $z_1 \bar{z}_2 / z_3$; (e) $z_1^2 z_2^2 - 2z_3^2$.

6.9 (Section 6.3). Find the modulus and principal argument of each of the following complex numbers:
(a) $z_1 = -2 + 2i$; (b) $z_2 = 4 - 4\sqrt{3}i$; (c) $z_3 = -5i$;
(d) $z_4 = -3$; (e) $z_5 = 3 + 4i$.

6.10 (Section 6.2). Let $z = x + iy$. Express each of the following equations in the complex variable z in real form in terms of x and y. Sketch, and identify in each case, the corresponding curve in the Argand diagram:
(a) $z\bar{z} = 1$; (b) Im $z = 2$;
(c) $|z - a| = 1$, where a is a complex number;
(d) $(z - \bar{z})^2 = -8(z + \bar{z})$;
(e) $|z - 1| + |z + 1| = 4$;
(f) Arg $z = \frac{1}{4}\pi$ (see Section 6.3);
(g) $|z| = \arg z$.

6.11 (Section 6.4). Express the following complex numbers in exponential form with principal arguments:
(a) $-1 + i$; (b) -2; (c) $-3i$;
(d) $7 - 7i\sqrt{3}$; (e) $(1 - i)(1 + i\sqrt{3})$;
(f) $\dfrac{1 - i}{1 + \sqrt{3}}$; (g) e^{2+i}; (h) $(1 + i)\,e^{2i}$;
(i) $(1 - i\sqrt{3})^9$; (j) $\dfrac{(1 + i)^4}{2 - 2i}$.

6.12 (Section 6.3). Using Euler's formula (6.8) for $e^{\pm i\theta}$, obtain the trigonometric identities for $\cos(\theta_1 \pm \theta_2)$ and $\sin(\theta_1 \pm \theta_2)$.

6.13 (Section 6.2). Using the parallelogram rule, sketch the locations in the Argand diagram, for general complex numbers z_1 and z_2, the following points: $z_1 + z_2$, $\bar{z}_1 + \bar{z}_2$, $z_1 - z_2$, $\bar{z}_1 + z_2$, $z_1 - \bar{z}_2$.

6.14 Let $f(\theta) = \cos\theta + i\sin\theta$. Verify that
$$\frac{d^2 f(\theta)}{d\theta^2} = -f(\theta).$$
Show that it is still true if
$$f(\theta) = a\cos\theta + b\sin\theta,$$
where a and b are arbitrary complex numbers.

6.15 Prove that $\tan ia = i\tanh a$, where a is a real number.

6.16 Find all the complex roots of the following equations:
(a) $\cosh z = 1$; (b) $\sinh z = 1$;
(c) $e^z = -1$; (d) $\cos z = \sqrt{2}$.

6.17 The **logarithm** of a complex number $z = r\,e^{i\theta}$ is defined by
$$\log z = \ln r + i\theta,$$
which will be a *multivalued* function because of the term $i\theta$. The **principal value** of the logarithm is denoted by Log z (note the capital letter L), and is defined by
$$\text{Log } z = \ln r + i\theta,$$
where $\pi < \theta \leqslant \pi$.
(a) Find the principal value Log$(1 + i\sqrt{3})$, and indicate its location on the Argand diagram.
(b) Find all roots of the equation $\log z = \pi i$.
(c) Express Log(ei) in standard form.
(d) Show that $e^{\log z} = z$.

6.18 If $z \neq 0$ and c are complex numbers, then z^c is defined by
$$z^c = e^{c \ln z}$$
(see Problem 6.17).
(a) Express 2^i in standard form.
(b) Find the principal value of i^i.
(c) Find all complex roots of $z^i = -1$.

6.19 (Section 6.4). Find all complex solutions of $z^5 = -1$, and sketch their locations on the Argand diagram.

6.20 (Section 6.5). Find the modulus, argument, and real and imaginary parts of each of the following complex numbers:
(a) $2\,e^{3+2i}$; (b) $4\,e^i$; (c) $5\,e^{\cos\frac{1}{4}\pi + i\sin\frac{1}{4}\pi}$; (d) e^{1+i}.

6.21 (Section 6.5). An oscillation in a system is given by $x = 0.04\,e^{-0.01t}\sin 12t$. Write this in the form
$$x = \text{Re}(c\,e^{\alpha + i\beta}).$$

6.22 (Section 6.5). The current in a branch of a circuit is given by
$$i(t) = c\,e^{-0.05t}\sin(0.4t + 0.5).$$
Write this in the form of the real part of a complex function.

6.23 A function $f(z)$, where $z = x + iy$, is known as a **function of complex variable** z. Find the real and imaginary parts of the following functions in terms of x and y:
(a) z^2; (b) $z + 2z^2 + 3z^3$; (c) $\sin z$;
(d) $\cos z$; (e) $e^z \cos z$; (f) e^{z^2}.

6.24 Let $w = f(z)$, where $z = x + iy$ and $w = u + iv$ are complex variables. If $f(z) = z^2$, find u and v in terms of x and y. The relation represents a **mapping** between two Argand diagrams. What curves do the hyperbolas $x^2 - y^2 = 1$ and $xy = 1$ map into the (u, v) plane?

6.25 Show that the **mapping**

$$w = z + \frac{c}{z},$$

where $z = x + iy$ and $w = u + iv$ and c is a real number, maps the circle $|z| = 1$ in the z plane into an ellipse in the w plane, and find its equation.

6.26 (Section 6.7). Show that

$$\cos^6\theta = \tfrac{1}{32}(\cos 6\theta + 6 \cos 4\theta + 15 \cos 2\theta + 10),$$

and find $\sin^6\theta$.

6.27 The damped oscillation of a vibrating block is given by

$$x = \mathrm{Re}\, z, \qquad z = e^{(-0.2+0.5i)t},$$

in terms of the time t. Find x, and determine the values of t where x is zero. Find the velocity of the block
(a) as dx/dt;
(b) as $\mathrm{Re}\, dz/dt$;
and confirm that the answers are the same.

6.28 Given that $2 + i$ is a solution of the equation

$$z^4 - 2z^3 - z^2 + 2z + 10 = 0,$$

find the other solutions.

6.29 Find the sum of the series

(a) $1 - \sin\theta + \dfrac{\sin 2\theta}{2!} - \dfrac{\sin 3\theta}{3!} + \cdots$;

(b) $1 + 2 \cos\theta + \dfrac{2^2 \cos 2\theta}{2!} + \dfrac{2^3 \cos 3\theta}{3!} + \cdots$.

PART II

Matrix algebra and vectors

7

Matrix algebra

CONTENTS

7.1 Matrix definition and notation

In many applications in physics and engineering it is useful to be able to represent and manipulate data in tabular or array form. An array which obeys certain algebraic rules of operation is known as a **matrix**. Capital letters are usually used to denote matrices. Thus

$$A = \begin{bmatrix} 1 & 2 & -1 \\ 0 & 3 & -4 \end{bmatrix}$$

is a matrix with two **rows** and three **columns**. The individual terms are known as **elements**: the element in the second row and third column is -4. This matrix is said to be of **order** 2×3, or a 2×3 matrix. A general $m \times n$ matrix, one with m rows and n columns, can be represented by the notation

$$A = \begin{bmatrix} a_{11} & a_{12} & \dots & a_{1n} \\ a_{21} & a_{22} & \dots & a_{2n} \\ \vdots & \vdots & \ddots & \vdots \\ a_{m1} & a_{m2} & \dots & a_{mn} \end{bmatrix} = [a_{ij}] \quad (1 \le i \le m, 1 \le j \le n),$$

where a_{ij} is the element in the ith row and jth column of A; or by

$$A = [a_{ij} : i = 1, \ldots, m; j = 1, \ldots, n],$$

or simply

$$A = [a_{ij}]$$

for brevity, if it is clear in context that the matrix is $m \times n$.

A 1×1 matrix is simply a number: for example, $[-5] = -5$. Matrices which have either one row or column are known as **vectors**. Thus

$$[1.3 \;\; -1.1 \;\; 2.9 \;\; 4.6] \quad \text{and} \quad \begin{bmatrix} -1.1 \\ 6.5 \\ -2.0 \end{bmatrix}$$

are respectively **row** and **column vectors**.

A matrix in which the number of rows equals the number of columns is called a **square matrix**: if $m \neq n$ then the matrix is said to be **rectangular**.

7.2 Rules of matrix algebra

We need to define consistent algebraic rules for manipulating matrices, such concepts as addition, multiplication, etc. As we shall see, these rules have their origins in the representation of linear equations and linear transformations, but for the present we simply state them as a list of rules.

1. *Equality.* Two matrices can only be equated if they are of the same order: that is, if they each have the same number of rows and the same number of columns. They are then said to be **equal** if the corresponding elements are equal. Thus if

$$A = \begin{bmatrix} a & b \\ c & d \end{bmatrix} \quad \text{and} \quad B = \begin{bmatrix} e & f \\ g & h \end{bmatrix},$$

then $A = B$ if and only if $a = e$, $b = f$, $c = g$, and $d = h$. In general, if $A = [a_{ij}]$ and $B = [b_{ij}]$ are both $m \times n$ matrices, then $A = B$ if and only if $a_{ij} = b_{ij}$ for $i = 1, 2, \ldots, m$ and $j = 1, 2, \ldots, n$.

Example 7.1 *Solve the equation $A = B$ when*

$$A = \begin{bmatrix} x & 1 & 2 \\ 0 & x^2 - y & 3 \end{bmatrix}, \qquad B = \begin{bmatrix} 1 & 1 & 2 \\ 0 & 2 & 3 \end{bmatrix}.$$

Since A and B must have the same elements, if follows that $x = 1$ and $x^2 - y = 2$. Hence, $y = x^2 - 2 = -1$. The solution is $x = 1$, $y = -1$.

---

2. *Multiplication by a constant.* Let k be a constant or **scalar**. By the product kA we mean the matrix in which **every element of A is multiplied by k**. Thus, if

$$A = \begin{bmatrix} 2.0 & 1.5 & 3.1 \\ -1.2 & 3.0 & -4.6 \end{bmatrix},$$

and $k = 10$, then

$$kA = \begin{bmatrix} 10 \times 2.0 & 10 \times 1.5 & 10 \times 3.1 \\ 10 \times -1.2 & 10 \times 3.0 & 10 \times -4.6 \end{bmatrix}$$

$$= \begin{bmatrix} 20 & 15 & 31 \\ -12 & 30 & -46 \end{bmatrix}.$$

Equally, we can 'factorize' a matrix. Thus

$$A = \begin{bmatrix} 5 & 25 & -30 \\ 10 & 15 & -5 \end{bmatrix} = 5\begin{bmatrix} 1 & 5 & -6 \\ 2 & 3 & -1 \end{bmatrix}.$$

3. *Zero matrix.* Any matrix in which every element is zero is called a **zero** or **null matrix**. If A is a zero matrix, we can simply write $A = 0$.

4. *Matrix sums and differences.* The sum of two matrices A and B has meaning only if A and B are of the same order, in which case $A + B$ is defined as the matrix C whose elements are the sums of the corresponding elements in A and B. We write $C = A + B$. Thus, if $A = [a_{ij}]$ and $B = [b_{ij}]$ are both $m \times n$ matrices, then

$$C = A + B = [a_{ij} + b_{ij}].$$

Example 7.2   *If*

$$A = \begin{bmatrix} 1 & 3 \\ 2 & 2 \\ 3 & 1 \end{bmatrix} \quad and \quad B = \begin{bmatrix} -4 & -6 \\ -5 & -5 \\ -6 & -4 \end{bmatrix},$$

then find $A + B$, $B + A$, and $A + 2B$.

We have

$$A + B = \begin{bmatrix} 1 - 4 & 3 - 6 \\ 2 - 5 & 2 - 5 \\ 3 - 6 & 1 - 4 \end{bmatrix} = \begin{bmatrix} -3 & -3 \\ -3 & -3 \\ -3 & -3 \end{bmatrix} = -3\begin{bmatrix} 1 & 1 \\ 1 & 1 \\ 1 & 1 \end{bmatrix} \quad \text{(by Rule 2)}.$$

Also

$$B + A = \begin{bmatrix} -4 + 1 & -6 + 3 \\ -5 + 2 & -5 + 2 \\ -6 + 3 & -4 + 1 \end{bmatrix} = \begin{bmatrix} -3 & -3 \\ -3 & -3 \\ -3 & -3 \end{bmatrix} = A + B.$$

Example 7.2 *continued*

Further

$$A + 2B = \begin{bmatrix} 1 & 3 \\ 2 & 2 \\ 3 & 1 \end{bmatrix} + 2\begin{bmatrix} -4 & -6 \\ -5 & -5 \\ -6 & -4 \end{bmatrix}$$

$$= \begin{bmatrix} 1 & 3 \\ 2 & 2 \\ 3 & 1 \end{bmatrix} + \begin{bmatrix} -8 & -12 \\ -10 & -10 \\ -12 & -8 \end{bmatrix} \quad \text{(by Rule 2)}$$

$$= \begin{bmatrix} -7 & -9 \\ -8 & -8 \\ -9 & -7 \end{bmatrix}.$$

As the second sum suggests, the commutative property of the real numbers, namely $a_{ij} + b_{ij} = b_{ij} + a_{ij}$, implies the **commutative property of matrix addition**, that is

$$A + B = B + A.$$

The difference of two matrices is written as $A - B$ which is interpreted as $A + (-1)B$, using Rule 2 for the multiplication of B by the number -1, and then Rule 4 for the sum of A and $(-1)B$. In practice, we simply take the difference of corresponding elements.

Example 7.3   *Find $A - B$ and $2A - 3B$, if*

$$A = \begin{bmatrix} 1 & -1 & 2 \\ 0 & 2 & -3 \end{bmatrix}, \quad B = \begin{bmatrix} 2 & -2 & -3 \\ 1 & 0 & -1 \end{bmatrix}.$$

We have

$$A - B = \begin{bmatrix} 1 & -1 & 2 \\ 0 & 2 & -3 \end{bmatrix} - \begin{bmatrix} 2 & -2 & -3 \\ 1 & 0 & -1 \end{bmatrix}$$

$$= \begin{bmatrix} 1-2 & -1+2 & 2+3 \\ 0-1 & 2-0 & -3+1 \end{bmatrix} = \begin{bmatrix} -1 & 1 & 5 \\ -1 & 2 & -2 \end{bmatrix}.$$

Also

$$2A - 3B = 2\begin{bmatrix} 1 & -1 & 2 \\ 0 & 2 & -3 \end{bmatrix} - 3\begin{bmatrix} 2 & -2 & -3 \\ 1 & 0 & -1 \end{bmatrix}$$

$$= \begin{bmatrix} 2 & -2 & 4 \\ 0 & 4 & -6 \end{bmatrix} - \begin{bmatrix} 6 & -6 & -9 \\ 3 & 0 & -3 \end{bmatrix}$$

$$= \begin{bmatrix} -4 & 4 & 13 \\ -3 & 4 & -3 \end{bmatrix}.$$

The rules of arithmetic as applied to the elements of matrices lead to the following results for matrices for which addition can be defined:

(a) $A + (B + C) = (A + B) + C$ (**associative law of addition**).

In other words, the order of addition of matrices is immaterial.

(b) $k(A + B) = kA + kB$, $(k + l)A = kA + lA$.

5. *Matrix multiplication.* We now need to define the concept of the product of two matrices. Not all matrices can be multiplied: they must have the right shape, or be **conformable** for multiplication to be defined. The product of A and B, in this order, is written as AB (no product sign is used), but it is only defined if the **number of columns in A equals the number of rows in B**. The product BA might not exist, and if it does, it will not in general be equal to AB.

Let us look at the case where A is a 1×3 matrix, which is a row vector, and B is a 3×1 matrix, which is a column vector, given by

$$A = [a_{11} \ a_{12} \ a_{13}], \qquad B = \begin{bmatrix} b_{11} \\ b_{21} \\ b_{31} \end{bmatrix}.$$

The product AB is defined as the 1×1 matrix C given by

$$AB = [a_{11} \ a_{12} \ a_{13}] \begin{bmatrix} b_{11} \\ b_{21} \\ b_{31} \end{bmatrix} = [a_{11}b_{11} + a_{12}b_{21} + a_{13}b_{31}] = C. \tag{7.1}$$

Here, the single remaining element is the sum of the products of corresponding elements from the row in A and the column in B. Thus the product of a 1×3 matrix and a 3×1 matrix is a 1×1 matrix (or simply on ordinary number). This is known as a **row-on-column** operation.

Suppose now that A is a 2×3 matrix and that B a 3×2 matrix which are given by

$$A = \begin{bmatrix} a_{11} & a_{12} & a_{13} \\ a_{21} & a_{22} & a_{23} \end{bmatrix}, \qquad B = \begin{bmatrix} b_{11} & b_{12} \\ b_{21} & b_{22} \\ b_{31} & b_{32} \end{bmatrix}.$$

The product AB is now a 2×2 matrix C given by

$$\begin{aligned} AB &= \begin{bmatrix} a_{11} & a_{12} & a_{13} \\ a_{21} & a_{22} & a_{23} \end{bmatrix} \begin{bmatrix} b_{11} & b_{12} \\ b_{21} & b_{22} \\ b_{31} & b_{32} \end{bmatrix} \\ &= \begin{bmatrix} a_{11}b_{11} + a_{12}b_{21} + a_{13}b_{31} & a_{11}b_{12} + a_{12}b_{22} + a_{13}b_{32} \\ a_{21}b_{11} + a_{22}b_{21} + a_{23}b_{31} & a_{21}b_{12} + a_{22}b_{22} + a_{23}b_{32} \end{bmatrix} \\ &= C. \end{aligned} \tag{7.2}$$

Note that each row in A 'operates' on each column in B giving four elements in the 2×2 matrix C.

Example 7.4   *Find AB if*

$$A = \begin{bmatrix} 1 & -1 & 0 \\ 2 & 1 & -3 \end{bmatrix}, \qquad B = \begin{bmatrix} 0 & 3 \\ 1 & -1 \\ -2 & 4 \end{bmatrix}.$$

We have

$$AB = \begin{bmatrix} 1 & -1 & 0 \\ 2 & 1 & -3 \end{bmatrix} \begin{bmatrix} 0 & 3 \\ 1 & -1 \\ -2 & 4 \end{bmatrix}$$

$$= \begin{bmatrix} 1 \times 0 + (-1) \times 1 + 0 \times (-2) & 1 \times 3 + (-1) \times (-1) + 0 \times 4 \\ 2 \times 0 + 1 \times 1 + (-3) \times (-2) & 2 \times 3 + 1 \times (-1) + (-3) \times 4 \end{bmatrix}$$

$$= \begin{bmatrix} -1 & 4 \\ 7 & -7 \end{bmatrix}.$$

We can use a *summation* (Section 1.15) notation to condense the expanded sums of products which occur in matrices. The sum of a string of numbers, say, $c_1 + c_2 + c_3$ can be expressed as

$$\sum_{i=1}^{3} c_i,$$

where i runs through all the integers from the lower limit on i under the Σ symbol, to the upper limit above. Thus, for example,

$$\sum_{i=3}^{8} h_i = h_3 + h_4 + h_5 + h_6 + h_7 + h_8.$$

We can also use the summation notation with the double-suffix notation, as in

$$\sum_{i=1}^{4} h_{i6} = h_{16} + h_{26} + h_{36} + h_{46}.$$

The product given by (7.1) can be written

$$AB = [a_{11}b_{11} + a_{12}b_{21} + a_{13}b_{31}] = \left[\sum_{j=1}^{3} a_{1j}b_{j1} \right] = \sum_{j=1}^{3} a_{1j}b_{j1}.$$

Similarly the elements in the square matrix (7.2) can be expressed as

$$AB = \left[\sum_{k=1}^{3} a_{ik}b_{kj} : i,j = 1, 2 \right].$$

This example gives a clue to the general expression for the product of an $m \times n$ matrix A and an $n \times p$ matrix B. Remember that the number of columns in A must always equal the number of rows in B for the product to be defined. Thus, the row-on-column definition of the product is the $m \times p$ matrix

$$AB = \left[\sum_{k=1}^{n} a_{ik} b_{kj} : i = 1, \ldots, m; \, j = 1, \ldots, p \right].$$

Multiplication rule

The element in the ith row and jth column of the product consists of the row-on-column product of the ith row in A and the jth column in B.

(7.3)

Example 7.5 *If A is a 5×4 matrix, B is a 4×5 matrix, and C is a 6×4 matrix, which of the following products are defined: AB, BA, AC, CB, (AB)C, (CB)A?*

AB is a 5×5 matrix.

BA is a 4×4 matrix.

AC is not defined since A has four columns and C has six rows.

CB is a 6×5 matrix.

$(AB)C$ is not defined since AB is a 5×5 and C is a 6×4 matrix.

CB is a 6×5 matrix; hence $(CB)A$ is a 6×4 matrix.

One conclusion which can be inferred from the previous example is that **matrix multiplication does not commute**; that is, in general, $AB \neq BA$. As the previous example indicates, one or both products may not be defined; when both are defined, AB and BA may be of different order; and, even when both are defined and of the same order, AB is generally not equal to BA. So we must be careful about the order of multiplication. In the product AB, we say that A is **multiplied on the right** by B, or that B is **multiplied on the left** by A. The expressions 'A **postmultiplied** by B' and 'B **premultiplied** by A' are also used. Statements such as 'A is multiplied by B' can be ambiguous without carefully stating how the product occurs.

Example 7.6 *If*

$$A = \begin{bmatrix} 1 & -1 & 0 \\ 3 & -2 & -1 \end{bmatrix}, \qquad B = \begin{bmatrix} 1 & 2 \\ 1 & 2 \\ 1 & 2 \end{bmatrix},$$

calculate AB and BA.

We have

$$AB = \begin{bmatrix} 1 & -1 & 0 \\ 3 & -2 & -1 \end{bmatrix}\begin{bmatrix} 1 & 2 \\ 1 & 2 \\ 1 & 2 \end{bmatrix} = \begin{bmatrix} 0 & 0 \\ 0 & 0 \end{bmatrix} = 0,$$

and

$$BA = \begin{bmatrix} 1 & 2 \\ 1 & 2 \\ 1 & 2 \end{bmatrix}\begin{bmatrix} 1 & -1 & 0 \\ 3 & -2 & -1 \end{bmatrix} = \begin{bmatrix} 7 & -5 & -2 \\ 7 & -5 & -2 \\ 7 & -5 & -2 \end{bmatrix}.$$

This example illustrates the point that AB can be a zero matrix without either A or B or BA being zero. Also, as a consequence, $A(B - C) = 0$ does not necessarily imply $B = C$.

We state the following results concerning sums and products, but proofs are omitted:

(a) $A(B + C) = AB + AC$ (distributive law of addition),
(b) $A(BC) = (AB)C$ (associative law of multiplication),

provided that the products are defined.

7.3 Special matrices

We define and give properties of several special matrices. Some properties apply to rectangular matrices; others are specific to square matrices.

The **transpose** of any matrix is one in which the rows and columns are interchanged. Thus the first row becomes the first column, the second row the second column, and so on. We denote the transpose of A by A^T. Hence,

$$\text{if} \quad A = \begin{bmatrix} a_{11} & a_{12} \\ a_{21} & a_{22} \\ a_{31} & a_{32} \end{bmatrix} \quad \text{then} \quad A^T = \begin{bmatrix} a_{11} & a_{21} & a_{31} \\ a_{12} & a_{22} & a_{32} \end{bmatrix}.$$

The 3×2 matrix A becomes the 2×3 matrix A^T.

Example 7.7 *Find the transposes of A, B, A + B^T, and AB, where*

$$A = \begin{bmatrix} 1 & 2 \\ 0 & 1 \\ -1 & 1 \end{bmatrix}, \qquad B = \begin{bmatrix} 3 & -1 & 0 \\ 1 & 2 & -2 \end{bmatrix}.$$

Confirm that $(AB)^T = B^T A^T$.

We see that

$$A^T = \begin{bmatrix} 1 & 0 & -1 \\ 2 & 1 & 1 \end{bmatrix}, \qquad B^T = \begin{bmatrix} 3 & 1 \\ -1 & 2 \\ 0 & -2 \end{bmatrix},$$

$$(A + B^T)^T = \left(\begin{bmatrix} 1 & 2 \\ 0 & 1 \\ -1 & 1 \end{bmatrix} + \begin{bmatrix} 3 & 1 \\ -1 & 2 \\ 0 & -2 \end{bmatrix} \right)^T = \begin{bmatrix} 4 & 3 \\ -1 & 3 \\ -1 & -1 \end{bmatrix}^T$$

$$= \begin{bmatrix} 4 & -1 & -1 \\ 3 & 3 & -1 \end{bmatrix}.$$

Also, note that

$$A^T + (B^T)^T = \begin{bmatrix} 1 & 0 & -1 \\ 2 & 1 & 1 \end{bmatrix} + \begin{bmatrix} 3 & -1 & 0 \\ 1 & 2 & -2 \end{bmatrix}$$

$$= \begin{bmatrix} 4 & -1 & -1 \\ 3 & 3 & -1 \end{bmatrix} = (A + B^T)^T.$$

$$(AB)^T = \left(\begin{bmatrix} 1 & 2 \\ 0 & 1 \\ -1 & 1 \end{bmatrix} \begin{bmatrix} 3 & -1 & 0 \\ 1 & 2 & -2 \end{bmatrix} \right)^T = \begin{bmatrix} 5 & 3 & -4 \\ 1 & 2 & -2 \\ -2 & 3 & -2 \end{bmatrix}^T$$

$$= \begin{bmatrix} 5 & 1 & -2 \\ 3 & 2 & 3 \\ -4 & -2 & -2 \end{bmatrix},$$

$$B^T A^T = \begin{bmatrix} 3 & 1 \\ -1 & 2 \\ 0 & -2 \end{bmatrix} \begin{bmatrix} 1 & 0 & -1 \\ 2 & 1 & 1 \end{bmatrix} = \begin{bmatrix} 5 & 1 & -2 \\ 3 & 2 & 3 \\ -4 & -2 & -2 \end{bmatrix}.$$

Hence $(AB)^T = B^T A^T$.

1. *Properties of the transpose.* Provided that the sum $A + B$ and product AB are defined for two matrices A and B, the last example points to the following two results concerning transposes:
(a) $(A + B)^T = A^T + B^T$;
(b) $(AB)^T = B^T A^T$.

2. *Symmetric matrices.* A **square matrix** is said to be **symmetric** if $A = A^T$. Since rows and columns are interchanged in the transpose, this is equivalent to $a_{ij} = a_{ji}$ for all elements if $A = [a_{ij}]$. Symmetric

matrices are easy to recognize since their elements are reflected in the **leading diagonal**, the diagonal string of elements from the top left to the bottom right of the matrix. Thus

$$A = \begin{bmatrix} 1 & 3 & -2 \\ 3 & 2 & 4 \\ -2 & 4 & -1 \end{bmatrix}$$

is a 3×3 **symmetric matrix**.

A square matrix A for which $A = -A^T$ is said to be **skew-symmetric**. Note that, if A is any square matrix, then $A + A^T$ is symmetric and $A - A^T$ is skew-symmetric. The elements along the leading diagonal of a skew-symmetric matrix must all be zero. Thus

$$\begin{bmatrix} 0 & 1 & 2 \\ -1 & 0 & -3 \\ -2 & 3 & 0 \end{bmatrix}$$

is skew-symmetric.

3. *Row and column vectors.* As we defined them is Section 7.1, a row vector is a matrix with one row, and a column vector is one with one column. For vectors, we usually use bold-faced small letters and write, for example,

$$\boldsymbol{a} = \begin{bmatrix} a_1 \\ a_2 \\ \vdots \\ a_n \end{bmatrix}, \qquad \boldsymbol{b} = [b_1 \ b_2 \ \dots \ b_n].$$

The transpose of a row vector is a column vector and vice versa. If A is an $m \times n$ matrix, then $A\boldsymbol{a}$ is a column vector with m rows.

Example 7.8 *If*

$$A = \begin{bmatrix} 1 & -1 & 2 \\ 3 & 1 & -4 \\ -1 & 2 & 1 \end{bmatrix}, \qquad \boldsymbol{x} = \begin{bmatrix} x \\ y \\ z \end{bmatrix}, \qquad \boldsymbol{d} = \begin{bmatrix} 2 \\ 1 \\ -1 \end{bmatrix},$$

find the set of equations for x, y, z represented by $A\boldsymbol{x} = \boldsymbol{d}$.

The matrix equation in full is

$$\begin{bmatrix} 1 & -1 & 2 \\ 3 & 1 & -4 \\ -1 & 2 & 1 \end{bmatrix} \begin{bmatrix} x \\ y \\ z \end{bmatrix} = \begin{bmatrix} 2 \\ 1 \\ -1 \end{bmatrix},$$

Example 7.8 *continued*

or

$$\begin{bmatrix} x - y + 2z \\ 3x + y - 4z \\ -x + 2y + z \end{bmatrix} = \begin{bmatrix} 2 \\ 1 \\ -1 \end{bmatrix}.$$

The set of **linear equations** for x, y, z is

$$x - y + 2z = 2,$$
$$3x + y - 4z = 1,$$
$$-x + 2y + z = -1.$$

We shall say more about the solutions of linear equations in Chapter 8.

4. *Diagonal matrices.* A square matrix all of whose elements off the leading diagonal are zero is called a **diagonal matrix**. Thus, if $A = [a_{ij}]$ is an $n \times n$ matrix, then A is diagonal if $a_{ij} = 0$ for all $i \neq j$. Hence

$$A = \begin{bmatrix} 1 & 0 & 0 \\ 0 & -2 & 0 \\ 0 & 0 & 3 \end{bmatrix}$$

is an example of a 3×3 diagonal matrix.

A diagonal matrix is obviously symmetric. If A and B are diagonal matrices of the same order then $A + B$ and AB are also both diagonal.

5. *Identity matrix.* The diagonal matrix with all diagonal elements 1 is called the **identity** or **unit matrix** I_n. Hence, the 3×3 identity is

$$I_3 = \begin{bmatrix} 1 & 0 & 0 \\ 0 & 1 & 0 \\ 0 & 0 & 1 \end{bmatrix}. \tag{7.4}$$

(If there is no confusion likely to arise, I_n or I_3 are simply replaced by the universal symbol I.) The reason for the definition becomes clear if we multiply a 3×3 matrix by I_3. If A is a general 3×3 matrix, then

$$AI_3 = \begin{bmatrix} a_{11} & a_{12} & a_{13} \\ a_{21} & a_{22} & a_{23} \\ a_{31} & a_{32} & a_{33} \end{bmatrix} \begin{bmatrix} 1 & 0 & 0 \\ 0 & 1 & 0 \\ 0 & 0 & 1 \end{bmatrix} = \begin{bmatrix} a_{11} & a_{12} & a_{13} \\ a_{21} & a_{22} & a_{23} \\ a_{31} & a_{32} & a_{33} \end{bmatrix} = A.$$

Similarly $I_3 A = A$.

A need not be square: provided that the products are defined, $AI = A$ and $IA = A$ for the appropriate identity matrix in each case.

6. *Powers of matrices*. If A is a square matrix of order $n \times n$, then we write AA as A^2, AA^2 as A^3, and so on.

If A is diagonal, as in

$$A = \begin{bmatrix} d_1 & 0 & 0 \\ 0 & d_2 & 0 \\ 0 & 0 & d_3 \end{bmatrix},$$

then

$$A^2 = \begin{bmatrix} d_1^2 & 0 & 0 \\ 0 & d_2^2 & 0 \\ 0 & 0 & d_3^2 \end{bmatrix}, \qquad A^3 = \begin{bmatrix} d_1^3 & 0 & 0 \\ 0 & d_2^3 & 0 \\ 0 & 0 & d_3^3 \end{bmatrix}, \qquad \text{etc.}$$

In particular $I_m^n = I_m$ for all positive integers n.

7.4 The inverse matrix

If A and B are *square matrices*, each of order $n \times n$, which satisfy the equations

$$AB = BA = I_n,$$

then B is called the **inverse** of A. We say *the* inverse because, if a matrix B exists with this property, then it is uniquely determined by A (although we shall not prove this here). We write $B = A^{-1}$ (*not* $B = I/A$). Since the definition is 'symmetric', it follows that A is the inverse of B, that is $A = B^{-1}$. The inverse matrix defines 'division' for matrices, but analogies with numbers must not be taken too far. It is a particularly useful operation since it enables us to manipulate matrix equations. Thus, if $AB = C$, and the inverse of B exists, then we can solve the equation and find A as $A = CB^{-1}$.

How do we find the inverse? Does it always exist? Let us look first at the case in which A is a 2×2 matrix, and consider the equation

$$Ax = d,$$

where

$$A = \begin{bmatrix} a_{11} & a_{12} \\ a_{21} & a_{22} \end{bmatrix}, \qquad x = \begin{bmatrix} x_1 \\ x_2 \end{bmatrix}, \qquad d = \begin{bmatrix} d_1 \\ d_2 \end{bmatrix}.$$

Thus

$$\begin{bmatrix} a_{11} & a_{12} \\ a_{21} & a_{22} \end{bmatrix} \begin{bmatrix} x_1 \\ x_2 \end{bmatrix} = \begin{bmatrix} d_1 \\ d_2 \end{bmatrix},$$

or

$$a_{11}x_1 + a_{12}x_2 = d_1, \qquad\qquad (7.5)$$

$$a_{12}x_1 + a_{22}x_2 = d_2. \qquad\qquad (7.6)$$

These are **linear equations** in the unknowns x_1 and x_2: we shall say more about their solution in Chapter 12. Eliminate x_2 by multiplying (7.5) by a_{22}, (7.6) by a_{12}, and by subtracting the two equations so that

$$(a_{11}a_{22} - a_{21}a_{12})x_1 = a_{22}d_1 - a_{12}d_2.$$

Similarly, elimination of x_1 leads to

$$-(a_{11}a_{22} - a_{21}a_{12})x_2 = a_{21}d_1 - a_{11}d_2.$$

Provided that $a_{11}a_{22} - a_{21}a_{12} \neq 0$, it follows that

$$x_1 = \frac{a_{22}d_1 - a_{12}d_2}{a_{11}a_{22} - a_{21}a_{12}}, \qquad x_2 = \frac{-a_{21}d_1 + a_{11}d_2}{a_{11}a_{22} - a_{21}a_{12}}.$$

We can now express the solution of the pair of equations (7.5), (7.6) in matrix form:

$$x = \begin{bmatrix} x_1 \\ x_2 \end{bmatrix} = \frac{1}{a_{11}a_{22} - a_{21}a_{12}} \begin{bmatrix} a_{22}d_1 - a_{12}d_2 \\ -a_{21}d_1 + a_{11}d_2 \end{bmatrix} = Cd,$$

where

$$C = \frac{1}{\det A} \begin{bmatrix} a_{22} & -a_{12} \\ -a_{21} & a_{11} \end{bmatrix}, \quad \text{with} \quad \det A = a_{11}a_{22} - a_{21}a_{12}.$$

If $Ax = d$ is multiplied on the left by the inverse A^{-1}, then

$$A^{-1}Ax = I_2 x = x = A^{-1}d.$$

Hence A^{-1} can be identified with the matrix C. In other words,

$$A^{-1} = C = \frac{1}{\det A} \begin{bmatrix} a_{22} & -a_{12} \\ -a_{21} & a_{11} \end{bmatrix}. \qquad\qquad (7.7)$$

It is worth remembering the rule for 2×2 matrices by which A^{-1} can be constructed from A.

Rule for 2×2 inverse

Suppose that $\det A \neq 0$. Then to form the inverse, the diagonal elements a_{11} and a_{22} are interchanged, the signs are changed for the other two elements, and the matrix is divided by $\det A$.

(7.8)

The number $\det A = a_{11}a_{22} - a_{21}a_{12}$ is known as the **determinant** of A. It may also be written directly in terms of the corresponding matrix as

$$\det A = \det \begin{bmatrix} a_{11} & a_{12} \\ a_{21} & a_{22} \end{bmatrix}.$$

The following is a common **notation** for a determinant:

$$\det A = \begin{vmatrix} a_{11} & a_{12} \\ a_{21} & a_{22} \end{vmatrix}.$$

The determinant is a function of the corresponding matrix, but is a number – *not* a matrix. If $\det A = 0$, then the matrix has no inverse. It is then said to be **singular**; if A has an inverse, then A is said to be **non-singular**. We shall say more about determinants in the next chapter.

Example 7.9 *Decide whether A, where*

$$A = \begin{bmatrix} 1 & 3 \\ -1 & 4 \end{bmatrix},$$

is singular or not. If it is non-singular, find its inverse.

Here $a_{11} = 1$, $a_{12} = 3$, $a_{21} = -1$, and $a_{22} = 4$. Hence

$$\det A = \begin{vmatrix} 1 & 3 \\ -1 & 4 \end{vmatrix} = 1 \times 4 - (-1) \times 3 = 4 + 3 = 7.$$

Since $\det A \neq 0$, then A is non-singular. Its inverse is, by the rule above,

$$A^{-1} = \tfrac{1}{7} \begin{vmatrix} 4 & -3 \\ 1 & 1 \end{vmatrix}.$$

Example 7.10 *If*

$$A = \begin{bmatrix} 1 & 3 \\ -1 & 4 \end{bmatrix}, \qquad B = \begin{bmatrix} 1 & 2 \\ 1 & -1 \end{bmatrix},$$

find A^{-1}, B^{-1}, and $(AB)^{-1}$.

Always check the determinants first. Here

$$\det A = 4 - (-3) = 7, \qquad \det B = -1 - 2 = -3.$$

These are not zero, so A and B have inverses:

$$A^{-1} = \tfrac{1}{7} \begin{bmatrix} 4 & -3 \\ 1 & 1 \end{bmatrix}, \qquad B^{-1} = -\tfrac{1}{3} \begin{bmatrix} -1 & -2 \\ -1 & 1 \end{bmatrix}.$$

Also

$$AB = \begin{bmatrix} 1 & 3 \\ -1 & 4 \end{bmatrix}\begin{bmatrix} 1 & 2 \\ 1 & -1 \end{bmatrix} = \begin{bmatrix} 4 & -1 \\ 3 & -6 \end{bmatrix}.$$

Example 7.10 *continued*

Thus, $\det(AB) = -24 + 3 = -21$, and

$$(AB)^{-1} = -\tfrac{1}{21}\begin{bmatrix} -6 & 1 \\ -3 & 4 \end{bmatrix}.$$

Note that

$$B^{-1}A^{-1} = -\tfrac{1}{21}\begin{bmatrix} -1 & -2 \\ -1 & 1 \end{bmatrix}\begin{bmatrix} 4 & -3 \\ 1 & 1 \end{bmatrix} = -\tfrac{1}{21}\begin{bmatrix} -6 & 1 \\ -3 & 4 \end{bmatrix} = (AB)^{-1}.$$

This last result suggests the following correct rule for the inverse of the product of two square matrices, namely

$$(AB)^{-1} = B^{-1}A^{-1}.$$

For the inverse of a 3×3 matrix we can adopt the same approach as for the 2×2 case by eliminating $x_1, x_2, \ldots$ successively between the set of equations $Ax = d$ or

$$a_{11}x_1 + a_{12}x_2 + a_{13}x_3 = d_1,$$
$$a_{21}x_1 + a_{22}x_2 + a_{23}x_3 = d_2, \qquad (7.9)$$
$$a_{31}x_1 + a_{32}x_2 + a_{33}x_3 = d_3.$$

The result is

$$x = A^{-1}d,$$

where

$$A^{-1} = \frac{1}{\det A}\begin{bmatrix} a_{22}a_{33} - a_{32}a_{23} & -(a_{12}a_{33} - a_{32}a_{13}) & a_{12}a_{23} - a_{22}a_{13} \\ -(a_{21}a_{33} - a_{31}a_{23}) & a_{11}a_{33} - a_{31}a_{13} & -(a_{11}a_{23} - a_{21}a_{13}) \\ a_{21}a_{32} - a_{31}a_{22} & -(a_{11}a_{32} - a_{31}a_{12}) & a_{11}a_{22} - a_{21}a_{12} \end{bmatrix}$$

$$(7.10)$$

with

$$\det A = a_{11}(a_{22}a_{33} - a_{32}a_{23}) - a_{12}(a_{21}a_{33} - a_{31}a_{23})$$
$$+ a_{13}(a_{21}a_{32} - a_{31}a_{22}), \qquad (7.11)$$

provided that $\det A \neq 0$. Again $\det A$ is known as the determinant of A and is denoted by

$$\det A = \begin{vmatrix} a_{11} & a_{12} & a_{13} \\ a_{21} & a_{22} & a_{23} \\ a_{31} & a_{32} & a_{33} \end{vmatrix}.$$

Equation (7.10) gives the inverse matrix, as can be verified by calculation of the products AA^{-1} and $A^{-1}A$. Even for 3×3 matrices, the formula for the inverse is quite complicated. If $\det A = 0$, then the matrix is singular.

Determinants, which have arisen in the content of inverse matrices, have important properties particularly with regard to their evaluation. They will be discussed in more detail in the next chapter.

Example 7.11 *Verify by direct multiplication that*

$$A = \begin{bmatrix} 0 & 1 & 1 \\ -1 & 1 & 1 \\ 1 & -1 & 1 \end{bmatrix}$$

has the inverse

$$B = \tfrac{1}{2}\begin{bmatrix} 2 & -2 & 0 \\ 2 & -1 & -1 \\ 0 & 1 & 1 \end{bmatrix}.$$

Check the matrix product BA:

$$BA = \tfrac{1}{2}\begin{bmatrix} 2 & -2 & 0 \\ 2 & -1 & -1 \\ 0 & 1 & 1 \end{bmatrix}\begin{bmatrix} 0 & 1 & 1 \\ -1 & 1 & 1 \\ 1 & -1 & 1 \end{bmatrix} = \begin{bmatrix} 1 & 0 & 0 \\ 0 & 1 & 0 \\ 0 & 0 & 1 \end{bmatrix} = I_3.$$

Hence $B = A^{-1}$.

Note that we need only verify that either $BA = I_3$ or $AB = I_3$, not both. If $BA = I_3$, then $AB = I_3$, and vice versa.

Example 7.12 *Using formula (7.10) find A^{-1}, where*

$$A = \begin{bmatrix} 2 & 1 & 0 \\ 1 & -1 & 5 \\ -1 & -1 & 2 \end{bmatrix}.$$

We first find $\det A$ using (7.11):

$$\begin{aligned}
\det A &= 2 \times [(-1) \times 2 - (-1) \times 5] - 1 \times [1 \times 2 - (-1) \times 5] \\
&\quad + 0 \times [1 \times (-1) - (-1) \times (-1)] \\
&= 2 \times 3 - 1 \times 7 \\
&= -1.
\end{aligned}$$

Thus, from (7.10),

$$A^{-1} = \frac{1}{-1}\begin{bmatrix} -1 \times 2 - (-1) \times 5 & -[1 \times 2 - (-1) \times 0] & 1 \times 5 - (-1) \times 0 \\ -[1 \times 2 - (-1) \times 5] & 2 \times 2 - (-1) \times 0 & -(2 \times 5 - 1 \times 0) \\ 1 \times (-1) - (-1) \times (-1) & -[2 \times (-1) - (-1) \times (-1)] & 2 \times (-1) - 1 \times (-1) \end{bmatrix}$$

$$= -\begin{bmatrix} 3 & -2 & 5 \\ -7 & 4 & -10 \\ -2 & 1 & -3 \end{bmatrix} = \begin{bmatrix} -3 & 2 & -5 \\ 7 & -4 & 10 \\ 2 & -1 & 3 \end{bmatrix}.$$

This is the 'formula' method of finding the inverse of a 3×3 matrix, but it is not an efficient procedure numerically. There are better methods using row operations which will be explained in Chapter 12.

Problems

7.1 (Section 7.1). The matrix $A = [a_{ij}]$ is given by

$$A = \begin{bmatrix} 1 & 2 & 3 \\ -1 & 0 & 1 \\ 2 & -2 & 4 \\ 1 & 5 & -3 \end{bmatrix}.$$

Identify the elements a_{13} and a_{31}.

7.2 (Section 7.2). Solve the equation $A = B$, where

$$A = \begin{bmatrix} 1 & -2 \\ 3 & 1 \\ -1 & 2 \end{bmatrix}, \qquad B = \begin{bmatrix} 1 & x \\ y-x & 1 \\ -1 & 2 \end{bmatrix},$$

for x and y.

7.3 (Section 7.2). Given that

$$A = \begin{bmatrix} 1 & 2 & -3 \\ -1 & 0 & 4 \end{bmatrix}, \qquad B = \begin{bmatrix} 2 & -1 & 3 \\ 4 & 1 & 2 \end{bmatrix},$$

find the matrices $A + B$, $A - B$, and $2A - 3B$.

7.4 (Section 7.2). Given that

$$A = \begin{bmatrix} 1 & 3 & 0 \\ 2 & 1 & 1 \end{bmatrix}, \quad B = \begin{bmatrix} 1 & 0 \\ 2 & 1 \\ -1 & -1 \end{bmatrix}, \quad C = \begin{bmatrix} 2 & 1 \\ -1 & 1 \\ 0 & 1 \end{bmatrix},$$

verify the distributive law $A(B + C) = AB + AC$ for the three matrices.

7.5 (Section 7.2). Let

$$A = \begin{bmatrix} -1 & 2 & -1 \\ 2 & 3 & 1 \end{bmatrix}, \quad B = \begin{bmatrix} -1 & 0 \\ 1 & 2 \\ 3 & -1 \end{bmatrix}, \quad C = \begin{bmatrix} 1 & 1 \\ -1 & 2 \end{bmatrix}.$$

Verify the associative law $A(BC) = (AB)C$ for these matrices.

7.6 (Section 7.2). Let

$$A = \begin{bmatrix} 4 & 2 \\ 2 & 1 \end{bmatrix}, \qquad B = \begin{bmatrix} -2 & -1 \\ 4 & 2 \end{bmatrix}.$$

Show that $AB = 0$, but that $BA \neq 0$.

7.7 (Section 7.3). Let

$$A = \begin{bmatrix} 2 & 1 & 3 \\ 1 & -1 & 2 \\ -2 & 1 & 1 \end{bmatrix}.$$

Find a matrix C such that $A + C$ is the identity matrix I_3. Deduce that $AC = CA$. Find AC, and hence the matrix $A^2 + C^2$.

7.8 (Section 7.3). A general $n \times n$ matrix is given by

$$A = [a_{ij}].$$

Show that $A + A^T$ is a symmetric matrix, and that $A - A^T$ is skew-symmetric.

Express the matrix

$$A = \begin{bmatrix} 2 & 1 & 3 \\ -2 & 0 & 1 \\ 3 & 1 & 2 \end{bmatrix}$$

as the sum of a symmetric matrix and a skew-symmetric matrix.

7.9 (Section 7.3). Let

$$A = \begin{bmatrix} 1 & 3 \\ -1 & 2 \\ 0 & 1 \end{bmatrix}.$$

Write down A^T, and find the products AA^T and A^TA.

7.10 (Section 7.3). If

$$A = \begin{bmatrix} 1 & -1 & 2 \\ 3 & 0 & 1 \\ -1 & 2 & -3 \end{bmatrix}, \qquad x = \begin{bmatrix} x \\ y \\ z \end{bmatrix}, \qquad d = \begin{bmatrix} 2 \\ 0 \\ -1 \end{bmatrix},$$

write down the set of equations defined by $Ax = d$. Confirm that the same set of equations is given by $x^TA^T = d^T$.

7.11 (Section 7.3). Let

$$A = \begin{bmatrix} 1 & 0 & 0 \\ a & -1 & 0 \\ b & c & 1 \end{bmatrix}.$$

Find A^2. For what relation between a, b, and c is $A^2 = I_3$? In this case, what is the inverse matrix of A? What is the inverse matrix of A^{2n-1} (n a positive integer)?

7.12 If

$$A = \begin{bmatrix} 2 & 0 & 1 \\ 2 & -2 & 2 \\ 0 & 4 & -4 \end{bmatrix}, \qquad B = \begin{bmatrix} 0 & \frac{1}{2} & \frac{1}{4} \\ 1 & -1 & -\frac{1}{4} \\ 1 & -1 & -\frac{1}{2} \end{bmatrix},$$

find the products AB and BA, and confirm that B is the inverse of A.

7.13 Let

$$A = \begin{bmatrix} 2 & 0 & 1 \\ 2 & -2 & 2 \\ 0 & 4 & 1 \end{bmatrix}.$$

Find the powers A^2 and A^3, and verify that

$$A^3 - A^2 - 12A = -12I_3.$$

Hence find the inverse matrix A^{-1} by multiplying the equation on both sides by A^{-1}.

7.14 (Section 7.4). Using the rule for inverses of 2×2 matrices, write down the inverses of:

(a) $\begin{bmatrix} 1 & 1 \\ 2 & -1 \end{bmatrix}$;   (b) $\begin{bmatrix} 2 & 3 \\ -7 & 11 \end{bmatrix}$;

(c) $\begin{bmatrix} 1 & 0 \\ 0 & -2 \end{bmatrix}$;   (d) $\begin{bmatrix} 10 & -7 \\ 8 & 0 \end{bmatrix}$;

(e) $\begin{bmatrix} -99 & 100 \\ 97 & 98 \end{bmatrix}$.

7.15 (Section 7.4). The sparsely filled matrix A is given by

$$A = \begin{bmatrix} 0 & 1 & 0 & 0 \\ 0 & 0 & 1 & 0 \\ 1 & 0 & 0 & 0 \\ 0 & 0 & 0 & 1 \end{bmatrix}.$$

Thinking about the row-on-column rule for matrix multiplication, can you guess the columns in the inverse matrix A^{-1}? How would this rule generalize to the matrix

$$A = \begin{bmatrix} 0 & a & 0 & 0 \\ 0 & 0 & b & 0 \\ c & 0 & 0 & 0 \\ 0 & 0 & 0 & d \end{bmatrix}?$$

7.16 (Section 7.4). Write down the set of equations given by $Ax = d$, where

$$A = \begin{bmatrix} 0 & 1 & 1 \\ 1 & -2 & 2 \\ 1 & 0 & 1 \end{bmatrix}, \quad x = \begin{bmatrix} x \\ y \\ z \end{bmatrix}, \quad d = \begin{bmatrix} 6 \\ 3 \\ -9 \end{bmatrix}.$$

Find A^{-1} and calculate the product $A^{-1}d$. What is the solution of the equation?

7.17 (Section 7.4). If A and B are both $n \times n$ matrices with A non-singular, show that

$$(A^{-1}BA)^2 = A^{-1}B^2A.$$

Let $A = \begin{bmatrix} 1 & 2 \\ -1 & 1 \end{bmatrix}$ and $B = \begin{bmatrix} 1 & 2 \\ -1 & 0 \end{bmatrix}$. Calculate $A^{-1}B^4A$.

7.18 (Section 7.4). For interpolation purposes for given data, it is required that the parabola $y = a + bx + cx^2$ should pass through the three points with coordinates (x_1, y_1), (x_2, y_2), and (x_3, y_3) in the (x, y) plane. Show that the matrix equation for the constants a, b, and c can be written as

$$\begin{bmatrix} 1 & x_1 & x_1^2 \\ 1 & x_2 & x_2^2 \\ 1 & x_3 & x_3^2 \end{bmatrix} \begin{bmatrix} a \\ b \\ c \end{bmatrix} = \begin{bmatrix} y_1 \\ y_2 \\ y_3 \end{bmatrix}.$$

Verify that the inverse of the 3×3 matrix on the left is

$$\begin{bmatrix} \dfrac{x_2 x_3}{(x_2 - x_1)(x_3 - x_1)} & \dfrac{x_3 x_1}{(x_3 - x_2)(x_1 - x_2)} & \dfrac{x_1 x_2}{(x_1 - x_3)(x_2 - x_3)} \\[2mm] -\dfrac{x_2 + x_3}{(x_2 - x_1)(x_3 - x_1)} & -\dfrac{x_3 + x_1}{(x_3 - x_2)(x_1 - x_2)} & -\dfrac{x_1 + x_2}{(x_1 - x_3)(x_2 - x_3)} \\[2mm] \dfrac{1}{(x_2 - x_1)(x_3 - x_1)} & \dfrac{1}{(x_3 - x_2)(x_1 - x_2)} & \dfrac{1}{(x_1 - x_3)(x_2 - x_3)} \end{bmatrix},$$

provided that certain conditions are met. What are they, and what implications have they for the given points in the plane? Find a, b, and c in terms of the given data. Find the equation of the parabola through the points $(-2, 0)$, $(1, -2)$, $(3, 4)$.

7.19 The elements in a 3×3 matrix $A = [a_{ij}]$ are given by the rule

$$a_{ij} = (-j)^i - ij.$$

Write down the matrix A. Calculate $\det A$ and the inverse of A.

7.20 If

$$A = \begin{bmatrix} 2 & 1 & 3 \\ 1 & -1 & 2 \\ 1 & 2 & 1 \end{bmatrix},$$

show that $A^3 - 2A^2 - 9A = 0$, but that $A^2 - 2A - 9I_3 \neq 0$. Does the inverse of A exist?

7.21 An nth-order square matrix A satisfies $A^2 = A$ and $A \neq I_n$. Show that
(a) $\det A = 0$;
(b) $(I_n + A)^{-1} = I_n - \frac{1}{2}A$;
(c) $(I_n + A)^m = I_n + (2^m - 1)A$ for any positive integer m.

7.22 Let $A_1 = \begin{bmatrix} x_1 & y_1 \\ -y_1 & x_1 \end{bmatrix}$, $A_2 = \begin{bmatrix} x_2 & y_2 \\ -y_2 & x_2 \end{bmatrix}$. Calculate

$A_1 + A_2$, A_1A_2, A_2A_1, A_1^{-1}. Compare your results with $z_1 + z_2$, z_1z_2, and $1/z_1$, where $z_1 = x_1 + iy_1$ and $z_2 = x_2 + iy_2$ are complex numbers (see Chapter 6). Consider the possibility of developing further parallels, such as to $|z|$ and e^z.

8 Determinants

CONTENTS

8.1   The determinant of a square matrix

As we saw in (7.8) (Section 7.4), certain combinations of elements from a square matrix appear as the denominator in the construction of the inverse matrix. If this number, called the **determinant** of the matrix, turns out to be zero, then the matrix is singular and no inverse exists. Here we look at the definition of the determinant of a matrix and its properties. Special emphasis will be placed on the 2×2 and 3×3 determinants which suggest generalizations to higher-order cases.

Given the matrix

$$A = \begin{bmatrix} a_{11} & a_{12} \\ a_{21} & a_{22} \end{bmatrix},$$

then the determinant of A is denoted and defined by

Expansion of 2×2 determinant

$$\det A = \begin{vmatrix} a_{11} & a_{12} \\ a_{21} & a_{22} \end{vmatrix} = a_{11}a_{22} - a_{21}a_{12}.$$

(8.1)

(The notation $|A|$ is also used extensively for the determinant.) For the 3×3 matrix

$$A = \begin{bmatrix} a_{11} & a_{12} & a_{13} \\ a_{21} & a_{22} & a_{23} \\ a_{31} & a_{32} & a_{33} \end{bmatrix},$$

its determinant is (see (7.11)) defined as

$$\det A = \begin{vmatrix} a_{11} & a_{12} & a_{13} \\ a_{21} & a_{22} & a_{23} \\ a_{31} & a_{32} & a_{33} \end{vmatrix} = a_{11}a_{22}a_{33} - a_{11}a_{32}a_{23} - a_{12}a_{21}a_{33} + a_{12}a_{31}a_{23} + a_{13}a_{21}a_{32} - a_{13}a_{31}a_{22}.$$

(8.2)

In (8.2) there are six terms, each of which is the product of three elements. Each term contains three elements, each from a different row and column. In other words, there are never two elements in any term from the same row or column. It can be seen that there must be just $3 \times 2 \times 1 = 6$ terms of this form, because three elements can be chosen from row 1, two from the two remaining elements in row 2, and one element from row 3.

Each term is prefixed by either $+1$ or -1. This is decided according to the following rule. Write each term in the form

$$a_{1j_1} a_{2j_2} a_{3j_3},$$

in which the first suffixes are in consecutive increasing order. Examine the **permutations** of the second suffixes $j_1 j_2 j_3$ (see Section 1.17). The permutation is said to be even (odd) if it has an even (odd) number of **inversions**. An inversion occurs whenever a larger integer precedes a smaller one. Thus the permutation 132 is odd, since 3 precedes 2, but 312 is even since there are two inversions because 3 precedes 1 and 2. If the number of permutations is even, then a + sign is attached; if the number is odd, then a − sign is attached. This rule can be extended to a determinant of any order.

While this expansion of the determinant says something about the structure of the determinant, it is not really a practical rule for evaluating determinants. Returning to (8.2), we can rewrite det A as

$$\det A = a_{11}(a_{22}a_{33} - a_{32}a_{23}) - a_{12}(a_{21}a_{33} - a_{31}a_{23}) + a_{13}(a_{21}a_{32} - a_{31}a_{22}).$$

The terms in brackets are themselves 2×2 determinants. Thus

Expansion of 3×3 determinant

$$\det A = \begin{vmatrix} a_{11} & a_{12} & a_{13} \\ a_{21} & a_{22} & a_{23} \\ a_{31} & a_{32} & a_{33} \end{vmatrix}$$

$$= a_{11} \begin{vmatrix} a_{22} & a_{23} \\ a_{32} & a_{33} \end{vmatrix} - a_{12} \begin{vmatrix} a_{21} & a_{23} \\ a_{31} & a_{33} \end{vmatrix} + a_{13} \begin{vmatrix} a_{21} & a_{22} \\ a_{31} & a_{32} \end{vmatrix}.$$

$$(8.3)$$

This expression is called an **expansion by the top row**. The term associated with a_{11}, namely

$$C_{11} = \begin{vmatrix} a_{22} & a_{23} \\ a_{32} & a_{33} \end{vmatrix},$$

is known as the **cofactor** of a_{11}. The cofactor of an element of A is obtained by deleting the row and column through the element and writing down the determinant of the elements of the remaining

2×2 submatrix with a $+$ or $-$ sign attached. The cofactors of a_{12} and a_{13} are

$$C_{12} = - \begin{vmatrix} a_{21} & a_{23} \\ a_{31} & a_{33} \end{vmatrix}, \qquad C_{13} = \begin{vmatrix} a_{21} & a_{22} \\ a_{31} & a_{32} \end{vmatrix},$$

where the signs attached should be noted. In the same way the cofactors of the elements in the second and third rows are defined as follows:

$$C_{21} = - \begin{vmatrix} a_{12} & a_{13} \\ a_{32} & a_{33} \end{vmatrix}, \qquad C_{22} = \begin{vmatrix} a_{11} & a_{13} \\ a_{31} & a_{33} \end{vmatrix}, \qquad C_{23} = - \begin{vmatrix} a_{11} & a_{12} \\ a_{31} & a_{32} \end{vmatrix},$$

$$C_{31} = \begin{vmatrix} a_{12} & a_{13} \\ a_{22} & a_{23} \end{vmatrix}, \qquad C_{32} = - \begin{vmatrix} a_{11} & a_{13} \\ a_{21} & a_{23} \end{vmatrix}, \qquad C_{33} = \begin{vmatrix} a_{11} & a_{12} \\ a_{21} & a_{22} \end{vmatrix}.$$

The signs associated with the cofactors alternate, starting with a $+$ at the top left as we move across or down from the top left-hand corner as shown:

$$\begin{vmatrix} + & - & + \\ - & + & - \\ + & - & + \end{vmatrix}.$$

The sign associated with C_{ij} is $+$ if $i+j$ is even and $-$ if $i+j$ is odd, and can be expressed as $(-1)^{i+j}$.

For example, if det A is expanded by the third column, then

$$\det A = a_{13}C_{13} + a_{23}C_{23} + a_{33}C_{33}.$$

Example 8.1   *Let*

$$A = \begin{bmatrix} 1 & -1 & 0 \\ 2 & 3 & -2 \\ 1 & -1 & 1 \end{bmatrix}.$$

Evaluate det A *by expanding by the first row. Find the cofactors* C_{13}, C_{23}, C_{33} *of the elements in the third column. Calculate*

$$a_{13}C_{13} + a_{23}C_{23} + a_{33}C_{33},$$

and verify that it also equals det A.

By (8.3),

$$\det A = 1 \times \begin{vmatrix} 3 & -2 \\ -1 & 1 \end{vmatrix} - (-1) \times \begin{vmatrix} 2 & -2 \\ 1 & 1 \end{vmatrix} + 0 \times \begin{vmatrix} 2 & 3 \\ 1 & -1 \end{vmatrix}$$

$$= (3 - 2) + (2 + 2) = 5.$$

The cofactors are (with due regard to the sign convention)

Example 8.1 *continued*

$$C_{13} = \begin{vmatrix} 2 & 3 \\ 1 & -1 \end{vmatrix} = -5, \qquad C_{23} = -\begin{vmatrix} 1 & -1 \\ 1 & -1 \end{vmatrix} = 0, \qquad C_{33} = \begin{vmatrix} 1 & -1 \\ 2 & 3 \end{vmatrix} = 5.$$

Hence expansion by the third column gives

$$a_{13}C_{13} + a_{23}C_{23} + a_{33}C_{33} = 0 \times (-5) + (-2) \times 0 + 1 \times 5 = 5,$$

which is the same as det A. (See also Rule 4 below.)

Example 8.2 *Evaluate the determinant*

$$\det A = \begin{vmatrix} 1 & 2 & k \\ 2 & -1 & 3 \\ -1 & 4 & -2 \end{vmatrix},$$

for any k. Find the value of k for which the determinant is zero.

Expanding by the first row gives

$$\det A = \begin{vmatrix} 1 & 2 & k \\ 2 & -1 & 3 \\ -1 & 4 & -2 \end{vmatrix} = 1 \times \begin{vmatrix} -1 & 3 \\ 4 & -2 \end{vmatrix} - 2 \times \begin{vmatrix} 2 & 3 \\ -1 & -2 \end{vmatrix} + k \times \begin{vmatrix} 2 & -1 \\ -1 & 4 \end{vmatrix}$$

$$= 1 \times (2 - 12) - 2 \times (-4 + 3) + k \times (8 - 1)$$

$$= -10 + 2 + 7k = -8 + 7k.$$

Hence det $A = 0$ if $k = \frac{8}{7}$.

The notion of cofactors generalizes to higher-order determinants. The alternating-sign rule applies from the top left-hand corner. For example, a 4×4 determinant has 16 cofactors, each of which is a 3×3 determinant.

8.2 Properties of determinants

We list here some properties of determinants. Many of them are useful in evaluating determinants. We shall not aim for complete generality but illustrate the rules mainly in the 3×3 case. However, the rules have obvious generalizations to higher orders.

1. det A^{T} = det A, *where A^{T} is the transpose of A (see Section 7.3).*

The determinants of a square matrix and its transpose are equal since

$$\det A^{\mathrm{T}} = \begin{vmatrix} a_{11} & a_{21} & a_{31} \\ a_{12} & a_{22} & a_{32} \\ a_{13} & a_{23} & a_{33} \end{vmatrix} = \begin{aligned} & a_{11}a_{22}a_{33} - a_{11}a_{23}a_{32} - a_{21}a_{12}a_{33} + a_{21}a_{13}a_{32} \\ & + a_{31}a_{12}a_{23} - a_{31}a_{13}a_{22}, \end{aligned}$$

and all terms in this expansion can be identified with those in (8.2). Hence det A^{T} = det A.

Example 8.3   *Evaluate*

$$\det A = \begin{vmatrix} 1 & 28 & -29 \\ 0 & 1 & -4 \\ 0 & -2 & 5 \end{vmatrix}.$$

Since the determinant has two zeros in the first column, it is advantageous to use Rule 1. The determinant of the transpose of A is given by

$$\det A^{\mathrm{T}} = \begin{vmatrix} 1 & 0 & 0 \\ 28 & 1 & -2 \\ -29 & -4 & 5 \end{vmatrix},$$

which now has two zeros in the first row. Hence the expansion by the top row becomes particularly easy:

$$\det A^{\mathrm{T}} = 1 \times \begin{vmatrix} 1 & -2 \\ -4 & 5 \end{vmatrix} = 5 - 8 = -3.$$

2. *If every element of any single row or column of the matrix A is multiplied by a scalar k, then the determinant of this matrix is k* $\det A$.

(Note: this rule is different from Rule 2, Section 7.2, for matrices.)

This is a self-evident result, since just one element from every row and column appears in every term. Thus, by (8.2), if every element of the second row in A is multiplied by k, then

$$\begin{vmatrix} a_{11} & a_{12} & a_{13} \\ ka_{21} & ka_{22} & ka_{23} \\ a_{31} & a_{32} & a_{33} \end{vmatrix} = \begin{aligned} &a_{11}ka_{22}a_{33} - a_{11}ka_{23}a_{32} - a_{12}ka_{21}a_{33} \\ &+ a_{12}ka_{23}a_{31} + a_{13}ka_{21}a_{32} - a_{13}ka_{22}a_{31} \end{aligned}$$
$$= k \det A.$$

By putting $k = 0$ in this result, note that any determinant must have zero value if all the elements of any row or column are zeros.

Example 8.4   *Evaluate the determinant*

$$\Delta = \begin{vmatrix} -1 & 99 & 1 \\ 2 & 33 & -2 \\ 3 & 55 & 1 \end{vmatrix}.$$

Since the second column obviously has a factor of 11, then we can remove this factor from the second column before expansion. Thus, by Rule 2,

$$\Delta = 11 \times \begin{vmatrix} -1 & 9 & 1 \\ 2 & 3 & -2 \\ 3 & 5 & 1 \end{vmatrix} = 11 \times \left((-1) \times \begin{vmatrix} 3 & -2 \\ 5 & 1 \end{vmatrix} - 9 \times \begin{vmatrix} 2 & -2 \\ 3 & 1 \end{vmatrix} + 1 \times \begin{vmatrix} 2 & 3 \\ 3 & 5 \end{vmatrix} \right)$$

$$= 11 \times [-(3 + 10) - 9 \times (2 + 6) + (10 - 9)]$$
$$= 11 \times (-13 - 72 + 1) = -924.$$

3. *If B is obtained from A by interchanging two rows (or columns) then* $\det B = -\det A$.

Suppose, for example, that rows 1 and 3 are interchanged, so that

$$A = \begin{bmatrix} a_{11} & a_{12} & a_{13} \\ a_{21} & a_{22} & a_{23} \\ a_{31} & a_{32} & a_{33} \end{bmatrix}, \qquad B = \begin{bmatrix} a_{31} & a_{32} & a_{33} \\ a_{21} & a_{22} & a_{23} \\ a_{11} & a_{12} & a_{13} \end{bmatrix}.$$

Then, by analogy with (8.2), the expansion of B by its first row is given by

$$\det B = a_{31}\begin{vmatrix} a_{22} & a_{23} \\ a_{12} & a_{13} \end{vmatrix} - a_{32}\begin{vmatrix} a_{21} & a_{23} \\ a_{11} & a_{13} \end{vmatrix} + a_{33}\begin{vmatrix} a_{21} & a_{22} \\ a_{11} & a_{12} \end{vmatrix}$$

$$= a_{31}a_{22}a_{13} - a_{31}a_{23}a_{12} - a_{32}a_{21}a_{13} + a_{32}a_{23}a_{11}$$
$$+ a_{33}a_{21}a_{12} - a_{33}a_{22}a_{11}.$$

These are the same terms as those present in (8.2) except that the sign of every term is changed. Therefore in this case

$$\det B = -\det A.$$

The same is true whichever row or column pairs are exchanged.
The rule applies to a determinant of any order.

Example 8.5 *Evaluate the determinant*

$$\Delta = \begin{vmatrix} 1 & 2 & 1 & 2 \\ 0 & 2 & 0 & 0 \\ -1 & 3 & 0 & 4 \\ -1 & 2 & 0 & -1 \end{vmatrix}.$$

There are several ways of approaching the evaluation of this determinant since the second row and third column each have three zeros. It is obviously advantageous to have as many zeros as possible in the top row. With this in view, interchange rows 1 and 2 using Rule 3:

$$\Delta = -\begin{vmatrix} 0 & 2 & 0 & 0 \\ 1 & 2 & 1 & 2 \\ -1 & 3 & 0 & 4 \\ -1 & 2 & 0 & -1 \end{vmatrix}.$$

Expanding by row 1, remembering the sign rule for cofactors:

$$\Delta = 2 \times \begin{vmatrix} 1 & 1 & 2 \\ -1 & 0 & 4 \\ -1 & 0 & -1 \end{vmatrix}.$$

Now successively use Rule 1 and interchange rows with columns, and then Rule 3 and interchange the new rows 1 and 2:

Example 8.5 *continued*

$$\Delta = 2 \times \begin{vmatrix} 1 & -1 & -1 \\ 1 & 0 & 0 \\ 2 & 4 & -1 \end{vmatrix}$$

$$= 2 \times \begin{vmatrix} 1 & -1 & -1 \\ 1 & 0 & 0 \\ 2 & 4 & -1 \end{vmatrix} = (-2) \times \begin{vmatrix} 1 & 0 & 0 \\ 1 & -1 & -1 \\ 2 & 4 & -1 \end{vmatrix} = (-2) \times \begin{vmatrix} -1 & -1 \\ 4 & -1 \end{vmatrix}$$

$$= (-2) \times (1 + 4) = -10.$$

4. *Expansion by any row or column.*

From (8.2), by grouping the terms differently, we can write, for example,

$$\det A = a_{31}(a_{12}a_{23} - a_{13}a_{22}) - a_{32}(a_{11}a_{23} - a_{13}a_{21}) + a_{33}(a_{11}a_{22} - a_{12}a_{21})$$

$$= a_{31} \begin{vmatrix} a_{12} & a_{13} \\ a_{22} & a_{23} \end{vmatrix} - a_{32} \begin{vmatrix} a_{11} & a_{13} \\ a_{21} & a_{23} \end{vmatrix} + a_{33} \begin{vmatrix} a_{11} & a_{12} \\ a_{21} & a_{22} \end{vmatrix}$$

$$= a_{31}C_{31} + a_{32}C_{32} + a_{33}C_{33}.$$

in terms of cofactors. Here the elements a_{31}, a_{32}, a_{33} constitute the third row, and we call this the expansion of det A by the third row.

It can be shown that the expansion can be written down similarly using any row or column. Thus

$$\det A = a_{12}C_{12} + a_{22}C_{22} + a_{32}C_{32}$$

is an expansion by the second column.

Example 8.6   *Evaluate* det A, *where*

$$A = \begin{bmatrix} 1 & 3 & 0 & 1 \\ 1 & 0 & 0 & 2 \\ -1 & 2 & 2 & 4 \\ 2 & 1 & 0 & -1 \end{bmatrix}.$$

Since column 3 contains three zeros, expand by this column. The cofactor of the element in row 3, column 3, is the 3×3 determinant obtained from A by deleting the third row and third column in A. It is associated with a + sign. Hence

$$\det A = 2 \begin{vmatrix} 1 & 3 & 1 \\ 1 & 0 & 2 \\ 2 & 1 & -1 \end{vmatrix} \quad \text{(remember the sign rule)}$$

$$= 2 \left(-1 \times \begin{vmatrix} 3 & 1 \\ 1 & -1 \end{vmatrix} - 2 \times \begin{vmatrix} 1 & 3 \\ 2 & 1 \end{vmatrix} \right) \quad \text{(expanding by row 2)}$$

$$= 2(4 + 10) = 28.$$

5. *If two rows (or columns) of A are identical, then* $\det A = 0$.

This is a direct consequence of Rule 3. Interchange the two identical rows (columns). The determinant looks the same, but its value is now $-\det A$. Hence $\det A = -\det A$, which implies that $\det A = 0$.

Also, as a consequence of Rule 2, it follows that, if the corresponding elements of two rows (columns) are in the same ratio, then the value of the determinant is zero. Thus, for example,

$$\begin{vmatrix} 99 & 18 & 63 \\ 11 & 2 & 7 \\ -2 & 3 & 4 \end{vmatrix} = 9 \begin{vmatrix} 11 & 2 & 7 \\ 11 & 2 & 7 \\ -2 & 3 & 4 \end{vmatrix} \quad \text{(by Rule 2)}.$$

$$= 0 \quad \text{(by Rule 4)}.$$

6. *If the matrix B is constructed from A by adding k times one row (or column) to another row (column) then* $\det B = \det A$: *in other words, any number of such operations on rows and on columns has no effect on the value of det A.*

For our standard matrix A, consider the matrix B which is obtained from A by adding k times the elements in the first row to the elements in the third row. Thus

$$\det B = \begin{vmatrix} a_{11} & a_{12} & a_{13} \\ a_{21} & a_{22} & a_{23} \\ a_{31} + ka_{11} & a_{32} + ka_{12} & a_{33} + ka_{13} \end{vmatrix}$$

$$= (a_{31} + ka_{11})C_{31} + (a_{32} + ka_{12})C_{32} + (a_{33} + ka_{13})C_{33}$$
$$\text{(expanding by row 3)}$$

$$= a_{31}C_{31} + a_{32}C_{32} + a_{33}C_{33} + k(a_{11}C_{31} + a_{12}C_{32} + a_{13}C_{33})$$

$$= \det A + k \begin{vmatrix} a_{11} & a_{12} & a_{13} \\ a_{21} & a_{22} & a_{23} \\ a_{11} & a_{12} & a_{13} \end{vmatrix}$$

$$= \det A,$$

since the second determinant vanishes by Rule 5 having two rows with the same elements.

Note that

$$a_{11}C_{31} + a_{12}C_{32} + a_{13}C_{33} = 0$$

that is, in its general form, **the sum of the products of the elements of one row (or column) and the cofactors of the elements of *another* row (column) is zero.** This follows since the left-hand side must arise from a matrix with two identical rows (columns).

Rule 5 is a particularly useful rule for simplifying the elements in a determinant before expansion and evaluation. We illustrate a number of these points in the next example.

Example 8.7 *Evaluate*

$$\Delta = \begin{vmatrix} 2 & 99 & -99 \\ 999 & 1000 & 1001 \\ 1000 & 1001 & 998 \end{vmatrix}.$$

Usually we use the rules (particularly 6) either to introduce zeros into the matrix or to reduce the size of elements as far as possible. It is important to list the operations in order to make the sequence of operations intelligible. For this purpose we identify the current rows by $r_1, r_2, \ldots$, and the current columns by $c_1, c_2, \ldots$. Denote the new rows and columns which have been changed by $r'_1, r'_2, \ldots$ and $c'_1, c'_2, \ldots$ respectively. There are many ways of approaching the evaluation of Δ. A first step in this example could be to add column 3 (c_3) to column 2 (c_2) since this produces a zero at the top of column 2. This operation is represented by $c'_2 = c_2 + c_3$, and we list the operations on the right-hand side as we proceed. The second operation is to subtract the new row 3 from the new row 2. A decision is taken at each step in the light of the new matrix. By Rule 5, these operations do not affect the value of Δ. Hence

$$\Delta = \begin{vmatrix} 2 & 99 & -99 \\ 999 & 1000 & 1001 \\ 1000 & 1001 & 998 \end{vmatrix}$$

$$= \begin{vmatrix} 2 & 0 & -99 \\ 999 & 2001 & 1001 \\ 1000 & 1999 & 998 \end{vmatrix} \quad (c'_2 = c_2 + c_3)$$

$$= \begin{vmatrix} 2 & 0 & -99 \\ -1 & 2 & 3 \\ 1000 & 1999 & 998 \end{vmatrix} \quad (r'_2 = r_2 - r_3)$$

$$= \begin{vmatrix} 2 & 0 & -99 \\ -2 & 2 & 2 \\ 0.5 & 1999 & -1.5 \end{vmatrix} \quad \left(\begin{aligned} c'_1 &= c_1 - \tfrac{1}{2}c_2 \\ c'_3 &= c_3 - \tfrac{1}{2}c_2 \end{aligned} \right)$$

$$= \begin{vmatrix} 2 & 0 & -93 \\ -2 & 2 & -4 \\ 0.5 & 1999 & 0 \end{vmatrix} \quad (c'_3 = c_3 + 3c_1)$$

$$= 2(4 \times 1999) - 93[(-2 \times 1999) - 1]$$

$$= 387\,899.$$

Note that while $r'_2 = r_2 + kr_3$ does not affect the value of the determinant, $r'_2 = kr_2 + r_3$ will change its value by a factor k.

7. *If A and B are square matrices of the some order, then* $\det AB = \det A \det B$. (We shall not prove this here.)

8.3 The adjoint and inverse matrices

We can now rewrite the formula for the inverse given in Section 7.4, using cofactors. The *transposed* matrix of cofactors given by

$$\text{adj } A = \begin{bmatrix} C_{11} & C_{21} & C_{31} \\ C_{12} & C_{22} & C_{32} \\ C_{13} & C_{23} & C_{33} \end{bmatrix} \tag{8.4}$$

is known as the **adjoint** of A; the term **adjugate** is also used. Hence the inverse matrix of A given by eqn. (7.10) becomes

$$A^{-1} = \frac{\text{adj } A}{\det A},$$

in terms of the adjoint and determinant of A.

We can alternatively confirm by direct matrix multiplication that $\text{adj } A/\det A$ is the inverse of A. Thus, using (8.4),

$$A\frac{\text{adj } A}{\det A} = \begin{bmatrix} a_{11} & a_{12} & a_{13} \\ a_{21} & a_{22} & a_{23} \\ a_{31} & a_{32} & a_{33} \end{bmatrix}\begin{bmatrix} C_{11} & C_{21} & C_{31} \\ C_{12} & C_{22} & C_{32} \\ C_{13} & C_{23} & C_{33} \end{bmatrix}\frac{1}{\det A}$$

$$= \frac{1}{\det A}\begin{bmatrix} a_{11}C_{11} + a_{12}C_{12} + a_{13}C_{13} & a_{11}C_{21} + a_{12}C_{22} + a_{13}C_{23} & a_{11}C_{31} + a_{12}C_{32} + a_{13}C_{33} \\ a_{21}C_{11} + a_{22}C_{12} + a_{23}C_{13} & a_{21}C_{21} + a_{22}C_{22} + a_{23}C_{23} & a_{21}C_{31} + a_{22}C_{32} + a_{23}C_{33} \\ a_{31}C_{11} + a_{32}C_{12} + a_{33}C_{13} & a_{31}C_{21} + a_{32}C_{22} + a_{33}C_{23} & a_{31}C_{31} + a_{32}C_{32} + a_{33}C_{33} \end{bmatrix}$$

$$= \frac{1}{\det A}\begin{bmatrix} \det A & 0 & 0 \\ 0 & \det A & 0 \\ 0 & 0 & \det A \end{bmatrix} = I_3.$$

This confirmation uses the results that the sum of the products of the elements of one row and their cofactors is the value of the determinant whilst the sum of the products of one row and the cofactors of another row is zero.

Example 8.8 *Find the inverse of*

$$A = \begin{bmatrix} 1 & 2 & -1 \\ 0 & 1 & -1 \\ 1 & -1 & -2 \end{bmatrix}.$$

We evaluate $\det A$ first. Thus

$$\det A = 1 \times (-2 - 1) - 2 \times (0 + 1) - 1 \times (0 - 1) = -4.$$

Since $\det A \neq 0$, the inverse exists. The cofactors are

$$C_{11} = -3, \quad C_{12} = -1, \quad C_{13} = -1,$$
$$C_{21} = 5, \quad\ \ C_{22} = -1, \quad C_{23} = 3,$$
$$C_{31} = -1, \quad C_{32} = 1, \quad\ \ C_{33} = 1.$$

Hence

$$A^{-1} = -\tfrac{1}{4}\begin{bmatrix} -3 & 5 & -1 \\ -1 & -1 & 1 \\ -1 & 3 & 1 \end{bmatrix}.$$

The definition of the adjoint generalizes to matrices of higher order. However, the adjoint of a 4×4 matrix contains 16 3×3 determinants, which is about the limit of hand calculations unless the determinant is only sparsely filled with nonzero elements or can be reduced to such a determinant. Such computations become a fertile source of errors. There are computer packages available which will quickly perform the arithmetic operations for determinants of reasonable size.

Determinant, adjoint, and inverse for 3×3 matrices

(a) *Determinant of A*

$$\det A = a_{11} \begin{vmatrix} a_{22} & a_{23} \\ a_{32} & a_{33} \end{vmatrix} - a_{12} \begin{vmatrix} a_{21} & a_{23} \\ a_{31} & a_{33} \end{vmatrix} + a_{13} \begin{vmatrix} a_{21} & a_{22} \\ a_{31} & a_{32} \end{vmatrix};$$

(b) *Adjoint or adjugate of A*

$$\text{adj } A = \begin{bmatrix} C_{11} & C_{21} & C_{31} \\ C_{12} & C_{22} & C_{32} \\ C_{13} & C_{23} & C_{33} \end{bmatrix};$$

(c) *Inverse of A*

$$A^{-1} = \frac{\text{adj } A}{\det A}.$$

(8.5)

Problems

8.1 Evaluate the following determinants.

(a) $\begin{vmatrix} 1 & 2 \\ -1 & 3 \end{vmatrix};$

(b) $\begin{vmatrix} 1 & 0 & 1 \\ 0 & 1 & 0 \\ 1 & 0 & 1 \end{vmatrix};$

(c) $\begin{vmatrix} 1 & -1 & 2 \\ 3 & 1 & -1 \\ 2 & 1 & -1 \end{vmatrix};$

(d) $\begin{vmatrix} 2 & 1 & 0 & -1 \\ 0 & 0 & 2 & 0 \\ 3 & -1 & 2 & 1 \\ 0 & 1 & -1 & 1 \end{vmatrix};$

(e) $\begin{vmatrix} 0 & 1 & 0 & 0 & 0 \\ 1 & 0 & 0 & 0 & 0 \\ 0 & 0 & 0 & 0 & 1 \\ 0 & 0 & 1 & 0 & 0 \\ 0 & 0 & 0 & 1 & 0 \end{vmatrix};$

(f) $\begin{vmatrix} 2 & 1 & 0 & 0 & 0 \\ 1 & 2 & 1 & 0 & 0 \\ 0 & 1 & 2 & 1 & 0 \\ 0 & 0 & 1 & 2 & 1 \\ 0 & 0 & 0 & 1 & 2 \end{vmatrix}.$

8.2 Without evaluating the following determinants, explain why they are all zero:

(a) $\begin{vmatrix} 2 & 3 & 4 \\ 4 & 6 & 8 \\ 1 & -1 & 2 \end{vmatrix};$

(b) $\begin{vmatrix} -1 & 2 & 3 \\ 3 & 1 & -2 \\ -2 & -3 & -1 \end{vmatrix};$

(c) $\begin{vmatrix} a & b & c \\ b & c & a \\ a-b & b-c & c-a \end{vmatrix};$

(d) $\begin{vmatrix} 1 & 1 & 1 \\ 3 & 0 & 0 \\ 5 & 0 & 0 \end{vmatrix}.$

8.3 Given that

$$\Delta = \begin{vmatrix} a & b & c \\ b & c & a \\ c & a & b \end{vmatrix},$$

what is the value of

$$\begin{vmatrix} a^3 & ab & ac^2 \\ ab & c & ac \\ ac & a & bc \end{vmatrix}$$

in terms of Δ?

8.4 Simplify first and then evaluate the following determinants

(a) $\begin{vmatrix} 99 & 100 & 200 \\ 98 & 102 & 199 \\ -1 & 2 & 3 \end{vmatrix}$;

(b) $\begin{vmatrix} 77 & 84 & 55 \\ 75 & 87 & 57 \\ 1 & -2 & 3 \end{vmatrix}$;

(c) $\begin{vmatrix} 2 & -1 & 1 \\ 99 & 98 & 55 \\ 200 & 197 & 111 \end{vmatrix}$;

(d) $\begin{vmatrix} 87 & 84 & 83 & 81 \\ 77 & 76 & 77 & 75 \\ 54 & 53 & 52 & 54 \\ -43 & -44 & -46 & -4 \end{vmatrix}$.

8.5 Explain why the determinant

$$\Delta = \begin{vmatrix} 1 & 1 & 1 \\ a & b & c \\ a^2 & b^2 & c^2 \end{vmatrix}$$

has factors $b - c$, $c - a$, and $a - b$. Express the value of Δ as the product of factors.

8.6 Factorize the determinant

$$\Delta = \begin{vmatrix} 1 & 1 & 1 \\ a & b & c \\ a^3 & b^3 & c^3 \end{vmatrix}.$$

8.7 Explain, using one of the rules for determinants, why the equation

$$\begin{vmatrix} x & y & 1 \\ a_1 & b_1 & 1 \\ a_2 & b_2 & 1 \end{vmatrix} = 0$$

represents the equation of the straight line through the points (a_1, b_1) and (a_2, b_2) in the (x, y) plane. If X_1 and X_2 are the cofactors of x and y, what is the slope of this line in terms of the cofactors? Using this method, find the equation of the straight line through the points: (a) $(1, -1)$ and $(2, 3)$; (b) $(-1, 0)$ and $(4, -1)$.

8.8 Find the value of a which makes the determinant

$$\begin{vmatrix} 1 & 1 & -1 \\ 1 & a & 2 \\ -1 & 1 & 2 \end{vmatrix}$$

equal to zero.

8.9 Explain why

$$\begin{vmatrix} x & 2 & -2 \\ 2 & x & 3 \\ x & -1 & x \end{vmatrix} = 0$$

will be at most a cubic equation in x, but that

$$\begin{vmatrix} 1 & 1 & 2 \\ 3 & x & 2 \\ x & 1 & x \end{vmatrix} = 0$$

will be at most a quadratic equation in x. Solve both equations, and find all roots including any complex ones.

8.10 Show that

$$\begin{vmatrix} a_{11} + b_{11} & a_{12} + b_{12} & a_{13} + b_{13} \\ a_{21} & a_{22} & a_{23} \\ a_{31} & a_{32} & a_{33} \end{vmatrix}$$

$$= \begin{vmatrix} a_{11} & a_{12} & a_{13} \\ a_{21} & a_{22} & a_{23} \\ a_{31} & a_{32} & a_{33} \end{vmatrix} + \begin{vmatrix} b_{11} & b_{12} & b_{13} \\ a_{21} & a_{22} & a_{23} \\ a_{31} & a_{32} & a_{33} \end{vmatrix}$$

8.11 The determinant

$$\begin{vmatrix} a_{11} + b_{11} & a_{12} + b_{12} & a_{13} + b_{13} \\ a_{21} + b_{21} & a_{22} + b_{22} & a_{23} + b_{23} \\ a_{31} + b_{31} & a_{32} + b_{32} & a_{33} + b_{33} \end{vmatrix}$$

is required as the sum of determinants each of which has just as or bs in columns. How many determinants are there in the sum? If the determinant is $n \times n$, how many determinants would there be in the sum?

8.12 Show that

$$\begin{vmatrix} 1 & a_1 - b_1 & a_1 + b_1 \\ 1 & a_2 - b_2 & a_2 + b_2 \\ 1 & a_3 - b_3 & a_3 + b_3 \end{vmatrix} = 2 \begin{vmatrix} 1 & a_1 & b_1 \\ 1 & a_2 & b_2 \\ 1 & a_3 & b_3 \end{vmatrix}.$$

8.13 Let D_n be the $n \times n$ **tridiagonal** determinant defined by

$$D_n = \begin{vmatrix} 2 & 1 & 0 & \cdots & 0 \\ 1 & 2 & 1 & \ddots & 0 \\ 0 & 1 & 2 & \ddots & \vdots \\ \vdots & \ddots & \ddots & \ddots & 1 \\ 0 & 0 & \cdots & 1 & 2 \end{vmatrix}.$$

Show that

$$D_n = 2D_{n-1} - D_{n-2}.$$

If $Q_n = D_n - D_{n-1}$, deduce that

$$Q_n = Q_{n-1} = \ldots = Q_3 = 1.$$

Show that $D_n = n + 1$.

8.14 Find all values of x for which

$$\begin{vmatrix} x & a & b & c \\ a & x & b & c \\ a & b & x & c \\ a & b & c & x \end{vmatrix}$$

is zero.

8.15 Let A and B be the two 2×2 matrices

$$A = \begin{bmatrix} a_{11} & a_{12} \\ a_{21} & a_{22} \end{bmatrix}, \qquad B = \begin{bmatrix} b_{11} & b_{12} \\ b_{21} & b_{22} \end{bmatrix}.$$

Write down det A and det B.

(a) Find the product AB and its determinant det AB. Confirm that

$$\det AB = \det A \det B.$$

Show also that

$$\det (A^2) = (\det A)^2.$$

(b) Write down A^T and find its determinant det A^T. Confirm that

$$\det A^T = \det A.$$

(c) Find A^{-1} and det A^{-1}, assuming that det $A \neq 0$. Confirm that

$$\det A^{-1} = 1/\det A.$$

(d) Show that

$$\det \text{adj } A = \det A.$$

(These formulae for 2×2 matrices suggest generalizations for $n \times n$ matrices. Thus, for two $n \times n$ matrices A and B,

$$\det AB = \det A \det B,$$

$$\det A^n = (\det A)^n,$$

$$\det A^T = \det A,$$

$$\det A^{-1} = 1/\det A,$$

$$\det \text{adj } A = (\det A)^{n-1}.$$

We shall not attempt to prove these formulae here.)

8.16 If

$$A = \begin{bmatrix} 1 & 2 & -1 \\ 0 & 1 & 2 \\ 1 & 3 & -1 \end{bmatrix},$$

$$B = \begin{bmatrix} 1 & 2 & -1 \\ 0 & 3 & 1 \\ 2 & 1 & 3 \end{bmatrix},$$

calculate det A, det B, det AB, A^T, det A^T, adj A, det adj A, A^{-1}, and det A^{-1}. Confirm the results conjectured at the end of the previous problem.

8.17 The elements in a 3×3 matrix $A = [a_{ij}]$ are given by the formula

$$a_{ij} = \alpha j + (-1)^i 2^j \quad (i, j = 1, 2, 3).$$

Show that det $A = 0$ for all real α.

Elementary operations with vectors

9.1 Displacement along an axis

Figure 9.1 shows an x axis with origin at O and a scale indicated. The positive direction for x is from left to right. Two points are marked, P at $x = x_P = -2$ and Q at $x = x_Q = 1.5$. The distance between two points is always expressed as a positive number, so in this case

distance from P to Q, or from Q to $P = PQ$ or $QP = 2 + 1.5$

$$= 3.5 \text{ units.}$$

Fig. 9.1

If we are told that $x_P = -2$, and that the distance between P and Q is 3.5 units, this does not tell us where Q is: x_Q might be either 1.5 or -5.5. We need a way to express, as a single piece of information, both the distance PQ and whether Q lies to the right or left of P.

This is done by attaching a plus or minus sign to the distance. We use **plus** if Q as viewed from P is in the **positive direction of the x axis** (to the right in this case), and **minus** if Q is in the **negative direction** (to the left in this case). This quantity is called the **displacement of Q relative to P**, or the displacement of Q from P, and is defined in terms of x_Q and x_P by

displacement of Q from $P = x_Q - x_P$.

In this case the displacement of Q from P is equal to $1.5 - (-2) = 3.5$. This is positive, showing that Q is to the right of P. By the same rule,

displacement of P from $Q = x_P - x_Q = (-2) - 1.5 = -3.5$.

The minus sign indicates that P is to the left to Q.

Example 9.1   *A pedestrian wanders up and down the high street, which extends east and west. Starting at the bus stop, she strolls 80 m east, 25 m west, 50 m east, then races 100 m west, at which point the returning bus drives off. Where was she, relative to the bus stop, at this time?*

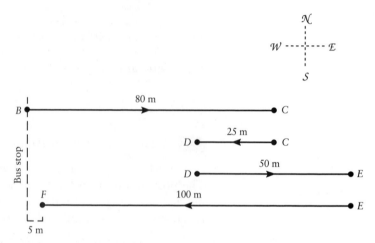

Fig. 9.2

There is no difficulty about this question: Fig. 9.2 shows that she ends up east of the bus stop with 5 more metres to go. Notice how natural it is to count one direction as positive and the other as negative. We shall formalize this, because can get useful illustrations about handling displacements from this problem.

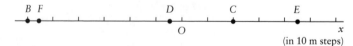

Fig. 9.3

In Fig. 9.3 we have drawn an east-pointing axis x. The origin is at O (it will make no difference where it is) and the bus stop is at B. The direction changes at C, D, E, and the end-point is F.

We want to find the displacement of F from B. This is defined by $x_F - x_B$. Write it in the form

$$x_F - x_B = (x_F - x_E) + (x_E - x_D) + (x_D - x_C) + (x_C - x_B)$$

which is identically true because x_E, x_D, and x_C cancel out. The quantities in the brackets are relative displacements: for example, $(x_D - x_C)$ represents the displacement of D relative to C.

Example 9.1 *continued*

The data of the problem consists of these displacements; all we have to do is to get the signs right. For example, since we chose the positive direction to be east, and the movement from C to D is west, the displacement of D from C is -25. By substituting all the information we obtain

$$x_F - x_B = 80 + (-25) + 50 + (-100) = 5.$$

Since this is positive, she ends up 5 m east of the bus stop.

We did not need to know the actual coordinates of any of the points B to F; the position of the origin O makes no difference to relative displacements.

Relative displacement along a line

Definition: Given an axis Ox, and points P and Q,

Displacement of Q from $P = x_Q - x_P$

$\qquad\qquad\qquad\qquad\qquad = -(\text{displacement of } P \text{ from } Q)$.

(a) The value of $x_Q - x_P$ is unaffected by changing the origin of x.

(b) Addition of displacements:

$$x_D - x_A = (x_D - x_C) + (x_C - x_B) + (x_B - x_A),$$

identically (there may be any number of intermediate points).

(c) The order in which the displacements take place does not affect the final displacement.

$\hfill$ **(9.1)**

9.2 Displacement vectors in two dimensions

We shall extend the idea of displacement into two dimensions. Suppose that a ferry boat stationed at a port A is instructed to proceed in three stages to a destination D. The instructions are:

(a) go 50 km east;

(b) continue 20 km north;

(c) continue 20 km north west to D.

The navigator can plot the route as in Fig. 9.4. The axes point east and north for convenience, and the origin O may be anywhere we please. The three stages are drawn to scale, with their directions indicated by the arrowheads.

Each instruction prescribes a **displacement in two dimensions** relative to the initial point of each stage. A notation for the displacements is

$$\overline{AB}, \overline{BC}, \overline{CD}.$$

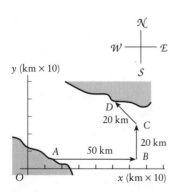

Fig. 9.4

The bar emphasizes that they have particular directions (e.g. A to B). The arrows in the figures indicate **displacement vectors**.

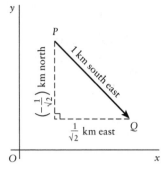

Fig. 9.5

Any displacement vector can be described by **a pair of numbers,** called its **components.** The two components are the **displacements in the x and y directions** that take place during the two-dimensional displacement. For example, Fig. 9.5 shows the displacement vector corresponding to the instruction:

Proceed 1 km south east from P.

The same point Q is arrived at by saying:

Proceed to a point $1/\sqrt{2}$ km east and $(-1/\sqrt{2})$ km north of P.

The numbers $1/\sqrt{2}$ and $(-1/\sqrt{2})$ are the components of the displacement vector $\overline{PQ}$.

In general, if $\overline{PQ}$ represents a displacement vector, and P and Q have coordinates (x_P, y_P) and (x_Q, y_Q), then

component of $\overline{PQ}$ in the x direction $= x_Q - x_P$,

component of $\overline{PQ}$ in the y direction $= y_Q - y_P$.

We may then write $\overline{PQ}$ in **component form:**

$$\overline{PQ} = (x_Q - x_P, y_Q - y_P).$$

Suppose we have a chain of successive displacements. In the ferry boat problem the chain consists of $\overline{AB}, \overline{BC}, \overline{CD}$. The final displacement, of D relative to A, is $\overline{AD}$. In component form

$$\overline{AD} = (x_D - x_A, y_D - y_A).$$

The final components can be broken up into the successive stages:

$$x_D - x_A = (x_D - x_C) + (x_C - x_B) + (x_B - x_A),$$
$$y_D - y_A = (y_D - y_C) + (y_C - y_B) + (y_B - y_A).$$

Since $\overline{CD} = (x_D - x_C, y_D - y_C)$, and so on, it is reasonable to write

$$\overline{AD} = \overline{CD} + \overline{BC} + \overline{AB}.$$

The components are ordinary numbers, so we can **change the order in which displacements are added,** and write instead

$$\overline{AD} = \overline{AB} + \overline{BC} + \overline{CD},$$

or reorder them in any other way: the boat will still arrive at the same point. This is true for any number of displacements.

Example 9.2 *Figure 9.6 shows the track of the ferry boat again.*
(a) *Find the displacement vector $\overline{AD}$ in component form.*
(b) *Express $\overline{AD}$ in terms of its length $\overline{AD}$ and the angle θ it makes with the positive x axis.*

(a) In component form,
$$\overline{AB} = (50, 0),\ \overline{BC} = (0, 20),\ \overline{CD} = (-20/\sqrt{2}, 20/\sqrt{2})$$

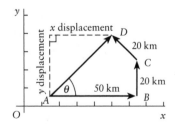

Fig. 9.6

Example 9.2 *continued*

(in km units) and suppose that

$$\overline{AD} = (X, Y).$$

Then, adding the individual x and y components,

$$X = 50 + 0 - 20/\sqrt{2} = 50 - 20/\sqrt{2},$$

and $\quad Y = 0 + 20 + 20/\sqrt{2} = 20 + 20/\sqrt{2},$

(b) From Fig. 9.6,

$$\text{length } AD = \sqrt{(X^2 + Y^2)} = [(50 - 20/\sqrt{2})^2 + (20 + 20/\sqrt{2})^2]^{\frac{1}{2}} = 49.5 \text{ (km)}.$$

$$\theta = \arctan \frac{Y}{X} = \arctan 0.952 = 43.6°.$$

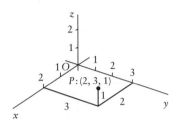

Fig. 9.7

9.3 Axes in three dimensions

From now on we shall consider both two- and three-dimensional situations. To locate points in a plane, two axes are needed. For three-dimensional space, introduce a third axis Oz perpendicular to the other two and drawn through the origin O in the direction shown in Fig. 9.7. These axes are indicated briefly by $Oxyz$. The position of any point P is then specified by a triplet of **coordinates**, (x, y, z), determined by reference to the three axes, Ox, Oy, Oz. For the point P in Fig. 9.7, $x = 2, y = 3,$ and $z = 1,$ and we indicate P by writing $P : (2, 3, 1)$.

There was a choice of two possible directions for Oz, as shown in Fig. 9.8. These two sets of axes cannot be superposed no matter how we turn them about: they are mirror images of each other, like a right shoe and a left shoe. The axes shown in Fig. 9.8a are called **right-handed axes** (left-handed axes, Fig. 9.8b, are seldom used).

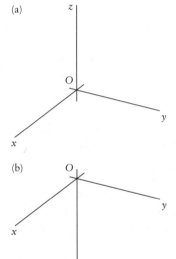

Fig. 9.8 (a) Right-handed axes. (b) Left-handed axes.

9.4 Vectors in two and three dimensions

A displacement vector is a case of a physical quantity which has a **magnitude** and a **direction**, and which follows a certain set of rules similar to those in ordinary algebra. Velocity, acceleration, and force are other examples. Such a quantity is called a **vector quantity**, and can be depicted in terms of **directed line segments** similar to those we used in Section 9.2 for displacements. The rules which follow apply to directed line segments. We shall illustrate later how the rules also apply to other vectors, such as forces.

1. *Components and magnitude.* Figure 9.9 shows a vector placed in a set of axes. Its **initial point** is $P : (x_P, y_P, z_P)$ and its **end-point** is $Q : (x_Q, y_Q, z_Q)$. We denote it either by $\overline{PQ}$, where the bar stresses the direction, P to Q; or (more often) by a single letter, say

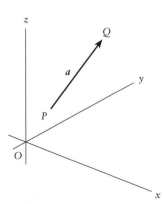

Fig. 9.9 A typical vector $\overline{PQ}$, or *a*.

a (in heavy print) or $\underline{a}$ (underlined when handwritten).

The **components** of a in the x, y, and z directions respectively are a_1, a_2, and a_3, where

$$a_1 = x_Q - x_P, \ a_2 = y_Q - y_P, \ a_3 = z_Q - z_P. \tag{9.2}$$

We write

$$\overline{PQ}, \text{ or } a = (a_1, a_2, a_3). \tag{9.3}$$

The **length** or **magnitude** of a is denoted by PQ (no bar) or QP, or $|a|$ or $|\overline{PQ}|$ or a. By Pythagoras's theorem the length PQ is given by

$$\sqrt{[(x_Q - x_P)^2 + (y_Q - y_P)^2 + (z_Q - z_P)^2]},$$

so

$$PQ \text{ or } |a| \text{ or } |\overline{PQ}| = \sqrt{(a_1^2 + a_2^2 + a_3^2)}. \tag{9.4}$$

This is always a **positive** number.

2. *Equality of two vectors.* We say that $a = b$ if their **components are equal**:

$$a_1 = b_1, \ a_2 = b_2, \ a_3 = b_3.$$

This is equivalent to saying that $a = b$ if they have the **same magnitude and direction**. Instead of saying 'in the same direction', we may say '**parallel** and with the same **sense**'.

The vectors shown in Fig. 9.10a are all called equal although they are in different places. Figure 9.10b shows four vectors in the form of a parallelogram, but only two letters, a and b, are needed to label it. Figure 9.10c shows two vectors which are parallel but have opposite senses, or directions.

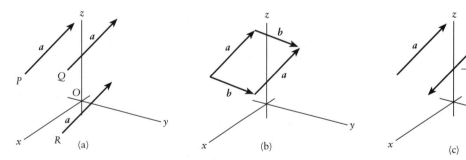

Fig. 9.10 (a) and (b) illustrate equality of vectors. (c) The vectors a and $-a$ have opposite senses.

3. *Multiplication by a positive or negative number.* If k is a real number, then

$$ka = (ka_1, ka_2, ka_3). \tag{9.5}$$

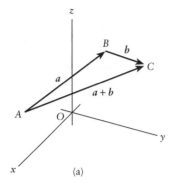

(a)

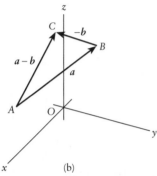

(b)

Fig. 9.11 The triangle rule.

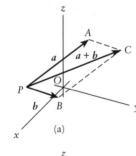

(a)

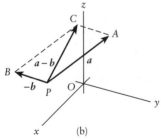

(b)

Fig. 9.12 The parallelogram rule.

Therefore $k\boldsymbol{a}$ is $|k|$ times as long as $\boldsymbol{a}$. If k is positive then $k\boldsymbol{a}$ is in the same direction as $\boldsymbol{a}$, and if k is negative it is in the direction opposite to $\boldsymbol{a}$.

The vector $(-\boldsymbol{a})$ means the same as $(-1)\boldsymbol{a}$:

$$-\boldsymbol{a} = (-a_1, -a_2, -a_3), \tag{9.6}$$

which has the same length as $\boldsymbol{a}$ and the opposite direction (Fig. 9.10c).

4. *Addition and subtraction.*

$$\begin{aligned} \boldsymbol{a} + \boldsymbol{b} &= (a_1, a_2, a_3) + (b_1, b_2, b_3) \\ &= (a_1 + b_1, a_2 + b_2, a_3 + b_3), \end{aligned} \tag{9.7}$$

so the sum of two vectors is obtained by adding the corresponding components (and similarly for any number of vectors).

This is equivalent geometrically to the **triangle rule**, illustrated in Fig. 9.11a. Choose any point A as the starting point, then draw $\boldsymbol{a}$ followed by $\boldsymbol{b}$, as if they were successive displacements from A. The definition says that

$$\boldsymbol{a} + \boldsymbol{b} = \overline{AB} + \overline{BC} = \overline{AC}, \tag{9.8}$$

where AC is the third side of the triangle.

Sometimes the **parallelogram rule**, illustrated in Fig. 9.12a, is more convenient. In this case, draw the vectors $\boldsymbol{a}$ and $\boldsymbol{b}$ out from the same point P, then complete the parallelogram $PACB$. The diagonal vector $\overline{PC}$ is equal to $\boldsymbol{a} + \boldsymbol{b}$, because

$$\overline{AC} = \overline{PB} = \boldsymbol{b}$$

and so by the triangle rule applied to PAC

$$\boldsymbol{a} + \boldsymbol{b} = \overline{PA} + \overline{AC} = \overline{PC}.$$

For subtraction,

$$\boldsymbol{a} - \boldsymbol{b} = \boldsymbol{a} + (-\boldsymbol{b}), \tag{9.9}$$

which is illustrated in Fig. 9.11b using the triangle rule and in Fig. 9.12b using the parallelogram rule.

Also,

$$\begin{aligned} \boldsymbol{a} - \boldsymbol{a} &= (a_1, a_2, a_3) + (-a_1, -a_2, -a_3) \\ &= (0, 0, 0). \end{aligned}$$

This is the zero vector, denoted by $\boldsymbol{0}$ (or $\underline{0}$, if handwritten).

5. *Brackets and rearrangement of sums of vectors.* Addition involves only the addition of the x, y, and z components separately. Since the *components* are ordinary numbers, we may **change the order** in which they are added; for two vectors we have

$$\boldsymbol{a} + \boldsymbol{b} = \boldsymbol{b} + \boldsymbol{a}. \tag{9.10}$$

We can also use **brackets** in the usual way:

$$a + (b + c) = (a + b) + c. \tag{9.11}$$

These are like the rules of ordinary algebra. A more complicated example is

$$(a + b) - (c + d) = (b - c) - (d - a).$$

6. *Vectors in two dimensions*. All of the foregoing definitions and properties apply equally to vectors in **two dimensions**. All that is necessary is to delete the z component. Thus, in two dimensions, if $a = (a_1, a_2)$ and $b = (b_1, b_2)$, then

$$a + b = (a_1 + b_1, a_2 + b_2).$$

Example 9.3   *Find $|a - b|$ when $a = (a_1, a_2)$ and $b = (b_1, b_2)$.*

$$a - b = (a_1 - b_1, a_2 - b_2);$$
$$|a - b| = \text{magnitude of } a - b$$
$$= \sqrt{[(a_1 - b_1)^2 + (a_2 - b_2)^2]}.$$

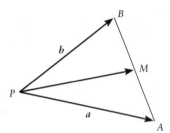

Fig. 9.13 M is the midpoint of AB.

Example 9.4   *M is the midpoint of the side AB of the triangle PAB (Fig. 9.13). Put $\overline{PA} = a$ and $\overline{PB} = b$. (a) Express the vector $\overline{PM}$ in terms of a and b. (b) Deduce that the diagonals of a parallelogram bisect each other. (You can think of this in two dimensions, but it applies equally in three.)*

(a) $\overline{PM} = \overline{PB} + \overline{BM}$ (triangle rule)

$\qquad = \overline{PB} + \frac{1}{2}\overline{BA}$ (M is the midpoint)

$\qquad = b + \frac{1}{2}\overline{BA}.$ \hfill **(i)**

Also $\overline{BA} = \overline{BP} + \overline{PA}$ (triangle rule; note the direction of $\overline{BP}$)

$\qquad = -\overline{PB} + \overline{PA}$ (see (9.6))

$\qquad = -b + a.$ \hfill **(ii)**

Substitute for BA in (i):

$\overline{PM} = b + \frac{1}{2}(-b + a)$

$\qquad = \frac{1}{2}(a + b)$ (after rearrangement).

(b) In Fig. 9.14 we have added a fourth vertex D to form a parallelogram. The point N is the midpoint of PD. Then

$\overline{PN} = \frac{1}{2}\overline{PD}$

$\qquad = \frac{1}{2}(a + b)$ (parallelogram rule).

Therefore, from the result in (a),

$\overline{PN} = \overline{PM},$

so the midpoints of PD and BA coincide.

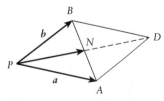

Fig. 9.14 N is the midpoint of PD.

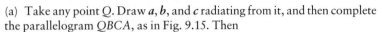

Example 9.5 *(In two dimensions.) In the (x, y) plane, a and b are two vectors which are not parallel, and c is another vector.*
(a) *Prove that $c = \lambda a + \mu b$, where λ and μ are constants.*
(b) *Find λ and μ when $a = (1, 1)$, $b = (2, 0)$, and $c = (3, 4)$.*

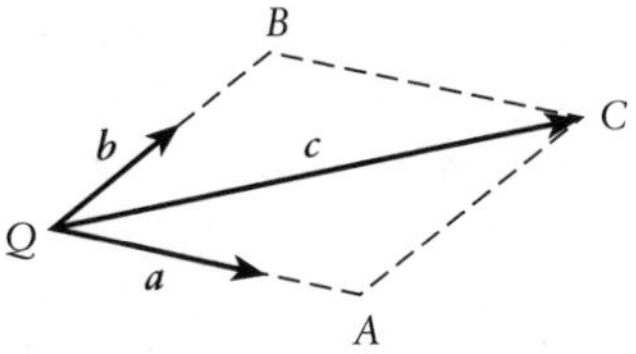

(a) Take any point Q. Draw a, b, and c radiating from it, and then complete the parallelogram $QBCA$, as in Fig. 9.15. Then
$$c = \overline{QC} = \overline{QA} + \overline{QB} \quad \text{(parallelogram rule)}.$$
But $\overline{QA}$ and $\overline{QB}$ point respectively in the directions of a and b so they are equal to certain (unique) multiples of a and b:
$$\overline{QA} = \lambda a \quad \text{and} \quad \overline{QB} = \mu b,$$
say. Therefore
$$c = \lambda a + \mu b.$$
(b) $a = (1, 1)$, $b = (2, 0)$, and $c = (3, 4)$, so from (a)
$$(1, 1) = \lambda(2, 0) + \mu(3, 4).$$
The individual components on the two sides must match, so
$$1 = 2\lambda + 3\mu,$$
and $\quad 1 = 0 + 4\mu$.
The solution is $\mu = \frac{1}{4}$ and $\lambda = \frac{1}{8}$, so
$$c = \tfrac{1}{8}a + \tfrac{1}{4}b.$$

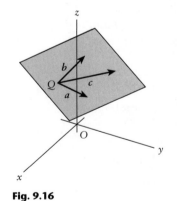

Fig. 9.15 Vectors in a plane: $c = \lambda a + \mu b$.

The result in Example 9.5a is important, and it extends to three dimensions as follows:

Relation between three coplanar vectors

a and b are two non-parallel, nonzero vectors with the same initial point Q, and c is any other vector at Q, in the same plane as a and b. Then
$$c = \lambda a + \mu b,$$
where λ and μ are certain (unique) constants. **(9.12)**

Figure 9.16 shows the three vectors a, b, c in their common plane. We can use the same argument as in the previous example. (It is not actually necessary for the vectors a, b, and c to be in the same plane and emerge from the same point to start with: it is sufficient for them merely to be *parallel to the same plane*, so that we can translate them to the positions in Fig. 9.16.) Then the argument in Example 9.5 follows.

Fig. 9.16

9.5 Relative velocity

In this section we shall assume that all the *velocities are constant*.

Velocity has magnitude and direction, so we can depict it by a directed line segment whose length is proportional to the **speed** (always a positive number), and which points in the right direction. But to decide whether velocity can be treated as a vector (i.e. whether it obeys the rules in Section 9.4) we need to say what addition of velocities is to mean physically.

Typically, addition of velocities is concerned with combining **relative velocities**. For example, if an escalator is moving at 0.5 m s⁻¹ relative to the wall, and a passenger is walking up at 1 m s⁻¹ relative to the escalator, then the actual velocity of the passenger is $1 + 0.5 = 1.5$ m s⁻¹ relative to the wall.

Since relative velocities are relative displacements per unit time, velocity vectors obey the same rules as displacement vectors. Take a set of axes Ox, Oy, Oz which are to be regarded as **fixed axes**. They might be fixed relative to the earth's surface, or relative to the directions of distant stars. Let

v_P = velocity of a point P relative to the fixed axes,

v_Q = velocity of a point Q relative to the fixed axes,

v_{QP} = velocity of Q relative to P.

Then the velocity v_{QP} of a point Q as observed from P, in terms of the velocities v_Q and v_P observed from the fixed axes, is given by

velocity of Q relative to P = velocity of Q − velocity of P,

or $v_{QP} = v_Q - v_P.$ (9.13)

....

Example 9.6   (*Figure 9.17*) *A river of width 0.2 km flows with uniform speed 3 km h⁻¹ from west to east. A boat sets off from a point S on the south bank, wishing to land at a point N on the north bank directly opposite S. It can travel at a speed of 5 km h⁻¹ relative to the water. In what direction should it point in order to arrive at N by a straight line route? How long does it take?*

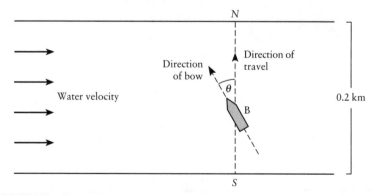

Fig. 9.17

The true path of the boat (i.e. as seen from the land, or relative to fixed axes) is not along the direction it is pointing, because it is also being carried

Example 9.6 *continued*

downstream. However, viewed from axes which travel along with the water, it does go in the direction it is pointing, at an apparent speed of 5 km h⁻¹. To visualize this, imagine there is a dense fog, so that the banks cannot be seen and the pilot is not aware of the current.

With B denoting 'boat' and W denoting 'water', put

v_B = velocity of B relative to fixed axes (direction north, magnitude, or speed, unknown);

v_{BW} = velocity of B relative to the water W (speed 5 km h⁻¹, in the unknown direction it is pointing);

v_W = velocity of the water W relative to fixed axes (direction east, speed 3 km h⁻¹).

We also know from (9.13) that these are connected by

$$v_{BW} = v_B - v_W,$$

or $v_B = v_{BW} + v_W$.

This information gives Fig. 9.18.

(a) From Fig. 9.18, the boat is directed at $\theta = \arcsin \frac{3}{5} = 36.9°$.

(b) Pythagoras's theorem gives the magnitude of v_B:

$$|v_B| = \sqrt{(5^2 - 3^2)} = 4 \text{ km h}^{-1}.$$

Therefore the time taken is $0.2/4 = 0.05$ h $= 3$ minutes.

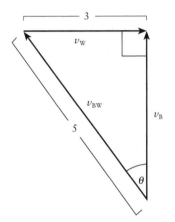

Fig. 9.18

9.6 Position vectors and vector equations

In Fig. 9.19 P is the point with coordinates $(2, 3, 1)$. The vector $\overline{OP}$, or r, which has its **initial point at the origin** of coordinates, O, is called the **position vector** of P. The **components** of r, or $\overline{OP}$, are then equal to the **coordinates** of P, so

$$\overline{OP} = r = (2, 3, 1)$$

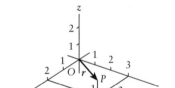

Fig. 9.19 Position vector, $r = \overline{OP}$, of the point P.

in this case. Position vectors are often distinguished from ordinary vectors by using the letter r. Apart from their being attached to the origin, the rules for position vectors are the same as for ordinary vectors.

This device enables us to specify, for example, the point at which a force acts, without mixing up vectors with coordinates in the same calculation. It also allows us to do coordinate geometry in vector terms, by obtaining **vector equations** describing curves and surfaces in terms of the position vector $r = (x, y, z)$.

Example 9.7 (*Two dimensions.*) *A circle has radius c, and its centre C at the point* (a, b). *(a) Obtain a vector equation for the circle. (b) Deduce the ordinary cartesian equation.*

(a) The circle is shown in Fig. 9.20. P is any point (x, y) on its circumference, so its position vector in component form is

$$r = (x, y).$$

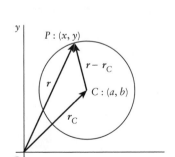

Fig. 9.20

Example 9.7 *continued*

The centre C has position vector r_C, where
$$r_C = (a, b).$$
Also, $\overline{CP} = r - r_C$. The length of $\overline{CP}$ must be constant and equal to c, so
$$|r - r_C| = c. \tag{i}$$
This is the vector equation required.

(b)   To turn (i) into x, y form write r and r_C in component form:
$$r - r_C = (x, y) - (a, b) = (x - a, y - b).$$
The length of this vector is given by
$$|r - r_C| = \sqrt{[(x - a)^2 + (y - b)^2]}.$$
Therefore, after squaring both sides in (i), we get
$$(x - a)^2 + (y - b)^2 = c^2,$$
which is the usual form for the equation of a circle. (This is not an efficient way of obtaining it, of course. We are simply checking that (i) makes sense.)

Example 9.8   *Three points, A, B, and C (which do not lie in a straight line), have position vectors **a**, **b**, and **c**. (a) Obtain a **parametric vector equation** for the plane through the points A, B, C. (b) Deduce **parametric cartesian** (i.e. x, y, z) equations for the plane in the case where the points are $A : (1, 2, 1)$, $B : (2, 2, 0)$, $C : (2, 1, 2)$. (c) Deduce the ordinary cartesian equation for this plane by eliminating the parameters occurring in (b).*

(a) Figure 9.21 shows the points A, B, C, and their position vectors. The point $P : (x, y, z)$ with position vector r is any point in the plane through A, B, and C. By the triangle rule
$$\overline{BA} = a - b, \overline{BC} = c - b, \overline{BP} = r - b.$$

By using the result (9.12), which relates any three coplanar vectors, we obtain
$$\overline{BP} = \lambda \overline{BA} + \mu \overline{BC},$$

or $r - b = \lambda(a - b) + \mu(c - b),$ 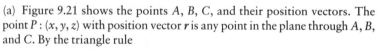 (i)

where λ, μ are two constants which depend on the position of P. We find every point r in the plane by letting the parameters λ, μ run through all possible values between $-\infty$ and $+\infty$, so (i) is a parametric vector equation for the plane through A, B, C.

(b) Since r, a, b, c are position vectors, their components are given by the coordinates of P, A, B, C, so eqn (i) becomes
$$(x, y, z) - (2, 2, 0) = \lambda[(1, 2, 1) - (2, 2, 0)] + \mu[(2, 1, 2) - (2, 2, 0)]$$
$$= \lambda(-1, 0, 1) + \mu(0, -1, 2).$$

Take the vector $(2, 2, 0)$ over to the right-hand side, and then match the x, y, z components separately:
$$x = 2 - \lambda, y = 2 - \mu, z = \lambda + 2\mu \tag{ii}$$
where λ and μ may take any values. These are cartesian parametric equations for the plane.

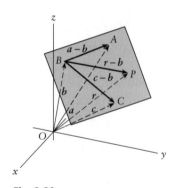

Fig. 9.21

Example 9.8 *continued*

(c) We obtain an x, y, z equation by eliminating λ and μ from the equations (ii). From the first two equations we have

$$\lambda = 2 - x \quad \text{and} \quad \mu = 2 - y.$$

Substitute these into the third equation of (ii):

$$z = (2 - x) + 2(2 - y),$$

which is the same as

$$x + 2y + z = 6. \tag{iii}$$

If A, B, and C do not lie on a straight line, the equation of the plane through them will always be like Example 9.8(iii):

> **Equation of a plane**
> The general equation of a plane is
> $$ax + by + cz = d,$$
> where a, b, c, d are constants. **(9.14)**

Example 9.9   *(Three dimensions.) Two points, A and B, have position vectors **a** and **b**. (a) Obtain a parametric vector equation for the straight line joining A and B. (b) Deduce parametric cartesian (i.e. x, y, z) equations for the case where the points are A : (2, 2, −1) and B : (0, 1, −2). (c) By eliminating the parameter between the equations in (b), find cartesian equations for this line.*

(a) Figure 9.22 shows the points A and B and their position vectors a and b. The point $P : (x, y, z)$ with position vector r represents any point on the line joining AB. Also,

$$\overline{AB} = b - a, \text{ and } \overline{AP} = r - a.$$

$\overline{AP}$ is some multiple, λ say, of $\overline{AB}$:

$$\overline{AP} = \lambda \overline{AB},$$

or $r - a = \lambda(b - a).$

Therefore

$$r = (1 - \lambda)a + \lambda b. \tag{i}$$

This is the required parametric vector equation, with λ as the parameter. As λ increases from $-\infty$ to $+\infty$, P traces out the straight line passing through A and B.

(b) Since r, a, and b are position vectors, their components are the same as the coordinates of P, A, and B:

$$r = (x, y, z), a = (2, 2, -1), b = (0, 1, -2).$$

Substitute these into (i):

$$(x, y, z) = (1 - \lambda)(2, 2, -1) + \lambda(0, 1, -2)$$
$$= (2 - 2\lambda, 2 - \lambda, -1 - \lambda).$$

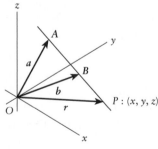

Fig. 9.22

Example 9.9 *continued*

Now match the x, y, z components on both sides:
$$x = 2 - 2\lambda,\ y = 2 - \lambda,\ z = -1 - \lambda. \tag{ii}$$
These are parametric cartesian equations, in which the parameter ranges from $-\infty$ to $+\infty$.

(c) In order to get rid of the parameter λ in (ii), write them successively in the form
$$\lambda = \frac{x-2}{-2},\quad \lambda = \frac{y-2}{-1},\quad \lambda = \frac{z+1}{-1}.$$
Since the three fractions are equal (equal to the current value of λ) we obtain the relation between x, y, z which holds on the line:
$$\frac{x-2}{-2} = \frac{y-2}{-1} = \frac{z+1}{-1}$$
which simplifies to
$$-\tfrac{1}{2}x + 1 = -y + 2 = -z - 1. \tag{iii}$$

The shape of the result (iii) of Example 9.9 might strike you as being peculiar. It really consists of two simultaneous equations, representing two planes which intersect along the required line AB. The expression cannot be reduced to a single equation. The general case will be given in Chapter 10.

Example 9.10   *Given the straight line*
$$2x - 2 = y + 1 = -2z, \tag{i}$$
(a) *Find any one point on the line.* (b) *Find a parametric equation for the line.* (c) *Find the coordinates of the point where the line crosses the plane*
$$x - y + z = 0. \tag{ii}$$

(a) Put, for example, $x = 1$. Then from (i), $2x - 2 = y + 1$, so when $x = 1$, $y = -1$. Also from (i), $2x - 2 = -2z$, so $z = 0$. Therefore, the point $(1, -1, 0)$ lies on the line. (Other values of x lead to other points.)

(b) Proceeding as in (a), put $x = \lambda$, where λ may take any value. Then we find that
$$y = 2\lambda - 3 \quad \text{and} \quad z = -\lambda + 1.$$
Therefore a set of parametric equations is
$$x = \lambda,\ y = 2\lambda - 3,\ z = -\lambda + 1. \tag{iii}$$

(c) From (ii) and (iii), at the point where the line meets the plane the value of λ must be given by
$$0 = x - y + z = \lambda - (2\lambda - 3) + (-\lambda + 1).$$
At this point $\lambda = 2$, so from (i) again, the line meets the plane at
$$x = 2,\ y = 1,\ z = -1.$$
(Alternatively, solve the equations (i) and (ii) simultaneously.)

9.7 Unit vectors and basis vectors

A vector of **unit magnitude** is called a **unit vector**. For example,

$a = (-\frac{2}{7}, \frac{3}{7}, \frac{6}{7})$ is a unit vector since

$a = |a| = \sqrt{[(-\frac{2}{7})^2 + (\frac{3}{7})^2 + (\frac{6}{7})^2]} = 1.$

The vector $(1, 0, 0)$ is a unit vector; it points is the direction of the x axis, since if it is drawn as a position vector it would join the origin to the point 1 unit along the x axis. Similarly, $(0, 1, 0)$ and $(0, 0, 1)$ are unit vectors in the y and z directions respectively. These vectors have the **special symbols** $\hat{\imath}, \hat{\jmath}$, and $\hat{k}$, and are called **basis vectors** for the given coordinates:

$$\hat{\imath} = (1, 0, 0), \hat{\jmath} = (0, 1, 0), \hat{k} = (0, 0, 1). \tag{9.15}$$

(They are sometimes spoken of as 'i-hat', and so on.) Figure 9.23 shows them as position vectors.

Any vector can be expressed in terms of $\hat{\imath}, \hat{\jmath}$, and $\hat{k}$. Suppose that $a = (a_1, a_2, a_3)$ in component form. Then

$$a = (a_1, 0, 0) + (0, a_2, 0) + (0, 0, a_3)$$
$$= a_1(1, 0, 0) + a_2(0, 1, 0) + a_3(0, 0, 1) = a_1\hat{\imath} + a_2\hat{\jmath} + a_3\hat{k}.$$

The components become the coefficients of $\hat{\imath}, \hat{\jmath}$, and $\hat{k}$.

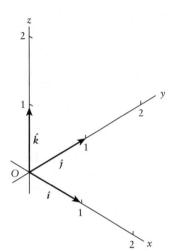

Fig. 9.23 Basis vectors, $\hat{\imath}, \hat{\jmath}, \hat{k}$.

Example 9.11 *Let $a = 2\hat{\imath} + 3\hat{\jmath} - \hat{k}$ and $b = \hat{\imath} - 3\hat{k}$. Express the vector x in the equation $3a + 2x = b$ in terms of $\hat{\imath}, \hat{\jmath}, \hat{k}$.*

In the usual way, we find that
$$x = \tfrac{1}{2}(b - 3a) = \tfrac{1}{2}b - \tfrac{3}{2}a$$
$$= \tfrac{1}{2}(\hat{\imath} - 3\hat{k}) - \tfrac{3}{2}(2\hat{\imath} + 3\hat{\jmath} - \hat{k}) = -\tfrac{5}{2}\hat{\imath} - \tfrac{9}{2}\hat{\jmath}.$$
The components of x are therefore $(-\tfrac{5}{2}, -\tfrac{9}{2}, 0)$.

If a is any vector, then the vector $\hat{a}$ (called 'a-hat')

$$\hat{a} = a/|a|,$$

obtained by dividing a by its own length (or magnitude), is a **unit vector in the direction of a** (we can say 'the direction of a is $\hat{a}$').

Example 9.12 *Obtain the unit vector $\hat{F}$ pointing in the direction of the force $F = 2\hat{\imath} - 3\hat{\jmath} - 6\hat{k}$.*

$$|F| = \sqrt{[2^2 + (-3)^2 + (-6)^2]} = \sqrt{49} = 7.$$
Therefore, the unit vector pointing in the same direction is
$$\hat{F} = (2\hat{\imath} - 3\hat{\jmath} - 6\hat{k})/7 = \tfrac{2}{7}\hat{\imath} - \tfrac{3}{7}\hat{\jmath} - \tfrac{6}{7}\hat{k},$$
or, in component form,
$$\hat{F} = (\tfrac{2}{7}, -\tfrac{3}{7}, -\tfrac{6}{7}).$$

Unit vectors

A unit vector is a vector of unit magnitude. The unit vector in the direction of *a* is denoted by $\hat{a}$ (*a*-hat).

(a) If *a* is any vector, then

$$\hat{a} = a/|a|.$$

(b) The vectors $\hat{\imath}, \hat{\jmath}, \hat{k}$ (basis vectors) are the unit vectors in directions Ox, Oy, Oz. If $a = (a_1, a_2, a_3)$ is any vector, then

$$a = a_1\hat{\imath} + a_2\hat{\jmath} + a_3\hat{k}.$$

(For two dimensions, use only $\hat{\imath}$ and $\hat{\jmath}$.)   **(9.16)**

Example 9.13   *Find the point Q where the straight line joining $A : (2, 3, 1)$ and $B : (1, 2, 2)$ intersects the plane $x + y + z = 0$.*

The position vectors of A and B, in terms of $\hat{\imath}, \hat{\jmath}, \hat{k}$, are $a = 2\hat{\imath} + 3\hat{\jmath} + \hat{k}$ and $b = \hat{\imath} + 2\hat{\jmath} + 2\hat{k}$ respectively. Let $r = x\hat{\imath} + y\hat{\jmath} + z\hat{k}$ be the position vector of a general point on the line AB. Then from Example 9.9a, the parametric equation of AB is

$$r = (1 - \lambda)a + \lambda b = (1 - \lambda)(2\hat{\imath} + 3\hat{\jmath} + \hat{k}) + \lambda(\hat{\imath} + 2\hat{\jmath} + 2\hat{k}).$$

After collecting terms in $\hat{\imath}, \hat{\jmath}$, and $\hat{k}$ on the right, this becomes

$$x\hat{\imath} + y\hat{\jmath} + z\hat{k} = (2 - \lambda)\hat{\imath} + (3 - \lambda)\hat{\jmath} + (1 + \lambda)\hat{k}. \tag{i}$$

Match the coefficients of $\hat{\imath}, \hat{\jmath}, \hat{k}$ on either side of (i); then

$$x = 2 - \lambda, y = 3 - \lambda, z = 1 + \lambda. \tag{ii}$$

The intersection point Q is on the plane, $x + y + z = 0$, so

$$(2 - \lambda) + (3 - \lambda) + (1 + \lambda) = 0.$$

Therefore $\lambda = 6$ at Q. Put this value back into (ii):

$$x = -4, y = -3, z = 7.$$

The position vector of Q is therefore

$$-4\hat{\imath} - 3\hat{\jmath} + 7\hat{k}.$$

Problems can be worked through with the vectors given either in component form or in $\hat{\imath}, \hat{\jmath}, \hat{k}$ form, whichever is convenient.

9.8 Tangent vector, velocity, and acceleration

Suppose that the coordinates of a point P depend on a parameter t (which might stand for time). Then we can write

$$r(t) = x(t)\hat{\imath} + y(t)\hat{\jmath} + z(t)\hat{k}.$$

As t runs from the value $t = a$ to $t = b$, where $b > a$, P follows a curve from A to B, as in Fig. 9.24.

Consider two points, P and Q, close together on the curve, where the parameter values are t and $t + \delta t$ respectively. The corresponding position vectors are $r(t)$ and $r(t + \delta t)$. By the triangle rule,

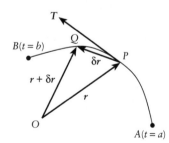

Fig. 9.24 Tangent vector T.

$$\overline{PQ} = r(t + \delta t) - r(t) = \delta r.$$

Now consider the vector T defined by

$$T = \lim_{\delta t \to 0} [r(t + \delta t) - r(t)]/\delta t = \lim_{\delta t \to 0} \delta r/\delta t.$$

This is like an ordinary derivative, so we denote this vector by

$$T = dr(t)/dt. \tag{9.17}$$

Notice also that this is equivalent to

$$T = dr/dt = \hat{\imath}\, dx/dt + \hat{\jmath}\, dy/dt + \hat{k}\, dz/dt,$$

since $\hat{\imath}, \hat{\jmath}, \hat{k}$ are constant.

As δt approaches zero, δr, and therefore $\delta r/\delta t$, become more and more nearly tangential to the curve. Therefore T is a **tangent vector** to the curve at P. To decide which way T points, consider the case when δt is positive. Then δr **points in the direction of increasing** t, so the tangent vector T must also point in this direction.

Derivative of $r(t)$

$r(t) = \hat{\imath}x(t) + \hat{\jmath}y(t) + \hat{k}z(t)$, where t is a parameter, represents a curve. The vector T given by

$$T = dr/dt = \hat{\imath}\, dx/dt + \hat{\jmath}\, dy/dt + \hat{k}\, dz/dt$$

is a tangent to the curve, in the direction of increasing t. $\qquad$ **(9.18)**

If the parameter t stands for time, then dr/dt is the **definition** of the **velocity** $v(t)$ of P, and dv/dt represents its vector **acceleration**:

Velocity and acceleration vectors

If a point P has position vector $r(t)$, then

$\qquad$ velocity $v(t) = dr/dt$,

$\qquad$ acceleration $a(t) = dv/dt \quad$ or $\quad d^2r/dt^2$.

Also the *speed* $= |v(t)|$. $\qquad$ **(9.19)**

Notice that velocity and acceleration are not generally parallel.

Example 9.14 *(Motion in the (x, y) plane.) The position vector of a point P is given by*

$$r(t) = \hat{\imath}c \cos \omega t + \hat{\jmath}c \sin \omega t,$$

where c and ω are positive constants. Find (a) *the velocity $v(t)$ and the speed of P;* (b) *the acceleration $a(t)$ of P.*

(a) $|r(t)| = c\sqrt{[\cos^2\omega t + \sin^2\omega t]} = c$, so P is moving around a circle of radius c in the (x, y) plane.

$$v = dr/dt = -\hat{\imath}c\omega \sin \omega t + \hat{\jmath}c\omega \cos \omega t.$$

Example 9.14 *continued*

The direction of v is tangential to the circle, by (9.18). By putting, say, $t=0$ we obtain $v = \hat{j}\omega c$, and since $c, \omega > 0$, this shows the motion to be anticlockwise. Also, speed $= |v| = c\omega$.

(The speed is constant, but the velocity is not, because its direction is changing.)

(b) $a = \mathrm{d}v/\mathrm{d}t = -\hat{i}c\omega^2 \cos \omega t - \hat{j}c\omega^2 \sin \omega t$

$\qquad = -\omega^2(\hat{i}c \cos \omega t + \hat{j}c \sin \omega t) = -\omega^2 r.$

The acceleration is therefore directed towards the centre of the circle (perhaps unexpectedly).

9.9 Motion in polar coordinates

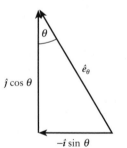

Fig. 9.25 Polar unit vectors $\hat{e}_r$ and $\hat{e}_\theta$.

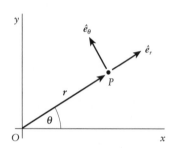

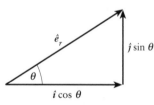

Fig. 9.26

Suppose that two-dimensional polar coordinates r, θ are appropriate to the geometry of an application. Figure 9.25 shows a point $P: (x, y)$ and its polar coordintes r, θ. There are also two *unit* vectors $\hat{e}_r$ and $\hat{e}_\theta$ associated with $P: \hat{e}_r$ in the direction of θ constant with r increasing, and $\hat{e}_\theta$ in the direction of r constant with θ increasing. The position vector of P is r, given by

$$\overline{OP} = r = r\hat{e}_r. \tag{9.20}$$

The unit vectors $\hat{e}_r$ and $\hat{e}_\theta$ *vary in direction* according to the value of θ, and are therefore functions of θ. They are related to the basis vectors $\hat{i}$ and $\hat{j}$ as in Fig. 9.26. By the triangle rule,

$$\hat{e}_r = \hat{i} \cos \theta + \hat{j} \sin \theta, \qquad \hat{e}_\theta = -\hat{i} \sin \theta + \hat{j} \cos \theta. \tag{9.21}$$

We shall need their derivatives with respect to θ:

$$\mathrm{d}\hat{e}_r/\mathrm{d}\theta = -\hat{i} \sin \theta + \hat{j} \cos \theta = \hat{e}_\theta \tag{9.22}$$

and $\quad \mathrm{d}e_\theta/\mathrm{d}\theta = -\hat{i} \cos \theta - \hat{j} \sin \theta = -\hat{e}_r. \tag{9.23}$

Now suppose that P is moving along a curved path. Then r and θ are functions of time, t, so we can write $r(t)$, $\theta(t)$ for its polar coordinates, and consider their derivatives with respect to t. There is a useful **dot notation for time derivatives** which saves a lot of writing – it works in the same way as the dash notation, (4.1):

> **Dot notation for time derivatives**
>
> If $x(t)$ represents a function of t, then $\dot{x}$ stands for $\mathrm{d}x/\mathrm{d}t$, $\ddot{x}$ stands for $\mathrm{d}^2x/\mathrm{d}t^2$, and so on.
>
> (9.24)

By using the chain rule, and writing $\dot{\theta}$ for $\mathrm{d}\theta/\mathrm{d}t$, we obtain from (9.22) and (9.23) the time variation of $\hat{e}_r$ and $\hat{e}_\theta$:

$$\mathrm{d}\hat{e}_r/\mathrm{d}t = \dot{\theta}\hat{e}_\theta \qquad \text{and} \qquad \mathrm{d}\hat{e}_\theta/\mathrm{d}t = -\dot{\theta}\hat{e}_r. \tag{9.25}$$

This result is used in the following example.

Example 9.15 *The polar coordinates of a point moving in a plane are $r(t)$, $\theta(t)$, where t is time. Find the polar components* (a) *of its velocity and* (b) *of its acceleration.*

(a) The position vector is $r(t) = r(t)\hat{e}_r$. The velocity v is dr/dt:

$$v(t) = dr/dt = d(r\hat{e}_r)/dt.$$

Both r and $\hat{e}_r$ depend on θ, so we use the product rule for differentiation:

$$v = \dot{r}\hat{e}_r + r\,d\hat{e}_r/dt = \dot{r}\hat{e}_r + r\dot{\theta}\hat{e}_\theta \qquad \text{(i)}$$

by (9.25). Therefore the **radial** velocity component is $\dot{r}$ and the **transverse** component is $r\dot{\theta}$.

(b) The acceleration is dv/dt, given by

$$dv/dt = (d/dt)(\dot{r}\hat{e}_r + r\dot{\theta}\hat{e}_\theta) \quad \text{(from (i))}$$

$$= \ddot{r}\hat{e}_r + \dot{r}\frac{d\hat{e}_r}{dt} + \frac{d(r\dot{\theta})}{dt}\hat{e}_\theta + r\dot{\theta}\frac{d\hat{e}_\theta}{dt}$$

$$= \ddot{r}\hat{e}_r + \dot{r}\dot{\theta}\hat{e}_\theta + (\dot{r}\dot{\theta} + r\ddot{\theta})\hat{e}_\theta - r\dot{\theta}^2\hat{e}_r$$

$$= (\ddot{r} - r\dot{\theta}^2)\hat{e}_r + (r\ddot{\theta} + 2\dot{r}\dot{\theta})\hat{e}_\theta.$$

Therefore the radial component of acceleration is $\ddot{r} - r\dot{\theta}^2$, and the transverse component is $r\ddot{\theta} + 2\dot{r}\dot{\theta}$.

Problems

9.1 Sketch the two-dimensional displacement vectors $\overline{PQ}$ and $\overline{QP}$, and state their x and y components, when the coordinates of P and Q are as follows.
(a) $P: (-2, 3), Q: (3, 0)$, (b) $P: (3, 4), Q: (2, 1)$,
(c) $P: (0, 1), Q: (-1, -2)$, (d) $P: (-1, -1), Q: (0, 0)$.

9.2 (a) to (h) represent two-dimensional displacement vectors expressed in terms of their x, y components. For each one obtain the length and the angle of inclination θ to the *positive direction* of the x axis in the range $-180°$ to $180°$.
(a) $(3, 0)$, (b) $(0, 2)$, (c) $(-1, 1)$ (d) $(1, 1)$,
(e) $(-1, -1)$, (f) $(-3, 4)$, (g) $(-3, -4)$, (h) $(-2, 1)$.
(Made sure that these angles are in the right quadrant by means of a rough sketch.)

9.3 Obtain the components of the vectors a in (a) to (d), where L is the magnitude and θ the angle made with the positive direction of the x axis ($-180° < \theta \leqslant 180°$):
(a) $L = 2$, $\theta = 45°$, (b) $L = 3$, $\theta = 120°$, (c) $L = 3$, $\theta = 60°$, (d) $L = 3$, $\theta = -150°$.

9.4 Two ships, S_1 and S_2, set off from the same point Q. Each follows a route given by successive displacement vectors. In axes pointing east and north, S_1 follows the path to B via $\overline{QA} = (2, 4)$, and $\overline{AB} = (4, 1)$. S_2 goes to E via $\overline{QC} = (3, 3)$, $\overline{CD} = (1, 1)$, and $\overline{DE} = (2, -3)$. Find the displacement vector $\overline{BE}$ in component form, the distance BE, and the final bearing of S_2 seen from S_1.

9.5 Find the distances between the pairs of points whose coordinates are: (a) $(0, 0, 0)$ and $(1, 2, 3)$, (b) $(1, 2, 3)$ and $(3, 2, 1)$, (c) $(1, 0, -1)$ and $(-1, 1, 0)$.

9.6 State the projections on the three axes of the vector $\overline{PQ}$ when P is the point $(1, 2, 1)$ and Q is $(2, 3, 3)$.

9.7 Find $2a$, $3b$, and $2a - 3b$ when
(a) $a = (1, 2, 1), b = (2, 1, 2)$,
(b) $a = (3, 2, 3), b = (1, 1, 2)$,
(c) $a = (6, 3, 1), b = (4, 2, 1)$.
  How do you recognize that $2a - 3b$ is parallel to the (x, y) plane in (b), and parallel to the z axis in (c)?

9.8 Sketch a diagram to show that if A, B, C are any three points, then $\overline{AB} + \overline{BC} + \overline{CA} = 0$. Formulate a similar result for any number of points.

9.9 Sketch a diagram to show that if A, B, C, D are any four points, then $\overline{CD} = \overline{CB} + \overline{BA} + \overline{AD}$. Formulate a similar result for any number of points.

9.10 $Oxyz$ and $QXYZ$ are two sets of axes with origins at O and Q respectively. QX is parallel to Ox and has the same sense (positive direction), and similarly for QY and QZ. The frame $QXYZ$ is said to be a **translation** (a motion without rotation) of the frame $Oxyz$.

Suppose that $\overline{OQ} = (2, -1, 3)$. (a) Find the coordinates of the point P in $QXYZ$ if it has coordinates $x = 5, y = 2, z = -3$ in $Oxyz$. (b) Find the equation of the sphere $x^2 + y^2 + z^2 = 1$ in terms of X, Y, and Z.

9.11 $ABCD$ is any quadrilateral in three dimensions. Prove that if P, Q, R, S are the midpoints of AB, BC, CD, DA respectively, then $PQRS$ is a parallelogram.

9.12 ABC is a triangle, and P, Q, R are the midpoints of the respective sides BC, CA, AB. Prove that the medians AP, BQ, CR meet at a single point G (called the centroid of ABC; it is the centre of mass of a uniform triangular plate).

9.13 Show that the vectors $\overline{OA} = (1, 1, 2), \overline{OB} = (1, 1, 1)$, and $\overline{OC} = (5, 5, 7)$ all lie in one plane. Show that the same is true if $\overline{OA} = (a, a, p), \overline{OB} = (b, b, q), \overline{OC} = (c, c, r)$, where a, b, c, p, q, r may stand for any numbers. Explain this result geometrically.

9.14 A glider is moving with a velocity $v = (40, 30, 10)$ relative to the air and is blown by the wind which has velocity relative to the earth of $w = (5, -10, 0)$. Find the velocity of the glider relative to the earth.

9.15 The captain of a boat at night can tell that it is moving relative to the sea with velocity $(5, 4) \text{ km h}^{-1}$, and by observation of lights on shore its true velocity is found to be $(4, 1)$. What is the velocity of the current?

9.16 A cyclist rides north along a straight road at 10 km h^{-1}. The wind appears to come from the west. If she increases her speed to 20 km h^{-1} then the wind appears to blow from the north west. Determine the speed and direction of the wind.

9.17 A ship travels south with speed u and the apparent wind direction is from the east. Another travels west with speed $2u/\sqrt{3}$, and the apparent wind direction is from $30°$ east of north. Find the true wind velocity.

9.18 r is the position vector $(2, 3, 1)$, and $a = (1, 1, 2)$ is a general vector. R is the position vector defined by $R = a + 2r$. Find the coordinates of the terminal point of R.

9.19 Find the angle θ, where $0 \leqslant \theta \leqslant 180°$, made by the position vector r with the *positive directions* of the axes Ox, Oy, Oz in the following cases: (a) $r = (1, 0, 0)$, (b) $r = (0, 1, 1)$, (c) $r = (0, 0, -1)$, (d) $r = (1, 1, 1)$, (e) $r = (1, 1, -1)$.

9.20 $P : (1, 1, 0), Q : (1, 1, 1)$, and $R : (1, 2, 1)$ are three of the vertices of a parallelogram with sides PQ and PR. Use vector methods to find the coordinates of (a) the fourth vertex, S, (b) the midpoint of PS, (c) the midpoint of QR. Show that (b) and (c) have the same coordinates (it is where the diagonals intersect).

Find the midpoints A, B, C, D of the four sides PR, RS, SQ, QP respectively. Show that $ABCD$ is a parallelogram.

9.21 Show that the points $A : (1, 2, -1), B : (3, 3, -2)$, and $C : (-3, 0, 1)$ are collinear (lie on a straight line), by considering the vectors $\overline{AB}$ and $\overline{AC}$ (or any other two combinations of A, B, and C). (a) Find which point is between the other two. (b) Find any other point on the line. (c) Show that the points $x = 2\lambda + 1, y = \lambda + 2, z = -\lambda - 1$, where λ is a parameter which may take any value, all lie on the line (these are parametric equations for the line).

9.22 Two points A and B have position vectors a and b respectively. In terms of a and b find the position vectors of the following points on the straight line passing through A and B: (a) the midpoint C of AB; (b) a point U between A and B for which $AU/UB = 1/3$; (c) a point V for which $AV/VB = 1/3$, but for which V does not lie between A and B.

9.23 Suppose that λ is a number such that $0 < \lambda < 1$. Find two points, U and V, on the line through A and B such that (a) $AU/UB = \lambda$ and U is between A and B. (b) $AV/BV = \lambda$ and V is not between A and B. (c) What is the case if $\lambda > 1$?

9.24 (a) Obtain a vector parametric equation for the straight line which passes through the point $(1, 4, 2)$ and is parallel to the line joining the points $(2, 3, 4)$ and $(1, 2, 3)$. (b) As in Example 9.9, deduce a pair of simultaneous cartesian equations for the line. (c) Obtain the points where the line intersects the (x, y) plane and the (y, z) plane. (c) By using these two points, obtain another pair of cartesian equations for the line.

9.25 Suppose that P has position vector r, and $r = \lambda a + (1 - \lambda)b$, where λ is a parameter, and A, B are points with a, b as position vectors. Show that P describes a straight line. Indicate on a diagram the relative positions of A, B, P when $\lambda < 0$, $0 < \lambda < 1$, and $\lambda > 1$.

9.26 Find the cartesian equation of the planes passing through the following points: (a) $(1, 0, 1)$, $(0, 1, 0)$, $(0, 0, 1)$, (b) $(0, 0, 0)$, $(1, 2, -1)$, $(2, 2, 2)$.

9.27 Find the shortest distance from the origin of the line given in vector parametric form by $r = a + tb$, where $a = (1, 2, 3)$, $b = (1, 1, 1)$, and t is the parameter. (Hint: use a calculus method, with t as the independent variable.)

9.28 For each of the following cases find a unit vector which has the same direction as a, and a unit vector which has the opposite direction. (a) $a = (3, 4, 3)$, (b) $a = 2\hat{\imath} + 3\hat{\jmath} + 6\hat{k}$, (c) $a = (-1, -1, 2)$, (d) $a = \hat{\imath} - 2\hat{\jmath} + \hat{k}$, (e) $3\hat{\imath} - 6\hat{\jmath} + 3\hat{k}$.

9.29 Express in terms of $\hat{\imath}, \hat{\jmath}, \hat{k}$ the vectors whose initial and terminal points are respectively given by the following position vectors: (a) $\hat{\imath} + \hat{\jmath} + \hat{k}$ and $-2\hat{\imath} + 3\hat{\jmath} + 5\hat{k}$, (b) $\hat{\imath} + 2\hat{\jmath} - \hat{k}$ and $3\hat{\imath} - \hat{\jmath} - 2\hat{k}$. Find the length of the vector in each case.

9.30 Show that the line joining the points with position vectors $\hat{\imath} - \hat{\jmath} + 2\hat{k}$ and $2\hat{\imath} - 2\hat{\jmath} - 3\hat{k}$ intersects the z axis.

9.31 A set of two-dimensional position vectors is given by $r = a\hat{\imath} + b\hat{\jmath}$, where $|a| + |b| < 1$. Describe the shape of the region which includes all the points with these position vectors.

9.32 A set of position vectors is given by $r = a\hat{\imath} + b\hat{\jmath} + c\hat{k}$, where $|a| + |b| + |c| < 1$. Describe the shape of the region which includes all the points with these position vectors.

9.33 Suppose that a weightless framework supports N particles, which have masses m_i and are located at points with position vectors r_i where $i = 1, 2, 3, \ldots, N$. You may assume that the centre of mass is at the point with position vector $\tilde{r}$, where

$$\tilde{r} = \sum m_i r_i / \sum m_i.$$

Find the centre of mass of three particles of masses 1 kg, 2 kg, and 3 kg at the points $\hat{\imath} + \hat{\jmath} + 2\hat{k}$, $-2\hat{\imath} + 3\hat{\jmath} - 5\hat{k}$, and $3\hat{\jmath} + 2\hat{k}$.

9.34 Obtain a parametric vector equation for the line which is parallel to $\hat{\imath} + 2\hat{\jmath} - \hat{k}$ and which passes through the point with position vector $\hat{\imath} + \hat{\jmath} + \hat{k}$. Find the point of intersection of this line with the plane $x - y + z = -2$.

9.35 An aircraft flying with constant speed V is circling horizontally at height H above an airfield which lies in the (x, y) plane. Its motion is in the clockwise direction when viewed from below. The centre of its circular path is at $P\hat{\imath} + H\hat{k}$, and at time $t = 0$ it is at the point $(P + R)\hat{\imath} + H\hat{k}$. Find the position vector for the aircraft at time t.

9.36 a and b are two position vectors. Find in terms of a and b a position vector which bisects the angle between them.

9.37 An aircraft A is flying along a path given by the position vector $0.41\hat{\imath} + 148t\hat{\jmath} + 0.99\hat{k}$, where t is the time in hours, and distance is in km. Another aircraft, B, takes off from an airfield at the origin O at time $t = 0$ and follows the path given by the position vector $100t\hat{\imath} + 250t\hat{\jmath} + 250t\hat{k}$. (a) Show that A and B are moving along straight lines at constant speeds, and find the speeds. (b) Show that a near miss between A and B will occur, and find the time that this happens.

9.38 Two moving points A and B have position vectors $r_A(t) = x_A(t)\hat{\imath} + y_A(t)\hat{\jmath} + z_A(t)\hat{k}$ and $r_B(t) = x_B(t)\hat{\imath} + y_B(t)\hat{\jmath} + z_B(t)\hat{k}$ respectively, which depend on the time t. (a) Show that the velocity of B relative to A is $dr_B(t)/dt - dr_A(t)/dt$. (b) Suppose that the two points are $A : (t, - t^2, t)$ and $B : (t^3, 2t^2, 1 + 3t)$. Find the velocity of B relative to A and the velocity of A relative to B. (c) Find the time t at which the relative speed is a minimum.

9.39 A particle describes an elliptical plane path with position vector $r = \hat{\imath}a \cos \omega t + \hat{\jmath}b \sin \omega t$, where t is time and ω, a, b are constants. Show that the acceleration is always directed towards the centre.

9.40 The position vector of a particle is given in polar coordinates by $r = \sec t$, $\theta = t$. Sketch the path for $0 \leqslant t < \frac{1}{2}\pi$. Find the radial and transverse components of acceleration.

9.41 The position vector of a particle P is given by

$$r = \hat{\imath}a \cos \omega t \sin \nu t + \hat{\jmath} \sin \omega t \sin \nu t + \hat{k}a \cos \nu t,$$

where a, ω, ν are constants and t is time. Show that P moves on a sphere of radius a. Find the velocity of the particle and show that its magnitude is $a(\nu^2 + \omega^2 \sin^2 \nu t)^{\frac{1}{2}}$. Deduce that the minimum speed occurs at the highest and lowest points of the sphere, and find where the maximum occurs.

10 The scalar product

CONTENTS

10.1 The scalar product of two vectors

Suppose that in component form

$$a = (a_1, a_2, a_3) \quad \text{and} \quad b = (b_1, b_2, b_3).$$

The **dot product** or **scalar product** of a and b is denoted by a dot and is *defined* by

$$a \cdot b = a_1 b_1 + a_2 b_2 + a_3 b_3.$$

(It is necessary to write the dot, because there is also another form of product, called the vector product.) The dot product is not a vector, but an ordinary number, or a **scalar** quantity. Some simple properties are:

Scalar or dot product

Definition: Let $a = (a_1, a_2, a_3)$ and $b = (b_1, b_2, b_3)$.
Then

$$a \cdot b = a_1 b_1 + a_2 b_2 + a_3 b_3.$$

(a) $a \cdot b = b \cdot a$ (commutative property).
(b) $a \cdot (b + c) = a \cdot b + a \cdot c$ (distributive property).
(c) Connection with the magnitude $|a|$:

$$a \cdot a = a_1^2 + a_2^2 + a_3^2 = |a|^2.$$

(For two dimensions, omit the third component.)   **(10.1)**

Example 10.1   *Find* $(a - b) \cdot (a + b)$ *when* $a = (-1, 0, 1)$ *and* $b = (2, 3, 2)$.

$a - b = (-3, -3, -1)$ and $a + b = (1, 3, 3)$.
Therefore

$$(a - b) \cdot (a + b) = (-3 \times 1) + (-3 \times 3) + (-1 \times 3) = -15.$$

Example 10.2   *Prove that* $(a - b) \cdot (a + b) = |a|^2 - |b|^2$.

Use the rules in (10.1) to proceed as in ordinary algebra:

$$(a - b) \cdot (a + b) = (a - b) \cdot a + (a - b) \cdot b = a \cdot a - b \cdot a + a \cdot b - b \cdot b$$
$$= a \cdot a - b \cdot b = |a|^2 - |b|^2 \quad \text{(by (10.1c))}.$$

10.2 The angle between two vectors

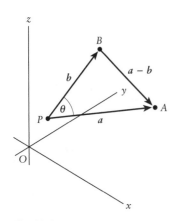

Fig. 10.1

In Fig. 10.1 we show two vectors, a and b, in three dimensions. Their initial points coincide at P. By the angle θ between a and b we mean the angle θ in the plane of a and b as shown: the angle chosen is the one which is **in the range 0° to 180°** (i.e. we refer to the internal angle, and do not use negative angles).

By the triangle rule

$$\overline{BA} = a - b,$$

and the lengths of the sides of the triangle ABP are given by

$$|\overline{PA}| = |a|, |\overline{PB}| = |b|, |\overline{BA}| = |a - b|.$$

The cosine rule (Appendix B) says that

$$BA^2 = PA^2 + PB^2 - 2PA \cdot PB \cos \theta,$$

or $\quad |a - b|^2 = |a|^2 + |b|^2 - 2|a||b| \cos \theta.$ \hfill (10.2)

But from (10.1c), putting $a - b$ in place of a,

$$|a - b|^2 = (a - b) \cdot (a - b) = a \cdot a + b \cdot b - 2a \cdot b,$$

or $\quad |a - b|^2 = |a|^2 + |b|^2 - 2a \cdot b.$ \hfill (10.3)

By comparing (10.2) with (10.3) we obtain

$$a \cdot b = |a||b| \cos \theta,$$

or $\quad \cos \theta = a \cdot b / |a||b|.$

Angle between two vectors

Let $\theta, 0° \leqslant \theta \leqslant 180°$, be the angle between the directions of a and b. Then

(a) $a \cdot b = |a||b| \cos \theta.$
(b) $\cos \theta = a \cdot b / |a||b|,$

or $\theta = \arccos(a \cdot b / |a||b|)$ (a calculator gives this angle uniquely in the range 0° to 180°). \hfill (10.4)

If a and b are not at the same point to start with we may still refer to θ as being the angle between them. The result (10.4b) can also be written in the form $\cos \theta = \hat{a} \cdot \hat{b}$ where $\hat{a}$ and $\hat{b}$ are the *unit* vectors in the directions of a and b.

Example 10.3   *Given three points* $A : (1, 1, 1)$, $B: (3, 2, 3)$, *and* $C : (0, -1, 1)$, *find the angle* θ *between* $\overline{CA}$ *and* $\overline{CB}$.

Put $\overline{CA} = a$, $\overline{CB} = b$.

Then $a = (1, 1, 1) - (0, -1, 1) = (1, 2, 0)$,

$\qquad b = (3, 2, 3) - (0, -1, 1) = (3, 3, 2)$.

$\qquad |a| = \sqrt{[1^2 + 2^2 + 0^2]} = \sqrt{5}$ and $|b| = \sqrt{[3^2 + 3^2 + 2^2]} = \sqrt{22}$.

$\qquad a \cdot b = (1 \times 3) + (2 \times 3) + (0 \times 2) = 9$.

From (10.4),

$$\cos \theta = \frac{a \cdot b}{|a||b|} = \frac{9}{\sqrt{110}} = 0.858.$$

Finally $\theta = 30.9°$

10.3 Perpendicular vectors

Cases when vectors are perpendicular or **orthogonal** are particularly important. The condition is that $\cos \theta = 0$.

Example 10.4   *Show that the vectors* $a = (1, 2, 3)$ *and* $b = (-5, 1, 1)$ *are perpendicular.*

We have

$$\cos \theta = \frac{a \cdot b}{|a||b|}$$

and

$\qquad a \cdot b = (1, 2, 3) \cdot (-5, 1, 1) = -5 + 2 + 3 = 0$.

Therefore

$\qquad \theta = 90°$, by (10.4).

From (10.4) the condition for two vectors to be perpendicular may be expressed as follows:

> **Perpendicular vectors**
>
> If a and b are nonzero vectors, they are perpendicular if
> $$a \cdot b = a_1 b_1 + a_2 b_2 + a_3 b_3 = 0.$$
> $\hfill$ (10.5)

The basis vectors $\hat{\imath}, \hat{\jmath}, \hat{k}$ are perpendicular, so

$$\hat{\imath}\cdot\hat{\jmath}=\hat{\jmath}\cdot\hat{k}=\hat{k}\cdot\hat{\imath}=0. \tag{10.6}$$

Also, they have unit magnitude, so by (10.1c),

$$\hat{\imath}\cdot\hat{\jmath}=\hat{\jmath}\cdot\hat{\jmath}=\hat{k}\cdot\hat{k}=1. \tag{10.7}$$

Suppose that $a = (a_1, a_2, a_3)$ in component form. Then

$$\hat{\imath}\cdot a = \hat{\imath}\cdot(a_1\hat{\imath}+a_2\hat{\jmath}+a_3\hat{k})$$
$$= a_1\hat{\imath}\cdot\hat{\imath}+a_2\hat{\imath}\cdot\hat{\jmath}+a_3\hat{\imath}\cdot\hat{k} = a_1,$$

from (10.6) and (10.7). The component a_1 is therefore picked out by scalar multiplication by $\hat{\imath}$. Similarly,

$$\hat{\jmath}\cdot a = a_2, \quad \hat{k}\cdot a = a_3.$$

We can therefore write any vector in the form

$$a = (\hat{\imath}\cdot a)\hat{\imath} + (\hat{\jmath}\cdot a)\hat{\jmath} + (\hat{k}\cdot a)\hat{k}.$$

(Remember that $\hat{\imath}\cdot a, \hat{\jmath}\cdot a$, and $\hat{k}\cdot a$ are ordinary numbers.)

Scalar products of $\hat{\imath}, \hat{\jmath}, \hat{k}$

(a) $\hat{\imath}\cdot\hat{\imath}=\hat{\jmath}\cdot\hat{\jmath}=\hat{k}\cdot\hat{k}=1;$
 $\hat{\imath}\cdot\hat{\jmath}=\hat{\jmath}\cdot\hat{k}=\hat{k}\cdot\hat{\imath}=0.$

(b) The components of any vector a are given by
 $a_1=\hat{\imath}\cdot a, a_2=\hat{\jmath}\cdot a, a_3=\hat{k}\cdot a.$ (10.8)

Example 10.5 *Find the numbers α, β, and γ which make the vectors*
$$a = \alpha\hat{\imath}+\hat{\jmath}+2\hat{k}, b = \hat{\imath}+\beta\hat{\jmath}-\hat{k}, c = \hat{\imath}-\hat{\jmath}+\gamma\hat{k}$$
mutually perpendicular.

We require that $a\cdot b = b\cdot c = c\cdot a = 0$.
$$a\cdot b = (\alpha\hat{\imath}+\hat{\jmath}+2\hat{k})\cdot(\hat{\imath}+\beta\hat{\jmath}-\hat{k}) = \alpha+\beta-2 = 0,$$
$$b\cdot c = (\hat{\imath}+\beta\hat{\jmath}-\hat{k})\cdot(\hat{\imath}-\hat{\jmath}+\gamma\hat{k}) = 1-\beta-\gamma = 0,$$
$$c\cdot a = (\hat{\imath}-\hat{\jmath}+\gamma\hat{k})\cdot(\alpha\hat{\imath}+\hat{\jmath}+2\hat{k}) = \alpha-1+2\gamma = 0.$$
Therefore α, β, γ must satisfy

$$\alpha+\beta \quad = 2, \tag{i}$$
$$-\beta-\gamma = -1, \tag{ii}$$
$$\alpha \quad +2\gamma = 1. \tag{iii}$$

Substitute α from (i) and γ from (ii) into (iii) to give

$$(2-\beta)+2(1-\beta) = 1,$$

so that $\beta = 1$. From (ii), $\gamma = 1-\beta = 0$, and from (i), $\alpha = 2-\beta = 1$. Therefore the required vectors are

$$a = \hat{\imath}+\hat{\jmath}+2\hat{k}, b = \hat{\imath}+\hat{\jmath}-\hat{k}, c = \hat{\imath}-\hat{\jmath}.$$

(a)

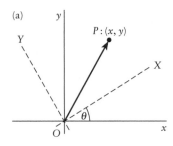

(b)

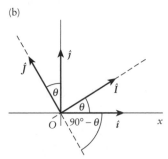

Fig. 10.2 (a) Change of axes in two dimensions. (b) The associated unit vectors.

10.4 Rotation of axes in two dimensions

In Fig. 10.2a, P is a point which has coordinates (x, y) in the axes Ox, Oy. OX, OY is another set of axes, rotated relatively to the first set by an angle θ. The positive direction for θ is anticlockwise, and θ may lie in the range $\pm 180°$ so as to cover all possibilities, like a polar angle. The unit basis vectors in the axes OX, OY are $\hat{I}$ and $\hat{J}$ respectively. The problem is to find the coordinates (X, Y) of P in the new axes.

We can express $\hat{i}$ and $\hat{j}$ in terms of $\hat{I}$ and $\hat{J}$. From Fig. 10.2b, their components in the X, Y axes are

$$\hat{i} = (\cos\theta, -\sin\theta), \quad \hat{j} = (\sin\theta, \cos\theta).$$

Therefore

$$\hat{i} = \hat{I}\cos\theta - \hat{J}\sin\theta,$$

and $\hat{j} = \hat{I}\sin\theta + \hat{J}\cos\theta.$

The position of P in space does not change when we change axes, so in terms of the new axes

$$\begin{aligned} X\hat{I} + Y\hat{J} &= x\hat{i} + y\hat{j} \\ &= x(\hat{I}\cos\theta - \hat{J}\sin\theta) + y(\hat{I}\sin\theta + \hat{J}\cos\theta) \\ &= (x\cos\theta + y\sin\theta)\hat{I} + (-x\sin\theta + y\cos\theta)\hat{J}. \end{aligned}$$

Finally, by equating the coefficients of $\hat{I}$ and $\hat{J}$, we obtain the result (10.9a):

> **Rotation of axes in two dimensions**
>
> Given axes inclined at θ as in Fig. 10.2a, the coordinates x, y and X, Y are related by
>
> (a) $X = x\cos\theta + y\sin\theta,$
> $Y = -x\sin\theta + y\cos\theta.$
> (b) $x = X\cos\theta - Y\sin\theta,$
> $y = X\sin\theta + Y\cos\theta.$
>
> (10.9)

The inverse relation (10.9b) can be obtained by solving the equations in (10.9a) for x and y; or by interchanging x, y and X, Y in (a) and putting $(-\theta)$ in place of θ.

10.5 Direction cosines

Figure 10.3 shows a position vector $r = \overline{OP}$, where P is the point (a, b, c), so

$$\overline{OP} = r = (a, b, c).$$

The angles between r and $\hat{i}, \hat{j}, \hat{k}$ respectively (chosen for definiteness between $0°$ and $180°$, as for θ in Section 10.2) are α, β, γ. These angles

Fig. 10.3 Angles made by $\overline{OP}$ with the axes.

specify the **direction** of r uniquely. It is convenient to use not the angles themselves, but their cosines, which are normally indicated by l, m, n:

$$l = \cos \alpha, m = \cos \beta, n = \cos \gamma.$$

These are called the **direction cosines** of r, and also specify the direction of r uniquely.

Referring to Fig. 10.3,

$$|r| = \sqrt{(a^2 + b^2 + c^2)}.$$

Also

$$l = \cos \alpha = a/|r|, m = \cos \beta = b/|r|, n = \cos \gamma = c/|r|.$$

Therefore

$$l^2 + m^2 + n^2 = \cos^2\alpha + \cos^2\beta + \cos^2\gamma$$
$$= (a^2 + b^2 + c^2)/|r|^2 = 1.$$

The vector given in component form by

$$(\cos \alpha, \cos \beta, \cos \gamma) = (l, m, n)$$

is therefore the **unit vector which specifies the direction of** r.

Now let s be a vector having any magnitude and location, but pointing in the same direction as r. Then s and r have the same inclinations α, β, γ to the axes, and $\cos \alpha, \cos \beta, \cos \gamma$, the direction cosines of s, are the same. To summarize:

Direction cosines l, m, n of any vector s

If the angles between s and Ox, Oy, Oz, are α, β, γ, respectively, in the range $0°$ to $180°$, then

$$l = \cos \alpha, m = \cos \beta, n = \cos \gamma$$

are the direction cosines of s.

(a) Any vector parallel to s with the same sense has the same direction cosines l, m, n.
(b) $l^2 + m^2 + n^2 = \cos^2\alpha + \cos^2\beta + \cos^2\gamma = 1$.
(c) $\hat{s} = (l, m, n)$ is a unit vector in the direction of s. **(10.10)**

Example 10.6 *Obtain the direction cosines of the vector* $s = \hat{\imath} + 2\hat{\jmath} - 2\hat{k}$. *Find the angles between s and the coordinate axes.*

The components of s are $(1, 2, -2)$, so its length is given by

$$s = \sqrt{[1^2 + 2^2 + (-2^2]} = 3.$$

Therefore the unit vector $\hat{s}$ has components

$$l = \tfrac{1}{3}, m = \tfrac{2}{3}, n = -\tfrac{2}{3}.$$

The corresponding angles in the range $0°$ to $180°$ are $\alpha = \arccos \tfrac{1}{3} = 70.5°$, $\beta = \arccos \tfrac{2}{3} = 48.2°$, $\gamma = \arccos(-\tfrac{2}{3}) = 131.8°$.

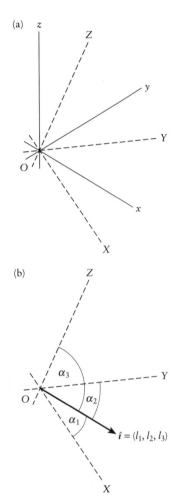

Fig. 10.4 (a) Change of axes in three dimensions. (b) Angles between $\hat{\imath}$ and the X, Y, Z axes.

10.6 Rotation of axes in three dimensions

Figure 10.4a shows two sets of axes $Oxyz$ and $OXYZ$, with the same origin O. The basis vectors are respectively $\hat{\imath}, \hat{\jmath}, \hat{k}$ and $\hat{I}, \hat{J}, \hat{K}$. We shall show how to change from one set of axes to the other, as we did in Section 10.4 in two dimensions.

The three components of any *unit* vector are equal to its three direction cosines. $\hat{I}, \hat{J},$ are $\hat{K}$ are unit vectors, so in the axes $Oxyz$ let them be given in terms of their direction cosines by

$$\hat{I} = (l_1, m_1, n_1), \hat{J} = (l_2, m_2, n_2), \hat{K} = (l_3, m_3, n_3). \tag{10.11a}$$

By inverting our view of the two sets of axes, we can also specify the components of $\hat{\imath}, \hat{\jmath}, \hat{k}$ in the axes $OXYZ$:

$$\hat{\imath} = (l_1, l_2, l_3), \hat{\jmath} = (m_1, m_2, m_3), \hat{k} = (n_1, n_2, n_3) \tag{10.11b}$$

(this is illustrated for the case of $\hat{\imath}$ in Fig. 10.4b).

Next, suppose that a *fixed point P* has position vector

$$r = (x, y, z) = x\hat{\imath} + y\hat{\jmath} + z\hat{k}$$

in the axes $Oxyz$. We need to find the components of r in the axes $OXYZ$. By substituting

$$\hat{\imath} = l_1\hat{I} + l_2\hat{J} + l_3\hat{K}, \text{ etc.}$$

from (10.11b) into r, we obtain

$$r = (l_1 x + m_1 y + n_1 z)\hat{I} + (l_2 x + m_2 y + n_2 z)\hat{J} + (l_3 x + m_3 y + n_3 z)\hat{K}.$$

The $OXYZ$ coordinates of the point P are therefore

$$(l_1 x + m_1 y + n_1 z, l_2 x + m_2 y + n_2 z, l_3 x + m_3 y + n_3 z). \tag{10.12a}$$

The inverse relation is obtained in a similar way. Given a fixed point Q, with coordinates (X, Y, Z) in the axes $OXYZ$ and position vector R, then

$$R = (X, Y, Z) = X\hat{I} + Y\hat{J} + Z\hat{K}.$$

Now use (10.11a) to show that R is given in the axes $Oxyz$ by

$$R = (l_1 X + l_2 Y + l_3 Z)\hat{\imath} + (m_1 X + m_2 Y + m_3 Z)\hat{\jmath} \\ + (n_1 X + n_2 Y + n_3 Z)\hat{k}.$$

The coordinates of Q in $Oxyz$ are therefore

$$(l_1 X + l_2 Y + l_3 Z, m_1 X + m_2 Y + m_3 Z, n_1 X + n_2 Y + n_3 Z). \tag{10.12b}$$

In matrix form, the coordinates in the two systems are related as follows:

> **Rotation of axes; three dimensions**
>
> $\hat{I} = (l_1, m_1, n_1), \hat{J} = (l_2, m_2, n_2), \hat{K} = (l_3, m_3, n_3)$ are the basis vectors for axes $OXYZ$, referred to axes $Oxyz$ (the components being direction cosines). Then
>
> (a) $\begin{bmatrix} X \\ Y \\ Z \end{bmatrix} = \begin{bmatrix} l_1 & m_1 & n_1 \\ l_2 & m_2 & n_2 \\ l_3 & m_3 & n_3 \end{bmatrix} \begin{bmatrix} x \\ y \\ z \end{bmatrix}$
>
> (b) $\begin{bmatrix} x \\ y \\ z \end{bmatrix} = \begin{bmatrix} l_1 & l_2 & l_3 \\ m_1 & m_2 & m_3 \\ n_1 & n_2 & n_3 \end{bmatrix} \begin{bmatrix} X \\ Y \\ Z \end{bmatrix}.$
>
> (10.13)

The matrix of direction cosines in (b) is the inverse of the matrix in (a).

Example 10.7 *In axes Ox, Oy, Oz, $\hat{I} = (\frac{1}{3}, -\frac{2}{3}, \frac{2}{3})$, $\hat{J} = (\frac{2}{3}, -\frac{1}{3}, -\frac{2}{3})$. $\hat{K} = (\frac{2}{3}, \frac{2}{3}, \frac{1}{3})$ are perpendicular unit vectors which are basis vectors for a new set of axes. Find the new coordinates of the point $P : (-3, -3, 3)$.*

From (10.13), the new coordinates are

$$X = \tfrac{1}{3}(-3) - \tfrac{2}{3}(-3) + \tfrac{2}{3}(3) = \ \ 3,$$
$$Y = \tfrac{2}{3}(-3) - \tfrac{1}{3}(-3) - \tfrac{2}{3}(3) = -3,$$
$$Z = \tfrac{2}{3}(-3) + \tfrac{2}{3}(-3) + \tfrac{1}{3}(3) = -3.$$

Example 10.8 (a) *Confirm that the matrices*

$$\begin{bmatrix} l_1 & m_1 & n_1 \\ l_2 & m_2 & n_2 \\ l_3 & m_3 & n_3 \end{bmatrix} \text{ and } \begin{bmatrix} l_1 & l_2 & l_3 \\ m_1 & m_2 & m_3 \\ n_1 & n_2 & n_3 \end{bmatrix}$$

are inverse matrices, where $\hat{I} = (l_1, m_1, n_1), \hat{J} = (l_2, m_2, n_2)$, and $\hat{K} = (l_3, m_3, n_3)$ are mutually perpendicular unit vectors.

(b) *Find the equation of the plane $3x + 3y + 3z = 1$ in the new axes, using the basis vectors given in Example 10.7.*

(a) Multiply the two matrices. The diagonal elements are

$$l_1^2 + l_2^2 + l_3^2, \quad m_1^2 + m_2^2 + m_3^2, \quad n_1^2 + n_2^2 + n_3^2,$$

all of which are equal to unity since $\hat{I}, \hat{J}, \hat{K}$ are unit vectors. The other elements have the typical form

$$l_1 l_2 + m_1 m_2 + n_1 n_2 = (l_1, m_1, n_1) \cdot (l_2, m_2, n_2).$$

All of these are zero because $\hat{I}, \hat{J}, \hat{K}$ are mutually perpendicular. Therefore, the product is the unit matrix.

(b) In this case we need x, y, z in terms of X, Y, Z. The equations corresponding to (10.13b) are

Example 10.8 *continued*

$$x = l_1 X + l_2 Y + l_3 Z = \tfrac{1}{3}X + \tfrac{2}{3}Y + \tfrac{2}{3}Z,$$
$$y = m_1 X + m_2 Y + m_3 Z = -\tfrac{2}{3}X - \tfrac{1}{3}Y + \tfrac{2}{3}Z,$$
$$z = n_1 X + n_2 Y + n_3 Z = \tfrac{2}{3}X - \tfrac{2}{3}Y + \tfrac{1}{3}Z.$$

In the new coordinates,

$$3x + 3y + 3z = (X + 2Y + 2Z) + (-2X - Y + 2Z) + (2X - 2Y + Z)$$
$$= X - Y + 5Z.$$

Therefore, the plane has the new equation

$$X - Y + 5Z = 1.$$

10.7 Direction ratios and coordinate geometry

In ordinary three-dimensional coordinate geometry the **inclination of a straight line** is specified without distinguishing between the two possible directions along the line. The method used (in vector terms) is equivalent to specifying the three **components of any vector s that is parallel to the line**. The length of s, and its direction forwards or backwards along the line, are immaterial. If

$$s = p\hat{\imath} + q\hat{\jmath} + r\hat{k} = (p, q, r)$$

is parallel to the line, then the triplet of numbers p, q, r is called a set of **direction ratios** for the line. Alternatively, if AB is any segment of the straight line, then the projections of AB on to Ox, Oy, Oz are a set of direction ratios for the line.

Any multiple of p, q, r, say $\lambda p, \lambda q, \lambda r$, is also a set of direction ratios for the line, because it corresponds to a parallel vector $s_1 = \lambda p\hat{\imath} + \lambda q\hat{\jmath} + \lambda r\hat{k}$. For example, if $s = 2\hat{\imath} + 3\hat{\jmath} + 6\hat{k}$ is parallel to a given line, then 2, 3, 6 and 6, 9, 18 are both sets of direction ratios for the line. So are $-2, -3, -6$, corresponding to the vector $(-1)s$, although it points in the opposite direction.

By putting $\lambda = \pm\tfrac{1}{7}$ we obtain the direction ratios $\pm\tfrac{2}{7}, \pm\tfrac{3}{7}, \pm\tfrac{6}{7}$ corresponding to the unit vectors $\pm\hat{s}$. These are also direction cosines for $\pm s$, from which the angles made with the directions of the axes can be obtained.

Example 10.9   *Find the angles made with Ox, Oy, and Oz by a line with direction ratios $2, 3, -6$.*

Put $s = 2\hat{\imath} + 3\hat{\jmath} - 6\hat{k}$: this is parallel to the line. Since $|s| = 7$, the corresponding unit vector $\hat{s}$ is given by

$$\hat{s} = \tfrac{1}{7}s = \tfrac{2}{7}\hat{\imath} + \tfrac{3}{7}\hat{\jmath} - \tfrac{6}{7}\hat{k} = \hat{\imath}\cos\alpha + \hat{\jmath}\cos\beta - \hat{k}\cos\gamma,$$

where $\cos\alpha$, $\cos\beta$, $\cos\gamma$ are its direction cosines. Therefore, the inclination of the line is specified by the angles

$$\alpha = \arccos\tfrac{2}{7} = 73.4°, \quad \beta = \arccos\tfrac{3}{7} = 64.6°, \quad \gamma = \arccos(-\tfrac{6}{7}) = 149°.$$

Example 10.10 (*Two dimensions.*) *Find a set of direction ratios for the straight line* $y = 2x + 1$.

We are looking for any vector which is parallel to the line. The points $A : (0, 1)$ and $B : (1, 3)$ lie on the line, so the vector $s = \overline{AB}$ given by

$$s = \overline{OB} - \overline{OA} = (\hat{i} + 3\hat{j}) - \hat{j} = \hat{i} + 2\hat{j}$$

is parallel to the line. Therefore one set of direction ratios is given by the numbers 1, 2.

Example 10.11 (*Two dimensions.*) *Find parametric and cartesian equations for the straight line through the point* $A : (a, b)$, *which has direction ratios* p, q.

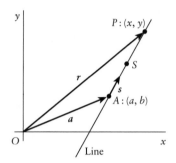

Fig. 10.5

In Fig. 10.5, A is the point with position vector $a = a\hat{i} + b\hat{j}$, and $s = p\hat{i} + q\hat{j}$. P is a general point on the line with position vector $r = x\hat{i} + y\hat{j}$, and $s = p\hat{i} + q\hat{k} = \overline{AS}$.

$$r = \overline{OA} + \overline{AP},$$

and $\overline{AP}$ is some multiple of s, say:

$$\overline{AP} = \lambda s.$$

Therefore

$$r = a + \lambda s, \tag{i}$$

where λ is a parameter. This is a parametric vector equation for the line.
By equating corresponding components we have

$$x = a + \lambda p, \ y = b + \lambda q, \tag{ii}$$

and these are parametric cartesian equations.
Now eliminate the parameter between the equations (ii):

$$(x - a)/p = (y - b)/q.$$

This is a cartesian equation, which could be reduced to the standard form $y = mx + c$.

Direction ratios of a straight line

Definition: if $p\hat{i} + q\hat{j} + r\hat{k}$ is parallel to the line, then p, q, r (or any multiple $\lambda p, \lambda q, \lambda r$) is a set of direction ratios for the line.

(a) The angles α, β, γ made with Ox, Oy, Oz are obtained from the equations

$$\cos \alpha = p/k, \ \cos \beta = q/k, \ \cos \gamma = r/k,$$

where $k = \sqrt{(p^2 + q^2 + r^2)}$.
(For two dimensions, suppress the third component.) **(10.14)**

10.8 Properties of a plane

Figure 10.6 shows a plane which passes through a given point $A : (a_1, a_2, a_3)$, and is perpendicular to a line CD having direction ratios p, q, r. We shall obtain equations for the plane.

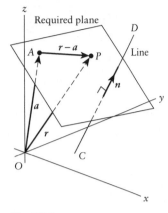

Fig. 10.6

The position vector of A is $\boldsymbol{a}$ given by

$$a = \overline{OA} = a_1\hat{\boldsymbol{i}} + a_2\hat{\boldsymbol{j}} + a_3\hat{\boldsymbol{k}}. \tag{10.15}$$

From (10.14) the vector $\boldsymbol{n}$ given by

$$n = p\hat{\boldsymbol{i}} + q\hat{\boldsymbol{j}} + r\hat{\boldsymbol{k}} \tag{10.16}$$

is parallel to the line CD, so $\boldsymbol{n}$ is also perpendicular to the plane ($\boldsymbol{n}$ is called a **normal to the plane**). $P : (x, y, z)$ represents an arbitrary point on the plane, with position vector $\boldsymbol{r}$ given by

$$r = \overline{OP} = x\hat{\boldsymbol{i}} + y\hat{\boldsymbol{j}} + z\hat{\boldsymbol{k}}. \tag{10.17}$$

By the triangle rule, $\overline{AP} = \boldsymbol{r} - \boldsymbol{a}$, and $\boldsymbol{n}$ must be perpendicular to $\overline{AP}$; therefore

$$\boldsymbol{n} \cdot (\boldsymbol{r} - \boldsymbol{a}) = 0,$$

or $\quad \boldsymbol{n} \cdot \boldsymbol{r} = \boldsymbol{n} \cdot \boldsymbol{a}. \tag{10.18}$

This is a vector equation for the plane.

By substituting for $\boldsymbol{n}$, $\boldsymbol{r}$, and $\boldsymbol{a}$ from (10.15, 16, and 17) we obtain the cartesian equation

$$px + qy + rz = pa_1 + qa_2 + ra_3.$$

Now suppose we start with an equation in the form

$$ax + by + cz = d, \tag{10.19}$$

such as $2x + 7y - 5z = 3$. We shall show how it can be written in the form (10.18). Put

$$r = x\hat{\boldsymbol{i}} + y\hat{\boldsymbol{j}} + z\hat{\boldsymbol{k}}, \quad \text{and} \quad p = a\hat{\boldsymbol{i}} + b\hat{\boldsymbol{j}} + c\hat{\boldsymbol{k}}.$$

Then (10.19) can be written in the form

$$\boldsymbol{p} \cdot \boldsymbol{r} = d. \tag{10.20}$$

Now let $A : (a_1, a_2, a_3)$ be any point we like that satisfies the equation (10.19), and put $\boldsymbol{a} = a_1\hat{\boldsymbol{i}} + a_2\hat{\boldsymbol{j}} + a_3\hat{\boldsymbol{k}}$. From (10.20), this means that

$$\boldsymbol{p} \cdot \boldsymbol{a} = d.$$

Therefore (10.19) can be written

$$\boldsymbol{p} \cdot \boldsymbol{r} = \boldsymbol{p} \cdot \boldsymbol{a}. \tag{10.21}$$

This is like (10.18). Therefore (10.19) represents a plane, the plane passes through the point with position vector $\boldsymbol{a}$, and $\boldsymbol{p}$ is perpendicular to the plane:

> **Vector equation of a plane**
>
> (a) A vector equation for a plane through the point a perpendicular to a vector n is $n \cdot r = n \cdot a$. Also, $n \cdot r = \text{constant}$ represents a plane perpendicular to n.
> (b) $ax + by + cz = d$ always represents a plane.
> (c) $p = a\hat{\imath} + b\hat{\jmath} + c\hat{k}$ is perpendicular to the plane $ax + by + cz = d$.
>
> (10.22)

Example 10.12 *Show that the plane $3x - 2z = 1$ is parallel to the y axis.*

By (10.22c) the vector $p = 3\hat{\imath} - 2\hat{k}$ is perpendicular to the plane. Also

$$\hat{\jmath} \cdot p = \hat{\jmath} \cdot (3\hat{\imath} - 2\hat{k}) = 3\hat{\jmath} \cdot \hat{\imath} - 2\hat{\jmath} \cdot \hat{k} = 0,$$

so p is perpendicular to $\hat{\jmath}$. Therefore the plane is parallel to $\hat{\jmath}$.

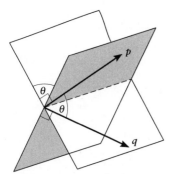

Fig. 10.7 The angle between the planes, θ, is equal to the angle between the normals p and q.

Example 10.13 *Find the angle of intersection between the two planes $2x + 3y + 4z = 5$ and $2x - 6y - 3z = 0$.*

From (10.22c), the vector $p = 2\hat{\imath} + 3\hat{\jmath} + 4\hat{k}$ is a normal to the first plane (i.e. it is perpendicular to it) and $q = 2\hat{\imath} - 6\hat{\jmath} - 3\hat{k}$ is a normal to the second plane.

   From Fig. 10.7, one of the angles between the two planes is equal to the standard angle θ (with $0° \leqslant \theta \leqslant 180°$) between the two normals, p and q. By (10.4a)

$$p \cdot q = |p||q| \cos \theta,$$

or $-26 = \sqrt{29} \times 7 \cos \theta$.

Therefore $\cos \theta = -26/(7\sqrt{29})$, so $\theta = 133.6°$.

Example 10.14 *Show that the planes $2x + 2y - z = 10$ and $3x - 2y + 2z = 0$ are perpendicular.*

The planes are perpendicular if their normal vectors are perpendicular. Taking the equations in order, by (10.22c) the vectors

$$p = 2\hat{\imath} + 2\hat{\jmath} - \hat{k} \quad \text{and} \quad q = 3\hat{\imath} - 2\hat{\jmath} + 2\hat{k}$$

are normal to the planes. Then

$$p \cdot q = 6 - 4 - 2 = 0,$$

so the planes are perpendicular.

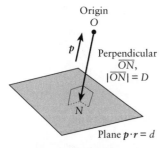

Fig. 10.8 The distance from the origin O to a plane.

   To find the distance D of the plane $ax + by + cz = d$ from the origin, consider Fig. 10.8. Drop a perpendicular $\overline{ON}$ from the origin O to the plane at N. The equation of the plane may be written

$$p \cdot r = d$$

where $p = a\hat{\imath} + b\hat{\jmath} + c\hat{k}$. Since N is a point on the plane,

$$p \cdot \overline{ON} = d$$

or, after dividing by $|p|$,

$$\hat{p} \cdot \overline{ON} = d/|p|,$$

where $\hat{p} = p/|p|$ is the corresponding unit vector. Also $\overline{ON} = \pm D\hat{p}$ depending on its sense, so

$$\pm D\hat{p} \cdot \hat{p} = d/|p|.$$

But $\hat{p} \cdot \hat{p} = 1$, and by taking the modulus we find D:

$$D = |d|/|p| = |d|/\sqrt{(a^2 + b^2 + c^2)}. \qquad (10.23)$$

Now let Q, position vector q, be any point, distance D_Q from the plane. Move the origin to Q, and let R denote the new general position vector measured from Q. Since $R = r - q$, the new equation of the plane is $p \cdot (R + q) = d$, or $p \cdot R = d - p \cdot q$. Therefore $d - p \cdot q$ is to be put in place of d in (10.23).

Distance of a point from a plane $ax + by + cz = d$
Put $a\hat{\imath} + b\hat{\jmath} + c\hat{k} = p$. Then
(a) Distance D of O from the plane:
$$D = |d|/|p| = |d|/\sqrt{(a^2 + b^2 + c^2)}.$$
(b) Distance D_Q of a point Q, position vector q:
$$D_Q = |p \cdot q - d|/|p|.$$

(10.24)

10.9 General equation of a straight line

Figure 10.9 shows the straight line through a point $A : (a_1, a_2, a_3)$ with position vector a. Its inclination is specified by direction ratios p, q, r. The vector

$$s = p\hat{\imath} + q\hat{\jmath} + r\hat{k}$$

is parallel to the line by (10.14), and is shown with its initial point at A, so that it lies along the line. $P : (x, y, z)$ is any point on the line, and has position vector r.

By the triangle rule

$$r = \overline{OA} + \overline{AP} = a + \overline{AP}.$$

$\overline{AP}$ is always some multiple λ of s:

$$\overline{AP} = \lambda s,$$

where λ is a parameter which may take any value, so finally

$$r = a + \lambda s$$

is a vector parametric equation for the line.

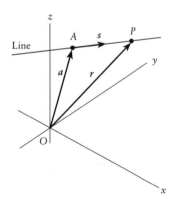

Fig. 10.9

In components this becomes

$$x = a_1 + \lambda p, \, y = a_2 + \lambda q, \, z = a_3 + \lambda r,$$

and these are parametric cartesian equations for the line.

Provided that none of p, q, r is zero the parameter λ can be eliminated by rearranging the equations

$$(x - a_1)/p = (y - a_2)/q = (z - a_3)/r. \tag{10.25}$$

This is the **cartesian equation of a straight line**, since any line passes through some point A and has some direction p, q, r. *This expression is not unique*, because (a_1, a_2, a_3) and p, q, r are not unique.

The equation (10.25) really consists of two simultaneous equations: for example, the pair

$$(x - a_1)/p = (y - a_2)/q \quad \text{and} \quad (y - a_2)/q = (z - a_3)/r.$$

These are the equations of two planes, and the line is their line of intersection.

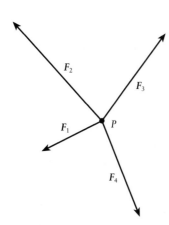

Fig. 10.10 Forces acting at a point P.

10.10 Forces acting at a point

The magnitude and direction of a force acting at a point in a body can be depicted by a directed line segment. This pictorial possibility does not automatically mean that forces behave like vectors: it must be established that the rules for combining vectors listed in Section 10.3 parallel the *experimental facts* of mechanics.

The analogy is a little different according to the physical situation. For the simplest case, Fig. 10.10 shows several forces, $F_1, F_2, \ldots$, acting on the same point P. P might be a single particle, or a single point fixed in a large body. It is ultimately an experimental fact that the forces have the same physical effect as a single force F, called the **resultant** of the forces shown, which acts at the same point and is obtained by vector addition:

$$F = F_1 + F_2 + \cdots. \tag{10.25}$$

A zero force has zero effect. Together with (10.25), this gives the condition for **equilibrium of a particle** under the influence of several forces $F_1, F_2, \ldots$: that the resultant force F must be zero, or

$$F = F_1 + F_2 + \cdots = 0. \tag{10.26}$$

The **magnitude** of a force (expressed in units such as newtons) is denoted by $|F|$, and is proportional to the length of the arrow that represents it. The component of a force in an arbitrary direction is illustrated in Fig. 10.11. Suppose the direction is indicated by the *unit* vector $\hat{s}$. Then

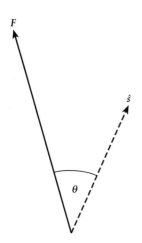

Fig. 10.11

$$\text{component of } F \text{ in direction } \hat{s} = F \cdot \hat{s} = |F| \cos \theta, \tag{10.27}$$

where θ is conveniently given a value between $0°$ and $180°$. Notice that if θ is between $90°$ and $180°$, then the component is negative. This agrees with the definition of vector components in the $\hat{\imath}, \hat{\jmath}, \hat{k}$ directions that we used before. The process of obtaining a component of F in a certain direction is often spoken of as **resolving F** in that direction.

Example 10.15   *Find the component of the force $F = 3\hat{\imath} + \hat{\jmath} + \hat{k}$ in the direction of the vector $s = 2\hat{\imath} + 3\hat{\jmath} + 6\hat{k}$.*

The unit direction vector s is given by
$$\hat{s} = s/|s| = (2\hat{\imath} + 3\hat{\jmath} + 6\hat{k})/7 = \tfrac{2}{7}\hat{\imath} + \tfrac{3}{7}\hat{\jmath} + \tfrac{6}{7}\hat{k}.$$
The component of F in this direction is given by
$$F \cdot \hat{s} = (3\hat{\imath} + \hat{\jmath} + \hat{k}) \cdot (\tfrac{2}{7}\hat{\imath} + \tfrac{3}{7}\hat{\jmath} + \tfrac{6}{7}\hat{k}) = 15/7.$$

In **two dimensions,** if the **components of a force F in any two non-parallel directions are zero, then F must be zero** (and conversely, of course). For suppose the angles made with the two directions are θ and ϕ, and they do not differ by $0°$ or $180°$. If

$$|F|\cos\theta = 0 \quad \text{and} \quad |F|\cos\phi = 0,$$

then $|F|$ must be zero, so $F = 0$. (One direction is not sufficient, since F might be perpendicular to that direction.) This principle, of 'resolving in two directions', is used frequently to solve problems. The following is a simple example.

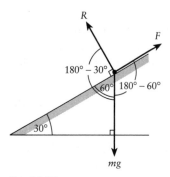

Fig. 10.12

Example 10.16   *Figure 10.12 represents a particle P at rest on a rough inclined plane of inclination $30°$. The forces acting on P are the force of gravity downwards of magnitude mg where g is the gravitational constant, the normal (perpendicular) reaction of the plane R, and the frictional force F. Find R and F.*

The arrows indicate provisional directions for the vectors R and F. The scalar quantities R and F attached to the arrows stand for the uknown *components of R and F in the assumed directions*, and these might not be postive numbers. This convention provides a safety net, for suppose we have, say, guessed the direction of F wrongly, and that it actually acts down the plane rather than up it. The mistake will do no harm, because F will simply turn out to be a negative number in our answer. This is a conventional way of lettering diagrams in mechanics.

  It is easiest to resolve in the assumed directions of F and R:

  in direction of F: $0 = F + mg\cos(180° - 60°)$

which is the same as

  $0 = F - mg\cos 60° = F - \tfrac{1}{2}mg$ (i)

  in direction of R: $0 = R + mg\cos(180° - 30°)$

Example 10.16 *continued*

which is the same as

$$0 = R - mg \cos 30° = R - \frac{\sqrt{3}}{2}mg. \tag{ii}$$

Therefore

$$F = \tfrac{1}{2}mg \quad \text{and} \quad R = \frac{\sqrt{3}}{2}mg.$$

(You would usually go straight for the commonsense way of writing the components given by (i) and (ii), avoiding the cosines of large angles.)

10.11 Curvature in two dimensions

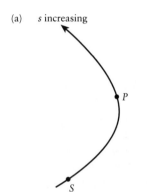

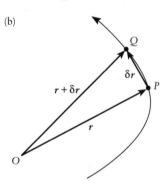

(a) *s increasing*

(b)

Fig. 10.13 The approach to a tangent at P.

In Fig. 10.13a, S is a fixed point on an arc and P is any other point on it, with position vector $r = x\hat{\imath} + y\hat{\jmath} + z\hat{k}$. A positive direction along the arc is indicated. P is then determined by specifying a number s, where

$$|s| = \text{arc-length } \widehat{SP},$$

and s is positive or negative according to whether P is on the positive or negative side of S. The parameter s is a kind of coordinate for P, measured along the arc. Indicate the dependence of r on s by writing $r(s)$ (compare $r(t)$ in Section 9.8). Given a particular vector function $r(s)$, the curve can in principle be reconstructed, although it is usually a complicated matter.

Figure 10.13b shows the vector $\overrightarrow{PQ} = \delta r$, where P has parameter value s and Q has parameter value $s + \delta s$. According to (9.18), the vector dr/ds is tangential to the arc at P, and points in the direction of increasing s. Also, in this case, when δs is small,

$$|\delta r| \approx |\delta s|,$$

approximately, and so

$$|dr(s)/ds| = \lim_{\delta s \to 0} |\delta r/\delta s| = 1. \tag{10.28}$$

Therefore, in the case when the parameter used is s, dr/ds is a **unit tangent vector**, which we can write as $\hat{t}$, pointing in the direction of increasing s.

Since $\hat{t}$ is a unit vector,

$$\hat{t} \cdot \hat{t} = 1.$$

Therefore, by using the product rule to differentiate,

$$\tfrac{d}{ds}(\hat{t} \cdot \hat{t}) = \hat{t} \cdot d\hat{t}/ds + (d\hat{t}/ds) \cdot \hat{t} = 0,$$

or $\quad 2(\hat{t} \cdot d\hat{t}/ds) = 0.$

Therefore, $d\hat{t}/ds$ is perpendicular to $\hat{t}$.

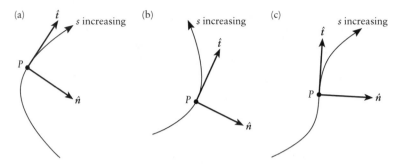

Fig. 10.14 (a) $\kappa > 0$, curve is concave viewed from the side of $\hat{n}$. (b) $\kappa < 0$, the curve is convex viewed from the side of $\hat{n}$. (c) $\kappa = 0$, a point of inflection.

Draw a **unit normal** $\overline{PN} = \hat{n}$ to the curve at P as in the diagrams of Fig. 10.14. As we walk along the curve in the direction of $\hat{t}$, the direction of $\hat{n}$ is *towards the right*. Since $\hat{t}$ and $d\hat{t}/ds$ are perpendicular, $d\hat{t}/ds$ must be a certain multiple, κ say, of $\hat{n}$:

$$d\hat{t}/ds = \kappa\hat{n}. \tag{10.29}$$

The three cases in Fig. 10.14 relate to the sign of κ. In Fig. 10.14a, the curve is **concave as viewed from the side of** $\hat{n}$, implying that if we make a small increase in s, then $\delta\hat{t}$ points in the direction of $\hat{n}$; therefore κ is positive. In Fig. 10.14b, the curve is **convex as viewed from the side of** $\hat{n}$; and in the same way it follows that κ is negative. In the case of a **point of inflection** (Fig. 10.14c), κ is zero.

The number κ is called the **curvature** of the curve at P. The greater is $|\kappa|$, the more sharply the curve is turning. The positive quantity ρ given by

$$\rho = 1/|\kappa|$$

is its **radius of curvature** at P. This is the radius of the circle that best fits the curve at P. We will not prove this, but illustrate it in the following example.

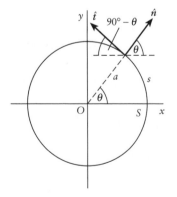

Fig. 10.15

Example 10.17 *Obtain expressions for $\hat{t}$ and $d\hat{t}/ds$ for the case of a circle of radius a with centre at the origin, and confirm that $\rho = a$* (Fig. 10.15).

Measure s from the point S; then

$$s = a\theta.$$

The unit tangent vector $\hat{t}$ has components $(-\sin\theta, \cos\theta)$, so

$$\hat{t} = -\hat{\imath}\sin\theta + \hat{\jmath}\cos\theta.$$

To differentiate with respect to s use the chain rule with $d\theta/ds = 1/a$ (because $ds/d\theta = a$)

$$\frac{d\hat{t}}{ds} = \frac{1}{a}(-\hat{\imath}\cos\theta - \hat{\jmath}\sin\theta)$$

Example 10.17 *continued*

(observe that $t \cdot dt/ds = 0$). The unit normal $\hat{n}$ in the right-hand direction has components $(\cos \theta, \sin \theta)$, so

$$\hat{n} = \hat{i} \cos \theta + \hat{j} \sin \theta.$$

Evidently

$$d\hat{t}/ds = -\frac{1}{a}\,\hat{n},$$

so $\kappa = -1/a$ (consistent with a curve that is convex viewed from the side of the normal n). Also

$$\rho = 1/|\kappa| = a,$$

the radius of the circle.

Problems

10.1 Obtain the scalar products of the pairs of vectors given in component form by: (a) $(2, 2, 1)$ and $(3, 1, 2)$, (b) $(2, -3, 2)$ and $(-2, 3, -1)$, (c) $(2, 2, -3)$ and $(-1, 1, -2)$, (d) $(2, 3, 4)$ and $(1, -2, 1)$, (e) $(p - q, p + q, p)$ and $(p + q, q, -p - q)$.

10.2 (Two dimensions). Obtain the scalar products of the pairs of vectors given in component form by (a) $(2, 3)$ and $(3, 4)$, (b) $(1, 0)$ and $(0, 1)$, (c) $(5, 6)$ and $(0, -4)$, (d) $(2, 3)$ and $(3, -2)$.

10.3 Prove that $|a + b|^2 + |a - b|^2 = 2(|a|^2 + |b|^2)$. (Hint: see eqn (10.1c).) Sketch the vectors $a, b, a + b, a - b$ on one diagram in order to obtain a geometrical theorem from this result. (There are two possible theorems, depending on what diagram you draw.)

10.4 Let $a = (2, -3, 4)$ and $b = (-1, -2, 3)$, or in the alternative form $a = 2\hat{i} - 3\hat{j} + 4\hat{k}$ and $b = -\hat{i} - 2\hat{j} + 3\hat{k}$. Evaluate $a \cdot b$, (a) using the first form, (b) using the second form with (10.8).

10.5 Given that $a = \hat{i} + 2\hat{j} - \hat{k}$ and $b = \hat{i} + 3\hat{j} + \hat{k}$, evaluate the following scalar products:
(a) $a \cdot b$, (b) $(a - b) \cdot (a + b)$,
(c) $(a - b) \cdot (a - b)$,
(d) $a \cdot a + 2a \cdot b + b \cdot b$,
(e) $(a \cdot a)a - (b \cdot b)b$.

10.6 Find the angles, in the range $0°$ to $180°$, between the pairs of vectors (a) $\hat{i} + \hat{j} + \hat{k}$ and $\hat{i} + \hat{j}$, (b) $\hat{i} - \hat{j} + \hat{k}$ and $\hat{i} + \hat{j}$, (c) $2\hat{i} - \hat{j} + 3\hat{k}$ and $\hat{i} + 3\hat{j} + 2\hat{k}$.

10.7 (Two dimensions). Find the angle θ $(0° \leqslant \theta \leqslant 180°)$ between the pairs of vectors: (a) $3\hat{i} + 4\hat{j}$ and $4\hat{i} - 3\hat{j}$, (b) $\hat{i} - 2\hat{j}$ and $2\hat{i} - \hat{j}$, (c) $\hat{i} - 2\hat{j}$ and $-6\hat{i} + 3\hat{j}$.

10.8 Find the angle between one of the edges of a cube and a diagonal line through one end.

10.9 A circular cone has its vertex at the origin and its axis in the direction of the unit vector $\hat{a}$. The half-angle at the vertex is α. Show that the position vector r of a general point on its surface satisfies the equation

$$\hat{a} \cdot r = |r| \cos \alpha.$$

Obtain the cartesian equation when $\hat{a} = (\frac{2}{7}, -\frac{3}{7}, -\frac{6}{7})$ and $\alpha = 60°$.

10.10 $A : (2, 2, -1)$, $B : (0, 1, 1)$, $C : (-1, 2, 0)$ are three points. Find the angles in the triangle ABC.

10.11 Confirm the fact that $a \cdot b = \frac{1}{4}(|a + b|^2 - |a - b|^2)$. (Hint: it is easier to start with the right-hand side.) Test the result using any two vectors. Deduce a simple geometrical theorem by sketching $a, b, a + b, a - b$ all on the same diagram: there are two theorems to be had, depending on whether you think of the triangle or the parallelogram rule.

10.12 Show that the component of a vector F in the direction of another vector a is given by $F \cdot a/|a|$. Find the components of $F = (8, 15, 9)$ in the directions of the three vectors a, b, c, where $a = (2, 3, 6)$, $b = (0, 3, 4)$, and $c = (2, 2, 1)$. Express F in the form $F = \lambda a + \mu b + \nu c$, where λ, μ, ν are constants.

10.13 Show that the vectors $a = \hat{\imath} + 3\hat{\jmath} + 4\hat{k}$ and $b = -2\hat{\imath} + 6\hat{\jmath} - 4\hat{k}$ are perpendicular. Obtain any vector $c = c_1\hat{\imath} + c_2\hat{\jmath} + c_3\hat{k}$ which is perpendicular to a and b, and derive from it two unit vectors (their senses will be opposite).

10.14 Let $a = \hat{\imath} + \hat{\jmath} - \hat{k}$ and $b = 2\hat{\imath} - \hat{\jmath} + 2\hat{k}$. Find the angle (in the range 0° to 180°) between a and b, and construct any vector perpendicular to a and b.

10.15 Find the value of λ such that the vectors $(\lambda, 2, -1)$ and $(1, 1, -3\lambda)$ are perpendicular.

10.16 Determine numbers α, β, γ which ensure that the vectors $a = (\alpha, 2, -3)$, $b = (-1, 2\beta, 2)$, and $c = (2, 1, -3\gamma)$ are mutually perpendicular.

10.17 The points $A : (1, 0, 0)$, $B : (0, 1, 0)$, $C : (0, 1, 1)$, and $D : (0, y, z)$ are the vertices of a tetrahedron. Find y and z such that ABD is an equilateral triangle and $\widehat{BCD}$ is a right angle.

10.18 (Change of axes in two dimensions). Oxy and OXY are two sets of right-handed axes with the same origin O. OX is reached from Ox by an anticlockwise rotation 45°. (a) Obtain the X, Y coordinates of a point P whose coordinates in Oxy are $(2, 2)$. (b) Find the values of x and y for the point Q for which $X = 1$, $Y = -1$. (c) Find the equation of the circle $(x-1)^2 + y^2 = 1$ in the axes OXY.

10.19 Find the lengths, and the direction cosines l, m, n, of the following vectors. (a) $\hat{\jmath}$, (b) $\hat{\imath} + \hat{\jmath} + \hat{k}$, (c) $\hat{\imath} - 2\hat{\jmath} - 2\hat{k}$, (d) $\hat{\imath} - \hat{\jmath} + \hat{k}$, (e) $\hat{\imath} - \hat{\jmath} - \hat{k}$, (f) $2\hat{\imath} + 3\hat{\jmath} + 6\hat{k}$, (g) $\hat{\imath} - 2\hat{\jmath} - 2\hat{k}$, (h) $3\hat{k}$, (i) $-3\hat{k}$.

10.20 (Change of axes). (a) Show that the vectors with components in $Oxyz$ given by $X = (6, 15, 10)/19$, $Y = (15, -10, 6)/19$, $Z = (10, 6, -15)/19$ are mutually perpendicular unit vectors. (b) A sketch will show that $[X, Y, Z]$ is a right-handed system, so it defines a new set of right-handed axes $OXYZ$. Write down the change-of-axes matrices in (10.13a) and (10.13b). (c) Find the coordinates of the point $x = 1$, $y = 2$, $z = 2$ in the new axes. (d) Express the equation of the plane $x + y + z = 0$ in the new coordinates.

10.21 The following are sets of direction ratios p, q, r for a straight line. Obtain two possible sets of direction cosines in each case. (a) 3, 4, 12; (b) 6, -10, 15.

10.22 A swarm of particles expands through all space. The velocity $v(t)$ of the particle with position vector $r(t)$ at time t in a given set of axes is equal to $f(t)r$. Show that the rule is the same when the velocity is measured relative to any given particle.

10.23 The angles made by a vector a and the positive directions of the axes Ox, Oz are 45° and 30° respectively. Find the angles that a may make with Oy.

10.24 The following are sets of direction ratios p, q, r for a straight line. Obtain two sets of direction cosines, describing unit vectors parallel to the line, for each csae. (a) 3, 4, 12; (b) 6, -10, 15.

10.25 (b) Find any constant vector parallel to the line given parametrically by $x = 1 - \lambda$, $y = 2 + 3\lambda$, $z = 1 + \lambda$. (Hint: see eqn (10.25).) (b) Find the equation of the plane which is perpendicular to line in (a) and which passes through the origin. (Hint: see eqn (10.22b).) (c) Find the equation of the plane such that the line in (a) lies in the plane, and the plane passes through the origin. (Hint: the new plane must be perpendicular to the plane in (b).)

10.26 Find the angle θ, in the range $0° \leqslant \theta \leqslant 90°$, between the pairs of planes given as follows: (a) $2x - 3y + z = 2$ and $x - y = 0$, (b) $x + y + z = 0$ and $z = 0$. (Hint: consider the normals.)

10.27 The vector equations of two planes are $a \cdot r = u$ and $b \cdot r = v$, where a and b are constant vectors and u and v are constants. What is the vector relation between a and b for the planes to be perpendicular? (Hint: see 10.22(b).) Obtain any plane perpendicular to $x + y + z = 0$.

10.28 (a) Show that the planes $ax + by + cz = d$, where a, b, c are fixed and d may take any value, are all parallel. (b) Show that the straight line through O and perpendicular to the plane $2x + y - z = 2$ has the parametric equation $r = \lambda(2, 1, -1)$, where λ is a parameter. (c) For (b), find the point at which the line intersects the plane, and deduce its length (this is the distance of the plane from the origin). (d) Find the distance between the plane in (b) and the plane $2x + y - z = 1$.

10.29 Let $p \cdot r = d$, where $p = (a, b, c)$ and $r = (x, y, z)$, be a plane, and Q be a point with position vector q. Show that the distance of Q from the plane is equal to $|p \cdot q - d|/|p|$.

Deduce the distance of the point $(1, 1, 2)$ from the plane $x + 2y - 4z + 3 = 0$.

10.30 $A: (0, -1, 3)$, $B: (1, 0, 3)$, and $C: (0, 0, 5)$ are three points. Let P_1 be the plane through A, perpendicular to $-\hat{\jmath} + \hat{k}$, and P_2 be the plane through A, B, and C.
(a) Find equations for P_1 and P_2. (b) Obtain the angle between P_1 and P_2. (c) Determine the perpendicular distance from the origin O to P_1. (d) Show that the line of intersection, L, of P_1 and P_2 meets the line OD, where D is the point $(1, 4, -4)$. (e) Determine the point of intersection of L with OD.

10.31 (Two dimensions). A straight line has a unit normal $\hat{n}$, and $\hat{s}$ points along the line. Let F_s and F_n be the component vectors of a vector F in the directions of $\hat{s}$ and $\hat{n}$ respectively, so that $F = F_s + F_n$. Show that $F = F_s - (F \cdot \hat{n})\hat{n}$. (Hint: see (10.8c).)
Find F_s and F_n when $F = \hat{\imath} - 3\hat{\jmath}$ and the straight line is given by $2x - 3y = 1$.

10.32 (Two dimensions). A mirror M_1 stands upright on a table (sketch it as a straight line M_1 throught the origin O in the (x, y) plane). $\hat{s}$ is a unit vector along M pointing away from O, and $\hat{n}$ is the unit normal vector to M pointing to the left of $\hat{s}$.
(a) A ray of light in the plane, with direction vector $\hat{u}$, falls on the mirror and is reflected in the direction $\hat{u}_1$. By considering its *vector* components in the directions of $\hat{s}$ and $\hat{n}$ show that

$\hat{u}_1 = -\hat{u} + 2(\hat{u} \cdot \hat{s})\hat{s}$.

(b) Find $\hat{u}_1$ when $\hat{u} = -(\hat{\imath} + \hat{\jmath})/\sqrt{2}$, and the mirror lies along the line $y = 0$. (c) Suppose there are two mirrors, M_1 and M_2, forming a wedge of angle 60° in the sector $x > 0$, $y > 0$, M_1 being along $y = 0$ and M_2 along $y/x = \sqrt{3}$. A ray enters the wedge in the direction $\hat{u} = -\hat{\imath} \cos\theta - \hat{\jmath} \sin\theta$. Use the result in (a) to find the direction of the twice-reflected ray.

10.33 (a) Find any two points on the line of intersection of the planes $x + y + z = 2$ and $2x + y - 2z = 1$ (e.g. one point is obtained by starting with $x = 0$). (b) Obtain a parametric vector equation for the line of intersection. (c) Deduce cartesian equations of the form (10.25) for the straight line. (Notice that the equations in (a), taken together, already define the line in cartesian form, but the form (10.25) is more informative since it contains the direction ratios.)

10.34 Obtain, in parametric form, the line of intersection of the planes $2x + 3y - z = 1$ and $x + y + z = 0$. Deduce the standard form (10.25).

10.35 Find direction ratios for the line of intersection of the planes $2x + 3y - 2z = 1$ and $x - 3y + 2z = 2$. (Notice that the line cannot be represented in the form (10.25).)

10.36 Three points are given by $A: (-1, -2, 1)$, $B: (-1, -2, 0)$, and $C: (-1, 0, 3)$. Let P_1 be the plane through B with normal vector $n_1 = \hat{\jmath} + \hat{k}$ and P_2 the plane through C with normal vector $n_2 = 2\hat{\imath} - \hat{\jmath} + 3\hat{k}$. Show that the line AC is perpendicular to P_1.

10.37 Given two planes, $a_1x + b_1y + c_1z = d_1$ and $a_2x + b_2y + c_2z = d_2$, show that any solution p, q, r of the simultaneous equations

$a_1p + b_1q + c_1r = 0, a_2p + b_2q + c_2r = 0$

is a set of direction ratios for the line of intersection.

10.38 (Perspective drawing). An observer's eye E is at the point $\hat{\imath} + \hat{\jmath} + \hat{k}$, and views objects through a plane screen which has the equation $r \cdot (1.1\hat{\imath} + 1.1\hat{\jmath} + \hat{k}) = 1$. Q is a general point on an object behind the screen, and its position vector is $r = x\hat{\imath} + y\hat{\jmath} + z\hat{k}$. Find the coordinates of the apparent position of Q on the screen. (Hint: find the equation of the line EQ; then find where it cuts the screen.)

10.39 An ellipse is given parametrically by $r = \hat{\imath}a \cos t + \hat{\jmath}b \sin t$, where a and b are constants and t is the parameter, with $-\pi < t \leq \pi$ (in radians). Show that $\delta s^2 \approx \delta x^2 + \delta y^2$, where s represents arc-length. Deduce that

$ds/dt = (a^2 \sin^2 t + b^2 \cos^2 t)^{\frac{1}{2}}$.

Find the unit tangent vector, a unit normal, the curvature, and the radius of curvature at the points where $t = 0$, $\frac{1}{4}\pi$, and $\frac{1}{2}\pi$.

10.40 A plane curve has the equation $y = f(x)$, and the position vector r of a point on the curve can be represented by

$r = x\hat{\imath} + f(x)\hat{\jmath}$

using x as the parameter. Show that the unit tangent vector to the curve is

$$\hat{t} = \frac{dr}{ds} = \frac{dr}{dx}\frac{dx}{ds}$$

$$= \frac{\hat{\imath} + f'\hat{\jmath}}{\sqrt{(1 + f'^2)}}.$$

Show that the curvature κ of the curve at any point is given by

$$k = \frac{-f''}{(1 + f'^2)^{3/2}}.$$

Find the curvature along

(a) the parabola $y = x^2$;
(b) the cosine curve $y = \cos x$.

11 Vector product

11.1 Vector product

A second form of product finds applications in problems about moments, angular velocity, and in other circumstances that involve rotation.

The **vector product**, or **cross product**, is denoted by a bold multiplication sign as in $a \times b$, or a caret sign as in $a \wedge b$. Its definition is:

Vector product $a \times b$

(a) $a \times b = (a_2 b_3 - a_3 b_2)\hat{\imath} - (a_1 b_3 - a_3 b_1)\hat{\jmath} + (a_1 b_2 - a_2 b_1)\hat{k},$

which can be written as a determinant (see (8.3)):

(b) $a \times b = \begin{vmatrix} \hat{\imath} & \hat{\jmath} & \hat{k} \\ a_1 & a_2 & a_3 \\ b_1 & b_2 & b_3 \end{vmatrix} = \hat{\imath}\begin{vmatrix} a_2 & a_3 \\ b_2 & b_3 \end{vmatrix} - \hat{\jmath}\begin{vmatrix} a_1 & a_3 \\ b_1 & b_3 \end{vmatrix} + \hat{k}\begin{vmatrix} a_1 & a_2 \\ b_1 & b_2 \end{vmatrix}.$

$$(11.1)$$

Example 11.1   *Find the vector products $a \times b$ and $b \times a$, where $a = 2\hat{\imath} - \hat{\jmath} + 3\hat{k}$ and $b = -\hat{\imath} + 2\hat{\jmath} + 4\hat{k}$.*

From (11.1a),

$$a \times b = \{[(-1) \times 4] - (3 \times 2)\}\hat{\imath} - \{(2 \times 4) - [3 \times (-1)]\}\hat{\jmath}$$
$$+ \{(2 \times 2) - [(-1) \times (-1)]\}\hat{k}$$
$$= -10\hat{\imath} - 11\hat{\jmath} + 3\hat{k}.$$

In evaluating $b \times a$, we exchange the a and b components in the expression (11.1a), so the sign of each of the three bracketed terms changes. Therefore

$$b \times a = -a \times b = 10\hat{\imath} + 11\hat{\jmath} - 3\hat{k}.$$

(Alternatively, we interchange the last two rows in the determinant form (11.1b), which changes its sign by Section 8.2, Rule 3.)

Algebraic manipulations are governed by the following rules:

Algebraic properties of $a \times b$

(a) $a \times b = -b \times a$ (the vector product does not commute).

(b) $a \times (b + c) = a \times b + a \times c$ (distributive law).

(c) $a \times (\lambda b) = \lambda a \times b$ where λ is any number.

(d) $a \times b = 0$ if b and a are parallel: in particular, $a \times a = 0$. **(11.2)**

These are proved as follows:

(a) If a and b are interchanged, then the three brackets in (11.1a) change sign.

(b) Put $b_1 + c_1$, $b_2 + c_2$, $b_3 + c_3$ in place of b_1, b_2, b_3 in (11.1a), and separate the groups of terms involving b and c. This is also a property of determinants.

(c) This follows immediately from (11.1a): λ is a factor throughout.

(d) a and b are parallel so $b = \lambda a$, where λ is some number. Therefore $a \times b = \lambda a \times a$ (from (11.2c)). If we now put $b_1 = a_1$ etc. into (11.1a), we obtain $a \times a = 0$, so $a \times b = 0$.

The unit vectors $\hat{i}$, $\hat{j}$, $\hat{k}$ are simply related by the cross product:

Vector products of $\hat{i}$, $\hat{j}$, $\hat{k}$

(a) $\hat{i} \times \hat{j} = \hat{k}$, $\quad \hat{j} \times \hat{k} = \hat{i}$, $\quad \hat{k} \times \hat{i} = \hat{j}$.

(b) $\hat{j} \times \hat{i} = -\hat{k}$, $\quad \hat{k} \times \hat{j} = -\hat{i}$, $\quad \hat{i} \times \hat{k} = -\hat{j}$. **(11.3)**

Notice that for the group in (11.3a), the cyclic order $\hat{i}$, $\hat{j}$, $\hat{k}$, $\hat{i}$, $\hat{j}$, ... is maintained, and for the group in (11.3b) there is a different **cyclic order** $\hat{j}$, $\hat{i}$, $\hat{k}$, $\hat{j}$, $\hat{i}$, ... To prove, for example, that $\hat{i} \times \hat{j} = \hat{k}$, put $\hat{i} = (1, 0, 0)$ and $\hat{j} = (0, 1, 0)$ into the definition (11.1b) (or into (11.1a) if you are not sure about determinants). Then we obtain

$$\hat{i} \times \hat{j} = \begin{vmatrix} \hat{i} & \hat{j} & \hat{k} \\ 1 & 0 & 0 \\ 0 & 1 & 0 \end{vmatrix} = 0\hat{i} + 0\hat{j} + 1\hat{k}$$

$$= \hat{k}.$$

The group (11.3b) follows by the change-of-order rule, (11.2a).

11.2 Nature of the vector $p = a \times b$

Firstly we show that $p = a \times b$ is perpendicular to both a and b. This is equivalent to proving that if we move a and b to emerge from a common point Q (see Fig. 11.1a) then p, or $a \times b$, is perpendicular to the plane containing a and b.

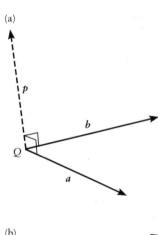

(a)

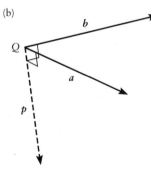

(b)

Fig. 11.1 It will be shown later that the direction of p is that given in (a).

Using the definition of $a \times b$,

$$a \cdot p = a \cdot (a \times b)$$
$$= (a_1\hat{\imath} + a_2\hat{\jmath} + a_3\hat{k}) \cdot [(a_2b_3 - a_3b_2)\hat{\imath} - (a_1b_3 - a_3b_1)\hat{\jmath} + (a_1b_2 - a_2b_1)\hat{k}]$$
$$= a_1(a_2b_3 - a_3b_2) - a_2(a_1b_3 - a_3b_1) + a_3(a_1b_2 - a_2b_1)$$
$$= 0.$$

Therefore, by (10.5), p is perpendicular to a. Similarly, p is perpendicular to b.

However, so far as we can tell from this argument, p might point in either of two directions, as suggested by the diagrams in Fig. 11.1a and b. We want to distinguish between them, and the distinction is similar to the distinction between right- and left-handed axes (compare Fig. 9.8). One way to recognize a right-handed system follows:

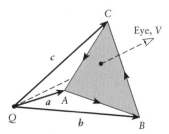

Fig. 11.2 Test for a right-handed system of vectors $[a, b, c]$. Viewed through the triangle, the vertices A, B, C follow in anticlockwise order.

Test for a right-handed system

(See Fig. 11.2.) Place $a = \overline{QA}$, $b = \overline{QB}$, $c = \overline{QC}$, at a common point Q. View Q from any point V on the *opposite side* of the triangle ABC from Q. Then

(a) $[a, b, c]$, in that order, is a right-handed system if the direction of the circuit A to B to C is seen from V as anticlockwise. Otherwise, $[a, b, c]$ is left-handed.

(b) If $[a, b, c]$ is right-handed, then (maintaining the cyclical order) $[b, c, a]$ and $[c, a, b]$ are right-handed. The others are left-handed.

(c) A set of axes is right-handed if $[\hat{\imath}, \hat{\jmath}, \hat{k}]$ is right-handed.   (11.4)

It is essential to place V on the *opposite* side of the triangle ABC from Q, otherwise the apparent direction of the circuit is reversed. In Fig. 11.1a, the system $[a, b, p]$ is right-handed, and in Fig. 11.1b, $[a, b, p]$ is left-handed.

Returning to the cross product $p = a \times b$, where the vectors all emerge from Q, set up a special set of right-handed axes Qx, Qy, Qz, as in Fig. 11.3. The axes satisfy the following conditions:

(i) Qx is in the direction of a.

(ii) Qy is in the plane of a and b, perpendicular to Qx. It is directed so that the y component of b is positive.

(iii) The direction of Qz makes the axes right-handed.

The unit vectors are $\hat{\imath}, \hat{\jmath}, \hat{k}$. From the conditions (i) and (ii), with the usual notation,

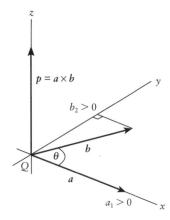

Fig. 11.3

$$p = \begin{vmatrix} \hat{\imath} & \hat{\jmath} & \hat{k} \\ a_1 & 0 & 0 \\ b_1 & b_2 & 0 \end{vmatrix} = a_1b_2\hat{k}.$$ (11.5)

Since, according to (i) and (ii), a_1 and b_2 are positive, p is in the direction of $\hat{k}$, and the test (11.4a) shows that

$[a, b, p]$ is a right-handed system. (11.6)

Therefore, Fig. 11.1a is the correct one, and Fig. 11.1b gives the direction of p incorrectly.

Moreover (see Fig. 11.3),

$b_2 = |b| \sin \theta$

(since $0° \leqslant \theta \leqslant 180°$, the sign of b_2 is positive as required). Also

$a_1 = |a|.$

Therefore, from (11.5),

$$p = a \times b = \hat{k}|a||b| \sin \theta, \tag{11.7}$$

which specifies p in a simple way.

Properties of $p = a \times b$

(a) p is perpendicular to a and b, in the direction making $[a, b, p]$ right-handed.

(b) $|p| = |a||b| \sin \theta$, where θ is the angle in the range $0°$ to $180°$ between the directions of a and b. (11.8)

The properties of $a \times b$ in (11.8) *depend only on the magnitude and direction* of a and b. We are bound to find the same results whatever axes we use to obtain them: the axes we actually used were chosen only to simplify the algebra. Therefore, we have shown that the cross product is **invariant with respect to changes in axes** (provided that we confine ourselves to right-handed axes; left-handed axes would produce $(-p)$).

Invariance of $a \times b$

$a \times b$ is invariant with respect to changes from one right-handed set of axes to another. (11.9)

Other **invariants** are the length and direction of a vector a, and therefore the vector a itself: its components are different in different axes, but the physical vector we are talking about does not change. The **scalar product**, $a \cdot b = a_1 b_1 + a_2 b_2 + a_3 b_3$, is also invariant; that is to say, it has the same numerical value in any right-handed axes: this value is equal to $|a||b| \cos \theta$ and so does not change.

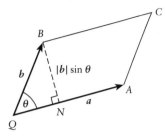

Fig. 11.4 The area of a parallelogram.

Example 11.2   *Let $a = \overline{QA}$ and $b = \overline{QB}$ be two vectors from Q, representing two sides of a parallelogram. Show that the area of the parallelogram is equal to $|a \times b|$.*

Complete the parallelogram as shown in Fig. 11.4. Construct a perpendicular BN on to QA. Then

Area $QACB$ = base $QA \times$ height BN

$$= |a||b| \sin \theta = |a \times b| \quad \text{(by (11.8))}.$$

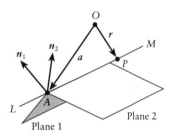

Fig. 11.5

Example 11.3   *Two planes have normals n_1 and n_2 respectively, and pass through a point A with position vector a. Obtain a vector parametric equation for their line of intersection.*

Figure 11.5 shows the two planes and their line of intersection LM, which contains the point A. $P : (x, y, z)$ with position vector r is a general point on the line.

Let p be any vector parallel to LM. Then $\overline{AP}$ is always a multiple of p, so

$$r - a = \lambda p \tag{i}$$

where λ is a parameter.

We may choose p to be given by

$$p = n_1 \times n_2; \tag{ii}$$

it is perpendicular to n_1 and n_2 by (11.8a), so it is parallel to LM. Therefore, from (i) and (ii),

$$r = a + \lambda n_1 \times n_2 \tag{iii}$$

is a parametric vector equation for the line.

11.3   The scalar triple product

The scalar quantity $a \cdot (b \times c)$ is called a **scalar triple product**. It has the following properties:

Properties of the triple scalar product

(a) $a \cdot (b \times c) = \begin{vmatrix} a_1 & a_2 & a_3 \\ b_1 & b_2 & b_3 \\ c_1 & c_2 & c_3 \end{vmatrix}$.

(b) $a \cdot (b \times c) = b \cdot (c \times a) = c \cdot (a \times b)$

(c) $a \cdot (c \times b) = b \cdot (a \times c) = c \cdot (b \times a)$
$= -a \cdot (b \times c).$

(d)   If any two vectors are equal or parallel, $a \cdot (b \times c) = 0$.

(e)   It is invariant for right-handed axes.

(11.10)

The proofs are as follows:

(a) Put $a = (a_1, a_2, a_3)$ and so on. Then

$$a \cdot (b \times c) = (a_1, a_2, a_3) \times (b_2 c_3 - b_3 c_2, b_3 c_1 - b_1 c_3, b_1 c_2 - b_2 c_1)$$
$$= a_1(b_2 c_3 - b_3 c_2) - a_2(b_1 c_3 - b_3 c_1) + a_3(b_1 c_2 - b_2 c_1)$$
$$= \begin{vmatrix} a_1 & a_2 & a_3 \\ b_1 & b_2 & b_3 \\ c_1 & c_2 & c_3 \end{vmatrix}.$$

(b) In the three permutations

$$a \cdot (b \times c), \, b \cdot (c \times a), \, c \cdot (a \times b),$$

the cyclic order $a, b, c, a, b, \ldots$ is maintained. The determinants for $b \cdot (c \times a)$ and $c \cdot (a \times b)$ are each obtained from $a \cdot (b \times c)$ by means of *two* row interchanges, and this leaves the determinant unaltered (see Section 8.2, Rule 3). Therefore they are all equal.

(c) Compare these three products with those in (b), and recall that $b \times c = -c \times b$ etc.

(d) If b is parallel to c, then $b \times c = 0$ from (11.2d). If a is parallel to b or c, then use the same argument on one of the equivalent permutations in (11.10b).

(e) Its value remains the same in any right-handed axes because the cross and dot products have this property (see (11.9)).

The **brackets in the triple scalar product** are not strictly necessary and are often omitted, because the alternative bracketing $(a \cdot b) \times c$ would be meaningless.

A **parallelepiped** is the three-dimensional analogue of a parallelogram and its volume can be expressed as a scalar triple product. Figure 11.6 shows the parallelepiped which has the vectors $\overline{QA} = a$, $\overline{QB} = b$, $\overline{QC} = c$ as three adjacent sides.

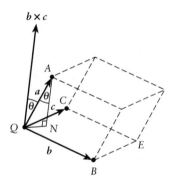

Fig. 11.6

Drop a perpendicular AN on to the plane $QBEC$. Then

volume = area $QBEC \times$ height AN.

But from Example 11.2, since $QBEC$ is a parallelogram,

area $QBEC = |b \times c|$.

Since $b \times c$ is perpendicular to the plane of b and c, $\overline{AN}$ and $b \times c$ are parallel, so

height $AN = QA \cos \theta = |a| \cos \theta.$

Therefore

volume $= |a| |b \times c| \cos \theta$
$= |a \cdot (b \times c)|$

from (10.2).

> **Volume of a parallelepiped**
> If the adjacent sides at a vertex Q are $\overline{QA} = a, \overline{QB} = b, \overline{QC} = c$, then
>
> $$\text{volume} = |a \cdot (b \times c)|.$$
>
> (11.11)

Vectors are said to be **coplanar** if, when drawn from the same point, they lie in the same plane. The condition for this is:

> **Coplanar vectors**
> Three nonzero vectors a, b, c at the same point are coplanar if, and only if,
>
> $$a \cdot (b \times c) = 0.$$
>
> (11.12)

(If they are not at the same point, then this is the condition that they should be *parallel to a common plane*.) The result follows from (11.11): the volume of the corresponding parallelepiped is zero.

Example 11.4   *Show that the points $A : (1, 2, 2)$, $B : (3, 4, 5)$, $C : (-1, 0, -1)$ lie on a plane through the origin.*

Suppose the three points A, B, C have position vectors a, b, c. To show a, b, c are coplanar, evaluate $a \cdot (b \times c)$:

$$a \cdot (b \times c) = (1, 2, 2) \cdot [(3, 4, 5) \times (-1, 0, -1)]$$

$$= \begin{vmatrix} 1 & 2 & 2 \\ 3 & 4 & 5 \\ -1 & 0 & -1 \end{vmatrix}$$

$$= 1 \begin{vmatrix} 4 & 5 \\ 0 & -1 \end{vmatrix} - 2 \begin{vmatrix} 3 & 5 \\ -1 & -1 \end{vmatrix} + 2 \begin{vmatrix} 3 & 4 \\ -1 & 0 \end{vmatrix}$$

$$= -4 - 2 \times 2 + 2 \times 4 = 0.$$

Therefore, A, B, C, and O are all in the same plane, so the points A, B, C are on a plane through the origin.

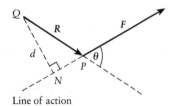

Line of action

Fig. 11.7

11.4   Moment of a force

Suppose that, in three dimensions, a force F is acting at a point P in a body (Fig. 11.7) and Q is any point. Then the magnitude M of the **moment or torque about the point Q** which is exerted by F is defined to be

$$M = |F|d,$$

where d is the length of the perpendicular QN from Q to the line of action of F. In Fig. 11.7, $\overline{QP} = R$, and θ is the angle between F and R, with $0 \leqslant \theta \leqslant 180°$. Then

$$d = |R| \sin \theta,$$

so

$$M = |F||R| \sin \theta. \tag{11.13}$$

This equation suggests a connection with the vector $R \times F$. Define a **vector** M by

$$M = R \times F \tag{11.14}$$

(note that R comes first in the product). Then by (11.8b),

$$|M| = |R \times F| = |R||F| \sin \theta,$$

which is the same as (11.13). We call M the **vector moment about the point Q of F acting at P.** M is perpendicular to the plane of R and F, in the direction making $[R, F, M]$ right-handed.

Example 11.5 *A force $F = \hat{\imath} - \hat{\jmath} + 2\hat{k}$ acts at $P : (1, 2, 1)$. Find its vector moment M about the point $Q : (2, 1, 1)$.*

In these axes the position vectors of P and Q are

$$p = \hat{\imath} + 2\hat{\jmath} + \hat{k}, q = 2\hat{\imath} + \hat{\jmath} + \hat{k},$$

so

$$R = \overline{QP} = p - q = -\hat{\imath} + \hat{\jmath}.$$

The moment M is given by

$$M = R \times F = \begin{vmatrix} \hat{\imath} & \hat{\jmath} & \hat{k} \\ -1 & 1 & 0 \\ 1 & -1 & 2 \end{vmatrix} = 2\hat{\imath} + 2\hat{\jmath}.$$

Example 11.6 *A force $F = \hat{\imath} - \hat{\jmath}$ (force units) acts at $P : (1, 2, 0)$. Find its vector moment about the origin O.*

O, P, and F all lie in the (x, y) plane, so the physical problem is two-dimensional. The Oz axis points towards you, out of the page, in Fig. 11.8.
The vector moment M is given by

$$M = R \times F = \begin{vmatrix} \hat{\imath} & \hat{\jmath} & \hat{k} \\ 1 & 2 & 0 \\ 1 & -1 & 0 \end{vmatrix} = -3\hat{k}.$$

Thus M is parallel to Oz and its z component is -3. Figure 11.8 shows the negative sign corresponds to F having a clockwise influence on a wheel turning about the point O.

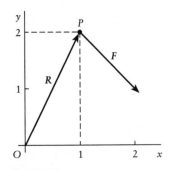

Fig. 11.8

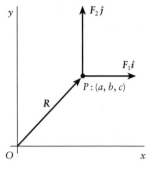

Fig. 11.9

Example 11.7 (*Generalizes Example* 11.6.) *A force* $F = F_1\hat{i} + F_2\hat{j}$
acts at $P : (a, b, 0)$. *Find its vector moment about the origin.*

We have

$$M = R \times F = \begin{vmatrix} \hat{i} & \hat{j} & \hat{k} \\ a & b & 0 \\ F_1 & F_2 & 0 \end{vmatrix} = (F_2 a - F_1 b)\hat{k}.$$

This situation is also (physically) two-dimensional; the z direction in Fig. 11.9 would only be needed in order to display M. The expression illustrates the separate clockwise and anticlockwise contributions respectively of F_1 and F_2.

The scalar triple product

$$\hat{s} \cdot M = \hat{s} \cdot (R \times F) \tag{11.15}$$

represents the **component of M in the direction of a unit vector $\hat{s}$.** Its physical significance concerns **torque or moment about an axis** (rather than about a point), as follows.

Figure 11.10 shows an **axis of rotation** AA' (in three dimensions) passing through a point Q and parallel to a unit vector $\hat{s}$. A force F acts at P. Q' is any other point on AA', and P' is any point on the line of action of F.

Put

$$\overline{QP} = R \quad \text{and} \quad \overline{Q'P'} = R'.$$

Then

$$\hat{s} \cdot (R' \times F') = \hat{s} \cdot [(\overline{Q'Q} + R + \overline{PP'}) \times F].$$

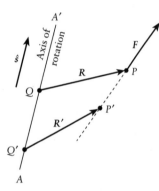

Fig. 11.10 Moment of F about an axis parallel to s:
$\hat{s} \cdot (R' \times F) = \hat{s} \cdot (R \times F)$.

But QQ' is parallel to $\hat{s}$, and PP' is parallel to $\hat{F}$, so by (11.10d) these make no contribution to the triple scalar product, and we obtain

$$\hat{s} \cdot (R' \times F) = \hat{s} \cdot (R \times F).$$

Thus *any* point on the axis and *any* point on the line of action of F may be put into the triple scalar product without affecting its value.

The freedom given by this result allows us to choose Q' and P' such that $Q'P' = R'$ is *perpendicular both to $\hat{s}$ and to the line of action of F*, as in Fig. 11.11. Put

$$|R'| = |\overline{Q'P'}| = d;$$

this is the distance between the two skew lines.

Next, construct a set of coordinate axes at origin Q'. Let $Q'x$ be in the direction of R', $Q'z$ in the direction of $\hat{s}$, and $Q'y$ perpendicular to $Q'x$ and $Q'y$ in the direction necessary for the axes to be right-handed. The unit vectors are $\hat{i}, \hat{j}, \hat{k}$.

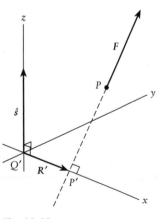

Fig. 11.11

Express F in terms of its components in directions $\hat{\imath}, \hat{\jmath}, \hat{k}$. The $\hat{\imath}$ component is zero since F is perpendicular to $Q'x$, so

$$F = F_2\hat{\jmath} + F_3\hat{k}.$$

Put $\hat{s}\cdot M = M$, say. Then

$$M = \hat{s}\cdot(R'\times F) = F\cdot(\hat{s}\times R') \text{ (by (11.10b))}.$$

Also

$$\hat{s}\times R' = (|\hat{s}||R'|\sin 90°)\hat{\jmath}$$
$$= |R'|\hat{\jmath} = d\hat{\jmath}.$$

Therefore

$$M = F\cdot d\hat{\jmath} = (F_2\hat{\jmath} + F_3\hat{k})\cdot(d\hat{\jmath})$$
$$= F_2 d. \tag{11.16}$$

The expression (11.16) corresponds to what we should expect about the turning effect of F about the given axis. There is no contribution for F_3 because $F_3\hat{k}$ is parallel to the axis of rotation, and F_1 is zero in these axes. What remains is $F_2\hat{\jmath}$, which is perpendicular to the axis of rotation $Q'z$, and d is the perpendicular distance of F from it.

For this reason the scalar quantity $M = \hat{s}\cdot(R\times F)$ is called the **moment of F about an axis of rotation AA'**, as in Fig. 11.10. Dropping the dashed quantities, the unit vector $\hat{s}$ is the direction AA', and $R = \overline{QP}$, where Q is any point on AA' and P any point on the line of action of F (M being independent of the choice of these points, by (11.15)).

11.5 Vector triple product

The vector

$$w = a\times(b\times c) \tag{11.17}$$

is called a **vector triple product**. The vector w is perpendicular to $b\times c$, but $b\times c$ is perpendicular to b and c, so b, c, and w are parallel to the same plane. Therefore (see (9.12)), it must be possible to express w in the simple form $w = \lambda b + \mu c$. The required relation is:

Vector triple product
$$a\times(b\times c) = (a\cdot c)b - (a\cdot b)c. \tag{11.18}$$

To prove (11.18), translate a, b, c to a common point Q, and set up axes Qx, Qy, Qz as in Fig. 11.12, such that Qz is in the direction of a. Then

$$a = a_3\hat{k},$$

so $\quad a\cdot b = a_3 b_3 \quad$ and $\quad a\cdot c = a_3 c_3.$ $\tag{11.19}$

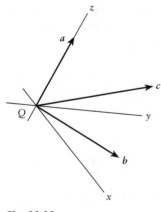

Fig. 11.12

Remember that $\hat{k}\times\hat{i}=\hat{j}$, $\hat{k}\times\hat{j}=-\hat{i}$, and $\hat{k}\times\hat{k}=0$. In these axes,

$$w = a\times(b\times c)$$
$$= a_3\hat{k}\times[(b_2c_3 - b_3c_2)\hat{i} - (b_1c_3 - b_3c_1)\hat{j} + (b_1c_2 - b_2c_1)\hat{k}]$$
$$= a_3(b_2c_3 - b_3c_2)\hat{j} + a_3(b_1c_3 - b_3c_1)\hat{i}$$
$$= a_3c_3(b_1\hat{i} + b_2\hat{j}) - a_3b_3(c_1\hat{i} + c_2\hat{j}).$$

The third components of b and c ($b_3\hat{k}$ and $c_3\hat{k}$) are missing in the brackets: to make them appear, add to the right-hand side the term

$$a_3b_3c_3\hat{k} - a_3b_3c_3\hat{k}\text{ (which} = 0).$$

After bringing them into the brackets we get

$$w = a_3c_3(b_1\hat{i} + b_2\hat{j} + b_3\hat{k}) - a_3b_3(c_1\hat{i} + c_2\hat{j} + c_3\hat{k})$$
$$= (a\cdot c)b - (a\cdot b)c.$$

Example 11.8 *Find $a\times(b\times c)$ when $a = \hat{i}+\hat{j}$, $b = 2\hat{i}-\hat{j}$, and $c = \hat{i}+\hat{j}+\hat{k}$.*

$$a\cdot b = (\hat{i}+\hat{j})\cdot(2\hat{i}-\hat{j}) = 2 - 1 = 1;$$
$$a\cdot c = (\hat{i}+\hat{j})\cdot(\hat{i}+\hat{j}+\hat{k}) = 1 + 1 = 2.$$

Therefore

$$a\times(b\times c) = (a\cdot c)b - (a\cdot b)c = 2b - c$$
$$= 2(2\hat{i}-\hat{j}) - (\hat{i}+\hat{j}+\hat{k}) = 3\hat{i} - 3\hat{j} - \hat{k}.$$

(The product could also be worked out directly.)

Problems

11.1 In component form let $a = (1, -2, 2)$, $b = (3, -1, -1)$, and $c = (-1, 0, -1)$. Evaluate the following:
(a) $a\times b$ (b) $b\times a$
(c) $a\times a$ (d) $a\cdot(b\times c)$
(e) $c\cdot(a\times b)$ (f) $b\cdot(a\times c)$
(g) $(a\times b)\cdot b$ (h) $a\times(a\times b)$
(i) $(c\times b)\times a$.

11.2 Given two planes, $r\cdot n_1 = d_1$, $r\cdot n_2 = d_2$, show that the plane through the origin perpendicular to their line of intersection is given by $r\cdot(n_1\times n_2) = 0$.

11.3 (a) Use the vector product to obtain a vector parametric equation for the straight line which is perpendicular to the vectors b and c, and passes through the point with position vector a. (b) Obtain the equation to the line when $a = \hat{i}+2\hat{j}+\hat{k}$, $b = \hat{i}-\hat{j}$, and $c = \hat{j}+\hat{k}$.

11.4 Show that the vector $a\times u$, where $a = (a_1, a_2, a_3)$ and u is any vector, is parallel to the plane $a_1x + a_2y + a_3z = d$. Obtain two vectors parallel to the plane $2x - 3y - z = 1$.

11.5 Under what conditions will $a\times b = 0$?

11.6 Show that the vectors $a = 2\hat{i} + 3\hat{j} + 6\hat{k}$ and $b = 6\hat{i} + 2\hat{j} - 3\hat{k}$ are perpendicular. Find a vector c which is perpendicular to b and c and such that $[a, b, c]$ is a right-handed set.

11.7 (a) The vertices of a triangle are A, B, C, with position vectors a, b, c. Show that the area of triangle ABC is given by $\frac{1}{2}|b\times c + c\times a + a\times b|$. (Hint: see Example 11.2.) (b) A second triangle has vertices at $a + \lambda(b - c)$, b, c, where λ is a scalar. Show that the areas of the two triangles are the same. What simple geometrical result does the equality exhibit? (c) Find the area of the triangle whose vertices are at $\hat{\imath} - 2\hat{\jmath} - \hat{k}$, $\hat{\imath} - \hat{\jmath} + 2\hat{k}$, $\hat{\imath} + 2\hat{\jmath} - \hat{k}$.

11.8 A, B, C are three points which do not lie on a straight line, and D is another point. Put $\overline{AB} = b$, $\overline{AC} = c$, and $\overline{AD} = d$. Show that the distance of D from the plane passing through A, B, C is equal to $|d \cdot (b \times c)|/|b \times c|$.

11.9 Show that, if QA, QB, QC are adjacent edges of a rectangular parallelepiped with coordinates

$$Q : (x_0, y_0, z_0), \qquad A : (x_1, y_1, z_1), \qquad B : (x_2, y_2, z_2),$$
$$C : (x_3, y_3, z_3),$$

then its volume is given by the *modulus* of the determinant

$$\begin{vmatrix} x_1 - x_0 & x_2 - x_0 & x_3 - x_0 \\ y_1 - y_0 & y_2 - y_0 & y_3 - y_0 \\ z_1 - z_0 & z_2 - z_0 & z_3 - z_0 \end{vmatrix}.$$

11.10 (Oblique coordinates). (a) Let a, b, c be three non-coplanar vectors, and v be any vector. Show that v can be expressed as

$$v = Xa + Yb + Zc,$$

where X, Y, Z are constants given by

$$X = v \cdot (b \times c)/D,$$
$$Y = v \cdot (c \times a)/D,$$
$$Z = v \cdot (a \times b)/D,$$

where

$$D = a \cdot (b \times c).$$

(Hint: start by forming, say, $v \cdot (a \times b)$. Equation (11.10d) gets rid of two terms.) (b) Check the formulae for the case $a = (1, 1, 0)$, $b = (1, 1, 0)$, $c = (1, 0, 1)$. $v = (1, 1, 1)$, by solving the three equations obtained by splitting the vector equation into components.

11.11 (Cramer's rule). In Problem 11.10, write the vector equation $v = Xa + Yb + Zc$ in the form of three simultaneous equations involving the components of a, b, c. Now write the formulae for X, Y, Z in determinant form. This is known as **Cramer's rule** (see Section 12.1), for solving any three simultaneous equations provided $D \neq 0$.

11.12 (a) Show that if three vectors a, b, c are non-coplanar and v is any vector, then constants X, Y, Z can be found such that $v = Xb \times c + Yc \times a + Za \times b$. (Hint: start by forming $a \cdot v$ from this expression.) (b) Find X, Y, Z if $v = 2\hat{\imath} + \hat{\jmath} - 2\hat{k}$, $a = \hat{\imath} - \hat{\jmath}$, $b = \hat{\imath} + 2\hat{\jmath}$, $c = \hat{\jmath} - 2\hat{k}$.

11.13 The equations $r = a + \lambda u$ and $r = b + \mu v$, where λ and μ are parameters, represent two skew lines L_1 and L_2 (straight lines which do not intersect). (a) Write down a vector w which is perpendicular to both L_1 and L_2. (b) Show that values of λ, μ, and ν can be found so that

$$(a + \lambda u) + \nu w = (b + \mu v),$$

and explain why this implies that there actually exists a straight line L_3 which joins L_1 and L_2 and is perpendicular to both. (c) For the case when $a = -\hat{\imath}$, $u = \hat{k}$, $b = \hat{\imath} - \hat{\jmath}$, $v = \hat{\imath} + \hat{\jmath} + \hat{k}$, find the values of λ, μ, ν. Deduce the points where L_3 meets L_1 and L_2. Find an equation for L_3, and the perpendicular distance between L_1 and L_2.

11.14 Find the vector moments M of the given forces F acting at the points P as specified. Make sketches, indicating the direction of M.
(a) $F = (2, 0, 0)$ at $P : (0, 3, 0)$. Find M about the origin O.
(b) $F = (2, 0, 0)$ at $P : (0, 3, 0)$. Find M about $Q : (0, 0, 3)$.
(c) $F = (2, 0, 0)$ at $P : (0, -3, 0)$. Find M about $Q : (0, 0, 3)$.

11.15 A force F of magnitude 4 acts at the point $(1, -1, 2)$ in the direction of $\hat{\imath} - 2\hat{\jmath} - 2\hat{k}$. Find the vector moment M of F, (a) about the origin, (b) about the point $(-2, 1, 2)$. (c) Find its component about the y axis, taken in the direction of $\hat{\jmath}$ (i.e. $\hat{s} = \hat{\jmath}$, in the text).

11.16 Find the moment M about the axes specified, where the force is $F = (2, 0, 0)$ acting at $P : (0, 3, 0)$. (Note that the sense of the axis needs to be specified. If the sense is reversed, then the sign of $\hat{s} \cdot (R \times F)$ changes.)
(a) The z axis, taken in the positive direction. (b) The z axis, taken in the negative direction. (c) The x axis, in the positive direction. (d) The y axis, in the positive direction. (e) The axis through the origin, direction $\hat{s} = (1/\sqrt{3}, 1/\sqrt{3}, 1/\sqrt{3})$.

11.17 Find the magnitude of the moment M of the force $F = (1, 1, 2)$ acting at $P : (2, -3, 1)$, about the axis $\overline{AB}$, where $A : (2, 3, 2)$ and $B : (1, 1, 1)$. Verify directly that the component of F in the direction of AB makes no contribution. Show that the component of F along any line joining P to the axis AB makes no contribution.

11.18 A fixed force F acts at a fixed point P with position vector r. An axis passes through the origin, but it can be adjusted so as to take any direction $\hat{s}$. Show that the magnitude $|M|$ of the moment M about the axis is a maximum when $\hat{s}$ is perpendicular to the plane containing r and F. (Hint: remember $a \cdot b = |a||b| \cos \theta$ in the usual notation.) Under what conditions is $|M|$ a minimum, and what is its value?

11.19 A rigid lamina in the (x, y) plane rotates at ω radians per second about the z axis in axes $Oxyz$, in the manner of a wheel on an axle. (a) Show that if r and θ are the polar coordinates of any point P, then the velocity of P is given by $v = -\hat{i}\omega r \sin\theta + \hat{j}\omega r \cos\theta$.

(b) Show that v may be written $v = \boldsymbol{\omega} \times r$, where $\boldsymbol{\omega} = \omega \hat{k}$ ($\boldsymbol{\omega}$ is called the angular velocity vector in two dimensions).

(c) Choose any point Q which travels round with the lamina, and let $QXYZ$ be another set of axes which remain parallel to $Oxyz$. Show that, viewed relative to $QXYZ$, any point P has velocity V given by $V = \boldsymbol{\omega} \times R$, where R is its position vector in $OXYZ$.

11.20 A point of a rigid body is fixed at the origin of coordinates O. It rotates about O with angular velocity $\boldsymbol{\omega}$ (i.e. at any instant the body is rotating at a rate $|\omega|$ rad s^{-1} about the line in the instantaneous direction of the vector $\boldsymbol{\omega}$).

Explain why every point of the body is moving perpendicularly to $\boldsymbol{\omega}$. Show that the velocity v of any point is given by $v = \boldsymbol{\omega} \times r$. (Hint: compare Problem 11.19.)

Find the matrix S such that $v = S\boldsymbol{\omega}$. Show that $|v|^2 = \boldsymbol{\omega}^T S^T S \boldsymbol{\omega}$, and that

$$S^T S = \begin{bmatrix} y^2 + z^2 & -xy & -zx \\ -xy & x^2 + z^2 & -yz \\ -zx & -yz & y^2 + z^2 \end{bmatrix}.$$

11.21 Supposing that a and b emerge from the same point, show geometrically that $a \times (a \times b)$ and $b \times (a \times b)$ are in the plane of a and b.

11.22 If $v = (a \times b) \times (c \times d)$, then v can be written in either of the forms $v = pc + qd$ or $v = ma + nb$. Justify this expectation geometrically, then obtain the constants by using eqn (11.18).

11.23 Prove that

$$a \times (b \times c) + b \times (c \times a) + c \times (a \times b) = 0.$$

11.24 (a) Find a vector which is perpendicular to n and in the plane of n and b, where n and b are any two vectors. (b) Show that the straight line $r = b + \mu n \times [(a - b) \times n]$, where μ is a parameter, passes through the point with position vector b, and meets the straight line given parametrically by $r = a + \lambda n$ in a right angle.

11.25 You are given two planes, $r \cdot n_1 = d_1$, $r \cdot n_2 = d_2$. Show that the point on their line of intersection that is closest to the origin has the position vector

$$\alpha(n_1 \times n_2) \times n_1 + \beta(n_1 \times n_2) \times n_2$$

where α and β are certain constants. Obtain a formula for the constants.

11.26 A particle P of mass m and position vector $r(t)$ moves with velocity $v(t)$ under the action of a single force F. A point Q at $q(t)$ has velocity $u(t)$. The moment of momentum (or angular momentum) $H(t)$ of P about Q is defined by

$$H = (r - q) \times (mv).$$

Show that $dH/dt = [(r - q) \times F] - mu \times v$. Deduce that if $u(t) = 0$, then $dH/dt = M$, where M is the moment of F about the point Q.

12 Linear algebraic equations

CONTENTS

12.1 Cramer's rule

An equation of the form

$$a_{11}x_1 + a_{12}x_2 + \cdots + a_{1n}x_n = 0,$$

where $a_{11}, a_{12}, \ldots, a_{1n}$ are constants, is an example of a **linear equation** in the unknowns $x_1, x_2, \ldots, x_n$. If several such linear equations are to be satisfied simultaneously, they can be solved by **eliminating** variables between them. For example, if x_1 and x_2 satisfy the two linear equations

$$x_1 + 2x_2 = 4,$$
$$x_1 - \ x_2 = 1,$$

then x_1 can be eliminated by subtracting the equations to give

$$2x_2 + x_2 = 4 - 1, \text{ or } 3x_2 = 3, \text{ or } x_2 = 1.$$

Similarly x_2 can be eliminated by adding twice the second equation to the first leaving

$$x_1 + 2x_1 = 4 + 2, \text{ or } 3x_1 = 6, \text{ or } x_1 = 2.$$

Hence the solution is $x_1 = 2$, $x_2 = 1$.

In general, if

$$a_{11}x_1 + a_{12}x_2 = d_1,$$
$$a_{21}x_1 + a_{22}x_2 = d_2,$$

the elimination of x_2 and x_1 by this process leads to the solution

$$x_1 = \frac{d_1 a_{22} - d_2 a_{12}}{a_{11}a_{22} - a_{12}a_{21}}, \quad x_2 = \frac{d_2 a_{11} - d_1 a_{21}}{a_{11}a_{22} - a_{12}a_{21}},$$

provided that the denominator $a_{11}a_{22} - a_{12}a_{21}$ is not zero. From Section 8.1 in the chapter on determinants, these ratios can be recognized as ratios of determinants:

$$x_1 = \frac{\begin{vmatrix} d_1 & a_{12} \\ d_2 & a_{22} \end{vmatrix}}{\begin{vmatrix} a_{11} & a_{12} \\ a_{21} & a_{22} \end{vmatrix}}, \quad x_2 = \frac{\begin{vmatrix} a_{11} & d_1 \\ a_{21} & d_2 \end{vmatrix}}{\begin{vmatrix} a_{11} & a_{12} \\ a_{21} & a_{22} \end{vmatrix}}.$$

This formula is known as **Cramer's rule**. As we shall see later, if the denominator is zero then the two equations can have no solutions or an infinity of solutions.

Elimination can be applied to equations with more unknowns as the following example illustrates.

Example 12.1   *Solve the equations*

$$x_1 - 2x_2 + x_3 = -4, \tag{i}$$
$$2x_1 + x_2 - x_3 = -1, \tag{ii}$$
$$x_1 + 3x_2 + 2x_3 = 7. \tag{iii}$$

Eliminate x_1 between (i) and (ii), and between (ii) and (iii). Thus 2(i) − (ii) gives

$$-5x_2 + 3x_3 = -7, \tag{iv}$$

whilst (ii) − 2(iii) gives

$$-5x_2 - 5x_3 = -15. \tag{v}$$

Now eliminate x_2 between (iv) and (v) by subtraction:

$$8x_3 = 8, \text{ or } x_3 = 1.$$

Rather than eliminate again to find the other unknowns, we can substitute back $x_3 = 1$ in (iv), say, so that

$$-5x_2 + 3 = -7, \text{ or } x_2 = 2.$$

Finally substitute $x_2 = 2$ and $x_3 = 1$ into (i):

$$x_1 - 4 + 1 = -4, \text{ or } x_1 = -1.$$

Hence the full solution is $x_1 = -1, x_2 = 2, x_3 = 1$.

For more unknowns and equations, this approach becomes increasingly laborious and prone to errors.

A **matrix equation**

$$Ax = d \tag{12.1}$$

defines a set of linear equations (we referred to them previously in connection with the inverse matrix in Section 7.4). In general, A will be an $m \times n$ matrix, while x and d are n and m column vectors respectively. Usually, but not always, we are interested in the case where the number of unknowns in the equations equals the number of equations. In other words, there is neither a surplus of unknowns nor equations.

In this case we have an $n \times n$ or **square matrix**, and this is the normal situation in applications. For example, the set of equations defined by

$$A = \begin{bmatrix} 1 & 2 & 1 \\ -2 & 3 & -1 \\ 1 & 4 & -2 \end{bmatrix}, \qquad x = \begin{bmatrix} x_1 \\ x_2 \\ x_3 \end{bmatrix}, \qquad d = \begin{bmatrix} 1 \\ -7 \\ -7 \end{bmatrix}$$

is

$$x_1 + 2x_2 + x_3 = 1, \qquad (12.2a)$$

$$-2x_1 + 3x_2 - x_3 = -7, \qquad (12.2b)$$

$$x_1 + 4x_2 - 2x_3 = -7. \qquad (12.2c)$$

Consider now the case in which A is an arbitrary square matrix. If the inverse of A exists, then multiplication of (12.1) on the left by A^{-1} leads to the solution vector

$$x = A^{-1}d = \frac{\text{adj } A}{\det A} d,$$

using the formula for the inverse given in Equation (8.5c): adj A is the adjoint of A. Let $n = 3$; then, for our standard matrix,

$$A = \begin{bmatrix} a_{11} & a_{12} & a_{13} \\ a_{21} & a_{22} & a_{23} \\ a_{31} & a_{32} & a_{33} \end{bmatrix},$$

we have

$$x = \begin{bmatrix} x_1 \\ x_2 \\ x_3 \end{bmatrix} = \frac{\text{adj } A}{\det A} d$$

$$= \frac{\text{adj } A}{\det A} \begin{bmatrix} d_1 \\ d_2 \\ d_3 \end{bmatrix} = \frac{1}{\det A} \begin{bmatrix} C_{11}d_1 + C_{21}d_2 + C_{31}d_3 \\ C_{12}d_1 + C_{22}d_2 + C_{32}d_3 \\ C_{13}d_1 + C_{23}d_2 + C_{33}d_3 \end{bmatrix},$$

where $C_{11}, C_{12}, \ldots$ are the cofactors of $a_{11}, a_{12}, \ldots$ (see Section 8.1). Thus, comparison of elements in the vectors leads to

$$x_1 = \frac{1}{\det A}(C_{11}d_1 + C_{21}d_2 + C_{31}d_3) = \frac{1}{\det A} \begin{vmatrix} d_1 & a_{12} & a_{13} \\ d_2 & a_{22} & a_{23} \\ d_3 & a_{32} & a_{33} \end{vmatrix},$$

$$x_2 = \frac{1}{\det A} \begin{vmatrix} a_{11} & d_1 & a_{13} \\ a_{21} & d_2 & a_{23} \\ a_{31} & d_3 & a_{33} \end{vmatrix}, \qquad x_3 = \frac{1}{\det A} \begin{vmatrix} a_{11} & a_{12} & d_1 \\ a_{21} & a_{22} & d_2 \\ a_{31} & a_{32} & d_3 \end{vmatrix}.$$

This is **Cramer's rule** for 3 equations in 3 unknowns. It is systematic in that, for x_1, the determinant in the numerator has the first column of A replaced by d; for x_2, the second column is replaced by d; and so on. The generalization to n linear equations in n unknowns is fairly clear from this formula. It is a useful theoretical result, but not generally a recommended method of solving more than four equations in four unknowns. High-order determinant evaluation is complicated.

12.2 Elementary row operations

Short of using computer software, the simplest method of solving equations involves systematic elimination. Consider the three equations

$$x_1 + 2x_2 + x_3 = 1, \tag{12.3a}$$
$$-2x_1 + 3x_2 - x_3 = -7, \tag{12.3b}$$
$$x_1 + 4x_2 - 2x_3 = -7. \tag{12.3c}$$

We can perform three **elementary row operations** on linear equations which do not affect the solution. They are

> **Elementary row operations**
>
> (i) any equation can be multiplied by a nonzero constant,
> (ii) any two equations can be interchanged,
> (iii) any equation can be replaced by the sum of itself and any multiple of another equation.
>
> These operations do not alter the solutions. (12.4)

Step 1. Eliminate x_1 from (12.3b,c) by adding multiples of (12.3a) from (12.3b) and (12.3c):

$$x_1 + 2x_2 + x_3 = 1,$$
$$7x_2 + x_3 = -5 \quad (\mathbf{r}_2' = \mathbf{r}_2 + 2\mathbf{r}_1),$$
$$2x_2 - 3x_3 = -8 \quad (\mathbf{r}_3' = \mathbf{r}_3 - \mathbf{r}_1).$$

The required operations between the equations are listed on the right.

Step 2. We now proceed to eliminate x_2 from (12.3c) using a multiple $\frac{2}{7}$ of the new row 2. Hence

$$x_1 + 2x_2 + x_3 = 1,$$
$$7x_2 + x_3 = -5,$$
$$-\tfrac{23}{7}x_3 = -\tfrac{46}{7} \quad (\mathbf{r}_3' = \mathbf{r}_3 - \tfrac{2}{7}\mathbf{r}_2).$$

Step 3. Using Rule (i) above, reduce the coefficients of x_2 and x_3 in the second and third equations above to 1:

$$x_1 + 2x_2 + x_3 = 1,$$
$$x_2 + \tfrac{1}{7}x_3 = -\tfrac{5}{7} \quad (\mathbf{r}_2' = \tfrac{1}{7}\mathbf{r}_2),$$
$$x_3 = 2 \quad (\mathbf{r}_3' = -\tfrac{7}{23}\mathbf{r}_3).$$

Step 4. Starting from the third equation, we can now solve the equations by **back substitution**. Since $x_3 = 2$, from the second equation,

$$x_2 = -\tfrac{5}{7} - \tfrac{1}{7}x_3 = -\tfrac{5}{7} - \tfrac{1}{7} \times 2 = -1,$$

and from the first equation,

$$x_1 = 1 - 2x_2 - x_3 = 1 + 2 - 2 = 1.$$

Thus the solution is

$$x_1 = 1, \qquad x_2 = -1, \qquad x_3 = 2.$$

The method is known as **Gaussian elimination**.

In fact, we need not write down the equations for x_1, x_2, x_3 at each stage, since all the information in (12.3) is given by the 3×4 matrix

$$\begin{bmatrix} 1 & 2 & 1 & 1 \\ -2 & 3 & -1 & -7 \\ 1 & 4 & -2 & -7 \end{bmatrix},$$

which is known as the **augmented matrix** for the system of equations: the fourth column consists of the constants on the right-hand sides of (12.3a,b,c). The elementary operations referred to previously become **elementary row operations on the matrix**. We can reproduce the steps above by the following more compact procedure:

$$\begin{bmatrix} \underline{1} & 2 & 1 & 1 \\ -2 & 3 & -1 & -7 \\ 1 & 4 & -2 & -7 \end{bmatrix} \to \begin{bmatrix} 1 & 2 & 1 & 1 \\ 0 & \underline{7} & 1 & -5 \\ 0 & 2 & -3 & -8 \end{bmatrix} \quad \begin{pmatrix} \mathbf{r}_2' = \mathbf{r}_2 + 2\mathbf{r}_1 \\ \mathbf{r}_3' = \mathbf{r}_3 - \mathbf{r}_1 \end{pmatrix}$$

$$\to \begin{bmatrix} 1 & 2 & 1 & 1 \\ 0 & 7 & 1 & -5 \\ 0 & 0 & -\tfrac{23}{7} & -\tfrac{46}{7} \end{bmatrix} \quad (\mathbf{r}_3' = \mathbf{r}_3 - \tfrac{2}{7}\mathbf{r}_2)$$

$$\to \begin{bmatrix} 1 & 2 & 1 & 1 \\ 0 & 1 & \tfrac{1}{7} & -\tfrac{5}{7} \\ 0 & 0 & 1 & 2 \end{bmatrix} \quad \begin{pmatrix} \mathbf{r}_2' = \tfrac{1}{7}\mathbf{r}_2 \\ \mathbf{r}_3' = -\tfrac{7}{23}\mathbf{r}_3 \end{pmatrix},$$

where the arrow '$\to$' means 'is transformed into'. The final matrix is said to be in **echelon** form: that is, it has zeros below the diagonal elements starting from the top left. We can now solve the equations by back substitution as before.

The elements underlined are known as **pivots** and they must be nonzero. They are used to clear the elements in the column below them. If any pivot turns out to be zero as the method progresses, then that equation or row is replaced by the first row *below* which has a nonzero coefficient in the column. If there are no further nonzero coefficients, then the pivot moves across to the next column.

It is now possible to complete the Gaussian elimination by using further row operations on the echelon matrix. Thus, continuing from the echelon form above

$$
\begin{bmatrix} 1 & 2 & 1 & 1 \\ 0 & 1 & \frac{1}{7} & -\frac{5}{7} \\ 0 & 0 & 1 & 2 \end{bmatrix} \rightarrow \begin{bmatrix} 1 & 2 & 0 & -1 \\ 0 & 1 & 0 & -1 \\ 0 & 0 & 1 & 2 \end{bmatrix} \quad \begin{pmatrix} \mathbf{r}_1' = \mathbf{r}_1 - \mathbf{r}_3 \\ \mathbf{r}_2' = \mathbf{r}_2 - \frac{1}{7}\mathbf{r}_3 \end{pmatrix}
$$

$$
\rightarrow \begin{bmatrix} 1 & 0 & 0 & 1 \\ 0 & 1 & 0 & -1 \\ 0 & 0 & 1 & 2 \end{bmatrix} \quad (\mathbf{r}_1' = \mathbf{r}_1 - 2\mathbf{r}_2),
$$

where the pivots are underlined again. The final matrix now represents the solution set $x_1 = 1$, $x_2 = -1$, $x_3 = 2$.

Example 12.2   *Using Gaussian elimination and back substitution, solve the set of equations*

$$
\begin{aligned}
x_1 + x_2 + 2x_3 &= 4, \\
2x_1 + 2x_2 + x_3 - x_4 &= -1, \\
x_2 + x_3 + x_4 &= 6, \\
x_2 - x_3 + 2x_4 &= 5.
\end{aligned}
$$

We first perform the pivotal row operations on the augmented matrix as follows:

$$
\begin{bmatrix} 1 & 1 & 2 & 0 & 4 \\ 2 & 2 & 1 & -1 & -1 \\ 0 & 1 & 1 & 1 & 6 \\ 0 & 1 & -1 & 2 & 5 \end{bmatrix} \rightarrow \begin{bmatrix} 1 & 1 & 2 & 0 & 4 \\ 0 & 0 & -3 & -1 & -9 \\ 0 & 1 & 1 & 1 & 6 \\ 0 & 1 & -1 & 2 & 5 \end{bmatrix} \quad (\mathbf{r}_2' = \mathbf{r}_2 - 2\mathbf{r}_1)
$$

$$
\rightarrow \begin{bmatrix} 1 & 1 & 2 & 0 & 4 \\ 0 & 1 & 1 & 1 & 6 \\ 0 & 0 & -3 & -1 & -9 \\ 0 & 1 & -1 & 2 & 5 \end{bmatrix} \quad (\mathbf{r}_2 \leftrightarrow \mathbf{r}_3)
$$

$$
\rightarrow \begin{bmatrix} 1 & 1 & 2 & 0 & 4 \\ 0 & 1 & 1 & 1 & 6 \\ 0 & 0 & -3 & -1 & -9 \\ 0 & 0 & -2 & 1 & -1 \end{bmatrix} \quad (\mathbf{r}_4' = \mathbf{r}_4 - \mathbf{r}_2)
$$

$$
\rightarrow \begin{bmatrix} 1 & 1 & 2 & 0 & 4 \\ 0 & 1 & 1 & 1 & 6 \\ 0 & 0 & -3 & -1 & -9 \\ 0 & 0 & 0 & \frac{5}{3} & 5 \end{bmatrix} \quad (\mathbf{r}_4' = \mathbf{r}_4 - \frac{2}{3}\mathbf{r}_3)
$$

$$
\rightarrow \begin{bmatrix} 1 & 1 & 2 & 0 & 4 \\ 0 & 1 & 1 & 1 & 6 \\ 0 & 0 & 1 & \frac{1}{3} & 3 \\ 0 & 0 & 0 & 1 & 3 \end{bmatrix} \quad \begin{pmatrix} \mathbf{r}_3' = -\frac{1}{3}\mathbf{r}_3 \\ \mathbf{r}_4' = \frac{3}{5}\mathbf{r}_4 \end{pmatrix}.
$$

Example 12.2 *continued*

(Note the row change $r_2 \leftrightarrow r_3$ because of the zero pivot.) Back substitution now gives

$$x_4 = 3, \quad x_3 = 3 - \tfrac{1}{3}x_4 = 2, \quad x_2 = 6 - x_3 - x_4 = 1, \quad x_1 = 4 - x_2 - 2x_3 = -1.$$

12.3 The inverse matrix by Gaussian elimination

Matrix multiplication is a row-on-column operation (see Section 7.2), so any elementary row operation applied simultaneously to both A and I in the identity $AA^{-1} = I$ maintains the equality. To obtain A^{-1} from A, apply to both sides of the identity a sequence of row operations that transform A into I, so that on the left AA^{-1} becomes A^{-1}. The parallel transformation of I gives A^{-1} explicitly. In other words:

> **Matrix inversion**
> Use elementary row operations to transform A into the identity I, and use the same operations to transform I into A^{-1}.   **(12.5)**

Suppose that we require the inverse of

$$A = \begin{bmatrix} 0 & 1 & 0 & 2 \\ 1 & 0 & 1 & 0 \\ 0 & 1 & 0 & 1 \\ 1 & 0 & 2 & 0 \end{bmatrix}.$$

We reduce A to I_4 and perform the same row operations on I_4. Thus, we can write down the steps **in parallel** as follows:

$$A = \begin{bmatrix} 0 & 1 & 0 & 2 \\ 1 & 0 & 1 & 0 \\ 0 & 1 & 0 & 1 \\ 1 & 0 & 2 & 0 \end{bmatrix} \qquad\qquad I_4 = \begin{bmatrix} 1 & 0 & 0 & 0 \\ 0 & 1 & 0 & 0 \\ 0 & 0 & 1 & 0 \\ 0 & 0 & 0 & 1 \end{bmatrix}$$

$$\rightarrow \begin{bmatrix} \underline{1} & 0 & 1 & 0 \\ 0 & 1 & 0 & 2 \\ 0 & 1 & 0 & 1 \\ 1 & 0 & 2 & 0 \end{bmatrix} (r_1 \leftrightarrow r_2) \qquad \rightarrow \begin{bmatrix} 0 & 1 & 0 & 0 \\ 1 & 0 & 0 & 0 \\ 0 & 0 & 1 & 0 \\ 0 & 0 & 0 & 1 \end{bmatrix}$$

$$\rightarrow \begin{bmatrix} 1 & 0 & 1 & 0 \\ 0 & \underline{1} & 0 & 2 \\ 0 & 1 & 0 & 1 \\ 0 & 0 & 1 & 0 \end{bmatrix} (r_4' = r_4 - r_1) \qquad \rightarrow \begin{bmatrix} 0 & 1 & 0 & 0 \\ 1 & 0 & 0 & 0 \\ 0 & 0 & 1 & 0 \\ 0 & -1 & 0 & 1 \end{bmatrix}$$

$$\rightarrow \begin{bmatrix} 1 & 0 & 1 & 0 \\ 0 & 1 & 0 & 2 \\ 0 & 0 & 0 & -1 \\ 0 & 0 & 1 & 0 \end{bmatrix} \ (r_3' = r_3 - r_2) \qquad \rightarrow \begin{bmatrix} 0 & 1 & 0 & 0 \\ 1 & 0 & 0 & 0 \\ -1 & 0 & 1 & 0 \\ 0 & -1 & 0 & 1 \end{bmatrix}$$

$$\rightarrow \begin{bmatrix} 1 & 0 & 1 & 0 \\ 0 & 1 & 0 & 2 \\ 0 & 0 & 1 & 0 \\ 0 & 0 & 0 & -1 \end{bmatrix} \ (r_3 \leftrightarrow r_4) \qquad \rightarrow \begin{bmatrix} 0 & 1 & 0 & 0 \\ 1 & 0 & 0 & 0 \\ 0 & -1 & 0 & 1 \\ -1 & 0 & 1 & 0 \end{bmatrix}$$

$$\rightarrow \begin{bmatrix} 1 & 0 & 1 & 0 \\ 0 & 1 & 0 & 2 \\ 0 & 0 & 1 & 0 \\ 0 & 0 & 0 & 1 \end{bmatrix} \ (r_4' = -r_4) \qquad \rightarrow \begin{bmatrix} 0 & 1 & 0 & 0 \\ 1 & 0 & 0 & 0 \\ 0 & -1 & 0 & 1 \\ 1 & 0 & -1 & 0 \end{bmatrix}$$

$$\rightarrow \begin{bmatrix} 1 & 0 & 1 & 0 \\ 0 & 1 & 0 & 0 \\ 0 & 0 & 1 & 0 \\ 0 & 0 & 0 & 1 \end{bmatrix} \ (r_2' = r_2 - 2r_4) \qquad \rightarrow \begin{bmatrix} 0 & 1 & 0 & 0 \\ -1 & 0 & 2 & 0 \\ 0 & -1 & 0 & 1 \\ 1 & 0 & -1 & 0 \end{bmatrix}$$

$$\rightarrow \begin{bmatrix} 1 & 0 & 0 & 0 \\ 0 & 1 & 0 & 0 \\ 0 & 0 & 1 & 0 \\ 0 & 0 & 0 & 1 \end{bmatrix} \ (r_1' = r_1 - r_3) \qquad \rightarrow \begin{bmatrix} 0 & 2 & 0 & -1 \\ -1 & 0 & 2 & 0 \\ 0 & -1 & 0 & 1 \\ 1 & 0 & -1 & 0 \end{bmatrix}$$

$$= I_4 \qquad\qquad = A^{-1}.$$

We conclude that

$$A^{-1} = \begin{bmatrix} 0 & 2 & 0 & -1 \\ -1 & 0 & 2 & 0 \\ 0 & -1 & 0 & 1 \\ 1 & 0 & -1 & 0 \end{bmatrix}.$$

12.4 Compatible and incompatible sets of equations

Not all sets of equations have solutions. For example, the simultaneous equations

$$x + y = 1,$$
$$x + y = 2,$$

clearly have no solutions. On the other hand the equations

$$x + 2y = 1,$$
$$2x + 4y = 2,$$

have the infinity of solutions

$$x = 1 - 2\lambda, \quad y = \lambda$$

for any real number λ. For two equations in two unknowns the occurrences of unique solutions are easy to detect: for higher-order sets of equations these possibilities are not so obvious.

Consider the set of equations

$$x + y - \ z = 3,$$
$$3x - y + 3z = 5,$$
$$x - y + 2z = 2.$$

We can sense that there might be a problem by first evaluating the determinant of the coefficients of x, y, and z. Thus

$$\begin{vmatrix} 1 & 1 & -1 \\ 3 & -1 & 3 \\ 1 & -1 & 2 \end{vmatrix} = \begin{vmatrix} 1 & 1 & -1 \\ 4 & 0 & 2 \\ 2 & 0 & 1 \end{vmatrix} \quad \begin{pmatrix} \mathbf{r}_2' = \mathbf{r}_2 + \mathbf{r}_1 \\ \mathbf{r}_3' = \mathbf{r}_3 + \mathbf{r}_1 \end{pmatrix}$$

$$= -\begin{vmatrix} 4 & 2 \\ 2 & 1 \end{vmatrix} = 0.$$

Thus Cramer's rule will fail, although there still may be solutions. We can determine whether solutions exist more readily by using Gaussian elimination. In this case the application of row operations on the augmented matrix leads to

$$\begin{bmatrix} 1 & 1 & -1 & 3 \\ 3 & -1 & 3 & 5 \\ 1 & -1 & 2 & 2 \end{bmatrix} \rightarrow \begin{bmatrix} 1 & 1 & -1 & 3 \\ 0 & -4 & 6 & -4 \\ 0 & -2 & 3 & -1 \end{bmatrix} \quad \begin{pmatrix} \mathbf{r}_2' = \mathbf{r}_2 - 3\mathbf{r}_1 \\ \mathbf{r}_3' = \mathbf{r}_3 - \mathbf{r}_1 \end{pmatrix}$$

$$\rightarrow \begin{bmatrix} 1 & 1 & -1 & 3 \\ 0 & -4 & 6 & -4 \\ 0 & 0 & 0 & 1 \end{bmatrix} \quad (\mathbf{r}_3' = \mathbf{r}_3 - \tfrac{1}{2}\mathbf{r}_2),$$

which is the echelon form for this set of equations. However, row 3 is impossible to be satisfied. Hence these equations can have no solutions.

On the other hand, consider the following set:

$$x + y - \ z = 1,$$
$$3x - y + 3z = 5,$$
$$x - y + 2z = 2$$

(this is the previous set with one change to the first equation). Gaussian elimination now gives

$$\begin{bmatrix} 1 & 1 & -1 & 1 \\ 3 & -1 & 3 & 5 \\ 1 & -1 & 2 & 2 \end{bmatrix} \rightarrow \begin{bmatrix} 1 & 1 & -1 & 1 \\ 0 & -4 & 6 & 2 \\ 0 & -2 & 3 & 1 \end{bmatrix} \quad \begin{pmatrix} \mathbf{r}_2' = \mathbf{r}_2 - 3\mathbf{r}_1 \\ \mathbf{r}_3' = \mathbf{r}_3 - \mathbf{r}_1 \end{pmatrix}$$

$$\rightarrow \begin{bmatrix} 1 & 1 & -1 & 1 \\ 0 & -4 & 6 & 2 \\ 0 & 0 & 0 & 0 \end{bmatrix} \quad (\mathbf{r}_3' = \mathbf{r}_3 - \tfrac{1}{2}\mathbf{r}_2).$$

Row 3 is now consistent, and row 2 is $-4y + 6z = 2$. Hence

$$y = -\tfrac{1}{4}(2 - 6z)$$

and, from row 1,

$$x = 1 - y + z = \tfrac{3}{2} - \tfrac{1}{2}z.$$

Thus z can take any value, say λ, so the full solution set is

$$\begin{bmatrix} x \\ y \\ z \end{bmatrix} = \begin{bmatrix} \tfrac{3}{2} - \tfrac{1}{2}\lambda \\ -\tfrac{1}{2} + \tfrac{3}{2}\lambda \\ \lambda \end{bmatrix}$$

for any value of λ. It can be seen in this case that there exists an infinite number of solutions, a different one for each different value of λ.

Geometrically, in three dimensions, it can be seen why equations can have a unique solution, no solution, or an infinite set of solutions. Any equation such as

$$ax + by + cz = d$$

represents a plane in $\mathbb{R}^3$. Three equations represent three planes, and we need only visualize how they might intersect or not. The coordinates of any point of intersection of the planes is the solution of the equations. The three diagrams in Fig. 12.1 show how three planes can intersect in a single point, no point, or a line of points.

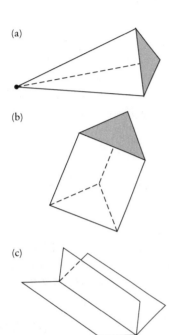

(a)

(b)

(c)

Fig. 12.1 (a) Unique solution, (b) no solution, (c) line of solutions.

Example 12.3   *Determine the complete sets of values for a and b which make the equations*

$$x - 2y + 3z = 2,$$
$$2x - y + 2z = 3,$$
$$x + y + az = b$$

have (i) a unique solution, (ii) no solutions, (iii) an infinite set of solutions.

Reduce the augmented matrix to echelon form using pivots to clear each column successively:

$$\begin{bmatrix} 1 & -2 & 3 & 2 \\ 2 & -1 & 2 & 3 \\ 1 & 1 & a & b \end{bmatrix} \rightarrow \begin{bmatrix} 1 & -2 & 3 & 2 \\ 0 & 3 & -4 & -1 \\ 0 & 3 & a-3 & b-2 \end{bmatrix} \begin{pmatrix} \mathbf{r}_2' = \mathbf{r}_2 - 2\mathbf{r}_1 \\ \mathbf{r}_3' = \mathbf{r}_3 - \mathbf{r}_1 \end{pmatrix}$$

$$\rightarrow \begin{bmatrix} 1 & -2 & 3 & 2 \\ 0 & 3 & -4 & -1 \\ 0 & 0 & a+1 & b-1 \end{bmatrix} \ (\mathbf{r}_3' = \mathbf{r}_3 - \mathbf{r}_2).$$

We can now interpret the echelon matrix.

Example 12.3 *continued*

(i) If $a \neq -1$, then z has the unique solution

$$z = \frac{b - 1}{a + 1}.$$

Also y and x can be found by back substitution.

(ii) If $a = -1$, and $b \neq 1$ then row 3 will lead to an inconsistency. Hence there are no solutions of the equations.

(iii) If $a = -1$, and $b = 1$ then row 3 implies $z = \lambda$ for any number λ. Also x and y can be found by back substitution.

This example illustrates the advantage of Gaussian elimination over the formula-based method of Cramer's rule. The Gaussian method can still be used if the number of equations differs from the number of unknowns. Consider the following example.

Example 12.4 *Investigate all solutions of*

$$\begin{aligned}
x + y - \ z &= 1 \\
3x - y + 3z &= 5, \\
x - y + 2z &= 2, \\
x \quad\quad + \ z &= 3.
\end{aligned}$$

There are 3 unknowns and 4 equations. The augmented matrix is

$$\begin{bmatrix} 1 & 1 & -1 & 1 \\ 3 & -1 & 3 & 5 \\ 1 & -1 & 2 & 2 \\ 1 & 0 & 1 & 3 \end{bmatrix} \rightarrow \begin{bmatrix} 1 & 1 & -1 & 1 \\ 0 & -4 & 6 & 2 \\ 0 & -2 & 3 & 1 \\ 0 & -1 & 2 & 2 \end{bmatrix} \begin{pmatrix} r_2' = r_2 - 3r_1 \\ r_3' = r_3 - r_1 \\ r_4' = r_4 - r_1 \end{pmatrix}$$

$$\rightarrow \begin{bmatrix} 1 & 1 & -1 & 1 \\ 0 & -4 & 6 & 2 \\ 0 & 0 & 0 & 0 \\ 0 & 0 & \frac{1}{2} & \frac{3}{2} \end{bmatrix} \begin{pmatrix} r_3' = r_3 - \frac{1}{2}r_2 \\ r_4' = r_4 - \frac{1}{4}r_2 \end{pmatrix}$$

$$\rightarrow \begin{bmatrix} 1 & 1 & -1 & 1 \\ 0 & -4 & 6 & 2 \\ 0 & 0 & \frac{1}{2} & \frac{3}{2} \\ 0 & 0 & 0 & 0 \end{bmatrix} \quad (r_3 \leftrightarrow r_4).$$

Row 4 is consistent, while row 3 implies $z = 3$. Then y and x can be found by back substitution in rows 2 and 1. Confirm that $y = 4$ and $x = 0$.

On the other hand, there may be more variables than equations, as in the following example. (Inconsistency is still possible.)

Example 12.5   *Show that the following equations are inconsistent:*

$$x_1 + x_2 + x_3 + 2x_4 = 1,$$
$$x_1 - 2x_2 + 3x_3 - x_4 = 4,$$
$$3x_1 - 3x_2 + 7x_3 \qquad = 7.$$

Proceed as before, and successively reduce the augmented matrix by pivots. Thus

$$\begin{bmatrix} \underline{1} & 1 & 1 & 2 & 1 \\ 1 & -2 & 3 & -1 & 4 \\ 3 & -3 & 7 & 0 & 7 \end{bmatrix} \rightarrow \begin{bmatrix} 1 & 1 & 1 & 2 & 1 \\ 0 & -\underline{3} & 2 & -3 & 3 \\ 0 & -6 & 4 & -6 & 4 \end{bmatrix} \quad \begin{pmatrix} \mathbf{r}_2' = \mathbf{r}_2 - \mathbf{r}_1 \\ \mathbf{r}_3' = \mathbf{r}_3 - 3\mathbf{r}_1 \end{pmatrix}$$

$$\rightarrow \begin{bmatrix} 1 & 1 & 12 & 1 & 1 \\ 0 & -3 & 2 & -3 & 3 \\ 0 & 0 & 0 & 0 & -2 \end{bmatrix} \quad (\mathbf{r}_3' = \mathbf{r}_3 - 2\mathbf{r}_2).$$

Since $0 \neq 2$, row 3 indicates an inconsistency, so the equations are incompatible.

12.5 Homogeneous sets of equations

Any set of equations $Ax = 0$ is known as a **homogeneous** set; it is a set of linear equations with zero right-hand sides. Clearly, the equations always have the so-called **trivial** solution $x = 0$, but there may exist **non-trivial** solutions. What are the conditions for their existence? Consider the following example.

Example 12.6   *Find the value of a for which the following equations have non-trivial solutions:*

$$x + y + z = 0,$$
$$x + 2y \qquad = 0,$$
$$x - 3y + az = 0.$$

Proceed in the usual way using Gaussian reduction. Thus

$$\begin{bmatrix} 1 & 1 & 1 & 0 \\ 1 & 2 & 0 & 0 \\ 1 & -3 & a & 0 \end{bmatrix} \rightarrow \begin{bmatrix} \underline{1} & 1 & 1 & 0 \\ 0 & 1 & -1 & 0 \\ 0 & -4 & a-1 & 0 \end{bmatrix} \quad \begin{pmatrix} \mathbf{r}_2' = \mathbf{r}_2 - \mathbf{r}_1 \\ \mathbf{r}_3' = \mathbf{r}_3 - \mathbf{r}_1 \end{pmatrix}$$

$$\rightarrow \begin{bmatrix} 1 & 1 & 1 & 0 \\ 0 & \underline{1} & -1 & 0 \\ 0 & 0 & a-5 & 0 \end{bmatrix} \quad (\mathbf{r}_3' = \mathbf{r}_3 + 4\mathbf{r}_2).$$

Non-trivial solutions exist if, and only if, $a = 5$. Put $z = \lambda$, any number. Then by back substitution

$$y = z = \lambda, \quad x = -y - z = -2\lambda$$

for any λ. The solution is non-trivial if $\lambda \neq 0$.

If A is a square matrix, then Cramer's rule (Section 12.1) implies that

$$x \det A = 0.$$

Hence x is a nonzero column vector only if $\det A = 0$. Further, it can be proved that if $\det A = 0$ then there is an infinite number of solutions. We therefore have the following test:

> **Homogeneous equations $Ax = 0$, where A is square.**
>
> If $\det A = 0$, there is an infinite number of non-trivial solutions.
> If $\det A \neq 0$, the only solution is $x = 0$. **(12.6)**

Example 12.7 *Find all conditions on the constants a, b, and c in order that*

$$\begin{aligned}
x + \ y + \ z &= 0, \\
ax + \ by + \ cz &= 0, \\
a^2 x + b^2 y + c^2 z &= 0
\end{aligned}$$

should have non-trivial solutions. Find the solutions in the cases (a) $a = 1, b = 1, c = 2$; (b) $a = 1, b = 1, c = 1$.

This system of equations will have non-trivial solutions for x, y, and z if, and only if,

$$D = \begin{vmatrix} 1 & 1 & 1 \\ a & b & c \\ a^2 & b^2 & c^2 \end{vmatrix} = 0.$$

Thus

$$D = \begin{vmatrix} 1 & 0 & 0 \\ a & b - a & c - a \\ a^2 & b^2 - a^2 & c^2 - a^2 \end{vmatrix} \quad \begin{pmatrix} c'_2 = c_2 - c_1 \\ c'_3 = c_3 - c_1 \end{pmatrix}$$

$$= (b - a)(c - a) \begin{vmatrix} 1 & 0 & 0 \\ a & 1 & 1 \\ a^2 & b + a & c + a \end{vmatrix}$$

$$= (b - a)(c - a)(c + a - b - a)$$

$$= (b - c)(c - a)(a - b).$$

Hence, non-trivial solutions exist if $b = c$, $c = a$, or $a = b$.

(a) $(a = 1, b = 1, c = 2)$ The equations become

$$\begin{aligned}
x + y + \ z &= 0, \\
x + y + 2z &= 0, \\
x + y + 4z &= 0.
\end{aligned}$$

Example 12.7 *continued*

The augmented matrix is

$$\begin{bmatrix} 1 & 1 & 1 & 0 \\ 1 & 1 & 2 & 0 \\ 1 & 1 & 4 & 0 \end{bmatrix} \rightarrow \begin{bmatrix} 1 & 1 & 1 & 0 \\ 0 & 0 & 1 & 0 \\ 0 & 0 & 3 & 0 \end{bmatrix} \begin{pmatrix} \mathbf{r}_2' = \mathbf{r}_2 - \mathbf{r}_1 \\ \mathbf{r}_3' = \mathbf{r}_3 - \mathbf{r}_1 \end{pmatrix}$$

$$\rightarrow \begin{bmatrix} 1 & 1 & 1 & 0 \\ 0 & 0 & 1 & 0 \\ 0 & 0 & 0 & 0 \end{bmatrix} \quad (\mathbf{r}_3' = \mathbf{r}_3 - 3\mathbf{r}_2).$$

Row 2 implies $z = 0$, while row 1 implies $x = -y$. Let $y = \lambda$, say. Then the solution set is

$$\begin{bmatrix} x \\ y \\ z \end{bmatrix} = \begin{bmatrix} -\lambda \\ \lambda \\ 0 \end{bmatrix} = \begin{bmatrix} -1 \\ 1 \\ 0 \end{bmatrix} \lambda,$$

for any λ.

(b) $(a = 1, b = 1, c = 1)$ Applying Gaussian elimination, we find that

$$\begin{bmatrix} 1 & 1 & 1 & 0 \\ 1 & 1 & 1 & 0 \\ 1 & 1 & 1 & 0 \end{bmatrix} \rightarrow \begin{bmatrix} 1 & 1 & 1 & 0 \\ 0 & 0 & 0 & 0 \\ 0 & 0 & 0 & 0 \end{bmatrix} \begin{pmatrix} \mathbf{r}_2' = \mathbf{r}_2 - \mathbf{r}_1 \\ \mathbf{r}_3' = \mathbf{r}_3 - \mathbf{r}_1 \end{pmatrix}$$

Hence, we are left with row 1, which implies

$$x + y + z = 0.$$

Let $z = \lambda$ and $y = \mu$. Then $x = -\lambda - \mu$. Hence the solution is

$$\begin{bmatrix} x \\ y \\ z \end{bmatrix} = \begin{bmatrix} -\lambda - \mu \\ \mu \\ \lambda \end{bmatrix} = \begin{bmatrix} -1 \\ 0 \\ 1 \end{bmatrix} \lambda + \begin{bmatrix} -1 \\ 1 \\ 0 \end{bmatrix} \mu$$

for any λ and any μ. Note that this is a two-parameter solution set.

12.6 Gauss–Seidel iterative method of solution

The method of Gaussian elimination, described in Section 12.2, is not a practical approach, by hand, for a large system with perhaps 30 equations in 30 unknowns. Whatever method is employed, the equations will have to be solved by computer. But decisions about what scheme may be the best for a given set of equations are not always easy. There are many direct iterative methods in addition to the row-operation method described in Section 12.2. Two such methods will be briefly explained here.

Consider the equations

$$3x_1 + x_2 + x_3 = -1,$$
$$-x_1 + 4x_2 + x_3 = -8,$$
$$2x_1 + x_2 + 5x_3 = -14.$$

Write the equations as

$$x_1 = \tfrac{1}{3}(-x_2 - x_3 - 1), \tag{12.7}$$

$$x_2 = \tfrac{1}{4}(x_1 - x_3 - 8), \tag{12.8}$$

$$x_3 = \tfrac{1}{5}(-2x_1 - x_2 - 14), \tag{12.9}$$

where x_1, x_2, and x_3 are now the subjects of the three equations.

To start the iteration choose initial values for x_2 and x_3, say $x_2^{(0)} = 0$ and $x_3^{(0)} = 0$, without thinking about the equations. Calculate $x_1^{(1)}$ from eqn (12.7) as

$$x_1^{(1)} = \tfrac{1}{3}(-x_2^{(0)} - x_3^{(0)} - 1). \tag{12.10}$$

Use $x_1^{(1)}$ in (12.8) with $x_3^{(0)}$. Thus

$$x_2^{(1)} = \tfrac{1}{4}(x_1^{(1)} - x_3^{(0)} - 8). \tag{12.11}$$

Finally, use the update $x_2^{(1)}$ in (12.9) to find $x_3^{(1)}$:

$$x_3^{(1)} = \tfrac{1}{5}(2x_1^{(1)} - x_2^{(1)} - 14). \tag{12.12}$$

Hence we have calculated a new approximate solution given by $x_1^{(1)}$, $x_2^{(1)}$, $x_3^{(2)}$. Now repeat the calculations starting with $x_2^{(1)}$ and $x_3^{(1)}$ as the new initial values to obtain $x_1^{(2)}$, $x_2^{(2)}$, $x_3^{(3)}$. The output from these **iterations** is shown in the following table.

i	0	1	2	3	4	5	6	7
$x_1^{(i)}$	–	−0.3333	1.1110	1.0570	1.0030	0.9984	0.9997	1.0000
$x_2^{(i)}$	0	−2.0830	−1.1600	−0.9825	−0.9926	−0.9997	−1.0000	−1.0000
$x_3^{(i)}$	0	−2.2500	−3.0130	−3.0260	−3.0030	−2.9990	−3.0000	−3.0000

All solutions are quoted to 4 decimal places. It can be seen that the exact solution, which is $x_1 = 1$, $x_2 = -1$, $x_3 = -3$, can be achieved to this accuracy after seven steps for this example. This is known as the **Gauss–Seidel scheme** for numerical solution of linear equations.

An alternative method without updating given by

$$x_1^{(1)} = \tfrac{1}{3}(-x_2^{(0)} - x_3^{(0)} - 1),$$

$$x_2^{(1)} = \tfrac{1}{4}(x_1^{(0)} - x_3^{(0)} - 8),$$

$$x_3^{(1)} = \tfrac{1}{5}(-2x_1^{(0)} - x_2^{(0)} - 14)$$

is known as **Jacobi's method**. However, convergence to the exact solution is slower with this scheme.

These methods do not always converge to the solution. For example, the Gauss–Seidel scheme applied to the system (12.2) fails since the iterates continue to increase in size. It can be shown that the Gauss–Seidel scheme converges if the magnitude of each leading diagonal element exceeds the sum of the magnitudes of the remaining elements in the same row of the matrix of coefficients. This is the case for the system given by (12.8a,b,c). Here the matrix of coefficients is

$$\begin{bmatrix} 3 & 1 & 1 \\ -1 & 4 & 1 \\ 2 & 1 & 5 \end{bmatrix}.$$

Each of the diagonal elements dominates the remaining elements in that row, since

$$3 \geqslant 1 + 1 = 2, \qquad 4 \geqslant |-1| + 1 = 2, \qquad 5 \geqslant 2 + 1 = 3.$$

This property of the system of equations is known as **diagonal dominance**. If the matrix is not diagonally dominant, then the scheme may or may not converge. Usually a few steps will indicate whether this is likely to be the case.

The schemes for both these methods can be expressed in matrix form as follows. Let the system of equations

$$Ax = d,$$

where $A = [a_{ij}]$ is an $n \times n$ matrix. Let

$$A = A_L + D + A_U,$$

where A_L, D, and A_U are respectively the lower triangular, diagonal, and upper triangular matrices given by

$$A_L = \begin{bmatrix} 0 & 0 & \cdots & 0 \\ a_{21} & 0 & \cdots & 0 \\ a_{31} & a_{32} & \cdots & 0 \\ \vdots & \vdots & \vdots & \vdots \\ a_{n1} & a_{n2} & \cdots & 0 \end{bmatrix}, \quad D = \begin{bmatrix} a_{11} & 0 & \cdots & 0 \\ 0 & a_{22} & \cdots & 0 \\ \vdots & \vdots & \vdots & \vdots \\ 0 & 0 & \cdots & a_{nn} \end{bmatrix},$$

$$A_U = \begin{bmatrix} 0 & a_{12} & a_{13} & \cdots & a_{1n} \\ 0 & 0 & a_{23} & \cdots & a_{2n} \\ \vdots & \vdots & \vdots & \vdots & \vdots \\ 0 & 0 & 0 & \cdots & 0 \end{bmatrix}.$$

The matrix equation becomes

$$A_L x + D x + A_U x = d.$$

It is easy to find x from an equation of the form $Dx = \cdots$, since D is a diagonal matrix with a simple inverse. This is the matrix which is updated by Jacobi's method. Assuming that $x^{(0)} = [x_1^{(0)}, \dots, x_n^{(0)}]^T$ is the given initial estimate, the approximate solution at step r will be computed from

$$Dx^{(r)} = -A_U x^{(r-1)} - A_L x^{(r-1)} + d.$$

On the other hand, in the Gauss–Seidel scheme, we take advantage of the observation that $x_1^{(r)}$, $x_2^{(r)}$, $\dots$ are successively computed from rows 1, 2, $\dots$. Hence they can be used in the rows that follow. Thus the Gauss–Seidel iterations are given by

$$Dx^{(r)} = -A_L x^{(r)} - A_U x^{(r-1)} + d.$$

Problems

12.1 (Section 12.1). Solve the following systems of linear equations using Cramer's rule:

(a)
$$x_1 \quad\quad + x_3 = 1,$$
$$x_2 - x_3 = 3,$$
$$2x_1 + x_2 \quad\quad = -1;$$

(b)
$$x_1 + 7x_2 + \quad x_3 = 1,$$
$$x_2 - \quad x_3 = 3,$$
$$2x_1 + \quad x_2 + 10x_3 = -1;$$

(c)
$$x_1 + 5x_2 - x_3 = 1,$$
$$-3x_1 + x_2 - x_3 = 1,$$
$$3x_1 + x_2 + x_3 = -3;$$

(d)
$$x_1 + \quad x_2 + \quad x_3 = 1,$$
$$ax_1 + bx_2 + cx_3 = d,$$
$$a^2 x_1 + b^2 x_2 + c^2 x_3 = d^2;$$

(e)
$$x_1 + 5x_2 \quad\quad + 2x_4 = 1,$$
$$- 3x_2 \quad\quad - x_4 = 1,$$
$$3x_2 + x_3 + \quad x_4 = 1,$$
$$2x_2 + x_3 + \quad x_4 = 2.$$

12.2 (Section 12.1). The currents i_1, i_2, i_3 (in amps) flow in parts of a circuit which contains a variable resistor of resistance R (in ohms). The equations for the currents are given by

$$4i_1 - \quad i_2 - \quad i_3 = 12,$$
$$-i_1 + Ri_2 \quad\quad = 24,$$
$$i_1 \quad\quad + 5i_3 = -12,$$

in terms of the voltages on the right-hand side. For design reasons, the current i_3 should be 2 amps. How many ohms should the resistance R be?

12.3 (Section 12.4). Show that the following sets of equations are inconsistent.

(a)
$$x_1 + 2x_2 + \quad x_3 = 3,$$
$$x_1 - 3x_2 + 2x_3 = 4,$$
$$5x_1 + 5x_2 + 6x_3 = 1;$$

(b)
$$x_1 + x_2 + \quad x_3 \quad\quad = 2,$$
$$x_1 \quad\quad + \quad x_3 + 2x_4 = 3,$$
$$x_1 + x_2 \quad\quad + \quad x_4 = 4,$$
$$- x_2 + 2x_3 \quad\quad = 2;$$

(c)
$$x_1 + \quad x_2 \quad\quad\quad\quad = 1,$$
$$x_2 + \quad x_3 \quad\quad\quad = 1,$$
$$x_3 + \quad x_4 \quad\quad = 1,$$
$$x_4 + \quad x_5 = 1,$$
$$x_1 + 3x_2 + 5x_3 + 7x_4 + 4x_5 = 1.$$

12.4 (Section 12.4). Determine the complete set of values for a and b that make the equations

$$x + \quad y - \quad z = 2,$$
$$2x + 3y + \quad z = 3,$$
$$5x + 7y + az = b$$

have (i) a unique solution, (ii) no solutions, (iii) an infinite set of solutions.

12.5 (Section 12.4). Investigate all solutions of the system

$$x - \quad y + 2z = 1,$$
$$x + \quad y + 3z = 2,$$
$$x + 2y - \quad z = 3,$$
$$x - 2y + 6z = 0.$$

12.6 (Section 12.4). Show that the following equations are inconsistent.

$$x_1 + \quad x_2 + \quad x_3 - x_4 = 10,$$
$$x_1 - \quad x_2 - \quad x_3 \quad\quad = 1,$$
$$4x_1 - 2x_2 - 2x_3 - x_4 = 5.$$

12.7 (Section 12.2). Solve the following equations by Gaussian elimination.

$$x_2 + 2x_3 - \quad x_4 = 11,$$
$$x_1 + x_2 + \quad x_3 + \quad x_4 = 1,$$
$$2x_1 + x_2 - \quad x_3 + 4x_4 = 0,$$
$$x_1 - x_2 + \quad x_3 - 2x_4 = 2.$$

12.8 (Section 12.4). Find the value of a for which the linear equations

$$ax - \quad y + 2z = 1,$$
$$x + 2y - az = 2,$$
$$4x + \quad y - 2z = 2$$

have no solutions.

12.9 (Section 12.3). Find the inverses of the following matrices.

(a)
$$\begin{bmatrix} 6 & -3 & 6 \\ 3 & 6 & 6 \\ -12 & -3 & 6 \end{bmatrix};$$

(b)
$$\begin{bmatrix} 1 & -1 & 2 \\ 1 & 2 & 1 \\ -4 & -1 & 2 \end{bmatrix};$$

(c) $\begin{bmatrix} 2 & -1 & 2 & 0 \\ 1 & 0 & -1 & 2 \\ 0 & 0 & -1 & 2 \\ -1 & 0 & 1 & 0 \end{bmatrix}$; (d) $\begin{bmatrix} 1 & 1 & 0 & 0 & 0 & 0 \\ 0 & 1 & 0 & 0 & 0 & 0 \\ 0 & 0 & 1 & 0 & 0 & 0 \\ 0 & 0 & 0 & 1 & 0 & 0 \\ 0 & 0 & 0 & 0 & 1 & 1 \\ 0 & 0 & 0 & 0 & 0 & 1 \end{bmatrix}$;

(e) $\begin{bmatrix} 1 & 0 & 0 & 0 & 0 \\ 1 & 1 & 0 & 0 & 0 \\ 1 & 1 & 1 & 0 & 0 \\ 1 & 1 & 1 & 1 & 0 \\ 1 & 1 & 1 & 1 & 1 \end{bmatrix}$.

12.10 Show that

$$\begin{bmatrix} 1 & 0 & 1 & 0 & 1 \\ 0 & 1 & 0 & 1 & 0 \\ 1 & 0 & 1 & 0 & 1 \\ 0 & 1 & 0 & 1 & 0 \\ 1 & 0 & 1 & 0 & 1 \end{bmatrix}$$

is a singular matrix.

12.11 (Section 12.1). The four planes

$$6x - 3y - z = -3,$$
$$2x - y + 5z = 15,$$
$$y + z = 1,$$
$$2x + y - z = 1$$

are the faces of a tetrahedron. Find the coordinates of all its vertices.

12.12 A light source is situated at the point $P : (3, 2, 2)$. A triangle has the points $A : (1, 1, 1)$, $B : (1, 0, 1)$, $C : (2, 1, 1)$ as vertices. Find the coordinates of the vertices of the shadow of the triangle on the coordinate planes $x = 0$, $y = 0$, and $z = 0$.

12.13 The parabola

$$y = \alpha + \beta x + \gamma x^2$$

is required which passes through the three points (x_1, y_1), (x_2, y_2), and (x_3, y_3). When solutions exist, find α, β, and γ, and discuss the cases where there are no solutions.

12.14 Find all values of the constants λ and μ in order that the equations

$$x + y + z = 4,$$
$$x - y + z = 2,$$
$$2x + y - \lambda z = \mu$$

may have (a) just one solution, (b) no solutions, (c) an infinite set of solutions.

12.15 (Section 12.3). For each of the sets of equations below, set up the augmented matrix and, using elementary row operations, decide on the consistency of the equations. If they are consistent, obtain all solutions in each case.

(a) $x + y + z = 3,$
$3x + 5y + z = -1,$
$x + 2y = 0;$

(b) $y + z = 1,$
$x + y + 2z = 3,$
$x - y = 1;$

(c) $x + 2y + z = 4,$
$x + y = -1,$
$3x + 4y - z = 12.$

12.16 (Section 12.5). Find all solutions of the determinant equation

$$\begin{vmatrix} 1 - k & 2 & -1 \\ 2 & 1 - k & -1 \\ -1 & -1 & 2 - k \end{vmatrix} = 0.$$

What are the values of k for which the following set of equations has non-trivial solutions?

$$(1 - k)x + 2y - z = 0,$$
$$2x + (1 - k)y - z = 0,$$
$$-x - y + (2 - k)z = 0.$$

12.17 (Section 12.5). Show that

$$\begin{vmatrix} a^2 + t & ab & ca \\ ab & b^2 + t & bc \\ ca & bc & c^2 + t \end{vmatrix} = t^2(t + a^2 + b^2 + c^2).$$

For what values of t do the equations

$$(1 + t)x + 2y + 3z = 0,$$
$$2x + (4 + t)y + 6z = 0,$$
$$3x + 6y + (9 + t)z = 0$$

have non-trivial solutions? Find all solutions in each case.

12.18 (Section 12.5). For what values of k do the equations

$$kx_1 + 4x_2 - x_3 + 3x_4 = 0,$$
$$4x_1 + kx_2 - x_3 + 3x_4 = 0,$$
$$4x_1 - x_2 + kx_3 + 3x_4 = 0,$$
$$4x_1 - x_2 + 3x_3 + kx_4 = 0$$

have non-trivial solutions?

12.19 (Section 12.4). Show that the equations

$$x_1 + 2x_2 + 3x_3 \qquad = 4,$$
$$2x_1 + 3x_2 + 8x_3 - x_4 = 20,$$
$$2x_1 + 5x_2 + 4x_3 + x_4 = 5$$

are inconsistent.

12.20 (Section 12.3). Find the inverses of

$$\begin{bmatrix} 1 & \lambda & 0 \\ 0 & 1 & \lambda \\ 0 & 0 & 1 \end{bmatrix} \text{ and } \begin{bmatrix} 1 & 0 & 0 \\ \mu & 1 & 0 \\ 0 & \mu & 1 \end{bmatrix}.$$

Hence find the inverse of

$$\begin{bmatrix} 1 + \lambda\mu & \lambda & 0 \\ \mu & 1 + \lambda\mu & \lambda \\ 0 & \mu & 1 \end{bmatrix}.$$

Find the inverse of

$$\begin{bmatrix} 13 & 3 & 0 \\ 4 & 13 & 3 \\ 0 & 4 & 1 \end{bmatrix}.$$

12.21 (Section 12.5). Express the determinant

$$\begin{vmatrix} 1 & 1 & 1 \\ a^2 & b^2 & c^2 \\ a(b+c) & b(c+a) & c(a+b) \end{vmatrix}$$

as the product of factors.

Obtain the values of a, b, and c for which non-trivial solutions of

$$x + a^2 y + a(b+c)z = 0,$$
$$x + b^2 y + b(c+a)z = 0,$$
$$x + c^2 y + c(a+b)z = 0$$

exist. Find the complete solution in the case $a + b = -c$.

12.22 (Section 12.6). Using the Gauss–Seidel iterative scheme solve the system of equations:

$$3x_1 + x_2 + x_3 \qquad = 5,$$
$$6x_2 - 2x_3 + 3x_4 = 6,$$
$$x_1 \qquad + 4x_3 - 2x_4 = 1,$$
$$x_2 + 2x_3 - 4x_4 = 2.$$

Show that the iterations converge to a solution accurate to 4 significant figures within 11 steps, starting from $(0, 1, 0, 0)$. Confirm also that the matrix of coefficients is diagonally dominant.

12.23 (Section 12.6). Show that the Gauss–Seidel scheme fails for the system

$$x_1 - 2x_2 + x_3 = 4,$$
$$x_1 - x_2 - x_3 = 1,$$
$$2x_1 + 3x_2 - 4x_3 = 4$$

(the matrix in this case is not diagonally dominant).

12.24 (Section 12.6). Show that one row of the matrix of coefficients fails to be dominant in the system

$$6x_1 - x_2 + x_3 = 2,$$
$$3x_1 + 2x_2 + x_3 = 1,$$
$$x_1 - x_2 + 4x_3 = 5.$$

However, confirm that the Gauss–Seidel scheme delivers a solution accurate to 4 significant figures after 10 iterations starting at $(0, 0, 1)$.

12.25 For comparison purposes with the Gauss–Seidel method, solve the equations (12.7)–(12.9), namely

$$x_1 = \tfrac{1}{3}(-x_2 - x_3 - 1),$$
$$x_2 = \tfrac{1}{4}(x_1 - x_3 - 8),$$
$$x_3 = \tfrac{1}{5}(-2x_1 - x_2 - 14),$$

using the Jacobi method. How many steps are required to achieve the same accuracy as that in the table in Section 12.6, that is to 5 significant figures?

13 Eigenvalues and eigenvectors

CONTENTS

13.1   Eigenvalues of a matrix

With any square matrix A, we can associate a set of homogeneous linear equations $Ax = 0$. As we saw in Section 12.5 of the previous chapter, such a set of equations will only have a non-trivial solution set if $\det A = 0$. Consider now the $n \times n$ set of equations

$$Ax = \lambda x, \quad \text{or} \quad (A - \lambda I_n)x = 0,$$

where λ is a parameter, and I_n is the unit matrix (see Section 7.3(5)). In order for these equations to have non-trivial solutions, we must have

$$\det(A - \lambda I_n) = 0. \tag{13.1}$$

This can only be satisfied if λ takes certain values. These are called the **eigenvalues** of the matrix A, and the equation they satisfy (eqn (13.1)) is called the **characteristic equation** of A. The characteristic equation is a polynomial equation, of degree n in λ. We usually list the eigenvalues as λ_1, λ_2, and so on.

Example 13.1   *Find the eigenvalues of*

$$A = \begin{bmatrix} 1 & 3 \\ 2 & 2 \end{bmatrix}.$$

The eigenvalues of A are given by the determinant equation

$$\det(A - \lambda I_2) = \begin{vmatrix} 1 - \lambda & 3 \\ 2 & 2 - \lambda \end{vmatrix} = 0,$$

Example 13.1 *continued*

which can be expanded into

$$(1 - \lambda)(2 - \lambda) - 6 = 0, \quad \text{or} \quad \lambda^2 - 3\lambda - 4 = 0.$$

This factorizes into $(\lambda - 4)(\lambda + 1) = 0$: hence the eigenvalues are $\lambda_1 = -1, \lambda_2 = 4$.

Example 13.2 *Find the eigenvalues of*

$$A = \begin{bmatrix} 2 & -2 \\ 1 & 4 \end{bmatrix}.$$

In this case

$$\det(A - \lambda I_2) = \begin{vmatrix} 2 - \lambda & -2 \\ 1 & 4 - \lambda \end{vmatrix} = (2 - \lambda)(4 - \lambda) + 2$$

$$= \lambda^2 - 6\lambda + 10 = 0,$$

and the quadratic equation has the roots

$$\lambda = \tfrac{1}{2}[6 \pm \sqrt{(36 - 40)}] = 3 \pm i.$$

Thus real matrices can have **complex eigenvalues**.

Example 13.3 *Find the eigenvalues of*

$$A = \begin{bmatrix} 1 & 2 & 1 \\ 2 & 1 & 1 \\ 1 & 1 & 2 \end{bmatrix}.$$

Here

$$\det(A - \lambda I_3) = \begin{vmatrix} 1 - \lambda & 2 & 1 \\ 2 & 1 - \lambda & 1 \\ 1 & 1 & 2 - \lambda \end{vmatrix}$$

$$= \begin{vmatrix} 4 - \lambda & 4 - \lambda & 4 - \lambda \\ 2 & 1 - \lambda & 1 \\ 1 & 1 & 2 - \lambda \end{vmatrix} \quad (\mathbf{r}_1' = \mathbf{r}_1 + \mathbf{r}_2 + \mathbf{r}_3)$$

$$= (4 - \lambda) \begin{vmatrix} 1 & 1 & 1 \\ 2 & 1 - \lambda & 1 \\ 1 & 1 & 2 - \lambda \end{vmatrix}$$

$$= (4 - \lambda) \begin{vmatrix} 1 & 0 & 0 \\ 2 & -1 - \lambda & -1 \\ 1 & 0 & 1 - \lambda \end{vmatrix} \quad \begin{pmatrix} \mathbf{c}_2' = \mathbf{c}_2 - \mathbf{c}_1 \\ \mathbf{c}_3' = \mathbf{c}_3 - \mathbf{c}_1 \end{pmatrix}$$

$$= (4 - \lambda)(-1 - \lambda)(1 - \lambda),$$

$$= 0,$$

if $\lambda = 4$ or ± 1. Hence the eigenvalues are

$$\lambda_1 = 4, \quad \lambda_2 = 1, \quad \lambda_3 = -1.$$

> **Eigenvalues**
>
> The eigenvalues of the $n \times n$ square matrix A are the solutions λ of the determinant equation
>
> $$\det(A - \lambda I_n) = 0.$$
>
> (13.2)

13.2 Eigenvectors

Associated with each eigenvalue λ of A, there will be an infinite number of non-trivial solutions of the equation $(A - \lambda I_n)x = 0$.

These are called the **eigenvectors of** A corresponding to the eigenvalue λ, and are generally denoted in this text by s. Thus, if λ is an eigenvalue of A, then there will exist a corresponding eigenvector $s \neq 0$, which is a non-trivial solution of

$$(A - \lambda I_n)s = 0.$$

The solutions of this set of linear equations can be found by Gaussian elimination.

Example 13.4 *Find the eigenvectors of*

$$A = \begin{bmatrix} 1 & 3 \\ 2 & 2 \end{bmatrix}.$$

From Example 13.1, the eigenvalues are $\lambda_1 = 4$ and $\lambda_2 = -1$. Let the corresponding eigenvectors be

$$s_1 = \begin{bmatrix} a_1 \\ b_1 \end{bmatrix}, \qquad s_2 = \begin{bmatrix} a_2 \\ b_2 \end{bmatrix}.$$

Thus $(A - \lambda_1 I_2)s_1 = 0$ becomes

$$\begin{bmatrix} 1-4 & 3 \\ 2 & 2-4 \end{bmatrix} \begin{bmatrix} a_1 \\ b_1 \end{bmatrix} = \begin{bmatrix} 0 \\ 0 \end{bmatrix}, \quad \text{or} \quad \begin{cases} -3a_1 + 3b_1 = 0 \\ 2a_1 - 2b_1 = 0 \end{cases}.$$

Solution is easy in this case, and the solutions can be expressed as $a_1 = b_1 = \alpha$ for any α. If we put $\alpha = 1$, then an eigenvector is

$$s_1 = \begin{bmatrix} 1 \\ 1 \end{bmatrix}.$$

Any nonzero value of α will give an eigenvector; we usually choose a convenient value for the parameter to give one solution. The others are multiples of this.

Similarly $(A - \lambda_2 I_2)s_2 = 0$ becomes

$$\begin{bmatrix} 1+1 & 3 \\ 2 & 2+1 \end{bmatrix} \begin{bmatrix} a_2 \\ b_2 \end{bmatrix} = 0, \quad \text{or} \quad \begin{cases} 2a_2 + 3b_2 = 0 \\ 2a_2 + 2b_2 = 0 \end{cases}.$$

The eigenvectors for this case are

$$s_2 = \begin{bmatrix} \beta \\ -\frac{2}{3}\beta \end{bmatrix},$$

Example 13.4 *continued*

for any nonzero β. As before, we choose a particular value of β which makes the eigenvector specific and simple. In this case we could put $\beta = 3$ to give the eigenvector

$$s_2 = \begin{bmatrix} 3 \\ -2 \end{bmatrix}.$$

Example 13.5   *Find the eigenvectors of*

$$A = \begin{bmatrix} 1 & 2 & 1 \\ 2 & 1 & 1 \\ 1 & 1 & 2 \end{bmatrix}.$$

The eigenvalues of A are $\lambda_1 = 4$, $\lambda_2 = 1$, $\lambda_3 = -1$ (see Example 13.3). Let the corresponding eigenvectors be

$$s_i = \begin{bmatrix} a_i \\ b_i \\ c_i \end{bmatrix} \quad (i = 1, 2, 3).$$

In each case, we need to solve $(A - \lambda_i I_3)s_i = 0$. If $\lambda_1 = 4$, then

$$-3a_1 + 2b_1 + c_1 = 0,$$
$$2a_1 - 3b_1 + c_1 = 0,$$
$$a_1 + b_1 - 2c_1 = 0.$$

Gaussian elimination leads to

$$\begin{bmatrix} -3 & 2 & 1 & 0 \\ 2 & -3 & 1 & 0 \\ 1 & 1 & -2 & 0 \end{bmatrix} \rightarrow \begin{bmatrix} -3 & 2 & 1 & 0 \\ 0 & -\frac{5}{3} & \frac{5}{3} & 0 \\ 0 & \frac{5}{3} & -\frac{5}{3} & 0 \end{bmatrix} \quad \begin{pmatrix} r_2' = r_2 + \frac{2}{3}r_1 \\ r_3' = r_3 + \frac{1}{3}r_1 \end{pmatrix}$$

$$\rightarrow \begin{bmatrix} -3 & 2 & 1 & 0 \\ 0 & -\frac{5}{3} & \frac{5}{3} & 0 \\ 0 & 0 & 0 & 0 \end{bmatrix} \quad (r_3' = r_3 + r_2).$$

By back substitution, if $c_1 = \alpha$, then $b_1 = c_1 = \alpha$, and $a_1 = \frac{1}{3}(2b_1 + c_1) = \alpha$. Thus, with $\alpha = 1$, an eigenvector is

$$s_1 = \begin{bmatrix} 1 \\ 1 \\ 1 \end{bmatrix}.$$

The other eigenvectors corresponding to λ_1 are simply multiples of s_1. Using the same procedure shows that the two eigenvectors corresponding respectively to λ_2 and λ_3 can be chosen to be

$$s_2 = \begin{bmatrix} -1 \\ -1 \\ 2 \end{bmatrix}, \quad s_2 = \begin{bmatrix} 1 \\ -1 \\ 0 \end{bmatrix}.$$

Example 13.6 *Find the eigenvalues and eigenvectors of*

$$A = \begin{bmatrix} 1 & 2 & -1 \\ 1 & 2 & -1 \\ 2 & 2 & -1 \end{bmatrix}.$$

In this example,

$$
\begin{aligned}
\det(A - \lambda I_3) &= \begin{vmatrix} 1-\lambda & 2 & -1 \\ 1 & 2-\lambda & -1 \\ 2 & 2 & -1-\lambda \end{vmatrix} \\
&= \begin{vmatrix} -\lambda & \lambda & 0 \\ 1 & 2-\lambda & -1 \\ 2 & 2 & -1-\lambda \end{vmatrix} \quad (\mathbf{r}_1' = \mathbf{r}_1 - \mathbf{r}_2) \\
&= \begin{vmatrix} -\lambda & 0 & 0 \\ 1 & 3-\lambda & -1 \\ 2 & 4 & -1-\lambda \end{vmatrix} \quad (\mathbf{c}_2' = \mathbf{c}_2 + \mathbf{c}_1) \\
&= -\lambda[(3-\lambda)(-1-\lambda) + 4] \\
&= -\lambda(\lambda - 1)^2.
\end{aligned}
$$

This particular matrix has an eigenvalue 0 and a **repeated** eigenvalue 1. How does this affect the eigenvectors? Let the eigenvectors be, for $\lambda_1 = 0$ and $\lambda_2 = 1$,

$$s_i = \begin{bmatrix} a_i \\ b_i \\ c_i \end{bmatrix} \quad (i = 1, 2).$$

For $\lambda_1 = 0$,

$$
\begin{aligned}
a_1 + 2b_1 - c_1 &= 0, \\
a_1 + 2b_1 - c_1 &= 0, \\
2a_1 + 2b_1 - c_1 &= 0.
\end{aligned}
$$

Hence $a_1 = 0$, $b_1 = \alpha$, $c_1 = 2\alpha$, for any α. An eigenvector is

$$s_1 = \begin{bmatrix} 0 \\ 1 \\ 2 \end{bmatrix}.$$

For $\lambda_2 = 1$,

$$
\begin{aligned}
2b_2 - c_2 &= 0, \\
a_2 + b_2 - c_2 &= 0, \\
2a_2 + 2b_2 - 2c_2 &= 0.
\end{aligned}
$$

If we let $b_2 = \beta$, then $c_2 = 2\beta$ and $a_2 = c_2 - b_2 = \beta$. Hence we can associate with $\lambda_2 = 1$ the eigenvector

$$s_2 = \begin{bmatrix} 1 \\ 1 \\ 2 \end{bmatrix},$$

by putting $\beta = 1$. There are only two independent eigenvectors in this example.

Note that **if A has a zero eigenvalue, then A must be a singular matrix** since $\det A = 0$. And conversely, if A is singular, then A has at least one zero eigenvalue.

The matrix in Example 13.6 has two eigenvalues (one repeated) and two eigenvectors. The meaning of this reduced eigenvector set will be illustrated in the context of coordinate transformations in Section 13.4. As the next example illustrates, a matrix can have a repeated eigenvalue but still retain a full set of independent eigenvectors.

Example 13.7 *Find the eigenvalues and eigenvectors of*

$$A = \begin{bmatrix} 3 & 0 & -1 \\ 0 & 1 & 0 \\ 2 & 0 & 0 \end{bmatrix}.$$

Thus

$$\det(A - \lambda I_3) = \begin{vmatrix} 3-\lambda & 0 & -1 \\ 0 & 1-\lambda & 0 \\ 2 & 0 & -\lambda \end{vmatrix}$$

$$= (3-\lambda)(1-\lambda)(-\lambda) + (-1)(-2)(1-\lambda)$$

$$= (1-\lambda)[-3\lambda + \lambda^2 + 2] = -(\lambda-2)(\lambda-1)^2.$$

Let $\lambda_1 = 2$ and $\lambda_2 = 1$ with corresponding eigenvectors

$$s_i = \begin{bmatrix} a_i \\ b_i \\ c_i \end{bmatrix} \quad (i = 1, 2).$$

For $\lambda_1 = 2$,

$$a_1 \quad - \ c_1 = 0,$$
$$-b_1 \qquad = 0,$$
$$2a_1 \quad - 2c_1 = 0.$$

We can let $b_1 = 0$, $c_1 = \alpha$, $a_1 = \alpha$. Hence we can choose

$$s_1 = \begin{bmatrix} 1 \\ 0 \\ 1 \end{bmatrix}.$$

For $\lambda_2 = 1$,

$$2a_2 - c_2 = 0,$$
$$0 = 0,$$
$$2a_2 - c_2 = 0.$$

If $a_2 = \beta$, then $c_2 = 2\beta$ but b_2 can then take any value γ, say. Hence, the eigenvector set is

$$s_2 = \begin{bmatrix} \beta \\ \gamma \\ 2\beta \end{bmatrix} = \beta \begin{bmatrix} 1 \\ 0 \\ 2 \end{bmatrix} + \gamma \begin{bmatrix} 0 \\ 1 \\ 0 \end{bmatrix},$$

Example 13.7 *continued*

this is it contains *two* parameters β and γ. The choices of $\beta = 1$ with $\gamma = 0$, and $\beta = 0$ with $\gamma = 1$, say, give two independent eigenvectors

$$\begin{bmatrix} 1 \\ 0 \\ 2 \end{bmatrix} \quad \text{and} \quad \begin{bmatrix} 0 \\ 1 \\ 0 \end{bmatrix}.$$

Unlike the previous example, three independent eigenvectors are associated with this matrix even though the matrix has only two eigenvalues. We shall take up this point again in connection with the diagonalization of matrices.

> **Eigenvectors**
>
> The eigenvectors of a square matrix A are the non-trivial solutions s of the homogeneous equations
>
> $(A - \lambda_r I_n)s = 0$, for each eigenvalue λ_r.   (13.3)

13.3 Linear dependence

It is useful in mathematics to gather, in a collection or set, elements which have common features. For example, we might consider the set of all integers, the set of all fractions, or the set of all real numbers. In a similar way, we can gather all $m \times n$ matrices. They all obey certain rules, and are said to form a vector space. We shall not consider the general case here, but restrict ourselves to the set of *all $m \times 1$* column vectors: this set is called an *m*-dimensional **vector space** $\mathcal{V}_m$. These vectors obey the rules of matrix algebra. Thus if

$$s_1 = \begin{bmatrix} a_1 \\ a_2 \\ \vdots \\ a_m \end{bmatrix}, \qquad s_2 = \begin{bmatrix} b_1 \\ b_2 \\ \vdots \\ b_m \end{bmatrix},$$

then s_1 and s_2 belong to $\mathcal{V}_m$, and so does $\alpha s_1 + \beta s_2$ for any constants α and β.

An important set of vectors in $\mathcal{V}_m$ is the set of **base vectors**

$$e_1 = \begin{bmatrix} 1 \\ 0 \\ 0 \\ \vdots \\ 0 \end{bmatrix}, \quad e_2 = \begin{bmatrix} 0 \\ 1 \\ 0 \\ \vdots \\ 0 \end{bmatrix}, \quad \dots, \quad e_m = \begin{bmatrix} 0 \\ 0 \\ 0 \\ \vdots \\ 1 \end{bmatrix}.$$

Any vector in $\mathcal{V}_m$ can be expressed as a **linear combination** of these vectors. Thus

$$s_1 = \begin{bmatrix} a_1 \\ a_2 \\ \vdots \\ a_m \end{bmatrix} = a_1 e_1 + a_2 e_2 + \cdots + a_m e_m.$$

The set of vectors $\{e_1, e_2, \ldots, e_m\}$ is said, therefore, to form a **basis** of $\mathcal{V}_m$. None of the vectors $e_1, e_2, \ldots, e_m$ can be expressed as a linear combination of the others, so that they are said to be **linearly independent**. A set of n column vectors $s_1, s_2, \ldots, s_n$ is said to be **linearly dependent** if there exist constants $\alpha_1, \alpha_2, \ldots, \alpha_n$, **not all zero**, such that

$$\alpha_1 s_1 + \alpha_2 s_2 + \cdots + \alpha_n s_n = 0.$$

If the above equation holds *only* when $\alpha_1 = \alpha_2 = \cdots = \alpha_n = 0$, then the vectors are linearly independent. It can be proved that any set of m linearly independent vectors form a basis of the vector space $\mathcal{V}_m$.

Example 13.8 *Show that the column vectors*

$$a_1 = (1, 1, 0)^T, \, a_2 = (1, 0, 1)^T, \, a_3 = (0, 1, 1)^T$$

form a basis in three dimensions.

We must test whether

$$xa_1 + ya_2 + za_3 = 0$$

has nonzero solutions for x, y, z. The equations in full are

$$\begin{aligned} x + y \quad &= 0, \\ x \quad + z &= 0, \\ y + z &= 0. \end{aligned}$$

The determinant of the coefficients is

$$D = \begin{vmatrix} 1 & 1 & 0 \\ 1 & 0 & 1 \\ 0 & 1 & 1 \end{vmatrix} = -2 \neq 0.$$

By (12.6) the only solution is $x = y = z = 0$. The vectors are therefore linearly independent and can form a basis.

By a similar argument it can be shown that

$$b_1 = (1, 1, 0)^T, \, b_2 = (1, 0, -1)^T, \, b_3 = (0, 1, 1)^T$$

are linearly dependent and therefore cannot form a basis.

13.4 Diagonalization of a matrix

We will take a constructive approach to this problem for a 3×3 matrix. Consider the matrix of Examples 13.3 and 13.5, namely

$$A = \begin{bmatrix} 1 & 2 & 1 \\ 2 & 1 & 1 \\ 1 & 1 & 2 \end{bmatrix}$$

which has the eigenvalues $\lambda_1 = 4$, $\lambda_2 = 1$, $\lambda_3 = -1$ and eigenvectors

$$s_1 = \begin{bmatrix} 1 \\ 1 \\ 1 \end{bmatrix} \qquad s_2 = \begin{bmatrix} -1 \\ -1 \\ 2 \end{bmatrix} \qquad s_3 = \begin{bmatrix} 1 \\ -1 \\ 0 \end{bmatrix}.$$

Construct a matrix C which has these eigenvectors as its columns:

$$C = [s_1 \quad s_2 \quad s_3] = \begin{bmatrix} 1 & -1 & 1 \\ 1 & -1 & -1 \\ 1 & 2 & 0 \end{bmatrix}.$$

The columns are independent, so C is nonsigular. Then

$$AC = A[s_1 \quad s_2 \quad s_3] = [As_1 \quad As_2 \quad As_3] = [\lambda_1 s_1 \quad \lambda_2 s_2 \quad \lambda_3 s_3],$$

the last equality holding since the eigenvector s_i is defined as a nonzero solution of $As_i = \lambda_i s_i$. This may be written as

$$AC = [s_1 \quad s_2 \quad s_3]D = CD,$$

where

$$D = \begin{bmatrix} \lambda_1 & 0 & 0 \\ 0 & \lambda_2 & 0 \\ 0 & 0 & \lambda_3 \end{bmatrix},$$

where D is a diagonal matrix with eigenvalue elements. If we premultiply this equation by C^{-1}, then

$$C^{-1}AC = C^{-1}CD = I_3D = D.$$

Therefore, the operation $C^{-1}AC$ has **diagonalized** the matrix A. In the example,

$$C^{-1} = \begin{bmatrix} 1 & -1 & 1 \\ 1 & -1 & -1 \\ 1 & 2 & 0 \end{bmatrix}^{-1} = \begin{bmatrix} \frac{1}{3} & \frac{1}{3} & \frac{1}{3} \\ -\frac{1}{6} & -\frac{1}{6} & \frac{1}{3} \\ \frac{1}{2} & -\frac{1}{2} & 0 \end{bmatrix}.$$

Finally, it can be checked that

$$C^{-1}AC = \begin{bmatrix} \frac{1}{3} & \frac{1}{3} & \frac{1}{3} \\ -\frac{1}{6} & -\frac{1}{6} & \frac{1}{3} \\ \frac{1}{2} & -\frac{1}{2} & 0 \end{bmatrix} \begin{bmatrix} 1 & 2 & 1 \\ 2 & 1 & 1 \\ 1 & 2 & 0 \end{bmatrix} \begin{bmatrix} 1 & -1 & 1 \\ 1 & -1 & -1 \\ 1 & 2 & 0 \end{bmatrix}$$

$$= \begin{bmatrix} 4 & 0 & 0 \\ 0 & 1 & 0 \\ 0 & 0 & -1 \end{bmatrix} = D.$$

It might appear at first sight that there is not a unique answer for D the eigenvectors are not uniquely defined. However, if C is replaced by

$$C_1 = [k_1 s_1, \quad k_2 s_2, \quad k_3 s_3],$$

where k_1, k_2, k_3 are any constants for which $k_1 \neq 0, k_2 \neq 0, k_3 \neq 0$, and

$$C = \begin{bmatrix} u_1 \\ u_2 \\ u_3 \end{bmatrix},$$

where u_1, u_2, u_3 are the rows in C, then

$$C_1^{-1} = \begin{bmatrix} u_1/k_1 \\ u_2/k_2 \\ u_3/k_3 \end{bmatrix}.$$

The factors k, k^{-1} from C, C^{-1} cancel, so $C^{-1}AC = D$.

Example 13.9 *Use the eigenvalues and eigenvectors of*

$$\begin{bmatrix} 3 & 0 & -1 \\ 0 & 1 & 0 \\ 2 & 0 & 0 \end{bmatrix}$$

obtained in Example 13.7 *to construct a transformation which diagonalizes A, and verify that the diagonalized matrix is*

$$D = \begin{bmatrix} 2 & 0 & 0 \\ 0 & 1 & 0 \\ 0 & 0 & 1 \end{bmatrix}.$$

From Example 13.7, we see that A has the eigenvalues $\lambda_1 = 2$ and $\lambda_2 = \lambda_3 = 1$. However, we can associate two linearly independent eigenvectors with the repeated eigenvalue. Thus, we can define C by

$$C = [s_1 \quad s_2 \quad s_3] = \begin{bmatrix} 1 & 1 & 0 \\ 0 & 0 & 1 \\ 1 & 2 & 0 \end{bmatrix}.$$

Example 13.9 *continued*

Its inverse is

$$C^{-1} = \begin{bmatrix} 2 & 0 & -1 \\ -1 & 0 & 1 \\ 0 & 1 & 0 \end{bmatrix}.$$

Finally it can be verified that

$$C^{-1}AC = \begin{bmatrix} 2 & 0 & -1 \\ -1 & 0 & 1 \\ 0 & 1 & 0 \end{bmatrix} \begin{bmatrix} 3 & 0 & -1 \\ 0 & 1 & 0 \\ 2 & 0 & 0 \end{bmatrix} \begin{bmatrix} 1 & 1 & 0 \\ 0 & 0 & 1 \\ 1 & 2 & 0 \end{bmatrix}$$

$$= \begin{bmatrix} 2 & 0 & 0 \\ 0 & 1 & 0 \\ 0 & 0 & 1 \end{bmatrix} = D.$$

Following the remarks just before this example, with $k_1 = 2$, $k_2 = 3$, $k_3 = -1$,

$$C_1 = [2s_1 \quad 3s_2 \quad -s_3] = \begin{bmatrix} 2 & 3 & 0 \\ 0 & 0 & -1 \\ 2 & 6 & 0 \end{bmatrix}$$

would equally well be an acceptable matrix in the diagonalization.

Example 13.10 *Find a transformation which diagonalizes the matrix*

$$A = \begin{bmatrix} 2 & -2 \\ 1 & 4 \end{bmatrix}.$$

From Example 13.2, the eigenvalues are $\lambda_1 = 3 + i$, $\lambda_2 = 3 - i$. Corresponding eigenvectors are

$$s_1 = \begin{bmatrix} -1 + i \\ 1 \end{bmatrix}, \qquad s_2 = \begin{bmatrix} -1 - i \\ 1 \end{bmatrix}.$$

The eigenvalues and eigenvectors are complex valued but this does not affect the method. The matrix C becomes

$$C = [s_1 \quad s_2] = \begin{bmatrix} -1 + i & -1 - i \\ 1 & 1 \end{bmatrix}.$$

Its inverse is

$$C^{-1} = \frac{1}{\det C} \begin{bmatrix} 1 & 1 + i \\ -1 & -1 + i \end{bmatrix} = \frac{1}{2i} \begin{bmatrix} 1 & 1 + i \\ -1 & -1 + i \end{bmatrix}.$$

Finally, check that

$$C^{-1}AC = \frac{1}{2i} \begin{bmatrix} 1 & 1 + i \\ -1 & -1 + i \end{bmatrix} \begin{bmatrix} 2 & -2 \\ 1 & 4 \end{bmatrix} \begin{bmatrix} -1 + i & -1 - i \\ 1 & 1 \end{bmatrix}$$

$$= \begin{bmatrix} 3 + i & 0 \\ 0 & 3 - i \end{bmatrix}.$$

> **Diagonalizing a matrix**
>
> To diagonalize a matrix A:
>
> (i) find the eigenvalues of A;
> (ii) find n linearly independent eigenvectors s_n of A (if they exist);
> (iii) construct the matrix C of eigenvectors;
> (iv) calculate the inverse C^{-1} of C;
> (v) compute $C^{-1}AC$.
>
> **(13.4)**

Not all matrices can be diagonalized in this way. In Example 13.6 where

$$A = \begin{bmatrix} 1 & 2 & -1 \\ 1 & 2 & -1 \\ 2 & 2 & -1 \end{bmatrix},$$

we can associate only two linearly independent eigenvectors with the eigenvalue 0 and the repeated eigenvalue 1, and no diagonalizing matrix C can be constructed.

13.5 Powers of matrices

The transformation C of the previous section can be used to obtain a formula for calculating powers of square matrices. This follows since it is a simple matter to find powers of diagonal matrices. Thus if

$$D = \begin{bmatrix} \lambda_1 & 0 & 0 \\ 0 & \lambda_2 & 0 \\ 0 & 0 & \lambda_3 \end{bmatrix},$$

then

$$D^2 = \begin{bmatrix} \lambda_1 & 0 & 0 \\ 0 & \lambda_2 & 0 \\ 0 & 0 & \lambda_3 \end{bmatrix} \begin{bmatrix} \lambda_1 & 0 & 0 \\ 0 & \lambda_2 & 0 \\ 0 & 0 & \lambda_3 \end{bmatrix}$$

$$= \begin{bmatrix} \lambda_1^2 & 0 & 0 \\ 0 & \lambda_2^2 & 0 \\ 0 & 0 & \lambda_3^2 \end{bmatrix},$$

and, in general,

$$D^n = \begin{bmatrix} \lambda_1^n & 0 & 0 \\ 0 & \lambda_2^n & 0 \\ 0 & 0 & \lambda_3^n \end{bmatrix}.$$

In the previous section we showed that, if a 3×3 matrix A has three linearly independent eigenvectors, then we can find a matrix C such that

$$AC = CD,$$

where D is a diagonal matrix, its elements consisting of the eigenvalues of A. Thus by multiplying on the right by C^{-1} we find that

$$A = CDC^{-1}.$$

Hence

$$A^2 = CDC^{-1}CDC^{-1} = CDI_3DC^{-1} = CD^2C^{-1},$$

since $C^{-1}C = I_3$. Continuing this process, we find that

$$A^3 = A^2A = CD^2C^{-1}CDC^{-1} = CD^3C^{-1},$$

and, in general,

$$A^n = CD^nC^{-1}.$$

Example 13.11   *Find a formula for A^n, where*

$$A = \begin{bmatrix} 1 & 2 & 1 \\ 2 & 1 & 1 \\ 1 & 1 & 2 \end{bmatrix}.$$

(See Examples 13.3 and 13.5 and Section 13.5.)

The eigenvalues of A are $\lambda_1 = 4$, $\lambda_2 = 1$, $\lambda_3 = -1$; and the diagonalizing transformation, with its inverse, is

$$C = \begin{bmatrix} 1 & -1 & 1 \\ 1 & -1 & -1 \\ 1 & 2 & 0 \end{bmatrix}, \qquad C^{-1} = \begin{bmatrix} \frac{1}{3} & \frac{1}{3} & \frac{1}{3} \\ -\frac{1}{6} & -\frac{1}{6} & \frac{1}{3} \\ \frac{1}{2} & -\frac{1}{2} & 0 \end{bmatrix}.$$

Hence

$$A^n = CD^nC^{-1} = \begin{bmatrix} 1 & -1 & 1 \\ 1 & -1 & -1 \\ 1 & 2 & 0 \end{bmatrix} \begin{bmatrix} 4 & 0 & 0 \\ 0 & 1 & 0 \\ 0 & 0 & -1 \end{bmatrix}^n \begin{bmatrix} \frac{1}{3} & \frac{1}{3} & \frac{1}{3} \\ -\frac{1}{6} & -\frac{1}{6} & \frac{1}{3} \\ \frac{1}{2} & -\frac{1}{2} & 0 \end{bmatrix}$$

$$= \begin{bmatrix} 1 & -1 & 1 \\ 1 & -1 & -1 \\ 1 & 2 & 0 \end{bmatrix} \begin{bmatrix} 4^n & 0 & 0 \\ 0 & 1^n & 0 \\ 0 & 0 & (-1)^n \end{bmatrix} \begin{bmatrix} \frac{1}{3} & \frac{1}{3} & \frac{1}{3} \\ -\frac{1}{6} & -\frac{1}{6} & \frac{1}{3} \\ \frac{1}{2} & -\frac{1}{2} & 0 \end{bmatrix}$$

$$= \begin{bmatrix} 4^n & -1 & (-1)^n \\ 4^n & -1 & -(-1)^n \\ 4^n & 2 & 0 \end{bmatrix} \begin{bmatrix} \frac{1}{3} & \frac{1}{3} & \frac{1}{3} \\ -\frac{1}{6} & -\frac{1}{6} & \frac{1}{3} \\ \frac{1}{2} & -\frac{1}{2} & 0 \end{bmatrix}$$

$$= \frac{4^n}{3} \begin{bmatrix} 1 & 1 & 1 \\ 1 & 1 & 1 \\ 1 & 1 & 1 \end{bmatrix} + \frac{1}{6} \begin{bmatrix} 1 & 1 & -2 \\ 1 & 1 & -2 \\ -2 & -2 & 4 \end{bmatrix} + \frac{(-1)^n}{2} \begin{bmatrix} 1 & -1 & 0 \\ -1 & 1 & 0 \\ 0 & 0 & 0 \end{bmatrix}.$$

Example 13.12 *Let*

$$P = \begin{bmatrix} 1 - \alpha & \alpha \\ \beta & 1 - \beta \end{bmatrix},$$

where $0 < \alpha, \beta < 1$. *Find* P^n *and* $\lim_{n \to \infty} P^n$.

The matrix P is an example of a **row-stochastic** matrix: that is, all elements are non-negative and the sum of the elements in each row is 1. The eigenvalues of P are given by

$$\begin{vmatrix} 1 - \alpha - \lambda & \alpha \\ \beta & 1 - \beta - \lambda \end{vmatrix} = 0.$$

Hence

$$(1 - \alpha - \lambda)(1 - \beta - \lambda) - \alpha\beta = 0,$$

or

$$\lambda^2 - \lambda(2 - \alpha - \beta) + 1 - \alpha - \beta = 0.$$

The roots $\lambda_1 = 1$, $\lambda_2 = 1 - \alpha - \beta = p$, say. Choose the corresponding eigenvectors

$$s_1 = \begin{bmatrix} 1 \\ 1 \end{bmatrix}, \qquad s_2 = \begin{bmatrix} -\alpha \\ \beta \end{bmatrix}.$$

Let

$$C = [s_1 \quad s_2] = \begin{bmatrix} 1 & -\alpha \\ 1 & \beta \end{bmatrix}.$$

Its inverse is given by

$$C^{-1} = \frac{1}{\alpha + \beta} \begin{bmatrix} \beta & \alpha \\ -1 & 1 \end{bmatrix}.$$

Thus

$$P^n = CD^nC^{-1} = \begin{bmatrix} 1 & -\alpha \\ 1 & \beta \end{bmatrix} \begin{bmatrix} 1 & 0 \\ 0 & p^n \end{bmatrix} \begin{bmatrix} \beta & \alpha \\ -1 & 1 \end{bmatrix} \frac{1}{\alpha + \beta}$$

$$= \frac{1}{\alpha + \beta} \begin{bmatrix} 1 & -\alpha p^n \\ 1 & \beta p^n \end{bmatrix} \begin{bmatrix} \beta & \alpha \\ -1 & 1 \end{bmatrix}$$

$$= \frac{1}{\alpha + \beta} \begin{bmatrix} \beta + \alpha p^n & \alpha - \alpha p^n \\ \beta - \beta p^n & \alpha + \beta p^n \end{bmatrix}$$

$$= \frac{1}{\alpha + \beta} \begin{bmatrix} \beta & \alpha \\ \beta & \alpha \end{bmatrix} + \frac{p^n}{\alpha + \beta} \begin{bmatrix} \alpha & -\alpha \\ -\beta & \beta \end{bmatrix}.$$

Since $0 < \alpha < 1$ and $0 < \beta < 1$, it follows that

$$p = 1 - \alpha - \beta < 1 \quad \text{and} \quad p = 1 - \alpha - \beta > 1 - 1 - 1 = -1,$$

that is $|p| < 1$. As $n \to \infty$, then $p^n \to 0$ and

$$P^n \to \frac{1}{\alpha + \beta} \begin{bmatrix} \beta & \alpha \\ \beta & \alpha \end{bmatrix}.$$

> **Powers of a square matrix**
> To find the power A^n of a diagonalizable matrix A:
> (i) find the eigenvalues and eigenvectors of A;
> (ii) construct a matrix C of eigenvectors such that $C^{-1}AC = D$ where D is the diagonal matrix of eigenvalues;
> (iii) the required answer is
> $$A^n = CD^nC^{-1}.$$
> **(13.5)**

13.6 Quadratic forms

Suppose that $x = [x_1, x_2, \dots, x_n]^T$, an n-dimensional column vector with elements $x_1, x_2, \dots, x_n$. Any polynomial function of these elements in which every term is of degree 2 in them is known as a **quadratic form**. Thus, if $n = 3$, then

$$x_1^2 + 8x_1x_2 + x_2^2 + 6x_2x_3 + x_3^2$$

is an example of a quadratic form. Quadratic forms can always be expressed as a matrix product of the form

$$x^{\mathrm{T}}Ax.$$

The example above can be written as

$$[x_1 \quad x_2 \quad x_3] \begin{bmatrix} 1 & 4 & 0 \\ 4 & 1 & 3 \\ 0 & 3 & 1 \end{bmatrix} \begin{bmatrix} x_1 \\ x_2 \\ x_3 \end{bmatrix}.$$
(13.6)

In this representation, A is required to be a *symmetric matrix*. Non-symmetric representations are possible with, for example, in the above

$$A = \begin{bmatrix} 1 & 0 & 0 \\ 8 & 1 & 2 \\ 0 & 4 & 1 \end{bmatrix},$$

but the symmetric form is adopted throughout this section.

Let us find the eigenvalues of the symmetric matrix in (13.6) in the usual way by solving

$$\begin{vmatrix} 1-\lambda & 4 & 0 \\ 4 & 1-\lambda & 3 \\ 0 & 3 & 1-\lambda \end{vmatrix} = 0.$$

Hence

$$(1-\lambda)[(1-\lambda)^2 - 9] - 4 \cdot 4 \cdot (1-\lambda) = 0$$

or

$$(1-\lambda)[(1-\lambda)^2 - 25] = 0.$$

It follows that the eigenvalues are $\lambda_1 = 1, \lambda_2 = -4, \lambda_3 = 6$. It can be shown by the methods previously explained that corresponding eigenvectors are

$$s_1 = \begin{bmatrix} 3 \\ 0 \\ -4 \end{bmatrix}, \quad s_2 = \begin{bmatrix} -4 \\ 5 \\ -3 \end{bmatrix}, \quad s_3 = \begin{bmatrix} 4 \\ 5 \\ 3 \end{bmatrix}.$$

If a and b are two column vectors, and

$$a^{\mathrm{T}}b = 0,$$

then a and b are said to be **orthogonal**. If we examine the eigenvectors $s_1, s_2,$ and s_3 above, then it is easy to see that

$$s_1^{\mathrm{T}}s_2 = [3 \quad 0 \quad -4]\begin{bmatrix} -4 \\ 5 \\ -3 \end{bmatrix} = -12 + 0 + 12 = 0,$$

and similarly that $s_2^{\mathrm{T}}s_3 = 0$ and $s_3^{\mathrm{T}}s_1 = 0$. Thus the three eigenvectors are **mutually orthogonal**: regarded as ordinary vectors in the sense of Chapter 9, they are **mutually perpendicular**.

It will be shown that this property of the eigenvalues follows from the symmetry of the matrix of the quadratic form. However, we first show that the eigenvectors of a symmetric matrix must be real numbers.

Theorem 13.1 *If A is a symmetric real matrix, then its eigenvalues are real.*

Proof. Suppose that $\lambda = \alpha + i\beta$ is an eigenvalue. Since the left-hand side of the equation $\det(A - \lambda I_n) = 0$ is a real polynomial in λ, it must also have an eigenvalue $\bar{\lambda} = \alpha - i\beta$. Let s and $\bar{s}$ be the eigenvectors corresponding to λ and its conjugate $\bar{\lambda}$. Thus

$$As = \lambda s, \qquad A\bar{s} = \bar{\lambda}\bar{s}. \tag{13.7}$$

Since A is symmetric, it follows that $(A\bar{s})^{\mathrm{T}} = \bar{s}^{\mathrm{T}}A^{\mathrm{T}} = \bar{s}^{\mathrm{T}}A$, and we can replace (13.7) by

$$As = \lambda s, \qquad \bar{s}^{\mathrm{T}}A = \bar{\lambda}\bar{s}^{\mathrm{T}}. \tag{13.8}$$

Multiply the first equation in (13.8) on the left by $\bar{s}^{\mathrm{T}}$, and the second equation on the right by s. Thus

$$\bar{s}^{\mathrm{T}}As = \lambda\bar{s}^{\mathrm{T}}s, \qquad \bar{s}^{\mathrm{T}}As = \bar{\lambda}\bar{s}^{\mathrm{T}}s.$$

Elimination of $\bar{s}^{\mathrm{T}}As$ leads to

$$(\lambda - \bar{\lambda})\bar{s}^{\mathrm{T}}s = 0. \tag{13.9}$$

To show that $\bar{s}^{\mathrm{T}}s \neq 0$, put $s^{\mathrm{T}} = (a_1, \dots, a_n)$. Then, since $\bar{a}_n a_n = |a_n|^2$,

$$\bar{s}^{\mathrm{T}}s = [\bar{a}_1 \quad \bar{a}_2 \dots \bar{a}_n]\begin{bmatrix} a_1 \\ a_2 \\ \vdots \\ a_n \end{bmatrix} = |a_1|^2 + |a_2|^2 + \cdots + |a_n|^2 > 0.$$

From (13.9), it follows that $\lambda = \bar{\lambda}$ or $\alpha + i\beta = \alpha - i\beta$, from which we conclude that $\beta = 0$. Therefore λ is real.

Theorem 13.2   *If A is a symmetric matrix, then the eigenvectors associated with two distinct eigenvalues are orthogonal.*

Proof. Let λ_1 and λ_2 be the distinct eigenvalues, and s_1 and s_2 their corresponding eigenvectors. Then

$$As_1 = \lambda_1 s_1, \qquad As_2 = \lambda_2 s_2.$$

Transpose the second equation so that the equations become

$$As_1 = \lambda_1 s_1, \qquad s_2^T A = \lambda_2 s_2^T,$$

since A is symmetrical. Multiply the first equation by s_2^T on the left, and the second equation by s_1 on the right. Hence

$$s_2^T As_1 = \lambda_1 s_2^T s_1, \qquad s_2^T As_1 = \lambda_2 s_2^T s_1.$$

Eliminate $s_2^T As_1$ between these equations, leaving

$$\lambda_1 s_2^T s_1 = \lambda_2 s_2^T s_1, \quad \text{or} \quad (\lambda_1 - \lambda_2)s_2^T s_1 = 0.$$

Since $\lambda_1 \neq \lambda_2$ it follows that $s_2^T s_1 = 0$; that is, s_1 and s_2 are orthogonal.

13.7   Positive-definite matrices

A quadratic form $x^T Ax$ is said to be **positive-definite** if $x^T Ax > 0$ for all $x \neq 0$. If this is true, we simply describe the matrix A as **positive-definite**.

Any quadratic form can be written as $x^T Ax$ where A is *symmetric*.

Consider the particular case in which A is a 3×3 symmetric matrix. Let $\lambda_1, \lambda_2, \lambda_3$ be its eigenvalues, with corresponding eigenvectors s_1, s_2, s_3 *which are chosen so that they are all unit vectors,* that is $s_1^T s_1 = s_2^T s_2 = s_3^T s_3 = 1$.

As we saw in Section 13.5, we can diagonalize A by using the matrix

$$C = [s_1 \quad s_2 \quad s_3],$$

so that

$$C^{-1}AC = D = \begin{bmatrix} \lambda_1 & 0 & 0 \\ 0 & \lambda_2 & 0 \\ 0 & 0 & \lambda_3 \end{bmatrix}.$$

For a symmetric matrix, the eigenvectors are orthogonal (Theorem 13.2). Hence

$$\begin{aligned} s_1^T C &= s_1^T [s_1 \quad s_2 \quad s_3] \\ &= [s_1^T s_1 \quad s_1^T s_2 \quad s_1^T s_3] \\ &= [1 \quad 0 \quad 0], \end{aligned}$$

since s_1 is a unit vector. In a similar way,

$$s_2^T C = [0 \quad 1 \quad 0], \qquad s_3^T C = [0 \quad 0 \quad 1].$$

Hence, if we construct a matrix with s_1^T, s_2^T, s_3^T as its rows, then

$$C^T C = \begin{bmatrix} s_1^T \\ s_2^T \\ s_3^T \end{bmatrix} C = \begin{bmatrix} 1 & 0 & 0 \\ 0 & 1 & 0 \\ 0 & 0 & 1 \end{bmatrix} = I_3.$$

In other words, **the transpose of C is equal to the inverse of C:**

$$C^T = \begin{bmatrix} s_1^T \\ s_2^T \\ s_3^T \end{bmatrix}$$

is the inverse of C, that is $C^T = C^{-1}$. Square matrices with this property are said to be **orthogonal** matrices.

Suppose that we now define a transformation by $x = CX$, where C is an orthogonal matrix. Then, in terms of X, the quadratic form becomes

$$
\begin{aligned}
x^T A x &= (CX)^T A CX \\
&= X^T C^T A CX \\
&= X^T DX = \lambda_1 X_1^2 + \lambda_2 X_2^2 + \lambda_3 X_3^2.
\end{aligned}
$$

It follows from this result, for 3×3 matrices, and by implication for higher order, that a **quadratic form is positive-definite if and only if all its eigenvalues are positive.**

Example 13.13   *Find an orthogonal matrix C which transforms the quadratic form $x^T A x$ where*

$$A = \begin{bmatrix} 3 & -1 & 0 \\ -1 & 3 & 0 \\ 0 & 0 & 1 \end{bmatrix}$$

into a diagonal quadratic form $X^T DX$.

The eigenvalues of A are given by $\det(A - \lambda I_3) = 0$, where

$$
\begin{aligned}
\det(A - \lambda I_3) &= \begin{vmatrix} 3 - \lambda & -1 & 0 \\ -1 & 3 - \lambda & 0 \\ 0 & 0 & 1 - \lambda \end{vmatrix}, \\
&= [(3 - \lambda)^2 - 1](1 - \lambda), \\
&= (\lambda - 2)(\lambda - 4)(1 - \lambda).
\end{aligned}
$$

Hence the eigenvalues are $\lambda_1 = 1, \lambda_2 = 2, \lambda_3 = 4$. Since all the eigenvalues are positive, it follows that the quadratic form is positive-definite. The corresponding eigenvectors are

$$s_1 = \begin{bmatrix} 0 \\ 0 \\ 1 \end{bmatrix}, \qquad s_2 = \begin{bmatrix} 1/\sqrt{2} \\ 1/\sqrt{2} \\ 0 \end{bmatrix}, \qquad s_3 = \begin{bmatrix} -1/\sqrt{2} \\ 1/\sqrt{2} \\ 0 \end{bmatrix}.$$

Hence the required orthogonal matrix C is

$$C = [s_1 \quad s_2 \quad s_3] = \begin{bmatrix} 0 & 1/\sqrt{2} & -1/\sqrt{2} \\ 0 & 1/\sqrt{2} & 1/\sqrt{2} \\ 1 & 0 & 0 \end{bmatrix}.$$

The relation between the coordinates (x, y, z) and (X, Y, Z) of a point fixed in space in the transformation

$$x = CX = [s_1 \quad s_2 \quad s_3]X,$$

where the eigenvectors s_1, s_2, s_3 are orthogonal unit vectors, can be seen as follows. Put $X = 1$, $Y = 0$, $Z = 0$, which is a point on the X axis. Since

$$X = \begin{bmatrix} 1 \\ 0 \\ 0 \end{bmatrix},$$

it follows that the corresponding point in the x frame is $x = s_1$. In other words the elements (a_1, b_1, c_1) of s_1 are the coordinates in the x space of the point $A_1 : (1, 0, 0)$ in the X space. Similarly, the elements of s_2 and s_3 are respectively the coordinates of $A_2 : (0, 1, 0)$ and $A_3 : (0, 0, 1)$ in the X space (see Fig. 13.1).

We know that the eigenvectors are mutually orthogonal, that is $s_i^T s_j = 0$ $(i \neq j)$. We want to show that this implies that the new axes $OXYZ$ are also mutually perpendicular. Consider the triangle OA_1A_2: we want to show that $\widehat{A_1OA_2}$ is a right angle, so that the triangle is subject to Pythagoras's theorem:

$$A_1A_2^2 - OA_1^2 - OA_2^2$$
$$= (a_1 - a_2)^2 + (b_1 - b_2)^2 + (c_1 - c_2)^2 - (a_1^2 + b_1^2 + c_1^2) - (a_2^2 + b_2^2 + c_2^2)$$
$$= -2(a_1a_2 + b_1b_2 + c_1c_2) = -2s_1^T s_2 = 0,$$

since the eigenvectors are unit vectors and orthogonal. Hence, by the theorem of Pythagoras, $\widehat{A_1OA_2}$ is a right angle. Similarly, the other angles $\widehat{A_2OA_3}$ and $\widehat{A_3OA_1}$ are right angles. Hence the new axes are mutually perpendicular. It can be shown that $\det C = \pm 1$. If $\det C = 1$, then the X coordinates can be obtained from the x coordinates by a rotation about the origin O. If $\det C = -1$, then a reflection and rotation are required.

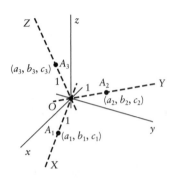

Fig. 13.1 Orthogonal mapping between axes. The coordinates $(a_1, b_1, c_1), (a_2, b_2, c_2), (a_3, b_3, c_3)$ are measured in the x space.

Example 13.14   *Show that*

$$C = \tfrac{1}{3}\begin{bmatrix} 1 & 2 & 2 \\ 2 & 1 & -2 \\ 2 & -2 & 1 \end{bmatrix}$$

is an orthogonal matrix. If $x = CX$, what does the point $x = 1$, $y = 2$, $z = -1$ map into in the (X, Y, Z) coordinates?

In this example,

$$s_1 = \tfrac{1}{3}\begin{bmatrix} 1 \\ 2 \\ 2 \end{bmatrix}, \qquad s_2 = \tfrac{1}{3}\begin{bmatrix} 2 \\ 1 \\ -2 \end{bmatrix}, \qquad s_3 = \tfrac{1}{3}\begin{bmatrix} 2 \\ -2 \\ 1 \end{bmatrix}.$$

Example 13.14 *continued*

Clearly

$$s_1^T s_1 = \tfrac{1}{9}[1 \quad 2 \quad 2]\begin{bmatrix} 1 \\ 2 \\ 2 \end{bmatrix} = 1.$$

Similarly $s_2^T s_2 = 1$ and $s_3^T s_3 = 1$. Also

$$s_1^T s_2 = \tfrac{1}{9}[1 \quad 2 \quad 2]\begin{bmatrix} 2 \\ 1 \\ -2 \end{bmatrix}$$

$$= \tfrac{1}{9}[1 \times 2 + 2 \times 1 + 2 \times (-2)] = 0.$$

Similarly, $s_2^T s_3 = 0$ and $s_3^T s_1 = 0$. We need to invert the transformation so that

$$X = C^{-1}x = C^T x$$

$$= \begin{bmatrix} s_1^T \\ s_2^T \\ s_3^T \end{bmatrix}\begin{bmatrix} 1 \\ 2 \\ -1 \end{bmatrix} = \tfrac{1}{3}\begin{bmatrix} 1 \times 1 + 2 \times 2 + 2 \times (-1) \\ 2 \times 1 + 1 \times 2 + (-2) \times (-1) \\ 2 \times 1 + (-2) \times 2 + 1 \times (-1) \end{bmatrix}$$

$$= [1 \quad 2 \quad -1]^T.$$

Hence $X = 1$, $Y = 2$, $Z = -1$.

13.8 An application to a vibrating system

Positive-definite matrices occur frequently in applications. For example, consider the system consisting of two particles of equal mass m and three equal springs stretched in a straight line between two supports as shown in Fig. 13.2. Suppose that, in equilibrium, the springs are unstretched, each of length a. The mechanical system vibrates longitudinally so that the displacements of the particles are x and y as shown.

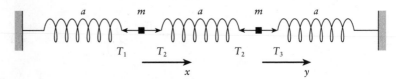

Fig. 13.2 Longitudinal oscillations.

If a spring is stretched or compressed from equilibrium by a length x, then its potential energy stored is $\tfrac{1}{2}kx^2$ where k is a constant known as the **stiffness** of the spring, which measures its reaction to being stretched or compressed. The total potential energy of the system is

$$V = \tfrac{1}{2}kx^2 + \tfrac{1}{2}k(y - x)^2 + \tfrac{1}{2}y^2.$$

Note that the extension of the middle spring is $y - x$. Thus

$$V = \tfrac{1}{2}kx^2 + \tfrac{1}{2}ky^2 - kyx + \tfrac{1}{2}kx^2 + \tfrac{1}{2}ky^2$$
$$= kx^2 - kxy + ky^2$$
$$= \tfrac{1}{2}x^{\mathrm{T}}Kx,$$

where

$$x = \begin{bmatrix} x \\ y \end{bmatrix}, \qquad K = \begin{bmatrix} 2k & -k \\ -k & 2k \end{bmatrix}.$$

The eigenvalues of K are given by $\det(K - \lambda I_2) = 0$, that is

$$\begin{vmatrix} 2k - \lambda & -k \\ -k & 2k - \lambda \end{vmatrix} = 0, \quad \text{or} \quad (2k - \lambda)^2 - k^2 = 0.$$

Hence, the eigenvalues are $\lambda_1 = k$ and $\lambda_2 = 3k$, which are both positive, implying that the potential energy is a positive-definite quadratic form. This is not surprising, since we might expect the potential energy to take a minimum value in equilibrium. The corresponding eigenvectors are

$$s_1 = \frac{1}{\sqrt{2}} \begin{bmatrix} 1 \\ 1 \end{bmatrix}, \qquad s_2 = \frac{1}{\sqrt{2}} \begin{bmatrix} 1 \\ -1 \end{bmatrix},$$

normalized as unit vectors. The matrix of eigenvectors, C, is given by

$$C = [s_1 \quad s_2] = \frac{1}{\sqrt{2}} \begin{bmatrix} 1 & 1 \\ 1 & -1 \end{bmatrix}.$$

The transformation $x = CX$ introduces the coordinates $X^{\mathrm{T}} = (X, Y)$ in which

$$V = \tfrac{1}{2}X^{\mathrm{T}} \begin{bmatrix} k & 0 \\ 0 & 3k \end{bmatrix} X = \tfrac{1}{2}(kX^2 + 3kY^2).$$

(X, Y) are known as the **normal coordinates** of the system, and are related to x and y by

$$\begin{bmatrix} x \\ y \end{bmatrix} = C \begin{bmatrix} X \\ Y \end{bmatrix} = \begin{bmatrix} (X + Y)/\sqrt{2} \\ (X - Y)/\sqrt{2} \end{bmatrix}.$$

Normal coordinates are often more convenient coordinates to use.

For the same problem, we can also derive the equations of motion of each particle. If T_1, T_2, and T_3 are the tensions in each of the springs, then applying Newton's law (force equals mass times acceleration) for each particle gives the **differential equations**

$$T_2 - T_1 = m\ddot{x}, \tag{13.10}$$

$$T_3 - T_2 = m\ddot{y}. \tag{13.11}$$

where $\ddot{x}$ and $\ddot{y}$ stand for d^2x/dt^2 and d^2y/dt^2 respectively. The tension in a spring is k times the extension, by Hooke's law, where k is the stiffness of the spring. Thus

$$T_1 = kx, \qquad T_2 = k(y - x), \qquad T_3 = -ky.$$

Substitution into (13.10) and (13.11) yields

$$-2kx + ky = m\ddot{x}, \tag{13.12}$$

$$kx - 2ky = m\ddot{y}. \tag{13.13}$$

In matrix form, these equations can be combined into the vector equation

$$\ddot{x} + Ax = 0,$$

where

$$\ddot{x} = \begin{bmatrix} \ddot{x} \\ \ddot{y} \end{bmatrix}, \qquad A = \begin{bmatrix} 2k/m & -k/m \\ -k/m & 2k/m \end{bmatrix}.$$

If we use the normal coordinates, then $x = CX$ implies

$$C\ddot{X} + ACX = 0.$$

Multiply this equation on the left by $C^{-1} = C^T$:

$$\ddot{X} + C^T ACX = 0,$$

or

$$\ddot{X} + DX = 0, \tag{13.14}$$

where

$$D = \begin{bmatrix} \lambda_1 & 0 \\ 0 & \lambda_2 \end{bmatrix} = \begin{bmatrix} k/m & 0 \\ 0 & 3k/m \end{bmatrix}.$$

Equation (13.14) now separates into the two differential equations

$$\ddot{X} + (k/m)X = 0,$$
$$\ddot{Y} + 3(k/m)Y = 0,$$

which, unlike (13.12) and (13.13), are now no longer *simultaneous* equations, but uncouple into two equations which can be solved separately and independently for X and Y. We say more about the solution of differential equations in Chapter 18.

Problems

13.1 (Sections 13.1, 2). Find the eigenvalues and eigenvectors of the following matrices:

(a) $\begin{bmatrix} 2 & 3 \\ 4 & 6 \end{bmatrix}$;   (b) $\begin{bmatrix} 6 & 3 \\ 2 & 7 \end{bmatrix}$;   (c) $\begin{bmatrix} 2 & 1 \\ 4 & 6 \end{bmatrix}$;

(d) $\begin{bmatrix} 1 & 1 \\ 4 & 5 \end{bmatrix}$;   (e) $\begin{bmatrix} 1 & 2 \\ 14 & 5 \end{bmatrix}$;   (f) $\begin{bmatrix} 2 & -2 \\ 4 & 6 \end{bmatrix}$.

13.2 (Section 13.1). Show that the eigenvalues of the *symmetric* matrix

$$A = \begin{bmatrix} a & b \\ b & c \end{bmatrix},$$

where a, b, and c are real numbers, are real.

13.3 (Section 13.1). Find the eigenvalues of

$$A = \begin{bmatrix} 6 & 3 \\ 2 & 7 \end{bmatrix}$$

(see Problem 13.1b). Find the inverse of A and find its eigenvalues. What relationship, would you guess, exists between the eigenvalues of A and those of A^{-1}? Find the eigenvalues of A^2. How do they relate to those of A?

13.4 (Sections 13.1, 2). Find the eigenvalues and eigenvectors of

(a) $\begin{bmatrix} 1 & 1 & 2 \\ 1 & 2 & 1 \\ 2 & 1 & 1 \end{bmatrix}$;   (b) $\begin{bmatrix} 2 & 1 & 2 \\ 1 & 2 & 2 \\ 2 & 1 & 2 \end{bmatrix}$;

(c) $\begin{bmatrix} 2 & 0 & 0 \\ 0 & 2 & 2 \\ 0 & 2 & -1 \end{bmatrix}$;   (d) $\begin{bmatrix} 6 & 5 & 5 \\ 5 & 6 & 5 \\ 5 & 5 & 6 \end{bmatrix}$.

13.5 (Sections 13.1, 2). Find the eigenvalues and eigenvectors of

$$\begin{bmatrix} 1 & 2 & 0 & 0 \\ 3 & 2 & 0 & 0 \\ 0 & 0 & 3 & 1 \\ 0 & 0 & 1 & 3 \end{bmatrix}.$$

13.6 (Sections 13.1, 2). Show that

$$A = \begin{bmatrix} 1 & 0 & 0 \\ 0 & 2 & 2 \\ 0 & 2 & 5 \end{bmatrix}$$

has a repeated eigenvalue. Find the corresponding eigenvectors. How many linearly independent eigenvectors are there?

13.7 (Sections 13.1, 2). Show that the matrix

$$A = \begin{bmatrix} -1 & -1 & a+1 \\ a+1 & -a & -1 \\ -a & a+1 & -a \end{bmatrix}$$

has a zero eigenvalue. For design reasons, a second eigenvalue must be 3. For what values of a does this occur? Find the third eigenvalue in each case.

13.8 (Sections 13.1, 2). A matrix is said to be **idempotent** if $A^2 = A$. Explain why all eigenvalues of A must be either 0 or 1. Show that

$$A = \begin{bmatrix} 1 & 0 & 0 \\ 0 & 3 & 6 \\ 0 & -1 & -2 \end{bmatrix}$$

is idempotent. Find the eigenvalues and eigenvectors of A and A^2 and confirm the above result.

13.9 (Sections, 13.1, 2). Let

$$A = \tfrac{1}{2} \begin{bmatrix} 1 & 1 & 1 & 1 \\ 1 & 1 & -1 & -1 \\ 1 & -1 & 1 & -1 \\ 1 & -1 & -1 & 1 \end{bmatrix}.$$

Show that $A^2 = I_4$. Explain why the eigenvalues of A must be either 1 or −1. Can A be diagonalized?

13.10 Find the eigenvalues $\lambda_1, \lambda_2, \lambda_3$ of

$$A = \begin{bmatrix} 1 & 2 & 1 \\ 2 & 1 & 1 \\ 1 & 1 & 2 \end{bmatrix}.$$

The **trace** of a square matrix is the sum of the elements in the leading diagonal. Thus if $B = [b_{ij}]$ is an $n \times n$ matrix, then

$$\text{trace } B = b_{11} + b_{22} + \cdots + b_{nn}.$$

Confirm for A above that trace $A = \lambda_1 + \lambda_2 + \lambda_3$. Also verify that det $A = \lambda_1 \lambda_2 \lambda_3$.

13.11 (Section 13.3). Show that the vectors

$$s_1 = \begin{bmatrix} 1 \\ 2 \\ 1 \end{bmatrix}, \qquad s_2 = \begin{bmatrix} 2 \\ -1 \\ 3 \end{bmatrix}, \qquad s_3 = \begin{bmatrix} 4 \\ 3 \\ 5 \end{bmatrix}$$

are linearly dependent.

13.12 (Section 13.5). Let

$$A = \begin{bmatrix} -4 & 1 & -2 \\ 2 & -2 & 1 \\ 0 & 1 & 0 \end{bmatrix}.$$

Find the eigenvalues of A and a set of corresponding eigenvectors. Hence construct a matrix C which makes $C^{-1}AC$ a diagonal matrix.

13.13 Find a matrix C which diagonalizes the matrix

$$A = \begin{bmatrix} 1 & 8 \\ 2 & 1 \end{bmatrix}.$$

13.14 (Section 13.5). Find a matrix C which diagonalizes

$$A = \begin{bmatrix} 2 & 0 & 0 \\ 0 & 2 & 2 \\ 0 & 2 & -1 \end{bmatrix}.$$

Verify that $C^{-1}AC = D$, where D is the diagonal matrix of eigenvalues.

13.15 Using the diagonalization result

$$C^{-1}AC = D$$

for a matrix A which has n linearly independent eigenvectors, show that

$$\det A = \lambda_1 \lambda_2 \dots \lambda_n,$$

where $\lambda_1, \lambda_2, \dots, \lambda_n$ are the eigenvalues of A. (Hint: use the result $\det AB = \det A \det B$ for square matrices.)

13.16 (Section 11.5). Find the eigenvalues and eigenvectors of the row-stochastic matrix.

$$A = \begin{bmatrix} \frac{1}{4} & \frac{1}{2} & \frac{1}{4} \\ \frac{1}{2} & \frac{1}{4} & \frac{1}{4} \\ \frac{1}{4} & \frac{1}{4} & \frac{1}{2} \end{bmatrix}.$$

Find a formula for A^n. How does A behave as $n \to \infty$?

13.17 Show that

$$A = \begin{bmatrix} 1 & 0 & 0 \\ 0 & \cos\alpha & -\sin\alpha \\ 0 & \sin\alpha & \cos\alpha \end{bmatrix}$$

is an orthogonal matrix. Describe the mapping defined by

$$X = Ax.$$

Which set of points remains unaffected by the mapping?

13.18 (Section 13.7). Show that

$$A = \tfrac{1}{2}\begin{bmatrix} 1 & -1 & 1 & -1 \\ 1 & -1 & -1 & 1 \\ 1 & 1 & -1 & -1 \\ 1 & 1 & 1 & 1 \end{bmatrix}$$

is an orthogonal matrix.

13.19 Show that, in the transformation

$$\begin{bmatrix} X \\ Y \end{bmatrix} = \begin{bmatrix} \cos\alpha & -\sin\alpha \\ \sin\alpha & \cos\alpha \end{bmatrix}\begin{bmatrix} x \\ y \end{bmatrix},$$

the angle between the two sets of axes is α. What do the axes of x and y become in the (X, Y) plane?

13.20 Show that the nonzero eigenvalues of the skew-symmetric matrix

$$A = \begin{bmatrix} 0 & a & b \\ -a & 0 & c \\ -b & -c & 0 \end{bmatrix}$$

are imaginary for a, b, c real.

13.21 Let

$$A = \begin{bmatrix} 1 & 2 & 1 \\ 2 & 1 & 1 \\ 1 & 1 & 2 \end{bmatrix}.$$

Show that

$$\det(A - \lambda I_3) = -\lambda^3 + 4\lambda^2 + \lambda - 4.$$

Verify that

$$-A^3 + 4A^2 + A - 4I_3 = 0.$$

In other words, the matrix A satisfies its own characteristic equation. This is known as the **Cayley–Hamilton** theorem, and holds generally for square matrices. Use the result to find the inverse matrix A^{-1}.

13.22 Find the eigenvalues and eigenvectors of

$$A = \begin{bmatrix} 5 & -1 & -3 & 3 \\ -1 & 5 & 3 & -3 \\ -3 & 3 & 5 & -1 \\ 3 & -3 & -1 & 5 \end{bmatrix}.$$

Construct a matrix C such that $C^{-1}AC$ is the diagonal matrix of eigenvalues. Write down $\det A$.

13.23 (Section 13.6). Express the following quadratic forms in the form x^TAx, where A is a 3×3 symmetric matrix:

(a) $x_1^2 + x_2^2 + x_3^2 + 4x_1x_2 - 4x_1x_3 + 4x_2x_3$;
(b) $x_1x_2 - x_1x_3 + x_2x_3$.

Find eigenvalues of A in each case, and find also a matrix C which transforms each into the form $\lambda_1 X_1^2 + \lambda_2 X_2^2 + \lambda_3 X_3^2$.

13.24 (Section 13.6). Which of each of the following quadratic forms is positive-definite?
(a) $4x_1^2 + x_2^2 - 4x_1x_2$;
(b) $x_1^2 + x_2^2 + 2x_3^2 + 2x_2x_3 + 2x_3x_1 + 4x_1x_2$;
(c) $6x_1^2 + 2x_2^2 - x_3x_1$.

13.25 (Section 13.8). Consider three particles, each of mass m, and four equal springs stretched in a straight line between fixed supports distance $4a$ apart by four springs each with unstretched lengths a (as in Fig. 13.2, but with three particles). Consider longitudinal oscillations of the systems and let x, y, z be the extensions of the springs. Assuming Hooke's law with stiffness k for the tension in each spring, show that x, y, z satisfy the differential equations

$$k(-2x + y) = m\ddot{x},$$

$$k(x - 2y + z) = m\ddot{y},$$

$$k(y - 2z) = m\ddot{z}.$$

Express the equations in the matrix form

$$\ddot{x} + Ax = 0.$$

Find the eigenvalues and eigenvectors of A. Construct a matrix C such that $C^{T}AC$ is diagonal. Obtain differential equations for the normal coordinates X, Y, Z.

13.26 Let

$$A = \begin{bmatrix} 0 & 1 & 0 \\ 0 & 0 & 1 \\ 1 & 0 & 0 \end{bmatrix}.$$

Calculate A^2 and A^3. Find a general formula for A^n.

Show that A has two complex eigenvalues, and find the corresponding eigenvectors. Construct a matrix C which diagonalizes A and find a formula for A^n. Compare this result with the *ad hoc* method above.

13.27 Let A be a square matrix, and let S represent the sum of the powers of A from A up to A^n:

$$S = A + A^2 + A^3 + \cdots + A^n.$$

By multiplying the equation by A and subtraction, show that

$$S = A(I + A^n)(I - A)^{-1},$$

and state any cases for which this method fails.

If

$$A = \begin{bmatrix} 1 & 3 \\ 2 & 2 \end{bmatrix}$$

(see Examples 13.1 and 13.4), find a formula for A^m and the sum

$$S = \sum_{m=1}^{n} A^m.$$

13.28 Let

$$A = \begin{bmatrix} 1 & 2 & 1 \\ 2 & 1 & 1 \\ 1 & 1 & 2 \end{bmatrix}.$$

Find the eigenvalues of A and confirm that

$$s_1 = \begin{bmatrix} 2 \\ 2 \\ 2 \end{bmatrix}, \quad s_2 = \begin{bmatrix} -1 \\ -1 \\ 2 \end{bmatrix}, \quad s_3 = \begin{bmatrix} -3 \\ 3 \\ 0 \end{bmatrix}$$

are eigenvectors of A. Construct the matrix C and verify that

$$D = C^{-1}AC,$$

where D is the diagonal matrix of eigenvalues. (This is a reworking of the problem at the beginning of Example 13.5, but with different eigenvectors.)

13.29 Given an $m \times n$ matrix

$$A = \begin{bmatrix} a_{11} & a_{12} & \cdots & a_{1n} \\ a_{21} & a_{22} & \cdots & a_{2n} \\ \vdots & \vdots & \ddots & \vdots \\ a_{m1} & a_{m2} & \cdots & a_{mn} \end{bmatrix},$$

the *determinant* of any square $r \times r$ submatrix of A obtained by eliminating $(m-r)$ rows and $(n-r)$ columns in A is called an rth-order **minor** of A. Since the submatrix must be square it follows that $1 \leqslant r \leqslant \min(m, n)$ ('min' means the smaller of m or n unless they are equal, in which case $\min(m, n) = m = n$). For example, if

$$A = \begin{bmatrix} 1 & 2 & 3 & 4 \\ 10 & 11 & 12 & 5 \\ 9 & 8 & 7 & 6 \end{bmatrix},$$

then

$$B = \begin{vmatrix} 1 & 2 & 4 \\ 10 & 11 & 5 \\ 9 & 8 & 6 \end{vmatrix} = -80 \quad \text{and} \quad C = \begin{vmatrix} 11 & 5 \\ 8 & 6 \end{vmatrix} = 26$$

are examples respectively of third-order and second-order minors of A. Find the remaining third-order minors of A.

13.30 (See Problem 13.29.) A matrix is said to have **rank** r if at least one of its rth-order minors is not zero whilst *every* $(r+1)$th-order minor is zero. Find the ranks of the following matrices:

(a) $\begin{bmatrix} 1 & 2 & 3 \\ 3 & 4 & 5 \\ 6 & 7 & 8 \end{bmatrix}$; (b) $\begin{bmatrix} 3 & 2 & 1 \\ 1 & 2 & 3 \\ 2 & 1 & 3 \end{bmatrix}$;

(c) $\begin{bmatrix} 1 & 2 & 3 & 2 \\ 1 & 3 & 4 & 5 \\ 2 & 3 & 5 & 1 \end{bmatrix}$.

13.31 (See Problem 13.30.) The method for checking the rank of a matrix by calculating minors can be a lengthy procedure. An alternative approach uses the elementary row operations of Section 12.2. Given a matrix A, row operations are applied to reduce A to echelon form, from which it is easier to test its rank. This is justified since it can be proved (but not here) that elementary row operations do not change the rank of a matrix. Express the matrix in Problem 13.30c in echelon form and check its rank.

13.32 Consider again Examples 13.6 and 13.7. Let the matrices in these examples be defined, respectively, by

$$A_1 = \begin{bmatrix} 1 & 2 & -1 \\ 1 & 2 & -1 \\ 2 & 2 & -1 \end{bmatrix} \quad \text{and} \quad A_2 = \begin{bmatrix} 3 & 0 & -1 \\ 0 & 1 & 0 \\ 2 & 0 & 0 \end{bmatrix}.$$

Both these matrices have a repeated eigenvalue. Find the ranks of the matrices $\lambda I_3 - A_1$ and $\lambda I_3 - A_2$ for all the eigenvalues. Confirm that if λ is the repeated eigenvalue of A_2, then the rank of $\lambda I_3 - A_2$ is 1, and there are two eigenvectors associated with this eigenvalue: the vector space defined by this eigenvalue has *dimension 2*. (In general, for any eigenvalue λ, the rank of $\lambda I_n - A$ indicates the dimension of the vector space associated with λ: if the root is r-fold and the rank of $\lambda I_n - A$ is s, where s must satisfy $n - 1 \leqslant s \leqslant n - r$, then the dimension of the vector space of λ is $n - s$. If the eigenvalue is unique, then the vector space has dimension 1; in other words, there is just one eigenvector associated with the eigenvalue, and if $r = 2$, there could be one or two eigenvectors depending on the rank, and so on.)

Find the eigenvalues of

$$A = \begin{bmatrix} 2 & 1 & 0 & 0 \\ 0 & -1 & 0 & 2 \\ 0 & 0 & 1 & 0 \\ 0 & 0 & 2 & 1 \end{bmatrix}.$$

Find the rank of A. What are the dimensions of the vector spaces associated with each eigenvalue?

Integration and differential equations

14 Antidifferentiation and area

CONTENTS

14.1 Reversing differentiation

Compare the following two problems:

Problem A: $\dfrac{d}{dx} \sin x = f(x)$; what is $f(x)$?

Problem B: $\dfrac{d}{dx} F(x) = \cos x$; what is $F(x)$?

For Problem A we know already that

$$f(x) = \cos x.$$

This provides *one* answer to Problem B, which is solved by

$$F(x) = \sin x.$$

Since $\cos x$ is the **derivative** of $\sin x$, we say that $\sin x$ is an **antiderivative** of $\cos x$ (we say *an* antiderivative because it is not the only one; for example, $\sin x + 1$ is also an antiderivative).

The **antidifferentiation** question in Problem B can be expressed in various ways; for example,

(a) What must be differentiated to get cos x?

(b) What curves have slope equal to cos x at every point?

(c) Find y as a function of x if $dy/dx = \cos x$.

Finding antiderivatives is the opposite or **inverse** process to that of finding derivatives.

The following examples show that **a function $f(x)$ has an infinite number of antiderivatives**: there is an infinite number of functions whose derivatives are $f(x)$. However, they are all very simple variants on a single function.

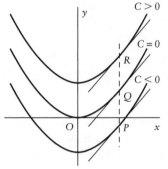

Fig. 14.1

Example 14.1 *Find y as a function of x if $dy/dx = 2x$.*

One solution is $y = x^2$, because its derivative is $2x$. But the derivatives of $x^2 + 3$, $x^2 - \frac{1}{2}$, and so on are also equal to $2x$. In fact

$$y = x^2 + C$$

is an antiderivative of $2x$ for any constant C.

Some of these solutions are shown in Fig. 14.1. Different choices for C just shift the graph bodily up or down parallel to itself. Therefore, at any particular value of x, such as is represented by the vertical line PQR, the slopes are all the same, independently of the value of C.

Evidently the same thing will happen whatever function we start with: if we find one solution, we can add constants to obtain more.

Example 14.2 *Find a collection of antiderivatives of $\sin 2x$.*

We want y such that $dy/dx = \sin 2x$. If we differentiate a cosine we get something involving a sine, so first of all test whether $y = \cos 2x$ is close to being an antiderivative of $\sin 2x$. We find that $dy/dx = -2 \sin 2x$. This contains an unwanted factor (-2). It can be eliminated by choosing instead

$$y = \frac{1}{-2} \cos 2x = -\tfrac{1}{2} \cos 2x,$$

for then we have $dy/dx = -\tfrac{1}{2}(-2 \sin 2x) = \sin 2x$, which is right. Therefore, one antiderivative is $-\tfrac{1}{2} \cos 2x$, and the rest are of the form

$$y = -\tfrac{1}{2} \cos 2x + C \quad \text{(C is any constant)}.$$

Example 14.3 *Solve the equation $dy/dx = e^{-3x}$ (that is to say, find a collection of antiderivatives of e^{-3x}).*

Try $y = e^{-3x}$; then $dy/dx = -3\,e^{-3x}$. To avoid the unwanted factor (-3) we should have taken

$$y = \frac{1}{(-3)} e^{-3x} = -\tfrac{1}{3} e^{-3x}.$$

From this we construct an infinite collection of antiderivatives:

$$-\tfrac{1}{3} e^{-3x} + C \quad \text{(C any constant)}.$$

It can be proved that the above process, of finding a particular anti-derivative of a function and adding constants, generates **all possible antiderivatives** for that function.

Antiderivatives of $f(x)$

A function $F(x)$ is called an **antiderivative** of $f(x)$ if

$$\frac{\mathrm{d}}{\mathrm{d}x} F(x) = f(x).$$

If $F(x)$ is any particular antiderivative of $f(x)$, then **all the antiderivatives** are given by

$$F(x) + C,$$

where C can be any constant. (Therefore, **any two antiderivatives differ by a constant**.)

(14.1)

An antiderivative of a function is also more usually called an **indefinite integral** of the function, and the process of getting it is called **integration**. If you know the term already, it is perfectly safe to use it. We shall change over to it in Chapter 15.

Example 14.4   *Find all the antiderivatives of* x^3.

We firstly have to find any y which fits the equation $\mathrm{d}y/\mathrm{d}x = x^3$. Differentiation reduces a power of x by unity, so try $y = x^4$:

$\mathrm{d}y/\mathrm{d}x = 4x^3$.

The factor 4 is unwanted; we needed $\frac{1}{4}x^4$ to give x^3. Therefore all antiderivatives are given by

$y = \frac{1}{4}x^4 + C,$

where C is any constant.

Sums of terms and constant multipliers are treated in the same way as in differentiation: the multipliers stay as multipliers and each term is treated separately, as in the next example.

Example 14.5   *Obtain all the antiderivatives of* $2\,\mathrm{e}^{-3x} - \frac{1}{2}x^3 + 2$.

From the previous examples, one antiderivative of e^{-3x} is $-\frac{1}{3}\,\mathrm{e}^{-3x}$, and one for x^3 is $\frac{1}{4}x^4$. Also, one antiderivative of 2 is obviously $2x$. Therefore one antiderivative of the given expression is

$2(-\frac{1}{3}\,\mathrm{e}^{-3x}) - \frac{1}{2}(\frac{1}{4}x^4) + 2x,$

and all its antiderivatives are of the form

$-\frac{2}{3}\,\mathrm{e}^{-3x} - \frac{1}{8}x^4 + 2x + C,$

where C is any constant.

The following two examples show the importance in practice of including the constant C.

Example 14.6   *A point is at $x = 2$ on the x axis at time $t = 0$, then moves with velocity $v = t - t^2$. Find where it is at time $t = 3$.*

Velocity is the rate at which displacement x changes with time: $v = dx/dt$. In this case

$$v = dx/dt = t - t^2.$$

Therefore x is some antiderivative of $t - t^2$. All of its antiderivatives are included in

$$x = \tfrac{1}{2}t^2 - \tfrac{1}{3}t^3 + C,$$

where C is any constant.

To find what value C must take in this case, we obviously have to take the starting point into consideration: $x = 2$ when $t = 0$. To obtain the value of C, substitute these values into our expression:

$$2 = 0 - 0 + C.$$

Therefore $C = 2$, so the position at any time is given by

$$x = \tfrac{1}{2}t^2 - \tfrac{1}{3}t^3 + 2.$$

Finally, when $t = 3$, we have $x = -\tfrac{5}{2}$.

Example 14.7   *Find the equation of the curve which passes through the point $(\pi, -1)$ and whose slope is given by $dy/dx = \sin 2x$.*

Since the required y is an antiderivative of $\sin 2x$, the equation of the curve must take the form

$$y = -\tfrac{1}{2} \cos 2x + C,$$

where C is *some* (not 'any') constant. Since also we know that the curve passes through the point $x = \pi$, $y = -1$, we must require

$$-1 = -\tfrac{1}{2} \cos 2\pi + C = -\tfrac{1}{2} + C,$$

so $C = -\tfrac{1}{2}$. Finally the required curve is

$$y = -\tfrac{1}{2} \cos 2x - \tfrac{1}{2}.$$

Example 14.8   *Obtain the antiderivatives of $(3x - 2)^3$.*

As in the earlier examples, we try to guess the structure of y, given that $dy/dx = (3x - 2)^3$. There is not much to go on, so try an analogy with x^3; it would lead us to try something like $y = (3x - 2)^4$. To check this, differentiate using the chain rule with $u = 3x - 2$ and $y = u^4$:

$$\frac{dy}{dx} = 4(3x - 2)^3 \cdot 3 = 12(3x - 2)^3.$$

The factor 12 is unwanted; we really needed $y = \tfrac{1}{12}(3x - 2)^4$. Therefore all the antiderivatives are given by $y = \tfrac{1}{12}(3x - 2)^4 + C$.

The technique used in the previous example can be used for functions like $(ax + b)^n$, e^{ax+b}, $\cos(ax + b)$, and $\sin(ax + b)$. However, it would not work in this simple way for a function such as $(2x^2 - 3)^2$ or $\sin(2x^2 - 3)$: the antiderivative of $(2x^2 - 3)^2$ is *not* equal to $\frac{1}{3}(2x^2 - 3)^3$, because x^2 is present rather than x (try it, using the chain rule).

14.2   Constructing a table of antiderivatives

Since antidifferentiation is the inverse of differentiation, any table of derivatives can be read backwards in order to provide antiderivatives. Suppose that two typical entries in a table of derivatives are as follows:

Given function	**Derivative**
$F(x)$	$f(x) = \dfrac{\mathrm{d}}{\mathrm{d}x} F(x)$
$\sin ax$	$a \cos ax$
e^{ax}	$a\,\mathrm{e}^{ax}$

By interchanging the columns and modifying the headings, we get two entries in a possible table of antiderivatives:

Given function	**One antiderivative**
$f(x)$	$F(x)$
$a \cos ax$	$\sin ax$
$a\mathrm{e}^{ax}$	e^{ax}

However, these entries are not yet in the form we should like them. For example, for the first entry we would prefer to have $\cos ax$ in the left column, instead of $a \cos ax$. Therefore divide both entries by the constant a, remember to introduce the arbitrary constant C to register *all* the antiderivatives, and we have a more convenient table:

Given function	**Antiderivatives**
$f(x)$	$F(x)$
$\cos ax$	$\dfrac{1}{a} \sin ax + C$
e^{ax}	$\dfrac{1}{a} \mathrm{e}^{ax} + C$

By such means the short table (14.2) is produced. To verify any entry, differentiate the function in the right-hand column; the result should be the entry on the left. The letter C stands for 'any constant' or 'an arbitrary constant'.

A short table of antiderivatives	
Given function	**Antiderivatives**
$f(x)$	$F(x)$
a (constant)	$ax + C$
* x^m (unless $m = -1$)	$\dfrac{1}{m+1} x^{m+1} + C$
** x^{-1} $\left(\text{i.e. } \dfrac{1}{x}\right)$	$\begin{cases} \ln x + C & \text{if } x > 0 \\ \ln(-x) + C & \text{if } x < 0 \end{cases}$ or $\ln\|x\| + C \ (x \neq 0)$
e^{ax}	$\dfrac{1}{a} e^{ax} + C$
$\cos ax$	$\dfrac{1}{a} \sin ax + C$
$\sin ax$	$-\dfrac{1}{a} \cos ax + C$

$$(14.2)$$

Notice particularly the two starred entries. The formula * covers most cases, but it does not produce antiderivatives of the function x^{-1} (i.e. of $1/x$). Here $m = -1$, so the entry on the right becomes infinite and therefore meaningless. Therefore the antiderivatives of x^{-1} must be given by some different formula, and this is shown under **. All we have to do is to verify the formula ** as in the following example. (The **modulus** or **absolute value** notation $|x|$ is explained in Section 1.1.)

Example 14.9 *Confirm that the antiderivatives of x^{-1} (i.e. $1/x$) are given by $\ln x + C$ if x is positive and by $\ln(-x) + C$ if x is negative, and that $\ln |x| + C$ covers both cases.*

(Remember that $\ln x$ does not have a meaning if x is negative or zero.) All we have to do to verify the correctness of the formulae is to differentiate the proposed antiderivatives. Since $(d/dx) \ln x = x^{-1}$, the result is right when x is positive.

Suppose now that x is negative. Then $-x$ is positive, so $\ln(-x)$ has a meaning. Using the chain rule (3.3) with $u = -x$,

$$\frac{d}{dx} \ln(-x) = \frac{1}{-x}(-1) = \frac{1}{x},$$

so the second result is confirmed.

But (see Section 1.1) $|x| = x$ if $x > 0$ and $|x| = -x$ when $x < 0$, so $\ln |x|$ is an antiderivative whether x is positive or negative.

Example 14.10 *Find the antiderivatives of* $(2x-3)^{-1}$.

The power $m=-1$ is the starred case in (14.2), so try $y=\ln(2x-3)$, supposing initially that $2x-3>0$. Then

$$\frac{dy}{dx}=\frac{2}{2x-3}, \quad \text{or} \quad 2(2x-3)^{-1}.$$

The unwanted factor 2 will not appear if we try again with $y=\frac{1}{2}\ln(2x-3)$ Also $2x-3$ might be negative, so we introduce a modulus sign. Finally we have

$$y=\tfrac{1}{2}\ln|2x-3|+C.$$

14.3 Signed area generated by a graph

Figure 14.2 shows the graph of a function $y=f(x)$ between $x=a$ and $x=b$, in which we assume that **the x and y scales are the same**. Divide the range as shown into N sections so that in any section y is either positive only, or negative only.

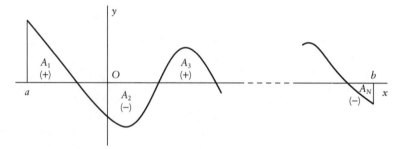

Fig. 14.2

Let $A_1, A_2, \ldots$ denote the **geometrical areas** of these segments, and A the sum of these. *Geometrical* area is always positive, so $A_1, A_2, \ldots$ are all **positive numbers**. Then

$$A=A_1+A_2+A_3+\cdots+A_N \tag{14.3}$$

is naturally called 'the geometrical area between the curve and the x axis'.

We require a different quantity, $\mathcal{A}$, called the **signed area** between the curve and the x axis. This is defined by

$$\mathcal{A}=A_1-A_2+A_3-\cdots-A_N. \tag{14.4}$$

In forming $\mathcal{A}$, we use the rule: **If y is positive, the contribution takes a positive sign; if y is negative, the contribution takes a negative sign.** This quantity has a far more useful range of applications than has geometrical area. For example, suppose that a point is moving on a straight line; then the signed displacement from its starting point is equal to the *signed* area of its velocity–time graph.

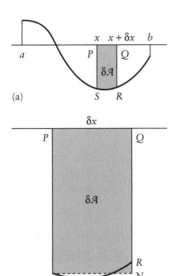

(a)

(b)

Fig. 14.3

We show how to calculate the signed area $\mathcal{A}$ of the graph of $y = f(x)$ between two given points, $x = a$ and $x = b$ (Fig. 14.3a). Let $\mathcal{A}(x)$ represent the signed area between a and a variable point with coordinate x (Fig. 14.3a). Increase x by a small step δx; the signed area from a to $x + \delta x$ is $\mathcal{A}(x + \delta x)$. The change in signed area, $\delta\mathcal{A} = \mathcal{A}(x + \delta x) - \mathcal{A}(x)$ (positive or negative), is equal to the signed area of $PQRS$ in Figs 14.3a and b. This is very nearly equal to the signed area of the rectangle $PQNS$ in Fig. 14.3b (in this case the required sign is negative) so

$$\delta\mathcal{A} \approx f(x)\,\delta x$$

which automatically takes the right sign. Therefore

$$\frac{\delta\mathcal{A}}{\delta x} \approx f(x).$$

Now let $\delta x \to 0$; '$\approx$' becomes '$=$', and $\delta\mathcal{A}/\delta x$ becomes $\mathrm{d}\mathcal{A}/\mathrm{d}x$, so that

$$\frac{\mathrm{d}\mathcal{A}}{\mathrm{d}x} = f(x). \tag{14.5}$$

From (14.5) $\mathcal{A}(x)$ must be one of the antiderivatives of $f(x)$. To find which one, choose *any particular antiderivative* and call it $F(x)$. Then $\mathcal{A}(x)$ can differ from $F(x)$ only by a constant, k say, so

$$\mathcal{A}(x) = F(x) + k. \tag{14.6}$$

To determine the value of k, use the fact that $\mathcal{A}(x) = 0$ at $x = a$, because the starting point is then the same as the end-point; that is to say,

$$\mathcal{A}(a) = 0.$$

Therefore, from (14.6)

$$\mathcal{A}(a) = 0 = F(a) + k,$$

or $\quad k = -F(a), \tag{14.7}$

a known quantity, since we selected the antiderivative $F(x)$ of $f(x)$ ourselves. The required area $\mathcal{A}$ between a and b is given by

$$\mathcal{A} = \mathcal{A}(b) = F(b) - F(a),$$

by putting $x = b$ into (14.6), with (14.7) as the value of k.

The signed area $\mathcal{A}$ of $f(x)$ between $x = a$ and b
$$\mathcal{A} = F(b) - F(a),$$
where $F(x)$ is *any antiderivative* of $f(x)$.

$\tag{14.8}$

In practice we naturally use the simplest antiderivative, in which the C in the table is zero. But any nonzero choice of C will cancel out and disappear, since it will be present in both $F(a)$ and $F(b)$.

Example 14.11 *Find the signed area of $y = x^2$ from $x = -1$ to $x = 2$.*

(This happens to be the same as the geometrical area, because y is never negative.) Here $a = -1$ and $b = 2$. Also, the simplest antiderivative of x^2 is

$$F(x) = \tfrac{1}{3}x^3.$$

Therefore, from (14.8),

$$\mathcal{A} = F(b) - F(a) = \tfrac{1}{3}(2)^3 - \tfrac{1}{3}(-1)^3 = 3.$$

There is a special notation, the **square-bracket notation**, which we shall use generally from now onward.

> **Square-bracket notation**
>
> $[F(x)]_a^b$ stands for $F(b) - F(a)$.
>
> (14.9)

Example 14.12 *Find* (a) *the signed area, and* (b) *the geometrical area, between $y = \sin x$ and the x axis from $x = 0$ to $x = 2\pi$.*

(a) $f(x) = \sin x$, so $F(x) = -\cos x$ is an antiderivative. From (14.8) and (14.9), with $a = 0$ and $b = 2\pi$, the signed area $\mathcal{A}$ is given by

$$\mathcal{A} = [-\cos x]_0^{2\pi} = -[\cos x]_0^{2\pi} = -(\cos 2\pi - \cos 0) = 0,$$

as is expected from Fig. 14.4: the positive and negative sections cancel.

(b) The geometrical area A can be obtained by splitting the range into a positive section 0 to π, and a negative section from π to 2π (see Fig. 14.4). The negatively signed section π to 2π must have its sign reversed in order to give the *geometrical* area:

$A = $ [geometrical area of 1st loop] + [geometrical area of 2nd loop]

  $= $ [signed area of 1st loop] − [signed area of 2nd loop].

This is equal to

$$[F(x)]_0^{\pi} - [F(x)]_{\pi}^{2\pi} = [-\cos x]_0^{\pi} - [-\cos x]_{\pi}^{2\pi}$$

$$= (-\cos \pi + \cos 0) - (-\cos 2\pi + \cos \pi)$$

$$= (1 + 1) - (-1 + (-1)) = 2 + 2 = 4.$$

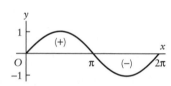

Fig. 14.4

Problems

Note: In case you have already met the term 'indefinite integral', the term 'antiderivative' has the same meaning.

14.1 Obtain all the antiderivatives of the following functions, and check their correctness by differentiating your results.

(a) x^5; $3x^4$; $2x^3$; $\frac{1}{3}x^2$; $6x$; $f(x) = 3$; $f(x) = 0$.

(b) $-\frac{1}{2}x^{-3}$; $2x^{-2}$; $3x^{-1}$ when $x > 0$ (if in doubt, see (14.2)).

(c) $x^{\frac{3}{2}}$; $x^{\frac{1}{2}}$; $x^{-\frac{1}{2}}$; $x^{\frac{4}{3}}$; $x^{-\frac{1}{3}}$.

(d) $1/x^2$ (write as x^{-2}); $1/x^4$; $1/x$ when $x < 0$ (see (14.2)).

(e) $\sqrt{x} (= x^{\frac{1}{2}})$; $1/\sqrt{x}$; $1/x^{\frac{3}{2}}$.

(f) $3x$; $\frac{1}{2}x^2$; $1/(3x^2)$; $3/(4x^{\frac{1}{4}})$.

(g) e^x; e^{-x}; $5e^{2x}$; $e^{-\frac{1}{2}x}$; $3e^{-2x}$.

(h) $\cos x$; $\cos 3x$; $\sin x$; $\sin 3x$;

(i) $1 - 3x$; $1 + 2x - 3x^2$; $3x^4 - 4x^2 + 5$.

(j) $x(x + 1)$ (expand by removing the brackets); $(1 + 2x)(1 - 2x)$; $(x + 1)^2$; $(1 + x)(1 - 1/x)$; $x^2(x + x^2)$.

(k) $(x + 1)/x$ (turn it into the sum of two terms); $(2\sqrt{x} - 1)/\sqrt{x}$ (put $\sqrt{x} = x^{\frac{1}{2}}$ and $1/\sqrt{x} = x^{-\frac{1}{2}}$, then simplify as the sum of two terms); $(x + 1)^2/x^3$.

(l) $e^x + e^{-x}$; $2e^{2x} - 3e^{3x}$; $e^{\frac{1}{2}x}(1 + e^{-\frac{1}{2}x})$; $1/e^{2x} (= e^{-2x})$; $(e^{2x} - e^{-2x})/e^{2x}$.

(m) $2 \cos 2x$; $3 \sin \frac{1}{2}x - 4 \cos \frac{1}{3}x$; $2 + \sin 2x$.

14.2 Find all the antiderivatives of the following by trial and error, as explained in the text. Confirm your answers by differentiation.

(a) $(x + 1)^3$ (start by trying $(x + 1)^4$); $(3x + 1)^3$; $(3x - 8)^3$.

(b) $(1 - x)^4$; $(8 - 3x)^{\frac{1}{2}}$; $(1 - x)^{\frac{1}{3}}$.

(c) $(2x + 1)^{-2}$; $(1 - x)^{-\frac{1}{2}}$; $2/(3x + 1)^3$; $1/[4(1 - x)^{\frac{1}{4}}]$.

(d) $2 \cos(3x - 2)$ (try first $\sin(3x - 2)$); $3 \sin(1 - x)$; $2 \sin(2 - 3x)$.

14.3 (See Example 14.10.) Find the antiderivatives of the following.

(a) $1/(x + 1)$; $1/(x - 1)$; $3/(3x - 2)$; $2/(5x - 4)$.

(b) $1/(1 - x)$; $1/(4 - 5x)$.

(c) $x/(x + 1)$ (it can be written as $1 - 1/(x + 1)$).

(d) $(x + 1)/(x - 1)$ (compare (c)).

14.4 Use the identities $\cos^2 A = \frac{1}{2}(1 + \cos 2A)$, $\sin^2 A = \frac{1}{2}(1 - \cos 2A)$, and $\sin A \cos A = \frac{1}{2} \sin 2A$ to get rid of the squares and products in the following expressions, and in that way obtain the antiderivatives.

(a) $\cos^2 x$; $\sin^2 x$; $\sin x \cos x$.

(b) $3 \cos^2 2x$; $\sin^2 3x$; $\sin 2x \cos 2x$.

(c) $\cos^4 x$ (you will have to use the identities twice).

14.5 (a) Show that $(d/dx)(x e^x) = e^x + x e^x$. By rearranging the terms, show that the antiderivatives of $x e^x$ are $e^x(x - 1) + C$ (use the fact that e^x can be written as $(d/dx) e^x$). Confirm the result by differentiation.

(b) Differentiate $x^2 e^x$. By rearranging the terms and using the result in (a), find the antiderivatives of $x^2 e^x$.

14.6 Use the result (14.8) to obtain the signed areas between the given graphs and the x axis. By roughly sketching the graphs of the functions for which you obtain zero, explain this fact.

(a) $y = x$, $0 \leq x \leq 2$;

(b) $y = x$, $-1 \leq x \leq 1$;

(c) $y = -x^2$, $0 \leq x \leq 1$;

(d) $y = \cos x$, $-\pi \leq x \leq \pi$;

(e) $y = \cos x - 1$, $0 \leq x \leq 2\pi$;

(f) $y = x^{-1}$, $-2 \leq x \leq -1$ (note that x is negative in this range);

(g) $y = \sin 3x$, $0 \leq x \leq \frac{2}{3}\pi$;

(h) $y = 1/(1 - x)$, $2 \leq x \leq 3$ (note: $1 - x$ is negative over this range, so make sure you understand Example 14.10; alternatively, write $1/(1 - x) = -1/(x - 1)$).

14.7 Obtain the geometric area between the graph and the x axis in each of the following cases. It is necessary to treat each positive or negative section separately.

(a) $y = -3$, $0 \leq x \leq 1$ (this is negative all the way);

(b) $y = x^3$, $-1 \leq x \leq 1$;

(c) $y = 4 - x^2$, $-1 \leq x \leq 3$;

(d) $y = \cos x$, $0 \leq x \leq 2\pi$.

14.8 Find the most general function which satisfies the following equations.

(Note: $\dfrac{d^2 x}{dt^2} = \dfrac{d}{dt}\dfrac{dx}{dt}$, $\dfrac{d^3 x}{dt^3} = \dfrac{d}{dt}\dfrac{d^2 x}{dt^2}$, etc. Work in several steps, finding the next lowest derivative in each step.)

(a) $\dfrac{d^2 x}{dt^2} = 0$; (b) $\dfrac{d^2 x}{dt^2} = t$;

(c) $\dfrac{d^2 x}{dt^2} = \sin t$;

(d) $\dfrac{d^3 x}{dt^3} = 0$; (e) $\dfrac{d^3 x}{dt^3} = \cos t$;

(f) $\dfrac{d^2 x}{dt^2} = g$ (g is a constant);

(g) $\dfrac{d^4 y}{dx^4} = w_0$ (w_0 is constant; this relates to the displacements $y(x)$ of a bending beam).

15 The definite and indefinite integral

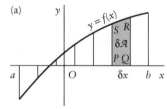

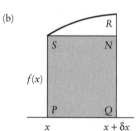

Fig. 15.1

15.1 Signed area as the sum of strips

Consider signed area from another point of view. Figure 15.1a represents the graph of a function $y = f(x)$ between $x = a$ and $x = b$. Since we are going to talk about area, assume that the **x and y scales are the same**. Divide the interval a to b into N small equal steps each of width

$$\delta x = \frac{b - a}{N}.$$

To any step PQ there is a signed area element $PQRS$ which we call δA. The total signed area A is equal to the sum of all these:

$$A = \sum_{x=a}^{x=b} \delta A$$

which signifies 'the sum of all the elements δA between a and b'. The typical area element $PQRS$ is shown magnified in Fig. 15.1b. When δx is small, the signed area δA is nearly that of the shaded rectangle $PQNS$. Therefore, for δx small, we have

$$A = \sum_{x=a}^{x=b} \delta A \approx \sum_{x=a}^{x=b} f(x)\, \delta x.$$

When $\delta x \to 0$ (with N increasing correspondingly), the approximation approaches perfection and we have

> **Signed area as a sum**
>
> Signed area $\mathcal{A}$ of $y = f(x)$, $x = a$ to b:
>
> $$\mathcal{A} = \lim_{\delta x \to 0} \sum_{x=a}^{x=b} f(x)\,\delta x.$$
>
> (15.1)

15.2 Numerical illustration of the sum formula

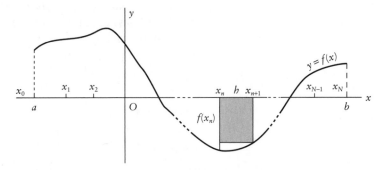

Fig. 15.2

We shall specify the sum in (15.1) in more detail, with the idea of obtaining a specific **algorithm** for actually calculating such a sum on a computer. In Fig. 15.2 we show the graph $y = f(x)$. There are N equal subdivisions: we shall call the **length of a subdivision** h rather than δx as in Fig. 15.1, since this is conventional when making numerical calculations:

$$h = \frac{b - a}{N}.$$

The points of subdivision are x_0 to x_N:

$$x_0 = a, \quad x_1 = a + h, \quad x_2 = a + 2h, \quad \dots, \quad x_{N-1} = a + Nh = b$$

which can be expressed as

$$x_n = a + nh \quad \text{for } n = 0, 1, 2, \dots, N.$$

Then the area of the nth approximating rectangle in (15.1) is $f(x_n)h$, and the approximating sum in (15.1) becomes

$$\mathcal{A} \approx f(a)h + f(a + h)h + \cdots + f(a + (N - 1)h)h$$

$$= h \sum_{n=0}^{N-1} f(x_n) \quad \text{where } x_n = a + nh, \text{ with } n = 0, 1, 2, \dots, N - 1.$$

> **Computation of approximating sums (rectangle rule)**
>
> If $y = f(x)$ with range $x = a$ to b, the signed-area approximation with N subdivisions is
>
> $$\mathcal{A} \approx h \sum_{n=0}^{N-1} f(x_n),$$
>
> where $h = (b - a)/N$ and $x_n = a + nh$.
>
> **(15.2)**

When we take larger and larger N, and smaller and smaller h correspondingly, we expect that *the approximation will approach the exact value*. The following example illustrates this for the very simple case of the signed area associated with a straight line. The **algorithm** (15.2) is very easy to program on a computer for any function $f(x)$. It is called the **rectangle rule**.

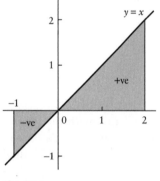

Fig. 15.3

Example 15.1   *Calculate the sum in* (15.2) *when* $f(x) = x$, $a = -1$, $b = 2$, *for* $N = 30, 300, 3000, \dots$, *showing how the results approach the exact value* 1.5.

The graph $y = x$ in Fig. 15.3 is a straight line, from which it is very easy to see that the *signed* area is exactly 1.5. The computed results are as follows ($b - a = 3$, and so $h = 3/N$):

N	30	300	3000	30 000	...
h	0.1	0.01	0.001	0.0001	...
$\mathcal{A} \approx$	1.05	1.455	1.4955	1.499 55	...

The approximations are approaching 1.5, though very slowly. We shall see in Section 16.3 how to improve such calculations.

15.3 The definite integral and area

The expression $\displaystyle\lim_{\delta x \to 0} \sum_{x=a}^{x=b} f(x)\,\delta x$ of (15.1), which is equal to the signed area, has a very important brief notation:

> **Definite-integral notation**
>
> $$\lim_{\delta x \to 0} \sum_{x=a}^{x=b} f(x)\,\delta x \text{ is denoted by } \int_a^b f(x)\,\mathrm{d}x.$$
>
> **(15.3)**

(Historically, a large letter S for 'sum' used to be printed instead of Σ: the sign $\int$ is really just an extended letter S.) The expression $\int_a^b f(x)\,\mathrm{d}x$ is called a **definite integral**, to be read: 'the integral of $f(x)\,\mathrm{d}x$ from a to b'. Here $f(x)$ is the **integrand**, or the function to be **integrated**. The letter x is the name of the **variable of integration**, while a is the **lower limit** and b the **upper limit** for the **integration** process.

We already found a way to obtain the signed area by using an antiderivative: see (14.8). In the new notation, (14.8) is expressed as follows.

> **Signed area expressed as a definite integral**
>
> The signed area $\mathcal{A}$ of $f(x)$ between $x = a$ and b is given by
>
> $$\int_a^b f(x)\,dx = [F(x)]_a^b = F(b) - F(a)$$
>
> where $F(x)$ is any antiderivative of $f(x)$. (15.4)

Notice that in a **definite integral any letter can be used for the variable of integration**, because the letter itself disappears in the course of evaluation; for example,

$$\int_0^1 x\,dx = [\tfrac{1}{2}x^2]_0^1 = [\tfrac{1}{2}x^2]_{x=0}^{x=1} = \tfrac{1}{2};$$

$$\int_0^1 t\,dt = [\tfrac{1}{2}t^2]_0^1 = [\tfrac{1}{2}t^2]_{t=0}^{t=1} = \tfrac{1}{2}; \quad \text{and so on.}$$

Consequently, the letter used is called a **dummy variable**. We should not choose a letter already being used for something else.

15.4 The indefinite-integral notation

The symbol

$$\int f(x)\,dx,$$

with no limits of integration specified, is called an **indefinite integral of** $f(x)$, and has exactly the **same meaning as the word 'antiderivative'** that we have used up until now, and which we have denoted by $F(x)$.

> **Indefinite-integral notation**
>
> $\int f(x)\,dx$, with no limits specified, stands for any antiderivative of $f(x)$. (15.5)

The expression $\int_a^b f(x)\,dx$ is called a definite integral because it takes a definite value: it represents a specified signed area and there is no arbitrary constant on the right. However, an indefinite integral $\int f(x)\,dx$ does not stand for a number; it represents an antiderivative, which is a function. This function is to be written in terms of the current variable of integration, so the name of the variable is usually significant. Also there will be a disposable, or arbitrary, constant, in the usual way. For example,

$$\int x^2 \, dx = \tfrac{1}{3}x^3 + C, \qquad \int e^{2t} \, dt = \tfrac{1}{2}e^{2t} + C,$$

$$\int \cos u \, du = \sin u + C$$

and so on, where C is a constant. In some problems we shall assign or discover a definite value for C; in others we might want to keep C as an arbitrary constant in order to express every possible antiderivative (hence, 'indefinite' integral).

Example 15.2   *Find the signed area A associated with the graph $y = 3 \, e^{2x}$ from $x = 1$ to $x = 3$ using the new notation.*

We shall need an antiderivative $F(x)$ (i.e. an indefinite integral) of $3 \, e^{2x}$. Using the notation (15.5), we may write

$$F(x) = \int 3 \, e^{2x} \, dx = \tfrac{3}{2} \, e^{2x}$$

(for this purpose any antiderivative will do, so we have put $C = 0$). Then, from (15.4),

$$A = \int_1^3 3 \, e^{2x} \, dx = [\tfrac{3}{2} \, e^{2x}]_1^3$$

$$= \tfrac{3}{2}[e^{2x}]_1^3 = \tfrac{3}{2}(e^6 - e^2).$$

In the last example, we might as well have written

$$\int_1^3 3 \, e^{2x} \, dx = \left[\int 3 \, e^{2x} \, dx \right]_1^3$$

in the first place, without ever introducing $F(x)$. If we do this, we get another version of (15.4) which it is often convenient to use:

Signed area, using the notation for definite and indefinite integrals

The signed area of $f(x)$ from a to b is

$$\int_a^b f(x) \, dx = \left[\int f(x) \, dx \right]_a^b,$$

where $\int f(x) \, dx$ is any indefinite integral (antiderivative) of $f(x)$.

(15.6)

15.5   Integrals unrelated to area

Integrals arise constantly in applications, but only seldom is there any direct connection with area. The following Example starts by giving information that seems to have nothing to do with area. However, we show that the problem can be thought of in terms of an area, and therefore can be solved in terms of a definite integral.

Example 15.3   *A small object P is pushed steadily along the x axis from $x = 0$ to $x = 1$, against a resistive force $f(x) = x^2$. Find the work done against the resistance.*

Divide the range $x = 0$ to 1 into a large number of short steps of length δx. In general, if the resistive force is constant the work done over a distance is (force) × (distance moved). Although the force on P is not constant, over a short distance δx the work δW done by the applied force is given approximately by

$$\delta W \approx f(x)\,\delta x = x^2\,\delta x.$$

The total work W is given by

$$W = \sum_{x=0}^{x=1} \delta W \approx \sum_{x=0}^{x=1} x^2\,\delta x.$$

Letting $\delta x \to 0$, we obtain exactly

$$W = \lim_{\delta x \to 0} \sum_{x=0}^{x=1} x^2\,\delta x. \tag{15.7}$$

But this expression matches eqn (15.1): *it represents the signed area of the curve $y = x^2$ between $x = 0$ and* 1. Consequently we can say immediately that

$$W = \int_0^1 x^2\,\mathrm{d}x = [\tfrac{1}{3}x^3]_0^1 = \tfrac{1}{3}. \tag{15.8}$$

You should think very carefully about the step from (15.7) to (15.8), because it can be generalized to apply to any similar problem. Suppose that there arises, in any context whatever, a sum of the type

$$\lim_{\delta x \to 0} \sum_{x=a}^{x=b} f(x)\,\delta x.$$

Then *such a sum can always be interpreted as representing a certain signed area* (namely the signed area of $y = f(x)$ between $x = a$ and $x = b$) so *it can always be represented by the definite integral $\int_a^b f(x)\,\mathrm{d}x$*. We do not have to repeat this argument every time we encounter such a sum; from now on, we call on the general statement:

> **The limit of a sum represented by a definite integral**
>
> In all cases
>
> $$\lim_{\delta x \to 0} \sum_{x=a}^{x=b} f(x)\, \delta x = \int_a^b f(x)\, dx.$$
>
> The integral is then evaluated using (15.4) or (15.6).   **(15.9)**

The variable occurring need not be denoted by x, as the following examples demonstrate.

Example 15.4   *An object is driven along a straight line with velocity $v(t) = e^{\frac{1}{2}t}$ between times $t = 0$ and $t = 2$. There is a resistive force $g(v) = 3v^2$, where v is velocity. Find the total work done against the resistance.*

In a short interval between times t and $t + \delta t$, the distance travelled, δx, is given approximately by

$$\delta x \approx v(t)\, \delta t = e^{\frac{1}{2}t}\, \delta t.$$

The work δW done in this time interval is approximated by

$$\delta W \approx g(v)\, \delta x = 3v^2\, \delta x$$
$$\approx 3v^2(e^{\frac{1}{2}t}\, \delta t) = 3\, e^t(e^{\frac{1}{2}t}\, \delta t) = 3\, e^{\frac{3}{2}t}\, \delta t.$$

Therefore the total work W required is given by

$$W = \lim_{\delta t \to 0} \sum_{t=0}^{t=2} 3\, e^{\frac{3}{2}t}\, \delta t$$

$$= \int_0^2 3\, e^{\frac{3}{2}t}\, dt = 2[e^{\frac{3}{2}t}]_0^2 = 2(e^3 - 1).$$

Example 15.5   *During a rainy period extending from $t = 0$ to $t = 10$ days, the rainfall rate r from moment to moment in units of centimetres per day is found to be $r(t) = \frac{3}{5}t - \frac{3}{50}t^2$. Find the total depth of rainfall, R, for the period.*

Take a short time interval from t to $t + \delta t$ (expressed as a fraction of a day). During this period, the rainfall δR is given approximately by

$$\delta R \approx r(t)\, \delta t = (\tfrac{3}{5}t - \tfrac{3}{50}t^2)\, \delta t$$

('approximately' because the rate of fall r varies a little even through a short time). The total rainfall from $t = 0$ to 10 days is equal to the sum of all the contributions as the steps δt tend to zero (while becoming proportionately more numerous):

$$R = \lim_{\delta t \to 0} \sum_{t=0}^{t=10} (\tfrac{3}{5}t - \tfrac{3}{50}t^2)\, \delta t = \int_0^{10} (\tfrac{3}{5}t - \tfrac{3}{50}t^2)\, dt \quad \text{(by (15.9))}$$

$$= [\tfrac{3}{5}(\tfrac{1}{2}t^2) - \tfrac{3}{50}(\tfrac{1}{3}t^3)]_0^{10} = \tfrac{3}{10}(10)^2 - \tfrac{1}{50}(10)^3 = 10 \quad \text{(cm).}$$

Example 15.6 *Suppose that, in Example 15.5, the rainfall rate is given by* $r(t) = t^{\frac{1}{2}} e^{-t}$ *(cm per day). Obtain the total rainfall R between t = 0 and 10 days.*

Proceeding as before, the total rainfall is given by

$$R = \int_0^{10} t^{\frac{1}{2}} e^{-t} \, dt.$$

We cannot find an indefinite integral to enable R to be evaluated. However, we know from (15.9) that R *must be equal to the area under the* (r, t) *graph*, which can be *computed numerically* by using the numerical method of eqn (15.2).

Divide the range $t = 0$ to 10 into N strips (so that $\delta t = 10/N$), then the approximation corresponding to (15.2) becomes

$$R \approx \frac{10}{N} \sum_{n=0}^{N-1} t_n^{\frac{1}{2}} e^{-t_n}, \quad \text{where} \quad t_n = n\left(\frac{10}{N}\right).$$

The following computed values show how the exact result is approached when we take N larger and larger:

N	5	10	100	1000
δx	2.00	1.00	0.10	0.100
R	0.4701	0.7070	0.8796	0.8859

The exact answer is $0.886\,07\ldots$.

We shall show in Chapter 16 that we are not tied to eqn (15.2) for calculating signed area, but can find far better computing formulae.

15.6 Improper integrals

If a definite integral has an **infinite range**, or the **integrand becomes infinite** at some point in its range, the integral is said to be **improper**. Usually these present no particular problem.

Example 15.7 *Evaluate* $\int_0^\infty e^{-2x} \, dx.$

Putting $\int e^{-2x} \, dx = -\frac{1}{2} e^{-2x}$, we have

$$\int_0^\infty e^{-2x} \, dx = -\frac{1}{2} [e^{-2x}]_0^\infty = -\frac{1}{2}(0 - 1) = \frac{1}{2}.$$

Example 15.8 *Evaluate* $\int_1^\infty \frac{dx}{x^2}.$

$$\int_1^\infty x^{-2} \, dx = [-x^{-1}]_1^\infty = [0 - (-1)] = 1.$$

In Examples 15.7 and 15.8 we have (see Fig. 15.4) two cases of an infinitely long figure which encloses a finite area. This cannot always happen, even if the integrand goes to zero when $x \to \infty$.

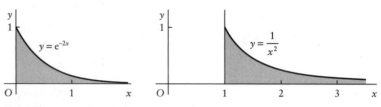

Fig. 15.4

Example 15.9   *Consider* $\displaystyle\int_1^{\infty} \frac{dx}{x}$.

We have

$$\int_1^{\infty} x^{-1}\, dx = [\ln x]_1^{\infty}.$$

The logarithm becomes infinite as x becomes infinite, so the integral is meaningless. The function x^{-1} does not tend to zero fast enough to keep the area finite as we extend the range to infinity.

The case when the **integrand becomes infinite** at some point in its range has similar features:

Example 15.10   *Consider* (a) $\displaystyle\int_0^1 x^{-\frac{1}{2}}\, dx$; (b) $\displaystyle\int_0^1 x^{-1}\, dx$.

Notice that $x^{-\frac{1}{2}}$ and x^{-1} are infinite at $x = 0$.

(a) $\displaystyle\int_0^1 x^{-\frac{1}{2}}\, dx = 2[x^{\frac{1}{2}}]_0^1 = 2[1 - 0] = 2.$

Therefore the integral gives no problem; it is again a case of an infinitely extended figure (extended in the y direction this time) containing a finite area.
(b) On the other hand,

$$\int_0^1 x^{-1}\, dx = [\ln x]_0^1,$$

and the integral this time is infinite, because $\ln 0$ is $(-\infty)$.

There are improper integrals which do not work out for a different reason:

Example 15.11 *Consider the integrals*

$$\text{(a)} \int_0^X \cos x \, dx, \quad \text{(b)} \int_0^\infty \cos x \, dx.$$

(a) We have

$$\int_0^X \cos x \, dx = [\sin x]_0^X = \sin X.$$

So long as X is finite, there is therefore no problem.

(b) However, for $\int_0^\infty \cos x \, dx$ we would, straightforwardly, have a term $\sin \infty$ to interpret. The only sensible meaning that we could attach to $\sin \infty$ is that it stands for $\lim_{x \to \infty} \sin X$. But $\sin X$ has no definite limit as $X \to \infty$; it goes up and down between ± 1 for ever.

Improper integrals which give a definite finite result are said to **converge**. If not, they are said to **diverge**.

15.7 Integration of complex functions: a new type of integral

To differentiate or integrate a function containing the 'imaginary' element i, simply treat i like an ordinary real constant. Thus, for example,

$$\frac{d}{dx} e^{ix} = i e^{ix}$$

and

$$\int e^{ix} \, dx = \frac{1}{i} e^{ix} + C = -i e^{ix} + C,$$

where C is an arbitrary constant (which in this context we would allow to be itself a complex number). Suppose that a and b are *real* numbers, and that

$$c = a + ib.$$

Then

$$\frac{d}{dx} e^{cx} = c \, e^{cx},$$

and

$$\int e^{cx} \, dx = \frac{1}{c} e^{cx} + C. \tag{15.10}$$

We use (15.10) with $c = a + ib$ to work out two integrals U and V which frequently occur in practice:

$$U = \int e^{ax} \cos bx \, dx \quad \text{and} \quad V = \int e^{ax} \sin bx \, dx.$$

Omit the arbitrary constant C for the moment. Observe that

$$U + iV = \int e^{ax} \cos bx \, dx + i \int e^{ax} \sin bx \, dx$$

$$= \int e^{ax}(\cos bx + i \sin bx) \, dx = \int e^{ax} \, e^{ibx} \, dx$$

$$= \int e^{(a+ib)x} \, dx$$

$$= \frac{1}{a + ib} e^{(a+ib)x} \quad \text{(from (15.10))}$$

$$= \frac{a - ib}{a^2 + b^2} e^{ax}(\cos bx + i \sin bx)$$

$$= \frac{1}{a^2 + b^2} e^{ax}[(a \cos bx + b \sin bx) + i(-b \cos bx + a \sin bx)].$$

Equate this last expression to $U + iV$: the real and imaginary parts must separately be equal; so, after introducing the arbitrary constant, we have

(a) $\int e^{ax} \cos bx \, dx$

$$= \frac{1}{a^2 + b^2} e^{ax}(a \cos bx + b \sin bx) + C,$$

(b) $\int e^{ax} \sin bx \, dx$

$$= \frac{1}{a^2 + b^2} e^{ax}(-b \cos bx + a \sin bx) + C.$$

(15.11)

The integrals can be expressed more simply in terms of a phase angle. Put

$$\frac{a}{(a^2 + b^2)^{\frac{1}{2}}} = \cos \phi \quad \text{and} \quad \frac{b}{(a^2 + b^2)^{\frac{1}{2}}} = -\sin \phi$$

into (15.11a), and

$$\frac{-b}{(a^2 + b^2)^{\frac{1}{2}}} = \cos \theta \quad \text{and} \quad \frac{a}{(a^2 + b^2)^{\frac{1}{2}}} = -\sin \theta$$

into (15.11b). Notice also that ϕ and θ must therefore be related by

$$\theta = \phi - \tfrac{1}{2}\pi.$$

Then (15.11) becomes

> (a) $\displaystyle \int e^{ax} \cos bx \, dx = \frac{1}{(a^2 + b^2)^{\frac{1}{2}}} e^{ax} \cos(bx + \phi) + C,$
>
> (b) $\displaystyle \int e^{ax} \sin bx \, dx = \frac{1}{(a^2 + b^2)^{\frac{1}{2}}} e^{ax} \cos(bx + \phi - \tfrac{1}{2}\pi) + C$
>
> where $\cos\phi = a/(a^2 + b^2)^{\frac{1}{2}}$, $\sin\phi = -b/(a^2 + b^2)^{\frac{1}{2}}$.
>
> **(15.12)**

Example 15.12   *Evaluate* $I = \displaystyle\int_0^\infty e^{-x} \cos 2x \, dx.$

Equation (15.11a) or (15.12a) can be used directly with $a = -1$, $b = 2$. Howev , we will go through the working from first principles, but express the argument differently. Remember that

$$e^{\alpha + i\beta} = e^\alpha \, e^{i\beta} = e^\alpha(\cos\beta + i\sin\beta);$$

then we have $e^{-x} \cos 2x = \text{Re } e^{(-1+2i)x}$. Therefore

$$I = \int_0^\infty e^{-x} \cos 2x \, dx = \text{Re} \int_0^\infty e^{(-1+2i)x} \, dx$$

$$= \text{Re} \left[\frac{1}{-1 + 2i} e^{(-1+2i)x} \right]_0^\infty$$

$$= \text{Re} \left(0 - \frac{1}{-1 + 2i} \right) = \text{Re} \frac{1 + 2i}{5} = \tfrac{1}{5}.$$

15.8 The area analogy for a definite integral

A signed area can be represented as a definite integral as in (15.4). Conversely, any definite integral $\int_a^b f(x) \, dx$, whatever it represents, can be interpreted as representing the signed area of the graph $y = f(x)$ between a and b. The connection with area means that we have a picture of an integral which can often give useful information without the need to evaluate the integral, which might in any case be impossible. One example of this is the simple numerical method described in Section 15.2. We restate the connection, calling it the **area analogy**.

> **The area analogy**
>
> The definite integral $\int_a^b f(x) \, dx$ always represents the signed area of the graph $y = f(x)$ from $x = a$ to b.
>
> **(15.13)**

The following section illustrates the use of the principle (15.13): it will also be referred to in later chapters.

15.9 Symmetric integrals

In this section we shall use t instead of x, and x instead of y, so that we consider functions of the form

$$x = f(t).$$

The reason is that time t is commonly the physical variable in contexts where these techniques are found useful.

Sometimes, by using the area analogy (15.13), the graph of the integrand of $\int_a^b f(t)\,dt$ makes it obvious that the value of a definite integral is zero. Figure 15.5 shows some simple cases.

The range of the integrals on the two sides of the origin are equal, and because of the special symmetry the positive and negative contributions cancel out. Such functions are called **odd functions**, or **functions odd about the origin**. They have the following property (see Section 1.4):

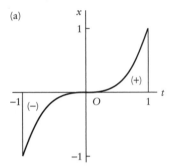

(a)

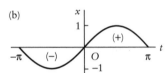

(b)

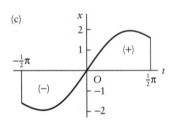

(c)

Fig. 15.5 (a) $x = t^3$; $\int_{-1}^{1} t^3\,dt = 0$.
(b) $x = \sin t$; $\int_{-\pi}^{\pi} \sin t\,dt = 0$.
(c) $x = t + \sin 2t$;
$\int_{-\frac{1}{2}\pi}^{\frac{1}{2}\pi}(t + \sin 2t)\,dt = 0$.

> **Odd functions** $f(t)$
> satisfy the condition $f(-t) = -f(t)$.
> (15.14)

Some basic odd functions are

$$t, t^3, t^5, \dots, \text{ and their reciprocals,}$$

$$\sin at, \sin^3 at, \dots, \text{ and } \tan at, \tan^3 at, \dots, \text{ where } a \text{ is constant.}$$

> **Symmetrical integrals over functions which are odd about the origin**
> If $f(-t) = -f(t)$ then $\displaystyle\int_{-c}^{c} f(t)\,dt = 0$.
> (15.15)

Another useful class are **even functions**, which are symmetrical about the x axis:

> **Even functions** $f(t)$
> $f(t)$ is even if $f(-t) = f(t)$.
> (15.16)

Some basic even functions are

$$t^2, t^4, t^6, \dots, \text{ and their reciprocals,}$$

$$\cos at \text{ and } \cos^n at \ (a \text{ is a constant}),$$

(a)

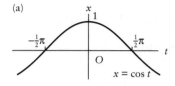

(b)

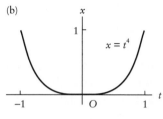

(c)

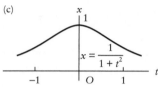

Fig. 15.6

and even powers of any odd function, such as $\tan^6 at$. It is also useful to realize that

(odd function) × (even function) = (odd function),

(odd function) × (odd function) = (even function).

Some even functions are shown in Fig. 15.6.

Example 15.13 *Show that the following integrals are zero*:

(a) $\displaystyle\int_{-\frac{1}{2}\pi}^{\frac{1}{2}\pi} t^4 \sin 3t \, dt$; (b) $\displaystyle\int_{-\pi}^{\pi} t^5 \cos 3t \cos \tfrac{1}{2}t \, dt$;

(c) $\displaystyle\int_{-1}^{1} (e^{2t} - e^{-2t}) \, dt$.

(a) The function t^4 is even and $\sin 3t$ is odd, so the integrand is odd. Since the range is symmetrical about the origin, the integral is zero.

(b) t^5 is odd, $\cos 3t$ is even, and $\cos \tfrac{1}{2}t$ is even, so the integrand is odd and the integral is zero as in (a).

(c) $e^{2t} - e^{-2t}$ is odd (put $-t$ in place of t in the function – it just changes its sign). Therefore the integral is zero.

If we integrate an even function between $\pm c$, the graph shows that we get equal contributions from both sides of the origin, which gives the following result.

> **Symmetrical integrals of even functions**
>
> If $f(t)$ is even, then $\displaystyle\int_{-c}^{c} f(t) \, dt = 2 \int_{0}^{c} f(t) \, dt$.
>
> **(15.17)**

These ideas may also be useful if there is special symmetry about some point other than the origin.

Example 15.14 *Show that* $\displaystyle\int_{0}^{2\pi} \cos^3 t \, dt = 0$.

(a)

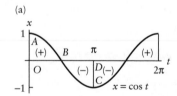

(b)

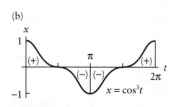

Fig. 15.7

In the graph of $x = \cos t$ (Fig. 15.7) the parts *OBA* and *DBC* are congruent, and similarly for the other pair of divisions; in fact all four divisions are congruent. For the graph of $x = \cos^3 t$ the shape is changed, but the four pieces remain congruent and retain their original sign. The resulting cancellation gives zero for the integral.

15.10   Definite integrals having variable limits

Integrals of the following type occur rather frequently in applications:

$$I(x) = \int_c^x f(t)\ \mathrm{d}t, \tag{15.18}$$

where c is a constant. Although $I(x)$ depends on x, this is still a *definite integral* because it has limits of integration; no arbitrary constant occurs. Notice that we **avoid using x as the variable of integration when a limit of integration involves x:** the same letter would be serving two totally different purposes. Therefore we have changed the variable of integration to t.

Suppose that $F(t)$ represents any particular indefinite integral, or antiderivative, of $f(t)$. Then, as always,

$$I(x) = [F(t)]_c^x = F(x) - F(c). \tag{15.19}$$

Since $F(c)$ is constant,

$$\frac{\mathrm{d}I(x)}{\mathrm{d}x} = \frac{\mathrm{d}}{\mathrm{d}x}\int_c^x f(t)\ \mathrm{d}t = \frac{\mathrm{d}F(x)}{\mathrm{d}x} = f(x).$$

Similarly we can obtain

$$\frac{\mathrm{d}}{\mathrm{d}x}\int_x^c f(t)\ \mathrm{d}t = -f(x).$$

Now consider the more complicated case

$$K(x) = \int_{u(x)}^{v(x)} f(t)\ \mathrm{d}t = F(v(x)) - F(u(x)).$$

By using the chain rule, we have

$$\frac{\mathrm{d}}{\mathrm{d}x}F(v(x)) = \frac{\mathrm{d}F(v)}{\mathrm{d}v}\frac{\mathrm{d}v(x)}{\mathrm{d}x} = f(v(x))\frac{\mathrm{d}v(x)}{\mathrm{d}x},$$

with a similar result for $F(u(x))$, and finally we have the results

---

Differentiation of integrals

(a) $\dfrac{\mathrm{d}}{\mathrm{d}x}\displaystyle\int_{u(x)}^{v(x)} f(t)\ \mathrm{d}t = f(v(x))\dfrac{\mathrm{d}v(x)}{\mathrm{d}x} - f(u(x))\dfrac{\mathrm{d}u(x)}{\mathrm{d}x}.$

(b) (Special cases):

$$\frac{\mathrm{d}}{\mathrm{d}x}\int_c^x f(t)\ \mathrm{d}t = f(x) \quad \text{and} \quad \frac{\mathrm{d}}{\mathrm{d}x}\int_x^c f(t)\ \mathrm{d}t = -f(x).$$

$$\tag{15.20}$$

(The results (15.20b) are simply (15.20a) in the respective cases $v(x) = x$, $u(x) = c$, or $v(x) = c$, $u(x) = x$.) It is worth noticing that (15.20) does not require you to integrate anything!

Example 15.15 *Obtain dI/dx when $I(x) = \displaystyle\int_{x^2}^{3x^2} e^t \, dt$.*

Here $f(t) = e^t$, $u(x) = x^2$, $v(x) = 3x^2$.

Therefore $\dfrac{dI(x)}{dx} = e^{3x^2} 6x - e^{x^2} 2x = 2x(3\,e^{3x^2} - e^{x^2})$.

Equation (15.19) shows that $I(x) = \int_c^x f(t)\, dt$ is an antiderivative of $f(x)$. It might be thought that, by choosing various values for c, we could reproduce *all* the antiderivatives $F(x)$ corresponding to $f(x)$. However, this expectation is only sometimes correct.

Example 15.16 *Let $f(x) = \cos x$, with antiderivatives $\sin x + C$, where C is an arbitrary constant. Demonstrate that for the integral $\int_c^x f(t)\, dt$ it is not possible to find a value of c which will reproduce the antiderivative $\sin x + 1000$.*

We have

$$\int_c^x \cos t \, dt = [\sin t]_c^x = \sin x - \sin c.$$

But $-1 \leqslant \sin c \leqslant 1$ no matter what value of c we take, so we could never make the integral equal to $\sin x + 1000$.

Example 15.17 *The function shown in Fig. 15.8a is described by $f(x) = 1$ when $0 \leqslant x < 1$, $f(x) = -1$ when $1 \leqslant x < 2$, and $f(x) = 0$ when $x \geqslant 2$. Sketch a graph of the function*

$$I(x) = \int_0^x f(t)\, dt.$$

Range $0 \leqslant x < 1$. $I(x) = \displaystyle\int_0^x 1 \, dt = x$. (i)

Range $1 \leqslant x < 2$. In this range, $f(t)$ is described by a different expression, -1 instead of 1, so we must split up the integral:

$$I(x) = \int_0^x f(t)\, dt = \int_0^1 f(t)\, dt + \int_1^x f(t)\, dt$$

$$= 1 + \int_1^x (-1)\, dt \quad \text{(after using (i) at } x = 1)$$

$$= 1 - (x - 1) = 2 - x.$$ (ii)

(a)

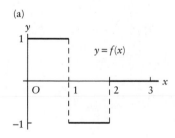

(b)

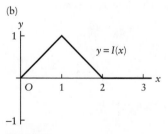

Fig. 15.8

Example 15.17 *continued*

Range x > 2. Split the integral again:

$$I(x) = \int_0^2 f(t)\, dt + \int_2^x f(t)\, dt$$

$$= 0 + \int_0^x 0\, dt = 0,$$ (iii)

where we used the value of (ii) at $x = 2$. The resulting graph of $I(x)$ is shown in Fig. 15.8b.

Problems

15.1 Sketch each of the following curves; then express the signed areas under them firstly as the sums of strips, as in (15.1), and secondly as definite integrals, as in (15.3); and finally evaluate them by (15.4).
(a) $y = x^3, -1 \leqslant x \leqslant 2$;   (b) $y = x^5, -1 \leqslant x \leqslant 1$;
(c) $y = \sin x, -\pi \leqslant x \leqslant 0$;   (d) $y = e^{-2x}, 0 \leqslant x \leqslant 1$.

15.2 Evaluate the following indefinite integrals (remember the arbitrary constant).

(a) $\int x^{\frac{1}{2}}\, dx$;   (b) $\int (x+1)^{\frac{1}{2}}\, dx$;   (c) $\int e^{\frac{1}{2}x}\, dx$;

(d) $\int \sin x\, dx$;   (e) $\int (\cos x - 2\sin 2x)\, dx$;

(f) $\int t^{-\frac{1}{2}}\, dt$;   (g) $\int \cos 2u\, du$;   (h) $\int 3e^{-\frac{1}{2}y}\, dy$;

(i) $\int (1 + 3t^2 - 2t)\, dt$;   (j) $\int (1 + 4\cos 4w)\, dw$;

(k) $\int (-x)^{\frac{1}{2}}\, dx$ when x is negative (you will have to experiment to find a valid antiderivative).

15.3 Evaluate the following definite integrals.

(a) $\int_{-1}^{1} x^3\, dx$;   (b) $\int_{-1}^{1} x^2\, dx$;   (c) $\int_0^1 dx$;

(d) $\int_0^4 x^{\frac{1}{2}}\, dx$;   (e) $\int_{-1}^{1} (1 - 3x + 2x^2)\, dx$;

(f) $\int_1^2 (x^{-3} + x^{-2})\, dx$;   (g) $\int_1^2 x^{-2}\, dx$;

(h) $\int_{-2}^{-1} x^{-1}\, dx$ (take care: the x values are negative);

(i) $\int_{-2}^{-1} (-x)^{\frac{1}{2}}\, dx$ (see the remark in Problem 15.2k);

(j) $\int_0^1 e^{-3x}\, dx$;   (k) $\int_0^{\frac{1}{4}\pi} \sin 4x\, dx$;

(l) $\int_0^{2\pi} \sin \frac{1}{2}x\, dx$;   (m) $\int_0^{2\pi} \cos \frac{1}{2}x\, dx$.

15.4 Evaluate the following integrals, using the notation of (15.4).

(a) $\int_0^1 x(x^2 + x + 1)\, dx$;   (b) $\int_{-1}^{1} (x-1)(x+1)\, dx$;

(c) $\int_0^2 x(x^2 - 1)\, dx$;   (d) $\int_1^2 \dfrac{x + x^2}{x^3}\, dx$;

(e) $\int_1^2 \dfrac{t(t+1)}{t^{\frac{1}{2}}}\, dt$;   (f) $\int_1^4 \dfrac{\sqrt{u} - 1}{u}\, du$;

(g) $\int_{-1}^0 \dfrac{dw}{2w + 3}$;   (h) $\int_{-2}^{-1} \dfrac{x}{x-1}\, dx$;

(i) $\int_0^{\pi} \cos^2 3t\, dt$ $(\cos^2 A = \frac{1}{2}(1 + \cos 2A))$.

15.5 Evaluate the following infinite integrals.

(a) $\int_1^{\infty} e^{-3t}\, dt$;   (b) $\int_0^{\infty} e^{-\frac{1}{2}v}\, dv$;   (c) $\int_1^{\infty} \dfrac{dx}{x^3}$;

(d) $\int_0^{\infty} \dfrac{dx}{(2x + 3)^2}$;   (e) $\int_0^1 \dfrac{ds}{s^{\frac{1}{4}}}$;   (f) $\int_1^2 \dfrac{dt}{(t-1)^{\frac{1}{2}}}$;

(g) $\int_0^{\infty} e^{-2t} \sin 3t\, dt$ (see Section 15.7);

(h) $\int_0^{\infty} e^{-\frac{1}{4}t} \cos 2t\, dt$ (see Section 15.7).

15.6 The **mean** or **average value** of $f(t)$ over an interval $0 \leqslant t \leqslant T$ is the quantity $\frac{1}{T}\int_0^T f(t)\,dt$. Find the mean values of the following over the intervals given.

(a) $f(t) = t, 0 \leqslant t \leqslant 1$;   (b) $f(t) = t, -1 \leqslant t \leqslant 1$;

(c) $f(t) = \sin t, 0 \leqslant t \leqslant \pi$;   (d) $f(t) = \sin t, 0 \leqslant t \leqslant 2\pi$;

(e) $f(t) = t^{-2}, 1 \leqslant t \leqslant T$;   (f) $f(t) = e^{-t}\cos t, 0 \leqslant t \leqslant 2\pi$;

(g) $f(t) = e^{-2t}\sin t, 0 \leqslant t < \infty$;

(h) $f(t) = 1 - e^{-t}$ (work out the mean value over $0 \leqslant t \leqslant T$ for several increasing values of T, and deduce the value of the mean over $0 \leqslant t < \infty$; if you put $T = \infty$ into the integral directly, it turns out to be infinite, or 'diverges' (see Section 15.6), so no conclusion can be drawn from this approach).

(i) $f(t) = t^{-1}, 1 \leqslant t < \infty$ (it is necessary to follow the procedure in the previous question, for the same reason).

15.7 Use the even/odd properties of the integrands (see Section 15.9) to prove the following results.

(a) $\int_{-\pi}^{\pi} \sin^4 t\,dt = 2\int_0^{\pi} \sin^4 t\,dt$;

(b) $\int_{-1}^{1} \dfrac{t^3}{(1+t^4)}\,dt = 0$;

(c) $\int_{-\pi}^{\pi} \dfrac{t\cos t}{1+t^2}\,dt = 0$;   (d) $\int_{-\frac{1}{2}\pi}^{\frac{1}{2}\pi} t^2\sin(t^3)\,dt = 0$.

15.8 (Computational: see Section 15.2 and Examples 15.1 and 15.6.) Write a simple program based on the algorithm (15.2) to evaluate a definite integral $\int_a^b f(x)\,dx$. Assume that you have a subroutine for evaluating $f(x)$, and that you input a, b, and N (the number of subdivisions); also either a permissible error E, or a parameter M which determines the number of iterations. If you use E, the process might be written to print out when two successive iterations are within E of each other. Check the correctness of the program by using a function such as x^2 as integrand.

Estimate the values of the following integrals.

(a) $\int_1^2 \dfrac{e^{-x}}{x}\,dx$;   (b) $\int_0^{\pi} \sin x^2\,dx$;   (c) $\int_0^1 \cos e^{-x}\,dx$.

15.9 (Computational). (a) Convince yourself that $e^{-x^2} < e^{-x}$ when $x > 1$. Use the area analogy (15.13) to show that, if $b > 1$, then $\int_b^{\infty} e^{-x^2}\,dx < \int_b^{\infty} e^{-x}\,dx$. Deduce that, if E is a positive number and $E < 1$, then $\int_b^{\infty} e^{-x^2}\,dx < E$ for $b > -\ln E$.

(b) Use the program written for Problem 15.8 to evaluate the *improper* integral $\int_0^{\infty} e^{-x^2}\,dx$ to within 2 decimal places, in the following way. You have to stop the

integral somewhere: the program cannot deal with $b = \infty$. Take a permissible error $E = 0.001$, say, to leave some leeway. Referring to (a), choose $b > -\ln E$, and compute the integral $\int_0^b e^{-x^2}\,dx$. The part of the original integrand between b and infinity will then be negligible.

15.10 (Section 15.10). Find dI/dx where $I(x)$ is given by the following integrals.

(a) $\int_0^x t^2\,dt$;   (b) $\int_0^x \sin^5 t\,dt$;   (c) $\int_0^x \dfrac{e^t}{1+t}\,dt$;

(d) $\int_0^{e^x} t\ln t\,dt$;   (e) $\int_{\sqrt{x}}^{\sqrt{x+1}} \sin(t^2)\,dt$.

15.11 (See Example 15.17: it is necessary to split up the integrals.) Obtain $\int_0^x f(t)\,dt$ where $f(x)$ is defined by the following.

(a) $f(x) = \begin{cases} 0 & \text{if } x < -1 \\ x & \text{if } -1 \leqslant x \leqslant 1 \\ 0 & \text{if } x > 1 \end{cases}$;consider positive and negative values of x.

(b) $f(x) = \begin{cases} x & \text{if } 0 \leqslant x < 1 \\ 2 - x & \text{if } 1 \leqslant x \leqslant \frac{3}{2} \\ 0 & \text{if } x > \frac{3}{2} \end{cases}$; consider positive x only.

15.12 An 'RL' circuit has a constant current I_0 flowing, produced by a constant applied voltage. A switch cuts off the voltage and closes the circuit again at time $t = t_0 > 0$. For $t > t_0$, the current is given by $I(t) = I_0 e^{-R(t-t_0)/L}$. Obtain expressions for $Q(t)$ for $t \geqslant 0$, where

$$Q(t) = \int_0^t I(u)\,du.$$

15.13 A function $f(x)$ is defined by

$$f(x) = \begin{cases} x^2, & 0 \leqslant x \leqslant 1, \\ 2 - x, & 1 < x \leqslant 2. \end{cases}$$

Sketch the graph of $y = f(x)$ for $0 \leqslant x \leqslant 2$. Find the area under the curve between $x = 0$ and $x = 2$.

15.14 Evaluate

$$\int_0^2 \frac{dx}{|x - 1|^{2/3}}.$$

Note that the integrand is infinite at $x = 1$. The result will be the sum of two improper integrals on $0 \leqslant x < 1$ and $1 < x \leqslant 2$.

16

Applications involving the integral as a sum

CONTENTS

Reminder: The short table (12.2) provides all the indefinite integrals (antiderivatives) required for this chapter.

16.1 Examples of integrals arising from a sum

The examples which follow show typical cases where integrals arise from sums of the type (15.9).

Example 16.1 *The tension T in an elastic string is given by $T = 0.01x$ (kg m s^{-2}), where x is the extension beyond the natural length. Find the work done on the string to stretch it 2 metres beyond its natural length.*

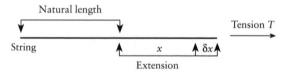

Fig. 16.1

To stretch it from extension x to $x + \delta x$ (Fig. 16.1), the work δW required is approximated by

$$\delta W \approx \text{force} \times \text{distance} = T\,\delta x = 0.01x\,\delta x.$$

The total work W is approximated by

$$W = \sum_{x=0}^{x=2} \delta W \approx \sum_{x=0}^{x=2} 0.01x\,\delta x.$$

Now let $\delta x \to 0$. Then we obtain

$$W = \lim_{\delta x \to 0} \sum_{x=0}^{x=2} 0.01x\,\delta x = \int_0^2 0.01x\,\mathrm{d}x \quad \text{(from (15.9))}$$

$$= \left[\int 0.01x\,\mathrm{d}x \right]_0^2 = 0.01[\tfrac{1}{2}x^2]_0^2 = 0.02 \quad (\text{kg m}^2\,\text{s}^{-2}).$$

Example 16.2 *A car runs from rest to rest in 1 hour, its velocity v being given by $v = 200t(1 - t)$ (in kilometres per hour). The rate of fuel consumption, f (in litres per kilometre), is related to the velocity by $f = 10^{-4} v^2$. Find (a) the distance travelled and (b) the amount of fuel used.*

(a) In time δt it travels a distance δx, where

$$\delta x \approx v \, \delta t.$$

The total displacement x (which is equal to the distance travelled since v is always positive) is therefore

$$x = \lim_{\delta t \to 0} \sum_{t=0}^{t=1} v \, \delta t = \int_0^1 v \, dt$$

$$= \int_0^1 200t(1 - t) \, dt = 200 \int_0^1 (t - t^2) \, dt$$

$$= 200[\tfrac{1}{2}t^2 - \tfrac{1}{3}t^3]_0^1 = 33\tfrac{1}{3} \quad (\text{km}).$$

(b) In distance δx, it uses an amount of fuel δF approximated by

$$\delta F \approx f \, \delta x \approx f v \, \delta t = 10^{-4} v^2 (v \, \delta t) = 800t^3(1 - t)^3 \, \delta t.$$

The total fuel used, F, is given by

$$F = \lim_{\delta t \to 0} \sum_{t=0}^{t=1} 800t^3(1 - t)^3 \, \delta t = \int_0^1 800t^3(1 - t)^3 \, dt,$$

(by comparing (15.9) with t in place of x)

$$= 800 \int_0^1 (t^3 - 3t^4 + 3t^5 - t^6) \, dt$$

$$= 800[\tfrac{1}{4}t^4 - \tfrac{3}{5}t^5 + \tfrac{1}{2}t^6 - \tfrac{1}{7}t^7]_0^1 = 5.71 \quad (\text{litres}).$$

Example 16.3 *The straight line $y = (r/h)x$, between $x = 0$ and $x = h$, is rotated around the x axis to sweep out a solid cone of height h and circular base radius r. Obtain an expression for its volume.*

Divide the interval OH into a large number of equal small steps δx (Fig. 16.2) Consider the step PQ between x and $x + \delta x$. This identifies a thin slice of the cone, like a slice of bread. Its volume δV is nearly that of a cylinder of radius y and thickness δx, so

$$\delta V \approx \pi y^2 \, \delta x.$$

The total volume is obtained by adding all the δV and then letting the slices tend to zero thickness (at the same time becoming proportionately more numerous):

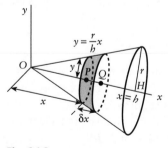

Fig. 16.2

$$V = \lim_{\delta x \to 0} \sum_{x=0}^{x=h} \pi y^2 \, \delta x = \int_0^h \pi y^2 \, dx$$

$$= \int_0^h \frac{\pi r^2}{h^2} x^2 \, dx = \frac{\pi r^2}{h^2} [\tfrac{1}{3}x^3]_0^h$$

$$= \frac{\pi r^2}{h^2} \frac{h^3}{3} = \tfrac{1}{3}\pi r^2 h.$$

The **volume** of any **solid of revolution** between $x = a$ and $x = b$, formed by rotating a **profile** $y = f(x)$ around the x axis, can be found in exactly the same way:

Volume of a solid of revolution around the x axis

For a profile $y = f(x)$, $a \leqslant x \leqslant b$,

the volume $V = \displaystyle\int_a^b \pi y^2 \, \mathrm{d}x.$

(16.1)

Example 16.4 *Find the geometrical area enclosed between the curves $y = 2x^2 - 1$ and $y = x^2$.*

This problem is complicated if we have to think all the time about the difference between *signed* and *geometrical* area as in Chapter 15.

Here it will be done in a different way. Divide the interval $-1 \leqslant x \leqslant 1$ into short steps of length δx and consider the area elements indicated in Fig. 16.3. They are nearly rectangular, and the **geometrical** (positive) area δA of each is given by

$$\delta A \approx |x^2 - (2x^2 - 1)| \, \delta x = (-x^2 + 1) \, \delta x$$

(we may drop the modulus signs since $-x^2 + 1 \geqslant 0$ in the given range).

The total geometrical area A is therefore given by

$$A = \lim_{\delta x \to 0} \sum_{x=-1}^{x=1} (-x^2 + 1) \, \delta x = \int_{-1}^{1} (-x^2 + 1) \, \mathrm{d}x$$

$$= [-\tfrac{1}{3}x^3 + x]_{-1}^1 = (-\tfrac{1}{3} + 1) - (\tfrac{1}{3} - 1) = \tfrac{4}{3}.$$

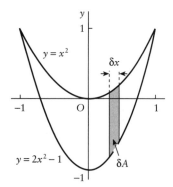

$y = x^2$

$y = 2x^2 - 1$

Fig. 16.3

16.2 Geometrical area in polar coordinates

In Fig. 16.4, $\widehat{AB}$ represents part of a curve which is described in polar coordinates by

$$r = f(\theta), \qquad \alpha \leqslant \theta \leqslant \beta.$$

Form a new, non-rectangular type of area element δA by dividing the θ range, $\theta = \alpha$ to $\theta = \beta$, into small angular steps $\delta\theta$, expressed in **radians**. (We use A rather than $\mathcal{A}$ because, in polar coordinates, we always regard r as being positive, and we shall count the area elements as positive.) A typical area element has the shape OPQ.

When $\delta\theta$ is small, OPQ has very nearly the same area as a narrow *circular* sector of radius r and angle $\delta\theta$ radians. Its area is therefore a fraction $\delta\theta/2\pi$ of a complete circle of radius r and area πr^2:

$$\delta A \approx \frac{\delta\theta}{2\pi}\pi r^2 = \tfrac{1}{2}r^2 \, \delta\theta.$$

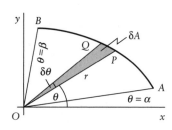

Fig. 16.4

The total area is obtained by adding all the elements and letting $\delta\theta$ tend to zero:

$$A = \lim_{\delta\theta\to 0} \sum_{\theta=\alpha}^{\theta=\beta} \tfrac{1}{2}r^2\,\delta\theta = \int_\alpha^\beta \tfrac{1}{2}r^2\,\mathrm{d}\theta,$$

where $r = f(\theta)$.

Area of a sector in polar coordinates
For a sector of $r = f(\theta)$, with $\alpha \leqslant \theta \leqslant \beta$,

$$A = \tfrac{1}{2}\int_\alpha^\beta r^2\,\mathrm{d}\theta.$$

(16.2)

Example 16.5 *Find the area of the loop of the curve $r = 3\sin 2\theta$ in the first quadrant.*

For the loop shown in Fig. 16.5, the range of θ is $0 \leqslant \theta \leqslant \tfrac{1}{2}\pi$. Thus in (16.2)

$$f(\theta) = 3\sin 2\theta, \qquad \alpha = 0, \qquad \beta = \tfrac{1}{2}\pi.$$

The area is therefore given by

$$A = \tfrac{1}{2}\int_0^{\frac{1}{2}\pi} (3\sin 2\theta)^2\,\mathrm{d}\theta = \tfrac{9}{2}\int_0^{\frac{1}{2}\pi}\sin^2 2\theta\,\mathrm{d}\theta.$$

But, for any angle B, $\sin^2 B = \tfrac{1}{2}(1-\cos 2B)$; so

$$A = \tfrac{9}{4}\int_0^{\frac{1}{2}\pi}(1-\cos 4\theta)\,\mathrm{d}\theta$$

$$= \tfrac{9}{4}[\theta - \tfrac{1}{4}\sin 4\theta]_0^{\frac{1}{2}\pi} = \tfrac{9}{8}\pi.$$

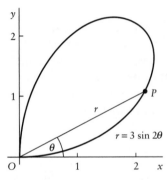

Fig. 16.5

16.3 The trapezium rule

Practical problems often give rise to integrals which the investigator cannot evaluate or find in a dictionary of integrals. Indeed, sometimes integrals which are very simple-looking cannot, *in principle*, be expressed in terms of ordinary 'formulae' at all. However, numerical approximations to definite integrals can usually be obtained to any required degree of accuracy by using numerical methods in conjunction with a computer. We will mention some very simple methods that call directly on the **area analogy** (15.13), which we repeat here:

The area analogy
The definite integral $\int_a^b f(x)\,\mathrm{d}x$ is equal to the signed area between $y = f(x)$ and the x axis from $x = a$ to $x = b$.

(16.3)

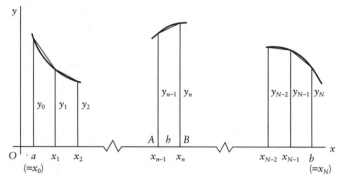

Fig. 16.6

In Examples 15.1 and 15.4, we illustrated the use of the area analogy (15.13) using as the area approximation the sum in (15.1), which had been introduced only for the purpose of establishing the principle. It only gives close approximations if we use very small step lengths; but, now that the area analogy is established, we can look for approximation methods that will be more efficient.

An improved area approximation is shown in Fig. 16.6, where the curve $y = f(x)$ is 'fitted' by a polygonal curve. The approximation to the area of each strip individually is obviously better in general than we would get from a rectangle. Divide the interval $x = a$ to $x = b$ into N steps. We shall denote the length of each step by h (instead of δx, because h is conventional in numerical analysis). Then

$$h = \frac{b - a}{N}.$$

Number the $N + 1$ points of division $0, 1, 2, \ldots, N$: the x values are

$$x_0 \ (=a), x_1, x_2, \ldots, x_{N-1}, x_N \ (=b).$$

and the y values $y_0, y_1, y_2, \ldots, y_{N-1}, y_N$.

Each of the approximating area elements is a trapezium. The signed area $\delta \mathcal{A}_n$ of the nth area element is given by

$$\delta \mathcal{A}_n \approx \tfrac{1}{2}(y_{n-1} + y_n)h = \frac{b - a}{2N}(y_{n-1} + y_n).$$

The total area $\mathcal{A}$ is approximated by the sum of these:

$$\mathcal{A} \approx \sum_{n=1}^{N} \frac{b - a}{2N}(y_{n-1} + y_n)$$

$$= \frac{b - a}{2N}[(y_0 + y_1) + (y_1 + y_2) + \cdots + (y_{N-1} + y_N)]$$

$$= \frac{b - a}{N}[\tfrac{1}{2}y_0 + (y_1 + y_2 + \cdots + y_{N-1}) + \tfrac{1}{2}y_N].$$

This is called the **trapezium rule**.

Trapezium rule

$$\int_a^b f(x)\,dx \approx \frac{b-a}{N}\left[\tfrac{1}{2}y_0 + (y_1 + y_2 + \cdots + y_{N-1}) + \tfrac{1}{2}y_N\right].$$

The interval is divided into N equal steps:

$x_0\,(=a), x_1, \ldots, x_N\,(=b)$ are the division points; and
$y_n = f(x_n)\ (n = 0, 1, 2, \ldots, N)$.

(16.4)

In the following example, we compare the trapezium rule (16.4) with the **rectangle rule** (13.2), which we can recast for comparison as

$$\int_a^b f(x)\,dx \approx \frac{b-a}{N}(y_0 + y_1 + \cdots + y_{N-1}).$$

Example 16.6 *Compare the efficiency of the trapezium rule (16.4) with the rectangle rule (15.2) for approximating to $\int_0^1 e^{-x}\,dx$.*

We set out the results in the following table.

N	10	100	1000
$h = (b-a)/N$	0.1	0.01	0.001
Rectangle rule	0.66	0.635	0.6324
Trapezium rule	0.632 657	0.632 125	0.632 120

The exact value is 0.632 120 5... . For three-decimal accuracy, the rectangle rule requires about 1000 divisions and the trapezium rule only about 12. There are many formulae which are far more efficient than even the trapezium rule, one of the best of these, for combining simplicity with accuracy, being Simpson's rule (see Problem 16.21). You should look at books on numerical analysis for others.

16.4 Centre of mass, moment of inertia

Suppose that there are N particles attached to a weightless plane sheet (Fig. 16.7), the nth particle being at $P : (x_n, y_n)$ and having mass m_n, where $n = 1, 2, \ldots, N$.

Let $G : (\bar{x}, \bar{y})$ be the **centre of mass**. It is the balancing point of the assembly, the point such that the total moment of the particles about any axis through G is zero. Consider in particular the axes CGD and AGB, parallel to the y *and* x axes and passing through G. Then

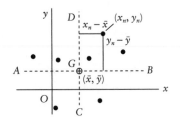

Fig. 16.7

$$\sum_{n=1}^{N} m_n(x_n - \bar{x}) = 0, \qquad \sum_{n=1}^{N} m_n(y_n - \bar{y}) = 0. \qquad (16.5)$$

These can be written

$$\sum_{n=1}^{N} m_n x_n - \bar{x}\sum_{n=1}^{N} m_n = 0, \qquad \sum_{n=1}^{N} m_n y_n - \bar{y}\sum_{n=1}^{N} m_n = 0.$$

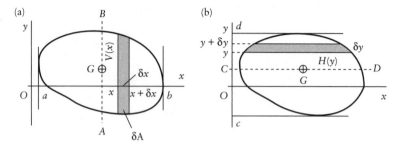

Fig. 16.8

Let $\displaystyle\sum_{n=1}^{N} m_n = M$, the total mass; then these equations give

$$\bar{x} = \frac{1}{M}\sum_{n=1}^{N} m_n x_n, \qquad \bar{y} = \frac{1}{M}\sum_{n=1}^{N} m_n y_n.$$

If instead of a number of particles there is a solid plate, then this too has a balancing point. Assume that the **plate is uniform** so that its mass per unit area, μ (Greek mu), is the same everywhere on it.

We also assume that the shape of the plate is such that no vertical or horizontal line cuts across the boundary more than twice: once going in and again going out. If the shape does not have this property, then the process as explained here has to be modified.

Suppose that the centre of mass G is at $(\bar{x}, \bar{y})$. Divide the area into narrow vertical strips of width δx (Fig. 16.8a). Let the total length, or height, of a representative strip as shown be $V(x)$. Then its geometrical area δA is nearly equal to $V\,\delta x$, and its mass δm is nearly $\mu V\,\delta x$. Therefore the moment about a vertical axis AB through $G : (\bar{x}, \bar{y})$ is approximately given by $(x - \bar{x})\,\delta m \approx (x - \bar{x})\,\mu V\,\delta x$. The sum of all the elementary moments must be zero, since G is the mass centre. So, in the limit as δx tends to 0, we have

$$\lim_{\delta x \to 0}\sum_{x=a}^{x=b}(x - \bar{x})V(x)\mu\,\delta x = 0,$$

where $x = a$ and $x = b$ represent the extreme left and right limits of the plate. Since μ and $\bar{x}$ are constants, this is the same as

$$\mu\lim_{\delta x \to 0}\sum_{x=a}^{x=b} xV(x)\,\delta x = \mu\bar{x}\lim_{\delta x \to 0}\sum_{x=a}^{x=b} V(x)\,\delta x = \mu A\bar{x},$$

where A is the area of the plate, equal to $\displaystyle\lim_{\delta x \to 0}\sum_{x=a}^{x=b} V(x)\,\delta x$. Cancelling μ, we obtain

$$\bar{x} = \frac{1}{A}\lim_{\delta x \to 0}\sum_{x=a}^{x=b} xV(x)\,\delta x = \frac{1}{A}\int_a^b xV(x)\,dx.$$

Similarly, by dividing the y axis into steps δy, and considering the moments of horizontal strips of length $H(y)$ (see Fig. 16.8b) about a horizontal axis CD through G, we obtain

$$\bar{y} = \frac{1}{A}\int_c^d yH(y)\,\mathrm{d}y,$$

where $y=c$ and $y=d$ are the extreme lower and upper limits of the plate.

In these expressions, all reference to mass has gone (μ is no longer present). Therefore the centre of mass of a **uniform plate** is also called the **centroid** of the figure representing the plate, and it depends only on its shape and size.

In fact the moments about *every* line through G are zero, not simply the moments about AB and CD parallel to the x and y axes that we used to find G.

Centre of mass of a uniform convex plate, or centroid of a convex area, $G : (\bar{x}, \bar{y})$

$$\bar{x} = \frac{1}{A}\int_a^b xV(x)\,\mathrm{d}x; \quad \bar{y} = \frac{1}{A}\int_c^d yH(y)\,\mathrm{d}y,$$

where A is the area. Here, respectively, $V(x)$ and $H(y)$ are the lengths of the vertical and horizontal strips, and $x = a, b$ (resp. $y = c, d$) are the extreme horizontal (resp. vertical) boundaries of the figure.

(16.6)

Example 16.7   *Find the position of the centroid or centre of mass of an isosceles triangle of height h and base b.*

Choose axes which make the job as simple as possible. In this case, use the axes shown in Fig. 16.9.

From the *symmetry* of the isosceles triangle about the x axis, the centroid must lie on this axis, so $\bar{y} = 0$ without any calculations.

The sides have equations

$$y = \pm\frac{b}{2h}x;$$

therefore the length of the strip at x is given by

$$V(x) = \frac{b}{h}x.$$

Also the area A of the triangle is given by

$$A = \tfrac{1}{2}bh.$$

Therefore by (16.6),

$$\bar{x} = \frac{2}{bh}\int_0^h x\left(\frac{b}{h}x\right)\mathrm{d}x = \frac{2}{h^2}\int_0^h x^2\,\mathrm{d}x = \tfrac{2}{3}h.$$

(In these coordinates, $\bar{x}$ is independent of the base length b.)

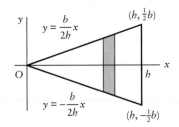

Fig. 16.9

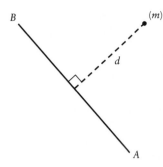

Fig. 16.10

The **moment of inertia** is important for problems in mechanics involving rotation: it plays a part similar to that of mass in non-rotational problems. The moment of inertia of a single particle of mass m about any axis AB is defined to be md^2, where d is its perpendicular distance from AB (see Fig. 16.10). For the moment of inertia of an assemblage of particles, the individual contributions are added. For a solid plate the contributions of small area elements are likewise added, as if they were particles, and in the limit we obtain a definite integral. It is important to select axes and suitably shaped area elements to make a particular problem manageable.

Example 16.8 *Find the moment of inertia I of a uniform rectangular plate ABCD about the edge AB when AB = 2, BC = 6, and the mass per unit area is 2.*

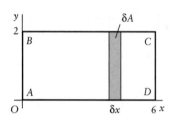

Fig. 16.11

Set up axes parallel to the sides, and area elements which are vertical strips of height 2 and width δx, as shown in Fig. 16.11. The axis of rotation is the y axis. The mass δm of each strip is given by

$$\delta m = \text{(surface density)} \times \text{(area)} = 2\,\delta A = 2 \times 2 \times \delta x = 4\,\delta x.$$

The moment of inertia of the strip distance x from the y axis is therefore

$$x^2\,\delta m = 4x^2\,\delta x.$$

The total moment of inertia I is given by

$$I = \lim_{\delta x \to 0} \sum_{x=0}^{x=6} 4x^2\,\delta x = 4\int_0^6 x^2\,\mathrm{d}x = 4[\tfrac{1}{3}x^3]_0^6 = 288.$$

Example 16.9 *Find an expression for the moment of inertia I of an isosceles triangle ABC about its base AB, when AB = b, its height is h, and its mass is M.*

The axes and the representative strip at x are shown in Fig. 16.12. The equations of BC and AC are

$$y = \pm\left(-\frac{b}{2h}x + \tfrac{1}{2}b\right)$$

respectively, so the length $V(x)$ to be assigned to the strip is

$$V(x) = -\frac{b}{h}x + b,$$

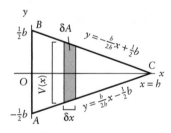

Fig. 16.12

and the area δA is approximated by

$$\delta A \approx \left(-\frac{b}{h}x + b\right)\delta x.$$

Since the plate is uniform, the mass per unit area is (total mass)/(area), or $M/\tfrac{1}{2}bh$, so the mass element δm is approximated by

$$\delta m \approx \frac{M}{\tfrac{1}{2}bh}\left(-\frac{b}{h}x + b\right)\delta x = \frac{2M}{h}\left(1 - \frac{x}{h}\right)\delta x.$$

Example 16.9 *continued*

Therefore the moment of inertia I is given by

$$I = \lim_{\delta x \to 0} \sum_{x=0}^{x=h} x^2\, \delta m = \lim_{\delta x \to 0} \sum_{x=0}^{x=h} x^2 \frac{2M}{h}\left(1 - \frac{x}{h}\right) \delta x$$

$$= \frac{2M}{h} \int_0^h x^2\left(1 - \frac{x}{h}\right) dx = \frac{2M}{h} \int_0^h \left(x^2 - \frac{1}{h}x^3\right) dx$$

$$= \frac{2M}{h}\left[\tfrac{1}{3}x^3 - \frac{1}{4h}x^4\right]_0^h = \frac{2M}{h}\frac{h^3}{12} = \tfrac{1}{6}Mh^2.$$

Example 16.10 *Find the moment of inertia of a circular disc of radius R and mass M about an axis through its centre and perpendicular to the plane of the disc.*

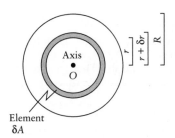

Element
δA

Fig. 16.13

The usual (x, y) coordinates are not natural to this problem. In Fig. 16.13, the polar coordinate r ranges from 0 to R. Break this range into ring-shaped steps as shown, the representative ring or **annulus** having inner radius r and thickness δr. These constitute the area elements δA.

We have $\delta A \approx 2\pi r\, \delta r$, and the mass per unit area is $M/\pi R^2$, so that the mass of the ring δm is approximately

$$\delta m \approx \frac{M}{\pi R^2} 2\pi r\, \delta r.$$

The moment of inertia of the ring must be equal to that of a suitable distribution of closely spaced particles along its circumference. The contribution of each of these imaginary particles to the moment of inertia of the ring is equal to its mass times r^2. Since r is constant on the ring, its moment of inertia δI is equal to the total mass of the ring times r^2:

$$\delta I \approx r^2\, \delta m \approx \frac{2M}{R^2} r^3\, \delta r.$$

Finally

$$I = \lim_{\delta r \to 0} \sum_{r=0}^{r=R} \frac{2M}{R^2} r^3\, \delta r = \frac{2M}{R^2} \int_0^R r^3\, dr$$

$$= \frac{2M}{R^2} \cdot \tfrac{1}{4}R^4 = \tfrac{1}{2}MR^2.$$

Problems

(Units are kilogram, metre, second (SI units) where they are unstated.)

16.1 The resistance R of a compression spring is given by $R = 100x + 1000x^2$, where x is the displacement from its natural length. Find the work done in compressing it through a distance of 0.01.

16.2 The velocity v of a point moving along the x axis is $v = 20 - 10t$, where t is the time. The displacement x taking place in a short time δt is approximated by $\delta x \approx v\, \delta t$. Express the displacement which takes place between $t = 2$ and $t = 4$ as a definite integral, and evaluate it. What is its x coordinate at $t = 4$ if it was at $x = 3$ when $t = 2$?

16.3 Each of the following curves is the profile of a solid of revolution which has the x axis as its central axis. Find the volume in each case (see (16.1) – you should briefly go through the whole argument until you understand it, not simply quote the formula):

(a) $y = e^{-x}, 0 \leqslant x \leqslant 1$;   (b) $y = 1/x, 1 \leqslant x \leqslant 2$;

(c) $y = x(1-x), 0 \leqslant x \leqslant 1$;   (d) $y = \sin x, 0 \leqslant x \leqslant \pi$;

(e) $y = x^3, -1 \leqslant x \leqslant 1$ (the fact that x^3 is negative over part of its range does not have to be taken into account: the volume elements are always positive, unlike area elements);

(f) $y = x(1-x), 0 \leqslant x \leqslant 2$ (see the note in (e));

(g) $y = x^{-1}, 1 \leqslant x < \infty$ (contrast Example 15.9, for area);

(h) $y = x^{\frac{1}{4}}, 0 \leqslant x \leqslant 1$.

16.4 Show that the volume of a sphere of radius R is $\frac{4}{3}\pi R^3$. (A sphere is a solid of revolution.)

16.5 (a) Find the volume of the ellipsoid obtained by rotating the elliptical profile $x^2/a^2 + y^2/b^2 = 1$ about the x axis.

(b) If the x and y scales of the profile ellipse in (a) are contracted or expanded by suitable factors, it becomes a unit circle. Deduce from this fact the formula for the volume of the ellipsoid of revolution.

16.6 The curve $y = \frac{1}{2}x$ between $y = 1$ and $y = 2$ is rotated about the y axis to profile a vertical spindle, or truncated cone. Find its volume.

16.7 A uniform beam AB of length L has mass m per unit length. It is cemented horizontally at A into a wall at the end A. Sum the moments about A of elements of length δx, form a definite integral, and so find the moment supporting the beam at A.

16.8 A 'beam' in the shape of a circular spindle made of material of density 500 is fixed to a vertical wall at the end A with its axis of symmetry horizontal. Its cross-sectional area (perpendicular to its axis) is $4 \times 10^{-4}(1 + 0.4x^2)$, where x is measured from A. Its length is 1. Find the moment at A required to support it under gravity.

Suppose that the data are the same, except that the cross-section is square, or possibly irregular in shape. Does this affect the answer? Suppose that the axis is bent, but that x still measures the perpendicular distance from the wall: is the calculation affected?

16.9 A narrow tube of length 10 cm and cross-section 0.1 cm^2 contains a chemical solution, with concentration $c(x) = 0.04 \, e^{-\frac{1}{4}x}$ g cm^{-3}, where x is the distance from one end. Find the total mass of solute in the tube.

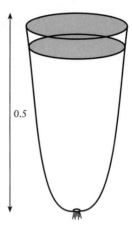

Fig. 16.14

16.10 The water clock in Fig. 16.14 has depth 0.5 m, and its profile is given by $r(h) = 0.39h^{\frac{1}{4}}$, where $r(h)$ is the redius at height h from the outlet in the bottom. The size of the outlet hole is such as to drain the water at a rate given by

$$\frac{dV}{dt} = -0.003h^{\frac{1}{2}} \text{ m}^3 \text{ h}^{-1},$$

where V is the volume of water remaining. Show that the water level falls at a uniform rate, and find how long it runs. (Consider the change δh in level which occurs in a short time δt.)

16.11 An alternating current $i = i_0 \cos \omega t$ flows through a resistor R. The instantaneous rate of heat generation is Ri^2 heat units per unit time. Find the heat generated in a complete cycle of the current, that is in a period $2\pi/\omega$. Does it make any difference at what instant you regard the period as starting? (To carry out the integration, you will need the identity $\cos^2 A = \frac{1}{2}(1 + \cos 2A)$.)

16.12 Find the geometric area enclosed between the curves $y = -x$ and $y = x(x - 1)$ on the interval $0 \leqslant x \leqslant 2$, by considering vertical strips between the curves of width δx.

16.13 Find the geometric area enclosed between the curves $y = -x$ and $y = x^3$ between $x = -1$ and $x = 1$ by considering vertical strips of width δx connecting the curves. (Be careful about signs: these curves cross.)

16.14 For the angular ranges specified, sketch the curves given in polar coordinates below and find the sectorial areas.

(a) $r = \theta, 0 \leqslant \theta \leqslant 2\pi$ (a spiral arc);
(b) $r = 2 \cos \theta, -\frac{1}{2}\pi \leqslant \theta \leqslant \frac{1}{2}\pi$ (a circle);
(c) $r = e^{\theta/2\pi}, 0 \leqslant \theta \leqslant \pi$ (spiral arc);
(d) $r = \sin 2\theta, 0 \leqslant \theta \leqslant \frac{1}{2}\pi$.
(Remember the identities $\cos^2 A = \frac{1}{2}(1 + \cos 2A)$, $\sin^2 A = \frac{1}{2}(1 - \cos 2A)$.)

16.15 The end of a water trough is a rectangle of height H and width L. Find the total force and moment on the end when the trough is full. (The pressure, meaning the force per unit area acting perpendicularly on any surface, at depth y is $\rho g y$, where ρ is density and g the gravitational constant.)

16.16 Determine the position of the centre of mass of a symmetrical cone of circular cross-section which has height H and base radius R.

16.17 Find the moment of inertia of a rectangle, having sides a and b, about an axis through its centre, parallel to the sides of length b.

16.18 Obtain the moment of inertia of an isosceles triangle of height H and base B about an axis through its vertex which is (a) parallel to the base, and (b) perpendicular to the base.

16.19 Use the trapezium rule (16.4) to evaluate the following integrals to 1% accuracy. (The exact value can be obtained by evaluating the integrals in the usual way.)

(a) $\displaystyle\int_0^1 e^{\frac{1}{2}x}\, dx;$ (b) $\displaystyle\int_0^\pi \sin x\, dx;$ (c) $\displaystyle\int_{-\frac{1}{2}\pi}^{\frac{1}{2}\pi} \cos x\, dx.$

16.20 The following integrals are either difficult or impossible to evaluate directly. Estimate them by using the trapezium rule (16.4). (Since you cannot know the exact answer in advance, you can proceed by running the program using increasingly fine divisions until you get no change in some predetermined number of decimal places.)

(a) $\displaystyle\int_0^{\frac{1}{2}\pi} \sin^{\frac{1}{2}}x\, dx;$ (b) $\displaystyle\int_0^1 e^{-x^2} dx;$

(c) $\displaystyle\int_1^2 \frac{e^x\, dx}{1+x^3};$ (d) $\displaystyle\int_1^2 \frac{\sin x}{x}\, dx.$

16.21 The following is called **Simpson's rule** for numerical integration. It results from splitting the points of division into successive groups of three, then exactly fitting the corresponding groups of points on the graph by second-degree polynomials. For this purpose, N **must be an even number**:

$$\int_a^b y\, dx \approx \frac{b-a}{3N}(y_0 + 4y_1 + 2y_2 + 4y_3 + 2y_4 + \cdots + 4y_{N-1} + y_N).$$

Show that $\int_0^1 e^{-x^2} dx$ is given correctly to 4 decimal places by using only four subdivisions. Compare the trapezium rule and the rectangle rule.

16.22 Consider the curve $y = f(x)$ for $a \leqslant x \leqslant b$. Show that the arc-length δs associated with a short step δx is given by $\delta s = [(\delta x)^2 + (\delta y)^2]^{\frac{1}{2}}$. Deduce that the total length s of the curve is given by

$$s = \int_a^b \left[1 + \left(\frac{dy}{dx}\right)^2\right]^{\frac{1}{2}} dx.$$

This type of integral is usually impossible to evaluate explicitly, but can be done numerically. Compute the lengths of the following curves. (Try the trapezium rule, Simpson's rule of Problem 16.21, and an integrating routine from a software package if you know how to use it: the interest lies in comparing them.)
(a) $y = \sin x, 0 \leqslant x \leqslant 1$;
(b) $y = x^2, 0 < x < 2$;
(c) $y = e^x, -1 < x < 1$;
(d) $y = (1 - x^2)^{\frac{1}{2}}, 0 \leqslant x \leqslant 1$ (a semicircle, so it can be done directly).

16.23 A curve is given in polar coordinates (r, θ) by $r = f(\theta)$. Show that the arc-length δs associated with a small change $\delta\theta$ in θ is given by

$$\delta s = [(\delta x)^2 + (\delta y)^2]^{\frac{1}{2}}$$
$$= [r^2(\delta\theta)^2 + (\delta r)^2]^{\frac{1}{2}},$$

using the relations $x = r \cos \theta, y = r \sin \theta$. Hence show that the total length of the curve between $\theta = \alpha$ and $\theta = \beta$ is

$$s = \int_\alpha^\beta \{[f(\theta)]^2 + [f'(\theta)]^2\}^{\frac{1}{2}} d\theta.$$

The cardioid (see also Problem 1.31a) is given by $r = a(1 + \cos \theta)$ in polar form. Find the length of its perimeter.

17 Systematic techniques for integration

17.1 Substitution method for $\int f(ax+b)\,dx$

Consider the indefinite integral

$$\int (3x-2)^3\,dx.$$

We carried out this integration in Example 14.8 by starting with a guess that the result will resemble $(3x-2)^4$. We now describe a method less dependent on trial and error.

We shall take up a clue suggested by the chain rule procedure (Section 3.3). Put

$$3x-2 = u. \tag{17.1}$$

Then the integral becomes

$$\int u^3\,dx.$$

Unfortunately this is *not* equal to $\frac{1}{4}u^4 + C$, because dx, not du, is present: the variable of integration is still x. Thinking in terms of an integral as a sum, δx is not the same size as δu; in fact from (17.1) $\delta u = 3\,\delta x$, which suggests what to do with the new integral.

From (17.1), $du/dx = 3$, which we write as

$$dx = \tfrac{1}{3}\,du.$$

Put this into the integral, and it works through straightforwardly:

$$\int (3x-2)^3\,dx = \int u^3(\tfrac{1}{3}\,du) = \tfrac{1}{3}\int u^3\,du = \tfrac{1}{12}u^4 + C.$$

Now use (17.1) to change back to x:

$$\int (3x - 2)^3 \, dx = \tfrac{1}{12}(3x - 2)^4 + C,$$

and this is correct. In checking its correctness by differentiation, we use the chain rule with $u = 3x - 2$, and find we are simply reversing the order of the operations that we just went through.

Example 17.1   *Use a substitution to obtain* $\displaystyle\int \frac{dx}{2x - 1}$.

Try

$\quad u = 2x - 1$.

We shall need to express dx in terms of u. Since $du/dx = 2$, we have

$\quad dx = \tfrac{1}{2} \, du$.

The integral therefore becomes, in terms of u,

$$\int \frac{dx}{2x - 1} = \int \frac{(\tfrac{1}{2} \, du)}{u} = \tfrac{1}{2} \ln |u| + C$$

$$= \tfrac{1}{2} \ln |2x - 1| + C.$$

Example 17.2   *Evaluate* $\displaystyle\int \sin(3x + 2) \, dx$.

Put

$\quad u = 3x + 2$,

then $du/dx = 3$, so $du = 3 \, dx$, or $dx = \tfrac{1}{3} \, du$. The integral becomes

$$\int \sin(3x + 2) \, dx = \int \sin u \cdot (\tfrac{1}{3} \, du)$$

$$= (-\tfrac{1}{3} \cos u) + C = -\tfrac{1}{3} \cos(3x + 2) + C.$$

The essence of the matter is that the **change of variable** or **substitution** led to a simpler integral than the one we started with. In general, for integrals of this type, we have the following result.

Type $\displaystyle\int f(ax + b) \, dx$

Put $u = ax + b$; then $\dfrac{du}{dx} = a$, or $dx = \dfrac{1}{a} du$. The integral transforms to $\dfrac{1}{a} \displaystyle\int f(u) \, du$

$$(17.2)$$

It is worth while to try this substitution in more general cases, even if it is not obvious that a simplification will take place.

Example 17.3   *Evaluate* $\int x(2x-1)^3\,\mathrm{d}x$.

This is not quite of the form (17.2) because of the presence of the loose x. Nevertheless, put

$u = 2x - 1,$

with the object of simplifying at least the most complicated part. Then

$\mathrm{d}u = 2\,\mathrm{d}x, \quad \text{or} \quad \mathrm{d}x = \tfrac{1}{2}\,\mathrm{d}u\,.$

We also need to express x in terms of u, using $u = 2x - 1$:

$x = \tfrac{1}{2}(u + 1).$

Now we have

$$\int x(2x-1)^3\,\mathrm{d}x = \int \tfrac{1}{2}(u+1)u^3(\tfrac{1}{2}\,\mathrm{d}u)$$

$$= \tfrac{1}{4}\int (u^4 + u^3)\,\mathrm{d}u = \tfrac{1}{20}u^5 + \tfrac{1}{16}u^4 + C$$

$$= \tfrac{1}{20}(2x-1)^5 + \tfrac{1}{16}(2x-1)^4 + C.$$

Do not miss the possibility of making a substitution in **simple cases**. For example:

$$\int e^{-3x}\,\mathrm{d}x: \text{put } u = -3x, \mathrm{d}x = -\tfrac{1}{3}\,\mathrm{d}u;$$

$$\int \sin 3x\,\mathrm{d}x: \text{put } u = 3x, \mathrm{d}x = \tfrac{1}{3}\,\mathrm{d}u;$$

$$\int \frac{1+x}{1-x}\,\mathrm{d}x: \text{put } u = 1-x, \mathrm{d}x = -\mathrm{d}u.$$

17.2 Substitution method for $\int f(ax^2 + b)x\,\mathrm{d}x$

Example 17.4   *Evaluate* $\int x\,e^{x^2}\,\mathrm{d}x$.

Try putting

$u = x^2,$

with the objective of simplifying the unfamiliar-looking term e^{x^2}. It is then necessary to deal with x and $\mathrm{d}x$ in the integral. We have

$$\frac{\mathrm{d}u}{\mathrm{d}x} = 2x,$$

Example 17.4 *continued*

which we can write as $du = 2x\,dx$, or

$x\,dx = \frac{1}{2}\,du.$

In this way we have translated the whole group $(x\,dx)$ into terms of u, instead of having to deal separately with x and dx. Therefore

$$\int x\,e^{x^2}\,dx = \int e^{x^2}(x\,dx) = \int e^u(\tfrac{1}{2}\,du)$$

$$= \tfrac{1}{2}\int e^u\,du = \tfrac{1}{2}e^u + C = \tfrac{1}{2}e^{x^2} + C,$$

where C is an arbitrary constant. The correctness of the result can be checked by differentiating it.

Example 17.5 *Evaluate* $\displaystyle\int \frac{x\,dx}{3x^2 + 2}.$

Notice that the integral can be written in the form

$$\int \frac{1}{3x^2 + 2}\,(x\,dx).$$

The integrand contains a function of x^2 and the combination $x\,dx$ which appeared in Example 17.4. This suggests putting $u = x^2$ to give a simpler integral. However, we can do even better than this.

Put

$u = 3x^2 + 2.$

Then $du/dx = 6x$, so that

$x\,dx = \frac{1}{6}\,du.$

Therefore

$$\int \frac{1}{3x^2 + 2}\,(x\,dx) = \int \frac{1}{u}\,(\tfrac{1}{6}\,du) = \tfrac{1}{6}\int \frac{du}{u}$$

$$= \tfrac{1}{6}\ln|u| + C = \tfrac{1}{6}\ln(3x^2 + 2) + C,$$

where C is an arbitrary constant. The modulus sign in the logarithm was discarded because $3x^2 + 2$ is always positive.

The general result is as follows:

Integrals of type $\displaystyle I = \int xf(ax^2 + b)\,dx$

Put $u = ax^2 + b$; then $x\,dx = \dfrac{1}{2a}\,du$, so

$$I = \frac{1}{2a}\int f(u)\,du.$$

(17.3)

17.3 **Substitution method for** $\int \cos^m ax \sin^n ax \, dx$
     (m or n odd)

Example 17.6   *Evaluate* $\int \sin^3 x \cos x \, dx$.

Aim to simplify the worst term by putting

$u = \sin x.$

Then $\sin^3 x$ becomes u^3, and we must deal with $\cos x \, dx$. As always, begin
with $du/dx = \cos x$. Therefore

$du = \cos x \, dx,$

so, by good fortune, $\cos x \, dx$ appears in one piece. Then we have

$$\int \sin^3 x \cdot (\cos x \, dx) = \int u^3 \, du = \tfrac{1}{4}u^4 + C = \tfrac{1}{4}\sin^4 x + C,$$

with C arbitrary. (Check by differentiating.)

Example 17.7   *Evaluate* $\int \tan x \, dx$.

We have

$$\int \tan x \, dx = \int \frac{\sin x}{\cos x} \, dx = \int \frac{1}{\cos x}(\sin x \, dx).$$

This time, put

$u = \cos x,$

so that $du/dx = -\sin x$. From this we obtain

$du = -\sin x \, dx,$

so, apart from the sign, we have exactly the combination required for the
rest of the integrand. Then

$$\int \frac{1}{\cos x}(\sin x \, dx) = \int \frac{1}{u}(-du) = -\ln|u| + C = -\ln|\cos x| + C,$$

where C is arbitrary. This often appears as $\ln|\sec x| + C$ in tables of
integrals, since $\ln|\sec x| = \ln(1/|\cos x|) = -\ln|\cos x|$.

This technique can be used for products $\cos^m ax \sin^n ax$, when either
m or n (*or both*) are *odd numbers, either positive or negative*, and for
certain other cases as well.

Example 17.8   *Evaluate* $\int \cos^3 x \, dx$. *(This is the case $m = 3$, $n = 0$.)*

Write $\int \cos^3 x \, dx = \int \cos^2 x \cdot (\cos x \, dx)$, and put

$u = \sin x$

(not $\cos x$ as possibly expected). Then $du/dx = \cos x$, so that

$du = \cos x \, dx.$

Example 17.8 *continued*

The remaining part of the integrand is $\cos^2 x$, and we can transform this by writing

$$\cos^2 x = 1 - \sin^2 x = 1 - u^2.$$

Then we have

$$\int \cos^2 x \cdot (\cos x \, dx) = \int (1 - u^2) \, du = u - \tfrac{1}{3}u^3 + C = \sin x - \tfrac{1}{3}\sin^3 x + C,$$

where C is arbitrary.

You should try also the substitution $u = \cos x$. It leads to an integral in terms of u that is correct but worse than the original.

Example 17.9   *Evaluate* $I = \displaystyle\int \cos^3 2x \sin^3 2x \, dx.$

(Here $m = 3, n = 3$.) The technique requires us to decompose the term whose power is odd. Here both powers are odd, so either will do. We shall split the integrand like this:

$$I = \int \cos^3 2x \sin^2 2x (\sin 2x \, dx).$$

Put $u = \cos 2x$ so that

$$\sin 2x \, dx = -\tfrac{1}{2} \, du.$$

Since $\sin^2 2x = 1 - \cos^2 2x$, the integral becomes

$$I = \int u^3 (1 - u^2)(-\tfrac{1}{2} \, du)$$

$$= -\tfrac{1}{2} \int (u^3 - u^5) \, du = -\tfrac{1}{8}u^4 + \tfrac{1}{12}u^6 + C$$

$$= -\tfrac{1}{8} \cos^4 2x + \tfrac{1}{12} \cos^6 2x + C,$$

with C arbitrary.

The general rule is as follows.

Integrals of type $I = \displaystyle\int \cos^m ax \, \sin^n ax \, dx$, m or n an odd positive or negative integer

(a) If n is odd, put $I = \displaystyle\int \cos^m ax \, \sin^{n-1} ax \, (\sin ax \, dx)$;

then $u = \cos ax$, $\sin ax \, dx = -\dfrac{1}{a} \, du$, and $\sin^2 ax = 1 - \cos^2 ax$.

(b) If m is odd, write

$$I = \int \cos^{m-1} ax \, \sin^n ax (\cos ax \, dx);$$

then $u = \sin ax$, $\cos ax \, dx = \dfrac{1}{a} \, du$, and $\cos^2 ax = 1 - \sin^2 ax$.

(c) If n and m are both odd, use either (a) or (b).   (17.4)

17.4 Definite integrals and change of variable

For the previous examples involving indefinite integrals, we changed the variable to u, carried out the integration, and then expressed the result back in terms of x. For a *definite* integral, it is often more convenient to *express the limits of integration in terms of u*, as well as the integrand, and in that way work with u right up to the end. In the following example, both procedures are illustrated.

Example 17.10   *Evaluate* $I = \displaystyle\int_0^{\frac{1}{2}\pi} \cos^3x \, dx$ *in two ways.*

(a) (First finding an indefinite integral in terms of x.) As in Example 17.8, put $u = \sin x$, $du = \cos x \, dx$;

$$\int \cos^3x \, dx = \int (1 - u^2) \, du = u - \tfrac{1}{3}u^3$$

$$= \sin x - \tfrac{1}{3}\sin^3x,$$

taking the simplest case with $C = 0$. Then

$$I = [\sin x - \tfrac{1}{3}\sin^3x]_0^{\frac{1}{2}\pi} = (1 - \tfrac{1}{3}) - 0 = \tfrac{2}{3}.$$

(b) (Working with u throughout.) Put $u = \sin x$ into I. In order to express the limits of integration in terms of u, note that $u = 0$ when $x = 0$, and $u = 1$ when $x = \tfrac{1}{2}\pi$. Then (writing the limits so as to make them more explicit)

$$\int_{x=0}^{x=\frac{1}{2}\pi} \cos^3x \, dx = \int_{u=0}^{u=1} (1 - u^2) \, du$$

$$= [u - \tfrac{1}{3}u^3]_{u=0}^{u=1} = (1 - \tfrac{1}{3}) - 0 = \tfrac{2}{3}.$$

In Example 17.10b it would have been *wrong* to write the integral in the form

$$\int_0^{\frac{1}{2}\pi} (1 - u^2) \, du.$$

This would imply that we were going to put u equal to 0 and $\tfrac{1}{2}\pi$ after integrating.

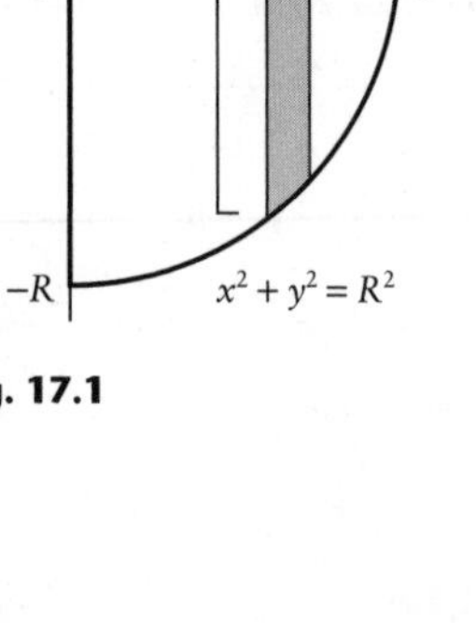

Example 17.11   *Find the centroid (centre of mass) of the uniform semicircular plate shown in Fig. 17.1.*

The symmetry shows that the centroid G lies on the x axis. From (17.6), the x coordinate of G is given by

$$\bar{x} = \frac{1}{\tfrac{1}{2}\pi R^2} \int_0^R V(x)x \, dx.$$

Since $x^2 + y^2 = R^2$, we have $V(x) = 2(R^2 - x^2)^{\frac{1}{2}}$, so that

Fig. 17.1

Example 17.11 *continued*

$$\bar{x} = \frac{4}{\pi R^2} \int_0^R (R^2 - x^2)^{\frac{1}{2}} x \, dx.$$

This is an integral of the type of (17.3). To simplify it, put $u = R^2 - x^2$, so that $du/dx = -2x$ and $x \, dx = -\frac{1}{2} \, du$. Also, $u = R^2$ when $x = 0$, and $u = 0$ when $x = R$. Therefore

$$\bar{x} = \frac{4}{\pi R^2} \int_{R^2}^0 u^{\frac{1}{2}}(-\tfrac{1}{2} \, du)$$

$$= -\frac{2}{\pi R^2} \, \tfrac{2}{3}[u^{\frac{3}{2}}]_{R^2}^0 = -\frac{4}{3\pi R^2}[0 - R^3] = \frac{4}{3\pi} R.$$

17.5 Occasional substitutions

Finding an advantageous new variable u is often a process of trial and error. Frequently the possible usefulness of a substitution is more easy to see in the form

$$x = f(u)$$

rather than u as a function of x as in the previous work.

Example 17.12 *Find a substitution to evaluate* $\displaystyle\int \frac{dx}{(1 - x^2)^{\frac{1}{2}}}.$

Try to simplify $(1 - x^2)^{\frac{1}{2}}$ first, hoping that dx will work out conveniently. To do this try

$$x = \sin u; \tag{17.5}$$

then $(1 - x^2)^{\frac{1}{2}} = (1 - \sin^2 u)^{\frac{1}{2}} = \cos u$. Also $dx/du = \cos u$, so

$$dx = \cos u \, du.$$

Therefore

$$\int \frac{dx}{(1 - x^2)^{\frac{1}{2}}} = \int \frac{\cos u \, du}{\cos u} = u + C = \arcsin x + C,$$

a result which can be confirmed from the table of derivatives in Appendix D. You might try putting $u = 1 - x^2$ instead: the resulting integral is different from, but no better than, the original.

Example 17.13 *From Example* 17.12, *we know that*

$$\int \frac{dx}{(1 - x^2)^{\frac{1}{2}}} = \arcsin x + C. \; \textit{Use this result to obtain} \int \frac{dx}{(4 - x^2)^{\frac{1}{2}}}.$$

Aim to convert $(4 - x^2)^{\frac{1}{2}}$ into something like $(1 - u^2)^{\frac{1}{2}}$, so as to be able to use the given result.

$$(4 - x^2)^{\frac{1}{2}} = 2(1 - \tfrac{1}{4}x^2)^{\frac{1}{2}} = 2[1 - (\tfrac{1}{2}x)^2]^{\frac{1}{2}},$$

↗

Example 17.13 *continued*

and make the substitution

$$u = \tfrac{1}{2}x.$$

Then $du/dx = \tfrac{1}{2}$, so that $dx = 2\,du$. Therefore

$$\int \frac{dx}{(4 - x^2)^{\frac{1}{2}}} = \int \frac{2\,du}{2(1 - u^2)^{\frac{1}{2}}} = \int \frac{du}{(1 - u^2)^{\frac{1}{2}}}$$

$$= \arcsin u + C = \arcsin \tfrac{1}{2}x + C.$$

Example 17.14   *From Appendix E,* $\displaystyle\int \frac{dx}{1 + x^2} = \arctan x + C.$

Use this result to evaluate $\displaystyle\int \frac{dx}{1 + 9x^2}.$

We want to transform $1 + 9x^2$ to a form close to $1 + u^2$, so put

$$u = 3x,$$

so that $dx = \tfrac{1}{3}\,du$. Then

$$\int \frac{dx}{1 + 9x^2} = \int \frac{\tfrac{1}{3}\,du}{1 + u^2} = \tfrac{1}{3}\arctan u + C = \tfrac{1}{3}\arctan 3x + C.$$

If the required integral does not seem to be similar to one that is already known, then one has in effect to *guess* a suitable substitution:

Example 17.15   *Evaluate* $\displaystyle\int \frac{\ln^3 x}{x}\,dx.$

We can simplify the logarithm (at the risk of extra complexity elsewhere) by putting

$$x = e^u$$

so that $\ln x = \ln e^u = u$. Since $dx/du = e^u$, we have $dx = e^u\,du$. Therefore

$$\int \frac{\ln^3 x}{x}\,dx = \int \frac{u^3}{e^u}\,e^u\,du = \int u^3\,du$$

$$= \tfrac{1}{4}u^4 + C = \tfrac{1}{4}\ln^4 x + C.$$

The general shape of the integrand sometimes suggests a substitution that is sure to simplify it. Suppose we notice that $f(x)$ takes the special form

$$f(x) = cg(u)\frac{du}{dx},$$

where c is a constant and u is a function of x. For example,

$$(x^4 + 1)^7 x^3 = \tfrac{1}{3} u^7 \frac{du}{dx};$$

where, in this case, $c = \tfrac{1}{4}$, $u(x) = x^4 + 1$, and $g(u) = u^7$. In the general case,

$$\int f(x)\, dx = c \int g(u) \frac{du}{dx}\, dx = c \int g(u)\, du.$$

(Any $f(x)$ can in principle be written in this form: the question is only whether it is easy to see how it breaks up.) Having observed the form of $u(x)$, the substitution should be made in the usual way.

Example 17.16 *Evaluate* $\displaystyle \int (x^{\frac{1}{2}} + 1)^{\frac{1}{3}} x^{-\frac{1}{2}}\, dx.$

The important thing is to spot that $(d/dx)(x^{\frac{1}{2}} + 1)$ is like the remaining factor, $x^{-\frac{1}{2}}$. This suggests that $u = x^{\frac{1}{2}} + 1$ is the right substitution. Specifically, put $u = x^{\frac{1}{2}} + 1$; then

$$\frac{du}{dx} = \tfrac{1}{2} x^{-\frac{1}{2}} \text{ and so } x^{-\frac{1}{2}}\, dx = 2\, du.$$

The integral becomes

$$2 \int u^{\frac{1}{3}}\, du = \tfrac{3}{2} u^{\frac{4}{3}} + C = \tfrac{3}{2}(x^{\frac{1}{2}} + 1)^{\frac{4}{3}} + C.$$

Some further special substitutions together with illustrative integrals are listed in Problem 17.23.

17.6 Partial fractions for integration

In Section 1.14, it was shown how a **rational function** $P(x)/Q(x)$, where $P(x)$ and $Q(x)$ are polynomials, $P(x)$ is of lower degree than $Q(x)$, and $Q(x)$ factorizes into real factors, can be expressed as the sum of simpler **partial fractions**. This provides a method for integrating rational functions.

Example 17.17 *Evaluate* $\displaystyle \int \frac{dx}{x^2 - 1}.$

By the methods of Section 1.14, we find that

$$\frac{1}{x^2 - 1} = \frac{1}{(x - 1)(x + 1)} = \tfrac{1}{2} \frac{1}{x - 1} - \tfrac{1}{2} \frac{1}{x + 1}.$$

Example 17.17 *continued*

Therefore

$$\int \frac{dx}{x^2 - 1} = \frac{1}{2} \int \frac{dx}{x - 1} - \frac{1}{2} \int \frac{dx}{x + 1}$$

$$= \frac{1}{2} \ln|x - 1| - \frac{1}{2} \ln|x + 1| + C.$$

Other equivalent forms are $\frac{1}{2} \ln\left|\frac{x - 1}{x + 1}\right| + C$, $\ln\left|\frac{x - 1}{x + 1}\right|^{\frac{1}{2}} + C$, and

$\frac{1}{2} \ln\left| B \frac{x - 1}{x + 1}\right|$, where C and B are arbitrary.

As a result of expanding in partial fractions we may encounter integrands of the type

$$\frac{cx + d}{px^2 + qx + r}$$

in which the equation $px^2 + qx + r = 0$ has no real roots (i.e. the denominator has no real factors). The following example shows how to evaluate them by '**completing the square**' in the denominator.

Example 17.18   *Evaluate* $I = \int \frac{(x + 1)\, dx}{x^2 + 4x + 8}$.

The quadratic form $x^2 + 4x + 8$ has no real factors. 'Completing the square' in the denominator consists of writing $x^2 + 4x + 8$ in the form $(x + a)^2 + b$. The first two terms, $x^2 + 4x$, can be written

$$x^2 + 4x = (x + 2)^2 - 4,$$

so

$$x^2 + 4x + 8 = (x + 2)^2 - 4 + 8 = (x + 2)^2 + 4.$$

The integral becomes

$$I = \int \frac{(x + 1)\, dx}{(x + 2)^2 + 4} = \frac{1}{4} \int \frac{(x + 1)\, dx}{[\frac{1}{2}(x + 2)]^2 + 1}.$$

Now put $u = \frac{1}{2}(x + 2)$, or $x = 2u - 2$, from which

$$dx = 2\, du.$$

Then

$$I = \frac{1}{2} \int \frac{2u - 1}{u^2 + 1}\, du = \int \frac{u}{u^2 + 1}\, du - \frac{1}{2} \int \frac{1}{u^2 + 1}\, du.$$

To evaluate the first integral, use the substitution $v = u^2 + 1$, as in Section 17.2; the second is a standard integral. We obtain

$$I = \frac{1}{2} \ln(x^2 + 4x + 8) - \frac{1}{2} \arctan \frac{1}{2}(x + 2) + C.$$

17.7 Integration by parts

This method is totally unrelated to the techniques we have so far described, and can be used to integrate special types of **product**. It is needed very frequently for obtaining fundamental general results.

Suppose that we are given any $u(x)$ and $v(x)$. Then, by the product rule (Section 3.1),

$$\frac{d}{dx}(uv) = u\frac{dv}{dx} + v\frac{du}{dx}.$$

Since both sides are equal, their indefinite integrals can only differ by a constant, so

$$\int \frac{d}{dx}(uv)\,dx = \int u\frac{dv}{dx}\,dx + \int v\frac{du}{dx}\,dx + B, \qquad (17.6)$$

where B is a constant. Look at the integral on the left. It means 'an antiderivative of $(d/dx)[u(x)v(x)]$'. But, from the definition (14.1), $u(x)v(x)$ is an antiderivative. Therefore (17.6) becomes

$$uv = \int u\frac{dv}{dx}\,dx + \int v\frac{du}{dx}\,dx + B.$$

Now rearrange the terms to obtain

$$\int u\frac{dv}{dx}\,dx = uv - \int v\frac{du}{dx}\,dx - B.$$

This is the formula for **integration by parts**. Replacing $-B$ by C:

Integration by parts

$$\int u\frac{dv}{dx}\,dx = uv - \int v\frac{du}{dx}\,dx + C,$$

where C is some constant.

(17.7)

It is not at first obvious how this complicated result could be of any use, but the point of it is that the right-hand integral might be simpler than the one on the left. The process was once called 'partial integration', because the uv part is already integrated out. (For the effect of missing out C, see Problem 17.19.)

Example 17.19 *Evaluate* $\int x\,e^x\,dx$ by integrating by parts.

First observe that the integrand consists of the **product** of two factors, x and e^x, both of which we can integrate and differentiate any number of times. We relate this fact to (17.7) by identifying them with u and dv/dx respectively: put

$$u = x \quad \text{and} \quad \frac{dv}{dx} = e^x. \tag{i}$$

Then

$$\frac{du}{dx} = 1 \quad \text{and} \quad v = \int e^x\,dx = e^x, \tag{ii}$$

where we have chosen v to be the simplest antiderivative of e^x. Nothing would ultimately be changed by introducing an arbitrary constant C into v: *any antiderivative will do* (see Problem 17.18).

Fill in the right-hand side of (17.7) by picking out u, v, du/dx from (i) and (ii), and introduce the constant C:

$$\int x\,e^x\,dx = x\,e^x - \int (e^x)(1)\,dx + C$$

$$= x\,e^x - \int e^x\,dx + C = x\,e^x - e^x + C,$$

where C is arbitrary.

We obtained a simplification because we chose x, rather than e^x, to be assigned to u. Since du/dx is simpler than u, it seemed possible that the right side of (17.7) might be simpler than the left. (To see what happens when we put $u = e^x$, $dv/dx = x$, see Example 17.21.)

As in Example 17.19, you should always write out stages (i) and (ii) in full and do the subsequent working in full, or you will make mistakes.

Example 17.20 *Evaluate* $\int x \cos 2x\,dx$.

Put $u = x$, $dv/dx = \cos 2x$. Then

$$\frac{du}{dx} = 1, \quad v = \int \cos 2x\,dx = \tfrac{1}{2}\sin 2x.$$

Substituting these functions into the right-hand side of (17.7):

$$\int x \cos 2x\,dx = x(\tfrac{1}{2}\sin 2x) - \int (\tfrac{1}{2}\sin 2x)(1)\,dx + C$$

$$= \tfrac{1}{2}x \sin 2x - \tfrac{1}{2}(-\tfrac{1}{2}\cos 2x) + C$$

$$= \tfrac{1}{2}x \sin 2x + \tfrac{1}{4}\cos 2x + C.$$

Example 17.21 For $\int x\,e^x\,dx$ (see Example 17.19), try the effect of assigning x and e^x to u and dv/dx the 'wrong way round'.

In Example 17.19, we successfully put $u = x$ and $dv/dx = e^x$. Now try instead

$$u = e^x, \qquad \frac{dv}{dx} = x,$$

then

$$\frac{du}{dx} = e^x, \qquad v = \int x\,dx = \tfrac{1}{2}x^2.$$

The integration-by-parts formula becomes

$$\int x\,e^x\,dx = e^x(\tfrac{1}{2}x^2) - \int (\tfrac{1}{2}x^2)\,e^x\,dx + C$$

$$= \tfrac{1}{2}x^2\,e^x - \tfrac{1}{2}\int x^2\,e^x\,dx + C,$$

which is a true result, but the transformed integral is worse than the original.

Sometimes it is not immediately obvious that the method can be made to work, as in the following.

Example 17.22 Evaluate $\int \ln x\,dx$.

Write $\ln x = (\ln x)(1)$, so that the integral becomes

$$\int (\ln x)(1)\,dx.$$

We can now put $u = \ln x$ and $dv/dx = 1$, so that

$$\frac{du}{dx} = \frac{1}{x}, \qquad v = x.$$

Then

$$\int (\ln x)(1)\,dx = (\ln x)(x) - \int (x)\left(\frac{1}{x}\right)dx + C$$

$$= x\ln x - \int dx + C = x\ln x - x + C,$$

where C is an arbitrary constant.

The **integrals of other inverse functions**, such as arcsin x and arctan x, respond to the same technique.

Example 17.23   *Evaluate* $\int x^2 \sin x \, dx$.

It is necessary in this problem to integrate by parts twice. Put

$$u = x^2, \qquad \frac{dv}{dx} = \sin x;$$

then

$$\frac{du}{dx} = 2x, \qquad v = -\cos x.$$

From (17.7),

$$\int x^2 \sin x \, dx = -x^2 \cos x + 2 \int x \cos x \, dx + C.$$

Integrate the integral on the right by parts; put

$$u = x, \qquad \frac{dv}{dx} = \cos x,$$

so that $du/dx = 1$ and $v = \sin x$. From (17.7), we obtain finally

$$\int x^2 \sin x \, dx = -x^2 \cos x + 2x \sin x + 2 \cos x + C.$$

17.8 Integration by parts: definite integrals

The integration-by-parts formula (17.7) expresses a relation between indefinite integrals, or antiderivatives. Suppose that we have a **definite integral** of the form

$$\int_a^b u \frac{dv}{dx} \, dx,$$

which we expect to integrate by parts. Then, from (17.7),

$$\int_a^b u \frac{dv}{dx} \, dx = \left[uv - \int v \frac{du}{dx} \, dx \right]_a^b$$

The operation $[\dots]_a^b$ applies to the two terms separately, so we have:

Integration by parts (definite integrals)

$$\int_a^b u \frac{dv}{dx} \, dx = [uv]_a^b - \int_a^b v \frac{du}{dx} \, dx.$$

(17.8)

This can sometimes considerably simplify the working, especially if more than one integration by parts is needed.

Example 17.24 *Evaluate* $\displaystyle\int_0^{\frac{1}{2}\pi} x^2 \sin x \, dx.$

As in Example 17.23, put $u = x^2$ and $dv/dx = \sin x$. Then

$$\frac{du}{dx} = 2x, \qquad v = -\cos x.$$

From (17.8),

$$\int_0^{\frac{1}{2}\pi} x^2 \sin x \, dx = [x^2(-\cos x)]_0^{\frac{1}{2}\pi} - \int_0^{\frac{1}{2}\pi} (-\cos x)(2x) \, dx$$

$$= 2\int_0^{\frac{1}{2}\pi} x \cos x \, dx,$$

because the bracketed term is zero; we did not have to wait to the end of the calculation to see it go. To evaluate the remaining integral, integrate by parts again, putting $u = x$ and $dv/dx = \cos x$; we have

$$\frac{du}{dx} = 1, \qquad v = \sin x.$$

Use (17.8) again:

$$2\int_0^{\frac{1}{2}\pi} x \cos x \, dx = 2\left([x \sin x]_0^{\frac{1}{2}\pi} - \int_0^{\frac{1}{2}\pi} \sin x \, dx \right)$$

$$= 2(\tfrac{1}{2}\pi + [\cos x]_0^{\frac{1}{2}\pi}) = 2[\tfrac{1}{2}\pi + (0 - 1)]$$

$$= \pi - 2.$$

The following result is important for Chapter 24, and involves the use of (17.8):

$$\int_0^\infty e^{-t} t^N \, dt = N!$$

when $N = 0, 1, 2, 3, \ldots$.
(0! is defined to be 1.)

(17.9)

Here N! stands for the **factorial** (see Section 1.17):

$$N! = N(N-1)(N-2) \ldots 3 \cdot 2 \cdot 1.$$

The symbol 0!, which is apparently arbitrarily given the value 1, does not fit this pattern; it should be regarded at this stage as being just a useful convention. The related **gamma function** $\Gamma(N) = (N-1)!$ is used in statistics in Section 41.6.

To prove (17.9), let k represent any of the numbers $0, 1, 2, \ldots$, and write

$$\int_0^\infty e^{-t} t^k \, dt = F(k),$$

to indicate the integral's dependence on the parameter k; for example, $\int_0^\infty e^{-t} t^3 \, dt$ is denoted by $F(3)$. Notice in particular that

$$F(0) = \int_0^\infty e^{-t} \, dt = [-e^{-t}]_0^\infty = 1. \tag{17.10}$$

For $k = 1, 2, \ldots$, integrate by parts. Put $u = t^k$ and $dv/dt = e^{-t}$; then

$$F(k) = \int_0^\infty e^{-t} t^k \, dt = [t^k(-e^{-t})]_0^\infty - \int_0^\infty (kt^{k-1})(-e^{-t}) \, dt$$

$$= k \int_0^\infty e^{-t} t^{k-1} \, dt$$

(the bracket is zero because $k \geqslant 1$). The integral is $F(k-1)$, so

$$F(k) = kF(k-1) \quad \text{for } k = 1, 2, \ldots . \tag{17.11}$$

By integrating by parts, we have reduced the degree of t by unity. We could evaluate $F(N)$, where N is given, by integrating by parts again and again until we reach $F(0)$, given as 1 by (17.10). But we do not have to integrate by parts any more: eqn (17.11) does it for us. Put $k = 0, 1, 2, 3, \ldots$, successively: we obtain

$F(0) = 1$ (by (17.10)),
$F(1) = 1F(0)$ (by (17.11)) $= 1 \cdot 1 = 1!$,
$F(2) = 2F(1)$ (by (17.11)) $= 2(1!) = 2!$,
$F(3) = 3F(2)$ (by (17.11)) $= 3(2!) = 3!$,

and so on (each line uses the result of the previous line). So, if we are given N, we shall reach $F(N)$ after N lines, and find that

$F(N) = N!$.

The argument above can be expressed in a different way. Using (17.11) repeatedly we have

$F(N) = NF(N-1) = N(N-1)F(N-2) = \cdots ,$

until we arrive at $F(0)$, which is 1, and we are left with $N!$ on the right.

Equation (17.11) is an example of a **reduction formula**, by which an integral can be systematically reduced, one step at a time, to progressively simpler integrals. (See Problems 17.14, 17.15, 17.16.)

Differentiating with respect to a parameter

(A fuller account of this topic is given in Section 28.8.) Suppose that we wish to integrate a function which, besides the variable of integration, contains a **parameter** (i.e. a general constant which may take any of a range of values). For example,

$$\int_0^\infty x\,e^{-\alpha x}\,dx$$

is such an integral, the parameter being α. This may be written

$$I(\alpha) = \int_0^\infty xe^{-\alpha x}\,dx = -\int_0^\alpha \frac{d}{d\alpha}(e^{-\alpha x})\,dx,$$

the derivative being *with respect to* α (not to x, which is treated like a constant for the purpose of the differentiation). It can be shown, as in Section 28.8, that the operator $d/d\alpha$ can be taken outside the integral sign, so that we have

$$I(\alpha) = \int_0^\infty x\,e^{-\alpha x}\,dx = -\frac{d}{d\alpha}\int_0^\infty e^{-\alpha x}\,dx = -\frac{d}{d\alpha}\left(\frac{1}{\alpha}\right) = \frac{1}{\alpha^2}.$$

In cases when we can foresee that the original integrand can be written in the form

$$\frac{d}{d\alpha} \text{ of something that we can integrate with respect to } x,$$

this procedure enables the original integral to be worked out. The following two examples further illustrate the procedure; the method can also be used for indefinite integrals.

Example 17.25 *Evaluate the indefinite integral*

$$\int x^\alpha \ln x\,dx = I(\alpha).$$

We observe that

$$\frac{d}{d\alpha}(x^\alpha) = \frac{d}{d\alpha}(e^{\alpha \ln x}) = e^{\alpha \ln x}\ln x = x^\alpha \ln x.$$

Therefore

$$I(\alpha) = \int \frac{d}{d\alpha}(x^\alpha)\,dx = \frac{d}{d\alpha}\int x^\alpha\,dx$$

$$= \frac{d}{d\alpha}\left(\frac{1}{(\alpha+1)}x^{\alpha+1}\right) = -\frac{1}{(\alpha+1)^2}x^{\alpha+1} + \frac{1}{(\alpha+1)}x^{\alpha+1}\ln x,$$

apart from a constant of integration.

Example 17.26   *Show that*

$$\int_0^\infty \frac{dx}{(x^2+1)^2} = \frac{1}{2}.$$

There is no parameter in the integral, so we shall introduce one and put $\alpha = 1$ at the end. Define $I(\alpha)$ by

$$I(\alpha) = \int_0^\infty \frac{dx}{(x^2+\alpha^2)^2}.$$ (i)

Observe that

$$\frac{d}{d\alpha}\left(\frac{1}{x^2+\alpha^2}\right) = -2\alpha\frac{1}{(x^2+\alpha^2)^2} \quad \text{or} \quad \frac{1}{(x^2+\alpha^2)^2} = -\frac{1}{2\alpha}\frac{d}{d\alpha}\left(\frac{1}{x^2+\alpha^2}\right).$$

Then

$$I(\alpha) = -\frac{1}{2\alpha}\int_0^\infty \frac{d}{d\alpha}\left(\frac{1}{x^2+\alpha^2}\right)dx = -\frac{1}{2\alpha}\frac{d}{d\alpha}\int_0^\infty \frac{dx}{x^2+\alpha^2}.$$ (ii)

But

$$\int_0^\infty \frac{dx}{x^2+\alpha^2} = \left[\frac{1}{\alpha}\arctan\left(\frac{x}{\alpha}\right)\right]_0^\infty = \frac{1}{\alpha}.$$ (iii)

Put (iii) into (ii); we obtain

$$I(\alpha) = -\frac{1}{2\alpha}\frac{d}{d\alpha}\left(\frac{1}{\alpha}\right) = \frac{1}{2\alpha^2}.$$ (iv)

By putting $\alpha = 1$ into (iv) we obtain from (i)

$$\int_0^\infty \frac{dx}{(x^2+1)^2} = \frac{1}{2},$$

as requested (though (iv) is a more general result).

Problems

17.1 (Section 17.1). Obtain $\int f(x)\,dx$ when the $f(x)$ are as follows.
(a) $\sin 3x$;   (b) $\cos 4x$;   (c) e^{-3x};
(d) $(1+x)^{10}$;   (e) $(1-x)^9$;
(f) $(3-2x)^5$;   (g) $(1+2x)^n$;
(h) $x(x-1)^4$;   (i) $(1-x)^{\frac{1}{2}}$;
(j) $(2x-3)^{-\frac{1}{2}}$ for $x > \frac{3}{2}$;
(k) $1/(3x+2)^2$;   (l) $1/(1-x)^4$;
(m) $1/(1+x)$;   (n) $1/(2x+3)$;
(o) $x/(1-x)^2$;   (p) $(1+x)/(1-x)$;
(q) $x/(x-1)^{\frac{1}{2}}$ for $x > 1$;
(r) $\cos(1-2x)$;   (s) $\sin(2x-3)$.

17.2 (Section 17.1). Evaluate the following indefinite integrals.

(a) $\int (2t-5)^5\,dt$;   (b) $\int \sin\frac{1}{2}(3t-1)\,dt$;

(c) $\int \frac{1}{(2w+1)^2}\,dw$;   (d) $\int e^{-3r}\,dr$;

(e) $\int (-t)^{\frac{1}{2}}\,dt$ if $t < 0$;   (f) $\int \frac{s\,ds}{(1-s)^3}$;

(g) $\int \cos(\omega t - \phi)\,dt$.

17.3 (Section 17.2). Obtain $\int f(x)\,dx$ when the $f(x)$ are as follows.

(a) $x\,e^{-x^2}$; (b) $x\sin x^2$;
(c) $x\cos x^2$; (d) $x\cos(x^2+3)$;
(e) $x\cos(1-3x^2)$; (f) $x(x^2-1)^4$;
(g) $x(3x^2+4)^3$; (h) $x/(1+2x^2)$;
(i) $x^3(1-x^2)^3$ (note: $x^3=xx^2$);
(j) $x/(1+x^2)$; (k) $x/(3x^2-2)$.

17.4 (Section 17.3). Find $\int f(x)\,dx$ where the $f(x)$ are as follows.

(a) $\sin x\cos x$; (b) $\sin^2 x\cos x$; (c) $\sin^2 2x\cos 2x$;
(d) $\cos^2 x\sin x$; (e) $\cos^2 3x\sin 3x$; (f) $\sin^3 x\cos x$;
(g) $\cot 2x$; (h) $\tan\frac{1}{2}x$; (i) $(\sin^3 x)/\cos x$;
(j) $\sin^3 x\,(=\sin^2 x\sin x)$; (k) $\tan^3 x$ (compare (j));
(l) $\cos^3 x$ (compare (j)).

17.5 Evaluate the following definite integrals by using any necessary substitutions.

(a) $\displaystyle\int_{-1}^{1}(1+x)^7\,dx$; (b) $\displaystyle\int_{-1}^{1}(1-\tfrac{1}{2}x)^7\,dx$;

(c) $\displaystyle\int_{0}^{1}x(1-x^2)^3\,dx$; (d) $\displaystyle\int_{0}^{1}\frac{x\,dx}{2x+3}$;

(e) $\displaystyle\int_{-3}^{-2}\frac{dx}{1+x}$ (note: $x<-1$); (f) $\displaystyle\int_{3}^{4}\frac{dx}{2-3x}$;

(g) $\displaystyle\int_{0}^{1}x^3(1-x^2)^3\,dx$; (h) $\displaystyle\int_{0}^{\frac{1}{4}\pi}\tan t\,dt$;

(i) $\displaystyle\int_{\frac{1}{12}\pi}^{\frac{1}{6}\pi}\cot 3w\,dw$; (j) $\displaystyle\int_{0}^{\frac{1}{2}\pi}\sin u\cos u\,du$;

(k) $\displaystyle\int_{0}^{\pi}(\sin v)^{\frac{1}{2}}\cos v\,dv$; (l) $\displaystyle\int_{-\frac{1}{2}\pi}^{\frac{1}{2}\pi}\cos^3\theta\,d\theta$;

(m) $\displaystyle\int_{0}^{\frac{1}{2}\pi}\sin 2t\,dt$; (n) $\displaystyle\int_{-\frac{1}{2}\pi/\omega}^{\frac{1}{2}\pi/\omega}\cos(\omega t+\phi)\,dt$.

17.6 Use the identities $\cos^2 A=\frac{1}{2}(1+\cos 2A)$, $\sin^2 A=\frac{1}{2}(1-\cos 2A)$, or $\sin A\cos A=\frac{1}{2}\sin 2A$ to evaluate the following.

(a) $\displaystyle\int_{0}^{\pi}\sin^2 t\,dt$; (b) $\displaystyle\int_{0}^{\pi}\cos^2 t\,dt$;

(c) $\displaystyle\int_{0}^{\frac{1}{2}\pi}\sin^2 2t\,dt$; (d) $\displaystyle\int_{0}^{\frac{1}{2}\pi}\cos^2\tfrac{1}{2}t\,dt$;

(e) $\displaystyle\int_{-\pi}^{\pi}\sin^2 3t\cos 3t\,dt$; (f) $\displaystyle\int_{0}^{\pi}\cos^4 u\,du$.

17.7 Use the substitutions suggested to evaluate $\int f(x)\,dx$ for the following $f(x)$. (In several of the questions the identity $1+\tan^2 A=1/\cos^2 A$ is needed. You may also have to refer to the table, Appendix E.)

(a) $\ln x/x$ (put $x=e^u$);
(b) $x(1-x^2)^{\frac{1}{2}}$ (try (i) $u=1-x^2$, (ii) $x=\sin u$);
(c) $1/(e^x+e^{-x})$ (put $u=e^x$);
(d) $1/(1-x^2)^{\frac{1}{2}}$ (try (i) $x=\sin u$, (ii) $x=\cos u$; why do the results seem to be different?);
(e) $\tan^2 x$ (put $u=\tan x$);
(f) $1/x^2(1+x^2)$ (put $x=1/u$, followed by another process);
(g) $1/(1+x^2)$ (put $x=\tan u$);
(h) $1/\cos^2 x$ (put $u=\tan x$);
(i) $\dfrac{1}{t^{\frac{1}{2}}(1+t)}$ (put $t=u^2$);
(j) $\dfrac{1}{t^2}\sin\dfrac{1}{t}$ (put $t=1/u$);
(k) $(1-x^2)^{\frac{1}{2}}$ (put $x=\sin u$);
(l) $1/(1+x^2)^{\frac{1}{2}}$ (put $x=\tan u$).

17.8 Use partial fractions to evaluate $\int f(x)\,dx$ for the following $f(x)$.

(a) $1/(x^2-4)$; (b) $1/x(x+2)$;
(c) $1/x^2(x-1)$; (d) $x/(2x+1)(x+1)$;
(e) $(x+1)/(4x^2-9)$; (f) $1/x(x^2+1)$;
(g) $x/(2x^2+3x+1)$; (h) $1/x^2(2x+1)$;
(i) $1/\cos x$ (first put $u=\sin x$);
(j) $1/\sin x$ (first put $u=\cos x$).

17.9 Obtain $\int f(x)\,dx$ for each of the following $f(x)$, noting that they take the form $cg(u)\,du/dx$ (see the remark at the end of Section 17.5), so that (a), for example, will respond to the substitution $u=x^3-1$.

(a) $x^2(x^3-1)^5$; (b) $(x-1)(x^2-2x+3)^{-1}$;
(c) $1/(x\ln^2 x)$; (d) $x^{\frac{1}{2}}(3x^{\frac{3}{2}}+2)^{\frac{1}{2}}$;
(e) $(e^x-e^{-x})/(e^x+e^{-x})$; (f) $1/x^{\frac{1}{2}}(x^{\frac{1}{2}}+1)$;
(g) $x^2/(x^3+1)$.

17.10 Use integration by parts (Section 17.7) to obtain $\int f(x)\,dx$ for each of the following $f(x)$.

(a) $x\,e^{-x}$; (b) $x\,e^{3x}$; (c) $x\,e^{-3x}$; (d) $x\cos x$;
(e) $x\sin x$; (f) $x\cos\frac{1}{2}x$; (g) $x\sin 2x$;
(h) $x(1-x)^{10}$; (i) $x\ln x$; (j) $x^n\ln x$, $n\neq-1$;
(k) $(\ln x)/x$ (the method might seem to have failed; but look again).

17.11 Use integration by parts (see Example 17.22), writing the integrand as $f(x)(1)$, to obtain $\int f(x)\,dx$ for each of the following $f(x)$.

(a) $\ln^2 x$; (b) $\arcsin x$; (c) $\arccos x$; (d) $\arctan x$.

17.12 To evaluate $\int f(x)\,dx$ for the following $f(x)$, integrate by parts twice; then look closely at your result. (If it does not work out you have probably made a mistake with a sign.) Compare your results with (15.11).
(a) $e^x \sin x$; (b) $e^{-x} \sin x$ (c) $e^{-x} \cos x$.

17.13 (Integration by parts: definite integrals, Section 17.8). Evaluate the following.

(a) $\displaystyle\int_0^{\frac{1}{2}\pi} x \cos x\,dx$;

(b) $\displaystyle\int_0^{\pi} x \cos 2x\,dx$;

(c) $\displaystyle\int_0^{\pi} x^2 \cos x\,dx$;

(d) $\displaystyle\int_0^{\infty} e^{-x} \sin x\,dx$ (integrate by parts twice);

(e) $\displaystyle\int_0^{\infty} e^{-x} \cos x\,dx$ (integrate by parts twice);

(f) $\displaystyle\int_1^2 \frac{\ln x\,dx}{x}$;   (g) $\displaystyle\int_0^1 \arcsin x\,dx$;

(h) $\displaystyle\int_{-1}^1 \arccos x\,dx$; (i) $\displaystyle\int_0^1 \arctan x\,dx$;

(j) $\displaystyle\int_1^2 \ln x\,dx$.

17.14 (Compare (17.9).) Denote $\int_0^1 x^k\,e^x\,dx$ by $F(k)$ for $k = 0, 1, 2, \ldots$. Integrate by parts to obtain the reduction formula

$$F(k) = e - kF(k-1)$$

(provided that k is positive). By applying it four times, show that

$$F(4) = \int_0^1 x^4\,e^x\,dx = -15e + 24F(0) = 9e - 24.$$

17.15 Denote $\int_0^{\frac{1}{2}\pi} \cos^k x\,dx$ by $F(k)$ when $k = 0, 1, 2, \ldots$. Integrate by parts to show that

$$F(k) = \frac{k-1}{k}\,F(k-2)\quad \text{for } k = 2, 3, \ldots.$$

Evaluate $F(0)$ and $F(1)$. Use the reduction formula repeatedly, together with $F(0)$ and $F(1)$, to evaluate

$$\int_0^{\frac{1}{2}\pi} \cos^4 x\,dx \text{ and } \int_0^{\frac{1}{2}\pi} \cos^5 x\,dx.$$

17.16 Follow the lines of Problems 17.14 and 17.15 to obtain the following reduction formulae, and to integrate the special cases given. The letter k is an integer as specified for each case.

(a) Let

$$F(k) = \int_1^2 (\ln x)^k\,dx \ (k \geqslant 0).$$

Show that $F(k) = 2(\ln 2)^k - kF(k-1)$ for $k \geqslant 1$, and evaluate $\int_1^2 (\ln x)^3\,dx$.

(b) Let $F(k) = \int_0^\pi x^k \sin x\,dx \ (k \geqslant 0)$. Integrate by parts twice to show that

$$F(k) = \pi^k - k(k-1)F(k-2)$$

for $k \geqslant 2$. Evaluate $\int_0^\pi x^4 \sin x\,dx$ and $\int_0^\pi x^5 \sin x\,dx$.

(c) Obtain a reduction formula for

$$F(k) = \int_0^{\frac{1}{2}\pi} \sin^k x\,dx,$$

and use it to evaluate $\int_0^{\frac{1}{2}\pi} \sin^4 x\,dx$ and $\int_0^{\frac{1}{2}\pi} \sin^5 x\,dx$.

17.17 (Change of variable etc.). Denote the integral $\int_1^c \dfrac{dx}{x}$ for $c > 0$ by $F(c)$. Deduce the properties (a) to (d) below. $F(c)$ is obviously equal to $\ln c$, but *do not use any of the known properties of the logarithm*; pretend that this is the first time you have ever seen the integral.
(a) $F(a^{-1}) = -F(a)$ if $a > 0$. (Hint: put $c = a^{-1}$ in the definition; then change the variable to u where $u = x^{-1}$.)
(b) $F(ab) = F(a) + F(b)$ if a and $b > 0$.
(c) $F(a/b) = F(a) - F(b)$, where a and $b > 0$.
(d) $F(a^n) = nF(a)$ if $a > 0$ and n has any value.

17.18 (Integration by parts). It is stated in Example 15.19 that, in obtaining v from dv/dx, we may take any antiderivative (so naturally we always take the simplest one, with $C = 0$ in the tables). Confirm that this is true for Example 17.19, in which $u = x$ and $dv/dx = e^x$, by choosing $v(x) = e^x + A$ instead.
    Prove that the truth of (15.7) is always unaffected by the choice of antiderivative for $v(x)$.

17.19 (Integration by parts: an apparent paradox). Consider the following calculation.

$$\int x^{-1}\,dx = \int x^{-1}(1)\,dx$$

$$= x^{-1}x - \int (-x^{-2})x\,dx = 1 + \int x^{-1}\,dx.$$

Therefore $0 = 1$. How is this to be resolved?

17.20 Verify the following moments of inertia I about the axis stated (Section 16.4):

(a) thin circular disc, mass m, radius a, about a diameter: $I = \frac{1}{4}ma^2$;

(b) solid uniform sphere, mass m, radius a, about a diameter: $I = \frac{2}{5}ma^2$;

(c) thin spherical shell, mass m, radius a, about a diameter: $I = \frac{2}{3}ma^2$;

(d) thin rectangle, mass m, side lengths $2a$ and $2b$, about a diagonal: $I = \frac{1}{3}m(a^2 + b^2)$;

(e) solid uniform cone, mass m, base radius a, height h, about its axis: $I = \frac{3}{10}ma^2$.

17.21 Assume that

$$\int e^{-at} \cos bt \, dt = A \, e^{-at} \cos bt + B \, e^{-at} \sin bt + C,$$

where A, B, and C are constants. By differentiating this expression and matching both sides, obtain the constants A and B in terms of a and b. Compare your result with eqn (15.11).

17.22 Evaluate the indefinite integral

$$I(\alpha) = \int x^2 e^{-\alpha x} \, dx,$$

using the technique of differeniating under the integral sign. (Hint: $(d^2/d\alpha^2)(e^{-\alpha x}) = x^2 e^{-\alpha x}$.)

17.23 (Some additional special substitutions). Evaluate the following integrals starting with the substitution suggested (further substitutions may be required: the table of integrals in Appendix E may also be helpful):

(a) $\displaystyle\int \frac{dx}{x\sqrt{(x^2 - a^2)}}$, $x = a/u$;

(b) $\displaystyle\int \frac{dx}{x\sqrt{(a^2 - x^2)}}$, $x = a/u$;

(c) $\displaystyle\int \frac{dx}{a^2 \sin^2 x + b^2 \cos^2 x}$, $u = (a \tan x)/b$;

(d) $\displaystyle\int \frac{dx}{\sin x}$, $u = \tan \frac{1}{2}x$;

(e) $\displaystyle\int \frac{dx}{3 + 5 \cos x}$, $u = \tan \frac{1}{2}x$;

(f) $\displaystyle\int \frac{dx}{5 \cosh x + 4 \sinh x}$, $u = \tanh \frac{1}{2}x$;

(g) $\displaystyle\int \sec x \, dx$, $u = \sec x + \tan x$;

(h) $\displaystyle\int_0^4 \frac{dx}{1 + \sqrt{x}}$, $x = u^2$;

(i) $\displaystyle\int x(1 + x)^{\frac{1}{3}} \, dx$, $x = u^3 - 1$;

(j) $\displaystyle\int_1^2 \frac{(x^2 + 1) \, dx}{x\sqrt{(x^4 + 7x^2 + 1)}}$, $u = x - \dfrac{1}{x}$;

(k) $\displaystyle\int_0^4 \sqrt{(1 + \sqrt{x})} \, dx$, $u = 1 + \sqrt{x}$.

17.24 If $p(x)$ is a polynomial of degree n, show that

$$\int e^x p(x) \, dx = e^x[p(x) - p'(x) + p''(x) - \cdots + (-1)^n p^{(n)}(x)] + C.$$

Hence evaluate

$$\int_0^1 e^x(x^3 - 2x^2 + x - 2) \, dx.$$

What is the formula for

$$\int_0^1 e^{-x} p(x) \, dx?$$

What is the value of the infinite integral

$$\int_0^\infty e^{-x} p(x) \, dx?$$

17.25 Find the centroid of the uniform plate bounded by the parabola $y^2 = 4ax$ and the straight line $x = h$ $(a, h > 0)$.

18 Unforced linear differential equations with constant coefficients

18.1 Differential equations and their solutions

Suppose that we have a problem in which a quantity x that we are studying depends on the time t; that is to say, x is a function of t, which we will write as $x(t)$. From the physics and geometry of the problem we can often obtain an *indirect* relation between x and t, called an *equation* for x. The equation might be an ordinary algebraic equation such as $x^2 + 2xt = 1$, but it might contain $\mathrm{d}x/\mathrm{d}t$ or $\mathrm{d}^2x/\mathrm{d}t^2$, as in the equation $\mathrm{d}^2x/\mathrm{d}t^2 = g$ for a falling body, where g is the gravitational acceleration. This is a simple example of a **differential equation**, and we can solve it by the methods of earlier chapters (compare Problem 14.8f).

The equation

$$\frac{\mathrm{d}x}{\mathrm{d}t} = 3x$$

is also a differential equation, but we do not yet know how to find an explicit solution for x in terms of t. Obviously not just anything will do; if for instance we try $x = t^2$ it does not work, because then $\mathrm{d}x/\mathrm{d}t = 2t$, but $3x = 3t^2$, and these are quite different.

A clue is given by interpreting the equation: it says that a quantity x always grows at a rate proportional to the amount of x already present. This is a property of the exponential function (see Section 1.10), so we might try exponential functions of t. In fact,

$$x = \mathrm{e}^{3t}$$

solves the equation, because then $\mathrm{d}x/\mathrm{d}t = 3\,\mathrm{e}^{3t}$, and this is equal to $3x$, as required. However, it is not the only solution, because

$$x = A\,e^{3t},$$

where A is any constant, also solves the equation.

In general, a **differential equation for x as a function of t** is an equation involving at least the first derivative dx/dt as well as, possibly, x and t separately. Some examples are

$$\frac{dx}{dt} + 2xt = 1, \quad \frac{d^2x}{dt^2} + \frac{dx}{dt} + x = 0, \quad \frac{d^3x}{dt^3} = \frac{x^2}{t^2}.$$

In such equations, t is called the **independent variable** and x the **dependent variable**. An equation is called **first-order, second-order,** and so on, according to the **order of the highest derivative** in it: dx/dt, d^2x/dt^2, and so on.

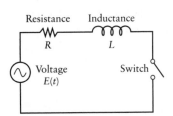

Resistance   Inductance

R L

Voltage   Switch
$E(t)$

Fig. 18.1

Problems in science and engineering are often most easily formulated in terms of differential equations. Suppose for example that in the RL circuit of Fig. 18.1 the switch is closed at time $t = 0$, and that subsequently the voltage applied is $E(t)$. Then the current $x(t)$ is found by solving the differential equation

$$L\frac{dx}{dt} + Rx = E(t).$$

Here we have collected all the terms that involve x (including dx/dt) on the left side and have put the term that does not involve x, namely $E(t)$, on the right. This is the conventional arrangement. The term independent of x which comes on the right is then called the **forcing term**, the reason being obvious in this case, since $E(t)$ drives the circuit.

The differential equation with the same left-hand side, but with a **zero forcing term** on the right, plays a key role in obtaining solutions of the original equation. Such equations are called **unforced differential equations**, or sometimes **homogeneous equations**, and are the subject of this chapter. Also, for the present, we shall further restrict ourselves to **linear equations with constant coefficients**, which have the form:

Linear unforced differential equations with constant coefficients

(a) First-order:

$$\frac{dx}{dt} + cx = 0 \ (c \text{ constant}).$$

(b) Second-order:

$$\frac{d^2x}{dt^2} + b\frac{dx}{dt} + cx = 0 \ (b, c \text{ constants}).$$

(18.1)

These are called **linear** because there are no squares, products, etc., involving x and its derivatives. Such equations have comparatively simple characteristics. The simplest instance of all is

$$\frac{dx}{dt} = 0.$$

It has solutions $x = A$, where A is any constant. There is therefore **an infinity of solutions**, and we must expect this to be true in more general cases too.

A **solution of a differential equation** is any function $x(t)$ which fits, or satisfies, the equation. This is illustrated in the next two examples.

Example 18.1   *For the differential equation* $dx/dt + 2x = 0$, *verify that* (a) $x = e^{2t}$ *is not a solution,* (b) $x = 2\,e^{-2t}$ *is a solution.*

(a) Test $x = e^{2t}$. Then $dx/dt = 2\,e^{2t}$ and so

$$\frac{dx}{dt} + 2x = 2\,e^{2t} + 2\,e^{2t} = 4\,e^{2t}.$$

This is not zero, so e^{2t} is not a solution.

(b) Test $x = 2\,e^{-2t}$. Then $\dfrac{dx}{dt} = -4\,e^{-2t}$ and so

$$\frac{dx}{dt} + 2x = -4\,e^{-2t} + 4\,e^{-2t} = 0.$$

The zero value is what the equation requires, so $2\,e^{-2t}$ is a solution.

Incidentally, we can confirm in the same way that $x = A\,e^{-2t}$, where A is any constant, is always a solution. We have

$$\frac{dx}{dt} + 2x = -2A\,e^{-2t} + 2A\,e^{-2t} = 0,$$

as it should be. This is the infinity of solutions we were expecting.

Example 18.2   *Verify that the following functions are solutions of the second-order equation* $d^2x/dt^2 + 4x = 0$: (a) $x = \cos 2t$, (b) $x = \sin 2t$, (c) $x = A\cos 2t + B\sin 2t$, *where A and B are any constants.*

Note that 'verify' means 'try out': you are not expected to show how the solutions were obtained.

(a) If $x = \cos 2t$, then $dx/dt = -2\sin 2t$, and $d^2x/dt^2 = -4\cos 2t$. Therefore

$$\frac{d^2x}{dt^2} + 4x = -4\cos 2t + 4\cos 2t = 0$$

as required.

(b) Similarly, if $x = \sin 2t$, then

$$\frac{d^2x}{dt^2} + 4x = -4\sin 2t + 4\sin 2t = 0.$$

Example 18.2 *continued*

(c) Confirmation is straightforward, but the underlying reason why the previous solutions can be combined into a new solution in this way is made clearer by organizing the calculation as follows.

$$\frac{d^2 x}{dt^2} + 4x = \frac{d^2}{dt^2}(A \cos 2t + B \sin 2t) + 4(A \cos 2t + B \sin 2t)$$

$$= A\left(\frac{d^2}{dt^2} \cos 2t + 4 \cos 2t\right) + B\left(\frac{d^2}{dt^2} \sin 2t + 4 \sin 2t\right),$$

by rearranging the terms. We already know that the two bracketed expressions are zero, so the whole expression is zero as required.

The separation of $d^2 x/dt^2 + 4x$ into an 'A' part and a 'B' part in this way is possible only because the equation is *linear*.

18.2 Solving first-order linear unforced equations

Consider the equation

$$\frac{dx}{dt} + cx = 0 \quad (c \text{ a fixed constant}). \tag{18.2}$$

If we write it in the form

$$\frac{dx}{dt} = (-c)x,$$

it can be seen to describe the variation of a quantity $x(t)$ which decays (if c is positive) or grows (if c is negative) at a rate proportional to the amount of x already present. From Section 1.10, we know that exponential functions have this property. We shall therefore test for solutions of the form

$$x(t) = A e^{mt} \tag{18.3}$$

where A and m are unknown constants which we shall try to adjust to fit the equation. From (18.3),

$$\frac{dx}{dt} + cx = Am e^{mt} + cA e^{mt} = A(m + c) e^{mt}.$$

This quantity must be *zero for all values of t in* order to fit the differential equation (18.2). Ignoring the possibility $A = 0$, which gives us the so-called **trivial solution** $x(t) = 0$, we must have

$$m = -c,$$

and in that case it does not matter what value is given to A. We have therefore found a collection of solutions $x(t) = A e^{-ct}$, where A is an arbitrary constant. It can be proved that there are no other solutions, and so we call the solutions we have found the **general solution** of the equation.

The general solution of

$$\frac{dx}{dt} + cx = 0$$

where c is a given constant, is

$$x = A\,e^{-ct},$$

where A is any constant.

(18.4)

Example 18.3   *Find the general solution of* $dx/dt - 4x = 0$.

We will rework the theory. Look for solutions of the form $x = A\,e^{mt}$:

$$\frac{dx}{dt} - 4x = Am\,e^{mt} - 4A\,e^{mt} = A\,e^{mt}(m - 4).$$

This is zero for all time if $m = 4$, whatever the value of A. Therefore the general solution (which includes the trivial one mentioned above) is

$$x = A\,e^{4t}, \quad \text{with } A \text{ an arbitrary constant.}$$

Figure 18.2 depicts several of these solutions, corresponding to various values of the arbitrary constant A.

Each value of A gives a different curve, and these **solution curves** fill the whole plane. Also the curves do not cross, so there is one and only one curve through every point. This corresponds to the fact that the slope dx/dt has one and only one value at every point, namely the value prescribed by the differential equation $dx/dt = 4x$ taken at the point. This is all strong evidence that we have found *all the solutions*. More is said about the graphical way of understanding differential equations in Chapters 22 and 23.

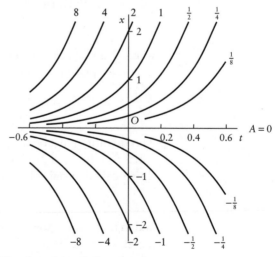

Fig. 18.2 The values of A are indicated on the curves.

Example 18.4 *Find all the solutions of* $3\dfrac{dx}{dt} + 2x = 0.$

We could carry out the full calculation as in the previous example. However, if instead we want to quote the *formula*, (18.4), we must first write the equation in the form

$$\frac{dx}{dt} + \tfrac{2}{3}x = 0.$$

Therefore $c = \tfrac{2}{3}$ (not 2), and the general solution is

$$x = A\,e^{-\frac{2}{3}t}, \quad \text{with } A \text{ any constant.}$$

It is worth while to memorize the formula (18.4).

In practical cases we do not usually need all the solutions, but only the one which satisfies some further condition of the problem. Frequently the condition supplied describes the condition prevailing at the start of the action, or at some other time, as in the following.

Example 18.5 *Find the solution of* $\dfrac{dx}{dt} - 4x = 0$ *for which* $x = 2$ *when* $t = 1$.

Other ways of saying this are 'find the solution curve which passes through the point $(1, 2)$', or 'find a solution $x(t)$ so that $x(1) = 2$'.

From Example 18.3, all the possible solutions are given by

$$x = A\,e^{4t}.$$

Since $x = 2$ when $t = 1$, we must have $2 = A\,e^4$. Therefore

$$A = 2\,e^{-4}$$

and the single solution picked out is

$$x = (2\,e^{-4})\,e^{4t} = 2\,e^{4(t-1)}.$$

An extra condition of this type is called an **initial condition**. It describes the **state of the system** at a given time. The differential equation together with its initial condition is called an **initial-value problem**.

Initial-value problem, first-order equation

(a) Differential equation: $\dfrac{dx}{dt} + cx = 0.$

(b) Initial condition: $x = x_0$ at $t = t_0$

(or $x(t_0) = x_0$), with x_0 and t_0 specified. (18.5)

18.3 Solving second-order linear unforced equations

For second-order differential equations of the type (18.1b), we use a similar technique.

Example 18.6 *Find some solutions of the equation*

$$\frac{d^2x}{dt^2} + \frac{dx}{dt} - 2x = 0.$$

We will look first for absolutely basic solutions. Test whether there are any solutions of the form $x(t) = e^{mt}$, where m is constant. Because $dx/dt = m\,e^{mt}$ and $d^2x/dt^2 = m^2\,e^{mt}$, we have

$$\frac{d^2x}{dt^2} + \frac{dx}{dt} - 2x = m^2\,e^{mt} + m\,e^{mt} - 2\,e^{mt}$$
$$= e^{mt}(m^2 + m - 2).$$

This is zero for all time if $m^2 + m - 2 = 0$, that is if

$$m = 1 \text{ or } -2.$$

This gives us two solutions, namely

$$x(t) = e^t \quad \text{and} \quad x(t) = e^{-2t}.$$

From this **basis**, we can obtain more solutions. Guided by Example 18.2c, we show that also

$$x(t) = A\,e^t + B\,e^{-2t},$$

where A and B are arbitrary constants, is a solution. By substituting into the equation and sorting the terms into those with coefficient A and those with coefficient B, we obtain

$$A\left(\frac{d^2}{dt^2}e^t + \frac{d}{dt}e^t - 2\,e^t\right) + B\left(\frac{d^2}{dt^2}e^{-2t} + \frac{d}{dt}e^{-2t} - 2\,e^{-2t}\right) = 0,$$

because e^t and e^{-2t} are known already to be solutions; so both of the bracketed expressions are zero.

This is the principle, but consider now the general case

$$\frac{d^2x}{dt^2} + b\frac{dx}{dt} + cx = 0.$$

Look for solutions of the form $x = e^{mt}$. Then

$$\frac{d^2x}{dt^2} + b\frac{dx}{dt} + cx = e^{mt}(m^2 + bm + c).$$

This will be zero for all t, as required by the differential equation, if

$$m^2 + bm + c = 0, \tag{18.6}$$

which is called the **characteristic equation**. Being quadratic, it may have two real solutions, exactly one real solution, or two complex solutions, depending on the coefficients. Consider the real cases first:

Roots m_1 and m_2 of the characteristic equation real and different
In this case,

$$x(t) = e^{m_1 t} \quad \text{and} \quad x(t) = e^{m_2 t}$$

are solutions of the differential equation, and from these we can construct a whole **family of solutions**

$$x(t) = A\, e^{m_1 t} + B\, e^{m_2 t},$$

where A and B are arbitrary. It can be proved that there are no more solutions: this gives the **general solution**. The pair of functions $(e^{m_1 t}, e^{m_2 t})$ is called a **basis** for the general solution.

Characteristic equation: unequal real roots

$$\frac{d^2 x}{dt^2} + b\frac{dx}{dt} + cx = 0; \text{ roots } m_1 \text{ and } m_2 \text{ of}$$

$m^2 + bm + c = 0$ real and different.

Basis of solutions: $e^{m_1 t}, e^{m_2 t}$.

General solution: $A\, e^{m_1 t} + B\, e^{m_2 t}$ (A, B arbitrary).   (18.7)

Example 18.7   *Find the general solution of* $2\dfrac{d^2 x}{dt^2} - \dfrac{dx}{dt} - x = 0.$

To correspond with the standard form, (18.7), we should have to write the equation in the form $d^2x/dt^2 - \frac{1}{2}dx/dt - \frac{1}{2}x = 0$, but there is no need to do this if we directly test for solutions of the form $x = e^{mt}$. The characteristic equation then takes the form $2m^2 - m - 1 = 0$, or $(2m + 1)(m - 1) = 0$, so that $m_1 = -\frac{1}{2}, m_2 = 1$. Therefore the basis for the general solution is the solution pair $(e^{-\frac{1}{2}t}, e^t)$, and the general solution is

$$x(t) = A\, e^{-\frac{1}{2}t} + B\, e^t, \quad A \text{ and } B \text{ arbitrary.}$$

Roots m_1 and m_2 of the characteristic equation are equal
Suppose that $m_1 = m_2 = m_0$, say. We have then only one function for our basis instead of two, and we might expect the general solution to be $A\, e^{m_0 t}$. However, all we know is that there is essentially only one solution of the form e^{mt} (ignoring simple multiples of e^{mt}), but we shall see in the next example that there is also a solution which is not of this form, namely

$$x(t) = t\, e^{m_0 t}. \tag{18.8}$$

We might therefore think there will be no end to it: if $t\, e^{m_0 t}$ is a solution, then why not $t^2\, e^{m_0 t}$, or some function of great complication? However, it can be proved that **every second-order linear differential equation has exactly two linearly independent solutions** (i.e. they are not just multiples of each other); also that **these form a basis of solutions**: we do not need any others to construct the most general solution. Formally:

> **Basis and general solution of**
>
> $$\frac{d^2x}{dt^2} + b\frac{dx}{dt} + cx = 0$$
>
> (a) There exist two linearly independent solutions.
> (b) If $u(t)$ and $v(t)$ are any two linearly independent solutions, these form a basis for the general solution; that is to say, the general solution is given by
>
> $$x(t) = Au(t) + Bv(t),$$
>
> where A and B are arbitrary constants.   **(18.9)**

Example 18.8   *Find the general solution of* $\dfrac{d^2x}{dt^2} + 4\dfrac{dx}{dt} + 4x = 0.$

The characteristic equation, formed by substituting $x(t) = e^{mt}$, is

$$m^2 + 4m + 4 = (m + 2)^2 = 0,$$

and the only value of m that we find is $m = -2$. It corresponds to the basic solution e^{-2t}.

The theorem (18.9) guarantees there is another independent solution, and it does not matter how we find it. Test the truth of (18.8), which proposes an independent solution having the form

$$x(t) = t\,e^{-2t}.$$

Then

$$\frac{dx}{dt} = (1 - 2t)\,e^{-2t},$$

and

$$\frac{d^2x}{dt^2} = (-4 + 4t)\,e^{-2t}.$$

Therefore

$$\frac{d^2x}{dt^2} + 4\frac{dx}{dt} + 4x = [(-4 + 4t) + 4(1 - 2t) + 4t]\,e^{-2t},$$

which is zero, so $x(t) = t\,e^{-2t}$ is a second solution, and it is independent of the first. By (18.9), the solution basis is therefore

$$(e^{-2t}, t\,e^{-2t}),$$

and the general solution is

$$x(t) = A\,e^{-2t} + Bt\,e^{-2t}, \quad A \text{ and } B \text{ arbitrary.}$$

The second solution always takes the same form (see Problem 18.8):

Characteristic equation: coincident roots

If $\dfrac{d^2x}{dt^2} + b\dfrac{dx}{dt} + cx = 0$, in which $b^2 - 4c = 0$ (for coincident

roots), and m_0 is the single solution of the characteristic
equation $m^2 + bm + c = 0$, then the solution basis is
$(e^{m_0t}, t\,e^{m_0t})$ and the general solution is $x(t) = A\,e^{m_0t} + Bt\,e^{m_0t}$
(A and B arbitrary constants). **(18.10)**

18.4 Complex roots of the characteristic equation

If $b^2 < 4c$, the roots m_1 and m_2 of the characteristic equation $m^2 + bm + c = 0$ for the differential equation $d^2x/dt^2 + b\,dx/dt + cx = 0$ are complex. Since they are roots of a quadratic equation, they must be **complex conjugate**, so put

$$m_1 = \alpha + i\beta, \qquad m_2 = \alpha - i\beta,$$

where α and β are real numbers. The corresponding functions

$$e^{(\alpha+i\beta)t} \quad \text{and} \quad e^{(\alpha-i\beta)t} \tag{18.11}$$

are genuine solutions of the differential equation. They are complex functions, so we call (18.11) a **complex basis** for solutions of the differential equation. If we are interested in complex as well as real solutions, then we can allow the arbitrary constants A and B to be complex as well, in an all-inclusive **general complex solution**

$$x(t) = A\,e^{(\alpha+i\beta)t} + B\,e^{(\alpha-i\beta)t}.$$

Suppose, however, that we want the general solution to consist only of real functions. Then a **basis for real solutions** can be got from (18.11) in the following way. By (6.8)

$$e^{(\alpha+i\beta)t} = e^{\alpha t}\,e^{i\beta t} = e^{\alpha t}\cos\beta t + i\,e^{\alpha t}\sin\beta t.$$

This function solves the differential equation, so its real and imaginary parts separately must also solve it. Therefore

$$(e^{\alpha t}\cos\beta t, e^{\alpha t}\sin\beta t)$$

is a **real basis for the general** (real) **solution**

$$x(t) = A\,e^{\alpha t}\cos\beta t + B\,e^{\alpha t}\sin\beta t, \tag{18.12}$$

where A and B are arbitrary (but real, of course). The second complex solution, $e^{(\alpha-i\beta)t}$, has the basis, $(e^{\alpha t}\cos\beta t, -e^{\alpha t}\cos\beta t)$, which leads to the same family of solutions, so we get nothing new by considering it.

Equation (18.12) can be written in a different form. Using the identity (1.19), we have

$$A \cos \beta t + B \sin \beta t = C \cos(\beta t + \varphi),$$

where C and φ are constants related to A and B. Therefore (18.12) can be written

$$x(t) = C \, e^{\alpha t} \cos(\beta t + \varphi).$$

Since A and B are arbitrary, so are C and φ.

Example 18.9 *Find the general solution of*

$$\frac{d^2 x}{dt^2} + 4x = 0.$$

The characteristic equation is $m^2 + 4 = 0$. Its solutions are $m = \pm 2i$. Therefore the complex solution basis is (e^{2it}, e^{-2it}). But

$$e^{2it} = \cos 2t + i \sin 2t,$$

and the real and imaginary parts give a basis for the real solutions:

$$(\cos 2t, \sin 2t).$$

Therefore the general solution is

$$x(t) = A \cos 2t + B \sin 2t \quad (A, B \text{ arbitrary}).$$

Example 18.10 *Find the general solution of*

$$\frac{d^2 x}{dt^2} + 2\frac{dx}{dt} + 2x = 0.$$

Setting $x = e^{mt}$ gives the characteristic equation $m^2 + 2m + 2 = 0$, so that $m = -1 \pm i$. Therefore

$$(e^{(-1+i)t}, e^{(-1-i)t})$$

is a basis for complex solutions. But

$$e^{(-1+i)t} = e^{-t}(\cos t + i \sin t),$$

whose real and imaginary parts are

$$e^{-t} \cos t, \quad e^{-t} \sin t.$$

These form the basis for the real solutions. The general solution is

$$x(t) = A \, e^{-t} \cos t + B \, e^{-t} \sin t.$$

If we chose instead to take the real and imaginary parts of $e^{(-1-i)t}$, we would obtain $(e^{-t} \cos t, -e^{-t} \sin t)$ as a basis. The minus sign will be absorbed into the *arbitrary* constant B: no new solutions appear.

The general solution method can be summed up as follows:

Characteristic equation: complex roots

$$\frac{d^2x}{dt^2} + b\frac{dx}{dt} + cx = 0, \text{when } m^2 + bm + c = 0 \text{ has complex roots}$$

$m_1, m_2 = \alpha \pm i\beta$ (i.e. $b^2 < 4c$).

Complex basis: $e^{(\alpha+i\beta)t}$, $e^{(\alpha-i\beta)t}$.

Real basis: $e^{\alpha t}\cos\beta t$, $e^{\alpha t}\sin\beta t$.

General solution:

(a) $x(t) = A\,e^{\alpha t}\cos\beta t + B\,e^{\alpha t}\sin\beta t$
     (A and B arbitrary);

     or

(b) $x(t) = C\,e^{\alpha t}\cos(\beta t + \phi)$ (C and ϕ arbitrary).   **(18.13)**

A very important case is when $b = 0$ and $c > 0$, illustrated by Example 18.9. In that case, $\alpha = 0$. In conventional notation, putting $c = \omega^2$, we obtain the following result:

Characteristic equation: special case

$$\frac{d^2x}{dt^2} + \omega^2 x = 0$$

Characteristic equation: $m^2 + \omega^2 = 0$; $m_1, m_2 = \pm i\omega$.

Complex basis: $e^{i\omega t}$, $e^{-i\omega t}$.

Real basis: $\cos\omega t$, $\sin\omega t$.

General solution: (a) $x(t) = A\cos\omega t + B\sin\omega t$, or
(b) $x(t) = C\cos(\omega t + \phi)$.   **(18.14)**

In the special case (18.14), the alternative solution form

$$x(t) = C\cos(\omega t + \phi)$$

shows that the solutions oscillate regularly, swinging above and below the t axis to an extent governed by the amplitude C. In the general case (18.13),

$$x(t) = C\,e^{\alpha t}\cos(\beta t + \phi),$$

the solutions oscillate, but the amplitude is governed by the factor $C\,e^{\alpha t}$. If α is positive, the oscillation constantly grows; if α is negative, it dies away to zero. This is fully discussed in Chapter 20, but Fig. 18.3 shows a particular case where α is negative.

The **damped unforced linear oscillator** is the simplest linear model of an oscillating mechanical or electrical system which has a small amount of friction or some other form of energy-loss mechanism (see Chapter 20 for a full discussion). In a customary notation the equation is

$$\frac{d^2x}{dt^2} + 2k\frac{dx}{dt} + \omega^2 x = 0.$$

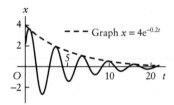

Fig. 18.3 Graph of $x(t) = 4\,e^{-0.2t}\cos(2t - 1)$.

The term $2k \, dx/dt$ expresses the energy-absorbing property. Assume $k^2 < \omega^2$.

The characteristic equation is $m^2 + 2km + \omega^2 = 0$, so that

$$m = -k \pm (k^2 - \omega^2)^{\frac{1}{2}} = -k \pm \mathrm{i}(\omega^2 - k^2)^{\frac{1}{2}},$$

since $k^2 < \omega^2$. From (18.13), $\alpha = -k$ and $\beta = (\omega^2 - k^2)^{\frac{1}{2}}$, so finally:

Damped linear oscillator

$$\frac{d^2x}{dt^2} + 2k\frac{dx}{dt} + \omega^2 x = 0 \text{ where } k^2 < \omega^2$$

General solution:

(a) $x(t) = A \, e^{-kt} \cos(\omega^2 - k^2)^{\frac{1}{2}}t + B \, e^{-kt} \sin(\omega^2 - k^2)^{\frac{1}{2}}t$

  (A and B arbitrary constants); or

(b) $x(t) = C \, e^{-kt} \cos[(\omega^2 - k^2)^{\frac{1}{2}}t + \phi]$

  (C and ϕ arbitrary).

  (18.15)

18.5 Initial conditions for second-order equations

The general solution of a second-order differential equation involves two arbitrary constants, and the solutions are therefore an order of magnitude more numerous than in the first-order case. Unlike the first-order case, the solution curves may cross – in fact, there is an infinite number of solution curves through any point on the (x, t) plane, as indicated in Fig. 18.4a.

To pick out a particular solution, we need to determine the two arbitrary constants. Two pieces of information are necessary. These may consist of **two initial conditions**, conditions which define the **state of the system at some starting time** t_0: the values of $x(t)$ and the slope dx/dt at $t = t_0$ are given (see Fig. 18.4b). For example, the equation $d^2x/dt^2 + \omega_0^2 x = 0$ describes the oscillations of a particle on a spring; the initial conditions tell us its position and velocity (i.e. its **state**) when it starts off. We then have an **initial-value problem**:

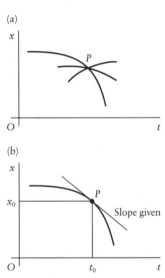

(a)

(b)

Fig. 18.4 (a) An infinite number of curves pass through each point. (b) Selection of a solution given P and the slope at P.

Initial-value problem

(i) Equation: $\dfrac{d^2x}{dt^2} + b\dfrac{dx}{dt} + cx = 0.$

(ii) Initial conditions:

$$x = x_0 \text{ and } \frac{dx}{dt} = x_1 \text{ at } t = t_0,$$

which may be expressed alternatively as

$$x(t_0) = x_0, \qquad x'(t_0) = x_1,$$

where x_0 and x_1 are given.

  (18.16)

Example 18.11 *Find the solution of* $d^2x/dt^2 + 4x = 0$ *for which* $x = 1$ *and* $dx/dt = 2$ *at* $t = 0$ *(i.e.* $x(0) = 1$, $x'(0) = 2$).

First we need all the solutions. From Example 18.9, these are $x(t) = A \cos 2t + B \sin 2t$, where A and B may take any values. Since $x = 1$ at $t = 0$,

$$1 = A + 0, \quad \text{so} \quad A = 1.$$

For the other condition, we first need $x'(t)$ in general:

$$x'(t) = -2A \sin 2t + 2B \cos 2t.$$

At $t = 0$, we are given that $x'(t) = 2$, so the last equation becomes

$$2 = 0 + 2B, \quad \text{or} \quad B = 1.$$

The required solution is therefore $x(t) = \cos 2t + \sin 2t$.

Problems

(For the 'dash' notation $x'(t) = dx/dt$ etc., see (4.1).)

18.1 Say which of the following equations are linear, unforced, with constant coefficients (i.e. can be rearranged to conform with (18.1a)).
(a) $x' = 3t$;   (b) $x' = \frac{1}{2}x$;   (c) $x' + tx = 0$;
(d) $3x' - 2x^2 = 0$;   (e) $x' - x = 0$;   (f) $x' = 0$;
(g) $\dfrac{x'}{x^2} = 3$;   (h) $\dfrac{dy}{dx} + \frac{1}{2}y = 1$;   (i) $\dfrac{1}{y}\dfrac{dy}{dx} = 2$;
(j) $L\dfrac{dI}{dt} + RI = 0$;   (k) $\dfrac{v' + v + v^2}{v' - v + v^2} = 1$.

18.2 Write down all the solutions of the following equations. Check one or two of them by substitution into the differential equation.
(a) $x' + 5x = 0$;   (b) $x' - \frac{1}{2}x = 0$;
(c) $x' - x = 0$;   (d) $x' + 3x = 0$;
(e) $3x' + 4x = 0$;   (f) $x' = 2x$;   (g) $x' = 3x$;
(h) $x'/x = -3$;   (i) $(x' + 1)/(x + 1) = 1$.

18.3 Solve the following initial-value problems.
(a) $x' + 2x = 0$, $x = 3$ when $t = 0$;
(b) $3x' - x = 0$, $x = 1$ when $t = 1$;
(c) $y' - 2y = 0$, $y = 2$ when $x = -3$;
(d) $x' + x = 0$, $x(-1) = 10$;
(e) $2y' - 3y = 0$, $y(0) = 1$;
(f) Find the curve whose slope at any point (x, y) is equal to $5x$, and which passes through the point $(1, -2)$.

18.4 Suppose that the generator in Fig. 18.1 is short-circuited and cut out at a moment when the current in the circuit is I_0. Find an expression for the current subsequently.

Show that the ratio L/R provides a measure of the time it takes for the current to die away.

18.5 A radioactive element disintegrates at a rate proportional to the amount of the original element still remaining. Show that if $A(t)$ represents the activity of the element at time t, then

$$\dfrac{dA}{dt} + kA = 0,$$

where k is a positive constant.
(a) Solve the initial-value problem for A if $A = A_0$ (given) at time $t = 0$.
(b) The time taken for the activity to drop to half of the starting value is called the half-life period. For uranium-232, it is found that 17.5% has decayed after 20 years. Show that its half-life period is about 72 years.

18.6 (*Götterdämmerung*). Once upon a time, rabbits in Elysium reached maturity instantly and bred with a birthrate of 20 rabbits per year per couple. No rabbit ever died. At the start of the experiment Zeus released 50 male and 50 female rabbits.

By treating the number of rabbits as a continuously varying quantity and considering the number born in a short time δt, construct a differential equation and then an initial-value problem for $R(t)$, the rabbit population. Find how many rabbits there were at the end of Year 4.

Appalled by this result and assisted by Pluto, Zeus launched another similar experiment, in which any rabbit was allowed to live for one year only. Construct the differential equation for the population. Did this alleviate the situation appreciably?

18.7 Obtain all solutions of the following equations. (The characteristic equations all have real roots, not necessarily distinct.)

(a) $x'' - 3x' + 2x = 0$;   (b) $x'' + x' - 2x = 0$;
(c) $x'' - x = 0$;   (d) $x'' - 4x = 0$;
(e) $3x'' - \frac{1}{4}x = 0$;   (f) $x'' - 9x = 0$;
(g) $x'' + 2x' - x = 0$;   (h) $x'' - 2x' - 2x = 0$;
(i) $2x'' + 2x' - x = 0$;   (j) $3x'' - x' - 2x = 0$;
(k) $x'' + 4x' + 4x = 0$;   (l) $x'' + 6x' + 9x = 0$;
(m) $4x'' + 4x' + x = 0$;   (n) $x'' = 0$.

18.8 Verify that, when the characteristic equation corresponding to $x'' + bx' + cx = 0$ has coincident roots, $m_1 = m_2 = m_0$, say, then the function $x(t) = t\,e^{m_0 t}$ provides a second solution for the basis of the general solution. (For coincident roots, $b^2 = 4c$.)

18.9 Solve the following initial-value problems.

(a) $x'' - 4x = 0$, $x(0) = 1, x'(0) = 0$;
(b) $x'' + x' - 2x = 0$, $x(0) = 0, x'(0) = 2$;
(c) $y'' - 4y' + 4y = 0$, $y(0) = 0, y'(0) = -1$;
(d) $y'' + 2y' + y = 0$, $y(1) = 0, y'(1) = 1$;
(e) $x'' - 9x = 0$, $x(1) = 1, x'(1) = 1$;
(f) $x'' - 4x' = 0$, $x(1) = 1, x'(1) = 0$.

18.10 Obtain all solutions of the following equations. (The roots of the characteristic equations are complex.)

(a) $x'' + x = 0$;   (b) $x'' + 9x = 0$;
(c) $x'' + \frac{1}{4}x = 0$;   (d) $x'' + \omega_0^2 x = 0$;
(e) $x'' + 2x' + 2x = 0$;   (f) $y'' - 2y' + 2y = 0$;
(g) $y'' + y' + y = 0$;   (h) $2x'' + 2x' + x = 0$;
(i) $3x'' + 4x' + 2x = 0$;   (j) $3x'' - 4x' + 2x = 0$.

18.11 Solve the following initial-value problems.

(a) $x'' + x = 0$, $x(0) = 0, x'(0) = 1$;
(b) $x'' + 4x = 0$, $x(0) = 1, x'(0) = 0$;
(c) $x'' + \omega_0^2 x = 0$, $x(0) = a, x'(0) = b$;
(d) $x'' + 2kx' + x = 0$, $x(0) = 0, x'(0) = b$ for the cases $k^2 > 1$, $k^2 < 1$, and $k^2 = 1$.

(Use the A, B form: finding the constants C and ϕ in (18.14b) for an initial-value problem can be comparatively difficult.)

18.12 The approximate equation for small swings of a pendulum is

$$\frac{d^2\theta}{dt^2} + \frac{g}{l}\theta = 0$$

where θ is the inclination from the vertical (in radians), l is the length, and g the gravitational acceleration. The pendulum is held still at an angle α, and is then passively released. Find the subsequent motion.

18.13 The pendulum in Problem 18.12 is hanging at rest; then the bob is given a small velocity v in the direction of θ increasing. Find the subsequent motion.

18.14 If there is a little friction in the pendulum of Problem 18.12, the equation of motion takes the form

$$\frac{d^2\theta}{dt^2} + K\frac{d\theta}{dt} + \frac{g}{l}\theta = 0,$$

where K is an additional positive constant which takes account of the friction (assumed to be proportional to the angular velocity). In a particular case (SI units), $g = 9.7$, $l = 20$, $K = 0.066$. The pendulum is at rest at first, hanging freely. It is then pushed so as to give the bob a velocity of 1 metre per second. Find the subsequent motion.

18.15 Consider the third-order differential equation

$$\frac{d^3 y}{dx^3} - y = 0.$$

Proceed by analogy with the method of Section 18.3: by substituting $y = e^{mx}$, and obtaining a characteristic equation for m (a cubic equation), find three distinct basic solutions of this type. By introducing arbitrary constants A, B, C, find as wide a variety of solutions as you can (in fact, this is the general solution).

18.16 By proceeding as in Problem 18.15, find a wide variety of solutions of the equation

$$\frac{d^3 y}{dx^3} + y = 0.$$

18.17 By proceeding with the equation

$$\frac{d^4 y}{dx^4} - y = 0$$

as in Problem 18.15, obtain the collection of solutions

$$y(x) = A\,e^x + B\,e^{-x} + C\cos x + D\sin x,$$

where A, B, C, D are arbitrary constants.

18.18 A tapered concrete column of height H metres is to support a statue of mass M (i.e. weight Mg force units, where g is the gravitational acceleration) at the top. Pressure (force per unit area) may not exceed P. Show that the most economical construction for the column is for its cross-sectional area $A(y)$, where y is distance above the ground, to satisfy the equation

$$A(y) = \frac{Mg}{P} + \frac{\rho g}{P}\int_y^H A(u)\,du,$$

where ρ is the density of concrete. By differentiating this expression (see Section 15.10), obtain a differential equation for $A(y)$, and an initial condition for the equation, and solve it.

19 Forced linear differential equations

CONTENTS

19.1 Particular solutions for standard forcing terms

Consider the equation

$$\frac{d^2x}{dt^2} + b\frac{dx}{dt} + cx = f(t).$$

The function $f(t)$ is called the **forcing term**; it represents physically the external **input** to the physical system that the equation describes, and the system will respond with an **output** $x(t)$ which depends on the input $f(t)$.

If $f(t)$ is an **exponential function** $K\,e^{\alpha t}$, a **sine** or **cosine function** $K\sin\beta t$ or $K\cos\beta t$, or a **polynomial**, then we can find an individual **particular solution** by trial.

Example 19.1 *Find a particular solution of*

$$\frac{d^2x}{dt^2} + \frac{dx}{dt} - 2x = 3\,e^{2t}.$$

Try for a solution containing the same exponential as that on the right-hand side of the equation:

$$x(t) = p\,e^{2t},$$

where p is some constant – not an arbitrary constant, but one whose value we shall settle by substitution: only one value will do. Then

$$\frac{dx}{dt} = 2p\,e^{2t}, \quad \text{and} \quad \frac{d^2x}{dt^2} = 4p\,e^{2t},$$

so that

$$\frac{d^2x}{dt^2} + \frac{dx}{dt} - 2x = 4p\,e^{2t} + 2p\,e^{2t} - 2p\,e^{2t}$$
$$= e^{2t}(4+2-2)p = 4p\,e^{2t}.$$

Example 19.1 *continued*

This must equal the given right-hand side, $3 \, e^{2t}$ *for all values of t*, which is only possible if $4p = 3$, or

$p = \frac{3}{4}$.

Therefore one particular solution is

$x(t) = \frac{3}{4} \, e^{2t}$.

There are many other solutions, as we shall see in Section 19.4, but they are all based on this particular solution.

Example 19.2   *Find a particular solution of $\dfrac{d^2x}{dt^2} + 4x = 2 \cos 3t$.*

Guess that there might be a solution of the form

$x = p \cos 3t$.

Then

$\dfrac{dx}{dt} = -3p \sin 3t, \quad \text{and} \quad \dfrac{d^2x}{dt^2} = -9p \cos 3t,$

so that

$\dfrac{d^2x}{dt^2} + 4x = -9p \cos 3t + 4p \cos 3t = -5p \cos 3t.$

This must be the same as the right-hand side of the equation in order for the guessed function to be a solution, so

$-5p \cos 3t = 2 \cos 3t.$

Therefore $p = -\frac{2}{5}$, and the required solution is

$x(t) = -\frac{2}{5} \cos 3t.$

In most cases when the right-hand side is a sine or cosine it will not work so simply, as is illustrated in the following example.

Example 19.3   *Find a particular solution of*

$\dfrac{d^2x}{dt^2} + \dfrac{dx}{dt} - 2x = 2 \cos 3t.$

The form $p \cos 3t$ cannot be made to fit this equation because the dx/dt term on the left produces a $\sin 3t$ term, making the left and right sides impossible to match for all t. Try instead

$x = p \cos 3t + q \sin 3t.$

Then

$\dfrac{dx}{dt} = -3p \sin 3t + 3q \cos 3t,$

and

$\dfrac{d^2x}{dt^2} = -9p \cos 3t - 9q \sin 3t.$

↗

Example 19.3 *continued*

Therefore

$$\frac{d^2x}{dt^2} + \frac{dx}{dt} - 2x = (-9p + 3q - 2p)\cos 3t + (-9q - 3p - 2q)\sin 3t$$
$$= (-11p + 3q)\cos 3t + (-3p - 11q)\sin 3t$$
$$= 2\cos 3t$$

which must match the right-hand side of the equation for all t.

The only way to satisfy this condition is to require both

$$-11p + 3q = 2, \qquad -3p - 11q = 0.$$

The solution of these two simultaneous equations for p and q is

$$p = -\tfrac{11}{65}, \qquad q = \tfrac{3}{65}.$$

Therefore a particular solution is

$$x = -\tfrac{11}{65}\cos 3t + \tfrac{3}{65}\sin 3t.$$

It will be necessary in nearly all cases to take both sine and cosine terms into account.

The case when $f(t)$ is a **constant** often occurs.

Example 19.4 *Find a particular solution of* $\dfrac{d^2x}{dt^2} - 2\dfrac{dx}{dt} + 4x = 3.$

Test whether there is a constant solution

$$x(t) = p \quad \text{(a constant)}.$$

By substituting this in the differential equation, we get

$$0 + 0 + 4p = 3,$$

so that $p = \tfrac{3}{4}$, and the particular solution is just $x(t) = \tfrac{3}{4}$, which is obvious after it has been worked out.

If the right-hand side is a **polynomial** then the solution will be a polynomial.

Example 19.5 *Find a solution of* $\dfrac{d^2x}{dt^2} - \dfrac{dx}{dt} + x = 3 + 2t^2.$

Try a solution of the form $x(t) = p + qt + rt^2$, where p, q, and r are constants. It is normally necessary to try a polynomial of the same degree as the forcing term, which in this case has degree 2, and to include all the lower-degree terms in the trial solution.

Since

$$\frac{dx}{dt} = q + 2rt, \quad \text{and} \quad \frac{d^2x}{dt^2} = 2r,$$

we must have

$$2r - (q + 2rt) + (p + qt + rt^2) = 3 + 2t^2.$$

Example 19.5 *continued*

Match up the coefficients of the three powers of t; we find that

$$r = 2, \qquad -2r + q = 0, \qquad 2r - q + p = 3.$$

The equations are easy to solve and lead to the solution

$$x(t) = 3 + 4t + 2t^2.$$

Particular solutions of

$$\frac{d^2x}{dt^2} + b\frac{dx}{dt} + cx = f(t) \text{ and } \frac{dx}{dt} + cx = f(t)$$

(a) $f(t) = K\,e^{\alpha t}$: try a solution $x(t) = p\,e^{\alpha t}$.
(b) $f(t) = K \cos \beta t$ or $K \sin \beta t$: try a solution

$$x(t) = p \cos \beta t + q \sin \beta t.$$

(c) $f(t)$ is a polynomial of degree N: try a polynomial
of the same degree, with all its terms present.

(19.1)

There are **exceptional cases** where these substitutions have to be modified. For example, $d^2x/dt^2 = t$ has a polynomial solution of degree 3, not degree 1. These cases are treated in Section 19.3.

If the forcing term on the right-hand side consists of the sum of several constituent terms, then obtain a particular solution for each one, and add them, as in the following example.

Example 19.6   *Obtain a particular solution of*

$$\frac{d^2x}{dt^2} + 4x = 1 + e^{-t}.$$

Solve $\dfrac{d^2x_1}{dt^2} + 4x_1 = 1$ for $x_1(t)$ and $\dfrac{d^2x_2}{dt^2} + 4x_2 = e^{-t}$ for $x_2(t)$; then

$$x(t) = x_1(t) + x_2(t)$$

will be a particular solution of the original equation.

For $x_1(t)$, try for a constant solution $x_1(t) = p$: it is found that $p = \frac{1}{4}$, so $x_1(t) = \frac{1}{4}$.

For $x_2(t)$, try $x_2(t) = q\,e^{-t}$ (following the method of Example 19.1). The substitution gives $q\,e^{-t} + 4q\,e^{-t} = e^{-t}$, so that $q = \frac{1}{5}$, and the solution is $x_2(t) = \frac{1}{5}\,e^{-t}$.

Therefore, a particular solution $x(t)$ of the original equation is

$$x(t) = x_1(t) + x_2(t) = \tfrac{1}{4} + \tfrac{1}{5}\,e^{-t}.$$

The method just described is another consequence of the linearity of the class of equations considered. It is also called the **superposition principle**.

These methods apply equally to **first-order equations**.

Example 19.7 *Obtain a particular solution of $\dfrac{dx}{dt} + x = 3 \cos 2t$.*

Remembering Example 19.3, we expect the solution will have to contain both cosine and sine terms, so try

$$x(t) = p \cos 2t + q \sin 2t.$$

The substitution gives $(p + 2q) \cos 2t + (-2p + q) \sin 2t = 3 \cos 2t$, so that $p = \frac{3}{5}$, $q = \frac{6}{5}$, and the solution is

$$x(t) = \tfrac{3}{5} \cos 2t + \tfrac{6}{5} \sin 2t.$$

19.2 Harmonic forcing term by using complex solutions

In Example 19.3, we solved a second-order equation with the term $\cos 2t$ on the right by choosing constants p and q so that the expression $p \cos 2t + q \sin 2t$ would fit the equation. We shall explain another important method for obtaining solutions, which derives the required real solutions from complex solutions of a related equation.

First of all, consider the general differential equation

$$\frac{d^2X}{dt^2} + b\frac{dX}{dt} + cX = a\,e^{i\beta t}, \tag{19.2}$$

where b, c, a, and β are real constants, and i is the complex element. Since the forcing term is an exponential, we shall test for a particular solution of (19.2) having the form

$$X(t) = P\,e^{i\beta t}$$

as in Example 19.1 (but this time we must expect that P will be a complex constant).

Example 19.8 *Find a particular (complex) solution of the complex differential equation $\dfrac{d^2X}{dt^2} + \dfrac{dX}{dt} + X = 3\,e^{2it}$.*

Look for a solution of the form $X(t) = P e^{2it}$. To find P, substitute this expression into the left-hand side of the differential equation:

$$(2i)^2 P e^{2it} + (2i)P e^{2it} + P e^{2it} = P(-4 + 2i + 1)\,e^{2it} = P(-3 + 2i)\,e^{2it}.$$

This must be the same as the right-hand side of the equation, $3\,e^{2it}$, for all values of t. Therefore, $P(-3 + 2i) = 3$, so

$$P = \frac{3}{-3 + 2i} = \frac{3(-3 - 2i)}{(-3)^2 + 2^2} = -\tfrac{3}{13}(3 + 2i).$$

Therefore $X(t) = -\tfrac{3}{15}(3 + 2i)\,e^{2it}$ is a particular solution. When expanded, it becomes

$$X(t) = -\tfrac{3}{15}(3 + 2i)(\cos 2t + i \sin 2t)$$
$$= -\tfrac{3}{15}(3 \cos 2t - 2 \sin 2t) + i[-\tfrac{3}{15}(2 \cos 2t + 3 \sin 2t)].$$

Consider next the real equation for $x(t)$:

$$\frac{d^2x}{dt^2} + b\frac{dx}{dt} + cx = a \cos \beta t, \tag{19.3}$$

where b, c, a, β are all real. We know that

$$\cos \beta t = \text{Re } e^{i\beta t}$$

(see (6.8)). Therefore, if we can find a particular solution $X(t)$ of the complex equation (19.2), its real part will solve the corresponding real equation (19.3).

Example 19.9   *Find a particular solution of the equation*

$$\frac{d^2x}{dt^2} + \frac{dx}{dt} - 2x = 2 \cos 3t.$$

This is the same problem as Example 19.3, reworked so that the methods can be compared. Since $\cos 3t = \text{Re } e^{3it}$, the corresponding complex equation for $X(t)$ is

$$\frac{d^2X}{dt^2} + \frac{dX}{dt} - 2X = 2 e^{3it}.$$

To find a particular solution of this new equation, try $X(t) = P e^{3it}$:

$$\frac{d^2X}{dt^2} + \frac{dX}{dt} - 2X = 9i^2P\, e^{3it} + 3iP\, e^{3it} - 2P\, e^{3it}$$
$$= (9i^2 + 3i - 2)P\, e^{3it} = (-11 + 3i)P\, e^{3it}.$$

This must equal $2 e^{3it}$ for all values of t, so

$$P = \frac{2}{-11 + 3i} = \frac{2(-11 - 3i)}{(-11)^2 + 3^2} = -(\tfrac{11}{65} + \tfrac{3}{65}i).$$

Therefore we have a complex solution of the complex equation:

$$X(t) = -(\tfrac{11}{65} + \tfrac{3}{65}i)\, e^{3it}$$
$$= -(\tfrac{11}{65} + \tfrac{3}{65}i)(\cos 3t + i \sin 3t).$$

For $x(t)$, we require only the real part of this expression:

$$x(t) = \text{Re } X(t)$$
$$= -\tfrac{11}{65} \cos 3t + \tfrac{3}{65} \sin 3t,$$

which is what we obtained in Example 19.3 for the same problem.

In the case when the right-hand side of the equation has the form $a \sin \omega t$, the calculation is the same, but the imaginary part of the complex solution must be extracted instead of the real part. The following example demonstrates also how right-hand sides of the form

$$a\, e^{\alpha t} \cos \beta t, \quad a\, e^{\alpha t} \sin \beta t$$

can be handled in the same way.

Example 19.10 *Find a solution of* $\dfrac{d^2x}{dt^2} + x = e^{-2t}\sin 3t.$

Use the fact that

$$e^{-2t}\sin 3t = \text{Im}(e^{-2t}\,e^{3it}) = \text{Im}\,e^{(-2+3i)t}.$$

Therefore, consider the *corresponding complex equation*

$$\frac{d^2X}{dt^2} + X = e^{(-2+3i)t}.$$

To find a solution, try the form

$$X(t) = P\,e^{(-2+3i)t}.$$

We find in the usual way that

$$(-2+3i)^2 P\,e^{(-2+3i)t} + P\,e^{(-2+3i)t} = e^{(-2+3i)t}$$

for all values of t. Therefore

$$P = \frac{1}{(-2+3i)^2 + 1} = \frac{-1}{4(1+3i)} = -\tfrac{1}{40}(1-3i)$$

and

$$X(t) = -\tfrac{1}{40}(1-3i)\,e^{(-2+3i)t}.$$

If we take the imaginary part of $X(t)$, we obtain a solution of the original equation:

$$x(t) = \text{Im}[-\tfrac{1}{40}(1-3i)(e^{-2t}\,e^{3it})]$$
$$= -\tfrac{1}{40}\,e^{-2t}(-3\cos 3t + \sin 3t).$$

The same result could be obtained by substituting

$$x(t) = p\,e^{-2t}\cos 3t + q\,e^{-2t}\sin 3t,$$

but this would be a very laborious and error-prone process.

The method is particularly advantageous when the coefficients are general constants. The following equation will be important in Chapter 20.

Example 19.11 *Find a particular solution of*

$$\frac{d^2x}{dt^2} + 2k\frac{dx}{dt} + \omega_0^2 x = a\cos\omega t, \quad \textit{where } a > 0.$$

Since $\cos\omega t = \text{Re }e^{i\omega t}$, first find a solution of

$$\frac{d^2X}{dt^2} + 2k\frac{dX}{dt} + \omega_0^2 X = a\,e^{i\omega t}.$$

By substituting $X(t) = P\,e^{i\omega t}$, we find that

$$P = a/[(\omega_0^2 - \omega^2) + i(2k\omega)].$$

It is easier if we put P into polar coordinates: $P = |P|\,e^{i\phi}$, where

$$|P| = |a/[(\omega_0^2 - \omega^2) + i(2k\omega)]|$$
$$= a/[(\omega_0^2 - \omega^2)^2 + (2k\omega)^2]^{\frac{1}{2}},$$

Example 19.11 *continued*

and

$$\phi = \arg[(\omega_0^2 - \omega^2) - \mathrm{i}(2k\omega)],$$

since $a > 0$. Then we obtain

$$x(t) = \mathrm{Re}(P\,\mathrm{e}^{\mathrm{i}\omega t}) = \frac{a\cos(\omega t + \phi)}{[(\omega_0^2 - \omega^2)^2 + (2k\omega)^2]^{\frac{1}{2}}},\qquad(19.4)$$

where ϕ is the polar angle of the point $((\omega_0^2 - \omega^2), -2k\omega)$ on an Argand diagram.

Particular solution of $\dfrac{\mathrm{d}^2x}{\mathrm{d}t^2} + b\dfrac{\mathrm{d}x}{\mathrm{d}t} + cx = f(t)$

(a) $f(t) = a\cos\beta t$ or $a\sin\beta t$.
Put $X(t) = P\,\mathrm{e}^{\mathrm{i}\beta t}$ to solve

$$X'' + bX' + cX = a\,\mathrm{e}^{\mathrm{i}\beta t}.$$

Then $x(t) = \mathrm{Re}\,X(t)$ or $\mathrm{Im}\,X(t)$, corresponding to $\cos\beta t$ or $\sin\beta t$ respectively.
(b) $f(t) = a\,\mathrm{e}^{\alpha t}\cos\beta t$ or $a\,\mathrm{e}^{\alpha t}\sin\beta t$.
Solve $X'' + bX' + cX = a\,\mathrm{e}^{(\alpha+\mathrm{i}\beta)t}$, and continue as in (a).   **(19.5)**

19.3 Particular solutions: exceptional cases

There are exceptional cases for each of the three rules (19.1), when the suggested substitution does not give any result because the trial function delivers zero when it is substituted into the left-hand side. This means (as with the similar exceptional case of a single solution of the characteristic equation, (18.10)) that the trial solution must have a different form.

The most important exception is the case of the equation

$$\frac{\mathrm{d}^2x}{\mathrm{d}t^2} + \beta^2x = a\cos\beta t \quad (\text{or } a\sin\beta t).$$

Note that β occurs on both sides of the equation. This is a special case of Example 19.11 in which $k = 0$ and $\omega^2 = \omega_0^2 = \beta^2$. The rule in (19.1) suggests substituting

$$x(t) = p\cos\beta t + q\sin\beta t,$$

and choosing p and q so that the two sides match. But we already know that this is a solution of the unforced equation $\mathrm{d}^2x/\mathrm{d}t^2 + \beta^2x = 0$, and the inevitable zero that we get on making the substitution cannot be matched to $a\cos\beta t$ on the right.

In this case, the solution is quite different. The following results can be confirmed by direct substitution.

Particular solutions: two exceptional cases

(a) $\dfrac{d^2x}{dt^2} + \beta^2 x = a \cos \beta t$: solution

$$x(t) = \frac{a}{2\beta} t \sin \beta t.$$

(b) $\dfrac{d^2x}{dt^2} + \beta^2 x = a \sin \beta t$: solution

$$x(t) = -\frac{a}{2\beta} t \cos \beta t.$$

(19.6)

Example 19.12 *Find a particular solution of* $\dfrac{d^2x}{dt^2} + 9x = 5 \sin 3t.$

Here, $x = p \cos 3t$ and $x = q \sin 3t$ both give $d^2x/dt^2 + 9x = 0$, so the standard solution form does not work. From (19.6), with $\beta = 3$ and $a = 5$, the required solution is

$$x(t) = -\frac{5}{2 \times 3} t \cos 3t = -\tfrac{5}{6} t \cos 3t.$$

This solution is sketched in Fig. 19.1. Unlike the ordinary sine- and cosine-type solutions, it grows indefinitely. Such solutions have an important physical significance described in Chapter 20.

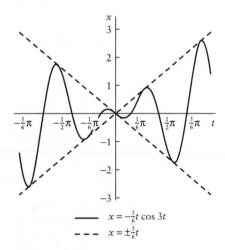

$$x = -\tfrac{5}{6}t \cos 3t$$
$$x = \pm\tfrac{5}{6}t$$

Fig. 19.1

There are other exceptional cases that are not so frequently encountered; some examples are given among the problems at the end of the chapter.

19.4   The general solution of forced equations

Consider the equation

$$\frac{d^2x}{dt^2} - x = -2\cos t. \tag{19.7}$$

A particular solution, $x_p(t)$, say, is

$$x_p(t) = \cos t. \tag{19.8}$$

From earlier experience, we should expect other solutions. In order to find some, consider what happens when we substitute various functions $x(t)$ in the expression

$$\frac{d^2}{dt^2}x(t) - x(t). \tag{19.9}$$

For example, when we put $x(t) = \cos t$, we obtain

$$\frac{d^2}{dt^2}\cos t - \cos t = -2\cos t,$$

as demanded by (19.7).

Suppose now that we can find another function, $x(t) = x_c(t)$ say, which produces *zero* out of (19.9). For example, $x(t) = x_c(t) = e^t$ makes $d^2x/dt^2 - x$ equal to zero. It is then obvious that if we put

$$x(t) = x_p(t) + x_c(t) = \cos t + e^t$$

into (19.9), we again obtain $(-2\cos t)$ on the right: that is to say, we have found another solution of (19.7).

But we already know, from Chapter 18, *all the functions* $x_c(t)$ that give zero when they are put into (19.9): they are the solutions of the equation

$$\frac{d^2x}{dt^2} - x = 0, \tag{19.10}$$

and are given by

$$x(t) = x_c(t) = A\,e^t + B\,e^{-t},$$

where A and B are any constants. Therefore

$$x(t) = \cos t + A\,e^t + B\,e^{-t} \tag{19.11}$$

is always a solution of (19.7). The differential equation (19.10) is called the **unforced equation corresponding to the original equation** (19.7), and its solutions $x(t)$ are called the **complementary functions** of the problem (they complement or extend the particular solution of (19.7) that we obtained).

To show that we have obtained all possible solutions of (19.7), take the particular solution $\cos t$ that we obtained, and suppose that $x_p(t)$ is any other solution of (19.7). Evidently the function $x(t) = x_p(t) - \cos t$ satisfies (19.10), so $x(t)$ must be a complementary function. Therefore

$$x_p(t) = \cos t + \text{(a complementary function)},$$

so $x_p(t)$ must be one of the solutions already expressed by (19.11). Therefore (19.11) is the **general solution** of (19.7). Exactly the same argument would have applied in the general case:

General solution of $\dfrac{d^2x}{dt^2} + b\dfrac{dx}{dt} + cx = f(t)$

(i) Obtain any particular solution, $x_p(t)$.
(ii) Obtain all the solutions $Ax_{c1}(t) + Bx_{c2}(t)$ of the corresponding unforced equation

$$\frac{d^2x}{dt^2} + b\frac{dx}{dt} + cx = 0$$

(the complementary functions).
The sum of these gives the general solution:

$$x(t) = x_p(t) + Ax_{c1}(t) + Bx_{c2}(t).$$

(19.12)

The theory and method is exactly the same for **linear equations** of the first order, and of any order, whether the coefficients are constant or not.

Example 19.13 *Find the general solution of* $\dfrac{d^2x}{dt^2} + 4x = 3\cos 5t.$

Particular solution $x_p(t)$. Looking forward into the calculation, it can be seen that the solution needs no $\sin 5t$ term. Therefore try

$$x_p(t) = p\cos 5t.$$

The substitution into the equation gives

$$p(-25\cos 5t) + 4p\cos 5t = 3\cos 5t$$

for all t, so $p = -\frac{1}{7}$. Therefore

$$x(t) = -\tfrac{1}{7}\cos 5t.$$

Complementary functions $x_c(t)$. We require the solutions $x_c(t)$ of the corresponding unforced equation $d^2x_c/dt^2 + 4x_c = 0$. Try for solutions of the form $x_c(t) = p\,e^{mt}$. The substitution produces the characteristic equation $m^2 + 4 = 0$. Therefore $m = \pm 2i$, so a pair of solutions

$$(e^{2it}, e^{-2it})$$

constitutes a complex basis. To get a real basis, choose either one, say e^{2it}, and find its real and imaginary parts. These are

$$\cos 2t, \quad \sin 2t,$$

and this is the required real basis. Therefore, all the complementary functions are given by

$$x_c(t) = A\cos 2t + B\sin 2t \quad (A \text{ and } B \text{ arbitrary constants}).$$

General solution. This is the sum of the two:

$$x(t) = -\tfrac{1}{7}\cos 5t + A\cos 2t + B\sin 2t.$$

As explained in Section 19.3, the straightforward trial method for the complementary functions fails if the forcing term on the right is already a complementary function, so it can be a useful tactic to look at the complementary functions first. The following example contains this feature, and is also an initial-value problem.

Example 19.14   (a) *Obtain the general solution of*

$$\frac{d^2x}{dt^2} + 4x = 3 + 2\cos 2t.$$

(b) *Find the particular solution for which $x = 0$ and $dx/dt = 0$ when $t = 0$.*

(a) *Complementary functions $x_c(t)$.* These are the solutions of $d^2x_c/dt^2 + 4x_c = 0$. We found them in Example 19.13: they are

$$x_c(t) = A\cos 2t + B\sin 2t \quad \text{with A and B arbitrary.}$$

Particular solution $x_p(t)$. There are two terms on the right, so find a particular solution for each term separately, and add them.

Therefore for a solution, $x_{p1}(t)$ say, of $d^2x_{p1}/dt^2 + 4x_{p1} = 3$, we can obviously take

$$x_{p1}(t) = \tfrac{3}{4}.$$

Corresponding to the other term, we need a solution, $x_{p2}(t)$ say, of $d^2x_{p2}/dt^2 + 4x_{p2} = 2\cos 2t$. We should normally expect a solution of the form $p\cos 2t + q\sin 2t$. However, looking at the complementary functions we found, this function is already a complementary function; so we have the exceptional case (19.6), which gives

$$x_{p2}(t) = \tfrac{1}{2}t\sin 2t.$$

Therefore a particular solution of the original equation is

$$x_p(t) = x_{p1}(t) + x_{p2}(t) = \tfrac{3}{4} + \tfrac{1}{2}t\sin 2t.$$

General solution.

$$x(t) = A\cos 2t + B\sin 2t + \tfrac{3}{4} + \tfrac{1}{2}t\sin 2t.$$

(b) *Initial-value problem.* We require also dx/dt:

$$\frac{dx}{dt} = -2A\sin 2t + 2B\cos 2t + \tfrac{1}{2}\sin 2t + t\cos 2t.$$

The initial conditions prescribe $x(0) = 0$, or

$$A + \tfrac{3}{4} = 0,$$

and $x'(0) = 0$, or

$$2B = 0.$$

Therefore $A = -\tfrac{3}{4}$ and $B = 0$, so the required particular solution is

$$x(t) = -\tfrac{3}{4}\cos 2t + \tfrac{3}{4} + \tfrac{1}{2}t\sin 2t.$$

19.5 First-order linear equations with a variable coefficient

So far, the coefficient c in the equation $dx/dt + cx = f(t)$ has been a constant. We shall now suppose c to be variable; call it $g(t)$:

$$\frac{dx}{dt} + g(t)x = f(t). \tag{19.13}$$

The equation is of linear type (no squares, products, etc., between terms involving x are present), and the idea of obtaining a general solution by adding complementary functions to any particular solution still holds good. However, it is nearly impossible to guess suitable trial functions so we need a new approach to finding solutions.

If we could express the left-hand side $dx/dt + g(t)x$ of the equation as

$$\frac{d}{dt} \text{(something)},$$

then the equation would be easy to solve. This cannot be done, but we can instead do the next best thing. This is to obtain a certain function $I(t)$, called an **integrating factor**, such that

$$I(t)\left(\frac{dx}{dt} + g(t)x\right) = \frac{d}{dt}[I(t)x] \tag{19.14}$$

identically (i.e. for every function $x(t)$ and for all values of t).

The following example shows the meaning of this idea and the way it is used.

Example 19.15   (a) *Show that $I(t) = e^t$ is an integrating factor for the expression $dx/dt + x$.* (b) *Use it to find the general solution of the equation $dx/dt + x = e^{2t}$.*

(a) We shall confirm (19.14), that

$$e^t\left(\frac{dx}{dt} + x\right) = \frac{d}{dt}(e^t x). \tag{i}$$

Work from the right-hand side of (i). Differentiate the product $e^t x$:

$$\frac{d}{dt}(e^t x) = e^t \frac{dx}{dt} + e^t x = e^t\left(\frac{dx}{dt} + x\right),$$

which is the same as the left-hand side of (i), so e^t is an integrating factor.
(b) Multiply both sides of the differential equation by e^t:

$$e^t\left(\frac{dx}{dt} + x\right) = e^t e^{2t} = e^{3t}.$$

↗

Example 19.15 *continued*

Because of the result in (a), we can write this as

$$\frac{d}{dt}(e^t x) = e^{3t}.$$

Therefore

$$e^t x = \int e^{3t}\, dt = \tfrac{1}{3}\, e^{3t} + A \quad (A \text{ arbitrary}),$$

or

$$x = \tfrac{1}{3}\, e^{2t} + A\, e^{-t}.$$

To find a general expression for an integrating factor, refer back to the definition (19.14); the integrating factor $I(t)$ is chosen so that

$$I(t)\left(\frac{dx}{dt} + g(t)x\right) = \frac{d}{dt}[I(t)x].$$

This is the same as

$$I(t)\frac{dx}{dt} + I(t)g(t)x = I(t)\frac{dx}{dt} + x\frac{dI(t)}{dt},$$

or

$$I(t)g(t) = \frac{dI(t)}{dt}$$

(after cancelling $I(t)\dfrac{dx}{dt}$, and dividing through by x). This can be written

$$\frac{1}{I(t)}\frac{dI(t)}{dt} = g(t)$$

or

$$\frac{d \ln I(t)}{dt} = g(t).$$

Therefore

$$\ln I(t) = \int g(t)\, dt,$$

or

$$I(t) = e^{\int g(t)\, dt}.$$

(In the case of Example 19.15, we had $g(t) = 1$, and the present formula gives $I(t) = e^{\int dt} = e^{t+C}$; the choice $C = 0$ gives the integrating factor suggested – any other choice would do.)

Integrating factor for the equation

$$\frac{dx}{dt} + g(t)x = f(t)$$

Put $I(t) = e^{\int g(t)\,dt}$;

then $I(t)\left(\dfrac{dx}{dt} + g(t)x\right) \equiv \dfrac{d}{dt}[I(t)x].$

(19.15)

Solution of $\dfrac{dx}{dt} + g(t)x = f(t)$

Multiply both sides by $I(t)$ (see (19.15)): the equation becomes

$$\frac{d}{dt}[I(t)x(t)] = I(t)f(t);$$

then $I(t)x(t) = \displaystyle\int I(t)f(t)\,dt + C$, giving $x(t)$.

(19.16)

Example 19.16 *Find the general solution of* $\dfrac{dx}{dt} - \dfrac{1}{t}x = t^3, \text{for } t > 0.$

Here $g(t) = -1/t$. Then $\int g(t)\,dt = C - \ln t$ (we need only consider t positive), so that

$$I(t) = e^{-\ln t} = t^{-1},$$

where we have chosen $C = 0$ for convenience. Multiply both sides by $I(t) = t^{-1}$:

$$t^{-1}\left(\frac{dx}{dt} - \frac{1}{t}x\right) = t^{-1}t^3 = t^2.$$

By (19.15), this can be written

$$\frac{d}{dt}(t^{-1}x) = t^2.$$

Therefore

$$t^{-1}x = \int t^2 \, dt = \tfrac{1}{3}t^3 + C,$$

so that

$$x(t) = \tfrac{1}{3}t^4 + Ct.$$

The solution obviously falls into the shape

particular solution + complementary function.

We should have gained nothing by considering negative t, or by adding an arbitrary constant, when working out $\int (1/t)dt$: we only need *any* integrating factor, not all possible ones.

Notice particularly that, in the examples, we did not need to calculate or check the truth of a statement like

$$t^{-1}\left(\frac{dx}{dt} - \frac{1}{t}x\right) = \frac{d}{dt}(t^{-1}x).$$

We already know that t^{-1} is an integrating factor, and this is the very property that an integrating factor is designed to possess.

Be prepared to recognize this type of equation in disguised form, or when different letters are involved; for example,

$$\frac{dy}{dx} = \frac{x + y}{x + 1}$$

is the same as

$$\frac{dy}{dx} - \frac{1}{x + 1}y = \frac{x}{x + 1}.$$

Problems

19.1 Find a particular solution of each of the following equations by trial as in Section 19.1.
(a) $x' + x = 3 e^{2t}$;
(b) $x' - 3x = t^3 + 1$;
(c) $2x' + 3x = t + 3 e^t$;
(d) $x'' + x = 3 e^{2t}$;
(e) $x'' - \frac{1}{4}x = 2 e^t + 3 e^{-t}$;
(f) $x'' - 2x' + x = 3$;
(g) $x'' + 4x' - x = 3t^2 - t$;
(h) $x'' - x = 2 \cos t$;
(i) $2x'' + 3x = 2 \sin 3t$;
(j) $2x'' + x' = \sin t - \cos t$;
(k) $x'' + 2x' + x = \cos 2t$;
(l) $\dfrac{d^2y}{dx^2} - y = 1 - 3 e^{2x}$;
(m) $\dfrac{d^2y}{dx^2} - \dfrac{dy}{dx} + 2y = 3 \sin 2x$.

19.2 Use the method of Section 19.2 to find a particular solution of the following.
(a) $x'' - x = 3 \cos 2t$;
(b) $x'' + x = 2 \sin 3t$;
(c) $x'' + 2x' + x = 3 \sin t$;
(d) $x'' - x' - x = 3 \cos t$;
(e) $2x'' + x' + 2x = 2 \cos 2t$;
(f) $3x'' + 2x' + x = 2 \sin 2t$;
(g) $x'' - 4x = e^{-t} \cos t$ (note: $e^{-t} \cos t = \mathrm{Re}\, e^{(-1+i)t}$);
(h) $x'' - 4x = 3 e^t \sin 2t$ (note: $e^t \sin 2t = \mathrm{Im}\, e^{(1+2i)t}$);
(i) Show that a solution of $x'' + x' + 4x = 5 \cos 3t$ is $(5/\sqrt{34}) \cos(3t + \phi)$, where $\phi = \arctan \frac{3}{5}$.

19.3 The following differential equations are examples of the exceptional cases treated in Section 19.3. Find a particular solution in each case.
(a) $x'' + x = 3 \cos t$;
(b) $x'' + 4x = 3 \sin 2t$;
(c) $x'' + 4x = 1 + 3 \cos 2t$;
(d) $\dfrac{d^2y}{dx^2} + 9y = 2 \sin 3x$;
(e) $\dfrac{d^2y}{dx^2} - 2\dfrac{dy}{dx} + 2y = e^x \cos x$.

19.4 The following are exceptional cases of types not described in Section 19.3. Find a particular solution for each.
(a) $x'' - x = e^t$; try a solution of the form pte^t.
(b) $x'' - 2x' + x = e^t$; try a solution of the form $pt^2 e^t$. (In this case, both e^t and $t e^t$ are complementary functions, so the form in (a) will not work.)
(c) Consider the simple differential equation $d^2x/dt^2 = t$. A first try with the form $pt + q$ suggested by (19.1) does not lead to a result. Try polynomials of higher degree than 1.
(d) $\dfrac{d^2y}{dx^2} + \dfrac{dy}{dx} = x$. The absence of a term in y causes the second-degree trial function $px^2 + qx + r$ to fail. Try a third-degree polynomial instead.
(e) $x'' - 2x' + 2x = e^t \cos t$. Try $t e^t (p \cos t + q \sin t)$, or modify the complex-number approach of Section 19.2 to obtain a particular solution.

(f) First-order equations also have exceptional cases. Consider the equation $\dfrac{dy}{dx} - y = e^x$. (If you have read as far as Section 19.5, you can also handle it by using an integrating factor.)

19.5 Find the general solution of the following equations.
(a) $x'' + 9x = 3\,e^{2t}$;
(b) $x'' - 4x = 2\,e^{-t}$;
(c) $4x'' - x = 1 + 3\cos 2t$;
(d) $\dfrac{d^2y}{dx^2} + 2\dfrac{dy}{dx} + 2y = 3$;
(e) $x'' - 2x' + 2x = 3\sin 2t$;
(f) $4x'' - 2x' - 2x = 3t^2$;
(g) $x'' + x' = 2 - 3\,e^{-t}\cos t$;
(h) $2x'' + x' - x = \frac{1}{2}t + 3\,e^{-t}$;
(i) $\dfrac{d^2y}{dx^2} + y = 1 + 2\,e^{3x} + x^2$;
(j) $\dfrac{d^2y}{dx^2} + 2\dfrac{dy}{dx} + y = 3\cos 2x + \sin 2x$;
(k) $\dfrac{d^2y}{dx^2} + 4\dfrac{dy}{dx} + 5y = e^{-x}\sin x$.

19.6 Use an integrating factor (Section 19.6) to find the general solution of the following equations.
(a) $x' - 3x = 0$;
(b) $x' + 2x = 3$;
(c) $x' - 2tx = t$;
(d) $x' - t^{-1}x = t + t\,e^{-t}$;
(e) $x' - t^{-1}x = t - 1$;
(f) $tx' - 2x + 3 = 0$;
(g) $\dfrac{dy}{dx} + \dfrac{1}{x+1}y = \sin x$
(you will need to use integration by parts to perform the integration);
(h) $3\dfrac{dy}{dx} + \dfrac{1}{x}y = x$;
(i) $(x-1)\dfrac{dy}{dx} - y = (x-1)^2$;
(j) $x' - \dfrac{1}{t}x = \ln t$;
(k) $tx' - x = 1 + t$;
(l) $\dfrac{dy}{dx} = \dfrac{x+y}{x+1}$;
(m) $x' + x\cos t = \cos t$;
(n) $x\dfrac{dy}{dx} = \dfrac{1-y}{1-x}$;
(o) $(1 - t^2)x' + tx = t$.

19.7 Show that the general solution of
$$\frac{dy}{dx} + \frac{1}{x}y = f(x)$$
is given by
$$y(x) = \frac{1}{x}\int xf(x)\,dx + \frac{C}{x},$$
where C is any constant. Find the solution of the equation
$$\frac{dy}{dx} + \frac{1}{x}y = \ln x$$
for $x > 0$, for which $y = 0$ when $x = 1$.

19.8 (a) Use an integrating factor to show that the general solution of
$$\frac{dy}{dx} + y = f(x)$$
is
$$y(x) = e^{-x}\int e^x f(x)\,dx + C\,e^{-x},$$
where C is an arbitrary constant.
(b) Show that the particular solution for which $y = y_0$ when $x = 0$ is given by
$$y(x) = y_0\,e^{-x} + e^{-x}\int_0^x e^u f(u)\,du.$$

19.9 (Newton cooling). An object is heated or cooled above or below the ambient air temperature T_0. Under certain physical assumptions, the body temperature T satisfies the equation
$$dT/dt = -k(T - T_0),$$
where k is a positive constant. Find the general solution of the equation.
The body is at $100°C$ in an atmosphere at $40°C$. After 3 minutes, its temperature is $85°C$. Find the value of k, and determine when the body will reach $60°C$.

20 Harmonic functions and the harmonic oscillator

CONTENTS

20.1 Harmonic oscillations

Consider the equation

$$\frac{d^2x}{dt^2} + \omega^2 x = 0,$$

where we assume $\omega > 0$. Its solutions (see (18.14b)) are

$$x(t) = C \cos(\omega t + \phi) \tag{20.1}$$

where C and ϕ are any constants. We can write

$$x(t) = C \cos(\omega t + \phi) = C \cos \omega[t + (\phi/\omega)].$$

Therefore the graph of (20.1) is merely the graph $x = C \cos \omega t$ shifted, or **translated**, a distance ϕ/ω along the t axis; to the left if ϕ is positive, to the right if ϕ is negative. Sine functions are included in the collection (20.1), because $\sin \omega t = \cos(\omega t - \frac{1}{2}\pi)$. These functions are spoken of generally as **harmonic functions**.

In applications it is usual to adjust ϕ so that

$$C > 0 \quad \text{and} \quad -\pi < \phi \leqslant \pi \tag{20.2}$$

which can always be done without changing the function values described by the expression (20.1). We say then that the function (20.1) is in **standard form**.

Example 20.1 *Express* $-2\cos(3t - \frac{7}{3}\pi)$ *in standard form.*

Note that $\cos(A + \pi) = -\cos A$. Therefore

$$-2\cos(3t - \tfrac{7}{3}\pi) = 2\cos(3t - \tfrac{7}{3}\pi + \pi) = 2\cos(3t - \tfrac{4}{3}\pi).$$

We now have positive C, but ϕ is still out of range according to (20.2). To bring it within range increase it by 2π, which alters nothing:

$$2\cos(3t - \tfrac{4}{3}\pi) = 2\cos(3t - \tfrac{4}{3}\pi + 2\pi) = 2\cos(3t + \tfrac{2}{3}\pi),$$

which is now in standard form.

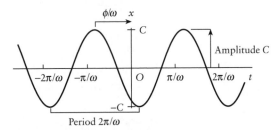

Fig. 20.1 $x = C\cos(\omega t + \phi);\ c > 0,\ -\pi < \phi \leqslant \pi.$

The features of the function $x(t) = C\cos(\omega t + \phi)$ are shown in Fig. 20.1. Assume that the expression is in standard form (20.2). The graph swings between $\pm C$, and C is its **amplitude**. It is periodic (see Section 1.6), repeating itself at intervals of length $2\pi/\omega$, which is its minimum **period**. The number of complete oscillations per unit time is the **frequency** (e.g. in cycles per second, or hertz units), and

$$\text{Frequency} = (\text{period})^{-1} = \omega/2\pi. \tag{20.3}$$

The parameter ω is **angular frequency**, often shortened merely to 'frequency'. The parameter ϕ is the **phase** or **phase angle**. As explained above, ϕ/ω represents the distance that the graph $x = C\cos\omega t$ has to be shifted to coincide with (20.1).

Frequently the independent variable represents **length** x instead of time t, as in a form such as $y = C\cos(\omega x + \phi)$. Then $2\pi/\omega$ is called **wavelength** rather than 'period', and $\omega/2\pi$ the **wave number** rather than 'frequency'.

Graphs of harmonic functions are often displayed by plotting x against the dimensionless variable

$$\tau = \omega t$$

(τ is the Greek letter 'tau') rather than against t. Thus τ will be the name of the new time-like axis, so that

$$x = C\cos(\tau + \phi),$$

which has period 2π in the variable τ. The $x,\ \tau$ graph is drawn in Fig. 20.2.

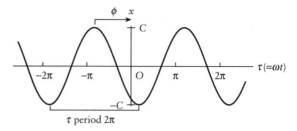

Fig. 20.2

The new graph has τ period 2π: it repeats itself when τ increases by 2π. It is the same as the graph of $x = C \cos \tau$ displaced through an interval in τ of length ϕ. Expressed in terms of the period or wavelength in Fig. 20.2, it is clear that $\phi = \pi$ represents a displacement of half a wavelength, $\phi = \frac{1}{2}\pi$ represents a displacement of a quarter of a wavelength, and so on.

20.2 Phase difference: lead and lag

Suppose that two oscillations have the same angular frequency ω, but are out of step because they have a different phase:

$$x_1(t) = C_1 \cos(\omega t + \phi_1), \qquad x_2(t) = C_2 \cos(\omega t + \phi_2).$$

Then they are said to be **out of phase** by an angle $\phi_2 - \phi_1$, or $\phi_1 - \phi_2$. More specifically, the following terminology is widely used in science and engineering applications:

> **Phase difference; lead and lag**
>
> $x(t) = C_1 \cos(\omega t + \phi_1)$ and $y(t) = C_2 \cos(\omega t + \phi_2)$ are harmonic functions in standard form with the same circular frequency ω. If $\phi_1 > \phi_2$, then x is said to **lead** y, or y is said to **lag** x, by an angle $\phi_1 - \phi_2$.
>
> (20.4)

The reason for these terms is illustrated in Example 20.2.

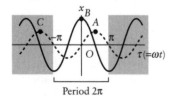

Fig. 20.3 —— $v = v_0 \cos \tau$;
------ $i = (v_0/\omega L) \cos(\tau - \frac{1}{2}\pi)$. The period inspected is symmetrical about the chosen feature at B.

Example 20.2   *If a voltage $v = v_0 \cos \omega t$ is applied to a coil having self-inductance L, the resulting current $i = \dfrac{v_0}{\omega L} \cos(\omega t - \frac{1}{2}\pi)$, so that v leads i, or i lags v, by $\frac{1}{2}\pi$ (or 90°). Illustrate the sense of the terms graphically for this case.*

The curves in Fig. 20.3 are plotted against the variable $\tau = \omega t$ and represent

$$v = v_0 \cos \tau \qquad \text{and} \qquad i = \frac{v_0}{\omega L} \cos(\tau - \frac{1}{2}\pi).$$

↗

Example 20.2 *continued*

Choose one of the curves, say the v curve, and select a prominent feature, say the maximum at B. Now search an interval within $\pm\pi$ of B (that is to say, within half a period on either side of B) for the corresponding feature of i. This is the maximum of i at A.

Now, as we move from left to right (time increasing) through the interval, B appears before A – that is to say, at a quarter period ($\frac{1}{2}\pi$) earlier than A. This will be true for any feature of v within its own symmetrical corresponding interval of $\pm\pi$. It is equivalent to saying that, when the two variables to be compared are in standard form, the one with the greater phase leads, and the other lags, by the phase difference (taken positively).

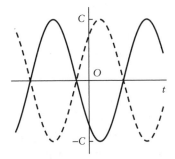

Fig. 20.4 ——— $x = C\cos(\omega t + \phi)$;
------ $x = C\cos(\omega t + \phi \pm \pi)$.

In Example 20.2 it is **essential to limit the search to the prescribed single period**. Otherwise we could argue (see Fig. 20.3) that because, say, C appears before B, therefore i leads v, which is contrary to the definition. (Any basic period of length 2π will give the same priority.) Also, notice that if one entity leads another it does not in the least imply that the first is to be taken as the cause of the second.

Suppose that two oscillations, having the same amplitude and frequency, differ in phase by π so they are displaced by half a period. If the oscillations are added together, there is total cancellation, as shown in Fig. 20.4. The following example shows what happens when the phase difference is less extreme.

Example 20.3 *Two waves described by $C\cos\omega t$ and $C\cos(\omega t + \phi)$ are superimposed (added). Show that the result is a harmonic wave of the same frequency, and show how the amplitude varies as ϕ varies between $\pm\pi$.*

From Appendix B the sum can be written

$$C[\cos\omega t + \cos(\omega t + \phi)] = 2C\cos\tfrac{1}{2}\phi\cos(\omega t + \tfrac{1}{2}\phi).$$

This is a harmonic oscillation with angular frequency ω, phase $\frac{1}{2}\phi$, and amplitude $2C\cos\frac{1}{2}\phi$. As ϕ goes from $-\pi$ through zero to π, the amplitude goes from zero (cancellation) through the value $2C$ and back to zero.

This type of superposition is of importance in describing **interference** and **diffraction** phenomena. If the amplitudes of the components are not the same, a similar calculation applies (see Problem 20.4 at the end of this chapter).

20.3 Physical models of a differential equation

Figure 20.5a shows a piston of mass m running in a cylinder, controlled by a spring which obeys Hooke's law (a linear spring) and has stiffness s, acted on by an external force $F(t)$. The displacement

(a)

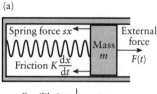

(b)

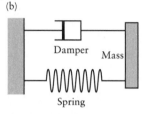

Fig. 20.5 (a) Mass–spring system. The arrows indicate the actual direction of the forces when $F(t)$, sx, and $K\,dx/dt$ take positive values. (b) Schematic representation: the spring and the frictional element must be in parallel.

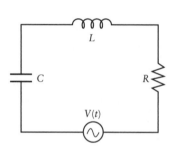

Fig. 20.6

of the piston from its equilibrium position is $x(t)$. Assume also that there is a frictional resistance proportional to the velocity:

$$\text{frictional resistance} = K\frac{dx}{dt}, \quad (K > 0).$$

The equation of motion, force equals mass times acceleration, becomes

$$F(t) - K\frac{dx}{dt} - sx = m\frac{d^2x}{dt^2},$$

$$\text{or} \quad \frac{d^2x}{dt^2} + \frac{K}{m}\frac{dx}{dt} + \frac{s}{m}x = \frac{F(t)}{m}. \tag{20.5}$$

This equation is of the type discussed in Chapter 19:

$$\frac{d^2x}{dt^2} + b\frac{dx}{dt} + cx = f(t), \tag{20.6}$$

in which

$$b = \frac{K}{m}, \quad c = \frac{s}{m}, \quad \text{and} \quad f(t) = \frac{F(t)}{m}.$$

Figure 20.6 represents an LCR circuit driven by a voltage source of zero impedance, $V(t)$. If Q is the charge on the capacitor, then

$$L\frac{d^2Q}{dt^2} + R\frac{dQ}{dt} + \frac{1}{C}Q = V(t),$$

$$\text{or} \quad \frac{d^2Q}{dt^2} + \frac{R}{L}\frac{dQ}{dt} + \frac{1}{LC}Q = \frac{1}{L}V(t). \tag{20.7}$$

Again, this is an equation of the type (20.6), with

$$x = Q, \quad b = \frac{R}{L}, \quad c = \frac{1}{LC}, \quad f(t) = \frac{1}{L}V(t).$$

These two physical systems serve as **models** of the differential equation (20.6). They are also models of each other, for by choosing the same values of b and c and the same forcing term $f(t)$ the circuit would serve as a precise **analogue** of the piston and mimic its behaviour exactly. A vast number of systems share the governing equation (20.6), at least approximately. Such a system is called a **linear oscillator.**

20.4 Free oscillations of a linear oscillator

Suppose that in the piston system there is no external force acting, so that $F(t) = 0$ for all t. We shall choose a conventional notation that simplifies the algebra a little. Equation (20.6) will be written

$$\frac{d^2x}{dt^2} + 2k\frac{dx}{dt} + \omega_0^2 x = 0 \tag{20.8}$$

(in which we have put $K/m = 2k$, $s/m = \omega_0^2$, $F(t) = 0$). This equation describes the **free oscillations** of the mass–spring system.

The parameter k is a measure of the amount of friction in the system. We shall consider the case when k is 'small'. This is not very meaningful because k is not dimensionless, so we could change our units so as to make it as large as we wished. The only thing that makes sense is to compare it with another parameter having the same dimensions. We specify that

$$k^2 < \omega_0^2. \tag{20.9}$$

We have already worked out this problem (see (18.15)). The solutions of (20.8) subject to (20.9) are given by

$$x(t) = C\,e^{-kt}\cos[(\omega_0^2 - k^2)^{\frac{1}{2}}t + \phi], \tag{20.10}$$

where C and ϕ are arbitrary. These are called the **free oscillations** or **natural oscillations** of the system represented by the equation.

If the friction, or so-called **damping**, is zero then $k = 0$ and the equation for the free oscillations becomes

$$\frac{d^2x}{dt^2} + \omega_0^2 x = 0 \tag{20.11}$$

with solutions

$$x(t) = C\cos(\omega_0 t + \phi) \tag{20.12}$$

which are harmonic functions with circular frequency ω_0.

The friction, or damping, changes (20.12) into (20.10). The frequency is changed from ω_0 to $(\omega_0^2 - k^2)^{\frac{1}{2}}$, which is a small change if k is small, and the regular oscillations of (20.12) are caused to die away through the factor e^{-kt} in (20.10). The general effect is shown in Fig. 20.7. We say that the oscillation **decays exponentially** down to zero, when all the initial energy is used up on friction. This is **weak damping** and the oscillation is said to be **underdamped**.

If $k^2 > \omega_0^2$, then there is a comparatively large amount of friction, and the form of the solution is different from (20.10) (see Fig. 20.7b). There are no oscillations; the $x(t)$ curve dies away without crossing the t axis more than once, as in a **dead-beat** electrical instrument or shock absorber. This is the case of **heavy damping** or an **overdamped oscillation**.

(a)

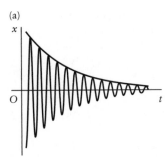

(b)

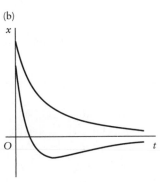

Fig. 20.7

20.5 Forced oscillations and transients

Return to the equation (20.5) for the mass–spring system with a non-zero external force $F(t)$ acting. As before, put $K/m = 2k$, $s/m = \omega_0^2$, and $F(t)/m = f(t)$, so that we get

$$\frac{d^2x}{dt^2} + 2k\frac{dx}{dt} + \omega_0^2 x = f(t),$$

which is a forced equation of the type considered in Chapter 19.

We shall consider only the case when $f(t) = K \cos \omega t$:

$$\frac{d^2x}{dt^2} + 2k\frac{dx}{dt} + \omega_0^2 x = K \cos \omega t, \tag{20.13}$$

and suppose as before that the friction (or resistance) is 'small':

$$k^2 < \omega_0^2.$$

The mass in the piston system is now subject to competing stimuli. Left to itself it would oscillate as in (20.10) with circular frequency $(\omega_0^2 - k^2)^{\frac{1}{2}}$, and finally come to rest. However, the forcing term is trying to make it oscillate with a different circular frequency ω. The result is described by the general solution of (20.13). This is equal to the sum of a particular solution (already worked out in (19.4)) and the complementary functions which are the free oscillations given in (18.15):

General solution of the forced linear oscillator equation
$(k^2 < \omega_0^2)$

$$\frac{d^2x}{dt^2} + 2k\frac{dx}{dt} + \omega_0^2 x = K \cos \omega t$$

$$x(t) = \frac{K}{[(\omega_0^2 - \omega^2)^2 + 4k^2\omega^2]^{\frac{1}{2}}} \cos(\omega t + \Phi)$$

$$+ C e^{-kt} \cos[(\omega_0^2 - k^2)^{\frac{1}{2}}t + \phi],$$

in which Φ is the polar angle of the point $(\omega_0^2 - \omega^2, -2k\omega)$, and C and ϕ are arbitrary.

(20.14)

The structure of (20.14) is very important: the general features are summarized in (20.15) below.

Forced oscillations of a linear oscillator

(A) The forced oscillation (first term of (20.14)) coexists with a free oscillation (second term). The free oscillation proceeds as if no forcing term were present.

(B) The term representing the forced oscillation is harmonic, with the same frequency as the forcing term, but a different phase and amplitude. The term is invariable; initial conditions can have no effect on it since it contains no adjustable constants.

(C) The free oscillation term adjusts to any initial conditions by means of the constants C and ϕ.

(D) If k is positive, that is to say if there is any friction (or resistance in the case of a circuit), the free oscillations die away to zero due to the factor e^{-kt}. Therefore all solutions ultimately settle into the same steady oscillation, independently of the initial conditions.

(20.15)

On account of (D) the free oscillation is called a **transient** oscillation, and may show itself, for example, by a brief irregularity in the voltage or current upon switching an electrical apparatus.

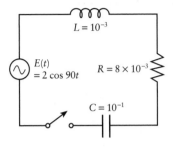

$L = 10^{-3}$

$E(t) = 2 \cos 90t$

$R = 8 \times 10^{-3}$

$C = 10^{-1}$

Fig. 20.8

Example 20.4 *The circuit shown in Fig. 20.8 is initially quiescent and uncharged. Find the charge $Q(t)$ on the capacitor after switching the circuit on.*

We shall rework the problem from first principles. The equation is

$$10^{-3}\frac{d^2Q}{dt^2} + 8 \times 10^{-3}\frac{dQ}{dt} + 10Q = 2 \cos 90t,$$

or $\quad \dfrac{d^2Q}{dt^2} + 8\dfrac{dQ}{dt} + 10^4 Q = 2 \times 10^3 \cos 90t.$

Complementary functions Q_c (natural oscillation). The characteristic equation is $m^2 + 8m + 10^4 = 0$, so that $m = -4 + 99.92i$ and the complementary functions are $Q_c = B\,e^{-4t}\cos(99.92t + \phi)$, where B and ϕ are arbitrary.

Particular solution Q_p (forced oscillation). Look for a solution to the corresponding complex equation

$$\frac{d^2X}{dt^2} + 8\frac{dX}{dt} + 10^4 X = 2 \times 10^3\, e^{90it},$$

and take its real part. By trying a solution of the form $X(t) = P\,e^{90it}$ we obtain $P = 0.9205 - 0.3488i$. In polar coordinates this becomes $P = 0.9843\, e^{-0.3622i}$. The corresponding complex solution, in polar coordinates, is

$X(t) = 0.9843\, e^{(90t - 0.3622)i}$.

Therefore the particular solution is

$Q_p(t) = \mathrm{Re}\, X(t) = 0.9843 \cos(90t - 0.3622)$.

The general solution. This is

$Q(t) = 0.984 \cos(90t - 0.3622) + B\,e^{-4t}\cos(99.92t + \phi)$.

Initial conditions. At $t = 0$, Q and dQ/dt are zero. After obtaining dQ/dt and substituting $t = 0$ into Q and dQ/dt, we obtain the equations

$B \cos\phi = -0.9204$,

$4B \cos\phi + 99.92B \sin\phi = 31.389$.

The solution is $B = 0.9851$, $\phi = 2.777$, so $Q(t)$ is given by

$Q(t) = 0.9843 \cos(90t - 0.3622) + 0.9851\, e^{-4t}\cos(99.92t + 2.777)$.

Figure 20.9 shows the individual contributions of the two terms.

(a)

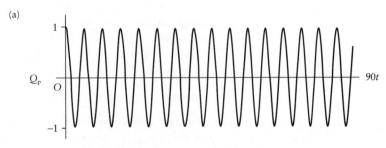

Fig. 20.9 (a) Forced oscillation, $Q_p = 0.9843 \cos(90t - 0.3622)$.

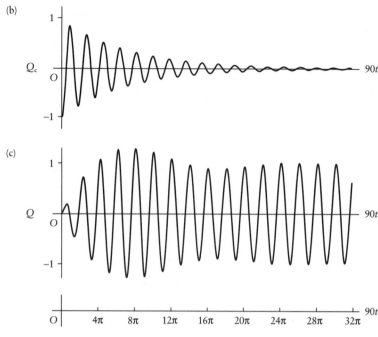

Fig. 20.9 (b) Transient, $Q_c = 0.9851\,e^{-4t}\cos(99.92t + 2.777)$. (c) Total oscillation, $Q = Q_p + Q_c$.

20.6 Resonance

Return to eqn (20.14) for the linear oscillator and its solutions and examine the forced oscillation, which is all that is left after the transient has died away. Its amplitude, A say, is given by

$$A = \frac{K}{[(\omega_0^2 - \omega^2)^2 + 4k^2\omega^2]^{\frac{1}{2}}}.$$

Different values for the forcing frequency ω will produce different amplitudes; some values of ω will be more effective than others in generating a large amplitude.

Regard ω_0 and k as representing the fixed characteristics of some kind of system, and consider an experiment in which we try to excite it with a controllable input $K \cos \omega t$, keeping K constant but trying various values of ω. The amplitude A will be greatest when $(\omega_0^2 - \omega^2)^2 + 4k^2\omega^2 = g(\omega)$, say, is a minimum with respect to the variable ω. It is found by solving $dg/d\omega = 0$ (see Problem 20.15), that is, the minimum occurs when

$$\omega^2 = \omega_0^2 - 2k^2, \quad (k^2 < \tfrac{1}{2}\omega^2).$$

When ω^2 takes this value the amplitude A will take its greatest possible value for the given K and ω, given by

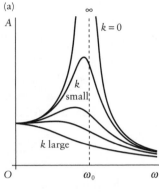

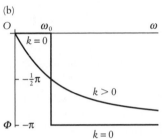

Fig. 20.10

$$A = \frac{K}{2k(\omega_0^2 - k^2)^{\frac{1}{2}}}.$$

Figure 20.10 shows schematically how the amplitude A, and also the phase Φ in (20.14), vary with forcing frequency ω. Different curves are obtained according to the amount of friction or damping (or resistance in the case of a circuit) in the system, measured by the size of k; as the damping decreases, the maximum increases. When the condition for a maximum is satisfied, the system is said to be in a state of **resonance**.

> **Resonating system**
>
> $$\frac{d^2x}{dt^2} + 2k\frac{dx}{dt} + \omega_0^2 x = K \cos \omega t.$$
>
> Forced amplitude $A = \dfrac{K}{[(\omega_0^2 - \omega^2)^2 + 4k^2\omega^2]^{\frac{1}{2}}}.$
>
> Resonant frequency $\omega^2 = \omega_0^2 - 2k^2.$
>
> Resonance amplitude $\dfrac{K}{2k(\omega_0^2 - k^2)^{\frac{1}{2}}}$
>
> (20.16)

A physical feeling for the buildup of a large amplitude can be obtained by thinking of a child being pushed on a swing by two people, one on either side of the swing. The method is to push the swing the way it wants to go, and not to work against it. This is best done by pushing it, forward and backward alternately, when it is at the bottom of its path. The driving frequency is then the same as the natural frequency of the swing. The driving cycle is a quarter of a period out of phase with the swing's cycle, because the force is a maximum when the displacement is a minimum. In terms of (20.16), k is assumed small, so that $\omega^2 = \omega_0^2$ very nearly, and the phase difference Φ is nearly $\frac{1}{2}\pi$ or a quarter of a period, the forcing term leading the response by this amount.

Suppose next that there is **zero friction**,

$$k = 0,$$

so that

$$\frac{d^2x}{dt^2} + \omega_0^2 x = K \cos \omega t, \tag{20.17}$$

and from (20.16) the forced amplitude A is

$$A = \frac{K}{\omega_0^2 - \omega^2}. \tag{20.18}$$

The natural frequency of this system is exactly ω_0. When ω (the forcing frequency) gets close to ω_0, the amplitude A can become very large, approaching infinity as ω approaches ω_0: see Fig. 20.10a.

When $\omega = \omega_0$ the equation becomes

$$\frac{d^2x}{dt^2} + \omega_0^2 x = K \cos \omega_0 t, \tag{20.19}$$

and apparently $A = \infty$. This result cannot be said to describe a steady solution of (20.19), but must be reconcilable with (20.19) in some way. In fact it is the 'exceptional case' of eqn (19.6a), and has a solution

$$x(t) = \frac{K}{2\omega_0} t \sin \omega_0 t. \tag{20.20}$$

This particular solution conveniently satisfies the initial conditions

$$x(0) = 0, \qquad x'(0) = 0, \tag{20.21}$$

that is to say, the conditions for initial quiescence. It therefore represents a system without friction and in a state of resonance, which starts up from rest. Its oscillations grow steadily to infinity due to the factor t in (20.20). The equation does not have any solutions corresponding to steady forced oscillations, such as we found earlier in systems having even a small amount of friction.

20.7 Nearly linear systems

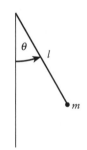

Fig. 20.11

Consider the pendulum of Fig. 20.11. It consists of a weightless rod of length l, pivoted at the top and carrying a point mass m at the lower end. It makes an angle $\theta(t)$ with the vertical. The equation of motion is

$$\frac{d^2\theta}{dt^2} + \frac{g}{l} \sin \theta = 0, \tag{20.22}$$

where g is the gravitational constant.

This equation is **nonlinear** since $\sin \theta$ is not of the form $a\theta + b$, so the methods of Chapter 19 do not apply to it. However, the Taylor series for $\sin \theta$ begins

$$\sin \theta = \theta - \tfrac{1}{6}\theta^3 + \cdots$$

(θ in radians), so provided that θ remains small enough we can approximate $\sin \theta$ by

$$\sin \theta \approx \theta.$$

The error is about 10% when $\theta = 45°$, and 0.1% at 5°. Put this into (20.22); we obtain the **approximate linearized equation**

$$\frac{d^2\theta}{dt^2} + \frac{g}{l}\theta = 0. \tag{20.23}$$

The general solution is $\theta(t) = C \cos[(g/l)^{\frac{1}{2}}t + \phi]$. The values of C and ϕ will depend on how it was set going; the initial conditions amount to prescribing the position and angular velocity at $t = 0$. However, C must be small for the approximation to be justified.

Exactly linear equations are uncommon. Most frequently they occur as the result of a simplifying approximation such as we carried out for the pendulum. Usually some function in the equation is linearized at the expense of a restriction on the dependent variable.

Example 20.5 *A mass m is fixed at the midpoint of a piece of elastic having natural length l and stiffness s. The elastic is stretched between two points a distance $L > l$ apart. Find the period of small lateral vibrations.*

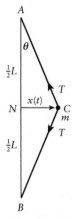

Fig. 20.12 *A* and *B* are fixed a distance *L* apart where $L > l$; the mass *m* at *C* is displaced from equilibrium at *N* by a distance $x(t)$.

(See Fig. 20.12.) The extension e of the branch AC is $AC - \frac{1}{2}l$, so the tension T in either branch is

$$T = se = s(AC - \tfrac{1}{2}l).$$

The total restoring force F is $2T \sin \theta$, and

$$\sin \theta = NC/AC = x/AC,$$

so $F = 2s(AC - \tfrac{1}{2}l)x/AC = 2s[1 - l/(2AC)]x.$

The equation of motion is

$$m\frac{\mathrm{d}^2x}{\mathrm{d}t^2} = -F,$$

so we must put AC in terms of x. Now

$$AC = (\tfrac{1}{4}L^2 + x^2)^{\frac{1}{2}},$$

so the equation will be nonlinear. However, if the oscillations which we expect are of small amplitude compared with L, we can put

$$AC \simeq \tfrac{1}{2}L$$

with an error of something like $2x^2/L^2$, and the approximation to the restoring force becomes

$$F \simeq 2s(1 - l/L)x.$$

The equation of motion becomes approximately

$$m\frac{\mathrm{d}^2x}{\mathrm{d}t^2} = -2s(1 - l/L)x,$$

or $\dfrac{\mathrm{d}^2x}{\mathrm{d}t^2} + \dfrac{2s}{m}(1 - l/L)x = 0.$

This is the linearized equation, good for small amplitudes. It has solutions

$$x(t) = C \cos\left\{ \left[\frac{2s}{m}\left(1 - \frac{l}{L}\right)\right]^{\frac{1}{2}} t + \phi \right\},$$

where C and ϕ are arbitrary. The approximate period is

$$2\pi \bigg/ \left[\frac{2s}{m}\left(1 - \frac{l}{L}\right)\right]^{\frac{1}{2}}.$$

It is interesting to consider the case when the string is *unstretched* in the equilibrium position, so that $L = l$. The resulting equation cannot be straightforwardly linearized. This case is treated in Example 23.2.

20.8 Stationary and travelling waves

The simplest type of wave motion is a **harmonic oscillation**: a periodic time variation of the form $C \cos(\omega t + \phi)$. However, physical vibrations arising in connection with subjects such as sound, elasticity, radar, X-ray analysis, optics, and many others involve one or more space variables x, y, z as well as time t. Oscillations are taking place everywhere in a region, and the variations in phase and amplitude from point to point are all-important for many applications.

Case (i) *Stationary waves in one space dimension*
Figure 20.13 shows a **sinusoidal** or **harmonic** wave in one space dimension, exemplified by the displacement of the string on a musical instrument when a perfectly pure tone is played. At every moment the shape of the string is sinusoidal in z, but each ordinate oscillates harmonically from moment to moment. The points N on the string where the displacement is zero (**nodes**), and the points A (**antinodes**), are *fixed points*. The motion is called a **standing wave** or a **stationary wave**. The ordinates go up and down as we watch, but there is no overall motion along the z axis.

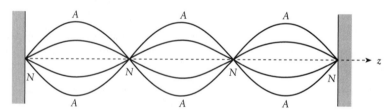

Fig. 20.13 A standing wave on a stretched string. The nodes N and antinodes A remain fixed in position.

The most general expression for a stationary sinusoidal wave is

$$u(t, x) = A \cos\left[2\pi\left(\frac{t}{T}\right) + \phi\right]\cos\left[2\pi\left(\frac{z}{\lambda}\right) + \alpha\right], \qquad (20.24)$$

where u is the displacement, z the coordinate along the string, λ the wavelength, T the period of the oscillation, $A > 0$ the amplitude, and ϕ and α arbitrary phase angles. Sines can be turned into cosines by increasing the phase by $\frac{1}{2}\pi$, so that (20.24) covers sines as well as cosines. We can express (20.24) in terms of **angular frequency** ω and **wave number** k by putting

$$T = \frac{2\pi}{\omega}, \quad \lambda = \frac{2\pi}{k}; \qquad (20.25a)$$

the frequency f in cycles per second (f Hertz) is given by

$$f = \frac{1}{T} = \frac{\omega}{2\pi}. \tag{20.25b}$$

Therefore

$$u(t, x) = A\cos(\omega t + \phi)\cos(kz + \alpha). \tag{20.26}$$

Case (ii) *Travelling waves in one space dimension*
Figure 20.14 illustrates another type of wave motion. It may be thought of in terms of a very long, taut string extending along the z axis to the right. A steady wave motion moving to the right can be initiated by wobbling the left-hand end of the string. It is called a **travelling** or **progressive wave**. The general form of a progressive harmonic wave is given by

$$u(t, z) = A\cos(\omega t - kz + \phi). \tag{20.27}$$

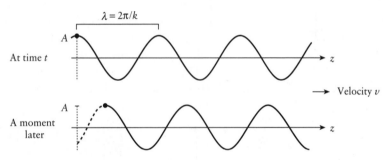

Fig. 20.14 A travelling wave $A\cos(\omega t - kz + \phi)$ moving with phase velocity $v = \omega/k$.

The shape of $u(t, z)$ at any *fixed moment* $t = t_0$ takes the form

$$u(t, z) = A\cos(-kz + B),$$

where $B = \omega t_0 + \phi$, a constant. Therefore, the graph of u against z maintains a constant shape for all t_0, but it is translated bodily along the z axis by a distance depending on t_0. To find the velocity of the translatory motion we may track the motion of any feature of the graph as time increases: we could use, say, any maximum of (20.27). A maximum occurs for values of t and z connected by $\omega t - kz + \phi = 0$; that is, at the moving point

$$z = \frac{\omega t}{k} + \frac{\phi}{k}.$$

From (20.25a,b) and (20.26), the velocity v of the wave along the z axis is therefore given by

$$v = \frac{\omega}{k} = \frac{\lambda}{T} = \lambda f. \tag{20.28}$$

The velocity of a sinusoidal wave is called the **phase velocity**: it is the velocity of a point for which the phase maintains a constant value; for example, following an antinode, as above, or a node. A more direct way of justifying the equation $v = \lambda f$ is as follows. The number of waves crossing the fixed point P per second is equal to the frequency f. Therefore, the length of the **wave train** crossing P per unit time is $f \times wavelength$, which is the velocity v.

The wavelength λ and the frequency f cannot be assigned arbitrarily in a physical problem (and the same applies to ω and k) since they are connected by (20.28), and the velocity of propagation v is determined by the physical medium (even if it varies with the frequency).

Case (iii) *Plane waves in three dimensions along the z axis*
Set up right-handed axes as in Fig. 20.15, and consider a disturbance (such as pressure in a sound wave) $u(t, x, y, z)$, described by

$$u(t, x, y, z) = A \cos(\omega t - kz + \phi), \tag{20.29}$$

in which A, ω, k, ϕ are constants. Although this looks similar to (20.27), its meaning is different; eqn (20.29) defines u at every point in the x, y, z space. The values of u are independent of x and y; that is to say, the value of u over any fixed plane perpendicular to the z axis, such as Σ in Fig. 20.15, is *uniform* at any particular moment, though this value varies with time t. The waves are therefore called **plane waves**.

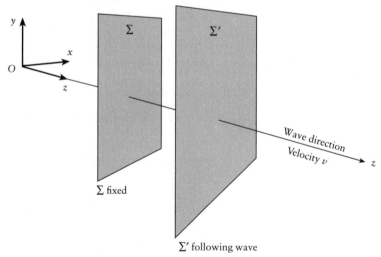

Fig. 20.15 A plane travelling wave. The disturbance u is uniform on the fixed plane Σ. It is uniform and also constant over Σ', which moves with velocity v.

The z axis is one **ray** of the three-dimensional wave; along it the situation is the same as in Case (ii) above. The value of u over Σ', moving with velocity v, is equal to the value on the z axis, so any plane Σ' (Fig. 20.15) that follows a given constant value of u must move to the right with speed $v = \omega/k = \lambda/T$. To sum up these results:

> **Plane wave travelling along the z axis**
>
> $$u(t, x, y, z) = A \cos(\omega t - kz + \phi),$$
>
> with amplitude $A > 0$, circular frequency ω, wave number k, phase angle ϕ, velocity $v = \omega/k$.
>
> **(20.30)**

Case (iv) *Plane waves in any direction*
Suppose that $\hat{s}$ is any unit vector (i.e. a vector of unit length; see Section 9.7). It can be used to indicate a definite direction in space. Equation (20.30) describes a plane wave travelling along the z axis; we shall verify that a plane wave travelling in the direction $\hat{s}$ is given by:

> **Plane wave $u(t, r)$ travelling in direction $\hat{s}$**
>
> $$u(t, r) = A \cos(\omega t - k\hat{s} \cdot r + \phi),$$
>
> where $\hat{s}$ is a unit vector in the direction of propagation, and r the position vector of any point of observation (x, y, z).
>
> **(20.31a)**

In (20.31a), $\hat{s} \cdot r$ represents the scalar or 'dot' product (see Section 10.1). If α, β, γ are the angles made by $\hat{s}$ with the positive directions of the x, y, z axes, then

$$\hat{s} = \hat{i} \cos \alpha + \hat{j} \cos \beta + \hat{k} \cos \gamma.$$

The components are direction cosines, so that

$$|\hat{s}|^2 = \cos^2\alpha + \cos^2\beta + \cos^2\gamma = 1$$

automatically, and

$$\hat{s} \cdot r = x \cos \alpha + y \cos \beta + z \cos \gamma. \tag{20.31b}$$

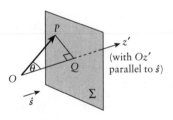

Fig. 20.16

To prove the result (20.31a), see Fig. 20.16. $P : (x, y, z)$ is a representative point. Σ is the plane which passes through P, and is perpendicular to the unit vector $\hat{s}$. OQ is perpendicular to Σ at Q, and passes through the origin O. Extend OQ to form a new coordinate axis Qz': this is parallel to $\hat{s}$ and in the same sense. Then (see (10.14))

$$\hat{s} \cdot r = |\hat{s}||r| \cos \theta = 1 \times OP \cos \theta = OQ = z',$$

where z' is the Oz' coordinate of Q (and P). The formula becomes

$$u(t, r) = A \cos(\omega t - k\hat{s} \cdot r + \phi) = A \cos(\omega t - kz' + \phi).$$

This expression is the same as (20.30), but refers to an axis Oz' parallel to $\hat{s}$, in place of the axis Oz. It therefore represents a plane travelling wave of the type (20.30) propagated in the direction of $\hat{s}$.

20.9 Compound oscillations; beats

Consider the superposition of two sinusoidal oscillations, $u_1(t)$ and $u_2(t)$. We shall limit the discussion to oscillations having the same amplitude and zero phase angle, but different frequencies, so that

$$u(t) = u_1(t) + u_2(t), \tag{20.32a}$$

with

$$u_1(t) = A \cos \omega_1 t, \qquad u_2(t) = A \cos \omega_2 t. \tag{20.32b}$$

The function $u(t)$ is not necessarily periodic. If ω_1/ω_2 is an *irrational number*, such as $\sqrt{2}$ (see Section 1.1), then $u(t)$ is *not periodic*. If ω_1/ω_2 is a *rational number*, let

$$\frac{\omega_1}{\omega_2} = \frac{p}{q},$$

where p and q are integers and p/q is in its lowest terms. In that case $u(t)$ is *periodic*, with period T given by

$$T = \frac{2\pi p}{\omega_1} \quad \left(\text{which also} = \frac{2\pi q}{\omega_2} \right). \tag{20.33}$$

Evidently T may be very much larger than the periods $2\pi/\omega_1$ and $2\pi/\omega_2$ of the individual components because of the possibly large size of the factors p and q (see Problem 20.22).

Express the difference between the component frequencies in (20.32) by $\Delta\omega$:

$$\omega_2 = \omega_1 + \Delta\omega, \tag{20.34}$$

and use the trigonometric identity from Appendix B(d):

$$u(t) = A \cos \omega_1 t + a \cos \omega_2 t$$
$$\equiv 2A \cos \tfrac{1}{2}(\omega_2 - \omega_1)t \cos \tfrac{1}{2}(\omega_1 + \omega_2)t$$
$$= (2A \cos \tfrac{1}{2}\Delta\omega_t)\cos \tfrac{1}{2}(\omega_1 + \omega_2)t. \tag{20.35}$$

This expression consists of an oscillation $\cos\tfrac{1}{2}(\omega_1 + \omega_2)t$, with angular frequency equal to the average of the component frequencies, **modulated** by an **amplitude function** $B(t)$ given by

$$B(t) = 2A \cos \tfrac{1}{2}\Delta\omega t. \tag{20.36}$$

If $\Delta\omega$ is fairly small compared with $\tfrac{1}{2}(\omega_1 + \omega_2)$, then $B(t)$ is a slowly varying function compared with $\cos\tfrac{1}{2}(\omega_1 + \omega_2)t$, and (20.35) takes on the appearance of Fig. 20.17c.

Figure 20.17 shows the components $u_1(t)$ and $u_2(t)$, and the composite function $u(t)$, together with the functions $\pm B(t)$, which form the profile of a stream of **wave packets** made up of faster oscillations.

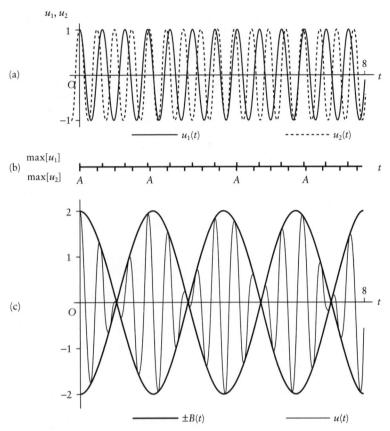

Fig. 20.17 Here $A = 1$, $\omega_1 = 10$, $\omega_2 = 13.1$. (a) $u_1 = A \cos \omega_1 t$, with $u_2 = A \cos \omega_2 t$. (b) Phase reinforcement of u_1, u_2 near points A. (c) $u_1 + u_2$ and $\pm B(t) = \pm 2A \cos \frac{1}{2} \Delta \omega t$.

These wave packets are called **beats**. The beats arise from a kind of interference: where $u_1(t)$ and $u_2(t)$ are nearly in phase (see Fig. 20.17b) they reinforce each other so that $u(t)$ is large; where they are opposed $u(t)$ is small (see Fig. 20.17b). Despite appearances, the beats will not in general contain an exact number of complete cycles of $u(t)$: in this case the period of $u(t)$ is about 31 beats long (see Problem 20.22b). The period and frequency of the *beats* (as distinct from the amplitude function $B(t)$) are defined to be equal to the period and frequency of the wave packets; therefore

Beat period

$$T_B = \tfrac{1}{2} \text{ period of } B(t) = 2\pi/(\Delta\omega).$$

Beat frequency

$$F_B = 2 \times \text{frequency of } B(t) = \Delta\omega/(2\pi). \tag{20.37}$$

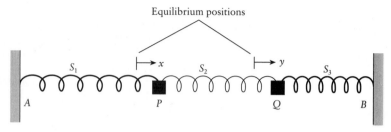

Fig. 20.18 Two loosely-coupled oscillating masses display beats.

If the wave concerned is a **sound wave**, the tone that is detected by the ear corresponds to the pulse or beat frequency rather than to that of the underlying frequencies f_1 and f_2, or of the function $B(t)$. In cases where f_1 and f_2 are large compared with the frequency f_B of the beats, the underlying rapid oscillation is sometimes referred to as a **carrier wave**, and the beats correspond to a **signal**.

For a mechanical example of the occurrence of beats, see Fig. 20.18. Unit particles P and Q are connected to fixed points A, B through springs S_1, S_2, S_3 of natural lengths l, with $l \leqslant \frac{1}{3}AB$; the stiffness of each of the springs S_1 and S_3 is K, and that of the connecting spring S_2 is k. The displacements of P, Q from equilibrium are $x(t)$, $y(t)$ respectively. The particles oscillate along the line AB, and their equations of motion are

$$\ddot{x} + (K+k)x = ky, \qquad \ddot{y} + (K+k)y = kx. \tag{20.38}$$

A related mechanical example was also considered in Section 13.8 in the chapter on eigenvalues.

It can be checked by substitution that a particular pair of solutions is given by

$$x(t) = \cos(t\sqrt{K}) + \cos[t\sqrt{(K+2k)}], \tag{20.39a}$$

$$y(t) = \cos(t\sqrt{K}) - \cos[t\sqrt{(K+2k)}]. \tag{20.39b}$$

As in (20.36b) we obtain the corresponding amplitude functions, say B_x and B_y:

$$B_x(t) = 2 \cos\tfrac{1}{2}[\sqrt{(K+2k)} - \sqrt{K}]t, \tag{20.40a}$$

and

$$\begin{aligned} B_y(t) &= -2 \sin\tfrac{1}{2}[\sqrt{(K+2k)} - \sqrt{K}]t \\ &= 2 \cos\{\tfrac{1}{2}[\sqrt{(K+2k)} - \sqrt{K}]t - \tfrac{1}{2}\pi\}, \end{aligned} \tag{20.40b}$$

which define the beats.

The particles P and Q therefore behave similarly, only their phases being different, so we shall only look in detail at the beats $B_x(t)$ corresponding to the motion $x(t)$ of P (eqn (20.40a)). To obtain an approximation, we shall assume that k is *small compared with K*. Put

$$\sqrt{K} = \omega_1, \qquad \sqrt{(K+2k)} = \omega_2.$$

Then

$$\Delta\omega = \omega_2 - \omega_1 = \sqrt{(K + 2k)} - \sqrt{K} = \sqrt{K}\{\sqrt{[1 + (2k/K)]} - 1\}$$
$$\approx k/\sqrt{K} \ll 1$$

(using the first term of the binomial theorem (Section 1.18 or 5.4) to approximate to $\sqrt{[1 + (2k/K)]}$). Therefore $\Delta\omega \approx k/\sqrt{K}$ in (20.36). The displacement $x(t)$ of particle P has beat period T_B given by (20.37):

$$T_B = \frac{2\pi}{\Delta\omega} \approx \frac{2\pi\sqrt{K}}{k}.$$

Beats with the same period also occur in $y(t)$, but are out of phase with those of $x(t)$ by half a beat period. A fixed stock of free mechanical energy is handed back and forth between P and Q: when one is vibrating vigorously, the other has only a small amplitude, and P and Q alternate in this respect. The same phenomenon occurs when two pendulum bobs are coupled by a weak spring.

20.10 Travelling waves; beats

Consider the superposition of two sinusoidal travelling waves, $u_1(t, z)$ and $u_2(t, z)$, having the same amplitude and zero phase angle, but different frequencies, so that

$$u(t, z) = u_1(t, z) + u_2(t, z), \tag{20.41a}$$

where

$$u_1(t, z) = A\cos(\omega_1 t - k_1 z), \quad u_2(t, z) = A\cos(\omega_2 t - k_2 z). \tag{20.41b}$$

By the identity from Appendix B(d), $u(t, z)$ may be written as

$$u(t, z) = 2A\cos[\tfrac{1}{2}(\omega_2 - \omega_1)t - \tfrac{1}{2}(k_2 - k_1)z]\cos[\tfrac{1}{2}(\omega_2 + \omega_1)t \\ - \tfrac{1}{2}(k_2 + k_1)z]. \tag{20.42}$$

In this section we visualize graphs of u plotted against z as in Fig. 20.19, for different values of time t.

Suppose firstly that *the phase velocity v is constant for all sinusoidal waves.* Put

$$\omega_2 = \omega_1 + \Delta\omega, \quad k_2 = k_1 + \Delta k. \tag{20.43}$$

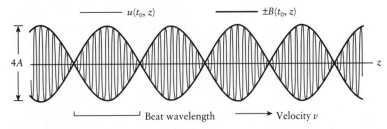

Fig. 20.19 $u(t, z)$ against z at a particular time.

The wave numbers and angular velocities are connected by (20.28):

$$vk_1 = \omega_1, \quad vk_2 = \omega_2, \quad v\Delta k = \Delta\omega.$$

Equation (20.42) becomes

$$u(t, z) = 2A \cos\tfrac{1}{2}\Delta\omega[t - (z/v)]\cos\tfrac{1}{2}(\omega_1 + \omega_2)[t - (z/v)]. \qquad (20.44)$$

This wave has the same form as the oscillation discussed in the previous section (eqn (20.36b)), except that we have $t - (z/v)$ in place of t. Plotted against z, as in Fig. 20.19, the wave travels unchanged at speed v. There is a carrier wave with wavelength $\lambda_C = 2\pi/[\tfrac{1}{2}(k_1 + k_2)]$, multiplied by a beat function $B(t, z)$:

$$B(t, z) = 2A \cos(\tfrac{1}{2}t\Delta\omega - \tfrac{1}{2}z\Delta k).$$

In terms of k the beat wavelength λ_B is

$$\lambda_B = \tfrac{1}{2}(\text{wavelength of } B(t, z)) = \tfrac{1}{2}[2\pi/(\tfrac{1}{2}\Delta k)] = 2\pi/(\Delta k)$$

and the beat frequency f_B is

$$f_B = 2(\text{frequency of } B(t, z)) = 2[(\tfrac{1}{2}\Delta\omega)/(2\pi)] = \Delta\omega/(2\pi).$$

The beats travel with the phase velocity v:

$$\text{beat velocity} = \tfrac{1}{2}\Delta\omega/(\tfrac{1}{2}\Delta k) = v\Delta k/(\Delta k) = v.$$

This is to be expected, since the components u_1 and u_2 have equal velocity v, and therefore remain in a constant phase relationship, reinforcing and cancelling each other over segments that remain in step as the waves travel. The theory of beats in travelling waves is related to **frequency modulation** in radio transmission.

20.11 Dispersion; group velocity

There are media and special situations where *the velocity of a travelling sinusoidal wave varies with its frequency* (or, equivalently, with its wavelength). Such waves are called **dispersive waves**: light waves are dispersive, leading to their spectral decomposition upon entering a refractive medium.

In a dispersive medium, one component wave will overtake the other; therefore u_1 and u_2 will not maintain a constant phase relationship as the wave travels, and the velocity associated with the beats is affected. Suppose that two dispersive waves, $u_1(t, z)$ and $u_2(t, z)$, have different angular frequencies ω_1 and ω_2, and phase velocities v_1 and $v_2 \neq v_1$ (whose values will depend on ω_1 and ω_2). Refer back to (20.44): if ω_2 and ω_1 are fairly close then distinct beats occur (in both time t and space z). Their profile is determined by the curves $\pm B(t, z)$, where

$$B(t, z) = 2A \cos(\tfrac{1}{2}t\Delta\omega - \tfrac{1}{2}z\Delta k), \qquad (20.45)$$

in which $\Delta\omega = \omega_2 - \omega_1$, $\Delta k = k_2 - k_1$. This beat profile represents a wave travelling along the z axis with a velocity v_g, called the **group velocity**, where

$$v_g = \frac{\frac{1}{2}\Delta\omega}{\frac{1}{2}\Delta k} = \frac{\omega_2 - \omega_1}{k_2 - k_1}. \tag{20.46}$$

In general v_g will differ from the phase velocities v_1 and v_2 of the constituent waves u_1 and u_2, given by $v_1 = \omega_1/k_1$, $v_2 = \omega_2/k_2$. In so-called **anomalous** cases, v_g can even exceed them; a signal may travel faster than the phase velocity of the constituent waves.

If the wave number k for a sinusoidal wave in a medium (whether dispersive or not) is prescribed, then the value of the angular frequency ω is also settled by $\omega = kv$. We can therefore regard both ω and v as functions of k only, and write

$$\omega(k) = kv(k). \tag{20.47}$$

A graph of $v(k)$ against k would contain all we need to know about the behaviour of v to enable wave interactions of any degree of complexity to be computed. (We could instead work with ω rather than k as the independent variable, or λ, or f (see Problem 20.18), but in any case only *one parameter* is needed to specify the variation of v.)

We shall relate this observation to the group velocity problem just discussed, for cases where *the wave number Δk of the beats is small compared with the wave number k of the carrier wave*. In this case the beats will be very distinct. From (20.45), (20.46), and (20.47)

$$\frac{\Delta\omega}{\Delta k} = v_g = \frac{\omega_2 - \omega_1}{k_2 - k_1} = \frac{k_2 v(k_2) - k_1 v(k_1)}{k_2 - k_1}. \tag{20.48}$$

The form of this equation suggests that we can approximate to v_g by an expression involving the derivative of $kv(k)$. Suppose that

$$\Delta k = k_2 - k_1 \to 0, \quad \text{or} \quad k_2 \to k_1.$$

For simplicity, write k in place of k_1; then from (20.47)

$$\frac{\Delta\omega}{\Delta k} \to \frac{\mathrm{d}(kv(k))}{\mathrm{d}k} = v(k) + \frac{k\,\mathrm{d}v(k)}{\mathrm{d}k}.$$

Therefore, for any value of k, and Δk small enough we have

Group velocity approximation

$$v_g \approx v + k\frac{\mathrm{d}v}{\mathrm{d}k},$$

or

$$\text{group velocity} \approx \text{phase velocity} + k\frac{\mathrm{d}v}{\mathrm{d}k}. \tag{20.49}$$

We have assumed Δk to be 'small', but in a physical context one would like to know when Δk is small enough for the approximation to be useful. It is sufficient that the dimensionless quantity $\Delta k/k$ should be small (see Problems 20.25 and 20.26).

20.12　The Doppler effect

Suppose that a steady, plane sound wave is being emitted from a large plane membrane vibrating with frequency f cycles per second, which is moving from left to right with velocity $u < v$, where v is the sound velocity relative to the medium (see Fig. 20.20). E and E' show the positions of the membrane at times t_0 and $t_1 > t_0$. There is a fixed observation point at P. The wave front that is at P at time t_0 travels with phase velocity v to the point Q at t_1. Speaking broadly, if the emitter is chasing the waves, then a given number of waves occupy a shorter length ahead of the emitter than if it were stationary. Therefore the motion of E reduces the wavelength λ. The frequency f of the emitter and the phase velocity v (relative to the medium, which we take to be stationary) are fixed, so that the frequency of arrival of waves at any *fixed* point P must be greater than f.

To examine the effect quantitatively, see Fig. 20.20. In the following, u may be positive, as shown, or negative (corresponding to E moving oppositely to the wave direction). Put

$$EP = L, \quad t_1 - t_0 = \Delta t.$$

Then at time t_1, E has moved to E' and the wave front to Q, so that

$$E'Q = L - u\Delta t + v\Delta t = L + (v - u)\Delta t.$$

Let the wavelength be λ. Then EP contains L/λ wavelengths, and $E'Q$ contains

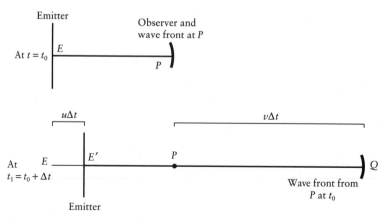

Fig. 20.20 A plane vibrating membrane at E is moving forward with velocity $u < v$.

$$\frac{E'Q}{\lambda} = \frac{L + (v - u)\Delta t}{\lambda} \text{ wavelengths.}$$

In the interval Δt, $f\Delta t$ new waves have been generated, so that

$$\frac{L + (v - u)\Delta t}{\lambda} - \frac{L}{\lambda} = f\Delta t, \quad \text{or} \quad \frac{(v - u)\Delta t}{\lambda} = f\Delta t.$$

Therefore

$$\lambda = \frac{v - u}{f}. \tag{20.50}$$

The fixed receiver at P records, say, f_P waves per second, and these are travelling at the normal phase velocity v, so from (20.28) and (20.50),

$$f_P = \frac{v}{\lambda} = \frac{vf}{v - u} = \frac{f}{1 - (u/v)}. \tag{20.51}$$

If $u > 0$ (E moving in the direction of v) then $f_P > f$. If $u < 0$ (E moving oppositely in the direction of v) then $f_P < f$.

The effect is observable if a vehicle with a siren speeds past an observer at a point P: as it passes there is a sudden lowering of the pitch. If the speed of the vehicle is u, the frequency drop Δf_P is given by

$$\Delta f_P = \frac{f}{1 - (u/v)} - \frac{f}{1 + (u/v)}, \tag{20.52}$$

which is approximately equal to $2uf/v$ if u/v is small. The so-called 'red shift' in astronomy, by which the velocity of a receding galaxy can be estimated from the change towards longer wavelengths (lower frequencies) in its spectrum, is explained on the same lines.

Problems

20.1 Express the following in standard amplitude–phase form $C \cos(\omega t + \phi)$, with $C > 0$ and $-\pi < \phi \leqslant \pi$.
(a) $3 \cos(3t + \frac{3}{2}\pi)$; (b) $3 \cos(\omega t - 3\pi)$; (c) $2 \sin 3t$;
(d) $3 \sin(2t + \frac{1}{2}\pi)$; (e) $-3 \cos(2t - \frac{1}{2}\pi)$;
(f) $-4 \cos(2t + \frac{1}{4}\pi)$; (g) $-\sin t$;
(h) $3 \cos 2t + 4 \sin 2t$; (i) $\cos 2t + \cos(2t - \pi)$;
(j) $\cos(2t - \frac{3}{2}\pi) - \cos(2t + \frac{3}{2}\pi)$.

20.2 State whether x leads or lags y in the following cases, and by how much.
(a) $x = 4 \cos 3t$, $y = 3 \cos(3t - \frac{1}{2}\pi)$.
(b) $x = 2 \cos(2t + \frac{1}{4}\pi)$, $y = 3 \cos(2t + \frac{9}{2}\pi)$.
(c) $x = -3 \cos 2t$, $y = 4 \cos 2t$.
(d) $x = \cos 3t$, $y = \sin 3t$.
(e) $x = 2 \cos 3t$, $y = \cos(3t - \frac{9}{4}\pi)$.

20.3 Obtain the free oscillations of the following in the form $C \cos(\omega t + \phi)$. State (i) the natural frequency if the damping coefficient is put to zero; (ii) the frequency that actually occurs in the cosine term of the solution; (iii) the number of complete cycles needed for the amplitude to drop to 0.1 of its value at $t = 0$.
(a) $x'' + 20x' + (2.5 \times 10^5)x = 0$.
(b) $x'' + 0.5x' + 4x = 0$.
(c) $x'' + 0.15x' + 3x = 0$.
(d) $x'' + x' + 20x = 0$.

20.4 Express $A \cos \omega t + B \sin(\omega t + \frac{1}{4}\pi)$ in the standard form $C \cos(\omega t + \phi)$ when (a) $A = 3^{\frac{1}{2}}, B = 1$; (b) $A = 3^{\frac{1}{2}}$, $B = -1$; (c) $A = -3^{\frac{1}{2}}, B = 1$; (d) $A = -3^{\frac{1}{2}}, B = -1$.

20.5 (a) Show that the maxima and minima of $x(t) = C \, e^{-kt} \cos(\omega t + \phi)$ occur at times T_N given by

$$\omega T_N + \phi = -\arctan\left(\frac{k}{\omega}\right) + N\pi,$$

where N is any integer.

(b) Show that the values of $x(t)$ at these points are given by

$$x(T_N) = \frac{(-1)^N \omega C \, e^{-kT_N}}{(\omega^2 + k^2)^{\frac{1}{2}}}.$$

20.6 Consider an expression of the form

$$x(t) = e^{-t/T} g(t),$$

where T is a constant, and $g(t)$ itself does not have any term in it like $e^{\pm kt}$ (e.g. $g(t)$ might be a constant, or $\cos t$, or even t^3, but it must not be, for example, $e^{-2t} \cos t$). Then T is called the time constant of $f(t)$.

(a) State the time constant for Q_c in Example 20.4.

(b) Describe how T provides a measure of the rate of exponential decay of $x(t)$, rather like the half-life period of a radioactive substance.

20.7 (Heavy damping). Find the general solution of the equation

$$x'' + 2kx' + \omega^2 x = 0$$

when $k^2 > \omega^2$. Describe the general character of the solutions, contrasting them with the case when $k^2 < \omega^2$.

20.8 Solve the equation

$$x'' + 10x' + 24x = 0$$

subject to the initial conditions $x(0) = -3$, $x'(0) = 20$. Show that the solution curve crosses the t axis only once, at the point $t = \ln 2$.

20.9 ('Critical damping'). Find the general solution of the equation

$$x'' + 2kx' + \omega^2 x = 0$$

for the case when $k^2 = \omega^2$.

20.10 The following equation could represent the damped vertical motion of a mass supported by a spring and subjected to an external periodic force:

$$x'' + x' + 36x = 10 \cos \omega t, \quad \text{for } t > 0,$$

the system being in equilibrium under no force for $t \leqslant 0$.

(a) Find the period of the free (damped) oscillations. Show that any free oscillations stimulated at startup are reduced by a factor of about 14 after five periods of oscillation.

(b) Obtain expressions in terms of ω for the amplitude and phase of the forced oscillation.

(c) Find the condition for resonance.

(d) Plot curves of amplitude and phase against ω for a range $4 \leqslant \omega \leqslant 8$.

20.11 A particle rolls to and fro under gravity at the bottom of a parabolic cylinder having vertical cross-section $y = ax^2$. There is negligible friction. The equation of motion in terms of horizontal displacement x is then

$$x'' + 2ax(g + 2ax'^2)/(1 + 4a^2 x^2) = 0.$$

Show that for small oscillations the period is $2^{\frac{1}{2}} \pi/(ag)^{\frac{1}{2}}$.

20.12 A particle is balanced at the topmost point, $x = y = 0$, of an inverted parabolic cylinder whose shape is described by $y = -ax^2$, y being measured vertically upward. Its equation of motion is

$$x'' + 2ax(2ax'^2 - g)/(1 + 4a^2 x^2) = 0.$$

By linearizing the equation show that, if the particle is slightly disturbed, it starts to move away from its initial position $(0, 0)$ at an increasing rate. (This condition is called **unstable equilibrium.**)

20.13 The equation for the displacement $x(t)$ of an electrical circuit fixed on springs and influenced by a current-carrying conductor is

$$x'' + 4[x - 2/(3 - x)] = 0.$$

(a) Show that there are two positions x at which the circuit could theoretically be in equilibrium. ('**Equilibrium**' means that

$$x(t) = \text{constant}$$

is a solution of the equation.)

(b) Call the equilibrium positions $x = a$ and $x = b$. To investigate the state of affairs near $x = a$, put

$$x = a + u$$

into the equation, so as to obtain an equation for $u(t)$, which is the distance from a. Then do the same thing near $x = b$ by putting

$$x = b + v,$$

where v is distance from b. Tidy the equations as far as possible.

(c) Suppose that u in one case, and v in the other, are small, and linearize the equations in each case.

(d) Show that in one case small oscillations take place, but that in the other the displacement tends to increase. (One is called a **stable** equilibrium state, the other **unstable**.)

20.14 A particle moves in a plane under a central attractive force γ/r^α per unit mass, where r and θ are its plane polar coordinates relative to an origin in the attracting body. Its equation of motion can be expressed in the form

$$\frac{d^2u}{d\theta^2} + u - \frac{\gamma}{H^2}u^{\alpha-2} = 0,$$

where $u = r^{-1}$ and H is its (constant) angular momentum per unit mass.

Show that the equation has a *constant* solution $u = u_0$, which is equivalent to a circular orbit. Does it stay close to this orbit if its position u is slightly changed from u_0, while H keeps its original value? (Hint: put $u = u_0 + x$ and linearize the equation for small x. You may assume that for small values of x/u (see (5.4d))

$$(u_0 + x)^{\alpha-2} \approx u_0^{\alpha-2}\left(1 + (\alpha - 2)\frac{x}{u_0}\right).$$

20.15 Given the expression for the forced amplitude A in eqn (20.16), deduce the expressions for the resonant frequency and resonant amplitude.

20.16 (a) A plane sound wave has period $\frac{1}{250}$ s and wavelength 1.2 m. Obtain the speed of sound in the medium.

(b) There is a broadcasting station with tuning frequency close to 100 MHz. Obtain the corresponding wavelength. (The speed of electromagnetic waves can be taken to be 3×10^8 m s^{-1} in round figures.)

20.17 Show that the stationary plane wave

$$u(t, x) = A \cos[(2\pi t/T) + \phi]\cos[(2\pi z/\lambda) + \alpha]$$

(see eqn (20.24)) is equivalent to the superposition of two plane waves travelling in opposite directions.

20.18 $u = \cos(\omega t - kz + \phi)$ represents a travelling plane wave. Express u in terms of (i) period T and wavelength λ; (ii) frequency f and wavelength λ; (iii) circular frequency ω and phase velocity v.

20.19 Show that the travelling wave $u = \cos(\omega t - kz + \phi)$ is equivalent to the superposition of two stationary waves.

20.20 $u(t, z) = A \cos(4500t - 3z)$ represents a travelling wave. Obtain the phase velocity v, the period t, the frequency f, and the wavelength λ.

20.21 Obtain a general expression for a plane harmonic wave $u(t, x, y, z)$ of angular frequency ω, travelling in a medium of wave velocity v, in the direction making equal (acute) angles with the axes Ox, Oy, Oz.

20.22 (a) Prove that $u(t, z) = A \cos \omega_1 t + A \cos \omega_2 t$ is periodic if, and only if, ω_1/ω_2 is a rational number (see Section 1.1), and confirm eqn (20.33) for the period.

(b) Obtain the period T of the oscillation cos $10t$ + cos $13.1t$ (shown in Fig. 20.17). Compare T with the period T_B of the beats, and with the periods T_1 and T_2 of the constituent oscillations.

20.23 (a) Obtain the sum of the two oscillations given by $A_1 \cos(\omega t + \phi_1)$ and $A_2 \cos(\omega t + \phi_2)$, in the form $A \cos(\omega t + \phi)$.

(b) Obtain the sum of two travelling waves $A_1 \cos(\omega t - kz + \phi_1)$ and $A_2 \cos(\omega t - kz + \phi_2)$ in the form of another travelling wave.

20.24 (a) A beam of light falls perpendicularly upon a surface, and is reflected without change of amplitude or phase. What is the nature of the combined wave?

(b) Consider separately the effects of a change in phase and a change in amplitude upon reflection.

20.25 Two superposed plane waves, u_1 and u_2, travel in the z direction through a dispersive medium in which the phase velocity v is regarded as a function of wavelength λ. They have the same amplitude A and phase angle zero, but different wave numbers (and consequently different angular frequencies ω). Show that $v_g = v - \lambda \, dv/d\lambda$, where v_g is the group velocity. (Hint: start with $v_g = \Delta\omega/\Delta k$.)

20.26 In the notation of Section 20.10, prove the identities

$$\left(1 + \frac{\Delta f}{f}\right)\left(1 + \frac{\Delta\lambda}{\lambda}\right) = 1 + \frac{\Delta v}{v}$$

and

$$\left(1 + \frac{\Delta k}{k}\right)\left(1 + \frac{\Delta v}{v}\right) = 1 + \frac{\Delta w}{w}.$$

(These relations are exact, but show that when small values are being considered the natural variables to use are $\Delta f/f$ etc.)

20.27 A fire truck speeds along a highway at 100 km h^{-1}, sounding its siren at a frequency of 350 cycles per second. Obtain the drop in pitch noticed by an observer standing on the sidewalk as it goes past.

21

Steady forced oscillations: phasors, impedance, transfer functions

CONTENTS

21.1 Phasors

We shall consider circuits driven by an applied *harmonically alternating* voltage, with resistances placed so that any free oscillations set up by switching on the circuit die away, leaving only a periodic forced oscillation, as described in Section 20.5. It is only this remaining, **steadily-oscillating state** that is discussed here.

Let $x(t)$ represent any variable in the circuit, such as the current in a particular branch. If the frequency of the applied voltage is $\omega/2\pi$, then all these possible variable $x(t)$ share the **same frequency** $\omega/2\pi$ once the transients have died away, though in general the **phases and amplitudes of different variables are different.** Here we adopt the standardized amplitude/phase form of (20.2), assuming that

$$x(t) = c\cos(\omega t + \phi), \quad \text{with } c > 0 \text{ and } -\pi < \phi \leqslant \pi. \tag{21.1}$$

We can write $x(t)$ in a complex form instead:

$$x(t) = \mathrm{Re}(c\,\mathrm{e}^{\mathrm{i}(\omega t + \phi)}) = \mathrm{Re}(c\,\mathrm{e}^{\mathrm{i}\phi}\,\mathrm{e}^{\mathrm{i}\omega t}).$$

The complex coefficient $c\,\mathrm{e}^{\mathrm{i}\phi}$ that multiplies $\mathrm{e}^{\mathrm{i}\omega t}$ is called the **phasor corresponding to** $x(t)$, and it is independent of time t. **Every variable** $x(t)$ **will have its own phasor**, but the factor $\mathrm{e}^{\mathrm{i}\omega t}$ is the same for each one. In a circuit, ϕ and c usually depend on ω, so the values of the corresponding phasors will depend on ω.

Corresponding to each variable denoted by a lowercase letter, we use a bold capital letter to denote the phasor. This style is traditional, and emphasizes that phasors, being complex numbers, can be treated as vectors in the Argand diagram. Typically, the variables are the voltages between any two nodes of a circuit, and the currents in any branch.

> **Phasor of a harmonic oscillation**
>
> The phasor of $x(t) = c \cos(\omega t + \phi)$ is the complex number
> $X = c\,e^{i\phi}$.
>
> (21.2)

In engineering applications, phasors $c\,e^{i\phi}$ are sometimes written in the form $c\underline{/\phi}$, and ϕ may be expressed in degrees. Thus, if $X = 3\,e^{-\frac{1}{4}\pi i}$, we can write

$$X = 3\underline{/-\tfrac{1}{4}\pi} = 3\underline{/-45^\circ}.$$

The two numbers displayed are the polar coordinates of the point which represents the phasor on an Argand diagram, in this case the point $(3/\sqrt{2}, -3/\sqrt{2})$ corresponding to

$$X = 3\cos(-45^\circ) + i3\sin(-45^\circ) = \frac{3}{\sqrt{2}} - i\frac{3}{\sqrt{2}}.$$

It is often convenient to express a phasor in the form $a + ib$ rather than in the polar form $c\,e^{i\phi}$.

Example 21.1   *Find the phasor of $x(t) = -3\cos(2t + \tfrac{1}{2}\pi)$.*

In standard form (21.1), $x(t) = 3\cos(2t - \tfrac{1}{2}\pi)$. The phasor X is therefore given by $X = 3\,e^{-\frac{1}{2}\pi i}$ or $3\underline{/-90^\circ}$.

Example 21.2   *Given that the prevailing angular frequency is $\omega = 10^4$, find the functions $x(t)$ having the following phasors.*
(a) $X = 1/(-1 + i)$, (b) $X = (1 - \sqrt{3}i)/(-1 + i)$.

(a) Put X into polar form:

$$X = \frac{1}{-1 + i} = \frac{-1 - i}{(-1)^2 + 1^2} = -\tfrac{1}{2} - \tfrac{1}{2}i = \frac{1}{\sqrt{2}}\,e^{-\frac{3}{4}\pi i}.$$

Therefore

$$x(t) = \frac{1}{\sqrt{2}}\cos(10^4 t - \tfrac{3}{4}\pi).$$

(b) $1 - \sqrt{3}i = 2\,e^{-\frac{1}{3}\pi i}$ (as can be seen by putting the point $1 - \sqrt{3}i$ on an Argand diagram). Therefore, using (a),

$$X = (2\,e^{-\frac{1}{3}\pi i})\left(\frac{1}{\sqrt{2}}\,e^{-\frac{3}{4}\pi i}\right) = \sqrt{2}\,e^{-\frac{13}{12}\pi i}.$$

The phase $(-\tfrac{13}{12}\pi)$ is out of the standard range (21.1), so add 2π to it, leaving X unchanged. We obtain $X = \sqrt{2}\,e^{\frac{11}{12}\pi i}$, so

$$x(t) = \sqrt{2}\cos(10^4 t + \tfrac{11}{12}\pi).$$

Example 21.3   *Let $x(t) = \sqrt{3}\cos \omega t - \sin \omega t$. Find the corresponding phasor.*

Take the terms separately:

$$\sqrt{3}\cos \omega t = \mathrm{Re}(\sqrt{3}\,\mathrm{e}^{\mathrm{i}\omega t});$$
$$\sin \omega t = \cos(\omega t - \tfrac{1}{2}\pi) = \mathrm{Re}(\mathrm{e}^{-\frac{1}{2}\pi\mathrm{i}}\,\mathrm{e}^{\mathrm{i}\omega t}).$$

Combining them, we obtain

$$x(t) = \mathrm{Re}[(\sqrt{3} - \mathrm{e}^{-\frac{1}{2}\pi\mathrm{i}})\,\mathrm{e}^{\mathrm{i}\omega t}].$$

Therefore

$$X = \sqrt{3} - \mathrm{e}^{-\frac{1}{2}\pi\mathrm{i}} = \sqrt{3} - (-\mathrm{i})$$
$$= 2\,\mathrm{e}^{\frac{1}{6}\pi\mathrm{i}} \text{ or } 2\underline{/30°}.$$

21.2   Algebra of phasors

As seen in Example 21.3, when oscillations associated with the same value of ω combine by addition, so do their phasors. Suppose, for instance, that $u(t)$ and $v(t)$ have the same angular frequency ω, and that their phasors are U and V. Then $u(t) = \mathrm{Re}(U\,\mathrm{e}^{\mathrm{i}\omega t})$ and $v(t) = \mathrm{Re}(V\,\mathrm{e}^{\mathrm{i}\omega t})$, so

$$u(t) + v(t) = \mathrm{Re}[(U + V)\,\mathrm{e}^{\mathrm{i}\omega t}],$$

whose phasor is $U + V$. The addition holds similarly if there are more terms present.

Addition principle for phasors

If $u(t), v(t), \dots$ have a common frequency, and
$z(t) = u(t) + v(t) + \cdots$, then $Z = U + V + \cdots$, where
$Z, U, V, \dots$ are the corresponding phasors.

(21.3)

Differentiation and integration give important results. If $x(t) = \mathrm{Re}(X\,\mathrm{e}^{\mathrm{i}\omega t})$, where X is the phasor, then $\mathrm{d}x/\mathrm{d}t = \mathrm{Re}(\mathrm{i}\omega X\,\mathrm{e}^{\mathrm{i}\omega t})$, so that the phasor of $\mathrm{d}x/\mathrm{d}t$ is $\mathrm{i}\omega X$. Differentiate again, and a further factor $\mathrm{i}\omega$ is introduced, so that the phasor of $\mathrm{d}^2 x/\mathrm{d}t^2$ is $(\mathrm{i}\omega)^2 X$, and so on. For $\int x(t)\,\mathrm{d}t$, we find in the same way that the phasor is $X/\mathrm{i}\omega$. The additive arbitrary constant in $\int \mathrm{e}^{\mathrm{i}\omega t}\mathrm{d}t$ has been put to zero because, in normal use, all the variables that occur oscillate.

Phasors of derivatives and integrals

Variable:	x	$\dfrac{\mathrm{d}x}{\mathrm{d}t}$	$\dfrac{\mathrm{d}^2 x}{\mathrm{d}t^2}$	$\displaystyle\int x\,\mathrm{d}t$
Phasor:	$X = c\,\mathrm{e}^{\mathrm{i}\phi}$	$\mathrm{i}\omega X$	$-\omega^2 X$	$\dfrac{1}{\mathrm{i}\omega}X$

(21.4)

Example 21.4   *Obtain the phasor of the expressions*

(a) $L\dfrac{d^2q}{dt^2} + R\dfrac{dq}{dt} + \dfrac{q}{C}$, (b) $L\dfrac{di}{dt} + \dfrac{1}{C}\displaystyle\int i\,dt$, *in terms of the phasors*

Q *of* $q(t)$ *and* I *of* $i(t)$. (L, R, *and* C *are circuit constants, and the prevailing frequency is* ω.)

(a) From (21.4) and the addition principle (21.3), the phasor is

$$L(i\omega)^2 Q + R(i\omega)Q + (1/C)Q = [(1/C - L\omega^2) + iR\omega]Q.$$

(b) The phasor is

$$L(i\omega)I + (1/C\,i\omega)I = i(L\omega - 1/C\omega)I.$$

Example 21.5   *Find the steady-state solution of*

$$\frac{d^2x}{dt^2} + 8\frac{dx}{dt} + 10^4 x = 2\times 10^3 \cos 90t.$$

This is equivalent to the circuit equation in Example 20.5, with $x(t)$ in place of $q(t)$. The prevailing value of ω is 90. Let X be the phasor of $x(t)$. The phasor of the right-hand side is 2×10^3, so by using (21.4) we obtain

$$[(90i)^2 + 8(90i) + 10^4]X = 2\times 10^3,$$

or

$$(1900 + 720i)X = 2\times 10^3,$$

from which X can be found.

$$X = \frac{2\times 10^3}{1900 + 720i} = \frac{1}{0.95 + 0.36i}$$

$$= \frac{1}{1.0159\,e^{0.3622i}} = 0.984\,e^{-0.362i}.$$

Therefore

$$x(t) = \text{Re}[0.984\,e^{-0.362i}\,e^{90it}]$$
$$= 0.984\cos(90t - 0.362),$$

as we found in Example 20.4 for the forced oscillation.

21.3 Phasor diagrams

Complex numbers can be represented by vectors in an Argand diagram (see Section 6.2), and they are added in the same way as the corresponding vectors. Phasors are just complex numbers, so they **add like vectors** too. This fact can be used to show pictorially how a number of superposed oscillations which are not in phase with each other contribute to the sum. The diagrams concerned are called **phasor diagrams**.

(a) Imaginary axis

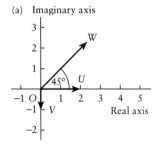

(b) Imaginary axis

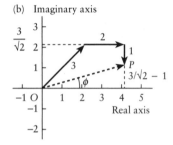

Fig. 21.1 (a) Argand diagram showing U, V, W. (b) The sum $U + V + W = \overline{OP}$.

Example 21.6   Let $u(t) = 2 \cos 10t$, $v(t) = \cos(10t - \frac{1}{2}\pi)$, and $w(t) = 3 \cos(10t + \frac{1}{4}\pi)$. Find $p(t) = u(t) + v(t) + w(t)$ by means of a phasor diagram.

The phasors corresponding to u, v, and w are $U = 2$, $V = e^{-\frac{1}{2}\pi i}$, and $W = 3 e^{\frac{1}{4}\pi i}$. In the polar-coordinate notation they are $U = 2\underline{/0°}$, $V = 1\underline{/-90°}$, $W = 3\underline{/45°}$. They are shown as position vectors in Fig. 21.1a, and in Fig. 21.1b they are strung together as usual for addition. The vector $\overline{OP}$ can be measured off from the diagram, or calculated using the dimensions shown. We have

$$|\overline{OP}| = [(3/\sqrt{2} + 2)^2 + (3/\sqrt{2} - 1)^2]^{\frac{1}{2}} = 4.27,$$

$$\phi = \arctan \frac{3/\sqrt{2} - 1}{3/\sqrt{2} + 2} = 0.479 \text{ (radians)}.$$

Therefore $p(t) = 4.27 \cos(10t + 0.479)$.

21.4 Phasors and complex impedance

In the following table, an electric current

$$i(t) = c \cos(\omega t + \phi)$$

with phasor

$$I = c\, e^{i\phi}$$

is caused to pass through a resistor, an inductor, and a capacitor, separately. The resulting voltage drop $v(t)$ associated with each is shown, together with its phasor V. It is the unique steadily oscillating state that is being described by the phasors.

	Resistor	Inductor	Capacitor
Voltage drop:	$v = Ri$	$v = L\dfrac{di}{dt}$	$v = \dfrac{1}{C}\displaystyle\int i\, dt$
Voltage phasor V:	$Rc\, e^{i\phi} = RI$	$i\omega L I$	$\dfrac{1}{i\omega C} I$
Voltage phase:	ϕ (in phase)	$\phi + \frac{1}{2}\pi$ (v leads i)	$\phi - \frac{1}{2}\pi$ (v lags i)

(21.5)

A similar table can be constructed if the voltage rather than current is prescribed. The entries can be read from the table above; for example, if the phasor of the voltage applied to an inductor is V, the phasor of the resulting current is $V/i\omega L$.

Discussion of circuits in terms of phasors is said to take place in the **frequency domain**, rather than the **time domain** associated with the differential equations of the circuits.

Each of the three cases in the table can be written in the form

$$V = ZI,$$

where Z is either R, $i\omega L$, or $(i\omega C)^{-1}$. The quantity Z is called the **complex impedance** of these elements. There is a plain analogy with Ohm's law for direct current through a resistance. We have

> **Complex impedance Z**
>
> Resistor $Z = R$
> Inductor $Z = i\omega L$
>
> Capacitor $Z = \dfrac{1}{i\omega C}.$
>
> (21.6)

By stringing elements of this type together in series and parallel, we can form composite units. The combined unit has a complex impedance which is the sum of the complex impedances of the individual elements:

Example 21.7 *Show that the complex impedance Z of two elements in series, whose complex impedances are Z_1 and Z_2, is given by*

$$Z = Z_1 + Z_2.$$

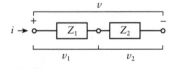

Fig. 21.2

Suppose that the impedance of the unit is Z; we mean by this that, if V is the phasor of the voltage drop across the unit and I is the phasor of the current through it (see Fig. 21.2), then

$$V = ZI.$$

From Fig. 21.2, $v = v_1 + v_2$; therefore, by (21.3), the corresponding phasors satisfy

$$V = V_1 + V_2.$$

But i, and therefore I, is the same for Z_1 and Z_2, so

$$V_1 = Z_1 I, \qquad V_2 = Z_2 I.$$

Therefore $V = Z_1 I + Z_2 I = ZI$, or

$$Z = Z_1 + Z_2.$$

If the two impedances are in parallel, the analogy with Ohm's law again exists:

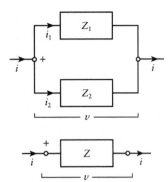

Fig. 21.3 Two impedances in parallel and their combined impedance.

Example 21.8 *Show that the complex impedance Z of any two elements Z_1 and Z_2 in parallel is given by $\dfrac{1}{Z} = \dfrac{1}{Z_1} + \dfrac{1}{Z_2}$.*

From Fig. 21.3, $i = i_1 + i_2$; so, by (21.3), $I = I_1 + I_2$. The voltage drop is the same for both branches, so

$$I_1 = V/Z_1, \quad I_2 = V/Z_2, \quad I = V/Z.$$

Therefore $I = (1/Z_1 + 1/Z_2)V = (1/Z)V$, from which the result follows.

It is easy to extend these two results to encompass more elements, and therefore we have the following general result.

Complex impedance Z of series and parallel circuits

(a) Impedances $Z_1, Z_2, \ldots$, in series:

$$Z = Z_1 + Z_2 + \cdots.$$

(b) Impedances $Z_1, Z_2, \ldots$, in parallel:

$$\frac{1}{Z} = \frac{1}{Z_1} + \frac{1}{Z_2} + \cdots.$$

(21.7)

The analogy with resistive circuits, evident from these formulae, goes much further. The general rules which govern voltages and currents in a passive linear circuit are Kirchhoff's laws: (i) that the algebraic sum of the voltages around any closed circuit is zero; (ii) that the resultant current entering any junction is zero. There is also a linear voltage/current relation for each branch. In terms of phasors and complex impedances for a circuit in a state of steady harmonic oscillation, these conditions become the following.

Kirchhoff's laws

Around any closed circuit, $\sum V = 0$.

At any junction, $\sum I = 0$.

On any branch, $V = ZI$.

(21.8)

These rules have the same form as the rules for resistive direct-current circuits, with V, I, and Z appearing in them in place of v, i, and R. It follows that **general rules applicable to DC circuits may be borrowed** for the purpose of the circuits we have been considering. Such rules are the Wheatstone bridge rules, Thévenin's theorem, and the structure of equivalent circuits. However, the restriction to steady harmonic oscillation must be remembered: many circuits can be made to 'balance' like a Wheatstone bridge for steady oscillations, but not for more general disturbances.

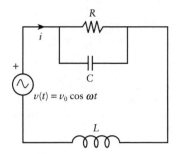

Fig. 21.4

Example 21.9 *Find the steady alternating current in the circuit shown in Fig.* 21.4.

The unit comprising R and C consists of two complex impedances in parallel, R and $(i\omega C)^{-1}$. If Z is the combined impedance, then

$$\frac{1}{Z} = \frac{1}{R} + \frac{1}{(i\omega C)^{-1}},$$

which gives

$$Z = \frac{R}{1 + i\omega RC}.$$

Z is in series with the other impedance, $i\omega L$, so the impedance of the circuit is given by

$$Z = \frac{R}{1 + i\omega RC} + i\omega L = \frac{R(1 - \omega^2 LC) + i\omega L}{1 + i\omega RC}.$$

Since $I = V/Z$, and $V = v_0$, we obtain

$$I = \frac{v_0(1 + i\omega RC)}{R(1 - \omega^2 LC) + i\omega L} = \frac{v_0(1 + \omega^2 R^2 C^2)^{\frac{1}{2}}}{[R^2(1 - \omega^2 LC)^2 + \omega^2 L^2]^{\frac{1}{2}}} e^{i(\phi_1 - \phi_2)},$$

where

$$\tan \phi_1 = \omega RC, \qquad \tan \phi_2 = \frac{\omega L}{R(1 - \omega^2 LC)}.$$

Finally,

$$i(t) = \text{Re}(I\,e^{i\omega t})$$

$$= \frac{v_0(1 + \omega^2 R^2 C^2)^{\frac{1}{2}}}{[R^2(1 - \omega^2 LC)^2 + \omega^2 L^2]^{\frac{1}{2}}} \cos(\omega t + \phi_1 - \phi_2).$$

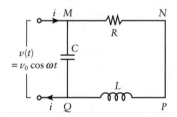

Fig. 21.5

Example 21.10 *Find the steady alternating current entering the circuit shown in Fig.* 21.5.

The phasor of the voltage source is $V = v_0$. By (21.6) the impedance of $MNPQ$ is $R + i\omega L$, and that of MQ is $1/i\omega C$. These are in parallel, so by (21.7) the impedance Z of the circuit viewed between M and Q is given by

$$\frac{1}{Z} = \frac{1}{R + i\omega L} + \frac{1}{1/i\omega C}.$$

Therefore

$$I = \frac{V}{Z} = v_0\left(\frac{1}{R + i\omega L} + i\omega C\right).$$

The simplest way to get an expression for $i(t)$ is to treat the two terms in the brackets on the right separately (though this does not give the answer in standard form). We obtain

$$i(t) = \frac{v_0}{(R^2 + \omega^2 L^2)^{\frac{1}{2}}} \cos\left(\omega t - \arctan\frac{\omega L}{R}\right) + v_0\omega C \cos(\omega t + \tfrac{1}{2}\pi).$$

(a)

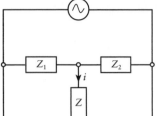

$v = v_0 \cos \omega t$

(b)

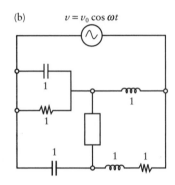

$v = v_0 \cos \omega t$

Fig. 21.6

Example 21.11 (*Balanced bridge circuit.*) (a) *For Fig. 21.6a, show that* (i) *if* $i(t) = 0$, *then* $Z_1/Z_2 = Z_3/Z_4$, (ii) *if* $Z_1/Z_2 = Z_3/Z_4$, *then* $i(t) = 0$. (b) *Check that* $i(t) = 0$ *in the circuit of Fig. 21.6b.*

(a) The analogy (21.8) between resistive and general circuits for steady harmonic oscillations enables us to borrow ordinary Wheatstone-bridge theory, substituting current and voltage phasors and complex impedances for the usual constant currents, voltages, and resistances. We can therefore say immediately that the circuit is balanced ($i(t) = 0$) if, and only if,

$$Z_1/Z_2 = Z_3/Z_4.$$

(b) Z_1 consists of a capacitor and resistor in parallel; so, by (21.6) and (21.7),

$$\frac{1}{Z_1} = \frac{1}{(i\omega)^{-1}} + \frac{1}{1} \quad \text{or} \quad Z_1 = \frac{1}{1 + i\omega}.$$

Also

$$Z_2 = i\omega, \qquad Z_3 = \frac{1}{i\omega}, \qquad Z_4 = 1 + i\omega.$$

Therefore

$$\frac{Z_1}{Z_2} = \frac{1}{i\omega(1 + i\omega)} \quad \text{and} \quad \frac{Z_3}{Z_4} = \frac{1}{i\omega(1 + i\omega)};$$

so, from (a), $i(t) = 0$ and the bridge is balanced.

21.5 Transfer functions in the frequency domain

Consider the circuit of Fig. 21.7, in which the applied voltage is $v_1(t)$, with phasor $V_1 = c_1 \, e^{i\phi_1}$. Suppose that the voltage drop $v_2(t)$ across R_2 has a phasor $V_2 = c_2 \, e^{i\phi_2}$.

Consider the ratio of these two phasors, denoting it by G_{12} (the letter G usually stands for **voltage gain**):

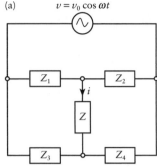

Fig. 21.7

$$G_{12} = \frac{V_2}{V_1} = \frac{c_2 \, e^{i\phi_2}}{c_1 \, e^{i\phi_1}} = \frac{c_2}{c_1} e^{i(\phi_2 - \phi_1)}.$$

Then (assuming that $c_1, c_2 > 0$,)

$$|G_{12}| = \frac{c_2}{c_1},$$

which is the ratio of the peak voltages, or amplitudes, of $v_2(t)$ and $v_1(t)$. The argument (polar angle) of G_{12} is the phase difference between them. If instead we are interested in the current $i_2(t)$ through R_2 produced by $v_1(t)$, then we need the ratio

$$Z_{12} = V_1/I_2,$$

where I_2 is the phasor of $i_2(t)$. This quantity, a voltage divided by a current, is called a **transfer impedance**. Alternatively, we could consider the ratio

$$Y_{21} = I_2/V_1,$$

in which Y_{21} is called a **transfer admittance** (whose parallel is conductance in DC theory).

In general, the ratio of an **output** (such as a current in a selected branch) to an **input** (such as a voltage driving a network) is called a **transfer function in the frequency domain**. A different class of transfer functions is discussed in Chapter 25, on Laplace transforms.

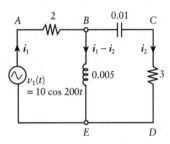

Fig. 21.8

Example 21.12   *Find the transfer impedance $Z_{12} = V_1/I_2$ for the circuit of Fig. 21.8 when the prevailing angular frequency ω is 200.*

The currents indicated take account of Kirchhoff's second rule (21.8), that the sum of the currents entering a junction is zero. The first law expressed in terms of the phasors (see (21.8)), that the sum of the voltage drops round closed circuits is zero, gives for the circuits *ABCDEA* and *BCDEB* respectively:

$$2I_1 + \left(\frac{1}{200 \times 0.01\mathrm{i}} + 3\right)I_2 = V_1,$$

and

$$\left(\frac{1}{200 \times 0.01\mathrm{i}} + 3\right)I_2 - (200 \times 0.005\mathrm{i})(I_2 - I_1) = 0.$$

After simplification, these become

$$2I_1 + (3 - \tfrac{1}{2}\mathrm{i})I_2 = V_1,$$
$$\mathrm{i}I_1 + (3 - \tfrac{3}{2}\mathrm{i})I_2 = 0.$$

The solution for I_2 is

$$I_2 = \frac{\mathrm{i}}{-\frac{11}{2} + 6\mathrm{i}} V_1.$$

The transfer function required is

$$Z_{12} = V_1/I_2 = 6 + \tfrac{11}{2}\mathrm{i} = 8.14\,\mathrm{e}^{0.74\mathrm{i}}.$$

The amplitude of $i_2(t)$ is given by

$$|I_2| = |V_1|/|Z_{12}| = 10/8.14 = 1.23.$$

Its phase is

$$(\text{phase of } V_1) - (\text{phase of } Z_{12}) = 0 - 0.74 = -0.74.$$

The current leads the voltage by this amount.

The methods described in this chapter were invented in the late nineteenth century to assist engineers working with alternating current to interpret and make calculations on their circuits. So long as only steady harmonic oscillations had to be considered, there was no need to solve differential equations: only algebraic equations are involved and these are much simpler to manipulate. Since that time, the methods have been extensively developed so as to permit computer calculation for circuits of any degree of complexity, using matrix algebra, graph theory, and other sophisticated techniques. In Section 24.16, another method for algebrizing circuit equations is described, using Laplace transforms.

21.6 Phasors and waves; complex amplitude

The use of phasors is not restricted to electrical circuits. Phase–amplitude variation from point to point in space, rather than between one circuit element and another, determines the diffraction, interference, and scattering properties of electromagnetic and other types of wave motion. It is not possible to describe the *physical* content of these subjects in this book, but only to indicate how phasors can be useful in such a context. They are employed in Sections 27.10 to 27.12 for diffraction problems.

(i) *Phasors and simple oscillations*

Firstly we shall slightly generalize Example 20.3 (the case of two superposed out-of-phase *oscillations* having the same frequency) by allowing the amplitudes to be different. In the notation we shall use, we take $u(t)$ to be given by

$$u(t) = A_1 \cos(\omega t + \phi_1) + A_2 \cos(\omega t + \phi_2), \tag{21.9a}$$

and assume for simplicity that $A_1 > 0$, $A_2 > 0$.

As in Section 21.1, $u(t)$ may be written as the real part of a complex function:

$$u(t) = \mathrm{Re}[A_1 e^{i(\omega t + \phi_1)} + A_2 e^{i(\omega t + \phi_2)}] = \mathrm{Re}[e^{i\omega t}(A_1 e^{i\phi_1} + A_2 e^{i\phi_2})]. \tag{21.9b}$$

The phasors of the two oscillations are U_1 and U_2, given by

$$U_1 = A_1 e^{i\phi_1}, \qquad U_2 = A_2 e^{i\phi_2}. \tag{21.10}$$

The phasor U of the composite oscillation is therefore given by

$$U = U_1 + U_2, \tag{21.11a}$$

and then

$$u(t) = \mathrm{Re}[U e^{i\omega t}]. \tag{21.11b}$$

Figure 21.9 shows a phasor diagram illustrating vectorial addition of the phasors. The real and imaginary parts of U, needed to evaluate (21.11), are equal to the components of the vector resultant shown. (Clearly, the greater the number of wave components, the greater the advantage of using phasors.)

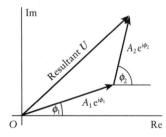

Fig. 21.9 Phasor diagram for $u(t)$, eqn (21.9a).

(ii) Complex amplitude and travelling waves

Here we take as an example the superposition of two *plane travelling waves*, u_1 and u_2 (see Chapter 20) which have the same direction, amplitude, and frequency, but *different phases*. Propagation is in the direction of the z axis. The variable t is to be thought of as a stopwatch time; we shall suppose the watch is switched on at the moment $t = 0$ when the origin of z is at a wave maximum. We lose no generality by this, and it simplifies the algebra. The composite wave is given (see Appendix B(d)) by

$$u(t, x, y, z) = u_1 + u_2 = A_0 \cos(\omega t - kz) + A_0 \cos(\omega t - kz + \phi)$$
$$= 2A_0 \cos \tfrac{1}{2}\phi \cos(\omega t - kz + \tfrac{1}{2}\phi). \tag{21.12}$$

This is a travelling wave having amplitude $2A_0 \cos \tfrac{1}{2}\phi$ and phase $\tfrac{1}{2}\phi$.
  Alternatively, put

$$A_0 \, e^{-ikz} = U_1, \qquad A_0 \, e^{-i(kz+\phi)} = U_2. \tag{21.13}$$

Then U_1, U_2, called the **complex amplitudes** of u_1 and u_2, behave like phasors. Since $u = u_1 + u_2$,

$$u = \mathrm{Re}[(U_1 + U_2) \, e^{i\omega t}] = \mathrm{Re}[U \, e^{i\omega t}],$$

where U is the complex amplitude of u.

  Figure 21.10 shows the phasor diagram for obtaining U on the plane $z = 0$, from which the amplitude and phase of $u(t, x, y, 0)$ are readily obtainable and agree with (21.12). On any other plane $z = z_0$, the phase angles become $-kz_0$ and $\tfrac{1}{2}\phi - kz_0$: the phasor diagram is rotated clockwise, bodily about the origin, through an angle $(-kz_0)$.

(a)

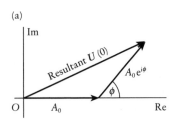

(b)

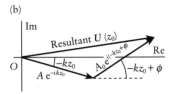

Fig. 21.10 Phasor diagram for (21.12). (a) $z = 0$, (b) $z = z_0$.

(iii) Intensity of a plane wave

Travelling waves carry energy along with them. A measure of this property is the energy being transported across a fixed wave front per unit time, per unit area: that is, power per unit area. The **time average** of this quantity is called the **intensity** of the wave. For the plane sinusoidal wave

$$u(t, x, y, z) = A \cos(\omega t - kt + \phi),$$

the instantaneous rate of transport can be shown to be proportional to u^2, so we require the time average of u^2. The motion is periodic (i.e. it is repetitive) so it is sufficient to average u^2 over a single period $2\pi/\omega$:

$$\text{time average of } u^2 = \frac{\omega}{2\pi} \int_0^{2\pi/\omega} A^2 \cos^2(\omega t - kz + \phi) \, dt$$

$$= A^2 \frac{\omega}{4\pi} \int_0^{2\pi/\omega} [1 + \cos 2(\omega t - kz + \phi)] \, dt = \tfrac{1}{2}A^2.$$

Therefore

$$\text{intensity } I = KA^2, \tag{21.14}$$

where K is a constant for the medium. In the case of optics, I is directly related to the brightness of an image on a screen. (We adopt standard practice by describing a light beam by means of a scalar wave.)

Now suppose that u is expressed in the form $u = \text{Re}[U\,e^{i\omega t}]$. Then

$$U = U_1 + U_2,$$

where

$$U_1 = A_0\,e^{-ikz} \quad \text{and} \quad U_2 = A_0\,e^{i(-kz+\phi)}.$$

Also $|U| = A > 0$, so that, from (21.14),

$$I = K|U|^2, \tag{21.15}$$

where K is constant. Intensities are not simply additive:

$$|U|^2 = |U_1 + U_2|^2 \geqslant |U_1|^2 + |U_2|^2.$$

As with a particle on a spring, doubling the amplitude involves four times as much energy.

(iv) *Interference of two inclined plane waves in two dimensions*
Consider two identical plane light waves u_1 and u_2, inclined to each other at an angle γ, and overlapping in a region. One propagates along the z direction and the other parallel to the (y, z) plane in the direction $\hat{s}$ with direction cosines $(0, \sin \gamma, \cos \gamma)$ (see eqn (20.31) and Fig. 21.11). They impinge on a vertical screen, which for simplicity we take to be the plane $z = 0$.

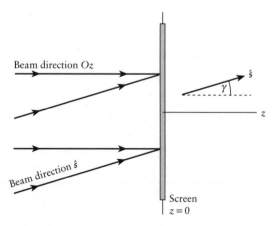

Fig. 21.11 Two light beams, in direction Oz and $\hat{s}$, interfere on arrival at a screen.

Let

$$u_1(t, y, z) = A_0 \cos(\omega t - kz), \tag{21.16a}$$

$$u_2(t, y, z) = A_0 \cos[\omega t - k(y \sin \gamma + z \cos z\gamma)], \tag{21.16b}$$

from (20.31a). The wave arriving at the screen $z = 0$ is then the resultant of the two waves

$$u_1(t, y, 0) = A_0 \cos \omega t, \quad u_2(t, y, 0) = A_0 \cos(\omega t - ky \sin \gamma). \tag{21.17}$$

By using the identity for the sum of two cosines (Appendix B(d)), we obtain

$$u_1 + u_2 = 2A_0 \cos \tfrac{1}{2}(ky \sin \gamma) \cos(\omega t - \tfrac{1}{2}ky \sin \gamma). \tag{21.18}$$

This represents a pattern of oscillatory disturbance on the screen having amplitude $2A_0 \cos(\tfrac{1}{2}ky \sin \gamma)$ and phase $(-\tfrac{1}{2}ky \sin \gamma)$, where both depend on the vertical coordinate y. It arises as the result of **interference** between the incoming waves where they meet on the screen.

On the screen $z = 0$ the phasors of the incoming waves are

$$U_1 = A_0, \qquad U_2 = A_0 e^{i(-ky \sin\gamma)}.$$

The phasor diagram for the equation $U = U_1 + U_2$ is shown in Fig. 21.12. The triangle OPQ is isosceles, so

$$\widehat{POQ} = \widehat{PQO} = -\tfrac{1}{2}ky \sin \gamma \quad \text{and}$$

$$OQ = ON + NQ = 2A_0|\cos(\tfrac{1}{2}ky \sin \gamma)|.$$

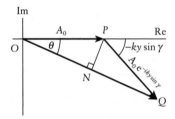

Fig. 21.12 Phasor diagram for interference of two beams on a screen. $U_1 = \overline{OP}, U_2 = \overline{PQ}, U = \overline{OQ}$.

Therefore the polar components of U are, as in (21.18),

$$|U| = OQ = 2A_0|\cos(\tfrac{1}{2}ky \sin \gamma)|, \quad \arg(U) = \theta = -\tfrac{1}{2}ky \sin \gamma.$$

In order to express the variations in brightness of the resulting pattern over the screen, we need the *intensity* of the combined wave (see (iii), above). By (21.15), this depends only on $|U|^2$, rather than U itself. From the phasor diagram (or from (21.18))

$$|U|^2 = |U_1 + U_2|^2 = 4A_0^2 \cos^2(\tfrac{1}{2}ky \sin \gamma) = 2A_0^2[1 + \cos(ky \sin \gamma)].$$

From (21.15), the intensity distribution I on the screen is therefore proportional to

$$1 + \cos(yk \sin \gamma). \tag{21.19}$$

The constant term in (21.19) corresponds to a constant average background illumination. The term $\cos(yk \sin \gamma)$ describes the varying brightness of the **interference fringes**. On arrival at the screen the beams reinforce or cancel each other at regular intervals in the y direction, due to changes in their phase difference $ky \sin \gamma$. In principle, a photographic plate on the (x, y) plane would show a pattern of brighter and darker interference fringes parallel to the x axis and equally spaced by an amount $2\pi/(k \sin \gamma)$ in the y direction.

Problems

21.1 Write down the phasors X corresponding to the oscillations $x(t)$ given below, in polar and $a + ib$ form.
(a) $2\cos(10t + \frac{1}{2}\pi)$; (b) $-2\cos(10t + \frac{1}{2}\pi)$;
(c) $3\sin\omega t$; (d) $-4\sin(3t - \frac{1}{4}\pi)$.

21.2 Write the following phasors X in polar form, and give the corresponding oscillations $x(t)$ when the angular frequency is ω.
(a) $1 - i$; (b) $2i$; (c) $-3i$; (d) $-\sqrt{2} - \sqrt{2}i$;
(e) $-2\sqrt{3} - 2i$; (f) $-1 - \sqrt{3}i$; (g) $1/(1 - 2i)$;
(h) $i/(1 - 2i)$; (i) $(2 + 3i)/(2 - 3i)$; (j) $1/(3i) + 2i$.

21.3 Write down the phasors corresponding to the following oscillators. State the amplitude and phase.
(a) $\cos 2t + \cos(2t - \frac{1}{4}\pi)$; (b) $\cos 3t - \sin 3t$;
(c) $\sin 3t + 2\cos 3t$.

21.4 Either algebraically, or by calculation or measurement based on a phasor diagram, give the phasors of the following functions.
(a) $-\cos 2t + \cos(2t + \frac{1}{4}\pi) + \cos(2t - \frac{1}{2}\pi)$;
(b) $\cos 1760t - 3\cos(1760t - \frac{1}{2}\pi) + \cos(1760t + \frac{1}{2}\pi)$.

21.5 Show that the point on an Argand diagram corresponding to $c\,e^{i(\omega t + \phi)}$ moves on a circle, centre the origin and radius c, with constant angular velocity ω, and that its projection on the x axis is given by $x = c\cos(\omega t + \phi)$. (A phase diagram such as Fig. 21.1 is a snapshot of conditions at $t = 0$ in such a representation: the whole diagram rotates unaltered.)

21.6 Obtain the complex impedance of the following circuit branches.

(a)
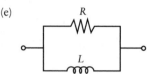

(b)

(c)

(d)
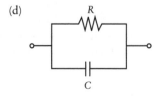

(e)
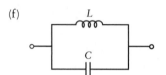

(f)
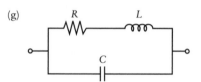

(g)
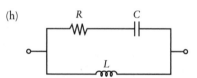

(h)
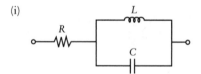

(i)
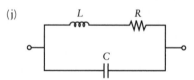

(j)
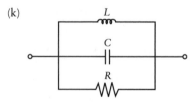

(k)
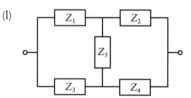

(l)

Fig. 21.13

21.7 A voltage $v = 2 \cos \omega t$ is applied across each of the circuits in Problem 21.6. Find the amplitude and phase of the current passing through the branches.

21.8 In Fig. 21.14, numerical values are given to the complex impedances (the standard units are ohms, although the quantities may be complex). A voltage with phasor V_0 is applied. Obtain the phasor V_1 as indicated, the corresponding voltage gain V_1/V_0, and the transfer impedance V_0/I_1.

(a)

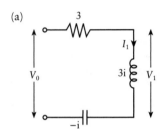

(b)

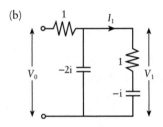

(c)

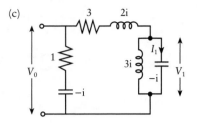

Fig. 21.14

21.9 Sketch phasor diagrams for the following cases and in each case calculate (or measure) to obtain the sum:
(a) $\cos 10t + 2 \cos(10t + 0.3)$;
(b) $\cos 10t + 2 \sin(10t + 10.2)$;
(c) $\cos 10t + 3 \cos(10t - 0.2)$;
(d) $\sin 20t - 3 \cos(20t + 0.75)$;
(e) $2 \cos(50t + 0.4) + \sin(50t + 0.3) - 3 \cos(50t - 0.5)$.

21.10 Use phasor diagrams for the following problems.
(a) Given axes x, y, z, obtain the general form for a plane electromagnetic wave $u(t, x, y, z)$, travelling in the direction that makes equal (acute) angles with the positive directions of the three axes. Investigate the form of the wave on a screen placed in the (x, y) plane.
(b) The wave in (a) is crossed by another identical plane wave, which travels in the direction of the z axis. Obtain the interference pattern on a photographic plate on the plane $z = 0$.

22 Graphical, numerical, and other aspects of first-order equations

CONTENTS

22.1 Graphical features of first-order equations

In this section we shall use x for the independent variable (instead of t), and y as the dependent variable (instead of x), and consider differential equations of the form

$$\frac{\mathrm{d}y}{\mathrm{d}x} = f(x, y), \tag{22.1}$$

where $f(x, y)$ is unrestricted. If $f(x, y)$ happens to take the form $g(x) + h(x)y$, the equation is **linear** and can be handled by the method of Section 19.5; otherwise none of the methods so far discussed will work.

However, we can always obtain a rough picture of the **solution curves** by using a simple fact. Choose any point $x = a$, $y = b$, on the (x, y) plane. Then eqn (22.1) says that the slope of the solution curve which passes through (a, b) must be equal to $f(a, b)$, and this has a definite numerical value that we can work out. So take a large number of points (a, b) on the (x, y) plane. For each of them, work out $f(a, b)$, and draw through the point a short line whose slope is equal to $f(a, b)$, as is done in Fig. 22.1a for the special case $f(x, y) = xy$. These are called **direction indicators**. Given enough of these, it is possible to draw a **family of curves** which follow their directions smoothly, as in Fig. 22.1b. Each of the curves represents a solution of (22.1), because its slope, or derivative, is correctly reproduced at each point on it. The picture is called a **lineal-element diagram** or a **direction field**. The technique can in principle be used for first-order equations however complicated they may be.

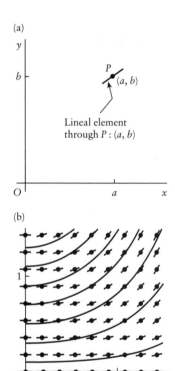

(a)

Lineal element
through $P : (a, b)$

(b)

Fig. 22.1 Lineal-element diagram indicating solution curves for $\mathrm{d}y/\mathrm{d}x = xy$.

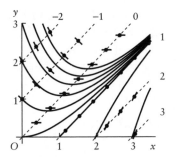

Fig. 22.2 Solution curves of $dy/dx = x - y$ in the first quadrant. ---------- isoclines, values of K indicated; ——— solution curves.

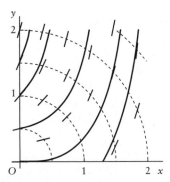

Fig. 22.3 Pattern of solution curves of $dy/dx = x^2 + y^2$ in the first quadrant. ---------- isoclines, ——— solution curves.

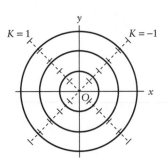

Fig. 22.4 The solution curves of $dy/dx = -x/y$. ---------- isoclines, ——— solution curves.

Rather than to place the direction indicators at grid points as in Fig. 22.1, it is often easier to look for curves, called **isoclines**, along which the slope is constant, as in the following example.

Example 22.1   *Sketch the solution curves of $\dfrac{dy}{dx} = x - y$.*

Here dy/dx takes *constant* values K on the isoclines $x - y = K$, or $y = x - K$. For example, $dy/dx = 0$ on $y = x$, $dy/dx = 1$ on $y = x - 1$, and so on. If we draw the line $y = x - K$, then the indicators along it are all parallel, with slope K, so it is easier to draw a large number of them. Figure 22.2 is constructed in this way. (This equation is in fact linear with constant coefficients, its solutions being $y = x - 1 + C e^{-x}$.)

Example 22.2   *Sketch the solution curves of $\dfrac{dy}{dx} = x^2 + y^2$.*

The isoclines are the circles $x^2 + y^2 = K$ (see Fig. 22.3), on each of which the slope is equal to K (which must be a positive number here).

Closed curves can occur, as in the following example.

Example 22.3   *Sketch the solution curves of $\dfrac{dy}{dx} = -\dfrac{x}{y}$.*

The isoclines having slope K are the radial straight lines $-x/y = K$ (see Fig. 22.4), or

$$y = -\frac{1}{K}x.$$

Thus, for example, if $K = -1$, the corresponding isocline is $y = x$, and solutions cut this straight line with slope -1 as shown in the figure. On $y = 0$, the slope K must be infinite so the direction indicators are vertical.

The method illustrates why there is always an infinite number of solutions: there will be **a single solution curve through every point where $f(x, y)$ has a definite value.** The type of exception that might arise is illustrated by the case $f(x, y) = (xy)^{\frac{1}{2}}$, which only has a meaning when x and y have the same sign; there are solutions in only the first and third quadrants of the (x, y) plane. Again, there can be points from which several solution curves emanate: this occurs at the origin in Example 22.3, where $f(x, y)$ takes the indeterminate form $0/0$; nothing can be taken for granted at such a point, but elsewhere the curves do not intersect.

To prescribe a point (a, b) through which a curve must pass is equivalent to imposing an **initial condition** on the solution: the corresponding initial condition would read 'Find the solution for which $y = b$ when $x = a$'. Therefore, it can be seen that even when an equation is not linear an initial condition of this sort will give exactly one solution, points where $f(a, b)$ is indeterminate being excepted.

22.2 The Euler method for numerical solution

For the equation

$$\frac{\mathrm{d}y}{\mathrm{d}x} = f(x, y),$$

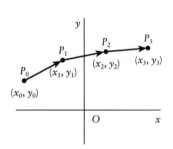

Fig. 22.5 Step-by-step use of direction indicators along a particular solution curve.

consider an adaptation of the graphical method described in the previous section. Start at any point $P_0 : (x_0, y_0)$, and draw an indicator with slope $f(x_0, y_0)$ from P_0 to $P_1 : (x_1, y_1)$ a short distance away (see Fig. 22.5). Then P_1 will lie close to the solution curve through P_0. Do the same thing starting with P_1, and so on, continuing as far as is necessary. It is also possible to proceed backwards from P_0. Provided that the steps are small enough, it seems likely that $P_0, P_1, P_2, \ldots$ will be close to the solution curve through P_0, so we have an approximate solution to the initial-value problem: Find the solution of $\mathrm{d}y/\mathrm{d}x = f(x, y)$ for which $y = y_0$ when $x = x_0$.

Obviously, to draw the solution curve in this way is not really practicable; but we need not actually draw it, because the same process can be carried out numerically as follows.

As shown in Fig. 22.6, choose a small, constant step length h in x for going from point to point: $P_0 \rightarrow P_1 \rightarrow P_2 \rightarrow \ldots$, where the vectors $\overline{P_0 P_1}, \overline{P_1 P_2}, \overline{P_2 P_3}, \ldots$ point in the direction of the indicators at their respective starting points $P_0, P_1, P_2, \ldots$. Corresponding to the x steps, call the y steps $k_1, k_2, k_3, \ldots$:

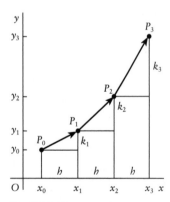

Fig. 22.6 Three steps in the numerical solution of $\mathrm{d}y/\mathrm{d}x = f(x, y)$, starting at the point $P_0 : (x_0, y_0)$.

$$y_1 = y_0 + k_1, \quad y_2 = y_1 + k_2, \quad y_3 = y_2 + k_3, \quad \ldots ,$$

where

$$k_1 = hy'(x_0) = hf(x_0, y_0), \quad k_2 = hy'(x_1) = hf(x_1, y_1), \quad \ldots .$$

Therefore

for P_1: $\quad x_1 = x_0 + h, \quad y_1 = y_0 + hf(x_0, y_0);$

for P_2: $\quad x_2 = x_1 + h, \quad y_2 = y_1 + hf(x_1, y_1);$

and, in general, with $n = 1, 2, 3, \ldots$, in turn,

for P_n: $\quad x_n = x_{n-1} + h, \quad y_n = y_{n-1} + hf(x_{n-1}, y_{n-1}).$

We expect that the points will be close to the solution curve. This is the **Euler method** for approximating to the solution of the differential equation.

> **Euler method for initial-value problems**
>
> Differential equation: $\dfrac{dy}{dx} = f(x, y)$.
>
> Initial condition: $y = b$ when $x = a$.
>
> Approximate solution: Put $x_0 = a$, $y_0 = b$; then
> $x_n = x_{n-1} + h$, $y_n = y_{n-1} + f(x_{n-1}, y_{n-1})h$
> for $n = 1, 2, \ldots$ successively.
>
> (22.2)

This recipe, or **algorithm**, describes a step-by-step repetitive process, or **iteration**; essentially the same thing has to be done over and over again. The procedure which produces a new (x, y) from the preceding (x, y) is called a **recurrence relation** (compare Newton's method for solving equations in Chapter 4). Such a process is easy to program on a computer, and Fig. 22.7 is the skeleton of a flow diagram. The program should contain a method for stopping itself when x has gone far enough; also, since small intervals h are usually necessary, it is useful to include a means of recording only the results for preset values of x to avoid voluminous output.

Fig. 22.7 Flow diagram for the initial-value problem.

Example 22.4 *Use the Euler method to obtain a solution of the initial-value problem*

$$\frac{dy}{dx} = xy^2, \text{ with } y = 1 \text{ at } x = 0,$$

between $x = 0$ and $x = 1$. Compare the result with the exact solution $y = (1 - \frac{1}{2}x^2)^{-1}$ when steps of $h = 0.2, 0.1, 0.01$, and 0.001 are adopted.

The following results are obtained.

x	0	0.2	0.4	0.6	0.8	1.0
Exact y	1.0000	1.0204	1.0870	1.2195	1.4706	2.0000
$h = 0.2$	1.0000	1.0000	1.0400	1.1265	1.2788	1.5405
$h = 0.1$	1.0000	1.0100	1.0623	1.1687	1.3601	1.7129
$h = 0.01$	1.0000	1.0193	1.0843	1.2139	1.4576	1.9618
$h = 0.001$	1.0000	1.0203	1.0867	1.2189	1.4693	1.9960

Euler's method is very simple; but it is usually good enough to provide reasonable accuracy over a finite range, provided that small enough intervals are used. The simplest way of checking accuracy is to experiment with successively smaller intervals h, noting when further reduction in h does not change the values of y obtained at the number of decimal places required. Several problems on these lines are given at the end of the chapter.

There exist, however, far more sophisticated algorithms which will give great accuracy over long ranges without having to use minute values of h (which can introduce problems of its own). The computer programs for such methods can be found in libraries of computer routines. For example, the software Mathematica has a program for the numerical solution and plotting of initial-value problems: see the projects in Chapter 42. The theoretical side of the subject is called numerical analysis; mathematical theory makes it possible, for example, to estimate the size of interval required without carrying out trials.

22.3   Nonlinear equations of separable type

The equation

$$\frac{dy}{dx} = \frac{y^2}{x^2}$$

is nonlinear (note the y^2 term), and none of the theory of Chapters 18 and 19 can be adapted to solve it. Write it in the form

$$\frac{dy}{y^2} = \frac{dx}{x^2}.$$

On the left only y appears, and on the right only x appears. The form looks like an invitation to integrate both sides:

$$\int \frac{dy}{y^2} = \int \frac{dx}{x^2} + C,$$

so $-1/y = -1/x + C$. Therefore

$$y = \frac{x}{1 - Cx},$$

where C is arbitrary. You should consider checking that these really are solutions by substituting into the equation. Notice that there is no sign of the complementary functions and particular solutions found for linear equations: an arbitrary constant C does occur, but it is imbedded deep in the expression. The solution curves are shown in Fig. 22.8: each curve has its individual asymptotes, namely the lines $y = -C^{-1}$, $x = C^{-1}$.

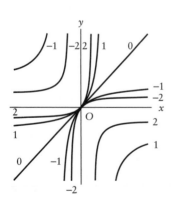

Fig. 22.8 Solution curves $y = x/(1 - Cx)$ for $dy/dx = y^2/x^2$. Values of C indicated on the curves.

The method is called **separation of variables**. It can be applied to equations which are **separable**, that is to say, ones that can be arranged in the form

$$\frac{dy}{dx} = g(x)h(y),$$

where the right-hand side is the product of two terms, one a function of x only, and the other a function of y only. Alternatively, you might see it more easily as an equation which can be put into the form

$$Y(y)\, dy = X(x)\, dx,$$

$X(x)$ and $Y(y)$ being functions respectively of x only and y only. For example, the equations

$$\frac{dy}{dx} = x^{\frac{1}{2}}y^{\frac{1}{2}}, \qquad \frac{dy}{dx} = e^x \sin y, \qquad \frac{y}{x}\frac{dy}{dx} = \cos(y^2)$$

are of the right type.

Separation of variables

Equation type: $\dfrac{dy}{dx} = g(x)h(y)$.

Separate the terms: $\dfrac{dy}{h(y)} = g(x)\, dx$.

Integrate: $\displaystyle\int \dfrac{dy}{h(y)} = \int g(x)\, dx + C$, so that y is expressed as function of x (usually an implicit function). C may take a range of values.   (22.3)

Example 22.5   *Find solutions of the equation* $y\dfrac{dy}{dx} = \cos x.$

This can be written $y\, dy = \cos x\, dx$. By integrating both sides, we obtain $\frac{1}{2}y^2 = \sin x + C$, giving $y = \pm 2^{\frac{1}{2}}(\sin x + C)^{\frac{1}{2}}$, where C is *only to a certain extent* arbitrary. It cannot be completely arbitrary; for example, if $C = -100$, then $\sin x + C$ will *always* be negative (because $-1 \leqslant \sin x \leqslant 1$), so the square root never has a real value. We must have $C > -1$ to get any real solution. If $-1 < C < 1$ there are regularly-spaced intervals on which $\sin x + C > 0$, giving the oval curves in Fig. 22.9. If $C > 1$ their $\sin x + C > 0$ for all x, giving the wavy phase paths.

↗

Example 22.5 *continued*

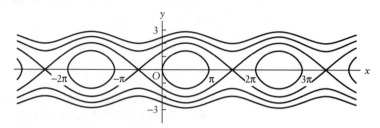

Fig. 22.9 Solution curves for the equation

$$y\frac{\mathrm{d}y}{\mathrm{d}x} = \cos x,$$

which are the functions $y = \pm 2^{\frac{1}{2}}(\sin x + C)^{\frac{1}{2}}$.

Example 22.6   *Find solutions of* $\dfrac{\mathrm{d}y}{\mathrm{d}x} = \dfrac{y(x + 1)}{x(y + 1)}$.

After separating, we have

$$\int \frac{1 + y}{y}\,\mathrm{d}y = \int \frac{1 + x}{x}\,\mathrm{d}x + C,$$

or

$$\int\left(1 + \frac{1}{y}\right)\mathrm{d}y = \int\left(1 + \frac{1}{x}\right)\mathrm{d}x + C.$$

Therefore $y + \ln|y| = x + \ln|x| + C$, or $y\,\mathrm{e}^y = Ax\,\mathrm{e}^x$, where A is arbitrary.

We cannot further reduce this 'solution' to express y explicitly in terms of x. The 'answer' is nearly as obscure as the original equation. More intelligible information about the solutions could be obtained by using the graphical or numerical methods of Sections 22.1 and 22.2.

The separation-of-variables technique requires initiative, even in the simplest cases, as can be seen from the following example.

Example 22.7   *Find solutions of* $\dfrac{\mathrm{d}y}{\mathrm{d}x} = 2y^{\frac{1}{2}}$.

After separating, we have $\frac{1}{2}\displaystyle\int y^{-\frac{1}{2}}\,\mathrm{d}y = \int \mathrm{d}x$, or

$$y^{\frac{1}{2}} = x + C. \tag{22.4}$$

To express y in terms of x, square both sides. We get

$$y = (x + C)^2. \tag{22.5}$$

This represents a family of parabolas, as shown in Fig. 22.10a. But it cannot be right: the curves cross at every point, although $\mathrm{d}y/\mathrm{d}x$ has only one value at any point. In fact, since $y^{\frac{1}{2}} > 0$, only the positive value of $y'(x)$ is legitimate, and this gives the right-hand branches, shown in Fig. 22.10b.

We also lost a solution, namely $y(x) = 0$. This is connected with the fact that, for this solution, we in effect divided by zero when we first separated the equation.

(a)

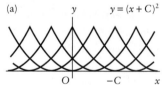

(b)

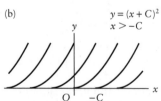

Fig. 22.10 (a) $y = (x + C)^2$ for various C. (b) The solutions of the differential equation, consisting only of the right-hand branches of the parabolas. (Note: $y(x) = 0$ is also a valid solution.)

The production of pseudosolutions and the non-appearance of certain **singular solutions** in the final formula, as in Example 22.7, is a problem which constantly arises in nonlinear differential equations.

22.4 Differentials and the solution of first-order equations

In this section, it is important to distinguish between an **identity** such as $d(y^2)/dx = 2y\, dy/dx$, which is true for any $y(x)$ (it is just a special case of the chain rule), and an **equation** such as $dy/dx = xy$, which will only be true for special functions $y(x)$.

Take an identity such as

$$\frac{d}{dx} x^2 = 2x, \tag{22.6a}$$

and consider another way of writing it:

$$d(x^2) = 2x\, dx. \tag{22.6b}$$

It is as if we formally multiply (22.6a) by dx to obtain (22.6b). Conversely, if we divide (22.6b) by dx, we recover (22.6a). We have already used this process to help to change the variable in an integral in Section 17.1.

Now consider a more complicated identity, obtained from the product rule for differentiation ($y(x)$ represents any function of x):

$$\frac{d}{dx}(xy) = y + x\frac{dy}{dx}. \tag{22.7a}$$

The parallel expression of the same identity, obtained as before, is

$$d(xy) = y\, dx + x\, dy. \tag{22.7b}$$

Given either one of them, we can immediately construct the other, so we shall regard such pairs of expressions as being simply different ways of writing the same thing. In effect this is what we did when carrying out the separation-of-variables process for differential equations in Section 22.3, and we are leading up to a generalization of this method.

In general, a **differential expression** or **differential form** has the shape

$$P(x, y)\, dx + Q(x, y)\, dy, \tag{22.8}$$

where $P(x, y)$ and $Q(x, y)$ are two functions of x and y. In (22.6b), we had $P(x, y) = 2x$ and $Q(x, y) = 0$; in (22.7b), we had $P(x, y) = y$ and $Q(x, y) = x$. The symbols on the left of (22.6b) and (22.7b), $d(x^2)$ and $d(xy)$, are called the **differentials** of x^2 and xy respectively.

The table (22.9) (below) gives a list of useful identities written in the usual form and the differential form for comparison.

Standard form	Differential form
$\dfrac{d}{dx}(C) = 0$ (C constant)	$dC = 0$
$\dfrac{d}{dx}(x^2) = 2x$	$d(x^2) = 2x\,dx$
$\dfrac{d}{dx}(y^2) = 2y\dfrac{dy}{dx}$	$d(y^2) = 2y\,dy$
$\dfrac{d}{dx}(xy) = x\dfrac{dy}{dx} + y$	$d(xy) = y\,dx + x\,dy$
$\dfrac{d}{dx}\left(\dfrac{y}{x}\right) = \dfrac{1}{x^2}\left(x\dfrac{dy}{dx} - y\right)$	$d\left(\dfrac{y}{x}\right) = -\dfrac{1}{x^2}(y\,dx - x\,dy)$
$\dfrac{d}{dx}\left(\dfrac{x}{y}\right) = \dfrac{1}{y^2}\left(y - x\dfrac{dy}{dx}\right)$	$d\left(\dfrac{x}{y}\right) = \dfrac{1}{y^2}(y\,dx - x\,dy)$
$\dfrac{d}{dx}\left(\ln\dfrac{y}{x}\right) = \dfrac{1}{xy}\left(x\dfrac{dy}{dx} - y\right)$	$d\left(\ln\dfrac{y}{x}\right) = -\dfrac{1}{xy}(y\,dx - x\,dy)$

$$(22.9)$$

Differential forms can be manipulated. For example:

(i) $2x\,dx - y\,dy = d(x^2) - d(\tfrac{1}{2}y^2) = d(x^2 - \tfrac{1}{2}y^2).$

(ii) $(x + y)\,dx + x\,dy = x\,dx + (y\,dx + x\,dy)$
$$= d(\tfrac{1}{2}x^2) + d(xy) = d(\tfrac{1}{2}x^2 + xy).$$

(iii)   If $u(x)$ and $v(x)$ are two functions, then
$$d(uv) = u\,dv + v\,du,$$

which is the product rule for derivatives in differential form. These results are all identities; that is, true for all functions $y(x), u(x), v(x)$.

Example 22.8   *Put* $d(x^3 + x\sin y)$ *into the form* $P\,dx + Q\,dy$.

We have
$$d(x^3 + x\sin y) = d(x^3) + d(x\sin y),$$
for the first term, $(d/dx)x^3 = 3x^2$, and in differential form this becomes $d(x^3) = 3x^2\,dx$. For the second term, we can use the product rule in the form (iii) above (or write it in standard form first):
$$d(x\sin y) = \sin y\,dx + x\,d(\sin y)$$
followed by the chain rule
$$d(x\sin y) = \sin y\,dx + x\cos y\,dy.$$
Finally,
$$d(x^3 + x\sin y) = (3x^2 + \sin y)\,dx + (x\cos y)\,dy,$$
so that, in (22.8), $P(x, y) = 3x^2 + \sin y$ and $Q(x, y) = x\cos y.$

First-order **differential equations** can be written alternatively as differential forms. The simplest is the equation

$$\frac{dy}{dx} = 0,$$

which has solutions $y(x) = C$, where C is any constant. In differential form, the equation becomes

$$dy = 0,$$

and the solutions are compatible with the first entry in the table, (22.9): $dC = 0$ when C is a constant.

Example 22.9 *Find solutions of the equation* $\dfrac{dy}{dx} = -\dfrac{x^3}{y^2}$.

In differential form, this becomes

$$x^3\,dx + y^2\,dy = 0.$$

But

$$x^3\,dx + y^2\,dy = d(\tfrac{1}{4}x^4) + d(\tfrac{1}{3}y^3) = d(\tfrac{1}{4}x^4 + \tfrac{1}{3}y^3).$$

This will be zero if

$$\tfrac{1}{4}x^4 + \tfrac{1}{3}y^3 = C,$$

where C is, in this case, any constant. The equation is in fact separable; you should compare this with the process in Section 22.3.

Example 22.10 *Find solutions of* $\dfrac{dy}{dx} = \dfrac{x - y}{x + y}$.

In differential form, this becomes

$$0 = (x - y)\,dx - (x + y)\,dy$$
$$= x\,dx - y\,dx - x\,dy - y\,dy.$$

Try to rearrange it so that recognizable forms appear:

$$0 = x\,dx - y\,dy - (y\,dx + x\,dy)$$
$$= d(\tfrac{1}{2}x^2) - d(\tfrac{1}{2}y^2) - d(xy)$$
$$= d(\tfrac{1}{2}x^2 - \tfrac{1}{2}y^2 - xy).$$

This differential will be zero as required if

$$\tfrac{1}{2}x^2 - \tfrac{1}{2}y^2 - xy = C,$$

where C is the 'variable constant', or **parameter**, which will generate a whole family of solutions.

In the previous example, the terms were rearranged in a search for a group like $y\,dx + x\,dy$ that would simplify, in that case, to $d(xy)$. If a differential form can be expressed *identically* (that is to say, for all $y(x)$) in the form of a single differential

$$P(x, y)\,dx + Q(x, y)\,dy \equiv dF(x, y), \tag{22.10}$$

where $F(x, y)$ is a fixed function of x and y, then it is called a **perfect differential form**, or a **perfect differential**. Usually this is impossible. For example, consider the differential form $y \, dx$. It can be proved that there *does not exist* any fixed function $F(x, y)$ such that

$$y(x) \, dx = dF(x, y(x)) \quad \text{for every } y(x)$$

(try looking for one).

To solve differential equations by this method, we search for a perfect differential in order to be able to conclude with the steps

$$\text{`} dF(x, y) = 0; \quad \text{therefore } F(x, y) = C\text{'},$$

as in the Examples. If a perfect differential is not already present, we might be able to produce one by multiplying through by a suitable function, called an **integrating factor** for the expression. For example, $y \, dx - x \, dy$ is not a perfect differential; but, from (22.10),

$$\frac{1}{x^2}(y \, dx - x \, dy) = d\left(-\frac{y}{x}\right),$$

so the new expression is a perfect differential.

Perfect differential forms

Let y be an arbitrary function of x. Then $P(x, y) \, dx + Q(x, y) \, dy$ is a perfect differential if it can be written as

$$P(x, y) \, dx + Q(x, y) \, dy \equiv dF(x, y),$$

where $F(x, y)$ is a fixed function of x and y.

(22.11)

Integrating factor for differential forms

A function $I(x, y)$ is an integrating factor for the differential form $P \, dx + Q \, dy$ if $I(P \, dx + Q \, dy)$ is a perfect differential.

(22.12)

It can be proved that every differential form has an integrating factor, but only occasionally is it easy to see one.

Example 22.11 *Find a family of solutions of* $x\dfrac{dy}{dx} = y$.

(This is a linear equation and it is also separable, so we have two other methods for solving it.) In differential form:

$$y \, dx - x \, dy = 0,$$

and we cannot do anything with the left-hand side as it stands. However, the remark above suggests we multiply by $1/x^2$, obtaining

$$0 = (1/x^2)(y \, dx - x \, dy) = d(-y/x).$$

↗

Example 22.11 *continued*

Therefore

$$-y/x = C, \quad \text{or} \quad y = -Cx,$$

are the solutions, as is easily confirmed.

There are other possibilities; for example (see (22.9)), we might divide by y^2 or xy. In the end these lead to the same set of solutions.

Note that, in Example 22.11, the equation is linear:

$$\frac{\mathrm{d}y}{\mathrm{d}x} + \left(-\frac{1}{x}\right)y = 0,$$

and so we can alternatively use the method of Section 19.5. An 'integrating factor' $I(x)$ is also used there: it is

$$I(x) = \mathrm{e}^{-\int x^{-1}\mathrm{d}x} = \mathrm{e}^{-\ln x + C} = 1/x$$

(where we choose $C = 0$ for simplicity), which is different from, though related to, the ones which work for the differential form $y\,\mathrm{d}x - x\,\mathrm{d}y$ above.

Example 22.12 *Find a set of solutions of* $x\dfrac{\mathrm{d}y}{\mathrm{d}x} = y + y^2 x.$

Equivalently, $y\,\mathrm{d}x - x\,\mathrm{d}y + y^2 x\,\mathrm{d}x = 0$. The first two terms cannot be written as $\mathrm{d}F(x, y)$. The table (22.9) offers three integrating factors, x^{-2}, y^{-2}, and $(xy)^{-1}$, to choose from. It is, however, also necessary to be able to manage the remaining term, $y^2 x\,\mathrm{d}x$, after multiplying by the integrating factor, so we choose y^{-2}, which gives

$$0 = (1/y^2)(y\,\mathrm{d}x - x\,\mathrm{d}y) + x\,\mathrm{d}x$$
$$= \mathrm{d}(x/y) + \mathrm{d}(\tfrac{1}{2}x^2) = \mathrm{d}(x/y + \tfrac{1}{2}x^2).$$

Therefore $x/y + \tfrac{1}{2}x^2 = C$, or $y = x/(C - \tfrac{1}{2}x^2)$, are solutions.

22.5 Change of variable in a differential equation

Occasionally we can find a **change of variable**, or **substitution**, which will simplify a differential equation. Some general types are given here for illustration.

> **Equations not involving y**
>
> If a differential equation contains
>
> $$x, \quad \mathrm{d}y/\mathrm{d}x, \quad \mathrm{d}^2y/\mathrm{d}x^2, \quad \dots,$$
>
> but not y, substitute the independent variable $w = \mathrm{d}y/\mathrm{d}x$ in place of y, producing an equation for w of lower order. **(22.13)**

To see how this works, consider the following Example.

Example 22.13   *Find solutions of* $\dfrac{d^2y}{dx^2} + \left(\dfrac{dy}{dx}\right)^2 = 0.$

The variable y is not independently present, so put

$$w = \frac{dy}{dx}.$$

Then $d^2y/dx^2 = dw/dx$, so that the equation becomes

$$\frac{dw}{dx} + w^2 = 0.$$

This is a separable equation, and the method of Section 22.3 gives

$$-\int \frac{dw}{w^2} = \int dx, \quad \text{or} \quad \frac{1}{w} = x + A,$$

where A is constant, so that

$$w = \frac{1}{x + A}.$$

For the second stage, remember that $w = dy/dx$, so we have

$$\frac{dy}{dx} = \frac{1}{x + A}.$$

Therefore

$$y = \ln|x + A| + B,$$

where A and B are constants which we see, in retrospect, may be chosen entirely arbitrarily.

Sometimes it is possible to change the independent variable y into something else to obtain a more manageable equation:

Equation of the form $\dfrac{dy}{dx} = f\left(\dfrac{y}{x}\right)$

Change to a new dependent variable v by

$$v = y/x$$

and solve the resulting separable equation.
(To make the change write $y = xv$, so that

$$dy/dx = x\,dv/dx + v.)$$

                                                                        (22.14)

Example 22.14 *Find solutions of $\dfrac{dy}{dx} = \dfrac{3y - x}{3x - y}$.*

This equation can be written in the form

$$\frac{dy}{dx} = \frac{3y/x - 1}{3 - y/x},$$

which has the form $f(y/x)$, so change the dependent variable from y to $v = y/x$. To obtain dy/dx in terms of v, write $y(x) = xv(x)$. Then $dy/dx = x\,dv/dx + v$, and in terms of v the equation becomes

$$x\frac{dv}{dx} + v = \frac{3v - 1}{3 - v}, \quad \text{or} \quad x\frac{dv}{dx} = \frac{v^2 - 1}{3 - v}.$$

This new equation is separable (Section 22.3) (a separable equation will always be obtained at this stage). Following (22.3):

$$\int \frac{1}{x}dx = \int \frac{3 - v}{v^2 - 1}dv$$

$$= \int \left(\frac{1}{v - 1} - \frac{2}{v + 1}\right)dv.$$

Therefore $\ln|[(v - 1)/(v + 1)^2]| = \ln|x| + C$, where C is an arbitrary constant. After returning to y and simplifying, we have

$$(y - x)/(y + x)^2 = c, \tag{22.15}$$

where $c = \pm e^C$. The solution curves are shown in Fig. 22.11, plotted by working directly from the differential equation and using a numerical method (see Section 22.2).

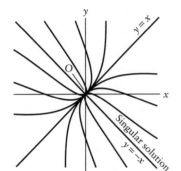

Fig. 22.11 Solution curves for

$$\frac{dy}{dx} = \frac{3y - x}{3x - y}.$$

Notice the special solutions $y = \pm x$.

The methods of separation of variables, differentials, substitutions, etc., used to solve nonlinear equations are rather hazardous. In Example 22.14 there are two solutions which do not appear in (22.15), represented by the straight line segments $y = -x$ for $x < 0$ and $x > 0$ in Fig. 22.11 and corresponding to the limiting case of infinite c. Therefore (22.15) is not a truly general solution. These extra **singular solutions** can be found independently in this case by trying the form $y = mx$ in the equation and solving the resulting quadratic for m.

Singular solutions are important in the theory of vibrations, population problems, and other nonlinear fields. In general, they represent limiting cases of the ordinary solutions; in this case, $y = x$ is an **envelope** of the ordinary solutions: that is, a curve that is tangential to each ordinary solution. This was the case with the singular solution $y(x) = 0$ of Example 22.7.

The following Example illustrates another way of transforming a differential equation. In a mechanical context, the transformation of the derivative d^2x/dt^2 involved is called the **energy transformation**.

Example 22.15   *The acceleration of a vehicle is constrained by its velocity:* $d^2x/dt^2 = Kv^{-2}$, *where* $v = dx/dt$ *and* K *is a constant. Find the velocity as a function of distance if* $x(0) = 0$ *and* $v(0) = 0$.

Transform the acceleration by the chain rule, using x as the intermediate variable:

$$\frac{d^2y}{dt^2} = \frac{dv}{dt} = \frac{dv}{dx}\frac{dx}{dt} = v\frac{dv}{dx} = \tfrac{1}{2}\frac{d(v^2)}{dx}.$$

The differential equation now relates v to x:

$$\tfrac{1}{2}\frac{d(v^2)}{dx} = Kv^{-2}.$$

We could put $v^2 = u$ and separate, but for the sake of adventurousness we shall leave v^2 as it is and turn the equation upside down:

$$2\frac{dx}{d(v^2)} = K^{-1}v^2.$$

Now integrate both sides, courageously using v^2 as the variable:

$$2x = K^{-1}\int v^2\, d(v^2) = \tfrac{1}{2}K^{-1}(v^2)^2 + C,$$

or

$$v^4 = 4Kx - 2CK.$$

The initial condition $x(0) = v(0) = 0$ then gives

$$v(x) = (4Kx)^{\frac{1}{4}}.$$

$x(t)$ is obtainable by solving the separable equation $v = dx/dt = (4kx)^{\frac{1}{4}}$.

Problems

In these problems, y' means dy/dx.

22.1 Sketch a lineal-element diagram for the solution curves of each of the following.
(a) $y' = -y$; (b) $y' = x - y$;
(c) $y' = x/y$; (d) $y' = xy$;
(e) $y' = -y/x$; (f) $y' = y/x$;
(g) $y' = (x - 1)y$; (h) $y' = \dfrac{1}{x^2 + y^2}$;
(i) $y' = \dfrac{1}{x^2 + y^2 - 1}$; (j) $y' = (1 - y^2)^{\frac{1}{2}}$;
(k) $y' = (y/x)^{\frac{1}{2}}$. Make sure that not all your curves lie in the first quadrant.

22.2 (Computational). Use the Euler method to compute approximate solutions to the following initial-value problems. Try various values of the step h. Compare the results with the exact solutions provided.

(a) $y' = -\tfrac{1}{2}y$ with $y = 1$ at $x = 0$, over the range $0 \leqslant x \leqslant 2$. (The exact solution is $y = e^{-\frac{1}{2}x}$.)
(b) $y' = -x/y$ with $y = -1$ at $x = -1$, over the range $-1 \leqslant x \leqslant 1$. (The exact solution is $y = -(2 - x^2)^{\frac{1}{2}}$. Try to extend your results forward and backward (using negative h) to the range $-2 \leqslant x \leqslant 2$.)
(c) $y' = (1 - y^2)^{\frac{1}{2}}$, with $y = 0$ at $x = 0$, over the range $0 \leqslant x \leqslant \tfrac{1}{2}\pi$. (The exact solution is $y = \sin x$.)

22.3 (Computational). Use the Euler method to calculate a few representative solution curves in the following cases. Each curve will have a different initial condition. You should follow each curve forwards, and probably backwards as well (negative h), sufficiently far to get a clear idea of how it is behaving. It might be advantageous to use smaller intervals h over some sections than over others.
(a) $y' = y(x + 1)/x(y + 1)$. This refers to Example 22.6, with the 'solutions' $y\, e^y = Ax\, e^x$, where A is an arbitrary constant.

(b) $y' = 2y^{\frac{1}{2}}$. This refers to Example 22.7.

(c) $y' = (y/x)^{\frac{1}{2}}$ has solution curves in the first and *third* quadrants. Sketch a lineal-element diagram to obtain the broad pattern, then compute a few representative curves. (The general solution is

$$|y| = (|x|^{\frac{1}{2}} - C)^{\frac{1}{2}} \quad \text{for} \quad |x|^{\frac{1}{2}} > C.)$$

22.4 (Separation of variables). Obtain solutions of the following equations.

(a) $y' = x/y$;   (b) $y' = 2x/y$;

(c) $y' = x/(y+2)$;   (d) $y' = (x+3)/(y+2)$;

(e) $y' = x^2/y^2$;   (f) $y' = -x^2/y^2$;

(g) $y' = y^2/x^2$;   (h) $y' = -y^2/x^2$;

(i) $2xy' = y^2$;   (j) $yy' + x = 1$;

(k) $\dfrac{dx}{dt} = 3t^2x^3$;

(l) $(\sin x)\dfrac{dx}{dt} = t$;

(m) $e^{x+y}\dfrac{dy}{dx} = 1$;

(n) $(1 + x^2)\dfrac{dy}{dx} + (1 + y^2) = 0$, with $y(0) = -1$.

22.5 Show that the solution of the initial-value problem

$$\frac{dy}{dx} = -\frac{x}{y},$$

where $y = 1$ when $x = 2$, is obtained from the equation

$$\int_2^x u \, du = -\int_1^y v \, dv.$$

Generalize this technique to apply to the initial-value problem

$$\frac{dy}{dx} = g(x)h(y)$$

where $y = b$ when $x = a$.

22.6 Solve the following equations and sketch the solution curves. Take care to avoid spurious solutions as in Example 22.7. Look out for solutions you might have lost in the process: these are usually suggested by the sketch.

(a) $x\dfrac{dy}{dx} = 2y^{\frac{1}{2}}$;

(b) $\dfrac{dy}{dx} = xy^{\frac{1}{2}}$;

(c) $\dfrac{dy}{dx} = (1 - y^2)^{\frac{1}{2}}$;

(d) $x\dfrac{dy}{dx} = (1 - y^2)^{\frac{1}{2}}$.

22.7 (Differential method). Obtain a family of solutions of the following equations. (Usually they must be left in implicit form.)

(a) $\dfrac{dy}{dx} = \dfrac{2x - y}{x + 2y}$ (check whether there is also a solution of the form $y = mx$);

(b) $\dfrac{dy}{dx} = \dfrac{y}{y^2 - x}$;   (c) $\dfrac{dy}{dx} = \dfrac{x^2 - y}{x + y}$;

(d) $\dfrac{dy}{dx} = \dfrac{2x - y}{x - 2y}$;   (e) $\dfrac{dy}{dx} = \dfrac{x - 2xy}{x^2 - y}$;

(f) $\dfrac{dy}{dx} = \dfrac{3x^2}{3y^2 + 1}$;

(g) $\dfrac{dy}{dx} + \dfrac{2xy}{x^2 - 1} = 0$ (this is also a linear equation);

(h) $(1 - \sin y)\dfrac{dy}{dx} + \cos x = 0$;

(i) $(1 + 3\,e^{3y})\dfrac{dy}{dx} = 2\,e^{2x} - 1$;

(j) $(e^{x+y} + 1)\dfrac{dy}{dx} + (e^{x+y} - 1) = 0$;

(k) $\dfrac{dy}{dx} = \dfrac{1 + \cos x \sin y}{1 + \sin x \cos y}$.

22.8 (Differential method). Solve the following equations. Some of these need an integrating factor (see eqn (22.12)) such as the ones suggested.

(a) $\dfrac{dy}{dx} = \dfrac{y}{x}\dfrac{y - 2x}{x - 2y}$ (check also for solutions of the for $y = mx$);

(b) $\dfrac{dy}{dx} = \dfrac{y(1 - x^2)}{x(1 + x^2)}$ (divide by x^2);

(c) $\dfrac{dy}{dx} = \dfrac{y^2}{y^2 - 1}$ (divide by y^2);

(d) $\dfrac{dy}{dx} = \dfrac{y(y - 1)}{y^2 - x}$ (divide by y^2);

(e) $\dfrac{dy}{dx} = \dfrac{y(x^2 + y^2 - y)}{x(x^2 + y^2)}$ (divide by x^2y^2);

(f) $\dfrac{dy}{dx} = \dfrac{y}{x}\dfrac{x^3 - y}{x^3 + y}$ (show that this reduces to $x^3y^2\,d(x/y) = y\,d(xy)$; now put $u = xy$ and $v = x/y$).

22.9 The 'logistic equation' $dP/dt = aP - bP^2$, where $a > 0$ and $b > 0$, represents the growth of a population $P(t) > 0$ in which unrestricted growth is prevented by the term $-bP^2$ representing pressure on the means of subsistence. Solve the equation, sketch the solution curves, and show that in all circumstances $P(t) \to a/b$ as t tends to infinity.

22.10 (Computational). A population $P(t)$ of protozoa is assumed to increase according to the equation $dP/dt = aP - bP^2$, where a and b are constants. Starting with 10 protozoa, they are observed to increase by 150% per day while the numbers are still low, and to reach a fairly steady level ($dP/dt = 0$) of 25 000 after a few weeks. Find an approximation to a and b.

Use a numerical method to compute the population curve for the first 10 days.

Compare the curve obtained from the law $dP/dt = aP - bP^4$.

22.11 (See Example 22.15.) (a) A falling body of mass m is subject to air resistance equal to Kv^α, where v is its speed of fall: its equation of motion is then

$$\frac{d^2x}{dt^2} = g - \frac{K}{m}\left(\frac{dx}{dt}\right)^\alpha,$$

where g is the gravitational acceleration and x represents its position measured vertically downwards. Without solving the equation, show that the limiting speed of fall is equal to $(mg/K)^{1/\alpha}$.

(b) Substitute $v = dx/dt$ to obtain an equation for v^2 of the form

$$\frac{d(v^2)}{dt} = 2\left(g - \frac{K}{M}(v^2)^{\frac{1}{2}\alpha}\right).$$

(c) Assume that (in mks units) $K = 4$, $m = 80$, $\alpha = 1.2$, $g = 10$, and that the mass is dropped from rest. Use Euler's method (22.2) to obtain v^2, and hence v, over a sufficient distance to compare with the limiting speed of fall.

22.12 (See Section 17.5.) Solve the following (implicitly) by putting $y = xw$.
(a) $dy/dx = (x^2 - xy + y^2)/xy$;
(b) $dy/dx + (x^2 + y^2)/xy = 0$;
(c) $dy/dx + (x - y)/(3x + y) = 0$;
(d) $dy/dx = 2xy/(3x^2 - 4y^2)$;
(e) $dy/dx + 2(2x^2 + y^2)/xy = 0$.

22.13 Show that, if the substitution $w = y^{1-n}$ is made in the equation $y' + g(x)y = h(x)y^n$ (called the Bernoulli equation), we obtain the linear equation $w' + (1 - n)g(x)w = (1 - n)h(x)$.

Use this result to solve (a) $y' + y = y^4$, (b) $y' + y = y^{-\frac{1}{2}}$.

22.14 The equation

$$d^2y/dx^2 + (b/x)\,dy/dx + (c/x^2)y = 0$$

is called **equidimensional** (the dx^2, $x\,dx$, and x^2 in the denominators are considered to have the same dimensions). It is a linear equation with zero on the right, so

we expect a general solution of the form $Ay_1(x) + By_2(x)$, where A and B are arbitrary. Show how a basis of solutions $(y_1(x), y_2(x))$ can be obtained in two ways:

(a) Look for solutions having the form $y = x^M$, where M is an unknown constant. Note that M might be complex; in that case, to obtain *real* solutions, use

$$x^{\alpha+i\beta} = e^{(\alpha+i\beta)\ln x} = e^{\alpha\ln x}e^{i\beta\ln x}$$

$$= x^\alpha[\cos(\beta\ln x) + i\sin(\beta\ln x)],$$

as in Section 18.4.

(b) Change the independent variable to t, where $t = \ln x$, or $x = e^t$. The new equation has constant coefficients.

(c) Use either method of find the solutions of the equations
(i) $d^2y/dx^2 - (2/x)\,dy/dx + (2/x^2)y = 0$;
(ii) $d^2y/dx^2 - (1/x)\,dy/dx + 1/x^2 = 0$;
(iii) $d^2y/dx^2 + (3/x)\,dy/dx + (2/x^2)y = 0$.

22.15 (Computational). A boat enters a river at O, and tries to reach the point A on the other bank, directly opposite O and distant H from O, by keeping its bow pointed towards A, at an angle θ from OA (see Fig. 22.12). The speed of the boat in still water is V, and the uniform stream speed is $v < V$.

Show that, when the boat is at (x, y),

$$dx/dt = V\cos\theta, \quad dy/dt = v - V\sin\theta$$

where t is time. By dividing one equation by the other find a differential equation for y in terms of x. Given the values $v = 1$ m s^{-1}, $V = 4$ m s^{-1}, $H = 30$ m, compute the path of the boat. (As you approach close to A, you will encounter a problem with dy/dx.)

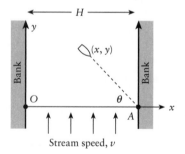

Fig. 22.12

22.16 (Computational). As in Problem 22.15, but construct a differential equation for a stream having a parabolic distribution of velocity, greatest in the middle and zero at the banks, of the form

$$v(x) = ax(H - x).$$

Put in plausible values for V, v, H, and a, and compute the path.

22.17 A mouse M enters a room at O and rushes to its hole at H with speed v, pursued by the cat C, who starts from B at the same moment as the mouse appears (see Fig. 22.13). The cat runs with speed $V > v$, always directly towards the mouse. Show that

$$dr/dt = v \cos \theta - V \quad \text{and} \quad d\theta/dt = -(v \sin \theta)/r,$$

where r and θ are polar coordinates for the cat relative to the (moving) mouse. Construct a differential equation

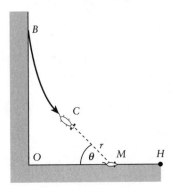

Fig. 22.13

for r in terms of θ, and solve it (it is really only a question of integration).

22.18 A satellite of mass m takes off vertically with speed V at time $t = 0$ from the surface of a planet of radius a. Assume that it is only influenced by the gravitational pull of the planet. If r is the distance of the satellite from the centre of the planet at time t, then, by Newton's law of gravitation, its equation of motion is

$$m\frac{d^2r}{dt^2} = -\frac{\gamma Mm}{r^2},$$

where M is the mass of the planet and γ is the gravitational constant. Using the identity

$$\frac{d^2r}{dt^2} = v\frac{dv}{dr}, \quad \text{where} \quad v = \frac{dr}{dt},$$

solve the first-order differential equation in v and r to obtain

$$\frac{1}{2}(v^2 - V^2) = \gamma\left(\frac{1}{r} - \frac{1}{a}\right).$$

Confirm that the **escape velocity** (i.e. the velocity above which the satellite will not return to the planet) from the surface of the planet is $\sqrt{(2\gamma M/a)}$.

Nonlinear differential equations and the phase plane

CONTENTS

However many methods may be invented for solving differential equations, there will always remain equations beyond their scope. But this does not mean that nothing can be done with them. The important van der Pol equation, which models a type of electrical oscillator,

$$\frac{d^2x}{dt^2} + c(x^2 - 1)\frac{dx}{dt} + x = 0,$$

where $c > 0$, cannot be solved explicitly. However, there are still comparatively simple methods which enable us to demonstrate its really important feature, which is that every solution, no matter how the device is started off, settles down into the same regular periodic oscillation.

Techniques enabling such conclusions to be drawn without actually solving the equation are called **qualitative methods**. This chapter outlines a way of looking at differential equations which is at the basis of many of these techniques. Qualitative methods do not consist of a collection of fixed results, and tend to be exploratory. Therefore computation is important. In the final section, a simple computing method is described which is easy to program but is effective enough to analyse realistic physical and biological models.

We shall take t (time) as the independent variable. For derivatives with respect to time, we use the conventional **dot notation** (just like the dash notation (4.1)):

$$\dot{x} = \frac{dx}{dt}, \qquad \ddot{x} = \frac{d^2x}{dt^2}.$$

23.1 Autonomous second-order equations

Let the independent variable be t and the dependent variable x. We shall only discuss equations which can be written in the form

$$\ddot{x}(t) = Q(x(t), \dot{x}(t)),$$

in which t **does not appear independently under Q on the right-hand side.** Such equations are called **autonomous**. For example, the equation $\ddot{x} - x\dot{x} + 1 = 0$ is autonomous, but $t\ddot{x} - x\dot{x} + 1 = 0$ is not autonomous. If startup conditions at $t = t_0$ are supplied so as to specify an **initial-value problem**

$$\ddot{x} = Q(x, \dot{x}), \qquad x(t_0) = x_0, \qquad \dot{x}(t_0) = y_0 \qquad (23.1)$$

(there is a special reason for using the symbol y_0 here), we expect that the initial conditions will select exactly one solution.

Suppose that the equation represents an electrical system, and that a graph of x against t for $t \geqslant t_0$ can be plotted automatically. If we find the clock in the plotter has been wrongly set, then it will not make the graph unusable; only its starting time t_0 will be wrong. Similarly, if we do one experiment starting at $t = 8.00$ h and repeat it at $t = 13.00$ h, the graphs plotted will be the same shape although the starting times are different (see Fig. 23.1). Intuition suggests that for autonomous equations, namely those in which t does not occur independently, it will not be a local clock time t_0 assigned to startup that counts, but the 'stopwatch' time elapsed from startup, $t - t_0$.

This intuition is correct. The mathematical reason is that a change of time scale from t to $t - t_0$ does not change the form of the differential equation, so the same phenomena follow. Put

$$T = t - t_0, \quad \text{and write} \quad x(t) = X(T).$$

Then $dX/dT = dx/dt$ and $d^2X/dT^2 = d^2x/dt^2$. Also $t = t_0$ becomes $T = 0$, so that the new initial-value problem is

$$\ddot{X} = Q(X, \dot{X}), \qquad X(0) = x_0, \qquad \dot{X}(0) = y_0. \qquad (23.2)$$

The equation is unchanged, but the starting time is assigned the value zero. Suppose (23.2) is solved in terms of T. Restore t by putting $T = t - t_0$. The solution $x(t)$ of (23.1) is then a function only of $t - t_0$, so it depends only on the time elapsed from startup.

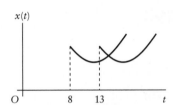

$x(t)$

O 8   13 t

Fig. 23.1 Two experiments, with a device described by an autonomous equation, which start at different times.

Example 23.1 *Solve the initial-value problems $\ddot{x} + \omega^2 x = 0$, with $x(t_0) = x_0$ and $\dot{x}(t_0) = y_0$.*

The general solution is $x(t) = A \cos \omega t + B \sin \omega t$ (see (18.14)). The process of finding A and B from the equations obtained by substituting the expressions for $x(t)$ and $\dot{x}(t)$ into the initial conditions is quite complicated (try it). Instead, put

$$T = t - t_0, \qquad x(t) = X(T).$$

↗

Example 23.1 *continued*

Then

$$\ddot{X} + \omega^2 X = 0, \qquad X(0) = x_0, \quad \dot{X}(0) = y_0.$$

The solution of this system is simple:

$$X(T) = x_0 \cos \omega T + \omega^{-1} y_0 \sin \omega T.$$

Put $T = t - t_0$; then the required solution is

$$x(t) = x_0 \cos \omega(t - t_0) + \omega^{-1} y_0 \sin \omega(t - t_0),$$

that is to say, x is a function only of the elapsed time $t - t_0$.

Autonomous second-order equations

The solution of the initial-value problem

$$\ddot{x} = Q(x, \dot{x})$$

with $x(t_0) = x_0$ and $\dot{x}(t_0) = y_0$ is a function only of elapsed time $t - t_0$.

(23.3)

23.2 Constructing a phase diagram for $(x, \dot{x})$

Consider again the initial-value problem of Example 23.1:

$$\ddot{x} + \omega^2 x = 0, \qquad x(t_0) = x_0, \quad \dot{x}(t_0) = y_0. \qquad (23.4)$$

This problem could arise in connection with a mass oscillating on a spring. The initial conditions imply that the position and velocity are prescribed at the start, $t = t_0$. If asked what the system was doing at $t = t_0$, specification of the position and velocity seems to constitute an adequate description of its state. It is in fact a perfect description, since it is exactly what is required to determine the whole future of the system. It is therefore reasonable to call the pair of numbers

$$(x_0, y_0)$$

the **state of the system** at t_0.

Subsequently the system moves smoothly through a succession of states: x and $\dot{x}$ will vary in time. Catch the system at any moment t_1; then the state $(x(t_1), \dot{x}(t_1))$ serves as fresh initial conditions for all the subsequent motion, but there can never be any conflict with what was predicted from the original initial conditions. It is the **succession of states** which is the subject of this chapter; the precise time that the states occur takes a secondary place.

To track the succession of states $(x(t), \dot{x}(t))$ for the initial-value problem (23.4), we could in principle begin by finding its solution. The solution (Example 23.1) is

$$x(t) = x_0 \cos \omega t + \omega^{-1} y_0 \sin \omega t,$$

so

$$\dot{x}(t) = -\omega x_0 \sin \omega t + y_0 \cos \omega t.$$

In effect, these equations specify the states **parametrically** (with parameter t). However, the expressions do not clearly reveal the association of x with $\dot{x}$, which is what we actually *observe* from moment to moment. Moreover, we need a method for equations we cannot solve.

We shall take a different route to the states $(x, \dot{x})$ which does not require that we solve the differential equation. Write

$$\dot{x} = y. \tag{23.5a}$$

Since $\ddot{x} = -\omega^2 x$, and $\ddot{x} = (d/dt)\dot{x}$, we have

$$\dot{y} = -\omega^2 x. \tag{23.5b}$$

These two **simultaneous *first-order*** differential equations are equivalent to the second-order differential equation (23.4). Divide (23.5b) by (23.5a), and use (3.8):

$$\frac{dy}{dt} \Big/ \frac{dx}{dt} = \frac{dy}{dx} = -\omega^2 \frac{x}{y}. \tag{23.6a}$$

Time has disappeared from the problem, and we now have a single first-order equation connecting x and y (i.e. connecting x and $\dot{x}$). Separate the variables (see Section 22.3) in (23.6a), and we obtain

$$\int y \, dy = -\omega^2 \int x \, dx,$$

or

$$\omega^2 x^2 + y^2 = C, \tag{23.6b}$$

where C is a positive, but otherwise arbitrary, constant.

The motive for introducing y as the symbol for $\dot{x}$ now becomes clear. Set up a pair of axes x and y as in Fig. 23.2. This framework is called the $(x, \dot{x})$ **phase plane**. The solution (23.6b) represents the family of ellipses displayed in the figure, and is called an $(x, \dot{x})$ **phase diagram for the differential equation** $\ddot{x} + \omega^2 x = 0$.

Any point on the diagram, say $A : (x_0, y_0)$, represents an initial state. If we follow the curve passing through A, we obtain the sequence of states for the corresponding solution. Equation (23.6a) alone does not tell us which way to travel along the curve; for this purpose, we momentarily resurrect t. The arrows indicate the directions that correspond to time going forwards rather than backwards. We defined y by

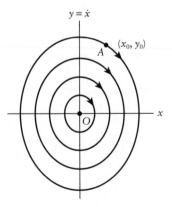

Fig. 23.2 $(x, \dot{x})$ phase diagram for $\ddot{x} + \omega^2 x = 0$, displaying $y = \dot{x}$ against x.

$$\frac{dx}{dt} = \dot{x} = y. \tag{23.7}$$

If we are in the **upper half plane**, then $y > 0$, so dx/dt is positive. Therefore $x(t)$ **is increasing**, and the directive arrow points **from left to right**. By a similar argument with $y < 0$ we find that in the **lower half plane** the arrow points **from right to left**. We must follow the arrow. Supplied with arrows, the state curves are called **phase paths**, or **trajectories**, or **orbits** for the differential equation $\ddot{x} + \omega^2 x = 0$.

Starting from $A : (x_0, y_0)$, follow the phase path. In going round, we can pick out a new feature in passing: that $\dot{x}$ is zero when x is at a minimum or maximum, and vice versa. Eventually we get back to A, renewing the initial state. Continue to follow the path around, duplicating the first circuit; the succession of states is repeated time after time.

This repetition does not itself establish that this is a truly *periodic* process. When we meet A again at the end of the first circuit, it is at a later time, t_1 say, so the initial conditions for the original equation are to this extent changed. Even though the system must follow the same path, perhaps it takes twice as long to go round the second time. However, from the discussion in Section 23.1, the time to complete any circuit, or to go repeatedly between any two fixed points on the circuit, is invariable because the equation is *autonomous*. This argument does not depend on what equation we started with, so we can say in general: **any closed phase path represents a periodic oscillation.**

Finally notice in Fig. 23.2 the bullet at the origin. This point represents a true solution, namely

$$x(t) = 0, \qquad y(t) = 0.$$

It is a special case of an **equilibrium point**, meaning a **constant solution**

$$x(t) = k, \qquad \dot{x}(t) = 0,$$

where k is a constant. Equilibrium points are of great importance in phase diagrams. A zero solution does not mean the same as no solution. If the constant k is zero, this solution is called 'trivial'; but this does not lessen its significance. An equilibrium point surrounded by closed curves is called a **centre**. It represents periodic oscillations about equilibrium. (Note, however, that oscillations of different amplitudes do not usually have the same period).

Since we chose a simple case, we have not discovered anything we did not know already, so consider the following example.

Example 23.2 *Sketch an* $(x, \dot{x})$ *phase plane for the equation* $\ddot{x} + cx^3 = 0$, *where* $c > 0$.

This represents small *lateral* oscillations $x(t)$ of a mass attached to the middle of an elastic string that is fixed at the ends and is *unextended* when $x = 0$. It is the same as Example 20.5, with $l = L$, and $c = 4s/ml$. We can regard explicit solutions as being unobtainable.

↗

Example 23.2 *continued*

Put $y = \dot{x}$. Then $\ddot{x} = \dot{y}$, and we have two first-order equations, together equivalent to the original equation:

$$\dot{x} = y, \qquad \dot{y} = -cx^3,$$

Therefore

$$\frac{dy}{dx} = \frac{dy}{dt} \bigg/ \frac{dx}{dt} = -c\frac{x^3}{y}.$$

By separating the variables, we obtain

$$\int y \, dy = -c \int x^3 \, dx,$$

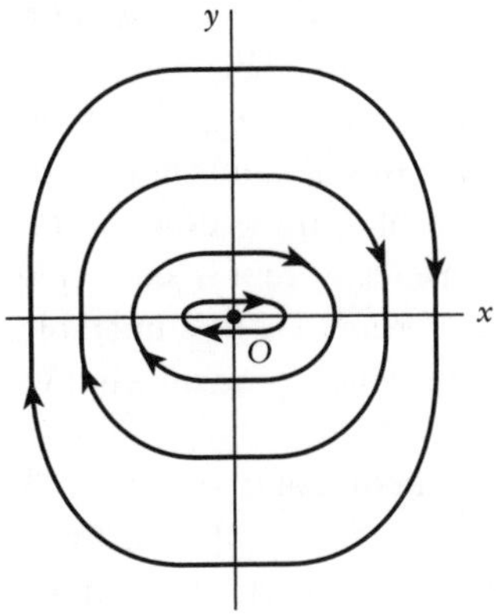

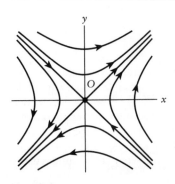

Fig. 23.3

so that $\frac{1}{2}y^2 = -\frac{1}{4}cx^4 + C$, where C is an arbitrary constant. Therefore

$$y = \pm(c/2)^{\frac{1}{2}}(A - x^4)^{\frac{1}{2}},$$

where A is arbitrary. For any $A > 0$, the phase path consists of two curves, one for $y > 0$ and one for $y < 0$. On both curves $y = 0$ where $x = \pm A^{\frac{1}{4}}$. The curves join smoothly at these two points so that the phase paths are closed curves. The family of phase paths is shown in Fig. 23.3.

The phase diagram of Fig. 23.3 consists entirely of closed curves; so, by exactly the same reasoning as in Example 23.1, we can deduce that every solution of the differential equation is a periodic oscillation. (However, we cannot say that they all have the same period: in fact they do not.) This phase diagram has therefore revealed an important fact about an equation that we could not solve.

23.3 $(x, \dot{x})$ phase diagrams for other linear equations; stability

For the moment, we shall stay with equations that are familiar from Chapter 16.

Example 23.3 *Construct the phase diagram for* $\ddot{x} - \omega^2 x = 0$.

Put $\dot{x} = y$; then $\dot{y} = \omega^2 x$. To eliminate t, form

$$\frac{\dot{y}}{\dot{x}} = \frac{dy}{dx} = \omega^2 \frac{x}{y}.$$

Separate the variables:

$$\int y \, dy = \omega^2 \int x \, dx,$$

or

$$y^2 - \omega^2 x^2 = A,$$

where A is arbitrary. This represents the family of hyperbolas in Fig. 23.4, having asymptotes $y = \pm\omega x$. The directions of the arrows follow the rule in Section 23.2: left to right in the upper half plane.

Fig. 23.4

Differential equations usually describe the behaviour of some circuit, machine, ecosystem, or something else in the real world, and our concern is with interpreting the phase diagrams in such a way as to bring to light features of practical importance. One important question is that of **stability of equilibrium** in a system. In our phase diagrams, the question turns into that of the **stability of an equilibrium point**, such as the origin in Examples 23.1 to 23.3.

In practice, systems are always subject to small external disturbances and internal fluctuations. For a system to work, it is important that small causes give small effects, and that the effects do not commence to grow catastrophically. In that case, a system is said to be **stable** with respect to small disturbances; otherwise it is **unstable**. The precise criterion for tolerable behaviour will depend on what we require from the particular system.

An equilibrium point surrounded by a structure of curves resembling those shown in Fig. 23.4 is called a **saddle**. It could hardly be classed as anything but an **unstable equilibrium point**. Apart from two special directions, if equilibrium is disturbed – even by a hairsbreadth – the system will find itself on one of the hyperbolas, and so it will be swept further and further away from equilibrium.

On the other hand a **centre**, exemplified in Figs 23.2 and 23.3, would often be called a **stable equilibrium point**. If equilibrium is disturbed by a small amount, the system does not go wild; it simply oscillates around its equilibrium position. However, a vehicle which behaved like that would be regarded as very unstable. Subject to a continual battering, it would vibrate objectionably, and the vibrations would never die away; therefore the stronger sort of stability illustrated in the following two examples is preferable.

Example 23.4   *Construct a phase diagram for $\ddot{x} + \frac{1}{4}\dot{x} + x = 0$ (weak damping).*

The equivalent pair of first-order equations is

$$\dot{x} = y, \qquad \dot{y} = -\tfrac{1}{4}y - x. \tag{i}$$

Therefore

$$\frac{dy}{dx} = \frac{-\tfrac{1}{4}y - x}{y}. \tag{ii}$$

This is hard to solve. To produce a phase diagram, we could solve the original equation for $x(t)$ by the methods of Chapter 18, and them obtain $y = \dot{x}(t)$. These are parametric equations for (x, y) curves. However, this would not be in the spirit of this chapter, because almost never are we able to solve the original equation. Instead, the Euler method of Section 22.2 can be used to obtain solution curves for (ii). (As we shall see later, it is easier to work from (i) using Section 23.8.) We obtain the pattern of spiral curves surrounding the origin shown in Fig. 23.5.

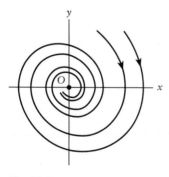

Fig. 23.5

Example 23.4 is a case of a linear oscillator with small damping, discussed in Section 18.4. The phase paths show that, from any starting point, the origin is approached via a sequence of diminishing spirals. Therefore any initial disturbance from equilibrium dies away. The equilibrium point is called a **stable spiral**. For the equation

$$\ddot{x} - \tfrac{1}{4}\dot{x} + x = 0,$$

the pattern of curves gives an outgoing **unstable spiral**.

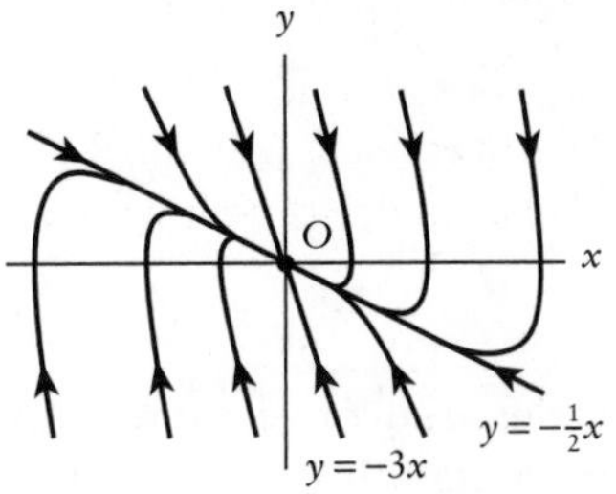

Example 23.5   *Construct a phase diagram for* $2\ddot{x} + 7\dot{x} + 3x = 0$.

This is a case of heavy damping (Section 20.4). We obtain

$$\dot{x} = y, \qquad \dot{y} = -\tfrac{7}{2}y - \tfrac{3}{2}x,$$

or

$$\frac{\mathrm{d}y}{\mathrm{d}x} = \frac{-\tfrac{7}{2}y - \tfrac{3}{2}x}{y}.$$

The phase paths, calculated numerically, are as shown in Fig. 23.6.

Fig. 23.6

In Fig. 23.6, the origin is called a **stable node**. All solutions fall straight into the origin without any oscillations: the system is **deadbeat**. Notice the structure of the node. There are two straight line solutions to the equation

$$\frac{\mathrm{d}y}{\mathrm{d}x} = \frac{-\tfrac{7}{2}y - \tfrac{3}{2}x}{y},$$

which can be found by trying for solutions of the form $y = mx$. Then $\mathrm{d}y/\mathrm{d}x = m = (-\tfrac{7}{2}m - \tfrac{3}{2})/m$, or $2m^2 + 7m + 3 = 0$. Therefore $m = -\tfrac{1}{2}$ or $m = -3$, and the two linear solutions are $y = -\tfrac{1}{2}x, y = -3x$. The divide the plane into four sectors which contain curved phase paths. Each of the curves has the property that it is tangential to $y = -\tfrac{1}{2}x$ at the origin, and parallel to $y = -3x$ at infinity. This behaviour is characteristic of nodes arising from linear equations, and the mutual tangency at the origin is common to all nodes, even those arising from nonlinear equations.

The technique for second-order differential equations can be summed up as follows.

> **$(x, \dot{x})$ phase plane for $\ddot{x} = Q(x, \dot{x})$**
>
> (a) *Phase path equations*:
> $$\dot{x} = y, \qquad \dot{y} = Q(x, y).$$
> (b) *The direction* of a phase path (x, y) is left to right if $y > 0$, and right to left if $y < 0$.
> (c) *Equilibrium points* (constant solutions) are at $(x, 0)$, where x is any solution of $Q(x, 0) = 0$.
> (d) *Alternative equation* $dy/dx = Q(x, y)/y$. (Shows that different phase paths may meet only at equilibrium points or other points where $Q(x, y)/y$ is undefined, and that paths cross the x axis at right angles.)
>
> **(23.8)**

23.4 The pendulum equation

The equation for a pendulum (Fig. 23.7) consisting of a light rod AB of length l freely pivoted at A and carrying a mass at B is

$$\ddot{x} + \omega^2 \sin x = 0,$$

where x is the angle of inclination and $\omega^2 = g/l$; here g is the gravitational acceleration. This equation can be solved, but only with difficulty, and by using recondite functions. From (23.7), the equations for the phase paths are

$$\dot{x} = y, \qquad \dot{y} = -\omega^2 \sin x. \tag{23.9}$$

The equilibrium points solve the equation $\sin x = 0$, so

$$x = 0, \pm\pi, \pm2\pi, \ldots, \quad \text{with} \quad y = 0. \tag{23.10a}$$

If

$$x = 0, \pm2\pi, \pm4\pi, \ldots, \tag{23.10b}$$

the pendulum is hanging vertically from its pivot in equilibrium. These values of x all represent the same *observed* state, though on the phase plane they correspond to different points. Similarly the values

$$x = \pm\pi, \pm3\pi, \ldots \tag{23.10c}$$

represent a state which is not usually thought of in connection with a pendulum: the pendulum rod is perched vertically upwards (and insecurely) on its pivot A.

Consider $x = 0$ as being representative of the freely hanging state, the equilibrium points (23.10b). See what happens when the displacement from $x = 0$ is small, but the pendulum is no longer in equilibrium. We can then put

$$\sin x \approx x.$$

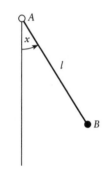

Fig. 23.7

The original equation becomes $\ddot{x} + \omega^2 x = 0$ (approximately), with solutions $x = C\cos(\omega t + \phi)$, where C (small) and ϕ are arbitrary. This is the familiar condition of small, isochronous oscillations. We have already solved the same problem for the phase plane in Section 23.2 and we found a centre: the family of ellipses shown in Fig. 23.2:

$$\omega^2 x^2 + y^2 = C,$$

where C is an arbitrary non-negative constant. This family will be repeated (for small C) around $x = \pm 2\pi, \pm 4\pi, \ldots$ in a progressively developing phase diagram; see Fig. 23.8.

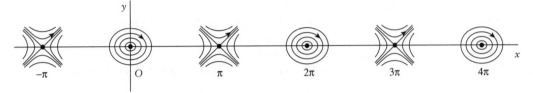

Fig. 23.8

Next, consider the case when the pendulum stands vertically: we choose as representative of (23.10c) the case $x = \pi$. To find what happens when the state is slightly displaced from the point $(\pi, 0)$, put

$$x = \pi + X,$$

where the new variable X is going to be small. Then

$$\sin x = \sin(\pi + X) = \sin\pi\cos X + \cos\pi\sin X$$
$$= -\sin X \approx -X.$$

From (23.8), with X instead of $\sin x$, the (approximate) equation for the displacement X from the equilibrium point is

$$\ddot{X} - \omega^2 X = 0.$$

From Example 23.3, this is equivalent to a saddle point on the phase plane at $X = 0$ (i.e. at $x = \pi$), and all the other equilibrium points in (23.10c) will have the identical structure, which implies instability.

The state of affairs around the equilibrium points is shown in Fig. 23.8. The rest of the phase diagram could be computed from (23.9), but in this case it is not difficult to sketch it in its entirely as in Fig. 23.9. From (23.9),

$$\frac{dy}{dx} = \frac{\omega^2 \sin x}{y}.$$

By separating the variables, we obtain the equations of the paths, which can be written in the form

$$y = \pm\sqrt{2}\omega(\cos x - A)^{\frac{1}{2}}, \tag{23.11}$$

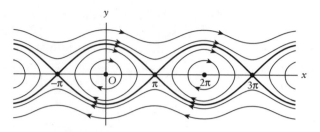

Fig. 23.9 Phase diagram $(x, \dot{x})$ for the pendulum equation $\ddot{x} + \omega^2 \sin x = 0$, given by $y = \pm\sqrt{2}\omega(\cos x - A)^{\frac{1}{2}}$ with $A \leqslant 1$. The figure extends with period 2π. The undulating curves (for $A < -1$) represent a whirling motion. The separatrices correspond to $A = -1$. There are centres at $x = 0, \pm 2\pi, \ldots$, and saddles at $x = \pm\pi, \pm 3\pi, \ldots$.

where A is (to an extent) arbitrary. Since $\cos x$ has period 2π, the repetitious nature of Fig. 23.8 is explained. Notice that $(\cos x - A)^{\frac{1}{2}}$ is real only when $\cos x \geqslant A$. Therefore $A \leqslant 1$. With that limitation, there are two main ranges of A which give significantly different patterns of (x, y) curves: $-1 \leqslant A \leqslant 1$ and $A < -1$. The centres correspond to $A = 1$, and the special curves joining the saddles, called the **separatrices**, correspond to $A = -1$. Notice the regular whirling motions which occur if $y = \dot{x}$ is large enough.

23.5 The general phase plane

There exists a great field of problems which, right from the start, take the form of simultaneous first-order differential equations:

$$\dot{x} = P(x, y), \qquad \dot{y} = Q(x, y). \tag{23.12}$$

Example 23.6   *A community of foxes and rabbits lives in uneasy harmony on an island. The rabbit population is $x(t)$, and they eat grass. The fox population is $y(t)$; they eat rabbits. Construct a differential equation model for the population variation.*

In a short time δt there is a rabbit population increase $ax\,\delta t$ $(a > 0)$ due to births and natural deaths, and a decrease $-bxy\,\delta t$ due to meetings with foxes, the frequency of which we suppose to be jointly proportional to the population densities of these animals. The net change in time δt is therefore

$$\delta x = ax\,\delta t - bxy\,\delta t.$$

Divide by δt and let $\delta t \to 0$; we obtain

$$\frac{\mathrm{d}x}{\mathrm{d}t} = ax - bxy. \tag{i}$$

For the foxes, assume that a shortage of prey causes a death rate c from starvation, offset by a fecundity factor dxy among those who get something to eat. Then, in time δt,

$$\delta y = -cy\,\delta t + dxy\,\delta t.$$

↗

Example 23.6 *continued*

Divide by δt and let $\delta t \to 0$:

$$\frac{dy}{dt} = -cy + dxy. \tag{ii}$$

(i) and (ii) form a **simultaneous (nonlinear) first-order system**:

$$\dot{x} = ax - bxy, \qquad \dot{y} = -cy + dxy, \tag{23.13}$$

with $a, b, c, d, > 0$.

We shall now show how important characteristics of the solutions $x(t)$ and $y(t)$ of the equations (23.13) resulting from this example can be revealed on a **general phase plane** by plotting y against x. (Notice that $\dot{x}$ is no longer equal to y, so this is not the same as the $(x, \dot{x})$ phase plane that we had before.) The pair of values (x, y) will be called a **state**. Although we do not prove it, the values of x and y at a particular t constitute **initial conditions** determining the solution for all subsequent time $t > t_0$, and because the equations are **autonomous**, the solutions are functions only of $t - t_0$.

Firstly, look for any **constant solutions** ($x =$ constant, $y =$ constant). These must satisfy the differential equations: that is to say

$$0 = x(a - by) \quad \text{and} \quad 0 = -y(c - dx).$$

Therefore $(x, y) = (0, 0)$ and $(x, y) = (c/d, a/b)$ are the constant solutions: on the (x, y) phase plane, they are the **equilibrium points**.

As before, divide the two equations, obtaining

$$\frac{dy}{dt} \Big/ \frac{dx}{dt} = \frac{dy}{dx} = -\frac{(c - dx)y}{(a - by)x}. \tag{23.14}$$

This is a separable equation; after separation it becomes

$$\int \left(\frac{a}{y} - b \right) dy = -\int \left(\frac{c}{x} - d \right) dx,$$

or

$$a \ln y - by + c \ln x - dx = C. \tag{23.15}$$

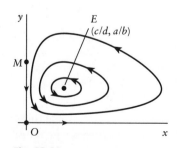

Fig. 23.10

Equation (23.15) represents the closed curves shown in Fig. 23.10. It is possible to see in advance that they are closed: the reason will be given shortly.

The direction arrows on the figure do not obey the rule (23.8b) for the $(x, \dot{x})$ phase plane. Each case has to be treated separately. The principle is easy: we have to find the **direction at a single point**, and the directions elsewhere are settled by **continuity of direction**: we expect adjacent curves to have the same direction. We might take the point $M : (0, m)$ in Fig. 23.10. At this point, the second equation of (23.13) gives $y = -cm < 0$, so y is decreasing at M. Once this direction is settled, the directions on the other curves follow by continuity.

There is a **centre** at the equilibrium point $E : (c/d, a/b)$. If the rabbit/fox populations take the values at E, the equations predict that the state will be permanent. A bad season for grass, or a disease amongst the foxes, will put the population state somewhere else, and thereafter the populations will undergo periodic oscillations. If foxes feast and thrive, rabbits languish, eventually starving the foxes; therefore rabbits prosper again; and so on.

The equilibrium point at O is unstable. If rabbits are introduced into a desert island paradise the population increase indefinitely, following the x axis arrow. If foxes are introduced to control the rabbits, a great periodic cycle is set up which goes on for as long as nothing else changes. Clearly the model is imperfect; but, provided that we have a program which will plot general phase paths, the complexity of the model is really a matter of no importance.

Earlier we said that the implicit equation (23.15) for the phase paths gives closed curves. It is useful to be able to recognize this feature.

> **Condition that $f(x) + g(y) = C$ (C arbitrary) represents a centre**
>
> If $f(x)$ has a minimum at $x = \alpha$, and $g(y)$ has a minimum at $y = \beta$, then there is an equilibrium point at (α, β) which is locally a centre (can also substitute 'maximum' for 'minimum' in both places).   **(23.16)**

To understand this you might need to look forward at Section 28.1. In three dimensions, x, y, z, the surface $z = f(x) + g(y)$ is bowl-shaped, with a minimum or maximum at (α, β). The paths reflect curves cut out by intersection with the horizontal planes $z = C$. The functions $c \ln x - dx$ and $a \ln y - by$ have maxima at $x = c/d$, $y = a/b$.

For a general system

$$\dot{x} = P(x, y), \qquad \dot{y} = Q(x, y),$$

the equilibrium points are where

$$P(x, y) = Q(x, y) = 0,$$

and might therefore appear anywhere in the phase plane, not just on the x axis as with the $(x, \dot{x})$ plane. On $Q(x, y) = 0$, $\dot{y} = 0$ so that phase paths cut this curve parallel to the x axis. Similarly on $P(x, y) = 0$, $\dot{x} = 0$ so that paths cut this curve parallel to the y axis. Between these curves the slopes of the paths will be either positive or negative depending on the sign of $Q(x, y)/P(x, y)$. The following statements recall the main features encountered in this section.

> **General phase plane**
> (a) *Phase path equations*: $\dot{x} = P(x, y)$, $\dot{y} = Q(x, y)$.
> (b) *Equilibrium points*: the solutions of $P(x, y) = Q(x, y) = 0$.
> (c) *Phase path direction*: find the direction at one point and use continuity for other paths.
> (d) *Alternative equation*: $dy/dx = Q(x, y)/P(x, y)$. **(23.17)**

23.6 Approximate linearization

In connection with the pendulum problem of Section 23.4, we were able to analyse equilibrium points by using a linear approximation valid near the points. In the general case, suppose that $\dot{x} = P(x, y)$, $\dot{y} = Q(x, y)$, and that (k, l) is an equilibrium point:

$$P(k, l) = Q(k, l) = 0. \tag{23.18}$$

We shall obtain a linear approximation to $P(x, y)$ and $Q(x, y)$ valid near (k, l). Put

$$x = k + X, \qquad y = l + Y, \tag{23.19}$$

where we suppose that X and Y are small. Then, because of (23.18), the approximations will take the form

$$P(x, y) = aX + bY, \qquad Q(x, y) = cX + dY \tag{23.20}$$

(with no constant term present).

Equations (23.20) should provide information about the phase paths near the equilibrium point $(x, y) = (k, l)$, which has become $(X, Y) = (0, 0)$, and tell us at least whether they are stable or unstable. This is often true, but not always: if, to take an extreme case, $a = b = c = d = 0$, then we would hardly want to rely on it.

The single equation corresponding to (23.17d) which connects X and Y in the approximation is

$$\frac{dY}{dX} = \frac{cX + dY}{aX + bY}. \tag{23.21}$$

The algebra involved in classifying this equation with respect to the coefficients is very complicated, and we merely summarize the results, omitting some special cases.

> **Equilibrium point $(0, 0)$ of the linear system** $\dot{X} = aX + bY$, $\dot{Y} = cX + dY$
>
> Put $p = a + d$, $q = ad - bc$, $\Delta = p^2 - 4q$.
> (a) If $q > 0$ and $\Delta > 0$: a node ($p < 0$, stable; $p > 0$, unstable.)
> (b) If $q > 0$ and $\Delta < 0$: a spiral ($p < 0$, stable; $p > 0$, unstable.)
> (c) If $q < 0$: a saddle.
> (d) If $p = 0$ and $q > 0$: a centre.
> (e) Path directions: investigate one point. **(23.22)**

Since these equations are **linear**, the phase diagram centred on $(0, 0)$ is **self-similar**: the pattern of paths is the same if viewed centrally through a microscope or seen over an immense field, so we are not restricted to small x and y.

In applying (23.22), do not be too ready to decide that the *original* equations have a centre just because the linear ones do: the small difference on changing back from the linear approximation may be all that is necessary to change a centre into a spiral.

Example 23.7 *Classify the equilibrium points of the system*

$$\dot{x} = x - y, \qquad \dot{y} = 1 - xy.$$

The equilibrium points are where $x - y = 0$, $1 - xy = 0$; that is, at $(1, 1)$ and $(-1, -1)$.

Near $(1, 1)$. Put $x = 1 + X$, $y = 1 + Y$. Then $x - y = X - Y$, and

$$1 - xy = 1 - (1 + X)(1 + Y) \approx -X - Y$$

for X and Y small. Therefore, in (23.20),

$$a = 1, \quad b = -1, \quad c = -1, \quad d = -1;$$

so $p = 0$, $q = -2$, $\Delta = 8$. According to (23.22), this is a saddle point (which is an unstable equilibrium point).

Near $(-1, -1)$. Put $x = -1 + X$, $y = -1 + Y$; then we obtain

$$x - y = X - Y, \qquad 1 - xy \approx -X - Y,$$

so that $a = 1, b = -1, c = 1, d = 1$. Therefore $q = 2 > 0, p = 2 > 0, \Delta = -4 < 0$; so, by (23.22), the point is an unstable spiral. The phase diagram is shown in Fig. 23.11. Note that the paths have zero slope on $xy = 1$ and infinite slope on $y = x$. See the remarks before (23.17).

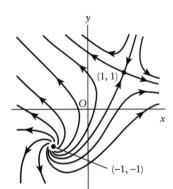

Fig. 23.11

23.7 Limit cycles

Spirals, centres, etc., occur for both linear and nonlinear systems, but a **limit cycle** is a feature only of nonlinear systems. When it occurs it usually represents the most important phenomenon in the phase plane. The following example includes a limit cycle.

Example 23.8 *Sketch a phase diagram for*

$$\ddot{x} + (x^2 + \dot{x}^2 - 1)\dot{x} + x = 0.$$

Put

$$\dot{x} = y, \qquad \dot{y} = (1 - x^2 - y^2)y - x. \tag{i}$$

It is possible to express the phase paths in polar coordinates r, θ:

$$r^2 = x^2 + y^2 \quad \text{and} \quad \tan\theta = y/x.$$

Differentiate these equations with respect to t:

$$r\dot{r} = x\dot{x} + y\dot{y},$$
$$\dot{\theta}/\cos^2\theta = (x\dot{y} - y\dot{x})/x^2. \tag{ii}$$

Example 23.8 *continued*

Substitute (i) into (ii): remember that $\dot{x} = y$, and put $x = r \cos\theta$ and $y = r \sin\theta$ as necessary. Then $r(t)$ and $\theta(t)$ are found to satisfy

$$\dot{r} = -r(r^2 - 1)\sin^2\theta, \tag{iii}$$

$$\dot{\theta} = -1 - (r^2 - 1)\sin\theta\cos\theta. \tag{iv}$$

A particular solution of (iii) and (iv) is $r = 1$, with $\dot{\theta} = -1$. This indicates a path consisting of the circle $r = 1$, followed around in the clockwise direction with unit angular velocity.

Also, from (iii),

$$\dot{r} \begin{cases} > 0 & \text{if } r < 1, \\ < 0 & \text{if } r > 1, \end{cases} \tag{v}$$

so the circle is approached from points inside by means of expanding spirals, and from points outside by contracting spirals. The phase diagram is shown in Fig. 23.12.

If we start from any initial conditions except for the equilibrium point $(0, 0)$, the system settles down gradually to the regular oscillation represented by the circle. This behaviour has a physical explanation. The 'coefficient' $x^2 + \dot{x}^2 - 1$, although variable, serves the purpose of a damping coefficient. Outside the circle, when $x^2 + y^2 - 1 > 0$ (remember $y = \dot{x}$), energy is lost and the paths tend to drift inwards. When $x^2 + y^2 - 1 < 0$ there is negative damping; energy is being supplied, so the amplitude of paths within the circle increases. For points on the circle $x^2 + y^2 - 1 = 0$ the damping is zero, so the motion is harmonic (the solutions are $x = \cos(t + \phi)$, with ϕ any constant), consistent with the circular path.

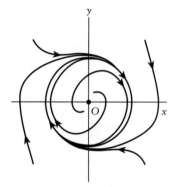

Fig. 23.12

The circular path $r = 1$ in Example 23.8 is an example of a **limit cycle**, which is defined generally as **an isolated closed phase path**. If the paths approach it spirally (in a broad sense) from both sides, it is called a **stable limit cycle**. It then represents a **stable oscillation**: if we disturb, or **perturb**, the oscillation by a small amount, it simply creeps back into the original oscillation. If the paths on one or both sides point away from the limit cycle, it is called **unstable** and is unlikely ever to be observed in practice.

To show how strange a limit cycle can be, we return to the van der Pol equation for the special case $\ddot{x} + 10(x^2 - 1)\dot{x} + x = 0$. Figure 23.13 shows its limit cycle in the $(x, \dot{x})$ phase plane together with the solution represented by the limit cycle.

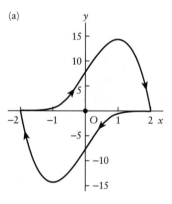

(b)

Fig. 23.13 (a) Limit cycle for $\ddot{x} + 10(x^2 - 1)\dot{x} + x = 0$. (b) The solution $x(t)$ corresponding to the limit cycle.

23.8 A numerical method for phase paths

We shall show a numerical method, related to Euler's method of Section 22.2, for plotting phase paths of the system

$$\dot{x}(t) = P(x(t), y(t)), \qquad \dot{y}(t) = Q(x(t), y(t)). \tag{23.23}$$

Essentially we use t as a parameter in a step-by-step solution. Start from an initial point $P : (x_0, y_0)$ at time t_0. The choice of t_0 does not affect the path constructed because the equations are autonomous. Take short time steps of length h. Then we proceed from point to point in the diagram;

$$P_0 : (x_0, y_0) \rightarrow P_1 : (x_1, y_1) \rightarrow P_2 : (x_2, y_2) \rightarrow \cdots .$$

Since, approximately,

$$x_{n+1} - x_n = h\dot{x}(t_n) = hP(x_n, y_n),$$

and similarly for $y_{n+1} - y_n$, the rule for getting from P_n to P_{n+1} is as follows.

> **Euler's method for $\dot{x} = P(x, y)$, $\dot{y} = Q(x, y)$**
>
> $$x_{n+1} = x_n + hP(x_n, y_n), \qquad y_{n+1} = y_n + hQ(x_n, y_n).$$
>
> Compute for $n = 0, 1, 2, \ldots$ successively.   (23.24)

This process gives rise to rather unevenly spaced points on a phase path, widely spaced when P and Q are large, and very closely spaced near an equilibrium point, where P and Q are inevitably small. However, they have the advantage that, if necessary, regular **time** indications can be marked on the path while it is being computed.

If **evenly spaced** points are wanted, the parameter can be changed from time t to arc-length s. We have $\delta s^2 = \delta x^2 + \delta y^2$; so

$$\frac{ds}{dt} = \left[\left(\frac{dx}{dt} \right)^2 + \left(\frac{dy}{dt} \right)^2 \right]^{\frac{1}{2}} = (P^2 + Q^2)^{\frac{1}{2}}.$$

Therefore

$$\frac{dx}{ds} = \frac{dx}{dt} \bigg/ \frac{ds}{dt} = \frac{P}{(P^2 + Q^2)^{\frac{1}{2}}} \quad \text{and} \quad \frac{dy}{ds} = \frac{Q}{(P^2 + Q^2)^{\frac{1}{2}}}$$

are equivalent equations for the path, in terms of arc-length s. This gives the following method.

> **To compute the paths of the system $\dot{x} = P(x, y)$, $\dot{y} = Q(x, y)$, at evenly spaced points**
>
> Apply (23.24) to the equivalent system
>
> $$\frac{dx}{ds} = \tilde{P}(x, y), \qquad \frac{dy}{ds} = \tilde{Q}(x, y),$$
>
> where
>
> $$\tilde{P} = P/(P^2 + Q^2)^{\frac{1}{2}}, \ \tilde{Q} = Q/(P^2 + Q^2)^{\frac{1}{2}}.$$
>
> The step length h is the distance along the path.   (23.25)

Problems

Many of the problems involve computation. The method of Section 23.8 is sufficient, but a high-accuracy computer library routine would allow the use of much larger values of h, and therefore be more efficient.

23.1 (Computation). Practise computing a phase diagram in the following cases. Information is given for checking, but imagine that you do not have it. The equilibrium points are at $(0, 0)$. Take different starting points; you may have to work backwards as well as forwards, by changing the sign of h.

(a) $\dot{x} = y, \dot{y} = -4x$. (A centre, $4x^2 + y^2 = C$. If your paths do not nearly close, try a smaller interval h.)

(b) $\dot{x} = y, \dot{y} = x$. (A saddle, $x^2 - y^2 = C$. Find the asymptotes by trying $y = mx$ in the equations: they are $y = \pm x$. It is difficult to make sense of the diagram without this information.)

(c) $\dot{x} = y, \dot{y} = -2x - 3y$. (Stable node. Find the two solutions $y = -x$, $y = 2x$, which are radial straight lines as in (b). These represent *four* paths since they are interrupted by the origin.)

(d) $\dot{x} = y, \dot{y} = -3x - y$. (Stable spiral.)

(e) $\dot{x} = y, \dot{y} = -2x + y$. (Unstable spiral.)

(f) Recompute (a), marking off a time scale on each of the paths, showing intervals in t of around 0.3.

(g) Recompute (b) with a time scale as in (f).

(h) $\dot{x} = y, \dot{y} = -2y$. A different type: what is the second-order equation that it comes from?

23.2 Sketch the phase paths for the following equations by first solving for them: form dy/dx and separate the variables.

(a) $\dot{x} = y, \dot{y} = x$; (b) $\dot{x} = x, \dot{y} = y$;
(c) $\dot{x} = -y, \dot{y} = x$; (d) $\dot{x} = -x, \dot{y} = y$;
(e) $\dot{x} = 2y, \dot{y} = x$; (f) $\dot{x} = -2y, \dot{y} = x$.

23.3 Solve the following by using the **energy transformation** $d^2x/dt^2 = \frac{1}{2}d(\dot{x}^2)/dx$ (Example 20.15), and sketch the $(x, \dot{x})$ phase diagrams.

(a) $\ddot{x} = e^x$;

(b) $\ddot{x} + \dot{x}^2 + x = 0$ (the transformed equation is linear in y^2);

(c) $\ddot{x} - 8x\dot{x} = 0$;

(d) $\ddot{x} = e^x - e^{-x}$ (the Poisson–Boltzmann equation).

23.4 Classify the equilibrium point $(0, 0)$ for each of the following linear equations by using (23.22). Sketch the phase diagram: in cases where it is appropriate you should first obtain the radial straight paths $y = mx$ by substitution. State which are unstable.

(a) $\dot{x} = x - 5y, \dot{y} = x - y$;
(b) $\dot{x} = x + y, \dot{y} = x - 2y$;
(c) $\dot{x} = -4x + 2y, \dot{y} = 3x - 2y$;
(d) $\dot{x} = x + 2y, \dot{y} = 2x + 2y$;
(e) $\dot{x} = 4x - 2y, \dot{y} = 3x - y$;
(f) $\dot{x} = 2x + 3y, \dot{y} = -3x - 3y$.

23.5 For the equations given: find any equilibrium points; obtain a linear approximation at each equilibrium point by the method of Section 23.6; classify it from (23.22) (finding the straight line paths in the case of nodes and saddles); and put the sketches on a phase diagram. Guess how the diagram away from the equilibrium points is filled in (isoclines, Section 17.1, might help here). Then turn to Problem 23.6.

(a) $\dot{x} = x - y, \dot{y} = x + y - 2xy$;
(b) $\dot{x} = 1 - xy, \dot{y} = (x - 1)(y + 1)$;
(c) $\dot{x} = x - y, \dot{y} = x^2 - 1$;
(d) $\ddot{x} + x - x^3 = 0$ (with $\dot{x} = y$);
(e) $\dot{x} = 4x - 2xy, \dot{y} = -2y + xy$, for $x > 0$ and $y > 0$ (foxes and rabbits, Example 23.6: classify $(0, 0)$ as if x and y could be negative).

23.6 (Computational). Check some of the phase diagrams you sketched in Problem 23.5 by computing representative phase paths. Look out for separatices, which end at equilibrium points.

23.7 Sketch possible phase diagrams from the information given. If a phase path ends in mid air, or if you have a closed curve without an equilibrium point inside, then there is something wrong. There are often several possibilities: for example, a path might either join two equilibrium points or split, forming two branches going to infinity. Suppose that the only equilibrium points at a finite distance are those given in the following cases.

(a) centre at $(0, 0)$, saddle at $(1, 0)$;
(b) centre at $(0, 0)$, saddles at $(\pm 1, 0)$;
(c) unstable node at $(0, 0)$, stable node at $(1, 0)$;
(d) centres at $(\pm 1, 0)$.

23.8 (Computational). Obtain a phase diagram for the following (in some of these the linear approximation point is zero, so it gives no information):

(a) $\ddot{x} + |\dot{x}|\dot{x} + x = 0$; (b) $\ddot{x} + |\dot{x}|\dot{x} + x^3 = 0$;
(c) $\dot{x} = x^4 - x^2$; (d) $\dot{x} = 2xy, \dot{y} = y^2 - x^2$;
(e) $\dot{x} = 2xy, \dot{y} = x^2 - y^2$;
(f) $\ddot{x} + \dot{x}(x^2 + \dot{x}^2) + x = 0$ (notice that the origin is a spiral, although the linear approximation has a centre – see the remark following (23.22)).

23.9 From the Taylor series (5.4b), $\sin x \approx x - \frac{1}{6}x^3$ for small x, so the pendulum equation (23.8) is approximated by $\ddot{x} + \omega^2(x - \frac{1}{6}x^3) = 0$ (the Duffing equation). Sketch or compute the phase diagram, and comment on the differences from Fig. 23.9, for the exact equation.

23.10 (Computational). For a modified form of the predator–prey problem (compare Example 23.6), in a special case, the equations are

$$\dot{x} = 4x - 2xy - x^2, \qquad \dot{y} = -2y + xy - 2y^2.$$

The additional terms in x^2 and y^2 are meant to account for competition for resources among rabbits and among foxes. Use a linear approximation at the equilibrium points in order to classify them, then compute the phase diagram.

23.11 A model for $H(t)$ hosts supporting $P(t)$ dangerous parasites is $\dot{H} = (a - bP)H$, $\dot{P} = (c - dP/H)P$, where a, b, c, d, are positive. Analyse the system in the (H, P) plane.

23.12 Figure 23.14 represents a spring of stiffness s and natural length l, pivoted at A at a height h above a smooth wire CD. At B is a lead m, attached to the spring and sliding on the wire. The equation of motion is

$$\ddot{x} + \frac{s}{m}\left(1 - \frac{l}{(h^2 + x^2)^{\frac{1}{2}}}\right)x = 0.$$

Classify the equilibrium points when $l < h$, $l = h$, and $l > h$.

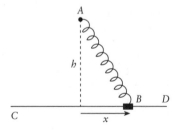

Fig. 23.14

23.13 (Computational). Solve Problem 23.12 modified so that there is friction between the bead and the wire equal to $k\dot{x}$. Classify the equilibrium points and construct the phase diagrams.

23.14 (Computational). Construct phase diagrams for the equation $\ddot{x} + k\dot{x} - x + x^2 = 0$. Consider various values of k.

23.15 (Computational). Construct a phase diagram for the following equations. (They each contain a limit cycle.)

(a) $\ddot{x} + \frac{1}{2}(x^2 + \dot{x}^2 - 1)\dot{x} + x = 0$;

(b) $\ddot{x} + \frac{1}{5}(x^2 - 1)\dot{x} + x = 0$;

(c) $\ddot{x} + \frac{1}{5}(\frac{1}{3}\dot{x}^2 - 1)\dot{x} + x = 0$;

(d) $\ddot{x} + 5(x^2 - 1)\dot{x} + x = 0$.

23.16 As in Problem 23.7, sketch phase diagrams for the general (x, y) phase plane compatible with the following information. The equilibrium points and limit cycles specified are the only ones allowed.

(a) $(0, 0)$ is a spiral and $x^2 + y^2 = 1$ is a stable limit cycle.

(b) $(0, 0)$ is a spiral, $x^2 + y^2 = 1$ a stable limit cycle, and $x^2 + y^2 = 4$ another limit cycle.

(c) $(\pm 1, 0)$ are saddles, $(0, 0)$ is a centre, and $x^2 + y^2 = 4$ is a stable limit cycle.

(d) $(\pm 1, 0)$ are centres, $(0, 0)$ is a saddle, and $x^2 + y^2 = 4$ is a stable limit cycle.

(e) $(0, 0)$ is a centre; the only closed path with $x^2 + y^2 > 1$ is the stable limit cycle $x^2 + y^2 = 4$.

23.17 Show that, in polar coordinates, the system

$$\dot{x} = -y + x(1 - x^2 - y^2),$$
$$\dot{y} = x + y(1 - x^2 - y^2)$$

becomes

$$\dot{r} = r(1 - r^2), \qquad \dot{\theta} = 1.$$

By investigating the sign of $\dot{r}$, explain why the system has just one limit cycle, which is stable. Sketch the phase diagram.

23.18 Find the locations of all the equilibrium points of

$$\dot{x} = (x^2 + y^2 - 1)y, \qquad \dot{y} = -(x^2 + y^2 - 1)x.$$

Explain why the circle $x^2 + y^2 = 1$ does not represent periodic motion.

23.19 Verify that the differential equation

$$\ddot{x} + \left(1 - x^2 - \frac{\dot{x}^2}{\omega^2}\right)\dot{x} + \omega^2 x = 0$$

has the particular solution $x = \cos \omega(t - t_0)$ for any t_0. What is the corresponding phase path in the $(x, y = \dot{x})$ plane? Put further details on a sketch of the phase diagram.

23.20 Locate all equilibrium points of the system

$$\dot{x} = (x^2 - 1)y, \qquad \dot{y} = (y^2 - 1)x,$$

and sketch its phase diagram.

23.21 The linear system given by (23.22), namely

$$\dot{x} = ax + by,$$

$$\dot{y} = cx + dy,$$

can be expressed in matrix form as

$$\dot{x} = Ax,$$

where

$$x = \begin{bmatrix} x \\ y \end{bmatrix}, \qquad A = \begin{bmatrix} a & b \\ c & d \end{bmatrix}.$$

Try an exponential solution in the form

$$x = C e^{\lambda t},$$

where C is a constant column vector, and show that λ must satisfy

$$\det(A - \lambda I_2) = 0.$$

In other words, the solutions for λ are the *eigenvalues* of A (see Chapter 13). If λ_1 and λ_2 are *distinct* eigenvalues, show that the general solution is

$$x = C_1 e^{\lambda_1 t} + C_2 e^{\lambda_2 t}.$$

What is the solution if $\lambda_1 = \lambda_2$?

Write down the roots of the quadratic equation, and discuss how the possible cases (e.g. real roots, imaginary roots, complex roots, etc.) fit in with the centre, saddle, node, and spiral, as classified in (23.22).

24

The Laplace transform

CONTENTS

24.1 The Laplace transform

Suppose that $f(t)$ is a specified function, and that s is a real positive parameter (that is to say, a supplementary variable). Then the integral

$$\int_0^\infty e^{-st} f(t)\, dt = F(s)$$

is called the **Laplace transform** of $f(t)$: the integral **transforms** $f(t)$ into another function $F(s)$.

For example, suppose that $f(t) = e^{2t}$ and $s > 2$. Then

$$F(s) = \int_0^\infty e^{-st} e^{2t}\, dt = \int_0^\infty e^{-(s-2)t}\, dt = \left[-\frac{1}{s-2}\, e^{-(s-2)t}\right]_0^\infty$$

$$= -\frac{1}{s-2}(e^{-\infty} - e^0) = -\frac{1}{s-2}(0 - 1) = \frac{1}{s-2}.$$

This result is true only if $s > 2$; otherwise the integral is infinite. We shall always assume that s is large enough to ensure that the integrals we encounter remain finite, or **converge** (see Section 15.6).

We also use the symbol L to stand for the 'Laplace transform of'. We have just proved that

$$F(s) = L\{e^{2t}\} = \frac{1}{s-2}.$$

Laplace transform of $f(t)$

$$L\{f(t)\} = F(s) = \int_0^\infty e^{-st} f(t)\, dt.$$

(24.1)

Another, very useful, notation is to indicate a transformed function by a **tilde sign**: $L\{f(t)\} = \tilde{f}(s)$, $L\{x(t)\} = \tilde{x}(s)$, and so on.

The letter p is often used for the parameter instead of s, especially in mainly theoretical texts.

24.2 Laplace transforms of t^n, $e^{\pm t}$, $\sin t$, $\cos t$

(a) *Positive, whole-number powers t^n, $n = 0, 1, 2, \dots$.*

$$L\{t^n\} = \int_0^\infty e^{-st} t^n\, dt.$$

Simplify the integral by substituting $u = st$, so that

$$t = \frac{1}{s}u \quad \text{and} \quad dt = \frac{1}{s}du.$$

Provided that s is **positive**, the limits of integration $t = 0$ and ∞ correspond to $u = 0$ and ∞ respectively. Therefore

$$L\{t^n\} = \int_0^\infty e^{-u}\left(\frac{u}{s}\right)^n \frac{du}{s} = \frac{1}{s^{n+1}} \int_0^\infty e^{-u} u^n\, du = \frac{n!}{s^{n+1}}$$

for $n = 0, 1, 2, \dots$ (from the standard integral, (17.9) for the factorial). Note that $0!$ is to be interpreted as being equal to 1.

Laplace transform of powers

$$L\{t^n\} = \frac{n!}{s^{n+1}} \quad \text{for } n = 0, 1, 2, \dots.$$

Special cases:

$$L\{1\} = \frac{1}{s}, \quad L\{t\} = \frac{1}{s^2}, \quad L\{t^2\} = \frac{2!}{s^3}, \quad L\{t^3\} = \frac{3!}{s^4}.$$

(24.2)

Example 24.1   *Find the Laplace transform F(s) of f(t) when*

$$f(t) = 1 - t + \frac{1}{2!}t^2 - \frac{1}{3!}t^3.$$

Composite expressions are dealt with in the following way.

$$F(s) \text{ or } L\{f(t)\} = L\left\{1 - t + \frac{1}{2!}t^2 - \frac{1}{3!}t^3\right\},$$

$$= L\{1\} - L\{t\} + \frac{1}{2}L\{t^2\} - \frac{1}{3!}L\{t^3\}$$

(which follows from the fact that each $L\{\cdots\}$ stands for the integral (24.1))

$$= \frac{1}{s} - \frac{1}{s^2} + \frac{1}{2!}\frac{2!}{s^3} - \frac{1}{3!}\frac{3!}{s^4},$$

$$= \frac{1}{s} - \frac{1}{s^2} + \frac{1}{s^3} - \frac{1}{s^4}.$$

(b) *Exponential* $e^{\pm t}$.

$$L\{e^{\pm t}\} = \int_0^\infty e^{-st} e^{\pm t} \, dt = \int_0^\infty e^{-(s\mp 1)t} \, dt$$

$$= -\frac{1}{s \mp 1}[e^{-(s\mp 1)t}]_0^\infty.$$

$s \mp 1$ are both positive if we take $s > 1$, in which case

$$L\{e^{\pm t}\} = -\frac{1}{s \mp 1}(0 - 1) = \frac{1}{s \mp 1}.$$

$$L\{e^t\} = \frac{1}{s - 1}, \qquad L\{e^{-t}\} = \frac{1}{s + 1}. \tag{24.3}$$

(c) *Sine and cosine.*

$$L\{\cos t\} = \frac{s}{s^2 + 1}, \qquad L\{\sin t\} = \frac{1}{s^2 + 1}. \tag{24.4}$$

Since $\cos t + i \sin t = e^{it}$, both of these can be verified at the same time by working out $L\{e^{it}\}$ and then separating the real and imaginary parts:

$$L\{e^{it}\} = \int_0^\infty e^{-st} e^{it} \, dt = \int_0^\infty e^{-(s-i)t} \, dt$$

$$= -\frac{1}{s - i}[e^{-(s-i)t}]_0^\infty = -\frac{1}{s - i}(0 - 1)$$

(since s is positive)

$$= \frac{1}{s-i} = \frac{s+i}{s^2+1}.$$

Therefore, as in (24.4),

$$\int_0^\infty e^{-st} \cos t \, dt = \text{Re} \int_0^\infty e^{-st} e^{it} \, dt = \frac{s}{s^2+1};$$

$$\int_0^\infty e^{-st} \sin t \, dt = \text{Im} \int_0^\infty e^{-st} e^{it} \, dt = \frac{1}{s^2+1}.$$

Example 24.2 *Find the Laplace transform of $3t^2 + 2\,e^{-t} - 5 \cos t$.*

$$L\{3t^2 + 2\,e^{-t} - 5\cos t\} = 3L\{t^2\} + 2L\{e^{-t}\} - 5L\{\cos t\}$$

$$= 3\frac{2!}{s^3} + 2\frac{1}{s+1} - 5\frac{s}{s^2+1}$$

$$= \frac{6}{s^3} + \frac{2}{s+1} - \frac{5s}{s^2+1},$$

from (24.3) and (24.5).

24.3 Scale rule; shift rule; factors t^n and e^{kt}

The following rules make it easy to derive more complicated transforms from the basic ones of Section 24.2.

> **Scale rule**
>
> If $L\{f(t)\} = F(s)$, and $k > 0$, then
>
> $$L\{f(kt)\} = \frac{1}{k}\,F\!\left(\frac{s}{k}\right).$$
>
> (24.5)

The proof is as follows.

$$L\{f(kt)\} = \int_0^\infty e^{-st}\,f(kt)\,dt.$$

Change the variable by putting $u = kt$, so that $t = u/k$ and $dt = du/k$. The limits of integration $t = 0$ and ∞ go into $u = 0$ and ∞ respectively because $k > 0$. Therefore

$$L\{f(kt)\} = \int_0^\infty e^{-s(u/k)} f(u)\left(\frac{du}{k}\right) = \frac{1}{k}\int_0^\infty e^{-(s/k)u} f(u)\,du = \frac{1}{k}\,F\!\left(\frac{s}{k}\right),$$

since $F(s) = \int_0^\infty e^{-su} f(u)\,du$.

The following are special cases.

If k is any constant, positive or negative, then

(a) $L\{e^{kt}\} = \dfrac{1}{s - k}$,

(b) $L\{\cos kt\} = \dfrac{s}{s^2 + k^2}$,

(c) $L\{\sin kt\} = \dfrac{k}{s^2 + k^2}$.

(24.6)

These are proved from the definition of L in (24.1), or as follows.

(a) Suppose that m is a positive number; then combining (24.3) with the scale rule (24.5) gives

$$L\{e^{\pm mt}\} = \frac{1}{m} \cdot \frac{1}{s/m \mp 1} = \frac{1}{s \mp m}.$$

The result (24.6a) therefore holds good for both positive and negative k.

(b) From (24.4),

$$L\{\cos t\} = \frac{s}{s^2 + 1}.$$

Therefore, by the scale rule (24.5), if $k > 0$,

$$L\{\cos kt\} = \frac{1}{k} \frac{s/k}{(s/k)^2 + 1} = \frac{s}{s^2 + k^2}.$$

This is true also if k is negative, since it is equal to $\int_0^\infty e^{-st} \cos kt \, dt$ (see (24.1)).

(c) is similar to (b).

Example 24.3   *Find the Laplace transform of* $\cos(3t + \tfrac{1}{4}\pi)$.

$\cos(3t + \tfrac{1}{4}\pi) = \cos \tfrac{1}{4}\pi \cos 3t - \sin \tfrac{1}{4}\pi \sin 3t = (\cos 3t - \sin 3t)/\sqrt{2}$.
Therefore by (24.6)

$$L\{\cos(3t + \tfrac{1}{4}\pi)\} = \frac{1}{\sqrt{2}}\left(\frac{s}{s^2 + 9} - \frac{3}{s^2 + 9}\right) = \frac{1}{\sqrt{2}} \frac{s - 3}{s^2 + 9}.$$

Suppose that we know the Laplace transform $F(s)$ of a function $f(t)$ already. Then the Laplace transform of $e^{kt} f(t)$ can immediately be written down

$$L\{e^{kt} f(t)\} = \int_0^\infty e^{-st} e^{kt} f(t) \, dt = \int_0^\infty e^{(s-k)t} f(t) \, dt.$$

But $\int_0^\infty e^{-st} f(t)\, dt = F(s)$, which is supposed to be known, and here we have $s - k$ in place of s. Therefore

$$L\{e^{kt} f(t)\} = F(s - k).$$

Shift rule (multiplication by e^{kt})

If $L\{f(t)\} = F(s)$ and k is any constant, then

$$L\{e^{kt} f(t)\} = F(s - k). \tag{24.7}$$

The shift rule is so called because the transform function $F(s)$ is 'shifted' a distance k along the s axis by the presence of the factor e^{kt}.

Example 24.4 *Find $L\{e^{-3t} \sin 2t\}$.*

From (24.6),

$$L\{\sin 2t\} = \frac{2}{s^2 + 4}.$$

By the shift rule (24.7) with $k = -3$, we deduce that

$$L\{e^{-3t} \sin 2t\} = \frac{2}{(s + 3)^2 + 4} = \frac{2}{s^2 + 6s + 13}.$$

Example 24.5 *Find $L\{t^3 e^{4t}\}$.*

From (24.2), $L\{t^3\} = 3!/s^4$. The shift rule with $k = 4$ gives

$$L\{e^{4t} t^3\} = \frac{3!}{(s - 4)^4}.$$

There is a rule similar to (24.7) by which we can find the Laplace transform of $t^n f(t)$ when the transform of $f(t)$ is known:

Multiplication by t^n

If $L\{f(t)\} = F(s)$, and n is a positive integer, then

$$L\{t^n f(t)\} = (-1)^n \frac{d^n F(s)}{ds^n}. \tag{24.8}$$

The simplest way to prove this is to start with the right-hand side. Since

$$\int_0^\infty e^{-st} f(t)\, dt = F(s),$$

then

$$\frac{dF(s)}{ds} = \frac{d}{ds} \int_0^\infty e^{-st} f(t) \, dt = \int_0^\infty \frac{d(e^{-st})}{ds} f(t) \, dt = \int_0^\infty (-t \, e^{-st}) f(t) \, dt$$

$$= -\int_0^\infty e^{-st} (tf(t)) \, dt = -L\{tf(t)\}.$$

Every time we differentiate, another factor t and another multiplication by -1 appear, which takes us to (24.8).

Example 24.6   *Find $L\{t \cos 3t\}$.*

Since, by (24.6b),

$$F(s) = L\{\cos 3t\} = \frac{s}{s^2 + 9},$$

then, by (24.8),

$$L\{t \cos 3t\} = -\frac{d}{ds} \frac{s}{s^2 + 9} = -\frac{9 - s^2}{(s^2 + 9)^2} = \frac{s^2 - 9}{(s^2 + 9)^2}.$$

Note the two following special cases, which occur frequently.

$$L\{t \cos kt\} = \frac{s^2 - k^2}{(s^2 + k^2)^2},$$

$$L\{t \sin kt\} = \frac{2ks}{(s^2 + k^2)^2}.$$

(24.9)

Example 24.7   *Find $L\{t^3 e^{-3t}\}$ (a) by using the shift rule, (b) by using (24.8), (c) by working directly from the definition of the Laplace transform.*

(a) From (24.2),

$$L\{t^3\} = \frac{6}{s^4}.$$

Therefore, using the shift rule (24.7a) with $k = -3$,

$$L\{e^{-3t} t^3\} = \frac{6}{(s + 3)^4}.$$

(b) From (24.6) with $k = -3$

$$L\{e^{-3t}\} = \frac{1}{s + 3}.$$

From (24.8) with $n = 3$,

$$L\{t^3 e^{-3t}\} = (-1)^3 \frac{d^3}{ds^3} \frac{1}{s + 3} = (-1)^3 \frac{(-1)(-2)(-3)}{(s + 3)^4} = \frac{6}{(s + 3)^4}.$$

Example 24.7 *continued*

(c) From the definition, (24.1),

$$\mathcal{L}\{t^3\, \mathrm{e}^{-3t}\} = \int_0^\infty \mathrm{e}^{-st}\, t^3\, \mathrm{e}^{-3t}\, \mathrm{d}t = \int_0^\infty \mathrm{e}^{-(s+3)t}\, t^3\, \mathrm{d}t.$$

From (24.2), this is equal to

$$\frac{3!}{(s+3)^4}.$$

24.4 Inverting a Laplace transform

Given a function $f(t)$, we obtain its transform $F(s)$ by using the definition (24.1). Alternatively, if a function $F(s)$ is presented, then we can try to recover the function $f(t)$, from which $F(s)$ is obtained. This second question is the **inverse problem** for the Laplace transform – to find '?' in the equation

$$\mathcal{L}\{?\} = F(s).$$

It can be proved that there is only one answer to this problem. The process of finding $f(t)$ from $F(s)$ is called **inversion** of $F(s)$.

The notation

$$f(t) \leftrightarrow F(s)$$

is a useful notation which underlines the two-way correspondence between $f(t)$ and $F(s)$.

We can open up a 'dictionary' for this purpose, as we did for derivatives and integrals. The most important results we have so far are given in the table (24.10) below.

$f(t)$ for $t > 0$		$F(s)$
t^n	$(n = 0, 1, \dots)$,	$\dfrac{n!}{s^{n+1}}$
$\dfrac{1}{(m-1)!}\, t^{m-1}$	$(m = 1, 2, \dots)$	$\dfrac{1}{s^m}$
e^{kt}	(any k)	$\dfrac{1}{s-k}$
$\cos kt$	(any k)	$\dfrac{s}{s^2 + k^2}$
$\sin kt$	(any k)	$\dfrac{k}{s^2 + k^2}$
$\dfrac{1}{k}\sin kt$	(any $k \neq 0$)	$\dfrac{1}{s^2 + k^2}$

(24.10)

A much fuller table which also includes the various rules can be found in Appendix F. Remember that everything we do with Laplace transforms refers to $t \geqslant 0$ only: the defining integral (24.1) calls only on values of $t \geqslant 0$.

Partial fractions are often useful for inverting transforms.

Example 24.8   *Given the transform $1/s(s + 1)$, find the original.*

In partial fractions,

$$\frac{1}{s(s + 1)} = \frac{1}{s} - \frac{1}{s + 1}.$$

From the table above,

$$\frac{1}{s} \leftrightarrow 1 \quad \text{and} \quad \frac{1}{s + 1} \leftrightarrow e^{-t},$$

so that

$$\frac{1}{s(s + 1)} \leftrightarrow 1 - e^{-t}.$$

Example 24.9   *Invert the Laplace transform*

$$\frac{s + 1}{s(s^2 + 4)}.$$

The partial-fraction rules require the form

$$\frac{s + 1}{s(s^2 + 4)} = \frac{A}{s} + \frac{Bs + C}{s^2 + 4}.$$

When the constants are determined by the method of Section 1.14, we find that $A = \frac{1}{4}, B = -\frac{1}{4}, C = 1$, so that

$$\frac{s + 1}{s(s^2 + 4)} = \frac{\frac{1}{4}}{s} + \frac{-\frac{1}{4}s + 1}{s^2 + 4}$$

$$= \frac{\frac{1}{4}}{s} - \frac{1}{4}\frac{s}{s^2 + 4} + \frac{1}{s^2 + 4}.$$

From (24.2),

$$\frac{1}{s} \leftrightarrow 1.$$

From the table (24.10),

$$\frac{s}{s^2 + 4} \leftrightarrow \cos 2t, \qquad \frac{1}{s^2 + 4} \leftrightarrow \tfrac{1}{2} \sin 2t.$$

Therefore

$$\frac{s + 1}{s(s^2 + 4)} \leftrightarrow \tfrac{1}{4} - \tfrac{1}{4} \cos 2t + \tfrac{1}{2} \sin 2t.$$

Example 24.10 *Invert the Laplace transform*

$$\frac{3s + 2}{s^2 + 2s + 2}.$$

The quadratic denominator does not have real factors, so partial fractions are not available. Instead we complete the square:

$$s^2 + 2s + 2 = (s + 1)^2 - 1 + 2 = (s + 1)^2 + 1.$$

We aim to write the whole expression in terms of $s + 1$ so that we can apply the shift rule (24.7). So put also

$$3s + 2 = 3(s + 1) - 3 + 2 = 3(s + 1) - 1,$$

and the transform becomes

$$\frac{3(s + 1) - 1}{(s + 1)^2 + 1}.$$

If we had s instead of $s + 1$, we could invert the transform:

$$\frac{3s - 1}{s^2 + 1} = \frac{3s}{s^2 + 1} - \frac{1}{s^2 + 1} \leftrightarrow 3 \cos t - \sin t.$$

Therefore, by the shift rule with $k = -1$,

$$\frac{3(s + 1) - 1}{(s + 1)^2 + 1} \leftrightarrow e^{-t}(3 \cos t - \sin t).$$

24.5 Laplace transforms of derivatives

Suppose that $L\{f(t)\} = F(s)$. Then the Laplace transforms of $df(t)/dt$, $d^2f(t)/dt^2$, ... can be expressed in terms of $F(s)$.

In the definition,

$$L\left\{\frac{df(t)}{dt}\right\} = \int_0^\infty e^{-st} \frac{df(t)}{dt} dt.$$

Integrate the right-hand side by parts. Using the notation of Section 17.7, put

$$u = e^{-st}, \qquad \frac{dv}{dt} = \frac{df(t)}{dt},$$

so that

$$\frac{du}{dt} = -s \, e^{-st}, \qquad v = f(t).$$

Then

$$L\left\{\frac{df(t)}{dt}\right\} = \int_0^\infty e^{-st} \frac{df(t)}{dt} dt = [e^{-st} f(t)]_0^\infty - \int_0^\infty (-s \, e^{-st}) f(t) \, dt$$

$$= 0 - e^0 f(0) + s \int_0^\infty e^{-st} f(t) \, dt = -f(0) + sL\{f(t)\}.$$

In other words, if $L\{f(t)\} = F(s)$, then

$$L\left\{\frac{df(t)}{dt}\right\} = sF(s) - f(0).$$ (24.11)

(Note that it is $f(0)$, not $F(0)$, that arises here.)

We can use (24.11) again and again to obtain

$$L\left\{\frac{d^2 f(t)}{dt^2}\right\} = L\left\{\frac{d}{dt}\frac{df(t)}{dt}\right\}$$

and higher derivatives successively, from which we obtain the sequence:

Laplace transform of derivatives

If $L\{f(t)\} = F(s)$, then

$$L\left\{\frac{df(t)}{dt}\right\} = sF(s) - f(0),$$

$$L\left\{\frac{d^2 f(t)}{dt^2}\right\} = s^2F(s) - sf(0) - f'(0),$$

$$L\left\{\frac{d^3 f(t)}{dt^3}\right\} = s^3F(s) - s^2f(0) - sf'(0) - f''(0),$$

and so on.   (24.12)

Example 24.11 *Obtain the transform of the expression*

$$\frac{d^2x}{dt^2} + 2\frac{dx}{dt} + 3x,$$

when $x = 4$ and $dx/dt = 5$ at $t = 0$.

Put $L\{x(t)\} = X(s)$. Then

$$L\left\{\frac{d^2x}{dt^2} + 2\frac{dx}{dt} + 3x\right\} = L\left\{\frac{d^2x}{dt^2}\right\} + 2L\left\{\frac{dx}{dt}\right\} + 3L[x]$$

$$= s^2X - sx(0) - x'(0) + 2[sX - x(0)] + 3X$$

$$= s^2X - 4s - 5 + 2(sX - 4) + 3X$$

$$= (s^2 + 2s + 3)X - 4s - 13.$$

24.6 Application to differential equations

The results (24.12) enable initial-value problems for differential equations to be solved.

Example 24.12 *Find the solution of*

$$\frac{dx}{dt} + 2x = e^{-t}$$

for which x = 3 when t = 0.

Since

$$\frac{dx}{dt} + 2x = e^{-t},$$

it is also true that

$$L\left\{\frac{dx}{dt}\right\} + 2L\{x\} = L\{e^{-t}\}.$$

Write

$$L\{x(t)\} = X(s).$$

By (24.12) the transformed equation becomes

$$sX - 3 + 2X = \frac{1}{s+1}$$

(where we put $x(0) = 3$ as specified by the initial condition). The transform $X(s)$ of $x(t)$ is therefore given by

$$X(s) = \frac{3s+4}{(s+1)(s+2)} = \frac{1}{s+1} + \frac{2}{s+2} \leftrightarrow x(t) = e^{-t} + 2\,e^{-2t},$$

which is the required solution.

It can be seen that the terms involving $f(0)$, $f'(0)$, … in (24.12), far from being merely a nuisance, are exactly what is required to translate a differential equation together with initial conditions into a simpler problem in ordinary algebra. We do not have to match up arbitrary constants with the initial conditions; these conditions are built into the transformed equations.

In many physical situations, we want to know what happens when an inactive or **quiescent** system is 'switched on'. In such cases, we have **zero initial conditions** at $t = 0$. For a system described by a second-order differential equation, we assume by this that the variable and its first derivative are initially set to zero.

Example 24.13 *A system is described by the equation*

$$\frac{d^2x}{dt^2} + 2\frac{dx}{dt} + 4x = 1.$$

It is initially quiescent and is then switched on. Find the subsequent time variation of x.

We have $x(0) = x'(0) = 0$. Let

$$x(t) \leftrightarrow X(s).$$

Example 24.13 *continued*

Then the equation transforms to

$$s^2X + 2sX + 4X = \frac{1}{s}$$

(notice the $1/s$) so that

$$X = \frac{1}{s(s^2 + 2s + 4)} = \frac{1}{4}\frac{1}{s} - \frac{1}{4}\frac{s+2}{s^2 + 2s + 4}.$$

The quadratic has no real factors; therefore the second term is rewritten in the manner of Example 24.10:

$$X = \frac{1}{4}\frac{1}{s} - \frac{1}{4}\frac{(s+1) + 1}{(s+1)^2 + 3}$$

$$= \frac{1}{4}\frac{1}{s} - \frac{1}{4}\left(\frac{s+1}{(s+1)^2 + 3} + \frac{1}{(s+1)^2 + 3}\right).$$

To invert the last two terms: from (24.10)

$$\frac{s}{s^2 + 3} \leftrightarrow \cos \sqrt{3}t, \qquad \frac{1}{s^2 + 3} = \frac{1}{\sqrt{3}} \sin \sqrt{3}t.$$

By using the shift rule (24.7) with $k = -1$, we obtain

$$\frac{s+1}{(s+1)^2 + 3} \leftrightarrow e^{-t} \cos \sqrt{3}t, \qquad \frac{1}{(s+1)^2 + 3} \leftrightarrow \frac{1}{\sqrt{3}} e^{-t} \sin \sqrt{3}t.$$

Therefore

$$x(t) = \tfrac{1}{4} - \tfrac{1}{4}(e^{-t} \cos \sqrt{3}t + \tfrac{1}{3}\sqrt{3}\, e^{-t} \sin \sqrt{3}t).$$

Example 24.14 *Solve the equation*

$$\frac{d^2x}{dt^2} + \omega_0^2 x = a \cos \omega_0 t,$$

with $x(0) = x'(0) = 0$.

If we put $\mathcal{L}\{x(t)\} = X(s)$, then the equation transforms into

$$s^2X + \omega_0^2 X = \frac{as}{s^2 + \omega_0^2},$$

so that

$$X = \frac{as}{(s^2 + \omega_0^2)^2}.$$

We can read off the inverse from (24.9) with $k = \omega_0$:

$$x(t) = \frac{a}{2\omega_0} t \sin \omega_0 t.$$

This equation is one of the exceptional resonant types discussed in Section 19.3. The advantage of using the Laplace transform is easy to see.

Example 24.15 *Solve the simultaneous equations*

$$\frac{dx}{dt} = x - y, \qquad \frac{dy}{dt} = x + y,$$

with the initial conditions $x(0) = 1$, $y(0) = 0$.

Let $L\{x(t)\} = X(s)$ and $L\{y(t)\} = Y(s)$. Then the transformed equations, including initial conditions, are

$$sX - 1 = X - Y, \qquad sY = X + Y.$$

Therefore

$$(1 - s)X - Y = -1,$$
$$X + (1 - s)Y = 0.$$

By solving these equations, we obtain

$$X = \frac{-1 + s}{s^2 - 2s + 2}, \qquad Y = \frac{1}{s^2 - 2s + 2}.$$

The denominators, $s^2 - 2s + 2$, have no real factors, so use the method of Example 24.10 to rewrite these expressions as

$$X = \frac{s - 1}{(s - 1)^2 + 1}, \qquad Y = \frac{1}{(s - 1)^2 + 1},$$

so that the shift rule (24.7) can be used to invert them. By (24.10),

$$\frac{s}{s^2 + 1} \leftrightarrow \cos t, \qquad \frac{s}{s^2 + 1} \leftrightarrow \sin t.$$

Therefore, by the shift rule with $k = 1$,

$$x(t) = e^t \cos t, \qquad y(t) = e^t \sin t.$$

24.7 The unit function and the delay rule

The Heaviside **unit function** H(t) (or U(t)) was introduced in Section 1.4. Here is a reminder of its definition:

Unit function H(t)

$$H(t) = \begin{cases} 0 & \text{when } t < 0, \\ 1 & \text{when } t \geq 0. \end{cases}$$

(24.13)

It is shown again in Fig. 24.1a. Figures 24.1b–e show how it can be used to describe various **step functions** and **switching functions**.

For example, the composition of the three segments of Fig. 24.1e is specified by:

$$e^t[H(t - 1) - H(t - 2)] = \begin{cases} e^t(0 - 0) = 0 & \text{if } t < 1, \\ e^t(1 - 0) = e^t & \text{if } 1 \leq t < 2, \\ e^t(1 - 1) = 0 & \text{if } t \geq 2. \end{cases}$$

Related Laplace transforms are given as follows.

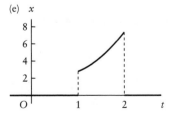
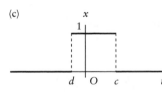
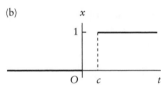
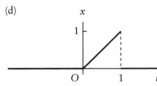
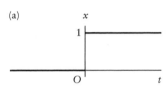

Fig. 24.1 (a) $x = H(t)$,
(b) $x = H(t - c)$,
(c) $x = H(t - d) - H(t - c)$,
(d) $x = t[H(t) - H(t - 1)]$,
(e) $x = e^t[H(t - 1) - H(t - 2)]$.

> **Laplace transform for the unit function**
>
> $$L\{H(t)\} = \frac{1}{s}, \qquad L\{H(t-c)\} = \frac{e^{-cs}}{s} \quad (c \text{ positive}).$$
>
> (24.14)

The various combination rules such as the shift rule (24.7) work for $H(t)$ in the same way as for smooth functions $f(t)$.

Example 24.16   *Find $L\{f(t)\}$ when $f(t) = e^t[H(t-1) - H(t-2)]$.*

This is the function shown in Fig. 24.1e. Then, from the definition,

$$L\{f(t)\} = \int_0^\infty e^{-st} e^t[H(t-1) - H(t-2)] \, dt,$$

$$= \int_1^2 e^{-(s-1)t} \, dt = -\frac{1}{s-1}[e^{-(s-1)t}]_1^2$$

$$= -\frac{1}{s-1}(e^{-2(s-1)} - e^{-(s-1)}).$$

Alternatively, we could use the shift rule (24.7), though it has no particular advantage.

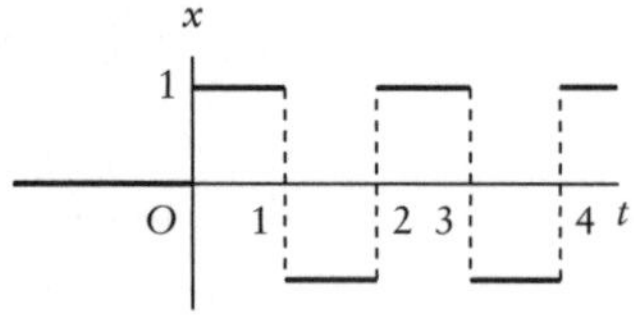

Fig. 24.2

Example 24.17   *Find the Laplace transform of the function shown in Fig. 24.2.*

By considering the segments one at a time and using Fig. 24.1c, we have
$$x(t) = [H(t) - H(t-1)] - [H(t-1) - H(t-2)]$$
$$+ [H(t-2) - H(t-3)] - \cdots,$$
$$= H(t) - 2H(t-1) + 2H(t-2) - 2H(t-3) + \cdots.$$

From (24.14)

$$H(t-n) \leftrightarrow \frac{e^{-ns}}{s}.$$

Therefore

$$L\{x(t)\} = \frac{1}{s} - \frac{2}{s}(e^{-s} - e^{-2s} + e^{-3s} - \cdots).$$

The brackets contain an infinite geometric series with first term e^{-s} and common ratio $-e^{-s}$. Therefore

$$L(x(t)) = \frac{1}{s} - \frac{2}{s}\frac{e^{-s}}{1 + e^{-s}}$$

$$= \frac{1 - e^{-s}}{s(1 + e^{-s})}.$$

(a)

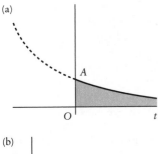

(b)

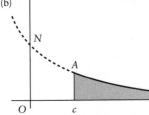

Fig. 24.3 (a) Graph of $g(t)$.
(b) Graph of $g(t-c)H(t-c)$.

Suppose that we have a function $g(t)$ which has a meaning for all positive t, such as $g(t) = e^{-t}$. Its Laplace transform is $G(s) = \int_0^\infty e^{-st} g(t) \, dt$. All values of $g(t)$ for t positive are called on to contribute to this integral, but none of its values for negative t are called upon (Fig. 24.3a).

Now translate the function a distance c (positive) to the right as in Fig. 24.3b. The new graph represents $g(t-c)$. It brings with it a section NA which originally corresponded to negative values of t. We cannot expect that the Laplace transform of this new function $g(t-c)$ can be expressed in terms of $G(s)$, because none of these t values played any part in the calculation of $G(s)$.

Therefore we cut out the section NA by considering not $g(t-c)$, but $g(t-c)H(t-c)$, which is shaded in Fig. 24.3b, and is congruent to the shaded part of Fig. 24.3a.

Then

$$L\{g(t-c)H(t-c)\} = \int_0^\infty e^{-st} g(t-c)H(t-c) \, dt$$

$$= \int_c^\infty e^{-st} g(t-c) \, dt.$$

Put $t - c = u$, so that $t = u + c$ and $dt = du$. The integral becomes

$$\int_0^\infty e^{-s(u+c)} g(u) \, du = e^{-sc} \int_0^\infty e^{-su} g(u) \, du = e^{-sc} G(s).$$

This is the **second shift rule**, or the **delay rule**, so called because $g(t-c)H(t-c)$ does not start until $t = c$.

> **Delay rule**
> If $G(s) \leftrightarrow g(t)$ and $c > 0$, then
> $$e^{-cs} G(s) \leftrightarrow g(t-c)H(t-c). \tag{24.15}$$

It is most often useful in inverting a Laplace transform.

Example 24.18 *Find the inverse Laplace transform of e^{-2s}/s^2.*

Put $G(s) = 1/s^2$. Then

$$G(s) = \frac{1}{s^2} \leftrightarrow g(t) = t.$$

By the delay rule,

$$\frac{e^{-2s}}{s^2} = e^{-2s} G(s) \leftrightarrow (t-2)H(t-2),$$

a function which suddenly takes off from zero at $t = 2$.

Example 24.19   *Find the inverse Laplace transform of*

$$\frac{e^{-2(s+1)}}{(s+1)(s+2)}.$$

Put

$$G(s) = \frac{1}{(s+1)(s+2)} = \frac{1}{s+1} - \frac{1}{s+2} \leftrightarrow g(t) = e^{-t} - e^{-2t}.$$

We require the inverse transform of $e^{-2(s+1)} G(s)$. By the delay rule with $c = 2$, this is given by

$$e^{-2(s+1)} G(s) = e^{-2} e^{-2s} G(s)$$
$$\leftrightarrow e^{-2}(e^{-(t-2)} - e^{-2(t-2)})H(t-2) = (e^{-t} - e^{-2t+2})H(t-2).$$

Example 24.20   *Solve the differential equation*

$$\frac{dx}{dt} + 2x = f(t)$$

with $x(0) = 0$, where (Fig. 24.4)

$$f(t) = \begin{cases} 0 & when\ t < 1, \\ e^{-t} & when\ 1 \leqslant t \leqslant 2, \\ 0 & when\ t > 2. \end{cases}$$

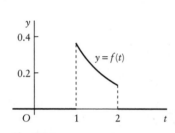

Fig. 24.4

Let $L\{x(t)\} = X(s)$. We need

$$L\{f(t)\} = \int_{1}^{2} e^{-st} e^{-t}\, dt = \int_{1}^{2} e^{-(s+1)t}\, dt$$

$$= \frac{1}{s+1}(e^{-(s+1)} - e^{-2(s+1)}) = F(s),$$

say. The transformed equation is then

$$sX + 2X = F(s), \quad \text{or} \quad X = F(s)/(s+2).$$

Therefore

$$X = (e^{-(s+1)} - e^{-2(s+1)})\frac{1}{(s+1)(s+2)} = (e^{-(s+1)} - e^{-2(s+1)})\left(\frac{1}{s+1} - \frac{1}{s+2}\right)$$

$$= e^{-1}e^{-s}\left(\frac{1}{s+1} - \frac{1}{s+2}\right) - e^{-2}e^{-2s}\left(\frac{1}{s+1} - \frac{1}{s+2}\right).$$

Apply the delay rule with $c = 1$ and $c = 2$, noting that

$$\frac{1}{s+1} - \frac{1}{s+2} \leftrightarrow e^{-t} - e^{-2t}.$$

We obtain

$$x(t) = e^{-1}(e^{-(t-1)} - e^{-2(t-1)})H(t-1) - e^{-2}(e^{-(t-2)} - e^{-2(t-2)})H(t-2)$$
$$= (e^{-t} - e^{1-2t})H(t-1) - (e^{-t} - e^{2-2t})H(t-2).$$

Both terms are zero before 'switch-on' at $t = 1$. Between $t = 1$ and 2, only the first term contributes. For $t > 2$ both terms are present, the second being stimulated by 'switching off'.

Problems

The dot notation, $\dot{x} = dx/dt$, $\ddot{x} = d^2x/dt^2$, etc., is used in some of the questions.

24.1 Write down $L\{x(t)\}$, where $x(t)$ is as follows.
(a) e^t;   (b) $4e^{-t}$;   (c) $3e^t - e^{-t}$;
(d) $3t^2 - 1$;   (e) $\frac{1}{2}t^3 + 2t^2 - 3$;   (f) $3 + 2t^4$;
(g) $3\sin t - \cos t$;   (h) $2(\cos t - \sin t)$;
(i) $1 + \dfrac{1}{1!}t + \dfrac{1}{2!}t^2 + \cdots + \dfrac{1}{n!}t^n$ (you get a geometric

series; see Section 1.16).

24.2 (Scale rule). Find $L\{x(t)\}$ for the following cases of $x(t)$.
(a) e^{3t};   (b) $1 - 2e^{-2t}$;
(c) $\sin \omega t$;   (d) $\cos \omega t$;
(e) $3\cos 2t - 2\sin 2t$;
(f) $\cos^2 t$ (express it in terms of $\cos 2t$);
(g) $\sin^2 t$ (see (f)).

24.3 (See Section 24.3.) Find $L\{x(t)\}$ in the following cases of $x(t)$.
(a) $t^2 e^t$ (easiest to start with t^2);
(b) $t e^{-2t}$;   (c) $t^2 e^{-t}$;
(d) $e^{2t} \cos t$;   (e) $e^{-t} \sin t$;
(f) $e^t \sin 3t$;   (g) $e^{-2t} \sin 3t$;
(h) $e^{-3t} \cos 2t$; (i) $t \cos 3t$;
(j) $t \sin 3t$;   (k) $t^2 \sin t$;
(l) $t^4 e^{-t}$ (compare the three methods: (i) start with t^4 and use the shift rule, (ii) start with e^{-t} and use (24.8), (iii) work directly from the definition (24.1)).

24.4 Obtain the Laplace transform for $t \sin kt$ by differentiating that of $\cos kt$ with respect to k.

24.5 Invert the following Laplace transforms.
(a) $1/s^2$;   (b) $1/s$;
(c) $3/(2s)$;   (d) $3/s^5$;
(e) $1/(s-3)$;   (f) $1/(s+4)$;
(g) $3/(2s-1)$;   (h) $2/(2-3s)$;
(i) $1/[s(s-1)]$;   (j) $1/(s^2+s-1)$;
(k) $s/(s^2-1)$;   (l) $(2s-1)/(s^2-1)$;
(m) $s/(s^2+1)$;   (n) $1/(s^2+4)$;
(o) $(2s-1)/(s^2+4)$;
(p) $(2s-1)/[s(s-1)]$;
(q) $(s^2-1)/[s(s-1)(s+2)(s+3)]$;
(r) $s/(s-1)(s^2+1)$;   (s) $1/(s-1)^3$;
(t) $(2s+1)/(s^2-2s+2)$;
(u) $s/[(s^2+1)(s^2+4)]$.

24.6 Find the Laplace transform of the following expressions involving $x(t)$, where $L\{x(t)\} = X(s)$.
(a) $\dot{x}(t)$, where $x(0) = 6$;
(b) $\dot{x}(t)$, where $x(0) = 0$;
(c) $\ddot{x}(t)$, where $x(0) = 3$, $\dot{x}(0) = 5$;
(d) $\ddot{x}(t)$, where $x(0) = 0$, $\dot{x}(0) = 0$;
(e) $2\ddot{x} + 3\dot{x} - 2x$, where $x(0) = 5$, $\dot{x}(0) = -2$;
(f) $3\ddot{x} - 5\dot{x} + x - 1$, where $x(0) = 0$, $\dot{x}(0) = 0$.

24.7 Use the Laplace transform to solve the following initial-value problems.
(a) $\ddot{x} + 3\dot{x} + 2x = 0$, $x(0) = 0, \dot{x}(0) = 1$;
(b) $\ddot{x} + \dot{x} - 2x = 0$, $x(0) = 3, \dot{x}(0) = 0$;
(c) $\ddot{x} + 4\dot{x} = 0$, $x(0) = x_0, \dot{x}(0) = y_0$;
(d) $\ddot{x} + \omega^2 x = 0$, $x(0) = c, \dot{x}(0) = 0$.
(e) $\ddot{x} + 2\dot{x} + 5x = 0$, $x(0) = 3, \dot{x}(0) = -3$;
(f) $d^4y/dx^4 - y = 0$, $y(0) = 1, y'(0) = 0, y''(0) = 0, y'''(0) = 0$ (use x instead of t as the variable in the Laplace transform).

24.8 Use the Laplace transform to solve the following initial-value problems.
(a) $\ddot{x} = 1 + t + e^t$, $x(0) = 0, \dot{x}(0) = 0$;
(b) $\ddot{x} + x = 3$, $x(0) = 0, \dot{x}(0) = 1$;
(c) $\ddot{x} + 2\dot{x} + 2x = 3$, $x(0) = 1, \dot{x}(0) = 0$;
(d) $\ddot{x} - x = e^{2t}$, $x(0) = 0, \dot{x}(0) = 1$;
(e) $\ddot{x} - x = t e^t$, $x(0) = 1, \dot{x}(0) = 1$;
(f) $\ddot{x} - 4x = 1 - e^{2t}$, $x(0) = 1, \dot{x}(0) = -1$;
(g) $\ddot{x} - 4x = e^{2t} + e^{-2t}$, $x(0) = 0, \dot{x}(0) = 0$;
(h) $\ddot{x} + \omega^2 x = C \cos \omega t$, $x(0) = x_0, \dot{x}(0) = y_0$;
(i) $\dddot{x} - 2\ddot{x} - \dot{x} + 2x = e^{-2t}$, $x(0) = 0, \dot{x}(0) = 0, \ddot{x}(0) = 2$ (look out for factors in the denominator of $X(s)$).

24.9 Solve the following simultaneous first-order differential equations, for the given initial values.
(a) $\dot{x} = x - y, \dot{y} = x + y$, $x(0) = 1, y(0) = 0$;
(b) $\dot{x} = 2x + 4y + e^{4t}, \dot{y} = x + 2y$, $x(0) = 1, y(0) = 0$;
(c) $\dot{x} = x - 4y, \dot{y} = x + 2y$, $x(0) = 2, y(0) = 1$.

24.10 Find the general solution of the following by putting $x(0) = A$, $\dot{x}(0) = B$, where A and B are arbitrary.
(a) $\ddot{x} + x = e^t$;
(b) $\ddot{x} - x = 3$;
(c) $\ddot{x} - 2\dot{x} + x = e^t$.

24.11 Find the general solution of $d^4y/dx^4 - y = e^x$, by putting $y(0) = A$, $y'(0) = B$, $y''(0) = C$, $y'''(0) = D$, where A, B, C, D are arbitrary. (Let the variable in the Laplace transform (24.1) be x instead of t.)

24.12 This is a system of first-order equations for $x_0(t), x_1(t), \ldots, x_n(t)$:

$$\dot{x}_0 = -\beta x_1, \qquad \dot{x}_r = \beta(x_{r-1} - x_r)$$

for $r = 1, 2, \ldots, n$. Solve them by using the Laplace transform, showing that

$$x_r = \frac{1}{r!}(\beta t)^r \, e^{-\beta t}.$$

24.13 Use the delay rule (24.15) to obtain the Laplace transform of $e^{-t}(t-2) \cos(t-2) H(t-2)$.

24.14 Find the functions which give rise to the following Laplace transforms:
(a) $e^{-2s}/(s+3)$;　(b) $(1 - s\,e^{-s})/(s^2 + 1)$;
(c) $e^{-2s}/(s-4)$;　(d) $s\,e^{-s}/[(s+1)(s+2)]$;
(e) $e^{-s}/[(s-1)(s^2 - 2s + 2)]$.

24.15 Solve the following differential equations assuming that the initial state is of quiescence $x(0) = \dot{x}(0) = 0$:
(a) $\ddot{x} + x = f(t)$, where

$$f(t) = \begin{cases} 1 & \text{for } 0 < t \leqslant 1, \\ 0 & \text{for } t > 1. \end{cases}$$

(b) $\ddot{x} - 4x = f(t)$, where

$$f(t) = \begin{cases} 1 & \text{for } 0 < t \leqslant 1, \\ 0 & \text{for } t > 1. \end{cases}$$

(c) $\ddot{x} - 4x = f(t)$, where

$$f(t) = \begin{cases} t & \text{for } 0 < t \leqslant 1, \\ 2 - t & \text{for } 1 < t \leqslant 2, \\ 0 & \text{for } t > 2. \end{cases}$$

(d) $\ddot{x} + x = f(t)$, where

$$f(t) = \begin{cases} \cos t & \text{for } 0 < t \leqslant \pi, \\ 0 & \text{for } t > \pi. \end{cases}$$

25 Laplace and *z* transforms: applications

CONTENTS

25.1   Division by *s* and integration

Multiplication by *s* is associated with differentiation (see (24.12)). Division by *s* is associated with integration, as follows.

> **Division rule**
>
> If $G(s) \leftrightarrow g(t)$, then $\dfrac{1}{s} G(s) \leftrightarrow \displaystyle\int_0^t g(\tau)\,\mathrm{d}\tau.$
>
> (25.1)

To prove this, put $(1/s)G(s) = F(s)$; then we must express $f(t)$ in terms of $g(t)$. Rewrite the relation between $F(s)$ and $G(s)$ in the form

$$sF(s) = G(s).$$

But from (22.12) we know that, in general,

$$\frac{\mathrm{d}f}{\mathrm{d}t} \leftrightarrow sF(s), \quad \text{provided that } f(0) = 0;$$

so then we have $\mathrm{d}f/\mathrm{d}t \leftrightarrow G(s)$. This is equivalent to the initial-value problem $\mathrm{d}f/\mathrm{d}t = g(t)$, with $f(0) = 0$. By integration we obtain

$$f(t) = \int_0^t g(\tau)\,\mathrm{d}\tau.$$

Example 25.1   *Find $f(t)$ when $F(s) = 1/[s(s^2 + 1)]$,*
(a) *by using partial fractions*, (b) *by using (25.1)*.

(a) $\dfrac{1}{s(s^2 + 1)} = \dfrac{1}{s} - \dfrac{s}{s^2 + 1} \leftrightarrow 1 - \cos t.$

(b)   In the notation of (25.1), put

$$G(s) = \frac{1}{s^2 + 1} \leftrightarrow \sin t.$$

Therefore

$$F(s) = \frac{1}{s(s^2 + 1)} = \frac{1}{s}\frac{1}{s^2 + 1}$$

$$\leftrightarrow \int_0^t \sin \tau \, d\tau = 1 - \cos t.$$

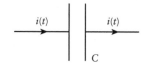

Fig. 25.1

Figure 25.1 shows a capacitor, of capacitance C, being charged by a current $i(t)$, the voltage drop across the plates being $v(t)$. Assume that *the capacitor is uncharged at $t = 0$*; then, at a later time t,

$$v(t) = \frac{1}{C}\int_0^t i(\tau)\, d\tau.$$

Therefore, according to (25.1), the relation between the Laplace transforms of $v(t)$ and $i(t)$ is

$$V(s) = \frac{1}{Cs}I(s). \tag{25.2}$$

We say that (25.2) describes the situation **in the s domain**, as we spoke of description in the frequency, or ω, domain in Section 21.3.

If the capacitor has a *nonzero initial charge q_0*, then

$$v(t) = \frac{1}{C}\left(\int_0^t i(\tau)\, d\tau + q_0\right).$$

Since $q_0 \leftrightarrow s^{-1}q_0$, this transforms into

$$V(s) = \frac{1}{C}\frac{1}{s}[I(s) + q_0]. \tag{25.3}$$

We shall not be concerned with this case.

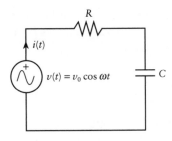

Fig. 25.2

Example 25.2 *The circuit shown in Fig. 25.2 is switched on at time $t = 0$. It is initially quiescent, and there is zero charge on the capacitor. Find the current for $t > 0$.*

The circuit equation is

$$v_0 \cos \omega t = Ri(t) + \frac{1}{C} \int_0^t i(\tau) \, d\tau.$$

Such an equation is called an **integral equation** for $i(t)$. The Laplace transform of the equation is

$$\frac{v_0 s}{s^2 + \omega^2} = RI(s) + \frac{1}{Cs}I(s),$$

so

$$I(s) = \frac{v_0}{R} \frac{s^2}{(s + 1/RC)(s^2 + \omega^2)}$$

$$= \frac{v_0}{R} \frac{1}{1 + (RC\omega)^2} \left(\frac{(RC\omega)^2 s}{s^2 + \omega^2} - \frac{RC\omega^2}{s^2 + \omega^2} + \frac{1}{s + 1/RC} \right)$$

after splitting into partial fractions. Therefore, for $t > 0$,

$$i(t) = \frac{v_0}{R} \frac{1}{1 + (RC\omega)^2} [(RC\omega)^2 \cos \omega t - RC\omega \sin \omega t + e^{-t/RC}].$$

The first two terms represent a steady forced oscillation and the final term is a transient.

25.2 The impulse function

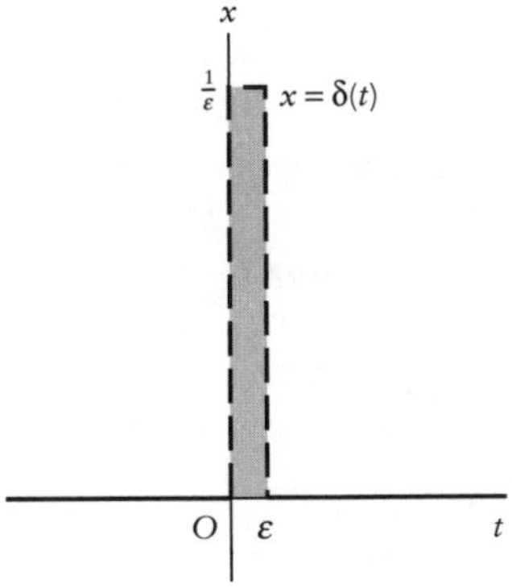

Figure 25.3 shows the graph of a function which is zero everywhere except for a tall, narrow rectangle with width ε and height $1/\varepsilon$, so that the area under the graph is equal to 1. Imagine that ε is a *very* small number, as small as we wish. This very tall and very narrow picture is a simplified version of the **impulse function or delta function**, usually denoted by $\delta(t)$. It is used in problems involving sudden and brief events, to represent (say) impulsive force between two bodies in collision; voltage from a lightning strike; or, if the variable is position rather than time, a point force.

Fig. 25.3

> **The impulse or delta function $\delta(t)$**
> Informal definition: $\delta(t) = 1/\varepsilon$ for $0 < t < \varepsilon$, and $\delta(t) = 0$ elsewhere, where ε is as small as is necessary. (25.4)

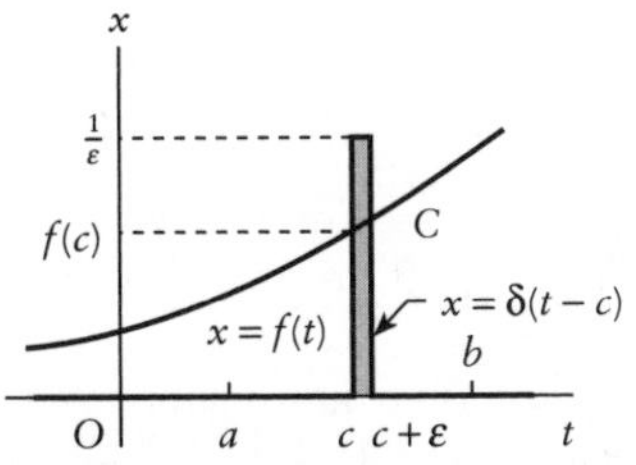

In Figure 25.4, $\delta(t)$ is moved to the right so as to be at $t = c$; the vertical strip therefore represents $\delta(t - c)$. An ordinary function $f(t)$ crosses it at C. Consider the integral

$$\int_a^b f(t) \, \delta(t - c) \, dt,$$

Fig. 25.4

where c lies between a and b. The integrand is zero except between c and $c + \varepsilon$; over this very narrow interval, $f(t)$ hardly changes from the value $f(c)$. Therefore (as closely as we wish)

$$\int_a^b f(t)\,\delta(t-c)\,\mathrm{d}t = \int_c^{c+\varepsilon} f(c)\varepsilon^{-1}\,\mathrm{d}t = f(c).$$

If c does not lie between a and b, then the integral is zero. The delta function is sometimes called a **sifting function** because of this property.

Sifting property of $\delta(t)$

$$\int_a^b f(t)\,\delta(t-c)\,\mathrm{d}t = \begin{cases} f(c) & \text{if } a \leqslant c < b, \\ 0 & \text{otherwise.} \end{cases}$$

(25.5)

We can obtain the Laplace transform of $\delta(t)$ from (25.5):

Laplace transform of $\delta(t-c)$

$$\mathcal{L}\{\delta(t-c)\} = \int_0^\infty \mathrm{e}^{-st}\,\delta(t-c)\,\mathrm{d}t = \mathrm{e}^{-cs},$$

for $c \geqslant 0$. In particular, $\mathcal{L}\{\delta(t)\} = 1$.

(25.6)

Example 25.3   *The equation $\mathrm{d}^2x/\mathrm{d}t^2 + \omega^2 x = f(t)$ represents the displacement x of a particle of unit mass on a spring of stiffness ω with external force $f(t)$. Find the motion for $t > 0$ if the particle is subjected to an impulse $I\,\delta(t-1)$ at time $t = 1$, assuming equilibrium at $t = 0$. (I has dimensions $[\text{force} \times \text{time}]$.)*

The equation is $\mathrm{d}^2x/\mathrm{d}t^2 + \omega^2 x = I\,\delta(t-1)$. Its transform is

$$s^2 X + \omega^2 X = I\,\mathrm{e}^{-s},$$

where $x(t) \leftrightarrow X(s)$. Therefore

$$X(s) = \frac{I}{s^2 + \omega^2}\,\mathrm{e}^{-s}.$$

We know that

$$\frac{1}{s^2 + \omega^2} \leftrightarrow \frac{1}{\omega}\sin \omega t;$$

so, by the delay rule (22.15), we have

$$X(s) = \frac{I}{s^2 + \omega^2}\,\mathrm{e}^{-s} \leftrightarrow x(t) = \frac{I}{\omega}\sin \omega(t-1)\mathrm{H}(t-1),$$

where H stands for the unit function (24.13). There is no motion until $t = 1$, when the impulse sets up free oscillations $(I/\omega)\sin \omega(t-1)$.

Example 25.4 *Find the current resulting from an impulsive voltage* $I_v \delta(t)$ *applied to the circuit of Fig. 25.5, the current being zero before application of the voltage. (The dimensions of* I_v *are* [*emf* × *time*].)

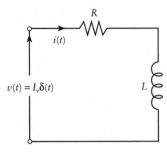

Fig. 25.5

The equation for the current is $L\, di/dt + Ri = I_v\, \delta(t)$. After transformation, with $i(0) = 0$, it becomes

$$LsI(s) + RI(s) = I_v.$$

Therefore

$$I(s) = \frac{I_v}{L(s + R/L)} \leftrightarrow i(t) = \frac{I_v}{L}\, e^{-Rt/L}.$$

The great, though brief, applied voltage gives only a finite current because of the counter-emf generated by the coil.

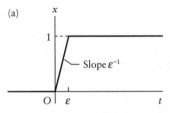

Fig. 25.6

The delta function can be regarded formally as the derivative of the unit function $H(t)$. As in Fig. 25.6a, smooth out the transition of $H(t)$, from zero to one, as *t* passes through the origin, by means of a sloping straight line segment. The derivative of this function is equal to zero outside the transition interval $(0, \varepsilon)$ and equal to ε inside it; this specifies $\delta(t)$ as in (25.4).

Connection between H(*t*) and δ(*t*)

$$\frac{dH(t)}{dt} = \delta(t).$$

(25.7)

This only conforms with the Laplace-transform derivative rule (24.12),

$$s\left(\frac{1}{s}\right) - H(0) = 1,$$

if we rather arbitrarily interpret $H(0)$ as being zero. It should be understood that certain weaknesses result from treating the impulse function very informally; the real justification for its use is an elaborate mathematical subject called **distribution theory**.

25.3 Impedance in the *s* domain

In the table (25.8) below three basic circuit elements are shown, together with their voltage-drop–current relations and the Laplace transforms of these relations, on the assumption that, at $t = 0$, **the current through the inductor and the charge on the capacitor are zero.** The expression '*s* domain' refers to transformed quantities.

	Resistor	Inductor	Capacitor
	$i(t)$ —W— $v(t)$	$i(t)$ —ᴍ— $v(t)$	$i(t)$ —╢— $v(t)$
Time domain:	$v(t) = Ri(t)$	$v(t) = L\dfrac{di(t)}{dt}$	$v(t) = \dfrac{1}{C}\displaystyle\int_0^t i(\tau)\,d\tau$
s domain:	$V(s) = RI(s)$	$V(s) = LsI(s)$	$V(s) = (1/Cs)I(s)$
Impedance $Z(s)$:	R	Ls	$1/Cs$

(25.8)

Table (25.8) should be compared with the table (25.5) for the case of steady forced oscillations of frequency $\omega/2\pi$. The **impedances** $Z(s)$ in the s plane are analogous to the complex impedances R, $i\omega L$, and $1/(i\omega C)$ of (21.6) for the steady case. One can pass from one to the other by substituting $i\omega$ for s, or $-is$ for ω. However, the s forms allow arbitrary inputs to the circuit to be considered.

Impedances combine in series and parallel in the same way as do complex impedances (see (21.7)) in the frequency domain, but it is to be remembered that they refer to **zero initial conditions only**.

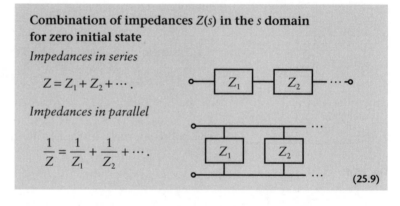

Combination of impedances $Z(s)$ in the s domain for zero initial state

Impedances in series

$$Z = Z_1 + Z_2 + \cdots .$$

Impedances in parallel

$$\frac{1}{Z} = \frac{1}{Z_1} + \frac{1}{Z_2} + \cdots .$$

(25.9)

Example 25.5 *The circuit shown in Fig. 25.7a is initially quiescent, with zero charge on the capacitor. The constant voltage v_0 is switched on at $t = 1$ and off at $t = 2$. Find the current $i(t)$.*

The corresponding s domain impedances are shown in Fig. 25.7b, in which the elements R and C are grouped. They are in parallel, so (25.8) and (25.9) give

$$\frac{1}{Z_1} = \frac{1}{R} + \frac{1}{(Cs)^{-1}} = \frac{1}{3} + \frac{s}{12} = \frac{s+4}{12}.$$

Hence

$$Z_1 = \frac{12}{s+4},$$

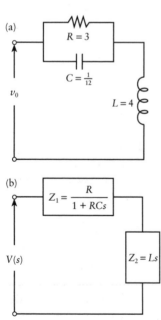

Fig. 25.7

Example 25.5 *continued*

and also $Z_2 = Ls = 4s$. Then Z for the whole circuit is given by

$$Z = 4s + \frac{12}{s + 4} = \frac{4(s + 1)(s + 3)}{s + 4}.$$

Therefore

$$I(s) = \frac{s + 4}{4(s + 1)(s + 3)} V(s).$$

Taking into account switch-on at $t = 1$ and switch-off at $t = 2$,

$$v(t) = v_0[H(t - 1) - H(t - 2)],$$

so

$$V(s) = v_0\left(\frac{1}{s}e^{-s} - \frac{1}{s}e^{-2s}\right).$$

Therefore

$$I(s) = \frac{v_0(s + 4)}{4s(s + 1)(s + 3)}(e^{-s} - e^{-2s})$$

$$= v_0\left(\frac{1}{3s} - \frac{3}{8}\frac{1}{s + 1} + \frac{1}{24}\frac{1}{s + 3}\right)(e^{-s} - e^{-2s}).$$

The first bracketed factor transforms back to

$$v_0(\tfrac{1}{3} - \tfrac{3}{8}e^{-t} + \tfrac{1}{24}e^{-3t}),$$

and, by using the delay rule (24.15) to deal with the exponentials,

$$i(t) = v_0(\tfrac{1}{3} - \tfrac{3}{8}e^{-(t-1)} + \tfrac{1}{24}e^{-3(t-1)})H(t - 1)$$

$$- v_0(\tfrac{1}{3} - \tfrac{3}{8}e^{-(t-2)} + \tfrac{1}{24}e^{-3(t-2)})H(t - 2).$$

Nothing happens until the system is switched on at $t = 1$, when the first term (only) is activated. At $t = 2$, when it is switched off, the second term comes in also; some current persists but it dies away to zero.

It must be emphasized that such a problem is considerably complicated when the initial conditions are not zero. For example, the expression (25.3) for an initially charged capacitor is not in the form of a voltage–impedance–current relationship. In such cases, it is necessary to start with the differential equations for individual branches.

25.4 Transfer functions in the *s* domain

The impedance $Z(s)$ which directly connects the current in a unit with the voltage drop across the same unit is a special case of a more general idea: to relate any two currents or voltages which occur in the network.

We suppose as before that we have a passive circuit consisting of linear resistors, capacitors, and inductors, and a single source of voltage which drives the circuit. Figure 25.8 represents such a network. We denote the driving voltage by $f(t)$, because much of what we say

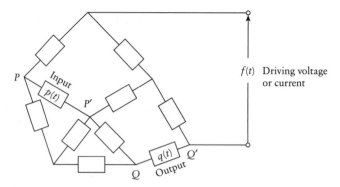

Fig. 25.8

can be taken over into mechanical and other systems. The unknown voltages and currents we call the **variables**.

Suppose that there are N currents and voltages to be determined, which we call $x_1(t)$, $x_2(t)$, ... , $x_N(t)$, with transforms $X_1(s)$, $X_2(s)$, ... , $X_N(s)$. A voltage $f(t)$, with transform $F(s)$, is applied somewhere in the network. Provided that the circuit is initially quiescent (zero initial conditions), each of the s-domain equations sufficient to determine the currents and voltages takes one of only two possible forms:

either $a_1X_1 + a_2X_2 + \cdots + a_NX_N = 0$

or $b_1X_1 + b_2X_2 + \cdots + b_NX_N = F$,

where the coefficients are functions of s. Therefore the transforms $X_1, X_2, \ldots, X_N$ are all proportional to F:

$X_n(s) = G_n(s)F(s)$,

for $n = 1, 2, \ldots, N$. The G_n are functions which **depend only on the circuit constants and not on the applied voltage**. They are called **transfer functions**. Nominate an arbitrary variable $p(t)$ in any branch as the **input**, and another variable $q(t)$ elsewhere as the **output**. The corresponding G_n are denoted by G_P and G_Q. The driving voltage $f(t)$ may serve as an input if we wish. Assuming that we start with **zero currents and charges**, which implies **zero initial conditions**, the transforms $P(s)$ and $Q(s)$ of $p(t)$ and $q(t)$ are related by

$Q(s)/P(s) = G_QF/G_PF = G_{PQ}(s)$,

say, where $G_{PQ}(s)$ is called the **transfer function from p to q**.

Transfer function $G_{PQ}(s)$ (zero initial conditions)

Let $p(t)$ (input) and $q(t)$ (output) be the voltage or current in any two branches. Then

$Q(s)/P(s) = G_{PQ}(s)$,

where $G_{PQ}(s)$ is the transfer function from p to q. G_{PQ} depends only on the circuit parameters.

(25.10)

Transfer functions which connect different types of variable are given various names and conventional symbols in literature on systems. For example, in the s domain, voltage ÷ current is **impedance**; current ÷ voltage is **admittance**; voltage ÷ voltage is **voltage gain**, and so on.

(a)

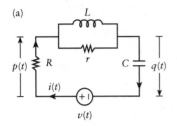

(b)

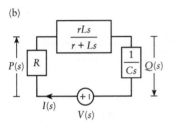

Fig. 25.9

Example 25.6 *Find the transfer function $G(s)$ from the voltage transform $P(s)$ over R, regarded as the input, and the voltage transform $Q(s)$ over C, regarded as the output, in Fig. 25.9a.*

Let the current $i(t)$ be as indicated. The impedances of the various groups are shown in Fig. 25.9b; these are in fact transfer functions between current and voltage for each unit. In terms of the transforms,

$$P(s) = RI(s), \qquad Q(s) = \frac{1}{Cs}I(s).$$

Therefore

$$G(s) = \frac{Q(s)}{P(s)} = \frac{1}{RCs}.$$

Thus

$$Q(s) = \frac{1}{RC}\frac{1}{s}P(s),$$

and so $q(t) = \dfrac{1}{RC}\displaystyle\int_0^t p(\tau)\,\mathrm{d}\tau$, as expected.

(a)   circuit A

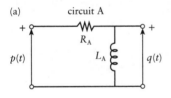

(b)   circuit A, s domain

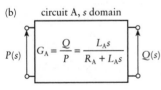

(c)

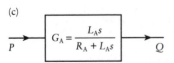

Fig. 25.10

Suppose now that we have a circuit such as the one in Fig. 25.10a, called Circuit A, where $p(t)$ is the input voltage and $q(t)$ the output voltage. Figure 25.10b schematizes the arrangement and specifies the transfer function $G(s) = Q(s)/P(s)$ between p and q.

We could also symbolize the dependence of q on p by the scheme in Fig. 25.10c. However, this figure suggests the beginnings of some kind of series arrangement: it looks as if we could attach another circuit to the original one without altering the transfer function, and so get an easy calculation for the combined circuit. This is not true in general, but sometimes it is a useful approximation.

To illustrate this question, we will append to Circuit A another Circuit B. It is shown in Fig. 25.11a, together with its s domain representation and its transfer function. In Fig. 25.11b, A and B are connected across MN; here p, q, r, and their transforms P, Q, R represent the actual voltages across the terminals indicated. The question is 'do the transfer functions written in the boxes still correctly give $Q(s)$ in terms of $P(s)$, and then $R(s)$ in terms of $Q(s)$?'

If an appreciable amount of current passes between Circuits A and B after attachment, then $Q(s)$ must change, so the true transfer functions of both of the circuits will be changed, and the changes will not compensate each other. In special circumstances, however,

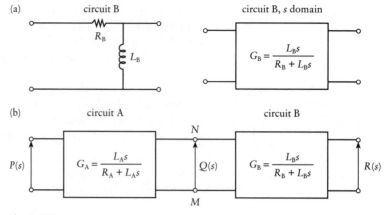

(a)

circuit B   circuit B, s domain

R_B

L_B $G_B = \dfrac{L_B s}{R_B + L_B s}$

(b)

circuit A   circuit B

N

$P(s)$ $G_A = \dfrac{L_A s}{R_A + L_A s}$ $Q(s)$ $G_B = \dfrac{L_B s}{R_B + L_B s}$ $R(s)$

M

Fig. 25.11

the circuits may behave almost independently, or can be made to do so by means of technical arrangements such as feedback.

Example 25.7   *The two circuits A and B shown in Fig. 25.12 are connected to form a composite circuit C. Show that*

$$G(s) \approx G_A(s) G_B(s)$$

(where the $G(s)$ are the transfer functions for the voltages shown) if $1/R$ is much smaller than $1/r + 1/r_1$.

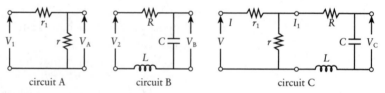

circuit A   circuit B   circuit C

Fig. 25.12

For Circuit A alone:

$$G_A(s) = \frac{V_A(s)}{V_1(s)} = \frac{r}{r + r_1}.$$

For Circuit B alone:

$$G_B(s) = \frac{V_B(s)}{V_2(s)} = \frac{\text{impedance of C}}{\text{total impedance}} = \frac{1}{Cs} \cdot \frac{1}{R + Ls + 1/Cs}.$$

Therefore

$$G_A(s) G_B(s) = \frac{1}{Cs} \cdot \frac{r}{r + r_1} \cdot \frac{1}{R + Ls + 1/Cs}.$$

For Circuit C, by following the voltage drops around closed subcircuits as usual, we get

$$V = r_1 I + r(I - I_1),$$
$$0 = (R + Ls + 1/Cs)I_1 - r(I - I_1),$$

Example 25.7 *continued*

from which

$$I_1 = \frac{Vr}{(r + r_1)(r + R + Ls + 1/Cs) - r^2},$$

which represents the current 'leaking' between A and B. Therefore

$$G_C(s) = \frac{V_C(s)}{V(s)} = \frac{1}{Cs} \cdot \frac{r}{(r + r_1)(r + R + Ls + 1/Cs) - r^2},$$

which we have to compare with $G_A(s)G_B(s)$ above. Rewrite $G_C(s)$ in the form

$$G_C(s) = \frac{1}{Cs} \cdot \frac{r}{r + r_1} \cdot \frac{1}{R + Ls + 1/Cs - rr_1/(r + r_1)}.$$

It can be seen that $G_C(s) \approx G_A(s)G_B(s)$ if $rr_1/(r + r_1)$ is much smaller than R. But

$$\frac{rr_1}{r + r_1} = 1 \Big/ \left(\frac{1}{r} + \frac{1}{r_1}\right),$$

and this is much smaller than R if $1/r + 1/r_1$ is much *greater* than $1/R$. The relation between the circuits could be represented in this case approximately by Fig. 25.13, as if they processed the voltage signals independently.

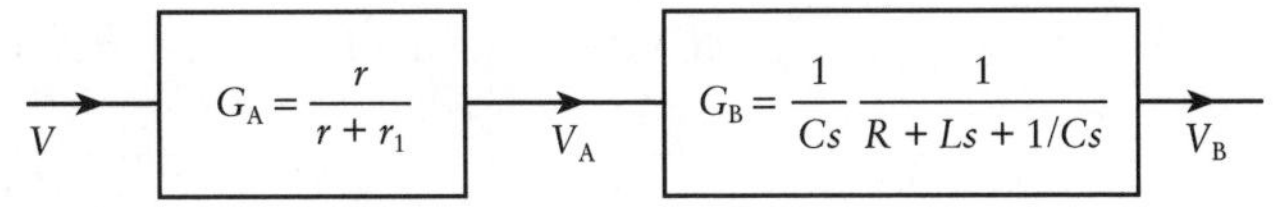

Fig. 25.13

Example 25.8 *Figure 25.14 shows a chain of three systems, which act independently upon their inputs according to the transfer functions $G_A(s)$, $G_B(s)$, and $G_C(s)$ indicated in the boxes. Find the transfer function $G(s)$ between $F(s)$ and $F_C(s)$. Find $f_C(t)$ when $f(t) = H(t)$, for zero initial conditions.*

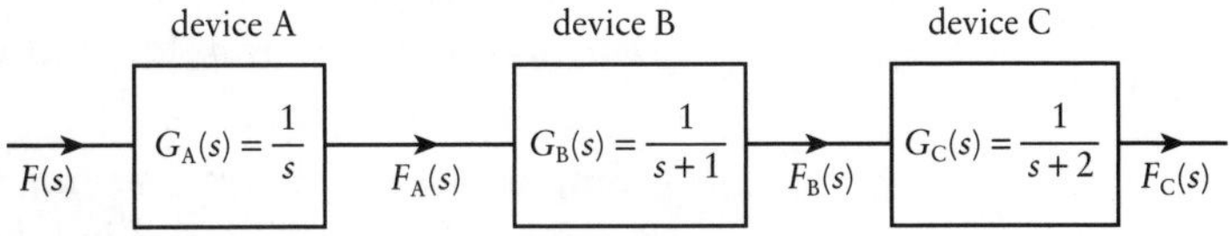

Fig. 25.14

We have

$$\frac{F_C}{F} = \frac{F_C}{F_B} \frac{F_B}{F_A} \frac{F_A}{F} = \frac{1}{s + 2} \frac{1}{s + 1} \frac{1}{s},$$

so

$$G(s) = \frac{1}{s(s + 1)(s + 2)}.$$

Example 25.8 *continued*

Now let $f(t) = H(t)$; then $F(s) = 1/s$. Therefore

$$F_C(s) = G(s)F(s) = \frac{1}{s(s+1)(s+2)}\frac{1}{s} = \frac{1}{s^2(s+1)(s+2)}.$$

In partial fractions,

$$F_C(s) = -\frac{3}{4}\frac{1}{s} + \frac{1}{2}\frac{1}{s^2} + \frac{1}{s+1} - \frac{1}{4}\frac{1}{s+2}.$$

Therefore $f(t) = -\frac{3}{4} + \frac{1}{2}t + e^{-t} - \frac{1}{4}e^{-2t}$ for $t > 0$.

It is possible to get an idea of important features of an output without going through the whole calculation:

Example 25.9   *In a particular system, the output $X(s)$ in the s domain is related to the input $F(s)$ by $X(s) = G(s)F(s)$. Find the general character of $x(t)$ if $f(t) = \cos 2t$, $G(s) = s/(s+1)(s^2 + 4s + 5)$.*

Since $f(t) \leftrightarrow s/(s^2 + 4)$, we have

$$X(s) = \frac{s}{(s+1)(s^2+4s+5)}\frac{s}{s^2+4}.$$

If we expanded this in partial fractions, we should have terms of the types

$$\frac{1}{s+1}, \frac{s}{(s+2)^2+1}, \text{ and } \frac{1}{(s+2)^2+1} \quad \text{(from } G\text{)},$$

and

$$\frac{s}{s^2+4} \text{ and } \frac{1}{s^2+4} \quad \text{(from } F\text{)}.$$

Therefore, in terms of time, we should obtain terms like

e^{-t}, $e^{-2t}\sin t$, and $e^{-2t}\cos t$ from G (which are transients),

$\cos 2t$ and $\sin 2t$ from F (a forced oscillation).

Finally we illustrate the relation between transfer functions in the s domain and complex transfer functions in the ω domain (Section 21.5).

Example 25.10   *The transfer function between an input $F(s)$ and an output $X(s)$ is $1/(s^2 + 1)$. Find the amplitude and phase of the steady forced oscillation produced by an input $f(t) = 3\sin 2t$.*

As pointed out in Section 25.3, the complex impedance is simply the s domain impedance with $i\omega$ substituted for s. The same is true for any transfer function. In the ω domain representation, the input and output will be represented by phasors $F(\omega) = 3e^{-\frac{1}{2}\pi i}$ and $X(\omega)$, corresponding to circular frequency $\omega = 2$ in this case. Then

$$X(\omega) = \frac{1}{(2i)^2 + 1}3e^{-\frac{1}{2}\pi i} = -e^{-\frac{1}{2}\pi i} = e^{\frac{1}{2}\pi i}.$$

The amplitude is the modulus of X, which is 1, and the phase is $\frac{1}{2}\pi$.

25.5 The convolution theorem

The following result enables us to interpret Laplace transforms which take the form of a **product** of two functions.

Convolution theorem

Suppose $F(s) = G(s)H(s)$, and

$$G(s) \leftrightarrow g(t), \quad H(s) \leftrightarrow h(t).$$

Then

$$f(t) = \int_0^t g(t-\tau)h(\tau)\,d\tau$$

(which is the same as $\int_0^t h(t-\tau)g(\tau)\,d\tau$). **(25.11)**

This result will be proved in Chapter 32, Example 32.12. For the present we shall verify that it is true in some special cases.

Example 25.11 *Find the inverse Laplace transform of*

$$F(s) = \frac{1}{(s+1)(s+2)}.$$

Put $F(s) = G(s)H(s)$, where

$$G(s) = \frac{1}{s+1}, \qquad H(s) = \frac{1}{s+2};$$

then $g(t) = e^{-t}$ and $h(t) = e^{-2t}$. The convolution theorem (25.11) gives

$$F(s) \leftrightarrow f(t) = \int_0^t e^{-(t-\tau)}\,e^{-2\tau}\,d\tau = \int_0^t e^{-t-\tau}\,d\tau$$

$$= \int_0^t e^{-t}\,e^{-\tau}\,d\tau = e^{-t}\int_0^t e^{-\tau}\,d\tau \qquad \text{(i)}$$

$$= e^{-t}(-e^{-t}+1) = e^{-t} - e^{-2t}.$$

This result can be confirmed by using partial fractions instead:

$$\frac{1}{(s+1)(s+2)} = \frac{1}{s+1} - \frac{1}{s+2}$$

$$\leftrightarrow e^{-t} - e^{-2t}.$$

Notice very carefully the distinction between t and τ in the integrals (25.11): τ is the variable of integration. **The variable t is a constant** so far as the integration process is concerned; so, for example, in eqn (i) point we took e^{-t} outside the integral sign.

Example 25.12   *Find the inverse transform of $1/[s(s^2+1)]$.*

In Example 25.1, we showed in two different ways that

$$\frac{1}{s(s^2+1)} \leftrightarrow 1 - \cos t.$$

To confirm that (25.11) gives the same result, put

$$G(s) = \frac{1}{s^2+1} \quad \text{and} \quad H(s) = \frac{1}{s},$$

say. Then (for $t > 0$) $g(t) = \sin t$ and $h(t) = 1$, so

$$g(t - \tau) = \sin(t - \tau) \quad \text{and} \quad h(\tau) = 1.$$

Therefore, by the convolution theorem,

$$F(s) \leftrightarrow \int_0^t \sin(t - \tau)1 \, d\tau = [\cos(t - \tau)]_{\tau=0}^t = 1 - \cos t,$$

as expected.

Example 25.13   *(See (25.11)). Confirm directly that*

$$\int_0^t g(t - \tau)h(\tau) \, d\tau = \int_0^t h(t - \tau)g(\tau) \, d\tau.$$

In the first integral, change the variable, putting

$$u = t - \tau.$$

Then (remember t is to be treated like a constant) $du = -d\tau$. Therefore

$$\int_0^t g(t - \tau)h(\tau) \, d\tau = \int_t^0 g(u)h(t - u)(-du) = \int_0^t h(t - u)g(u) \, du,$$

which is the integral required, merely using u instead of τ for the variable of integration.

Example 25.14   *Find an expression for the inverse transform of*

$$F(s) = \frac{1}{s+1} H(s)$$

in terms of $h(t)$, the inverse transform of $H(s)$.

Use the convolution theorem, (25.11), putting $G(s) = 1/(s+1)$. Then

$$g(t) = e^{-t}.$$

We therefore obtain from (25.11)

$$f(t) = \int_0^t e^{-(t-\tau)}h(\tau) \, d\tau,$$

or its alternative form

$$f(t) = \int_0^t e^{-\tau}h(t - \tau) \, d\tau.$$

25.6 General response of a system from its impulsive response

We shall take an electrical network as our example, though what we say applies to linear mechanical systems as well. Suppose that it is activated by an applied voltage $f(t)$ (regarded as the input). Focus on any particular one of the currents or voltages in the circuit, and call it $x(t)$ (the output). The transfer function between input and output will be called $G(s)$. We have then

$$X(s) = G(s)F(s). \tag{25.12}$$

Suppose that we conduct an experiment in which we excite the circuit by means of a voltage impulse $I_v \delta(t)$, and record the result (the dimensions of I_v are [emf × time]). Then

$$f(t) = I_v \delta(t), \quad \text{so that} \quad F(s) = I_v.$$

The current resulting from this special voltage (an impulsive input) will be called $x^*(t)$, with transform $X^*(s)$. Now put $F(s) = I_v$ and X^* for X into (25.12), and it becomes

$$X^*(s) = I_v G(s). \tag{25.13}$$

Such an experiment would therefore give us the corresponding transfer function $G(s)$ directly (we could even arrange for I_v to equal unity). Thus, even if the circuit is a 'black box' with its details unknown, we still know from (25.13) what to put into (25.12) for the case when $f(t)$ is any function at all:

$$X(s) = I_v^{-1} X^*(s)F(s).$$

Therefore, by the convolution theorem (25.11),

$$x(t) = I_v^{-1} \int_0^t x^*(t - \tau)f(\tau)\,\mathrm{d}\tau, \quad \text{or} \quad I_v^{-1} \int_0^t x^*(\tau)f(t - \tau)\,\mathrm{d}\tau.$$

This type of result applies to the other circuit variables such as voltages and charges, and to mechanical systems governed by linear differential equations. In terms of general outputs and inputs:

Output $x(t)$ from an input $f(t)$ to a quiescent linear system, in terms of the output $x^*(t)$ from an impulsive input $I \delta(t)$

$$x(t) = I^{-1} \int_0^t x^*(t - \tau)f(\tau)\,\mathrm{d}\tau,$$

or

$$x(t) = I^{-1} \int_0^t x^*(\tau)f(t - \tau)\,\mathrm{d}\tau. \tag{25.14}$$

Example 25.15   *The displacement $x^*(t)$ caused by an impulse $I\,\delta(t)$ applied to a certain mechanical linear system at rest is found to be $x^*(t) = \mathrm{e}^{-t} - \sin 2t$. Find the displacement $x(t)$ corresponding to an applied force $f(t) = \sin t$ starting at $t = 0$.*

We have

$$x(t) = I^{-1}\int_0^t [\mathrm{e}^{-(t-\tau)} - \sin 2(t-\tau)] \sin\tau\, \mathrm{d}\tau \quad \text{(from (25.14))}$$

$$= I^{-1}\mathrm{e}^{-t}\int_0^t \mathrm{e}^{\tau}\sin\tau\,\mathrm{d}\tau - I^{-1}\int_0^t \sin(2t - 2\tau)\sin\tau\,\mathrm{d}\tau$$

$$= I^{-1}\mathrm{e}^{-t}\int_0^t \mathrm{e}^{\tau}\sin\tau\,\mathrm{d}\tau - \tfrac{1}{2}I^{-1}\int_0^t [\cos(2t - 3\tau) - \cos(2t - \tau)]\,\mathrm{d}\tau,$$

by using the identity (1.17b). In the end, we find

$$x(t) = \tfrac{1}{2}I^{-1}\,\mathrm{e}^{-t} + \tfrac{1}{3}I^{-1}\sin 2t - \tfrac{1}{6}I^{-1}(3\cos t + \sin t)$$

for $t > 0$. The first term is a transient, and the second an induced free oscillation, and the third term represents the forced oscillation.

25.7 Convolution integral in terms of memory

An integral of the type

$$x(t) = \int_0^t g(t - \tau)f(\tau)\,\mathrm{d}\tau,$$

such as arose in the convolution theorem (25.11), is called a **convolution integral**. Typically, f acts as some kind of 'cause', such as a driving force or voltage, and $x(t)$ stands for a certain 'effect' produced.

Choose a time t for observation; then divide the interval $\tau = 0$ to $\tau = t$ into a large number of equal time steps $\delta\tau$. We have

$$x(t) = \int_0^t g(t - \tau)f(\tau)\,\mathrm{d}\tau \approx \sum_{\tau=0}^{\tau=t} g(t - \tau)f(\tau)\,\delta\tau.$$

Now choose any moment τ_1 between 0 and t: there was a force $f(\tau_1)$ applied at this moment, and its contribution to x at time $t > \tau_1$ is

$$g(t - \tau_1)f(\tau_1)\,\delta\tau.$$

The factor $g(t - \tau_1)$ takes into account the **time elapsed** between the cause and its effect – in some problems is would be appropriate to call $t - \tau_1$ the 'age' of $f(\tau_1)$ at the moment t of observation, and g an ageing factor. Depending on the type of problem, this factor might weaken or amplify the contribution of $f(\tau_1)$ to the integral as time t passes. The elapsed time is increased if either we take an earlier τ_1, or delay the time of observation by increasing t. Figure 25.15a shows a

(a)

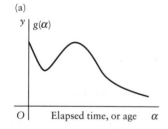

(b)

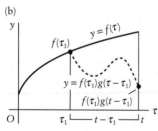

Fig. 25.15

representative function $g(\alpha)$, where α stands for 'age', and Fig. 25.15b illustrates its effect on the influence of f at time τ_1 on x at a later time t.

25.8 Discrete systems

Suppose we have a **system** or **processor**, which we shall generally think of as an electrical circuit. *All the time functions used are zero for $t < 0$.* The **input** will be denoted by $x(t)$ and the **output** by $y(t)$, and either may be referred to as a **signal**. The system is said to be **linear and time invariant** if there is a fixed transfer function $G(s)$ such that for all inputs and at all times $t \geq 0$ the input/output relation between the Laplace transforms of $x(t)$ and $y(t)$ has the form

$$Y(s) = G(s)X(s) \tag{25.15}$$

where

$$x(t) \leftrightarrow X(s) \quad \text{and} \quad y(t) \leftrightarrow Y(s),$$

subject to the condition of quiescence at $t = 0$. Thus $G(s)$ completely describes the effect of the circuit. By (25.11), the convolution theorem (25.15) is equivalent to

$$y(t) = \int_0^t x(\tau)g(t - \tau)\, d\tau \quad \text{or} \quad \int_0^t x(t - \tau)g(\tau)\, d\tau \tag{25.16}$$

where $g(t) \leftrightarrow G(s)$.

For the **impulsive input** $x(t) = x^*(t)$, where $x^*(t) = \delta(t)$, we have, by (25.5), $X^*(s) = 1$, so by (25.15), $Y^*(s) = G(s)$, or

$$g(t) = y^*(t), \quad \text{where} \quad g(t) \leftrightarrow G(s). \tag{25.17}$$

In other words, the interpretation of $g(t)$ is that it is *equal to the output from a unit delta-function input* at $t = 0$. This repeats the result (25.14).

So far in the chapter we have only considered circuits made up from the traditional elements, resistances, capacitances, and inductances, but there exists a far greater variety of basic units. We shall not describe the circuits which contain these new features, but only specify their properties.

Figure 25.16a shows a smooth signal $x(t)$ starting at $t = 0$. Imagine that this serves as the input to a circuit that picks out the values of $x(t)$ at times $t = 0, T, 2T, 3T, \ldots$, samples them over very short time intervals, and ignores the values of $x(t)$ in between, treating them as if they were zero. This process is indicated by the shaded strips in Fig. 25.16a. The device registers a sequence of values

$$\{x(0), x(T), x(2T), x(3T), \ldots\},$$

(a)

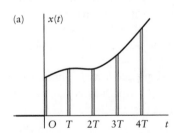

(b)

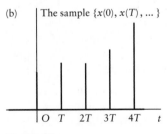

The sample $\{x(0), x(T), \ldots\}$

Fig. 25.16

called a **sample of** $x(t)$ **at equal intervals** T. In an actual instrument the output will consist of a succession of 'spikes' as in Fig. 25.16b. These can be thought of as brief puffs of energy generated by the circuit, which are equal in 'content' to the sequence of values above, so it is plausible to represent the sample, $y(t)$ say, by

$$y(t) = \sum_{k=0}^{K} x(kT)\, \delta(t - kT) \tag{25.18}$$

(where K may be infinite). Such a function is called **discrete**. The circuit works like the first stage of an analogue-to-digital converter.

Suppose next that we have a circuit which processes *discrete inputs* of interval T, and produces *discrete outputs* of interval T. Such circuits may amplify, or filter, or delay, or modify the input in a variety of ways. We then have a completely **discrete system**. The input $x(t)$ and output $y(t)$, and their Laplace transforms $X(s)$ and $Y(s)$, take the form

$$x(t) = \sum_{n=0}^{N} x_n\, \delta(t - nT), \quad X(s) = \sum_{n=0}^{N} x_n\, e^{-nTs}, \tag{25.19}$$

$$y(t) = \sum_{k=0}^{K} y_k\, \delta(t - kT), \quad Y(s) = \sum_{k=0}^{K} y_k\, e^{-kTs}, \tag{25.20}$$

where x_n and y_k are constants, and N and K may be infinite. We may alternatively express $x(t)$ and $y(t)$ in the form

$x(t) = \{x_0, x_1, x_2, \dots, x_N\}$, or simply as $\{x_n\}$;

$y(t) = \{y_0, y_1, y_2, \dots, y_K\}$, or as $\{y_k\}$.

Thus, $\{n + 3\}$ stands for $\{3, 4, 5, \dots\}$. In a case such as $\{1, 2, 0, 0, 0, 0, \dots\}$ we may further shorten it to $\{1, 2\}$.

Assume next that there exists a transfer function $G(s)$ so that $Y(s) = G(s)X(s)$. Let $g(t) \leftrightarrow G(s)$; then $g(t)$ is equal to the output resulting from the unit impulsive input

$x^*(t) = \delta(t)$

(or $x^*(t) = \{1\}$ or $\{1, 0, 0, 0, \dots\}$ in the sequence form). The device generates only discrete outputs. Then, $g(t)$, which is equal to the response to $x^*(t)$, must also have a discrete form:

$$g(t) = \sum_{m=0}^{M} g_m\, \delta(t - mT), \quad \text{so} \quad G(s) = \sum_{m=0}^{M} g_m\, e^{-mTs}. \tag{25.21}$$

Example 25.16 *A discrete circuit delays any incoming signal by an interval T (see Fig. 25.17a). (a) Obtain a transfer function G(s) by considering the response to a delta-function input. (b) Confirm that this transfer function delays an arbitrary discrete signal by an interval T, and that therefore the circuit is linear.*

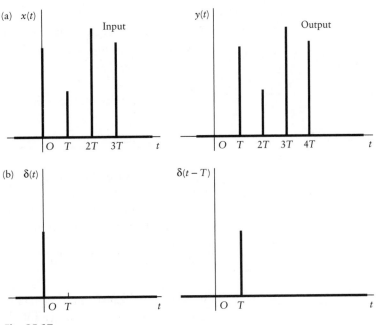

Fig. 25.17

(a) If the input is $x(t)$, then the output is $y(t) = x(t - T)$. Therefore, if $x(t) = \delta(t)$, the output is $\delta(t - T)$ as in Fig. 25.17b, and by (25.17), we must have $g(t) = \delta(t - T)$, so the transfer function is $G(s) = e^{-Ts}$.

(b) To check that this transfer function really works for a general discrete input, put $x(t) = \sum_{n=0}^{\infty} x_n \delta(t - nT)$. We then have

$$Y(s) = X(s)G(s) = X(s)\, e^{-Ts}.$$

By the delay rule (24.25),

$$y(t) = x(t - T)H(t - T),$$

where $H(t)$ is the unit function. Therefore $x(t)$ is delayed by an interval T.

In the general case when the transfer function takes the form $\{g_1, g_2, \ldots, g_M\}$, an input $x(t) = \delta(t)$, represented by $x(t) = \{1\}$, generates a string of impulses $\sum_{m=0}^{M} g_m \delta(t - mT)$, delayed by intervals mT, $m = 0$ to M. We shall look at this case in the next sections.

25.9 The z transform

In the previous section the only functions of s that appear are exponentials of the form e^{-nTs}, representing $\delta(t - Nt)$, where n is a positive integer or zero.

They may be written

$$e^{-nTs} = (e^{Ts})^{-n} = 1/(e^{Ts})^n.$$

The algebra connected with discrete systems is simplified by introducing a new variable z, defined by

$$z = e^{Ts}. \tag{25.22}$$

Then we may write for the transform of a typical discrete input $x(t)$:

$$X(s) = \sum_{n=0}^{N} x_n \, e^{-nTs} \equiv \sum_{n=0}^{N} \frac{x_n}{z^n}.$$

We shall reformulate the previous results in terms of z. Suppose we have a discrete signal $x(t) = \{x_0, x_1, x_2, \dots \}$ consisting of equally spaced impulses or samples with interval T, and $x(t) = 0$ for $t < 0$. Then the function $X(z)$ given by

$$X(z) = x_0 + \frac{x_1}{z} + \frac{x_2}{z^2} + \frac{x_3}{z^3} + \cdots \tag{25.23}$$

is called the z **transform** of $x(t)$. Given $x(t)$ we can write down the z transform. Conversely, given a suitable function $X(z)$, we can expand it by Taylor's theorem for large z in powers of z^{-1} in order to obtain the sequence of coefficients $\{x_0, x_1, x_2, \dots \}$ in (25.23), which defines $x(t)$. This sequence is called the **inverse transform** of $X(z)$.

Suppose that $\{x_n\}$ is supplied as input to a discrete linear system. The z transform of the output $y(t) = \{y_0, y_1, y_2, \dots \}$ is

$$\mathcal{Y}(z) = y_0 + \frac{y_1}{z} + \frac{y_2}{z^2} + \cdots. \tag{25.24}$$

We already know from (25.21) that if the circuit is linear it has a transfer function $G(s)$ which represents a similar sequence of impulsive terms. Therefore $g(t)$ has a z transform:

$$G(z) = g_0 + \frac{g_1}{z} + \frac{g_2}{z^2} + \cdots. \tag{25.25}$$

Finally, from (25.15) (since all we have done is to write a shorthand for e^{Ts}), the z transforms of output and input are related by

$$\mathcal{Y}(z) = G(z)X(z) \tag{25.26}$$

which is simply the product of two polynomials in powers of z^{-1}.

We have lost sight of T in these expressions, but we can always recover it by returning to time-domain or s-domain formulae by putting $z = e^{Ts}$. To summarize:

The *z* transform of a discrete signal

(a) If $x(t) = \{x_0, x_1, x_2, \ldots\}$, its z transform is

$$X(z) = x_0 + \frac{x_1}{z} + \frac{x_2}{z^2} + \cdots.$$

(b) The sequence $\{x_0, x_1, x_2, \ldots\}$ is called the inverse transform of $X(z)$.

(c) z is related to the Laplace transform by $z = e^{Ts}$.

(25.27)

The transfer function in terms of *z*

(a) The transfer function takes the form

$$G(z) = g_0 + \frac{g_1}{z} + \frac{g_2}{z^2} + \cdots.$$

(b) The input/output relation is

$$\mathcal{Y}(z) = G(z)X(z).$$

(c) The inverse of $G(z)$ is the response to an input $\delta(t)$, and has the form $\{g_0, g_1, g_2, \ldots\}$.

(25.28)

Example 25.17 *Obtain the z transform of the discrete signal* $x(t)$ *defined by the sequences* (a) $\{1\}$; (b) $x_n = 1$ *for* $n \geqslant 0$; (c) $x_n = 1$ *if n is even,* $x_n = 0$ *if n is odd.*

(a) $X(z) = 1 + \dfrac{0}{z} + \dfrac{0}{z^2} + \cdots = 1.$

(b) $X(z) = 1 + \dfrac{1}{z} + \dfrac{1}{z^2} + \cdots.$

This is an infinite geometric series with common ratio z^{-1} (it converges only if $|z| > 1$, but do not worry about this). From eqn (5.4a):

$$X(z) = \frac{1}{1 - z^{-1}} = \frac{z}{z - 1}.$$

(c) $X(z) = 1 + \dfrac{1}{z^2} + \dfrac{1}{z^4} + \cdots.$

The common ratio is z^{-2}, so by Section 5.4

$$X(z) = \frac{1}{1 - z^{-2}} = \frac{z^2}{z^2 - 1}.$$

Example 25.18   *Obtain the z transform of* $x(t) = \{1, 2, 3, \ldots\}$, *or* $\{n + 1\}$.

We see that

$$X(z) = 1 + \frac{2}{z} + \frac{3}{z^2} + \frac{4}{z^3} + \cdots.$$

To sum this series, multiply it by $1/z$:

$$\frac{1}{z}X(z) = \frac{1}{z} + \frac{2}{z^2} + \frac{3}{z^3} + \cdots.$$

Subtract the second expression from the first:

$$\left(1 - \frac{1}{z}\right)X(z) = 1 + \frac{1}{z} + \frac{1}{z^2} + \cdots = \frac{z}{z - 1}$$

(as in the previous Example). Therefore

$$X(z) = \frac{z}{z - 1} \bigg/ \left(1 - \frac{1}{z}\right) = \frac{z^2}{(z - 1)^2}.$$

Example 25.19   (a) *Obtain the inverse z transform of the function* $X(z) = z/(z - 2)$. (b) *Deduce the time function* $x(t)$ *which it represents.*

(a) We need to find the coefficients in the infinite series form for $X(z)$:

$$X(z) = x_0 + \frac{x_1}{z} + \frac{x_2}{z^2} + \cdots.$$

This is a Taylor expansion of $X(z)$ in powers of $1/z$ for large z (see Section 5.6). To obtain it, we start by expressing $X(z)$ in terms of $1/z$:

$$X(z) = \frac{z}{z - 2} = 1 \bigg/ \left(1 - \frac{2}{z}\right) = \left(1 - \frac{2}{z}\right)^{-1}.$$

The binomial expansion (5.4f), with $\alpha = -1$ and $x = -2/z$, gives

$$X(z) = 1 + \frac{2}{z} + \frac{2^2}{z^2} + \frac{2^3}{z^3} + \cdots.$$

Therefore the sequence of coefficients (i.e. the inverse) is $\{1, 2, 2^2, 2^3, \ldots\}$.

(b) The corresponding time function $x(t)$ is therefore

$$x(t) = \delta(t) + 2\delta(t - T) + 2^2\delta(t - 2T) + 2^3\delta(t - 3T) + \cdots.$$

Example 25.20   *The response of a discrete system to the input* $x(t) = \delta(t) + \delta(t - T)$ *is found to be* $y(t) = \delta(t) + 2\delta(t - T) + \delta(t - 2T)$. *Find* (a) *the z transfer function* $G(z)$, (b) *the Laplace transfer function* $G(s)$, (c) *the response to a unit impulse* $\delta(t)$.

(a) Put $X(z) = 1 + \dfrac{1}{z}$, $\mathcal{Y}(z) = 1 + \dfrac{2}{z} + \dfrac{1}{z^2}$, and $G(z) = g_0 + \dfrac{g_1}{z} + \dfrac{g_2}{z^2} + \cdots$ (for all we know at this stage, there might be an infinite number of terms in $G(z)$). Since $\mathcal{Y}(z) = G(z)X(z)$. Then

Example 25.20 *continued*

$$G(z) = \mathcal{Y}(z)/X(z) = \left(1 + \frac{2}{z} + \frac{1}{z^2}\right) \bigg/ \left(1 + \frac{1}{z}\right) = \left(1 + \frac{1}{z}\right)^2 \bigg/ \left(1 + \frac{1}{z}\right) = 1 + \frac{1}{z}.$$

(b) Restore *s* by putting $z = e^{Ts}$, where *T* is the spacing interval:

$G(s) = 1 + e^{-Ts}.$

(c) The impulse response is the inverse transform, $g(t)$, of $G(s)$:

$g(t) = \delta(t) + \delta(t - T),$

which can be obtained also from (a).

Example 25.21 *A smooth signal $x(t)$ is sampled at intervals T to produce the discrete signal $\{x(0), x(T), x(2T), \dots \}$. Obtain the z transform when (a) $x(t) = \cos \omega t$; (b) $x(t) = \sin \omega t$.*

The sample sequences in (a) and (b) are respectively $\{1, \cos \omega T, \cos 2\omega T, \dots \}$ and $\{0, \sin \omega T, \sin 2\omega T, \dots \}$. We can deal with both at the same time by remembering that $\cos n\omega T$ and $\sin n\omega T$ are respectively equal to the real and imaginary parts of $e^{in\omega T}$. Therefore, consider the sequence resulting from the complex input sequence

$$\{1, e^{i\omega T}, e^{2i\omega T}, \dots \},$$

which has the z transform

$$1 + \frac{e^{i\omega T}}{z} + \frac{e^{2i\omega T}}{z^2} + \dots = 1 + (e^{i\omega T}z^{-1}) + (e^{i\omega T}z^{-1})^2 + \dots .$$

This is an infinite geometric series with common ratio $e^{i\omega T}z^{-1}$, so its sum is equal to

$$1/(1 - e^{i\omega T}z^{-1}) = z/(z - e^{i\omega T}).$$

The complex conjugate of the denominator is $z - e^{-i\omega T}$, so write

$$\frac{z}{z - e^{i\omega T}} = \frac{z}{z - e^{i\omega T}} \frac{z - e^{-i\omega T}}{z - e^{-i\omega T}}$$

$$= \frac{z(z - e^{-i\omega T})}{z^2 - z(e^{i\omega T} + e^{-i\omega T}) + 1} = \frac{z(z - e^{-i\omega T})}{z^2 - 2z \cos \omega T + 1}.$$

The transforms of $\cos \omega T$ and $\sin \omega T$ are the real and imaginary parts respectively of this expression, so:

$$\text{transform of } \cos \omega T = \frac{z(z - \cos \omega T)}{z^2 - 2z \cos \omega T + 1};$$

$$\text{transform of } \sin \omega T = \frac{z \sin \omega T}{z^2 - 2z \cos \omega T + 1}.$$

Finally, we note the discrete form of the convolution theorem, (25.11), expressed in terms of *z*. For a discrete linear system there exists a transfer function $G(z)$ such that input and output are related by $\mathcal{Y}(z) = G(z)\mathcal{Y}(z)$, or

$$y_0 + \frac{y_1}{z} + \dots = \left(g_0 + \frac{g_1}{z} + \dots\right)\left(x_0 + \frac{x_1}{z} + \dots\right).$$

By matching the coefficients of inverse powers of z on both sides we obtain:

Discrete form of the convolution theorem

If $\mathcal{Y}(z) = G(z)X(z)$, then

$$y_0 = g_0 x_0,$$
$$y_1 = g_1 x_0 + g_0 x_1,$$
$$y_2 = g_2 x_0 + g_1 x_1 + g_0 x_2,$$

and so on. In general,

$$y_n = \sum_{r=0}^{n} g_r x_{n-r}.$$

(25.29)

The structure of these formulae resembles that of a convolution integral, with r in place of τ and n in place of t in (25.11).

25.10 Behaviour of z transforms in the complex plane

Suppose that a string of impulses represented by $x(t) = \{x_0, x_1, x_2, \dots\}$ is fed into a discrete processor. When the first impulse arrives at $t = 0$, it triggers the circuit to produce a scaled copy of the transfer function $G(z)$ in the time domain, a string of impulses given by $x_0 g(t) = \{x_0 g_0, x_0 g_1, x_0 g_2, \dots\}$. The second impulse is felt at $t = T$, and $G(z)$ forms another scaled copy of itself, $x_1 g(t - T)$, starting at $t = T$, and so on. These sequences overlap: the second one starts before the first has ended, and the output sequence consists of the sum of all the effects which are still present at $T, 2T, 3T, \dots$. This is illustrated in Fig. 25.18.

If $G(z)$ has an infinite number of terms, the effect of any input term will be present for ever after. This extension into the distant future of the influence of an individual piece of input resembles the presence of transients in systems governed by differential equations. Very long-term effects are usually undesirable; in particular, they should not increase as time goes on. Their increase or decrease is described by the rate of increase or decrease of the coefficients in the series

$$G(z) = g_0 + \frac{g_1}{z} + \frac{g_2}{z^2} + \cdots. \tag{25.30}$$

We shall illustrate how information about this question can be obtained by examining the behaviour of $G(z)$ when it is given in closed form, and the variable z is allowed to be complex.

We limit consideration to cases where $G(z)$ is a *rational function* of z:

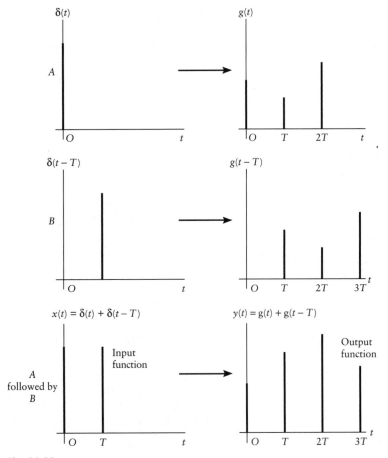

Fig. 25.18

$$G(z) = \frac{a_M z^M + a_{M-1} z^{M-1} + \cdots + a_0}{b_N z^N + b_{N-1} z^{N-1} + \cdots + b_0}. \tag{25.31}$$

We shall assume that $M < N$. Suppose that the a_m and b_n are all real numbers, and that the N solutions of the equation

$$b_N z^N + b_{N-1} z^{N-1} + \cdots + b_0 = 0 \tag{25.32}$$

are

$$z = z_1, z_2, z_3, \ldots, z_N.$$

For simplicity, we shall assume that these numbers are *all different*. The denominator of (25.31) then has N different factors of the form $(z - z_n)$, for $n = 1$ to N, so (25.31) can be written

$$G(z) = \frac{a_M z^M + a_{M-1} z^{M-1} + \cdots + a_0}{b_N (z - z_1)(z - z_2) \ldots (z - z_N)}. \tag{25.33}$$

Notice that $G(z)$ is *infinite* at the points $z_1, z_2, \ldots, z_N$. These points are called the **poles** of $G(z)$. Some of them may be *complex numbers*.

If so, they occur in pairs: if z_n is a solution of (25.32), then so it its complex conjugate $\bar{z}_n$. Equation (25.33) may now be expressed as the sum of partial fractions (now in general complex) as in Section 1.14:

$$G(z) = \frac{C_1}{z - z_1} + \frac{C_2}{z - z_2} + \cdots + \frac{C_N}{z - z_N} \tag{25.34}$$

since $M < N$, where C_1 to C_N are constants. A typical term has the form

$$\frac{C}{z - c}, \tag{25.35}$$

where c may be complex: if so, then C might be complex as well. This term is the source of a part of the discrete output signal $g(t)$ produced by an input $x(t) = \delta(t)$, and we shall see whether it generates an *increasing* or a *decreasing* output.

Suppose firstly that we find a pole at $z = c$ in (25.35), where c is a *real number*. Then C is also real, and

$$\frac{C}{z - c} = \frac{C}{z}\left(1 - \frac{c}{z}\right)^{-1} = \frac{C}{z} + \frac{Cc}{z^2} + \frac{Cc^2}{z^3} + \cdots .$$

In the time domain this corresponds to the sequence

$$\{C, Cc, Cc^2, \ldots \}.$$

If $|c| > 1$ the terms are *increasing* in magnitude, and the system is said to be **unstable**. If $|c| < 1$ they are *decreasing* in magnitude. The rate of increase or decrease is actually **exponential**, because

$$|Cc^n| = |C|\, e^{n \ln|c|}.$$

If $c = \pm 1$, then the output time sequence is nondecreasing, and unstable:

$$\{C, \pm C, C, \pm C, \ldots \}.$$

Next, suppose that c is *complex*. Then there is another pole at $z = \bar{c}$. Taking these together, we obtain a pair of complex conjugate terms, generating *real* coefficients:

$$\frac{C}{z - c} + \frac{\bar{C}}{z - \bar{c}} = 2\,\mathrm{Re}\frac{C}{z - c} = 2\,\mathrm{Re}\frac{C}{z}\frac{1}{1 - cz^{-1}}$$

$$= 2\,\mathrm{Re}\left(\frac{C}{z} + \frac{Cc}{z^2} + \frac{Cc^2}{z^3} + \cdots\right)$$

$$= \{2\,\mathrm{Re}\,(C), 2\,\mathrm{Re}\,(Cc), \ldots \}. \tag{25.36}$$

Evidently the *magnitude* (modulus) of the coefficients follows the same rule as before.

Each of the terms (23.34) in $G(z)$ contributes to $g(t)$ in a similar way. Therefore, the response $y(t)$ to a delta function input $x(t) = d(t)$ depends upon the poles c_n of $G(z)$ as follows:

Stability of a linear system

(i) If $|c_n| < 1$ for every pole c_n of $G(z)$, the response $y(t)$ dies
     away, so the system is stable.
(ii) If not, then the system is unstable.   **(25.37)**

We can interpret the complex poles more closely. Put

$$C = |C|\, e^{i\phi} \text{ and } c = |c|\, e^{i\omega}. \tag{25.38}$$

From (25.38)

$$Cc^n = |C|\,|c|^n\, e^{i(n\omega+\phi)}.$$

Therefore, for $n = 0, 1, 2, \ldots$,

$$2\, \text{Re}\,(Cc^n) = |C|\,|c|^n \cos(n\omega + \phi).$$

Put this into the time sequence (25.36). It becomes

$$\{2|C| \cos\phi,\, 2|C|\,|c|\, \cos(\omega+\phi),\, 2|C|\,|c|^2 \cos(2\omega+\phi),$$
$$2|C|\,|c|^3 \cos(3\omega+\phi),\, \ldots\}.$$

This sequence would be obtained by sampling at $t = 1, 2, 3, \ldots$ from
the *smooth* function

$$2|C|\,|c|^t \cos(\omega t + \phi)H(t), \tag{25.39}$$

so a picture of the progress of the discrete transient can be obtained
as in Fig. 25.19. Alternatively, this is equivalent to samples at $t = T$,
$2T, \ldots$ taken from

$$2|C|\,|c|^{t/T} \cos(\omega t/T + \phi)H(t). \tag{25.40}$$

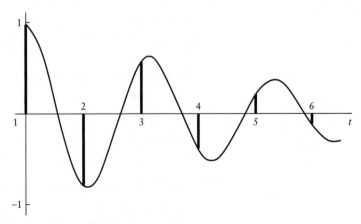

Fig. 25.19 Discrete transient of $C/(z-c)$. Suppose that $C = 0.5$ and $c = 0.8\, e^{2.9i}$.
Then $|C| = 0.5$, $|c| = 0.8$, $\phi = 0$, $\omega = 2.9$. The curve $y = (0.8)^t \cos(2.9t)$ and the impulsive
response to $\delta(t)$ are shown.

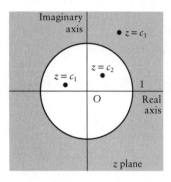

Fig. 25.20 The unit circle $|z| = 1$, and several poles of a transfer function $G(z)$. One of the poles is *outside* the circle so the circuit is unstable and a transient associated with this pole will grow exponentially.

In Fig. 25.20 we show an Argand diagram with the unit circle $|z| = 1$ indicated. This is used as a design tool to obtain a qualitative idea of how a proposed circuit will behave, and to modify its properties. We can find the poles (the points where $G(z)$ is infinite), and place them on the diagram. Poles within the circle promise transients which die away; if there is a pole outside, then a stimulus applied to the circuit will produce ever-increasing output, so the system will be unstable. Poles lying on the circle $|z| = 1$ produce transients which do not approach zero or infinity in magnitude. If the values associated with the circuit elements can be adjusted so that all the poles lie inside the unit circle, than we shall have a circuit for which all disturbances die away with time.

25.11 *z* transforms and difference equations

Systems can be constructed whose output y_{n+1} at time $t = (n + 1)T$ depends not only on the *input* up to that time, but also on the *preceding outputs*. This is achieved by delay elements which pick up each y_n at time nT, store it for a time T, then feed it back into the system so as to modify y_{n+1} in some way. A chain of delay elements can reach back further into the history of the outputs. In this way y_{n+1} may be related to the current input and earlier outputs by equations such as

$$y_{n+1} = y_n + x_n,$$

or

$$y_{n+2} = 2y_{n+1} - y_n + x_n,$$

for $n = 0, 1, 2, \ldots$. Such equations are called **difference equations** or **recurrence relations**. The equations above are called linear difference equations, because the terms in y only appear linearly. The circuits producing such equations are not necessarily linear in the sense we have used so far: in the sense of possessing a transfer function. Difference equations are treated more fully in Chapter 38; for the present we shall outline a connection with z transforms.

Example 25.22   *Obtain the sequence* $\{y_n\}$, *where*

$$y_{n+1} = y_n + x_n,$$

given that

$$y_0 = 3 \text{ and } \{x_n\} = \{1, 2, 3, \ldots\}.$$

This is easily done by simply counting.
   For $n = 0$: $y_1 = y_0 + x_0 = 3 + 1$.
   For $n = 1$: $y_2 = y_1 + x_1 = (3 + 1) + 2$.
   For $n = 2$: $y_3 = y_2 + x_2 = (3 + 1 + 2) + 3$, and so on. Evidently,

$$y_n = 3 + (1 + 2 + 3 + \cdots + n) = 3 + \tfrac{1}{2}n(n + 1)$$

by using a well-known formula (see Appendix A(f)).

Notice that we had to *prescribe* y_0: it was not given by the difference equation, and we could have assigned any value to it. It resembles the initial condition of a first-order differential equation.

Example 25.23 *Use z transforms to determine the stability of a feedback circuit which processes the digital signal $\{x_n\}$ according to the difference equation*

$$y_{n+2} = 3y_{n+1} - 2y_n + x_n,$$

where y_0, y_1, and the sequence $\{x_n\}$ are given.

It is usual to collect together the *y* terms on the left-hand side:

$$y_{n+2} - 3y_{n+1} + 2y_n = x_n \tag{i}$$

as with a differential equation. The sequences and their transforms are given by

$$\{x_n\} = \{x_0, x_1, x_2, \dots\},$$
$$X(z) = x_0 + x_1 z^{-1} + x_2 z^{-2} + \cdots;$$
$$\{y_n\} = \{y_0, y_1, y_2, \dots\},$$
$$\mathcal{Y}(z) = y_0 + y_1 z^{-1} + y_2 z^{-2} + \cdots;$$
$$\{y_{n+1}\} = \{y_1, y_2, y_3, \dots\},$$
$$\mathcal{Y}_1(z) = y_1 + y_2 z^{-1} + y_3 z^{-2} + \cdots;$$
$$\{y_{n+2}\} = \{y_2, y_3, y_4, \dots\},$$
$$\mathcal{Y}_2(z) = y_2 + y_3 z^{-1} + y_4 z^{-2} + \cdots.$$

By simply looking at the *z* series it can be seen that

$$\mathcal{Y}_1(z) = z\mathcal{Y}(z) - zy_0$$

and

$$\mathcal{Y}_2(z) = z\mathcal{Y}_1(z) - zy_1 = z^2\mathcal{Y}(z) - z^2 y_0 - zy_1.$$

Therefore the *z* transforms of the sequences obeying the relation (i) are connected by

$$(z^2\mathcal{Y}(z) - z^2 y_0 - zy_1) - 3(z\mathcal{Y}(z) - zy_0) + 2\mathcal{Y}(z) = X(z),$$

or

$$(z^2 - 3z + 2)\mathcal{Y}(z) - (z^2 - 3z)y_0 - zy_1 = X(z).$$

From this equation we obtain

$$\mathcal{Y}(z) = \frac{X(z) + (z^2 - 3z)y_0 + zy_1}{z^2 - 3z + 2} = \frac{X(z) + (z^2 - 3z)y_0 + zy_1}{(z-1)(z-2)}. \tag{ii}$$

The denominator is independent of y_0, y_1, and $\{x_n\}$, which are arbitrary. $\mathcal{Y}(z)$ has a pole at $z = 2$, so the results of the previous section predict that unless y_0, y_1, and $\{x_n\}$ are specially chosen, the output will grow exponentially.

In Example 25.23, eqn (ii), the denominator has the form

$$az^2 + bz + c$$

where a, b, c are the coefficients of y_{n+2}, y_{n+1}, and y_n respectively. The denominator alone determines the growth of transients, so there is really no need to work right through the problem if all we want is information about the stability. In fact, if $y_0 = y_1 = 0$, which would be a natural condition, the circuit has a transfer function equal to $1/(az^2 + bz + c)$, so the situation is exactly the same as in the previous section. Similar considerations apply to linear difference equations of any order.

Problems

25.1 Invert the transforms (a) $1/[s(s^2+1)]$, (b) $1/[s^2(s^2+1)]$, (c) $1/[s^3(s^2+1)]$, by using (25.1).

25.2 The equation for the current $i(t)$ in an RLC circuit for zero initial charge is

$$L\frac{di}{dt} + Ri + \frac{1}{C}\int_0^t i(\tau)\,d\tau = v(t).$$

(a) Solve this equation when $L = 2$, $R = 3$, $C = \frac{1}{3}$, $v(t) = 3\cos t$ in conveniently scaled units, for zero initial current and charge.
(b) Adapt the equation to the case when $v(t) = 0$ and there is an initial charge q_0 on the capacitor, and solve it, given that $i(0) = 0$.
(c) The circuit in (a) is quiescent with zero charge; then, at $t = t_0$, a voltage of 300 units acts in it for 0.01 time units. Approximate the applied voltage by a suitable impulse function, and solve the equation for $i(t)$.

25.3 The displacement $x(t)$ of a mass on a spring with velocity damping and external force $f(t)$ per unit mass reduces to the conventional form $\ddot{x} + 2k\dot{x} + \omega^2 x = f(t)$. The initial conditions are $x(0) = 1$, $\dot{x}(0) = 1$. An impulse I is applied at $t = t_0$. Find the solution for $t > 0$ for $k^2 > \omega^2$.

25.4 A light plank of length l rests across a crevasse, and sags under the weight of a mountaineer of mass M standing at the centre. The displacement $u(x)$, where x is measured from one end, is determined in general by $K\,d^4u/dx^4 = f(x)$, where K is constant and $f(t)$ is force per unit length along the plank. The **boundary conditions**, which say that the plank merely rests on its ends, are

$u(0) = u''(0) = u(l) = u''(l) = 0.$

Treat the mountaineer as a point force and solve the problem using Laplace transforms. (Hint: two

conditions are prescribed at $x = 0$, but four are needed: call the missing ones A, B. Find A and B by requiring $u(l) = u''(l) = 0$.)

25.5 Find the impedances of the circuits in Fig. 25.21.

(a)
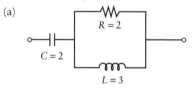
$R = 2$
$C = 2$
$L = 3$

(b)
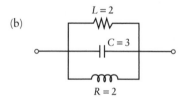
$L = 2$
$C = 3$
$R = 2$

(c)
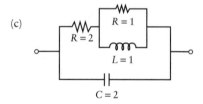
$R = 1$
$R = 2$
$L = 1$
$C = 2$

(d)
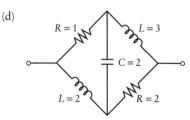
$R = 1$ $L = 3$
$C = 2$
$L = 2$ $R = 2$

Fig. 25.21

25.6 Find the transfer functions $V_2(s)/V_1(s)$ and $V_2(s)/I(s)$ in the circuits in Fig. 25.22.

(a)

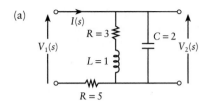

(b)

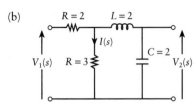

Fig. 25.22

25.7 Evaluate the convolution integral

$$\int_0^t g(\tau)h(t-\tau)\,d\tau,$$

or

$$\int_0^t h(\tau)g(t-\tau)\,d\tau,$$

in the following cases. Sometimes it might be easier to invert the corresponding Laplace transform (25.11).
(a) $g(t) = e^t, h(t) = 1$;   (b) $g(t) = 1, h(t) = 1$;
(c) $g(t) = e^t, h(t) = e^t$;   (d) $g(t) = e^{-t}, h(t) = t$;
(e) $g(t) = t, h(t) = \sin t$;   (f) $g(t) = \cos t, h(t) = t$;
(g) $g(t) = \sin 3t, h(t) = e^{-2t}$;
(h) $g(t) = \sin t, h(t) = \sin t$;
(i) $g(t) = t^4, h(t) = \sin t$;   (j) $g(t) = t^n, h(t) = t^m$.

25.8 Use the convolution theorem (25.11) to obtain an expression, in the form of an integral, for a particular solution of the following equations.

(a) $\dfrac{d^2x}{dt^2} + \omega^2 x = f(t)$;   (b) $\dfrac{d^2x}{dt^2} - \omega^2 x = f(t)$.

25.9 Use the convolution theorem (25.11) to find a solution $x(t)$ of the following **Volterra-type integral equations.**

(a) $\displaystyle\int_0^t x(\tau)(t-\tau)\,d\tau = t^4$;

(b) $x(t) = 1 + \displaystyle\int_0^t x(\tau)(t-\tau)\,d\tau$;

(c) $x(t) = \sin t + \displaystyle\int_0^t x(\tau)\cos(t-\tau)\,d\tau$.

25.10 By following a similar argument to that leading up to (25.14), show that

$$x(t) = \frac{d}{dt}\int_0^t x^{**}(\tau)f(t-\tau)\,d\tau,$$

where $x^{**}(t)$ represents the response of a quiescent 'black box' to a unit-function input $H(t)$, and $x(t)$ is its response from quiescence to an input $f(t)$.

Suppose that the transform $X^{**}(s)$ of the unit-function response is given by $1/(s-1)(s+2)$ in a particular case. Obtain the response from zero initial conditions to an input $H(t)\sin \omega t$.

25.11 (a) A student learning a language aims to memorize 50 new words a day, starting at $t = 0$. She is successful in this but, after a time lapse α, remembers only a fraction $e^{-0.01\alpha}$ of those learned at any time. Express the number $N(t)$ of words still in her vocabulary in terms of a convolution integral, and evaluate it.

(b) The student decides to increase the number of words available by attempting $50 + 0.1t$ words per day. Find the new $N(t)$, assuming the same initial success and the same rate of forgetting.

25.12 A population $p(t)$ for $t > 0$ develops as follows. The p_0 individuals in existence at $t = 0$ die out on average via a factor $e^{-\gamma t}$, so that at time t only about $p_0 e^{-\gamma t}$ are still in existence. For the rest, take any time $\tau < t$. The number born between τ and $\tau + \delta\tau$ is $bp(\tau)\,\delta\tau$, where b is the birthrate; these individuals die out through a factor $e^{-\beta(t-\tau)}$, where $\beta < \gamma$ and $t - \tau$ is the time elapsed from birth. Show that

$$p(t) = p_0 e^{-\gamma t} + b\int_0^t p(\tau)\,e^{-\beta(t-\tau)}\,d\tau,$$

and solve the equation.

25.13 A simple harmonic oscillator with displacement x is subject to a constant force F_0 for $0 < t < t_0$, and allowed to oscillate freely for $t > t_0$. If $H(t)$ is the Heaviside function its equation of motion is

$$m\ddot{x} + kx = F_0[H(t) - H(t-t_0)].$$

If the system starts from rest in equilibrium, show that the Laplace transform is

$$\mathcal{L}\{x(t)\} = \frac{F_0}{m}\frac{(1 - e^{-st})}{s(s^2 + \omega^2)},$$

where $\omega = \sqrt{(k/m)}$. Show that, for $0 < t < t_0$, the solution is

$$x(t) = \frac{F_0}{k}(1 - \cos \omega t),$$

and find the solution for $t > t_0$.

25.14 An equation of the form

$$\frac{dx(t)}{dt} = x(t-1) + t,$$

which relates the derivative at time t to the value of the function at an earlier time, is an example of a **differential delay equation**. If $x(t) = 0$ for $t \leqslant 0$, show that the Laplace transform of the solution is

$$\mathcal{L}\{x(t)\} = \frac{1}{s^2(s + e^{-s})} = \frac{1}{s^3(1 + e^{-s}/s)}.$$

Expand $1/(1 + e^{-s}/s)$ in powers of e^{-s}/s using a binomial expansion, and show that

$$x(t) = 2 \sum_{n=0}^{\lfloor t \rfloor} \frac{(t-n)^{n+2}}{(n+2)!},$$

where $\lfloor t \rfloor$ is the **integer floor function** (the largest integer less than or equal to t: for example, $\lfloor 2.3 \rfloor = 2$, $\lfloor 3 \rfloor = 3$, and $\lfloor -2.3 \rfloor = -3$).

25.15 The equation

$$2 \int_0^t \cos(t-u)x(u) \, du = x(t) - t$$

is an example of an **integral equation**. Note that the integral is of convolution type, which means that the Laplace transform of the equation is

$$2\mathcal{L}\{\cos t\}X(s) = X(s) - \frac{1}{s^2}.$$

Show that the solution is

$$x(t) = 2(t-1) \, e^t + t + 2.$$

25.16 The differential equation

$$\frac{d^2x}{dt^2} + t\frac{dx}{dt} - x = 0$$

does *not* have constant coefficients: the coefficient of dx/dt is t. Using the results (24.8) and (24.12), show that the transform of the differential equation subject to the conditions $x(0) = 0$ and $x'(0) = 1$ satisfies the first-order equation

$$-s\frac{dX(s)}{ds} + (s^2 - 2)X(s) = 1.$$

Verify that $X(s) = 1/s^2$ satisfies this equation, and hence obtain the required solution of the original equation.

25.17 Using the method outlined in Problem 25.16, solve the following variable-coefficient equations using Laplace transforms:
(a) $tx''(t) + (1-t)x'(t) - x(t) = 0$, $x(0) = x'(0) = 1$;
(b) $x''(t) + tx'(t) - 2x(t) = 2$, $x(0) = x'(0) = 0$;
(c) $tx''(t) - x'(t) + tx(t) = \sin t$, $x(0) = 1$, $x'(0) = 0$.

25.18 (Discrete systems, Section 25.8). The following signals are expressed in the sequence forms (25.30). Write the explicit form of $x(t)$ and its Laplace transform (25.19) in each case.
(a) $\{1, 2, 1, 0, 0, 0, \ldots\}$.
(b) $\{0, 1, 2, 3, \ldots\}$.
(c) $\{3\}$.
(d) $\{(-2)^n\}$.
(e) $\{0, 0, 3\}$.

25.19 The transfer functions $g(t)$ in the time domain, and inputs $x(t)$, are given below. Obtain the outputs $y(t)$ in each case.
(a) $g(t) = \{1, 1\}$, $x(t) = \{1, 1\}$.
(b) $g(t) = \{1, 1/2, 1/2^2, \ldots\}$, $x(t) = \{1, 1\}$.
(c) $g(t) = \{1, -1, 1, -1, \ldots\}$, $x(t) = \{0, 2, 2\}$.

25.20 Obtain the output $y(t)$, when the transfer function is $G(s) = 1/(1 - \frac{1}{3}e^{-Ts})$ and the Laplace transform of the input is $X(s) = e^{-Ts} + 2\,e^{-2Ts}$. (Hint: expand $G(s)$ in the form of an appropriate infinite series in powers of e^{-sT}.)

25.21 Obtain the z transforms corresponding to the various specifications that follow:
(a) $x(t) = \delta(t-T) + 2\delta(t-2T) - \delta(t-3T)$.
(b) $x(t) = \{1, -1, 1, -1, \ldots\}$.   (c) $x(t) = \{1/2^n\}$.
(d) $X(s) = e^{-Ts}/(1 - e^{-2Ts})$.

25.22 The following functions are sampled at interval T. Obtain the z transform of the (discrete) sampled functions ($H(t)$ is the unit function (1.13)).
(a) $tH(t)$.
(b) $e^{-t}H(t)$.
(c) $\cos \omega t \, H(t)$, when $T = \pi/2\omega$.
(d) $\sin \omega t$, when $T = \pi/2\omega$.

25.23 Obtain the z transforms of the transfer functions, $G(z)$, of various discrete, linear, systems which have been tested for the particular input $x(t)$ and output $y(t)$ as specified:
(a) $x(t) = \{1, 1\}$, $y(t) = \{1, -1\}$, and find the sequence for $g(t)$.
(b) $x(t) = \{1, 0, 0, 3\}$, $y(t) = \{1, 1\}$.
(e) $x(t) = \{1, -1\}$, $y(t) = \{1, 1\}$.
(d) $x(t) = \{1, 1, 1, \ldots\}$, $y(t) = \{1, 0, -1, 0, 1, 0, -1, 0, 1, \ldots\}$.
(e) $x(t) = \{1, 0, 1, 0, \ldots\}$,
$y(t) = \{1, 0, -1, 0, 1, 0, -1, 0, 1, \ldots\}$.

25.24 Prove that if the z transform of the discrete function given by $\{x_0, x_1, x_2, \ldots\}$ is $X(z)$, then the discrete transform of $y(t) = \{x_0, x_1 \, e^{-CT}, x_2 \, e^{-2CT}, \ldots\}$ is $X(c^{CT}z)$.

25.25 (a) Prove that if the z transform of the discrete function $x(t)$ defined by $\{x_0, x_1, x_2, \dots\}$ is $X(z)$, then the transform of $(0, x_0, x_1, \dots)$ is $(1/z)X(z)$.

(b) Deduce that the transform of $\{0, 0, \dots, 0, x_0, x_1, \dots\}$ (starting with N zeros) is $(1/z)^N X(z)$. (This is a time-delay rule for z transforms.)

25.26 Prove that if the z transform of $\{x_0, x_1, x_2, \dots\}$ is $X(z)$, then the transform of $\{x_N, x_{N+1}, x_{N+2}, \dots\}$ is

$$z^N X(z) - z^N x_0 - z^{N-1} x_1 - \cdots - z x_{N-1}.$$

(This resembles the differentiation rule for Laplace transforms, (24.12). Start the process with $N = 1$, then $N = 2$ etc., until the sequence becomes clear.)

25.27 The following represent transfer functions for discrete systems, $G(z)$. Find the poles, mark them on an Argand diagram as in Fig. 25.20, and state whether the systems are stable or not. Obtain the rate of growth or decay of their transients.

(a) $(z+1)/(z^2-4)$.

(b) $(z^2-z)/(4z^2-1)$.

(c) $1/(4z^2+1)$.

(d) $(z^3+1)/(2z^4+5z^2+2)$.

25.28 $\{x_n\}$ and $\{y_n\}$ represent inputs and outputs to discrete systems governed by the difference equations shown. Use z transforms to obtain the transforms $\mathcal{Y}(z)$ in terms of $X(z)$ and the initial values y_0 and y_1. State whether the systems are stable or not.

(a) $4y_{n+2} - y_n = x_n;\ y_0 = 1,\ y_1 = 2$.

(b) $y_{n+2} - 3y_{n+1} + 2y_n = 2x_n;\ y_0 = 0,\ y_1 = 1$.

(c) $2y_{n+2} + y_{n+1} + y_n = x_{n+1} - x_n;\ y_0 = 0,\ y_1 = 1$.

(d) $2y_{n+2} + 3y_{n+1} - y_n = x_n;\ y_0 = 1,\ y_1 = 1$.

26 Fourier series

CONTENTS

26.1   The composition of vibrations

If a note on a piano is played, firstly by pressing the key and then by plucking the string, the sounds produced are very different although the **pitch** or **fundamental frequency** heard is the same in both cases. The note produced by an instrument is not a pure tone or sinusoidal wave; it is a richer sound which contains other frequencies. These occur in different proportions when the same note is stimulated in different ways, or is sounded on different instruments.

A trained ear can detect some detail in these differences; the extra components can be distinguished and their pitch recognized, or they can be isolated by using resonators. The extra component frequencies of a note are all higher than the fundamental frequency, and related to it in a simple manner. If the fundamental frequency is f, then the **harmonics** present have frequencies

$$f, \quad 2f, \quad 3f, \quad 4f, \quad 5f, \quad \dots ,$$

the strength of the harmonics dropping off to zero as their frequency increases.

When these components are added, a profile for the composite wave is obtained. A particular note was found to have components as shown:

Order of harmonic:	1	2	3	4	5	…
Frequency:	f	$2f$	$3f$	$4f$	$5f$	…
Relative amplitude:	1.0	0.9	0.3	0.3	0.1	…

The shape and amplitude of the component harmonic waves, and of the composite wave, are shown in Fig. 26.1.

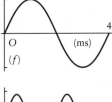

(f)

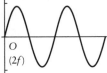

(2f)

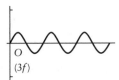

(3f)

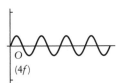

(4f)

(5f)

Compound sound

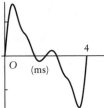

Fig. 26.1

By means of an electronic synthesizer the proportions in which harmonics occur can be controlled and a great variety of sound quality generated, from flute to drum. Given any particular fundamental frequency f, it is plausible that we could generate a sound wave of any preassigned quality (that is to say, of any shape) by adjusting the balance of the harmonics. This possibility is essentially what the theory of Fourier series is about, though in a wider context than that of sound waves.

26.2 Fourier series for a periodic function

The following symbols are used in connection with **periodic functions** (see Section 20.1):

f = frequency (cycles/time; if time is in seconds, the unit is the hertz);
ω = angular frequency (radians/time): $\omega = 2\pi f = 2\pi/T$;
T = period or wavelength: $T = 1/f = 2\pi/\omega$.

A typical **periodic** function $P(t)$ with period T is shown in Fig. 26.2. Any full-period interval may be chosen for discussion; suppose it is the interval between $t = -\pi/\omega$ and $t = \pi/\omega$.

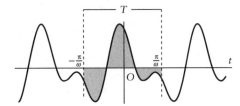

Fig. 26.2 A periodic function $P(t)$ with period $T = 2\pi/\omega$.

We shall express $P(t)$ over this interval, denoted by $[-\pi/\omega, \pi/\omega]$, as the sum of harmonic (sinusoidal) curves having frequencies f, $2f$, $3f, \ldots$, where $f = 1/T$; or, equivalently, angular frequencies ω, 2ω, $3\omega, \ldots$, where $\omega = 2\pi/T$. A **constant term** is also needed, since the average value of $P(t)$ will not generally be zero. **Both sine and cosine terms** are needed, because if we involve only sines or only cosines, the sum will have a symmetry, odd or even (Section 15.9), which $P(t)$ might not have. Then we expect that

Fourier series for a function of period T

$$P(t) = \tfrac{1}{2}a_0 + (a_1 \cos \omega t + b_1 \sin \omega t) + (a_2 \cos 2\omega t + b_2 \sin 2\omega t) + \cdots$$

$$= \tfrac{1}{2}a_0 + \sum_{n=1}^{\infty}(a_n \cos n\omega t + b_n \sin n\omega t),$$

where $\omega = 2\pi/T$.

(26.1)

Equation (26.1) is a **Fourier series** for $P(t)$, and the constants a_0; a_1, b_1; a_2, b_2; ... are its **Fourier coefficients**. It will be shown how to determine the coefficients in Section 26.4: the factor $\frac{1}{2}$ in the constant term $\frac{1}{2}a_0$ is introduced to simplify the working.

We have spoken in terms of the one-period range $t = -\pi/\omega$ to $t = \pi/\omega$, but every term on the right of (26.1) is periodic with the same period $T = 2\pi/\omega$ as $P(t)$. Therefore **the series will describe** $P(t)$ **for every value of** t, not merely for t in the interval between $\pm\pi/\omega$.

26.3 Integrals of periodic functions

We prove two results needed for the next section, in which the values of the coefficients in (26.1) are determined. Figure 26.3 represents a periodic function $P(t)$ with period T. Choose any value of t, say $t = t_0$, and compare the two integrals

$$\int_0^T P(t)\,dt \quad \text{and} \quad \int_{t_0}^{t_0+T} P(t)\,dt,$$

each of which is taken over a one-period interval of $P(t)$. The figure shows that the integrals are equal by virtue of the area analogy (15.13). The two shaded areas in Figs 26.3a, b are assembled from identical elements which are simply added up in a different order.

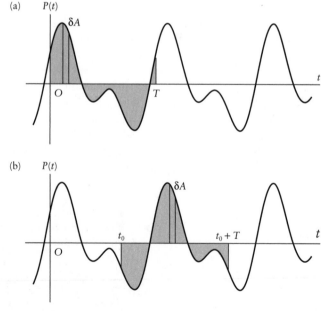

Fig. 26.3 Illustrating the area analogy for (a) $\int_0^T P(t)\,dt$ and (b) $\int_{t_0}^{t_0+T} P(t)\,dt$, where $P(t)$ has period T.

> **The integral over any one-period interval of a function $P(t)$ having period T,**
>
> $$\int_{t_0}^{t_0+T} P(t)\, dt,$$
>
> **does not depend on t_0.**
>
> (26.2)

Example 26.1 *Show that $\displaystyle\int_0^\pi \sin 2t\, \cos^2 t\, dt = 0.$*

The period of $\cos^2 t$ is π, because

$$\cos^2 t = \tfrac{1}{2}(1 + \cos 2t)$$

and the period of $\cos 2t$ is π. The period of $\sin 2t$ is also π. Therefore, by (26.2),

$$\int_0^\pi \sin 2t\, \cos^2 t\, dt = \int_{-\frac{1}{2}\pi}^{\frac{1}{2}\pi} \sin 2t\, \cos^2 t\, dt,$$

since the range $-\tfrac{1}{2}\pi$ to $\tfrac{1}{2}\pi$ also covers a period π. But the integrand is an odd function about the origin, so that the value of the last version is zero (Section 15.9).

The following special results can be proved by using the trigonometric identities in Appendix B which convert products to sums.

> **Trigonometric integrals over a one-period interval**
>
> (a) For n and $m = 0, 1, 2, \ldots$ with $n \neq m$,
>
> $$\int_{-\pi/\omega}^{\pi/\omega} \cos n\omega t\, \cos m\omega t\, dt = 0,$$
>
> $$\int_{-\pi/\omega}^{\pi/\omega} \sin n\omega t\, \sin m\omega t\, dt = 0,$$
>
> $$\int_{-\pi/\omega}^{\pi/\omega} \cos n\omega t\, \sin m\omega t\, dt = 0.$$
>
> (b) For $n = 1, 2, \ldots$
>
> $$\int_{-\pi/\omega}^{\pi/\omega} \cos^2 n\omega t\, dt = \int_{-\pi/\omega}^{\pi/\omega} \sin^2 n\omega t\, dt = \pi/\omega.$$
>
> For $n = 0$, we obtain
>
> $$\int_{-\pi/\omega}^{\pi/\omega} dt = 2\pi/\omega \quad \text{and} \quad \int_{-\pi/\omega}^{\pi/\omega} 0\, dt = 0.$$
>
> (c) The range $-\pi/\omega$ to π/ω may be replaced by any interval of length $2\pi/\omega$.
>
> (26.3)

26.4 Calculating the Fourier coefficients

From (26.1), we expect that any periodic function $P(t)$ having period $2\pi/\omega$ can be expressed in the form of a Fourier series

$$P(t) = \tfrac{1}{2}a_0 + \sum_{n=1}^{\infty} (a_n \cos n\omega t + b_n \sin \omega t). \tag{26.4}$$

To find a particular coefficient a_N, multiply both sides of (24.4) by $\cos N\omega t$:

$$P(t) \cos N\omega t = \tfrac{1}{2}a_0 \cos N\omega t$$

$$+ \sum_{n=1}^{\infty} (a_n \cos n\omega t \cos N\omega t + b_n \sin n\omega t \cos N\omega t).$$

Integrate both sides of this equation between $-\pi/\omega$ and π/ω:

$$\int_{-\pi/\omega}^{\pi/\omega} P(t) \cos N\omega t \, dt = \tfrac{1}{2}a_0 \int_{-\pi/\omega}^{\pi/\omega} \cos N\omega t \, dt$$

$$+ \sum_{n=1}^{\infty} \left(a_n \int_{-\pi/\omega}^{\pi/\omega} \cos n\omega t \cos N\omega t \, dt + b_n \int_{-\pi/\omega}^{\pi/\omega} \sin n\omega t \cos N\omega t \, dt \right). \tag{26.5}$$

(i) *The constant term*
Consider the case $N = 0$. According to (26.3a), all terms under the summation sign in (26.5) are zero; so, after putting $\cos N\omega t = \cos 0 = 1$, we are left with

$$\int_{-\pi/\omega}^{\pi/\omega} P(t) \, dt = \tfrac{1}{2}a_0 \int_{-\pi/\omega}^{\pi/\omega} dt = \frac{\pi}{\omega} a_0.$$

Therefore

$$a_0 = \frac{\omega}{\pi} \int_{-\pi/\omega}^{\pi/\omega} P(t) \, dt \tag{26.6}$$

which is equal to the *average value* of $f(t)$ over a period.

(ii) *The cosine terms*
Suppose that $N \neq 0$. By (26.3), all the integrals on the right of (26.6) are zero except the single one that involves a_N, so (26.6) reduces to

$$\int_{-\pi/\omega}^{\pi/\omega} P(t) \cos N\omega t \, dt = a_N \int_{-\pi/\omega}^{\pi/\omega} \cos^2 N\omega t \, dt = \frac{\pi}{\omega} a_N.$$

Therefore, for $N = 1, 2, 3, \ldots$,

$$a_N = \frac{\omega}{\pi} \int_{-\pi/\omega}^{\pi/\omega} P(t) \cos N\omega t \, \mathrm{d}t. \tag{26.7}$$

By comparing (26.7) with (26.6), it can be seen that a_0 and $a_1, a_2, \ldots$ are all given by the same formula. That is why the constant term in (26.4) is written as $\frac{1}{2}a_0$ instead of a_0.

(iii) *The sine terms*
To find b_N for $N = 1, 2, \ldots$, multiply (26.4) by $\sin N\omega t$ and integrate. In a similar way to that described above, we find that, for $N = 1, 2, 3, \ldots$,

$$b_N = \frac{\omega}{\pi} \int_{-\pi/\omega}^{\pi/\omega} P(t) \sin N\omega t \, \mathrm{d}t. \tag{26.8}$$

Since $P(t)$ is a known function, the integrals in (26.6), (26.7), and (26.8) can be evaluated to give all the coefficients in the Fourier series (26.4).

In the following summary, the letter n is used in place of N to simplify the form of the results.

Fourier series for periodic functions

Function: $P(t)$, period $T = 2\pi/\omega$,

Fourier series: $P(t) = \frac{1}{2}a_0 + \sum_{n=1}^{\infty}(a_n \cos n\omega t + b_n \sin n\omega t)$,

Fourier coefficients:

$$a_n = \frac{\omega}{\pi} \int_{-\pi/\omega}^{\pi/\omega} P(t) \cos n\omega t \, \mathrm{d}t \quad (n = 0, 1, 2, \ldots),$$

$$b_n = \frac{\omega}{\pi} \int_{-\pi/\omega}^{\pi/\omega} P(t) \sin n\omega t \, \mathrm{d}t \quad (n = 1, 2, \ldots)$$

(in place of the range of integration, $-\pi/\omega$ to π/ω, any other one-period interval may be used). $\tag{26.9}$

It can be seen also that since

$$\frac{1}{2}a_0 = \frac{\omega}{2\pi} \int_{-\pi/\omega}^{\pi/\omega} P(t) \, \mathrm{d}t = \frac{1}{T} \int_{-\frac{1}{2}T}^{\frac{1}{2}T} P(t) \, \mathrm{d}t,$$

the following is true:

Average value of $P(t)$

The average value of $P(t)$ over a one-period interval is equal to the constant term $\frac{1}{2}a_0$.   **(26.10)**

Notice the case of period 2π, which often occurs. In such cases $\omega = 2\pi/T = 1$:

Fourier series for functions $P(t)$ with period 2π

$$P(t) = \tfrac{1}{2}a_0 + \sum_{n=1}^{\infty}(a_n \cos nt + b_n \sin nt),$$

where

$$a_n = \frac{1}{\pi}\int_{-\pi}^{\pi} P(t)\cos nt \, dt,$$

$$b_n = \frac{1}{\pi}\int_{-\pi}^{\pi} P(t)\sin nt \, dt.$$

(The integrals may be taken over any one-period interval instead of $[-\pi, \pi]$.)   **(26.11)**

26.5 Examples of Fourier series

The actual calculation of Fourier coefficients requires attention to detail, especially in respect of a_0.

Example 26.2   *Find the Fourier series of the function $P(t)$ shown in Fig. 26.4.*

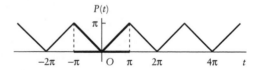

Fig. 26.4

The period is 2π, so that $\omega = 2\pi/2\pi = 1$. Choosing the interval $-\pi$ to π as the basis of the calculation yields

$$P(t) = \begin{cases} -t & \text{if } -\pi \leqslant t \leqslant 0, \\ t & \text{if } 0 \leqslant t \leqslant \pi. \end{cases}$$

Example 26.2 continued

The coefficients can be obtained from (26.11):

Coefficients b_n. $P(t)$ is an even function about the origin (see Section 15.9), and sin nt is odd; therefore $P(t)$ sin nt is odd. Hence the integrals defining b_n are all zero:

$$b_n = 0 \quad (n = 1, 2, \dots). \tag{i}$$

Coefficients a_n. Since $P(t)$ is even and cos nt is even, $P(t)$ cos nt is even; so (26.11) gives

$$a_n = \frac{2}{\pi} \int_0^\pi P(t) \cos nt \, dt = \frac{2}{\pi} \int_0^\pi t \cos nt \, dt$$

$$= \frac{2}{\pi} \left(\left[\frac{t \sin nt}{n} \right]_0^\pi - \int_0^\pi \frac{\sin nt}{n} \, dt \right), \tag{ii}$$

after integrating by parts.

At this point it is seen that $n = 0$, as before, is a case requiring separate treatment (basically because $\int \cos nt \, dt \neq n^{-1} \sin nt + C$ when $n = 0$). Postponing the question of $n = 0$, suppose firstly that $n = 1, 2, \dots$. The formula becomes

$$a_n = \frac{2}{\pi n^2} [\cos nt]_0^\pi = \frac{2}{\pi} \frac{(-1)^n - 1}{n^2}.$$

Therefore

$$a_n = \begin{cases} 0 & \text{if } n \text{ is even,} \\ -4/\pi n^2 & \text{if } n \text{ is odd.} \end{cases} \tag{iii}$$

We still have to find a_0, which is given by (26.11) as

$$a_0 = \frac{2}{\pi} \int_0^\pi t \, dt = \pi. \tag{iv}$$

Collect the coefficients from (i), (iii), and (iv) and put them back into the Fourier series:

$$P(t) = \tfrac{1}{2}\pi - \frac{4}{\pi} \left(\frac{\cos t}{1^2} + \frac{\cos 3t}{3^2} + \frac{\cos 5t}{5^2} + \cdots \right).$$

In Fig. 26.5, we show how $P(t)$ is gradually shaped as we take more and more terms of the Fourier series in Example 26.2. Here

$$P(t) = \tfrac{1}{2}\pi - \frac{4}{\pi} \left(\frac{\cos t}{1^2} + \frac{\cos 3t}{3^2} + \frac{\cos 5t}{5^2} + \cdots \right),$$

$$= 1.571 - 1.273 \cos t - 0.141 \cos 3t - 0.051 \cos 5t - \cdots.$$

Example 26.3 *Find the Fourier series for the function shown in Fig. 26.6.*

The period is $T = 2\pi$, so that $\omega = 1$ and the Fourier series is

$$P(t) = \tfrac{1}{2}a_0 + \sum_{n=1}^{\infty} (a_n \cos nt + b_n \sin nt).$$

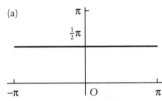

(a)

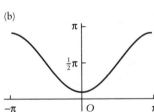

(b)

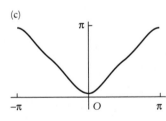

(c)

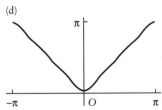

(d)

Fig. 26.5 (a) 1.571;
(b) $1.571 - 1.273 \cos t$;
(c) $1.571 - 1.273 \cos t - 0.141 \cos 3t$;
(d) $1.571 - 1.273 \cos t - 0.141 \cos 3t$
$- 0.051 \cos 5t$.

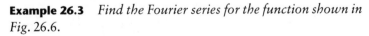

Example 26.3 *continued*

It makes no difference to the ease of calculation whether $-\pi$ to π or 0 to 2π is chosen as the basic interval. We will take 0 to 2π to remind you of the possibility. Then

$$P(t) = \begin{cases} t & (0 \leqslant t \leqslant \pi), \\ 0 & (\pi < t \leqslant 2\pi). \end{cases}$$

Coefficient a_n. From (26.11),

$$a_n = \frac{1}{\pi} \int_0^{2\pi} P(t) \cos nt \, dt = \frac{1}{\pi} \int_0^{\pi} t \cos nt \, dt.$$

Warned by Example 26.2, we deal first with the case $n = 1, 2, 3, \ldots$:

$$a_n = \frac{1}{\pi} \left[\frac{1}{n} t \sin nt + \frac{1}{n^2} \cos nt \right]_0^{\pi}$$

$$= \frac{1}{\pi} \left[\left(0 + \frac{1}{n^2} \cos n\pi \right) - \left(0 + \frac{1}{n^2} \right) \right] = \frac{1}{\pi n^2} [(-1)^n - 1].$$

The sequence has every even-order term zero:

$$a_1 = -\frac{2}{\pi}, \quad a_2 = 0, \quad a_3 = -\frac{2}{\pi 3^2}, \quad a_4 = 0, \quad a_5 = -\frac{2}{\pi 5^2}, \ldots.$$

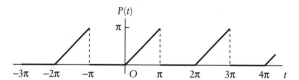

Fig. 26.6

The case $n = 0$ is again special:

$$a_0 = \frac{1}{\pi} \int_0^{2\pi} P(t) \, dt = \frac{1}{\pi} \left[\tfrac{1}{2} t^2 \right]_0^{\pi} = \tfrac{1}{2} \pi.$$

Coefficient b_n.

$$b_n = \frac{1}{\pi} \int_0^{2\pi} P(t) \sin nt \, dt = \frac{1}{\pi} \int_0^{\pi} t \sin nt \, dt = \frac{1}{\pi} \left[-\frac{t \cos nt}{n} + \frac{1}{n^2} \sin nt \right]_0^{\pi}$$

$$= \frac{1}{\pi} \left[-\frac{1}{n} (\pi \cos n\pi - 0) + \frac{1}{n^2} (0 - 0) \right] = -\frac{(-1)^n}{n}.$$

The series is difficult to write if the cosine and sine terms are kept together. By separating them, we obtain

$$P(t) = \tfrac{1}{4} \pi - \frac{2}{\pi} \left(\cos t + \frac{1}{3^2} \cos 3t + \frac{1}{5^2} \cos 5t + \cdots \right)$$

$$+ \left(\sin t - \frac{1}{2} \sin 2t + \frac{1}{3} \sin 3t - \cdots \right).$$

In Example 26.3, the function $P(t)$ jumps from π to zero at the points

$$t = \dots, -\pi, \pi, 3\pi, \dots .$$

To see what values are generated by the *series* at such points put, say, $t = \pi$ into the series we obtained. All the cosine terms become (-1) and all the sine terms are zero, so that at $x = \pi$ the series delivers

$$\frac{1}{4}\pi + \frac{2}{\pi}\left(1 + \frac{1}{3^2} + \frac{1}{5^2} + \cdots\right).$$

A few minutes with a calculator make it clear that this series for $P(\pi)$ cannot add up to π, and plainly it does not give zero either. In fact its sum is $\frac{1}{2}\pi$, half-way between these values. The general rule is as follows.

> **Fourier series at a jump in value of a function**
> The sum of a Fourier series at a jump is equal to the average of the two function values on either side. This is written as
> $$\tfrac{1}{2}[x(t_0^-) + x(t_0^+)].$$
> (26.12)

Figure 26.7 shows how the function is fitted by the series when the six terms up to $\cos 3t$ and $\sin 3t$ are taken.

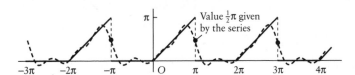

Fig. 26.7

26.6 Use of symmetry: sine and cosine series

In general, the Fourier series for a periodic function will contain both sine and cosine terms. However, the following results hold.

> **Even and odd functions $P(t)$**
> (a) If $P(t)$ is even about the origin, then
> $$b_1 = b_2 = \cdots = 0.$$
> (b) If $P(t)$ is odd about the origin, then
> $$a_0 = a_1 = a_2 = \cdots = 0.$$
> (26.13)

These results follow from (26.9) and (26.11), because $P(t) \sin n\omega t$ is odd if $P(t)$ is even, and $P(t) \cos n\omega t$ is odd if $P(t)$ is odd.

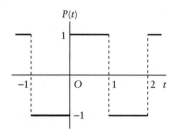

Fig. 26.8

Example 26.4   *Obtain the Fourier series for the* **switching function** *$P(t)$ shown in Fig. 26.8.*

The period T is 2, so that $\omega = \pi$. Choose the basic interval to be $t = -1$ to 1. On this interval,

$$P(t) = \begin{cases} -1 & \text{for } -1 \leqslant t < 0, \\ 1 & \text{for } 0 \leqslant t \leqslant 1. \end{cases}$$

Since $P(t)$ is odd about the origin,

$$a_0 = a_1 = a_2 = \cdots = 0.$$

For the b_n, from (26.9), since the integrands are even functions,

$$b_n = \frac{\pi}{\pi} \int_{-1}^{1} P(t) \sin n\pi t \; dt = 2 \int_{0}^{1} \sin n\pi t \; dt = -\frac{2}{n\pi} [\cos n\pi t]_0^1$$

$$= -\frac{2}{n\pi} [(-1)^n - 1]$$

for $n = 1, 2, \ldots$. The sequence b_n is therefore

$$b_1 = \frac{4}{\pi}, \quad b_2 = 0, \quad b_3 = \frac{4}{\pi} \frac{1}{3}, \quad b_4 = 0, \quad \ldots,$$

and the Fourier series is

$$P(t) = \frac{4}{\pi} \left(\sin \pi t + \frac{1}{3} \sin 3\pi t + \frac{1}{5} \sin 5\pi t + \cdots \right) = \frac{4}{\pi} \sum_{r=1}^{\infty} \frac{\sin(2r-1)\pi t}{2r-1}.$$

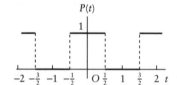

Fig. 26.9

Example 26.5   *Obtain the Fourier series for the switching function $P(t)$ shown in Fig. 26.9.*

The period is 2, so that $\omega = \pi$. Choose $[-1, 1]$ as the representative interval; then

$$P(t) = \begin{cases} 1 & \text{if } -\frac{1}{2} \leqslant t \leqslant \frac{1}{2}, \\ 0 & \text{elsewhere on the interval.} \end{cases}$$

Since $P(t)$ is an even function, $b_1 = b_2 = b_3 = \cdots = 0$. The coefficients a_n are given by

$$a_n = \int_{-1}^{1} P(t) \cos n\pi t \; dt = \int_{-\frac{1}{2}}^{\frac{1}{2}} \cos n\pi t \; dt = \frac{1}{n\pi} [\sin n\pi t]_{-\frac{1}{2}}^{\frac{1}{2}} = \frac{2}{n\pi} \sin \tfrac{1}{2} n\pi.$$

As we have seen before, a_0 gives trouble since this formula is meaningless when $n = 0$. We have, in fact,

$$a_0 = \int_{-\frac{1}{2}}^{\frac{1}{2}} 1 \; dt = 1.$$

Then

$$a_0 = 1, \quad a_1 = 2/\pi, \quad a_2 = 0, \quad a_3 = -2/3\pi, \quad a_4 = 0, \quad \ldots,$$

so that the odd-order coefficients alternate in sign.

Finally the series is

$$P(t) = \tfrac{1}{2} + \frac{2}{\pi} \left(\cos \pi t - \frac{1}{3} \cos 3\pi t + \frac{1}{5} \cos 5\pi t - \cdots \right)$$

$$= \tfrac{1}{2} - \frac{2}{\pi} \sum_{r=1}^{\infty} \frac{(-1)^r \cos(2r-1)\pi t}{2r-1}.$$

26.7 Functions defined on a finite range: half-range series

It is often necessary to obtain a Fourier-type series for a function which is of interest only over some finite interval, and whose natural extension, if any, is not necessarily periodic. The Fourier *series* is invariably periodic, so that it cannot fit a non-periodic function everywhere. For example, consider the problem of finding a Fourier series which will fit $f(t)$, where (Figure 26.10a)

$f(t) = t$ between $t = 0$ and π,

when our *only* concern is whether the series fits $f(t)$ between 0 and π, the behaviour of the series elsewhere being a matter of indifference.

Figure 26.10 illustrates a technique for producing such series. We hold on to the given function inside the interval of interest, but **extend it by means of an artificial function which is periodic.** This extended function will have a Fourier series of its own, and it will agree with $f(t)$ on the interval 0 to π.

In Fig. 26.10b we have extended the non-periodic function $f(t) = t$ on $0 \leqslant t \leqslant \pi$ to an artificial function $f_s(t)$ which has period 2π and is an odd function. Being odd, it has a Fourier series consisting of sine

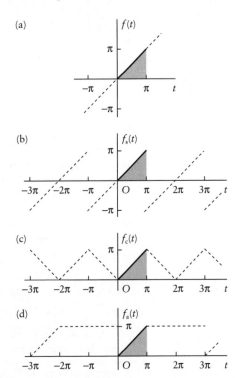

Fig. 26.10 (a) $f(t) = t$ on $0 \leqslant t \leqslant \pi$, with natural non-periodic extension.
(b) $f_s(t) = t$ on $0 \leqslant t \leqslant \pi$, has period 2π and is an odd function.
(c) $f_c(t) = t$ on $0 \leqslant t \leqslant \pi$, has period 2π and is an even function.
(d) $f_a(t) = t$ on $0 \leqslant t \leqslant \pi$, and has an arbitrary extension of period 3π.

terms only, and this odd extension will correctly reproduce $f(t)$ on $0 \leq t \leq \pi$.

Alternatively, Fig. 26.10c shows how to get a series of cosine terms by an *even extension* $f_c(t)$, keeping $f_c(t) = f(t)$ on $0 \leq t \leq \pi$.

Again, Fig. 26.10d shows a fairly arbitrary extension of period 3π, which will have a Fourier series containing both sine and cosine terms. Obviously there is an infinite number of possibilities, the most important being the so-called **half-range sine** and **cosine series**, corresponding to **odd and even extensions** respectively.

Example 26.6   *Obtain a Fourier sine series for $f(t) = t$ on the interval $0 \leq t \leq \pi$.*

Extend $f(t)$ on $0 \leq t \leq \pi$ as an odd function $f_s(t)$ with period 2π (not π) as shown in Fig. 26.10b. Then $\omega = 1$ in (26.9). Choose the interval $-\pi$ to π as basic. Then since $f_s(t)$ is odd, we know in advance from (26.13) that

$$f_s(t) = \sum_{n=1}^{\infty} b_n \sin nt$$

(sine terms only), where

$$b_n = \frac{1}{\pi} \int_{-\pi}^{\pi} f_s(t) \sin nt \, dt$$

$$= \frac{2}{\pi} \int_0^{\pi} f_s(t) \sin nt \, dt \quad \text{(since } f_s(t) \text{ is odd; see (15.17))}$$

$$= \frac{2}{\pi} \int_0^{\pi} f(t) \sin nt \, dt \quad \text{(since } f_s(t) \text{ agrees with } f(t) \text{ on } 0 \leq t \leq \pi)$$

$$= \frac{2}{\pi} \int_0^{\pi} t \sin nt \, dt = \frac{2}{\pi} \left[-\frac{1}{n} t \cos nt + \frac{1}{n^2} \sin nt \right]_0^{\pi}$$

$$= \frac{2}{\pi} \left(-\frac{\pi}{n} \cos n\pi \right) = \frac{2}{n} (-1)^{n+1}.$$

Therefore the required series is

$$\frac{2}{\pi} \left(\sin t - \frac{1}{2} \sin 2t + \frac{1}{3} \sin 3t - \cdots \right) = \frac{2}{\pi} \sum_{n=1}^{\infty} \frac{(-1)^{n+1} \sin nt}{n},$$

and this is equal to t on $0 \leq t < \pi$, but nowhere else. (By (26.12), the value delivered by the series at $t = \pi$ is zero.)

(a) **Half-range cosine series for $0 \leq t \leq \pi$**

$$f(t) = \tfrac{1}{2} a_0 + \sum_{n=1}^{\infty} a_n \cos nt, \qquad a_n = \frac{2}{\pi} \int_0^{\pi} f(t) \cos nt \, dt.$$

(b) **Half-range sine series for $0 \leq t \leq \pi$**

$$f(t) = \sum_{n=1}^{\infty} b_n \sin nt, \qquad b_n = \frac{2}{\pi} \int_0^{\pi} f(t) \sin nt \, dt.$$

(26.14)

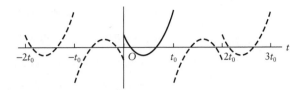

Fig. 26.11 $f(t), 0 \leqslant t \leqslant t_0$; odd extension, $f_s(t)$, period $2t_0$.

Suppose that, more generally, a *sine* series representing $f(t)$ for $0 \leqslant t \leqslant t_0$ is required (Fig. 26.11). Extend $f(t)$ to an *odd* function $f_s(t)$ having period $2t_0$. Then, in eqn (26.9),

$$\omega = 2\pi/2t_0 = \pi/t_0,$$

and, since $f_s(t)$ is odd,

$$f_s(t) = \sum_{n=1}^{\infty} b_n \sin(n\pi t/t_0),$$

where

$$b_n = \frac{1}{t_0} \int_{-t_0}^{t_0} f_s(t) \sin \frac{n\pi t}{t_0} \, dt$$

$$= \frac{2}{t_0} \int_0^{t_0} f(t) \sin \frac{n\pi t}{t_0} \, dt,$$

since $f_s(t) = f(t)$ on the interval 0 to t_0.

If the extension is carried out so as to produce an *even* periodic function, a similar calculation leads to a *cosine* expansion.

(a) **Half-range cosine series for** $0 \leqslant t \leqslant t_0$

$$f(t) = \tfrac{1}{2}a_0 + \sum_{n=1}^{\infty} a_n \cos \frac{n\pi t}{t_0}, \qquad a_n = \frac{2}{t_0} \int_0^{t_0} f(t) \cos \frac{n\pi t}{t_0} \, dt.$$

(b) **Half-range sine series for** $0 \leqslant t \leqslant t_0$

$$f(t) = \sum_{n=1}^{\infty} b_n \sin \frac{n\pi t}{t_0}, \qquad b_n = \frac{2}{t_0} \int_0^{t_0} f(t) \sin \frac{n\pi t}{t_0} \, dt.$$

(26.15)

26.8 Spectrum of a periodic function

Suppose that $P(t)$ is a periodic function with period T. The Fourier series has the form

$$P(t) = \tfrac{1}{2}a_0 + \sum_{n=1}^{\infty} (a_n \cos n\omega t + b_n \sin n\omega t)$$

where $\omega = 2\pi/T$. By the identity (1.18),

$$a_n \cos n\omega t + b_n \sin n\omega t = c_n \cos(n\omega t + \phi_n),$$

where ϕ_n is a phase angle, and

$$c_n = \sqrt{(a_n^2 + b_n^2)} \quad (n = 1, 2, \dots),$$

which is the (positive) amplitude or strength of the nth term. For completeness, we include

$$c_0 = \tfrac{1}{2}|a_0|.$$

The sequence $c_0, c_1, c_2, \dots$ is called the **spectrum** of $P(t)$. If the series consists only of cosine (or sine) terms, then correspondingly $c_n = \sqrt{(a_n^2)} = |a_n|$ (or $|b_n|$) – the **spectral components** are always positive or zero.

The spectrum can be displayed as if it were a physical spectrum. Figure 26.12 shows the spectrum of the function worked out in Example 26.5. The property which makes the spectrum a useful concept is that **the spectrum is independent of the time origin of t,** although the Fourier series itself is not. If

$$P(t) = \tfrac{1}{2}a_0 + \sum_{n=1}^{\infty} c_n \cos(n\omega t + \phi_n),$$

then the series for $P(t - t_0)$, whose graph is the same shape as $P(t)$ but moved to the right a distance t_0, is

$$P(t - t_0) = \tfrac{1}{2}a_0 + \sum_{n=1}^{\infty} c_n \cos[n\omega(t - t_0) + \phi_n].$$

The c_n remain the same, and only the phase angle changes. Therefore it is only the shape of $P(t)$ which determines its spectrum, not its clock-timing. For this reason the spectral or harmonic composition of a piano note is always the same, independently of what time of day the note is played.

It is important to realize that the spectrum refers only to a *complete periodic function*, and not to an isolated segment such as those discussed in Section 26.7. For the functions shown in Fig. 26.10, there correspond different Fourier series which have different spectra.

26.9   Obtaining one Fourier series from another

There exist 'dictionaries' of Fourier series, but the entries cannot match exactly all the functions required in practice. If the broad shape of the dictionary entry is the same as that of the function whose series is needed, then scaling or translation along the t axis, or along the axis of $P(t)$, might be all that is required. The transition can require more than one stage.

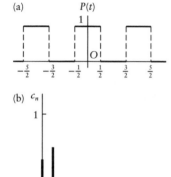

Fig. 26.12 (See Example 26.5.)
(a) $P(t) = \tfrac{1}{2} + (2/\pi)(\cos \pi t + \tfrac{1}{3}\cos 3\pi t + \cdots)$.
(b) Spectral components $\tfrac{1}{2}, 2/\pi, 2/(3\pi), 2/(5\pi), \dots$.

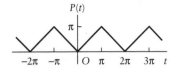

Fig. 26.13

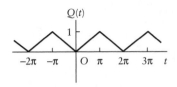

Fig. 26.14

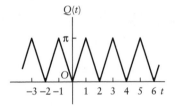

Fig. 26.15

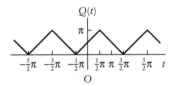

Fig. 26.16

The examples which follow are based on the standard form shown in Fig. 26.13, which was expressed as a Fourier series in Example 26.2:

$$P(t) = \tfrac{1}{2}\pi - \frac{4}{\pi}\left(\frac{\cos t}{1^2} + \frac{\cos 3t}{3^2} + \frac{\cos 5t}{5^2} + \cdots \right). \tag{26.16}$$

Example 26.7 *Find the Fourier expansion of the function $Q(t)$ shown in Fig. 26.14.*

This is the same as Fig. 26.13 except that the vertical dimension is reduced by a factor $1/\pi$. Therefore, from (26.16),

$$Q(t) = \tfrac{1}{2} - \frac{4}{\pi^2}\left(\cos t + \frac{1}{3^2}\cos 3t + \cdots \right).$$

Example 26.8 *Find the Fourier expansion of the function $Q(t)$ shown in Fig. 26.15.*

Here the t scale is changed by a factor π. We obtain, from (26.16),

$$Q(t) = \tfrac{1}{2}\pi - \frac{4}{\pi}\left(\cos \pi t + \frac{1}{3^2}\cos 3\pi t + \frac{1}{5^2}\cos 5\pi t + \cdots \right).$$

It is necessary to be careful here: it is not t/π but πt in the new series. Check the period: it is equal to 2, which is correct.

Example 26.9 *Find the Fourier expansion of $Q(t)$ in Fig. 26.16.*

The graph of $P(t)$ in Fig. 26.13 has been shifted a distance $\tfrac{1}{2}\pi$ to the left (see Fig. 26.16). Therefore

$$Q(t) = P(t + \tfrac{1}{2}\pi)$$

$$= \tfrac{1}{2}\pi - \frac{4}{\pi}\left(\cos(t + \tfrac{1}{2}\pi) + \frac{1}{3^2}\cos 3(t + \tfrac{1}{2}\pi) + \frac{1}{5^2}\cos 5(t + \tfrac{1}{2}\pi) + \cdots \right).$$

As n goes through the sequence 1, 3, 5, 7, ... , $\cos \tfrac{1}{2}n\pi = 0$ and $\sin \tfrac{1}{2}n\pi$ becomes the alternating sequence 1, −1, 1, −1, ... Therefore

$$Q(t) = \tfrac{1}{2}\pi - \frac{4}{\pi}\left(-\sin t + \frac{1}{3^2}\sin 3t - \frac{1}{5^2}\sin 5t - \cdots \right).$$

26.10 The two-sided Fourier series

Equations (26.9) define the Fourier series in terms of **circular frequency** ω, where $\omega = 2\pi/T$, and T is the period. For the rest of the chapter we shall instead use the **fundamental frequency** f_0 (complete cycles per unit time), since it will simplify the subsequent development of Fourier transforms in Chapter 27. We then have

$$T = \frac{1}{f_0} \quad \text{and} \quad \omega = 2\pi f_0.$$

In terms of f_0, (26.9) becomes

Fourier series in terms of frequency f_0

$x_P(t)$ a real or complex function with period $T = 1/f_0$.

(a) *Fourier series*

$$x_P(t) = \tfrac{1}{2}a_0 + \sum_{n=1}^{\infty}(a_n \cos 2\pi n f_0 t + b_n \sin 2\pi n f_0 t).$$

(b) *Coefficients*

$$a_n = 2f_0 \int_{\text{Period}} x_P(t) \cos 2\pi n f_0 t \, dt$$

$$b_n = 2f_0 \int_{\text{Period}} x_P(t) \sin 2\pi n f_0 t \, dt.$$

(26.17)

We shall now show that (26.17) may be reorganized into another shape, as follows:

The two-sided Fourier series

$x_P(t)$ a real or complex function with period $T = 1/f_0$.

(a) *Two-sided series*

$$x_P(t) = \sum_{n=-\infty}^{\infty} X_n e^{i2\pi n f_0 t}$$

(b) *Coefficients X_n*

$$X_n = f_0 \int_{\text{Period}} x_P(t) \, e^{-i2\pi n f_0 t} \, dt.$$

(26.18)

The coefficients are in general complex even if $x_P(t)$ is real, and the series runs from $n = -\infty$ to $n = \infty$.

To prove (26.18) we shall work backwards from it to arrive at (26.17). Start with (26.18a):

$$x_P(t) = X_0 + \sum_{n=1}^{\infty} X_n e^{i2\pi n f_0 t} + \sum_{n=-\infty}^{-1} X_n e^{i2\pi n f_0 t}$$

$$= X_0 + \sum_{n=1}^{\infty} X_n e^{i2\pi n f_0 t} + \sum_{n=1}^{\infty} X_{-n} e^{-i2\pi n f_0 t},$$

(26.19)

changing the counting index n to $(-n)$ in the final term.

From (26.18b), for n positive and negative,

$$X_n = f_0 \int_{\text{Period}} x_{\text{P}}(t)[\cos 2\pi n f_0 t - \text{i} \sin 2\pi n f_0 t]\, \text{d}t = \tfrac{1}{2}(a_n - \text{i}b_n),$$

$$(26.20)$$

where

$$\left.\begin{aligned} a_n &= 2f_0 \int_{\text{Period}} x_{\text{P}}(t) \cos 2\pi n f_0 t\, \text{d}t, \\ b_n &= 2f_0 \int_{\text{Period}} x_{\text{P}}(t) \sin 2\pi n f_0 t\, \text{d}t. \end{aligned}\right\}$$

$$(26.21)$$

Therefore

$$a_{-n} = a_n \quad \text{and} \quad b_{-n} = -b_n. \tag{26.22}$$

It follows from (26.20) and (26.22) that when $n \geqslant 0$, as in the sums (26.19),

$$X_n = \tfrac{1}{2}(a_n - \text{i}b_n), \qquad X_{-n} = \tfrac{1}{2}(a_{-n} - \text{i}b_{-n}) = \tfrac{1}{2}(a_n + \text{i}b_n), \qquad (26.23)$$

where a_n and b_n are the *same numbers as the coefficients in the original series* (26.17).

Finally, (26.19) becomes

$$x_{\text{P}}(t) = \tfrac{1}{2}a_0 + \sum_{n=1}^{\infty} [\tfrac{1}{2}(a_n - \text{i}b_n)\, \text{e}^{\text{i}2\pi n f_0 t} + \tfrac{1}{2}(a_n + \text{i}b_n)\, \text{e}^{-\text{i}2\pi n f_0 t}].$$

After using Euler's formula (6.8) for the exponentials, and carrying out the multiplications, the terms in which i appear cancel, and we are left with

$$x_{\text{P}}(t) = \tfrac{1}{2}a_0 + \sum_{n=1}^{\infty} (a_n \cos 2\pi n f_0 t + b_n \sin 2\pi n f_0 t),$$

which is the original form (26.17a). (Since $x_{\text{P}}(t)$ may be complex, so may a_n and b_n, so we should not shorten the final calculation by taking twice the real part of $\tfrac{1}{2}(a_n - \text{i}b_n)\, \text{e}^{\text{i}2\pi n f_0 t}$.)

The following properties sometimes save calculation:

Properties of X_n in the two-sided Fourier series

(a) $X_n = \tfrac{1}{2}(a_n - \text{i}b_n)$, and $X_{-n} = \tfrac{1}{2}(a_n + \text{i}b_n)$, where a_n and b_n are derived as in (26.17).

(b) If $x_{\text{P}}(t)$ is real, then $X_{-n} = \bar{X}_n$.

(c) If $x_{\text{P}}(t)$ is real and even, X_n is real and $X_{-n} = X_n$.

(d) If $x_{\text{P}}(t)$ is real and odd, X_n is pure imaginary and $X_{-n} = -X_n$.

$$(26.24)$$

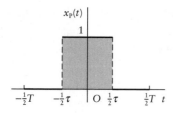

Fig. 26.17

Example 26.10 *Obtain the two-sided Fourier series for the function* $x_p(t)$, *having period* T, *of which a single period is shown in Fig. 26.17.*

In (26.18) $f_0 = 1/T$. Therefore

$$X_n = \frac{1}{T}\int_{-\frac{1}{2}T}^{\frac{1}{2}T} x_p(t)\,e^{-i2\pi nt/T}\,dt = \frac{1}{T}\int_{-\frac{1}{2}\tau}^{\frac{1}{2}\tau} e^{-i2\pi nt/T}\,dt$$

$$= \frac{1}{T}\frac{T}{(-i2\pi n)}[e^{-i2\pi nt/T}]_{-\frac{1}{2}\tau}^{\frac{1}{2}\tau} = \frac{-i}{2\pi n}(e^{i\pi nt/T} - e^{-i\pi nt/T})$$

$$= \frac{1}{\pi n}\sin(\pi n\tau/T) \quad \text{(from (6.10)).}$$

Finally

$$x_p(t) = \sum_{n=-\infty}^{\infty} \frac{1}{\pi n}\sin\frac{\pi n\tau}{T}\,e^{i\pi nt/T}.$$

Problems

26.1 Draw a sketch of the following odd 2π-periodic functions defined for $-\pi < t \le \pi$, and find a general formula for their Fourier coefficients:

(a) $f(t) = \begin{cases} -1 & (-\pi < t < 0), \\ 1 & (0 \le t \le \pi); \end{cases}$

(b) $f(t) = t \quad (-\pi < t \le \pi);$

(c) $f(t) = \begin{cases} -t^2 & (-\pi < t < 0), \\ t^2 & (0 \le t \le \pi); \end{cases}$

(d) $f(t) = \begin{cases} e^t - 1 & (-\pi < t < 0), \\ -(e^t - 1) & (0 \le t \le \pi); \end{cases}$

(e) $f(t) = \begin{cases} 1 & (-\pi < t \le -\frac{1}{2}\pi), \\ -1 & (-\frac{1}{2}\pi < t \le 0), \\ 1 & (0 < t \le \frac{1}{2}\pi), \\ -1 & (\frac{1}{2}\pi < t \le \pi). \end{cases}$

26.2 Draw a sketch of the following even 2π-periodic functions defined for $-\pi < t \le \pi$, and find a general formula for their Fourier coefficients:

(a) $f(t) = \begin{cases} -1 & (-\pi < t \le \frac{1}{2}\pi), \\ 1 & (-\frac{1}{2}\pi < t \le \frac{1}{2}\pi), \\ -1 & (\frac{1}{2}\pi < t \le \pi); \end{cases}$

(b) $f(t) = t^2;$

(c) $f(t) = \cos\frac{1}{2}t.$

26.3 Draw a sketch of the following 2π-periodic functions defined for $-\pi < t \le \pi$, and find a general formula for their Fourier coefficients:

(a) $f(t) = \begin{cases} 0 & (-\pi < t < 0), \\ t & (0 \le t \le \pi); \end{cases}$

(b) $f(t) = \begin{cases} t + \pi & (-\pi < t < 0), \\ t & (0 < t \le \pi). \end{cases}$

26.4 A half-rectified sine wave is given by the 2π-periodic function

$$f(t) = \begin{cases} 0 & (-\pi < t \le 0), \\ \sin t & (0 < t \le \pi). \end{cases}$$

Find the Fourier series of $f(t)$.

26.5 Show that the Fourier series of the 2π-periodic function

$$f(t) = \begin{cases} 0 & (-\pi < t \le 0), \\ 1 & (0 < t \le \pi), \end{cases}$$

is

$$\frac{1}{2} + \frac{2}{\pi}\left(\sin t + \frac{1}{3}\sin 3t + \frac{1}{5}\sin 5t + \cdots\right).$$

What value does the Fourier series take at $t = 0$?

By choosing a particular value of t, find the sum of the series

$$1 - \frac{1}{3} + \frac{1}{5} - \frac{1}{7} + \cdots.$$

26.6 A signal $F \sin t$ with amplitude $F > 0$ is fully rectified into $F|\sin t|$. Find the Fourier series of the rectified signal. What is the amplitude of its first harmonic?

26.7 A Fourier series is given by

$$\sum_{n=1}^{\infty} \frac{n + a}{n^3 + an + 3} \sin nt,$$

where a is a design parameter in the system. Find a in order that the leading harmonics $n = 1$ and $n = 2$ have amplitudes in the ratio $2 : 1$. What is the amplitude of the next harmonic?

26.8 The two 2π-periodic signals shown in Fig. 26.18 are added. Find the Fourier series of the combined signal. What value should F take in order that the leading harmonic should disappear?

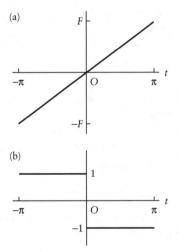

Fig. 26.18

26.9 A T-periodic function is defined by

$$Q(t) = \tfrac{1}{4}T^2 - t^2 \quad \text{for } -\tfrac{1}{2}T \leqslant t \leqslant \tfrac{1}{2}T.$$

Find the Fourier series of $Q(t)$. What is the error between the sum of the first four terms of the series and $Q(t)$ at (a) $t = 0$, (b) $t = \tfrac{1}{4}T$?

26.10 A 2π-periodic function is defined by

$$f(t) = \begin{cases} \beta t(\pi - t) & (0 < t \leqslant \pi), \\ \beta t(\pi + t) & (-\pi < t \leqslant 0). \end{cases}$$

Find the Fourier series for $f(t)$. What is the ratio of the amplitudes of the third and first harmonics? Compare the values of $f(t)$ and the Fourier series up to and including the coefficient b_3 at $t = \tfrac{1}{2}\pi$.

26.11 Find the Fourier series of the 2π-periodic function defined by $f(t) = t(\pi^2 - t^2)$ for $-\pi < t \leqslant \pi$. Find the derivative of $f(t)$ for $-\pi < t \leqslant \pi$ and find its Fourier series. Confirm that the derivative of the Fourier series of $f(t)$ obtained by term-by-term differentiation is the same series as the Fourier series of $f'(t)$.

Consider now the function $g(t) = t^3$ defined for $-\pi < t \leqslant \pi$. Find the Fourier series of $g(t)$ and $g'(t)$. Confirm that the derivative of the Fourier series of $g(t)$ is *not* the same series as the Fourier series of $g'(t)$.

Comparing the functions of $f(t)$ and $g(t)$, what feature of $g(t)$ do you think causes the problem with its differentiated Fourier series?

26.12 Sketch the rectified sine wave defined by

$$P(t) = \begin{cases} 0 & (-\pi \leqslant t \leqslant 0), \\ |\sin 2t| & (0 < t \leqslant \pi), \end{cases}$$

extended so as to have period 2π. Find its Fourier series. (The identities

$$\sin A \cos B = \tfrac{1}{2}[\sin(A + B) + \sin(A - B)],$$
$$\sin A \sin B = \tfrac{1}{2}[-\cos(A + B) + \cos(A - B)],$$

will be needed.)

26.13 Show that

$$t = 2 \sum_{n=1}^{\infty} \frac{(-1)^{n-1}}{n} \sin nt$$

for $-\pi < t \leqslant \pi$. Integrate the terms from $t = 0$ to $t = x$, and rearrange them to show that

$$x^2 = 4 \sum_{n=1}^{\infty} \frac{(-1)^{n-1}}{n^2} - 4 \sum_{n=1}^{\infty} \frac{(-1)^{n-1}}{n^2} \cos nx.$$

Now use (26.10) to establish the value of the constant term in this Fourier series.

26.14 From Problem 26.13, or by direct means, obtain the Fourier series valid for $-\pi \leq t \leq \pi$:

$$t^2 = \tfrac{1}{3}\pi^2 + 4\sum_{n=1}^{\infty} \frac{(-1)^n}{n^2}\cos nt.$$

By integrating all the terms in this expression from $t = 0$ to $t = x$, obtain a Fourier series for $x^3 - \pi^2 x$. (It is always valid to integrate a Fourier series in this way in order to obtain a new one, but differentiation of the terms does not always lead to a valid series.)

26.15 Obtain the Fourier series of the function having period T which is defined for $-\tfrac{1}{2}T \leq t < \tfrac{1}{2}T$ by

$$P(t) = \begin{cases} -2t & (-\tfrac{1}{2}T \leq t < 0), \\ 2t & (0 \leq t < \tfrac{1}{2}T). \end{cases}$$

Display its spectrum as in Fig. 26.12.

26.16 The function $f(t)$ is defined on the interval $0 \leq t \leq 1$ by $f(t) = 1$. Express $f(t)$ as a half-range Fourier series on $0 \leq t \leq 1$, (a) as a sine series, (b) as a cosine series. Sketch the sum of the *series* on $-\infty < t < \infty$ in both cases.

26.17 The function $f(t)$ is defined on $0 \leq t \leq 1$ by $f(t) = t$. Express $f(t)$ as a half-range Fourier series on $0 \leq t \leq 1$, (a) as a cosine series, (b) as a sine series.

26.18 Express $f(t) = \sin \omega t$ on $0 \leq t \leq \pi/\omega$ as a half-range cosine series. Sketch the sum of the *series* on $-\infty < t < \infty$.

26.19 Express $f(t) = \cos \omega t$ on $0 \leq t \leq \pi/\omega$ as a half-range sine series. Sketch the sum of the *series* on $-\infty < t < \infty$.

26.20 Express $f(t) = \cos t$ on $0 \leq t \leq 2\pi$ as a half-range sine series.

26.21 Express $f(t) = \cos t$ on $0 \leq t \leq 2\pi$ as a half-range cosine series.

26.22 Express the function $f(t)$, for $0 \leq t \leq \pi$, (a) as a half-range sine series, (b) as a half-range cosine series:

$$f(t) = \begin{cases} 1 & (0 \leq t < \tfrac{1}{2}\pi), \\ 0 & (\tfrac{1}{2}\pi \leq t \leq \pi). \end{cases}$$

26.23 The Fourier series for the function $P(t)$, period 2π, given by

$$P(t) = \begin{cases} -t & (-\pi \leq t \leq 0), \\ t & (0 \leq t \leq \pi), \end{cases}$$

is

$$P(t) = \tfrac{1}{2}\pi - \frac{4}{\pi}\left(\frac{\cos t}{1^2} + \frac{\cos 3t}{3^2} + \frac{\cos 5t}{5^2} + \cdots\right)$$

(see Example 26.2). Deduce from this the Fourier expansions of the following periodic functions.

(a) $Q(t)$, period 4, where

$$Q(t) = \begin{cases} -3t & (-2 \leq t \leq 0), \\ 3t & (0 \leq t \leq 2); \end{cases}$$

(b) $R(t)$, period 2, where

$$R(t) = \begin{cases} 1+t & (-1 \leq t \leq 0), \\ 1-t & (0 \leq t \leq 1). \end{cases}$$

(Sketch $R(t)$ to understand the connection with $P(t)$.)

(c) Check that $P(t)$, $Q(t)$, $R(t)$ have similar spectra.

26.24 The Fourier series for the function $P(t)$, period 2, given by

$$P(t) = \begin{cases} -1 & (-1 \leq t < 0), \\ 1 & (0 \leq t < 1), \end{cases}$$

for one period, is

$$P(t) = \frac{2}{\pi}\left(\sin \pi t + \frac{1}{3}\sin 3\pi t + \frac{1}{5}\sin 5\pi t + \cdots\right)$$

(see Example 26.4).

Deduce the expansion of the function $Q(t)$, period T, defined on one period by

$$Q(t) = \begin{cases} -a & (0 \leq t < \tfrac{1}{2}T), \\ a & (\tfrac{1}{2}T \leq t < T). \end{cases}$$

26.25 Find the Fourier series of the 2π-periodic saw-tooth wave defined by

$$f(t) = t \quad (-\pi \leq t \leq \pi).$$

Determine the *forced* part of the solution of the second-order differential equation

$$\frac{d^2 x}{dt^2} + \Omega^2 x = K \sin \omega t,$$

where $\omega \neq \pm\Omega$. Hence put together the periodic output of the forced system

$$\frac{d^2 x}{dt^2} + \Omega^2 x = f(t),$$

where $f(t)$ is the sawtooth wave above. For what values of Ω does the system exhibit resonance?

26.26 The Fourier series of a function with period T is given by

$$f(t) = \tfrac{1}{2}a_0 + \sum_{n=1}^{\infty}(a_n \cos \omega t + b_n \sin \omega t),$$

where $T = 2\pi/\omega$. Multiply both sides of the equation by $f(t)$ and integrate between $-\frac{1}{2}T$ and $\frac{1}{2}T$, to obtain **Parseval's identity**

$$\frac{2}{T} \int_{-\frac{1}{2}T}^{\frac{1}{2}T} f(t)^2 \, dt = \frac{1}{2}a_0^2 + \sum_{n=1}^{\infty} (a_n^2 + b_n^2).$$

(a) Let $T = \pi$, and

$$f(t) = \begin{cases} -1 & (-\frac{1}{2}\pi < t \leq 0), \\ 1 & (0 < t \leq \frac{1}{2}\pi). \end{cases}$$

Show that

$$\sum_{n=1}^{\infty} \frac{1}{(2n+1)^2} = \frac{\pi^2}{8}.$$

(b) Let $f(t) = t$ $(-\pi < t \leq \pi)$ be a 2π-periodic function. Find its Fourier series (see Problem 26.1b), and deduce the corresponding Parseval identity.

26.27 The function $f(t)$ with period T has the Fourier series

$$f(t) = \frac{1}{2}a_0 + \sum_{n=1}^{\infty} (a_n \cos n\omega t + b_n \sin n\omega t).$$

Find the Laplace transform of the function as the sum of a series of Laplace transforms of the trigonometric terms. Hence find the Laplace transform of the 2π-periodic function defined by

$$f(t) = \begin{cases} -t^2 & (-\pi < t < 0), \\ t^2 & (0 \leq t \leq \pi). \end{cases}$$

(See Problem 26.1c.)

26.28 A radio wave described by

$$x(t) = a \cos \omega t \cos \omega_0 t,$$

where ω_0 is very much greater than ω, represents a **carrier wave** subject to amplitude modulation by the comparatively slowly varying term $\cos \omega t$, which represents a musical note. Roughly sketch the general character of $x(t)$.

(a) If $\omega = 500$ and $\omega_0 = 100\,001$ (notice the 1 at the end), what is the period of $x(t)$?

(b) Let $\omega = p/q$ and $\omega_0 = r/s$, where p, q, r, s are whole numbers such that no two of them have a common divisor other than 1. What is the period? Express $x(t)$ as the sum of two waves with angular frequencies $\omega \pm \omega_0$ (these are called the sidebands). What is the Fourier cosine expansion based on this period?

(c) If you know about irrational numbers, show that $x_1(t) = \cos t \cos \sqrt{2}t$ never repeats itself *exactly*: it is not periodic.

26.29 (a) Prove that

$$\int_{-\frac{1}{2}T}^{\frac{1}{2}T} e^{i2\pi n f_0 t} e^{-i2\pi m f_0 t} \, dt = \begin{cases} T, & m = n, \\ 0, & m \neq n. \end{cases}$$

(b) Confirm that the expansion (26.18) is valid by multiplying both sides of (26.18a) by $e^{-i2\pi N f_0 t}$ and integrating the result over one period.

26.30 Obtain the two-sided Fourier series for the sawtooth function $x_p(t)$ defined by $x_p(t) = t/T$ for $0 < t < T$, together with its periodic extension of period T.

27 Fourier transforms

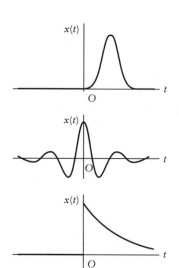

Fig. 27.1 Three non-periodic functions.

27.1 Sine and cosine transforms

Figure 27.1 shows examples of functions that are not periodic. **Non-periodic functions** can still be expressed in terms of harmonic functions (sines, cosines, and their complex exponential forms), but instead of an infinite series of harmonic terms associated with discrete frequencies $f_0, 2f_0, 3f_0, \ldots$, where f_0 is the fundamental frequency, an infinite *integral* over a *continuum* of frequencies is required.

A full derivation of such results is too complicated to give in this book, but representation by a continuous distribution of frequencies can be made plausible by regarding a non-periodic function as being equivalent to a periodic function with an infinite period. To illustrate this approach we shall consider a simple case.

Let $p_O(t)$ be a real-valued function for $-\infty < t < \infty$, periodic with period T, and odd (i.e. $p_O(-t) = -p_O(t)$). Further, suppose that it consists of a stream of discrete, equally spaced 'pulses', each of duration $\tau < T$, and has the value zero between them, as illustrated in Fig. 27.2. The function $p_O(t)$ can be represented by a **Fourier sine series** (see Section 26.6). Up to this point we have expressed Fourier series in terms of the circular frequency $\omega = (2\pi/T)$, but here we shall use the **fundamental frequency** $f_0 = 1/T$ instead. Then (26.13b) becomes

$$p_O(t) = \sum_{n=0}^{\infty} b_n \sin(2\pi n f_0 t), \tag{27.1}$$

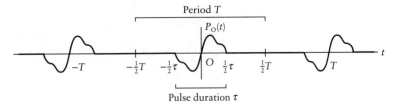

Fig. 27.2

where

$$b_n = 2f_0 \int_{-\frac{1}{2}T}^{\frac{1}{2}T} p_O(t) \sin(2\pi n f_0 t) \, \mathrm{d}t = 4f_0 \int_0^{\frac{1}{2}T} p_O(t) \sin(2\pi n f_0 t) \, \mathrm{d}t \quad (27.2)$$

(since the integrand is an even function).

We shall seek a representation of the fixed single pulse present in the interval $-\frac{1}{2}T < t < \frac{1}{2}T$ by letting $T \to \infty$. This pushes away to infinity the periodic copies of the central pulse, whilst leaving the central pulse unaffected. (In a physical context, as in passing a solitary pulse through an electrical filter, we should be likely to disregard extraneous pulses that arrive only every hour, or every month, or every century, as the period T is taken larger and larger.)

The series (27.1) becomes increasingly intractable as $T \gg \tau$ (meaning T 'is much greater than' τ); too many terms have to be taken in order to get a reasonable approximation to $p_O(t)$. However, we can recast the series as an integral, and this problem disappears. Put

$$\frac{b_n}{2f_0} = 2 \int_0^{\frac{1}{2}T} p_O(t) \sin(2\pi n f_0 t) \, \mathrm{d}t = X_s(nf_0) \quad (27.3)$$

from (27.2). As $T \to \infty$, $f_0 = 1/T \to 0$. Also, the successive frequency components nf_0 are separated by a distance f_0. Now, for f_0 small, write

$$f_0 = \delta f, \qquad n f_0 = f_n.$$

Equation (27.1) becomes

$$p_O(t) = 2\delta f \sum_{n=0}^{\infty} X_s(f_n) \sin(2\pi f_n t). \quad (27.4)$$

When $T \to \infty$ so that $\delta f \to 0$ the periodic copies of the central pulse are consigned to infinity, and we are left with a solitary pulse $x(t)$ given by

$$x(t) = \begin{cases} p_O(t), & -\frac{1}{2}\tau < t < \frac{1}{2}\tau; \\ 0, & \text{elsewhere.} \end{cases}$$

At the same time (by eqn (15.9)) the sum in (27.4) approaches an infinite integral, as does the finite integral in (27.3). We then have the symmetrical pair of relations:

Fourier sine transform

(a) $x(t) = 2 \int_0^\infty X_s(f) \sin(2\pi f t)\, df,$

where

(b) $X_s(f) = 2 \int_0^\infty x(t) \sin(2\pi f t)\, dt.$

(27.5)

The function $X_s(f)$ is called the **Fourier sine transform** of $x(t)$, or the **spectral density** or **frequency distribution function** corresponding to $x(t)$, in the context of sine transforms. All positive frequencies are represented. Notice that (27.5a) automatically defines $x(t)$ as an odd function if the context demands that we be concerned with the time range $-\infty < t < \infty$. The normal use for the sine transform is, however, for $t \geq 0$ only.

If we start with an *even* periodic chain of pulses $p_E(t)$ and its Fourier *cosine* series, we arrive similarly at the cosine transform pair:

Fourier cosine transform

(a) $x(t) = 2 \int_0^\infty X_c(f) \cos(2\pi f t)\, df,$

where

(b) $X_c(f) = 2 \int_0^\infty x(t) \cos(2\pi f t)\, dt.$

(27.6)

The equations (27.5a) and (27.6a) are also known as the **inverse transforms** of $X_s(f)$ and $X_c(f)$. As with Laplace transforms, they solve the problem: 'given a frequency distribution, obtain the corresponding time function'.

To arrive at the sine and cosine equations we assumed that the **signal** $x(t)$ consists of a pulse of finite extent τ. However, the results are true for suitably behaved functions having infinite extent, from $t = 0$ to $t = \infty$ (or from $t = -\infty$ to ∞ provided that they are appropriately odd or even functions). Thus $x(t) = e^{-t}$ for $t \geq 0$ has both a sine and a cosine transform for $t \geq 0$.

As in eqn (26.12) for Fourier series, the value attributed to $x(t)$ by (27.5) and (27.6) is the average of its values on either side of a **jump discontinuity** at $t = t_0$:

$$x(t) = \tfrac{1}{2}[x(t_0^-) + x(t_0^+)]$$

(27.7)

(in the notation of (26.12)).

Example 27.1 (a) *Obtain the cosine transform of the function* $x(t)$ *given by*

$$x(t) = \begin{cases} 1, & 0 \leqslant t \leqslant 1, \\ 0, & t > 1, \end{cases}$$

(*see Fig. 27.3a*) *and write down the inverse transform (without evaluating it).*
(b) *Deduce that*

$$\int_0^\infty \frac{\sin u}{u}\, du = \tfrac{1}{2}\pi.$$

(c) *Show that the value attributed to* $x(1)$, *a point of discontinuity of* $x(t)$, *conforms with eqn* (27.7).

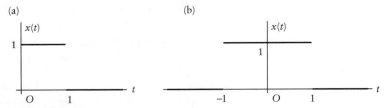

(a) (b)

Fig. 27.3

(a) From (27.6b)

$$X_c(f) = 2\int_0^\infty x(t)\,\cos(2\pi ft)\,dt = 2\int_0^1 \cos(2\pi ft)\,dt$$

$$= \frac{1}{\pi f}[\sin(2\pi ft)]_0^1 = \frac{\sin(2\pi f)}{\pi f}.$$

The inverse transform is

$$2\int_0^\infty \frac{\sin(2\pi f)}{\pi f}\cos(2\pi ft)\,df. \tag{i}$$

This is equal to $x(t)$ at points of continuity of $x(t)$.

(b) By putting $(-t)$ for t into (i), $x(t)$ is extended to an even function (i.e. $x(-t) = x(t)$) on $-\infty < t < \infty$: see Fig. 27.3b). The point $t = 0$ is therefore a point of continuity of $x(t)$ on $-\infty < t < \infty$, so that

$$x(0) = 2\int_0^\infty \frac{\sin 2\pi f}{\pi f}\,df = 1. \tag{ii}$$

Substitute $u = 2\pi f$, $df/f = du/u$, and we obtain from (ii) the standard integral

$$\int_0^\infty \frac{\sin u}{u}\,du = \tfrac{1}{2}\pi. \tag{iii}$$

(c) The point $t = 1$ marks a jump in value of $x(t)$ from 1 to 0. Equation (27.7) predicts that the integral (i) will deliver the value $x(1) = \tfrac{1}{2}(1 + 0) = \tfrac{1}{2}$. To confirm this, put $t = 1$ in eqn (i):

Example 27.1 *continued*

$$2 \int_0^\infty \frac{\sin 2\pi f}{\pi f} \cos 2\pi f \, df = \int_0^\infty \frac{\sin 4\pi f}{\pi f} \, df.$$

Put $u = 4\pi f$, $df/f = du/u$; then by using the result (iii):

$$\int_0^\infty \frac{\sin 4\pi f}{\pi f} \, df = \frac{1}{\pi} \int_0^\infty \frac{\sin v}{v} \, dv = \frac{1}{\pi} \cdot \tfrac{1}{2}\pi = \tfrac{1}{2},$$

as predicted.

27.2 The exponential Fourier transform

The sine and cosine transforms have limited usefulness for the applications in this chapter since they refer only to functions $x(t)$ that are odd or even, unless we restrict the interval of t to $t \geqslant 0$. Functions on $-\infty < t < \infty$ are in general neither odd nor even, and the following transform pair is free from these limitations:

> **The complex exponential Fourier transform pair**
>
> $x(t)$ is any well-behaved real or complex function on $-\infty < t < \infty$ such that $\int_{-\infty}^{\infty} |x(t)| \, dt$ exists. Then at points of continuity of $x(t)$
>
> (a) $\quad x(t) = \displaystyle\int_{-\infty}^{\infty} X(f) \, e^{2\pi i f t} \, df,$
>
> where
>
> (b) $\quad X(f) = \displaystyle\int_{-\infty}^{\infty} x(t) \, e^{-2\pi i f t} \, dt.$
>
> (27.8)

These formulae closely resemble eqns (26.18) for the two-sided (complex) Fourier series, and it is possible to calculate the transition from (26.18) to the exponential transform by the procedure described in the previous section. Alternatively, (27.8) can be obtained from the sine and cosine transforms.

The condition that $\int_{-\infty}^{\infty} |x(t)| \, dt$ should exist (i.e. converge) appears to be rather restrictive. For example, any function which does not tend to zero as $t \to \pm\infty$ is suspect. Besides functions like t and e^t, this condition would disqualify *all* periodic functions such as $\sin \omega t$. The imprecise term 'well-behaved' in (27.8) implies further unspecified restrictions. Here we shall only say that if $\int_{-\infty}^{\infty} |x(t)| \, dt$ exists, the only exclusions are functions having a degree of eccentricity rarely encountered in physical applications. Simple **jump discontinuities** in the value of $x(t)$ are allowed, and, as with Fourier series, eqn (27.8a) delivers a value at such points equal to the average of the values of $x(t)$ on either side of the jump. If there is a jump at $t = t_0$, then

$$x(t) = \tfrac{1}{2}[x(t_0^-) + x(t_0^+)]. \tag{27.9}$$

The scope of the Fourier transform is not paralysed by restrictions, and the examples in this chapter will show that the system is far more flexible than this discussion might suggest.

Example 27.2 *(Compare Example 26.10.) Find the spectral distribution function $X(f)$ of the signal $x(t)$ defined by*

$$x(t) = \begin{cases} 1, & -\tfrac{1}{2}\tau < t < \tfrac{1}{2}\tau, \\ 0, & elsewhere. \end{cases}$$

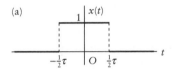

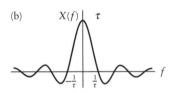

Fig. 27.4

Figure 27.4a shows $x(t)$. The spectral distribution function, or Fourier transform, of $x(t)$ is given by

$$X(f) = \int_{-\infty}^{\infty} x(t)\, e^{-2\pi i f t}\, dt = \int_{-\frac{1}{2}\tau}^{\frac{1}{2}\tau} e^{-2\pi i f t}\, dt$$

$$= \frac{1}{-2\pi i f}[e^{-2\pi i f t}]_{-\frac{1}{2}\tau}^{\frac{1}{2}\tau} = \frac{i}{2\pi f}(e^{-\pi i f \tau} - e^{\pi i f \tau})$$

$$= \frac{i}{2\pi f}(-2i)\sin \pi f \tau = \frac{1}{\pi f}\sin \pi f \tau.$$

The signal $x(t)$ is reconstructed from the spectral components $X(f)$ by

$$x(t) = \int_{-\infty}^{\infty} X(f)\, e^{2\pi i f t}\, df = \int_{-\infty}^{\infty} \frac{\sin \pi f \tau}{\pi f} e^{2\pi i f t}\, df.$$

Figure 27.4b shows the frequency distribution, which in this case is a real function.

The following properties are sometimes useful:

Properties of the exponential Fourier transform $X(f)$
If $x(t)$ is a real function, then:
(a) $X(-f) = \overline{X}(f)$.
(b) If $x(t)$ is even, $X(f)$ is real and even.
(c) If $x(t)$ is odd, $X(f)$ is pure imaginary, and odd. (27.10)

27.3 Short notations: alternative expressions

There are several conventional **notations** in common use in connection with the sine, cosine, and exponential Fourier transforms. For example:

$$\text{(a)} \quad \mathcal{F}[x(t)] = \int_{-\infty}^{\infty} x(t)\, e^{-2\pi i f t}\, dt \tag{27.11a}$$

(sometimes expressed as $\mathcal{F}[x](t)$) denotes the exponential Fourier transform of $x(t)$. $\mathcal{F}_s$ and $\mathcal{F}_c$ are used for the sine and cosine transforms respectively.

(b) $\mathcal{F}^{-1}[X(f)]$ denotes the inverse Fourier transform: it gives us back the originating function $x(t)$ if we know the frequency distribution $X(f)$:

$$\mathcal{F}^{-1}[X(f)] = \int_{-\infty}^{\infty} X(f)\, e^{2\pi i f t}\, df = x(t). \tag{27.11b}$$

(c)   The 'tilde' notation is often convenient:

$$\mathcal{F}[x(t)] = \tilde{x}(t).$$

Thus, if we have several time-dependent functions $u(t), v(t), \ldots$, their transforms may be written $\tilde{u}(f), \tilde{v}(f), \ldots$. Many books also use the notation $x(t) \leftrightarrow X(f)$.

More importantly, there are several **different versions** of the results (27.5), (27.6), and (27.8), all of which are widely used, so it is necessary to establish which one has been adopted in a particular piece of work. For example, the Fourier cosine transform and its inverse may appear in the form

$$G_c(\omega) = \frac{2}{\pi} \int_0^{\infty} x(t) \cos \omega t\, dt, \quad x(t) = \int_0^{\infty} G_c(\omega) \cos \omega t\, d\omega, \tag{27.12}$$

where ω can be interpreted as circular frequency ($\omega = 2\pi f_0$). Here we have preferred to present more symmetrical expressions, but all versions are equivalent to each other by means of a change of variable. There are similar variants of the exponential transform (27.8); in particular, the positive and negative signs in the complex exponents $\pm 2\pi i f t$ may be interchanged. See also Problem 27.7.

27.4   **Fourier transforms of some basic functions**

(Note: a more complete list of transforms is given in Appendix G.)

(a) *The top-hat function* $\Pi(t)$

The functions $\Pi(t)$ and $\Pi(t/\tau)$ are shown in Fig. 27.5. The transforms were found in Example 27.2:

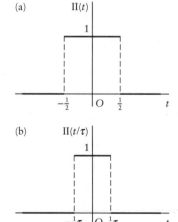

Fig. 27.5 (a) $\Pi(t)$, (b) $\Pi(t/\tau)$ (width τ).

Top-hat function

(a) $\mathcal{F}[\Pi(t)] = \dfrac{\sin \pi f}{\pi f}$.

(b) $\mathcal{F}[\Pi(t/\tau)] = \dfrac{\sin (\pi f \tau)}{\pi f}$.

(27.13)

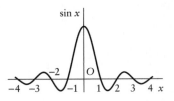

Fig. 27.6 sinc $x = \sin(\pi x)/(\pi x)$.

(b) *The function* sinc

The functions on the right of (27.13) are related to a standard function defined by

$$\text{sinc } x = \frac{\sin (\pi x)}{\pi x}.$$

Its graph is shown in Fig. 27.6. It is an even function, and it can be shown that the signed area under the curve is equal to unity.

The function sinc x

(a) sinc $x = \sin(\pi x)/\pi x$, sinc $0 = 1$.

(b) $\displaystyle\int_{-\infty}^{\infty} \text{sinc } x \, dx = 1.$

(27.14)

The transform of the top-hat functions (27.13) become

Fourier transform of $\Pi(t)$

(a) $\mathcal{F}[\Pi(t)] = \text{sinc } f.$
(b) $\mathcal{F}[\Pi(t/\tau)] = \tau \, \text{sinc } (\tau f).$

(27.15)

To find the Fourier transform of sinc t, start with (27.15b). Since $\tau \, \text{sinc } \tau f = \mathcal{F}[\Pi(t/\tau)]$, it follows that

$$\Pi(t/\tau) = \mathcal{F}^{-1}[\tau \, \text{sinc } (\tau f)],$$

$$= \int_{-\infty}^{\infty} \tau \, \text{sinc } (\tau f) \, e^{i2\pi ft} \, df.$$

Interchange the letters t and f, take the complex conjugate of the result to make the sign in the exponential negative, and put $1/\tau$ in place of τ. We obtain

$$\Pi(\tau f) = \int_{-\infty}^{\infty} \frac{1}{\tau} \text{sinc } \frac{t}{\tau} e^{-i2\pi ft} \, dt.$$

Multiply through by τ to obtain the results:

Fourier transform of sinc t

(a) $\mathcal{F}[\text{sinc } t] = \Pi(f).$
(b) $\mathcal{F}[\text{sinc}(t/\tau)] = \tau \Pi(\tau f).$

(27.16)

Equations (27.15) and (27.16) illustrate a general fact: that as the duration of a signal increases (e.g. as τ increases in (27.15b)), the effective frequency range tends to become narrower, and conversely.

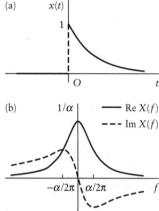

Fig. 27.7 (a) $x(t) = e^{-\alpha t}H(t)$.
(b) $X(f) = \mathcal{F}[e^{-\alpha t}H(t)]$.

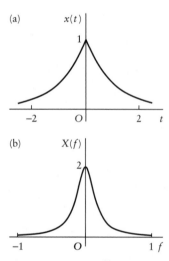

Fig. 27.8 (a) $x(t) = e^{-|t|}H(t)$.
(b) $X(f) = \mathcal{F}[e^{-|t|}]$.

(c) *A one-sided exponential function*
Consider the function in Fig. 27.7a defined by

$$x(t) = e^{-\alpha t}H(t),$$

where H(t) is the unit function (1.13) and α is positive.

$$\mathcal{F}[x(t)] = \int_{-\infty}^{\infty} x(t)\,e^{-i2\pi ft}\,dt = \int_{0}^{\infty} e^{-\alpha t}\,e^{-i2\pi ft}\,dt$$

$$= \int_{0}^{\infty} e^{-(\alpha+i2\pi f)t}\,dt = \frac{-1}{\alpha + i2\pi f}[e^{-(\alpha+i2\pi f)t}]_{0}^{\infty}$$

$$= \frac{1}{\alpha + i2\pi f}.$$

Therefore

$$\mathcal{F}[e^{-\alpha t}H(t)] = \frac{1}{\alpha + i2\pi f}. \tag{27.17}$$

Since $x(t)$ is neither even nor odd the spectral distribution is a complex function. Its real and imaginary parts are shown in Fig. (27.7b).

Example 27.3 *Find the Fourier transform of the function given by* $x(t) = e^{-|t|}$ (see Fig. 27.8a)

We have

$$X(f) = \int_{-\infty}^{\infty} e^{-|t|}\,e^{-i2\pi ft}\,dt$$

$$= \int_{-\infty}^{0} e^{t}\,e^{-i2\pi ft}\,dt + \int_{0}^{\infty} e^{-t}\,e^{-i2\pi ft}\,dt$$

$$= \frac{1}{1 - i2\pi f}[e^{(1-i2\pi f)t}]_{-\infty}^{0} + \frac{(-1)}{1 + i2\pi f}[e^{-(1+i2\pi f)t}]_{0}^{\infty}$$

$$= \frac{1}{1 - i2\pi f} + \frac{1}{1 + i2\pi f} = \frac{2}{1 + 4\pi^2 f^2}.$$

This function is shown in Fig. 27.8b.

27.5 Rules for manipulating transforms

The following rules enable new transforms to be obtained from known ones. The constants A, B, C, D, K are assumed to be real, but the signals $x(t)$ may be real or complex.

The proofs are left to the Problems. Most of them are obtained by writing down the appropriate Fourier integral or its inverse and then making a simple change of variable. The following Examples illustrate how these results are used.

		Signal $x(t)$	Transform $X(f) = \mathcal{F}[x(t)]$
(a)	Linearity	$Ax_1(t) + Bx_2(t)$	$AX_1(f) + BX_2(f)$
(b)	Time scaling	$x(At)$	$X(f/A)/|A|$
	Time reversal	$x(-t)$	$X(-f)$
(c)	Time delay	$x(t - B)$	$X(f)\, e^{-i2\pi Bf}$
(d)	Frequency scaling	$x(t/C)/|C|$	$X(Cf)$
(e)	Frequency shift	$x(t)\, e^{i2\pi Dt}$	$X(f - D)$
(f)	Modulation	$x(t) \cos 2\pi Kt$	$[X(f-K) + X(f+K)]/2$
		$x(t) \sin 2\pi Kt$	$[X(f-K) - X(f+K)]/(2i)$
(g)	Duality	$X(t)$	$x(-f)$
(h)	Differentiation	$dx(t)/dt$	$(i2\pi f)X(f)$
		$d^n x(t)/dt^n$	$(i2\pi f)^n X(f)$

$$(27.18)$$

Example 27.4   *Given that $\mathcal{F}[\Pi(t)] = \operatorname{sinc} f$, obtain $\mathcal{F}[\operatorname{sinc}(t/\tau)]$.*

Use the time-scaling rule (27.18b) with $A = 1/\tau$:

$$\Pi(t/\tau) \leftrightarrow \tau \operatorname{sinc}(f\tau).$$

Example 27.5   *Given that $\mathcal{F}[\Pi(t)] = \operatorname{sinc} f$, obtain $\mathcal{F}[\Pi(at + d)]$.*

We have

$$\Pi(at + d) = \Pi(a\{t + d/a\}).$$

From the time-scaling rule (27.18b).

$$\mathcal{F}[\Pi(at)] = (1/|a|)\, \operatorname{sinc}(f/a).$$

Then, using the time-delay rule (27.18c), with $B = -d/a$,

$$\mathcal{F}[\Pi(a\{t + d/a\})] = (1/|a|)\, \operatorname{sinc}(f/a)\, e^{i2\pi df/a}.$$

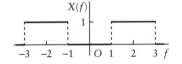

Fig. 27.9

Example 27.6   *Obtain the signal $x(t)$ produced by the spectral distribution $X(f)$ shown in Fig. 27.9.*

The two rectangular pulses are arrived at by extending the range of $\Pi(f)$ by a factor 2 to give $\Pi(t/2)$, then shifting this graph along the f axis a distance 2 to the left and 2 to the right to give

$$X(f) = \Pi(\tfrac{1}{2}\{f + 2\}) + \Pi(\tfrac{1}{2}\{f - 2\}).$$

From the frequency-scaling rule (27.18d) with $C = \tfrac{1}{2}$,

$$\Pi(\tfrac{1}{2}f) \leftrightarrow 2 \operatorname{sinc} 2t.$$

Then, by the frequency-shift rule (27.18e), with $K = \mp 2$,

$$X(f) \leftrightarrow (e^{i4\pi t} + e^{-i4\pi t}) \cdot 2 \operatorname{sinc} 2t$$

$$= 4 \cos 4\pi t \operatorname{sinc} 2t.$$

(The modulation rule (27.18f) could have been adopted for the final stage instead.)

Example 27.7   *Let*

$$x(t) = \begin{cases} -e^t, & t < 0, \\ e^{-t}, & t > 0. \end{cases} \tag{i}$$

Given that

$$x(t) \leftrightarrow -4\pi i f/(1 + 4\pi^2 f^2),$$

use the duality theorem (27.18g) *to deduce the Fourier transform of* $t/(1 + t^2)$.

Put

$$X(f) = -4\pi i f/(1 + 4\pi^2 f^2).$$

By the frequency-scaling rule (27.18d), with $C = 1/2\pi$,

$$X\left(\frac{f}{2\pi}\right) = -2if/(1 + f^2) \leftrightarrow 2\pi x(2\pi t).$$

Divide by $(-2i)$ and rename the functions obtained $Y(f)$ and $y(t)$:

$$f/(1 + f^2) = Y(f) \leftrightarrow i\pi x(2\pi t) = y(t). \tag{ii}$$

$x(t)$ is given, so we know the *inverse* transform of

$$Y(f) = f/(1 + f^2).$$

We need the transform of the identical *time* function given by

$$Y(t) = t/(1 + t^2).$$

The duality theorem (27.18g) tells how time may be exchanged for frequency in a *given* function. Applying it to $Y(f)$ in (ii) we have

$$Y(t) \leftrightarrow y(-f) = i\pi x(-2\pi f),$$

so we must substitute $(-2\pi f)$ for t into (i) (including the inequalities $t < 0$ and $t > 0$) giving

$$t/(1 + t^2) \leftrightarrow \begin{cases} i\pi \, e^{2\pi f}, & f < 0, \\ -i\pi \, e^{-2\pi f}, & f > 0. \end{cases}$$

Example 27.8   *(Sidebands) The voltage signal* $x(t) = v(t) \cos 2\pi f_0 t$ *represents an audiofrequency signal* $v(t)$ *used to modulate a carrier wave of high frequency* f_0. *Suppose that* $\mathcal{F}[v(t)] = V(f)$, *where f lies in the range* $-f_m < f < f_m < f_0$. *Use the modulation formula* (27.18f) *to illustrate the general nature of the spectral distribution function* $X(f) = \mathcal{F}[x(t)]$.

From (27.18f)

$$X(f) = \tfrac{1}{2}[V(f - f_0) + V(f + f_0)]. \tag{i}$$

$V(f - f_0)$ is zero unless

$$-f_m < f - f_0 < f_m,$$

that is unless

$$f_0 - f_m < f < f_0 + f_m, \tag{ii}$$

↗

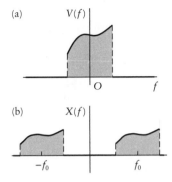

(a) $V(f)$

(b) $X(f)$

$-f_0$　　　f_0

Fig. 27.10 (a) Spectral distribution of $v(t)$. (b) Spectral distribution of $v(t) \cos 2\pi f_0 t$.

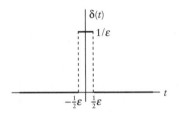

$\delta(t)$

$1/\varepsilon$

$-\frac{1}{2}\varepsilon$　$\frac{1}{2}\varepsilon$　　t

Fig. 27.11

Example 27.8 *continued*

and similarly $V(f + f_0)$ is zero unless

$$-f_0 - f_m < f < -f_0 + f_m. \tag{iii}$$

The intervals (ii) and (iii) do not overlap, since $f_m < f_0$. Therefore the spectral distribution (i) falls into two separate parts on opposite sides of the origin of f, as in Fig. 27.10. They are related to the sidebands of communication engineering. The two parts have the same shape, since their graphs consist of the graph of $V(f)$ moved through distances $\pm f_0$. (In general they would be complex, and even if they are real they will not generally correspond to two real signals.)

27.6 The delta function and periodic functions

The delta or impulse function $\delta(t)$ was defined in Section 25.5. It is convenient for the present purposes to modify slightly the definition given there. For the Laplace transform, which is concerned with $t \geq 0$ only, we constructed a rectangle of width ε and height $1/\varepsilon$, placing it on the interval $t = 0$ to $t = \varepsilon$. Here we are dealing with $-\infty < t < \infty$ and instead place it on the interval $t = -\frac{1}{2}\varepsilon$ to $t = \frac{1}{2}\varepsilon$ (see Fig. 27.11). Nothing else changes. The principal properties are:

The impulse or delta function $\delta(t)$

(a) Informal definition: $\delta(t) = 1/\varepsilon$　for $-\frac{1}{2}\varepsilon < t < \frac{1}{2}\varepsilon$, and $\delta(t) = 0$ elsewhere (allowing ε to be as small we wish).

(b) Sifting property: if $a < c < b$,

$$\int_a^b f(t)\, \delta(t - c)\, dt = f(c),$$

and the integral is otherwise zero.

(c) Fourier transform: by (b)

$$\mathcal{F}[\delta(t - c)] = \int_{-\infty}^{\infty} \delta(t - c)\, e^{-2\pi i f t}\, dt = e^{-2\pi i f c}. $$

$$(27.19)$$

The signal giving rise to $\delta(f - f_0)$ is given by the inverse transform:

$$\mathcal{F}^{-1}[\delta(f - f_0)] = \int_{-\infty}^{\infty} \delta(f - f_0)\, e^{i2\pi f t}\, df = e^{i2\pi f_0 t}.$$

Therefore

$$e^{i2\pi f_0 t} \leftrightarrow \delta(f - f_0),$$

and similarly

$$e^{-i2\pi f_0 t} \leftrightarrow \delta(f + f_0).$$

These are complex signals. But

$$\cos(2\pi f_0 t) = \tfrac{1}{2}(e^{i2\pi f_0 t} + e^{-i2\pi f_0 t});$$

so

$$\cos(2\pi f_0 t) \leftrightarrow \tfrac{1}{2}[\delta(f - f_0) + \delta(f + f_0)]. \tag{27.20a}$$

Similarly,

$$\sin(2\pi f_0 t) = \frac{1}{2i}(e^{i2\pi f_0 t} - e^{-i2\pi f_0 t}),$$

so

$$\sin(2\pi f_0 t) \leftrightarrow \frac{1}{2i}[\delta(f - f_0) - \delta(f + f_0)]. \tag{27.20b}$$

Therefore, the (real) cosine and sine functions having frequency f_0 are each associated with a pair of spectral lines, located at $f = \pm f_0$, as in Fig. 27.12.

The delta function is not at all a normal function. It belongs to a class of mathematical entities called **generalized functions**. They are essential in practical applications, since their use greatly simplifies what would otherwise be very difficult calculations. Generalized functions play a part similar to the symbol i in complex numbers: i is not an ordinary number, but in most ways it behaves like one.

There are apparent anomalies associated with generalized functions; for example, we have just obtained the Fourier transform of $\cos(2\pi f_0 t)$, but the normal definition of a Fourier transform (27.8b) does not work with a periodic function, because the integral does not approach a definite value when we apply the infinite limits of integration. Exact justification and interpretation of these questions are far beyond the scope of this book. You should regard relations such as (27.20) as being usually safe, and call on them as if you were using a dictionary, as did the original inventors of these methods.

In this sense we can obtain the Fourier transform of a general periodic function. The result is:

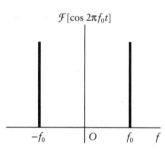

$\mathcal{F}[\cos 2\pi f_0 t]$

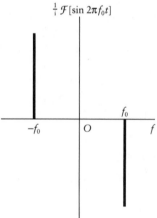

$\frac{1}{i}\mathcal{F}[\sin 2\pi f_0 t]$

Fig. 27.12

> **Transform of a periodic function**
>
> $x_P(t)$ is periodic with period T, and $f_0 = 1/T$.
>
> (a) $\mathcal{F}[x_P(t)] = \displaystyle\sum_{n=-\infty}^{\infty} X_n \delta(f - n f_0),$
>
> where
>
> (b) $X_n = f_0 \displaystyle\int_{\text{Period}} x_P(t)\, e^{-i2\pi n f_0 t}\, dt$
>
> $\hfill (27.21)$

The spectral frequency distribution consists of an infinite row of 'spikes' $\delta(f - nf_0)$ spaced at equal intervals f_0. These are weighted by X_n, which are just the two-sided Fourier *series* coefficients for the periodic function $x_P(t)$ given by (26.18b).

To prove the result (27.21), take the Fourier *series* representation (26.18a), and use (26.20a and b) to transform the cosines and sines in the series term by term. We obtain

$$\mathcal{F}[x_P(t)] = \sum_{n=-\infty}^{\infty} X_n \delta(f - nf_0),$$

which is (27.21a). The coefficients X_n are given by (26.18b), which is the same as (27.21b).

27.7 Convolution theorem for Fourier transforms

Suppose that we have a spectral distribution $X(f)$ which can be written as the product of two simpler functions $X_1(f)$ and $X_2(f)$, whose inverses we know:

$$X(f) = X_1(f)X_2(f), \tag{27.22}$$

where

$$x_1(t) \leftrightarrow X_1(f), \quad x_2(t) \leftrightarrow X_2(f). \tag{27.23}$$

The inverse transform of $X(f)$ is given by

$$x(t) = \int_{-\infty}^{\infty} e^{i2\pi ft} X_1(f)X_2(f) \, df \tag{27.24}$$

in which

$$X_1(f) = \int_{-\infty}^{\infty} e^{-i2\pi ft} x_1(t) \, dt = \int_{-\infty}^{\infty} e^{-2\pi i fu} x_1(u) \, du, \tag{27.25}$$

changing t to u, since t is already in use in (27.24). Substitute (27.25) into (27.24):

$$x(t) = \int_{-\infty}^{\infty} e^{i2\pi ft} \left(\int_{-\infty}^{\infty} e^{-i2\pi fu} x_1(u) \, du \right) X_2(f) \, df$$

$$= \int_{-\infty}^{\infty} \left(\int_{-\infty}^{\infty} e^{i2\pi f(t-u)} x_1(u) \, du \right) X_2(f) \, df$$

$$= \int_{-\infty}^{\infty} x_1(u) \left(\int_{-\infty}^{\infty} e^{i2\pi f(t-u)} X_2(f) \, df \right) du$$

after changing the order of integration (this process is justified in Section 32.1). The interior integral is equal to the inverse of $X_2(f)$ at time $(t - u)$, so it is equal to $x_2(t - u)$. Therefore

$$x(t) = \int_{-\infty}^{\infty} x_i(u) x_2(t - u) \, du. \qquad (27.26a)$$

If we had started by substituting for $X_2(t)$, we should obviously have arrived at

$$x(t) = \int_{-\infty}^{\infty} x_1(t - u) x_2(u) \, du, \qquad (27.26b)$$

so the two integrals on the right of (27.26a and b) are equal. This enables us to invert **products of spectral distributions**.

The integrals

$$\int_{-\infty}^{\infty} x_i(u) x_2(t - u) \, du \quad \text{or} \quad \int_{-\infty}^{\infty} x_1(t - u) x_2(u) \, du \qquad (27.27a)$$

(which are equal) are often written in the short **notation**

$$x_1(t) * x_2(t) \quad \text{or} \quad x_2(t) * x_1(t). \qquad (27.27b)$$

$x_1(t) * x_2(t)$ (or $x_2(t) * x_1(t)$) is called the **convolution of** $x_1(t)$ **and** $x_2(t)$. The result (27.26) is the **convolution theorem**. In the short notation:

> **Convolution theorem for Fourier transforms**
> Let $x_1(t) \leftrightarrow X_1(f)$ and $x_2(t) \leftrightarrow X_2(f)$. Then
> $$X_1(f) X_2(f) \leftrightarrow x_1(t) * x_2(t)$$
> where $x_1(t) * x_2(t)$ is given by (27.27).
>
> (27.28)

In the **convolution integrals** (27.27a) the variable of integration is u, and *t is to be treated like a parameter.*

Example 27.9 *Obtain* $x(t) = x_1(t) * x_2(t)$ *when* $x_1(t) = \Pi(t)$ *and* $x_2(t) = 1/(1 + t^2)$.

Write the convolution in the form

$$x(t) = \int_{-\infty}^{\infty} x_1(u) x_2(t - u) \, du = \int_{-\infty}^{\infty} \Pi(u) \frac{1}{1 + (t - u)^2} \, du.$$

Since $\Pi(u) = 0$ unless $-\frac{1}{2} < u < \frac{1}{2}$, the limits of integration become $\pm\frac{1}{2}$, so

$$x(t) = \int_{-\frac{1}{2}}^{\frac{1}{2}} \Pi(u) \frac{1}{1 + (t - u)^2} \, du = \int_{-\frac{1}{2}}^{\frac{1}{2}} \frac{1}{1 + (t - u)^2} \, du = \int_{t-\frac{1}{2}}^{t+\frac{1}{2}} \frac{dv}{1 + v^2}$$

(after putting $t - u = v$)

$$= \arctan(t + \tfrac{1}{2}) - \arctan(t - \tfrac{1}{2}) = \arctan\frac{4}{4t^2 + 3}.$$

At (i) in Example 27.9 the limits of integration were modified to take account of the fact that the integrand is zero except over the interval $-\frac{1}{2} < u < \frac{1}{2}$. In many typical cases it is quite difficult to establish the new limits. Consider, for example, the convolution of two identical pulses $\Pi(t)$:

$$x(t) = \Pi(t) * \Pi(t) = \int_{-\infty}^{\infty} \Pi(u)\Pi(t - u) \, du. \tag{27.29}$$

For different values of t, $\Pi(t - u)$ occupies a different position on the u axis. For certain ranges of t it partially overlaps $\Pi(u)$ from the left, or from the right, and for other ranges of t there is no overlap, as illustrated in Fig. 27.13.

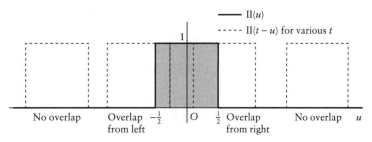

Fig. 27.13

To take this into account, set up a diagram as in Fig. 27.14, with axes t and u. The region in which $\Pi(t)\Pi(t - u)$ is nonzero is easy to find by carrying out the following construction.

(i) $\Pi(u)$ is *nonzero* only if $-\frac{1}{2} \le u \le \frac{1}{2}$. The edges of this region are the straight lines

$$u = -\tfrac{1}{2} \quad \text{and} \quad u = \tfrac{1}{2}.$$

Draw these and label them with the u values.

(ii) $\Pi(t - u)$ is *nonzero* only if $-\frac{1}{2} \le t - u < \frac{1}{2}$, or if

$$u - \tfrac{1}{2} \le t < u + \tfrac{1}{2}.$$

In terms of u, the edges of this region are therefore

$$u = t + \tfrac{1}{2} \quad \text{and} \quad u = t - \tfrac{1}{2}.$$

Draw these lines and label them.

(iii) The parallelogram enclosed by the four lines contains the u, t values for which the integrand (27.29) is *nonzero*. Next, draw a vertical line representing the current values of t as in Fig. 27.14. The effective limits of integration are represented by the points where the t line intersects the sides, and the u values are already written on these sides. (The limits of integration are therefore *different* functions of t for values of t on either side of a vertex.)

Other ways of interpreting convolution integrals will be found elsewhere, but this is by far the simplest way for working them out.

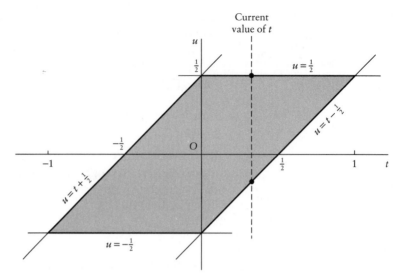

Fig. 27.14

It can be adapted for use whatever the nonzero intervals for the two functions may be. In practice it is used as follows:

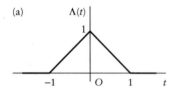

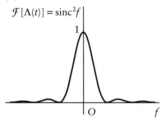

Fig. 27.15

Example 27.10   (a) *Show that* $\Pi(t) * \Pi(t) = \Lambda(t)$, *where* (Fig. 27.15a)

$$\Lambda(t) = \begin{cases} 1 + t, & -1 < t < 0, \\ 1 - t, & 0 < t < 1, \\ 0, & elsewhere. \end{cases}$$

(b) *Show that* $\mathcal{F}[\Lambda(t)] = \operatorname{sinc}^2 f$.

(a) Put $x(t) = \Pi(t) * \Pi(t)$, and use the diagram Fig. 27.14 as described in (iii) above.

If $t < -1$ or $t > 1$, there is no overlap, so $x(t) = 0$.

If $-1 \leqslant t \leqslant 0$, the limits of integration are from $u = -\frac{1}{2}$ to $t + \frac{1}{2}$, so

$$x(t) = \int_{-\frac{1}{2}}^{t+\frac{1}{2}} 1 \times 1 \, du = 1 + t.$$

If $0 \leqslant t \leqslant 1$, the limits of integration are from $u = t - \frac{1}{2}$ to $\frac{1}{2}$, so

$$x(t) = \int_{t-\frac{1}{2}}^{\frac{1}{2}} 1 \times 1 \, du = 1 - t.$$

Therefore the convolution is equal to $\Lambda(t)$, shown in Fig. 27.15a.

(b) From the convolution theorem, (27.28),

$$\mathcal{F}[\Pi(t) * \Pi(t)] = \mathcal{F}[\Pi(t)]\mathcal{F}[\Pi(t)] = \{\mathcal{F}[\Pi(t)]\}^2,$$

and

$$\mathcal{F}[\Pi(t)] = \operatorname{sinc} f$$

by (27.15). Therefore (see Fig. 27.15b)

$$\mathcal{F}[\Lambda(t)] = \operatorname{sinc}^2 f.$$

A more convenient way to express the triangle function is given in (27.30a):

Triangle function

(a) Definition

$$\Lambda(t) = \begin{cases} 1 - |t|, & -1 < t < 1; \\ 0, & \text{elsewhere.} \end{cases}$$

(b) Transform

$$\mathcal{F}[\Lambda(t)] = \text{sinc}^2 f.$$

(27.30)

27.8 The shah function

The generalized function $\text{III}_T(t)$ (pronounced 'shah') otherwise called a **Dirac comb** is defined by

$$\text{III}_T(t) = \sum_{n=-\infty}^{\infty} \delta(t - nT)$$

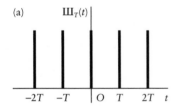

(a) $\text{III}_T(t)$

$-2T \quad -T \quad O \quad T \quad 2T \quad t$

(Fig. 27.16a). It is an even function consisting of an infinite string of equal 'spikes' (the delta functions) spaced at a constant interval T, one of them being at $t = 0$. Since it is periodic, its Fourier transform is given by (27.21a):

$$\mathcal{F}[\text{III}_T(t)] = \sum_{n=-\infty}^{\infty} X_n \delta(f - nf_0),$$

(27.31)

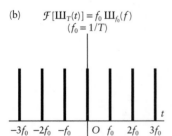

(b) $\mathcal{F}[\text{III}_T(t)] = f_0 \, \text{III}_{f_0}(f)$
$(f_0 = 1/T)$

$-3f_0 \quad -2f_0 \quad -f_0 \quad O \quad f_0 \quad 2f_0 \quad 3f_0$

Fig. 27.16

(a) $\text{III}_T(t) = \sum_{n=-\infty}^{\infty} \delta(t - nT)$.

(b) $\mathcal{F}[\text{III}_T(t)] = f_0 \sum_{n=-\infty}^{\infty} \delta(f - nf_0)$.

where $f_0 = 1/T$, and, from (27.21b),

$$X_n = f_0 \int_{\text{Period}} e^{-i2\pi nf_0 t} \text{III}_T(t) \, dt$$

$$= f_0 \sum_{n=-\infty}^{\infty} \int_{-\frac{1}{2}T}^{\frac{1}{2}T} e^{-i2\pi nf_0 t} \delta(t - nT) \, dt = f_0,$$

by the sifting rule (27.19b), since the only delta function within the period is the one where $n = 0$. Therefore, from (27.31),

$$\mathcal{F}[\text{III}_T(t)] = f_0 \text{III}_{f_0}(f).$$

It is shown in Fig. (27.16b).

The shah function

(a) $\text{III}_T(t) = \sum_{n=-\infty}^{\infty} \delta(t - nT).$

(b) $\mathcal{F}[\text{III}_T(t)] = f_0 \text{III}_{f_0}(f)$, where $f_0 = 1/T$.

(27.32)

Example 27.11   *The function $x(t)$ is zero when $t < -\frac{1}{2}T$ and $t > \frac{1}{2}T$. Show the convolution $y(t) = \text{Ш}_T(t) * x(t)$ is the periodic function, having period T, which agrees with $x(t)$ in the range $-\frac{1}{2}T < t < \frac{1}{2}T$.*

Write

$$\text{Ш}_T(t) * x(t) = \int_{-\infty}^{\infty} x(u)\text{Ш}_T(t-u)\,\mathrm{d}u$$

$$= \sum_{n=-\infty}^{\infty} \int_{-\infty}^{\infty} x(u)\,\delta(t-u-nT)\,\mathrm{d}u = \sum_{n=-\infty}^{\infty} x(t-nT),$$

using the sifting theorem (the critical points are where $t - u - nT = 0$). The term with $n = 0$ reproduces $x(t)$, which is zero outside the range $-\frac{1}{2}T$ to $\frac{1}{2}T$. The term with $n = 1$ slides that graph a distance T to the right, and we have a non-overlapping copy of $x(t)$ in the range $\frac{1}{2}T$ to $\frac{3}{2}T$, and so on. The general picture is shown in Fig. 27.17: $y(t)$ is a periodic copy of $x(t)$, with period T.

(a)

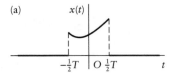

(b)

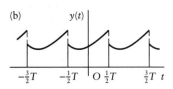

Fig. 27.17 (a) $x(t)$ (non-periodic). (b) $\text{Ш}_T(t) * x(t)$.

27.9 Energy in a signal: Rayleigh's theorem

The total energy E carried by a signal, from $t = -\infty$ to $t = \infty$, often takes the form

$$E = \int_{-\infty}^{\infty} |x(t)|^2 \,\mathrm{d}t.$$

This can be expressed in terms of the spectral distribution $X(f)$ as follows:

Rayleigh's theorem

$$\int_{-\infty}^{\infty} |x(t)|^2 \,\mathrm{d}t = \int_{-\infty}^{\infty} |X(f)|^2 \,\mathrm{d}f$$

of

$$\int_{-\infty}^{\infty} x(t)\bar{x}(t) \,\mathrm{d}t = \int_{-\infty}^{\infty} X(f)\bar{X}(f) \,\mathrm{d}f.$$

$$(27.33)$$

We have

$$E = \int_{-\infty}^{\infty} |x(t)|^2 \,\mathrm{d}t = \int_{-\infty}^{\infty} x(t)\bar{x}(t) \,\mathrm{d}t$$

$$= \int_{-\infty}^{\infty} x(t)\left(\int_{-\infty}^{\infty} \bar{X}(f)\,\mathrm{e}^{-\mathrm{i}2\pi ft} \,\mathrm{d}f\right)\mathrm{d}t$$

(after expressing $x(t)$ as the inverse transform of $X(f)$, and taking its complex conjugate). Now change the order of integration:

$$E = \int_{-\infty}^{\infty} \bar{X}(f) \left(\int_{-\infty}^{\infty} x(t) \, e^{-i2\pi ft} \, dt \right) df$$

$$= \int_{-\infty}^{\infty} \bar{X}(f) X(f) \, df = \int_{-\infty}^{\infty} |X(f)|^2 \, df.$$

Parseval's theorem extends this result for cases when the energy depends on two functions, $x(t)$ and $y(t)$, as in the case of current and voltage in circuits. It states that

$$\int_{-\infty}^{\infty} x(t) \bar{y}(t) \, dt = \int_{-\infty}^{\infty} X(f) \bar{Y}(f) \, df \tag{27.34}$$

27.10 Diffraction from a uniformly radiating strip

(Note: For the necessary background to waves and phasors see Sections 20.8 and 21.6.)

We shall illustrate a type of calculation which arises in **diffraction problems** in several branches of physics. In optics it occurs in Fraunhofer diffraction by a narrow slit, and there are similar problems in acoustics. Also there is a close connection with the theory of radiating antennas. We shall present the problem in an abstract way, since the process of tailoring it to a real situation involves additional physical considerations.

Consider the half-space $z \geqslant 0$, criss-crossed by travelling waves all having the same frequency f and wavelength λ. At every point P there is a disturbance $u(t, P)$ produced by superposition of all the rays passing through P, and interference between these rays determines the resultant amplitude and phase of the oscillation at P. Instead of using $u(t, P)$ we shall assign a phasor, or complex amplitude (see Section 21.6), $U(P)$ to every point, so that

$$u(t, P) = \text{Re}[U(P) \, e^{2\pi i ft}].$$

We need a preliminary result. Figure 27.18 show a ray directed along an arbitrary axis Oz. It has constant amplitude a. The disturbance is given by

Fig. 27.18

$$u(t, z) = a \cos\left[2\pi \left(\frac{t}{T} - \frac{z}{\lambda} \right) + \phi \right], \tag{27.35}$$

where T, λ, and ϕ are the period, wavelength, and a constant phase angle, and the wave velocity $v = \lambda/T$ is directed towards the right. Let Q and P be arbitrary *fixed* points on Oz. In an obvious notation

$$u(t, z_Q) = a \cos\left[2\pi\left(\frac{t}{T} - \frac{z_Q}{\lambda}\right) + \phi\right],$$

$$u(t, z_P) = a \cos\left[2\pi\left(\frac{t}{T} - \frac{z_P}{\lambda}\right) + \phi\right].$$

The corresponding phasors or complex amplitudes at Q and P are U_Q, U_P given by

$$U_Q = a\, e^{i[\phi - (2\pi i z_Q/\lambda)]} = a\, e^{i\phi_Q},$$

$$U_P = a\, e^{i[\phi - (2\pi i z_P/\lambda)]} = a\, e^{i\phi_P}.$$

The out-of-step behaviour of the oscillations at Q and P is defined by the phase difference

$$\phi_{QP} = \phi_P - \phi_Q = -2\pi(z_P - z_Q)/\lambda.$$

Therefore:

Phase change along a ray QP

$$U_P = U_Q\, e^{i\phi_{QP}} = U_Q\, e^{-2\pi i QP/\lambda},$$

where QP is the distance from Q to P. Therefore the phase of the complex amplitude decreases from Q to P by an amount equal to

$$2\pi \times \text{distance } QP \text{ measured in wavelengths.}$$

(27.36)

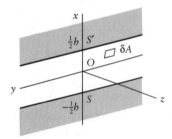

Fig. 27.19 Infinite radiating strip, width h, parallel to the y axis, radiating into $z > 0$, is typical radiating element.

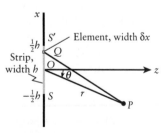

Fig. 27.20 Cross-section $y = 0$ of Fig. 27.19.

Figure 27.19 shows an infinite radiating strip in the (x, y) plane having width h, its central line along the y axis, and infinite length $-\infty < y < \infty$. Each infinitesimal element, or **source**, δA on the strip emits a harmonic wave spreading equally in all directions (i.e. it generates a **spherical wave**). We assume firstly that *the distribution of sources on the strip is uniform*: the contribution of any element of area δA is $\alpha\,\delta A$ where α is independent of position on the strip (one may imagine a uniform distribution of tiny, equal, hemispherical loudspeakers). Secondly, all the sources have the *same frequency and phase*: they are all oscillating in step.

Since the strip is infinitely long and the source distribution is independent of y, the problem is two-dimensional: the wave fields over all cross-sections $y = $ constant are identical. Figure 27.20 shows the cross-section $y = 0$, $z \geq 0$. P is a typical point distant r from O, and OP is inclined at θ to Oz (the positive direction for θ is *clockwise* here). Q is a typical elementary source at $(0, x)$, with $-h \leq x \leq h$, and width δx. The waves arriving at P from all points on the strip SS' interfere, and when the quantities h/λ, r/λ, are of the appropriate magnitude, a systematic distribution of intensity is observed as the angle of view θ varies between $\mp\frac{1}{2}\pi$, called a **diffraction pattern** or

angular spectrum. (Note: in the physics literature, this problem is usually referred to as one-dimensional.)

Firstly we shall determine the phases of the rays at P. For simplicity we take the common *phase of the sources to be zero*. Then by (27.36) the phase ϕ_{QP} of the complex amplitude component induced at P by the ray QP is given by

$$\phi_{QP} = -2\pi QP/\lambda. \tag{27.37}$$

To obtain an expression for the length QP: by the cosine rule (Appendix B(f)) applied to the triangle OQP

$$QP = [r^2 + x^2 - 2rx \cos(\tfrac{1}{2}\pi + \theta)]^{\frac{1}{2}}$$

$$= r\left(1 + \frac{2x \sin \theta}{r} + \frac{x^2}{r^2}\right)^{\frac{1}{2}}$$

$$= r(1+q)^{\frac{1}{2}}$$

say, where

$$q = \frac{2x \sin \theta}{r} + \frac{x^2}{r^2}.$$

It can be shown that if $h/r < \sqrt{5} - 1$ (which would always be so in practice), then $|q| < 1$ for all x in $-\tfrac{1}{2}h \leqslant x \leqslant \tfrac{1}{2}h$ and all θ in $-\tfrac{1}{2}\pi \leqslant \theta \leqslant \tfrac{1}{2}\pi$. In that case we can use the binomial theorem (5.4f) to approximate to $(1 + q)^{\frac{1}{2}}$. The first few terms are given by

$$(1 + q)^{\frac{1}{2}} = 1 + \tfrac{1}{2}q - \tfrac{1}{8}q^2 + \cdots .$$

Therefore

$$(1+q)^{\frac{1}{2}} = r\left[1 + \frac{1}{2}\left(\frac{2x \sin \theta}{r} + \frac{x^2}{r^2}\right) - \frac{1}{8}\left(\frac{2x \sin \theta}{r} + \frac{x^2}{r^2}\right)^2 + \cdots\right]$$

$$= r + x \sin \theta + r\left(\frac{1}{2}\frac{x^2}{r^2}\sin^2\theta - \frac{1}{2}\frac{x^3}{r^3}\sin \theta - \frac{1}{8}\frac{x^4}{r^4}\cdots\right)$$

$$= (r + x \sin \theta) + \frac{x^2}{2r}\left(\cos^2\theta - \frac{x}{r}\sin \theta - \frac{x^2}{4r^2} + \cdots\right). \tag{27.38}$$

We shall retain the linear expression $(r + x \sin \theta)$ as the approximation. However, we must ensure that *the error is much smaller than one wavelength λ.* Otherwise the subsequent calculation of the interference effect at P of all the sources together will be seriously affected. The error consists of the neglected group of terms in (27.38), so we require $\tfrac{1}{2}(x^2/r)\cos^2\theta \ll \lambda$, where $\ll$ means 'is very much less than', and merely indicates an order of magnitude. This is satisfied for all the x and θ values if

$$h^2/r \ll 8\lambda,\tag{27.39}$$

Finally we have

> **Phase change ϕ_{QP} along the ray QP (Fig. 27.19)**
> If $-\frac{1}{2}h \leqslant x \leqslant \frac{1}{2}h$ and $h^2/r \ll 8\lambda$, then
> $$\phi_{QP} = -2\pi(r + x \sin\theta)/\lambda,$$
> with an error $\ll 2\pi$ (radians).   **(27.40)**

Note that the approximation to ϕ_{QP} is *linear in $x \sin\theta$*.

Next we shall transform all distances into multiples of a wavelength λ. Put

$$r/\lambda = R, \qquad x/\lambda = X, \qquad h/\lambda = H.\tag{27.41}$$

These are natural variables for the problem; the physical outcome depends on the number of wavelengths in h, for example. If we double the wavelength we must double r, x, and h to preserve the same geometry. Since R, X, H are dimensionless: if the unit of length is changed, say from metres to angstrom units, these quantities are unaffected. Equation (27.40) becomes:

> **Phase change ϕ_{QP} along the ray QP**
> (Distances measured in wavelengths.)
> If $-\frac{1}{2}H \leqslant x \leqslant \frac{1}{2}H$ and $H^2/R \ll 8$, then
> $$\phi_{QP} = -2\pi(R + X \sin\theta),$$
> accurate to a small fraction of 2π.   **(27.42)**

Suppose that the amplitude of the source at Q is $a\delta X$ in the new units. We can allow to some extent for **attenuation** along the ray QP, provided that it depends effectively only on distance R. We approximate its contribution, $\delta_Q U$, to the complex amplitude U_P at P by putting

$$\delta U_Q = u(R)\, e^{-2\pi i X \sin\theta},\tag{27.43}$$

where, using (27.42),

$$u(R) = a(R)\, e^{-2\pi i R}\, \delta X,\tag{27.44}$$

and $a(R)$ also includes an attenuation factor.

The *resultant complex amplitude U_P* at P arising from all the sources on $-H \leqslant X \leqslant H$ is then given by

$$\begin{aligned}
U_P &= \lim_{\delta X \to 0} \sum \delta_Q U = u(R) \int_{-\frac{1}{2}H}^{\frac{1}{2}H} e^{-2\pi i X \sin\theta}\, dX \\
&= u(R) \int_{-\infty}^{\infty} \Pi(X/H)\, e^{-2\pi i X \sin\theta}\, dX,
\end{aligned}\tag{27.45}$$

where Π is the top-hat function (27.13).

We are interested only in the dependence of the expression (27.45) for U_P on the angle θ, and not on the variation with distance, or on the R-dependent part of the phase of U_P. We therefore define the **angular spectrum function** $F(\sin\theta)$ by casting off $u(R)$, so that for *constant R*:

$$F(S) = \int_{-\infty}^{\infty} \Pi(X/H)\, e^{-2\pi iXS}\, dX, \tag{27.46a}$$

where

$$S = \sin\theta. \tag{27.46b}$$

By comparing (27.46a) with (27.8b) (with X in place of t and S in place of f), it can be seen that $F(S)$ *is the Fourier transform of* $\Pi(X/H)$. Also, we can refer to (27.13b) to evaluate it (with H standing in place of τ). We obtain

$$F(S) = \frac{\sin(\pi SH)}{\pi S} = H\,\mathrm{sinc}(HS). \tag{27.47}$$

In terms of the original variables x, h, λ, θ, therefore, over a circular arc $r = constant$,

angular distribution of amplitude $\propto \mathrm{sinc}(h\sin\theta/\lambda)$. (27.48a)

This angular dependence is illustrated in Fig. 27.21b. The intensity distribution (see Section 21.4(iii)) is proportional to $|U_P|^2$, so that

angular spectrum of intensity $\propto \mathrm{sinc}^2(h\sin\theta/\lambda)$, (27.48b)

shown in Fig. 27.21c.

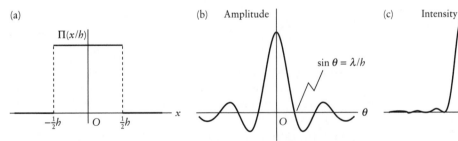

Fig. 27.21 (a) Source distribution $\propto \Pi(x/h)$. (b) Amplitude spectrum $\propto \mathrm{sinc}(h\sin(\theta/\lambda))$ (zeros at $\sin\theta = n\lambda/h$.) (c) Intensity spectrum $\propto \mathrm{sinc}^2(h\sin\theta/\lambda)$.

27.11 General source distribution and the inverse transform

Suppose now that the source distribution is not necessarily confined absolutely to a strip, and is not necessarily uniform. Let its complex amplitude be $e(x)$ where

$$e(x) = |e(x)|\,e^{i\phi(x)}, \quad -\infty < x < \infty. \tag{27.49}$$

By following exactly the same procedure, we obtain the angular spectrum

$$F(S) = \int_{-\infty}^{\infty} E(X)\, e^{-2\pi i S X}\, dX, \tag{27.50}$$

where $S = \sin\theta$, $X = x/\lambda$, and $E(X) = e(x)$. In principle, the source may be infinitely extended in the $\pm x$ directions, though realistically we shall assume $E(X)$ to be negligible beyond a certain range of values. Equation (27.50) is again the Fourier transform of the source distribution, and its inverse transform is given by

$$E(X) = \int_{-\infty}^{\infty} F(S)\, e^{2\pi i S X}\, dS. \tag{27.51}$$

Equation (27.51) suggests that we might be able to construct a source distribution $E(X)$ (that consists, for example, of a suitable array of antennas), having prescribed directional properties defined by a fairly arbitrary function of direction $F(S) = F(\sin\theta)$. However, there seems to be a difficulty, in that the inversion integral (27.51) requires a value of $F(S)$ at *every* value of S, $-\infty < S < \infty$. But in the physical world S stands for $\sin\theta$ with $-\frac{1}{2}\pi \leqslant \theta \leqslant \frac{1}{2}\pi$, or for $-1 \leqslant S \leqslant 1$. Outside of this range we cannot prescribe values of S in advance.

To meet this difficulty we shall add another approximation requirement to the small print of the theory. For 'well-behaved' functions (see the remarks following (27.8)), $|F(S)|$ defined by (27.50) approaches zero as $S \to \pm\infty$. If we can be confident that

$$\int_{-\infty}^{\infty} F(S)\, e^{2\pi i S X}\, dX \approx \int_{-1}^{1} F(S)\, e^{2\pi i S X}\, dS \tag{27.52}$$

to an acceptable degree of accuracy, then we may ignore the range $|S| > 1$ for the purpose of obtaining $E(X)$ from a given $F(S)$. A commonly arising physical situation that provides support for the approximation (27.52) involves radiation fields that are *strongly directional*, the diffracted rays being effectively confined to a fairly narrow range of θ. The radiation from a uniform strip (Section 27.10) is of this character if the dimensions are right.

27.12 Transforms in radiation problems

We continue to use the dimensionless variables (27.41) for simplicity of expression. The reformulation of earlier results (see Appendix G(a)) in terms of the new variables is obtained by the following correspondences:

Appendix G(b):	f		t	x	X	D	B	u
Current symbols:	$S(= \sin\theta)$		X	E	F	K	D	W

The results we shall be using, recast in the present notation, are:

> **Rules for radiation problems**
>
> | Rule | Source distribution | Amplitude spectrum |
> |---|---|---|
> | (a) Scaling | $E(AX)$ | $F(S/A)/|A|$ |
> | (b) Linear phase factor | $E(X)\,e^{iKX}$ | $F[S-(K/2\pi)]$ |
> | (c) Source displacement | $E(X-D)$ | $F(S)\,e^{-2\pi iDS}$ |
> | (d) Reciprocity (duality) | $F(X)$ | $E(-S)$ |
> | (e) Convolution | $g(X) * E(X)$ | $G(X)F(S)$ |
> | $\quad(G(s) = \mathcal{F}[g(X)])$ | | |
>
> (where $f(X) * g(X) = \int_{-\infty}^{\infty} f(W)g(X-W)\,\mathrm{d}W$
>
> $\qquad\qquad\qquad = \int_{-\infty}^{\infty} f(X-W)g(W)\,\mathrm{d}W).$
>
> (27.53)

We now give some examples showing the significance of these rules for radiation problems. It will be assumed that the estimates (27.42) and (27.52) apply where necessary. Notice that if the effective diffracted range of θ is small enough, S can be identified with θ for the purpose of visualizing the diffraction patterns that arise.

(i) *Change of scale*

Suppose that $\mathcal{F}[E(X)] = F(S)$ and $\mathcal{F}[E(AX)] = F_A(S)$, where $F(S)$ and $F_A(S)$ describe the respective angular spectra. If $0 < A < 1$, the graph of $E(AX)$ is obtained from that of $E(X)$ by *stretching* it uniformly by a factor $1/A > 1$ along the X axis. Then the scaling rule (27.3a) states that the graph of $F_A(S)$ is obtained by *contracting* the graph of $F(S)$ by a factor A. (The amplitude of $F_A(S)$ is affected by the $1/|A|$, but this does not affect the angular distribution, which is all we are interested in.) If $A > 1$, then $E(X)$ is contracted and $F(S)$ is stretched.

This is illustrated by the uniform source distribution shown in Fig. 27.20, where (in terms of X and H, which are measured in wavelengths)

$$E(X) = \Pi\left(\frac{X}{H}\right), \quad F(S) = H \operatorname{sinc} HS,$$

so that $A = 1/H$. The breadth of the central loop of $F(S)$ and its satellites is inversely proportional to H.

(ii) *Linear phase change across a radiating strip*

Let $E(X)$ and $E(X)e^{iKX}$ be two source distributions and $F(S)$ the angular spectrum of $E(X)$. Equation (27.53b) states that the spectrum of $E(X)\,e^{iKX}$ is equal to $F[S-(K/2\pi)]$. The angular displacement of the diffracted pattern (given in terms of S) is therefore shifted by a constant amount $K/(2\pi)$, the pattern itself remaining unchanged. The diffracted pattern 'swings' through an angle determined by $\Delta S = K/(2\pi)$, where $S = \sin\theta$ and ΔS is the change in S. Such a phase gradient can be induced across an antenna array to redirect the main beam.

(iii) *Displacement of the source*

Equation (27.53c) states that if we move the emitter bodily up the X axis by a distance D (wavelengths), then $F(S)$ becomes $F(S)\,e^{-2\pi iDS}$. This result may seem a little curious, since it is physically obvious that the new spectrum is simply the old spectrum moved up a distance D. However, θ is still being measured from the same origin, and this formula, easy to prove, condenses some awkward geometry. Note that this is the dual or reciprocal property (27.53d) corresponding to item (ii).

(iv) *Interference between two narrow uniform sources*

Figure 27.22 shows two identical and uniform radiating strips of width h placed symmetrically a distance d apart as measured between their centres. The source distribution $e(x)$ is given by

$$e(x) = a\Pi\left(\frac{x + \frac{1}{2}d}{h}\right) + a\Pi\left(\frac{x - \frac{1}{2}d}{h}\right),$$

where a is a constant. In terms of the dimensionless variables X, H, and $D = d/\lambda$,

$$E(X) = a\Pi\left(\frac{X + \frac{1}{2}D}{H}\right) + a\Pi\left(\frac{X - \frac{1}{2}D}{H}\right). \tag{27.54}$$

By (27.47),

$$\mathcal{F}[\Pi(X/H)] = F(S) = H\,\text{sinc}(HS).$$

Then by using the property (27.53c)

$$F(S) = aF(S)\,e^{\pi iDS} + aF(S)\,e^{-\pi iDS} = 2H\,\text{sinc}(HS)\cos(DS). \tag{27.55}$$

The zeros of $F(S)$ due to the term sinc HS are at $S = n\pi/H$ and those due to cos DS are at $S = (n + \frac{1}{2})\pi/D$, and they are interlaced. If $D > H$ (not necessarily hugely greater, but perhaps 10 times greater) an interference of the type shown in Fig. 27.23 is obtained for the

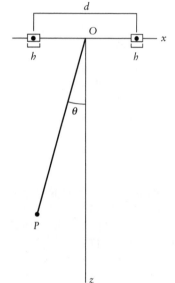

Fig. 27.22

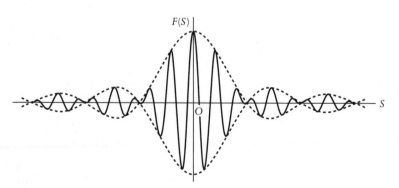

$F(S)$

S

O

Fig. 27.23 The angular spectrum of the arrangement in Fig. 27.21.

angular spectrum $F(S)$. The envelope is proportional to sinc(HS). The intensity spectrum is proportional to the square of this function, sinc$^2(HS)\cos^2(DS)$ (see Fig. 27.23). If $D \gg H$, the underlying fine-scale oscillation may be difficult to resolve instrumentally.

(v) Arrays of sources in terms of a convolution

Suppose (see Fig. 27.24) that we have an array of N identical radiating elements centred on the points $X = X_1, X_2, \ldots, X_N$. The nth element has the distribution $E_0(X - X_n)\Pi[(X - X_n)/H]$, where H is the constant width (in wavelengths). Assume that $X_1, X_2, \ldots, X_N$ are spaced so that the elements are non-overlapping. The overall source distribution $E(X)$ is then given by

$$E(X) = \sum_{n=1}^{N} E_0(X - X_n)\Pi\left(\frac{X - X_n}{H}\right). \tag{27.56}$$

We shall show that $E(X)$ can be expressed in the form of the convolution

$$E(X) = E_0(X)\Pi(X/H) * g(X), \tag{27.57a}$$

where

$$g(X) = \sum_{n=1}^{N} \delta(X - X_n), \tag{27.57b}$$

and δ represents the delta function (27.19). $g(x)$ is called the **distribution function for the array**. To prove (27.57): by the definition of the convolution (27.53e),

$$E_0(X)\Pi\left(\frac{X}{H}\right) * g(X) = \int_{-\infty}^{\infty} E_0(X')\Pi\left(\frac{X'}{H}\right)\sum_{n=1}^{N}\delta(X - X' - X_n)\,dX'$$

$$= \sum_{n=1}^{N}\int_{-\infty}^{\infty} E_0(X')\Pi\left(\frac{X'}{H}\right)\delta(X - X' - X_n)\,dX'$$

$$= \sum_{n=1}^{N}\int_{-\infty}^{\infty} E_0(X - X_n - w)\Pi\left(\frac{X - X_n - w}{H}\right)\delta(w)\,dw$$

(after putting $w = X - X_n - X'$)

$$= \sum_{n=1}^{N} E_0(X - X_n)\Pi\left(\frac{X - X_n}{H}\right)$$

(from the sifting property of the delta function (27.19b))

$$= E(X)$$

as required by (27.57a).

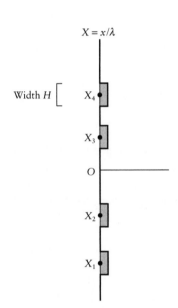

$X = x/\lambda$

Width H

X_4

X_3

O

X_2

X_1

Fig. 27.24 An array of sources.

Let the transforms of $g(x)$ and $E_0(X)\Pi(X/H)$ be given by

$$\mathcal{F}[g(X)] = G(S), \qquad \mathcal{F}[E_0(X)\Pi(X/H)] = F_0(S).$$

Then by (27.53e) and (27.57a)

$$\mathcal{F}[E(X)] = \mathcal{F}[E_0(X)\Pi(X/H) * g(X)] = G(S)F_0(S). \tag{27.58}$$

Therefore the spectrum of the array is equal to the transform of the array distribution function, multiplied by the spectrum of the single element centred on the origin.

Alternatively, we can obtain $G(S)$ explicitly:

$$G(S) = \int_{-\infty}^{\infty} \sum_{n=1}^{N} \delta(X - X_n) \, e^{-2\pi i X S} \, dX = \sum_{n=1}^{N} e^{-2\pi i X_n S}, \tag{27.59}$$

so that

$$F(S) = F_0(S) \sum_{n-1}^{N} e^{-2\pi i X_n S}. \tag{27.60}$$

Each displaced source

$$E_0(X - X_n)\Pi[(X - X_n)/H]$$

is subject to the displacement rule (27.53c), which introduces the factor $e^{2\pi i X_n S}$ into the spectrum of $E_0(X)\Pi(X/H)$, by (iii) above. Therefore we obtain the sum (27.60) more directly. This approach is simpler, but for more general cases convolution methods are more versatile.

Problems

27.1 Obtain the Fourier sine and cosine transforms of the function $x(t) = e^{-t}$ for $t \geqslant 0$. Find the value delivered by the inverse sine transform at $t = 0$. (Hint: $\cos(2\pi ft) + i \sin(2\pi ft) = e^{2\pi i ft}$.)

27.2 Show that the cosine transform of the function

$$x(t) = \begin{cases} 1 - t, & 0 \leqslant t \leqslant 1, \\ 0, & t > 1 \end{cases}$$

is $X_c(f) = \sin^2(\pi f)/(\pi^2 f^2)$. By considering the inverse transform show that

$$\int_0^\infty \frac{\sin^2 u}{u^2} \, du = \tfrac{1}{2}\pi.$$

27.3 Let $x(t) = e^{-t^2}, t \geqslant 0$. Use the procedure that follows to show that its Fourier cosine transform, $X_c(f)$, is given by $X_c(f) = \sqrt{\pi}\, e^{-\pi^2 f^2}$:

(i) Write the integral defining $X_c(f)$, and obtain dX_c/df by differentiating under the integral sign (see Sections 17.9 or 27.8).

(ii) Integrate by parts to obtain the differential equation

$$\frac{dX_c}{df} = -2\pi f X_c,$$

and obtain the general solution.

(iii) Use the fact (see Example 32.11) that

$$\int_0^\infty e^{-x^2} \, dx = \tfrac{1}{2}\sqrt{\pi}$$

to provide the initial condition $X_c(0) = \sqrt{\pi}$ for (ii), and deduce that $X_c(f) = \sqrt{\pi}\, e^{-\pi^2 f^2}$.

27.4 (a) Show that if $x(t)$ is an even function, then $\mathcal{F}[x(t)]$ is an even function of f, and takes the form of the cosine transform of $x(t)$. (Hint: split the range of integration into two parts, $-\infty$ to 0 and 0 to ∞.)

(b) Use the result of Problem 27.3 that the cosine transform of e^{-t^2} is $\sqrt{\pi}\,e^{-\pi^2 f^2}$, together with the scaling rule (27.18d) to find the cosine transform of e^{-at^2}.

27.5 Prove that if $x(t)$ is an odd function then $X(f)$ is a pure imaginary odd function. Show that $X(f)$ reduces to $iX_s(f)$, where X_s is the sine transform of x. Confirm that the inverse sine transform of $iX_s(f)$ is $x(t)$.

27.6 From Problem 27.3, the cosine transform of e^{-t^2} is $\sqrt{\pi}\,e^{-\pi^2 f^2}$. Use this result together with the scaling rule (27.17b) to prove that $\mathcal{F}[e^{-\pi t^2}] = e^{-\pi f^2}$.

27.7 By means of a change of variable show that an alternative form of the Fourier transform pair is given by

$$X(\omega) = \frac{1}{\sqrt{(2\pi)}} \int_{-\infty}^{\infty} x(t)\, e^{i\omega t}\, dt,$$

$$x(t) = \frac{1}{\sqrt{(2\pi)}} \int_{-\infty}^{\infty} X(\omega)\, e^{-i\omega t}\, dt.$$

27.8 Prove that if $x(t)$ is an even function then $\mathcal{F}[x(t)]$ is an even function of f. Use this fact to reduce the Fourier transform pair to a real form. (Hint: split the ranges of integration into two parts, $-\infty$ to 0 and 0 to ∞.)

27.9 Prove that if $x(t)$ is an odd function, then $X(f)$ is a pure imaginary odd function. Show that the Fourier transform pair can then be reduced to a pair of *real* equations.

27.10 Prove the time-scaling rule, (27.18b), and the time-delay rule, (27.10).

27.11 By (27.4), $\mathcal{F}[\Pi(t)] = \text{sinc}\, f$. (a) Use the time-delay rule (27.18c) to obtain the transform of

$$x(t) = \begin{cases} 1, & 0 < t < 1, \\ 0, & \text{elsewhere.} \end{cases}$$

(b) Confirm the result (a) by evaluating $\mathcal{F}[x(t)]$ directly.
(c) Use the time-delay rule and the time-scaling rule to obtain $\mathcal{F}[x(t)]$ where $b > \frac{1}{2}c$ and

$$x(t) = \begin{cases} -1, & -b - \frac{1}{2}c < t < -b + \frac{1}{2}c, \\ 1, & b - \frac{1}{2}c < t < b + \frac{1}{2}c, \\ 0, & \text{elsewhere.} \end{cases}$$

(Hint: sketch a diagram.)

27.12 Given that $\mathcal{F}[\Lambda(t)] = \text{sinc}^2 f$ (proved in Example (27.10)), where

$$\Lambda(t) = \begin{cases} 1 + t, & -1 < t < 0, \\ 1 - t, & 0 < t < 1, \\ 0, & \text{elsewhere,} \end{cases}$$

obtain (a) $\mathcal{F}[\Lambda(2t)]$; (b) $\mathcal{F}[\Lambda(2t - 3)]$.

27.13 (a) Prove the frequency-shift property, (27.18).
(b) Obtain $\mathcal{F}[x(t)e^{\pm i2\pi f_0 t}]$.
(c) From (b) deduce the modulation rules, (27.18), for $\mathcal{F}[x(t)\cos 2\pi f_0 t]$ and $\mathcal{F}[x(t)\sin 2\pi f_0 t]$.
(d) Obtain $\mathcal{F}[\Pi(\frac{1}{2}t)\cos 2\pi f_0 t]$ and $\mathcal{F}[\Pi(\frac{1}{2}t)\sin 2\pi f_0 t]$.

27.14 (a) Given that $\Lambda(t) \leftrightarrow \text{sinc}^2 f$, obtain $\mathcal{F}[\text{sinc}^2 t]$ either by using the duality rule (27.18), or by a direct method.
(b) Use the result (a), together with the time-shift and time-scaling rules, to find $\mathcal{F}[\Lambda(at + b)]$. ($\Lambda(t)$ is defined by

$$\Lambda(t) = \begin{cases} 1 + t, & -1 < t < 1, \\ 1 - t, & 0 < t < 1, \\ 0, & \text{elsewhere.} \end{cases}$$

27.15 (a) Prove the differentiation rule (27.18).
(b) Given that $e^{-|t|} \leftrightarrow 2/(1 + 4\pi^2 f^2)$, obtain $\mathcal{F}^{-1}[if/(1 + 4\pi^2 f^2)]$.

27.16 From the result $e^{-\alpha t}H(t) \leftrightarrow 1/(\alpha + i2\pi f)$, use the time-reversal rule to obtain $\mathcal{F}[e^{-\alpha|t|}]$, where $\alpha > 0$.

27.17 (a) Obtain

$$\mathcal{F}[e^{-\alpha t}\cos \beta t\, H(t)] \quad \text{and} \quad \mathcal{F}[e^{-\alpha t}\sin \beta t\, H(t)],$$

where $\alpha > 0$. (Hint: look at the table of simplifying rules (27.18) before trying to tackle these directly.)
(b) Obtain $\mathcal{F}[e^{-\alpha t}\cos(2\pi f_0 t + \phi)]$, where $\alpha > 0$.

27.18 Prove that

$$H(t) * \{x(t)\, H(t)\} = \int_0^t x(\tau)\, d\tau.$$

27.19 (a) Obtain $x_1(t) * x_2(t)$ when $x_1(t) = x_2(t) = e^{-t}H(t)$.
(b) Use your result together with the convolution theorem (27.28) to obtain the transform of a new function, $t\, e^{-t}$.
(c) Obtain $\mathcal{F}[t\, e^{-\alpha t}]$ from (b), where $\alpha > 0$.
(d) Obtain the same result as in (c) by noticing that

$$\frac{d}{d\alpha}(e^{-\alpha t}) = -t\, e^{-\alpha t}.$$

27.20 (a) Prove that

$$\Pi(t - \tfrac{1}{2}) * \Pi(t + \tfrac{1}{2}) = \Lambda(t).$$

(Hint: use the convolution theorem, (27.28).)

(b) Show that

$$\Pi(t - a) * \Pi(t - b) = \Lambda(t - a - b).$$

(c) Show that

$$\Pi(t) * \Pi(\tfrac{1}{2}t) = \begin{cases} 0, & t < -\tfrac{3}{2} \text{ and } t > \tfrac{3}{2}, \\ \tfrac{3}{2} + t, & -\tfrac{3}{2} < t < -\tfrac{1}{2}, \\ 1, & -\tfrac{1}{2} < t < \tfrac{1}{2}, \\ \tfrac{3}{2} - t, & \tfrac{1}{2} < t < \tfrac{3}{2}. \end{cases}$$

27.21 Show that the total energy in the signal $x(t) = e^{-\alpha t}H(t)$ ($\alpha > 0$) is equal to $1/2\alpha$. Show that the total energy due to the frequency range $-f_0 < f < f_0$ is equal to

$$\frac{1}{\pi \alpha} \arctan(2\pi f_0 / \alpha).$$

27.22 Prove the result of Example 27.15 by using the convolution theorem (27.28) together with the expression (27.31) for $\mathcal{F}[\text{Ш}_T(t)]$.

27.23 Use the Fourier transform to obtain a particular solution of the differential equation

$$\frac{d^2 x}{dt^2} - x = \frac{1}{1 + t^2},$$

in the form of a convolution integral.

27.24 (a) Given that $\mathcal{F}[\text{sinc } t] = \Pi(t)$, deduce that

$$\int_0^\infty \frac{\sin u}{u} \, du = \tfrac{1}{2}\pi \quad \text{and} \quad \int_0^\infty \text{sinc } u \, du = \tfrac{1}{2}.$$

(b) Given (see Problem 27.6) that $\mathcal{F}[e^{-\pi t^2}] = e^{-\pi f^2}$, deduce that $\mathcal{F}[t \, e^{-\pi t^2}] = \mathrm{i} f e^{-\pi f^2}$.

27.25 Use the convolution theorem and time-delay rule (27.17) to show that

$$\text{sinc } t * \text{sinc } t = \text{sinc } t$$

and, more generally, that

$$\text{sinc}(t - a) * \text{sinc}(t + a) = \text{sinc } t$$

27.26 The function $g_\tau(t)$ defined by

$$g_\tau(t) = \frac{1}{\tau} \int_{t - \frac{1}{2}\tau}^{t + \frac{1}{2}\tau} g(u) \, du$$

is called the **moving average** of g over a range of length τ. (The output from a recording instrument is often a moving average overs a short interval τ.)
(a) Show that $g_\tau(t) = \Pi(t/\tau) * g(t)$.
(b) Obtain the moving average $g_\tau(t)$ when $g(t) = \Pi(t)$, for values $\tau = \tfrac{1}{4}, \tfrac{3}{4}, 2$, and indicate their general nature by sketches.

27.27 in the manner of Example 27.11, interpret the convolution

$$\text{Ш}_T(t - a) * g(t - b)\Pi(\{t - b\}/\tau),$$

where $\tau \leqslant t$.

27.28 Obtain the function $h(t) = \Pi(t) * \Lambda(t)$ and its Fourier transform. Use the result to evaluate $\int_{-\infty}^{\infty} \text{sinc}^3 u \, du$. (Hint: the segments of $h(t)$ should join up continuously (check this), and the work is halved by noticing that $h(t)$ is even.)

27.29 Show that, viewed from any direction θ at a sufficiently large distance, the rays arriving from a uniform strip (Section 27.10) appear as a narrow *beam of simple plane waves*: that is, the phase across a perpendicular cross-section is constant. (In an optical case the beams may be brought to a focus, giving a diffraction pattern on a photographic plate.)

27.30 x, y, and z are any three (suitable) functions. Prove that
(a) $x(t) * \{Ay(t) + Bz(t)\} = Ax(t) * y(t) + Bx(t) * z(t)$.
(b) $x(t) * y(t) = y(t) * x(t)$.
(c) $x(t) * \{y(t) * z(t)\} = \{x(t) * y(t)\} * z(t)$ (i.e. the brackets may be omitted).

28 Differentiation of functions of two variables

CONTENTS

28.1 Functions of more than one variable

Quantities in nature usually depend on, or are functions of, more than one variable. The elevation H of land above sea level depends on two map coordinates x and y; so H is a function of the two variables x and y, and we write $H(x, y)$. If we want to take account of geological changes, then time t becomes a consideration, and in that case H is a function of three variables x, y, t, and we write $H(x, y, t)$. It is easy to produce examples involving many variables; for example, the distance between two points $P : (x_1, y_1, z_1)$ and $Q : (x_2, y_2, z_2)$ is a function of six variables. The state of the economy is a function of a multitude of variables. We alternatively speak of a function in one, two, three, … **dimensions**.

Suppose that a quantity z, called the **dependent variable**, depends on two **independent variables** x and y. The dependence can often be expressed by an explicit **formula** such as

$$z = x^3 + y^3, \qquad z = e^{x-2y}, \qquad z = |xy|,$$

and so on. To make statements which apply to all sorts of dependence we use the notation

$$z = f(x, y),$$

or $z = g(x, y)$ etc. The letter f on its own signifies a particular **function** or **process**: a computer subroutine, a particular formula, or a set of rules which will generate a single number z when two numbers x and y are fed to it in the right order.

Thus, if

$$f(x, y) = 2x + y^2,$$

then

$$f(3, -2) = (2 \times 3) + (-2)^2 = 10,$$
$$f(-2, 3) = [2 \times (-2)] + 3^2 = 5,$$
$$f(a, b) = 2a + b^2, \qquad f(u^2, v) = 2u^2 + v^2,$$
$$f(-x, x) = -2x + x^2, \qquad f(y, x) = 2y + x^2.$$

Notice the last one particularly: it is different from $f(x, y)$.

28.2 Depiction of functions of two variables

Consider the particular function

$$f(x, y) = x^2 + y^2.$$

Set up x, y, z axes; put

$$z = x^2 + y^2,$$

and proceed as if plotting a graph. Take a large number of pairs (x, y), work out z for each, then put the point (x, y, z) in the axes. For example, if $x = 1$ and $y = 2$, then $z = 5$ and we 'plot' the point $(1, 2, 5)$ as shown in Fig. 28.1a. For Fig. 28.1b, a great number of points is supposed to have been plotted. They cover a surface shaped like an inverted bowl.

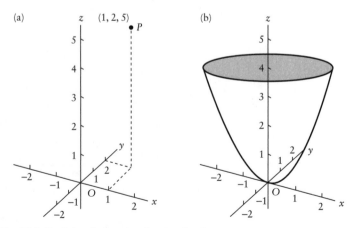

Fig. 28.1 Depicting the function $f(x, y) = x^2 + y^2$.

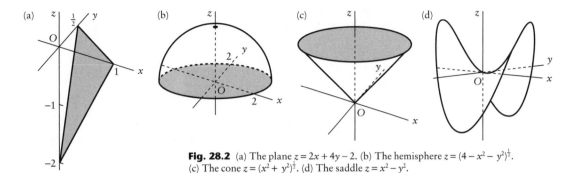

Fig. 28.2 (a) The plane $z = 2x + 4y - 2$. (b) The hemisphere $z = (4 - x^2 - y^2)^{\frac{1}{2}}$. (c) The cone $z = (x^2 + y^2)^{\frac{1}{2}}$. (d) The saddle $z = x^2 - y^2$.

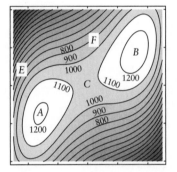

Fig. 28.3

Every function has a characteristic **surface** shape, which is the analogue in three dimensions of the graphs used for functions of a single variable. Some other functions are depicted in Fig. 28.2.

Another way of depicting a function is to sketch its **contour map** consisting of its **level curves**. Figure 28.3 shows a contour map of a patch of countryside. Along each contour the height is constant, and is indicated on the curve. The important features of the terrain are very easy to pick out; there are peaks at A and B, a pass at C (which is a 'saddle' as in Fig. 28.2d), valleys north west and south east of C and ascents north east and south west of C. At E the contours are close together, so the slope is steep, and at F the contours are widely spaced so the slope is comparatively gentle.

Consider again the function $f(x, y) = x^2 + y^2$ depicted in Fig. 28.1. The contour of height c is the circle

$$x^2 + y^2 = c,$$

where $c > 0$, which is a circle of radius $c^{\frac{1}{2}}$, as shown in Fig. 28.4a. This can be visualized as in Fig. 28.4b, as a horizontal slice of the surface $z = x^2 + y^2$ at height c, projected on to the (x, y) plane.

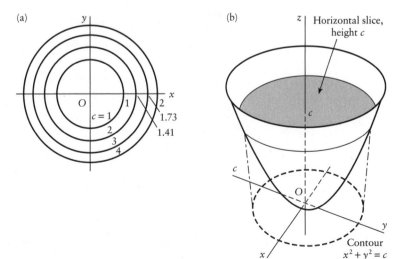

Fig. 28.4

Example 28.1   *Sketch the contours of the function xy.*

The contour of height c is given by the equation

$$z = xy = c,$$

or

$$y = c/x.$$

These curves are known as **rectangular hyperbolas**. By varying c, taking positive and negative values, the contour map or level curves of Fig. 28.5 are obtained.

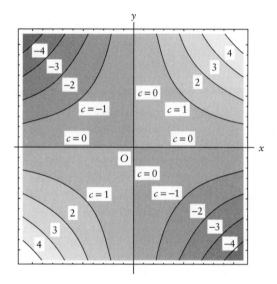

Fig. 28.5

28.3 Partial derivatives

Suppose that $z = f(x, y)$ represents the height above sea level of a piece of countryside.

In Fig. 28.6a, an observer stands at the point $P : (x, y)$, facing east, in the direction of the x axis. A short step forward takes the observer

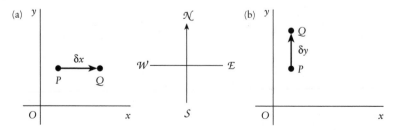

Fig. 28.6

to $Q : (x + \delta x, y)$, up or down a slope. The altitude changes by an amount

$$\delta z = f(x + \delta x, y) - f(x, y).$$

The **average slope** in this direction over the step length δx is $\delta z / \delta x$, so the slope at P facing the observer is given by

$$\lim_{\delta x \to 0} \frac{\delta z}{\delta x} = \lim_{\delta x \to 0} \frac{f(x + \delta x, y) - f(x, y)}{\delta x}.$$

Since the variable y is constant during the step, this is in effect an ordinary derivative, taken with respect to x *only*. However, it is customary to signal that another variable is present, which is done by using the special sign ∂ (still called 'dee') instead of the usual d for the derivative, writing

$$\frac{\partial f}{\partial x} \quad \text{or} \quad \frac{\partial z}{\partial x}$$

instead of df/dx or dz/dx. This is called the **partial derivative of** $f(x, y)$**, or of** z**, with respect to** x.

If the observer faces north and takes a step δy, as in Fig. 28.6b, then we obtain in the same way the slope $\partial f / \partial y$ or $\partial z / \partial y$ in the y direction.

Partial derivatives

If $z = f(x, y)$, then

$$\frac{\partial f}{\partial x} \text{ or } \frac{\partial z}{\partial x} = \lim_{\delta x \to 0} \frac{f(x + \delta x, y) - f(x, y)}{\delta x},$$

$$\frac{\partial f}{\partial y} \text{ or } \frac{\partial z}{\partial y} = \lim_{\delta y \to 0} \frac{f(x, y + \delta y) - f(x, y)}{\delta y}.$$

(28.1)

Example 28.2 *Find* $\partial z / \partial x$ *and* $\partial z / \partial y$ *at the point* $x = 1, y = 3$ *when*
$z = x^2 y + 2x^2 - 3y + 4$.

For $\partial z / \partial x$, y has the status of a constant for the purpose of the differentiation, so

$$\frac{\partial z}{\partial x} = 2xy + 4x - 0 + 0 = 2xy + 4x.$$

At the point $(1, 3)$, $\partial z / \partial x = 10$.

For $\partial z / \partial y$, x is treated as constant, so

$$\frac{\partial z}{\partial y} = x^2 + 0 - 3 + 0 = x^2 - 3.$$

At $(1, 3)$, $\partial z / \partial y = -2$.

We often need to indicate the particular point at which a derivative is to be evaluated, like the point $(1, 3)$ in the previous example. There are many notations in use for this purpose. We use

$$\left(\frac{\partial z}{\partial x}\right)_{(a,b)} \quad \text{and} \quad \left(\frac{\partial f}{\partial x}\right)_{(a,b)}$$

or

$$\left(\frac{\partial z}{\partial y}\right)_{P} \quad \text{and} \quad \left(\frac{\partial f}{\partial y}\right)_{P}$$

to mean the derivatives are to be evaluated at $P : (a, b)$. In this connection, the following definitions are equivalent to (28.1):

Partial derivatives at (a, b)

$$\left(\frac{\partial z}{\partial x}\right)_{(a,b)} \quad \text{or} \quad \left(\frac{\partial f}{\partial x}\right)_{(a,b)} = \lim_{x \to a} \frac{f(x, b) - f(a, b)}{x - a},$$

$$\left(\frac{\partial z}{\partial y}\right)_{(a,b)} \quad \text{or} \quad \left(\frac{\partial f}{\partial y}\right)_{(a,b)} = \lim_{y \to b} \frac{f(a, y) - f(a, b)}{y - b}.$$

$$(28.2)$$

Example 28.3   *Obtain* (a) $\dfrac{\partial}{\partial x}\left(\dfrac{x}{x + y}\right)$; (b) $\dfrac{\partial}{\partial y}\left(\dfrac{1}{(x^2 + y^2)^{\frac{1}{2}}}\right)$.

(a) We hold y constant and use the quotient rule (3.2):

$$\frac{\partial}{\partial x}\left(\frac{x}{x + y}\right) = \left((x + y)\frac{\partial x}{\partial x} - x\frac{\partial}{\partial x}(x + y)\right)\Big/(x + y)^2$$

$$= \frac{y}{(x + y)^2},$$

since $\partial x/\partial x = 1$, and y is constant.

(b) x is held constant. Use the chain rule (3.3), putting

$$u = x^2 + y^2, \qquad z = u^{-\frac{1}{2}};$$

then

$$\frac{\partial z}{\partial y} = \frac{dz}{du}\frac{\partial u}{\partial y}.$$

(We write $\partial u/\partial y$ instead of du/dy in the chain rule because both x and y are present in u, and x is being held constant.) Continuing, we have

$$\frac{\partial z}{\partial y} = (-\tfrac{1}{2}u^{-\frac{3}{2}})(2y) = -y(x^2 + y^2)^{-\frac{3}{2}}.$$

Example 28.4 *The potential function $V(x, t) = A\,e^{-qt}\sin k(x - ct)$ represents an attenuating wave travelling to the right along a cable with speed c. Here A, q, k, c are constants. Find*
(a) the rate of change of V with time t at any fixed point x;
(b) the 'potential gradient' $\partial V/\partial x$ along the wire at any moment.

(a) For $\partial V/\partial t$, use the product rule (3.1) with $u = A\,e^{-qt}$ and $v = \sin k(x - ct)$. We treat x as constant, so $\partial v/\partial t$ instead of dv/dt will be written into the product rule:

$$\frac{\partial V}{\partial t} = \frac{\partial(uv)}{\partial t}$$

$$= u\frac{\partial v}{\partial t} + v\frac{du}{dt}$$

$$= A\,e^{-qt}[-kc\cos k(x - ct)] + (-qA\,e^{-qt})\sin k(x - ct)$$

$$= -A\,e^{-qt}[kc\cos k(x - ct) + q\sin k(x - ct)].$$

(b) $\dfrac{\partial V}{\partial x} = A\,e^{-qt}\dfrac{\partial}{\partial x}\sin k(x - ct)$

$$= k\,A\,e^{-qt}\cos k(x - ct),$$

t being treated as constant.

It will be seen that **no new rules have to be learned** in order to obtain the partial derivatives of given functions. In fact you have always unconsciously carried out partial differentiation when differentiating expressions like $A\sin(\omega t + \phi)$, without worrying whether A, ω, ϕ were really constants or just to be treated as such while differentiating.

28.4 Higher derivatives

Having differentiated a function, we might want to differentiate it again. If $z = f(x, y)$, we can form $\dfrac{\partial z}{\partial x}$ and then $\dfrac{\partial}{\partial x}\left(\dfrac{\partial z}{\partial x}\right)$ or $\dfrac{\partial}{\partial y}\left(\dfrac{\partial z}{\partial x}\right)$, thus forming **second derivatives**, or derivatives higher than the second. There are four second derivatives, written as follows:

$$\frac{\partial}{\partial x}\left(\frac{\partial z}{\partial x}\right) = \frac{\partial^2 z}{\partial x^2}, \qquad \frac{\partial}{\partial y}\left(\frac{\partial z}{\partial x}\right) = \frac{\partial^2 z}{\partial y\,\partial x},$$

$$\frac{\partial}{\partial x}\left(\frac{\partial z}{\partial y}\right) = \frac{\partial^2 z}{\partial x\,\partial y}, \qquad \frac{\partial}{\partial y}\left(\frac{\partial z}{\partial y}\right) = \frac{\partial^2 z}{\partial y^2}.$$

Example 28.5 *Obtain the four second derivatives when*

$$z = x^3y + xy^2 + x + y^2 + 1.$$

The first derivatives are

$$\frac{\partial z}{\partial x} = 3x^2y + y^2 + 1, \qquad \frac{\partial z}{\partial y} = x^3 + 2xy + 2y.$$

Therefore

$$\frac{\partial^2 z}{\partial x^2} = \frac{\partial}{\partial x}\left(\frac{\partial z}{\partial x}\right) = 6xy,$$

$$\frac{\partial^2 z}{\partial y\,\partial x} = \frac{\partial}{\partial y}\left(\frac{\partial z}{\partial x}\right) = 3x^2 + 2y,$$

$$\frac{\partial^2 z}{\partial x\,\partial y} = \frac{\partial}{\partial x}\left(\frac{\partial z}{\partial y}\right) = 3x^2 + 2y,$$

$$\frac{\partial^2 z}{\partial y^2} = \frac{\partial}{\partial y}\left(\frac{\partial z}{\partial y}\right) = 2x + 2.$$

In the last example, we see that the **mixed derivatives** satisfy $\partial^2 z/\partial y\,\partial x = \partial^2 z/\partial x\,\partial y$. This is always true for normal functions, although the proof is difficult:

> **Mixed derivatives**
>
> For any function $f(x, y)$,
>
> $$\frac{\partial^2 f}{\partial y\,\partial x} = \frac{\partial^2 f}{\partial x\,\partial y}.$$
>
> In higher derivatives, the ∂x and ∂y in the denominator may be arranged in any order.
>
> **(28.3)**

For example, $\dfrac{\partial^3 f}{\partial x\,\partial y^2} = \dfrac{\partial^3 f}{\partial y^2\,\partial x} = \dfrac{\partial^3 f}{\partial y\,\partial x\,\partial y}$, and so on.

The next example shows how to manage a problem in notation. Often a function $f(x, y)$ is used in which the variables x and y only occur in a fixed combination $u = h(x, y)$, so that

$$f(x, y) = g(u), \quad \text{with} \quad u = h(x, y),$$

where g represents a general, unspecified, **function of a single variable**. To obtain a general formula for $\partial f/\partial x$ use the chain rule (3.3) (see also Example 4.2c):

$$\frac{\partial f}{\partial x} = \frac{\mathrm{d}g}{\mathrm{d}u}\frac{\partial u}{\partial x} = g'(u)\frac{\partial u}{\partial x} = \frac{\partial h}{\partial x}g'[h(x, y)].$$

It is a common mistake to write $\partial g/\partial x$ instead of $g'[h(x, y)]$ in this context, presumably misreading the chain rule. You must work out

$g'(u)$ first, before substituting $u = h(x, y)$. Thus suppose that $f(x, y) = g(5x - 3y)$; then

$$\frac{\partial f}{\partial x} = 5g'(5x - 3y) \text{ and } \frac{\partial f}{\partial y} = -3g'(5x - 3y).$$

If $z = g(u)$, where $u = h(x, y)$, then

$$\frac{\partial z}{\partial x} = \frac{dg}{du}\frac{\partial u}{\partial x} = g'(h(x, y))\frac{\partial h}{\partial x},$$

$$\frac{\partial z}{\partial y} = \frac{dg}{du}\frac{\partial u}{\partial y} = g'(h(x, y))\frac{\partial h}{\partial y}.$$

(28.4)

Example 28.6 *Prove that if $z = \phi(x - ct)$, where ϕ is any function, then*

$$\frac{\partial^2 z}{\partial x^2} = \frac{1}{c^2}\frac{\partial^2 z}{\partial t^2}.$$

Put $z = \phi(u)$ where $u = x - ct$. Then

$$\frac{\partial z}{\partial x} = \frac{d\phi}{du}\frac{\partial u}{\partial x} = \phi'(u).$$

By the chain rule again,

$$\frac{\partial^2 z}{\partial x^2} = \frac{\partial}{\partial x}\phi'(u) = \frac{d\phi'(u)}{du}\frac{\partial u}{\partial x} = \phi''(u).$$

Similarly

$$\frac{\partial z}{\partial t} = \frac{d\phi}{du}\frac{\partial u}{\partial t} = \phi'(u)(-c),$$

so

$$\frac{\partial^2 z}{\partial t^2} = \frac{\partial}{\partial t}[-c\phi'(u)] = \frac{d}{du}[-c\phi'(u)]\frac{\partial u}{\partial t} = (-c)^2\phi''(u).$$

Therefore

$$\frac{\partial^2 z}{\partial x^2} = \frac{1}{c^2}\frac{\partial^2 z}{\partial t^2}.$$

The equation

$$\frac{\partial^2 z}{\partial x^2} = \frac{1}{c^2}\frac{\partial^2 z}{\partial t^2}$$

in Example 28.6 is called the **wave equation** in one space dimension. It is a **partial differential equation** as contrasted with the **ordinary**

differential equations treated earlier in the book. We have verified that $\phi(x - ct)$ is always a solution, for any function ϕ. The general solution is

$$\phi(x - ct) + \psi(x + ct),$$

where ϕ and ψ are arbitrary functions. The general solution of partial differential equations involves arbitrary functions rather than the arbitrary constants which occur in ordinary differential equations: even the simple equation $\partial z / \partial x = 0$ has the general solution $z = f(y)$, where $f(y)$ is an arbitrary function.

28.5 Tangent plane and normal to a surface

The tangent plane to a surface $z = f(x, y)$ at a point Q on the surface plays the same role as the tangent line to a curve for functions of a single variable. The tangent plane is the plane that fits the surface near Q better than any other possible plane, as when a coin is pressed against a teapot at a particular point.

Suppose that the tangent plane at Q (Fig. 28.7) has the equation

$$z = Ax + By + C.$$

There are three constants to be determined, so we need three conditions to settle the values. The conditions it is reasonable to expect the tangent plane to satisfy are

(i) It must pass through Q; so $c = Aa + Bb + C$.

(ii) In the x direction at Q, the slope A of the *plane* must be equal to the slope of the *surface*; so

$$A = \left(\frac{\partial f}{\partial x}\right)_Q.$$

(iii) In the y direction at Q the slope B of the *plane* must be equal to the slope of the *surface*; so

$$B = \left(\frac{\partial f}{\partial y}\right)_Q.$$

Then the equation for the tangent plane becomes

$$z = \left(\frac{\partial f}{\partial x}\right)_Q x + \left(\frac{\partial f}{\partial y}\right)_Q y + \left[c - \left(\frac{\partial f}{\partial x}\right)_Q a - \left(\frac{\partial f}{\partial y}\right)_Q b\right],$$

or, more tidily,

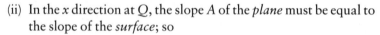

$$z - c = \left(\frac{\partial f}{\partial x}\right)_Q (x - a) + \left(\frac{\partial f}{\partial y}\right)_Q (y - b),$$

where the values of the coefficients are to be calculated using $z = f(x, y)$.

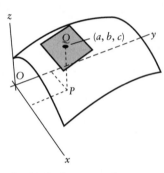

Fig. 28.7 The tangent plane to $z = f(x, y)$ at $Q : (a, b, c)$, where $c = f(a, b)$.

Tangent plane at $Q: (a, b, c)$ on the surface $z = f(x, y)$

$$z - c = \left(\frac{\partial f}{\partial x}\right)_{(a,b)} (x - a) + \left(\frac{\partial f}{\partial y}\right)_{(a,b)} (y - b).$$

(28.5)

Example 28.7 *Find the equation of the tangent plane at the point $Q: (2, 1, -2)$ on the sphere $x^2 + y^2 + z^2 = 9$.*

Recast the equation into the form $z = f(x, y)$, noticing that Q is on the lower half of the sphere:

$$z = -(9 - x^2 - y^2)^{\frac{1}{2}}.$$

Work out the coefficients first. The chain rule gives:

$$\frac{\partial f}{\partial x} = -(-2x) \cdot \tfrac{1}{2}(9 - x^2 - y^2)^{-\frac{1}{2}}, \quad \text{and} \quad \left(\frac{\partial z}{\partial x}\right)_{(2,1)} = 1;$$

$$\frac{\partial f}{\partial y} = -(-2y) \cdot \tfrac{1}{2}(9 - x^2 - y^2)^{-\frac{1}{2}}, \quad \text{and} \quad \left(\frac{\partial z}{\partial y}\right)_{(2,1)} = \tfrac{1}{2}.$$

Therefore the equation of the tangent plane at Q is

$$z - (-2) = 1(x - 2) + \tfrac{1}{2}(y - 1),$$

or

$$z = x + \tfrac{1}{2}y - \tfrac{9}{2}.$$

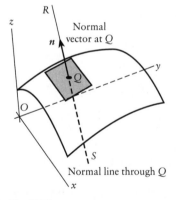

R

Normal vector at Q

n

z

y

O

Q

S

Normal line through Q

x

Fig. 28.8

A straight line SQR (Fig. 28.8) is said to be **normal** or perpendicular to the surface $z = f(x, y)$ at Q if it is perpendicular to its tangent plane at Q. The equation (28.5) for the tangent plane can be written in the form

$$\left(\frac{\partial f}{\partial x}\right)_Q x + \left(\frac{\partial f}{\partial y}\right)_Q y + (-1)z = C,$$

where C is a constant, so (see eqn (10.22)) a triplet of direction ratios for the line normal to the surface at Q is

$$\left(\left(\frac{\partial f}{\partial x}\right)_Q, \left(\frac{\partial f}{\partial y}\right)_Q, -1\right).$$

(28.6)

Example 28.8 *Find the cartesian (x, y, z) equation of the straight line normal to the surface $x^2 + y^2 + z^2 = 9$ at $(2, 1, -2)$.*

From Example 28.5 (which has the same data), the direction ratios in (28.6) are

$$1, \quad \tfrac{1}{2}, \quad -1.$$

Therefore the equation of the normal line at Q is

$$\frac{x - 2}{1} = \frac{y - 1}{\frac{1}{2}} = \frac{z + 2}{-1}.$$

The triplet of direction ratios in (28.6) can be regarded as the three components of a vector parallel to the normal line. Such a vector is called a **normal vector** at Q, and is denoted usually by $\boldsymbol{n}$:

$$\boldsymbol{n} = \left(\left(\frac{\partial z}{\partial x} \right)_Q, \left(\frac{\partial z}{\partial y} \right)_Q, -1 \right).$$

Any multiple of this vector is another normal vector, since it will be parallel to the same line. A normal vector placed at Q is shown in Fig. 28.8.

Normal vector $\boldsymbol{n}$ at Q : (a, b, c) where $c = f(a, b)$, on the surface $z = f(x, y)$

$$\boldsymbol{n} = \left(\frac{\partial z}{\partial x} \right)_Q \hat{\boldsymbol{i}} + \left(\frac{\partial z}{\partial y} \right)_Q \hat{\boldsymbol{j}} + (-1)\hat{\boldsymbol{k}}$$

$$= \left(\left(\frac{\partial z}{\partial x} \right)_Q, \left(\frac{\partial z}{\partial y} \right)_Q, -1 \right),$$

or any multiple of this vector. Its components are direction ratios of the normal line at Q.

           (28.7)

Example 28.9 *Find several vectors normal to the sphere $x^2 + y^2 + z^2 = 9$ at the point $(2, 1, -2)$ on the sphere.*

The data are again the same as in Example 28.5. The normal taken from (28.7) is $(1, \frac{1}{2}, -1)$. Another is $(-1, -\frac{1}{2}, 1)$, pointing in the opposite direction, while $(\frac{2}{3}, \frac{1}{3}, -\frac{2}{3})$ is a unit vector which is a normal.

28.6   Maxima, minima, and other stationary points

For a function of a single variable, a local maximum or minimum or a point of inflection occurs where the tangent line to the graph of the function is horizontal. For a function $f(x, y)$ of two variables, there are similar possibilities at **points where the tangent plane is horizontal**. Such points, or rather their (x, y) coordinates, are called **stationary points** of $f(x, y)$, because as we pass through them the function is momentarily neither increasing nor decreasing. Sometimes a stationary point is a local **minimum or maximum** as illustrated in Fig. 28.9a, b.

The condition for the tangent plane at Q on $z = f(x, y)$ to be horizontal is that the normal $\boldsymbol{n}$ at Q should be vertical, or parallel

(a)

(b)

(c)

(d)

Fig. 28.9 (a) A local minimum. (b) A local maximum. (c) A saddle. (d) A shoulder.

to the z axis. Therefore the x and y components of $\mathbf{n}$ in (28.7) must be zero:

$$\frac{\partial f}{\partial x} = 0, \qquad \frac{\partial f}{\partial y} = 0.$$

These constitute two simultaneous equations whose solutions (x, y) are the stationary points of $f(x, y)$.

Stationary points of $f(x, y)$
are at the solutions (x, y) of

$$\frac{\partial f}{\partial x} = 0, \qquad \frac{\partial f}{\partial y} = 0.$$

(28.8)

We shall usually describe a stationary point of $f(x, y)$ as being 'at $P : (x, y)$' rather than 'at $Q : (x, y, z)$ on $z = f(x, y)$'. If necessary, the corresponding value of z can be worked out after finding (x, y).

Example 28.10 *Find the stationary points of*

$$f(x, y) = \tfrac{1}{3}x^3 - xy^2 - 2y,$$

and the value of $f(x, y)$ there.

Since $\partial f/\partial x = x^2 - y^2$ and $\partial f/\partial y = -2xy - 2$, stationary points occur where

$$x^2 - y^2 = 0, \qquad xy + 1 = 0.$$

The first equation is equivalent to $y = \pm x$. Consider these alternatives *separately*:

If $y = x$, the second equation becomes $x^2 + 1 = 0$, which has no solution. Therefore reject $y = x$.

If $y = -x$, the second equation becomes $-x^2 + 1 = 0$, which has solutions $x = \pm 1$. Corresponding to these we have

$$y = -x = \mp 1.$$

Therefore there are two stationary points, $(1, -1)$ and $(-1, 1)$. The values of $f(x, y)$ at these points are

$$f(1, -1) = \tfrac{4}{3}, \qquad f(-1, 1) = -\tfrac{4}{3}.$$

A stationary point at (a, b) is a **local maximum** if $f(a, b)$ is **greater than** $f(x, y)$ at all points in its immediate locality; it is a **local minimum** if the words 'less than' are substituted for 'greater than'. On a contour map, a maximum or minimum shows its presence by being surrounded by closed contours as in Fig. 28.10b.

As with functions of a single variable, the test for a maximum or minimum involves second derivatives. The following test enables maxima, minima, and other stationary points to be distinguished in most cases, but we omit the proof, which is difficult.

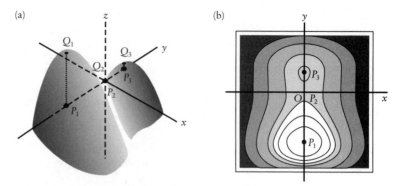

(a)

(b)

Fig. 28.10 (a) Local maxima at Q_1 and Q_3 and a saddle at Q_2.
(b) The contour map shows closed curves around the maxima.

Test for the character of a stationary point $P : (a, b)$ of $f(x, y)$

Suppose than $\partial f/\partial x = \partial f/\partial y = 0$ at P. Then P is

(a) a saddle if $\dfrac{\partial^2 f}{\partial x^2} \dfrac{\partial^2 f}{\partial y^2} - \left(\dfrac{\partial^2 f}{\partial x\, \partial y} \right)^2 < 0$ at P,

(b) a maximum if $\dfrac{\partial^2 f}{\partial x^2} \dfrac{\partial^2 f}{\partial y^2} - \left(\dfrac{\partial^2 f}{\partial x\, \partial y} \right)^2 > 0$

with $\dfrac{\partial^2 f}{\partial x^2} < 0 \left(\text{or } \dfrac{\partial^2 f}{\partial y^2} < 0 \right)$ at P,

(c) a minimum if $\dfrac{\partial^2 f}{\partial x^2} \dfrac{\partial^2 f}{\partial y^2} - \left(\dfrac{\partial^2 f}{\partial x\, \partial y} \right)^2 > 0$

with $\dfrac{\partial^2 f}{\partial x^2} > 0 \left(\text{or } \dfrac{\partial^2 f}{\partial y^2} > 0 \right)$ at P.

(d) If none of these apply, the point might be any type.   **(28.9)**

Example 28.11   *Find and classify the stationary points of*
$$f(x, y) = \tfrac{1}{3}x^3 + \tfrac{1}{3}y^3 - x^2 - y^2.$$

The stationary points are the solutions of $\partial f/\partial x = 0$, $\partial f/\partial y = 0$, or
$$x^2 - 2x = 0, \qquad y^2 - 2y = 0.$$
From the first, we obtain $x = 0$ or $x = 2$. From the second, $y = 0$ or $y = 2$. Therefore there are stationary points at $(0, 0)$, $(0, 2)$, $(2, 0)$, $(2, 2)$. To test them, we need the second derivatives at a general point:
$$\frac{\partial^2 f}{\partial x^2} = 2x - 2, \qquad \frac{\partial^2 f}{\partial y^2} = 2y - 2, \qquad \frac{\partial^2 f}{\partial x\, \partial y} = 0.$$
At $(0, 0)$, these become respectively $-2, -2, 0$. Then

Example 28.11 *continued*

$$\frac{\partial^2 f}{\partial x^2}\frac{\partial^2 f}{\partial y^2} - \left(\frac{\partial^2 f}{\partial x\,\partial y}\right)^2 = 4 > 0, \qquad \frac{\partial^2 f}{\partial x^2} = \frac{\partial^2 f}{\partial y^2} = -4 < 0.$$

Since the conditions of (28.9b) apply, the point is a maximum.

At $(0, 2)$ and $(2, 0)$,

$$\frac{\partial^2 f}{\partial x^2}\frac{\partial^2 f}{\partial y^2} - \left(\frac{\partial^2 f}{\partial x\,\partial y}\right)^2 = -4 < 0;$$

so, by (28.9a), both points are saddles.

At $(2, 2)$,

$$\frac{\partial^2 f}{\partial x^2}\frac{\partial^2 f}{\partial y^2} - \left(\frac{\partial^2 f}{\partial x\,\partial y}\right)^2 = 4 > 0, \qquad \frac{\partial^2 f}{\partial x^2} = \frac{\partial^2 f}{\partial y^2} = 4 > 0.$$

Therefore, by (28.9c), the point is a minimum.

28.7 The method of least squares

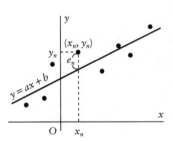

Fig. 28.11

Suppose that a succession of experiments is performed in which we vary one quantity x, such as voltage applied to a circuit, and measure the corresponding value of another variable y, say the resulting current. The values recorded for one or both of the variables might be subject to random errors of measurement; on a graph of the results, this will show up as scatter among the points, as in Fig. 28.11.

We might have reason to believe that the underlying relation between x and y is a straight line. There is no way of deducing this line with certainty, but the following method is often used to obtain a convincing straight line fit to the points.

Suppose that there are N points altogether; call them

$$(x_1, y_1), \quad (x_2, y_2), \quad \dots, \quad (x_N, y_N).$$

The general point is called (x_n, y_n). Figure 28.11 shows a candidate for the best-fitting straight line,

$$y = ax + b,$$

and we have to adjust the constants a and b to obtain a good fit. The vertical deviation e_n of a point (x_n, y_n) from the line is shown:

$$e_n = y_n - (ax_n + b).$$

The criterion we shall use to determine the best straight line is to choose a and b so that $\sum_{n=1}^{N} e_n^2$ is as small as possible; that is to say, we want to minimize

$$\sum_{n=1}^{N} e_n^2 = \sum_{n=1}^{N} (y_n - ax_n - b)^2 = f(a, b) \quad \text{(say)}.$$

Therefore a and b are the variables in this problem, and everything else has fixed values.

For a minimum, we require at least that

$$\frac{\partial f}{\partial a} = \frac{\partial f}{\partial b} = 0.$$

The derivatives are given by

$$\frac{\partial f}{\partial a} = \sum_{n=1}^{N} 2(-x_n)(y_n - ax_n - b) = 2\sum_{n=1}^{N}(ax_n^2 + bx_n - x_n y_n),$$

$$\frac{\partial f}{\partial b} = \sum_{n=1}^{N}(-2)(y_n - ax_n - b) = 2\sum_{n=1}^{N}(ax_n + b - y_n).$$

Noting that $\sum_{n=1}^{N} b = b + b + \cdots + b = Nb$, we find the conditions for a minimum as the following pair of simultaneous equations for a and b:

Method of least squares

To fit a straight line $y = ax + b$ to the N points (x_n, y_n) $(n = 1, 2, \ldots, N)$:

find a and b by solving the simultaneous equations

$$a\sum_{n=1}^{N} x_n^2 + b\sum_{n=1}^{N} x_n = \sum_{n=1}^{N} x_n y_n,$$

$$a\sum_{n=1}^{N} x_n + bN = \sum_{n=1}^{N} y_n.$$

(28.10)

We shall not prove that the stationary point of $f(a, b)$ found by this method is actually a minimum (see Problem 28.21).

Example 28.12   *Find the straight line which best fits the data:*

x_n	0.0	1.1	3.2	3.9	7.1	8.9
y_n	1.1	1.6	1.6	2.8	2.9	3.8

Here $N = 6$, and the coefficients in (28.10) are

$$\sum_{n=1}^{6} x_n = 24.2, \qquad \sum_{n=1}^{6} y_n = 13.8,$$

$$\sum_{n=1}^{6} x_n^2 = 156.28, \qquad \sum_{n=1}^{6} x_n y_n = 72.21.$$

The equations for a and b therefore become

$$156.28a + 24.2b = 72.21,$$
$$24.2a + 6b = 13.8.$$

By solving these we find that $a = 0.28$, $b = 1.16$, so the required line is $y = 0.28x + 1.16$.

The equations for a and b are sometimes **ill-conditioned**, meaning that the solutions are very sensitive to small changes in the coefficients. It is therefore advisable to retain all the significant figures given by the data while solving them, despite the fact that we know they already embody the errors of measurement.

28.8 Differentiating an integral with respect to a parameter

Suppose that we have an integral whose integrand contains a parameter α as well as the variable of integration – for example,

$$\int_0^1 e^{\alpha t}\,dt, \qquad \int_{-\infty}^{\infty} g(x)h(x+\alpha)\,dx, \qquad \int \frac{dx}{x+\alpha}.$$

We shall consider a definite integral, though the process works in the same way for indefinite integrals. Indicate the dependence on α in the general case by

$$I(\alpha) = \int_a^b f(t, \alpha)\,dt.$$

Then $dI(\alpha)/d\alpha$ can be obtained by the following rule:

> **Differentiating an integral with respect to a parameter**
>
> If $\displaystyle\int_a^b f(t, \alpha)\,dt = I(\alpha)$, then
>
> $$\frac{dI(\alpha)}{d\alpha} = \int_a^b \frac{\partial f(t, \alpha)}{\partial \alpha}\,dt.$$
>
> (28.11)

This process is also called **differentiation under the integral sign**. To prove (28.11), change α to $\alpha + \delta\alpha$; then $I(\alpha)$ changes to $I(\alpha + \delta\alpha)$. Put

$$I(\alpha + \delta\alpha) - I(\alpha) = \delta I(\alpha).$$

Then

$$\frac{\delta I(\alpha)}{\delta\alpha} = \frac{I(\alpha + \delta\alpha) - I(\alpha)}{\delta\alpha}$$

$$= \frac{1}{\delta\alpha}\left(\int_a^b f(t, \alpha + \delta\alpha)\,dt - \int_a^b f(t, \alpha)\,dt\right)$$

$$= \int_a^b \frac{f(t, \alpha + \delta\alpha) - f(t, \alpha)}{\delta\alpha}\,dt.$$

Now let $\delta\alpha \to 0$. Then $\delta I(\alpha)/\delta\alpha$ becomes $dI(\alpha)/d\alpha$, and the integrand becomes $\partial f(t, \alpha)/\partial\alpha$, which is the result (28.11).

Example 28.13   *Evaluate* $I(\alpha) = \displaystyle\int_0^\infty \dfrac{dt}{t^2 + \alpha^2}$, *where* $\alpha > 0$, *and use*

(28.11) *to evaluate* $J(\alpha) = \displaystyle\int_0^\infty \dfrac{dt}{(t^2 + \alpha^2)^2}$.

From Appendix E,

$$I(\alpha) = \int_0^\infty \frac{dt}{t^2 + \alpha^2} = [\alpha^{-1} \arctan(t/\alpha)]_0^\infty = \frac{\pi}{2\alpha}.$$

By (28.11),

$$\frac{dI}{d\alpha} = \int_0^\infty \frac{\partial}{\partial \alpha} \frac{1}{t^2 + \alpha^2}\, dt = \frac{d}{d\alpha} \frac{\pi}{2\alpha},$$

or

$$\int_0^\infty \frac{-2\alpha}{(t^2 + \alpha^2)^2}\, dt = -\frac{\pi}{2\alpha^2}.$$

Therefore

$$J(\alpha) = \int_0^\infty \frac{dt}{(t^2 + \alpha^2)^2} = \frac{\pi}{4\alpha^3}.$$

Problems

28.1 Sketch contour maps of the following functions:
(a) $2x - 3y + 4$;   (b) $-x + 2y - 1$;
(c) $(x - 1)(y - 1)$;   (d) $x^2 + \frac{1}{4}y^2 - 1$;
(e) $x^2 + 2x + y^2$ (complete the square in x);
(f) y/x;   (g) $y^2 - x^2$;   (h) y/x^3;
(i) $x^3 + 4y^2$;   (j) $y/(x + y)$.

28.2 By sketching rough contour maps, indicate the paths of steepest ascent (the paths on which z increases most rapidly), starting at the point $(1, 1)$:
(a) $z = 2x - 3y + 4$;   (b) $z = x - y$;
(c) $z = x^2 y^2$;   (d) $z = (x - 1)^2 + \frac{1}{4}(y - 1)^2$.

28.3 Obtain $\partial f/\partial x$ and $\partial f/\partial y$ at the point $(2, 1)$ for the following functions.
(a) $3x + 7y - 2$;   (b) $-2x + 3y + 4$;
(c) $2x^2 - 3y^2 - 2xy - x - y + 1$;
(d) $\frac{1}{8}x^3 + y^3 - 2y - 1$;   (e) $x^4 y^2 - 1$;
(f) $(x - 1)(y - 2)$;   (g) $1/(xy)$;
(h) x/y;   (i) $\dfrac{x - y}{x + y}$;   (j) $\dfrac{3}{x^2 + y^2}$;

(k) $(x^2 + y^2)^{\frac{1}{2}}$;   (l) $(2x - 3y + 2)^3$;   (m) $e^{x^2 + y^2}$;
(n) $\cos(x^2 - y^2)$;   (o) $\sin(x/y)$;   (p) $\arctan(y/x)$.

28.4 (a) Let $z = g(ax + by)$, where a and b are constants. Express $\partial z/\partial x$ and $\partial z/\partial y$ in terms of $g'(ax + by)$ (which means $g'(u)$ when u is subsequently put equal to $ax + by$). Check your result for the cases when $g(u) = \cos u$ and $g(u) = e^u$.

(b) Let $z = g(\sin xy)$. Express $\partial z/\partial x$ and $\partial z/\partial y$ in terms of x, y, and $g'(\sin xy)$. Check the result by differentiating $e^{\sin xy}$ directly.

(c) A certain physical quantity V is a function only of the radial coordinate r in plane polar coordinates: $V = g(r)$, where $r = (x^2 + y^2)^{\frac{1}{2}}$. Express $\partial V/\partial x$ and $\partial V/\partial y$, firstly in terms of x and y, then in terms of r and θ.

28.5 In plane polar coordinates (r, θ) in the first quadrant, $r = (x^2 + y^2)^{\frac{1}{2}}$ and $x = r \cos \theta$. Form $\partial r/\partial x$ and $\partial x/\partial r$, and show that

$$\frac{\partial r}{\partial x} \frac{\partial x}{\partial r} \neq 1.$$

By considering the meaning of the derivatives $\partial r/\partial x$ and $\partial x/\partial r$ near a particular point P in the manner of Fig. 28.6, show why it is not to be expected that the product should equal 1. (In the case of a single variable and ordinary derivatives, we often get true results by formally cancelling out symbols like dx, du, etc., as in the chain rule. This almost never works when more variables are present: see for example the next problem.)

28.6 (a) Let $z = \sin(x - y)$; show that $\dfrac{\partial z}{\partial x} \Big/ \dfrac{\partial z}{\partial y} = -1$.

(b) Let $z = g(x - y)$; show that $\dfrac{\partial z}{\partial x} \Big/ \dfrac{\partial z}{\partial y} = -1$.

28.7 Show that, if $z = g(x/y)$, then

$$x\frac{\partial z}{\partial x} + y\frac{\partial z}{\partial y} = 0,$$

and check the result in the case $z = \sin x/y$.

28.8 Find $\dfrac{\partial^2 f}{\partial x^2}, \dfrac{\partial^2 f}{\partial y^2}, \dfrac{\partial^2 f}{\partial y\,\partial x}, \dfrac{\partial^2 f}{\partial x\,\partial y}$ in each of the following cases (see (28.3)).
(a) $ax + by + c$; (b) $x^2 + 2y^2 + 3xy - x + 1$;
(c) $\sin(x - y)$; (d) y/x; (e) e^{2x+3y};
(f) $1/x + 1/y$; (g) $\sin 3x + \cos 2y$; (h) $(3x - 4y)^4$;
(i) $1/(x + y)$; (j) $\ln xy$; (k) $1/(x^2 + y^2)^{\frac{1}{2}}$.

28.9 Confirm that, if $r = (x^2 + y^2)^{\frac{1}{2}}$ and $z = \ln r$, then

$$\frac{\partial z}{\partial x} = \frac{x}{r^2} \quad \text{and} \quad \frac{\partial^2 z}{\partial x^2} = \frac{1}{r^2} - \frac{2x^2}{r^4}.$$

Show that $z = \ln r$ is a solution of the equation

$$\frac{\partial^2 z}{\partial x^2} + \frac{\partial^2 z}{\partial y^2} = 0.$$

(This is called Laplace's partial differential equation in two dimensions.)

28.10 Obtain the tangent plane and a normal vector for the following surfaces at the points given.
(a) $z = x^2 + y^2$ at $(1, 1, 2)$; (b) $z = xy$ at $(2, 2, 4)$;
(c) $z = x/y$ at $(2, 1, 2)$;
(d) $z = (29 - x^2 - y^2)^{\frac{1}{2}}$ at $(3, 4, 2)$;
(e) $z = x^2 + y^2 - 2x - 2y$ at $(1, 1, -2)$;
(f) $z = e^{xy}$ at $(0, 0, 1)$.

28.11 The two surfaces $z = x^2 + y^2$ and $z = x - y + 2$ intersect at the point $Q : (1, 1, 2)$. Find normal vectors at Q to each of the two surfaces, n_1 to the first and n_2 to the second. By considering the scalar product $n_1 \cdot n_2$, find the angle between the normals and hence the angle at which the surfaces cut at Q.

28.12 Find the stationary points of the following functions, and classify them using (28.9).
(a) $(x - 1)(y + 2)$; (b) $x^2 + y^2 - 2x + 2y$;
(c) $\frac{1}{3}x^3 - \frac{1}{3}y^3 - x + y + 3$; (d) $\cos x + \cos y$;
(e) $\ln(x^2 + x) + \ln(y^2 + y)$; (f) $e^{x^2+y^2-2x+2y}$;
(g) $xy + 1/x + 1/y$; (h) $x^3 + y^3 - 3xy + 1$;
(i) $\sin x + \sin y$; (j) $xy^2 - x^2y + x - y + 1$;
(k) $(x^2 - y^2) + 2xy$; (l) $(2 - x^2 - y^2)^2$;
(m) $x^4 + y^4 + y - x$;
(n) $x^4 + y^4$ (this eludes the test (28.9) – the point is obviously a minimum).

28.13 Classify the stationary point of $ax^2 + 2hxy + by^2$ at $(0, 0)$ for various relations between a, b, and h.

28.14 Find positive numbers a, b, c so that
(a) $a + b + c = 21$ and abc is a maximum.
(b) $abc = 64$ and $a + b + c$ is a minimum.

28.15 Find the *absolute maximum* value of $(2 - x^2 - y^2)^2$ in the 'box' $-1 \leqslant x \leqslant 2$, $-1 \leqslant y \leqslant 1$. (It will be necessary to investigate the function on the four edges of the box separately, since the *absolute* maximum will not be revealed by the conditions (28.9) if it is on the edges.)

28.16 Find the shortest distance between the straight lines $x = y = z$ and $2x = y = z + 2$, by using a simple parametrization of each line. (Use different letters for the two parameters: these will be the new variables for the minimization.)

28.17 N points $(x_1, y_1), (x_2, y_2), \ldots, (x_N, y_N)$ are given in a plane, and $P : (x, y)$ is a general point. Find P so that the sum of the squares of its distances from the N given points is as small as possible.

28.18 (a) A rectangular box with a lid must hold a given volume V, and have the smallest possible surface area. Show that it must be a cube. (Call the lengths of two of its sides x and y.)
(b) An open-topped rectangular box must have a given volume V and its surface area must be as small as possible. Find its dimensions.
(c) A circular-cylindrical box must have a fixed volume V and minimum surface area. Find its dimensions (i) if it has a lid, (ii) if it has no lid.
(d) A rectangular container is required to have total surface area S, and a volume as large as possible. Find its dimensions (i) if it has a lid, (ii) if it does not have a lid.

28.19 Find the straight line which best fits the experimental data in the sense of Section 28.7:

x	1	2	3	4	5
y	3.1	2.1	2.0	1.8	1.2

28.20 The population P of a fast-breeding rodent was observed over a period of 12 months, and the following estimates obtained:

t (months)	0	2	3	5	8	10	12
P (pop'n)	12	23	26	60	170	300	690

Assume that the underlying growth law takes the form (see Section 1.12)

$$P = A\,e^{bt},$$

where A and b are constants.

To estimate A and b, take the logarithm of this expression and treat $y = \ln P$ as a variable in the least-squares method of Section 28.7.

28.21 For the least-squares method of Section 28.7, use the test (28.9) to show that the values of a and b obtained do minimize the sum of squares. (This is, of course, rather obvious intuitively.)

28.22 Using Laplace transforms with respect to t, solve the partial differential equation

$$\frac{\partial z}{\partial t} + x\frac{\partial z}{\partial x} + z = 2x,$$

for $x > 0$ and $t > 0$, where $z(0, t) = 0$ and $z(x, 0) = 0$.

28.23 A grain silo of height $2a$ with a square floor on $z = 0$ has vertical sides given by $x = 0$, $x = a$, $y = 0$, $y = a$ in x, y, z space. Grain is poured into the silo and eventually settles with a surface given by

$$z = [2a^2 - (x - \tfrac{1}{2}a)^2 - y^2]/a.$$

Find the highest and lowest points of the surface of the grain in the silo. (Note: the lowest point(s) are not stationary points.)

28.24 If $z = f(x, y)$, how many nth-order partial derivatives of $f(x, y)$ are these of the form $\partial^n f/\partial x^r \partial y^{n-r}$ assuming that the order of differentiation is immaterial? How many would there be if the order did matter?

28.25 (A necessary condition for functional dependence.) By using eqn. (28.4), show that if $g(x) = H\{f(x, y)\}$, then

$$\Delta = \frac{\partial f}{\partial x}\frac{\partial g}{\partial y} - \frac{\partial f}{\partial y}\frac{\partial g}{\partial x} \equiv 0,$$

H being a function of a single variable.

By using (29.6), show that if $\Delta \equiv 0$, then the families of level curves of $f(x, y)$ are identical.

29

Functions of two variables: geometry and formulae

CONTENTS

29.1 The incremental approximation

It was explained in Section 28.5 that the tangent plane at a point is the plane that best fits a surface at the point. The formula for the tangent plane to a surface $z = f(x, y)$ at $Q : (a, b, c)$, where $c = f(a, b)$, is

$$z - c = \left(\frac{\partial f}{\partial x}\right)_{(a,b)} (x - a) + \left(\frac{\partial f}{\partial y}\right)_{(a,b)} (y - b)$$

(see (28.5)). We will set up new axes with origin at Q, parallel to the old ones, and call them $\delta x, \delta y, \delta z$ (see Fig. 29.1), anticipating that we shall be concerned with *small* distances from Q. Then

$$\delta x = x - a, \qquad \delta y = y - b, \qquad \delta z = z - c.$$

In the new coordinates, the equation of the tangent plane is

$$\delta z = \left(\frac{\partial f}{\partial x}\right)_{(a,b)} \delta x + \left(\frac{\partial f}{\partial y}\right)_{(a,b)} \delta y.$$

Now consider the quantity δf, where

$$\delta f = f(x, y) - f(a, b).$$

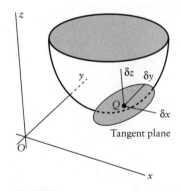

Fig. 29.1

This is the change in z *on the surface* $z = f(x, y)$ from its value at Q. The tangent plane is the best-fitting plane to the surface at Q, so the formula

$$f(x, y) - f(a, b) = \delta f \approx \left(\frac{\partial f}{\partial x}\right)_{(a,b)} \delta x + \left(\frac{\partial f}{\partial y}\right)_{(a,b)} \delta y$$

must give the **best-fitting linear approximation** to δf near $x = a, y = b$:

> **Best linear approximation to $f(x, y)$ near (a, b)**
>
> $$\delta f \approx \left(\frac{\partial f}{\partial x}\right)_{(a,b)} \delta x + \left(\frac{\partial f}{\partial y}\right)_{(a,b)} \delta y,$$
>
> where $\delta f = f(x, y) - f(a, b)$, $\delta x = x - a$, $\delta y = y - b$. **(29.1)**

You are more likely to **remember** the formula obtained by calling the general point (x, y) instead of (a, b), and putting z in place of f. Also the approximation will be good enough to be useful only when δx and δy are 'small' (how small will depend on circumstances):

> **Incremental approximation for $f(x, y)$ (mnemonic version)**
>
> For small enough increments δx and δy:
>
> $$f(x + \delta x, y + \delta y) - f(x, y) \approx \frac{\partial f}{\partial x} \delta x + \frac{\partial f}{\partial y} \delta y.$$
>
> If we put $z = f(x, y)$, this can be written
>
> $$\delta z = \frac{\partial z}{\partial x} \delta x + \frac{\partial z}{\partial y} \delta y \quad \text{(approximately)}.$$
>
> **(29.2)**

This will be the source of almost all our results from now on, but remember that $\partial z/\partial x$ and $\partial z/\partial y$ are *constants* given by the explicit formula (29.1).

Example 29.1 *Let $z = x^2 + 3y^2$. Find an approximation to δz in terms of δx and δy near the points* (a) $x = 2, y = 1$; (b) $x = 3, y = 2$; (c) $x = 0, y = 0$. (d) *Find the exact value of δz in case* (a) *and compare it with the approximate values in the three cases when δx and δy both take the values $0.1, 0.01$, and 0.001.*

In general, $\dfrac{\partial z}{\partial x} = 2x$ and $\dfrac{\partial z}{\partial y} = 6y$.

(a) At $(2, 1)$, $\partial z/\partial x = 4$ and $\partial z/\partial y = 6$. Therefore, from (29.2)

$\delta z = 4 \, \delta x + 6 \, \delta y$ approximately.

(b) At $(3, 2)$, $\partial z/\partial x = 6$ and $\partial z/\partial y = 12$; so

$\delta z = 6 \, \delta x + 12 \, \delta y$ approximately.

(c) At $(0, 0)$, $\partial z/\partial x = \partial z/\partial y = 0$; so the formula predicts $\delta z = 0$ approximately. The reason is that $(0, 0)$ is a stationary point, so z hardly changes when we move a short distance from $(0, 0)$.

(d) From (a), the approximation near $(2, 1)$ when $\delta x = \delta y = 0.1$ is

$\delta z = (4 \times 0.1) + (6 \times 0.1) = 1.0$.

The exact value is given by

$\delta z = f(2.1, 1.1) - f(2, 1) = 1.04$,

so the error in estimating δz is -4%. If $\delta x = \delta y = 0.01$, the error is -0.4%; if $\delta x = \delta y = 0.001$, it is -0.04%.

We see from (d) in the Example that the **approximation improves percentagewise as δx and δy get smaller**: it is not merely that the error decreases because δx, δy, δz all go to zero together. The following Example shows the reason for this.

Example 29.2 *Find the exact algebraic form of the error incurred by using (29.2) to estimate δz at (2, 1) when $z = x^2 + 3y^2$ (see Example 29.1a).*

Put $x = 2 + \delta x$ and $y = 1 + \delta y$. Then
$$\delta z = f(2 + \delta x, 1 + \delta y) - f(2, 1)$$
$$= (2 + \delta x)^2 + 3(1 + \delta y)^2 - 7$$
$$= (4\,\delta x + 6\,\delta y) + (\delta x^2 + 3\,\delta y^2).$$

The first two terms represent the linear approximation obtained in Example 29.1a. The remainder is the error incurred, the part we ignore in the approximation. The error consists only of **higher powers of δx and δy**, and this will always be the case. Therefore **the error is an order of magnitude smaller than the linear terms retained in the incremental approximation (29.2).**

29.2 Small changes and errors

The incremental approximation (29.1) or (29.2) can be used to estimate the effect of making **small changes** in the values of variables in a formula.

Example 29.3 *Estimate the change in the value of*
$$z = \frac{1}{(x^2 + y^2)^{\frac{1}{2}}}$$
when (x, y) change from (3, 4) to (3.1, 3.8).

Using (29.2), put $(x, y) = (3, 4)$, $\delta x = 0.1$, $\delta y = -0.2$. We require
$$\left(\frac{\partial z}{\partial x}\right)_{(3,4)} = [-x(x^2 + y^2)^{-\frac{3}{2}}]_{(3,4)} = -\tfrac{3}{125},$$
$$\left(\frac{\partial z}{\partial y}\right)_{(3,4)} = [-y(x^2 + y^2)^{-\frac{3}{2}}]_{(3,4)} = -\tfrac{4}{125}.$$

Therefore, approximately,
$$\delta z = (-\tfrac{3}{125})(0.1) + (-\tfrac{4}{125})(-0.2) = 0.004.$$

(The exact value of δz is 0.003 91 ....)

Example 29.4   *The period T of the swings of a pendulum is equal to $2\pi(l/g)^{\frac{1}{2}}$, where l is its length and g the gravitational constant. Estimate the error in calculating T if, instead of using closely correct values l = 1.015 and g = 9.812 in the formula, we use the rounded values l = 1 and g = 10.*

The formula corresponding to (29.12) is

$$\delta T \approx \frac{\partial T}{\partial l}\delta l + \frac{\partial T}{\partial g}\delta g.$$

Suppose for simplicity we decide to substitute the *rounded values* l = 1 and g = 10 into the coefficients: we obtain

$$\frac{\partial T}{\partial l} = (\pi l^{-\frac{1}{2}}g^{-\frac{1}{2}})_{(1,10)} = 0.993,$$

$$\frac{\partial T}{\partial g} = (-\pi l^{\frac{1}{2}}g^{-\frac{3}{2}})_{(1,10)} = -0.099.$$

Equation (29.2) then requires that we put

$$\delta l = \text{(true value)} - \text{(rounded value)} = 0.015,$$

$$\delta g = \text{(true value)} - \text{(rounded value)} = -0.188.$$

Then

$$\delta T \approx \text{(true value)} - \text{(rounded value)}$$

$$\approx (0.993)(0.015) + (-0.099)(-0.188) = 0.0335$$

But this is not the *error*: for that we need

$$\text{(error)} = \text{(rounded value)} - \text{(true value)} = -\delta T,$$

so the *error* is about −0.0335. (The exact error is −0.0339 ... .)

In the last example, we substituted the rounded (erroneous) values into $\partial T/\partial l$ and $\partial T/\partial g$, which led to a complication we might have avoided. However, usually there is no choice, the exact values being unknown. Let $z = f(x, y)$, and suppose that we want to estimate the error is z which could arise from using measured (i.e. approximate) values for x and y. The error Δx in x is

$$\Delta x = \text{(measured value of } x) - \text{(exact value of } x),$$

and similarly for Δy and Δz.

Usually we only know a **range of possible error**, not the errors themselves. For example, we might say that a parcel weighed 1430(±15) g, meaning that we think it is between 1415 and 1445 g. Therefore, the values of Δx and Δy are unknown, so the exact values of x and y are unknown, and are not available to go into (29.2) in place of (x, y). Instead, in such cases, take x, y to be convenient **reference values**, at which the derivates are evaluated.

$$x \text{ and } y = \text{(reference values)}.$$

To correspond with this, the definition of $\delta x, \delta y, \delta z$ in (29.2) requires

$\delta x, \delta y, \delta z = (\text{true values}) - (\text{reference values}),$

Therefore

$$\delta x = -\Delta x, \qquad \delta y = -\Delta y, \qquad \delta z = -\Delta z$$

go into (29.2). Every term has then a negative sign, so the formula in terms of $\Delta x, \Delta y, \Delta z$ has the same shape as the incremental formula:

Small-error formula

If $z = f(x, y)$, then

$$\Delta z = \frac{\partial z}{\partial x} \Delta x + \frac{\partial z}{\partial y} \Delta y \quad \text{(approximately)},$$

where x and y are reference values, and Δ stands for

error $= (\text{reference value}) - (\text{exact value})$.   **(29.3)**

This is used in the following way.

Example 29.5   *In a triangle ABC, the side BC has length a given by*

$$a = \frac{c \sin A}{\sin(A + B)}.$$

Suppose that $c = 10$ (exactly), and angles A and B are measured to $5°$ accuracy: $A = 45(\pm 5)°$, $B = 30(\pm 5)°$. Estimate the largest possible resulting error in a.

Put

$$a = f(A, B) = 10 \frac{\sin A}{\sin(A + B)}.$$

Then

$$\frac{\partial a}{\partial A} = 10 \frac{\sin(A + B) \cos A - \cos(A + B) \sin A}{\sin^2(A + B)} = 10 \frac{\sin B}{\sin^2(A + B)}.$$

Similarly

$$\frac{\partial a}{\partial B} = -10 \frac{\sin A}{\sin^2(A + B)}.$$

Choose as reference values $A = 45°$ and $B = 30°$ (for this seems to be the simplest choice). We get $\partial a/\partial A = 5.36$ and $\partial a/\partial B = -7.58$. The error formula (29.3) becomes

$$\Delta a = 5.36 \, \Delta A - 7.58 \, \Delta B$$

approximately, where ΔA and ΔB *must be measured in radians.*

The greatest possible magnitude of Δa occurs if ΔA and ΔB happen to have the opposite signs and their greatest possible magnitudes; that is, if $\Delta A = -\Delta B = \pm 0.087$ radians. In that case, $\Delta a = \pm 1.13$. Therefore

$$a = 7.32(\pm 1.13),$$

showing a possible error of 15%.

Example 29.6 *One solution of the equation $x^2 + bx + c = 0$ is $x = \frac{1}{2}[-b + (b^2 - 4c)^{\frac{1}{2}}]$. (a) Find an approximate expression for the error Δx arising from small errors Δb and Δc in b and c. (b) Estimate the maximum possible error in the solution x if b and c are rounded to one decimal to give $b \approx 3.1$, $c \approx 2.1$.*

(a) We have $\Delta x = (\partial x/\partial b)\, \Delta b + (\partial x/\partial c)\, \Delta c$, in which we must put

$$\frac{\partial x}{\partial b} = \tfrac{1}{2}[-1 + b(b^2 - 4c)^{-\frac{1}{2}}], \quad \frac{\partial x}{\partial c} = -(b^2 - 4c)^{-\frac{1}{2}}.$$

(b) Since b and c are rounded numbers, all that we know about them is that

$$b = 3.1(\pm 0.05), \qquad c = 2.1(\pm 0.05),$$

meaning that the error might be *anywhere* in the range indicated. Putting the reference values $b = 3.1$ and $c = 2.1$ into (a), we obtain

$$\frac{\partial x}{\partial b} = 0.909, \quad \frac{\partial x}{\partial c} = -0.909;$$

so, by (29.3),

$$\Delta x = 0.909\, \Delta b - 0.909\, \Delta c.$$

This takes its greatest possible magnitude when Δb and Δc take their maximum values and have *opposite* sign: that is, when

$$\Delta b = \pm 0.05, \qquad \Delta c = \mp 0.05.$$

In that case $\Delta x = \pm 0.909(0.05 + 0.05) = \pm 0.091$.

The value of x estimated from the rounded coefficients is $x = -1$. Although the rounding error is only at most 2.5%, the error in the solution could be as large as $\pm 8.3\%$.

29.3 The derivative in any direction

The plane in Fig. 29.2 in a **map** of a surface $z = f(x, y)$ with all detail omitted. At $P : (x, y)$ we see a slope $\partial z/\partial x$ if we look east, a slope $\partial z/\partial y$ looking north, and other slopes in other directions. We can find the slopes in other directions in terms of $\partial z/\partial x$ and $\partial z/\partial y$. It might seem that we could make the intermediate slopes equal to anything we liked, but if the surface at P is smooth enough to have a tangent plane, this is not so. In effect, the slopes we see are the slopes of the tangent plane in the various directions.

Consider the direction $\overline{PQ}$ which makes an angle θ with the positive x axis, the direction for positive angles being anticlockwise as with polar coordinates. Let the length $PQ = \delta s$, a short step, and let δx and δy be as shown. Then, by (29.2), the change in elevation in this direction is given approximately by

$$\delta z \approx \frac{\partial z}{\partial x}\delta x + \frac{\partial z}{\partial y}\delta y.$$

Divide by δs; we obtain

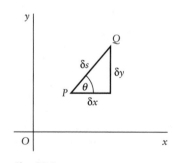

Fig. 29.2

$$\frac{\delta z}{\delta s} \approx \frac{\partial z}{\partial x}\frac{\delta x}{\delta s} + \frac{\partial z}{\partial y}\frac{\delta y}{\delta s} = \frac{\partial z}{\partial x}\cos\theta + \frac{\partial z}{\partial y}\sin\theta$$

from Fig. 29.2. Now let $\delta s \to 0$; the approximation becomes exact, and we have an expression for the slope in any direction. Using the notation for the **directional derivative**,

$$\lim_{\delta s \to 0}\frac{\delta z}{\delta s} = \frac{dz}{ds},$$

we have the following formula.

> **Directional derivative**
>
> The slope of $z = f(x, y)$ at P in direction θ:
>
> $$\frac{dz}{ds} = \left(\frac{\partial z}{\partial x}\right)_P \cos\theta + \left(\frac{\partial z}{\partial y}\right)_P \sin\theta.$$
>
> (29.4)

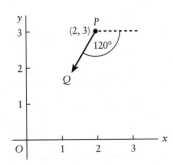

Fig. 29.3

Example 29.7 *Find the slope of the surface $z = xy + x^2$ at $P: (2, 3)$ in the direction $-120°$.*

The direction is shown on Fig. 29.3.

$$\left(\frac{\partial z}{\partial x}\right)_{(2,3)} = (y + 2x)_{(2,3)} = 7,$$

$$\left(\frac{\partial z}{\partial y}\right)_{(2,3)} = (x)_{(2,3)} = 2.$$

Also

$$\cos(-120°) = -\sin 30° = -\tfrac{1}{2}$$

and

$$\sin(-120°) = -\cos 30° = -\tfrac{1}{2}\sqrt{3};$$

so

$$\frac{dz}{ds} = 7(-\tfrac{1}{2}) + 2(-\tfrac{1}{2}\sqrt{3}) = -\tfrac{1}{2}(7 + 2\sqrt{3}).$$

Example 29.8 *The temperature distribution in a plate heated at the point $(0, 0)$ is given by $T = 1/(x^2 + y^2)^{\frac{1}{2}}$. (a) Find the temperature gradient at the point $(3, 3)$ in a direction of $45°$ to the positive x axis. (b) In polar coordinates, $T = 1/r$. Show that the result (a) is the same as $\partial T/\partial r$ taken at any point on the circle $r = 3\sqrt{2}$.*

(a) $$\left(\frac{\partial T}{\partial x}\right)_{(3,3)} = \left(-\frac{x}{(x^2 + y^2)^{\frac{3}{2}}}\right)_{(3,3)} = -\frac{1}{18\sqrt{2}},$$

$$\left(\frac{\partial T}{\partial y}\right)_{(3,3)} = \left(-\frac{y}{(x^2 + y^2)^{\frac{3}{2}}}\right)_{(3,3)} = -\frac{1}{18\sqrt{2}}.$$

Example 29.8 *continued*

Also cos $\theta = 1/\sqrt{2}$ and sin $\theta = 1/\sqrt{2}$. Therefore the temperature gradient at $(3, 3)$ in the given direction is

$$\frac{dT}{ds} = -\frac{1}{18}.$$

(b) $T = 1/r$, so $\partial T/\partial r = -1/r^2$. At the given point, $r = 3\sqrt{2}$, so the result is the same.

Example 29.9   *At any point on the plane $z = \sqrt{3}x - y + 4$, find
(a) an expression for the slope dz/ds in every direction, (b) the
directions in which dz/ds = 0, (c) the directions in which dz/ds is a
maximum and a minimum.*

(a) $\partial z/\partial x = \sqrt{3}$ and $\partial z/\partial y = -1$; and these are the same at every point. By (29.4),

$$\frac{dz}{ds} = \sqrt{3} \cos \theta - \sin \theta.$$

(b) $dz/ds = 0$ where $\sqrt{3} \cos \theta - \sin \theta = 0$, or tan $\theta = \sqrt{3}$. Therefore $\theta = 60°$ or $\theta = -120°$. These directions are opposed: see Fig. 29.4. They give the direction of the contour through any point.

(c) dz/ds is a maximum (direction of **steepest ascent**), or a minimum (**steepest descent**), in directions such that

$$\frac{d}{d\theta}\left(\frac{dz}{ds}\right) = 0,$$

or $-\sqrt{3} \sin \theta - \cos \theta = 0$, or tan $\theta = -1/\sqrt{3}$. Therefore $\theta = -30°$ or $\theta = 150°$, these directions being directly opposed: see Fig. 29.4. By considering the sign of

$$\frac{d^2}{d\theta^2}\left(\frac{dz}{ds}\right),$$

or just by thinking about it, it can be seen that the directions of steepest ascent and descent are as shown.

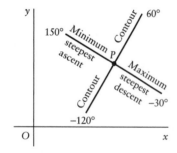

Fig. 29.4

In the last example, the directions of steepest ascent/descent at any point are perpendicular to the directions of the contours; we shall now show that this is true for all surfaces. On the contour map of $z = f(x, y)$, the slope in the direction θ at a point $P : (x, y)$ has the form (29.4):

$$\frac{dz}{ds} = A \cos \theta + B \sin \theta,$$

where A and B are the values of $\partial z/\partial x$ and $\partial z/\partial y$ at P. This is zero in the directions θ_1 where

$$\tan \theta_1 = -A/B.$$

The two directions θ_1 which satisfy this equation differ by π, so they indicate smooth passage of the contour through P. The gradient dz/ds is a maximum or minimum when

$$\frac{d}{d\theta}\left(\frac{dz}{ds}\right) = 0,$$

or in directions θ_2 where

$$\tan\theta_2 = B/A,$$

which give the directions of steepest ascent/descent. Since

$$\tan\theta_1 \tan\theta_2 = -1,$$

these directions are perpendicular, a fact known intuitively by any hill walker.

> **Steepest ascent/descent**
>
> At each point on the map of $z = f(x, y)$, the direction of steepest ascent/descent is perpendicular to the contour.   (29.5)

The two systems of curves, consisting of the contours and the curves which follow directions of steepest ascent or descent, are perpendicular wherever they cross, so they are called **orthogonal systems** of curves (see Section 30.4).

29.4   Implicit differentiation

An equation of the type

$$f(x, y) = c,$$

where c is a constant, describes a curve or curves in the (x, y) plane, since we can imagine solving it to obtain y as a function of x. For example, $x^2 + y^2 = 4$ represents the two semicircles $y = \pm(4 - x^2)^{\frac{1}{2}}$. Another interpretation is that the equation $f(x, y) = c$ describes the contour $z = c$ of a surface $z = f(x, y)$, projected into the (x, y) plane, as in Fig. 28.4b.

Although it is usually impossible in practice to solve for y in terms of x, it is always possible to obtain an expression for the slope dy/dx of the curve in terms of x and y. Choose any point $P : (x, y)$ on the curve (Fig. 29.5), and move along it a short distance to $Q : (x + \delta x, y + \delta y)$. Then dy/dx on the curve is given by

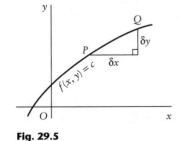

Fig. 29.5

$$\frac{dy}{dx} = \lim_{\delta x \to 0} \frac{\delta y}{\delta x}.$$

Since P and Q both lie on the curve, $\delta f = 0$; so the incremental approximation (29.2) gives

$$\frac{\partial f}{\partial x}\delta x + \frac{\partial f}{\partial y}\delta y \approx 0,$$

or

$$\frac{\delta y}{\delta x} \approx -\frac{\partial f}{\partial x}\bigg/\frac{\partial f}{\partial y}.$$

Now let $\delta x \to 0$. The '$\approx$' becomes '$=$', and $\delta y/\delta x$ becomes dy/dx, from which we obtain:

The implicit-differentiation formula

The slope of $f(x, y) = c$ at any point (x, y) on the curve is given by

$$\frac{dy}{dx} = -\frac{\partial f}{\partial x}\bigg/\frac{\partial f}{\partial y}.$$

(29.6)

The process is called **implicit differentiation** because $f(x, y) = c$ gives y in terms of x only 'implicitly', not explicitly.

Example 29.10   *Find an expression for* dy/dx *at a general point* (x, y) *on the circle* $x^2 + y^2 = 4$.

Here $f(x, y) = x^2 + y^2 = 4$, and so

$$\frac{\partial f}{\partial x} = 2x, \qquad \frac{\partial f}{\partial y} = 2y.$$

Therefore, by (29.6),

$$\frac{dy}{dx} = -\frac{2x}{2y} = -\frac{x}{y}$$

(provided that (x, y) is actually a point on the given circle).

In the last Example we would have obtained exactly the same result for the circle $x^2 + y^2 = 1$, or $x^2 + y^2 = 100$. It is the numerical values of x and y to be put in the right-hand side which will distinguish the circle under discussion from all the other circles. In fact the equation we obtained,

$$\frac{dy}{dx} = -\frac{x}{y},$$

can be thought of as a differential equation. Its solutions (obtained by the method of Section 22.3) are $x^2 + y^2 = C$, which includes the given circle and all the others as well.

Example 29.11 *Find* dy/dx *on the curve* $x^3y - xy^3 = 6$ *at the point* (2, 1).

(You can check that the point (2, 1) is really on the curve.) Putting $f(x, y) = x^3y - xy^3$, we have

$$\frac{\partial f}{\partial x} = 3x^2y - y^3, \qquad \frac{\partial f}{\partial y} = x^3 - 3xy^2.$$

Therefore, at any point (x, y) on the curve,

$$\frac{dy}{dx} = -\frac{3x^2y - y^3}{x^3 - 3xy^2}.$$

At (2, 1), the slope is

$$\left(\frac{dy}{dx}\right)_{(2,1)} = -\frac{\partial f/\partial x}{\partial f/\partial y} = -\frac{11}{2}.$$

(This is *not* a differential equation: it is a *numerical value* which holds at only a *single point*.)

The link with differential equations can be used in many ways, as in the following example.

Example 29.12 *Find the family of curves which is orthogonal (perpendicular) to the curves* $xy = C$.

The curves $xy = C$ are the contours of the function $f(x, y) = xy$, and the new family will be the curves of steepest ascent/descent on the contour map of xy.

The differential equation of the family $xy = C$ is, from the implicit-differentiation formula (29.6),

$$\frac{dy}{dx} = -\frac{y}{x}.$$

Wherever the new curves intersect with these they must cut at a right angle, so the product of their slopes at any intersection must be equal to -1. Therefore the new family must have the differential equation

$$\frac{dy}{dx} = \frac{x}{y}$$

because $(-y/x)_P (x/y)_P = -1$ at any point P. This equation can be solved by separating the variables (Section 22.3), which gives

$$\int y \, dy = \int x \, dx$$

or

$$y^2 - x^2 = B,$$

where B is an arbitrary constant. This is another family of hyperbolas. A small region of the (x, y) plane is shown in Fig. 29.6.

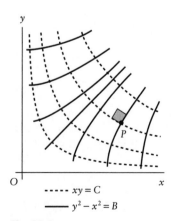

$\cdots\cdots$ $xy = C$

—— $y^2 - x^2 = B$

Fig. 29.6

29.5 Normal to a curve

The slope at any point P on the curve $f(x, y) = c$ is equal to

$$-\left(\frac{\partial f}{\partial x}\right)_P \bigg/ \left(\frac{\partial f}{\partial y}\right)_P \quad \text{(by (29.6))}.$$

We shall obtain a vector $\boldsymbol{n}$ perpendicular or **normal** to the curve at P. A straight line through P perpendicular to the curve must have slope

$$\left(\frac{\partial f}{\partial y}\right)_P \bigg/ \left(\frac{\partial f}{\partial x}\right)_P,$$

because the product of the slopes must be equal to -1. A vector with components (a, b) has slope b/a, so one **normal vector** $\boldsymbol{n}$ is

$$\boldsymbol{n} = \left(\left(\frac{\partial f}{\partial x}\right)_P, \left(\frac{\partial f}{\partial y}\right)_P\right).$$

Any multiple of this $\boldsymbol{n}$ is also a normal at the point. Dropping the suffix P, we have the following result.

Normal vector $\boldsymbol{n}$ at the point (x, y) on the curve $f(x, y) = c$

$$\boldsymbol{n} = \left(\frac{\partial f}{\partial x}, \frac{\partial f}{\partial y}\right) = \frac{\partial f}{\partial x}\hat{\boldsymbol{i}} + \frac{\partial f}{\partial y}\hat{\boldsymbol{j}}.$$

(29.7)

Example 29.13   *Find several normal vectors at the point $(2, 1)$ on the curve $x^2 + y^2 = 5$.*

Putting $f(x, y) = x^2 + y^2$, we have $\partial f/\partial x = 2x$ and $\partial f/\partial y = 2y$; so

$$\left(\frac{\partial f}{\partial x}\right)_{(2,1)} = 4, \qquad \left(\frac{\partial f}{\partial y}\right)_{(2,1)} = 2.$$

Therefore one vector normal to the circle at $(2, 1)$ is

$$\boldsymbol{n} = (4, 2)$$

and from this any number of other normal vectors can be constructed by taking multiples. For example, $(2, 1)$, $(-2, -1)$, and $(\frac{2}{\sqrt{5}}, \frac{1}{\sqrt{5}})$ are also normals, the last one being a **unit normal** (one having unit length), which is often important.

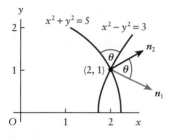

Fig. 29.7

Example 29.14   *Find the angle of intersection between the curves* $x^2 + y^2 = 5$ *and* $x^2 - y^2 = 3$ *at the point* (2, 1).

In the last Example, we showed that $n = (4, 2)$ is normal to $x^2 + y^2 = 5$ at (2, 1). Similarly the vector $n = (4, -2)$ is normal to $x^2 - y^2 = 3$ at the point. From Fig. 29.7, it can be seen that the acute angle θ between the normals is equal to one of the angles between the curves (the other is $\pi - \theta$). From (10.4),

$$\cos\theta = \frac{n_1 \cdot n_2}{|n_1||n_2|} = \frac{(4, 2)\cdot(4, -2)}{\sqrt{20}\sqrt{20}} = \frac{3}{5},$$

so $\theta = 53.1°$.

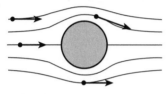

Fig. 29.8

29.6 Gradient vector in two dimensions

It is familiar that the value of a quantity such as pressure or temperature depends on, or is a function of, position (x, y). These are **scalar functions**: the values they take up are ordinary numbers. There are also vector quantities that depend on position. Figure 29.8 shows some streamlines for a fluid flowing over a long cylinder (assuming that the flow is always in the plane of the paper). The velocity v is a **vector** which varies from point to point, so we can write $v = v(x, y)$. Gravitational, magnetic, and electric fields are other instances of **vector functions of position** or **vector fields**.

Associated with any scalar function, there is an important vector function which arises as follows. We repeatedly produce formulae involving the pair of elements $\partial f/\partial x$ and $\partial f/\partial y$ in combinations of the form $U\,\partial f/\partial x + V\,\partial f/\partial y$, where U and V are constants or functions; for example, as in (29.7), (29.2), and (29.4). We can manipulate this pair as a unit by regarding

$$\frac{\partial f}{\partial x}\hat{\imath} + \frac{\partial f}{\partial y}\hat{\jmath} \quad \text{or} \quad \left(\frac{\partial f}{\partial x}, \frac{\partial f}{\partial y}\right)$$

as a vector function. We call this vector function the **gradient of** f and denote it by

$$\mathbf{grad}\, f \quad \text{or} \quad \nabla f$$

(∇ is pronounced 'del' or 'nabla'). We shall see that it works rather like an ordinary derivative, but in two dimensions; hence its name.

Alternatively we can regard the symbol **grad** or ∇ standing alone as an **operator** (compare d/dx): it operates on scalar functions $f(x, y)$, instructing us to carry out the operation $\hat{\imath}\,\partial/\partial x + \hat{\jmath}\,\partial/\partial y$ or $(\partial/\partial x, \partial/\partial y)$ on $f(x, y)$:

$$\mathbf{grad}\, f(x, y) = \left(\hat{\imath}\,\frac{\partial}{\partial x} + \hat{\jmath}\,\frac{\partial}{\partial y}\right) f(x, y).$$

> **Gradient in two dimensions**
>
> Given a scalar function $f(x, y)$, **grad** f or ∇f stands for
>
> $$\hat{i}\frac{\partial f}{\partial x} + \hat{j}\frac{\partial f}{\partial y} \quad \text{or} \quad \left(\frac{\partial f}{\partial x}, \frac{\partial f}{\partial y}\right).$$
>
> Alternatively, **grad** or ∇ stands for the operator
>
> $$\hat{i}\frac{\partial}{\partial x} + \hat{j}\frac{\partial}{\partial y} \quad \text{or} \quad \left(\frac{\partial}{\partial x}, \frac{\partial}{\partial y}\right).$$
>
> (29.8)

Example 29.15   Let $f(x, y) = x^2 + y^2$. *Obtain* (a) *the vector function* **grad** f; (b) *the value of* **grad** f *at the point* $(1, 2)$; (c) *an expression for the magnitude, or length, of* **grad** f *at* (x, y).

(a) **grad** $f = \dfrac{\partial f}{\partial x}\hat{i} + \dfrac{\partial f}{\partial y}\hat{j} = 2x\hat{i} + 2y\hat{j}$;

or we can use the alternative notations, and even the operator viewpoint:

$$\nabla f = \left(\frac{\partial}{\partial x}, \frac{\partial}{\partial y}\right)(x^2 + y^2) = (2x, 2y).$$

(b) At $x = 1$, $y = 2$, we have

   grad $f = (2, 4)$.

(c) The magnitude or length of a vector $v = (a, b)$ is $|v| = (a^2 + b^2)^{\frac{1}{2}}$, so $|\textbf{grad } f| = [(2x)^2 + (2y)^2]^{\frac{1}{2}} = 2(x^2 + y^2)^{\frac{1}{2}}$.

We can re-express some earlier results in terms of **grad**. For example, we may write (29.7) immediately as follows.

> **A normal vector** n at the point (x, y) on the curve $f(x, y) = c$ is
>
> $n = \textbf{grad } f(x, y)$.
>
> (29.9)

As we remarked earlier, expressions occurring in physical theory frequently take the form

$$U\frac{\partial f}{\partial x} + V\frac{\partial f}{\partial y},$$

where U and V may be constants, or various functions. Then we can write such expressions as a scalar ('dot') product by inventing a new vector function $S = U\hat{i} + V\hat{j}$:

> If $S = U\hat{i} + V\hat{j}$, then
>
> $$U\frac{\partial f}{\partial x} + V\frac{\partial f}{\partial y} = (U, V)\cdot\left(\frac{\partial f}{\partial x}, \frac{\partial f}{\partial y}\right) = S\cdot\textbf{grad } f.$$
>
> (29.10)

Now consider the directional-derivative formula (29.4), regarding it as representing the rate of change of $f(x, y)$ in the direction θ:

$$\frac{df}{ds} = \frac{\partial f}{\partial x} \cos \theta + \frac{\partial f}{\partial y} \sin \theta.$$

To recast this in the form of (29.10), we require the vector

$$\hat{\imath} \cos \theta + \hat{\jmath} \sin \theta.$$

This is a *unit vector* (i.e. it has length unity) because

$$(\cos^2\theta + \sin^2\theta)^{\frac{1}{2}} = 1;$$

so put

$$\hat{\imath} \cos \theta + \hat{\jmath} \sin \theta = \hat{s},$$

where $\hat{s}$ is a unit vector pointing in the desired direction, and (29.10) becomes

> **Directional derivative**
>
> In the direction of a unit vector $\hat{s}$, the rate of change of $f(x, y)$ is given by
>
> $$\frac{df}{ds} = \hat{s} \cdot \mathbf{grad}\, f;$$
>
> that is to say, df/ds is equal to the component of $\mathbf{grad}\, f$ in the direction of $\hat{s}$.
>
> **(29.11)**

Equation (29.11) can be written in a different way. If a and b are two vectors, then the angle between them, ϕ, can be obtained from the identity

$$\mathbf{a} \cdot \mathbf{b} = |\mathbf{a}|\,|\mathbf{b}| \cos \phi, \text{ with } 0 < \phi \leqslant \pi$$

(see (10.4)). If we put $\mathbf{a} = \hat{s}$ and $\mathbf{b} = \mathbf{grad}\, f$, and use the fact that $|\hat{s}| = 1$, we obtain an alternative form of (29.11).

> **Directional derivative (alternative form)**
>
> $$\frac{df}{ds} = |\mathbf{grad}\, f| \cos \phi,$$
>
> where ϕ is the interior angle between $\mathbf{grad}\, f$ and the required direction.
>
> **(29.12)**

By using (29.12), the perpendicularity of the directions of steepest ascent and the contours of $f(x, y)$, proved in Section 29.3, can be recovered.

Problems

29.1 Use the incremental approximation (29.1) or (29.2) to estimate the change δz due to changes δx and δy as specified, and check the percentage error by calculating the exact result.
(a) $z = x^2 + y^2$ at $(3, 1)$, $\delta x = 0.1$, $\delta y = 0.3$;
(b) $z = \sin xy$ at $(0.5, 1.2)$, $\delta x = 0.1$, $\delta y = -0.05$;
(c) $z = e^{x^2 + 3y^2}$ at $(1, 1)$, $\delta x = 0.1$, $\delta y = 0.2$;
(d) $z = 1/(x^2 + y^2)^{\frac{1}{2}}$ at $(2, 1)$, $\delta x = -0.2$, $\delta y = 0.1$.

29.2 Given $z = x^2 - y^2$ and two points $P : (1.0, 2.1)$ and $Q : (1.1, 2.0)$, (a) estimate the change in z in going from P to Q; (b) estimate the change in going from Q to P; (c) explain in general terms why the second estimate is not precisely the negative of the first.

29.3 (See Example 29.2.) Obtain the *exact algebraical form* of the error incurred in δf, where

$$\delta f = f(x + \delta x, y + \delta y) - f(x, y),$$

in using the approximation $\delta f \approx \dfrac{\partial f}{\partial x} \delta x + \dfrac{\partial f}{\partial y} \delta y$
(a) for $f(x, y) = xy$ near the point $(2, 1)$;
(b) for $f(x, y) = x/y$ near the point $(2, 1)$.

29.4 The relation between the object distance u, the image distance v, and the focal length f of a thin lens is

$$\frac{1}{u} + \frac{1}{v} = \frac{1}{f}.$$

Suppose that the measured values of u and v are $u = 0.31(\pm 0.01)$, $v = 0.56(\pm 0.03)$; calculate the greatest possible error in estimating f, and the corresponding percentage error.

29.5 A viscous liquid is forced through a tube of diameter $a = 10(\pm 0.05 \times 10^{-3})$ and length $l = 0.1$ under a pressure $p = 10(\pm 5 \times 10^4)$, and is found to pass fluid at a rate $v = 0.625 \times 10^{-9}$ per unit time. The viscosity η is given by the formula

$$\eta = \frac{\pi}{128} \frac{pd^4}{vl}.$$

Find the maximum error in the viscosity estimate.

29.6 One root of the equation $x^2 + bx + c = 0$ is $x = \frac{1}{2}[-b + (b^2 - 4c)^{\frac{1}{2}}]$. Suppose that $b = 20.4$ and $c = 95.5$. Estimate the percentage error in the root which would arise if these were rounded to $b = 20$, $c = 96$.

29.7 The area S of a triangle with base b and base angles A and C is given by

$$S = \frac{\frac{1}{2}b^2 \tan A \tan C}{\tan A + \tan C}.$$

Suppose that nominally $b = 2$, $A = 30°$, $C = 60°$, but that C is found to be too large by 5%. By what amount should A be changed so that S would be restored to the correct area?

29.8 A certain type of experiment to measure surface tension S requires the formula $S = ahr^3/p^2$, where a is a constant and h, r, and p are measured quantities. Take the logarithm of the formula to find the fractional change in $\delta S/S$ in S, in terms of simultaneous fractional changes in h, r, and p.

29.9 Find the directional derivative df/ds of each of the following functions according to the data. Also, for the given point, find the directions of the contour and the direction of steepest ascent.
(a) $f(x, y) = x^2 + y^2$ at $(1, 2)$, direction $\theta = 30°$;
(b) $f(x, y) = x^2y^2$ at $(2, 1)$, direction $\theta = -45°$;
(c) $f(x, y) = x^2y - xy^2 + 2$ at $(-1, 1)$, direction $\theta = 120°$;
(d) $f(x, y) = \sin xy$ at $(\frac{1}{2}, \pi)$, direction $\theta = -90°$;
(e) $f(x, y) = \cos(x^2 - y)$ at $(0, -\pi)$, direction $\theta = 0$;
(f) $f(x, y) = e^{x-y}$ at $(1, 1)$, direction $\theta = -45°$.

29.10 Find $\dfrac{dy}{dx}$ at the prescribed points on the curves given.
(a) $xy = 1$ at $(2, \frac{1}{2})$; (b) $x^2 + y^2 = 25$ at $(3, 4)$;
(c) $1/x - 1/y = \frac{1}{2}$ at $(1, 2)$; (d) $\frac{1}{10}x^2 + \frac{1}{15}y^2 = 1$ at $(2, 3)$;
(e) $x^3 + 2y^3 = 3$ at $(1, 1)$;
(f) $x^3y + 3x^2 - y^2 - 19 = 0$ at $(2, 1)$;
(g) $xy^2 - x^2y + 6 = 0$ at $(3, 2)$;
(h) $x^2 + y^2 = 4$ at $(2\cos\theta, 2\sin\theta)$;
(i) $x^2/a^2 + y^2/b^2 = 1$ at $(a\cos t, b\sin t)$;
(j) $x\cos y = y\sin x$ at $(\pi/2, 0)$;
(k) $y^2 - 4ax = 0$ at $(at^2, 2at)$.

29.11 The ideal-gas equation, for a fixed mass of gas is $PV = RT$, where R is a constant. There are three variables: P is pressure, V is volume, and T is absolute temperature. Show that

$$\left(\frac{\partial V}{\partial P}\right)_T = -\left(\frac{\partial T}{\partial P}\right)_V \bigg/ \left(\frac{\partial T}{\partial V}\right)_P.$$

(The notation $(\partial u/\partial v)_w$ means that the variable w is kept constant during differentiation when $u = g(v, w)$. Use (29.6).)

29.12 Find the cartesian equation of the tangent line at a point (x_1, y_1) on each of the following curves. (Find dy/dx first.)
(a) $x^2 + y^2 = a^2$; (b) $x^2/a^2 + y^2/b^2 = 1$;
(c) $a^2x^2 - b^2y^2 = c$; (d) $xy = 1$; (e) $x^{\frac{2}{3}} + y^{\frac{2}{3}} = 1$;
(f) $ax^2 + 2hxy + by^2 + 2gx + 2fy + c = 0$.

29.13 Suppose that the curves $f(x, y) = \alpha$ and $g(x, y) = \beta$ intersect at right angles at a point (a, b). Find dy/dx at the point for each curve and deduce that, at (a, b),

$$\frac{\partial f}{\partial x}\frac{\partial g}{\partial x} + \frac{\partial f}{\partial y}\frac{\partial g}{\partial y} = 0.$$

Use this result to confirm that, in the following cases, the two systems of curves are orthogonal (i.e. they always intersect at right angles). Here α and β are the parameters for the two systems – by varying them we obtain all the curves for the systems.
(a) $x^2 + y^2 = \alpha$, $y/x = \beta$; (b) $x^2 - y^2 = \alpha$, $xy = \beta$;
(c) $y^3 - x^3 = \alpha$, $1/y + 1/x = \beta$;
(d) $(x^2 + y^2)/x = \alpha$, $(x^2 + y^2)/y = \beta$.

29.14 Let (x, y) be any point on the curve $y^3 - x^3 = 1$. Find an expression for dy/dx at the point. Since this expression holds good for every point on the curve, it is a differential equation, having the given curve as one of its solution curves. Verify this by solving it, and obtain the other solutions.

29.15 (Numerical). Form the differential equation for the following families of curves, in which c is the parameter; then use the numerical solution method of Section 22.2 to obtain a contour map of the functions concerned.
(a) $x^2 + 2y^2 = c$, $c > 0$; (b) $x^2 + xy - y^3 = c$;
(c) $\dfrac{x^2 + y}{x + y^2} = c$; (d) $xy\,e^{-x} = c$.

29.16 Form the differential equation for each system of curves, and deduce the differential equation for the orthogonal (perpendicular) system. Solve it to obtain the orthogonal system.
(a) $y^2 - x^2 = c$; (b) $y^3 + x^3 = c$;
(c) $y^2 = cx$; (d) $e^y - e^x = c$.

29.17 Find the curves of steepest ascent from an arbitrary point (a, b) for each of the following functions.
(a) $\frac{1}{2}x^2 + y^2$; (b) x^3y^3; (c) $\frac{1}{2}y^2 - y - x^2$.

29.18 Implicit differentiation of y with respect to x can be carried out as follows when $f(x, y)$ is given explicitly. Consider $f(x, y) = x^2 + 2xy + y^2 = c$. Then, by differentiating this equation and treating y as a function of x, we obtain

$$2x + 2x\frac{dy}{dx} + 2y + 2y\frac{dy}{dx} = 0,$$

from which dy/dx can be found. Check that (29.6) gives the same result.

29.19 Find normal vectors to the curves below, and find the angle between them at the intersection given.
(a) $xy = 2$, $x^2 - y^2 = -3$, at intersection $(1, 2)$.
(b) $y = x^3$, $x^2 + \frac{1}{2}y^2 = 36$, at intersection $(2, 8)$.
(c) $x^2 + xy + y^2 = 3$, $x + y = 2$, at intersection $(1, 1)$; interpret your result geometrically.
(d) $ax^2 + 2hxy + by^2 + c = 0$ and

$$ax_0x + h(x_0 + x)(y_0 + y) + by_0y + c = 0,$$

at any point (x_0, y_0) which lies on the first curve.

29.20 Find d^2y/dx^2 on the following curves.
(a) $x^4 - y^4 = 1$; (b) $xy = 1$; (c) $xy\,e^{xy} = 1$.

29.21 Obtain **grad** f, where $f(x, y)$ is given by the following. Give its components, its direction, and its magnitude at the points specified.
(a) $1/(x + y)$ at $(1, -2)$; (b) y/x at $(2, 0)$;
(c) $y^2 - 3x^2 + 1$ at $(0, 0)$; (d) $1/x - 1/y$ at $(2, 1)$;
(e) $1/r$, where r is the polar coordinate, $r = (x^2 + y^2)^{\frac{1}{2}}$; confirm that the gradient vector points in a radial direction.

29.22 Use the gradient vector to obtain a *unit* vector perpendicular to the following curves at the points given
(a) $2x - 3y + 1 = 0$ at any point;
(b) $x^2 + y^2 = 5$ at $(2, 1)$;
(c) $x^2 + y^2 = r^2$ at (x_0, y_0) on the circle;
(d) $x^2/a^2 + y^2/b^2 = 1$ at (x_0, y_0) on the ellipse;
(e) $y = 3x^2 - 2$ at $(2, 10)$.

29.23 Use the property (29.9) to find the angle of intersection of the following curves at the point of intersection given.
(a) $y^2 - x^2 = -3$ and $x^3 - y^3 = 7$ at $(2, 1)$;
(b) $x^2y - xy^2 = 0$ and $x/y - y/x = 0$ at $(2, 2)$;
(c) $x^2 + y^2 + 2x - 4y + 4 = 0$ and $y = x^2 + 2x + 2$ at $(-1, 1)$; explain the result geometrically.

29.24 Use (29.12) to prove the results given in Section 29.3 for a general $f(x, y)$: that (a) the directions of most rapid increase and decrease through a point (x, y) are perpendicular to the direction of the contour through the point; (b) the maximum rate of increase from the point is equal to $|\textbf{grad } f|$ at the point.

30 Chain rules, restricted maxima, coordinate systems

CONTENTS

30.1 Chain rule for a single parameter

Suppose that x and y depend on, or are functions of, another variable t (say) which we call the **parameter**. It might represent time, for example. We shall write

$$x = x(t), \qquad y = y(t).$$

As t varies, the point (x, y) follows a curve of some sort which is said to be **defined parametrically**. The curve also has a characteristic **direction**, which is the direction the curve is described as t **is increasing**, and is indicated by an arrow. We then have a **directed path**.

Example 30.1 *Show that both of the following parametrizations define a unit semicircle, centred at the origin, in the upper half plane, traced anticlockwise:* (a) $x = \cos t, y = \sin t$, *where t increases from 0 to π;* (b) $x = -u, y = (1 - u^2)^{\frac{1}{2}}$, *where u increases from -1 to 1.*

(a) The shape of the curve is obtainable by eliminating t:

$$x^2 + y^2 = \cos^2 t + \sin^2 t = 1;$$

so the points lie on the unit circle. Also, as t increases from 0 to π, y is positive and x decreases from 1 to -1. The path is the upper semicircle from $(1, 0)$ to $(-1, 0)$, described in a single direction, as shown in Fig. 30.1a.

(b) $x^2 + y^2 = (-u)^2 + (1 - u^2) = 1$. As u increases from -1 to 1, y remains positive while x decreases from 1 to -1. The path is as in (a): see Fig. 30.1b.

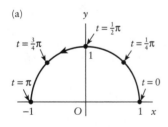

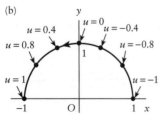

Fig. 30.1

Given a function $f(x, y)$ which can take values all over the (x, y) plane, the function

$$g(t) = f(x(t), y(t))$$

picks out only the values on the path $(x(t), y(t))$. As we move along this path, the function value varies, and we might be concerned with the rate at which it changes with t. (This is generally different from the rate at which $f(x, y)$ changes with *distance along the path*, which is equal to the directional derivative (29.4), and corresponds to using arc-length s as the parameter.)

To find df/dt, suppose that t increases from t to $t + \delta t$. Then, on the curve $(x(t), y(t))$, x changes from x to $x + \delta x$ and y to $y + \delta y$. Divide (29.2) (the incremental approximation) by δt:

$$\frac{\delta f}{\delta t} \approx \frac{\partial f}{\partial x}\frac{\delta x}{\delta t} + \frac{\partial f}{\partial y}\frac{\delta y}{\delta t}.$$

Let $\delta t \to 0$. Then '$\approx$' becomes '$=$', $\dfrac{\delta x}{\delta t} \to \dfrac{dx}{dt}$, and $\dfrac{\delta y}{\delta t} \to \dfrac{dy}{dt}$, and we have the **chain rule** (or **total derivative**):

Chain rule for one parameter

Given $f(x, y)$, $x = x(t)$ and $y = y(t)$,

$$\frac{df}{dt} = \frac{\partial f}{\partial x}\frac{dx}{dt} + \frac{\partial f}{\partial y}\frac{dy}{dt}$$

(and similarly with z in place of f, if we write $z = f(x, y)$). **(30.1)**

This expression is like the chain rule (3.3) for functions of a single variable with an extra term in it for the variable y. Partial derivative rather than ordinary derivative signs are then written as necessary.

Example 30.2 *Let $f(x, y) = xy - y^2$, $x = t^2$, $y = t^3$. (a) Find df/dt using the chain rule; (b) find df/dt by substitution.*

(a) $\dfrac{\partial f}{\partial x} = y$, $\dfrac{\partial f}{\partial y} = x - 2y$, $\dfrac{dx}{dt} = 2t$, $\dfrac{dy}{dt} = 3t^2$.

Therefore, by (30.1),

$$\frac{df}{dt} = \frac{\partial f}{\partial x}\frac{dx}{dt} + \frac{\partial f}{\partial y}\frac{dy}{dt} = y(2t) + (x - 2y)3t^2$$

$$= 2t^4 + (t^2 - 2t^3)3t^2 = 5t^4 - 6t^5.$$

(This expression can be written in various ways in terms of x and y, for example as $5x^2 - 6xy$, or $5yx^{\frac{1}{2}} - 6x^2y^{\frac{1}{3}}$. These all look very different, but they all take the same values since x and y are *connected* by the fact that (x, y) lies on the given curve.)

↗

Example 30.2 *continued*

(b) By substitution,

$$f(x(t), y(t)) = xy - y^2 = t^2 t^3 - (t^3)^2 = t^5 - t^6.$$

Therefore, as before,

$$\frac{df}{dt} = 5t^4 - 6t^5.$$

Example 30.3 *Prove the implicit-differentiation formula (29.6) by using the chain rule with x treated as the parameter.*

If $f(x, y) = c$, then there is a solution $y = y(x)$ for which

$$f(x, y(x)) = c$$

is *automatically* true for every value of x involved (i.e. it is an **identity**). Therefore

$$\frac{df(x, y(x))}{dx} = 0.$$

Comparing this with (30.1), the chain rule, we have x in place of t for the parameter. In terms of the chain rule, we therefore have

$$0 = \frac{\partial f}{\partial x}\frac{dx}{dx} + \frac{\partial f}{\partial y}\frac{dy}{dx} = \frac{\partial f}{\partial x} + \frac{\partial f}{\partial y}\frac{dy}{dx}.$$

From this we recover the implicit-derivative formula (29.6)

$$\frac{dy}{dx} = -\frac{\partial f}{\partial x}\bigg/\frac{\partial f}{\partial y}.$$

The chain rule is more useful for obtaining general results, as in Example 30.3, than in working out special instances such as Example 30.2.

30.2 Restricted maxima and minima: the Lagrange multiplier

Consider the simple function

$$f(x, y) = x + y.$$

This has no maxima, minima, or other stationary points, since $z = x + y$ represents an inclined plane. However, if we travel around the plane *on a particular path*, we are likely to encounter high points and low points, and points where we are momentarily travelling on the level. Suppose that we walk on the circular path $x^2 + y^2 = 1$, shown on a map in Fig. 30.2.

Then A is the highest point; this is where we were walking uphill but then turn downhill: this is a **local maximum point on the path**. If we plotted a graph of elevation against time, this point would show up as local maximum on the graph.

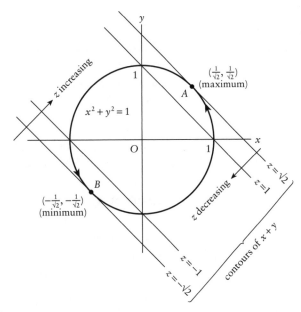

Fig. 30.2

The clue which reveals A to be a maximum is that *one of the contours of $x + y$ is a tangent to the path at A.* Those nearby contours that the path crosses are all lower than the one through A. Similarly, at B, there is a **local minimum** for the path.

This is an example of a **restricted stationary-point problem**, the 'restriction' being the condition that the only points considered are those that lie on a particular curve. A general statement of the problem is as follows.

> **Restricted stationary-point problem**
> Find the stationary points of $f(x, y)$ subject to the condition $g(x, y) = c$.　　　　　　(30.2)

Very simple problems of this type can be solved by an elementary method, as in the following example.

Example 30.4　*Find the maximum possible area a rectangle may have if the perimeter is restricted to length* 10 *units.*

Call the sides x and y. Then we require the maximum of the area A:
$$A = f(x, y) = xy \tag{i}$$
subject to the restriction on the perimeter P
$$P = g(x, y) = 2x + 2y = 10. \tag{ii}$$
From the perimeter equation (ii), we have $y = 5 - x$; so the area can be expressed in terms of x only:
$$A = x(5 - x).$$

Example 30.4 *continued*

This has a stationary point where $dA/dx = 0$, or

$5 - 2x = 0,$

that is at $x = \frac{5}{2}$. The perimeter equation (ii) gives correspondingly $y = \frac{5}{2}$, so the desired shape is a square of area $\frac{25}{4}$.

However, although the following problem looks very similar, there turns out to be a difficulty.

Example 30.5   *Find the maximalminima of $z = f(x, y) = x^2 - y^2$ on the circle $g(x, y) = x^2 + y^2 = 1$.*

On the given curve,

$y^2 = 1 - x^2.$ (i)

The values taken by $z = x^2 - y^2$ on this curve are given in terms of x by

$z = x^2 - (1 - x^2) = 2x^2 - 1.$

The stationary points of this function are where

$$\frac{d}{dx}(2x^2 - 1) = 0,$$

which is at $x = 0$. At $x = 0$, the curve equation (i) gives $y = \pm 1$, so we have found the points $A : (0, 1)$ and $A' : (0, -1)$. These are in fact minima, and they are shown on the path in Fig. 30.3a.

However, there are plainly two *maxima* also, at B and B', which are completely missed by the process above. We could have found them (but lost A and A') if we had substituted for x instead of y by means of $x^2 = 1 - y^2$. You can see the reason for losing A and A' if you sketch the function $2x^2 - 1$ between $x = \pm 1$. The maxima are at the ends, but cannot be found by differentiating; see also Example 4.8.

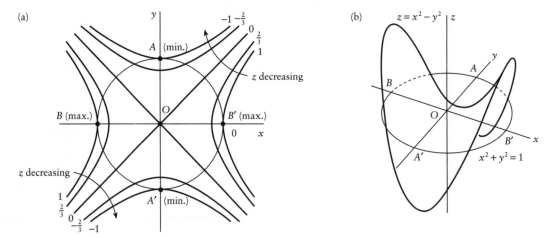

Fig. 30.3 (a) Contour map of $x^2 - y^2$, showing also the curve $x^2 + y^2 = 1$. Here A and A' are minima, and B and B' are maxima. (b) The path $x^2 + y^2 = 1$ in the (x, y) plane, with the corresponding values of $z = x^2 - y^2$ shown.

We can get over this difficulty by parametrizing the curve $g(x, y) = c$ as in the following example, which repeats Example 30.5.

Example 30.6 *Find the stationary points of $x^2 - y^2$ on the curve* $x^2 + y^2 = 1$.

Put

$$x = \cos t, \qquad y = \sin t, \quad 0 \leqslant t < 2\pi,$$

then the circle $C: x^2 + y^2 = 1$ is traced once, anticlockwise, starting and ending at $(1, 0)$. On C,

$$f(x, y) = \cos^2 t - \sin^2 t.$$

As we go along the path C, stationary points are encountered where

$$0 = \frac{df(x(t), y(t))}{dt} = 2 \cos t(-\sin t) - 2 \sin t \cos t$$

$$= -4 \sin t \cos t = -2 \sin 2t.$$

The solutions of this equation in the range $0 \leqslant t < 2\pi$ are $t = 0, \frac{1}{2}\pi, \pi, \frac{3}{2}\pi$, which correspond to the points $(1, 0)$, $(0, 1)$, $(-1, 0)$, $(0, -1)$. Therefore this approach successfully found all the stationary points on the path, which the method of Example 30.5 failed to do.

We shall now describe the **Lagrange-multiplier method** for solving the restricted stationary-value problem (30.2). This uses the parametric idea, but all reference to a parameter is eliminated eventually so that we do not have to invent a parametrization and then go through the resulting algebra.

Think of time t as a possible parameter, and $P : (x(t), y(t))$ as a point moving along the curve with velocity $(dx/dt, dy/dt)$. We shall imagine $g(x, y) = c$ is expressed parametrically so that (a) that path is traced exactly once as t moves through its range, and (b) dx/dt and dy/dt are never both zero together (if t is time, this means that the moving point P never pauses).

Then as $P : (x(t), y(t))$ moves along $g(x, y) = c$, the points Q where $df/dt = 0$ are the stationary points of $f(x(t), y(t))$. Therefore, by the chain rule (30.1),

$$\frac{\partial f}{\partial x}\frac{dx}{dt} + \frac{\partial f}{\partial y}\frac{dy}{dt} = 0.$$

To get rid of dx/dt and dy/dt, which are special to the particular parametrization chosen, we need another equation. On the curve, $g(x, y)$ has a constant value c, so $dg/dt = 0$ at every point including Q. Therefore by the chain rule,

$$\frac{\partial g}{\partial x}\frac{dx}{dt} + \frac{\partial g}{\partial y}\frac{dy}{dt} = 0.$$

These last two equations can be regarded as a pair of homogeneous algebraic equations for dx/dt and dy/dt. From Section 12.5, the equations have a non-trivial solution if and only if the determinant of the coefficients is zero, so at the (unknown) point Q

$$\frac{\partial f}{\partial x}\frac{\partial g}{\partial y} - \frac{\partial f}{\partial y}\frac{\partial g}{\partial x} = 0.$$

This can be written alternatively in the form

$$\frac{\partial f}{\partial x}\bigg/\frac{\partial g}{\partial x} = \lambda, \qquad \frac{\partial f}{\partial y}\bigg/\frac{\partial g}{\partial y} = \lambda, \qquad\qquad \text{(30.3a, b)}$$

where λ ('lambda') is a new unknown constant, called the **Lagrange multiplier** for the problem. We have lost some information here, because the condition $dg/dt = 0$ does not distinguish between one value of c and another, so we have to reassert the condition

$$g(x, y) = c. \qquad\qquad \text{(30.3c)}$$

Looking back, we have three unknowns: x and y (the coordinates of any stationary point Q) and λ, another constant. To determine these, there are three equations: (30.3a, b) and (30.3c). Finally, we summarize the method.

> **Lagrange-multiplier method for the restricted stationary-value problem**
>
> To find the stationary points of $f(x, y)$ subject to $g(x, y) = c$, solve the following equations for x, y, λ:
>
> $$g(x, y) = c, \qquad\qquad \text{(i)}$$
>
> $$\frac{\partial f}{\partial x} - \lambda\frac{\partial g}{\partial x} = 0, \qquad\qquad \text{(ii)}$$
>
> $$\frac{\partial f}{\partial y} - \lambda\frac{\partial g}{\partial y} = 0. \qquad\qquad \text{(iii)}$$
>
> (The value of λ can usually be discarded.)
>
> **(30.4)**

Notice that all reference to the parameter t has disappeared. There are many ways of proving (30.4), but this is probably the simplest for two dimensions. The problem is treated for three dimensions in Section 31.8.

Example 30.7   *Find the stationary points of $x^2 - y^2$ on the curve $x^2 + y^2 = 1$ (compare Examples 30.5 and 30.6).*

In (30.4), $f(x, y) = x^2 - y^2$ and $g(x, y) = x^2 + y^2 = 1$. The equations to be solved, in the order of (30.4), become

$$x^2 + y^2 = 1, \qquad\qquad \text{(i)}$$
$$2x - \lambda(2x) = 0 \quad \text{or} \quad (1 - \lambda)x = 0, \qquad\qquad \text{(ii)}$$
$$-2y - \lambda(2y) = 0 \quad \text{or} \quad (1 + \lambda)y = 0. \qquad\qquad \text{(iii)}$$

↗

Example 30.7 *continued*

From (ii), either $\lambda = 1$ or $x = 0$. Taking these possibilities in order:

If $\lambda = 1$, then (iii) gives $y = 0$; consequently (i) gives $x = \pm 1$.

Therefore we have found the points $(1, 0)$ and $(-1, 0)$ which we called B' and B in Example 30.5.

If $x = 0$, then (iii) gives $\lambda = -1$, and (i) gives $y = \pm 1$. We have therefore found the points $(0, 1)$ and $(0, -1)$ which we called A and A' in Example 30.5.

The equations obtained are often awkward to solve. It is best to be very systematic, not wandering aimlessly between the equations. Be careful not to overlook possibilities (such as that (ii) in Example 30.7 is solved by $\lambda = 1$); and check at the end that the solutions actually fit. The values found for λ do have a special significance in certain subjects but otherwise can be thrown away.

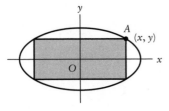

Fig. 30.4

Example 30.8 *Find the rectangle of maximum area which can be placed symmetrically in the ellipse $x^2 + 4y^2 = 1$ as shown in Fig. 30.4.*

Suppose that one of the vertices, say A, is at (x, y). We shall require that x and y be positive, since this is sufficient, given the symmetry. The area is equal to $4xy = f(x, y)$, while x and y are subject to $g(x, y) = x^2 + 4y^2 = 1$.

The three equations, taken in the order of (30.4), become

$$x^2 + 4y^2 = 1, \tag{i}$$

$$2y - \lambda x = 0, \tag{ii}$$

$$x - 2\lambda y = 0. \tag{iii}$$

Suppose that neither x nor y is zero (that could not give a maximum). Then, from (ii) and (iii),

$$\lambda = 2y/x = x/(2y), \tag{iv}$$

so $x = \pm 2y$. However, these must have the same sign for positive area, and we postulated that x and y should be positive. Therefore

$$x = 2y > 0; \tag{v}$$

so, from (iv) again,

$$\lambda = 1. \tag{vi}$$

Use (v) to substitute for x in (i): we get $8y^2 = 1$, or (rejecting negative values of y)

$$y = 1/(2\sqrt{2}),$$

and (v) gives correspondingly

$$x = 1/\sqrt{2}.$$

The sides have length $1/\sqrt{2}$ and $\sqrt{2}$, so the area is 1.

30.3 Curvilinear coordinates in two dimensions

Suppose that x and y are functions of **two parameters**, u and v. To indicate this, write

$$x = x(u, v), \qquad y = y(u, v).$$

This situation arises when we change coordinates from (x, y) to another system. For example, the equations

$$x = u \cos v, \qquad y = u \sin v,$$

represent polar coordinates, with u as the radial and v as the angular coordinate. Now hold v constant; put

$$v = \beta,$$

say, and let u vary. Then

$$x = u \cos \beta, \qquad y = u \sin \beta.$$

Here u is the only active parameter; as it varies, (x, y) traces a radial straight line. Suppose instead that u is held constant, say

$$u = \alpha;$$

then, as v varies, (x, y) follows the circle

$$x = \alpha \cos v, \qquad y = \alpha \sin v.$$

The point where the two curves intersect can be described either by

$$u = \alpha, \qquad v = \beta$$

in the new (polar) coordinates, or in the original coordinates by

$$x = \alpha \cos \beta, \qquad y = \alpha \sin \beta.$$

In general, if we have

$$x = x(u, v), \qquad y = y(u, v),$$

and vary u and v together in an arbitrary way, then the corresponding points (x, y) will completely cover some **area** in the (x, y) plane. If, however, we put $u = \alpha$ and vary v, then put $v = \beta$ and vary u, we obtain two curves in parametric form:

$$(x(\alpha, v), y(\alpha, v)) \quad \text{and} \quad (x(u, \beta), y(u, \beta)).$$

By choosing different values for α and β, we produce a net consisting of two independent systems of curves. This can serve as a new **coordinate system**.

(a)

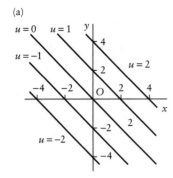

(b)

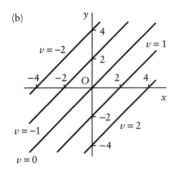

(c)

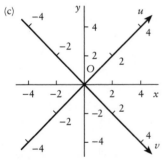

Fig. 30.5

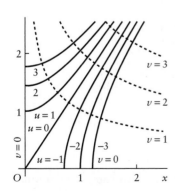

Fig. 30.6

Example 30.9 *Sketch the coordinate system defined by*

$$x = u + v, \qquad y = u - v.$$

Put $u = \alpha$ (constant) and vary v in the equations

$$x = \alpha + v, \qquad y = \alpha - v.$$

Eliminating the active parameter v between the two equations:

$$y = -x + 2\alpha,$$

which is a straight line. By taking different values of α, we obtain a system of parallel straight lines as in Fig. 30.5a.

Put $v = \beta$ and vary u:

$$x = u + \beta, \qquad y = u - \beta.$$

Therefore

$$y = x - 2\beta,$$

which gives another family of parallel straight lines, obtained by taking various values for the constant β, as in Fig. 30.5b.

The two families happen to be at right angles. Taken together, as in Fig. 30.5c, they form a left-handed system of cartesian coordinates (u, v) with origin at $x = 0$, $y = 0$.

New coordinates (u, v) can also be defined in the form

$$u = u(x, y), \qquad v = v(x, y).$$

For example, the system

$$u = (x^2 + y^2)^{\frac{1}{2}}, \qquad v = \arctan(y/x)$$

defines polar coordinates, u (radial) and v (angular). To sketch the curves corresponding to constant u or v, we put

$$\alpha = u(x, y) \quad \text{or} \quad \beta = v(x, y);$$

each of these gives the corresponding curve implicitly.

Example 30.10 *Sketch the coordinate system (u, v) described by*

$$u = y^2 - 2x^2, \qquad v = x^{\frac{1}{2}}y.$$

The curve $u = \alpha$ is obtained in terms of x and y by solving

$$\alpha = y^2 - 2x^2.$$

These curves (for various values of α) are in fact recognizable without solving, being a system of hyperbolas with asymptotes

$$y = \pm\sqrt{2}x.$$

The curves $v = \beta$ are given in $x^{\frac{1}{2}}y = \beta$, so $y = \beta/x^{\frac{1}{2}}$. The system is sketched in Fig. 30.6 for the first quadrant. Notice that $v = 0$ on both $x = 0$ and $y = 0$: the connection between (x, y) and (u, v) is not one-to-one over the whole (x, y) plane.

30.4 Orthogonal coordinates

Suppose that we have a (u, v) system of coordinates defined either by $x = x(u, v)$, $y = y(u, v)$, or by $u = u(x, y)$, $v = v(x, y)$, and the curves $u = \alpha$ and $v = \beta$ always intersect at right angles for any constants α and β. Then the (u, v) system is said to be an **orthogonal system of coordinates**. For example, polar coordinates are orthogonal. Coordinate systems which are not orthogonal are seldom used because of the complexity of the formulae connected with them. A test for orthogonality is the following.

Conditions for an orthogonal system of coordinates

The (u, v) system is orthogonal if

either (a) $u = u(x, y)$, $v = v(x, y)$, and

$$\frac{\partial u}{\partial x}\frac{\partial v}{\partial x} + \frac{\partial u}{\partial y}\frac{\partial v}{\partial y} = 0;$$

or (b) $x = x(u, v)$, $y = y(u, v)$, and

$$\frac{\partial x}{\partial u}\frac{\partial x}{\partial v} + \frac{\partial y}{\partial u}\frac{\partial y}{\partial v} = 0.$$

(30.5)

We prove this result as follows.

(a) Consider the curve from each family which passes through (x, y). According to (29.7), normal vectors to the two curves at (x, y) are $\mathbf{n}_1 = (\partial u/\partial x, \partial u/\partial y)$ and $\mathbf{n}_2 = (\partial v/\partial x, \partial v/\partial y)$ respectively. The curves meet in a right angle if their normals do so, and the condition for this is $\mathbf{n}_1 \cdot \mathbf{n}_2 = 0$, which is equivalent to the condition given in (30.5a).

(b) Consider the curves $u = \alpha$ and $v = \beta$ which pass through a point P which has new coordinates (α, β). Their parametric equations are

$$x = x(\alpha, v), \quad y = y(\alpha, v), \quad \text{for the curve } u = \alpha,$$

$$x = x(u, \beta), \quad y = y(u, \beta), \quad \text{for the curve } v = \beta.$$

Their slopes at P are respectively given by

$$\left(\frac{\mathrm{d}y}{\mathrm{d}x}\right)_P = \left(\frac{\partial y}{\partial v}\Big/\frac{\partial x}{\partial v}\right)_P \quad \text{and}$$

$$\left(\frac{\mathrm{d}y}{\mathrm{d}x}\right)_P = \left(\frac{\partial y}{\partial u}\Big/\frac{\partial x}{\partial u}\right)_P.$$

The condition for the curves to be perpendicular is that the product should equal -1, and this is equivalent to the result in (30.5b).

Example 30.11 *Confirm that the following coordinate systems (u, v) are orthogonal.* (a) $u = y^2 - 2x^2, v = x^{\frac{1}{2}}y$; (b) $x = 2uv, y = u^2 - v^2$.

For (a), use (30.5a). We have

$$\frac{\partial u}{\partial x} = -4x, \quad \frac{\partial v}{\partial x} = \tfrac{1}{2}x^{-\frac{1}{2}}y, \quad \frac{\partial u}{\partial y} = 2y, \quad \frac{\partial v}{\partial y} = x^{\frac{1}{2}},$$

so

$$\frac{\partial u}{\partial x}\frac{\partial v}{\partial x} + \frac{\partial u}{\partial y}\frac{\partial v}{\partial y} = -4x(\tfrac{1}{2}x^{-\frac{1}{2}}y) + 2y(x^{\frac{1}{2}}) = 0.$$

For (b), use (30.5b); notice how this condition is differently structured from (30.5a). We have

$$\frac{\partial x}{\partial u} = 2v, \quad \frac{\partial x}{\partial v} = 2u, \quad \frac{\partial y}{\partial u} = 2u, \quad \frac{\partial y}{\partial v} = -2v;$$

so

$$\frac{\partial x}{\partial u}\frac{\partial x}{\partial v} + \frac{\partial y}{\partial u}\frac{\partial y}{\partial v} = 2v(2u) + 2u(-2v) = 0.$$

30.5 The chain rule for two parameters

Suppose that we have a new set of coordinates defined by

$$x = x(u, v), \qquad y = y(u, v),$$

and a function $f(x, y)$: an arbitrary **function of position**. The function $f(x, y)$ can be expressed in terms of the new coordinates; for example, if

$$x = u^2 - v^2, \quad y = 2uv, \quad \text{and} \quad f(x, y) = x^2 + y^2,$$

then

$$f(x, y) = (u^2 - v^2)^2 + (2uv)^2 = (u^2 + v^2)^2$$

when evaluated at the same point.

If we put

$$z = f(x, y),$$

then the derivatives $\partial z/\partial u$ and $\partial z/\partial v$ indicate how z, or $f(x, y)$, changes as we follow the curves of constant v and constant u respectively. Consider the derivative

$$\frac{\partial z}{\partial u}$$

in which v is held constant, at $v = \beta$ say. *Since only u varies, we are able to adopt the single-variable chain rule* (30.1), *with u instead of t.* However, we must write $\partial x/\partial u$ and $\partial y/\partial u$ instead of dx/du

and $\mathrm{d}y/\mathrm{d}u$ in order to indicate that another variable v is present, although it is regarded as constant for the differentiation. We obtain the following.

Chain rule for two parameters

If $x = x(u, v)$, $y = y(u, v)$, $z = f(x, y)$, then

$$\frac{\partial z}{\partial u} = \frac{\partial z}{\partial x}\frac{\partial x}{\partial u} + \frac{\partial z}{\partial y}\frac{\partial y}{\partial u},$$

$$\frac{\partial z}{\partial v} = \frac{\partial z}{\partial x}\frac{\partial x}{\partial v} + \frac{\partial z}{\partial y}\frac{\partial y}{\partial v}.$$

(Or f may be written instead of z.)

(30.6)

Example 30.12   *Use the chain rule (30.6) to obtain $\partial z/\partial v$ where $x = u^2 - v^2$, $y = 2uv$, and $z = xy$; check the result by substitution.*

For the chain rule, we require

$$\frac{\partial z}{\partial x} = y, \quad \frac{\partial z}{\partial y} = x, \quad \frac{\partial x}{\partial v} = -2v, \quad \frac{\partial y}{\partial v} = 2u.$$

Then

$$\frac{\partial z}{\partial v} = \frac{\partial z}{\partial x}\frac{\partial x}{\partial v} + \frac{\partial z}{\partial y}\frac{\partial y}{\partial v} = -2yv + 2xu = 2u^3 - 6uv^2.$$

To check the result, write z in terms of u and v:

$$z = xy = (u^2 - v^2)2uv = 2u^3v - 2uv^3.$$

Therefore $\dfrac{\partial z}{\partial v} = 2u^3 - 6uv^2$, as before.

There is clearly no advantage in using the chain rule for a simple explicit case such as this. The use of such rules is to obtain general results as in the following examples.

Example 30.13   *Find expressions for $\partial z/\partial r$ and $\partial z/\partial\theta$ when $x = r\cos\theta$, $y = r\sin\theta$, and z is a function of position.*

To use (30.6), put (r, θ) in place of (u, v):

$$\frac{\partial z}{\partial r} = \frac{\partial z}{\partial x}\frac{\partial x}{\partial r} + \frac{\partial z}{\partial y}\frac{\partial y}{\partial r} = \cos\theta\,\frac{\partial z}{\partial x} + \sin\theta\,\frac{\partial z}{\partial y},$$

$$\frac{\partial z}{\partial\theta} = \frac{\partial z}{\partial x}\frac{\partial x}{\partial\theta} + \frac{\partial z}{\partial y}\frac{\partial y}{\partial\theta} = -r\sin\theta\,\frac{\partial z}{\partial x} + r\cos\theta\,\frac{\partial z}{\partial y}.$$

Example 30.14 *Find expressions for $\partial z/\partial x$ and $\partial z/\partial y$ in terms of $\partial z/\partial r$ and $\partial z/\partial \theta$, where $x = r \cos\theta$, $y = r \sin\theta$.*

The appropriate form for chain rule (30.6) will be

$$\frac{\partial z}{\partial x} = \frac{\partial z}{\partial r}\frac{\partial r}{\partial x} + \frac{\partial z}{\partial \theta}\frac{\partial \theta}{\partial x}, \qquad \frac{\partial z}{\partial y} = \frac{\partial z}{\partial r}\frac{\partial r}{\partial y} + \frac{\partial z}{\partial \theta}\frac{\partial \theta}{\partial y}.$$

To find $\partial r/\partial x$ etc., use the alternative form for polar coordinates:

$$r = (x^2 + y^2)^{\frac{1}{2}}, \qquad \theta = \arctan y/x;$$

then

$$\frac{\partial r}{\partial x} = \frac{x}{(x^2+y^2)^{\frac{1}{2}}} = \frac{r\cos\theta}{r} = \cos\theta;$$

$$\frac{\partial\theta}{\partial x} = \frac{1}{1+(y/x)^2}\left(-\frac{y}{x^2}\right) = -\frac{y}{x^2+y^2} = -\frac{r\sin\theta}{r^2} = -\frac{\sin\theta}{r}.$$

Therefore

$$\frac{\partial z}{\partial x} = \cos\theta\,\frac{\partial z}{\partial r} - \frac{\sin\theta}{r}\frac{\partial z}{\partial\theta}.$$

Similarly $\partial r/\partial y$ and $\partial\theta/\partial y$ can be calculated to give

$$\frac{\partial z}{\partial y} = \sin\theta\,\frac{\partial z}{\partial r} + \frac{\cos\theta}{r}\frac{\partial z}{\partial\theta}.$$

(These can also be obtained by treating the pair of expressions for $\partial z/\partial r$ and $\partial z/\partial\theta$ obtained in Example 30.13 as if they were a pair of simultaneous equations for $\partial z/\partial x$ and $\partial z/\partial y$, and solving them.)

Example 30.15 *Supposing that no further information is provided, simplify the expression*

$$\frac{\partial P}{\partial U}\frac{\partial U}{\partial M} + \frac{\partial P}{\partial V}\frac{\partial V}{\partial M}.$$

We may understand from the notation that

$$P = P(U, V).$$

The partial derivative notation $\partial U/\partial M$ and $\partial V/\partial M$ indicates that

$$U = U(M, \dots) \quad \text{and} \quad V = V(M, \dots),$$

at least one more variable being present: the expression does not tell us its name. The chain rule automatically simplifies the expression to

$$\frac{\partial P}{\partial U}\frac{\partial U}{\partial M} + \frac{\partial P}{\partial V}\frac{\partial V}{\partial M} = \frac{\partial P}{\partial M}.$$

Notice how the expressions in (30.6) are formed. Suppose that

$$P = P(U, V), \quad Q = Q(U, V), \quad U = (X, Y), \quad V = V(X, Y).$$

To form for example $\dfrac{\partial P}{\partial X}$, write

$$\frac{\partial P}{\partial X} = \frac{\partial P}{\partial X} + \frac{\partial P}{\partial X},$$

then fill in the spaces in the first term with ∂U and the second with ∂V.

Example 30.16 *Prove that if (x, y) and (u, v) are coordinates related by*

$$x = x(u, v) \quad and \quad y = y(u, v), \tag{i}$$

or alternatively by

$$u = u(x, y) \quad and \quad v = v(x, y), \tag{ii}$$

then

$$\begin{bmatrix} \dfrac{\partial x}{\partial u} & \dfrac{\partial x}{\partial v} \\ \dfrac{\partial y}{\partial u} & \dfrac{\partial y}{\partial v} \end{bmatrix} \begin{bmatrix} \dfrac{\partial u}{\partial x} & \dfrac{\partial u}{\partial y} \\ \dfrac{\partial v}{\partial x} & \dfrac{\partial v}{\partial y} \end{bmatrix} = \begin{bmatrix} 1 & 0 \\ 0 & 1 \end{bmatrix} = I_2.$$

In the first matrix, the relations (i) are implied, and in the second the relations (ii). By multiplying the matrices we obtain

$$\begin{bmatrix} \dfrac{\partial x}{\partial u}\dfrac{\partial u}{\partial x} + \dfrac{\partial x}{\partial v}\dfrac{\partial v}{\partial x} & \dfrac{\partial x}{\partial u}\dfrac{\partial u}{\partial y} + \dfrac{\partial x}{\partial v}\dfrac{\partial v}{\partial y} \\ \dfrac{\partial y}{\partial u}\dfrac{\partial u}{\partial x} + \dfrac{\partial y}{\partial v}\dfrac{\partial v}{\partial x} & \dfrac{\partial y}{\partial u}\dfrac{\partial u}{\partial y} + \dfrac{\partial y}{\partial v}\dfrac{\partial v}{\partial y} \end{bmatrix}.$$

Each of these elements has the right shape for the representation of a derivative by the chain rule (30.6), though the variable combinations occupying the various positions my seem unusual. The matrix becomes

$$\begin{bmatrix} \dfrac{\partial x}{\partial x} & \dfrac{\partial x}{\partial y} \\ \dfrac{\partial y}{\partial x} & \dfrac{\partial y}{\partial y} \end{bmatrix} = \begin{bmatrix} 1 & 0 \\ 0 & 1 \end{bmatrix}.$$

30.6 The use of differentials

Problems are sometimes made easier by working directly with the incremental approximation (29.2): if $z = f(x, y)$, then

$$\delta z \approx \frac{\partial z}{\partial x}\delta x + \frac{\partial z}{\partial y}\delta y.$$

This can be more fruitful than searching for a chain rule or other formula which will work. It is customary in certain applications, particularly in **thermodynamics**, to write this formula in the form

$$\mathrm{d}z = \frac{\partial z}{\partial x}\mathrm{d}x + \frac{\partial z}{\partial y}\mathrm{d}y,$$

in which '≈' becomes '=' and dx, dy, dz are put in place of δx, δy, δz. Such expressions can be manipulated in the same way as the differential forms described in Section 22.4 for functions of a single variable (the theory, however, is somewhat difficult). Here we shall adopt '=' for brevity, but retain δx etc.

Example 30.17   *Find a vector normal to the curve $f(x, y) = c$ at a point (x, y) on the curve. (Compare Section 29.5.)*

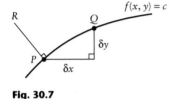

Fig. 30.7

Let P be (x, y) and Q a nearby point $(x + \delta x, y + \delta y)$ also on the curve (see Fig. 30.7). Put

$$z = f(x, y).$$

Then, since z is constant on the curve (it equals c),

$$\delta z = 0 = \frac{\partial z}{\partial x}\,\delta x + \frac{\partial z}{\partial y}\,\delta y,$$

where the derivatives are evaluated at P. This can be written

$$\left(\frac{\partial z}{\partial x}, \frac{\partial z}{\partial y}\right) \cdot (\delta x, \delta y) = 0.$$

But $(\delta x, \delta y) = \overline{PQ}$ is in the direction of the tangent at P (more and more nearly as PQ becomes smaller, of course), so $(\partial z/\partial x, \partial z/\partial y)$ is a vector in the direction of the normal, as we found in Section 29.6.

Example 30.18   *Show that the coordinate system (u, v) defined by*

$$x = 2uv \quad and \quad y = v^2 - u^2$$

is orthogonal.

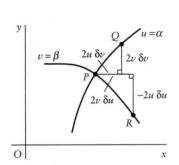

Fig. 30.8

We have to show that any two curves given respectively by $u = \alpha$ and $v = \beta$ intersect in a right angle, as in Fig. 30.8. If u and v are allowed to vary arbitrarily, the incremental formula gives

$$\delta x = 2v\,\delta u + 2u\,\delta v, \qquad \delta y = -2u\,\delta u + 2v\,\delta v. \tag{i}$$

But u does not vary on the curve $u = \alpha$, so $\delta u = 0$ and (i) becomes $\delta x = 2u\,\delta v$, $\delta y = 2v\,\delta v$.
The vector $\overline{PQ}$ points nearly in the direction of the tangent at P:

$$\overline{PQ} = (\delta x, \delta y) = (2u\,\delta v, 2v\,\delta v). \tag{ii}$$

Similarly, on the curve $v = \beta$, we have $\delta v = 0$; so

$$\delta x = 2v\,\delta u, \qquad \delta y = -2u\,\delta u. \tag{iii}$$

$\overline{PR}$ points in the direction of the tangent to $v = \beta$, and

$$\overline{PR} = (\delta x, \delta y) = (2v\,\delta u, -2u\,\delta u).$$

From (ii) and (iii), we have

$$\overline{PQ} \cdot \overline{PR} = (2u\,\delta v, 2v\,\delta v) \cdot (2v\,\delta u, -2u\,\delta u)$$
$$= 4uv\,\delta u\,\delta v - 4uv\,\delta u\,\delta v = 0,$$

so the curves intersect in a right angle.

Problems

30.1 Find a parametrization $(x(t), y(t))$ suitable for the following curves, specifying the range of t required to traverse the curve exactly once, in the anticlockwise direction if the curve is closed.
(a) $x^2 + y^2 = 25$;
(b) $\frac{1}{4}x^2 + \frac{1}{9}y^2 = 1$;
(c) $xy = 4$;
(d) $x^2 - y^2 = 1$ (try using the identity $1 + \tan^2 A = 1/\cos^2 A$);
(e) $\frac{1}{4}x^2 - \frac{1}{9}y^2 = 1$;
(f) $y^2 = 4ax$;
(g) $(x-1)^2 + (y-2)^2 = 9$;
(h) $2x - 5y + 2 = 0$.

30.2 For each of the following cases, obtain df/dt in terms of t by means of the chain rule (30.1).
(a) $f(x, y) = x^2 + y^2$, $x(t) = t$, $y(t) = 1/t$;
(b) $f(x, y) = x^2 - y^2$, $x(t) = \cos t$, $y(t) = \sin t$;
(c) $f(x, y) = xy$, $x(t) = 2\cos t$, $y(t) = \sin t$;
(d) $f(x, y) = x \sin y$, $x(t) = 2t$, $y(t) = t^2$;
(e) $f(x, y) = 4x^2 + 9y^2$, $x(t) = \frac{1}{2}\cos t$, $y(t) = \frac{1}{3}\sin t$.

30.3 Two athletes run around concentric circular tracks of radius r and R with speeds v and V respectively. They start on the same radial line. By using time as a parameter, find the rate of change with time of the distance between them and interpret any stationary points.

30.4 Use the Lagrange-multiplier method to solve the following problems.
(a) Find the maximum area of a rectangle having perimeter of length 10.
(b) Find the rectangle with area 9 which has the shortest perimeter.
(c) Find the stationary points of $x^2 + 2y^2$ subject to $x^2 + y^2 = 1$.
(d) Find the largest rectangle in the first quadrant of the (x, y) plane which has two of its sides along $x = 0$ and $y = 0$ respectively, and a vertex on the line $2x + y = 1$.
(e) Find the minimum distance of the straight line $x + 2y = 1$ from the point $(1, 1)$. (It is easier to consider the *square* of the distance.)
(f) Find the shortest distance from the origin to the curve $x^2 + 8xy + 7y^2 = 225$.
(g) With reference to Fig. 30.4, find the rectangle in the ellipse which has the minimum perimeter.
(h) Find the stationary points of $(x - y + 1)^2$ on $y = x^2$.
(i) Show that in general there are three normals to a parabola from any given point inside it.

30.5 Find the stationary points of $f(x, y)$ on $g(x, y) = c$
(i) by parametrizing the given path as in Example 30.6,
(ii) by using the Lagrange-multiplier technique, in each of the following cases.
(a) $f(x, y) = x^2 + y^2$ on $g(x, y) = xy = 1$;
(b) $f(x, y) = x^2 + y^2$ on $(x - 1)^2 + y^2 = 1$;
(c) $f(x, y) = x^2 + 4y^2$ on $x^2 + y^2 = 1$;
(d) $f(x, y) = 3x - 2y$ on $x^2 - y^2 = 4$;
(e) $f(x, y) = xy$ on $g(x, y) = x^2 + y^2 = 1$ (compare this with (a)).

30.6 Show by means of sketches that, for the restricted stationary-value problem, a stationary point can be expected at any point where the curve $g(x, y) = c$ is tangential to a contour of $f(x, y)$.

Use this observation to derive the Lagrange-multiplier principle. (Hint: consider the normals at the point of tangency; or use implicit differentiation to get expressions for the directions of the curves there.)

There are cases when a stationary point can occur although the curves are not tangential there. Try to identify these cases by sketching various possibilities. (Hint: they correspond to $\lambda = 0$.)

30.7 A change of coordinates from (x, y) to (u, v) is specified by each of the following. Show that the new coordinate system is orthogonal.
(a) $u = 2x + 3y$, $v = -3x + 2y$;
(b) $u = xy$, $v = x^2 - y^2$;
(c) $u = x^2 + 2y^2$, $v = y/x^2$;
(d) $u = xy^2$, $v = y^2 - 2x^2$;
(e) $u = x + 1/x + y^2/x$, $v = y - 1/y + x^2/y$;
(f) $x = 2u - v$, $y = u + 2v$;
(g) $x = u^2 - v^2$, $y = 2uv$;
(h) $x = u/(u^2 + v^2)$, $y = v/(u^2 + v^2)$;
(i) $x = u^2 - v^2$, $y = -2uv$.

30.8 Let $r(t)$ and $\theta(t)$ be polar coordinates which are functions of a parameter t.
(a) Express dx/dt and dy/dt in terms of dr/dt, $d\theta/dt$, r, and θ.
(b) Use (a) to obtain expressions for d^2x/dt^2 and d^2y/dt^2.
(c) Prove that

$$\cos\theta \frac{d^2x}{dt^2} + \sin\theta \frac{d^2y}{dt^2} = \frac{d^2r}{dt^2} - r\left(\frac{d\theta}{dt}\right)^2,$$

$$\cos\theta \frac{d^2y}{dt^2} - \sin\theta \frac{d^2x}{dt^2} = \frac{1}{r}\frac{d}{dt}\left(r^2 \frac{d\theta}{dt}\right).$$

(These two equations express the radial and tangential components of acceleration, given on the left, in terms of polar coordinates.)

30.9 Use the chain rule (30.6) to find $\partial f/\partial u$ and $\partial f/\partial v$ in terms of u and v in each of the following cases.

(a) $f(x, y) = 2x - y, x = uv, y = u^2 - v^2$;
(b) $f(x, y) = y/x, x = u + v, y = u - v$;
(c) $f(x, y) = y^2, x = u^2 + v^2, y = v/u$;
(d) $f(x, y) = (x - y)/(x + y), x = v, y = u - v$.

30.10 By using the chain rule (30.6) twice, obtain $\partial^2 f/\partial u^2$, $\partial^2 f/\partial v^2$, and $\partial^2 f/\partial u \, \partial v$ in each of the following cases.

(a) $f(x, y) = y/x, x = u + v, y = u - v$;
(b) $f(x, y) = x^2 + y^2, x = uv, y = u^2 - v^2$;
(c) $f(x, y) = y^2, x = uv, y = v$.

30.11 Find expressions for $\partial f/\partial u$, $\partial f/\partial v$, $\partial^2 f/\partial u^2$, $\partial^2 f/\partial v^2$, and $\partial^2 f/\partial u \, \partial v$ if

$f(x, y) = g(x^2 - y^2), x = u + v, y = u - v$.

(The expressions will involve the functions $g'(x^2 - y^2)$ etc.)

30.12 Let $w = w(u, v), u = u(x, y), v = v(x, y)$, where u and v are related in such a way that

$$\frac{\partial u}{\partial x} = \frac{\partial v}{\partial y}, \quad \frac{\partial u}{\partial y} = -\frac{\partial v}{\partial x}.$$

Prove that

$$\frac{\partial^2 u}{\partial x^2} + \frac{\partial^2 u}{\partial y^2} = 0, \quad \frac{\partial^2 v}{\partial x^2} + \frac{\partial^2 v}{\partial y^2} = 0.$$

Use the chain rule (30.6) to prove that

$$\frac{\partial^2 w}{\partial x^2} + \frac{\partial^2 w}{\partial y^2} = \left[\left(\frac{\partial u}{\partial x} \right)^2 + \left(\frac{\partial v}{\partial y} \right)^2 \right] \left[\frac{\partial^2 w}{\partial x^2} + \frac{\partial^2 w}{\partial y^2} \right].$$

30.13 Let r and θ be the usual polar coordinates, and $z = f(x, y)$; show that:

(a) $\left(\dfrac{\partial z}{\partial x} \right)^2 + \left(\dfrac{\partial z}{\partial y} \right)^2 = \left(\dfrac{\partial z}{\partial r} \right)^2 + \dfrac{1}{r^2} \left(\dfrac{\partial z}{\partial \theta} \right)^2$;

(b) $\dfrac{\partial^2 z}{\partial x^2} + \dfrac{\partial^2 z}{\partial y^2} = \dfrac{\partial^2 z}{\partial r^2} + \dfrac{1}{r} \dfrac{\partial z}{\partial r} + \dfrac{1}{r^2} \dfrac{\partial^2 z}{\partial \theta^2}$.

31 Functions of any number of variables

31.1 The incremental approximation; errors

For functions of three and more variables, simple pictorial representations are not available. Nevertheless many of the important formulae follow the pattern of the two-variable case, simply containing more terms of the same type. This follows from the incremental approximation (29.1), extended to three and more variables.

Suppose that $f(x, y, z, \dots)$ is any function of N ($\geqslant 3$) variables. The **partial derivatives**, $\partial f/\partial x, \partial f/\partial y, \partial f/\partial z, \dots$, have the same meaning as they did in Chapter 28: during differentiation, all the variables except the named one are treated as constants.

Higher derivatives are defined as with functions of two variables; for example,

$$\frac{\partial^3 f}{\partial x \, \partial y \, \partial z} = \frac{\partial}{\partial x} \frac{\partial}{\partial y} \frac{\partial f}{\partial z}.$$

It follows from the result for second derivatives (Equation (28.3)) that

$$\frac{\partial^3 f}{\partial x \, \partial y \, \partial z} = \frac{\partial^3 f}{\partial y \, \partial x \, \partial z} = \frac{\partial^3 f}{\partial z \, \partial y \, \partial x}$$

and so on: the derivatives may be taken in any order.

The **incremental approximation** has the same form as (29.1) and (29.2), simply containing further terms corresponding to the extra variables:

> **Incremental approximation for** $f(x, y, z, \dots)$
>
> For small enough increments $\delta x, \delta y, \delta z, \dots$:
> $$\delta f = f(x + \delta x, y + \delta y, z + \delta z, \dots) - f(x, y, z, \dots)$$
> $$\approx \frac{\partial f}{\partial x} \delta x + \frac{\partial f}{\partial y} \delta y + \frac{\partial f}{\partial z} \delta z + \cdots.$$
> If we put $w = f(x, y, z, \dots)$, this can be written
> $$\delta w \approx \frac{\partial w}{\partial x} \delta x + \frac{\partial w}{\partial y} \delta y + \frac{\partial w}{\partial z} \delta z + \cdots.$$
>
> (31.1)

To prove (31.1), the idea of a tangent plane is not available, so we must go directly for the linear approximation to the function. Put $w = f(x, y, z, \dots)$, and consider a fixed 'point' $P : (x, y, z, \dots)$ and another nearby point $Q : (x + \delta x, y + \delta y, z + \delta z, \dots)$. Then w changes to $w + \delta w$. We assume that the relation between δw and δx, $\delta y, \delta z, \dots$ is **close to linear** for small $\delta x, \delta y, \delta z, \dots$; that is to say,

$$\delta w = A \, \delta x + B \, \delta y + C \, \delta z + \cdots + \varepsilon, \tag{31.2}$$

where $A, B, \dots$ are certain constants and the **error ε is of a lower order of magnitude than the δ quantities** (compare (29.2) for two variables).

In order firstly to find A, vary only x, so that

$$\delta x \neq 0, \quad \delta y = \delta z = \cdots = 0.$$

Put these into (31.2) and divide by δx, giving

$$\frac{\delta w}{\delta x} = A + \frac{\varepsilon}{\delta x}.$$

Now let $\delta x \to 0$. Then $\delta w / \delta y \to \partial w / \partial x$, and (since ε is of lower order of magnitude than δx) $\varepsilon / \delta x \to 0$. Therefore

$$\frac{\partial w}{\partial x} = A.$$

Similarly $\partial w / \partial y = B$, and so on, which gives the result (31.1).

The incremental approximation (31.1) can be used to estimate **errors** as in Section 29.2.

> **Small-error formula**
>
> If $w = f(x, y, z, \dots)$, then (approximately)
> $$\Delta w = \frac{\partial w}{\partial x} \Delta x + \frac{\partial w}{\partial y} \Delta y + \frac{\partial w}{\partial z} \Delta z + \cdots,$$
> where $x, y, z, \dots$ stand for the measured values and Δ stands for
>
> error = (measured value) − (exact value).
>
> (31.3)

Example 31.1   *In a triangle ABC,* $\cos C = (c^2 - a^2 - b^2)/(2ab)$. *In a particular case, the measured side lengths are* $a = 3$, $b = 4$, *and* $c = 5.5$ *units. Possible errors of measurement lie between* ±0.1 *units. Find the error in estimating C in the worst case.* (*Compare Example 29.6.*)

For ease of differentiation put

$$w = \cos C = \frac{c^2}{2ab} - \frac{a}{2b} - \frac{b}{2a}.$$

Then

$$\Delta(\cos C) = \Delta w \approx \frac{\partial w}{\partial a} \Delta a + \frac{\partial w}{\partial b} \Delta b + \frac{\partial w}{\partial c} \Delta c,$$

where

$$\frac{\partial w}{\partial a} = -\frac{c^2}{2a^2 b} - \frac{1}{2b} + \frac{b}{2a^2},$$

and similarly for the other two derivatives. From the measurements,

$$\frac{\partial w}{\partial a} = -0.323, \qquad \frac{\partial w}{\partial b} = -0.388, \qquad \frac{\partial w}{\partial c} = 0.458.$$

Therefore $\Delta(\cos C) \approx -0.323 \, \Delta a - 0.388 \, \Delta b + 0.458 \, \Delta c$.
    To obtain ΔC from $\Delta(\cos C)$:

$$\Delta(\cos C) \approx \left(\frac{d}{dC} \cos C \right) \Delta C = (-\sin C) \, \Delta C,$$

where C is in *radians*. The value of C estimated from the cosine rule using the measured values is 1.350 radians, and $\sin 1.350 = 0.976$. Therefore

$$\Delta C \approx -0.334 \, \Delta a - 0.398 \, \Delta b + 0.469 \, \Delta c.$$

The magnitude of ΔC is a maximum if by chance the errors are

$$\Delta a = \mp 0.1, \qquad \Delta b = \mp 0.1, \qquad \Delta c = \pm 0.1,$$

and then

$$\Delta C \approx \pm 0.120.$$

which is about a 9% error.

31.2 Implicit differentiation

There is an analogy with the implicit-differentiation formula (29.6). Suppose that

$$f(x, y, z, \dots) = 0. \tag{31.4}$$

This condition implies that any one of the variables depends on, or is a function of, all the others. For example, if the variables are x, y, z, and r, and

$$x^2 + y^2 + z^2 - r^2 = 0,$$

then

$$y = \pm (r^2 - x^2 - z^2)^{\frac{1}{2}}.$$

Subject to (31.4) we can therefore talk about partial derivatives such as $\partial y/\partial x$: we think of y as being a function of the other variables, but with all the variables except x and y held constant.

Suppose that $(x, y, z, \dots)$ and $(x + \delta x, y + \delta y, z + \delta z, \dots)$ both satisfy condition (31.4). Then $\delta f = 0$ and the incremental approximation gives

$$\frac{\partial f}{\partial x}\delta x + \frac{\partial f}{\partial y}\delta y + \frac{\partial f}{\partial z}\delta z + \cdots \approx 0. \tag{31.5}$$

Suppose next that all the variables except x and y are kept constant, so that $\delta x \neq 0$ and $\delta y \neq 0$, but $\delta z = \cdots = 0$. Equation (31.5) becomes $(\partial f/\partial x)\,\delta x + (\partial f/\partial y)\,\delta y \approx 0$, so that

$$\frac{\delta y}{\delta x} \approx -\frac{\partial f}{\partial x}\bigg/\frac{\partial f}{\partial y}.$$

Now let $\delta x \to 0$ and the equation becomes (compare (29.61)

$$\frac{\partial y}{\partial x} = -\frac{\partial f}{\partial x}\bigg/\frac{\partial f}{\partial y}:$$

Implicit differentiation

If $f(x, y, z, \dots) = 0$, then

$$\frac{\partial y}{\partial x} = -\frac{\partial f}{\partial x}\bigg/\frac{\partial f}{\partial y}.$$

Any other two variables may be substituted for x and y. **(31.6)**

Example 31.2 *For a fixed mass of gas, an equation of the form $f(P, V, T) = 0$ holds (the 'equation of state'), where P, V, and T represent the pressure, volume, and temperature respectively. Show that*

(a) $\dfrac{\partial P}{\partial T}\dfrac{\partial T}{\partial V} = -\dfrac{\partial P}{\partial V}$, (b) $\dfrac{\partial P}{\partial T}\dfrac{\partial T}{\partial V}\dfrac{\partial V}{\partial P} = -1$.

The relation $f(P, V, T) = 0$ implies that any of P, V, or T is a function of the other two variables: $P = P(V, T)$, $V = V(T, P)$, and $T = T(P, V)$. If we put, say $P = P(V, T) = $ constant, then implicit differentiation, by (31.6), gives $\partial V/\partial T$ or $\partial T/\partial V$ in terms of $\partial f/\partial V$ and $\partial f/\partial T$ (where we are reminded of the 'constant P' condition by the partial derivative signs instead of $\mathrm{d}V/\mathrm{d}T$ and $\mathrm{d}T/\mathrm{d}V$). Similarly we obtain $\partial P/\partial T$, $\partial T/\partial P$, $\partial P/\partial V$, and $\partial V/\partial T$.

(a) $\dfrac{\partial P}{\partial T} = -\dfrac{\partial f}{\partial T}\bigg/\dfrac{\partial f}{\partial P}$ and $\dfrac{\partial T}{\partial V} = -\dfrac{\partial f}{\partial V}\bigg/\dfrac{\partial f}{\partial T}$

(from (31.6)). Therefore

$$\frac{\partial P}{\partial T}\frac{\partial T}{\partial V} = \frac{\partial f}{\partial V}\bigg/\frac{\partial f}{\partial P} = -\frac{\partial P}{\partial V} \quad \text{(using (31.6) again).}$$

(b) By repeating the process (a) with different variables,

$$\frac{\partial P}{\partial T}\frac{\partial T}{\partial V}\frac{\partial V}{\partial P} = \left(-\frac{\partial f}{\partial T}\bigg/\frac{\partial f}{\partial P}\right)\left(-\frac{\partial f}{\partial V}\bigg/\frac{\partial f}{\partial T}\right)\left(-\frac{\partial f}{\partial P}\bigg/\frac{\partial f}{\partial V}\right) = -1.$$

There are many more similar formulae obtainable by permuting P, T, V: these identities are important in the theory of **thermodynamics**. A helpful **notation**, encountered in physics texts, contains a reminder of the variables to be held constant during differentiation. If $f(P, V, T) = 0$, then the new notation puts

$$\left(\frac{\partial P}{\partial V} \right)_T$$

in place of $\partial P / \partial V$. The formula in Example 31.2a would become

$$\left(\frac{\partial P}{\partial T} \right)_V \left(\frac{\partial T}{\partial V} \right)_P = -\left(\frac{\partial P}{\partial V} \right)_T,$$

indicating that the symbols ∂T on the left cannot simply be 'cancelled' as with ordinary derivatives.

31.3 Chain rules

The **chain rule for a single parameter** t is obtained exactly as in the case of a single variable: divide (31.1) by δt and take the limit, to give the following formula.

> **Chain rule for one parameter**
> Given $f(x, y, z, \dots)$, where $x = x(t)$,
> $\qquad y = y(t), z = z(t), \dots,$
> $$\frac{df}{dt} = \frac{\partial f}{\partial x} \frac{dx}{dt} + \frac{\partial f}{\partial y} \frac{dy}{dt} + \frac{\partial f}{\partial z} \frac{dz}{dt} + \cdots$$
> (or with w in place of f if $w = f(x, y, z, \dots)$).
>
> (31.7)

Notice that $(x(t), y(t), z(t))$ defines a directed path in three dimensions.

In the case of more than one parameter, the results of Section 30.1 may be extended as follows.

> **Chain rule for more than one parameter**
> For a function $f(x, y, z, \dots)$, where $x, y, z, \dots$ are functions of parameters $u, v, \dots$, we have
> $$\frac{\partial f}{\partial u} = \frac{\partial f}{\partial x} \frac{\partial x}{\partial u} + \frac{\partial f}{\partial y} \frac{\partial y}{\partial u} + \frac{\partial f}{\partial z} \frac{\partial z}{\partial u} + \cdots,$$
> $$\frac{\partial f}{\partial v} = \frac{\partial f}{\partial x} \frac{\partial x}{\partial v} + \frac{\partial f}{\partial y} \frac{\partial y}{\partial v} + \frac{\partial f}{\partial z} \frac{\partial z}{\partial v} + \cdots,$$
> and so for any other parameters. (If $w = f(x, y, z, \dots)$ then w may be written in place of f.)
>
> (31.8)

31.4 The gradient vector in three dimensions

The **gradient vector function**, introduced for two dimensions in Section 29.6, extends to any number of dimensions, though we shall restrict consideration to three variables in this section.

In the equations we have obtained, such as (31.1) and (31.8), there repeatedly occurs the triplet of elements $\partial f/\partial x$, $\partial f/\partial y$, $\partial f/\partial z$, added, together with various multipliers. We can manipulate this group as a unit by regarding

$$\left(\frac{\partial f}{\partial x}, \frac{\partial f}{\partial y}, \frac{\partial f}{\partial z}\right), \quad \text{or} \quad \frac{\partial f}{\partial x}\hat{\imath} + \frac{\partial f}{\partial y}\hat{\jmath} + \frac{\partial f}{\partial z}\hat{k},$$

as a vector function – **the gradient of** f, now in three dimensions – and denote it by

grad f or ∇f,

as before. As in Section 29.6, we can also think of **grad** or ∇ standing alone as an operator: an instruction to carry out the process

$$\hat{\imath}\frac{\partial}{\partial x} + \hat{\jmath}\frac{\partial}{\partial y} + \hat{k}\frac{\partial}{\partial z}, \quad \text{or} \quad \left(\frac{\partial}{\partial x}, \frac{\partial}{\partial y}, \frac{\partial}{\partial z}\right),$$

on some scalar function $f(x, y, z)$. The definition is stated for reference as follows.

> **Gradient vector function (three dimensions)**
>
> For a scalar function $f(x, y, z)$:
>
> $$\textbf{grad } f \quad \text{or} \quad \nabla f = \left(\frac{\partial f}{\partial x}, \frac{\partial f}{\partial y}, \frac{\partial f}{\partial z}\right) = \hat{\imath}\frac{\partial f}{\partial x} + \hat{\jmath}\frac{\partial f}{\partial y} + \hat{k}\frac{\partial f}{\partial z}.$$
>
> Alternatively, **grad** or ∇ stands for the **operator**
>
> $$\left(\hat{\imath}\frac{\partial}{\partial x} + \hat{\jmath}\frac{\partial}{\partial y} + \hat{k}\frac{\partial}{\partial z}\right), \quad \text{or} \quad \left(\frac{\partial}{\partial x}, \frac{\partial}{\partial y}, \frac{\partial}{\partial z}\right).$$
>
>   (31.9)

Example 31.3 *Let* $f(x, y, z) = x^2 + y^2 + z^2$. *Obtain* (a) *the vector function* **grad** $f(x, y, z)$; (b) *the value of* **grad** $f(x, y, z)$ *at the point* $(1, 2, 3)$; (c) *an expression for the magnitude (or length) of* **grad** $f(x, y, z)$.

(a) **grad** $f(x, y, z) = \left(\dfrac{\partial f}{\partial x}, \dfrac{\partial f}{\partial y}, \dfrac{\partial f}{\partial z}\right) = (2x, 2y, 2z);$

or one can use the 'operator' idea and the other way of writing a vector:

$$\textbf{grad } f = \left(\hat{\imath}\frac{\partial}{\partial x} + \hat{\jmath}\frac{\partial}{\partial y} + \hat{k}\frac{\partial}{\partial z}\right)(x^2 + y^2 + z^2) = \hat{\imath}(2x) + \hat{\jmath}(2y) + \hat{k}(2z).$$

Example 31.3 *continued*

(b) At $x = 1$, $y = 2$, $z = 3$,

    grad $f = (2, 4, 6)$.

(c) The magnitude or length $|v|$ of a vector $v = (a, b, c)$ is $|v| = (a^2 + b^2 + c^2)^{\frac{1}{2}}$; so

$$|\mathbf{grad}\, f| = [(2x)^2 + (2y)^2 + (2z)^2]^{\frac{1}{2}} = 2(x^2 + y^2 + z^2)^{\frac{1}{2}}.$$

Expressions which occur in the theory frequently take the form

$$U\frac{\partial f}{\partial x} + V\frac{\partial f}{\partial y} + W\frac{\partial f}{\partial z}, \tag{31.10}$$

where U, V, and W may be constants, or various functions. If we put

$$\hat{i}U + \hat{j}V + \hat{k}W = S,$$

where S is another vector (compare Section 29.6), then we can write (31.10) in the form

$$U\frac{\partial f}{\partial x} + V\frac{\partial f}{\partial y} + W\frac{\partial f}{\partial z} = (U, V, W) \cdot \left(\frac{\partial f}{\partial x}, \frac{\partial f}{\partial y}, \frac{\partial f}{\partial z}\right) = S \cdot \mathbf{grad}\, f,$$

as in the following example.

Example 31.4 *Suppose that the concentration of plankton in the sea is $C(x, y, z, t)$. A whale travels on the path $x = x(t)$, $y = y(t)$, $z = z(t)$, where t is time. Show that, on the path of the whale,*

$$\frac{dC}{dt} = \frac{\partial C}{\partial t} + v \cdot \mathbf{grad}\, C,$$

where v is its velocity.

By the chain rule (31.7),

$$\frac{dC}{dt} = \frac{\partial C}{\partial x}\frac{dx}{dt} + \frac{\partial C}{\partial y}\frac{dy}{dt} + \frac{\partial C}{\partial z}\frac{dz}{dt} + \frac{\partial C}{\partial t},$$

after putting $dt/dt = 1$ into the final term. The whale's velocity is

$$v = \left(\frac{dx}{dt}, \frac{dy}{dt}, \frac{dz}{dt}\right),$$

so that

$$\frac{dC}{dt} = \frac{\partial C}{\partial t} + v \cdot \mathbf{grad}\, C.$$

(If the whale drifted with the motion of the sea, v would represent the velocity of the current. This case is related to the concept of **material derivative** in fluid mechanics. Instead of C there is a quantity such as the density or momentum of a particular piece of fluid, whose variation we follow as the fluid moves around.)

31.5 Normal to a surface

An equation of the form

$$g(x, y, z) = k$$

represents a **surface** in three dimensions, because we can imagine 'solving' the equation for z in order to obtain equivalent equation(s):

$$z = f(x, y).$$

Thus, if $x^2 + y^2 + z^2 = 1$, then $z = \pm(1 - x^2 - y^2)^{\frac{1}{2}}$. The normal to the surface can be expressed neatly as follows.

> **Normal (perpendicular) to a surface**
> Let P be any point on a **surface** $g(x, y, z) = k$. Then
> **grad** g, evaluated at P, is normal to the surface at P. (31.11)

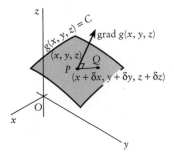

Fig. 31.1

(Compare (29.9), for the normal to a curve in two dimensions.) The proof is as follows. In Fig. 31.1, $P : (x, y, z)$ is the given point on the surface and $Q : (x + \delta x, y + \delta y, z + \delta z)$ is *any* nearby point on the surface. Then

$$g(x + \delta x, y + \delta y, z + \delta z) - g(x, y, z) = 0,$$

or

$$\delta g = 0.$$

Therefore, by the incremental formula (31.1),

$$0 = \frac{\partial g}{\partial x} \delta x + \frac{\partial g}{\partial y} \delta y + \frac{\partial g}{\partial z} \delta z$$

$$= (\mathbf{grad}\ g) \cdot (\delta x, \delta y, \delta z).$$

This shows that **grad** g is perpendicular to the vector $(\delta x, \delta y, \delta z)$. But $\overline{PQ}$ can be chosen to point in any direction from P in the surface, so the only possibility is that **grad** g is perpendicular to the surface itself at P.

We already know (from (28.7)) that a vector normal to a suface described in the form $z = f(x, y)$ is

$$\left(\frac{\partial f}{\partial x}, \frac{\partial f}{\partial y}, -1 \right).$$

This is reconciled with (31.11) if we write its equation in the form

$$g(x, y, z) = f(x, y) - z = 0.$$

31.6 Equation of the tangent plane

Suppose that a surface is specified in the form $g(x, y, z) = k$, and that the point $P : (x, y, z)$ is on the surface. The conditions that the tangent plane must satisfy are (a) it contains the point P; and (b) it is perpendicular to the normal vector **grad** $g(x, y, z)$ evaluated at P, as in (31.11). These conditions are satisfied by the equation

$$\left(\frac{\partial g}{\partial x}\right)_P (x - a) + \left(\frac{\partial g}{\partial y}\right)_P (y - b) + \left(\frac{\partial g}{\partial z}\right)_P (z - c) = 0. \tag{31.12}$$

It can be seen that the expression is zero when $x = a$, $y = b$, $z = c$. Also the coefficient vector

$$\left(\left(\frac{\partial g}{\partial x}\right)_P, \left(\frac{\partial g}{\partial y}\right)_P, \left(\frac{\partial g}{\partial z}\right)_P\right) = [\mathbf{grad}\ g]_P$$

is perpendicular to the plane. Therefore we may state the equation of the plane as follows.

Tangent plane to the surface $g(x, y, z) = k$ **at**
$P : (a, b, c)$

$$\left(\frac{\partial g}{\partial x}\right)_P (x - a) + \left(\frac{\partial g}{\partial y}\right)_P (y - b) + \left(\frac{\partial g}{\partial z}\right)_P (z - c) = 0.$$

(31.13)

31.7 Directional derivative in terms of gradient

The vector **grad** f contains the necessary information to calculate the rate of change of $f(x, y, z)$ in any direction. In Fig. 31.2, let $P : (a, b, c)$ be any point. Suppose that we require the rate of change with distance of $f(x, y, z)$ in the direction PR.

Choose a nearby point $Q : (x + \delta x, y + \delta y, z + \delta z)$ on PR, and put

$$PQ = \delta s = (\delta x^2 + \delta y^2 + \delta z^2)^{\frac{1}{2}}$$

(where δs is a standard symbol for a small element of distance). Then

$$\frac{\delta x}{\delta s} = \cos \alpha, \qquad \frac{\delta y}{\delta s} = \cos \beta, \qquad \frac{\delta z}{\delta s} = \cos \gamma,$$

where $\cos \alpha, \cos \beta, \cos \gamma$ are the **direction cosines** of PQ (Section 10.5). Now divide the incremental approximation (31.1) through by δs and take the limit as $\delta s \to 0$. We obtain an expression for the rate of change of $f(x, y, z)$ with distance in any direction:

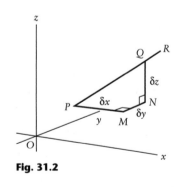

Fig. 31.2

> **Directional derivative in three dimensions**
>
> In the direction having direction cosines $(\cos \alpha, \cos \beta, \cos \gamma)$:
>
> $$\frac{\partial f}{\partial s} = \frac{\partial f}{\partial x} \cos \alpha + \frac{\partial f}{\partial y} \cos \beta + \frac{\partial f}{\partial z} \cos \gamma.$$
>
> (31.14)

In the two-dimensional version (29.4), the coefficients $\cos \theta$ and $\sin \theta$ are equal to the two-dimensional direction cosines, $\cos \theta$ and $\cos(\frac{1}{2}\pi - \theta)$, so (29.4) is compatible with (31.14).

The **direction cosines** $\cos \alpha, \cos \beta, \cos \gamma$ have the property $\cos^2\alpha + \cos^2\beta + \cos^2\gamma = 1$ (see Section 10.5), so they are the components of a **unit vector** $\hat{s}$ which points in the desired direction. Therefore (31.14) can be written differently:

> **Directional derivative in three dimensions in terms of the gradient**
>
> In the direction of the unit vector $\hat{s}$,
>
> $$\frac{\mathrm{d}f}{\mathrm{d}s} = \hat{s} \cdot \mathbf{grad}\, f,$$
>
> which is the component of **grad** f in direction $\hat{s}$.
>
> (31.15)

As in Section 29.6, the result (31.15) can be expressed in a third way. If a and b are two vectors, and ϕ is the angle between them, then $a \cdot b = |a||b| \cos \phi$. Putting $\hat{s}$ for a and **grad** f for b in (31.15), and using the fact that $|\hat{s}| = 1$, we obtain the next result.

> **Directional derivative in three dimensions**
>
> $$\frac{\mathrm{d}f}{\mathrm{d}s} = |\mathbf{grad}\, f| \cos \phi,$$
>
> where ϕ is the angle between **grad** f and the unit direction vector $\hat{s}$.
>
> (31.16)

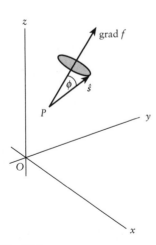

Fig. 31.3

Now take a function $f(x, y, z)$, and a point $P : (x_1, y_1, z_1)$ as in Fig. 31.3. By means of (31.16), we can explore the rate of variation of $f(x, y, z)$ in all directions, by pointing $\hat{s}$ in the required directions. The only thing that changes when we do this is the angle ϕ. It can be seen from (31.16) that: (i) if $\phi = \frac{1}{2}\pi$, then $\mathrm{d}f/\mathrm{d}s = 0$, which is consistent with $\hat{s}$ pointing tangentially to the surface $f(x, y, z) = f_P$, where f_P is the value of f at P; (ii) $\mathrm{d}f/\mathrm{d}s$ takes its maximum value $|\mathbf{grad}\, f|$, when $\phi = 0$. That is to say, **grad** f *points in the direction of most rapid increase of f and it is normal to the surface $f(x, y, z) = f_P$.*

It is worth noticing that, for a fixed angle ϕ, the unit vector $\hat{s}$ may point anywhere along the generators of a cone having axis **grad** f, as shown in Fig. 31.3. The directional derivative df/ds is the same in all these directions.

Example 31.5   *Let $f(x, y, z) = 4 - x^2 - \frac{1}{2}y^2 - \frac{1}{2}z^2$ represent the atmospheric concentration of a chemical which attracts insects. (a) Write down **grad** f at (x, y, z). (b) Find a unit vector $\hat{s}$ which points in the direction of most rapid rate of increase in $f(x, y, z)$ at the point $(1, 1, 1)$. (c) An insect sets off from $(1, 1, 1)$ and flies a short distance δs in the direction given by (b). Find its new coordinates (approximately).*

(a) $\mathbf{grad}\ f(x, y, z) = \left(\dfrac{\partial f}{\partial x}, \dfrac{\partial f}{\partial y}, \dfrac{\partial f}{\partial z} \right)$

$\qquad\qquad\qquad = (-2x, -y, -z),\quad \text{or}\quad -2x\hat{i} - y\hat{j} - z\hat{k}.$

(b) **grad** f always points in the required direction; at the point $(1, 1, 1)$, its components are $(-2, -1, -1)$. To obtain the corresponding *unit* vector $\hat{s}$, divide by the length $[(-2)^2 + (-1)^2 + (-1)^2]^{\frac{1}{2}} = \sqrt{6}$, obtaining $\hat{s} = (-2/\sqrt{6}, -1/\sqrt{6}, -1/\sqrt{6})$.

(c) The insect moves a distance δs from the point $P : (1, 1, 1)$ along $\hat{s}$ (see Fig. 31.4), so its *vector* displacement $\overline{PQ}$ is

$$\hat{s}\,\delta s = \left(-\frac{2}{\sqrt{6}}\delta s, -\frac{1}{\sqrt{6}}\delta s, -\frac{1}{\sqrt{6}}\delta s \right).$$

The components of this vector are the x, y, z displacements

$$\delta x = -\frac{2}{\sqrt{6}}\delta s, \qquad \delta y = -\frac{1}{\sqrt{6}}\delta s, \qquad \delta z = -\frac{1}{\sqrt{6}}\delta s.$$

The new coordinates are therefore

$$x = 1 - \frac{2}{\sqrt{6}}\delta s, \qquad y = 1 - \frac{1}{\sqrt{6}}\delta s, \qquad z = 1 - \frac{1}{\sqrt{6}}\delta s.$$

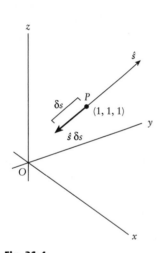

Fig. 31.4

Example 31.6   *Using Example 31.5c as a model, give a systematic method, suitable for computation, for approximating to the path of an insect which always flies in the direction of most rapidly increasing concentration.*

Figure 31.5 shows notionally the path of such an insect. It starts at $P_0 :$ (x_0, y_0, z_0), and the path consists of short steps of equal length h (instead of δs, for the purpose of programming). The progression is $P_0, P_1, P_2, \dots, P_n,$ $P_{n+1}, \dots$, with coordinates numbered (x_n, y_n, z_n) for $n = 0, 1, 2, \dots$. At each P_n, the insect moves in the unit-vector direction $\hat{s}_n$, where $s_n = \mathbf{grad}\ f(x, y, z)$ and, as in Example 31.5c,

$\qquad \hat{s}_n = [(\mathbf{grad}\ f)/|\mathbf{grad}\ f|]_{P_n}$

$\qquad\quad = (-2x_n, -y_n, -z_n)/(4x_n^2 + y_n^2 + z_n^2)^{\frac{1}{2}}$

$\qquad\quad = (a_n, b_n, c_n)\quad \text{(say)}.$ \hfill (31.17)

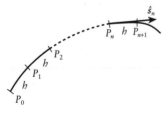

Fig. 31.5

Example 31.6 *continued*

For the general step from P_n to P_{n+1}, we obtain the small displacement components δx_n, δy_n, δz_n in the x, y, and z directions (in general these will differ from step to step):

$$(\delta x_n, \delta y_n, \delta z_n) = \hat{s}_n h = (a_n h, b_n h, c_n h),$$

from (31.17). Therefore

$$(x_{n+1}, y_{n+1}, z_{n+1}) = (x_n + \delta x_n, y_n + \delta y_n, z_n + \delta z_n)$$
$$= (x_n + a_n h, y_n + b_n h, z_n + c_n h). \qquad (31.18)$$

Equations (31.17) and (31.18), with the starting point (x_0, y_0, z_0) given, form a step-by-step process which is easy to computerize. The following table of the early stages was calculated with $h = 0.05$; the starting point in this case is the point $(1, 1, 1)$, where $f(x, y, z) = 4 - x^2 - \frac{1}{2}y^2 - \frac{1}{2}z^2$ as in Example 31.5.

n	x_n	y_n	z_n
0	1	1	1
1	0.959	0.980	0.980
2	0.919	0.959	0.959
3	0.878	0.938	0.938
4	0.839	0.917	0.917
5	0.799	0.895	0.895

If a surface is defined by the equation

$$f(x, y, z) = c,$$

where c is a constant, it is called a **level surface of the function** f (it is the analogy of a contour in the theory for functions of two variables). According to (31.11), therefore, we can say in different language:

> **Normal to a level surface of $f(x, y, z)$**
>
> **grad** f, evaluated at a point P, is perpendicular to the level surface of $f(x, y, z)$ through P. $\qquad$ **(31.19)**

It follows that the insect in Example 31.6 crosses perpendicularly all the level surfaces that it meets.

31.8 Stationary points

Stationary points (which include **maxima and minima**) are more difficult to discuss in more than two dimensions, since we no longer have the horizontal tangent plane to refer to. We should expect a stationary point of $f(x, y, z, \dots)$ to occur at any point Q where

$$\frac{\partial f}{\partial x} = \frac{\partial f}{\partial y} = \frac{\partial f}{\partial z} = \cdots = 0, \qquad (31.20)$$

since all our previous formulae have been merely extended versions of the two-dimensional case. To show that this criterion is the right one, choose *any* path through Q, and suppose that we describe it parametrically by

$$x = x(t), \quad y = y(t), \quad z = z(t), \quad \dots \quad .$$

Then, if (31.20) holds at Q, the chain rule (31.7) together with (31.20) gives

$$\frac{df}{dt} = \frac{\partial f}{\partial x}\frac{dx}{dt} + \frac{\partial f}{\partial y}\frac{dy}{dt} + \dots = 0.$$

Therefore a turning point of $f(x(t), y(t), z(t), \dots)$ is encountered at Q on *every* path passing through Q, and this is what we should wish to happen for the point Q to be described as stationary.

Stationary points of $f(x, y, z, \dots)$

The stationary points are the solutions $(x, y, z, \dots)$ of the equations

$$\frac{\partial f}{\partial x} = \frac{\partial f}{\partial y} = \frac{\partial f}{\partial z} = \dots = 0.$$

(31.21)

Example 31.7 *Find the stationary points of the function*

$$f(x, y, z) = x^2 + y^2 + z^2 - xy - 2yz - zx - z.$$

The conditions (31.21) become

$$\partial f / \partial x = 2x - y - z = 0,$$
$$\partial f / \partial y = -x + 2y - 2z = 0,$$
$$\partial f / \partial z = -x - 2y + 2z - 1 = 0.$$

By elimination the only solution of these equations is $x = -\frac{1}{2}, y = -\frac{5}{8}, z = -\frac{3}{8}$.

Restricted stationary-value problems (see Section 30.2) may occur in any number of dimensions. In three dimensions, the **restriction** may be either to **values of $f(x, y, z)$ on some given curve**, or to **values on some given surface**.

To help visualize a three-dimensional situation, suppose that a fish swims through a field of pollution of density $P = f(x, y, z)$. At some point in the sea the pollution is at an overall maximum, but this is of no concern to the fish if it does not swim through it. However, it will notice highs and lows along its own path even if there is nothing special about such points from an overall viewpoint. These are restricted maxima and minima on the fish's path. Suppose that the path of the fish is expressed parametrically:

$$x = x(t), \quad y = y(t), \quad z = z(t).$$

Then the stationary points peculiar to the path are where $df/dt = 0$. By the chain rule (31.7), these are the points where

$$\frac{\partial f}{\partial x}\frac{dx}{dt} + \frac{\partial f}{\partial y}\frac{dy}{dt} + \frac{\partial f}{\partial z}\frac{dz}{dt} = 0.$$

When written in terms of t, this is an equation giving the critical values of t. (We must be careful to avoid a parametrization such that $dx/dt = dy/dt = dz/dt = 0$ at some point on the path: at such a point, a non-existent stationary point would be predicted.)

Stationary points of $f(x, y, z)$ on the path $(x(t), y(t), z(t))$

The stationary points are the solutions of

$$\frac{\partial f}{\partial x}\frac{dx}{dt} + \frac{\partial f}{\partial y}\frac{dy}{dt} + \frac{\partial f}{\partial z}\frac{dz}{dt} = 0.$$

(31.22)

(In particular cases it might be easier to substitute $x(t)$, $y(t)$, $z(t)$ directly into $f(x, y, z)$ for the turning points of f with respect to t.)

It is more usual for restricted stationary-value problems to be formulated in a way that avoids parametric considerations. The **restriction to a surface** is the easier case. Instead of a fish in the body of the sea, consider a crab which confines itself to the undulating seabed described by an equation of the form

$$g(x, y, z) = c,$$

encountering there the local pollution, given throughout the sea by $f(x, y, z)$. The crab does not know about the rest of the sea, but as it moves around it will meet highs and lows (and other stationary points) unconnected with possibly more extreme pollution in the body of the sea. A stationary point will be found at a point Q on the surface $g(x, y, z) = c$ if

$$\frac{df}{ds} = 0 \quad \text{at } Q$$

in all directions $\hat{s}$ from Q which do not point into the body of the sea, but are **tangential to the surface** $g(x, y, z) = c$.

Figure 31.6 shows such a point Q, and various tangential directions denoted by unit vectors $\hat{s}$ pointing away from Q. From (31.15), one condition for a restricted stationary point is

$$\frac{df}{ds} = 0 = \hat{s} \cdot \mathbf{grad}\, f \quad \text{at } Q, \text{ for all such } \hat{s}. \tag{31.23}$$

In other words, **grad** f *must be perpendicular to the surface at Q* (ignoring for the moment the chance that **grad** f might be zero at Q). But, by (31.11), **grad** g is *always perpendicular to the surface*

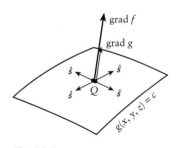

Fig. 31.6

$g(x, y, z) = c$; in particular at Q. Therefore **grad** f and **grad** g, evaluated at Q, are parallel vectors; so

grad $f = \lambda$ **grad** g at Q,

where λ is an (unknown) constant, called a **Lagrange multiplier** for the problem. By writing **grad** f and **grad** g in their components, we obtain

$$\frac{\partial f}{\partial x} - \lambda \frac{\partial g}{\partial x} = 0, \quad \frac{\partial f}{\partial y} - \lambda \frac{\partial g}{\partial y} = 0, \quad \frac{\partial f}{\partial z} - \lambda \frac{\partial g}{\partial z} = 0. \quad \text{(31.24a,b,c)}$$

We now have three equations for the four unknowns: (x, y, z) (the position of Q) and λ. To find another equation, notice that (31.24a,b,c) would be unaffected if we had $g(x, y, z)$ equal to some constant other than c, so it is necessary to reassert the particular surface:

$$g(x, y, z) = c. \quad \text{(31.24d)}$$

The very special case mentioned above, that (31.23) is satisfied alternatively if (by chance) **grad** $f = 0$ at Q, is still governed by the same equations. When they are solved, we should merely find that $\lambda = 0$. (We should not usually realize this in advance.) The case corresponds to the unrestricted stationary-point problem (see (31.20)), where the point found happens to lie in the specified surface.

Restricted stationary-point problem: Stationary points of $f(x, y, z)$ subject to $g(x, y, z) = c$

Solve for x, y, z, λ the equations

$$g = c, \quad \text{(i)}$$

$$\frac{\partial f}{\partial x} - \lambda \frac{\partial g}{\partial x} = 0, \quad \text{(ii)}$$

$$\frac{\partial f}{\partial y} - \lambda \frac{\partial g}{\partial y} = 0, \quad \text{(iii)}$$

$$\frac{\partial f}{\partial z} - \lambda \frac{\partial g}{\partial z} = 0. \quad \text{(iv)}$$

$$\text{(31.25)}$$

Example 31.8   *Find the stationary points of $f(x, y, z) = x^2 + y^2 + yz + zx$ on the hyperboloid $g(x, y, z) = x^2 + y^2 - z^2 = 1$.*

The four equations (31.25) are

$$x^2 + y^2 - z^2 = 1, \quad \text{(i)}$$

$$2x + z - 2\lambda x = 0, \quad \text{or} \quad (2 - 2\lambda)x + z = 0; \quad \text{(ii)}$$

$$2y + z - 2\lambda y = 0, \quad \text{or} \quad (2 - 2\lambda)y + z = 0; \quad \text{(iii)}$$

$$y + x - 2\lambda z = 0, \quad \text{or} \quad x + y + 2\lambda z = 0. \quad \text{(iv)}$$

↗

Example 31.8 *continued*

Equations (ii), (iii), and (iv) constitute a set of homogeneous linear algebraic equations for x, y, z. The only possibilities are *either* that $x = y = z = 0$, which is excluded since these values do not satisfy (i), *or* that the determinant of the coefficients is zero:

$$\det \begin{bmatrix} 2 - 2\lambda & 0 & 1 \\ 0 & 2 - 2\lambda & 1 \\ 1 & 1 & 2\lambda \end{bmatrix} = 0,$$

so that $(1 - \lambda)(2\lambda^2 - 2\lambda + 1) = 0$. The only real solution is

$$\lambda = 1.$$

The equations then become

$$z = 0, \quad z = 0, \quad x + y + 2z = 0,$$

or

$$z = 0, \quad y = -x. \tag{v}$$

Substitute for y in terms of x into (i). Then

$$2x^2 = 1, \quad \text{or} \quad x = \pm 1/\sqrt{2}.$$

Therefore, using (v) again gives the stationary points

$$(\pm 1/\sqrt{2}, \quad \mp 1/\sqrt{2}, 0).$$

For the corresponding problem of finding the **stationary points of** $f(x, y, z)$ **on a specified curve**, as for the fish problem discussed at the beginning of the section, the curve will be assumed to be specified by the intersection of two surfaces:

$$g(x, y, z) = c_1, \qquad h(x, y, z) = c_2.$$

The method of solution involves two Lagrange multipliers:

> **Restricted stationary-point problem: stationary points of**
> $f(x, y, z)$ **subject to** $g(x, y, z) = c_1$ **and** $h(x, y, z) = c_2$.
>
> Solve for x, y, z, λ, μ the equations
>
> $$g = c_1 \qquad h = c_2, \tag{i), (ii}$$
>
> $$\frac{\partial f}{\partial x} - \lambda \frac{\partial g}{\partial x} - \mu \frac{\partial h}{\partial x} = 0, \tag{iii}$$
>
> $$\frac{\partial f}{\partial y} - \lambda \frac{\partial g}{\partial y} - \mu \frac{\partial h}{\partial y} = 0, \tag{iv}$$
>
> $$\frac{\partial f}{\partial z} - \lambda \frac{\partial g}{\partial z} - \mu \frac{\partial h}{\partial z} = 0. \tag{v}$$
>
> $$\tag{31.26}$$

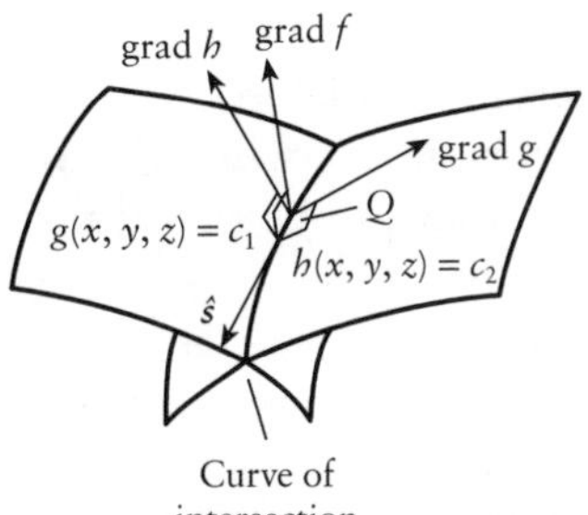

Fig. 31.7

We shall not give the proof in full. Briefly, the situation is shown in Fig. 31.7. Q is a stationary point on the curve of intersection and $\hat{s}$ a unit vector tangential to it at Q. Since

$$\frac{df}{ds} = \hat{s} \cdot \mathbf{grad}\, f = 0 \quad \text{at } Q,$$

grad f is perpendicular to $\hat{s}$ at Q. For the same reason as in the earlier case, **grad** g and **grad** h are also perpendicular to $\hat{s}$ at Q. Therefore the three vectors **grad** f, **grad** g, **grad** h all lie in the same plane (which is perpendicular to $\hat{s}$), so **grad** f can be expressed in terms of the other two vectors:

$$\mathbf{grad}\, f = \lambda\, \mathbf{grad}\, g + \mu\, \mathbf{grad}\, h,$$

where λ and μ are certain constants, the **Lagrange multipliers** for this problem. Then split this equation into its components to obtain (31.26).

Example 31.9   *Find the stationary points of* $x^2 + y^2 + z^2$ *on the curve of intersection of the vertical cylinder* $x^2 + y^2 = 1$ *with the plane* $x + y + z = 1$. *(This is an inclined ellipse.)*

Here $f(x, y, z) = x^2 + y^2 + z^2$, $g(x, y, z) = x^2 + y^2$, and $h(x, y, z) = x + y + z$. The equations to be solved become

$$x^2 + y^2 = 1, \tag{i}$$

$$x + y + z = 1, \tag{ii}$$

$$2x - \lambda 2x - \mu = 0, \quad \text{or} \quad 2x(1 - \lambda) = \mu, \tag{iii}$$

$$2y - \lambda 2y - \mu = 0, \quad \text{or} \quad 2y(1 - \lambda) = \mu, \tag{iv}$$

$$2z - \mu = 0. \tag{v}$$

From (iii) and (iv), *either* (a) $\lambda = 1$, so that $\mu = 0$, *or* (b) $\lambda \neq 1$, so that $x = y$. We consider these possibilities in order.

(a) *The case* $\lambda = 1$, $\mu = 0$. (We cannot deduce anything about x and y from (iii) and (iv) if this is true.) From (v) we obtain $z = 0$, so (i) and (ii) become

$$x^2 + y^2 = 1, \qquad x + y = 1.$$

The solutions are $x = 0$, $y = 1$, and $x = 1$, $y = 0$. Then we have found two solutions:

$$(0, 1, 0) \quad \text{and} \quad (1, 0, 0).$$

(b) *The case* $\lambda \neq 1$, $x = y$. From (i), $x = \pm 1/\sqrt{2}$, $y = \pm 1/\sqrt{2}$. Equation (ii) then gives $z = 1 - x - y = 1 \mp \sqrt{2}$. Thus we have two more solutions:

$$(1/\sqrt{2}, 1/\sqrt{2}, 1 - \sqrt{2}) \quad \text{and} \quad (-1/\sqrt{2}, -1/\sqrt{2}, 1 + \sqrt{2}).$$

For a **restricted stationary-point problem in N variables**, there may be up **to $N - 1$ restricting equations, or constraints, with the same number of Lagrange multipliers.** The equations to be solved then follow the pattern of (31.25) and (31.26).

The identification of a **maximum** or **minimum** is usually of most interest. The general question is difficult, but sometimes it is fairly obvious. For instance, in the previous example the values of f are restricted to a closed curve, so the values of f obtained make it clear that the points (a) give minima of f, and points (b) give maxima.

31.9 The envelope of a family of curves

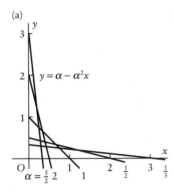

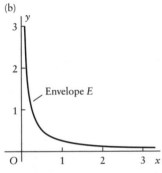

Fig. 31.8

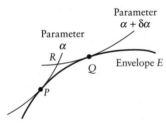

Fig. 31.9

Figure 31.8a shows the straight lines

$$y = \alpha - \alpha^2 x,$$

for several values of α, which we call the **parameter of the family** of straight lines. The 'boundary' of the family is starting to form itself into a curve E, which is sketched in Fig. 31.8b. The reason why the curve E is sharply defined is because all the straight lines are tangential to it, and therefore reinforce it along its length. The curve is called the **envelope of the family** $y = \alpha - \alpha^2 x$, where α is the **parameter of the family**.

The family does not have to consist of straight lines. Suppose that the family is described by

$$f(x, y, \alpha) = 0.$$

To find the envelope (Fig. 31.9) consider two close values of the parameter, α and $\alpha + \delta\alpha$, the corresponding curves of the family being

$$f(x, y, \alpha) = 0 \quad \text{and} \quad f(x, y, \alpha + \delta\alpha) = 0.$$

The intersection point R is the point where

$$f(x, y, \alpha) = f(x, y, \alpha + \delta\alpha) \quad (= 0).$$

Therefore, at the point R,

$$\frac{f(x, y, \alpha + \delta\alpha) - f(x, y, \alpha)}{\delta\alpha} = 0.$$

Now let $\delta\alpha \to 0$. Then R and Q come together at P on the envelope, and this equation becomes

$$\frac{\partial f(x, y, \alpha)}{\partial \alpha} = 0, \tag{31.27}$$

at P. Also P lies on the curve

$$f(x, y, \alpha) = 0. \tag{31.28}$$

(We had not so far used the fact that f is *zero* rather than some other constant.) If we eliminate α between (31.27) and (31.28), we obtain an equation in x and y which describes the envelope.

> **Envelope of the family of curves $f(x, y, \alpha) = 0$, where α is a parameter**
> The result of eliminating α between the equations $f(x, y, \alpha) = 0$
> and $\dfrac{\partial f}{\partial \alpha} = 0$ contains the envelope.
> $$\tag{31.29}$$

(The solution might also include the track of other peculiarities.)

Example 31.10   *Find the envelope of the family of straight lines*
$y = \alpha - \alpha^2 x$, *where α is the parameter. (See Fig. 31.8.)*

Let $f(x, y, \alpha) = y - \alpha + \alpha^2 x$. Then

$$\frac{\partial f}{\partial \alpha} = -1 + 2\alpha x = 0.$$

Therefore

$$\alpha = 1/(2x).$$ (i)

On the envelope, also

$$y - \alpha + \alpha^2 x = 0;$$ (ii)

so, from (i), $y - 1/(2x) + 1/(4x) = 0$, or

$$y = 1/(4x),$$

which is a rectangular hyperbola (see Fig. 31.8b).

Problems

31.1 Write down the incremental approximation for δf in the following cases.
(a) $f(x, y, z) = 2x + 3y^2 + 4z^2 - 3$;
(b) $f(x, y, t) = (x^2 + y^2)^{-\frac{1}{2}} e^{-t}$;
(c) $f(r, \theta, t) = e^{-t} r \cos \theta$;
(d) $f(x, y, z, t) = x^2 + y^2 + z^2 - t^2$;
(e) $f(x_1, y_1, x_2, y_2) = (x_1 - x_2)^2 + (y_1 - y_2)^2$;
(f) $f(x, y, z, t) = (1/r) e^{-(x^2+y^2)/t}$. Compare with the expression for δg when $g(r, t) = (1/r) e^{-r^2/t}$ in polar coordinates.

31.2 The distance d between two points (x_1, y_1, z_1) and (x_2, y_2, z_2) in a plane is given by $d^2 = (x_1 - x_2)^2 + (y_1 - y_2)^2 + (z_1 - z_2)^2$. Find approximately the change from $(1, 1, 2)$, $(1, 2, 1)$ to $(1.1, 0.9, 1.8)$, $(0.9, 2.1, 1.1)$.

31.3 R_1, R_2, R_3, R_4 are resistances in a circuit whose overall resistance is R, arranged so that

$$1/R = 1/R_4 + (R_1 + R_2)/(R_1 R_2 + R_2 R_3 + R_3 R_1).$$

Find an expression for δR in terms of δR_1, δR_2, δR_3, and δR_4.

Suppose that initially $R_1 = 3$, $R_2 = 10$, $R_3 = 5$, and $R_4 = 10$, and that R_1 becomes 3.2 and R_2 becomes 9.8. Estimate the change in R_3 necessary if R is to remain unaltered.

31.4 The equation $2x^3 - 3x - 45 = 0$ has a solution $x = 3$. Find an approximate solution to the equation $2.1x^3 - 2.9x - 47 = 0$.

31.5 Estimate the maximum possible error and the corresponding percentage error in w for the following cases.
(a) $w = yz + zx + xy$, $x = 2$ (± 0.1), $y = 3$ (± 0.2), $z = 1$ (± 0.1).
(b) $w = (x - y)(y - z)(z - x)$, $x = 1$ (± 0.1), $y = 2$ (± 0.1), $z = 3$ (± 0.1).
(c) $w = (x + y + z - t)^{-1}$, where it is known only that $x = 1.2$, $y = 2.9$, $z = 1.9$, and $t = 2.1$ after rounding to 1 decimal place. Compare with the exact maximum and percentage errors.

31.6 Estimate the maximum error and the maximum percentage error for the following.
(a) (For c) $c^2 = a^2 + b^2 - 2ab \cos A$ (the 'cosine rule' for a triangle ABC). Here $a = 2$ (± 0.1), $b = 4$ (± 0.1), $A = 135°$ ($\pm 2°$). (Note: $\Delta(c^2) \approx 2c \, \Delta c$.)
(b) (For d)

$$d^2 = (x_1 - x_2)^2 + (y_1 - y_2)^2 + (z_1 - z_2)^2,$$

where the measured values $(x_1, y_1, z_1) = (1, 2, 1)$ and $(x_2, y_2, z_2) = (2, 1, 1)$ have been rounded to 1 significant figure. (Note: $\Delta(d^2) \approx 2d \, \Delta d$.)
(c) (For A) The area of a triangle with sides a, b, c is given by

$$A = [s(s - a)(s - b)(s - c)]^{\frac{1}{2}},$$

where $s = \frac{1}{2}(a + b + c)$. Consider the case when $a = 2$, $b = 4$, $c = 3$, all with possible errors as large as ± 0.1. (You can substitute s directly into the formula for A or A^2, but it is easier algebraically to obtain two simultaneous equations, with numerical coefficients, involving ΔA, Δs, Δa, Δb, Δc.)

31.7 (Section 24.2). If $f(x, y, z, w) = c$ (a constant), then any of the four variables is a function of the other three. Use (31.6) to show that

(a) $\dfrac{\partial x}{\partial y} \dfrac{\partial y}{\partial x} = 1$;   (b) $\dfrac{\partial x}{\partial y} \dfrac{\partial y}{\partial z} = -\dfrac{\partial x}{\partial z}$.

(c) Simplify $\dfrac{\partial x}{\partial y} \dfrac{\partial y}{\partial z} \dfrac{\partial z}{\partial w} \dfrac{\partial w}{\partial x}$.

Test the truth of the results in the cases:
(i) $x + 2y + 3z + 4w = 5$;   (ii) $xy^2z^3w = 1$.

31.8 Assume that the following relations define z implicitly as a function of x and y. Write down the relation between δx, δy, δz at the points prescribed. Without solving for z, deduce $\partial z/\partial x$ and $\partial z/\partial y$ at the points.
(a) $2x - 3y + 4z = 1$ at points satisfying the condition;
(b) $x^2 + y^2 + z^2 = 14$ at $(1, 2, -3)$;
(c) $4x^3 + y^4 + 9z^3 - xyz^2 = 13$ at $(1, 1, 1)$;
(d) $x^2 - z^2 = 9$ at $x = 5$, $y = y_0$, $z = 4$ (this is a hyperbolic cylinder).

31.9 (a) Compare the result of using the chain rule (31.7) with that of direct substitution in order to find df/dt when $f(x, y, z) = xy/z$ and $x = t$, $y = 4t$, $z = 2t$.
(b) The same parametrization as in (a), but with the function $f(x, y, z) = \sin(xy/z)$.
(c) Obtain an expression for df/dt on the path in (a) when $f(x, y, z) = g(xy/z)$, g being any function, and confirm that it works with case (b). (Hint: express the result in terms of g'.)

31.10 Cylindrical coordinates r, θ, z are shown in Fig. 31.10. They are related to x, y, z by $x = r \cos \theta$, $y = r \sin \theta$, $z = z$.

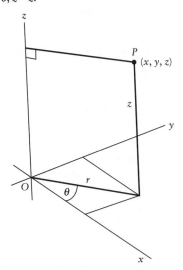

Fig. 31.10

(a) Given $f(x, y, z)$, use the chain rule (31.8) with r, θ, z as the parameters to express $\partial f/\partial r$, $\partial f/\partial \theta$, $\partial f/\partial z$ in terms of $\partial f/\partial x$, $\partial f/\partial y$, $\partial f/\partial z$.
(b) Regarding (a) as a pair of equations for $\partial f/\partial x$ and $\partial f/\partial y$, show that

$$\frac{\partial f}{\partial x} = \cos \theta \frac{\partial f}{\partial r} - \frac{\cos \theta}{r} \frac{\partial f}{\partial \theta} \quad \text{and}$$

$$\frac{\partial f}{\partial y} = \sin \theta \frac{\partial f}{\partial r} + \frac{\cos \theta}{r} \frac{\partial f}{\partial \theta}.$$

(c) The results (b) show that the differentiation operations $\partial/\partial x$ and $\partial/\partial r$ are equivalent respectively to the polar forms

$$\cos \theta \frac{\partial}{\partial r} - \frac{\sin \theta}{r} \frac{\partial}{\partial \theta} \quad \text{and}$$

$$\sin \theta \frac{\partial}{\partial r} + \frac{\cos \theta}{r} \frac{\partial}{\partial \theta}.$$

Use this fact to confirm that

$$\frac{\partial^2 f}{\partial x^2} + \frac{\partial^2 f}{\partial y^2} = \frac{\partial^2 f}{\partial r^2} + \frac{1}{r} \frac{\partial f}{\partial r} + \frac{1}{r^2} \frac{\partial^2 f}{\partial \theta^2}.$$

31.11 Obtain the vector function **grad** f for each of the following.
(a) $x + y + z$;
(b) $2x - 3y + 5z - 6$;
(c) $x^2 + y^2 + z^2$;
(d) $x^3 + 3z^3 - 1$ (in three dimensions);
(e) $x^2 - \frac{1}{4}y^2 + \frac{1}{9}z^2$;
(f) $1/r$, where $r = (x^2 + y^2 + z^2)^{\frac{1}{2}}$; confirm that the gradient vector points in the direction of the position vector (x, y, z).

31.12 Obtain a vector which is normal to the following surfaces at the points specified, and construct a unit vector from it:
(a) $x - 2y + z = 0$ at any point;
(b) $y^2 + z^2 = 2$ at any point;
(c) $x^2 + y^2 + z^2 = 9$ at $(2, 1, -2)$;
(d) $\frac{1}{4}x^2 + \frac{1}{9}y^2 + \frac{1}{16}z^2 = 3$ at $(2, 3, 4)$;
(e) $x^3y + zx^3 = 5$ at $(1, 2, 3)$;
(f) $\dfrac{1}{x} + \dfrac{1}{y} + \dfrac{1}{z} = 1$ at $(2, 3, 6)$;
(g) $(x^2 + 4y^2 - z^2)^{-1} = \frac{1}{16}$ at $(4, 1, 2)$.

31.13 By finding the gradient vectors, obtain the angle between the following surfaces at the point of intersection given:
(a) $x^2 + y^2 + z^2 = 9$, $x^2 - z^2 = 0$ at $(2, 1, 2)$;
(b) $x^2 - y^2 + z^2 = 1$, $2x - 3y + z + 1 = 0$ at $(2, 2, 1)$;
(c) $x^2 + y^2 - z^2 = 0$, $3x + 4y + 5z = 50$ at $(3, 4, 5)$. Explain the result.

31.14 (a) Find **grad** f for

$$f(x, y, z) = A\,e^{\alpha(2x^2+4y^2+z^2)^{\frac{1}{2}}},$$

where A and α are constants. Deduce that the vector $(2x, 4y, z)$ points in the direction of **grad** f.
(b) Let $f(x, y, z) = g[u(x, y, z)]$, where g and u are two other functions. Show that

$$\mathbf{grad}\,f = \left(g'(u)\frac{\partial u}{\partial x},\, g'(u)\frac{\partial u}{\partial y},\, g'(u)\frac{\partial u}{\partial z}\right),$$

and deduce that **grad** u points either in the same or in the opposite direction to **grad** f.

31.15 Write down expressions for the directional derivative of the following at the point (x, y, z), in terms of a unit direction vector $\hat{s}$.
(a) $x + 2y + 3z$;   (b) $x^2 - y^2 - 3z$;
(c) $(x-1)^3 + y^3 + z^3$.

31.16 Find df/ds for the following functions f, taken at the point $(2, 3, 2)$ in the direction $\hat{s} = (\frac{1}{4}\sqrt{2}, \frac{1}{4}\sqrt{2}, \frac{1}{2}\sqrt{3})$.
(a) $x - y + 2z$;   (b) $xy + yz + zx$;
(c) $(xy + yz + zx)^2$;
(d) $x^2 - y^2 + 5$ (in three dimensions: this represents a vertical cylinder).

31.17 The equations for two surfaces, $f(x, y, z) = a$, $g(x, y, z) = b$, where a and b are constants, together represent their curve of intersection, C. Show that the vector product **grad** $f \times$ **grad** g, evaluated at a point on C, points in the direction of C. Use this to find a unit vector $\hat{s}$ in the direction of C in the following cases:
(a) $2x + 3y - z = 1$, $x - y - z = 0$, at any common point.
(b) $x + y = 0$, $x - z = 0$, at any common point.
(c) $x^2 + y^2 + z^2 = 6$, $x - y + z = 0$, $(1, 2, 1)$.
(d) $x^2 + (y-1)^2 = 1$, $x^2 + (y-2)^2 = 4$, at $x = 0$, $y = 0$, and any value of z. Explain what is happening here.
(e) $xy + yz + zx = 3$, $x + y + z = 3$, at $(1, 1, 1)$.

31.18 Find the stationary points of the following functions with respect to all the variables named in f:
(a) $f(x, y, z) = x^2 + y^2 + z^2$;
(b) $f(x, y, z) = x^3 - 3x + y^3 - 3yz + 2z^2$;
(c) $f(x, y, z) = xy + yz + zx + y - z$;
(d) $f(x, y, z) = x/z + y/x + z/y$;
(e) $f(x, y, z, \lambda) = (x + y + z) - \lambda(x^2 + y^2 + z^2 - 1)$;
(f) $f(x, y, z) = x^4 + y^4 + z^4 - 2(x - y + z)^2$.

31.19 Find the stationary points of $x^2 + y^2 + z^2$ on the path

$$x = \cos t, \qquad y = \sin t, \qquad z = \sin\tfrac{1}{2}t,$$

where $0 < t < 4\pi$.

31.20 At the point (x, y, z) in the air, an insecticide maintains a concentration

$$s = C\exp\{-\alpha[2(x-1)^2 + 4y^2 + z^2]\},$$

where C is a constant. An insect is trying to escape by following the path of most rapid decrease in concentration.
(a) Show that, when it is at (x, y, z), its direction is that of the vector $(2(x-1), 4y, z)$.
(b) In a short interval of time δt, it moves to $(x + \delta x, y + \delta y, z + \delta z)$. Show that, approximately,

$$\frac{\delta x}{2(x-1)} = \frac{\delta y}{4y} = \frac{\delta z}{z}.$$

(c) By letting $\delta t \to 0$, show that its path is described by the two differential equations

$$\frac{dz}{dx} = \frac{z}{2(x-1)}, \qquad \frac{dz}{dy} = \frac{z}{4y}.$$

(Such simultaneous equations would often be written $dx/2(x-1) = dy/4y = dz/z$.)
(d) Show that the general solution of these equations, which expresses a path in space, can be written as

$$z = Ay^{\frac{1}{4}} = B(x-1)^{\frac{1}{2}},$$

where A and B are arbitrary constants.
(e) Assuming that the insect starts at $(0, 1, 1)$, find its path.

31.21 Use the Lagrange-multiplier technique of (31.25)–(31.26) to solve the following restricted stationary-point (SP) problems.
(a) SPs of $x + y + z$ subject to $1/x + 1/y + 1/z = 1$.
(b) SPs of xyz subject to $1/x + 1/y + 1/z = 1$.
(c) SPs of $x^2 + y^2 + z^2$ subject to $ax + by + cz = 1$.
(d) SPs of $xy + yz + zx$ subject to $xyz = 1$. (This corresponds to finding the rectangular block of given volume which has the smallest surface area.)
(e) The problem of the rectangular block of greatest volume which can be fitted into an ellipsoid leads to the problem: find the SPs of xyz subject to $x^2/a^2 + y^2/b^2 + z^2/c^2 = 1$.
(f) SPs of $x^2 + 4y^2 + z^2$ on the intersection of the two planes $x - y - 2z = 0$ and $z = 1$.
(g) SPs of $x^2 - y^2 - z^2$ on the straight line

$$\frac{x-1}{2} = \frac{y-2}{-1} = \frac{z-2}{3}.$$

(h) SPs of xyz subject to $xy + yz + zx = 1$. (Compare (d).)
(i) SPs of $x - y - 2z$ on the intersection of $z = 1$ with $x^2 + 4y^2 + z^2 = 6$. (Compare (f).)

31.22 (Numerical). (a) Write a program to carry out the numerical scheme suggested in Example 31.6 *in two dimensions* in order to obtain the curves along which a function $f(x, y)$ increases most rapidly. It may also be possible for you to display the curves on the screen.

(b) Use (a) to obtain a numerical solution of the following problems. Try a succession of decreasing step lengths h in order to ensure plotting accuracy.

(c) The altitude H of a part of a hill is given in km by

$$H = 0.5 - x^2 - 4y^2.$$

Start for example with $(x, y) = (2, 2)$, and go to the summit. (For comparison, the exact solution to the problem is $y = \frac{1}{8}x^4$; the summit is at the origin.)

(d) Plot the track of most rapid *descent* from the point $(3, 2)$ in the case where $H = \frac{1}{2} + x^2 - y^2$. (To descend, use negative h. The shape is a saddle: viewed from the origin, H increases east and west and decreases north and south.)

(e) In certain types of fluid flow in two dimensions, the velocity vector $v(x, y)$ is equal to the gradient of a single scalar **potential** function $\phi(x, y, z): v = \mathbf{grad}\ \phi$. A **stream-line** through any point is in the direction of v. Plot some streamlines on and outside of the circle $x^2 + y^2 = 1$ when

$$\phi(x, y) = x\left(1 + \frac{1}{x^2 + y^2}\right).$$

31.23 Often a function $f(x, y, z)$ takes the form of 'a function of a function': $w = f(x, y, z) = g(u(x, y, z))$. (An example is

$$w = f(x, y, z) = \sin xyz : u = xyz, w = \sin u.)$$

(a) Write down several examples of functions which can be regarded in this way.

(b) Show that

$$\frac{\partial f}{\partial x} = g'(u)\frac{\partial u}{\partial x}, \quad \frac{\partial f}{\partial y} = g'(u)\frac{\partial u}{\partial y}, \quad \frac{\partial f}{\partial z} = g'(u)\frac{\partial u}{\partial z}.$$

(You only need the one-variable chain rule (3.3).)

(c) Check the correctness of the formulae (b) in the cases when (i) $w = e^{x^2-y^2+z^2}$, (ii) $w = \sin(xy/z)$.

(d) Using the results (b), rewrite the chain rule (31.7) in the form appropriate to functions of the form $g(u(x, y, z))$.

(e) The path $x = \cos t, y = \sin t, z = t$ represents a helix whose axis is the z axis. Find an expression in terms of t for df/dt on the path when $f(x, y, z) = g(xy/z)$, where g is any function. Confirm the result for any simple case.

31.24 Often a function takes the form

$$f(u, v, w),$$

where u, v, and w are themselves functions of x, y, and z. Write down a version of the chain rule (31.8) which enables $\partial f/\partial x, \partial f/\partial y, \partial f/\partial z$ to be found. (In this case, x, y, and z function like parameters and u, v, and w like the principal variables.) Use this result to prove the following results:

(a) If $\phi = f(x - y, y - z, z - x)$, where f is any function, then

$$\frac{\partial\phi}{\partial x} + \frac{\partial\phi}{\partial y} + \frac{\partial\phi}{\partial z} = 0.$$

Check your result with the function

$$(x - y)(y - z)(z - x).$$

(b) If $\phi = f(y/x, z/x)$, where f is any function of *two* variables, then

$$x\frac{\partial\phi}{\partial x} + y\frac{\partial\phi}{\partial y} + z\frac{\partial\phi}{\partial z} = 0.$$

Try it out with the function $x/y + y/z + z/x$, noting that

$$\frac{y}{x}\bigg/\frac{z}{x} = \frac{y}{z}.$$

31.25 Let $f(x, y, z, t) = e^{i(k_1x+k_2y+k_3z-\omega t)}$, where i is the complex element $(i^2 = -1)$, and k_1, k_2, k_3, and ω are constants. Show that

$$\frac{\partial^2 f}{\partial x^2} + \frac{\partial^2 f}{\partial y^2} + \frac{\partial^2 f}{\partial z^2} = \frac{1}{c^2}\frac{\partial^2 f}{\partial t^2},$$

where $c = \omega/(k_1 + k_2 + k_3)$. (This is called the **wave equation** in three dimensions, and $f(x, y, z, t)$ is one of its solutions.)

Prove that $g(k_1x + k_2y + k_3z - \omega t)$, where g is any function of a single variable, is also a solution.

31.26 Find the envelopes of the following families.

(a) $y = \alpha + \alpha^2 x$ (parameter α);

(b) $y + \alpha^2 x = \alpha$ (parameter α);

(c) $\dfrac{x}{\alpha} + \dfrac{y}{1 - \alpha} = 1$ (parameter α);

(d) $x\cos\theta + y\sin\theta = 1$ (parameter θ).

31.27 The cross-sectional profile of a long cylindrical mirror is the semicircle $x^2 + y^2 = 1$ in the right-hand half plane. Rays from the left, parallel to the x axis, fall on the mirror.

(a) Show that the equation of the ray reflected from the point $(\cos\theta, \sin\theta)$ on the mirror is $x\sin 2\theta - y\cos 2\theta = \sin\theta$.

(b) By regarding θ as the parameter, show that the envelope of these reflected rays is given by $x^2 + y^2 = \frac{1}{4}(3y^{\frac{2}{3}} + 1)$. (In optics, this envelope is called the **caustic** of the reflected rays.)

31.28 Show that the envelope of the family of straight line's such that the length cut off between the x and y axes is L, a constant, is given by $x^{\frac{2}{3}} + y^{\frac{2}{3}} = L^{\frac{2}{3}}$. Sketch the curve (it has four segments).

32 Double integration

CONTENTS

32.1 Repeated integrals with constant limits

Before explaining how they arise, we show first how a **repeated integral** is written and evaluated. The following is an example of a repeated integral with **constant limits**:

$$I = \int_0^1 \int_0^2 (xy + y^2 - 1) \, dx \, dy.$$

There are two stages of integration; first with respect to x, then with respect to y, this being determined by the order in which dx and dy appear under the integral signs.

You are recommended to copy the following procedure step by step at first.

(i) Put brackets round the **inner integral**, which is the first to be evaluated:

$$I = \int_0^1 \left(\int_0^2 (xy + y^2 - 1) \, dx \right) dy.$$

(ii) Make it clear which variable connects with which limits of integration, by explicitly labelling them as shown:

$$I = \int_{y=0}^1 \left(\int_{x=0}^2 (xy + y^2 - 1) \, dx \right) dy.$$

(iii) Evaluate the inner integral with respect to the first variable (here x), **treating the other variable (y) as a constant**:

$$\int_{x=0}^{2} (xy + y^2 - 1)\, dx = [\tfrac{1}{2}x^2 y + y^2 x - x]_{x=0}^{2} = 2y + 2y^2 - 2.$$

This process eliminates the variable x.

(iv) Use the result of (iii) as the integrand of the **outer integral**:

$$I = \int_{y=0}^{1} (2y + 2y^2 - 2)\, dy = [y^2 + \tfrac{2}{3}y^3 - 2y]_{y=0}^{1} = -\tfrac{1}{3}.$$

This eliminates the variable y, so that the final **result is a definite number**. If you find you are left with an x or y in the result, then you have not followed the process correctly.

Example 32.1 *Evaluate the repeated integrals*

(a) $I = \displaystyle\int_{0}^{1}\int_{2}^{4} (xy + 1)\, dx\, dy,$ (b) $J = \displaystyle\int_{2}^{4}\int_{0}^{1} (xy + 1)\, dy\, dx.$

(a) $I = \displaystyle\int_{y=0}^{1} \left(\int_{x=2}^{4} (xy + 1)\, dx \right) dy.$

The inner integral becomes

$$\int_{x=2}^{4} (xy + 1)\, dx = [\tfrac{1}{2}x^2 y + x]_{x=2}^{4} = 8y + 4 - (2y + 2)$$
$$= 6y + 2.$$

This forms the integrand of the outer integral:

$$I = \int_{y=0}^{1} (6y + 2)\, dy = [3y^2 + 2y]_{y=0}^{1} = 5.$$

(b) Here the order of the symbols $\int_{x=2}^{4}$ and $\int_{y=0}^{1}$ has been reversed, and also the order of dx and dy. In other words, the same processes are to be carried out, but in the reverse order. The details, however, look different.
   We have

$$J = \int_{x=2}^{4} \left(\int_{y=0}^{1} (xy + 1)\, dy \right) dx.$$

The inner integral is with respect to y, and we treat x as a constant:

$$\int_{y=0}^{1} (xy + 1)\, dy \quad (x \text{ constant}).$$

This is equal to

$$[x(\tfrac{1}{2}y^2) + y]_{y=0}^{1} = \tfrac{1}{2}x + 1.$$

The outer integral becomes

$$J = \int_{x=2}^{4} (\tfrac{1}{2}x + 1)\, dx = 5,$$

which is the same as I in (a).

In this example, it makes no difference whether we integrate with respect to x or y first. Later we show that this is always true when the repeated integral has constant limits.

32.2 Examples leading to repeated integrals with constant limits

Figure 32.1a represents a heap of grain in a rectangular silo of length 8 m and breadth 4 m. The top surface of the grain is curved, with the equation

$$z = \tfrac{1}{32}x^2 + \tfrac{1}{16}y^2 + 2 \quad (0 \leqslant x \leqslant 8; 0 \leqslant y \leqslant 4),$$

and the problem is to find the volume, V say, of the grain.

Imagine the grain divided into thin vertical plane slices, parallel to the (x, z) plane, the thickness of a slice being δy. A typical slice is shown in Fig. 32.1a, and the value of y is constant on its faces. It is lifted out and displayed in elevation separately in Fig. 32.1b. The face area is given by

$$\text{Area } ABCD = \int_{x=0}^{8} (\tfrac{1}{32}x^2 + \tfrac{1}{16}y^2 + 2) \, dx,$$

in which y takes the current constant value. Therefore its volume, δV say, is given by

$$\delta V \approx \left(\int_{x=0}^{8} (\tfrac{1}{32}x^2 + \tfrac{1}{16}y^2 + 2) \, dx \right) \delta y.$$

When we take the sum of all the elements δV and let $\delta y \to 0$, we obtain in the usual way

$$V = \int_{y=0}^{4} \left(\int_{x=0}^{8} (\tfrac{1}{32}x^2 + \tfrac{1}{16}y^2 + 2) \, dx \right) dy.$$

The result has therefore taken the form of a repeated integral of the kind described in Section 32.1. In evaluating it, the inner integral gives the cross-sectional area of a slice on which y is held constant:

$$\int_{x=0}^{8} (\tfrac{1}{32}x^2 + \tfrac{1}{16}y^2 + 2) \, dx = [\tfrac{1}{96}x^3 + \tfrac{1}{16}y^2 x + 2x]_{x=0}^{8}$$

$$= \tfrac{64}{3} + \tfrac{1}{2}y^2.$$

Finally

$$V = \int_{y=0}^{4} (\tfrac{64}{3} + \tfrac{1}{2}y^2) \, dy = [\tfrac{64}{3}y + \tfrac{1}{6}y^3]_{y=0}^{4} = 96.$$

(a)

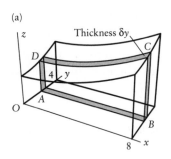

(b)

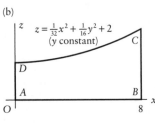

Fig. 32.1

It can be seen that, if we had taken the slices parallel to the (y, z) plane, the process would have led to the integral

$$V = \int_{x=0}^{8} \left(\int_{y=0}^{4} (\tfrac{1}{32}x^2 + \tfrac{1}{16}y^2 + 2) \, dy \right) dx.$$

The integrand is the same, and the result must be the same, when the integrations over x and y are carried out in the opposite order.

In general, and repeated integral with *constant limits*,

$$\int_{c}^{d} \int_{a}^{b} f(x, y) \, dx \, dy,$$

can be interpreted when $f(x, y)$ is positive as the volume of material in a box standing on the rectangle specified by $a \leqslant x \leqslant b, c \leqslant y \leqslant d$, when the material in it has depth $f(x, y)$. If $f(x, y)$ **is negative** over any part of this area, then it will obviously make a **negative contribution to the integral**. This **signed-volume analogy** is closely similar to the signed-area analogy (15.13). We can use the idea to say that the order of integration for repeated integrals having constant limits is immaterial:

Changing order of integration in a repeated integral with constant limits

$$\int_{c}^{d} \int_{a}^{b} f(x, y) \, dx \, dy = \int_{a}^{b} \int_{c}^{d} f(x, y) \, dy \, dx.$$

(32.1)

There is frequently an advantage to be had from changing the order of integration in this way.

Example 32.2 *Evaluate* $I = \int_{0}^{1} \int_{0}^{2} x \, e^{xy} \, dx \, dy.$

The inner integral is

$$\int_{x=0}^{2} x \, e^{xy} \, dx,$$

which, though not very difficult, does involve integration by parts. To avoid this, try the alternative order of integration:

$$I = \int_{x=0}^{2} \left(\int_{y=0}^{1} x \, e^{xy} \, dy \right) dx.$$

The inner integral, with x being treated as a constant, is

$$\int_{y=0}^{1} x \, e^{xy} \, dy = \left[x \frac{1}{x} e^{xy} \right]_{y=0}^{1} = e^x - 1.$$

Example 32.2 *continued*

Then

$$I = \int_0^2 (e^x - 1)\, dx = [e^x - x]_0^2 = e^2 - 3,$$

which is much simpler.

32.3 Repeated integrals over non-rectangular regions

Suppose now that the base of the silo, loaded with grain, has the triangular shape OPQ shown in Fig. 32.2; to take a definite instance, we could consider its depth $f(x, y)$ to be the same as before: $f(x, y) = \frac{1}{32}x^2 + \frac{1}{16}y^2 + 2$; but our discussion will hold for a general function f. Again we measure the volume V by summing the volumes of slices parallel to the x axis, having thickness δy.

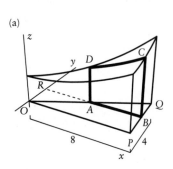

Figure 32.2b shows a typical slice lifted out and viewed in (x, z) axes in order to obtain its face area, and Fig. 32.2c shows the base of the silo in plan view; the slice chosen is along AB.

The slices all have different x values at their starting points, and these values depend on y, so **the limits of integration are not constant** in this case. In order to determine the range of integration of the slice at level y, it is necessary to refer to the triangular area in Fig. 32.2c, called the **region of integration** for this problem. The equation of the side OQ is

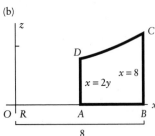

$$y = \tfrac{1}{2}x$$

for $0 \leqslant x \leqslant 8$. Since we need x in terms of y, we express this as

$$x = 2y,$$

and it is helpful to write it on OQ, as shown, together with the simpler information required for the other limits of integration.

The face area of the slice $ABCD$ at level y is therefore given by

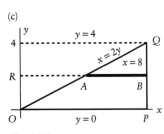

$$\text{area } ABCD = \int_{x=2y}^{8} f(x, y)\, dx.$$

Its volume δV is equal to (area $ABCD \times \delta y$), so

$$\delta V \approx \left(\int_{x=2y}^{8} f(x, y)\, dx \right) \delta y,$$

and finally the whole volume V is

Fig. 32.2

$$V = \int_0^4 \int_{2y}^{8} f(x, y)\, dx\, dy.$$

Notice that **the limits of integration have nothing to do with the integrand $f(x, y)$, but depend only on the shape of the region of integration** in the (x, y) plane (in this case the triangle OPQ in Fig. 32.2c). The limits of integration are the same no matter what the integrand.

Example 32.3 *Evaluate the integral $I = \int_0^1 \int_{2y}^2 (x + y) \, dx \, dy$.*

Write the integral

$$I = \int_{y=0}^1 \left(\int_{x=2y}^2 (x + y) \, dx \right) dy.$$

The inner integral is

$$\int_{x=2y}^2 (x + y) \, dx = [\tfrac{1}{2}x^2 + yx]_{x=2y}^2$$
$$= (2 + 2y) - (2y^2 + 2y^2) = 2 + 2y - 4y^2.$$

(Follow the calculation carefully.) Then

$$I = \int_{y=0}^1 (2 + 2y - 4y^2) \, dy = [2y + y^2 - \tfrac{4}{3}y^3]_0^1 = \tfrac{5}{3}.$$

Example 32.4 *Evaluate the integral $I = \int_0^2 \int_0^y xy \, dx \, dy$, and sketch the region of integration.*

Let

$$I = \int_{y=0}^2 \left(\int_{x=0}^y xy \, dx \right) dy.$$

The inner integral is

$$\int_{x=0}^y xy \, dx = [\tfrac{1}{2}x^2 y]_{x=0}^y = \tfrac{1}{2}y^3.$$

Therefore

$$I = \int_0^2 \tfrac{1}{2}y^3 \, dy = \tfrac{1}{8}[y^4]_0^2 = 2.$$

A sketch of the region of integration can be constructed in the following way. The region of integration consists of the points (x, y) which simultaneously satisfy

(i) $0 \leqslant y \leqslant 2$ and (ii) $0 \leqslant x \leqslant y$.

One way of finding these is to sketch the *boundaries* of the required region. These are the lines

$$y = 0, \quad y = 2, \quad \text{and} \quad x = 0, \quad x = y,$$

and they are shown on Fig. 32.3a. The region consists of any points which lie between both pairs of boundaries (i) and (ii). This is the triangle shown in Fig. 32.3.

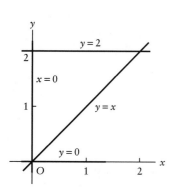

Fig. 32.3

Changing the order of integration for non-rectangular regions

If the region of integration is a *rectangle* with sides parallel to the axes, then (32.1) states that changing the order of integration simply involves performing the same operations in the opposite order. But if we do the same thing with the previous example, we get

$$\int_0^y \int_0^2 xy \, dy \, dx.$$

This is obviously nonsense: the answer contains y, whereas we ought to get the answer 2 again. In fact the new form means nothing at all.

If the region of integration is non-rectangular we have to write

$$\int\!\!\int f(x, y) \, dy \, dx$$

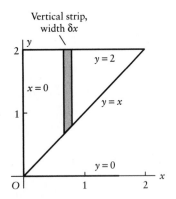

Vertical strip, width δx

$y = 2$

$x = 0$

$y = x$

$y = 0$

Fig. 32.4

and begin again, filling in the limits of integration so that we cover the same region. The interior integral is now with respect to y, so we must start with strips parallel to the y axis as shown in Fig. 32.4. Then the inner integral gives the contribution from the strip:

$$\delta x \int_{y=x}^2 f(x, y) \, dy.$$

The outer integral involves all x between 0 and 2, so finally we have

$$\int_0^2 \int_0^y f(x, y) \, dx \, dy = \int_0^2 \int_x^2 f(x, y) \, dy \, dx.$$

Each case has to be considered individually in this way.

Example 32.5 *Change the order of integration in the repeated integral*

$$\int_0^{\frac{1}{2}} \int_{-\sqrt{(1-4y^2)}}^{\sqrt{(1-4y^2)}} y \, dx \, dy$$

and so evaluate the integral.

Write the integral in the form

$$I = \int_{y=0}^{\frac{1}{2}} \left(\int_{x=-\sqrt{(1-4y^2)}}^{\sqrt{(1-4y^2)}} y \, dx \right) dy.$$

The limits of integration express the boundaries of the region of integrations:

$$x = (1 - 4y^2)^{\frac{1}{2}}, \quad x = -(1 - 4y^2)^{\frac{1}{2}}, \quad y = 0, \quad y = \tfrac{1}{2}.$$

Example 32.5 *continued*

These are shown in Fig. 32.5 (the curved part can be written $x^2 + 4y^2 = 1$: an ellipse with semi-axes equal to 1 and $\frac{1}{2}$). Figure 32.5a shows how the form given is obtained, by starting with **horizontal strips** which end at $x = \pm(1 - 4y^2)^{\frac{1}{2}}$.

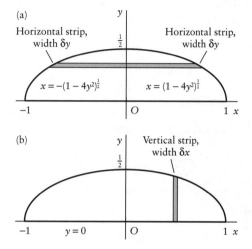

(a)

Horizontal strip, width δy

Horizontal strip, width δy

$x = -(1 - 4y^2)^{\frac{1}{2}}$ $x = (1 - 4y^2)^{\frac{1}{2}}$

(b)

Vertical strip, width δx

$y = 0$

Fig. 32.5

In Fig. 32.5b, the position with regard to **vertical strips** is shown, for which the inner integral will be over y. When the order of integration is changed by this means, we obtain

$$I = \int_{x=-1}^{1} \left(\int_{y=0}^{\frac{1}{2}\sqrt{(1-x^2)}} y \, dy \right) dx.$$

The inner integral is now

$$\int_{0}^{\frac{1}{2}\sqrt{(1-x^2)}} y \, dy = [\tfrac{1}{2}y^2]_0^{\frac{1}{2}\sqrt{(1-x^2)}} = \tfrac{1}{8}(1 - x^2).$$

Therefore

$$I = \tfrac{1}{8} \int_{-1}^{1} (1 - x^2) \, dx = \tfrac{1}{8}[x - \tfrac{1}{3}x^3]_{-1}^{1} = \tfrac{1}{6}.$$

It is left to you to try it in the original form; it is perfectly possible, but more complicated.

32.5 Double integrals

The repeated-integral notation is very informative and self-contained: all the information needed is contained in the integral. It even suggests the coordinates to be used, and gives explicitly the boundary of the region of integration. However, problems do not always fall easily into this form.

(b)

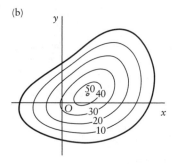

(b)

Region of
integration
$\mathcal{R}$

δA at P

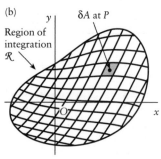

Fig. 32.6

Suppose we have a lake of any shape, as shown in Fig. 32.6a, whose depth is $f(x, y)$: notional contours are suggested. We want to find the volume of water in the lake.

Call the area covered by the lake the **region of integration** $\mathcal{R}$ for the problem. Construct a **mesh** on $\mathcal{R}$ consisting of small area elements δA as in Fig. 32.6b: the mesh may be quite arbitrary for the present purpose. A typical area element δA is at P. Below δA is a depth of water we shall denote by $f(P)$ (we shall not use $f(x, y)$ because cartesian coordinates might not be the ones we eventually want to use). The volume δV in the vertical column of water below δA at the point P is approximated by

$$\delta V \approx f(P)\, \delta A.$$

If we add up all the volume elements in the usual way, we obtain the total volume V. Denote this operation by

$$V = \sum_{\mathcal{R}} \delta V,$$

thus indicating a certain region of integration $\mathcal{R}$, which can be obtained by reference to the diagram, or might be specified separately in words or in some other way. Now let all the area elements δA tend to zero, while becoming more numerous in order to cover $\mathcal{R}$. We obtain

$$V = \sum_{\mathcal{R}} \delta V = \lim_{\delta A \to 0} \sum_{\mathcal{R}} f(P)\, \delta A.$$

It is natural to write this as some kind of integral as we did in one dimension (see Section 15.1). There are several notations; we shall write

$$V = \iint_{\mathcal{R}} f(P)\, \mathrm{d}A,$$

which is to be read: the **double integral of f over the region** $\mathcal{R}$. Unlike a repeated integral it does not give any clue as to how to evaluate it.

As a rule, the argument that gives rise to a double integral by way of a certain summation will not have anything to do with volume; but, as a result of the summation that it represents, the **signed-volume analogy** referred to in Section 32.2 will always hold good:

Double integral $I = \displaystyle\iint_{\mathcal{R}} f(P)\, \mathrm{d}A$ **and the signed-volume analogy**

(i) I stands for $\displaystyle\lim_{\delta A \to 0} \sum_{\mathcal{R}} f(P)\, \delta A$; $\mathcal{R}$ represents a given region in a plane; and δA is a typical area element of $\mathcal{R}$, taken at the point P. The summation is over all the elements δA of $\mathcal{R}$.

(ii) (Signed-volume analogy) Whatever its origin, the integral is numerically equal to the signed volume between a surface $z = f(x, y)$ and the plane $z = 0$, taken over the region $\mathcal{R}$. (Where z is negative the contribution counts as negative.) (32.2)

Example 32.6 *A flat plate occupying a region* $\mathcal{R}$ *is acted on at every point P on the plate by a variable normal stress* $\sigma(P)$ *per unit area. Express the total (resultant) force F on the plate as a double integral.*

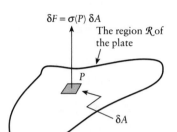

$\delta F = \sigma(P) \, \delta A$

The region $\mathcal{R}$ of the plate

P

δA

Fig. 32.7

The position is as in Fig. 32.7: $\mathcal{R}$ is the plate. The force δF acting on a typical area element δA at P is given by

$$\delta F \approx \sigma(P) \, \delta A.$$

Add up the contributions of all the elements covering $\mathcal{R}$, and take the limit as the mesh becomes finer and finer. We obtain

$$F = \lim_{\delta A \to 0} \sum_{\mathcal{R}} \sigma(P) \, \delta A = \iint_{\mathcal{R}} \sigma(P) \, dA.$$

Confronted with a double integral, one has to decide on what coordinate system to use (say cartesian or polar coordinates) so as to turn it into a repeated integral. In the following example, cartesian coordinates (x, y) are appropriate.

Example 32.7 *We have shown in Example 32.6 that the resultant force F on any flat plate* $\mathcal{R}$ *subject to a normal distribution of stress* $\sigma(P)$ *is given by* $F = \iint_{\mathcal{R}} \sigma(P) \, dA.$ *Find the force on a rectangular plate of sides 2 and 3 units when* $\sigma = 3(r^2 - 2)$, *where r is the distance from one of the corners.*

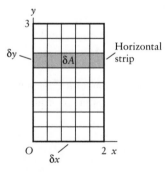

y

3

δy

δA

Horizontal strip

O δx 2 x

Fig. 32.8

This double-integral expression is perfectly general, applying to any plate, any distribution of force, and any coordinates. We have to reformulate the problem for this case. Place the rectangle as in Fig. 32.8, with the corner to which the data refer at the origin. A suitable mesh is the rectangular mesh, with δA having sides δx and δy: the area element is $\delta A = \delta x \, \delta y$. Also $\sigma = 3(x^2 + y^2 - 2)$.

We can add (which ultimately means integrate) the contributions

$$\delta F \approx 3(x^2 + y^2 - 2) \, \delta x \, \delta y$$

in any order that is convenient. Suppose we decide to add the contributions along each horizontal strip at level y, and then to add the results from the strips. Then from the strip at level y we obtain the contribution

$$\left(\int_{x=0}^{2} 3(x^2 + y^2 - 2) \, dx \right) \delta y,$$

after letting $\delta x \to 0$. When we add the contributions from all the strips and let $\delta y \to 0$, we have the repeated integral

$$F = \int_{0}^{3} \int_{0}^{2} 3(x^2 + y^2 - 2) \, dx \, dy = 3 \int_{0}^{3} (-\tfrac{4}{3} + 2y^2) \, dy = 42.$$

The following examples show the adaptability of the notation. In each case, $\mathcal{R}$ represents the region in question, with P a representative point of $\mathcal{R}$ and dA the corresponding element.

(i) *Area.* Area of $\mathcal{R}$: $\iint_{\mathcal{R}} dA$.

(ii) *Variable surface density.* Total mass of a thin flat plate, of variable mass $\sigma(P)$ per unit area: $\iint_{\mathcal{R}} \sigma(P)\, dA$.

(iii) *Moments.* Moment of (ii) about the x axis: $\iint_{\mathcal{R}} y\sigma(P)\, dA$.

(iv) *Moments of inertia.* Moment of inertia of (ii) about the y axis: $\iint_{\mathcal{R}} x^2\sigma(P)\, dA$.

(v) *Probability.* A function $f(x, y)$ is eligible to be a probability density function for random variables X and Y over a region $\mathcal{R}$ if $f(x, y) \geqslant 0$ and $\iint_{\mathcal{R}} f(x, y)\, dA = 1$.

The probability that (X, Y) lies in a subregion S is then $\iint_{S} f(x, y)\, dA$. (Here it is helpful to retain x and y: we are not obliged to use P if it does not have the right associations.)

(vi) *Vector resultant.* A force per unit area, $f(P)$ (stress), variable in direction and magnitude, is applied to the surface of a flat plate $\mathcal{R}$. The resultant force F is given by $F = \iint_{\mathcal{R}} f(P)\, dA$.

In order to interpret or evaluate the integral, we write f in its components $f = \hat{i}f_1 + \hat{j}f_2 + \hat{k}f_3$: the original double integral with a vector function as integrand is really three double integrals in one.

The resultant force $F \cdot \hat{s}$ in a fixed direction $\hat{s}$ is $\iint_{\mathcal{R}} f(P) \cdot \hat{s}\, dA$. This integrand $f \cdot \hat{s}$ is not a vector. It can be rewritten in any convenient way: for example, as $|f| \cos \theta$, where θ is the angle between f and $\hat{s}$.

32.6 Polar coordinates

If the boundary of $\mathcal{R}$ in a double integral is circular, or if $\mathcal{R}$ is a circular sector, it might be easiest to work it out using polar coordinates. However, when we do this, the integrand changes in an important way.

Figure 32.9a shows an annular sector $\mathcal{R}$ whose boundaries are specified by $r = a, r = b, \theta = \alpha, \theta = \beta$, where r and θ are polar coordinates. We want to evaluate

$$\iint_{\mathcal{R}} f(P)\, dA = \lim_{\delta A \to 0} \sum_{\mathcal{R}} f(P)\, \delta A, \qquad (32.3)$$

where P is a representative point of $\mathcal{R}$, and δA for the moment permits any kind of division of $\mathcal{R}$ into small area elements. We want to put everything in terms of polar coordinates. This process must include a suitable choice of elements δA, so that the summation, or integration, (32.3) can be carried out in an orderly way over the δA elements – the equivalent of 'strips' in (x, y) coordinates.

The mesh suitable for this purpose is also shown in Fig. 32.9a, and one of the area elements δA is shown in Fig. 32.9b. It is nearly a rectangle, with sides δr and $r\, \delta \theta$, so

(a)
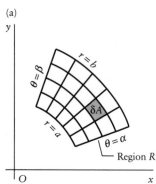

(b)
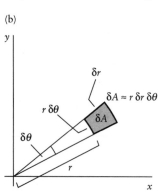

Fig. 32.9

$$\delta A \approx r \, \delta r \, \delta \theta.$$

The sum in (32.3) therefore becomes, in polar coordinates,

$$\lim_{\substack{\delta r \to 0 \\ \delta \theta \to 0}} \sum_{\mathcal{R}} f(r, \theta)(r \, \delta r \, \delta \theta).$$

The sum of the elements along the radial line θ is

$$\left(\int_{r=a}^{b} f(r, \theta) r \, dr \right) \delta \theta$$

after letting $\delta r \to 0$. Now add up the contributions from all these narrow sectors, ranging from $\theta = \alpha$ to $\theta = \beta$, and let $\delta \theta \to 0$. We obtain the repeated integral

$$\int_{\theta=\alpha}^{\beta} \left(\int_{r=a}^{b} f(r, \theta) r \, dr \right) d\theta.$$

Notice that this contains **an extra element r in the integrand**.

Double integral in polar coordinates, when the region of integration $\mathcal{R}$ is an annular sector

If the sector $\mathcal{R}$ is the region $a \leqslant r \leqslant b$, $\alpha \leqslant \theta \leqslant \beta$, then

$$\iint_{\mathcal{R}} f(P) \, dA = \int_{\alpha}^{\beta} \int_{a}^{b} f(r, \theta) r \, dr \, d\theta.$$

(32.4)

Example 32.8 *Find the volume V between the two planes $x + y + z = 4$ and $z = 0$, over the quadrant $0 \leqslant r \leqslant 1$, $0 \leqslant \theta \leqslant \frac{1}{2}\pi$.*

Here $\mathcal{R}$ is the region $0 \leqslant r \leqslant 1$, $0 \leqslant \theta \leqslant \frac{1}{2}\pi$ in the plane $z = 0$. Expressed as a double integral, the required volume V is

$$V = \iint_{\mathcal{R}} f(P) \, dA = \iint_{\mathcal{R}} z \, dA.$$

Here z is given by

$$z = 4 - x - y = 4 - r \cos \theta - r \sin \theta = f(r, \theta).$$

Then by (32.4),

$$V = \int_{0}^{\frac{1}{2}\pi} \int_{0}^{1} (4 - r \cos \theta - r \sin \theta) r \, dr \, d\theta$$

$$= \int_{\theta=0}^{\frac{1}{2}\pi} \left(\int_{r=0}^{1} (4r - r^2 \cos \theta - r^2 \sin \theta) \, dr \right) d\theta$$

$$= \int_{\theta=0}^{\frac{1}{2}\pi} [2r^2 - \tfrac{1}{3}r^3 \cos \theta - \tfrac{1}{3}r^3 \sin \theta]_{r=0}^{1} \, d\theta$$

$$= \int_{0}^{\frac{1}{2}\pi} (2 - \tfrac{1}{3} \cos \theta - \tfrac{1}{3} \sin \theta) \, d\theta = \pi - \tfrac{2}{3}.$$

Example 32.9   *A circular disc of radius 0.1 m has a surface charge density* $\sigma = 10^{-6}(1 + 10^3 r^3 \sin \frac{1}{2}\theta)$ *coulombs per square metre. Find the total charge.*

The total charge Q is given by $\iint_{\mathcal{R}} \sigma(P)\, dA$, where the region $\mathcal{R}$ is the disc $0 \leqslant r \leqslant 0.1, 0 \leqslant \theta \leqslant 2\pi$ (if in doubt, sketch it). Remembering that, in polar coordinates, $\delta A = r\, dr\, d\theta$ (or reading straight from (32.4)), we have

$$Q = \int_0^{2\pi} \int_0^{0.1} \sigma(r, \theta) r\, dr\, d\theta$$

$$= \int_0^{2\pi} \int_0^{0.1} 10^{-6}(1 + 10^3 r^3 \sin \tfrac{1}{2}\theta) r\, dr\, d\theta$$

$$= 10^{-6} \int_{\theta=0}^{2\pi} \left(\int_{r=0}^{0.1} (r + 10^3 r^4 \sin \tfrac{1}{2}\theta)\, dr \right) d\theta$$

$$= 10^{-6} \int_{\theta=0}^{2\pi} [\tfrac{1}{2}r^2 + \tfrac{1}{5}10^3 r^5 \sin \tfrac{1}{2}\theta]_{r=0}^{0.1}\, d\theta$$

$$= 10^{-8} \int_0^{2\pi} (\tfrac{1}{2} + \tfrac{1}{5} \sin \tfrac{1}{2}\theta)\, d\theta = 10^{-8}[\tfrac{1}{2}\theta - \tfrac{2}{5} \cos \tfrac{1}{2}\theta]_0^{2\pi}$$

$$= 3.94 \times 10^{-8}.$$

Since the repeated integral has constant limits, the same result would be obtained by integrating in the reverse order (see (32.1)).

Example 32.10   *The curve* $r = \cos \theta$ $(0 \leqslant \theta \leqslant \frac{1}{4}\pi)$, *together with the radii from the origin to its ends, forms the boundary of the region* $\mathcal{R}$, *and is shown in Fig. 32.10. Obtain* (a) *the area of* $\mathcal{R}$, *and* (b) *its moment about the y axis.*

(a) In general, the area of a region $\mathcal{R}$ is $\iint_{\mathcal{R}} dA$. In this case, we shall add up the contribution δA along radial sectors inclined at angle θ, one of which is shown, and then sum the results for all these sectors to obtain the total area A. We can indicate this, together with the range for r and θ, by writing

$$A = \sum_{\theta=0}^{\theta=\frac{1}{4}\pi} \sum_{r=0}^{r=\cos\theta} \delta A \approx \sum_{\theta=0}^{\theta=\frac{1}{4}\pi} \sum_{r=0}^{r=\cos\theta} (r\, \delta r\, \delta\theta).$$

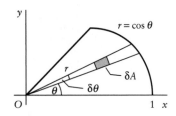

Fig. 32.10

When we let δr and $\delta \theta$ tend to zero, we have a repeated integral with a variable limit:

$$A = \int_0^{\frac{1}{4}\pi} \int_0^{\cos\theta} r\, dr\, d\theta = \int_{\theta=0}^{\frac{1}{4}\pi} [\tfrac{1}{2}r^2]_{r=0}^{\cos\theta}\, d\theta = \tfrac{1}{2} \int_0^{\frac{1}{4}\pi} \cos^2\theta\, d\theta = \tfrac{1}{16}\pi + \tfrac{1}{8}.$$

(b) The moment δM, about the y axis, of an area element δA is

$$\delta M \approx x\, \delta A;$$

so, as a double integral, the total moment M is given by

$$M = \iint_{\mathcal{R}} x\, dA.$$

Example 32.10 *continued*

In the same way as in (a), but with an extra factor

$$x = r \cos \theta,$$

we have

$$M = \int_0^{\frac{1}{4}\pi} \int_0^{\cos\theta} r^2 \cos\theta \, dr \, d\theta = \tfrac{1}{32}\pi + \tfrac{1}{12}.$$

32.7 Separable integrals

Suppose that we have a repeated integral I **with constant limits**, whose integrand $f(x, y)$ is the product of a function of x only with a function of y only:

$$I = \int_c^d \int_a^b f(x, y) \, dx \, dy = \int_c^d \int_a^b g(x)h(y) \, dx \, dy.$$

It is called a **separable integral** because of the following property.

The inner integral is

$$\int_a^b g(x)h(y) \, dx = h(y) \int_a^b g(x) \, dx,$$

since y is held constant. Therefore

$$I = \int_c^d h(y) \left(\int_a^b g(x) \, dx \right) dy.$$

The integral $\int_a^b g(x) \, dx$, once worked out, is just a constant, so we can take it out from under the y integral, obtaining

$$I = \int_a^b g(x) \, dx \int_c^d h(y) \, dy,$$

which is simply the product of two ordinary integrals. We have the following result:

> **Separable integrals**
> $$\int_c^d \int_a^b g(x)h(y) \, dx \, dy = \int_a^b g(x) \, dx \int_c^d h(y) \, dy.$$
> (32.5)

This can sometimes speed up the working when evaluating integrals, but the following example proves an important result by applying (32.5) the other way round.

Example 32.11   *Prove that* $\int_0^\infty e^{-x^2} \, dx = \frac{1}{2}\pi^{\frac{1}{2}}$.

Put $I = \int_0^\infty e^{-x^2} \, dx$.

The name given to the variable of integration in a definite integral is a matter of indifference, so we can equally well put

$$I = \int_0^\infty e^{-y^2} \, dy.$$

The product is I^2:

$$I^2 = \int_0^\infty e^{-x^2} \, dx \int_0^\infty e^{-y^2} \, dy.$$

By (32.5), this can be written as a repeated integral

$$I^2 = \int_0^\infty \int_0^\infty e^{-x^2} e^{-y^2} \, dx \, dy, \tag{32.6}$$

because this repeated integral is separable.

Regard x and y as cartesian coordinates. The region of integration $\mathcal{R}$ is the whole of the first quadrant $(0 \leqslant x < \infty;\ 0 \leqslant y < \infty)$ in Fig. 32.11. Now change to polar coordinates, putting

$$e^{-x^2} e^{-y^2} = e^{-(x^2+y^2)} \quad \text{and} \quad x^2 + y^2 = r^2.$$

The area element is $dA = r \, dr \, d\theta$, and the same region $\mathcal{R}$ is described in polars by

$$0 \leqslant r < \infty, \qquad 0 \leqslant \theta \leqslant \tfrac{1}{2}\pi.$$

Then

$$I^2 = \int_0^{\frac{1}{2}\pi} \int_0^\infty e^{-r^2} r \, dr \, d\theta$$

$$= \int_0^{\frac{1}{2}\pi} d\theta \int_0^\infty e^{-r^2} r \, dr \text{ (by (32.5) again)}$$

$$= \tfrac{1}{2}\pi \tfrac{1}{2}[e^{-r^2}]_{r=0}^\infty = \tfrac{1}{4}\pi.$$

Therefore $I = \frac{1}{2}\pi^{\frac{1}{2}}$.

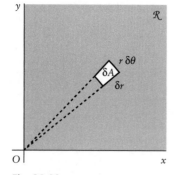

Fig. 32.11

Example 32.12   *Prove the convolution theorem for Laplace transforms (see (25.11)): that is, if F(s) and G(s) are the Laplace transforms of f(t) and g(t) respectively, then F(s)G(s) is the Laplace transform of $\int_0^t f(\tau)g(t-\tau) \, d\tau$.*

Consider the Laplace transform $P(s)$ of $\int_0^t f(\tau)g(t-\tau) \, d\tau$:

$$P(s) = \int_0^\infty e^{-st} \int_0^t f(\tau)g(t-\tau) \, d\tau \, dt$$

$$= \int_0^\infty \int_0^t e^{-st} f(\tau)g(t-\tau) \, d\tau \, dt.$$

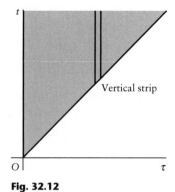

Fig. 32.12

Example 32.12 *continued*

The region of integration is the triangle in the (τ, t) plane shown in Fig. 32.12. Change the order of integration by summing vertical strips: we find that

$$P(s) = \int_0^\infty \int_\tau^\infty e^{-st} f(\tau) g(t - \tau) \, dt \, d\tau.$$

Now change the variable in the inner integral from t to u, where

$$u = t - \tau,$$

remembering that τ is constant in the inner integral. We obtain

$$P(s) = \int_0^\infty \int_0^\infty g(u) f(\tau) \, e^{-s(u+\tau)} \, du \, d\tau$$

$$= \int_0^\infty \int_0^\infty e^{-su} \, e^{-s\tau} \, g(u) f(\tau) \, du \, d\tau$$

$$= \int_0^\infty e^{-su} \, g(u) \, du \int_0^\infty e^{-s\tau} \, f(\tau) \, d\tau$$

(since the integral is separable)

$$= F(s) G(s).$$

32.8 General change of variable; the Jacobian determinant

Consider the integral

$$I = \iint_{\mathcal{R}} f(x, y) \, dx \, dy \tag{32.7}$$

where $\mathcal{R}$ is the region of integration in the (x, y) plane and the area elements δA are small rectangles of side δx and δy as in Fig. 32.13. The shape of $\mathcal{R}$ might suggest the use of another system of coordinates to evaluate I. The special case of polar coordinates was illustrated in Section 32.6.

Suppose that new coordinates u and v are defined by the relations

$$x = x(u, v); \qquad y = y(u, v); \tag{32.8}$$

where there is a one-to-one correspondence between (x, y) and (u, v). The objective is to put (32.7) entirely in terms of u and v.

Figure 32.14 shows a general point P at (x_P, y_P), or at $(u = u_P, v = v_P)$ in the new coordinates. The coordinate curves $u = u_P$ and $v = v_P$ through P are also shown.

Now let δu and δv represent *positive* small increments in u and v respectively. In Fig. 32.15(a) the two curves $u = u_P + \delta u$ and $v = v_P + \delta v$ are also shown. The area element $PQRS$, denoted by $\delta A'$, is of the type appropriate for the new coordinates, and when δu and δv are small, $PQRS$ is nearly a parallelogram, as indicated in Fig. 32.15(b).

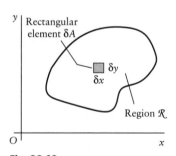

Fig. 32.13

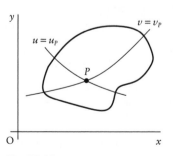

Fig. 32.14

(a) $u = u_P + \delta u$ $v = v_P + \delta v$

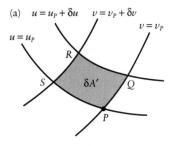

(b)

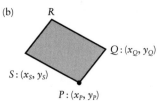

Fig. 32.15 (a) Area element $\delta A'$ for the u, v coordinates. (b) When δu and δv are small, $PQRS$ is nearly a parallelogram.

The area of the parallelogram $PQRS$ is given by

$$\delta A' = \left| \det \begin{bmatrix} x_Q - x_P & x_S - x_P \\ y_Q - y_P & y_S - y_P \end{bmatrix} \right|$$

where the verticals stand for the *modulus of the determinant* between them (see Problem 32.17).

The elements of the determinant are given approximately by

$$x_Q - x_P = x(u_P + \delta u, v_P) - x(u_P, v_P) = \frac{\partial x}{\partial u} \delta u,$$

$$x_S - x_P = x(u_P, v_P + \delta v) - x(u_P, v_P) = \frac{\partial x}{\partial v} \delta v,$$

$$y_Q - y_P = y(u_P + \delta u, v_P) - y(u_P, v_P) = \frac{\partial y}{\partial u} \delta u,$$

$$y_S - y_P = y(u_P, v_P + \delta v) - y(u_P, v_P) = \frac{\partial y}{\partial v} \delta v,$$

where the partial derivatives are evaluated at P. Therefore

$$\delta A' = \left| \det \begin{pmatrix} \dfrac{\partial x}{\partial u} & \dfrac{\partial x}{\partial v} \\ \dfrac{\partial y}{\partial u} & \dfrac{\partial y}{\partial v} \end{pmatrix} \delta u\, \delta v \right| = \left| \det \begin{pmatrix} \dfrac{\partial x}{\partial u} & \dfrac{\partial x}{\partial v} \\ \dfrac{\partial y}{\partial u} & \dfrac{\partial y}{\partial v} \end{pmatrix} \right| \delta u\, \delta v, \qquad (32.9)$$

since we required δu and δv to be positive.

The determinant which occurs in (32.9) is of wide importance. It is called the **Jacobian** of the transformation (32.8), and has the notation

$$\frac{\partial(x, y)}{\partial(u, v)}.$$

For brevity it is sometimes denoted simply by J.

> **The Jacobian determinant of the transformation**
> $x = x(u, v)$, $y = y(u, v)$
>
> $$J \quad \text{or} \quad \frac{\partial(x, y)}{\partial(u, v)} = \det \begin{pmatrix} \dfrac{\partial x}{\partial u} & \dfrac{\partial x}{\partial v} \\ \dfrac{\partial y}{\partial u} & \dfrac{\partial y}{\partial v} \end{pmatrix}$$
>
> (32.10)

From (32.8) and (32.9), remembering the modulus, we can therefore say:

Area $\delta A'$ of an element at P with sides in the directions of the u, v coordinate curves

$$\delta A' = \left| \frac{\partial(x, y)}{\partial(u, v)} \right| \delta u \, \delta v \quad \text{(or } |J| \, \delta u \, \delta v)$$

where $\delta u, \delta v$ are positive, and the Jacobian determinant is evaluated at P.

(32.11)

We can now rewrite the original integral (32.7) in terms of the new coordinates u and v:

To express a double integral in new coordinates

If $x = x(u, v)$, $y = y(u, v)$, then

$$\iint_{R} f(x, y) \, dx \, dy = \iint_{S} f(x(u, v), y(u, v)) \left| \frac{\partial(x, y)}{\partial(u, v)} \right| du \, dv,$$

where S is the region R transformed to the cartesian (u, v) plane.

(32.12)

The effect of making the change of variable is to change the integrand to something different. This is not surprising; a similar thing happens in the one-dimensional case. If

$$I = \int_{a}^{b} f(x) \, dx$$

and we change the variable by putting $x = x(u)$, the new factor $\dfrac{dx}{du}$ appears in the integrand.

The final step is to convert (32.12) into a repeated integral in terms of u and v, so that the integrations can be carried out.

Example 32.13 *Transform the integral $I = \displaystyle\iint_{R} (x^2 + y^2) \, dx \, dy$ into polar coordinates r, θ, where R is the region shown in Fig. 32.16.*

Here r and θ stand in place of u and v, and

$$x = r \cos \theta, \qquad y = r \sin \theta,$$

$$\frac{\partial x}{\partial r} = \cos \theta, \qquad \frac{\partial x}{\partial \theta} = -r \sin \theta,$$

$$\frac{\partial y}{\partial r} = \sin \theta, \qquad \frac{\partial y}{\partial \theta} = r \cos \theta.$$

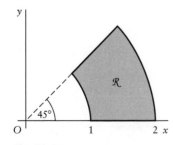

Fig. 32.16

Example 32.13 *continued*

Therefore,

$$\frac{\partial(x, y)}{\partial(r, \theta)} = \det\begin{pmatrix} \cos\theta & -r\sin\theta \\ \sin\theta & r\cos\theta \end{pmatrix} = r\cos^2\theta + r\sin^2\theta = r.$$

This is already positive, so the new area elements are given by

$$\delta A' = r\,\delta r\,\delta\theta,$$

as we found in Example 32.11. Also

$$f(x, y) = x^2 + y^2 = r^2.$$

Finally

$$I = \iint_S r(r^2\,dr\,d\theta) = \iint_S r^3\,dr\,d\theta.$$

This is to be read straightforwardly as a fresh double integral in variables called r and θ, *with rectangular area elements $\delta r\,\delta\theta$*. When we draw the diagram to find the shape of S, r and θ are to be *treated as cartesian coordinates* in axes labelled r and θ (see Fig. 32.17). In this frame, S is bounded by the straight lines $r = 1$, $r = 2$, $\theta = 0$, $\theta = \frac{1}{4}\pi$.

Expressed as a repeated integral, using (say) strips parallel to the r axis,

$$I = \int_0^{\frac{1}{4}\pi}\int_1^2 r^3\,dr\,d\theta = \int_1^2 r^3\,dr\int_0^{\frac{1}{4}\pi}d\theta \quad \text{(since the integral is separable)}$$

$$= [\tfrac{1}{4}r^4]_1^2\,[\theta]_0^{\frac{1}{4}\pi} = \tfrac{15}{16}\pi.$$

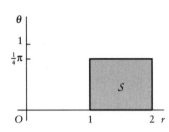

Fig. 32.17

Example 32.14 *Evaluate $I = \displaystyle\iint_{\mathcal{R}} (y^2 - x^2)\,dx\,dy$ over the square*

region in Fig. 32.18 by changing the variables to u, v, where $x = v - u$, $y = v + u$).

Figure 32.18 shows the x, y and the u, v equations of the sides of $\mathcal{R}$. We have

$$\frac{\partial(x, y)}{\partial(u, v)} = \det\begin{bmatrix} \dfrac{\partial x}{\partial u} & \dfrac{\partial x}{\partial v} \\ \dfrac{\partial y}{\partial u} & \dfrac{\partial y}{\partial v} \end{bmatrix} = \det\begin{bmatrix} -1 & 1 \\ 1 & 1 \end{bmatrix} = -2.$$

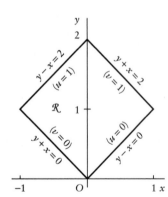

Fig. 32.18

Therefore

$$\delta A = \left|\frac{\partial(x, y)}{\partial(u, v)}\right|\delta u\,\delta v = 2uv\,\delta u\,\delta v.$$

In terms of u and v, $y^2 - x^2 = 4uv$, so

$$I = \iint_S 4uv(2\,du\,dv) = 8\iint_S uv\,du\,dv.$$

The corresponding region S in the (u, v) plane is shown in Fig. 32.19. Therefore

$$I = \int_0^1\int_0^1 uv\,du\,dv = 8\int_0^1 u\,du\int_0^1 v\,dv \quad \text{(since the integral is separable)}$$

$$= 8 \times \tfrac{1}{2} \times \tfrac{1}{2} = 2.$$

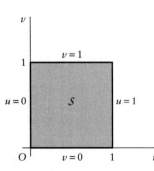

Fig. 32.19

Problems

32.1 Evaluate the following repeated integrals with constant limits.

(a) $\displaystyle\int_0^1\int_1^2 xy^2\,dx\,dy;$ (b) $\displaystyle\int_0^1\int_0^1 y\,e^{xy}\,dx\,dy;$

(c) $\displaystyle\int_c^d\int_a^b dx\,dy;$ (d) $\displaystyle\int_a^b\int_c^d dx\,dy;$

(e) $\displaystyle\int_c^d\int_a^b dy\,dx;$ (f) $\displaystyle\int_0^{\frac12}\int_0^{\frac12\pi} y\sin xy\,dx\,dy;$

(g) $\displaystyle\int_{-1}^1\int_{-1}^1 x^2\,dx\,dy;$ (h) $\displaystyle\int_1^2\int_0^1 x^2\,dy\,dx;$

(i) $\displaystyle\int_0^1\int_{-1}^1 (xy^2-x^2y)\,dx\,dy;$ (j) $\displaystyle\int_{-1}^1\int_0^1 (xy^2-x^2y)\,dy\,dx;$

(k) $\displaystyle\int_0^1\int_0^1 (x+y^2+1)^2\,dx\,dy;$

(l) $\displaystyle\int_0^{\frac12\pi}\int_0^{\frac12\pi}\cos(x+y)\,dy\,dx;$ (m) $\displaystyle\int_1^2\int_0^1 \frac{x}{y}\,dx\,dy.$

32.2 Find the signed volume between the given surfaces and the plane $z=0$ over the specified rectangular regions.
(a) $z=xy,\quad 0\leqslant x\leqslant 1, 0\leqslant y\leqslant 1;$
(b) $z=xy,\quad -1\leqslant x\leqslant 1, 0\leqslant y\leqslant 1$ (explain the result);
(c) $z=x+y,\quad -1\leqslant x\leqslant 2, -2\leqslant y\leqslant 1;$
(d) $z=-1,\quad a\leqslant x\leqslant b, c\leqslant y\leqslant d;$
(e) $z=2x-y+3,\quad 0\leqslant x\leqslant 1, 0\leqslant y\leqslant 1;$
(f) $z=1/(x+y),\quad 1\leqslant x\leqslant 2, 0\leqslant y\leqslant 1;$
(g) $z=(x+2y-1)^2,\quad -2\leqslant x\leqslant 1, -1\leqslant y\leqslant 1.$

32.3 In the following problems, the region of integration is not rectangular. In each case, sketch the region of integration and indicate a typical strip for the inner integral.

(a) $\displaystyle\int_0^1\int_0^y dx\,dy;$ (b) $\displaystyle\int_0^1\int_y^1 x^2y\,dx\,dy;$

(c) $\displaystyle\int_0^1\int_0^x x^2y\,dy\,dx$ (compare (b));

(d) $\displaystyle\int_0^1\int_0^y (x+y^2)^2\,dx\,dy;$ (e) $\displaystyle\int_0^1\int_{-y}^y y\,dx\,dy;$

(f) $\displaystyle\int_0^2\int_{-\frac12 y}^y y^2\sin xy\,dx\,dy;$ (g) $\displaystyle\int_0^2\int_0^{1-\frac12 x} x^2\,dy\,dx;$

(h) $\displaystyle\int_0^1\int_0^{\sqrt{(1-y^2)}} x\,dx\,dy;$ (i) $\displaystyle\int_0^1\int_0^{\sqrt{(1-x^2)}} x\,dy\,dx.$

32.4 Find the volume of the wedge-shaped object having one curved surface which is part of the cylinder $x^2+y^2=1$, and whose flat surfaces are $z=0$ and the plane $z=2y$. (Consider the simple wedge in $z\geqslant 0$ only.)

32.5 Reverse the order of integration in each of the following cases. It is *necessary* to sketch the region of integration and to indicate a typical strip corresponding to the new order of integration, as in Section 32.4.

(a) $\displaystyle\int_0^1\int_0^y f(x,y)\,dx\,dy;$ (b) $\displaystyle\int_0^1\int_y^1 f(x,y)\,dx\,dy;$

(c) $\displaystyle\int_1^2\int_0^{y+1} f(x,y)\,dx\,dy;$ (d) $\displaystyle\int_0^1\int_{-\sqrt{(1-y^2)}}^{\sqrt{(1-y^2)}} f(x,y)\,dx\,dy;$

(e) $\displaystyle\int_2^4\int_0^{\frac12 y} f(x,y)\,dx\,dy;$ (f) $\displaystyle\int_0^1\int_{x^3}^{x^2} f(x,y)\,dy\,dx;$

(g) $\displaystyle\int_0^1\int_{-1+x}^{1-x} f(x,y)\,dy\,dx$ (it becomes the sum of two integrals);

(h) $\displaystyle\int_{-1}^1\int_{1-\sqrt{(x^2-1)}}^{1+\sqrt{(x^2-1)}} f(x,y)\,dy\,dx.$

32.6 Change the order of integration in the following, and hence evaluate them. It is *necessary* to sketch the region of integration $\mathcal{R}$ and to indicate the strip corresponding to the new inner integral.

(a) $\displaystyle\int_0^{\frac12}\int_0^{\frac12\pi} x\sin xy\,dx\,dy;$ (b) $\displaystyle\int_1^2\int_0^{2(y-1)} x^2\,dx\,dy;$

(c) $\displaystyle\int_0^1\int_0^y x^2\,e^{xy}\,dx\,dy;$ (d) $\displaystyle\int_0^\infty\int_{\frac14 y^2}^{y^2} x^2y\,e^{-x^2y^2}\,dx\,dy;$

(e) $\displaystyle\int_{-1}^1\int_{-2}^2 y(1-x^2-y^2)^2\,dx\,dy;$

(f) $\displaystyle\int_1^2\int_0^1 \frac{y}{x^2+y^2}\,dx\,dy;$ (g) $\displaystyle\int_1^\infty\int_0^\infty \frac{y}{(x+y)^3}\,dx\,dy;$

(h) $\displaystyle\int_0^1\int_y^1 y(x^2-y^2)^{\frac12}\,dx\,dy;$

(i) $\displaystyle\int_0^2\int_{-y-1}^{y-1} x^2\,dx\,dy$ (the integral must be split into two parts);

(j) $\displaystyle\int_0^1\int_y^1 \frac{y\,dx\,dy}{(x^2-y^2)^{\frac12}}.$

32.7 The symbol $\iint_R f(P)\,dA$ represents a double integral taken over the region R, and dA is the area element at the point P in R (see Section 32.5). In the following cases, the region R is described, and $f(P)$ given in cartesian coordinates. Evaluate the integrals.

(a) R is the rectangle with corners at $(1, 1)$, $(2, 1)$, $(2, 4)$, and $(1, 4)$, and $f(P) = x^2 + y^2$.

(b) R is the equilateral triangle with vertices at $(0, -1)$, $(3^{\frac{1}{2}}, 0)$, and $(0, 1)$, and $f(P) = x$.

(c) R is the circle of radius 2, centred at the origin, and $f(P) = y^2$.

32.8 As in Problem 32.7, but polar coordinates are to be used for the evaluation (see Section 32.6). Remember the change in the area element; see (32.4).

(a) R is the disc $x^2 + y^2 \leqslant 1$, and $f(P) = x^2 + y^2$.

(b) R is the disc $x^2 + y^2 \leqslant 1$, and $f(P) = y^2$.

(c) R is the area whose boundary consists of the x axis between $x = 0$ and 2, the y axis between $y = 0$ and 2, and a quarter of the circle $x^2 + y^2 = 4$. Also $f(x, y) = xy$.

(d) R is the sector $1 \leqslant r \leqslant 2, 0 \leqslant \theta \leqslant \frac{1}{2}\pi$, and $f(P) = xy$.

(e) R is the disc $x^2 + y^2 \leqslant 4$, and $f(x, y) = \arctan(y/x)$.

(f) R is the first quadrant of the plane, and $f(x, y) = e^{-4(x^2+y^2)}$.

(g) Show that the volume of a sphere of radius a is $\frac{4}{3}\pi a^3$. (Consider the hemisphere
$$0 \leqslant z \leqslant (a^2 - x^2 - y^2)^{\frac{1}{2}}.)$$

(h) R is the half plane $y \geqslant 0$, and $f(P) = y\,e^{-(x^2+y^2)}$. (Hint: separate the integral: see (32.5).)

32.9 A circular hole of radius $\frac{1}{2}a$ is drilled through a sphere of radius a in such a way that the edge of the hole passes through the centre of the sphere. Let the equation of the sphere and the cylinder be $x^2 + y^2 + z^2 = a^2$ and $(x - \frac{1}{2}a)^2 + y^2 = \frac{1}{4}a^2$. If $x = r\cos\theta$ and $y = r\sin\theta$, show that the volume V_c of material removed (the section in the (x, y) plane is shown in Fig. 32.20) is given by

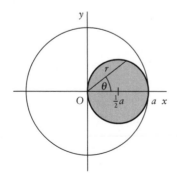

Fig. 32.20

$$V_c = 2\int_{-\frac{1}{2}\pi}^{\frac{1}{2}\pi}\int_0^a \sqrt{a^2 - r^2}\,r\,dr\,d\theta.$$

Hence find the volume of the remaining part of the sphere.

32.10 Find the Jacobian

$$J(u, v) = \frac{\partial(x, y)}{\partial(u, v)} = \begin{vmatrix} \dfrac{\partial x}{\partial u} & \dfrac{\partial x}{\partial v} \\ \dfrac{\partial y}{\partial u} & \dfrac{\partial y}{\partial v} \end{vmatrix}$$

of the following transformations:

(a) $x = u^2 - v^2, y = uv$;

(b) $x = u - v, y = 2v$;

(c) $u = 2x - y, v = x + 2y$;

(d) $x = u - e^{-v}, y = u - e^v$.

32.11 Find the Jacobian of the transformation $x = u/v$, $y = uv$. Let R be the region bounded by $y = 2x$, $y = x$, $xy = 1$, $xy = 8$. Express

$$\iint_R xy^2\,dx\,dy$$

as a repeated integral in the (u, v) plane, and evaluate it.

32.12 Sketch the region in the (x, y) plane bounded by the parabolas $y = x^2$, $y = 2x^2$, $x = y^2$, $x = 2y^2$. Find the Jacobian of the transformation given by

$$u = \frac{y}{x^2}, \qquad v = \frac{x}{y^2}.$$

Hence find the area bounded by the parabolas.

32.13 Evaluate

$$\iint_R x\,e^{x+y}\,dA,$$

where R is the region bounded by the square $|x| + |y| = 1$.

32.14 A plastic component is cut from a solid plastic rod which is cylindrical with cross-section bounded by the rhombus R, $y = 1 - \frac{1}{2}x$, $y = -1 - \frac{1}{2}x$, $y = 1 + \frac{1}{2}x$, $y = -1 + \frac{1}{2}x$ in the (x, y) plane, with the rod in the z direction. The ends of the component are shaped into the surfaces $z = x^2 + 2$ and $z = -(x^2 + 2)$. Find the volume

$$V = 2\iint_R (x^2 + 2)\,dA$$

of the component.

32.15 For the polar transformation $x = r \cos\theta$, $y = r \sin\theta$, show that

$$\frac{\partial(x, y)}{\partial(r, \theta)} = r.$$

Show that r and θ are given by

$$r = \sqrt{x^2 + y^2}, \qquad \tan\theta = \frac{y}{x}.$$

Show that

$$\frac{\partial(r, \theta)}{\partial(x, y)} = \frac{1}{r} = 1 \bigg/ \frac{\partial(x, y)}{\partial(r, \theta)}.$$

In fact under fairly general conditions, the Jacobian satisfies this inverse rule, which is helpful in some cases since it can avoid the inversion of transformations (see Example 30.16).

Find $\partial(u, v)/\partial(x, y)$ if $u = y/x^2$ and $v = x/y^2$ using this rule, and confirm that

$$\frac{\partial(x, y)}{\partial(u, v)} = \frac{1}{3u^2v^2}.$$

32.16 Find the Jacobian $J(u, v)$ of the transformation $u = x^2 - y^2$, $v = 2xy$. Draw a sketch of the region $\mathcal{R}$ in the (x, y) plane bounded by the curves $x^2 - y^2 = 1$, $x^2 - y^2 = 4$, $xy = 2$, $xy = 4$. By using the change of variable from (x, y) to (u, v), evaluate

$$\iint_{\mathcal{R}} (x^2 + y^2)\, dx\, dy.$$

32.17 Let $PQRS$ be a parallelogram with P at (x_P, y_P), Q at (x_Q, y_Q), and S at (x_S, y_S). Show that its area is given by the modulus of the determinant

$$\det\begin{bmatrix} x_Q - x_P & x_S - x_P \\ y_Q - y_P & y_S - y_P \end{bmatrix}.$$

(Hint: simplify the problem by placing P at the origin of coordinates.)

32.18 The following technique can be regarded as integrating under another integral sign with respect to a parameter (compare Section 17.9).

(a) Noting that

$$\int_a^b e^{-xy}\, dy = \frac{1}{x}(e^{-ax} - e^{-bx})$$

evaluate the integral

$$\int_0^\infty \frac{e^{-ax} - e^{-bx}}{x}\, dx,$$

where $a > 0$ and $b > 0$.

(b) In a similar way, evaluate

$$\int_{-\infty}^\infty \frac{\cos ax - \cos bx}{x^2}\, dx,$$

where a and b may take any values. In this problem the result depends on the signs of a and b. (You may assume that

$$\int_0^\infty \frac{\sin u}{u}\, du = \frac{1}{2}\sqrt{\pi}.)$$

33 Line integrals

CONTENTS

33.1 Illustrating a line integral

Consider the following scenario. The success of a museum is measured by two variables. These are the monthly income x from visitors, and the monthly income y from grants and donations. The variation is smoothed out so that they form a continuous record. The exhibitions director receives a variable monthly bonus I which rewards success and penalizes failure in promoting attendance: when attendance changes by a small amount δx, positive or negative, there is a change in bonus (up or down) of δI, where

$$\delta I \simeq f(x, y)\, \delta x. \tag{33.1}$$

Figure 33.1 charts the fortunes, or the **state** (x, y), of the museum over a period, starting at state A and arriving at state B, in the form of a curve joining A and B. Time does not register on this diagram, except that direction of development as time increases is indicated by the arrow. The directed curve is called the **path from A to B**, denoted by (AB) (there may be more letters in the brackets). Suppose that the bonus at the starting state A is I_A, and at the state B it is I_B. Then the problem is to find the change in bonus over the period, $I_{(AB)}$:

$$I_B - I_A = I_{(AB)}. \tag{33.2}$$

Divide up the path into many short segments such as PP' (Fig. 33.1). The increment δI over a typical segment is given by

$$(\delta I)_{\text{along } PP'} \approx f(x, y)\, \delta x.$$

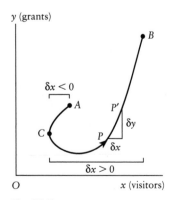

Fig. 33.1

Add the contributions of all the segments to obtain $I_{(AB)}$:

$$I_{(AB)} \approx \sum_{(AB)} f(x, y) \, \delta x. \tag{33.3}$$

Given a specific function $f(x, y)$ and a *specific path* (AB), $I_{(AB)}$ could in principle be computed by carrying out the summation (33.3) numerically, taking δx very small, and allowing for the fact that δx is *sometimes positive and sometimes negative*. We have to split up the path for this purpose: in Fig. 33.1, δx is negative along (AC) and positive along (CB). If the path is vertical along a section, then δx will be *zero*, and there will be a zero change in the bonus along this part despite the fact that y is changing.

In imagination, let $\delta x \to 0$. Then '$\approx$' becomes '$=$'. It is natural to write the result as a kind of integral:

$$I_{(AB)} = \lim_{\delta x \to 0} \sum_{(AB)} f(x, y) \, \delta x = \int_{(AB)} f(x, y) \, \mathrm{d}x, \tag{33.4}$$

where the notation reminds us that we take values of (x, y) which lie on (AB) and take account of the sign of δx at each point on the path.

The integral in (33.4) is called a **line integral**. It is not straightforwardly an ordinary integral because **the direction, left to right or right to left, at every point must be taken into account**. The director is losing money along (AC). In order to arrive at B from A many paths are possible. In general, a line integral $I_{(AB)}$ will depend on the total history, on what path has been followed between A and B, and we say that the integral is **path dependent**.

To show how to intepret (33.4) in terms of ordinary integrals, suppose that the bonus function $f(x, y)$ is given by

$$f(x, y) = x + \tfrac{1}{2}y \quad \text{so that} \quad \delta I = (x + \tfrac{1}{2}y) \, \delta x$$

(in suitable units). Suppose that the museum starts off at state A in Figs 33.2a,b; it thrives in Fig. 33.2a and declines in Fig. 33.2b.

Consider the case (a). The graph AB can be expressed in principle as a function of x, and the curve chosen for illustration is

$$y = x^2 - 7x + 15.$$

Then

$$I_{(AB)} = \int_{(AB)} (x + \tfrac{1}{2}y) \, \mathrm{d}x = \int_{(AB)} [x + \tfrac{1}{2}(x^2 - 7x + 15)] \, \mathrm{d}x.$$

But δx is **positive** all the way; so, regarded as the limit of the sum in (33.4), this is just an **ordinary integral**. After simplifying the integrand, we have

(a)
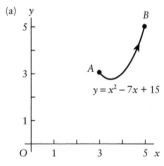

(b)
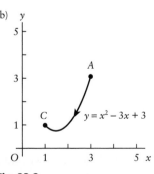

Fig. 33.2

$$I = \int_{x=3}^{5} \tfrac{1}{2}(x^2 - 5x + 15)\, dx = 11.33.$$

For the case (b), the equation of the curve from A to C is

$$y = x^2 - 3x + 3,$$

so that $I_{(AC)}$ becomes

$$I_{(AC)} = \int_{(AC)} \tfrac{1}{2}(x^2 - x + 3)\, dx.$$

In this case, however, the δx in the sum (33.4) are all *negative*: x is *decreasing*. To turn this into an ordinary integral, we have therefore to reverse the sign:

$$I_{(AC)} = -I_{(CA)} = -\int_{x=1}^{3} \tfrac{1}{2}(x^2 - x + 3)\, dx = -5.33. \qquad (33.5)$$

There is a reduction of bonus for bringing the museum to the edge of ruin.

While we are still observing things, notice first that, in connection with the sign change for negative δx on (AC) in (33.5), we can write

$$I = -\int_{1}^{3} \tfrac{1}{2}(x^2 + x + 3)\, dx = (+)\int_{3}^{1} \tfrac{1}{2}(x^2 + x + 3)\, dx.$$

In other words, we obtain the correct result by setting the x coordinate of the **starting point** as the **lower limit**, and that of the **end-point** as the **upper limit**, whether x is constantly decreasing or constantly increasing along the path, and this is a general result.

Lastly we compare the result for the parabolic path (AC) in Fig. 33.2 with a straight path from A to C whose equation is

$$y = x.$$

Then

$$\int_{(AC)} (x + \tfrac{1}{2}y)\, dx = \int_{3}^{1} \tfrac{3}{2}x\, dx = \tfrac{3}{2}[\tfrac{1}{2}x^2]_{3}^{1} = -6.$$

This is different from (33.5), so we must in general expect that line integrals will be path dependent.

The following summary generalizes the special case we have discussed.

The line integral $I_{(AB)} = \displaystyle\int_{(AB)} f(x, y)\, dx$

(a) Definition: $I_{(AB)} = \displaystyle\lim_{\delta x \to 0} \sum_{(AB)} f(x, y)\, \delta x$, where (x, y) takes values on the path (AB) from A to B.

(b) $I_{(AB)}$ is path dependent, and $I_{(BA)} = -I_{(AB)}$.

(c) If δx has constant sign on the path $y = y(x)$ from $C : (x_C, y_C)$ to $D : (x_D, y_D)$, then

$$\int_{(CD)} f(x, y) = \int_{x_C}^{x_D} f(x, y(x))\, dx.$$

(33.6)

(a)

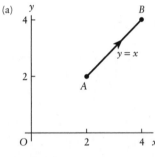

(b)

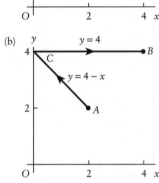

Fig. 33.3

Example 33.1 *Evaluate the two line integrals*

(a) $\displaystyle\int_{(AB)} xy\, dx$,   (b) $\displaystyle\int_{(ACB)} xy\, dx$,

shown in Figs 33.3a,b respectively.

(a) On (AB), $y = x$ and $\delta x > 0$. Here $A = (2, 2)$ and $B = (4, 4)$; so, by (33.6),

$$I_{(AB)} = \int_{(AB)} xy\, dx = \int_2^4 x^2\, dx = [\tfrac{1}{3}x^3]_2^4 = \tfrac{56}{3}.$$

(b) The path (ACB) has to be broken into two parts: (AC), on which $\delta x < 0$, and (CB), on which $\delta x > 0$, where $C = (0, 4)$. Then

$$I_{(ACB)} = \int_{(ACB)} xy\, dx = \int_{(AC)} xy\, dx + \int_{(CB)} xy\, dx.$$

On (AC), $y = 4 - x$; on (CD), $y = 4$. Therefore

$$I_{(ACB)} = \int_{(AC)} x(4 - x)\, dx + \int_{(CB)} 4x\, dx$$

$$= \int_2^0 (4x - x^2)\, dx + \int_0^4 4x\, dx$$

$$= [2x^2 - \tfrac{1}{3}x^3]_2^0 + 2[x^2]_0^4 = \tfrac{80}{3}.$$

Despite (33.6c), reduction to ordinary integrals over x is not usually the best way to evaluate line integrals, as will be seen later on.

33.2 General line integrals in two and three dimensions

Line integrals of the type

$$\int_{(AB)} g(x, y)\, dy$$

are to be understood in a similar way:

$$\int_{(AB)} g(x, y)\, \delta y \quad \text{means} \quad \lim_{\delta y \to 0} \sum_{(AB)} g(x, y)\, \delta y,$$

in which the sign of δy is **positive** on a segment along which y is **increasing** and **negative** on a segment where y is **decreasing**. We can similarly consider paths and functions in three dimensions:

$$\int_{(AB)} f(x, y, z)\, dx, \qquad \int_{(AB)} g(x, y, z)\, dy, \qquad \int_{(AB)} h(x, y, z)\, dz,$$

and string these types together to obtain a general integral

$$\int_{(AB)} (f\, dx + g\, dy + h\, dz).$$

In the following definition, (AB) is a directed path in three dimensions with representative point $P : (x, y, z)$, and f, g, h are any three functions, their values at P being denoted by f_P, g_P, h_P:

General line integrals

(a) $\displaystyle \int_{(AB)} (f\, dx + g\, dy + h\, dz) = \lim_{\delta x, \delta y, \delta z \to 0} \sum_{(AB)} (f_P\, \delta x + g_P\, \delta y + h_P\, \delta z)$

($\delta x, \delta y, \delta z$ are positive (negative) where x, y, z are increasing (decreasing)).

(b) $\displaystyle \int_{(AB)} (f\, dx + g\, dy + h\, dz) = -\int_{(BA)} (f\, dx + g\, dy + h\, dz).$

(c) For two dimensions, suppress the variable z.

(d) The above integrals generally depend on the path between A and B.

(33.7)

To organize such an integral in order to take account of the signs of $\delta x, \delta y, \delta z$ is often difficult. For example, if the path (AB) consists of an ellipse inclined to the three axes, each term must be broken into two sections, leading to six integrals in all, to ensure constancy of sign of $\delta x, \delta y,$ or δz along each. However, if a **parametric representation of the path** is adopted, the correct interpretation is obtained **automatically**.

Consider the integral with respect to x,

$$\int_{(AB)} f(x, y, z)\, dx,$$

where (AB) is parametrized as

$$x = x(t), y = y(t), z = z(t),$$

so that, as the parameter t increases or decreases from t_A at A to t_B at B, the path is traced exactly once in the right direction. Then, in the short interval from t to $t + \delta t$, the change in δx is approximated by

$$\delta x = \frac{dx}{dt} \delta t,$$

and δx automatically has the right sign. Now put $(dx/dt)\,\delta t$ into the defining sum in place of δx; correspondingly $(dx/dt)\,dt$ will go into the original integral in place of dx. After doing the same thing with the y and z integrals, we have the following result.

Parametric evaluation of line integrals

(a) If $x = x(t)$, $y = y(t)$, $z = z(t)$, and $P : (x, y, z)$ covers (AB) exactly once in the correct direction as t increases or decreases from t_A to t_B, then

$$\int_{(AB)} (f\,dx + g\,dy + h\,dz) = \int_{t_A}^{t_B} \left(f\frac{dx}{dt} + g\frac{dy}{dt} + h\frac{dz}{dt} \right) dt.$$

(b) For the two-dimensional case, omit z.

(33.8)

Example 33.2 *Evaluate* $I = \displaystyle\int_{(AB)} (x^2 + y)\,dy$, *where (AB) is the path shown in Fig. 33.4.*

On (AB), $y = -x$, so we can use $x = t$, $y = -t$, with t running from $t = 1$ to $t = -1$. This covers (AB) once in the right direction (it is like using x as the parameter). Then

$$x^2 + y = t^2 - t \quad \text{and} \quad \frac{dy}{dt} = -1,$$

so

$$I = -\int_1^{-1} (t^2 - t)\,dt$$

$$= -[\tfrac{1}{3}t^3 - \tfrac{1}{2}t^2]_1^{-1} = \tfrac{2}{3}.$$

It is immaterial what parametrization is used, so long as it satisfies the conditions in (33.8). The following example compares two parametrizations.

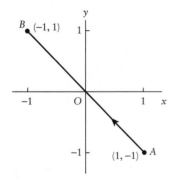

Fig. 33.4

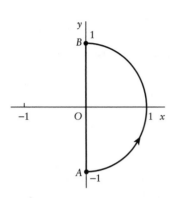

Fig. 33.5

Example 33.3   *Evaluate* $I = \displaystyle\int_{(AB)} (x\,\mathrm{d}y - y\,\mathrm{d}x)$, *where* (AB) *is the semicircle shown in Fig. 33.5, by means of two parametrizations:*

(a) $x = \cos t,\ y = \sin t\ for\ -\tfrac{1}{2}\pi \leqslant t \leqslant \tfrac{1}{2}\pi$;

(b) $x = (1 - t^2)^{\frac{1}{2}},\ y = t,\ for\ -1 \leqslant t \leqslant 1.$

(a) $x = \cos t,\ y = \sin t$; so

$$\frac{\mathrm{d}x}{\mathrm{d}t} = -\sin t \qquad \frac{\mathrm{d}y}{\mathrm{d}t} = \cos t.$$

Then

$$I = \int_{-\frac{1}{2}\pi}^{\frac{1}{2}\pi} [\cos t\ \cos t - \sin t(-\sin t)]\,\mathrm{d}t$$

$$= \int_{-\frac{1}{2}\pi}^{\frac{1}{2}\pi} [\cos^2 t + \sin^2 t]\,\mathrm{d}t$$

$$= \int_{-\frac{1}{2}\pi}^{\frac{1}{2}\pi} \mathrm{d}t = \pi.$$

(b) $x = (1 - t^2)^{\frac{1}{2}},\ y = t$; so

$$\frac{\mathrm{d}x}{\mathrm{d}t} = -t(1 - t^2)^{-\frac{1}{2}}, \qquad \frac{\mathrm{d}y}{\mathrm{d}t} = 1.$$

Then

$$I = \int_{-1}^{1} \{(1 - t^2)^{\frac{1}{2}} - t[-t(1 - t^2)^{-\frac{1}{2}}]\}\,\mathrm{d}t$$

$$= \int_{-1}^{1} \frac{\mathrm{d}t}{(1 - t^2)^{\frac{1}{2}}} = [\arcsin t]_{-1}^{1} = \pi.$$

Example 33.4   *Evaluate* $I = \displaystyle\int_{(AB)} (x\,\mathrm{d}x + y\,\mathrm{d}y + z\,\mathrm{d}z)$, *where* (AB) *is the path* $x = a\cos t,\ y = a\sin t,\ z = bt$ *between* $t = 0$ *and* 4π. *(* (AB) *is a* **helix** *along the z axis.)*

We have

$$\frac{\mathrm{d}x}{\mathrm{d}t} = -a\sin t, \qquad \frac{\mathrm{d}y}{\mathrm{d}t} = a\cos t, \qquad \frac{\mathrm{d}z}{\mathrm{d}t} = b,$$

so the integral becomes

$$I = \int_{0}^{4\pi} (-a^2 \cos t\ \sin t + a^2 \sin t\ \cos t + b^2 t)\,\mathrm{d}t$$

$$= b^2 \int_{0}^{4\pi} t\,\mathrm{d}t = 8b^2\pi^2.$$

33.3 Paths parallel to the axes

Sometimes it is necessary to evaluate line integrals along line segments which are parallel to the axes. In such cases the easiest approach is a direct one, as shown in the following example.

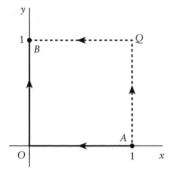

Fig. 33.6

Example 33.5 *(See Fig. 33.6.) Evaluate the line integral $\displaystyle\int_{(AB)} x\,\mathrm{d}y$ (a) over the path (AOB);   (b) over the path (AQB).*

In this method, we refer to the sums in the definition (33.7).

(a) On (AO), $\delta y = 0$ since y is constant; so $\displaystyle\int_{(AO)} x\,\mathrm{d}y = 0$.

On (OB), $x = 0$; so $\displaystyle\int_{(OB)} x\,\mathrm{d}y = 0$.

Therefore

$$\int_{(AOB)} x\,\mathrm{d}y = \int_{(AO)} x\,\mathrm{d}y + \int_{(OB)} x\,\mathrm{d}y = 0.$$

(b) On (AQ), $x = 1$; so $\displaystyle\int_{(AQ)} x\,\mathrm{d}y = \int_{(AQ)} 1\,\mathrm{d}y = 1$.

On (QB), y is constant; so $\delta y = 0$, and $\displaystyle\int_{(QB)} x\,\mathrm{d}y = 0$. Therefore

$$\int_{(AQB)} x\,\mathrm{d}y = \int_{(AQ)} x\,\mathrm{d}y + \int_{(QB)} x\,\mathrm{d}y = 1.$$

33.4 Path independence and perfect differentials

Despite the fact that the value of a line integral taken between two given points usually depends on the path chosen, there are many cases of physical importance for which the value is **independent of the path**: all paths between A and B lead to the same value. To show that such cases can exist, the following two examples show integrands for which the integral is independent of the path.

Example 33.6 *(In two dimensions). Show that $I = \displaystyle\int_{(AB)} (y\,\mathrm{d}x + x\,\mathrm{d}y)$ is independent of the path chosen from A to B.*

We can write the integrand in terms of a **perfect differential** (see Section 22.4):

$$y\,\mathrm{d}x + x\,\mathrm{d}y = \mathrm{d}(xy).$$

Now express the integral in the form

$$I = \int_{(AB)} (y\,\mathrm{d}x + x\,\mathrm{d}y) = \int_{(AB)} \mathrm{d}(xy). \tag{i}$$

Example 33.6 *continued*

From (33.7), the meaning of the integral is the limit of a certain sum, which can be recast in the form

$$\lim_{\delta x, \delta y \to 0} \sum_{(AB)} (y\,\delta x + x\,\delta y) = \lim_{\delta(xy) \to 0} \sum_{(AB)} \delta(xy). \tag{ii}$$

As we travel along (AB), the value of (xy) starts at $x_A y_A$, where the values are taken at A, then goes by steps $\delta(xy)$ until it attains the value $x_B y_B$. In other words,

$$\sum_{(AB)} \delta(xy) = x_B y_B - x_A y_A,$$

which is independent of what path connects A to B.

Example 33.7 *Prove that*

$$I = \int_{(AB)} [(y+z)\,dx + (z+x)\,dy + (x+y)\,dz]$$

is independent of the path from A to B.

We can use the idea of differentials in exactly the same way: it is suggested by the incremental approximation (31.1). If we put

$$f(x, y, z) = yz + zx + xy,$$

the incremental approximation gives

$$\delta f = (y+z)\,\delta x + (z+x)\,\delta y + (x+y)\,\delta z,$$

which parallels the corresponding statement involving differentials:

$$d(yz + zx + xy) = (y+z)\,dx + (z+x)\,dy + (x+y)\,dz.$$

In the same way as in Example 33.6, we have

$$I = \int_{(AB)} [(y+z)\,dx + (z+x)\,dy + (x+y)\,dz]$$

$$= \int_{(AB)} d(yz + zx + xy)$$

$$= (yz + zx + xy)_B - (yz + zx + xy)_A,$$

the suffices A and B meaning the values of the brackets at the end-points A and B. This is independent of the path.

In general, suppose that we can recognize the functions f, g, and h in the differential form

$$f(x, y, z)\,dx + g(x, y, z)\,dy + h(x, y, z)\,dz$$

as being expressible in terms of a single-valued function $S(x, y, z)$ in the following way:

$$f = \frac{\partial S}{\partial x}, \qquad g = \frac{\partial S}{\partial y}, \qquad h = \frac{\partial S}{\partial z}.$$

Then we can write $f\,dx + g\,dy + h\,dz$ as a perfect differential

$$f\,dx + g\,dy + h\,dz = \frac{\partial S}{\partial x}dx + \frac{\partial S}{\partial y}dy + \frac{\partial S}{\partial z}dz = dS.$$

This can be substituted into our integrals to yield

$$\int_{(AB)} (f\,dx + g\,dy + h\,dz) = \int_{(AB)} dS = S_B - S_A.$$

Provided that there is no ambiguity in the values to be assigned to S_B and S_A (this possibility is discussed in Section 33.10 below), this provides a way of evaluating the integral and possibly demonstrating path independence.

Evaluation of integrals over perfect differentials

If $f\,dx + g\,dy + h\,dz = dS$, and S is single-valued then

$$\int_{(AB)} (f\,dx + g\,dy + h\,dz) = S_B - S_A.$$

(33.9)

33.5 Closed paths

A closed path is one that returns to its starting point, so that B has the same coordinates as A, as in Figs 33.7a,b. We shall discuss only **simple closed paths**. These do not cross over themselves, as do the curves in Fig. 33.7b.

It is clear from the definition (33.7) that, when A and B are the same point, and the path is closed, their position on the curve will not affect the value of the integral. Consequently its coordinates are not usually stated; a closed path is indicated by a symbol such as C, and the integral is written

$$\int_C (f\,dx + g\,dy + h\,dz).$$

In three dimensions, the **direction** along C is specified by extra information, such as by an arrow on a sketch of the curve. However, in **two dimensions**, a convention operates: if it is not otherwise indicated, the **standard direction is anticlockwise**.

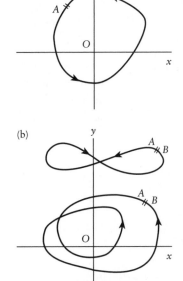

Fig. 33.7 (a) A simple closed path. (b) Closed paths which are not simple.

Example 33.8 *Evaluate* $I = \int_C (x\,dy - y\,dx),$ *where C is the ellipse* $x^2/a^2 + y^2/b^2 = 1$, *described in the standard direction.*

The ellipse can be parametrized for the anticlockwise direction by

$$x = a\cos t \quad \text{and} \quad y = b\sin t \quad (0 \leqslant t \leqslant 2\pi),$$

Example 33.8 *continued*

where we assume a and b to be positive. We had a choice for the range of t, because we can start at any point on the ellipse. For the choice 0 to 2π, the path starts and ends at $(a, 0)$. Then

$$\frac{\mathrm{d}x}{\mathrm{d}t} = -a \sin t, \qquad \frac{\mathrm{d}y}{\mathrm{d}t} = b \cos t;$$

so

$$I = \int_0^{2\pi} [a \cos t(b \cos t) - b \sin t(-a \sin t)] \, \mathrm{d}t$$

$$= ab \int_0^{2\pi} \mathrm{d}t = 2\pi ab.$$

The following result is sometimes useful:

A criterion for general path independence

If $\displaystyle\int_C (f \, \mathrm{d}x + g \, \mathrm{d}y + h \, \mathrm{d}z) = 0$ for every closed curve C,

then $\displaystyle\int_{(AB)} (f \, \mathrm{d}x + g \, \mathrm{d}y + h \, \mathrm{d}z)$ is path independent for

every A and B.

(33.10)

(a)
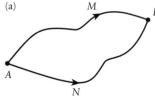

(b)
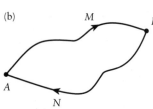

Fig. 33.8

To prove this, see Fig. 33.8a. Here A and B are *any* two points. While (AMB) and (ANB) are *any* two paths from A to B. If we reverse the direction of the path (ANB) we have Fig. 33.8b, which is a closed curve C. Suppose that we know the integral around every closed curve to be zero. Then

$$0 = \int_{(AMBNA)} (f \, \mathrm{d}x + g \, \mathrm{d}y + h \, \mathrm{d}z)$$

$$= \int_{(AMB)} (f \, \mathrm{d}x + g \, \mathrm{d}y + h \, \mathrm{d}z) + \int_{(BNA)} (f \, \mathrm{d}x + g \, \mathrm{d}y + h \, \mathrm{d}z)$$

$$= \int_{(AMB)} (f \, \mathrm{d}x + g \, \mathrm{d}y + h \, \mathrm{d}z) - \int_{(ANB)} (f \, \mathrm{d}x + g \, \mathrm{d}y + h \, \mathrm{d}z).$$

Therefore the integrals along (AMB) and (ANC) are equal.

Path independence for *any two particular points* is sufficient to ensure path independence between *all pairs of points*:

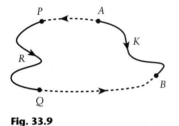

Fig. 33.9

The fixed points A and B are shown on Fig. 33.9, and (P, Q) is any other pair of points. (AKB), (AP), (QB), and (PRQ) are arbitrary paths joining the points specified by their brackets. Since $I_{(AB)}$ is independent of the path joint A and B,

$$I_{(APRQB)} = I_{(AKB)},$$

so

$$I_{(AP)} + I_{(PRQ)} + I_{(QB)} = I_{(AKB)},$$

or

$$I_{(PRQ)} = I_{(AKB)} - I_{(AP)} - I_{(QB)}.$$

But the right-hand side does not depend on which path was chosen for (PQ). Therefore $I_{(PQ)}$ is independent of the path joining P and Q, which proves the result.

33.6 Green's theorem

The following theorem connects a **two-dimensional line integral around a closed curve** C with a certain **double integral over the region** $\mathcal{A}$ **enclosed by** C, as shown in Fig. 33.10. The functions $P(x, y)$ and $Q(x, y)$ which occur are assumed to be 'smooth' in the region considered. 'Smoothness' has a technical meaning, that all the first derivatives of P and Q are continuous, but it will be enough for us to say that P and Q must have no undefined values, jumps or infinities on C and its interior $\mathcal{A}$ taken together. The theorem is:

Although the result is true in general, we shall prove it only for a curve like that in Fig. 33.10a, for which lines parallel to the axes cut the curve in at most two points.

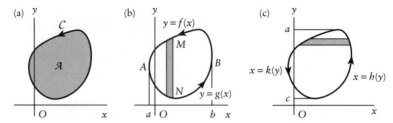

Fig. 33.10 (a) The diagram for Green's theorem. (b) For the integration of $\partial P/\partial y$. (c) For the integration of $\partial Q/\partial x$.

Consider $\displaystyle\iint_{\mathcal{A}} \frac{\partial P}{\partial y}\,\mathrm{d}\mathcal{A}$. We shall integrate it by vertical strips as in Fig. 33.10b. Suppose that the top part AMB of C, between $x = a$ and b, has the equation $y = f(x)$, and the lower part ANB is $y = g(x)$. Then

$$\iint_{\mathcal{A}} \frac{\partial P}{\partial y}\,\mathrm{d}\mathcal{A} = \int_{a}^{b} \int_{g(x)}^{f(x)} \frac{\partial P}{\partial y}\,\mathrm{d}y\,\mathrm{d}x$$

$$= \int_{a}^{b} [P(x, f(x)) - P(x, g(x))]\,\mathrm{d}x$$

$$= \int_{(AMB)} P(x, y)\,\mathrm{d}x - \int_{(ANB)} P(x, y)\,\mathrm{d}x$$

$$= -\int_{(BMA)} P(x, y)\,\mathrm{d}x - \int_{(ANB)} P(x, y)\,\mathrm{d}x$$

$$= -\int_{C} P(x, y)\,\mathrm{d}x. \tag{i}$$

Similarly, but by using horizontal strips as in Fig. 33.10c,

$$\iint_{\mathcal{A}} \frac{\partial Q}{\partial x}\,\mathrm{d}\mathcal{A} = \int_{c}^{d} \int_{k(y)}^{h(y)} \frac{\partial Q}{\partial x}\,\mathrm{d}x\,\mathrm{d}y = \int_{C} Q\,\mathrm{d}y. \tag{ii}$$

By subtracting (i) and (ii) we obtain the result required.

Example 33.9   *Show that if C is a simple closed curve, the geometrical area it encloses is equal to $\frac{1}{2}\int_{C}(x\,\mathrm{d}y - y\,\mathrm{d}x)$.*

Put $P = -y$ and $Q = x$ in Green's theorem (33.2):

$$\frac{1}{2}\int_{C}(-y\,\mathrm{d}x + x\,\mathrm{d}y) = \frac{1}{2}\iint_{\mathcal{A}}\left(\frac{\partial}{\partial x}(x) - \frac{\partial}{\partial y}(-y)\right)\mathrm{d}\mathcal{A} = \iint_{\mathcal{A}} \mathrm{d}\mathcal{A},$$

which is the geometrical area enclosed.

Green's theorem (33.12) enables us to produce a criterion by which path independence can be recognized:

> **A condition for path independence**
> If $\partial P/\partial y = \partial Q/\partial x$, then for any points A and B, $\int_{(AB)} (P \, dx + Q \, dy)$ is independent of the path (AB).
>
> **(33.13)**

To prove this, let C be any closed curve. Its interior is denoted by $\mathcal{A}$. Then, by Green's theorem,

$$\int_C (P \, dx + Q \, dy) = \iint_{\mathcal{A}} \left(\frac{\partial Q}{\partial x} - \frac{\partial P}{\partial y} \right) d\mathcal{A} = 0.$$

Therefore, by (33.10), we have path independence.

33.7 Line integrals and work

A particle follows a certain path (AB) in three-dimensional space under the action of various forces and its own inertia. Consider *one* of the forces, $F(x, y, z)$, which might be the contribution of a force field such as gravity, or a point force such as friction or the tension in a string. F is a **vector**, and it does not necessarily point along the path of the particle.

The path can be parametrized by using t, the time, as the parameter, and its position specified briefly by its position vector $r(t)$:

$$(x(t), y(t), z(t)) = r(t)$$

as in Fig. 33.11a. In a time interval δt, the particle moves from P to Q, and $r(t)$ changes to $r(t + \delta t)$, the change being denoted by δr. Figure 33.11a also shows the force F acting on the particle when it is at P. During the interval δt, the work δW done by F *alone* on the particle is approximated by

$$\delta W = (\text{component of } F \text{ in direction } PQ) \times (\text{distance } PQ)$$
$$= (|F| \cos \theta)|\delta r| = F \cdot \delta r.$$

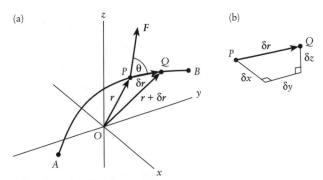

Fig. 33.11

The total work $W_{(AB)}$ done on the particle by F along the path (AB) is given by

$$W_{(AB)} = \sum_{(AB)} \delta W \approx \sum_{(AB)} F \cdot \delta r.$$

When the step length goes to zero the sum can be written as an integral:

$$W_{(AB)} = \int_{(AB)} F \cdot dr.$$

This integral has an ordinary meaning when it is written in (dx, dy, dz) form by splitting F and δr into their components (see Fig. 33.11b):

$$F = F_1 \hat{\imath} + F_2 \hat{\jmath} + F_3 \hat{k} \quad \text{and} \quad \delta r = \delta x\,\hat{\imath} + \delta y\,\hat{\jmath} + \delta z\,\hat{k}.$$

Then

$$F \cdot \delta r = F_1\,\delta x + F_2\,\delta y + F_3\,\delta z,$$

and

$$W_{(AB)} \approx \sum_{(AB)} (F_1\,\delta x + F_2\,\delta y + F_3\,\delta z).$$

Finally, taking the limit as δx, δy, δz approach zero, we obtain the exact result.

Work done by a force F along a path (AB)

$$W_{(AB)} = \int_{(AB)} F \cdot dr = \int_{(AB)} (F_1\,dx + F_2\,dy + F_3\,dz).$$

(For two dimensions, suppress z.)

(33.14)

Example 33.10   *A field of force F is constant everywhere. Show that the work done by F alone on a particle which moves from a fixed point A to a fixed point B is independent of the path followed.*

Put $F = a\hat{\imath} + b\hat{\jmath} + c\hat{k}$ where a, b, c are constants. Then, by (33.14), $W_{(AB)}$ is given by

$$W_{(AB)} = \int_{(AB)} (a\,dx + b\,dy + c\,dz) = \int_{(AB)} d(ax + by + cz)$$

$$= (ax + by + cz)_B - (ax + by + cz)_A,$$

which is a quantity independent of the path followed.

Example 33.11 (a) r is a position vector in axes x, y, z, and $r = |\mathbf{r}|$. Show that, in terms of differentials,

$$d(r^{-1}) = -(x/r^3)\,dx - (y/r^3)\,dy - (z/r^3)\,dz.$$

(b) The gravitational force $\mathbf{F}$ of the earth acting on a particle of mass m at a distance r from the earth's centre O is equal to $-m\gamma\mathbf{r}/r^3$ (γ constant; $\mathbf{F}$ is directed towards O and has magnitude $m\gamma/r^2$). Show that the work done by $\mathbf{F}$ when the particle moves between any two points A and B is equal to $m\gamma(r_B^{-1} - r_A^{-1})$ (it is path independent).

(a) $r = (x^2 + y^2 + z^2)^{\frac{1}{2}}$, so by (31.1)

$$\delta(r^{-1}) \approx \frac{\partial}{\partial x}(r^{-1})\,\delta x + \frac{\partial}{\partial y}(r^{-1})\,\delta y + \frac{\partial}{\partial z}(r^{-1})\,\delta z$$

$$= -\frac{x}{(x^2+y^2+z^2)^{\frac{3}{2}}}\,\delta x - \frac{y}{(x^2+y^2+z^2)^{\frac{3}{2}}}\,\delta y - \frac{z}{(x^2+y^2+z^2)^{\frac{3}{2}}}\,\delta z$$

$$= -(x/r^3)\,\delta x - (y/r^3)\,\delta y - (z/r^3)\,\delta z.$$

The corresponding differential relation is

$$d(r^{-1}) = -(x/r^3)\,dx - (y/r^3)\,dy - (z/r^3)\,dz.$$

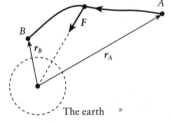

Fig. 33.12

(b) Refer to Fig. 33.12 and (33.14). The components of $\mathbf{F}$ are $(F_1, F_2, F_3) = (-m\gamma x/r^3, -m\gamma y/r^3, -m\gamma z/r^3)$, obtained by splitting $\mathbf{r}$ into its components. Therefore the work $W_{(AB)}$ done is

$$W_{(AB)} = \int_{(AB)} (F_1\,dx + F_2\,dy + F_3\,dz)$$

$$= m\gamma \int_{(AB)} \left(-\frac{x}{r^3}\,dx - \frac{y}{r^3}\,dy - \frac{z}{r^3}\,dz\right)$$

$$= m\gamma \int_{(AB)} d(r^{-1}) \quad \text{(from (a))}$$

$$= m\gamma(r_B^{-1} - r_A^{-1}).$$

Examples 33.10 and 33.11 illustrate cases where the work done by a force between two fixed points is independent of the path between them, but this is not a universal state of affairs: for example, it is not the case for the force field $(y, -x, 0)$.

33.8 Conservative fields

We have spoken in a general way of a field of force and its action on a particle. By a **particle** we mean an object small enough for its exact shape, physical constitution, state of rotation, and so on to be unimportant on the scale of the problem being considered; it behaves in the way we imagine a point should behave.

However, the magnitude of the force exerted upon it by gravity, electrostatic influence, etc., will still depend on the mass or charge assigned to the particle. We need a way to specify the strength of the force field itself, a **field intensity**, which is independent of what particle we put into it. When the field strength is specified, we should be able to deduce its effect on any particle.

This is not always quite straightforward, because the introduction of a new particle into (say) an electrostatic field might change the distribution of charge that constitutes the **source of the field**, so that in effect we would be putting the particle into a modified situation. The case is similar with gravity: if an asteroid enters the moon's gravitational field, the moon will respond by moving, and the field entered will change, if only by a little. For the purpose of defining field intensity, we imagine that somehow such an effect is prevented from taking place. Subject to this, we have the following definition.

> **Field intensity f_P at P**
>
> f_P is equal to the vector force that would act on a particle of unit mass (charge, etc.) at P if the sources are assumed to be unaffected by the particle.
>
> **(33.15)**

Therefore, if the gravitational field intensity is G_P at P, the force with which the field acts on a particle of mass m at P is mG_P. One can alternatively imagine a particle of extremely small mass μ to be introduced as a test particle. Then f_P will be equal to μ^{-1} times the force exerted on such a particle.

Consider the action of a field of intensity $f(x, y, z)$ on a **unit** particle which is travelling on a path (AB) (Fig. 33.13). We shall consider not the work done *by f on* the particle, but the **work done against the field by the particle**, which has the opposite sign. Denote this quantity generally by v. The work done against f in a step PQ is given by

$$\delta v \approx -f \cdot \delta r. \tag{33.16}$$

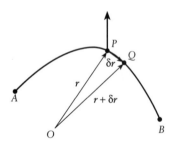

Fig. 33.13

The total work along the path is the limit of the sum of the δv, which can be expressed as a line integral as before:

$$v_{(AB)} = -\int_{(AB)} f \cdot dr = -\int_{(AB)} (f_1 \, dx + f_2 \, dy + f_3 \, dz),$$

where $f = (f_1, f_2, f_3)$.

The important case is when $v_{(AB)}$ is **independent of the path from A to B**, in which case the **field is said to be a conservative field**. In practice our 'field' will not usually consist of the whole of space; some space will be occupied by impenetrable bodies, or we might be interested only in the region $\mathcal{R}$ inside a metal cage. According to (33.11) it is only necessary to check path independence for any single pair of points (A, B) within this region. Therefore:

> **Conservative field in a region $\mathcal{R}$**
>
> Let A and B be two given points in $\mathcal{R}$. Then $f(x, y, z)$ is conservative in $\mathcal{R}$ if $v_{(AB)} = -\int_{(AB)} f \cdot dr$ is independent of the path in $\mathcal{R}$ from A to B. (Or equivalently if $\int_C f \cdot dr$ is zero for every closed path C in $\mathcal{R}$: see (33.10).)
>
> (33.17)

The constant field of Example 33.10 and the gravitation field of Example 33.11 are conservative.

33.9 Potential for a conservative field

Suppose that $f(x, y, z)$ is *conservative* in a region $\mathcal{R}$, and that A is a fixed point in $\mathcal{R}$. Since f is conservative, the integral

$$v_{(AP)} = -\int_{(AP)} f \cdot dr,$$

where P is another point in $\mathcal{R}$, is independent of the path (AP), and so its value depends only on the location (x, y, z) of P. Therefore we shall write

$$v_{(AP)} = V(x, y, z) \quad \text{or} \quad V_P, \tag{33.18}$$

in which we have suppressed the coordinates of A since they are constant.

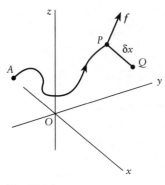

Fig. 33.14

In Fig. 33.14, suppose that (AP) is a fixed reference path from A to $P : (x, y, z)$. Let $Q : (x + \delta x, y, z)$ be a point close to P, displaced from it a distance δx in the x direction only. Choose a path (AQ) consisting of two parts: the selected path (AP) and a *straight line extension* (PQ) from P to Q. The choice of these paths rather than any others does not affect the values of V_P and V_Q since the field is conservative. Then, from (33.18),

$$v_{(AQ)} - v_{(AP)} = V_Q - V_P = V(x + \delta x, y, z) - V(x, y, z).$$

But also, from (33.16), $\delta v = -f \cdot \delta r$ with $\delta y = \delta z = 0$; so

$$v_{(AQ)} - v_{(AP)} \approx -f_1 \, \delta x,$$

where f_1 is the x component of f. Equating the last two results and dividing by δx, we obtain

$$f_1 = -[V(x + \delta x, y, z) - V(x, y, z)]/\delta x.$$

When $\delta x \to 0$, this becomes

$$f_1 = -\frac{\partial V}{\partial x},$$

and similarly

$$f_2 = -\frac{\partial V}{\partial y} \quad \text{and} \quad f_3 = -\frac{\partial V}{\partial z}.$$

Therefore

$$f = f_1 \hat{i} + f_2 \hat{j} + f_3 \hat{k}$$
$$= -\left(\frac{\partial V}{\partial x} \hat{i} + \frac{\partial V}{\partial y} \hat{j} + \frac{\partial V}{\partial z} \hat{k} \right),$$

or

$$f = -\text{grad } V. \tag{33.19}$$

We call V a **potential function for the field** f, or simply a **potential**. The single scalar function $V(x, y, z)$ contains all the information necessary to define the *three* scalar components of f: $f_1(x, y, z)$, $f_2(x, y, z), f_3(x, y, z)$. The point A is commonly taken to be at infinity: you might recognize the idea of 'the work required to bring a particle in from infinity' in mechanics. However, if we choose a different reference point A, it only changes V by an additive constant, and does not, therefore, affect the truth of (33.19); we get the same f whatever location A has. We sum up this result as follows.

> **Potential V of a conservative field f**
>
> If $f(x, y, z)$ is conservative in a region $\mathcal{R}$, then
>
> $$f = -\text{grad } V$$
>
> in $\mathcal{R}$, where V is a scalar potential function for f. Also V is defined in the region $\mathcal{R}$ by
>
> $$V_P = -\int_{(AP)} f \cdot d\mathbf{r},$$
>
> where A is a fixed point. $\tag{33.20}$

As an example of a potential, the gravitational field from a particle of mass M, namely $f = -M\gamma \mathbf{r}/r^3$, has the potential $V = -M\gamma/r$. This can be checked from the working of Example 33.10. The potential function V is equal to the work done against the field in moving a unit particle from a fixed point A to the current point P in cases when the field is conservative. Therefore the **potential energy** of a particle of mass m, relative to the reference point A at P, is equal to mV. Alternatively, V can be regarded as energy stored by the gravitational field, like energy stored in a spring.

33.10 Single-valuedness of potentials

There is a connection between the question of single-valuedness in a perfect differential and the conservative property of a force field. There exist fields $f(x, y, z)$ which have a potential, but are not conservative, because they do not satisfy the condition (33.17), that $\int_{(AB)} f \cdot d\mathbf{r}$ should be independent of path.

> **Potential field**
>
> If there is a scalar function V such that
>
> $f = -\text{grad } V$,
>
> then $f(x, y, z)$ is called a **potential field**. (33.21)

We can test whether such a field is conservative or not:

> **Condition for a potential field to be conservative**
>
> If $f = -\text{grad } V$, and V is single valued, then f is conservative. In this case, the work $v_{(AB)}$ in moving a unit particle from A to B against the field is equal to $V_B - V_A$. (33.22)

This is proved as follows:

$$v_{(AB)} = -\int_{(AB)} f \cdot dr = \int_{(AB)} (\text{grad } V) \cdot dr$$

$$= \int_{(AB)} \left(\hat{i} \frac{\partial V}{\partial x} + \hat{j} \frac{\partial V}{\partial y} + \hat{k} \frac{\partial V}{\partial z} \right) \cdot (\hat{i}\, dx + \hat{j}\, dy + \hat{k}\, dz)$$

$$= \int_{(AB)} \left(\frac{\partial V}{\partial x} dx + \frac{\partial V}{\partial y} dy + \frac{\partial V}{\partial z} dz \right) = \int_{(AB)} dV \quad \text{(see (31.1))}$$

$$= V_B - V_A.$$

Provided that $\int_{(AB)} dV$ is independent of the path from A to B, the value to be assigned to $V_B - V_A$ is unambiguous and we say that V is single valued. However, the values of V may depend not only on the position, *but also on the way in which the position was reached* (analogously to the time spent reaching a point on the other side of a road being dependent on whether you cross directly or via the underpass). For example, in the plane, let

$$V = \theta,$$

where θ is the polar angle traversed in reaching the current position, measured continuously from a given starting point. What do we mean by

$$\int_{(AB)} d\theta?$$

Figure 33.15 shows two paths from A to B: (ACB) goes from A to B more or less directly, and (ADB) circles the origin completely first. The definition of the integral is that

$$\int_{(AB)} d\theta = \lim_{\delta\theta \to 0} \sum_{(AB)} \delta\theta,$$

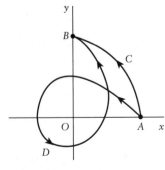

Fig. 33.15

where the summation is carried out by taking small steps along the path. On (ACB), θ passes smoothly from $\theta = 0$ to $\theta = \frac{1}{2}\pi$, so

$$\int_{(ACB)} dV = \int_{(ACB)} d\theta = \theta_B - \theta_A = \frac{1}{2}\pi.$$

But on (ADB), θ starts at $\theta_A = 0$ and increases smoothly through values $\frac{1}{2}\pi, \pi, \frac{3}{2}\pi, 2\pi$, to $\theta_B = \frac{5}{2}\pi$. Therefore

$$\int_{(ADB)} dV = \int_{(ADB)} d\theta = \theta_B - \theta_A = \frac{5}{2}\pi.$$

Therefore V in this case is path dependent. If the potential of a force field is given by

$$V = \theta,$$

where θ is the *traversed* polar angle, then the field is strictly not conservative: various paths from A to B involve different amounts of work by a unit particle moving in the field.

Example 33.12   *Show that the two-dimensional field*

$$f(x, y) = \frac{-y}{x^2 + y^2}\hat{\imath} + \frac{x}{x^2 + y^2}\hat{\jmath}$$

is not a conservative field.

Apart from physical constants this represents the circumferential magnetic field around a straight wire carrying a current, or the velocity field of a vortex. Put $r = x\hat{\imath} + y\hat{\jmath}$; then

$$f \cdot r = x\frac{-y}{x^2 + y^2} + y\frac{x}{x^2 + y^2} = 0.$$

Therefore the field is perpendicular to the radius vector at every point, as in Fig. 33.16. It is easy to confirm that

$$f = -\operatorname{grad} V,$$

where V is a (path-dependent) *continuous* function such that

$$\tan V = y/x.$$

Thus we may take $V = \theta$ as described in the case we just discussed. (We cannot write $V = \arctan y/x$, because this function is discontinuous across the y axis: it would have an infinite gradient there.) The figure makes it obvious that the field is not conservative: more work is done if you take a unit magnetic pole against the field 50 times around the origin in order to travel between two points than if you go directly.

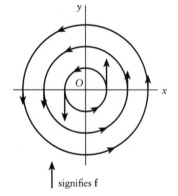

signifies f

Fig. 33.16

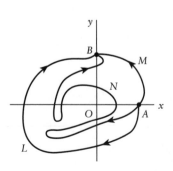

Fig. 33.17

The field in Example 33.12 is not conservative, but whole classes of paths *are* equivalent. Suppose that, as in Fig. 33.17, we have two paths, (AMB) and (ANB), which can be steadily deformed into each other

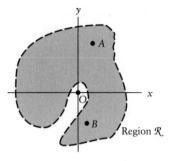

Fig. 33.18

(as if A and B were connected by a piece of elastic) *without passing over the origin*. Then these two paths are equivalent. In this case, θ starts at $\theta_A = 0$; although the value of θ wanders about on (ANB), increasing and decreasing, it still ends at the value $\theta_B = \frac{1}{2}\pi$, as on the path (AMB).

However, (AMB) *cannot* be deformed into the third path (ALB) *without passing over the origin*; by following it around, it can be seen that $\theta_B = -\frac{3}{2}\pi$ for this path.

Suppose that we confine consideration to a 'patch', or region $\mathcal{R}$ as in Fig. 33.18, which neither contains nor surrounds the origin O. Then, *within this region*, the field behaves as if it were conservative, because any path from A to B inside the region can be deformed into any other without crossing the origin. We could not tell, from experiments confined to $\mathcal{R}$, that the field is not conservative over the whole plane.

Problems

33.1 (Section 33.1). Evaluate the following line integrals where (AOB) is shown in Fig. 33.19a.

(a) $\displaystyle\int_{(AOB)} x \, dx$; (b) $\displaystyle\int_{(AOB)} y \, dx$; (c) $\displaystyle\int_{(AOB)} x^2 \, dx$.

33.2 Evaluate the following integrals; $\mathcal{P}$ represents the parabolic path (AOB) on $y^2 = x$, shown in Fig. 33.19b.

(a) $\displaystyle\int_{\mathcal{P}} x \, dx$; (b) $\displaystyle\int_{\mathcal{P}} y \, dx$; (c) $\displaystyle\int_{\mathcal{P}} x^2 \, dx$;

(d) $\displaystyle\int_{\mathcal{P}} (x+y) \, dy$; (e) $\displaystyle\int_{\mathcal{P}} xy^2 \, dy$; (f) $\displaystyle\int_{\mathcal{P}} (x \, dx + y \, dy)$;

(g) $\displaystyle\int_{\mathcal{P}} (\tfrac{1}{2} dx - y \, dy)$; (h) $\displaystyle\int_{\mathcal{P}} (y \, dx - x \, dy)$.

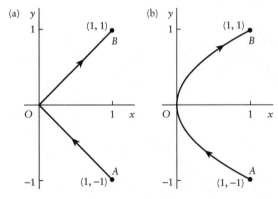

Fig. 33.19

33.3 (Section 33.2). Evaluate the following line integrals over the various paths $\mathcal{P}$, which are specified parametrically.

(a) $\displaystyle\int_{\mathcal{P}} xy^2 \, dx$; $\mathcal{P}$ is $x = t^2, y = t; 0 \leqslant t \leqslant 1$.

(b) $\displaystyle\int_{\mathcal{P}} (x \, dy - y \, dx)$; $\mathcal{P}$ is $x = \cos t, y = \sin t; 0 \leqslant t \leqslant \pi$.

(c) $\displaystyle\int_{\mathcal{P}} (z \, dx - x \, dy + y \, dz)$; $\mathcal{P}$ is $x = t + 1, y = t, z = 2t$;

$0 \leqslant t \leqslant 1$.

(d) $\displaystyle\int_{\mathcal{P}} (x^2 \, dx + y^2 \, dy + z^2 \, dz)$; $\mathcal{P}$ is $x = \cos t, y = \sin t, z = t$;

$0 \leqslant t \leqslant 2\pi$.

(e) Compare (c) when $\mathcal{P}$ joins the same two points, $(1, 0, 0)$ to $(2, 1, 2)$, but $x = t^2 + 1, y = 2t - t^2, z = 2t^2$; $0 \leqslant t \leqslant 1$.

33.4 (Section 33.2). The line integral $\int_{(AB)} f(x, y) \, dy$, where the path (AB) is described by the curve $y = k(x)$, can be written formally as

$$\int_{(AB)} f(x, k(x)) \frac{dk}{dx} \, dx.$$

Apply this formula to $\int_{(AB)} (x+y) \, dy$, taken over the parabolic path in Fig. 33.19b. Express it as the sum of two ordinary integrals over x. (This is like using x as the parameter in Section 33.1.)

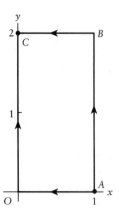

Fig. 33.20

33.5 (Section 33.3). The references are to Fig. 33.20.)

(a) $\int_{(ABC)} dx;$ (b) $\int_{(AOC)} dy;$

(c) $\int_{(ABC)} (x\,dy - y\,dx);$ (d) $\int_{(AOC)} (x\,dy - y\,dx);$

(e) $\int_{(ABC)} y\,dy;$ (f) $\int_{(AOC)} y\,dy;$

(g) $\int_{(ABC)} (y\,dx + x\,dy);$ (h) $\int_{(AOC)} (y\,dx + x\,dy).$

33.6 (Section 33.4). The integrands given are perfect differentials; $\mathcal{P}$ represents any path having the right direction which joins the two given points. Evaluate

(a) $\int_{\mathcal{P}} (x\,dx + y\,dy + z\,dz);$ $\mathcal{P}$ is $(-1, 1, -1)$ to $(1, -1, 1)$.

(b) $\int_{\mathcal{P}} (yz\,dx + zx\,dy + xy\,dz);$ $\mathcal{P}$ is $(0, 0, 0)$ to $(1, 1, 1)$.

(c) $\int_{\mathcal{P}} e^{x^2+y^2+z^2}(x\,dx + y\,dy + z\,dz);$ $\mathcal{P}$ is $(0, 0, 0)$ to $(1, 1, 1)$.

(d) $\int_{\mathcal{P}} [(y + z)\,dx + (z + x)\,dy + (x + y)\,dz];$ $\mathcal{P}$ is $(1, 1, 1)$ to $(0, 1, 0)$.

(e) $\int_{\mathcal{P}} [\cos(xy + yz + zx)]\,[(y + z)\,dx + (z + x)\,dy + (x + y)\,dz];$ P is $(1, 0, \pi)$ to $(0, \pi, 1)$.

(f) $\int_{\mathcal{P}} (xy^2\,dx + x^2y\,dy);$ $\mathcal{P}$ is $(1, 1)$ to $(2, 2)$.

33.7 (Section 33.5). Evaluate the following two-dimensional line integrals over the closed paths C given, the direction being anticlockwise.

(a) $\int_C (x^2\,dy - y^2\,dx);$ C is the circle $x^2 + y^2 = 4$.

(b) $\int_C \left(\dfrac{x}{y}\,dx + \dfrac{y}{x}\,dy\right);$ C is the ellipse $\frac{1}{4}x^2 + \frac{1}{9}y^2 = 1;$ use the parametrization $x = 2\cos\theta, y = 3\sin\theta$.

33.8 Evaluate the following (all the paths C are closed).

(a) $\int_C (y\,dx + z\,dy + x\,dz);$ C is $x = \sin t, y = \cos t, z = \sin t;$ $0 \leqslant t \leqslant 2\pi$.

(b) $\int_{(ABC)} (y\,dx + z\,dy + x\,dz);$ (ABC) is the triangle $A : (1, 0, 0), B : (0, 1, 0), C : (0, 0, 1)$.

(c) $\int_C (yz\,dx + zx\,dy + xy\,dz);$ C is any closed path.

33.9 Show that $\int_{(AB)} (yx^2\,dx + \frac{1}{3}x^3\,dy)$ is path independent between any two points A and B. Use this fact to evaluate the integral along the spiral path given in polar coordinates (r, θ) by $r = e^{\theta/2\pi}$ for $0 \leqslant \theta \leqslant \pi$.

33.10 Show that if $\int_{(AB)} (f\,dx + g\,dy)$ is independent of the path (AB) for every two points A and B, then the integral around every closed path is zero. (Hint: A and B may coincide.)

33.11 Show that if the variables are changed in a perfect differential form, it remains a perfect differential. Illustrate this by transforming the identity $y\,dx + dy = d(xy)$ into polar coordinates.

33.12 (Green's theorem, Section 33.6). Confirm the truth of Green's theorem (33.12) for some very simple cases for which you know you can work out both the line integral and the double integral involved.

33.13 (Green's theorem, Section 33.6). Check the correctness of the area formula, Example 33.9, by evaluating the line integral $\frac{1}{2}\int_C(x\,dy - y\,dx)$ taken around the following closed paths.
(a) The circle $x^2 + y^2 = 4$.
(b) The ellipse $\frac{1}{4}x^2 + \frac{1}{9}y^2 = 1$.
(c) The triangle with vertices $(-1, 0), (2, 0), (0, 4)$.

33.14 Find the area of the star-shaped region bounded by the curve $x^{\frac{2}{3}} + y^{\frac{2}{3}} = 1$, by parametrizing its equation as in Example 33.9.

33.15 The gravitation force F arising from a particle of mass M at the origin upon a particle of mass m at a point with position vector r is given by $F = -\gamma Mmr/r^3$. Find the work done by F on a particle which travels in from infinity to r.

33.16 Use Green's theorem with (33.10) to decide whether the following represent conservative fields (in two dimensions) or not in the stated regions.
(a) $(x^2 - y^2, 2xy)$; all x, y.
(b) $(\frac{1}{2} \ln(x^2 + y^2), \arctan(y/x))$; $x > 0$.

33.17 A force field has field intensity $f(x, y, z) = y\hat{\imath} + \hat{\jmath} + x\hat{k}$. Is f conservative? Find the work done against the field by a unit particle moving in a straight line from $(0, 0, 0)$ to $(1, 1, 1)$.

33.18 A force f is given by $f(x, y, z) = yz\hat{\imath} + xz\hat{\jmath} + xy\hat{k}$. Show that it is conservative. Find the work done against f along the path $x = \cos t$, $y = \sin t$, $z = \sin t \cos t$; $-\frac{1}{2} \le t \le \frac{1}{2}\pi$. Are you doing this the easiest way?

33.19 Prove that a force field f having the form $f = r^\alpha \hat{r}$, where α is any constant, r is distance from the origin, $\hat{r}$ is the *unit* position vector, and $\hat{r} = r/r$, is a conservative field. (Hint: start by putting $r = (x^2 + y^2 + z^2)^{\frac{1}{2}}$, and guess something that f might be the gradient of. If you cannot guess, then use the fact that grad $F(r) = \hat{r}(\partial F/\partial r)$.)

33.20 Generalize Problem 33.19 to a field $f = \hat{r}f(r)$. What is the potential of such a field?

33.21 Confirm that Green's theorem still holds for boundary C of the annular region A between the circles $x^2 + y^2 = 1$ and $x^2 + y^2 = 4$ for the line integral

$$\int_C [(2x - y^3) \, dx - xy \, dy].$$

What are the directions on C?

33.22 Show that $\int_C (5x^4y \, dx + x^5 \, dy) = 0$ holds for any closed curve C for which Green's theorem is true.

33.23 Sketch the curve given parametrically by

$$x = \cos t - \tfrac{1}{2} \sin 2t, \ y = \sin t; \ 0 \le t < 2\pi.$$

Using Green's theorem, find the area enclosed by the curve.

34 Vector fields: divergence and curl

CONTENTS

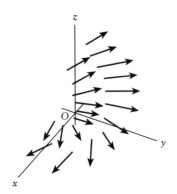

Fig. 34.1 Vector field.

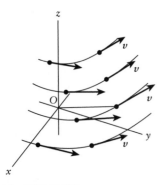

Fig. 34.2 Field lines.

34.1 Vector fields and field lines

Vector fields in two dimensions have already been encountered in Section 29.6. A vector field in three dimensions extends this notion to a vector with three components which are functions of position in space. In terms of cartesian components a vector field $F(x, y, z)$ will have the form

$$F(x, y, z) = F_1(x, y, z)\hat{\imath} + F_2(x, y, z)\hat{\jmath} + F_3(x, y, z)\hat{k}.$$

Vector fields abound in physical and engineering applications. Fluid velocity, gravitational forces, magnetic and electric fields are examples of vector fields. In time-varying applications the vector field and its components will also depend on a fourth variable, namely time, but here we shall concentrate only on the position variables.

At each point where the vector field is defined we can draw a vector. Figure 34.1 shows a region with a sample of local vectors drawn. Generally their magnitudes and directions will vary from point to point.

Assuming that the components of the vector field are smooth functions, we can associate with the vector field, **field lines** or **integral curves**, which are such that the vector field at any point is always tangential to a field line (Fig. 34.2). (The streamlines in Fig. 29.8 are field lines for a two-dimensional velocity field.) Suppose that a particular field line is given by the position vector $r = r(t)$, where t is any suitable parameter. Then its tangent is in the direction of dr/dt (see eqn (9.18)) which must be in the same direction as F:

$$\frac{dr}{dt} = \mu(t)F(x, y, z)$$

where $\mu(t)$ is some scalar function of the parameter t. Hence in component form

$$\frac{dx}{dt} = \mu(t)F_1(x, y, z), \quad \frac{dy}{dt} = \mu(t)F_2(x, y, z), \quad \frac{dz}{dt} = \mu(t)F_3(x, y, z).$$

Elimination of the unknown $\mu(t)$ leads to:

Equations for field lines

$$\frac{dx}{F_1(x, y, z)} = \frac{dy}{F_2(x, y, z)} = \frac{dz}{F_3(x, y, z)},$$

(34.1)

---

which are two simultaneous differential equations for x, y, and z in differential form (see Section 22.4). The solution is not always easy, but here is an example of a vector field whose field lines can be found.

Example 34.1 *Find the field lines of the vector field*

$$F = xy^2z\hat{\imath} + xz\hat{\jmath} + x\hat{k}.$$

Equation (34.1) becomes

$$\frac{dx}{xy^2z} = \frac{dy}{xz} = \frac{dz}{x},$$

which is equivalent to the two differential equations

$$\frac{dx}{dy} = y^2, \quad \frac{dy}{dz} = z,$$

which are both *separable* differential equations. Hence

$$\int dx = \int y^2 \, dy, \quad \text{or} \quad x = \tfrac{1}{3}y^3 + C_1,$$

(34.2)

and

$$\int dy = \int z \, dz, \quad \text{or} \quad y = \tfrac{1}{2}z^2 + C_2.$$

(34.3)

Equations (34.2) and (34.3) are two families of surfaces (both are cylindrical) and their curves of intersections are the field lines of F.

34.2 Divergence of a vector field

The **divergence** of a vector field F, denoted by div F, is a *scalar field* defined by:

Divergence of a vector field

$$\operatorname{div} F = \nabla \cdot F = \frac{\partial F_1}{\partial x} + \frac{\partial F_2}{\partial y} + \frac{\partial F_3}{\partial z}.$$

(34.4)

The notation $\boldsymbol{\nabla} \cdot \boldsymbol{F}$ emphasizes the del operator (Section 29.6)

$$\boldsymbol{\nabla} = \hat{\imath}\frac{\partial}{\partial x} + \hat{\jmath}\frac{\partial}{\partial y} + \hat{k}\frac{\partial}{\partial z}$$

again. Here $\boldsymbol{\nabla} \cdot \boldsymbol{F}$ is the 'scalar product' of the operator and the vector field.

Example 34.2 *Find the divergence of*

$\boldsymbol{F} = \sin(xy)\hat{\imath} + y\cos(z)\hat{\jmath} + xz\cos(z)\hat{k}.$

From the definition above

$$\text{div } \boldsymbol{F} = \frac{\partial}{\partial x}(\sin(xy)) + \frac{\partial}{\partial y}(y\cos z) + \frac{\partial}{\partial z}(zx\cos z)$$

$$= y\cos(xy) + (1+x)\cos z - zx\sin z.$$

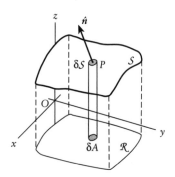

Fig. 34.3

34.3 Surface and volume integrals

Let S be a surface (Fig. 34.3), and let δS be an element of area on S. Suppose that $f(x, y, z)$ is a given function. The **surface integral** of $f(x, y, z)$ over the surface S, written as

$$\iint_S f(x, y, z)\, \mathrm{d}S$$

is the limit

$$\lim_{\delta S \to 0} \sum_S f(x, y, z)\, \delta S.$$

It is evaluated as follows.

Let the projection of S on to the (x, y) plane be $\mathcal{R}$, and let δA be the projection of the element δS. (We assume that any line parallel to the z axis cuts S in at most one point.) The element δA could be the rectangular element having area $\delta x\, \delta y$, in which case, for small δx and δy, δS would be approximately a parallelogram on the tangent plane at a point P within δS.

The relation between δS and δA in Fig. 34.3 depends on the unit normal $\hat{n}$ at P. Consider a vertical plane through P containing the vector $\hat{n}$ and $\hat{k}$ as shown in Fig. 34.4. Let θ be the smaller angle between $\hat{n}$ and $\hat{k}$, that is $0 \leqslant \theta \leqslant 180°$.

Then the length of any line element in δS perpendicular to the plane is unaltered by the projection, but line elements in δS lying in the plane are changed in length by projection by a factor $\cos\theta = |\hat{n} \cdot \hat{k}|$. Hence

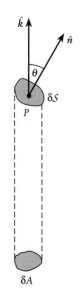

Fig. 34.4

$$\delta A = |\hat{n} \cdot \hat{k}|\, \delta S.$$

Thus

$$\iint_S f(x, y, z) \, dS = \iint_R f(x, y, z) \frac{dA}{|\hat{n} \cdot \hat{k}|} \tag{34.5}$$

which can be used as a definition of the surface integral.

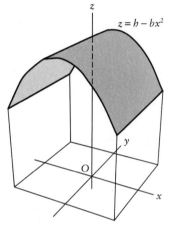

Fig. 34.5

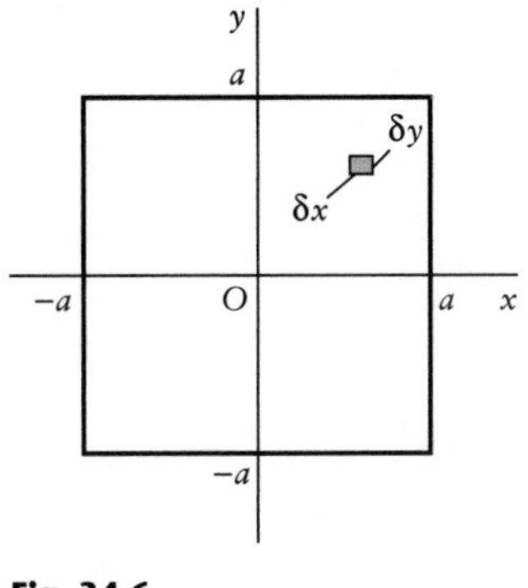

Fig. 34.6

Example 34.3 *The roof of a building has the cylindrical shape $z = h - bx^2$ over a square floor plan given by $|x| \leqslant a, |y| \leqslant a$, where $h > 2a^2b$ (see Fig. 34.5). Find the surface area of the roof.*

The surface area is given by

$$S = \iint_S \delta S.$$

In this case we use cartesian coordinates to define the element δA, which is the rectangle with sides parallel to the axes with lengths δx and δy. Thus $\delta A = \delta x \, \delta y$. The integration takes place over the square $|x| \leqslant a, |y| \leqslant a$ in the (x, y) plane (Fig. 34.6). We also require the *unit* normal $\hat{n}$. By (28.7) the unit normal will be

$$\hat{n} = \frac{(-2bx, 0, -1)}{\sqrt{4b^2x^2 + 1}}.$$

Hence

$$|\hat{n} \cdot \hat{k}| = \frac{1}{\sqrt{4b^2x^2 + 1}},$$

and by (34.5)

$$S = \int_{-a}^{a} \int_{-a}^{a} \sqrt{4b^2x^2 + 1} \, dx \, dy.$$

The repeated integral is *separable* (Section 32.7). Hence

$$S = \int_{-a}^{a} \sqrt{4b^2x^2 + 1} \, dx \int_{-a}^{a} dy$$

$$= 2a \int_{-a}^{a} \sqrt{4b^2x^2 + 1} \, dx.$$

The remaining integral can be evaluated using the substitution $x = (\sinh u)/(2b)$. The result is

$$S = \frac{a}{b}[2ab\sqrt{4a^2b^2 + 1} + \sinh^{-1}(2ab)].$$

If the surface S is given by $z = f(x, y)$, we can obtain a general cartesian formula for the surface area. A vector in the direction of the normal at any point on the surface is given (see Section 28.5) by

$$n = \left(-\frac{\partial f}{\partial x}, -\frac{\partial f}{\partial y}, 1\right)$$

where we have chosen n to be in the direction in which its $\hat{k}$ component is positive. This ensures a positive value for the area. A *unit* vector in the direction of n is

$$\hat{n} = \left(-\frac{\partial f}{\partial x}, -\frac{\partial f}{\partial y}, 1 \right) \Bigg/ \sqrt{\left[1 + \left(\frac{\partial f}{\partial x} \right)^2 + \left(\frac{\partial f}{\partial y} \right)^2 \right]}.$$

Hence

$$\hat{k} \cdot \hat{n} = 1 \Bigg/ \sqrt{\left[1 + \left(\frac{\partial f}{\partial x} \right)^2 + \left(\frac{\partial f}{\partial y} \right)^2 \right]},$$

and the surface area of S is therefore given by

$$\iint_S dS = \iint_{\mathcal{R}} \sqrt{\left[1 + \left(\frac{\partial f}{\partial x} \right)^2 + \left(\frac{\partial f}{\partial y} \right)^2 \right]} \, dx \, dy,$$

where $\mathcal{R}$ is the projection of S on to the (x, y) plane.

A **surface** in three dimensions is a two-dimensional object, which means that it can be represented by a position vector which is a **function of *two* parameters**. Remember that for a *curve* in three dimensions, the position vector is a function of a *single* parameter. Unlike the cartesian form $z = f(x, y)$, parametric equations enable the creation of much more complicated surfaces.

Parametric form of a surface

A surface can be represented by a position vector r as a function of two parameters u and v in the form

$$r(u, v) = x(u, v)\hat{i} + y(u, v)\hat{j} + z(u, v)\hat{k},$$

where $a \leqslant u \leqslant b, c \leqslant v \leqslant d$.

(34.6)

The parameters u and v are defined over a rectangle in the (u, v) plane. For example, for the surface

$$r = a \cos u \sin v \, \hat{i} + a \sin u \sin v \, \hat{j} + a \cos v \, \hat{k},$$

we can see that

$$|r| = \sqrt{[a^2 \cos^2 u \sin^2 v + a^2 \sin^2 u \sin^2 v + a^2 \cos^2 v]}$$
$$= \sqrt{[a^2(\cos^2 u + \sin^2 u) \sin^2 v + a^2 \cos^2 v]}$$
$$= a\sqrt{[\sin^2 v + \cos^2 v]} = a,$$

which means that the position vector r traces out a sphere of radius a, centre at the origin. We need to specify u and v to determine which part of the sphere is defined. For the whole surface, the parameters must range over the intervals $0 \leqslant u \leqslant 2\pi$ and $0 \leqslant v \leqslant \pi$.

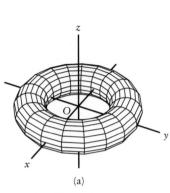
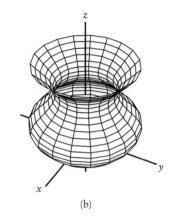

(a)   (b)

Fig. 34.7 (a) Torus; (b) a vase.

More complicated surfaces can be generated in this way, and their graphical representation has become easier using symbolic computer software (see Chapter 42, projects for this chapter). For example, the position vector r defined by

$$r = (3 + \cos v) \cos u\,\hat{\imath} + (3 + \cos v) \sin u\,\hat{\jmath} + \sin v\,\hat{k}$$

generates a **torus** (like the shape of a doughnut) with its axis in the $\hat{k}$ direction (Fig. 34.7a). The vase-shaped surface in Fig. 34.7b is generated by

$$r = (1 + a \sin bu) \cos v\,\hat{\imath} + (1 + a \sin bu) \sin v\,\hat{\jmath} + u\,\hat{k},$$

where $a = 0.3$, $b = 3.5$ for $0 \leqslant u \leqslant 2$ and $0 \leqslant v < 2\pi$.

Triple integrals or **volume integrals** can also be defined in vector calculus. By analogy with the double integral, the triple integral of a function f (either a scalar or vector field) over a three-dimensional region $\mathcal{V}$ is

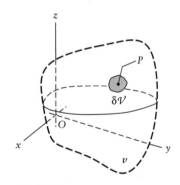

Fig. 34.8

$$I = \iiint_{\mathcal{V}} f(P)\,\mathrm{d}\mathcal{V},$$

where $\delta\mathcal{V}$ is an increment of volume and P is a point in $\delta\mathcal{V}$ (see Fig. 34.8). Its evaluation requires it to be converted into a repeated integral with three integrations.

Example 34.4 *A cube of metal occupying $|x| \leqslant a$, $|y| \leqslant a$, $|z| \leqslant a$ has a radial density distribution given by*

$$\rho(x, y, z) = \alpha + \beta(x^2 + y^2 + z^2),$$

where α and β are positive constants. Find the mass of the cube.

We choose a rectangular grid with volume element $\delta\mathcal{V} = \delta x\,\delta y\,\delta z$ (Fig. 34.9). The mass of this element is approximately

$$\rho\,\delta x\,\delta y\,\delta z = [\alpha + \beta(x^2 + y^2 + z^2)]\,\delta x\,\delta y\,\delta z.$$

↗

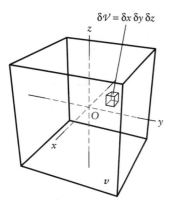

$$\delta \mathcal{V} = \delta x \, \delta y \, \delta z$$

Fig. 34.9

Example 34.4 *continued*

The total mass is therefore the *sum* or *integral* of these elements within the cube. The integral, with δx, δy, and δz parallel to the axes, sweeps out the interior of the cube if it is integrated in the x, y, and z directions, in turn, between $-a$ and a in each case. Hence the mass M of the cube is

$$M = \int_{-a}^{a} \int_{-a}^{a} \int_{-a}^{a} [\alpha + \beta(x^2 + y^2 + z^2)] \, dx \, dy \, dz.$$

This integral can now be evaluated as a repeated integral as follows:

$$M = \int_{-a}^{a} \int_{-a}^{a} [\alpha x + \beta(\tfrac{1}{3}x^3 + xy^2 + xz^2)]_{-a}^{a} \, dy \, dz$$

$$= \int_{-a}^{a} \int_{-a}^{a} [2\alpha a + 2\beta(\tfrac{1}{3}a^3 + ay^2 + az^2)] \, dy \, dz$$

$$= \int_{-a}^{a} [2\alpha ay + 2\beta(\tfrac{1}{3}a^3 y + \tfrac{1}{3}ay^3 + ayz^2)]_{-a}^{a} \, dz$$

$$= \int_{-a}^{a} [4\alpha a^2 + 4\beta(\tfrac{1}{3}a^4 + \tfrac{1}{3}a^4 + a^2 z^2)] \, dz$$

$$= [4\alpha^2 z + 4\beta(\tfrac{2}{3}a^4 z + \tfrac{1}{3}a^2 z^3)]_{-a}^{a}$$

$$= 8a^3(\alpha + \beta a^2).$$

34.4 The divergence theorem

The **divergence theorem** (due to Gauss) relates a volume integral to a surface integral. Let $\mathcal{V}$ be a region in three dimensions which is bounded by a smooth surface S. We shall prove the theorem in the restricted case in which any straight line parallel to any of the cartesian axes cuts S in at most two points. It will look something like the surface shown in Fig. 34.10. The theorem is:

> **Divergence theorem**
>
> Let S be a surface enclosing a region $\mathcal{V}$, and let F be a smooth vector field defined in $\mathcal{V}$. Then
>
> $$\iiint_{\mathcal{V}} \operatorname{div} F \, d\mathcal{V} = \iint_{S} F \cdot \hat{n} \, dS,$$
>
> where $\hat{n}$ is the unit normal to S drawn *outwards* from $\mathcal{V}$.   (34.7)

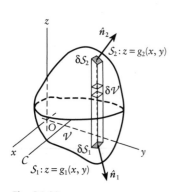

Fig. 34.10

Within the restrictions imposed on S we can divide S into two surfaces, an upper one S_2 with equation, say, $z = g_2(x, y)$ and a lower one S_1 with equation $z = g_1(x, y)$, the two surfaces meeting on the curve C. We shall use the cartesian increment $\delta x \, \delta y \, \delta z$ for $\delta \mathcal{V}$.

The divergence theorem is really the sum of three results. Suppose that $F = F_1\hat{\imath} + F_2\hat{\jmath} + F_3\hat{k}$, and consider first

$$I_3 = \iiint_\mathcal{V} \frac{\partial F_3}{\partial z}\,dx\,dy\,dz.$$

If $\mathcal{R}$ is the projection of C on to the (x, y) plane, then as a repeated integral,

$$I_3 = \iint_\mathcal{R} \left[\int_{z=g_1(x,y)}^{z=g_2(x,y)} \frac{\partial F_3}{\partial z}\,dz\right] dx\,dy$$

$$= \iint_\mathcal{R} [F_3(x, y, z)]_{z=g_1(x,y)}^{z=g_2(x,y)}\,dx\,dy$$

$$= \iint_\mathcal{R} [F_3(x, y, g_2(x, y)) - F_3(x, y, g_1(x, y))]\,dx\,dy.$$

From the previous section, noting carefully the directions of $\hat{n}_1$ and $\hat{n}_2$, the *outward* normals to S_1 and S_2, it follows that

$$dx\,dy = \hat{k}\cdot\hat{n}_2\,dS_2 \text{ on } S_2 \quad \text{but } dx\,dy = -\hat{k}\cdot\hat{n}_1\,dS_1 \text{ on } S_1,$$

since the angle between $\hat{k}$ and $\hat{n}_1$ is obtuse. Hence

$$I_3 = \iint_{S_2} F_3(x, y, g_2(x, y))\hat{k}\cdot\hat{n}_2\,dS_2 + \iint_{S_1} F_3(x, y, g_1(x, y))\hat{k}\cdot\hat{n}_1\,dS_1$$

$$= \iint_S F_3\hat{k}\cdot\hat{n}\,dS.$$

Similarly it can be shown that

$$I_1 = \iiint_\mathcal{V} \frac{\partial F_1}{\partial x}\,d\mathcal{V} = \iint_S F_1\hat{\imath}\cdot\hat{n}\,dS.$$

$$I_2 = \iiint_\mathcal{V} \frac{\partial F_2}{\partial y}\,d\mathcal{V} = \iint_S F_2\hat{\jmath}\cdot\hat{n}\,dS.$$

Addition of these results gives the divergence theorem:

$$I_1 + I_2 + I_3 = \iiint_\mathcal{V} \left(\frac{\partial F_1}{\partial x} + \frac{\partial F_2}{\partial y} + \frac{\partial F_3}{\partial z}\right) d\mathcal{V} = \iiint_\mathcal{V} \operatorname{div} F\,d\mathcal{V}$$

$$= \iint_S (F_1\hat{\imath} + F_2\hat{\jmath} + F_3\hat{k})\cdot\hat{n}\,dS$$

$$= \iint_S F\cdot\hat{n}\,dS.$$

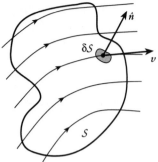

Fig. 34.11

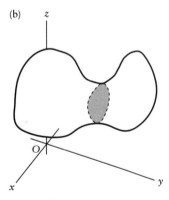

Fig. 34.12

The divergence theorem tells us something about the physical interpretation of the divergence of a vector field. In Fig. 34.11 the curves represent streamlines of the flow of an incompressible fluid, and v is the local velocity of the fluid. Consider *any* fixed closed surface S drawn in the flow. Then the outflow through an element of area δS on the surface will be $v \cdot \hat{n}\, \delta S$ per unit time. The total outflow through S will be

$$\iint_S v \cdot \hat{n}\, dS.$$

Assuming that fluid is neither being created nor destroyed within S, it follows that

$$\iint_S v \cdot \hat{n}\, dS = 0,$$

that is the net outflow is zero. By the divergence theorem it must be true that

$$\iiint_{\mathcal{V}} \operatorname{div} v \, d\mathcal{V} = 0$$

for every such surface S. Therefore

$$\operatorname{div} v = 0$$

throughout the flow. This is known as the **equation of continuity** in fluid dynamics. A vector field which satisfies div $v = 0$ is said to be **solenoidal**. Generally the divergence of a vector field at a point P measures the rate at which the vector field spreads out from P.

It is not difficult to generalize the divergence theorem to regions which have corners or parts of S parallel to an axis as in Fig. 34.12a, or to regions for which the two-point rule does not apply as in Fig. 34.12b. This region can be split into regions to each of which the divergence theorem applies, and the theorem applies to the whole region by addition. The surface integrals over the joins cancel out.

34.5 Curl of a vector field

The **curl** of a vector field

$$F(x, y, z) = F_1(x, y, z)\hat{\imath} + F_2(x, y, z)\hat{\jmath} + F_3(x, y, z)\hat{k}$$

is a vector field defined as follows:

Curl of a vector field

$$\text{curl } F = \left(\frac{\partial F_3}{\partial y} - \frac{\partial F_2}{\partial z}\right)\hat{i} + \left(\frac{\partial F_1}{\partial z} - \frac{\partial F_3}{\partial x}\right)\hat{j} + \left(\frac{\partial F_2}{\partial x} - \frac{\partial F_1}{\partial y}\right)\hat{k}$$

$$= \begin{vmatrix} \hat{i} & \hat{j} & \hat{k} \\ \dfrac{\partial}{\partial x} & \dfrac{\partial}{\partial y} & \dfrac{\partial}{\partial z} \\ F_1 & F_2 & F_3 \end{vmatrix}$$

(34.8)

This 'determinant' is a useful hybrid form which has unit vectors on the top row, operators on the second, and components on the third, and it is evaluated using the first row expansion rule for determinants. This rule is analogous to the determinant rule for the vector product given in Section 11.2. The del form is

$$\text{curl } F = \nabla \times F.$$

Example 34.5 *Find the curl of*

$$F = e^{xyz}\hat{i} + (x^2 + y)\hat{j} + xz\, e^y\hat{k}.$$

Using (34.8)

$$\text{curl } F = \begin{vmatrix} \hat{i} & \hat{j} & \hat{k} \\ \dfrac{\partial}{\partial x} & \dfrac{\partial}{\partial y} & \dfrac{\partial}{\partial z} \\ e^{xyz} & x^2 + y & xz\, e^y \end{vmatrix}$$

$$= \left(\frac{\partial}{\partial y}(xz\, e^y) - \frac{\partial}{\partial z}(x^2 + y)\right)\hat{i} + \left(\frac{\partial}{\partial z}(e^{xyz}) - \frac{\partial}{\partial x}(xz\, e^y)\right)\hat{j}$$

$$+ \left(\frac{\partial}{\partial x}(x^2 + y) - \frac{\partial}{\partial y}(e^{xyz})\right)\hat{k}$$

$$= xz\, e^y\hat{i} + (xy\, e^{xyz} - z\, e^y)\hat{j} + (2x - xz\, e^{xyz})\hat{k}.$$

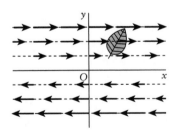

Fig. 34.13

We can interpret the curl of a vector field as follows. Consider a **rectilinear** fluid flow with velocity

$$v = \omega y\hat{i} \quad (\omega \text{ constant})$$

in which the flow is in one direction only. Imagine that we are looking down on the surface of the flow in Fig. 34.13.

The divergence of v is zero so that the flow satisfies the equation of continuity. Its curl is given by

$$\text{curl } v = \begin{vmatrix} \hat{i} & \hat{j} & \hat{k} \\ \dfrac{\partial}{\partial x} & \dfrac{\partial}{\partial y} & \dfrac{\partial}{\partial z} \\ \omega y & 0 & 0 \end{vmatrix} = -\omega\hat{k},$$

which is a vector perpendicular to the surface. The fluid as a whole does not appear to rotate, but a small leaf placed on the flow will rotate in a clockwise sense as it is carried along with the stream. For example, if it is placed so that $y > 0$ for all points on the leaf, then the points furthest from the x axis will be moving faster than those nearest. The spin, or local angular velocity, turns out to be

$$\tfrac{1}{2}\,\mathbf{curl}\,v = -\tfrac{1}{2}\omega\hat{k}$$

everywhere.

A vector field which satisfies $\mathbf{curl}\,v = 0$ is said to be **irrotational**. There are two important identities for special vector fields.

> **A conservative potential field** ϕ **is irrotational since**
> $$\mathbf{curl}\,\mathbf{grad}\,\phi = 0.$$
>
> (34.9)

> **The curl of a vector field** F **is solenoidal since**
> $$\mathrm{div}\,\mathbf{curl}\,F = 0.$$
>
> (34.10)

The verification of these results is straightforward. For the first one

$$\mathbf{curl}\,\mathbf{grad}\,\phi = \begin{vmatrix} \hat{i} & \hat{j} & \hat{k} \\ \dfrac{\partial}{\partial x} & \dfrac{\partial}{\partial y} & \dfrac{\partial}{\partial z} \\ \dfrac{\partial\phi}{\partial x} & \dfrac{\partial\phi}{\partial y} & \dfrac{\partial\phi}{\partial z} \end{vmatrix}$$

$$= \left(\frac{\partial^2\phi}{\partial y\,\partial z} - \frac{\partial^2\phi}{\partial z\,\partial y}\right)\hat{i} + \left(\frac{\partial^2\phi}{\partial z\,\partial x} - \frac{\partial^2\phi}{\partial x\,\partial z}\right)\hat{j}$$

$$+ \left(\frac{\partial^2\phi}{\partial x\,\partial y} - \frac{\partial^2\phi}{\partial y\,\partial x}\right)\hat{k}$$

$$= 0$$

assuming that scalar field is smooth enough to ensure that all the mixed partial derivatives cancel.

For the second result

$$\mathrm{div}\,\mathbf{curl}\,F = \frac{\partial}{\partial x}\left(\frac{\partial F_3}{\partial y} - \frac{\partial F_2}{\partial z}\right) + \frac{\partial}{\partial y}\left(\frac{\partial F_1}{\partial z} - \frac{\partial F_3}{\partial x}\right) + \frac{\partial}{\partial z}\left(\frac{\partial F_2}{\partial x} - \frac{\partial F_1}{\partial y}\right)$$

$$= \frac{\partial^2 F_3}{\partial x\,\partial y} - \frac{\partial^2 F_2}{\partial x\,\partial z} + \frac{\partial^2 F_1}{\partial y\,\partial z} - \frac{\partial^2 F_3}{\partial y\,\partial x} + \frac{\partial^2 F_2}{\partial z\,\partial x} - \frac{\partial^2 F_1}{\partial z\,\partial y}$$

$$= 0.$$

Example 34.6 *Show that the vector field*

$$\mathbf{F} = (y^2z + z + y\,e^{xy})\hat{\boldsymbol{i}} + (2xyz + x\,e^{xy})\hat{\boldsymbol{j}} + (xy^2 + x)\hat{\boldsymbol{k}}$$

is conservative. Find the scalar potential ϕ of $\mathbf{F}$.

We first check that **curl** $\mathbf{F} = 0$. Thus

$$\text{curl grad } \phi = \begin{vmatrix} \hat{\boldsymbol{i}} & \hat{\boldsymbol{j}} & \hat{\boldsymbol{k}} \\ \dfrac{\partial}{\partial x} & \dfrac{\partial}{\partial y} & \dfrac{\partial}{\partial z} \\ y^2z + z + y\,e^{xy} & 2xyz + x\,e^{xy} & xy^2 + x \end{vmatrix}$$

$$= \left(\frac{\partial}{\partial y}(xy^2 + x) - \frac{\partial}{\partial z}(2xyz + x\,e^{xy}) \right)\hat{\boldsymbol{i}}$$

$$+ \left(\frac{\partial}{\partial z}(y^2z + z + y\,e^{xy}) - \frac{\partial}{\partial x}(xy^2 + x) \right)\hat{\boldsymbol{j}}$$

$$+ \left(\frac{\partial}{\partial x}(2xyz + x\,e^{xy}) - \frac{\partial}{\partial y}(y^2z + z + y\,e^{xy}) \right)\hat{\boldsymbol{k}}$$

$$= \hat{\boldsymbol{i}}(2xy - 2xy) + \hat{\boldsymbol{j}}[(y^2 + 1) - (y^2 + 1)]$$
$$+ \hat{\boldsymbol{k}}[(2yz + e^{xy} + xy\,e^{xy}) - (2yz + e^{xy} + xy\,e^{xy})]$$

$$= 0.$$

We now need to find ϕ such that grad $\phi = \mathbf{F}$, that is

$$\frac{\partial \phi}{\partial x} = y^2z + z + y\,e^{xy},$$

$$\frac{\partial \phi}{\partial y} = 2xyz + x\,e^{xy},$$

$$\frac{\partial \phi}{\partial z} = xy^2 + x.$$

Integrate the *partial* derivatives with respect to x, y, and z to give:

$$\phi = \int (y^2z + z + y\,e^{xy})\,\mathrm{d}x + f(y, z) = xy^2z + xz + e^{xy} + f(y, z); \tag{34.11}$$

$$\phi = \int (2xyz + x\,e^{xy})\,\mathrm{d}y + g(z, x) = xy^2z + e^{xy} + g(z, x); \tag{34.12}$$

$$\phi = \int (xy^2 + x)\,\mathrm{d}z + h(x, y) = xy^2z + xz + h(x, y). \tag{34.13}$$

Here the 'constants of integration' become functions of the other two variables in each case since partial derivatives are being integrated. Finally, ϕ given by (34.11), (34.12), and (34.13) must all result in the same answer. This can be achieved by the choices

$$f(y, z) = C, \quad g(z, x) = xz + C, \quad h(x, y) = e^{xy} + C,$$

where C is any constant. Hence

$$\phi = xy^2z + xz + e^{xy} + C.$$

Note that potentials of conservative fields can only be found to within an additive constant.

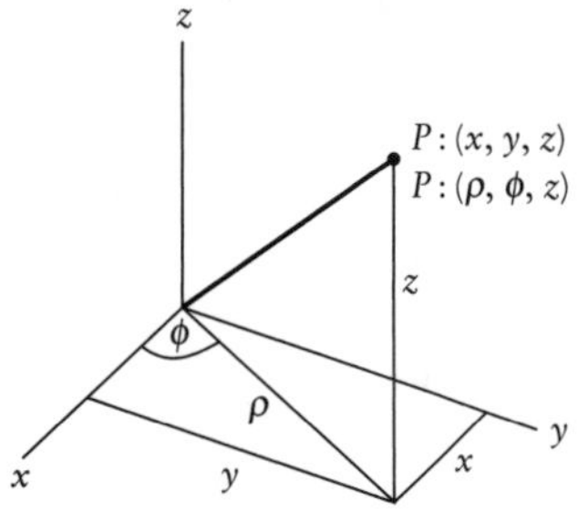

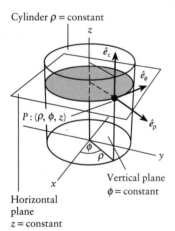

Fig. 34.14 Cylindrical polar coordinates.

Cylinder ρ = constant

$P : (\rho, \phi, z)$

Vertical plane
ϕ = constant

Horizontal
plane
z = constant

Fig. 34.15

34.6 Cylindrical polar coordinates

In many applications it is advantageous to use alternative three-dimensional coordinate systems. Usually the geometry of the application suggests an appropriate system, and one system which is suitable for problems involving cylinders uses cylindrical polarcoordinates (ρ, θ, z) (see Fig. 34.14) defined by:

Cylindrical polar coordinates ρ, ϕ, z

$$x = \rho \cos \phi, \qquad y = \rho \sin \phi, \qquad z = z$$

$$(0 \leqslant \rho < \infty, 0 \leqslant \phi < 2\pi, -\infty < z < \infty).$$

(34.14)

A point P can be viewed as lying at the intersection of three surfaces (Fig. 34.15): the cylinder ρ = a constant, the radial plane ϕ = a constant through the z axis, and the horizontal plane z = a constant. These surfaces meet at right angles at every point, and coordinate systems with this property are said to be **orthogonal**.

The point P can be represented by the position vector **r**, where

$$r = r(\rho, \phi, z) = \rho \cos \phi \, \hat{\imath} + \rho \sin \phi \, \hat{\jmath} + z \, \hat{k}.$$

Along the ρ-increasing line through P, ϕ and z are constant. The vector $\partial r/\partial \rho$, evaluated at P, is a tangent to this curve at P, pointing in the direction of increasing ρ. The corresponding unit vector in this direction is

$$\hat{e}_\rho = \frac{\partial r}{\partial \rho} \bigg/ \left| \frac{\partial r}{\partial \rho} \right| = \frac{1}{h_\rho} \frac{\partial r}{\partial \rho}$$

where $h_\rho = |\partial r/\partial \rho|$ is called the **scale factor** associated with ρ. In cylindrical coordinates

$$h_\rho = \left| \frac{\partial r}{\partial \rho} \right| = |\cos \phi \, \hat{\imath} + \sin \phi \, \hat{\jmath}|$$

$$= (\cos^2\phi + \sin^2\phi)^{\frac{1}{2}} = 1.$$

Similarly, $h_\phi = \rho$ and $h_z = 1$. Therefore the unit vectors in cylindrical polars are

$$\hat{e}_\rho = \frac{\partial r}{\partial \rho}, \qquad \hat{e}_\phi = \frac{1}{\rho} \frac{\partial r}{\partial \phi}, \qquad \hat{e}_z = \frac{\partial r}{\partial z}. \tag{34.15}$$

All partial derivatives are to be evaluated at P.

The **gradient** of a scalar function $U(x, y, z)$ can be expressed in terms of the unit vectors $\hat{e}_\rho, \hat{e}_\phi, \hat{e}_z$. Suppose that

$$\mathbf{grad}\, U = g_\rho \hat{e}_\rho + g_\phi \hat{e}_\phi + g_z \hat{e}_z,$$

where we require the components g_ρ, g_ϕ, g_z. Treating U as a function of ρ, ϕ, z, the incremental formula gives

$$\delta U = \frac{\partial U}{\partial \rho} \delta \rho + \frac{\partial U}{\partial \phi} \delta \phi + \frac{\partial U}{\partial z} \delta z. \tag{34.16}$$

Also, from (31.15), the directional derivative of U is

$$\frac{dU}{ds} = \hat{s} \cdot \mathbf{grad}\, U,$$

where $\hat{s}$ represents an arbitrary direction. Since $|\delta s| = |\delta r|$,

$$\hat{s} = \frac{dr}{ds} = \frac{\partial r}{\partial \rho} \frac{d\rho}{ds} + \frac{\partial r}{\partial \phi} \frac{d\phi}{ds} + \frac{\partial r}{\partial z} \frac{dz}{ds}$$

$$= h_\rho \frac{d\rho}{ds} \hat{e}_\rho + h_\phi \frac{d\phi}{ds} \hat{e}_\phi + h_z \frac{dz}{ds} \hat{e}_z.$$

Therefore,

$$\frac{dU}{ds} = h_\rho g_\rho \frac{d\rho}{ds} + h_\phi g_\phi \frac{d\phi}{ds} + h_z g_z \frac{dz}{ds},$$

or, expressed in increments,

$$\delta U = h_\rho g_\rho \, \delta \rho + h_\phi g_\phi \, \delta \phi + h_z g_z \, \delta z. \tag{34.17}$$

Compare (34.16) and (34.17), which are true for arbitrary $\delta \rho$, $\delta \phi$, δz. In turn, put one of $\delta \rho$, $\delta \phi$, δz to a nonzero value, and the other two to zero. We obtain

$$g_\rho = \frac{1}{h_\rho} \frac{\partial U}{\partial \rho}, \qquad g_\phi = \frac{1}{h_\phi} \frac{\partial U}{\partial \phi}, \qquad g_z = \frac{1}{h_z} \frac{\partial U}{\partial z},$$

so that $\mathbf{grad}\, U$ is given by

$$\mathbf{grad}\, U = \frac{1}{h_\rho} \frac{\partial U}{\partial \rho} \hat{e}_\rho + \frac{1}{h_\phi} \frac{\partial U}{\partial \phi} \hat{e}_\theta + \frac{1}{h_z} \frac{\partial U}{\partial z} \hat{e}_z$$

$$= \frac{\partial U}{\partial \rho} \hat{e}_\rho + \frac{1}{\rho} \frac{\partial U}{\partial \phi} \hat{e}_\phi + \frac{\partial U}{\partial z} \hat{e}_z.$$

The **divergence** and **curl** also have their cylindrical polar forms. For the vector field $\mathbf{F} = F_\rho \hat{e}_\rho + F_\phi \hat{e}_\phi + F_z \hat{e}_z$, these are

$$\mathrm{div}\, \mathbf{F} = \frac{1}{\rho} \left[\frac{\partial}{\partial \rho} (\rho F_\rho) + \frac{\partial}{\partial \phi} (F_\phi) + \frac{\partial}{\partial z} (\rho F_z) \right],$$

and

$$\text{curl } F = \frac{1}{\rho} \begin{vmatrix} \hat{e}_\rho & \rho\hat{e}_\phi & \hat{e}_z \\ \dfrac{\partial}{\partial \rho} & \dfrac{\partial}{\partial \phi} & \dfrac{\partial}{\partial z} \\ F_\rho & \rho F_\phi & F_z \end{vmatrix} = \left(\frac{1}{\rho} \frac{\partial F_z}{\partial \phi} - \frac{\partial F_\phi}{\partial z} \right) \hat{e}_\rho + \left(\frac{\partial F_\rho}{\partial z} + \frac{\partial F_z}{\partial \rho} \right) \hat{e}_\phi$$

$$+ \frac{1}{\rho} \left(\frac{\partial}{\partial \rho}(\rho F_\phi) - \frac{\partial}{\partial \phi}(F_\rho) \right) \hat{e}_z.$$

34.7 Curvilinear coordinates

In this section we simply summarize the generalizations of the results given in the previous section for cylindrical polar coordinates to **orthogonal curvilinear coordinates**. Suppose that the position vector r of a point is expressed in terms of the curvilinear coordinates u_1, u_2, u_3 so that

$$r = r(u_1, u_2, u_3) = x(u_1, u_2, u_3)\hat{i} + y(u_1, u_2, u_3)\hat{j} + z(u_1, u_2, u_3)\hat{k}.$$

Assume that the curvilinear coordinates are orthogonal, that is the surfaces u_1 = a constant, u_2 = a constant, u_3 = a constant meet at right angles at every point (Fig. 34.16). The unit vector $\hat{e}_1$ is in the direction of the curve along which the surfaces u_2 = constant and u_3 = constant meet, and it points in the direction of u_1 increasing. The other unit vectors are in the directions of the intersections of the other surface pairs as shown in Fig. 34.16.

The scale factors and unit vectors are given by:

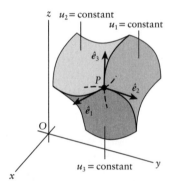

Fig. 34.16 Orthogonal curvilinear coordinates.

Scale factors, unit vectors

$$h_1 = \left| \frac{\partial r}{\partial u_1} \right|, \quad h_2 = \left| \frac{\partial r}{\partial u_2} \right|, \quad h_3 = \left| \frac{\partial r}{\partial u_3} \right|.$$

$$\hat{e}_1 = \frac{1}{h_1} \frac{\partial r}{\partial u_1}, \quad \hat{e}_2 = \frac{1}{h_2} \frac{\partial r}{\partial u_2}, \quad \hat{e}_3 = \frac{1}{h_3} \frac{\partial r}{\partial u_3}.$$

Elements of distance δs in the u_1, u_2, u_3 directions are respectively

$$h_1 \delta u_1, \quad h_2 \delta u_2, \quad h_3 \delta u_3.$$

(34.18)

We simply state the formulae for **grad**, div, and **curl** in general curvilinear coordinates without derivation. They are given by:

Gradient of U

$$\text{grad } U = \frac{1}{h_1} \frac{\partial U}{\partial u_1} \hat{e}_1 + \frac{1}{h_2} \frac{\partial U}{\partial u_2} \hat{e}_2 + \frac{1}{h_3} \frac{\partial U}{\partial u_3} \hat{e}_3.$$

(34.19)

Divergence of $F = F_1\hat{e}_1 + F_2\hat{e}_2 + F_3\hat{e}_3$

$$\text{div } F = \frac{1}{h_1 h_2 h_3}\left[\frac{\partial}{\partial u_1}(h_2 h_3 F_1) + \frac{\partial}{\partial u_2}(h_3 h_1 F_2) + \frac{\partial}{\partial u_3}(h_1 h_2 F_3)\right].$$

(34.20)

Curl in orthogonal curvilinear coordinates

$$\text{curl } F = \frac{1}{h_1 h_2 h_3}\begin{vmatrix} h_1\hat{e}_1 & h_2\hat{e}_2 & h_3\hat{e}_3 \\ \dfrac{\partial}{\partial u_1} & \dfrac{\partial}{\partial u_2} & \dfrac{\partial}{\partial u_3} \\ h_1 F_1 & h_2 F_2 & h_3 F_3 \end{vmatrix}.$$

(34.21)

Example 34.7 *In terms of (x, y, z), spherical polar coordinates (r, θ, ϕ) are given by*

$$r = x\hat{\imath} + y\hat{\jmath} + z\hat{k} = r\sin\theta\cos\phi\,\hat{\imath} + r\sin\theta\sin\phi\,\hat{\jmath} + r\cos\theta\,\hat{k},$$
$$r \geqslant 0, 0 \leqslant \theta \leqslant \pi, 0 \leqslant \phi < 2\pi.$$

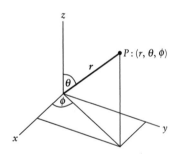

Fig. 34.17 Spherical polar coordinates (r, θ, ϕ).

The coordinates are shown in Fig. 34.17: the coordinates are orthogonal with coordinate surfaces of a sphere, r = constant, a vertical plane, ϕ = constant, and a cone, θ = constant. Find the scale factors of these curvilinear coordinates. Hence, obtain the gradient of the scalar field U, and the divergence of the vector field $F = F_r\hat{e}_r + F_\theta\hat{e}_\theta + F_\phi\hat{e}_\phi$ in spherical polar coordinates.

The scale factors are

$$h_r = \left|\frac{\partial r}{\partial r}\right| = |\sin\theta\cos\phi\,\hat{\imath} + \sin\theta\sin\phi\,\hat{\jmath} + \cos\theta\,\hat{k}|$$
$$= \sqrt{\sin^2\theta\,(\cos^2\phi + \sin^2\phi) + \cos^2\theta} = 1,$$

$$h_\theta = \left|\frac{\partial r}{\partial\theta}\right| = |r\cos\theta\cos\phi\,\hat{\imath} + r\cos\theta\sin\phi\,\hat{\jmath} - r\sin\theta\,\hat{k}|$$
$$= r\sqrt{\cos^2\theta(\cos^2\phi + \sin^2\phi) + \sin^2\theta} = r,$$

$$h_\phi = \left|\frac{\partial r}{\partial\phi}\right| = |-r\sin\theta\sin\phi\,i + r\sin\theta\cos\phi\,j| = r\sin\theta.$$

From (34.19)

$$\text{grad } U = \frac{\partial U}{\partial r}\hat{e}_r + \frac{1}{r}\frac{\partial U}{\partial\theta}\hat{e}_\theta + \frac{1}{r\sin\theta}\frac{\partial U}{\partial\phi}\hat{e}_\phi.$$

By (34.20)

$$\text{div } F = \frac{1}{r^2}\frac{\partial}{\partial r}(r^2 F_r) + \frac{1}{r\sin\theta}\frac{\partial}{\partial\theta}(\sin\theta\,F_\theta) + \frac{1}{r\sin\theta}\frac{\partial}{\partial\phi}(F_\phi).$$

Problems

34.1 Find the surface area of the spherical cap of height h whose equation for $z \geqslant 0$ is

$$z = \sqrt{[a^2 - x^2 - y^2]} - a + h, \quad (0 < h \leqslant a).$$

34.2 Evaluate the following triple integrals as repeated integrals:

(a) $\displaystyle\int_0^1 \int_0^x \int_y^{2y} x \, dx \, dy \, dz;$

(b) $\displaystyle\int_0^1 \int_0^z \int_0^{\sqrt{[1-y^2]}} x \, dx \, dy \, dz;$

(c) $\displaystyle\int_0^1 \int_0^z \int_{-\frac{1}{2}\sqrt{[1-y^2-z^2]}}^{\sqrt{[1-y^2-z^2]}} x^3 \, dx \, dy \, dz.$

34.3 It is intended to evaluate the integral

$$\iiint_V f(x, y, z) \, dx \, dy \, dz$$

as a repeated integral over the interior of the sphere $x^2 + y^2 + z^2 = a^2$ which lies in the first *octant* $x \geqslant 0, y \geqslant 0, z \geqslant 0$. Work out the limits of integration if the order of integration is x followed by y followed by z.

34.4 Show that the volume of the tetrahedron bounded by the coordinate planes $x = 0$, $y = 0$, $z = 0$ and the plane

$$\frac{x}{a} + \frac{y}{b} + \frac{z}{c} = 1, \qquad a > 0, b > 0, c > 0,$$

is $\frac{1}{6}abc$.

34.5 Find the area of the surface $z = x^2 + y$ for which $|x| \leqslant 1$ and $|y| \leqslant 1$.

34.6 Show that the vector field

$$F = (yz \, e^{xyz} - y \sin xy + z)\hat{\imath} + (xz \, e^{xyz} - x \sin xy)\hat{\jmath}$$
$$+ (xy \, e^{xyz} + x)\hat{k}$$

is irrotational. Find the scalar potential of F.

34.7 **Paraboloidal coordinates** (u, v, ϕ) are defined by

$$x = uv \cos \phi, \quad y = uv \sin \phi, \quad z = \frac{1}{2}(u^2 - v^2),$$

where $u \geqslant 0$, $v \geqslant 0$, and $0 \leqslant \phi < 2\pi$. Find the corresponding scale factors. Find also div F in paraboloidal coordinates.

34.8 Using the definitions of **grad**, div, and **curl**, verify the following identities:
(a) $\mathbf{grad}(UV) = U \, \mathbf{grad} \, V + V \, \mathbf{grad} \, U;$
(b) $\mathrm{div}(UF) = (\mathbf{grad} \, U) \cdot F + U \, \mathrm{div} \, F;$
(c) $\mathrm{div}(F \times G) = (\mathbf{curl} \, F) \cdot G - F \cdot (\mathbf{curl} \, G);$
(d) $\mathbf{curl} \, \mathbf{curl} \, F = \mathbf{grad}(\mathrm{div} \, F) - \mathrm{div} \, \mathbf{grad} \, F$. By div grad F is meant $\hat{\imath} \, \mathrm{div}(\mathbf{grad} \, F_1) + \hat{\jmath} \, \mathrm{div}(\mathbf{grad} \, F_2) + \hat{k} \, \mathrm{div}(\mathbf{grad} \, F_3)$.
(e) $\mathbf{grad}(F \cdot G) = F \times \mathbf{curl} \, G + G \times \mathbf{curl} \, F + (F \cdot \mathbf{grad})G + (G \cdot \mathbf{grad})F$.

34.9 Show that

$$\mathrm{div} \, \mathbf{grad} \, \phi = \frac{\partial^2 \phi}{\partial x^2} + \frac{\partial^2 \phi}{\partial y^2} + \frac{\partial^2 \phi}{\partial z^2}.$$

This is often written as $\nabla^2 \phi$. The *equation*

$$\nabla^2 \phi = 0$$

is known as **Laplace's equation**.

Show that $\phi = 1/\sqrt{x^2 + y^2 + z^2}$ is a solution of Laplace's equation.

34.10 Prove that
(a) $\mathrm{div}(F + G) = \mathrm{div} \, F + \mathrm{div} \, G;$
(b) $\mathbf{curl}(F + G) = \mathbf{curl} \, F + \mathbf{curl} \, G.$

34.11 Find the divergence of each of the following vector fields:
(a) $F = e^{xyz}\hat{\imath} + e^{y^2z}\hat{\jmath} + e^{xz}\hat{k};$
(b) $F = (xz - y)\hat{\imath} + yz\hat{\jmath} + 2xy\hat{k};$
(c) $F = (xz - y^2)\hat{\imath} + yz\hat{\jmath} + 2x^2y\hat{k}.$
Indicate any vector fields which are solenoidal.

34.12 Find the **curl** of each of the following vector fields:
(a) $F = e^{xyz}\hat{\imath} + e^{y^2z}\hat{\jmath} + e^{xz}\hat{k};$
(b) $F = (xz - y)\hat{\imath} + yz\hat{\jmath} + 2xy\hat{k};$
(c) $F = (2xy + yz)\hat{\imath} + (x^2 + xz)\hat{\jmath} + xy\hat{k}.$

34.13 A vector field v is both irrotational and solenoidal. Show that its scalar potential Φ satisfies Laplace's equation:

$$\nabla^2 \Phi = 0.$$

34.14 If $r = x\hat{\imath} + y\hat{\jmath} + z\hat{k}$ and r is its magnitude find
(a) $\mathrm{div}(r^2 r);$ (b) $\mathbf{curl}(r^3 r);$
(c) $\mathbf{grad} \, r^3;$ (d) $\mathrm{div}(r/r^3);$
(e) $\mathbf{curl}(r/r^2);$ (f) div $\mathbf{grad} \, r^3.$

34.15 Prove the identity

$$(v \cdot \text{grad})v = \tfrac{1}{2}\,\text{grad}\,v^2 - v \times \text{curl}\,v,$$

where $v = |v|$.

34.16 Show that Laplace's equation (see Problem 34.9) in cylindrical polar coordinates is given by

$$\frac{1}{\rho}\frac{\partial}{\partial \rho}\left(\rho\frac{\partial U}{\partial \rho}\right) + \frac{1}{\rho^2}\frac{\partial^2 U}{\partial \phi^2} + \frac{\partial^2 U}{\partial z^2} = 0.$$

If $U = f(\rho)$, that is U is independent of the other variables, show that f satisfies the ordinary differential equation

$$\rho f''(\rho) + f'(\rho) = 0.$$

Hence show that $f(\rho) = A + B\,\ln \rho$, where A and B are constants.

34.17 Show that Laplace's equation in spherical polar coordinates is given by

$$\frac{1}{r^2}\frac{\partial}{\partial r}\left(r^2\frac{\partial U}{\partial r}\right) + \frac{1}{r^2 \sin \theta}\frac{\partial}{\partial \theta}\left(\sin \theta\,\frac{\partial U}{\partial \theta}\right)$$

$$+ \frac{1}{r^2 \sin^2 \theta}\frac{\partial^2 U}{\partial \phi^2} = 0.$$

A solution with spherical symmetry is sought for U, that is with $U = f(r)$. Show that $f(r) = A + (B/r)$, where A and B are constants.

34.18 A vector field is given by $F = xy^2\hat{\imath} + xz\hat{\jmath} + xyz\hat{k}$. Let S be the surface of a cube bounded by the planes $x = \pm 1$, $y = \pm 1$, $z = \pm 1$. Use the divergence theorem to evaluate

$$\iint_S F \cdot \hat{n}\,dS,$$

where $\hat{n}$ is the outward normal to the cube.

34.19 Prove that

$$\iint_S \hat{n} \cdot \text{curl}\,F\,dS = 0$$

for any closed surface S for which the divergence theorem holds.

34.20 Suppose that F is a smooth vector field which equals the outward unit normal $\hat{n}$ on S. Use the divergence theorem to show that the surface area of S is given by

$$\iiint_V \text{div}\,F\,dV,$$

where V is the interior of S.

34.21 Let S be a closed surface surrounding a region V for which the divergence theorem holds. By using the vector field $r = x\hat{\imath} + y\hat{\jmath} + z\hat{k}$, show that the volume enclosed by S is

$$\frac{1}{3}\iint_S r \cdot \hat{n}\,dS,$$

where $\hat{n}$ is the outward normal to S.

Using this result verify that
(a) the volume of the sphere enclosed by $x^2 + y^2 + z^2 = a^2$ is $\tfrac{4}{3}\pi a^3$;
(b) the volume of a cone with vertex at the origin and plane base of area A in the plane $z = h$ is $\tfrac{1}{3}Ah$.

34.22 Let S be a closed surface surround a region V for which the divergence theorem holds. Let F be a vector field which satisfies div $F = 1$ in a region which contains V. Show that the volume enclosed by S is given by the formula

$$\iint_S F \cdot \hat{n}\,dS.$$

PART VI

Discrete mathematics

35 Sets

CONTENTS

35.1 Notation

We are often interested in grouping together objects that have common characteristics or features. We might be interested in the integers $1, 2, 3, 4$, or in all the integers. The set of all points in a plane would consist of pairs of numbers of the form (x, y), where x and y are coordinates which can take any real values. These examples all involve numbers, but the elements of sets can be other objects such as functions, or matrices, or Fourier series, or Laplace transforms, etc.

A **set** is a collection of objects or **elements**. The elements in the set can be defined by a rule or in any descriptive manner. Sets are usually denoted by capital letters such as S, A, B, X, etc., and their elements by lowercase letters such as s, a, b, x, etc. The elements in a set are listed between braces $\{ \dots \}$. If the set A consists of just two numbers 0 and 1, then we write

$$A = \{0, 1\}, \quad \text{or} \quad A = \{1, 0\}, \tag{35.1}$$

the order being a matter of indifference. We say that 0 and 1 are the **elements** or **members** of the set A, or **belong to** A. We write

$$0 \in A, \quad 1 \in A,$$

read as '0 belongs to the set A', etc. The number 2 does not belong to A, and we write

$$2 \notin A,$$

that is '2 does not belong to the set A'.

The set defined by (35.1) is the **binary** set which could represent the *on* and *off* states of a system. This could be the state of a light switch, for example.

Sets can be either **finite**, having a finite number of elements, or **infinite**, in which case the set contains an infinite number of elements. Thus the set given by (35.1) defines a finite set A, while

$$B = \{1, 2, 3, \dots \},$$

the list of all positive integers, defines an infinite set.

Some of the more common sets have their own special symbols:

> **Notation for sets of numbers**
>
> $\mathbb{R}$, the set of all real numbers
> $\mathbb{C}$, the set of all complex numbers
> $\mathbb{R}^+$, the set of all positive real numbers (excludes zero)
> $\mathbb{Z}$, the set of all integers (positive, negative, and zero)
> $\mathbb{N}^+$, the set of all positive integers
> $\mathbb{N}^-$, the set of all negative integers
> $\mathbb{Q}$, the set of all rational numbers (i.e. numbers of the form p/q where $q \neq 0$ and p are integers)
>
> (35.2)

Often the elements are defined by a rule rather than by a list or formula. We write the set as

$$S = \{x \,|\, x \text{ satisfies specified rules}\},$$

which can be translated as 'S is the set of values of x which satisfy the stated rules'. The rules occur after the vertical $|$. Thus

$$S = \{x \,|\, x \in \mathbb{N}^+ \text{ and } 2 \leqslant x \leqslant 8\}$$

is an alternative way of writing $S = \{2, 3, 4, 5, 6, 7, 8\}$. As another example,

$$S = \{x \,|\, x \in \mathbb{R} \text{ and } 0 \leqslant x \leqslant 1\}$$

is the closed interval $[0, 1]$.

35.2 Equality, union, and intersection

Two sets A and B are said to be **equal** if they contain exactly the same elements. If this is the case, we write

$$A = B.$$

For example,

$$A = \{1, 2, 3\}, \qquad B = \{3, 2, 1\}, \qquad C = \{3, 1, 2, 1\}$$

are all equal, that is $A = B = C$. The order of the elements is immaterial, and repeated elements are discounted.

In a given context, the set of all elements of interest is known as the **universal set**. It could be the set $\mathbb{R}$ (the set of real numbers), or the set of all complex numbers, but it will vary from application to application.

We now define how sets can be combined to create new sets. The **union** of two sets A and B is the set of all elements that belong to A, *or to B, or to both*. It is written as

$$A \cup B = \{x \mid x \in A \text{ or } x \in B \text{ or both}\},$$

and read as 'A union B'.

Example 35.1   *Find the union of*
$A = \{x \mid x \in \mathbb{R} \text{ and } 0 \leqslant x \leqslant 2\}$ *and*
$B = \{x \mid x \in \mathbb{R} \text{ and } 1 \leqslant x \leqslant 3\}$.

The elements in the union have to belong to one or other of the intervals $0 \leqslant x \leqslant 2$, or $1 \leqslant x \leqslant 3$, or to both. The interval $0 \leqslant x \leqslant 3$ contains all these numbers. Hence

$A \cup B = \{x \mid \mathbb{R} \text{ and } 0 \leqslant x \leqslant 3\}$.

The **intersection** of two sets A and B is the set $A \cap B$ that contains *all elements common to both A and B*. It is written and defined by

$$A \cap B = \{x \mid x \in A \text{ and } x \in B\}.$$

Example 35.2   *Find the intersection of the sets A and B in Example 35.1.*

The elements in the intersection have to belong simultaneously to both intervals, that is to the overlapping part of the intervals $[0, 2]$ and $[1, 3]$, which is $[1, 2]$. Thus

$A \cap B = \{x \mid x \in \mathbb{R} \text{ and } 1 \leqslant x \leqslant 2\}$.

In the definitions of $A \cup B$ and $A \cap B$ above, we can see that the logical operation 'or' is associated with union, while 'and' is associated with intersection.

If A and B have no elements in common, then A and B are said to be **disjoint**. The set with no elements is called the **empty set** and denoted by $\varnothing$. Thus, if A and B are disjoint, then $A \cap B = \varnothing$.

The **complement** of a set A is the set of all those elements which belong to the universal set U but do not belong to A. We denote this set by $\bar{A}$ (the notations A^c and A' are also frequently used). Hence, the complement of A is, assuming that $x \in U$,

$$\bar{A} = \{x \mid x \notin A\}.$$

We say that A is a **subset** of B, expressed as $A \subseteq B$, if every element of A also belongs to the set B. It follows that $A \subseteq U$ if $B \subseteq U$. If there are elements of B which are not in A, then A is called a **proper subset** of B and written $A \subset B$. The statement $A \subseteq B$ includes the possibility that $A = B$, while $A \subset B$ does not. If $A \subseteq B$ and $B \subseteq A$, then all elements in A are contained in B, and vice versa; in other words, $A = B$.

The sets of integers $\mathbb{Z}$ and rational numbers $\mathbb{Q}$ are proper subsets of the real numbers $\mathbb{R}$, that is

$$\mathbb{Z} \subset \mathbb{R}, \quad \text{and} \quad \mathbb{Q} \subset \mathbb{R}.$$

We can summarize the results as follows.

Set operations

(a) **Union:** $A \cup B = \{x \mid x \in A \text{ or } x \in B \text{ or both}\}$.
(b) **Intersection:** $A \cap B = \{x \mid x \in A \text{ and } x \in B\}$.
(c) **Complement:** $\bar{A} = \{x \mid x \notin A\}$.
(d) **Empty set:** $\varnothing$, the set with no elements.
(e) **Subset:** $A \subseteq B$ means that A is a **subset** of B.
(f) **Proper subset:** $A \subset B$ means that $A \subseteq B$ but $A \neq B$. **(35.3)**

35.3 Venn diagrams

Useful graphical views and interpretations of sets and operations on them can be provided by **Venn diagrams**. We represent sets by regions in the plane, with the interpretation that the region stands for those elements belonging to the given set. The diagrams are symbolic: the set $A = \{1, 2\}$, for example, could be represented by the circle as shown in Fig. 35.1. Usually, sets are represented by the interiors of circles, but any closed curves can be used. In a given context, all the sets are subsets of a certain **universal set** U, whose nature will differ according to the context.

If the universal set is represented by a rectangle, then a subset A of U is represented by the interior of a circle within the rectangle shown in Fig. 35.2. This is a Venn diagram for U and A. Remember that A could represent an infinite number of elements, or include just one element, or be the empty set $\varnothing$. The union, intersection, complement, and proper subset can be represented by the Venn diagrams shown in Fig. 35.3. The shaded regions indicate the elements defined by the operations.

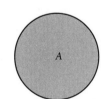

Fig. 35.1

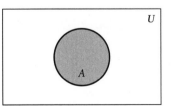

Fig. 35.2 Venn diagram for the universal set U and a set A.

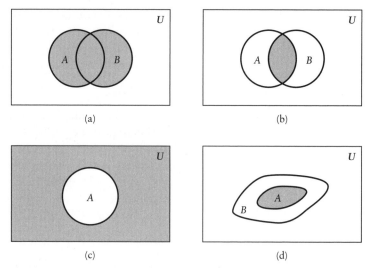

Fig. 35.3 (a) Union $A \cup B$. (b) Intersection $A \cap B$. (c) Complement $\bar{A}$.
(d) Proper subset $A \subset B$.

From the definitions of union, intersection, and complement, or from Venn diagrams, the following laws of the algebra of sets can be deduced:

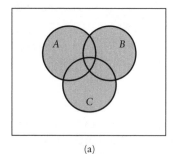

(a)

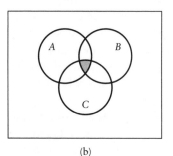

(b)

Fig. 35.4 (a) $(A \cup B) \cup C$ or $A \cup (B \cup C)$. (b) $(A \cap B) \cap C$ or $A \cap (B \cap C)$.

Algebra of sets
$$A \cup A = A, \qquad A \cap A = A.$$
Commutative laws:
$$A \cup B = B \cup A, \qquad A \cap B = B \cap A.$$
Associative laws (see Fig. 35.4):
$$(A \cup B) \cup C = A \cup (B \cup C),$$
$$(A \cap B) \cap C = A \cap (B \cap C).$$
Distributive laws:
$$A \cap (B \cup C) = (A \cap B) \cup (A \cap C),$$
$$A \cup (B \cap C) = (A \cup B) \cap (A \cup C). \tag{35.4}$$

Sets also satisfy the following identity and complementary laws:

Identity laws: $A \cup \varnothing = A, \qquad A \cap U = A.$
Complementary laws:
$$A \cup \bar{A} = U, \qquad A \cap \bar{A} = \varnothing, \qquad \bar{\bar{A}} = A. \tag{35.5}$$

For example, $\bar{A}$ consists of all elements that do not belong to A, and none that do; so there are no elements common to A and $\bar{A}$. Therefore $A \cap \bar{A} = \varnothing$.

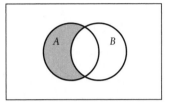

Fig. 35.5 Venn diagram for the difference $A \backslash B$ (shaded).

The **difference** of the sets A and B, written as $A \backslash B$, consists of the set of those elements that belong to A but do not belong to B. Thus

$$A \backslash B = \{x \mid x \in A \text{ and } x \notin B\} \quad \text{or} \quad \overline{A \cap B} \cap A.$$

(The notation $A - B$ is also used for $A \backslash B$.) Figure 35.5 shows a Venn diagram for $A \backslash B$.

Example 35.3 *Using Fig. 35.6 as the Venn diagram of two sets A and B, mark by shading the following sets:*

(a) $A \cup \bar{B}$, (b) $A \cap \bar{B}$, (c) $\bar{A} \cap \bar{B}$, (d) $\bar{A} \cup \bar{B}$, (e) $\overline{A \cup B}$, (f) $\overline{A \cap B}$.

Venn diagrams of the sets are shown in Fig. 35.7.

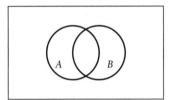

Fig. 35.6

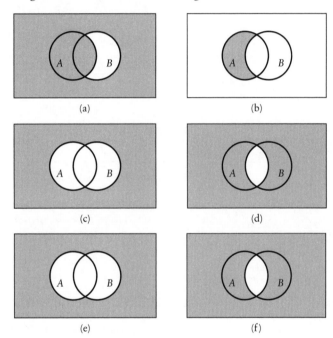

Fig. 35.7

The previous example confirms **de Morgan's laws**, which are

De Morgan's laws
$$\overline{A \cup B} = \bar{A} \cap \bar{B}, \qquad \overline{A \cap B} = \bar{A} \cup \bar{B}. \tag{35.6}$$

Example 35.4 *Using Fig. 35.8 as the Venn diagram of three sets A, B, and C, shade the following sets:*

(a) $(A \cap B) \cup C$, (b) $(A \cap B) \cap C$, (c) $(A \cap B) \cap (A \cap C)$, (d) $(A \cup B) \cup (A \cap C)$.

The required sets are shown in Fig. 35.9.

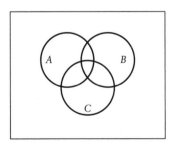

Fig. 35.8

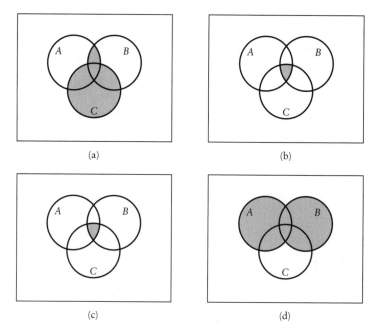

Fig. 35.9

Example 35.5 *Show that* $(A \cap B) \cup (A \cap \bar{B}) = A.$

By the distributive law (35.4),

$$(A \cap B) \cup (A \cap \bar{B}) = A \cap (B \cup \bar{B})$$
$$= A \cap U \quad \text{(by the complementary law)}$$
$$= A \quad \text{(by the identity law)}.$$

Example 35.6 *Show that* $(A \cup B) \cup (A \backslash B) = A \cup B.$

From Fig. 35.5, we can observe that $A \backslash B = A \cap \bar{B}$. Hence

$$(A \cup B) \cup (A \backslash B) = (A \cup B) \cup (A \cap \bar{B})$$
$$= A \cup (B \cup (A \cap \bar{B})) \quad \text{(associative law)}$$
$$= A \cup ((B \cup A) \cap (B \cup \bar{B})) \quad \text{(distributive law)}$$
$$= A \cup ((B \cup A) \cap U)$$
$$= A \cup (B \cup A) \quad \text{(identity law)}$$
$$= (B \cup A) \cup A \quad \text{(commutative law)}$$
$$= B \cup (A \cup A)$$
$$= B \cup A$$
$$= A \cup B.$$

Alternatively, and more intuitively, we may notice that, since $A \backslash B$ is a subset of A, it is therefore also a subset of $A \cup B$, and so adds nothing to $A \cup B$ when united with it.

Example 35.7 *In a manufacturing process, a product passes through three production stages and is given a quality check at all three stages, which it either passes or fails. Let P_i represent the set of products passing the quality check at stage i. Draw a Venn diagram of the process. Interpret the quality failures of the products in the sets given by $\bar{P}_1$, $P_2\backslash(P_1 \cup P_3)$, and $(P_1 \cup P_2) \cap P_3$. What set represents the completely satisfactory products?*

A production run of 1000 occurs, of which 8 fail all stages, 20 pass only stage P_1, 31 only stage P_2, and 17 only stage P_3; 814 pass stages P_1 and P_2, 902 stages P_2 and P_3, and 800 stages P_3 and P_1. Determine the final number which pass all quality checks.

$\bar{P}_1$ represents all products which fail the P_1 quality check.
$P_2\backslash(P_1 \cup P_3)$ represents those products which pass only P_2 stage.
$(P_1 \cup P_2) \cap P_3$ represents those products which are satisfactory at stages P_3 and P_1 or P_2. The set $P_1 \cap P_2 \cap P_3$ represents those products which are satisfactory at all stages.

The numbers associated with each subset of the universal set U are shown in Fig. 35.10. Since 8 fail all quality checks, then the number of elements in $P_1 \cup (P_2 \cup P_3)$ is 992. In the figure, k represents the number of products which pass all the quality checks. Hence $800 - k$, for example, represents those products which are satisfactory in stages P_1 and P_2, but fail in P_3. Thus $P_1 \cup P_2 \cup P_3$ contains

$$992 = 20 + 31 + 17 + (814 - k) + (902 - k) + (800 - k) + k$$

products. Hence $992 = 2584 - 2k$, and so

$$k = 796.$$

Of the 1000 products manufactured, 796 passed all the quality checks.

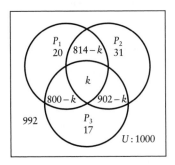

Fig. 35.10

In the previous Example, we are really interested in the *numbers* of elements in each of the sets. For example, the number of elements in U is 1000 and the number of elements in $P_2\backslash(P_1 \cup P_3)$, those products which pass only stage 2, is 31. We write

$$\mathrm{n}(U) = 1000, \qquad \mathrm{n}[P_2\backslash(P_1 \cup P_3)] = 31.$$

The number of elements in the set S is $\mathrm{n}(S)$: this number is known as the **cardinality** of S. Many sets can have **infinite cardinality**. For example, $\mathrm{n}(\mathbb{Q})$, where $\mathbb{Q}$ is the set of rational numbers, is an infinite number. We write $\mathrm{n}(\mathbb{Q}) = \infty$. The empty set $\varnothing$ has no elements: hence $\mathrm{n}(\varnothing) = 0$.

The following results apply to *finite sets*. If two finite sets A and B are disjoint, then they have no elements in common. It follows that

$$\mathrm{n}(A \cup B) = \mathrm{n}(A) + \mathrm{n}(B).$$

This result applies to any number of disjoint sets. It is clear that they must be disjoint, since otherwise elements would be counted more than once.

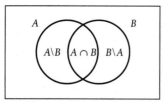

Fig. 35.11 Counting elements in the union of two sets.

This last result is also a useful method of counting elements when combined with a Venn diagram. Consider just two sets A and B as shown in Fig. 35.11. The sets representing each of the subsets in the Venn diagram $A \backslash B$, $A \cap B$, and $B \backslash A$ are shown in Fig. 35.11. Since these sets are disjoint, then we can obtain a formula for the number of elements in the union of A and B, namely

$$n(A \cup B) = n(A \backslash B) + n(A \cap B) + n(B \backslash A). \qquad (35.7)$$

For sets A and B separately,

$$n(A) = n(A \backslash B) + n(A \cap B), \qquad n(B) = n(B \backslash A) + n(A \cap B). \quad (35.8)$$

Elimination of $n(A \backslash B)$ and $n(B \backslash A)$ between (35.7) and (35.8) leads to the alternative result

$$n(A \cup B) = n(A) + n(B) - n(A \cap B).$$

For three finite sets A, B, and C the corresponding result is

$$n(A \cup B \cup C) = n(A) + n(B) + n(C) + n(A \cap B \cap C)$$
$$- n(B \cap C) - n(C \cap A) - n(A \cap B).$$

This result can be constructed from the Venn diagram.

Problems

35.1 (Section 35.1). List the elements in the following sets:
(a) $S = \{x \,|\, x \in \mathbb{N}^+ \text{ and } 3 \leqslant x \leqslant 10\}$;
(b) $S = \{x \,|\, x \in \mathbb{N}^+ \text{ and } -2 \leqslant x \leqslant 4\}$;
(c) $S = \{x \,|\, x \in \mathbb{Z} \text{ and } -2 \leqslant x \leqslant 4\}$;
(d) $S = \{x \,|\, x \in \mathbb{N}^+, \mathbb{N}^-, \text{ and } -2 \leqslant x \leqslant 4\}$;
(e) $S = \{1/x \,|\, x \in \mathbb{N}^+ \text{ and } 3 \leqslant x \leqslant 8\}$;
(f) $S = \{x^2 \,|\, x \in \mathbb{N}^+ \text{ and } |x| \leqslant 3\}$;
(g) $S = \{x + iy \,|\, x \in \mathbb{N}^+, y \in \mathbb{N}^+, 1 \leqslant x \leqslant 4, 2 \leqslant y \leqslant 5\}$.

35.2 (Section 35.3). Show on Venn diagrams the following sets:
(a) $A \cup \bar{B}$;
(b) $\bar{A} \cap \bar{B}$;
(c) $A \cap (B \cup C)$;
(d) $(A \cap B) \cup (B \cap C)$;
(e) $\overline{A \cap B}$;
(f) $(A \backslash B) \cap C$;
(g) $A \backslash (B \cap C)$;
(h) $\overline{(A \backslash B)} \cup (B \backslash C)$.

35.3 (Section 35.2). Determine the union $A \cup B$ of each of the following pairs of sets A and B:

(a) $A = \{x \,|\, x \in \mathbb{R} \text{ and } -1 \leqslant x \leqslant 2\}$,
    $B = \{x \,|\, x \in \mathbb{R} \text{ and } -1 \leqslant x \leqslant 4\}$;
(b) $A = \{x \,|\, x \in \mathbb{R} \text{ and } -1 \leqslant x < 0\}$, $B = \{x \,|\, x \in \mathbb{R} \text{ and } 0 < x < 1\}$;
(c) $A = \{1, 2, 3, 4\}$, $B = \{-4, -3, -2, -1\}$;
(d) $A = \{y \,|\, y = \cos x, x \in \mathbb{R}, \text{ and } 0 \leqslant x \leqslant \frac{1}{2}\pi\}$,
    $B = \{y \,|\, y = \sin x, x \in \mathbb{R}, \text{ and } -\frac{1}{2}\pi \leqslant x \leqslant \frac{1}{2}\pi\}$.

35.4 (Section 35.2). Determine the intersections $A \cap B$ of the following sets:
(a) $A = \{x \,|\, x \in \mathbb{R}, \text{ and } -2 \leqslant x \leqslant 1\}$,
    $B = \{x \,|\, x \in \mathbb{R}, \text{ and } -1 \leqslant x \leqslant 2\}$;
(b) $A = \{x \,|\, x \in \mathbb{N}^+ \text{ and } -5 \leqslant x \leqslant 2\}$, $B = \{x \,|\, x \in \mathbb{R}, \text{ and } -5 \leqslant x \leqslant 2\}$;
(c) $A = \{n \,|\, n = 1/m \text{ and } m \in \mathbb{N}^+\}$, $B = \{n \,|\, n = 1/m^2 \text{ and } m \in \mathbb{N}^+\}$;
(d) $A = \{x \,|\, x \in \mathbb{R} \text{ and } x^2 - 3x + 2 = 0\}$,
    $B = \{x \,|\, x \in \mathbb{R} \text{ and } 2x^2 + x - 3 = 0\}$;
(e) $A = \{x \,|\, x \in \mathbb{R} \text{ and } |x| \leqslant 2\}$,
    $B = \{x \,|\, x \in \mathbb{R} \text{ and } |x - 1| \leqslant 1\}$.

35.5 (Section 35.3). Construct a set formula for the shaded sets of Fig. 35.12:

(a)

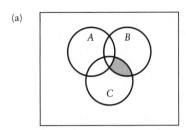

(b)

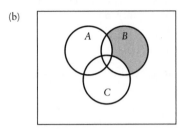

(c)

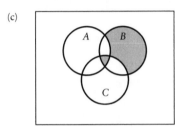

(d)

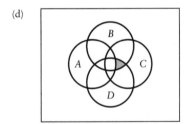

Fig. 35.12

35.6 The set S consists of products, each of which is given n pass/fail tests, numbered 1 to n. The set S_r, consists of those products that pass test r. What is the set of products that
(a) fails all tests,
(b) fails only test 1,
(c) fails some tests?

35.7 At Keele University, all first-year students must take three subjects of which at least one must be a science subject, and at least one must be a humanities or social science subject. Let A be the set of all first-year students in a given year, A_1 the set of students who take exactly one science subject, B_1 the set of students who take just one humanities subject, and B_2 the set of those who take two social science subjects. Draw a Venn diagram to represent the different sets of students classified by groups of subjects. Give set formulae for students who take
(a) just one social science subject,
(b) no humanities subject,
(c) one subject from each group.

35.8 (Section 35.3). The rules listed in (35.4) illustrate the **duality principle** which states that every statement involving sets which is true *for all sets* has a dual in which $\cup$ and $\cap$ are interchanged, and $\varnothing$ and U are interchanged everywhere.

Use Venn diagrams to establish the following:
(a) $(A\backslash B)\cap C=(A\cap C)\backslash B$;
(b) $A\cap(B\cup C)=(A\cap B)\cup(A\cap C)$.
What are their dual identities?

35.9 Three sets A, B, and C satisfy

$$A\cap B\cap C=(A\cap C)\cup(B\cap C).$$

Explain why the duality principle of Problem 34.8 does not apply. What condition of the duality principle is violated?

35.10 The **cartesian product** of two sets A and B is the set of all **ordered pairs** $\{(a, b)\}$, where $a\in A$ and $b\in B$. It is written as

$$A\times B=\{(a, b)\,|\,a\in A\text{ and }b\in B\}.$$

If $A=B$, then we write $A\times A=A^2$. Let $A=\{1, 2\}$ and $B=\{1, 2, 3\}$; write down all the elements in the sets $A\times B$, $B\times A$, A^2, and B^2.

35.11 The cartesian product extends to the products of three or more sets. Thus

$$A\times B\times C=\{(a, b, c)\,|\,a\in A\text{ and }b\in B\text{ and }c\in C\}.$$

Let $A=\{1, 2, 3\}$, $B=\{0, 1\}$, and $C=\{1, 2\}$. Write down all the elements in

$$A\times B\times C,\ A^2\times C,\ (A\cup B)\times C,\ (A\cap B)\times C.$$

35.12 At the end of a production process, 500 electrical components pass through three quality checks P, Q, and R. It is found that 38 components fail check P, 29 fail Q, 30 fail R, 7 fail P and Q, 5 fail Q and R, 8 fail R and P, and 3 fail all checks. Determine how many components:
(a) pass all checks,
(b) fail just one check,
(c) fail just two checks.

35.13 (Section 35.4). For three finite sets A, B, and C, show that the number of elements in the union of the sets is given by

$$n(A \cup B \cup C) = n(A) + n(B) + n(C)$$
$$+ n(A \cap B \cap C) - n(B \cap C)$$
$$- n(C \cap A) - n(A \cap B).$$

35.14 If A and B are two finite sets, explain why, for the cartesian product (defined in Problem 35.10 above),

$$n(A \times B) = n(A)n(B).$$

35.15 The menu in a restaurant contains three courses: 4 starters (set A), 5 main courses (set B), and 3 sweets (set C). Customers can choose either the full menu or, alternatively, a main course and a sweet. In terms of cartesian products what is the set of all possible meals (the answer is really a *set of pairs and triples*). For how many different orders can customers ask?

35.16 Given $A = \{1, 2, 3\}$, $B = \{3, 4\}$, and $C = \{2, 3, 4, 5\}$, find the elements in the sets $B \cup C$, $B \cap C$, and the cartesian products $A \times B$ and $A \times C$. Verify that

$$A \times (B \cup C) = (A \times B) \cup (A \times C),$$

$$A \times (B \cap C) = (A \times B) \cap (A \times C).$$

(This example suggests general results which are true for all sets.)

36 Boolean algebra: logic gates and switching functions

CONTENTS

36.1 Laws of Boolean algebra

We are now going to present some new operations between special entities. They have some analogies with ordinary addition and multiplication, and the symbols for them will be similar but not the same, since we need to emphasize that these are Boolean operations. The algebra involved is named after George Boole (1815–64) who first developed the modern ideas of symbolic logic. Boolean algebra has applications in logic and switching circuits.

Consider a set B which consists of just two elements 0 and 1, that is $B = \{0, 1\}$. We shall denote the **sum of two elements** a and b of B by $a \oplus b$ (the notations $\vee$, $\cup$, and +, and the alternative term **join** are also used); we denote the **product of the two elements** by $a * b$ (the notations $\wedge$, $\cap$, $\times$, and $\cdot$, or simply ab, and the alternative term **meet** are also in use) and the **complement** of a by $\bar{a}$ ($\sim a$ and $\neg a$ are used in logic). These **binary operations** applied to the members of B are defined to give the elements shown in Table 36.1.

Table 36.1   *Binary operations*

Sum			Product			Complement	
a	b	$a \oplus b$	a	b	$a * b$	a	$\bar{a}$
0	0	0	0	0	0	0	1
0	1	1	0	1	0	1	0
1	0	1	1	0	0		
1	1	1	1	1	1		

Thus, for example

$$0 \oplus 1 = 1, \quad 1 \oplus 1 = 1, \qquad 0 * 0 = 0, \quad 1 * 1 = 1, \qquad \bar{0} = 1, \quad \bar{1} = 0.$$

The elements of B are known as **Boolean variables**. We have restricted our set B to one with just two elements or binary digits, because this is the main application in circuits and computer design, but definitions can be interpreted for more general sets. A **Boolean algebra** is a set with the operations $\oplus$, $*$, and $^-$ defined on it, together with the following laws on any elements a, b, c which belong to B:

Commutative laws:
$$a \oplus b = b \oplus a, \qquad a * b = b * a;$$
Associative laws:
$$a \oplus (b \oplus c) = (a \oplus b) \oplus c, \qquad a * (b * c) = (a * b) * c;$$
Distributive laws:
$$a * (b \oplus c) = (a * b) \oplus (a * c),$$
$$a \oplus (b * c) = (a \oplus b) * (a \oplus c). \tag{36.1}$$

In addition, the set must contain distinct identity elements 0 and 1 for the operations $\oplus$ and $*$ respectively. For these elements we must have the **identity laws**

$$a \oplus 0 = a, \qquad a * 1 = a.$$

Finally, the **complement laws** must hold:

$$a \oplus \bar{a} = 1, \qquad a * \bar{a} = 0.$$

To summarize, we can say that a Boolean algebra consists of the collection

$$(B, \oplus, *, ^-, 0, 1),$$

in other words, a set B, the binary operations $\oplus$ and $*$, the complement $^-$, and the identity elements 0 and 1.

In our case $B = \{0, 1\}$, the **binary set**, which consists simply of identity elements. We can check that the definitions in Table 36.1 satisfy the laws in (36.1). They are essentially the laws of set operations with sum $\oplus$ and product $*$ replacing union $\cup$ and intersection $\cap$, and with 1 replacing the universal set U and 0 the empty set $\varnothing$.

Just as with sets, we can deduce further laws, some of which are included in (36.2):

Absorption laws:
$$a \oplus (a * b) = a, \qquad a * (a \oplus b) = a;$$
de Morgan's laws: $\overline{a \oplus b} = \bar{a} * \bar{b}, \qquad \overline{a * b} = \bar{a} \oplus \bar{b};$
Identity laws:
$$a \oplus 0 = a, \qquad a * 1 = a,$$
$$1 \oplus a = a \oplus 1 = 1, \qquad 0 * a = a * 0 = 0;$$
Reflexive law: $\bar{\bar{a}} = a.$
$$\tag{36.2}$$

Note that * *takes precedence over* ⊕ in the absence of brackets. Thus, in the first absorption law, $a \oplus a * b$ means $a \oplus (a * b)$; in the second absorption law, the brackets are essential.

We will prove one of the absorption laws to illustrate how proofs are approached in Boolean algebra.

Example 36.1 *Prove that* $a \oplus a * b = a$.

For all $a, b \in B$

$$a \oplus a * b = a * 1 \oplus a * b \quad \text{(identity law)}$$
$$= a * (1 \oplus b) \quad \text{(distributive law)}.$$

Now

$$1 \oplus b = (1 \oplus b) * 1 \quad \text{(identity law)}$$
$$= 1 * (b \oplus 1) \quad \text{(associative law)}$$
$$= (b \oplus \bar{b}) * (b \oplus 1) \quad \text{(complement law)}$$
$$= b \oplus \bar{b} * 1 \quad \text{(distributive law)}$$
$$= b \oplus \bar{b} \quad \text{(identity law)}$$
$$= 1 \quad \text{(complement law)}.$$

Finally

$$a \oplus a * b = a * 1 = a.$$

36.2 Logic gates and truth tables

Any expression made up from the elements of B and the operations ⊕, *, and ⁻ is known as a **Boolean expression**. For example,

$$a \oplus b, \quad a \oplus \bar{b}, \quad a \oplus \bar{a} * b,$$

are Boolean expressions. For the binary set, the elements 1 and 0 can represent 'on' or 'off' states in digital circuits. The basic components in a computer are **logic gates** which can produce an output from inputs. All the outputs and inputs can be in one of two states, usually either low voltage (0) or high voltage (1).

The fundamental Boolean operations of ⊕, *, and ⁻ correspond to devices known respectively as the OR **gate**, AND **gate**, and NOT **gate**. As with circuit components such as resistance and inductance, each has its own symbol.

The OR gate has two inputs and a single output represented by the symbol in Fig. 36.1. The output is $f = a \oplus b$. The inputs a and b can each take either of the values 0 or 1. Hence there are four possible inputs into the device as listed in Table 36.2. The final column f can be completed using the sum rule in Table 36.1. Then, if a is 'on' (1) and b is 'off' (0), the output f is 'on' (1). Table 36.2 is known as the **truth table** of the OR gate.

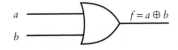

Fig. 36.1 The OR gate.

Table 36.2   *Truth table for the* OR *gate*

a	b	$f = a \oplus b$
0	0	0
0	1	1
1	0	1
1	1	1

The symbol and truth table for the AND **gate** are shown in Fig. 36.2 and Table 36.3. Again the device has two inputs and the single output $f = a * b$, the product of a and b.

Table 36.3   *Truth table for the* AND *gate*

a	b	$f = a * b$
0	0	0
0	1	0
1	0	0
1	1	1

Fig. 36.2 The AND gate.

Finally the NOT **gate** is shown in Fig. 36.3 with its truth table given as Table 36.4. The NOT gate has a single input and a single output which is the complement of its input.

Table 36.4   *Truth table for the* NOT *gate*

a	$f = \bar{a}$
0	1
1	0

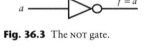

Fig. 36.3 The NOT gate.

There is further jargon associated with these gates. The output $a \oplus b$ is known as the **disjunction** of a and b, while $a * b$ is known as the **conjunction** of a and b, and $\bar{a}$ is called the **negation** of a.

These devices can be connected in series and parallel to create new logic devices, each of which will have its own truth table.

A series connection between a NOT gate and an AND gate is shown in Fig. 36.4a. The output $a * b$ of the AND gate becomes the input of the NOT gate which results in the output $\overline{a * b}$. This device is known as the NAND **gate**, and it has its own symbolic representation shown in Fig. 36.4b. Its truth table is given in Table 36.5.

A series connection between a NOT gate and an OR gate produces the NOR **gate** as shown in Fig. 36.5a. The output f is the complement of the sum of a and b. The NOR gate also has its own symbol shown in Fig. 36.5b. It has the truth table shown in Table 36.6.

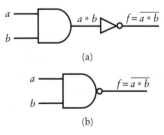

Fig. 36.4 The NAND gate.

Table 36.5 *Truth table for the* NAND *gate*

a	b	$f = \overline{a * b}$
0	0	1
0	1	1
1	0	1
1	1	0

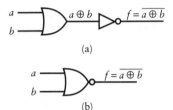

(a)

(b)

Fig. 36.5 The NOR gate.

Table 36.6 *Truth table for the* NOR *gate*

a	b	$f = \overline{a \oplus b}$
0	0	1
0	1	0
1	0	0
1	1	0

36.3 Logic networks

The five gates introduced in the previous section can be linked in series and parallel combinations to create further **logic networks**. Some examples are presented here.

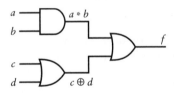

Fig. 36.6

Example 36.2 *Construct the Boolean expression for the output f of the device shown in Fig. 36.6.*

Starting from the left in Fig. 36.6, the upper AND gate produces an output $a * b$ and the lower OR gate has an output $c \oplus d$. These become the inputs into the OR gate on the right. Hence the final output is

$$f = a * b \oplus c \oplus d.$$

Since there are four inputs, the output f can be determined for each of the $2^4 = 16$ possible inputs.

Example 36.3 *Figure 36.7 shows a logical network with three inputs a, b, c, and four devices. Find a Boolean expression for the output f. Write down the truth table for the system.*

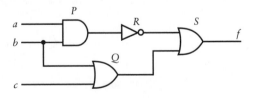

Fig. 36.7

Example 36.3 *continued*

The input b is the same in both devices P and Q. The output from the AND gate P is $a * b$, and the output from R is $\overline{a * b}$. The output from Q is $b \oplus c$. Hence the inputs $\overline{a * b}$ and $b \oplus c$ into S produce an output

$$f = \overline{a * b} \oplus b \oplus c.$$

The truth table for this network is given in Table 36.7.

Table 36.7

a	b	c	$\overline{a * b}$	$b \oplus c$	$\overline{a * b} \oplus (b \oplus c)$
0	0	0	0	1	1
0	0	1	0	1	1
0	1	0	0	1	1
0	1	1	0	1	1
1	0	0	0	1	1
1	0	1	0	1	1
1	1	0	1	0	1
1	1	1	1	0	1

Example 36.4 *Show that, using just the NOR gate, it is possible to build a logic network to model any Boolean expression.*

Given inputs a and b, we have to show that devices can be constructed using just NOR gates with outputs of $a \oplus b$, $a * b$, and $\bar{a}$. For inputs of a and b, the single NOR gate generates an output of $\overline{a \oplus b}$. Figure 36.8 shows three devices which simulate the required outputs.

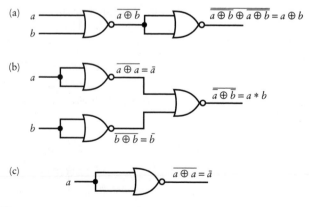

Fig. 36.8

Example 36.5　*Design a logic network using* OR, AND, *and* NOT *gates to reproduce the Boolean expression* $f = a * \bar{b} \oplus a$ *for inputs a and b.*

From input b we obtain $\bar{b}$ by a NOT gate. The inputs a and $\bar{b}$ are then fed into an AND gate to produce $a * \bar{b}$. Finally a spur from the a input and the $a * \bar{b}$ output are fed into an OR gate as shown in Fig. 36.9.

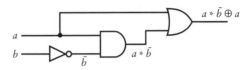

Fig. 36.9

36.4　The inverse truth-table problem

In this problem we attempt to recreate a Boolean expression for a given truth table. For example, Table 36.8 is a truth table for two inputs a and b. We illustrate a method for the construction of a Boolean expression which will generate this truth table. Pick out cases for which $f = 1$. For the case $a = 0$, $b = 1$, write down $\bar{a} * b$, and for $a = 1$, $b = 0$ write down $a * \bar{b}$, using in the products, the *complement of any zero element*. Now form the sum of the elements which produce $f = 1$. We obtain

$$f = \bar{a} * b \oplus a * \bar{b}, \tag{36.3}$$

Table 36.8

a	b	f
0	0	0
0	1	1
1	0	1
1	1	0

where it can be checked that, if $a = b$, then $f = 0$, and if a and b are not the same, then $f = 1$.

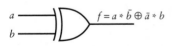

Fig. 36.10 The exclusive-OR gate.

This particular gate is known as the **exclusive-OR gate**, or EXOR **gate**, and has its own symbol shown in Fig. 36.10. This form of f obtained by the construction just described is known as the **disjunctive normal form**. By the definitions in Table 36.1, the construction *guarantees* a Boolean expression for any truth table.

The method can be applied to more complex truth tables. Table 36.9 shows an output for three inputs. The output 1 appears in rows 2, 4, 5, 7, 8. The disjunctive normal form for a corresponding

Table 36.9

a	b	c	f
0	0	0	0
0	0	1	1
0	1	0	0
0	1	1	1
1	0	0	1
1	0	1	0
1	1	0	1
1	1	1	1

Boolean expression is, following the rules for products of elements and their complements,

$$f = \bar{a} * \bar{b} * c \oplus \bar{a} * b * c \oplus a * \bar{b} * \bar{c} \oplus a * b * \bar{c} \oplus a * b * c.$$

Thus in row 2 write $\bar{a} * \bar{b} * c$, since $a = b = 0$ but c is 1, in row 4, $\bar{a} * b * c$, and so on. Check that f does give the required output. The disjunctive normal form always guarantees an answer, but it is not necessarily the simplest or most efficient in circuit architecture.

36.5 Switching circuits

A circuit of **on–off** switches can also be represented by Boolean expressions. For example, Fig. 36.11 shows a simple on–off switch in part of a circuit. Current flows if the switch S is in the *on* or closed position ($a = 1$), and does not flow if the switch is in the *off* or open position ($a = 0$). The variable a represents the state of the switch.

Fig. 36.11 On–off switch.

Consider two switches S_1 and S_2 in series (Fig. 36.12). Current only flows if both switches are closed, that is when $a_1 = 1$ and $a_2 = 1$, where a_1 and a_2 represent the states of the switches. Hence the truth table for the series switches is as shown in Table 36.10. Thus the state of current flow is given by $f = a * b$, the product of a and b.

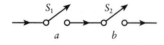

Fig. 36.12 Two switches in series.

Table 36.10 *Truth table for two switches in series*

a	b	f
0	0	0
0	1	0
1	0	0
1	1	1

Table 36.11 *Truth table for two switches in parallel*

a	b	f
0	0	0
0	1	1
1	0	1
1	1	1

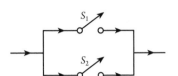

Fig. 36.13 Two switches in parallel.

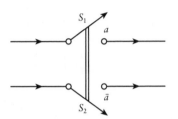

Fig. 36.14 Complement of a switch using a rigid tie.

Similarly two switches in parallel (Fig. 36.13) correspond to the sum of a and b. The truth table is given in Table 36.11. The final column indicates that $f = a \oplus b$.

The complement of a, the state of switch S_1, is another switch S_2 in the circuit which is always in the complementary state to S_1, off when S_1 is on and vice versa. It can be represented symbolically by Fig. 36.14, in which the switches S_1 and S_2 are joined by a rigid tie.

These devices are analogous to the gates of Section 36.3. For switching circuits, the Boolean expressions are often referred to as **switching functions**.

Example 36.6 *Find a switching function f for the system shown in Fig. 36.15.*

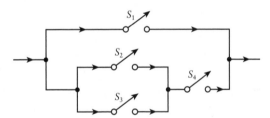

Fig. 36.15

Let a_1, a_2, a_3, a_4 represent respectively the states of each switch S_1, S_2, S_3, S_4. Since S_2 and S_3 are in parallel, their output will be $a_2 \oplus a_3$. This combined in series with a_4 will give an output of $(a_2 \oplus a_3) * a_4$. In turn, this is in parallel with S_1. Hence, the final output is

$$(a_2 \oplus a_3) * a_4 \oplus a_1.$$

Example 36.7 *A light on a staircase is controlled by two switches S_1 and S_2, one at the bottom of the stairs and one at the top. Switches can be separately 'up' or 'down'. If both switches are up, the light is off. Either switch changed to down switches the light on, and any subsequent change to a switch alters the state of the light. Design a truth table for the circuit.*

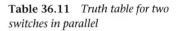

Example 36.7 *continued*

Table 36.12

Switch S_1	Switch S_2	Light	a_1	a_2	f
up	up	off	0	0	0
down	up	on	1	0	1
down	down	off	1	1	0
up	down	on	0	1	1

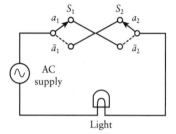

Fig. 36.16 Two-switch light control.

The truth table is shown in Table 36.12, where the state of S_i ($i = 1, 2$) is $a_i = 0$ when the switch is up (off) and $a_i = 1$ when the switch is down (on). The light on is $f = 1$, and the light off is $f = 0$. This truth table is the same as that for the exclusive-OR gate in Section 36.4. Hence, from (36.4), the circuit can be represented by the switching function

$$f = a_1 * \bar{a}_2 \oplus \bar{a}_1 * a_2.$$

The actual circuit is shown in Fig. 36.16, where S_1 and S_2 are one-pole two-way switches. At S_1, the state a_1 represents the switch 'up' and its complement $\bar{a}_1$ is the switch down. A similar state operates at S_2.

Problems

36.1 Read through Example 36.1. Now prove the other absorption law:

$$a * a \oplus b = a.$$

(Example 36.1 and this result illustrate the **duality principle**, which states that any theorem which can be proved in Boolean algebra implies another theorem with $*$ and $\oplus$ interchanged for the same elements.)

36.2 (Section 36.1). Prove the de Morgan result

$$\overline{a \oplus b} = \bar{a} * \bar{b},$$

by showing that $(a \oplus b) \oplus (\bar{a} * \bar{b}) = 1$. Explain how the duality result (Problem 36.1) gives the other de Morgan theorem.

36.3 (Section 36.1). Let B be the Boolean algebra with the two elements 0 and 1. For arbitrary $a, b \in B$, prove the following:
(a) $a * (\bar{a} \oplus b) = a * b$;
(b) $(a \oplus b) * (a \oplus \bar{b}) = a$;
(c) $(a \oplus b) * \bar{a} * \bar{b} = 0$.

36.4 (Section 36.1). Using the laws of Boolean algebra for the set with two elements 0 and 1, show that:
(a) $a * b \oplus a * \bar{b} = a$;   (b) $a \oplus \bar{a} * \bar{b} * c = a \oplus \bar{b} * c$.
Use the result to obtain the truth tables in each case.

36.5 (Section 36.4). In Problem 36.4b, it is shown that

$$a \oplus \bar{a} * \bar{b} * c = a \oplus \bar{b} * c.$$

Design two sequences of gates which give the same output for the inputs a, b, and c. The resultant gates are said to be **logically equivalent**.

36.6 (Section 36.4). Design a circuit of gates to produce the output

$$(a \oplus \bar{b}) * (a \oplus \bar{c}).$$

Construct the truth table for this Boolean expression.

36.7 (Section 36.1). Show that the Boolean expressions $(a \oplus b) * (\bar{a} \oplus b) \oplus a$ and $a \oplus b$ are equivalent.

36.8 (Section 36.1). Show that the following Boolean expressions are equivalent:
(a) $a \oplus b$;   (b) $a \oplus b * b$.

36.9 (Section 36.3). Find a Boolean expression f which corresponds to the truth table shown in Table 36.13.

Table 36.13

a	b	c	f
0	0	0	1
0	0	1	1
0	1	0	0
0	1	1	0
1	0	0	1
1	0	1	1
1	1	0	0
1	1	1	0

36.10 (Section 36.3). Construct Boolean expressions for the output f in the devices shown in Figs 36.17a–d. Construct the truth tables in each case.

(a)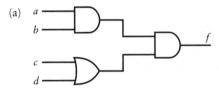

(b)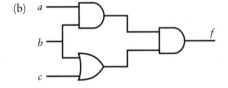

(c)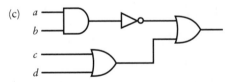

(c)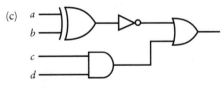

Fig. 36.17

36.11 Find the outputs f and g in the logic circuits shown in Fig. 36.18. This device can represent **binary addition** in which g is the 'carry' in the binary table shown in Table 36.14. The output g gives the '1' in the '10' in the binary sum $1 + 1 = 10$.

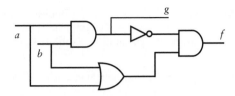

Fig. 36.18

Table 36.14

x	y	$x + y$
0	0	0
0	1	1
1	0	1
1	1	10

36.12 (Section 36.3). Reproduce the logic gate in Fig. 36.6 using just the NOR gate.

36.13 (Section 36.4). Using the disjunctive normal form, construct a Boolean expression f for the truth tables given in Tables 36.15 and 36.16.

Table 36.15

a	b	f
0	0	0
0	1	1
1	0	1
1	1	1

Table 36.16

a	b	c	f
0	0	0	1
0	0	1	0
0	1	0	0
0	1	1	1
1	0	0	1
1	0	1	0
1	1	0	1
1	1	1	0

36.14 (Section 36.3). Show that any Boolean expression can be modelled using just a NAND gate. (Hint: use a method similar to that explained in Example 36.4.)

36.15 (Section 36.4). Find switching functions for the switching circuits shown in Figs 36.19a,b.

36.16 A lecture theatre has three entrances and the lighting can be controlled from each entrance; that is, it can be switched on or off independently. The light is 'on' if the output f equals 1 and 'off' if $f = 0$. Let $a_i = 1$ $(i = 1, 2, 3)$ when switch i is up, and let $a_i = 0$ $(i = 1, 2, 3)$ when it is down. Construct a truth table for the state of the lighting for all states of the switches. Also specify a Boolean expression which will control the lighting.

(a)

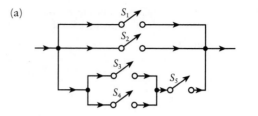

(b)

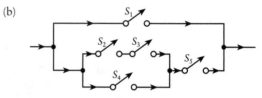

Fig. 36.19

37 Graph theory and its applications

CONTENTS

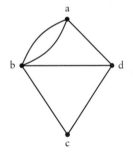

Fig. 37.1

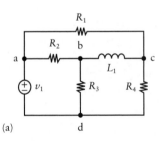

(a)

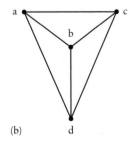

(b)

Fig. 37.2

37.1 Examples of graphs

A graph is a network or diagram composed of **points**, or **nodes** or **vertices**, joined together by lines or **edges**, each of which has a **vertex** at each end. Figure 37.1 shows a graph which has four vertices {a, b, c, d} and six edges {ab, ab, ad, bd, bc, cd}. Two vertices are not joined in this graph, namely a and c, while a and b are joined by two edges. Generally, it is not the shape of the graph which is important; it is usually the number and connection of the edges which is significant.

Here are some practical examples of situations and objects which can be usefully represented by graphs.

(i) *Electrical circuits*. Figure 37.2a shows an electrical circuit with three resistors R_1, R_2, and R_3, an inductor L, and a voltage source V_1. Each edge has just one component, and the joins between components are the vertices (the term **node** is frequently used in circuit theory) in the graph. Care has to be taken with the definition of nodes (see Section 37.6): they are not necessarily where three or more wires meet. This circuit has four vertices a, b, c, d, and it can be represented by the graph in Fig. 37.2b. The presence of a line or edge between two nodes in the graph indicates that there is a component between the nodes.

Figure 37.3 shows another circuit with six vertices in which the boxes indicate electrical components. The wires joining c to f and b to e cross over each other. In the design of printed circuits, it is useful to know whether the circuit can be redrawn so that no wires cross. Such a graph, with no edges crossing, is known as a **planar** graph. The graph in Fig. 37.2 is planar, but the graph of the circuit in Fig. 37.3 has no planar drawing: at least two edges will cross in any plane diagram of it.

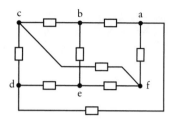

Fig. 37.3

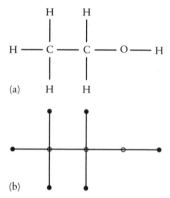

(a)

(b)

Fig. 37.4 Ethanol molecule.

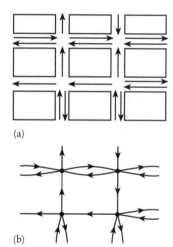

(a)

(b)

Fig. 37.5 (a) Traffic flow in a road grid, (b) Digraph representation of the roads in (a).

(ii) *Chemical molecules*. The molecule of ethanol can be represented by Fig. 37.4a. In its graph representation in Fig. 37.4b, the vertices represent **atoms** and the edges **bonds**. The number of bonds which meet at an atom is the **valency** of the atom. Thus carbon (C) has valency 4, oxygen (O) valency 2, and hydrogen (H) valency 1. Generally in graphs, the number of edges that meet at a vertex is known as the **degree** of the vertex.

(iii) *Road maps*. Road maps and street plans are graphs with roads as edges and junctions as vertices. However, most road networks include one-way streets. Hence graphs need to be modified to indicate directions in which movement or flow is permitted. Figure 37.5a shows a typical section of a street plan with some one-way streets. We have to associate directions with the edges as shown in the graph of the plan in Fig. 37.5b. Note that two-way streets now have *two directed edges* associated with them. This is an example of a **directed graph**, which is also known by the shortened term **digraph**.

(iv) *Shortest paths*. Figure 37.6 shows a digraph with **weights** associated with each edge. The graph could represent routes between towns S and F which pass through intermediate towns A, B, ... , the weights associated with each directed edge could stand for distances or times. This graph is shown as a digraph, but weights could be present without directions in some cases. We might be interested in this example in the shortest distance between the start (S) and the finish (F).

37.2 Definitions and properties of graphs

As we have seen, a **graph** is an object composed of vertices and edges with one vertex at each end of every edge. An edge which joins a vertex to itself is known as a **loop**. If two or more edges join the same two vertices then they are known as **multiple edges**. A graph with no loops or multiple edges is known as a **simple graph**. A graph with loops and/or multiple edges is known as a **multigraph**.

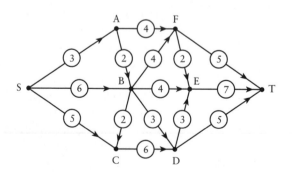

Fig. 37.6

A graph in which every vertex can be reached from every other vertex along a succession of edges is said to be **connected**. Otherwise the graph is said to be **disconnected**. A connected graph is in one piece; a disconnected graph is in two or more pieces.

The **degree** of a vertex x is the number of edges that meet there, denoted by $\deg(x)$. If, in a graph G, all the vertices have the same degree r, then G is said to be **regular of degree** r.

Example 37.1 *Find the degree of the vertices in the graph in Fig.* 37.1.

Three edges meet at the vertex a. Hence $\deg(a) = 3$. Four edges meet at b. Hence $\deg(b) = 4$. Similarly, $\deg(c) = 2$ and $\deg(d) = 3$.

A simple graph in which every vertex is joined to every other vertex by just one edge is called a **complete graph**.

Figure 37.7 shows some examples of the various graphs described above.

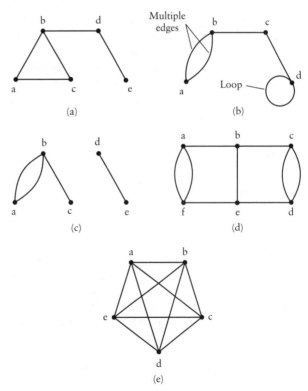

Fig. 37.7 (a) Connected simple graph. (b) Connected multigraph. (c) Disconnected multigraph. (d) Regular graph of degree 3. (e) Complete graph with five vertices: $\deg(a) = 4$.

Since every edge has a vertex at each end, it follows that the sum of all the vertex degrees equals twice the number of edges. This is known as the **handshaking lemma**. For example, from Example 37.1,

$$\deg(a) + \deg(b) + \deg(c) + \deg(d) = 3 + 4 + 2 + 3 = 12,$$

which is twice the number of edges in the graph shown in Fig. 37.1.

There are two immediate consequences of the handshaking lemma:

(i) the sum of all the vertex degrees in a graph is an even number;

(ii) the number of vertices of odd degree is even.

37.3 How many simple graphs are there?

Graphs can be described as **labelled**, in which case the vertices are distinguishable as in Fig. 37.8a or **unlabelled** as in Fig. 37.8b. If we look at graphs with just three vertices, there are eight labelled simple graphs as shown in Fig. 37.9, but there are just four distinct unlabelled graphs as shown in Fig. 37.10. In Fig. 37.9, the three labelled graphs with one edge will correspond to the one unlabelled graph in Fig. 37.10.

The number of labelled simple graphs with n vertices is fairly easy to calculate. Between any two vertices, there is the possibility of an edge. Any vertex can be joined to $n - 1$ other vertices. Since this will duplicate edges, there will be $\frac{1}{2}n(n - 1)$ possible edges. Each edge may be either present or not. Hence the number of possible combinations of present and absent edges will be $2^{\frac{1}{2}n(n-1)}$, which is the number of labelled graphs. Thus there must be $2^{\frac{1}{2}4(4-1)} = 2^6 = 64$ labelled graphs with four vertices; of these, 11 can be identified as unlabelled graphs.

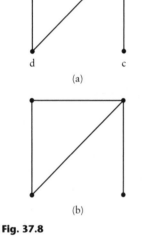

Fig. 37.8

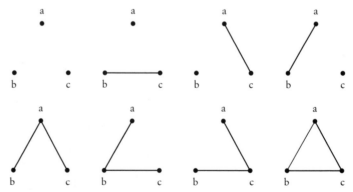

Fig. 37.9 Labelled graphs with three vertices.

Fig. 37.10 Unlabelled graphs with three vertices.

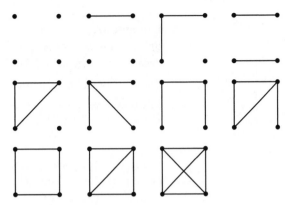

Fig. 37.11 All unlabelled graphs with four vertices.

The latter graphs are shown in Fig. 37.11. Of the 11 unlabelled graphs it can be seen that six are connected and four are regular.

For applications involving circuits, the main interest is in connected graphs. The numbers of the various categories of graphs up to $n = 7$ vertices are given in Table 37.1. It can be seen from the table that the number of unlabelled graphs is a considerable reduction on the labelled set, and that regular graphs are comparatively rare. The counting of unlabelled graphs does not follow from a simple formula.

Table 37.1

n	1	2	3	4	5	6	7
Labelled graphs	1	2	8	64	1024	32 768	2 097 152
Unlabelled graphs	1	2	4	11	34	156	1 044
Connected graphs	1	1	2	6	21	112	853
Regular graphs	1	2	2	4	3	8	6

37.4 Paths and cycles

Consider a graph G. Suppose we follow a succession of connected edges between two vertices a and z, along which there may be repeated edges and vertices. This is known as a **walk** between a and z. If all the edges walked are different (i.e. no edge is covered more than once but vertices may be visited more than once), then the walk defines what is known as a **trail**. A trail is said to be **closed** if the first and last vertices are the same. If all the vertices on a trail are different, except possibly the end pair, then the succession defines a **path**. A closed path is known as a **cycle** or **circuit**. For example, in Fig. 37.12, a–f–b–c–d is a path between a and d, but a–b–f–e–b–c–d is only a trail since vertex b is passed through twice. Also, a–b–c–d–e–f–a is an example of a cycle.

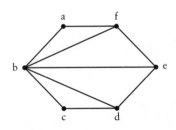

Fig. 37.12

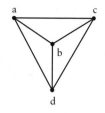

Fig. 37.13

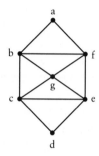

Fig. 37.14

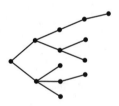

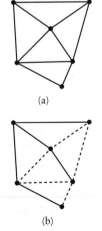

Fig. 37.15 An example of a tree.

Example 37.2   *Electrical circuits are usually such that every edge of their representative graph is part of a cycle. List all the distinct cycles in the circuit in Fig. 37.2(a).*

The graph of the circuit is repeated in Fig. 37.13. The complete list of cycles is:

3-edge cycles: a–b–c–a, a–b–d–a, a–d–c–a, b–d–c–b;
4-edge cycles: a–b–d–c–a, a–d–b–c–a, a–b–c–d–a.

Some graphs have special closed-path and cycle properties. A connected graph G is said to be **eulerian** if there exists a closed trail that includes every edge in G. A connected graph G is said to be **hamiltonian** if there exists a cycle that includes every vertex in G. The graph in Fig. 37.13 is hamiltonian but not eulerian. One hamiltonian cycle in its graph is a–b–d–c–a. Note that this cycle does not have to cover *every* edge in the graph.

The graph in Fig. 37.14 is both eulerian and hamiltonian. An eulerian trail is

a–b–c–d–e–f–g–e–c–g–b–f–a,

and a hamiltonian cycle is

a–b–c–d–e–g–f–a.

It can be shown that a graph is eulerian if and only if every vertex has even degree. This provides an easy test for the eulerian property of a graph.

37.5 Trees

A connected graph which has no cycles is known as a **tree**. An example of a tree is shown in Fig. 37.15. The edges in a tree are called **branches**.

Suppose that a graph G consists of the set $V(G)$ of vertices and the set $E(G)$ of edges. Then any graph whose vertices and edges are subsets of $V(G)$ and $E(G)$ respectively is called a **subgraph**. It is important to note that the subgraph must be a graph whose vertices and edges come from G; and only edges that join two vertices of the subgraph are permitted in the subset of $E(G)$.

Suppose that G is a connected graph.

> A **spanning tree** of G is a subgraph of G which is a tree and includes all vertices of G.

Figure 37.16a shows a connected graph G and Fig. 37.16b shows a spanning tree of G. Graphs can have many different spanning trees. The set of edges that are not part of the spanning tree (the broken edges in Fig. 37.16b) is known as the **cotree** and its edges are called **links**.

Fig. 37.16 (a) Connected graph. (b) The same graph with a spanning tree.

(a)

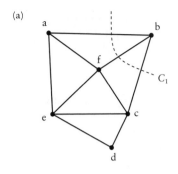

(b)

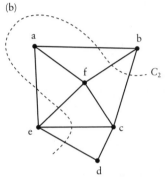

Fig. 37.17 A cutset of a graph.

Consider a tree with n vertices. Construct the tree from a chosen vertex by adding edges. Each edge added must introduce a new vertex, since otherwise a cycle would be created and the graph would no longer be a tree. Hence a tree with n vertices must have just $n - 1$ branches. It follows that a graph with n vertices must have a cotree with $e - n + 1$ links, where e is the number of edges of the graph.

We now introduce the **cutset**, by which we can disconnect a graph into two subgraphs which together contain all the original vertices, by removing a minimum set of edges in the graph.

Cutset

In a connected graph, a **cutset** is a set of edges (a) whose removal disconnects the graph into two subgraphs and (b) no proper subset of the cutset disconnects the graph.

In other words, there must be no redundancy in the cutset. Thus, for example in Fig. 37.17a, the broken line C_1, which removes the edges ba, bf, and bc, defines a cutset {ba, bf, bc}; but C_2 in Fig. 37.17b does not define a cutset, since the *subset* {ab, bf, bc} of edges disconnects the graph.

37.6 Electrical circuits: the cutset method

In this section we give a brief description of the representation of circuits by graphs, and show how Kirchhoff's laws can be applied to cutsets of the resulting graphs. Figure 37.18a shows a plan of a circuit with seven resistors, a voltage supply, and two capacitors. This particular circuit has 10 components and 10 edges. Note that A will be a vertex or **node** (a preferred term in circuits) but that the joins B, C, and D are not separate nodes but can be coalesced into a single node. The equivalent graph is shown in Fig. 37.18b: it has five nodes and 10 edges. Note that it is a multigraph with two nodes joined by two edges and two nodes joined by four edges.

A **circuit loop** in the circuit is a cycle in the graph.

Kirchhoff's laws have already been stated in eqn (21.8), but for convenience they are given again here in graph terms. They state (i) that *the algebraic sum of the voltages around any loop is zero*, and (ii) that *the algebraic sum of the currents entering any node is zero*.

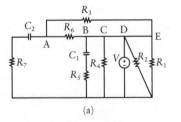

(a)

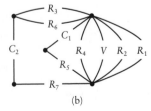

(b)

Fig. 37.18 A circuit and its graph.

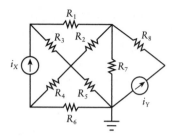

Fig. 37.19

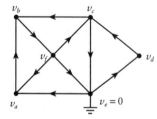

Fig. 37.20

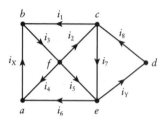

Fig. 37.21

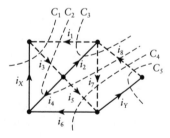

Fig. 37.22

In addition, for resistors we also have **Ohm's law** which states that *the voltage across a resistor is directly proportional to the current flowing through it*, that is

$$v \propto i \quad \text{or} \quad v = Ri,$$

where the constant R is measured in units called *ohms* (Ω). Figure 37.19 shows a circuit with two independent maintained current sources i_X and i_Y: the symbol of the circle enclosing an arrow represents a maintained current in the direction of the arrow.

The corresponding six-node digraph with currents $i_1, i_2, \ldots, i_8$ in the directions indicated is shown in Fig. 37.20. If any current turns out to be negative then its direction will be opposite to that shown.

Now introduce **nodal voltages** $v_a, v_b, \ldots, v_f$ as shown in Fig. 37.21. The use of nodal voltages means that effectively Kirchhoff's first law is automatically satisfied. The earthing at e makes $v_e = 0$ and other voltages can be measured relative to this zero ground potential.

This circuit has 13 unknowns: 8 currents and 5 nodal voltages. The problem with circuits is the selection of the minimum number of consistent equations from Kirchhoff's laws and Ohm's law sufficient to determine the unknowns.

The graph of this circuit is the same as that in Fig. 37.16a, and we shall use the same spanning tree as shown in Fig. 37.16b. In this graph, the number of nodes n is 6, the number of edges e is 10. Hence the cotree has, from the previous section, $e - n + 1 = 10 - 6 + 1 = 5$ links. Any cutset of the original graphs which contains one and only one branch of the spanning tree (the rest of the cutset consisting of links) is known as a **fundamental cutset** of the circuit. Hence we can associate five fundamental cutsets with the spanning tree in Fig. 37.16b. Five possible cutsets $C_1, C_2, \ldots, C_5$ are shown in Fig. 37.22.

By repeated use of Kirchhoff's second law to the nodes on one side of a cutset, it follows that the algebraic sum of the currents crossing the cutset must be zero. Hence the five cutset equations are:

$$C_1: i_1 - i_3 + i_X = 0, \tag{37.1}$$

$$C_2: i_1 - i_3 + i_4 + i_5 + i_7 - i_8 = 0, \tag{37.2}$$

$$C_3: i_i - i_2 + i_7 - i_8 = 0, \tag{37.3}$$

$$C_4: i_6 - i_5 - i_7 + i_8 = 0, \tag{37.4}$$

$$C_5: i_Y - i_8 = 0. \tag{37.5}$$

These equations must be independent since each one contains a current from a branch of the spanning tree which does not appear in any other equation. Further any non-fundamental cutset equation will be a linear combination of the five fundamental cutset equations. The number of branches in the spanning tree defines the number of independent equations.

We can also apply Ohm's law to each resistor (note that current flows from high to low potential). Thus the voltage difference across R_1 is $v_c - v_b$, so that

$$i_1 = (v_c - v_b)/R_1. \tag{37.6}$$

Similarly

$$i_2 = (v_f - v_c)/R_2, \tag{37.7}$$

$$i_3 = (v_b - v_f)/R_3, \tag{37.8}$$

$$i_4 = (v_f - v_a)/R_4, \tag{37.9}$$

$$i_5 = v_f/R_5, \tag{37.10}$$

$$i_6 = (-v_a)/R_6, \tag{37.11}$$

$$i_7 = v_c/R_7, \tag{37.12}$$

$$i_8 = (v_d - v_c)/R_8. \tag{37.13}$$

We can now substitute for the currents from (37.6) to (37.13) into (37.1) to (37.5) resulting in five linear equations to determine the nodal voltages v_a, v_b, v_c, v_d, v_f in terms of the known currents i_X and i_Y. The remaining currents can then be calculated from (37.6) to (37.13).

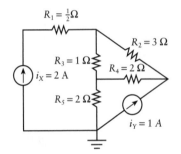

Fig. 37.23

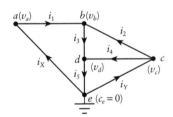

Fig. 37.24

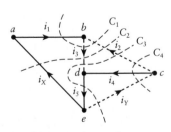

Fig. 37.25 Fundamental cutsets.

Example 37.3 *Using the cutset method, find all currents and nodal voltages in the circuit shown in Fig. 37.23.*

The circuit can be represented by a graph with five nodes (Fig. 37.24) with the currents i_1, i_2, i_3, i_4, i_5 in the directions shown.

A spanning tree with three links is shown in Fig. 37.25 together with cutsets C_1, C_2, C_3, C_4. Hence Kirchhoff's second law implies:

$$C_1: i_1 - i_3 + i_2 = 0, \tag{37.14}$$

$$C_2: i_X - i_3 + i_2 = 0, \tag{37.15}$$

$$C_3: -i_Y + i_5 - i_3 + i_2 = 0, \tag{37.16}$$

$$C_4: -i_Y + i_4 + i_2 = 0. \tag{37.17}$$

With $v_e = 0$, the currents in terms of the nodal voltages v_a, v_b, v_c, v_d are, by Ohm's law:

$$i_1 = (v_a - v_b)/R_1 = 2(v_a - v_b), \tag{37.18}$$

$$i_2 = (v_c - v_b)/R_2 = \tfrac{1}{3}(v_c - v_b), \tag{37.19}$$

$$i_3 = (v_b - v_d)/R_3 = v_b - v_d, \tag{37.20}$$

$$i_4 = (v_c - v_d)/R_4 = \tfrac{1}{2}(v_c - v_d), \tag{37.21}$$

$$i_5 = v_d/R_5 = \tfrac{1}{2}v_d. \tag{37.22}$$

Eliminate the currents in (37.14) to (37.17) using (37.18) to (37.22):

$$2v_a - \tfrac{10}{3}v_b + \tfrac{1}{3}v_c + v_d = 0, \tag{37.23}$$

$$\tfrac{4}{3}v_b - \tfrac{1}{3}v_c - v_d = 2, \tag{37.24}$$

$$-\tfrac{4}{3}v_b + \tfrac{1}{3}v_c - \tfrac{3}{2}v_d = 2, \tag{37.25}$$

$$-\tfrac{1}{3}v_b + \tfrac{5}{6}v_c - \tfrac{1}{2}v_d = 1. \tag{37.26}$$

Example 37.3 *continued*

These are linear equations which can be solved using the methods of Chapter 12. Computer algebra is also very useful in solving sets of equations of this type (see the computer algebra applications for Chapter 12 in Chapter 42). The answers are

$$v_a = 5\text{ V}, \qquad v_b = 4\text{ V}, \qquad v_c = 4\text{ V}, \qquad v_d = 2\text{ V}.$$

Since $v_c = v_b$, no current flows through the resistor on bc.

We can summarize the result for an earthed circuit which contains only resistors and current sources. Suppose that the representative graph of the circuit contains n nodes and e edges of which f contain known current sources. The curcuit will have $e - f$ unknown currents and $n - 1$ unknown nodal voltages giving $e - f + n - 1$ unknowns in total. Its spanning tree will have $n - 1$ edges which will lead to $n - 1$ fundamental cutset equations, and Ohm's law will apply to $e - f$ resistors. Hence we shall always have a consistent set of $e - f + n - 1$ equations to find the unknowns.

This result can be extended to circuits with current sources, voltage sources (batteries), and resistors. If the representative graph has n nodes and e edges of which f contain current sources and s maintained voltage sources, then the number of unknown currents will be $e - f$ and the number of unknown nodal voltages will be $n - 1 - s$ since the nodal voltage difference across a battery will be known. Hence the number of unknowns is $e - f + n - 1 - s$ which will satisfy $n - 1$ cutset equations and $e - f - s$ Ohm's laws.

37.7 Signal-flow graphs

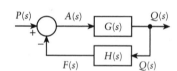

Fig. 37.26 Negative-feedback control system.

Figure 37.26 shows a **block diagram** of a **negative-feedback control system**. The input into the system is $P(s)$ and the output $Q(s)$. All operations are defined by their transfer functions (see Section 25.4). The boxes represent devices or controllers. The circle represents a sum operator, and the return sign on $F(s)$ indicates positive or negative feedback. The output signal $Q(s)$ is fed back into the input through $H(s)$, and it is a **negative feedback** which will reduce the output. In a later problem, we shall consider a device with a positive feedback. Thus the input into $G(s)$ is

$$A(s) = P(s) - F(s). \tag{37.27}$$

The boxes each produce outputs given by the transfer functions

$$Q(s) = G(s)A(s), \tag{37.28}$$

$$F(s) = H(s)Q(s). \tag{37.29}$$

We wish to find $Q(s)$ in terms of $P(s)$, $G(s)$, and $H(s)$, from the equations (37.27) to (37.29). Thus, from (37.28)

$$Q(s) = G(s)A(s) = G(s)[P(s) - F(s)],$$
$$= G(s)[P(s) - H(s)Q(s)].$$

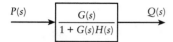

$P(s)$ $\rightarrow$ $\dfrac{G(s)}{1+G(s)H(s)}$ $\rightarrow$ $Q(s)$

Fig. 37.27 Block-reduced diagram for Fig. 37.26.

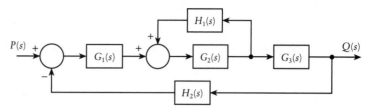

Fig. 37.28 A multiple-feedback control system.

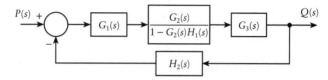

Fig. 37.29 First stage in the block reduction of the multiple-feedback control system.

Hence the output transfer function is

$$Q(s) = \frac{G(s)}{1+G(s)H(s)} P(s).$$

This is the closed-loop transfer function. The actual signal can be obtained by finding the inverse Laplace transform for $Q(s)$. Hence the system is equivalent to that shown in Fig. 37.27.

If the feedback reinforces the input signal it is called **positive feedback**. Figure 37.28 shows a multiple-feedback control system with a positive and a negative feedback. The output signal is given by

$$Q(s) = \frac{G_1(s)G_2(s)G_3(s)}{1 - G_2(s)H_1(s) + G_1(s)G_2(s)G_3(s)H_2(s)} P(s), \qquad (37.30)$$

which can be obtained by the method of **block-diagram reduction**. For example, the feedback through H_1 makes the system equivalent to that shown in Fig. 37.29. We can now combine the series devices which reduce the system to the negative-feedback control system considered at the beginning of this section. The details are omitted here.

This block-reduction method can get quite complicated for a complex feedback system. Instead of using block reduction in this way, represent the system by a **weighted digraph** as shown in Fig. 37.30, where the weights are the transfer functions – except that the edges representing the input and output are assigned weight 1 since they carry no devices. Also the negative feedback is replaced by $-H_2(s)$, to make sure that it reduces the input into $G_1(s)$. This is the **signal-flow graph** of the system. Let the inputs into the nodes be x_1, x_2, x_3, and x_4 as shown; then, for the positive-feedback cycle,

$$x_3 = G_2 x_2, \qquad x_2 = G_1 x_1 + H_1 x_3.$$

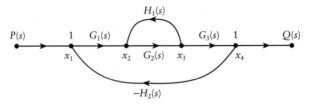

Fig. 37.30 Signal-flow graph for the multiple-feedback control system shown in Fig. 37.28.

(The argument (s) has now been dropped from the working.) Hence

$$x_3 = \frac{G_1 G_2 x_1}{1 - G_2 H_1}.$$

In other words, we can replace (a) by (b) in Fig. 37.31.

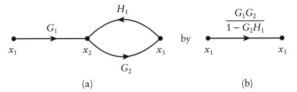

Fig. 37.31

There are other rules, and a complete list now follows for the replacements for subgraphs in the graph.

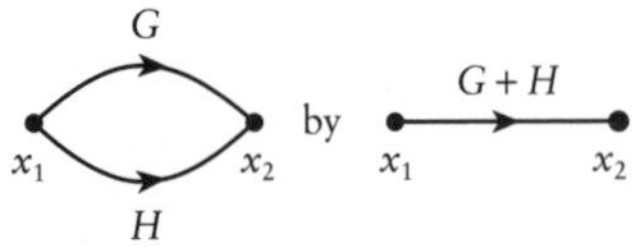

(a) *Multiple edges.* See Fig. 37.32. This follows since
$$x_2 = Gx_1 + Hx_1 = (G + H)x_1.$$
(b) *Edges in series.* See Fig. 37.33. This follows since
$$x_3 = Hx_2 = H(Gx_1) = HGx_1.$$

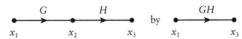

Fig. 37.33 Edges in series.

(c) *Cycles.* See Fig. 37.34. This follows since
$$x_3 = Hx_2 \quad \text{and} \quad x_2 = Gx_1 + Jx_3.$$
Assume that $HJ \neq 1$; otherwise there is infinite gain.

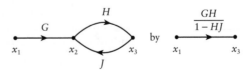

Fig. 37.34 Cycle.

Fig. 37.32 Multiple edges.

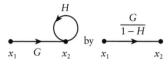

Fig. 37.35 Loop.

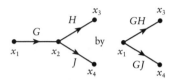

Fig. 37.36 Stem.

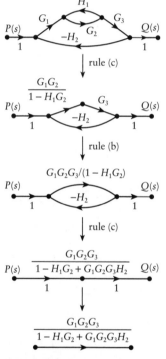

Fig. 37.37 Successive steps in the reduction of the signal-flow graph of the control system shown in Fig. 37.28.

(d) *Loops*. See Fig. 37.35. This follows since

$$x_2 = Gx_1 + Hx_2$$

with $H \neq 1$.

(e) *Stems*. See Fig. 37.36. This follows since

$$x_2 = Gx_1, \qquad x_3 = Hx_2 = HGx_1, \qquad x_4 = Jx_2 = JGx_1.$$

Apply these rules to the successive reduction of the feedback system in Fig. 37.30. The sequence of steps in the reduction of the signal-control graph to a single-edge graph is shown in Fig. 37.37. The weight of the final edge agrees with the output in eqn (37.38).

Essentially the operations in a signal-flow graph are those applied to a *weighted digraph* as illustrated in the following example.

Example 37.4 *Find the output–input relation in the signal-flow graph shown in Fig. 37.38.*

Applying rule (a) to the multiple edge, and rule (c) to the cycle, the graph is reduced to Fig. 37.39. Apply the series rule to the divided edges to give Fig. 37.40. Finally the multiple-edge and series rules give Fig. 37.41. Thus the output is given by

$$q = \frac{abd}{1 - bc} + he(g + f).$$

In the actual control system $a, b, c, \ldots$ will be transfer functions.

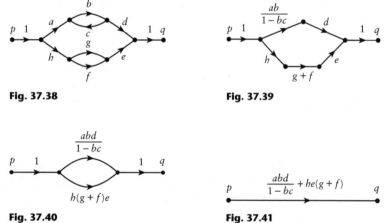

Fig. 37.38

Fig. 37.39

Fig. 37.40

Fig. 37.41

37.8 Planar graphs

As we remarked in Section 37.1, planar graphs are important in circuit design since planar circuits can be manufactured as a single board. A planar graph is a graph that can be drawn with no edges crossing or meeting except at vertices. The standard example of a

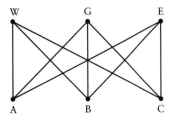

Fig. 37.42 Bipartite graph K$_{3,3}$.

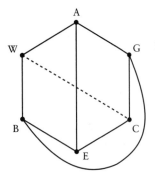

Fig. 37.43

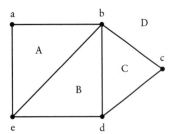

Fig. 37.44 A planar graph with fiver vertices, seven edges, and four faces.

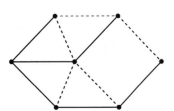

Fig. 37.45 A graph with a spanning tree.

simple application which cannot be represented by a planar graph is the delivery of three services, water (W), gas (G), and electricity (E), to three houses A, B, C (Fig. 37.42). This graph has no plane drawing. The reorganization of the graph in Fig. 37.43 shows the impossibility of this; if W and C are connected last then this edge must cross either AE or BG.

The graph in Fig. 37.42 is an example of a **bipartite graph** in which one set of vertices may be connected to another set of vertices, but not to vertices in the same set. If every vertex in one set is connected by one edge to every vertex in the other set then it is called a **complete bipartite graph**. If the sets have m and n vertices respectively, then the notation K$_{m,n}$ denotes the complete bipartite graph. Figure 37.42 shows the graph K$_{3,3}$ and this graph is not planar. Check that the graphs K$_{2,2}$ and K$_{2,3}$ are planar.

In planar graphs there is a relation between the numbers of vertices, edges, and faces. In a plane drawing of a graph, the plane is divided into regions called **faces**. One face is the region external to the graph. Figure 37.44 shows a planar graph with five vertices and seven edges, and with four faces: A, B, C, and the external face D.

A remarkable formula, due to Euler, links the numbers of vertices, edges, and faces of a graph.

> **Theorem** (Euler). Suppose that the graph G has a planar drawing, and let v be the number of vertices, e the number of edges, and f the number of faces of G. Then
> $$v - e + f = 2.$$

Proof. For the graph G, define a spanning tree (see, for example, Fig. 37.45). The spanning tree must have n vertices and $n - 1$ edges (see Section 37.5). It must also have just one face. Since

$$n - (n - 1) + 1 = 2,$$

Euler's formula holds for the spanning tree. Successively replace the other edges in the graph. Each time an extra edge is added, a face is divided and one extra face is added. However, algebraically, this cancels the additional edge in the accumulation to Euler's formula for the spanning tree. Hence

$$v - e + f = 2$$

for the reconstructed graph G.

A complete graph with n vertices is a simple graph in which every vertex is joined to every other vertex. For the n-vertex graph, it is denoted by K$_n$. The graphs of K$_2$, K$_3$, K$_4$, and K$_5$ are shown in Fig. 37.46. Of these graphs, K$_2$, K$_3$, and K$_4$ are planar, but K$_5$ and all succeeding complete graphs are not.

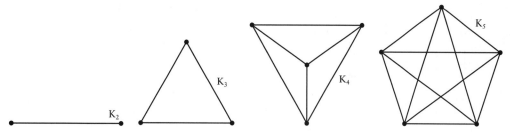

Fig. 37.46 The complete graphs K_n for $n = 2, 3, 4, 5$.

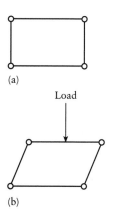

(a)

Load

(b)

Fig. 37.47 Single unbraced pin-jointed frame.

Fig. 37.48 Braced frame.

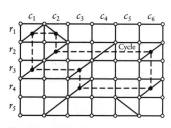

Fig. 37.49 5×6 framework.

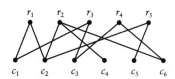

Fig. 37.50

The graphs $K_{3,3}$ and K_5 are the keys to tests for planarity of graphs, and whether it is possible to design, for example, a plane printed-circuit board to make the required connections between electronic components. It was proved by Kuratowski in 1930 that every non-planar graph contains subgraphs which are either $K_{3,3}$ or K_5, or $K_{3,3}$ or K_5 with additional vertices on their edges.

37.9 Further applications

Braced frameworks

Consider a frame which consists of four struts in the shape of a rect-angle (Fig. 37.47a) with pin joints at each corner. Without a diagonal tie the structure will not support a vertical load, but will collapse into a parallelogram as shown in Fig. 37.47b. The structure can be made rigid and load bearing by the insertion of a diagonal strut as in Fig. 37.48.

Consider now a pinjointed framework with $m \times n$ rectangular frames with some individual frames braced. How can we decide whether a particular framework is braced, that is no part of it can be sheared? And if it is braced, how many ties could be removed to leave a **minimum bracing**? The framework is similar to a vertical section of scaffolding or a steel-framed building, although in both cases the joins are bolted but can still need bracing to ensure rigidity.

Figure 37.49 shows a 5×6 framework with 11 braces as shown (braces can be diagonal struts in either direction). Label the *cell* rows $r_1, r_2, \ldots, r_5$ and the cell columns $c_1, c_2, \ldots, c_6$ as shown in Fig. 37.49. The framework will be represented by a *bipartite graph* (see Section 37.8) with the cell rows and columns as vertices. Arrange them in rows as shown in Fig. 37.50.

If a particular rectangular cell is braced then the identifying row and column vertices are joined by an edge. Thus the cell $r_1 c_1$ is braced so that an edge joins r_1 and c_1 in the bipartite graph. No edge joins r_1 and c_3 since this cell is not braced. The bipartite graph representing the framework is shown in Fig. 37.50. If the graph is *connected*, then the framework is braced since the shearing of any cell or group of cells is not then possible. The graph is connected in this case, and the framework is braced. Can any braces be removed in such a way that

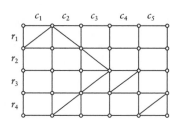
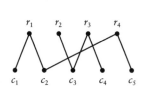
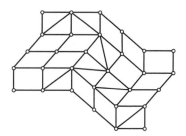

Fig. 37.51 An unbraced framework.

the framework is still braced? Any brace which is removed must not *disconnect* the graph. If the graph contains a *cycle* (Section 37.4) then any edge removed from the cycle will not disconnect the graph. This removal rule can be applied to each cycle in the graph. If, at the end of this process, there are no cycles remaining and the graph remains connected, then the framework is said to have a **minimum bracing**. The framework graph in Fig. 37.50 contains just one cycle, namely $r_1c_1r_3c_3r_4c_6r_2c_2r_1$ (see Fig. 37.49). *Any* edge can be removed from this cycle leaving a minimum bracing. The removal of any further edges will disconnect the graph.

If every cell is braced in a framework then the bipartite graph will be complete, and the framework will be seriously overbraced. You might note that a complete bipartite graph $K_{m,n}$ has mn edges but a minimum bracing for an $m \times n$ framework has $m + n - 1$ edges: for example, if $m = 5$ and $n = 6$ then $mn = 30$ whilst $m + n - 1 = 10$.

Figure 37.51 shows an unbraced 4×5 framework, its (disconnected) graph, and the same framework sheared.

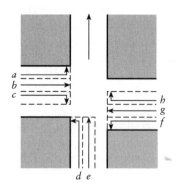

Phasing of traffic signals

Figure 37.52 shows a road junction with eight incoming lanes of traffic and a one-way exit. Suppose that each lane can be controlled by its own individual signal.

Fig. 37.52 Road junction.

One solution for traffic management would be to allow each lane to have a green signal in sequence with the remaining all on red, but this would be inefficient since obviously several lanes of traffic can move simultaneously without risk. How can an efficient phasing of the signals be designed?

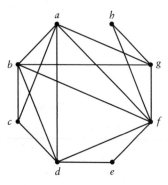

Label each incoming lane $a, b, c, \ldots, h$ as shown, and let these be vertices of a graph (Fig. 37.53). Starting, say, with a we decide which traffic lanes are *compatible* with a; that is, which lanes can also have green lights simultaneously without risk of a collision. Thus a and b are compatible, and we therefore join a and b by an edge. Lanes a and c are also compatible, and we therefore join a and b by an edge. Lanes a and c are also compatible, but a and e are not, and so on. The graph G in Fig. 37.53 shows which lanes are compatible, and is known as the **compatibility graph** for this junction.

Fig. 37.53

Table 37.2

| Time | Subgraph | | | |
	abcd	abfg	def	fgh
$0-\frac{1}{4}T$	green	red	red	red
$\frac{1}{4}T-\frac{1}{2}T$	red	green	red	red
$\frac{1}{2}T-\frac{3}{4}T$	red	red	green	red
$\frac{3}{4}T-T$	red	red	red	green

We now look for *complete subgraphs* (Section 37.8) in G. An edge is a complete subgraph (K_2), a triangle (K_3) is a complete subgraph with three vertices, K_4 with four vertices, and so on. We try to use the largest subgraphs in any *covering* of G, that is a list of subgraphs which includes *all* vertices. In G, *abcd*, *abdf*, and *abfg* are K_4 subgraphs, and there are a large number of triangles. For example, we can cover G by the set of subgraphs

$$\{abcd, abfg, def, fgh\}.$$

Generally, we include as many large subgraphs as possible. In this list it is better to use *fgh* rather than just *gh*: this could be chosen since f is included in other subgraphs.

Suppose that the period of the traffic signal sequence is T seconds with each lane having a green light for at least $\frac{1}{5}T$. There are four different traffic flows represented by the subgraphs. Suppose that each subgraph list of lanes has a green light for $\frac{1}{4}T$. The green/red phasing sequence is shown in Table 37.2.

The actual phasing lane by lane is shown in Fig. 37.54 where the solid line indicates the green light for a lane. For example, between $\frac{1}{4}T$ and $\frac{1}{2}T$, lanes a, b, e, f are on green with the others on red.

The **total waiting time** for the traffic at the junction is a measure of the efficiency of the timings and phases. Let t_a, t_b, t_c, ... be the waiting times of the lanes so that, from Fig. 37.54, we can see that $t_a = \frac{1}{2}T$, $t_b = \frac{1}{2}T$, $t_c = \frac{3}{4}T$, etc. Hence the total waiting time W_T is given by

$$W_T = t_a + t_b + \cdots + t_h$$
$$= \tfrac{1}{2}T + \tfrac{1}{2}T + \tfrac{3}{4}T + \tfrac{1}{2}T + \tfrac{3}{4}T + \tfrac{1}{4}T + \tfrac{1}{2}T + \tfrac{3}{4}T$$
$$= \tfrac{9}{2}T.$$

Can the waiting time be reduced within the time constraints by choosing either a different set of subgraphs to cover G, or a different sequence of timings? Figure 37.55 shows the same choice of subgraphs but with different timings. The result is a slightly shorter waiting time of $\frac{22}{5}T$.

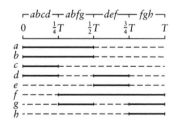

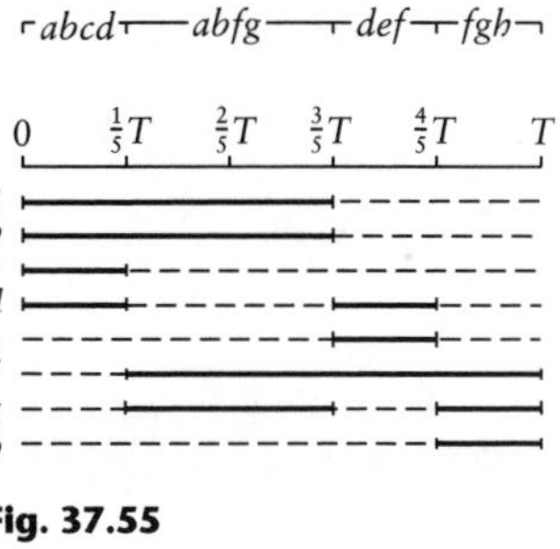

Fig. 37.54 Traffic phasing.

Fig. 37.55

Problems

37.1 (Section 37.2). Write down the degree of each vertex in the graph in Fig. 37.56.

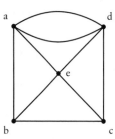

Fig. 37.56

37.2 (Section 37.2). Draw the complete graph with six vertices. How many edges does it have?

37.3 (Section 37.2). Sketch the 21 connected unlabelled graphs with five vertices. How many of them are planar?

37.4 Sketch the eight regular graphs with six vertices. How many of them are connected?

37.5 The **adjacency matrix** of a graph G with no loops is a vertex–vertex matrix, in which the element in the ith row and jth column is 0 if vertices i and j are not joined by an edge, and r if i and j are joined by r edges. Thus, if we number the vertices a, b, c, d as 1, 2, 3, 4 respectively, then the adjacency matrix of the graph in Fig. 37.1 is

$$A = \begin{bmatrix} 0 & 2 & 0 & 1 \\ 2 & 0 & 1 & 1 \\ 0 & 1 & 0 & 1 \\ 1 & 1 & 1 & 0 \end{bmatrix}.$$

Note that the leading diagonal has zeros if there are no loops. The adjacency matrix is a formula for the graph.

Evaluate A^2. What is the interpretation of the matrix in terms of the edges of G?

37.6 Draw the graphs defined by the following adjacency matrices:

(a) $A = \begin{bmatrix} 0 & 1 & 1 & 1 & 1 \\ 1 & 0 & 1 & 1 & 1 \\ 1 & 1 & 0 & 1 & 1 \\ 1 & 1 & 1 & 0 & 1 \\ 1 & 1 & 1 & 1 & 0 \end{bmatrix}$, (b) $A = \begin{bmatrix} 0 & 2 & 0 & 0 \\ 2 & 0 & 1 & 1 \\ 0 & 1 & 0 & 1 \\ 0 & 1 & 1 & 0 \end{bmatrix}$.

37.7 Write down the adjacency matrices of the graphs in Fig. 37.7. Note that a single loop introduces an element 1 into the appropriate position on the leading diagonal. What characterizes the matrix of a disconnected graph?

37.8 (Section 37.4). How many different cycles pass through a single vertex in a complete graph with four vertices?

37.9 (Section 37.4). List all trails between vertices a and f in the graph shown in Fig. 37.57. Identify which trails in the list are also paths.

37.10 (Section 37.4). Is the graph in Fig. 37.57 eulerian? If it is find an eulerian closed trail. Is it hamiltonian?

37.11 (Section 37.5). Construct a spanning tree for the graph shown in Fig. 37.57. Draw its cotree. Show that there is a spanning tree in which no vertex has degree more than two.

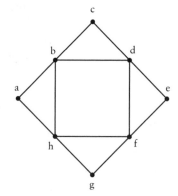

Fig. 37.57

37.12 Figure 37.58 shows a graph with seven vertices.
(a) Decide whether the graph is eulerian.
(b) Construct a spanning tree for the graph. How many branches does the tree have?
(c) Draw a cutset which disconnects the vertices a, b, g, f from the vertices c, d, e.

37.13 Figure 37.59 shows a digraph. How many trails are there between a and e? Which of them are also paths? Can you find a four-edge cycle?

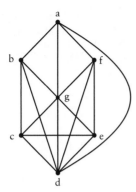

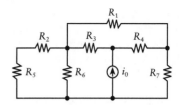

Fig. 37.60

Fig. 37.58

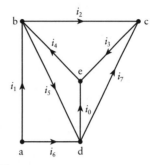

Fig. 37.61

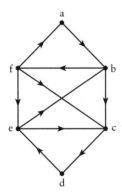

Fig. 37.59

in which the current i_1 passes through a resistor R_1 and so on. Define a spanning tree for the graph. How many fundamental cutsets are required? Write down the current equations associated with each of the cutsets. If $i_1 = 2$ A, a maintained current, $v_c = 0$ (earthed), and $R_k = 1\ \Omega$ for $k = 1, 2, \dots, 7$, find the remaining voltages v_a, v_b, v_c, v_e.

37.14 (Section 37.6). Figure 37.60 shows a circuit with an independent current source i_0. Represent the circuit by a graph. How many vertices does the graph have?

37.15 (Section 37.6). A circuit is represented by the graph shown in Fig. 37.61. The current i_0 is from an independent source, and all other edges contain a resistor

37.16 (Section 37.6). Figures 37.62a,b show two circuits with current sources and resistors. Figure 37.62c shows a circuit with current sources and a constant voltage source (battery). Use the cutset method to find the modal voltages and currents through the resistors.

37.17 Complete the block-reduction method for the multi-feedback control system shown in Fig. 37.64.

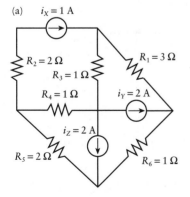

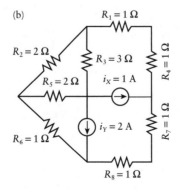

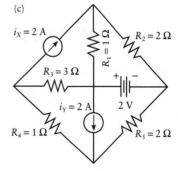

Fig. 37.62

37.18 (Section 37.5). Figure 37.63 shows a positive-feedback control system. If $P(s)$ is the system input, find its output $Q(s)$, and the transfer function of a single equivalent device.

37.19 (Section 37.7). Find the outputs in the systems shown in Figs 37.64a,b by progressively replacing parts of the system by equivalent devices until just one device remains. Find the transfer function of the resulting equivalent single device.

37.20 (Section 37.7). Reduce each of the signal-flow graphs in Figs 37.65a,b,c,d to an equivalent single edge, and (e) to a stem, and find the transfer function in each case.

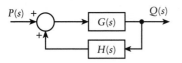

Fig. 37.63

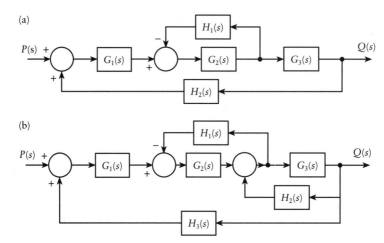

Fig. 37.64

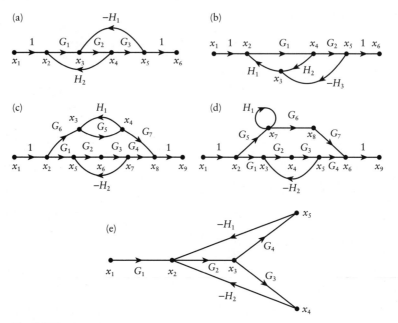

Fig. 37.65

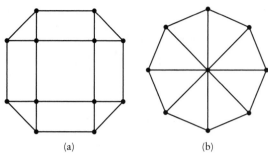

(a)   (b)

Fig. 37.66

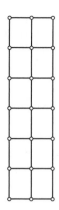

Fig. 37.68

37.21 (Section 37.8). Label the edges, vertices, and faces of the graphs shown in Figs 37.66a,b and verify Euler's formula.

37.22 (Section 37.8). Show that the bipartite graph $K_{2,3}$ has a planar representation.

37.23 (Section 37.8). The complete graph K_5 does not have a plane drawing. What is the minimum number of edge crossings in a plane representation of the graph?

37.24 (Section 37.1). List all the paths between S and T in the network given in Fig. 37.6, and hence find the shortest and longest paths. (This method of simply listing all paths can become very extensive for larger networks: efficient algorithms are really required to reduce the number of calculations.)

37.25 (Section 37.9). Show that the framework in Fig. 37.67 is overbraced. How many ties can be removed to leave a minimum bracing?

(a)

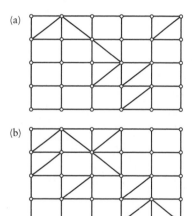

(b)

(c)

Fig. 37.69

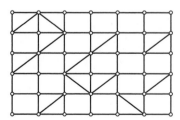

Fig. 37.67

37.26 (Section 37.9). How many ties will be needed to secure a minimum bracing for the framework shown in Fig. 37.68? Draw in a suitable set of ties for a minimum bracing.

37.27 (Section 37.9). Decide whether the frameworks shown in Fig. 37.69 are overbraced, have a minimum bracing, or are not braced.

37.28 (Section 37.9). The framework in Fig. 37.69c is required to be strengthened so that it is overbraced with each diagonal tie as an edge in at least one cycle in the associated bipartite graph. What is the minimum number of ties which must be added?

37.29 Figure 37.70 shows a junction with eight distinct lanes of traffic each controlled by a separate traffic signal. This is really a 'design and solve' problem. Here is one model: of the doubtful cases assume that lane a is compatible with both c and e, and that e is compatible with h. Draw the compatibility graph for this junction. List all complete subgraphs with four and three vertices. If the period of the traffic signal cycle is T and the subgraphs

   $\{abef, cdg, aeh\}$

are chosen with each allowed green for $\frac{1}{3}T$, calculate the total waiting time. Suppose that the subgraph $abef$ runs for $\frac{1}{2}T$ and the others for $\frac{1}{4}T$ each. How does this affect the total waiting time?

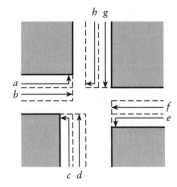

Fig. 37.70

38 Difference equations

CONTENTS

38.1 Discrete variables

In many applications, functions can only take discrete values – that is, they cannot (for various reasons) take a continuous spectrum of values. It is reasonable to model the temperature in a room by a function which varies continuously with time – most of the calculus in this book is concerned with such functions. On the other hand, the population size of a country can only take integer values. As births and deaths occur, the population size is *discontinuous* in time, and the graph of population size against time will be a **step function**. Between births and deaths the population number will be constant so that we are only concerned with changes which take place at these events. In this problem jumps occur at variable time intervals.

We can obtain discrete data from a continuous signal or function by sampling the signal at regular time steps rather than keeping a continuous record. This is often the situation in microprocessor-driven operations.

Let us start by considering a simple financial application which generates discrete values. In compound interest the sum of £P_0 is invested in an account to which interest accrues annually at a compound rate of $100I\%$. If £P_1 is the amount in the account at the end of the first year, then

$$P_1 = (1 + I)P_0. \tag{38.1}$$

Let £P_n be the sum after n years. Then, similarly

$$P_n = (1 + I)P_{n-1}. \tag{38.2}$$

This is an example of a **difference equation** or **recurrence relation**. It gives the values of P_n at the integer values 1, 2, … in terms of the immediately preceding value. Treating the variable as n, the **difference** in this case is 1. The notation $P(n)$ instead of P_n is often used to emphasize the function aspect of P but we have chosen the more economical subscript form P_n.

It is fairly easy to solve (38.2) by repeated application of the formula starting with (38.1). Thus

$$P_2 = (1+I)P_1 = (1+I)^2 P_0,$$
$$P_3 = (1+I)P_2 = (1+I)^3 P_0,$$

and so the formula

$$P_n = (1+I)^n P_0 \tag{38.3}$$

holds at least for values of n up to 3. Suppose that (38.3) holds for $n = k$. Then (38.2) implies that

$$P_{k+1} = (1+I)P_k = (1+I)^{k+1} P_0.$$

So the same formula holds for P_{k+1}. Hence, if the result is true for k then it is also true for $k + 1$. Equation (38.1) confirms that it is true for $k = 1$. It follows sequentially that it is true for $n = 2$, $n = 3$, and so on. (This method of proof is known as **induction**.)

Example 38.1   *£1000 is invested for 5 years at the following rates: (a) 5% annually; (b) $\frac{5}{12}$% calendar monthly; (b) $\frac{5}{365}$% daily (ignoring leap years). (c) Calculate the final amount in the account in each case.*

In each case the formula is

$$P_n = (1+I)^n P_0,$$

with $P_0 = 1000$, but the I and n differ.

(a) This is the original problem with $n = 5$ and $I = 0.05$. Hence

$$P_5 = (1 + 0.05)^5 \times 1000 = 1.05^5 \times 1000 = 1276.28$$

(in £, to the nearest penny).

(b) This account has 12 *compounding periods* each year, giving a total of 60 over the 5 years. Hence we require

$$P_{60} = \left(1 + \frac{0.05}{12}\right)^{60} \times 1000 = 1283.36.$$

(c) For the daily rate, there are $365 \times 5 = 1825$ compounding periods. Thus we require

$$P_{1825} = \left(1 + \frac{0.05}{365}\right)^{1825} \times 1000 = 1284.00.$$

There is a slight gain with increasing number of compounding periods.

Example 38.2 *The sum of £50 000 is borrowed over 25 years and 23 to be repaid in equal annual instalments, the interest on the outstanding balance in any year being 8%. Find the annual repayments. This scheme equalizes the repayments over the term of the loan.*

Let us solve the general mortgage-repayment problem. Suppose that $£P$ is borrowed over N years at an interest rate of $100I\%$ on the outstanding loan. Let the amount outstanding after m years be $£Q_m$. Thus $Q_0 = P$ and $Q_N = 0$. Let $£A$ be the annual repayment. Then this must include the interest on the debt still owed and capital repayment. Thus

$$A = IQ_{m-1} + (Q_{m-1} - Q_m),$$

or

$$Q_m - (1 + I)Q_{m-1} = -A \quad (m = 1, 2, \dots, N). \tag{38.4}$$

It can be done, but it is not quite so obvious now how to iterate Q_m from Q_0. It is sometimes helpful to look for any *constant* solutions of the difference equation. In this case, let $Q_m = C$ for $m = 1, 2, \dots, N$. Then

$$C - (1 + I)C = -A, \quad \text{or} \quad C = A/I.$$

This is known as a **fixed point** or an **equilibrium value** of the difference equation.

Put

$$Q_m = C + U_m = A/I + U_m$$

into (38.4). Then

$$(A/I + U_m) - (1 + I)(A/I + U_{m-1}) = -A.$$

Hence U_m satisfies

$$U_m - (1 + I)U_{m-1} = 0,$$

which is the difference equation (38.1) again. Thus

$$U_m = (1 + I)^m U_0 = (1 + I)^m (Q_0 - A/I),$$

and so

$$Q_m = A/I + U_m = A/I + (1 + I)^m (Q_0 - A/I).$$

Finally, the **boundary conditions** $Q_0 = P$ and $Q_N = 0$ imply

$$0 = A/I + (1 + I)^N (P - A/I).$$

Hence

$$A = \frac{IP(1 + I)^N}{(1 + I)^N - 1}.$$

For the data given, $P = 50\,000$, $I = 0.08$, $N = 25$, and consequently

$$A = \frac{0.08 \times 50\,000 \times 1.08^{25}}{1.08^{25} - 1} = 4683.94 \quad \text{(to 2 d.p.),}$$

which represents the annual payments in £ to the nearest penny. The total repayment over the 25 years is

$$AN = A \times 25 = 117\,098.47 \quad \text{(to 2 d.p.).}$$

In the first year, in £,

$$Q_1 = A/I + (1 + I)(P - A/I) = P(1 + I) - A = 49\,316.06$$

which indicates, as one might expect, that interest payments dominate in the early years.

38.2 Difference equations: general properties

Any equation of the form

$$u_n = f(u_{n-1}, u_{n-2}, \dots, u_{n-m}) \tag{38.5}$$

(where m is an integer $\geqslant 1$) for any successive sequence of integers n, which may terminate or not, is known as a **difference equation**. The term **discrete dynamical system** is also frequently used. Thus

$$u_n = 2u_{n-1} + 2, \tag{38.6}$$

$$u_n = 3u_{n-1} + 2u_{n-2} + n^2, \tag{38.7}$$

$$u_{n+1} = ku_n(1 - u_n) \tag{38.8}$$

are examples of difference equations.

The number m in (38.5) is known as the **order** of the difference equation: it is the difference between the largest and smallest subscripts attached to u, namely

$$n - (n - m) = m.$$

Thus (38.6) and (38.8) are first-order difference equations, while (38.7) is second-order. The sequence of integers attached to u can be translated (i.e. any integer can be added to the index) without affecting the difference equation. The difference equation

$$u_{n+2} = 3u_{n+1} + 2u_n + (n + 2)^2,$$

is the same as (38.7): n has been replaced by $n + 2$ throughout.

Given **initial conditions**, the successive terms are very easy to compute. For a first-order difference equation, we can assume that u_0 is given, but it could be any term, say u_r, which is taken as the initial condition. Generally, our aim is to find a **sequence** $\{u_n\}$ and a formula for u_n for $n \geqslant r$ which satisfies the difference equation.

The difference equation (38.8) (which is known as the **logistic equation**) with $k = 2$ is

$$u_{n+1} = 2u_n(1 - u_n). \tag{38.9}$$

Suppose that we put $u_0 = \frac{1}{4}$; then the sequence

$$u_1 = \frac{3}{8}, \quad u_2 = \frac{15}{32}, \quad u_3 = \frac{255}{512}, \quad u_4 = \frac{65\,535}{131\,072}, \quad \dots,$$

follows by successive substitution. This sequence of numbers is actually approaching the value $\frac{1}{2}$ as n increases. We can sketch the sequence by discrete values at integer values of x in the usual cartesian axes. The series of dots in Fig. 38.1 is a graphical representation of the sequence.

The implied limiting value of u_n as $n \to \infty$ for this particular sequence suggests that $u_n = \frac{1}{2}$ is a constant solution of the difference equation (38.9), and this can be confirmed. We can find all constant solutions by simply putting $u_n = u$ for all n. From (38.9), the constant solutions are given by

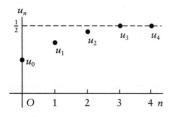

Fig. 38.1 Iterations of the sequence $u_{n+1} = 2u_n(1 - u_n)$ with $u_0 = \frac{1}{4}$.

$$u = 2u(1 - u), \qquad \text{or } 2u^2 - u = 0,$$

which implies that $u = 0$ and $u = \frac{1}{2}$ are solutions. These are also known as the **fixed points** or **equilibrium values** of the difference equation.

> **Fixed points or equilibrium values**
>
> For any first-order difference equation $u_{n+1} = f(u_n)$, its fixed points are given by solutions of
>
> $$u = f(u). \tag{38.10}$$

You might notice that the solutions of (38.9) vary qualitatively with the **initial value**, u_0. If $0 < u_0 < 1$, then $u_n \to \frac{1}{2}$ as $n \to \infty$; but, if $u_0 > 1$ or $u_0 < 0$, then u_n becomes unbounded for large n. We shall discuss the logistic equation further in Section 38.5.

For second-order difference equations, the same process gives equilibrium values. For example, if

$$u_{n+2} - 2u_{n+1} + 4u_n = 6,$$

then the equilibrium value is given by

$$u - 2u + 4u = 6, \quad \text{or} \quad u = 2.$$

On the other hand, the second-order difference equation (38.7) has no fixed points since

$$u - 3u - 2u - n^2 = -4u - n^2$$

which can never be zero for constant u.

38.3 First-order difference equations and the cobweb

An alternative method of representing solutions of difference equations graphically is the **cobweb**. Consider the first-order difference equation

$$u_{n+1} = f(u_n) = \tfrac{1}{2}u_n + 1.$$

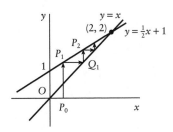

Fig. 38.2

Plot the lines $y = x$ and $y = \frac{1}{2}x + 1$ (Fig. 38.2). Select an initial value, say, $u_0 = \frac{1}{2}$, which corresponds to P_0 on the x axis. Then

$$u_1 = \tfrac{1}{2} \cdot \tfrac{1}{2} + 1 = \tfrac{5}{4},$$

which we can represent by $P_1 : (u_0, u_1)$ on the line $y = \frac{1}{2}x + 1$. Locate the point $Q_1 : (u_1, u_1)$ on $y = x$, which can be achieved by drawing P_1Q_1 parallel to the x axis. Next P_2 on $y = \frac{1}{2}x + 1$ can be found by drawing Q_1P_2 parallel to the y axis. Its ordinate must be u_2. Repeat the process by drawing lines between $y = x$ and $y = \frac{1}{2}x + 1$ using the same rules.

The usefulness of the method is that a graphical representation and interpretation of the solutions can be achieved by a simple line drawing. It is particularly helpful for finding fixed points and assessing their **stability**. We can see that this difference equation has a fixed point, which is **stable** since *all* cobwebs approach the fixed point.

Example 38.3   *Sketch a cobweb solution for*

$$u_{n+1} = -ku_n + k,$$

for (a) $k = \frac{1}{2}$, (b) $k = \frac{3}{2}$, (c) $k = 1$, *using the initial value* $u_0 = \frac{3}{4}$ *in each case.*

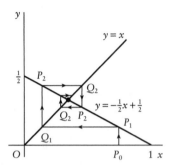

Fig. 38.3 Cobweb for $u_{n+1} = -\frac{1}{2}u_n + \frac{1}{2}$ with $u_0 = \frac{3}{4}$.

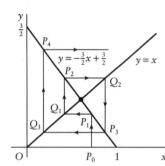

Fig. 38.4 Cobweb for $u_{n+1} = -2u_n + 2$ with $u_0 = \frac{3}{4}$.

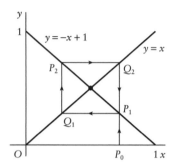

Fig. 38.5 Cobweb for $u_{n+1} = -u_n + 1$ with $u_0 = \frac{3}{4}$.

(a) Plot the lines $y = x$ and $y = -\frac{1}{2}x + \frac{1}{2}$. They intersect at the fixed point $(\frac{1}{3}, \frac{1}{3})$. Starting from $P_0 : (\frac{3}{4}, 0)$, the cobweb traces $P_0P_1Q_1P_2Q_2P_3 \ldots$ in Fig. 38.3. Evidently it approaches the fixed point as $n \to \infty$, indicating stability.

(b) The lines are $y = x$ and $y = -\frac{3}{2}x + \frac{3}{2}$. The fixed point is at $(\frac{3}{5}, \frac{3}{5})$, and the cobweb path is $P_0P_1Q_1P_2Q_2 \ldots$ in Fig. 38.4. The path moves away from the fixed point implying its instability.

(c) The lines are $y = x$ and $y = -x + 1$ with fixed point $(\frac{1}{2}, \frac{1}{2})$. The path starting at $P_0 : (\frac{3}{4}, 0)$ follows the rectangle $P_1Q_1P_2Q_2$, indicating **periodicity** (Fig. 38.5). This is true for any starting value except that of the fixed point itself.

Graphs of the sequences u_n versus n are shown in Fig. 38.6.

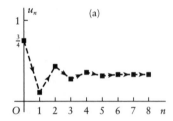

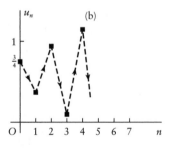

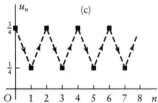

Fig. 38.6 Solutions of $u_{n+1} = -k(u_n+1)$ for (a) $k = \frac{1}{2}$, (b) $k = 2$, (c) $k = 1$.

The stability of the fixed point of the first-order linear difference equation can be summarized as follows.

Stability

The first-order difference equation $u_{n+1} = -ku_n + a$ has

(a) a stable fixed point if $-1 < k < 1$,
(b) a fixed point that is not stable if $|k| > 1$,
(c) a periodic fixed point if $k = 1$.

(38.11)

38.4 Constant-coefficient linear difference equations

Any difference equation of the form

$$u_n + a_{n-1}u_{n-1} + \cdots + a_{n-m}u_{n-m} = f(n),$$

where the a_i ($i = n - m, \ldots, n - 1$) are constants, is a constant-coefficient linear difference equation. We shall look in detail at the second-order case

$$u_{n+2} + 2au_{n+1} + bu_n = f(n), \tag{38.12}$$

where a and b are constants and $f(n)$ is a given function. The methods generalize in a fairly obvious way to higher-order systems.

There are many parallels between the difference equation (38.12) and second-order constant-coefficient equations (Chapters 18–19). The equation is said to be **homogeneous** if $f(n) = 0$, and **inhomogeneous** otherwise, just as in the case of second-order differential equations. However, this section is self-contained and reference back is not necessary. The general solution of the inhomogeneous case requires that of the homogeneous case: hence we start with the latter.

Homogeneous equations. We can see how to proceed by looking at the first-order constant-coefficient equation

$$u_{n+1} - cu_n = 0. \tag{38.13}$$

As can be seen from (38.2) or verified directly, the general solution of this equation is

$$u_n = Ac^n, \tag{38.14}$$

where A is any constant. Notice that we could equally well write

$$u_n = Ac^{n-1}, \quad \text{or} \quad u_n = Ac^{n+1}:$$

the result would be equally correct, although A would take different values for the same initial condition. The significant property of (38.13) and its solution (38.14) is that u_{n+1} is a constant multiple of u_n.

With this in view, we attempt to find solutions of

$$u_{n+2} + 2au_{n+1} + bu_n = 0 \tag{38.15}$$

in the form

$$u_n = p^n,$$

where p is a constant. Thus

$$u_{n+2} + 2au_{n+1} + bu_n = p^{n+2} + 2ap^{n+1} + bp^n$$
$$= (p^2 + 2ap + b)p^n = 0,$$

for all n, if $p = 0$ or

$$p^2 + 2ap + b = 0. \tag{38.16}$$

The case $p = 0$ leads to the self-evident *trivial* solution $u_n = 0$. We are interested in solutions of (38.16), which is known as the **characteristic equation** of (38.15).

There are various cases to consider. Suppose that the roots of (38.16) are the distinct numbers p_1 and p_2. Hence $u_n = p_1^n$ and $u_n = p_2^n$ are solutions of (38.15). Since this equation is homogeneous and *linear*, it follows that any linear combination of p_1^n and p_2^n is also a solution. We state this as follows.

> **Distinct roots**
>
> The general solution of $u_{n+2} + 2au_{n+1} + bu_n = 0$ for distinct roots p_1 and p_2 of $p^2 + 2ap + b = 0$ is
>
> $$u_n = Ap_1^n + Bp_2^n, \quad \text{for any constants } A \text{ and } B.$$
>
> **(38.17)**

Example 38.4 *Find the general solution of*

$$u_{n+2} - u_{n+1} - 6u_n = 0. \tag{38.18}$$

The characteristic equation of (38.18) is

$$p^2 - p - 6 = 0, \quad \text{or} \quad (p-3)(p+2) = 0.$$

The roots are $p_1 = 3$, $p_2 = -2$. Hence the general solution is

$$u_n = A \cdot 3^n + B(-2)^n.$$

Example 38.5 *Find the solution of*

$$u_{n+2} + 2u_{n+1} - 3u_n = 0$$

that satisfies $u_0 = 1$, $u_1 = 2$.

The characteristic equation is

$$p^2 + 2p - 3 = 0, \quad \text{or} \quad (p+3)(p-1) = 0.$$

The roots are $p_1 = -3$, $p_2 = 1$. Hence the general solution is

$$u_n = A(-3)^n + B \cdot 1^n = A(-3)^n + B.$$

From the initial conditions,

$$u_0 = 1 = A + B, \quad u_1 = 2 = -3A + B.$$

Hence $A = -\frac{1}{4}$ and $B = \frac{5}{4}$. The required solution is

$$u_n = -\frac{1}{4} \cdot (-3)^n + \frac{5}{4}.$$

The characteristic equation can have equal roots, which is a special case. Consider the difference equation

$$u_{n+2} - 2au_{n+1} + a^2 u_n = 0,$$

where $a \neq 0$. Its characteristic equation is

$$p^2 - 2ap + a^2 = 0, \quad \text{or} \quad (p-a)^2 = 0,$$

which has the repeated root $p = a$. One solution is Aa^n; but we require a second independent solution. Let $u_n = v_n a^n$. Then

$$0 = u_{n+2} - 2au_{n+1} + a^2 u_n = a^{n+2} v_{n+2} - 2a^{n+2} v_{n+1} + a^{n+2} v_n$$
$$= a^{n+2}(v_{n+2} - 2v_{n+1} + v_n).$$

We require therefore

$$v_{n+2} - 2v_{n+1} + v_n = 0$$

since $a \neq 0$. It can be verified that this equation has a solution $v_n = n$. Hence a further independent solution is $u_n = Bna^n$.

Equal roots

The general solution of $u_{n+2} - 2au_{n+1} + a^2u_n = 0$ is

$$u_n = (A + Bn)a^n. \tag{38.19}$$

Roots can also be complex. Consider the difference equation

$$u_{n+2} + 2u_{n+1} + 2u_n = 0.$$

Its characteristic equation is

$$p^2 + 2p + 2 = 0$$

with roots $p_1 = -1 + i$, $p_2 = -1 - i$. The method still works and the general solution becomes

$$u_n = A(-1 + i)^n + B(-1 - i)^n.$$

For a real-valued problem, the constants A and B will be complex conjugates which ensure that u_n is real. The solution can be cast in real form by using the polar forms (Section 6.3) of the complex numbers. In this case

$$-1 \pm i = \sqrt{2} \, e^{\pm \frac{3}{4}\pi i}.$$

Hence

$$\begin{aligned}
u_n &= A2^{\frac{1}{2}n} e^{\frac{3}{4}\pi in} + B2^{\frac{1}{2}n} e^{-\frac{3}{4}\pi in} \\
&= 2^{\frac{1}{2}n}[A(\cos \tfrac{3}{4}\pi n + i \sin \tfrac{3}{4}\pi n) + B(\cos \tfrac{3}{4}\pi n - i \sin \tfrac{3}{4}\pi n)] \\
&= 2^{\frac{1}{2}n}(C \cos \tfrac{3}{4}\pi n + D \sin \tfrac{3}{4}\pi n),
\end{aligned}$$

where $C = A + B$ and $D = (A - B)i$.

Complex roots, $\alpha \pm i\beta = r\,e^{\pm\theta i}$

The general complex solution of

$$u_{n+2} + 2au_{n+1} + bu_n = 0,$$

where $a^2 < b$, is

$$u_n = A(\alpha + i\beta)^n + B(\alpha - i\beta)^n.$$

The general real solution is

$$u_n = r^n(C \cos \theta n + D \sin \theta n). \tag{38.20}$$

Example 38.6 *Obtain the general solution of*

$$u_{n+2} + u_n = 0.$$

The characteristic equation is

$$p^2 + 1 = 0,$$

giving roots $p_1 = i$, $p_2 = -i$. Hence

$$u_n = Ai^n + B(-i)^n.$$

In polar form, $i = e^{\frac{1}{2}\pi i}$, $-i = e^{-\frac{1}{2}\pi i}$. Hence the real form of the solution is

$$u_n = C \cos \tfrac{1}{2}\pi n + D \sin \tfrac{1}{2}\pi n.$$

Inhomogeneous equations. The general inhomogeneous equation is

$$u_{n+2} + 2au_{n+1} + bu_n = f(n) \qquad\qquad (38.21)$$

(see (38.12)). Let $u_n = v_n + q_n$, where v_n is the *general solution* of the corresponding *homogeneous equation*. Substitute this form of u_n into (38.21):

$$(v_{n+2} + q_{n+2}) + 2a(v_{n+1} + q_{n+1}) + b(v_n + q_n) = f(n),$$

or

$$(v_{n+2} + 2av_{n+1} + bv_n) + (q_{n+2} + 2aq_{n+1} + bq_n) = f(n).$$

Since v_n satisfies the homogeneous equation, it follows that

$$q_{n+2} + 2aq_{n+1} + bq_n = f(n),$$

which means that q_n must be a **particular solution** of the inhomogeneous equation. As in differential equations, v_n is known as the **complementary function.**

We construct particular solutions by appropriate choices of functions usually containing adjustable parameters which are suggested by the form of the function $f(n)$. If a particular choice fails, then we reject it and try something else. For example, if $f(n) = k$, a constant, then we might try $q_n = C$. This will work provided that the homogeneous equation does not itself have a constant solution, a point which the next two examples illustrate.

Example 38.7   *Obtain the general solution of*

$$u_{n+2} - u_{n+1} - 6u_n = 4.$$

From Example 38.4, the complementary function is

$$v_n = 3^n A + (-2)^n B.$$

For the particular solution, we try $q_n = C$, since $f_n = 4$. Then

$$q_{n+2} - q_{n+1} - 6q_n - 4 = C - C - 6C - 4$$
$$= -6C - 4 = 0,$$

if $C = -\frac{2}{3}$. Hence $q_n = -\frac{2}{3}$, and the general solution is

$$u_n = 3^n A + (-2)^n B - \frac{2}{3}.$$

Note that the two unknown constants in the general solution occur in the complementary function.

Example 38.8   *Obtain the general solution of*

$$u_{n+2} + 2u_{n+1} - 3u_n = 4.$$

From Example 38.5, the complementary function is

$$v_n = (-3)^n A + B.$$

Example 38.8 *continued*

In this case we expect the choice $q_n = C$ to fail, since **it must make the left-hand side of the difference equation vanish.** When this happens, we try

$$q_n = Cn.$$

Then

$$q_{n+2} + 2q_{n+1} - 3q_n - 4 = C(n+2) + 2C(n+1) - 3Cn - 4$$
$$= 2C + 2C - 4 = 4C - 4 = 0,$$

if $C = 1$. Hence the general solution is

$$u_n = (-3)^n A + B + n.$$

Table 38.1 lists some simple forcing terms $f(n)$ with suggested forms of particular solution and alternatives containing parameters to be determined by direct substitution.

Table 38.1

$f(n)$	Trial solution q_n
k (a constant)	C; or Cn, if C fails; or Cn^2, if C and Cn fail; etc.
k^n	Ck^n; or Cnk^n, if Ck^n fails; etc.
n	$C_0 + C_1 n$
n^p (p an integer)	$C_0 + C_1 n + \cdots + C_p n^p$ (may need higher powers of n in special cases)
$\sin kn$ or $\cos kn$	$C_1 \cos kn + C_2 \sin kn$

Example 38.9 *Find the general solution of*

$$u_{n+2} - 4u_n = n.$$

The characteristic equation is

$$p^2 - 4 = 0, \quad \text{or} \quad (p-2)(p+2) = 0.$$

The roots are $p_1 = 2$, $p_2 = -2$. Hence the complementary function is

$$v_n = 2^n A + (-2)^n B.$$

For the particular solution, try (choosing from Table 38.1)

$$q_n = C_0 + C_1 n.$$

Then

$$q_{n+2} - 4q_{n+1} - n = C_0 + C_1(n+2) - 4C_0 - 4C_1 n - n$$
$$= (-3C_0 + 2C_1) + n(-3C_1 - 1).$$

The right-hand side vanishes for all n if

$$-3C_0 + 2C_1 = 0, \qquad -3C_1 - 1 = 0.$$

Hence $C_1 = -\frac{1}{3}$, $C_0 = 2C_1/3 = -\frac{2}{9}$, and the general solution is

$$u_n = 2^n A + (-2)^n B - \tfrac{2}{9} - \tfrac{1}{3}n.$$

38.5   The logistic difference equation

Consider again the logistic difference equation

$$u_{n+1} = \alpha u_n(1 - u_n), \tag{38.22}$$

where α is a **parameter** which will take various values. This *nonlinear* equation can model population growth of generations. If u_n represents the population size of generation n and α is the birthrate, then we might expect the population size of the next generation to be αu_n in the absence of any inhibiting factors such as lack of resources or overcrowding. If $\alpha > 1$, then the population model given by the first-order difference equation $u_{n+1} = \alpha u_n$ would imply that the population would grow to infinity, since the equation has the solution $u_n = \alpha^n u_0$. To counter this possibility, we can introduce a feedback term $-\alpha u_n^2$ which will tend to reduce population growth when the population is large.

Fixed points of the equation (38.22) occur where

$$u = \alpha u(1 - u);$$

that is, for $u = 0$ and $u = 1 - 1/\alpha$. We can adapt the cobweb method of Section 38.3 to this nonlinear difference equation by plotting graphs of the parabola $y = f(x) = \alpha x(1 - x)$ and the straight line $y = x$. Fixed points of the difference equation will occur where the line and the parabola intersect. The values of x at these points will be given by the solutions of

$$\alpha x(1 - x) = x, \quad \text{or} \quad x(\alpha x - 1 - \alpha) = 0.$$

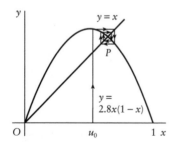

Fig. 38.7 Cobweb solution for $u_{n+1} = 2.8u_n(1 - u_n)$ showing a solution starting from $x = u_0$ approaching the fixed point at P.

The points are $x = 0$ and $P : x = 1 - 1/\alpha$. We shall only look at values of $\alpha > 1$, so that one fixed point is in the first quadrant, $x > 0, y > 0$. A cobweb solution for the case $\alpha = 2.8$ is shown in Fig. 38.7.

Notice that, for this choice of α, the fixed point P is *stable*; that is, the cobweb solution approaches P. The slope of the graph of $y = \alpha x(1 - x)$ at P determines the stability or instability of the solutions. The slope at P is $m = f'(x) = \alpha - 2\alpha x$, where $x = 1 - 1/\alpha$. Hence

$$m = \alpha - 2\alpha(1 - 1/\alpha) = -\alpha + 2.$$

As with the cobweb for two intersecting lines for the linear difference equation in Section 38.3, the fixed point P is **locally stable** if $m > -1$, in that all cobweb paths starting close to $1 - 1/\alpha$ approach the fixed point P as $n \to \infty$. This corresponds to $\alpha < 3$. Notice also that, if $1 < \alpha < 2$, then $y = x$ intersects the parabola $y = \alpha x(1 - x)$ between the origin and its maximum value. This follows since the maximum occurs at $x = \frac{1}{2}$ and $0 < 1 - 1/\alpha < \frac{1}{2}$ implies $1 < \alpha < 2$.

For $\alpha \geqslant 3$ the solutions become more complicated. The fixed point at the origin is unstable: hence there is no *stable* fixed point to which

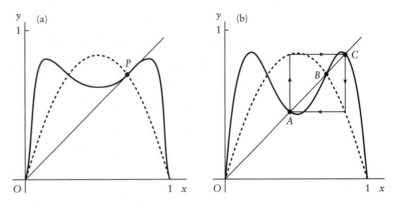

Fig. 38.8 (a) Graph of $y = f(f(x))$ for the critical case $\alpha = 3$. (b) Graph of $y = f(f(x))$ for $\alpha = 3.4$ showing fixed points O, A, B, C.

solutions can approach. We can obtain a clue as to what happens if we look at the function of a function given by

$$y = f(f(x)) = \alpha[\alpha x(1-x)][1 - \alpha x(1-x)]$$
$$= \alpha^2 x(1-x) - \alpha^3 x^2(1-x)^2.$$

When $\alpha = 3$, this curve intersects $y = x$ at $x = 0$ and at P only. This can be checked by noting that

$$x = \alpha^2 x(1-x) - \alpha^3 x^2(1-x)^2$$

can be written as

$$x(27x^3 - 54x^2 + 36x - 8) = 0, \quad \text{or} \quad x(3x-2)^3 = 0,$$

when $\alpha = 3$. Graphs of the curves $y = f(x)$ and $y = f(f(x))$ for $\alpha = 3$ are shown in Fig. 38.8a. The fixed point P is at $(\frac{2}{3}, \frac{2}{3})$. As α increases, two additional fixed points develop on the line $y = x$. Further graphs of the two functions $y = f(x)$ and $y = f(f(x))$ for $\alpha = 3.4$ are shown in Fig. 38.8(b), together with the line $y = x$. There are fixed points at O, A, B, C, of which A and C are stable. While A and C are fixed points of this equation, they can be associated with 2-**cycles** or **period**-2 solutions of the difference equation, as shown by the superimposed cobweb on $y = \alpha x(1-x)$ in Fig. 38.8b. This phenomenon is known as **period doubling**. It first appears when the fixed point P on $y = f(x)$ ceases to be stable at $\alpha = 3$. The 2-cycle then grows in 'amplitude' as α increases. The solution is said to **bifurcate** at $\alpha = 3$. This type of bifurcation, where the stable solution becomes suddenly unstable and throws off two stable solutions on either side, is an example of a **pitchfork bifurcation**, called this because of its fork-like appearance (see Fig. 38.9).

For general α, the fixed points of $y = f(f(x))$ occur where

$$x = \alpha^2 x(1-x) - \alpha^3 x^2(1-x)^2.$$

To solve this equation, put $1 - x = u$. Then u satisfies

$$1 - u = \alpha^2(1-u)u - \alpha^3(1-u)^2 u^2,$$

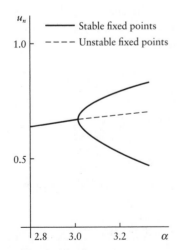

Fig. 38.9 Pitchfork bifurcation.

which can be written as $G(u) = 0$, where

$$G(u) = (u - 1)(\alpha^3 u^3 - \alpha^3 u^2 + \alpha^2 u - 1). \tag{38.23}$$

One obvious solution of $G(u) = 0$ is $u = 1$, while the cubic factor has the solution $u = 1/\alpha$ as can be verified. Hence

$$G(u) = (u - 1)(\alpha u - 1)[\alpha^2 u^2 + \alpha(1 - \alpha)u + 1].$$

At A and C, u satisfies

$$\alpha^2 u^2 + \alpha(1 - \alpha)u + 1 = 0. \tag{38.24}$$

Hence, at the two fixed points A and C,

$$\left.\begin{array}{c} x_1 \\ x_2 \end{array}\right\} = \frac{1 + \alpha \mp \sqrt{[(\alpha + 1)(\alpha - 3)]}}{2\alpha} \quad (\alpha > 3),$$

respectively, while $x = (\alpha - 1)/\alpha$ at B. Since

$$
\begin{aligned}
f(x_1) &= \alpha x_1(1 - x_1) \\
&= \tfrac{1}{2}[1 + \alpha - \sqrt{[(\alpha + 1)(\alpha - 3)]}] \\
&\quad \times \left[1 - \frac{1}{2\alpha}\{1 + \alpha - \sqrt{[(\alpha + 1)(\alpha - 3)]}\}\right] \\
&= \frac{1}{4\alpha}[\alpha + 1 - \sqrt{[(\alpha + 1)(\alpha - 3)]}] \times [\alpha - 1 + \sqrt{[(\alpha + 1)(\alpha - 3)]}] \\
&= \frac{1}{4\alpha}[\alpha^2 - \{1 - \sqrt{[(\alpha + 1)(\alpha - 3)]}\}^2] \\
&= \frac{1}{2\alpha}\{1 + \alpha + \sqrt{[(\alpha + 1)(\alpha - 3)]}\} = x_2,
\end{aligned}
$$

and similarly $f(x_2) = x_1$. Hence period doubling occurs, and transfers between x_1 and x_2 can take place around the square cobweb in Fig. 38.8b.

The fixed point A will remain stable if the absolute value of its slope is less than 1. The same condition will also apply at C. In fact the critical slope is -1, and we will find the value of α at which this occurs. We have

$$
\begin{aligned}
\frac{\mathrm{d}}{\mathrm{d}x} f(f(x)) &= \alpha^2 - 2\alpha^2 x - \alpha^3(2x - 6x^2 + 4x^3) \\
&= \alpha^2 - 2\alpha^2(1 + \alpha)x + 6\alpha^3 x^2 - 4\alpha^3 x^3.
\end{aligned} \tag{38.25}
$$

We require the value of α given by

$$\frac{\mathrm{d}}{\mathrm{d}x} f(f(x)) = -1, \tag{38.26}$$

or $4\alpha^3 x^3 - 6\alpha^3 x^2 + 2\alpha^2(1 + \alpha)x - \alpha^2 = 1$,

when x satisfies (38.24) with $u = 1 - x$, which is when

$$\alpha^2 x^2 - \alpha(1 + \alpha)x + \alpha + 1 = 0. \tag{38.27}$$

Remove the x^3 term from (38.26) by multiplying (38.27) by $4\alpha x$, and subtracting it from (38.26). Then

$$-2\alpha^2(\alpha - 2)x^2 + 2\alpha(\alpha - 2)(\alpha + 1)x - (1 + \alpha^2) = 0. \qquad (38.28)$$

Equations (38.27) and (38.28) must have the *same* roots in x. In each case, make the coefficient of x^2 equal to 1. The results for comparison are

$$x^2 - \frac{(\alpha + 1)}{\alpha}x + \frac{(\alpha + 1)}{\alpha^2} = 0,$$

$$x^2 - \frac{(\alpha + 1)}{\alpha}x + \frac{(\alpha^2 + 1)}{2\alpha^2(\alpha - 2)} = 0.$$

These equations have the same roots if

$$\frac{\alpha + 1}{\alpha^2} = \frac{(\alpha^2 + 1)}{2\alpha^2(\alpha - 2)}, \qquad (38.29)$$

or

$$\alpha^2 - 2\alpha - 5 = 0.$$

We are interested in values of $\alpha > 3$, so that the required root of (38.29) is $\alpha = 1 + \sqrt{6} = 3.449\ldots$. In fact the slopes at both A and C both become -1 for this value of α. Thus, for

$$3 < \alpha < 1 + \sqrt{6},$$

the 2-cycle solution is stable.

At $\alpha = 1 + \sqrt{6}$, the system bifurcates again into a 4-cycle or period-4 solution, which corresponds to the set of stable fixed points of $y = f(f(f(f(x))))$. A graph of this function for $\alpha = 3.54$ is shown in Fig. 38.10 together with the eight fixed points. The cycle doubles again at about $\alpha = 3.544, \ldots$ and so on. The intervals between the bifurcations of the period doubling rapidly decrease, until a limit is reached at about $\alpha = 3.570, \ldots$ beyond which **chaos** occurs. The iterations are no longer periodic for most values of α beyond this point, although there are some brief intervals of periodicity.

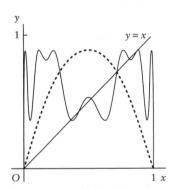

Fig. 38.10 Fixed points of $y = f(f(f(f(x))))$ for $\alpha = 3.54$, given by the intersection of the curve and the line $y = x$.

Logistic equation

$$u_{n+1} = f(u_n) = \alpha u_n(1 - u_n).$$

Fixed point for $\alpha > 0$, $x > 0$ at $x_0 = (\alpha - 1)/\alpha$.
Fixed point x_0 stable if

$$f'(x_0) = \alpha - 2\alpha x_0 = -\alpha + 2 > -1, \text{ that is if } \alpha < 3.$$

Period-2 solution: fixed points $(\alpha > 3)$

$$x_1, x_2 = \{1 + \alpha \mp \sqrt{[(\alpha + 1)(\alpha - 3)]}\}/(2\alpha).$$

Period-2 solution stable if $3 < \alpha < 1 + \sqrt{6}$. $\qquad$ **(38.30)**

The sequence of period-doubling bifurcations is known as the **Feigenbaum sequence**, and it has certain universal aspects in that it is not just a consequence of the logistic equation, but has common features with other difference equations which generate period doubling.

The simplest way to view the progressively complex behaviour is through a computer-drawn picture of the iterations of

$$u_{n+1} = \alpha u_n (1 - u_n)$$

for stepped increases in α starting at $\alpha = 2.8$ up to $\alpha = 3.8$, which covers the main area of interest. The result is shown in Fig. 38.11. The series of single dots for each α in $2.8 \leqslant \alpha \leqslant 3$ indicates the fixed point, which then bifurcates into a stable 2-cycle **attractor** for $3 < \alpha \leqslant 1 + \sqrt{6}$. This in turn bifurcates into a stable 4-cycle attractor at $\alpha = 1 + \sqrt{6}$ and so on. The effect of infinite period doubling is that the solution is ultimately non-periodic. The generally chaotic and noisy behaviour of the difference equation can clearly be seen in the large number of dots for larger values of α. These non-periodic sets are known as **strange attractors**. The successive iterates of the logistic equation wander about in a seemingly random but bounded manner, and never settle into a periodic solution. However, within the chaotic band of α values, there appear windows of periodic cycles. Problem 38.26, for example, confirms that there is a 3-cycle around $\alpha = 3.83$.

The logistic equation can be thought of as a relatively simple model example. Many similar nonlinear difference equations also exhibit similar period-doubling bifurcations and strange attractors.

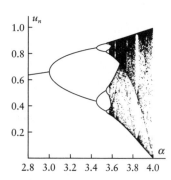

Fig. 38.11 Period doubling for the logistic equation for increasing α, followed by chaotic iterations beyond about $\alpha = 3.57$.

Problems

38.1 £1000 is invested over 10 years at an interest rate of 6% annually. Find the final total investment. What should the *monthly* interest rate be to achieve the same final total?

38.2 The sum of £50 000 is borrowed over 25 years and the money is repaid in equal annual instalments. The interest rate on the outstanding balance in any year is 10%. Find what the annual repayments would be. After 5 years, the interest rate is reduced to 9%.
(a) Find the required adjustment to the annual repayments for the loan to be repaid over the original term.
(b) If the repayments are not changed, by how much will the mortgage term be reduced?

38.3 Find the fixed points of the following difference equations:
(a) $u_{n+1} = u_n(2 - u_n)$;   (b) $u_{n+1} = u_n(1 + u_n)(2 - 3u_n)$;
(c) $u_{n+1} = \sin u_n$;   (d) $u_{n+1} = \frac{1}{2}\sin u_n$;   (e) $u_{n+1} = e^{u_n} - 1$.

38.4 Given the initial value u_0 in each case, calculate the sequence of terms up to u_5 for each of the following first-order difference equations:
(a) $u_{n+1} = 2u_n(3 - u_n)$, $u_0 = 1$;
(b) $u_{n+1} = 2u_n(1 - u_n)$, $u_0 = \frac{1}{2}$;
(c) $u_{n+1} = 3.2u_n(1 - u_n)$, $u_0 = \frac{1}{2}$;
(d) $u_{n+1} = 4u_n(1 - u_n)$, $u_0 = \frac{1}{2}$.

38.5 (Section 38.3). Sketch the cobweb solutions for the following first-order equations with the stated initial conditions, and discuss the stability of the fixed point:
(a) $u_{n+1} = \frac{1}{2}u_n + \frac{1}{2}$, $u_0 = \frac{1}{2}$ and $u_0 = \frac{3}{2}$;
(b) $u_{n+1} = 2u_n - 2$, $u_0 = \frac{1}{2}$ and $u_0 = \frac{3}{2}$;
(c) $u_{n+1} = -u_n + 2$, $u_0 = \frac{1}{2}$ and $u_0 = \frac{3}{4}$;
(d) $u_{n+1} = -\frac{1}{2}u_n + \frac{3}{2}$, $u_0 = \frac{1}{2}$ and $u_0 = \frac{3}{2}$;
(e) $u_{n+1} = -2u_n + 3$, $u_0 = \frac{1}{2}$ and $u_0 = \frac{3}{2}$.

38.6 The function $f(n)$ satisfies

$$f(n) = f(\tfrac{1}{2}n) + 1.$$

Put $n = 2^m$ and $g(m) = f(2^m)$, and show that

$$g(m) = g(m-1) + 1.$$

Hence find $f(n)$ given that $f(1) = 0$.

38.7 Use the method suggested in the previous problem to solve

$$f(n) = f(\tfrac{1}{3}n) + \tfrac{5}{8},$$

given the initial condition $f(1) = 0$.

38.8 (Section 38.3). Find the general solutions of the following difference equations:
(a) $u_{n+2} + 2u_{n+1} - 3u_n = 0$;
(b) $u_{n+2} - 9u_n = 0$;
(c) $u_{n+2} + 9u_n = 0$;
(d) $u_n - 4u_{n-1} + 5u_{n-2} = 0$;
(e) $u_{n+2} - 4u_{n+1} + 4u_n = 0$;
(f) $u_{n+3} - u_{n+2} + u_{n+1} - u_n = 0$;
(g) $u_{n+3} - u_n = 0$;
(h) $u_{n+3} - 3u_{n+2} + 3u_{n+1} - u_n = 0$;
(i) $u_{n+2} - u_{n+1} - u_n + u_{n-1} = 0$.

38.9 Express the solution of the initial-value problem

$$u_{n+2} - 6u_{n+1} + 13u_n = 0, \qquad u_0 = 0, \quad u_1 = 1,$$

in real form.

38.10 Find the difference equation satisfied by

$$u_n = A \cdot 2^n + B \cdot (-5)^n,$$

for all A and B.

38.11 Obtain particular solutions of the following inhomogeneous difference equations:
(a) $u_{n+2} + 2u_{n+1} - 3u_n = f(n)$, where
    (i) $f(n) = 2^n$; (ii) $f(n) = n$; (iii) $f(n) = 2$
    (iv) $f(n) = (-3)^n$.
(b) $u_{n+2} + 2u_{n+1} + 2u_n = f(n)$, where
    (i) $f(n) = 1$; (ii) $f(n) = n + 3$;
    (iii) $f(n) = \cos \tfrac{3}{4}\pi n$.
(c) $u_{n+3} - 3u_{n+2} + 3u_{n+1} + u_n = f(n)$, where
    (i) $f(n) = 1$; (ii) $f(n) = n$; (iii) $f(n) = n^2$.
(d) $u_{n+2} - 6u_{n+1} + 9u_n = f(n)$, where (i) $f(n) = 2^n$;
    (ii) $f(n) = 3$; (iii) $f(n) = 3^n$; (iv) $f(n) = n3^n$.

38.12 A ball bearing is dropped from a height $z = h_0$ on to a metal plate, and the coefficient of restitution between the ball and the plate is ε, where $0 < \varepsilon < 1$. Set up a difference equation for the maximum height reached after n impacts. Solve the equation. (Assume that a ball dropped from a

height h hits the plate with speed $v = \sqrt{(2gh)}$, where g is the acceleration due to gravity. The rebound speed of the ball is εv.) Instead of being stationary, the plate now oscillates so that it is moving upwards at a speed u (a constant) at the moment of each impact with the ball. Find the difference equation for h_n. Show that the difference equation has a fixed point and interpret its meaning.

38.13 $D_n(x)$ is the $n \times n$ determinant defined by

$$D_n(x) = \begin{vmatrix} 2x & 1 & 0 & \cdots & 0 \\ 1 & 2x & 1 & \cdots & 0 \\ \vdots & \vdots & \vdots & \vdots & \vdots \\ 0 & 0 & 0 & \cdots & 2x \end{vmatrix} \quad (n > 2),$$

$$D_2(x) = \begin{vmatrix} 2x & 1 \\ 1 & 2x \end{vmatrix}, \quad D_1(x) = 2x.$$

Show that

$$D_n(x) = 2xD_{n-1}(x) - D_{n-2}(x).$$

Solve the difference equation for $x \neq 1$ and $x = 1$.

38.14 Let $\{u_n\}$ $(n = 0, 1, \dots)$ be a sequence. The power series

$$f(u_n, x) = \sum_{n=0}^{\infty} u_n x^n$$

is known as the **generating function** of the sequence. Thus, for example, if $u_n = (-1)^n/n!$, then

$$f(u_n, x) = \sum_{n=0}^{\infty} \frac{(-1)^n}{n!} x^n = e^{-x},$$

which means that e^{-x} is the generating function of $\{u_n\}$.
    The generating function of $\{u_{n+1}\}$ is

$$f(u_{n+1}, x) = \sum_{n=0}^{\infty} u_{n+1}x^n = \frac{1}{x} \sum_{n=0}^{\infty} \frac{(-1)^{n+1}}{(n+1)!} x^{n+1}$$

$$= \frac{1}{x}\left(\sum_{n=0}^{\infty} \frac{(-1)^n}{n!} x^n - 1 \right) = \frac{1}{x}[f(u_n, x) - 1].$$

Consider the difference equation

$$u_{n+2} + u_{n+1} - 2u_n = 0, \qquad u_0 = 1, \quad u_1 = -2.$$

By taking the generating function of the equation, show that

$$f(u_n, x) = \frac{1}{1 + 2x}.$$

Using the binomial theorem find u_n.

38.15 A Fibonacci sequence is defined as a sequence in which any term is the sum of the two preceding terms. For the Fibonacci sequence starting with $u_1 = 1$, $u_2 = 2$, find and solve the difference equation for u_n.

38.16 Solve the initial-value difference equation

$$3u_{n+2} - 2u_{n+1} - u_n = 0, \qquad u_1 = 2, \quad u_2 = 1,$$

and show that $u_n \to \frac{5}{4}$ as $n \to \infty$.

38.17 A symmetric **random walk** takes place on the integer steps on the line between $x = 0$ and $x = N$. At any position $x = r$ $(1 \leqslant r \leqslant N - 1)$, the probability that the walker moves to either $x = r + 1$ or $x = r - 1$ at any stage is $\frac{1}{2}$. The probability u_k that the walker reaches $x = 0$ first, given an initial position $x = k$, satisfies the difference equation

$$u_k = \tfrac{1}{2}u_{k-1} + \tfrac{1}{2}u_{k+1}, \qquad u_0 = 1, \quad u_N = 0,$$

for $1 \leqslant k \leqslant N - 1$. Find u_k. What is the probability that the walker reaches $x = N$ first?

If d_k is the expected number of steps in the walk before it reaches 0 or N, then d_k satisfies

$$d_k = \tfrac{1}{2}(1 + d_{k+1}) + \tfrac{1}{2}(1 + d_{k-1}), \qquad d_0 = d_N = 0$$

for $1 \leqslant k \leqslant N - 1$. Find the expected duration of the walk.

38.18 Show that $u_n = n!$ is a solution of the second-order difference equation

$$u_{n+2} = (n + 2)(n + 1)u_n.$$

By using the substitution $u_n = v_n n!$, find a second independent solution.

38.19 Given that

$$s_n = \sum_{k=1}^{n} k^3,$$

find a first-order difference equation for s_n. Solve the equation to find a formula for the sum s_n.

38.20 Show that the difference equation

$$u_{n+2} + 2au_{n+1} + bu_n = 0$$

can be expressed as

$$z_{n+1} = Az_n,$$

where

$$z_n = \begin{bmatrix} u_n \\ v_n \end{bmatrix}, \qquad A = \begin{bmatrix} -2a & b \\ -1 & 0 \end{bmatrix}.$$

Deduce that

$$z_n = A^n z_0.$$

Consider the case with $a = 1$ and $b = -8$. Find the eigenvalues of A and use the methods of Section 13.5 to find a formula for A^n. Hence solve the difference equation for u_n in terms of u_0 and u_1.

38.21 (Section 38.5). Consider the logistic equation

$$u_{n+1} = \alpha u_n(1 - u_n).$$

Draw cobweb solutions starting at $u_0 = \frac{1}{2}$ for the cases $\alpha = 2.7$, $\alpha = 2.9$, and $\alpha = 3.3$. What do you infer about the stability of the fixed point in the first quadrant?

38.22 (Section 38.5). In the logistic equation $u_{n+1} = \alpha u_n(1 - u_n)$, for what positive values of α is the origin a stable fixed point?

38.23 (Section 38.5). Find the two stable values between which u_n ultimately oscillates in the logistic equation $u_{n+1} = 3.25u_n(1 - u_n)$.

38.24 Consider the difference equation

$$u_{n+1} = \alpha(\tfrac{1}{2} - |u_n - \tfrac{1}{2}|).$$

Sketch the function $y = f(x) = \alpha(\tfrac{1}{2} - |x - \tfrac{1}{2}|)$ for $\alpha = \frac{2}{3}$. Where are the equilibrium points of the difference equation for $\alpha > 1$? Show that the origin is stable if $\alpha < 1$, and unstable if $\alpha > 1$. What happens if $\alpha = 1$?

Sketch the graph of $y = f(f(x))$ for $\alpha = 2$. Show that there exists a 2-cycle and locate the periodic values of u_n.

38.25 Find the fixed points of

$$u_{n+1} = \alpha u_n(1 - u_n^3),$$

for all α. Determine the slope of $y = f(x) = \alpha x(1 - x^3)$ at the nonzero fixed point. Confirm that this fixed point is stable if $\alpha < \frac{5}{3}$ and unstable if $\alpha > \frac{5}{3}$. Sketch cobweb solutions for $\alpha = 1.2, 1.4, 1.8$.

38.26 By starting from $u_0 = 0.957\,417$, compute u_1, $u_2, \ldots, u_5$ for the difference equation

$$u_{n+1} = \alpha u_n(1 - u_n), \qquad \alpha = 3.83,$$

and confirm that the logistic equation appears to have a 3-cycle for this value of α.

38.27 Find the fixed points of the difference equation

$$u_{n+1} = \alpha u_n(1 - u_n)^2,$$

in the three cases (a) $\alpha = 9$, (b) $\alpha = 4$, (c) $\alpha = \frac{9}{4}$. Discuss the stability of the fixed points in each case.

38.28 Show that the special logistic equation

$$u_{n+1} = 4u_n(1 - u_n)$$

has the solution

$$u_n = \sin^2(2^n C\pi)$$

where C is any constant. This general solution includes closed-form chaotic solutions. For example, if $C = 1/\pi$, then

$$u_n = \sin^2(2^n)$$

which never repeats itself for $n = 0, 1, 2, \ldots$.

Probability and statistics

Probability

CONTENTS

39.1 Introduction

An **experiment** or **trial** is described as **random** if the result or outcome of the experiment is not predictable or contains uncertainty. The theory of probability is essential in the modelling and analysis of random experiments. In some aspects of life we expect and often hope that situations we meet behave in a predictable or **deterministic** manner. We expect water to freeze at 0°C under normal pressure; we expect the sun to rise at the appropriate time each day. For important safety reasons we expect an aircraft to have predictable characteristics in a wide range of sometimes extreme situations. However, the weather is largely unpredictable looking more than a week into the future. The distinction between random and deterministic has become less 'certain' in more recent times. Some physical phenomena such as the weather can be modelled by deterministic equations but still exhibit long-term, seemingly unpredictable, behaviour. Such systems, which display what is known as **chaos** (see Section 38.5 for a model difference equation with a chaotic output), show extreme sensitivity to small initial changes. Chaos is distinct from random behaviour but the outcome can show very similar manifestations.

If an experiment is repeatable we can count the occasions when a particular outcome occurs. (This only makes sense if the conditions surrounding the experiment do not change with time.) In repeating such an experiment many times, the proportion of favourable outcomes may achieve some regularity. We can measure this by calculating the **relative frequency** of this outcome defined by

$$\text{relative frequency} = \frac{\text{number of occurrences of the given outcome}}{\text{total number of experiments}}.$$

After a large number of experiments, this ratio may approach a steady value which is known as the **probability** of this particular outcome.

For example, the standard die has six faces numbered $1, 2, 3, 4, 5, 6$. After a large number of throws, we would expect the number 1 (or any other number) to appear on the upper face with a relative frequency of 1/6. Hence we expect that the probability of a 1 appearing is 1/6.

Many probabilities are based on data, past records, the 'degree of belief', the view of individuals, and so on. Horse races are usually not repeated so that there can be no relative frequency approach, but bookmakers and punters bet on the basis of the previous form of the horses, the state of the course, and the pattern of bets. Generally as the race approaches the bookmakers' odds reflect how the accumulation of bets has been distributed among the runners. Many outcomes will be assigned probabilities with at least some subjective element.

Probabilities are important in measuring risk, and there can be surprising results. From past data the earth receives a significant meteor impact every 100 years. The probability of a particular individual being killed by such an impact is very small but nonzero. However, the impact could be cataclysmic, which means that by some measures the probability of being killed by a meteor impact is greater than that arising from a plane crash. In engineering, as the reliability of components improves, the likelihood of failure becomes more remote, but might as a consequence have more serious implications if it does occur.

39.2 Sample spaces, events, and probability

The first task with our random experiment is to define the list, or **set, of all possible outcomes** which is known as the **sample space**. A simple example is the single spin of a coin, in which there are two possible outcomes with either a head or a tail showing. The outcomes can be denoted by H (for head) and T (for tail). The sample space S for this experiment has two **elements** H and T and we denote it in set terms by

$$S = \{H, T\}.$$

(Information about sets and set notation can be found in Chapter 35.) For the single throw of a fair die, the sample space has the six possible outcomes, namely 1, 2, 3, 4, 5, 6. Hence its sample space is

$$S = \{1, 2, 3, 4, 5, 6\}.$$

Some sample spaces have an infinite number of elements. Suppose we spin a coin until a tail appears. Any number of heads could appear before a tail. Hence the sample space is

$$S = \{0 \text{ head}, 1 \text{ head}, 2 \text{ heads}, 3 \text{ heads, and so on}\}.$$

However, the sample space is **countable**, that is the elements in the sample space can be matched against the positive integers. A sample space is said to be **discrete** if it contains a **finite** or **countably infinite** set of outcomes. A list of outcomes such as $\{2, 4, 6, 8, \dots\}$ would be countably infinite.

A collection of elements satisfying a common requirement in a sample space is known as an **event**. For a die the event could be the appearance of a particular number, say 5, an odd-number outcome, or any number less than 5. These events are respectively the sets

$$A_1 = \{5\}, \qquad A_2 = \{1, 3, 5\}, \qquad A_3 = \{1, 2, 3, 4\}:$$

they are *subsets* of the sample space; that is, in the notation of (35.3) $A \subseteq S$ in each case.

As we mentioned in the introduction, the probability of an event is the relative frequency that the event takes place in a large number of repetitions of the experiment. The probability of an event A is denoted by $P(A)$. For the single spin of a coin we expect heads and tails to be equally likely to occur. Thus

$$P(H) = \tfrac{1}{2}, \qquad P(T) = \tfrac{1}{2}.$$

We can also view this in a non-experimental way. If an event can occur in n different ways out of a total number of N possible ways, *all of which are equally likely*, then the probability of the event is n/N. For a fair coin a head can arise in one way from two equally likely ways. Hence $P(H) = \tfrac{1}{2}$.

For the die the probability that any individual number x is face up is given by $P(x) = 1/6$. The probability that a number less than 5 appears will be

$$P(A_3) = \frac{\text{number of ways in which numbers less than 5 occur}}{\text{total number of possible outcomes}}$$

$$= \frac{4}{6} = \frac{2}{3},$$

where $A_3 = \{1, 2, 3, 4\}$.

Example 39.1   *Two coins are spun. What is the probability that at least one head appears?*

It is essential in the solution to distinguish the coins, as, say, a and b. Thus if H_a is the event that coin a shows a head, T_a that a shows a tail, and so on, then the sample space has four elements:

$$S = \{(H_a, H_b), (H_a, T_b), (T_a, H_b), (T_a, T_b)\},$$

which are all equally likely. Thus

$$P((H_a, H_b)) = P((H_a, T_b)) = P((T_a, H_b)) = P((T_a, T_b)) = \tfrac{1}{4}.$$

The event A in the problem is the subset

$$A = \{(H_a, H_b), (H_a, T_b), (T_a, H_b)\},$$

which contains three of the four elements. Hence at least one head occurs with probability $P(A) = \tfrac{3}{4}$.

Example 39.2   *Two distinguishable dice a and b are rolled. What are the elements of the sample space? What is the probability that the sum of the face values of the two dice is 8? What is the probability that at least one 5 appears?*

We distinguish the outcome of each die separately, so that there are $6 \times 6 = 36$ possible outcomes for the pair. The sample space has 36 elements of the form (i, j) where i and j take all integer values 1, 2, 3, 4, 5, 6, and i is the outcome of die a and j is the outcome of b. The full list is

$$
\begin{aligned}
S = \{ & (1, 1), (1, 2), (1, 3), (1, 4), (1, 5), (1, 6), \\
& (2, 1), (2, 2), (2, 3), (2, 4), (2, 5), (2, 6), \\
& (3, 1), (3, 2), (3, 3), (3, 4), (3, 5), (3, 6), \\
& (4, 1), (4, 2), (4, 3), (4, 4), (4, 5), (4, 6), \\
& (5, 1), (5, 2), (5, 3), (5, 4), (5, 5), (5, 6), \\
& (6, 1), (6, 2), (6, 3), (6, 4), (6, 5), (6, 6) \},
\end{aligned}
$$

and they are all equally likely. If A_1 is the event that the sum of the dice is 8, then from the list

$$A_1 = \{(2, 6), (3, 5), (4, 4), (5, 3), (6, 2)\}$$

which occurs for 5 elements out of 36. Hence

$$P(A_1) = \tfrac{5}{36}.$$

The event that at least one 5 appears is the list

$$A_2 = \{(1, 5), (2, 5), (3, 5), (4, 5), (5, 1), (5, 2), (5, 3), (5, 4), (5, 5), (5, 6), (6, 5)\},$$

which has 11 elements. Hence

$$P(A_2) = \tfrac{11}{36}.$$

39.3   Sets and probability

Set notation is very helpful in representing sample spaces and events. This section uses the properties of sets and Venn diagrams explained in Chapter 35. Consider Example 39.2 again: this is the problem when two dice are rolled. In set terms the sample space S can be thought of as the universal set for this experiment. Suppose that we are interested in the event A_3 in which *either* the sum of the two dice is 8 (event A_1) *or* at least one 5 appears (event A_2), *or both*. This event is the *union* of the subsets of S, namely A_1 and A_2, represented by

$$A_3 = A_1 \cup A_2,$$

where

$$A_1 = \{(2, 6), (3, 5), (4, 4), (5, 3), (6, 2)\},$$
$$A_2 = \{(1, 5), (2, 5), (3, 5), (4, 5), (5, 1), (5, 2), (5, 3), (5, 4),$$
$$(5, 5), (5, 6), (6, 5)\}.$$

The event A_3 has 14 elements of which two are common to both A_1 and A_2. If A_4 is the event that both A_1 *and* A_2 occur, then A_4 is the *intersection* of A_1 and A_2, namely

$$A_4 = A_1 \cap A_2 = \{(3, 5), (5, 3)\}.$$

The two events are shown diagrammatically in Figure 39.1.

Remember that the *complement* of a set or event is denoted by $\bar{A}$, and the empty set by $\varnothing$.

(a)

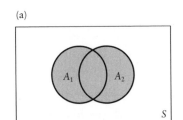

(b)

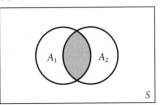

Fig. 39.1 (a) The event $A_3 = A_1 \cup A_2$, (b) The event $A_4 = A_1 \cap A_2$.

Example 39.3 *Suppose that A, B, C are three events in the sample space S. Write down the sets which represent the events that:*
(a) *A occurs, but B and C do not;* (b) *A, B, and C all occur.*

(a) The event that B or C occurs will be $B \cup C$. The event that neither B nor C occurs will be the complement $\overline{B \cup C}$. The required set will be the intersection of this event and A, namely

$$A \cap (\overline{B \cap C}).$$

By de Morgan's first law (35.6), this is equivalent to

$$A \cap (\bar{B} \cap \bar{C}),$$

which can be written unambiguously as $A \cap \bar{B} \cap \bar{C}$ by the associative law for intersection (35.4).

(b) Events B and C occur in the set $B \cap C$. Events A and $B \cap C$ occur in the event

$$A \cap (B \cap C) \quad \text{or} \quad A \cap B \cap C.$$

Two events are said to be **mutually exclusive** if they cannot occur together in a single **trial** (or experiment), which in set terms is equivalent to the two subsets of S being disjoint: that is, having no elements in common. Consider the following illustrative application of a single die, which is rolled and the score noted. An event of interest in a random experiment can be specified in many ways. A player could be interested in even or odd scores, the score 2 or not, or scores which are factors of 6 or not. In each case the sample space is divided into two disjoint sets or mutually exclusive events, together constituting an **exhaustive** (meaning that there are no outcomes which are not in at least one event) list of outcomes. For example, if A stands for the event of an even score, then $\bar{A}$ must represent an odd score. Thus

$$A \cap \bar{A} = \varnothing, \quad \text{and} \quad A \cup \bar{A} = S.$$

If A_i denotes the event of a score i where $i = 1, 2, \ldots, 6$ for the rolling of a die, then the events are mutually exclusive and exhaustive, and the sample space will be the union of these events:

$$S = A_1 \cup A_2 \cup \cdots \cup A_6.$$

Any union of events can be expressed in terms of the union of certain mutually exclusive events. For example, the union $A \cup B$ of two events A and B can be **partitioned** into the mutually exclusive events $A \cap \bar{B}$, $A \cap B$, and $\bar{A} \cap B$. Then

$$A \cup B = (\bar{A} \cap B) \cup (A \cap B) \cup (A \cap \bar{B}).$$

In another example, an event A in the sample space which also contains B can be divided as

$$A = (A \cap B) \cup (A \cap \bar{B}),$$

which can be interpreted as meaning that A can occur either with B or without B.

Suppose the sample space is partitioned into the n mutually exclusive and exhaustive events $A_1, A_2, \ldots, A_n$. If A is *any* event, then

$$A = (A \cap A_1) \cup (A \cap A_2) \cup \cdots \cup (A \cap A_n).$$

This means that, if A occurs, then it must occur as one, and only one, of the events $A_1, A_2, \ldots, A_n$. It might happen that $A \cap A_i = \varnothing$ for some intersections, but this does not matter.

Example 39.4   *In Example 39.2, express the sample space S and the events A_1 and A_2 in set terms.*

The sample space is given by

$$S = \{(i, j) \,|\, i, j = 1, 2, 3, 4, 5, 6\},$$

which has 36 elements since the dice are distinguishable and (i, j) is distinct from (j, i). The events A_1 and A_2 can be written

$$A_1 = \{(i, j) \,|\, i + j = 8\},$$
$$A_2 = \{(i, j) \,|\, \text{either } i = 5 \text{ or } j = 5 \text{ or both}\}.$$

The rules governing probability are as follows:

Axioms of probability

For every event A in a sample space S, the probability $P(A)$ must satisfy:

(a) $0 \leqslant P(A) \leqslant 1$;

(b) for the empty set (or non-event) and the sample space S:
$$P(\emptyset) = 0, \qquad P(S) = 1;$$

(c) for n mutually exclusive events $A_1, A_2, \ldots, A_n$,
$$P(A_1 \cup A_2 \cup \cdots \cup A_n) = P(A_1) + P(A_2) + \cdots + P(A_n). \qquad \textbf{(39.1)}$$

The rules can be interpreted as

(a) every probability must lie between 0 and 1;

(b) the probability of an impossible event is zero, and the probability of the occurrence of some element in a sample space is certain;

(c) the probability that one of a set of mutually exclusive events occurs is the sum of the probabilities of each event.

Example 39.5 *Two dice are rolled. What is the probability that a total score of 4 or 7 occurs?*

Let A_1 be the event of a score 4 and A_2 be the event of a score 7. These cannot occur together, so they must be mutually exclusive events. Hence by (39.1c) and the complete list of outcomes in Example 39.2,
$$P(A_1 \cup A_2) = P(A_1) + P(A_2) = \tfrac{3}{36} + \tfrac{6}{36} = \tfrac{1}{4}.$$

If two events A_1 and A_2 are *not* mutually exclusive then they must have elements of the sample space in common. Using partitioning, which was explained previously in this section, A_1, A_2, and therefore the union of A_1 and A_2 can be partitioned into unions of mutually exclusive events. Thus

$$A_1 = (A_1 \cap \bar{A}_2) \cup (A_1 \cap A_2),$$

$$A_2 = (\bar{A}_1 \cap A_2) \cup (A_1 \cap A_2),$$

$$A_1 \cup A_2 = (A_1 \cap \bar{A}_2) \cup (\bar{A}_1 \cap A_2) \cup (A_1 \cap A_2),$$

since $(A_1 \cap A_2) \cup (A_1 \cap A_2) = A_1 \cap A_2$. Hence by rule (c) in (39.1),

$$P(A_1) = P(A_1 \cap \bar{A}_2) + P(A_1 \cap A_2), \qquad (39.2)$$

$$P(A_2) = P(\bar{A}_1 \cap A_2) + P(A_1 \cap A_2), \qquad (39.3)$$

$$P(A_1 \cup A_2) = P(A_1 \cap \bar{A}_2) + P(\bar{A}_1 \cap A_2) + P(A_1 \cap A_2). \qquad (39.4)$$

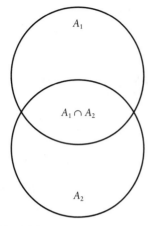

Fig. 39.2

Elimination of $P(A_1 \cap \bar{A}_2)$ and $P(\bar{A}_1 \cap A_2)$ between (39.2), (39.3), and (39.4) leads to:

Probability addition law

For two events which are not mutually exclusive:

$$P(A_1 \cup A_2) = P(A_1) + P(A_2) - P(A_1 \cap A_2).$$

(39.5)

Geometrically the result can be seen from Fig. 39.2 in which the intersection $A_1 \cap A_2$ is 'counted twice' in $P(A_1) + P(A_2)$.

Example 39.6 *If two dice are rolled, what is the probability that either the sum is 8 or at least one 5 appears?*

As we saw in Example 39.2, if A_1 is the event that the sum is 8 and A_2 the event that at least one 5 appears, then

$$P(A_1) = \tfrac{5}{36}, \qquad P(A_2) = \tfrac{11}{36}.$$

These are not mutually exclusive events because both events occur when the outcomes are $\{(5, 3)\}$ or $\{(3, 5)\}$. Therefore

$$A_1 \cap A_2 = \{(3, 5), (5, 3)\},$$

in which case

$$P(A_1 \cap A_2) = \tfrac{2}{36} = \tfrac{1}{18}.$$

Hence, by (39.5),

$$P(A_1 \cup A_2) = \tfrac{5}{36} + \tfrac{11}{36} - \tfrac{2}{36} = \tfrac{7}{18},$$

which means that the sum is 8 or at least one 5 appears with probability $\tfrac{7}{18}$.

39.4 Frequencies and combinations

In many applications the total number of elements in a sample space or in an event needs to be counted. Enumeration of outcomes can become a lengthy process. For example, suppose that an experiment consists of trials such as the spinning of k coins or the rolling of k dice. If there are n possible outcomes for each coin or die, then the same space has n^k possible outcomes. The rolling of four dice leads to a sample space with $6^4 = 1296$ elements.

As we saw in Section 39.1, probabilities can be obtained using relative frequency arguments. For the counting process which is needed, permutation and combination formulae are often useful. Section 1.17 provides a full account of permutations and combinations. We shall review the main results here.

A **permutation** is a particular *ordered selection*. The notation $_nP_r$ means *the number of ways in which r different items can be selected from n distinct items taking due regard of the order of selection*. If items are not replaced, the first item can be chosen in n ways leaving

$n-1$ items. Hence the second item can be chosen in $n-1$ ways. The first two items can be chosen in $n(n-1)$ different ways. Continuing this process r times we obtain

$$_nP_r = n(n-1)(n-2)\ldots(n-r+1) = \frac{n!}{(n-r)!}.$$

Example 39.7 *How many permutations of the letters a, b, c, d can be made if two are selected each time?*

In this example $n = 4$ and $r = 2$. Thus
$$_4P_2 = 4\cdot 3 = 12.$$
The full list of permutations is
$$\{ab, ac, ad, ba, bc, bd, ca, cb, cd, da, db, dc\}.$$

A **combination** is an *unordered selection*. The notation for a combination is $_nC_r$ which means the number of ways in which r *different items can be selected from n items without regard to order*. Among the $_nP_r$ permutations there are $r!$ which give the same combination, because the first position can be chosen in r ways, the second in $r-1$ ways, and so on. Thus

$$_nC_r = \frac{_nP_r}{r!} = \frac{n!}{(n-r)!r!}.$$

In the example above the items ab and ba are not distinguished in the combination and so on, so that two different letters may be chosen from four different letters in

$$_4C_2 = \frac{4!}{2!2!} = 6$$

ways.

Note that

$$_nC_r = \frac{n!}{(n-r)!r!} = \frac{n!}{(n-(n-r))!(n-r)!} = {_nC_{n-r}}.$$

An alternative notation for $_nC_r$ is

$$_nC_r = \binom{n}{r}.$$

(See eqn (1.44)). Notice also that the sequence $_nC_r$ $(r = 0, 1, 2, \ldots, n)$ generates the coefficients of the binomial series in $(a + b)^n$ (see Section 1.18 and Appendix A(c)). Special values are

$$_nC_0 = 1, \qquad _nC_n = 1.$$

Example 39.8   *How many different five-card hands can be dealt from a standard deck with 52 cards? What is the probability that a hand dealt at random consists of five spades?*

This is a *combination* problem, not a permutation one. Thus there are

$$_{52}C_5 = \frac{52!}{47!5!} = \frac{52 \cdot 51 \cdot 50 \cdot 49 \cdot 48}{1 \cdot 2 \cdot 3 \cdot 4 \cdot 5} = 2\,598\,960$$

different hands.

The number of different hands consisting of five spades is, since there are 13 spades in the pack,

$$_{13}C_5 = \frac{13!}{8!5!} = \frac{13 \cdot 12 \cdot 11 \cdot 10 \cdot 9}{1 \cdot 2 \cdot 3 \cdot 4 \cdot 5} = 1287.$$

To obtain the probability that a random five-card hand contains five spades we can use the counting argument, namely that out of the 2 598 960 equally likely different hands 1287 will have five spades. Hence

$$P(\text{five-card spade hand}) = \frac{1287}{2\,598\,960} \approx 0.0005,$$

which implies that about one hand in 2000 will have five spades.

Example 39.9   *A box contains 20 balls of which 7 are red(r), 5 are white(w), and 8 are black(b) balls. If three balls are drawn at random, without replacement, find the probability that*

(a) *two red balls and one black ball are drawn;*
(b) *one of each colour is drawn;*
(c) *one or more red balls are drawn;*
(d) *all are of the same colour.*

The total number of three-ball selections which can be made is

$$N = {}_{20}C_3 = 1140$$

for labelled balls. They are all equally likely to be drawn.

(a) The numbers of ways in which two red balls and one black ball can be drawn is

$$_7C_2 \times 8 = \frac{7 \cdot 6}{1 \cdot 2} \cdot 8 = 168.$$

Hence

$$P(2r \text{ and } 1b) = \frac{168}{N} = \frac{168}{1140} = \frac{14}{95} \approx 0.15.$$

(b) The number of ways in which one of each colour can be chosen is $7 \times 5 \times 8 = 280$ from a total of 1140. Hence

$$P(1r \text{ and } 1w \text{ and } 1b) = \frac{280}{1140} = \frac{14}{57} \approx 0.25.$$

Example 39.9 *continued*

(c) The number of ways in which no red ball is drawn is $_{13}C_3 = 286$ from the total of 1140. Hence the probability that a selection contains at least one red ball is

$$P(\geqslant 1r) = 1 - P(0r) = 1 - \frac{286}{1140} = \frac{854}{1140} = \frac{427}{570} \approx 0.75.$$

(d) Since the events are mutually exclusive, using (39.1c),

$$\begin{aligned}
P(3r \text{ or } 3w \text{ or } 3b) &= P(3r \cup 3w \cup 3b) \\
&= P(3r) + P(3w) + P(3b) \\
&= \frac{_7C_3 + _5C_3 + _8C_3}{_{20}C_3} \\
&= \frac{101}{1140} \approx 0.09.
\end{aligned}$$

39.5 Conditional probability

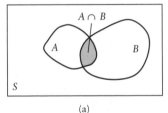

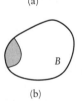

Fig. 39.3 (a) Both A and B occur in the shaded intersection $A \cap B$. (b) $P(A|B)$ refers to the new universal set B.

In many applications we are interested in an event A given that an event B occurs. The probability of A, **conditional** that B occurs, is written as $P(A|B)$. A Venn diagram showing the overlapping events is displayed in Fig. 39.3a. The probability $P(A|B)$ refers to the restricted set in Fig. 39.3b in which effectively the new universal set is B. In enumeration terms we can derive

$$\begin{aligned}
P(A|B) &= \frac{(\text{number of outcomes in } A \cap B)}{(\text{number of outcomes in } B)} \\
&= \frac{(\text{number of outcomes in } A \cap B)/(\text{number of outcomes in } S)}{(\text{number of outcomes in } B)/(\text{number of outcomes in } S)}.
\end{aligned}$$

Hence the formal definition is, assuming that $P(B) \neq 0$,

Conditional probability of A given B

$$P(A|B) = \frac{P(A \cap B)}{P(B)}.$$

(39.6)

Example 39.10 *Six cards are dealt from a well-shuffled deck of playing cards. Given that all six cards are black, find the probability that they are all of the same suit.*

Let A and B represent the following events:

   $A = \{$the cards are black$\}$, $B = \{$the six cards in the same suit$\}$.

Thus

   $A \cap B = \{$six black cards of the same suit$\}$.

Example 39.10 *continued*

Therefore

$$P(A \cap B) = \frac{\text{(number of combinations of six clubs or six spades)}}{\text{(number of combinations of six cards)}}$$

$$= \frac{2 \cdot {}_{13}C_6}{{}_{52}C_6}.$$

Also

$$P(B) = \frac{\text{(number of combinations of six black cards)}}{\text{(number of combinations of six cards)}} = \frac{{}_{26}C_6}{{}_{52}C_6}.$$

Hence the conditional probability that they are all of the same black suit is

$$P(A|B) = \frac{P(A \cap B)}{P(B)} = \frac{2 \cdot {}_{13}C_6}{{}_{52}C_6} \cdot \frac{{}_{52}C_6}{{}_{26}C_6}$$

$$= 2 \cdot \frac{13!}{6!7!} \cdot \frac{6!20!}{26!} = \frac{12}{805} \approx 0.015.$$

Note the following properties of conditional probabilities:

(i) $P(A|A) = 1$.

(ii) $P(A|B)P(B) = P(B|A)P(A)$.

The latter follows since $A \cap B = B \cap A$ and

$$P(A|B) = P(A \cap B)/P(B), \qquad P(B|A) = P(B \cap A)/P(A)$$

from definition (39.6).

Example 39.11 *A production line is supplied with the same component made by two different machines M_1 and M_2. It is known from samples of the outputs that the probability that a component from M_1 is not faulty is 0.91 and from M_2 is 0.85. Machine M_1 supplied 60% of the components and machine M_2 40%. Components are chosen at random and tested before the next stage of production. What is the probability that*

(a) *given that a component was made by M_2 it is not faulty?*
(b) *a component is not faulty?*

Let A_1, A_2, and B be the events

$A_1 = \{\text{component made by } M_1\}, \qquad A_2 = \{\text{component made by } M_2\},$
$B = \{\text{component not faulty}\}.$

From the 60%/40% supply we know that $P(A_1) = 0.6$ and that $P(A_2) = 0.4$. The known failure rates in M_1 and M_2 give the conditional probabilities $P(B|A_1) = 0.91$ and $P(B|A_2) = 0.85$.

↗

Example 39.11 *continued*

(a) The answer is $P(B|A_2) = 0.85$.

(b) Write the event B as $(B \cap A_1) \cup (B \cap A_2)$ which is still the event that the component is not faulty. Since $B \cap A_1$ and $B \cap A_2$ are mutually exclusive, it follows that

$$P[(B \cap A_1) \cup (B \cap A_2)] = P(B \cap A_1) + P(B \cap A_2)$$
$$= P(B|A_1)P(A_1) + P(B|A_2)P(A_2)$$
$$= 0.91 \times 0.6 + 0.85 \times 0.4 = 0.886,$$

using (39.6). Hence the probability of a non-faulty component is 0.89 approximately. In solving this problem we have encountered a new law in (b) called the law of total probability which will be discussed further in Section 39.7.

39.6 Independent events

The recognition of independence of events and data is crucial in probability and statistics. Two events are said to be **independent** if the occurrence of either event has no effect on the occurrence of the other. In terms of conditional probability this means that two events A and B are independent if and only if

$$P(B|A) = P(B) \quad \text{or} \quad P(A|B) = P(A). \tag{39.7a}$$

In that case

$$P(A \cap B) = P(A)P(B) \tag{39.7b}$$

by (39.6). The independence result (39.7b) generalizes for N independent events $A_1, A_2, \ldots, A_N$ to

$$P(A_1 \cap A_2 \cap \cdots \cap A_N) = P(A_1)P(A_2) \ldots P(A_N). \tag{39.8}$$

The following simple illustration shows the distinction between dependent and independent events. Two cards are chosen at random from a pack of 52 cards. In the first case, the first card is *replaced* before the second card is chosen. The events considered are

$A = \{$first card is an ace$\}$,

$B = \{$second card is an ace$\}$.

Then

$$P(A) = \tfrac{4}{52} = \tfrac{1}{13}, \quad \text{and} \quad P(B|A) = \tfrac{4}{52} = \tfrac{1}{13} = P(A).$$

In other words the events are independent.

On the other hand if there is *no replacement*, then

$$P(A) = \tfrac{1}{13} \quad \text{but} \quad P(B|A) = \tfrac{3}{51} \neq P(A),$$

indicating that A and B are not independent events.

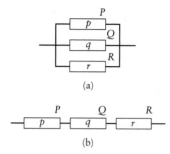

(a)

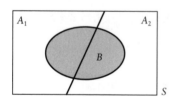

(b)

Fig. 39.4 (a) Components in parallel. (b) Components in series.

Example 39.12 *Figure 39.4 shows parts of two circuits which contain electrical components P, Q, and R placed in parallel and series. For the parallel case the circuit fails if all three components fail, but in the series case failure occurs if just one component fails. In some time interval the probabilities of failure of P, Q, and R are respectively p, q, and r. What are the probabilities of circuit breakdown in the two cases?*

Let A, B, and C be the events

$$A = \{P \text{ fails}\}, \qquad B = \{Q \text{ fails}\}, \qquad C = \{R \text{ fails}\},$$

where we assume that failures are independent events.

For the parallel case failure occurs if $A \cap B \cap C$ occurs. By (39.8)

$$P(A \cap B \cap C) = P(A)P(B)P(C) = pqr,$$

which means that the probability of failure is pqr.

For the series case failure occurs if the event $A \cup B \cup C$ occurs. Using (39.5) twice and (39.8)

$$
\begin{aligned}
P(A \cup B \cup C) &= P(A) + P(B \cup C) - P(A \cap (B \cup C)) \\
&= p + P(B \cup C) - P(A)P(B \cup C) \\
&= p + (1-p)(P(B) + P(C) - P(B \cap C)) \\
&= p + (1-p)(q + r - qr) \\
&= (p + q + r) - (qr + rp + pq) + pqr
\end{aligned}
$$

which is the probability of series failure.

39.7 Total probability

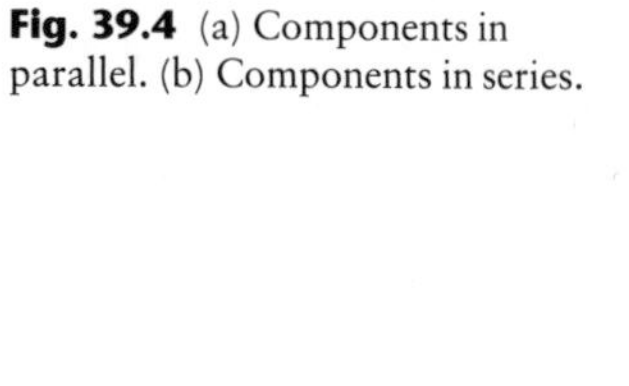

Suppose that a sample space is partitioned (see Section 38.3) into two events A_1 and A_2 which are mutually exclusive. In other words $A_1 \cap A_2 = \varnothing$ and $A_1 \cup A_2 = S$. Let B be an event in S (see Fig. 39.5). The sets $B \cap A_1$ and $B \cap A_2$ are mutually exclusive so that

$$P(B) = P(B \cap A_1) + P(B \cap A_2).$$

From the notion of conditional probability (39.6) we obtain:

Fig. 39.5

> **The law of total probability**
> For mutually exclusive events A_1 and A_2
> $$P(B) = P(B|A_1)P(A_1) + P(B|A_2)P(A_2).$$
> (39.9)

The result generalizes to the case in which S contains n mutually exclusive and exhaustive events $A_1, A_2, \ldots, A_n$. If B is an event in S, then

$$P(B) = \sum_{i=1}^{n} P(B|A_i)P(A_i).$$

Example 39.13 *A box contains 8 red and 13 black components. A machine draws components at random from the box and fits them into a circuit. What is the probability that the second component is red?*

Suppose now that components in the box are replaced with components of the same colour as they are used. What is the probability that the second component is red?

Define the event as follows:

$A_1 = \{\text{first component is red}\}$,

$A_2 = \{\text{first component is black}\}$,

$B = \{\text{second component is red}\}$.

Then

$$P(A_1) = \tfrac{8}{21}, \qquad P(A_2) = \tfrac{13}{21}.$$

Also

$$P(B\,|\,A_1) = \tfrac{7}{20}, \qquad P(B\,|\,A_2) = \tfrac{8}{20}.$$

Using (39.9), since A_1 and A_2 are mutually exclusive

$$P(B) = P(B\,|\,A_1)P(A_1) + P(B\,|\,A_2)P(B\,|\,A_2)$$

$$= \tfrac{7}{20}\cdot\tfrac{8}{21} + \tfrac{8}{20}\cdot\tfrac{13}{21} = \tfrac{8}{21}.$$

Hence the probability that the second component draw is red is 8/21, which is the same as $P(A_1)$. This suggests (correctly) that the probability that the second ball is red does *not depend on the colour of the first ball*.

The first solution was selection *without replacement*. In the second part of the question the components are replaced. In this case $P(B) = \tfrac{8}{21}$; in other words, with or without replacement, the probability that the second component is red is still 8/21.

39.8 Bayes' theorem

Suppose that the sample space S is the union of the mutually exclusive events A_1 and A_2. In this case $A_2 = \bar{A}_1$, and the notation suggests the generalization which follows. Suppose that an event B occurs. We ask the question: if B occurs, what is the probability that A_1 occurs? In other words, what is $P(A_1\,|\,B)$?

From the rule for conditional probability (39.6) we can deduce, since $B \cap A_1 = A_1 \cap B$, etc.,

$$P(B \cap A_1) = P(A_1\,|\,B)P(B) = P(B\,|\,A_1)P(A_1), \tag{39.10}$$

$$P(B \cap A_2) = P(A_2\,|\,B)P(B) = P(B\,|\,A_2)P(A_2). \tag{39.11}$$

From (39.9) we also have

$$P(B) = P(B\,|\,A_1)P(A_1) + P(B\,|\,A_2)P(A_2). \tag{39.12}$$

Elimination of $P(B)$ between (39.10) and (39.12) leads to:

> **Bayes' theorem**
>
> For mutually exclusive events A_1 and A_2
>
> $$P(A_1|B) = \frac{P(B|A_1)P(A_1)}{P(B)}$$
>
> $$= \frac{P(B|A_1)P(A_1)}{P(B|A_1)P(A_1) + P(B|A_2)P(A_2)}.$$
>
> (39.13)

Example 39.14 *It is known that 4% of a batch of components in a manufacturing process are faulty. Components are tested on the production line with 90% probability that a fault component is detected, but it is known that in 2% of the cases a component which is not faulty is nevertheless recorded as faulty. What is the probability that a component which is recorded as faulty is actually faulty?*

Let A_1 and A_2 be the events

   $A_1 = \{\text{component faulty}\}$,

   $A_2 = \{\text{component not faulty}\}$,

and let B be the event

   $B = \{\text{test indicates faulty}\}$.

Then

   $P(A_1) = 0.04,$ $P(A_2) = 0.96,$ $P(B|A_1) = 0.90,$ $P(B|A_2) = 0.02.$

We require the probability that the component is faulty given that the test recorded faulty, that is $P(A_1|B)$. By Bayes' theorem (39.13)

$$P(A_1|B) = \frac{P(B|A_1)P(A_1)}{P(B|A_1)P(A_1) + P(B|A_2)P(A_2)}$$

$$= \frac{0.9 \times 0.04}{0.9 \times 0.04 + 0.02 \times 0.96}$$

$$= 0.65.$$

If the sample space is partitioned by $\{A_i\}$ ($i = 1, 2, \dots, n$) then the generalized Bayes' theorem is

$$P(A_i|B) = \frac{P(B|A_i)P(A_i)}{\sum_{j=1}^{n} P(B|A_j)P(A_j)}.$$

Problems

39.1 How many elements do the following sample spaces contain?
(a) the spinning of five coins;
(b) the rolling of three dice;
(c) a coin and a die randomly thrown together;
(d) a dart thrown at a dartboard.

39.2 Two dice are rolled. What is the probability that the sum of the face values is 7? What is the probability that no 5 appears? What is the probability that the score is 7 or less?

39.3 Two dice are rolled and the scores noted. Write down the elements in the sample space. How many elements does the set have? Let A denote the event {the sum of the outcomes is 5}, and B denote the event {at least one die shows 4}.

Express the sets of these events in formula terms. List all the elements in A, B, $A \cup B$, and $A \cap B$.

39.4 Suppose that A, B, and C are three events of the sample space S. Write down the set formulae for the events:
(a) only B occurs,
(b) exactly one of A, B, or C occurs.

39.5 Suppose that a sample space S includes the events A and B. Show that the number of elements in $A \cup B$ can be expressed as

$$n(A \cup B) = n(A \cap B) + n(\bar{A} \cap B) + n(A \cap \bar{B})$$

(this is an alternative version of (35.7)).

Suppose two dice are rolled. Let A denote the event {the sum of the outcomes is 6} and B the event {both dice show the same number}. List the elements in $A \cap B$, $\bar{A} \cap B$, and $A \cap \bar{B}$, and find $n(A \cup B)$ using the formula above.

39.6 A card is drawn from a deck of 52 playing cards. If A is the event that an ace is drawn, B is the event that a heart is drawn, and C is the event that a black card is drawn, explain in terms of the cards drawn what the following events represent:
(a) $A \cap B$; (b) $A \cap C$; (c) $A \cup B$;
(d) $A \cup B \cup C$; (e) $A \backslash B$; (f) $\bar{A} \backslash \bar{B}$; (g) $\bar{A} \backslash \bar{C}$;
(h) $(A \cap B) \cup C$; (i) $(A \cap B) \cup (A \cap \bar{C})$.

39.7 Cards are drawn from a deck of 52 playing cards without replacement. What is the probability that

(a) the first card is a king?
(b) the first two cards are kings?
(c) the first card is a king, the second and third cards are not kings, and the fourth card is a king?

39.8 A well-shuffled deck of cards is cut twice randomly. What is the probability that two aces are shown? (This is a problem of selection *with* replacement.)

39.9 Evaluate the following permutations:
(a) $_5P_3$; (b) $_{10}P_4$; (c) $_7P_7$; (d) $_7P_1$.

39.10 How many different three-letter 'words' can be made up from the letters a, b, c, d, e with no repetition of letters?

39.11 How many five-digit numbers can be formed (numbers cannot start with 0) from 0, 1, 2, 3, 4, 5, 6, 7, 8, 9, if
(a) numbers are selected without replacement?
(b) any number of repetitions of numbers is allowed?
(c) without replacement but such that the number must be divisible by 5?

39.12 Calculate the following combinations:
(a) $_7C_3$; (b) $_{99}C_{96}$; (c) $_{11}C_5$.

39.13 Prove that

$$_{n-1}C_r + {_{n-1}C_{r-1}} = {_nC_r}.$$

39.14 Prove that

(a) $\displaystyle\sum_{r=0}^{n} {_nC_r} = 2^n$,

(b) $\displaystyle\sum_{r=0}^{n} {_nC_r} 3^r = 4^n$.

39.15 How many different four-card hands can be dealt from a deck of 52 playing cards? How many hands contain four cards of the same suit? What is the probability that a hand dealt randomly contains four cards from the same suit?

39.16 In the previous question investigate how the probabilities change for n-card hands ($1 \leqslant n \leqslant 13$) with n cards from the same suit.

39.17 A box contains 22 balls of which 7 are red, 9 are white, and 6 are black. Four balls are drawn at random from the box without replacement. Find the probability that
(a) three red balls and one white ball are drawn;
(b) the balls are red;
(c) the balls are all of the same colour;
(d) there is at least one ball of each colour.

39.18 A production line is supplied with the same component made by two different machines M_1 and M_2. It is known from samples of the outputs that the probability that a component from M_1 is not faulty is 0.89 and from M_2 is 0.83. Machine M_1 supplies 70% of the components and machine M_2 30%. Components are chosen at random and tested before the next stage of production. What is the probability that
(a) given that it was made by M_1, a component is not faulty?
(b) a component is not faulty?
(c) given that a component was faulty that was manufactured by M_2?

39.19 A production line is supplied with the same component made by three different machines M_1, M_2, and M_3. It is known from samples of the outputs that the probability that a component from M_1 is not faulty is 0.87, from M_2 is 0.84, and from M_3 is 0.91. Machine M_1 supplies 45% of the components, machine M_2 30%, and machine M_3 25%. Components are chosen at random and tested before the next stage of production. What is the probability that
(a) a component is not faulty?
(b) given that a component was faulty that it was manufactured by M_2?
(c) given that a component was faulty that it was made by M_1 or M_2?

39.20 Figure 39.6 shows part of a circuit with six components in a parallel and series combination. The probabilities of failures of components are p_1, p_2, p_3, q,

r_1, and r_2 as shown and are independent. What is the probability that this part of the circuit fails? If all components have the *same* probability of failure of 0.98, what is the probability that this part of the circuit fails? (Parallel and series failures are as in Example 39.12.).

39.21 It is known that in a batch of 100 microprocessors, 5 are defective.
(a) A microprocessor is chosen at random without replacement. What is the probability that it is defective?
(b) Two are chosen at random without replacement. What is the probability that both are defective?
(c) Two are chosen without replacement. Given that the first is defective, what is the probability that the second is also defective?

39.22 In the UK national lottery 6 numbered balls are selected at random from 49 balls numbered 1, 2, 3, ..., 49 without replacement. Prizes are given to those who correctly select three, four, five, or six numbers. Find the probability of winning in each case. A seventh bonus ball is also drawn from the remaining 43 balls and further prizes are given for those who correctly choose the bonus ball and any five of the six drawn numbers. Find the probability of winning in this case. What is the overall probability that a lottery ticket wins at least one prize?

39.23 A game is played in which n players each spins a coin and the outcome is examined. The game continues until the outcome is either $n - 1$ heads and 1 tail, or 1 head and $n - 1$ tails. The single player with the different outcome wins the coins from the other players. Show that the probability that the game ends at a given play is $n/2^{n-1}$, and that the probability that the game finishes at the ith play is given by the geometric distribution

$$\frac{n}{2^{n-1}}\left(1 - \frac{n}{2^{n-1}}\right)^{i-1}.$$

Find also the mean number of plays to the end of the game.

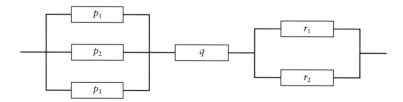

Fig. 39.6

40 Random variables and probability distributions

CONTENTS

40.1 Random variables

In experiments or trials in which the outcome is numerical, the outcomes are values of what is known as a **random variable**. For example, suppose that a coin is spun three times and we record the outcomes and ask: how many heads appear? Then the answer will be 0, 1, 2, or 3 heads. The sample space S, which lists all possible outcomes in trial, has eight elements given by

$$S = \{(HHH), (THH), (HTH), (HHT), (TTH), (THT),$$
$$(HTT), (TTT)\}.$$

The random variable X associated with the question is the number of heads obtained. Generally, the random variable X assigns a *number* to each event in the sample space S. This set of numbers is denoted by S_X. In this example

$$S_X = \{0, 1, 2, 3\},$$

which is a list of the possible numerical outcomes of the number of heads.

The random variable X can be thought of as a **function** or **mapping** from the sample space S to S_X which, since it is a set of real numbers, can be represented by points on a straight line. A representation of the mapping is displayed in Fig. 40.1 in which it is shown that the element s in S is mapped by X into the value $X(s)$ on the real line S_X. In the example s could be (HTT) giving $X(s) = 1$, but notice that THT and TTH will also map into $X(s) = 1$.

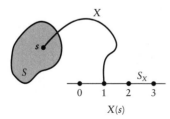

Fig. 40.1 Mapping of the random variable X from the sample space S onto the real line S_X.

In the example above, X is a **discrete random variable** since $X(s)$ is one of a finite set of numbers. In some cases the possible outcomes are infinite in number but can still be **counted**. For example, suppose that X is the random variable of the number of spins of a coin until a head appears. The list of possible outcomes is $\{1, 2, 3, \dots\}$ which is unbounded but countable, and X is still called a discrete random variable.

Obviously many random variables can be associated with the same experiment. In the example above where a coin is spun three times, a random variable Y, say, could be the number of tails observed.

40.2 Probability distributions

Let the random variable X take the values $x_1, x_2, \dots$ (depending on the context it is sometimes more convenient to start with $x_0, x_1, \dots$), where the set of numbers can be finite or infinite. In terms of a random variable we write probabilities as $P(X = x_i)$, which means the probability that the random variable X takes the value x_i, or we could consider $P(X < x_i)$, which is the probability that the random variable takes values strictly less than x_i, and so on. Often we denote $P(X = x_i)$ simply by the symbol p_i. The pairs (x_i, p_i) for $i = 1, 2, \dots$ define the **probability distribution** or **probability function** for the random variable X. Note that for any probability distribution of a discrete random variable we must have:

> **Probability distribution** $P(X = x_i) = p_i$
>
> (i) $0 \leqslant p_i \leqslant 1$;
> (ii) $\sum_{i=1}^{n} p_i = 1$, if X has n possible outcomes, or $\sum_{i=1}^{\infty} p_i = 1$
> if X has a countably infinite set of outcomes. **(40.1)**

A discrete probability distribution can be expressed in a table such as

$x_i =$	x_1	x_2	x_3	$\dots$
$p_i =$	p_1	p_2	p_3	$\dots$

For the coin spun three times in Section 40.1, the distribution would be

$x_i =$	0	1	2	3
$p_i =$	$\frac{1}{8}$	$\frac{3}{8}$	$\frac{3}{8}$	$\frac{1}{8}$

since each of the outcomes in the original S is equally likely.

Example 40.1   *A box contains six components of which two are defective. Components are selected at random without replacement until a defective component is chosen. Find the probability distribution of the number of components drawn from the box.*

Let X be the random variable (number of components withdrawn including the defective). Then

$$S_X = \{1, 2, 3, 4, 5\} = \{x_i\} \qquad (i = 1, 2, 3, 4, 5).$$

The probability

$$p_1 = P(X = x_1) = \tfrac{2}{6} = \tfrac{1}{3},$$

since there is a 2 in 6 chance of choosing a defective on the first selection. Also

$$P(X = x_2) = \tfrac{4}{6} \cdot \tfrac{2}{5} = \tfrac{4}{15},$$

since the probability of choosing a non-defective component at the first stage is $\tfrac{4}{6}$ which leaves two defective in the remaining five. Similarly

$$P(X = x_3) = \tfrac{4}{6} \cdot \tfrac{3}{5} \cdot \tfrac{2}{4} = \tfrac{1}{5}, \qquad P(X = x_4) = \tfrac{4}{6} \cdot \tfrac{3}{5} \cdot \tfrac{2}{4} \cdot \tfrac{2}{3} = \tfrac{2}{15},$$

$$P(X = x_5) = \tfrac{4}{6} \cdot \tfrac{3}{5} \cdot \tfrac{2}{4} \cdot \tfrac{1}{3} \cdot 1 = \tfrac{1}{15}.$$

The complete distribution is

$x_i =$	1	2	3	4	5
$p_i =$	$\tfrac{1}{3}$	$\tfrac{4}{15}$	$\tfrac{1}{5}$	$\tfrac{2}{15}$	$\tfrac{1}{15}$

The distribution can be represented graphically as shown in Fig. 40.2.

Fig. 40.2

40.3   The binomial distribution

Suppose a series of trials are independent, and have two possible outcomes which occur with probabilities p and $(1 - p)$. If p is constant throughout then these are known as **Bernoulli trials**. A simple example is the spinning of a coin. We could define a random variable X which takes the value 1 if a head appears and 0 if a tail appears. The probabilities that these occur is $\tfrac{1}{2}$ in each case. The terms *success* and *failure* are frequently used in this context, and generally Bernoulli trials apply to populations that naturally divide into pairs of alternatives, for example on/off, male/female, alive/dead, etc. With 1/0 representing success/failure, a Bernoulli sequence of trials might look like

$$1\,0\,0\,0\,1\,1\,1\,1\,0\,0\,1\,0\,1\,1\,0\,0\,\ldots.$$

If p is the probability of success at each trial and $q = 1 - p$ is the probability of failure, then the probability distribution of Bernoulli trials is

$x_i =$	0	1
$p_i =$	q	p

Let us consider a further distribution which can arise from Bernoulli trials. A series of independent Bernoulli trials takes place with the probability of success or failure of any given trial given by p or q where $p + q = 1$. Consider the probability distribution of i successes in a *fixed* number of trials n. In the notation of probability distributions

$$x_i = i \quad (i = 0, 1, 2, \dots, n).$$

Here is a particular sequence:

$$\underbrace{1\,1\,1\dots1}_{i \text{ times}} \ \underbrace{0\,0\,0\dots0}_{n-i \text{ times}}.$$

This sequence, in which there are i successes followed by $n - i$ failures, occurs with probability

$$p^i q^{n-i},$$

since the probability of a success followed by another success is $p \times p = p^2$, and so on. However, there are many sequences which have i successes (1) and $n - i$ failures (0), and the number of possible arrangements is $_nC_i$ (see Section 39.4 for an explanation of the combination notation). Including every arrangement, the probability of i successes in n trials is

$$_nC_i p^i q^{n-i}.$$

This is called the binomial distribution for X, the random variable of the number of successes in n trials.

> **Binomial distribution** which has the probability function
>
> $$P(X = x_i) = p_i = {_nC_i} p^i q^{n-i} = \frac{n! p^i q^{n-i}}{(n-i)! i!} \quad (i = 0, 1, 2, \dots, n).$$
>
> (40.2)

The binomial distribution contains two parameters, n the number of trials and p the probability of success. Since $q = 1 - p$, the first few entries in the distribution are

$x_i =$	0	1	2	3	$\dots$
$p_i =$	q^n	npq^n	$\dfrac{n(n-1)}{2!} p^2 q^{n-2}$	$\dfrac{n(n-1)(n-2)}{3!} p^3 q^{n-3}$	$\dots$

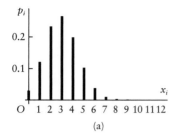

(a)

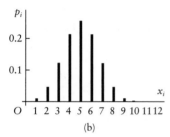

(b)

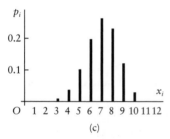

(c)

Fig. 40.3 Binomial distribution for $n = 10$ and (a) $p = 0.3$, (b) $p = 0.5$, (c) $p = 0.7$.

which are recognizably the first few terms in the binomial expansion of $(p + q)^n$ (see Appendix A(c)). Hence

$$\sum_{i=0}^{n} p_i = \sum_{i=0}^{n} \frac{n! p^i q^{n-i}}{(n-i)! i!} = (p + q)^n = 1,$$

since $p + q = 1$. This confirms that (40.2) does satisfy the key requirement for a probability distribution. Some **bar charts** for the binomial distribution are shown for $n = 10$ and $p = 0.3, 0.5, 0.7$ in Fig. 40.3.

Example 40.2   *Three dice are rolled simultaneously. What is the probability that two 5s appear with the third face showing a different number?*

Let the random variable X be the number of 5s which appear. Then

$$S_X = \{0, 1, 2, 3\}.$$

The outcomes from each die are independent with a 5 showing called a success and no 5 showing a failure. The probability that a single die shows a 5 is $\frac{1}{6}$. Hence X has a binomial distribution with parameters $n = 3$ and $p = \frac{1}{6}$. Hence, by (40.1),

$$P(X = 2) = {}_3C_2 (\tfrac{1}{6})^2 (\tfrac{5}{6}) = \tfrac{15}{216} \approx 0.069,$$

which is quite small. The other probabilities are

$$P(X = 0) = \tfrac{125}{216} \approx 0.579, \; P(X = 1) = \tfrac{75}{216} \approx 0.347, \; P(X = 3) = \tfrac{1}{216} \approx 0.005.$$

The odds for obtaining three 5s are 1 in 216.

40.4 Expected value and variance

The **expected value** or **mean** or **expectation** of a random variable is defined in terms of a **weighted average** of outcomes: the weighting is equal to the probability p_i with which x_i occurs. Thus if X is a random variable which can take the values x_1, x_2, ... with probabilities p_1, p_2, ... then

> **Expected value** or **mean** of X is defined by
>
> $$E(X) = \sum_i p_i x_i$$
>
> where the summation is over all i, either finite or countably infinite.
>   The symbol μ is often used for the expected value $E(X)$. **(40.3)**

For the binomial distribution (40.2) with parameters n and p, the expected value is (note that the distribution has $n + 1$ elements)

$$E(X) = \sum_{i=0}^{n} {}_nC_i p^i q^{n-i} i$$

$$= \sum_{i=1}^{n} \frac{n! p^i q^{n-i}}{(n-i)!(i-1)!} = np \sum_{i=1}^{n} \frac{(n-1)! p^{i-1} q^{n-i}}{(n-i)!(i-1)!}$$

$$= np \sum_{i=0}^{n-1} \frac{(n-1)! p^i q^{n-i-1}}{(n-1-i)! i!} = np(p+q)^{n-1}$$

$$= np,$$

using the binomial expansion (Appendix A(c)).

Example 40.3   *In Example* 40.2 *what is the expected value of the number of 5s which appear when three dice are rolled?*

From the definition of expected value and the results in the previous example,

$$E(X) = \sum_{i=0}^{3} P(X=i)i = \frac{125}{216} \cdot 0 + \frac{75}{216} \cdot 1 + \frac{15}{216} \cdot 2 + \frac{1}{216} \cdot 3 = \frac{108}{216} = \frac{1}{2}.$$

This result checks with $np = \frac{3}{6} = \frac{1}{2}$.

Random variables can be combined as in $X + Y$, and it is possible to consider functions of random variables $g(X)$. Expected values satisfy the following theorems (which will not be proved here, however). If c is a constant, X and Y are random variables, and $g(X)$ is a function of X, then

> **Rules for expected values**
> (i) $E(cX) = cE(X)$;
> (ii) $E(X + Y) = E(X) + E(Y)$;
> (iii) $E(XY) = E(X)E(Y)$ (if X and Y are independent);
> (iv) $E(g(X)) = \sum_{i=1}^{n} g(x_i) p_i$ (for a finite distribution).   (40.4)

Whilst the expected value of a random variable is a useful average it gives no idea of the *spread* of the distribution about the expected value. Two distributions can have the same mean but can have very different shapes in relation to the mean. A measure of the spread is the difference $X - E(X)$, the difference between the random variable and its mean. However, its expectation is always zero since, using (40.4)(i), (ii) above,

$$E(X - E(X)) = \sum_{i=1}^{n} (x_i - E(X)) p_i = \sum_{i=1}^{n} x_i p_i - E(X) \sum_{i=1}^{n} p_i$$

$$= E(X) - E(X) \cdot 1 = 0,$$

which is obviously not helpful as a measure of spread. Instead we choose the random variable $(X - E(X))^2$. Its expected value is known as the **variance** and is denoted by

Variance of a random variable

$$\text{Var}(X) = \sigma^2 = E[(X - E(X))^2] = E[(X - \mu)^2],$$

where $\mu = E(X)$.

(40.5)

Using (40.4), note that the variance can be expressed in the form

$$\text{Var}(X) = E(X^2 - 2\mu X + \mu^2) = E(X^2) - 2\mu E(X) + \mu^2$$
$$= E(X^2) - \mu^2$$

(40.6)

which is more convenient.

Since the units associated with the variance are squares, the symbol σ^2 is frequently used for variance so that in 'linear' terms the spread can be defined by $\sigma = \sqrt{\text{Var}(X)}$. This is known as

Standard deviation of the random variable X:

$$\sigma = \sqrt{\text{Var}(X)}.$$

(40.7)

Example 40.4   *Find the variance of the binomial distribution given by (40.2).*

Using (40.2) and (40.4)(iv)

$$\text{Var}(X) = E(X^2) - \mu^2 = \sum_{i=0}^{n} i^2 \, _nC_i p^i q^{n-i} - \mu^2,$$

$$= \sum_{i=1}^{n} \frac{i \cdot n! p^i q^{n-i}}{(n-i)!(i-1)!} - \mu^2.$$

As a device for summing the series we assume that p and q are independent parameters, and use the formula $E(X) = np(p + q)^{n-1}$ for the expected value of the binomial distribution. Thus

$$\sum_{i=1}^{n} \frac{i \cdot n! p^i q^{n-i}}{(n-i)!(i-1)!} - \mu^2 = p\frac{\partial}{\partial p}\left(\sum_{i=1}^{n} \frac{n! p^i q^{n-i}}{(n-i)!(i-1)!}\right) - \mu^2$$

$$= p\frac{\partial}{\partial p}(E(X)) - \mu^2$$

$$= p\frac{\partial}{\partial p}(np(p + q)^{n-1}) - \mu^2$$

$$= p[n(p + q)^{n-1} + n(n-1)p(p + q)^{n-2}] - n^2 p^2$$

$$= pn[1 + p(n-1)] - n^2 p^2 \quad \text{(since } p + q = 1\text{)}$$

$$= np(1 - p).$$

The following rules for variances can be proved:

> **Rules for variances**
> (i) $\mathrm{Var}(X + c) = \mathrm{Var}(X);$
> (ii) $\mathrm{Var}(cX) = c^2\,\mathrm{Var}(X);$
> (iii) $\mathrm{Var}(X + Y) = \mathrm{Var}(X) + \mathrm{Var}(Y)$
> (if X and Y are independent).
>
> **(40.8)**

40.5 Geometric distribution

Consider again a sequence of independent Bernoulli trials explained in Section 40.3 with in any trial a probability p of success (1) and a probability $q = 1 - p$ of failure (0). Suppose that we are interested in the *number of trials up to and including the first success*. Call this random variable X, and let $X = i$ correspond to the sequence

$$\underbrace{0\,0\,0\ldots 0}_{i-1\,\text{times}}\,1.$$

The probability of $i - 1$ successive failures is q^{i-1} so that

$$P(X = i) = p_i = q^i p \quad (i = 1, 2, \ldots\,).$$

Unlike the binomial distribution, this distribution has an *infinite* sample space. It defines

> **The geometric distribution**
> $$P(X = i) = p_i = (1 - p)^{i-1} p \quad (i = 1, 2, \ldots\,).$$
>
> **(40.9)**

Note that

$$\sum_{i=1}^{\infty} p_i = p \sum_{i=1}^{\infty} q^{i-1} = p(1 + q + q^2 + \ldots) = \frac{p}{1 - q} = 1,$$

using the formula for the sum of a geometric series (see Section 1.16). A bar chart of a geometric distribution with $p = 0.2$ is shown in Fig. 40.4.

The expected value of the random variable of the geometric distribution is

$$\mu = E(X) = \sum_{i=1}^{\infty} i p_i = \sum_{i=1}^{\infty} i p q^{i-1}$$

$$= p(1 + 2q + 3q^2 + \ldots)$$

$$= p[(1 + q + q^2 + \ldots) + q(1 + 2q + 3q^2 + \ldots)]$$

$$= \frac{p}{1 - q} + q\mu = 1 + q\mu.$$

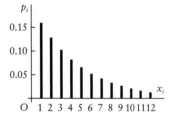

Fig. 40.4 Geometric distribution with $p = 0.2$.

Hence

$$\mu = \frac{1}{p}.$$

In a similar manner it can be shown that the variance is given by

$$\sigma^2 = \frac{1-p}{p^2}.$$

Example 40.5 *In a drug-testing programme, independent and sequential tests are conducted. Each test costs £500. The probability of success at each test is p. However, for each test after the first there is an additional cost per test of £200. What should p be greater than if the expected cost of the tests should not exceed £2000?*

Let X be the random variable of the number of tests up to and including the first success. We are actually interested in a random variable which is a function of X, namely the cost $C(X)$ which is given by

$$C(X) = 500X + 200(X-1) = 700X - 200.$$

Thus $C(1) = 500$, $C(2) = 1200$, $C(3) = 1900$, etc. Using (40.2), the expected value of $C(X)$ is

$$E(C(X)) = E(700X - 200) = 700E(X) - 200 = \frac{700}{p} - 200,$$

since X is a random variable with a geometric distribution. This expected cost is less than £2000 if

$$\frac{700}{p} - 200 < 2000 \quad \text{or} \quad \frac{700}{p} < 2200.$$

Hence the probability must satisfy the inequality $p > \frac{7}{22} \approx 0.32$.

40.6 Poisson distribution

Let X be a random variable which can take values 0, 1, 2, ... with probability

$$p_n = P(X = n) = \frac{\lambda^n e^{-\lambda}}{n!} \quad (n = 0, 1, 2, \dots).$$

This is a probability distribution since

$$\sum_{n=0}^{\infty} p_n = \sum_{n=0}^{\infty} \frac{\lambda^n e^{-\lambda}}{n!} = e^{-\lambda} e^{\lambda} = 1,$$

using the power series for e^{λ} (see eqn (5.4b)). This is known as the **Poisson distribution** with parameter λ. It occurs in problems in which discrete data accumulate, such as, for example, in the Geiger counter which records the number of radioactive particles which hit the instrument from a radioactive source. The distribution is appropriate for data arriving in a sequential random manner.

The Poisson distribution has mean

$$\mu = E(X) = \sum_{n=0}^{\infty} \frac{n\lambda^n e^{-\lambda}}{n!} = \sum_{n=1}^{\infty} \frac{\lambda^n e^{-\lambda}}{(n-1)!}$$

$$= \lambda \sum_{n=0}^{\infty} \frac{\lambda^n e^{-\lambda}}{n!} = \lambda e^{-\lambda} \sum_{n=0}^{\infty} \frac{\lambda^n}{n!} = \lambda e^{-\lambda} e^{\lambda} = \lambda$$

(using (5.4b)). Its variance is

$$\sigma^2 = \text{Var}(X) = E(X^2) - \mu^2 = \sum_{n=1}^{\infty} \frac{n^2 \lambda^n e^{-\lambda}}{n!} - \mu^2$$

$$= e^{-\lambda} \sum_{n=1}^{\infty} \frac{n\lambda^n}{(n-1)!} - \lambda^2 = e^{-\lambda} \lambda \frac{\mathrm{d}}{\mathrm{d}\lambda} [\lambda\, e^{\lambda}] - \lambda^2$$

$$= \lambda\, e^{-\lambda} (e^{\lambda} + \lambda\, e^{\lambda}) - \lambda^2 = \lambda.$$

Apart from being a distribution in its own right, the Poisson distribution is also a useful approximation to the binomial distribution (see Section 40.3) for large n. In the binomial term $_nC_i\, p^i q^{n-i}$ put the parameter $p = \lambda/n$. Then

$$_nC_i\, p^i q^{n-i} = \frac{n!}{(n-i)!\,i!} \left(\frac{\lambda}{n}\right)^i \left(1 - \frac{\lambda}{n}\right)^{n-i}$$

$$= \frac{n(n-1)\dots(n-i+1)}{i!} \left(\frac{\lambda}{n}\right)^i \left(1 - \frac{\lambda}{n}\right)^{n-i}$$

$$= \left[\frac{\left(1 - \dfrac{1}{n}\right)\dots\left(1 - \dfrac{i-1}{n}\right)}{i!\left(1 - \dfrac{\lambda}{n}\right)^i} \lambda^i \left(1 - \frac{\lambda}{n}\right)^n\right].$$

Now let $n \to \infty$. The term in the square brackets approaches $\lambda^i/i!$ as $n \to \infty$, whilst

$$\left(1 - \frac{\lambda}{n}\right)^n \to e^{-\lambda}.$$

(This limit can be obtained by putting $h = n/\lambda$ in the approximation for e given in Section 1.10.) Thus

$$_nC_i\, p^i q^{n-i} \to \frac{\lambda^i}{i!} e^{-\lambda}$$

as $n \to \infty$. This is a useful approximation as the following application illustrates.

Example 40.6 *Certain processors are known to have a failure rate of 1.2%. They are shipped in batches of 150. What is the probability that a batch has exactly one defective processor? What is the probability that it has two?*

We assume that the defects are independent. We use the binomial distribution with probability

$$p_i(n, p) = {}_nC_i p^i q^{n-i}$$

with $n = 150$ and $p = 0.012$ (failure of the component is 'success' in the binomial convention). Hence for $i = 1, 2$, a direct calculation gives

$$p_1(150, 0.012) = {}_{150}C_1(0.988)^{149}(0.012)^1 = 0.297\,891,$$

$$p_2(150, 0.012) = {}_{150}C_2(0.988)^{148}(0.012)^2 = 0.269\,549.$$

In this problem n is 'large', so that it is suitable for the Poisson approximation. The parameter λ for the corresponding Poisson distribution is given by

$$\lambda = np = 150 \times 0.012 = 1.8.$$

Hence the probability of one failure is

$$\frac{\lambda}{1!} e^{-\lambda} = 1.8\,e^{-1.8} = 0.297\,538,$$

and of two failures is

$$\frac{\lambda^2}{2!} e^{-\lambda} = \frac{(1.8)^2}{2} e^{-1.8} = 0.267\,784,$$

which show accuracy to 2 decimal places compared with the binomial distribution. This is more than sufficient in many applications. The Poisson approximation avoids the rounding errors which can occur in calculating probabilities raised to large powers.

40.7 Other discrete distributions

(a) *The Pascal or negative binomial distribution.* This is the distribution with function

$$p_i = {}_{i-1}C_{k-1} p^k (1 - p)^{i-k} \quad (i = k, k+1, \dots).$$

This distribution is an extension of the geometric distribution, and arises from the random variable which is the number of Bernoulli trials to achieve k successes where a success occurs with probability p. This is sometimes known as **inverse sampling**, since the number of successes k is specified in advance.

Its mean and variance are

$$\mu = \frac{k}{p}, \qquad \sigma^2 = \frac{k(1-p)}{p^2}.$$

(b) *Hypergeometric distribution.* Consider a box containing w white balls and b black balls. Suppose that n balls are chosen at random from the box *without replacement*. What is the probability that

i white balls are chosen? The i balls must be chosen from w, and the $n - i$ balls from b. Hence the number of possible samples is $_wC_i {_b}C_{n-i}$. By this counting method we obtain

$$P(X = i) = p_i = \frac{_wC_i {_b}C_{n-i}}{_{w+b}C_n},$$

where

$$i = \begin{cases} 0, 1, 2, \ldots, n & \text{if } n \leqslant w, \\ 0, 1, 2, \ldots, w & \text{if } n > w. \end{cases}$$

The function p_i defines the **hypergeometric distribution**. Its mean and variance are given by

$$\mu = \frac{nb}{w + b}, \qquad \sigma^2 = \frac{nwb(b + w + n)}{(w + b)^2(w + b - 1)}.$$

The same problem with replacement leads to the binomial distribution.

40.8 Continuous random variables and distributions

In many applications the discrete random variable which takes its values from a countable list is inappropriate. For example, the random variable X could be the time from, say, $t = 0$ until a light bulb fails. Whilst it would be possible to measure failure to the nearest hour and use a discrete random variable, it is often more convenient and more accurate to use a **continuous random variable**, which is defined for the continuous variable $t \geqslant 0$, and is no longer a countable list of values.

Instead of the sequence of probabilities $\{P(X = x_i)\} = \{p_i\}$, we define a **probability density function** (pdf) $f(x)$ over $-\infty < x < \infty$ which has the properties:

> **Probability density function** (pdf)
> (a) $f(x) \geqslant 0 \ (-\infty < x < \infty)$;
> (b) $\int_{-\infty}^{\infty} f(x) \, dx = 1$;
> (c) for any x_1, x_2 such that $-\infty < x_1 < x_2 < \infty$,
>
> $$P(x_1 \leqslant X \leqslant x_2) = \int_{x_1}^{x_2} f(x) \, dx.$$
>
> (40.10)

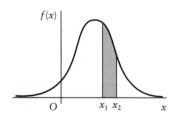

Fig. 40.5 Probability density function.

The random variable X can take any value of the continuous variable x. A graph of a density function $f(x)$ against x is shown in Fig. 40.5. By (a) the curve must never fall below the x axis, by (b) the area under the curve must be 1, and by (c) the probability that X lies between two values x_1 and x_2 is the shaded area under the graph. Unlike p_i, the pdf $f(x)$ is not itself a probability.

We can associate with the pdf a **cumulative distribution function** (**cdf**) $F(x)$ which is defined by

> **Cumulative distribution function** (cdf)
>
> $$F(x) = P(X \leqslant x) = \int_{-\infty}^{x} f(u)\, du.$$
>
> (40.11)

It represents the probability that $X \leqslant x$. By (40.10b) it follows that

$$F(x) \to 1 \quad \text{as } x \to \infty,$$

and

$$P(x_1 \leqslant x \leqslant x_2) = \int_{x_1}^{x_2} f(u)\, du = F(x_2) - F(x_1).$$

A typical cdf, which must be a non-decreasing function, is shown in Fig. 40.6.

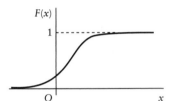

Fig. 40.6 Cumulative distribution function.

Example 40.7 *Let X be the random variable of time to failure of a light bulb measured from time $t = 0$. Assume that X has a pdf*

$$f(t) = \begin{cases} \alpha\, e^{-\alpha t} & t \geqslant 0 \\ 0 & t < 0, \end{cases}$$

where t is measured in hours. What is the probability that the bulb has failed at $t = 10$ hours? What is the probability that the light bulb fails between $t = 10$ hours and $t = 20$ hours?

Note that $f(t)$ is a pdf since $f(t) \geqslant 0$ and

$$\int_{-\infty}^{\infty} f(t)\, dt = \int_{0}^{\infty} \alpha\, e^{-\alpha t}\, dt = [-e^{-\alpha t}]_0^{\infty} = 1.$$

For the first question we require

$$P(X \leqslant 10) = \int_{0}^{10} \alpha\, e^{-\alpha t}\, dt = [-e^{\alpha t}]_0^{10} = 1 - e^{-10\alpha}.$$

For the second question

$$P(10 \leqslant X \leqslant 20) = \int_{10}^{20} e^{-\alpha t}\, dt = [-3^{-\alpha t}]_{10}^{20}$$

$$= e^{-10\alpha} - e^{-20\alpha} = e^{-10\alpha}(1 - e^{-10\alpha}).$$

Thus the light bulb fails before 10 hours with probability $1 - e^{-10\alpha}$, and between 10 hours and 20 hours with probability $e^{-10\alpha}(1 - e^{-10\alpha})$.

The pdf in the previous example is the **exponential distribution** which is frequently used in 'time to failure' problems. Its cdf is given by

$$F(x) = \begin{cases} \displaystyle\int_0^x \alpha \, e^{-\alpha u} \, du = 1 - e^{-\alpha x}, & x \geqslant 0; \\ 0, & x < 0. \end{cases}$$

Note that density functions do not have to be continuous: they can include jumps. Also if some event can only take place after a given time, say, then we put the density function equal to zero until that time.

40.9 Mean and variance of continuous random variables

By analogy with that for discrete random variables the **expected value** or the **mean**, and the **variance** of a continuous random variable X with pdf $f(x)$, are defined to be

> **Mean of continuous random variable:**
>
> $$\mu = E(X) = \int_{-\infty}^{\infty} x f(x) \, dx.$$
>
> **Variance of continuous random variable:**
>
> $$\sigma^2 = \text{Var}(X) = E((X - \mu)^2)$$
>
> $$= \int_{-\infty}^{\infty} (x - \mu)^2 f(x) \, dx.$$
>
> **(40.12)**

For the exponential distribution with pdf given by

$$f(t) = \begin{cases} \alpha \, e^{-\alpha t}, & t \geqslant 0, \\ 0, & t < 0, \end{cases}$$

its mean or expected value is

$$\mu = \int_{-\infty}^{\infty} t f(t) \, dt = \int_0^{\infty} \alpha t \, e^{-\alpha t} \, dt$$

$$= -\int_0^{\infty} t \frac{d}{dt} (e^{-\alpha t}) \, dt$$

$$= -[t \, e^{-\alpha t}]_0^{\infty} + \int_0^{\infty} e^{-\alpha t} \frac{d(t)}{dt} \, dt \quad \text{(integrating by parts)}$$

$$= 0 + \int_0^{\infty} e^{-\alpha t} \, dt = \frac{1}{\alpha}.$$

It can be shown similarly that

$$\sigma^2 = \frac{1}{\alpha^2}.$$

40.10 The normal distribution

The **normal distribution** with pdf defined by

> **Normal distribution, $N(\mu, \sigma^2)$,**
>
> $$f(x) = \frac{1}{\sigma\sqrt{2\pi}} e^{-(x-\mu)^2/2\sigma^2}, \quad -\infty < x < \infty,$$
>
> (40.13)

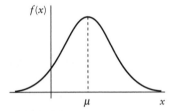

Fig. 40.7 A normal distribution.

is particularly important in many applications. It has a symmetrical bell-shaped distribution about its mean μ. Note also that σ in (40.13) is its standard deviation. A typical normal distribution is shown in Fig. 40.7. It can be verified that

$$\int_{-\infty}^{\infty} f(x)\,\mathrm{d}x = 1,$$

$$\int_{-\infty}^{\infty} x f(x)\,\mathrm{d}x = \mu,$$

$$\int_{-\infty}^{\infty} x^2 f(x)\,\mathrm{d}x = \sigma^2.$$

The normal distribution $N(\mu, \sigma^2)$ is a two-parameter distribution with its mean and standard deviation as parameters.

The **standardized normal distribution** is $N(0, 1)$ with pdf

$$f(z) = \frac{1}{\sqrt{2\pi}} e^{-\frac{1}{2}z^2}.$$

It has mean zero and standard deviation 1. Any normal random variable X with distribution $N(\mu, \sigma^2)$ can be 'standardized' by considering the random variable $Z = (X - \mu)/\sigma$. In the distribution (40.12) this is equivalent to the substitution $z = (x - \mu)/\sigma$. Thus $N(0, 1)$ has the density

$$\frac{1}{\sqrt{2\pi}} e^{-\frac{1}{2}z^2}.$$

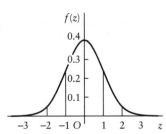

Fig. 40.8 The standard normal curve.

The **standard normal curve** representing $N(0, 1)$ is shown in Fig. 40.8. The standard deviations within 1, 2, 3 units of the mean are also shown in the figure. If Z is the corresponding random variable then the probability that Z lies within one standard deviation of the mean zero is the area under the curve between -1 and 1. Thus

$$P(-1 \leqslant Z \leqslant 1) = \frac{1}{\sqrt{2\pi}} \int_{-1}^{1} e^{-\frac{1}{2}z^2}\, dz = 0.6827,$$

but numerical integration is required to evaluate this integral. Tables of standard normal distributions can also be used to estimate the answer (see Appendix G).

Similarly

$$P(-2 \leqslant Z \leqslant 2) = \frac{1}{\sqrt{2\pi}} \int_{-2}^{2} e^{-\frac{1}{2}z^2}\, dz = 0.9545,$$

$$P(-3 \leqslant Z \leqslant 3) = \frac{1}{\sqrt{2\pi}} \int_{-3}^{3} e^{-\frac{1}{2}z^2}\, dz = 0.9973.$$

The last result implies that there is a 99.73% chance that a selected item lies within three standard deviations of the mean for the standardized normal distribution.

The importance of the normal distribution lies in the observation that in many measurements, which almost always involve random experimental errors, the distribution of the errors seems to be *normal* (see eqn (40.13)).

The cdf for the standardized normal distribution $N(0, 1)$ is

$$\Phi(z) = P(Z \leqslant z) = \frac{1}{\sqrt{2\pi}} \int_{z}^{\infty} e^{-\frac{1}{2}u^2}\, du,$$

whose values can be obtained from Appendix H. A graph of $\Phi(z)$ against z is shown in Fig. 40.9: it can be used to estimate probabilities for the normal distribution.

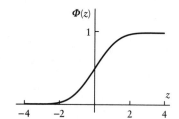

Fig. 40.9 Cumulative distribution function for the standardized normal distribution.

Example 40.8   *The mean height of 459 university students is 180 cm with a standard deviation of 4.2 cm. Assuming that the heights are normally distributed estimate the number of students who have heights greater than 200 cm, and the number who have heights between 175 cm and 185 cm.*

For this sample $\mu = 180$ and $\sigma = 4.2$. Hence the normal distribution $N(180, 17.64)$ is given by

$$\frac{1}{\sqrt{2\pi}} \frac{1}{4.2} e^{-(x-180)^2/35.28}.$$

We can obtain the corresponding standardized normal distribution by putting

$$z = \frac{x - 180}{17.64}.$$

Example 40.8 *continued*

If $x = 200$ then $z = (200 - 180)/17.64 = 1.13$. Hence

$$P(Z \geqslant 1.13) = 1 - F(1.13) = 1 - 0.87 = 0.13,$$

approximately (this can be read either from Fig. 40.8, or by using tables). Hence around $0.13 \times 459 \approx 59$ students will have heights in excess of 200 cm; similarly if $x = 195$, then $z = -0.28$, and if $x = 185$ then $z = 0.28$. Thus

$$P(-0.28 \leqslant Z \leqslant 0.28) = F(0.28) - F(-0.28) = F(0.28) - (1 - F(0.28))$$
$$= 2F(0.28) - 1 = 2 \times 0.61 - 1 = 0.22,$$

approximately. Hence it expected that about 101 students will have heights between 175 cm and 185 cm.

Problems

40.1 A biased coin is spun three times. The probability of a head appearing is 0.45 and of a tail 0.55. If X is the random variable of the number of heads shown, what is the sample space of X? What is the probability distribution of X?

Sketch a bar chart showing the probability distribution. What is the probability that X is greater than or equal to 1, that is $P(X \geqslant 1)$?

40.2 Explain why the sequence

$$p_j = \frac{1}{3}\left(\frac{1}{2^j} + \frac{1}{2^{j-1}}\right) \quad (j = 1, 2, 3, \dots)$$

can be interpreted as a probability distribution. If $P(X = j) = p_j$ find $P(X \geqslant 6)$.

40.3 The probability of success in a sequence of independent Bernoulli trials is $\frac{1}{3}$. If 12 trials take place calculate the probabilities of 0, 1, ... , 12 successes. Calculate also the mean and standard deviation of the random variable which is the number of successes.

40.4 The **uniform distribution** has the pdf

$$f(x) = \begin{cases} 1/(b-a) & a < x < b \\ 0 & \text{elsewhere.} \end{cases}$$

Sketch the graphs of the pdf and its cdf. Find the mean and standard deviation of the uniform distribution.

40.5 Prove that the variance of the geometric distribution $p_i = (1-p)^{i-1}p$, $(i = 1, 2, \dots)$, is $(1-p)/p^2$.

40.6 Components join a production assembly line in sequence. The probability that a particular component is faulty is 0.012. How many components (excluding the faulty component) will be expected to join the assembly line before a faulty one is encountered? What is the standard deviation of the number of components to failure?

40.7 A coin is spun until a tail a shown. What is the probability that eight heads appear before the first tail?

40.8 In a series of Bernoulli trials the probability of success is p. Let X be the random variable until r successes occur. For example, if 1 denotes success and 0 denotes failure then, in the sequence

1 0 0 1 1 0 0 0 1 1 1 0 0 1 0 0

7 successes will have occurred in 16 trials, that is $X = 16$ for $r = 7$ in this case. Show that

$$p_i = P(X = i) = \binom{i-1}{r-1} p^r (1-p)^{i-r}$$

for $i = r, r+1, r+2, \dots$. This is the negative binomial distribution. Confirm that

$$\sum_{i=r}^{\infty} p_i = 1.$$

Show that

$$E(X) = \frac{r}{p}, \qquad \text{Var}(X) = \frac{r(1-p)}{p^2}.$$

40.9 In a milk-bottling plant bottles are filled with milk and their weights checked. If a bottle is underweight or more than 4% overweight the production line is stopped and the problem investigated. Assume that a bottle fails randomly with the same probability p. What would be an appropriate distribution for this problem? On average it is found that breakdown occurs every 1503 bottles. What is the probability that an individual bottle fails the weight test?

40.10 Suppose that the random variable X has the exponential distribution with pdf

$$f(t) = \begin{cases} 1.5\, e^{-1.5t}, & t \geq 0, \\ 0, & t < 0. \end{cases}$$

Find the following probabilities
(a) $P(0 < X < 1)$; (b) $P(X < 0)$; (c) $P(X \geq 1)$;
(d) $P(X \leq 1)$; (e) $P(X < 2)$ or $P(X < 1)$.

40.11 Calls to a freephone information line are assumed to occur so that the times between calls are exponentially distributed with mean time of 20 minutes between calls. If X is the random variable of the time between calls, (a) What is the probability that there are no calls in a one-hour interval?
(b) What is the probability that there is at least one call within a 15-minute interval?

40.12 A geiger counter is an instrument for counting the number of radioactive particles emitted by a radioactive sample which strike the instrument. In a probability model of the counter, the random variable X, which is the number of radioactive particles detected in a given time interval, has a **Poisson distribution**

$$P(X = \lambda) = \frac{e^{-\lambda}\lambda^n}{n!} \qquad (n = 0, 1, 2, \ldots)$$

where λ is a parameter which characterizes the radio-activity of the sample. Show that the mean and variance of the probability distribution are both λ.

What is the probability that five or more hits occur in the time interval?

40.13 The random variable Z has a standardized normal distribution. Estimate the following probabilities:
(a) $P(Z \geq 0.8)$; (b) $P(Z \leq 0.7)$;
(c) $P(-0.5 \leq Z \leq 0.8)$.

40.14 A particular repetitive operation on a production line has a uniform distribution (see Problem 40.4) with pdf $f(t) = 0.1$ for $33 < t < 43$, where time is measured in seconds. What are the mean time and variance of the operation? On average what proportion of operations take longer than 40 seconds?

40.15 It is required in an application that

$$f(t) = \begin{cases} A(a^2 - t^2) & -a \leq t \leq a \\ 0 & \text{elsewhere} \end{cases}$$

should be a pdf. What should the parameter A be in terms of a? Find the variance of the distribution. What should A and a be for the distribution to have a standard deviation of 1?

40.16 The time to failure of catalytic converters in exhaust systems of cars is modelled by a normal random variable with mean of 1200 hours. If 95% of the converters are to last at least 1000 hours without failure, what is the maximum value which the standard deviation of the normal distribution can take?

40.17 The random variable of the time to failure in a batch of light bulbs is assumed to be exponentially distributed with mean time to failure of 500 hours. What is the probability that a light bulb is still functioning after 640 hours? A room is lit by four light bulbs which are not replaced as they fail. What is the probability that just two bulbs will still be working at 640 hours?

41 Descriptive statistics

CONTENTS

41.1 Representing data

Statistics is a subject concerned with the collection, analysis, and interpretation of data. Any method which seeks to interpret the data is a branch of statistical inference. The data set usually consists of a **random sample** from some larger set called a **population** and may be quite a small proportion of it. The objective is to make inferences about the population as a whole from a small sample. Hence if we want to find out what the mean salary of a population is, then a random sample of individuals is taken, and the mean of the resulting sample is used to estimate and make inferences about the unknown mean salary of the population. Generally the values in a sample are known as **variates**. From this process of sampling the aim is to infer properties about the whole population: this is known as **statistical inference**. Any quantity calculated from the sample is known as a **statistic,** and the corresponding (usually unknown) value in the population is known as a **parameter**.

Let us first look at graphical ways of representing the data. Table 41.1 shows the number of vehicles which cross an automatic census cable on a road on a particular day. The day is split into two-hour time slots from midnight to midnight. We can represent the data by a **histogram**, in which each two-hour slot is represented by a rectangle whose height is the frequency or the number of vehicles in this case, and whose width is the time interval, as shown in Fig. 41.1. We can draw a **frequency polygon** by joining the midpoints of the tops of the rectangles. If there is a large number of time intervals then the polygon may sometimes be replaced by a smooth curve fitted to the data.

Given a set of data, the design of histogram for the data is a matter of judgement. In the example, the 24 hours were divided into 12 two-hour time intervals, but we could have collected alternatively over

Table 41.1

Time interval	Number of vehicles
00:00–02:00	6
02:00–04:00	4
04:00–06:00	9
06:00–08:00	21
06:00–10:00	24
10:00–12:00	15
12:00–14:00	16
14:00–16:00	18
16:00–18:00	29
18:00–20:00	20
20:00–22:00	16
22:00–24:00	10

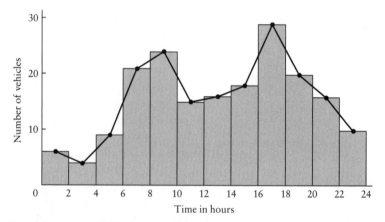

Fig. 41.1 Histogram of the data in Table 41.1.

24 one-hour intervals. The **intervals** are also known as **cells** or **bins**. The intervals should usually be of equal 'width'. Also the number of intervals should not be too large for the data. A working rule is that *the number of intervals may roughly increase like* $\sqrt{n}$*, where n is the number of observations.* In the example above there are 188 observed vehicles which according to the rule suggests about 14 intervals which is close to our choice of 12.

Here is another example. A 'snapshot' of vehicles on a short stretch of road is taken at the same hour on the same weekday for 59 occasions. Table 41.2 is a frequency table, which collocates the numbers of cars. The histogram is shown in Fig. 41.2.

The **sample mean** $\bar{x}$ of n observations $\{x_i\}$, where x_i occurs with frequency f_i, is defined by

Table 41.2

Number of vehicles (x_i)	Frequency (f_i)
0	12
1	15
2	13
3	8
4	5
5	3
6	2
7	1

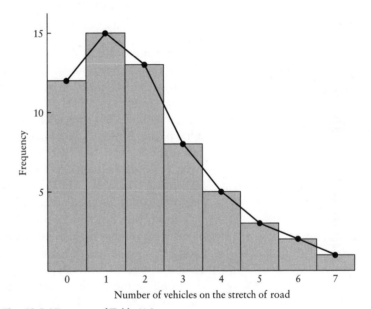

Fig. 41.2 Histogram of Table 41.2.

$$\bar{x} = \frac{\sum_{i=1}^{n} f_i x_i}{\sum_{i=1}^{n} f_i},$$

which is equivalent to the average, or **mean**, of the total set of observations since there must be $\sum_{i=1}^{n} f_i$ of them. For the traffic census in Table 41.2

$$\bar{x} = \frac{(0 \times 12) + (1 \times 15) + (2 \times 13) + (3 \times 8) + (4 \times 5) + (5 \times 3) + (6 \times 2) + (7 \times 1)}{12 + 15 + 13 + 8 + 5 + 3 + 2 + 1}$$

$$= \frac{119}{59} \approx 2.017.$$

The mean $\bar{x}$ of this sample will be an **estimate** for the true population mean. If the samples are not classified into categories then $f_i = 1$ and, as before,

$$\bar{x} = \frac{1}{n} \sum_{i=1}^{n} x_i,$$

where n is now the number of samples.

There are other measures of the central characteristics of samples. The **mode** is the value of x_i which occurs most often in a sample, and is therefore most likely to occur in other random samples. Thus in Table 41.2, the number 1 appears most often (15 times), so that the mode of these data is 1.

The central item in an ordered list of sample values is known as the **median**. Suppose that a list of examination marks is given by

Examination marks: 31, 36, 38, 39, 45, 46, 57, 60, 65, 65, 69, 72, 75, 79

in increasing order. If the sample has an odd number of items, say $2n + 1$, then the median is the $(n + 1)$th item: if the number is even, say $2n$, then it is defined to be the average of the nth and the $(n + 1)$th numbers in the ranking. In the list of examination marks above the median is $\frac{1}{2}(57 + 60) = 58.5$. The mode is 65 but it would be a number of no particular significance in this list, since the mode contains only two marks.

In Table 41.2, there are 59 numbers from 0 to 7 consisting of 12 zeros, 15 ones, etc. The median is the 30th number which will be one of the twos. Hence the median is 2.

The **box plot** displays graphically important features of data such as the median, the spread, and symmetry of the data, and is particularly useful in comparing different data sets, as for example in the results in a series of associated examination papers. Suppose that the examination marks in three papers are as percentages (each in increasing order) as shown in Table 41.3. We first find the median of the marks in each paper. Thus the medians are 59.5 for Paper 1, 61 for Paper 2, and 56 for Paper 3.

Suppose that the data contain $2n$ observations. Then the **first quartile** is the median of the n smallest observations, the **second**

Table 41.3

	Examination marks (0–100)
Paper 1 (16 results)	27, 40, 46, 48, 55, 55, 56, 58, 61, 63, 64, 66, 68, 69, 72, 78
Paper 2 (11 results)	30, 38, 39, 48, 58, 61, 64, 68, 69, 70, 81
Paper 3 (9 results)	26, 40, 43, 54, 56, 61, 62, 72, 74

Table 41.4

	First quartile	Median	Third quartile
Paper 1	51.5	59.5	67
Paper 2	43.5	61	68.5
Paper 3	43	56	62

quartile is the overall median of the data, and the **third quartile** the median of the n largest observations. If the data contain $2n + 1$ observations then the first quartile is the median of the $n + 1$ smallest observations and the third quartile the median of the $n + 1$ largest observations. (The second quartile is the overall median.) The quartiles divide the observations into four approximately equal numbers of observations.

For the examination marks paper by paper the quartiles are given in Table 41.4. The difference between the third and first quartiles is a measure of the spread of the data, and is known as the **interquartile range**.

Create a vertical scale 0–100 as shown in Fig. 41.3. For each paper, position a box such that its upper edge is level with the third quartile on the scale, and its lower edge is level with the first quartile. The line across the middle of the box is the median. Extend each box by a line to the extreme marks above and below the box. These lines are known as **whiskers**. Visually we can see how the average and spread of the marks compare. A compressed box indicates poor discrimination

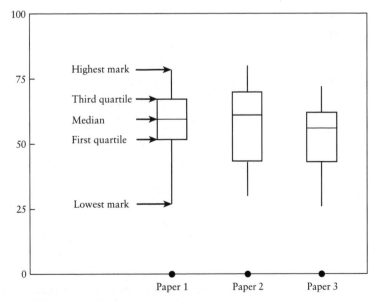

Fig. 41.3 Box plots for three examination papers.

in the marks, and long whiskers might indicate exceptional successes or failures (often known as **outliers**). Examiners may wish to take remedial action by scaling in the light of the comparative box plots if there are candidates in common among the papers.

41.2   Random samples and sampling distributions

One aim in statistics for a set of data is to fit a probability model to it so that inferences can be drawn concerning the data. This usually requires the selection of a probability distribution to model the data, often on the basis of minimal information. Having chosen the distribution, the parameter values of the distribution have to be estimated from the data. The set of data is the only hard information.

Consider a simple yes/no poll of the population in the UK. The question is asked: are you in favour of the UK joining the European Monetary Union (EMU)? The answer must be either 'yes' or 'no': 'don't know' responses are not permitted. The question could be put to the whole population in a **referendum** (at considerable expense), but if we are interested just in an opinion poll then the question could be put, say, to a random sample of 500 individuals chosen randomly from the population. (There is the question of how this can be achieved but we will not dwell on this polling problem.) Suppose that in the sample 56% say 'yes' and 44% say 'no'. We could conclude that the population is in favour but we could also ask how much weight should be attached to this poll result. Is it far enough away from the 50% critical value, for example, for a confident prediction?

More information might be obtained if we took a *number* of 500-person polls from the population, and examined the *distribution* of the random variable X representing the number of votes for EMU. This distribution is known as the **sampling distribution** of X. We could model the posing of the question to 500 individuals as a series of 500 independent Bernoulli trials which has a binomial distribution (see Section 40.3) with parameters $n = 500$ and an unknown probability p for the number of 'yes' votes.

For a single poll with n persons, the binomial distribution (the probability of i yes votes) is

$$\frac{n! p^i (1-p)^{n-i}}{(n-i)! i!} \qquad (i = 1, 2, \dots, n).$$

The mean of the binomial distribution, which is the mean number of yes votes, is np.

We can estimate p from each poll. Since the mean for the binomial distribution is np, it seems reasonable to estimate p as X/n. We shall denote an **estimator** for p from a sample by $\hat{p}$. The value of a random variable is known as an **estimate**. The symbol $\hat{p}$ is used for both the random variable and its value.

One test of whether we are looking at an appropriate measure of the probability p is the behaviour of the expected value of $\hat{p}$. Thus

$$E(\hat{p}) = E(X/n) = \frac{1}{n}E(X) = \frac{np}{n} = p, \tag{4.1}$$

since we are assuming a binomial distribution. Hence the expected value of the estimates gives the probability p, the mean of the sampling distribution of $\hat{p}$. Generally, if the expected value of the estimate equals the parameter being estimated, then the estimate is called **unbiased**; if this is not the case then the estimate is called **biased**.

The spread of the estimate can be found by calculating the expected value $E[(\hat{p} - p)^2]$. Then

$$E[(\hat{p} - p)^2] = E[\{\hat{p} - E(\hat{p})\}^2] = \text{Var}[\hat{p}]$$

by (40.5). Hence, the variance of the sampling distribution of $\hat{p}$ is given by

$$\begin{aligned}
\text{Var}[\hat{p}] &= \text{Var}\left[\frac{X}{n}\right] \\
&= \frac{1}{n^2}\text{Var}[X] \quad \text{(by (40.8(ii))} \\
&= \frac{np(1-p)}{n^2} = \frac{p(1-p)}{n},
\end{aligned} \tag{4.2}$$

for the binomial distribution. As we might expect, the variance of the sample means decreases with increasing sample size.

Given the *one* sample at the beginning of this section, the estimate for p is $p = 0.56$. The estimated variance of this single sample replacing p by $\hat{p}$ is

$$\begin{aligned}
\text{Var}[\hat{p}] &= \frac{\hat{p}(1-\hat{p})}{n} \\
&= \frac{0.56 \times 0.44}{500} \approx 0.005.
\end{aligned}$$

The corresponding **standard error** for $\hat{p}$ is $\sqrt{\text{Var}(\hat{p})} \approx 0.022$.

41.3 Sample mean and variance, and their estimation

The general sampling problem is as follows. Suppose that we take a sample of n observations $x_1, x_2, \ldots, x_n$ of a population where x_i is a value of a random variable X_i. The random variables (or **random sample**) $X_1, X_2, \ldots, X_n$ are assumed to be independent with the same density distributions. We assume that the sampling distribution can be modelled by a known probability distribution. Estimates, preferably unbiased, are required for the mean and variance of the underlying distribution.

An obvious choice for the mean is simply the **sample mean** which is the average value of the sample.

The **sample mean** is a random variable defined by

Sample mean for sample size n

$$\bar{X} = \frac{X_1 + X_2 + \cdots + X_n}{n}.$$

(41.3)

If $x_1, x_2, \ldots, x_n$ are *values* obtained in a particular sample then its mean is

$$\bar{x} = \frac{1}{n} \sum_{i=1}^{n} x_i.$$

What is the relation between the sample mean and the mean of the population? The expected value of $\bar{X}$ is

$$E(\bar{X}) = E\left(\frac{1}{n} \sum_{i=1}^{n} X_i \right) = \frac{1}{n} \sum_{i=1}^{n} E(X_i)$$

$$= \frac{1}{n} n\mu = \mu.$$

As we might expect, the expected value of the sample mean is the same as the mean of the population.

The variance of the sample mean is, by (40.6),

$$\mathrm{Var}(\bar{X}) = \mathrm{Var}\left(\frac{1}{n} \sum_{i=1}^{n} X_i \right) = \frac{1}{n^2} \sum_{i=1}^{n} \mathrm{Var}(X_i)$$

$$= \frac{n\sigma^2}{n^2} = \frac{\sigma^2}{n}$$

(41.4)

where σ^2 is the unknown variance of the population. Its standard deviation $\sigma/\sqrt{n}$ is known as the **standard error** of the sample mean.

We also need an estimate for the variance σ^2 of the population. We might choose

$$T^2 = \sum_{i=1}^{n} \frac{(X_i - \bar{X})^2}{n},$$

which is the variance of the sample, but is it unbiased? In other words does its expected value equal σ^2? The following algebra supplies the answer:

$$E(T^2) = E\left[\frac{1}{n}\sum_{i=1}^{n}(X_i - \bar{X})^2\right]$$

$$= E\left[\frac{1}{n}\sum_{i=1}^{n}[(X_i - \mu) - (\bar{X} - \mu)]^2\right]$$

$$= E\left[\frac{1}{n}\sum_{i=1}^{n}[(X_i - \mu)^2 - 2(X_i - \mu)(\bar{X} - \mu) + (\bar{X} - \mu)^2]\right]$$

$$= \frac{1}{n}E\left[\sum_{i=1}^{n}(X_i - \mu)^2 - n(\bar{X} - \mu)^2\right]$$

$$= \frac{1}{n}[n\sigma^2 - nE[(\bar{X} - \mu)^2]]$$

$$= \sigma^2 - \mathrm{Var}(\bar{X}) = \sigma^2 - \frac{\sigma^2}{n} = \frac{n-1}{n}\sigma^2$$

using (41.4) in the last line. In other words the expected value of the sample variance is *not* an unbiased estimator of the variance of the sampling distribution: there is a correction factor of $(n-1)/n$. A better statistic for an unbiased estimator of the variance is

> **Estimator for the sample variance**
>
> $$S^2 = \frac{\sum_{i=1}^{n}(X_i - \bar{X})^2}{n-1}.$$
>
> (41.5)

For large samples the difference between T^2 and S^2 is small but it can be significant for small sample sizes. The estimator is often known simply as the **sample variance**.

41.4 Central limit theorem

The normal distribution was introduced in Section 40.10, and its importance in the context of random errors was hinted at there. The pdf for a normal distribution with mean μ and variance σ^2 is (eqn (40.10))

$$f(x) = \frac{1}{\sigma\sqrt{2\pi}}e^{-(x-\mu)^2/2\sigma^2}.$$

The central limit theorem (which will not be proved here) states that if random samples are taken from a distribution with mean μ and standard deviation σ, then the sampling distribution of the random variable $\bar{X}$ of the sample mean will be normally distributed with mean μ and standard deviation $\sigma/\sqrt{n}$ as $n \to \infty$ whatever the original distribution of the X_i. Analytically this can be expressed as

Central limit theorem

$$\lim_{n\to\infty} P\left(\frac{\bar{X} - n\mu}{\sigma\sqrt{n}} \leq x\right) = \frac{1}{\sqrt{2\pi}} \int_{-\infty}^{x} e^{-\frac{1}{2}u^2}\, du.$$

(41.6)

In this result $(\bar{X} - \mu)\sqrt{n}/\sigma$ is the standardized random variable derived from $\bar{X}$, and for large n it is normally distributed.

As we have already stated the true significance of this result is that it is *independent* of the distribution of each X_i, which need not be normal.

This result can be illustrated in the case of the throwing of n dice in which the frequencies of average scores are kept. The probabilities can easily be computed (see Project 41.4 in Chapter 42) for small values of n. For example, if $n = 2$, then the possible average scores and the probabilities with which they occur are given in Table 41.5 and Fig. 41.4a. Graphs for $n = 2, 4, 6$ computed using a program to generate the bar charts are shown in Fig. 41.4. The bounding curve begins to show for $n = 6$ the familiar shape of the normal distribution.

Table 41.5

Average score	1	$\frac{3}{2}$	2	$\frac{5}{2}$	3	$\frac{7}{2}$	4	$\frac{9}{2}$	5	$\frac{11}{6}$	6
Probabilities	$\frac{1}{36}$	$\frac{2}{36}$	$\frac{3}{36}$	$\frac{4}{36}$	$\frac{5}{36}$	$\frac{6}{36}$	$\frac{7}{36}$	$\frac{8}{36}$	$\frac{9}{36}$	$\frac{10}{36}$	$\frac{11}{36}$

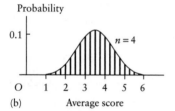

(a)   Average score

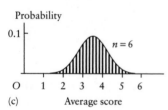

(b)   Average score

Probability

0.1 $n = 6$

O   1   2   3   4   5   6
(c)   Average score

Fig. 41.4 Probabilities versus average scores for rolling two, four and six dice.

Example 41.1   *A die is rolled 6000 times. The number T of times face 1 appears is counted. Find m_1 and m_2 in $P(m_1 < T < m_2)$ in order that T should lie within one standard deviation of its mean value 1000.*

Let X be the random variable that a 1 appears face up on the die. Then $E(X) = \frac{1}{6}$. Its variance is given by

$$\sigma^2 = \text{Var}(X) = E(X^2) - E(X)^2 = 1\cdot\frac{1}{6} - \frac{1}{36} = \frac{5}{36}.$$

By the central limit theorem

$$P\left(k_1 \leq \frac{T - 6000\cdot\frac{1}{6}}{\frac{\sqrt{5}}{6}\sqrt{6000}} \leq k_2\right) \approx \frac{1}{\sqrt{2\pi}} \int_{k_1}^{k_2} e^{-\frac{1}{2}u^2}\, du,$$

or

$$P\left(\frac{\sqrt{5}}{6}k_1\sqrt{6000} + 1000 \leq T \leq \frac{\sqrt{5}}{6}k_2\sqrt{6000} + 1000\right) \approx \frac{1}{\sqrt{2\pi}} \int_{k_1}^{k_2} e^{-\frac{1}{2}u^2}\, du.$$

Hence by the normal distribution table (Appendix G)

$$k_1 = -0.8413, \qquad k_1 = 0.8413.$$

Hence

$$m_1 = -0.8413\frac{\sqrt{5}}{6}\sqrt{6000} + 1000 \approx 976,$$

$$m_2 \approx 1024.$$

41.5 Regression

Suppose that we have a set of data in which one quantity is measured in relation to another quantity. For example, the fuel consumption of a car will vary with the speed of the car, or the weight of an individual will vary with the height of the person. We may wish to speculate as to what the relationship is between two (or more) quantities.

Suppose that a sample of measurements is taken, for example fuel consumption (y) for different speeds (x) of a car. This leads to the paired data $(x_1, y_1), (x_2, y_2), \ldots, (x_n, y_n)$, in which one or both variables may contain random errors. We can obtain an idea of the likely relationship between x and y by plotting the coordinates (x_i, y_i) as points in rectangular cartesian coordinates, giving what is known as a **scatter diagram**. Some examples are shown in Fig. 41.5. If we fit a curve to the data shown in the scatter diagrams, then we might guess a straight line **fit** to the data in Fig. 41.5a, and a curve in Fig. 41.5b, whereas in Fig. 41.5c, which shows data centred around a point, we might feel that no relationship exists between the variables. Often in scientific experiments the relationship between the variables can be inferred from some underlying theory although parameters may be unknown. For example, it might be known that the formula relating x and y is linear so that we need to find the best **straight line fit** to the data. For others we might need to guess the likely shape of the curve from the scatter of the data as in Fig. 41.5b.

In some data sets there can be errors in both measurements. In others, one variable known as the **controlled** or **independent variable** x is specified (measurements could be made at specified times which are known accurately) and y, which will contain random errors, is known as the **response** or **dependent variable**. In the fuel efficiency tests the speed of the car could be measured accurately (controlled variable), but the fuel consumption (response variable) might be affected by other factors (ambient temperature, engine tuning, etc.). On the other hand in the height/weight data the measurements could be accurate, although the weight could vary over time. There is unlikely to be a 'formula' relating height and weight (there may be other parameters involved) but nevertheless it is useful to have a working relation between the two for life tables used by insurance companies. The process of estimating the response variable from a set of controlled variables is known as **regression**.

If the hypothesis is that the data follow a straight line relationship then the model is known as a **linear regression** model. This regression model assumes that the random variable Y of the data $\{y_i\}$ is given by

$$Y = ax + b + \varepsilon,$$

where a and b are unknown parameters and ε is a random error with mean 0 and unknown variance σ^2. Note that the variance of Y is

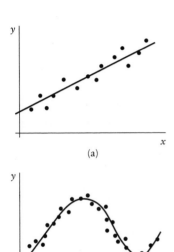

(a)

(b)

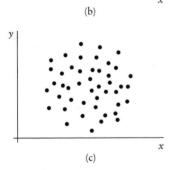

(c)

Fig. 41.5 Scatter diagrams.

$$\mathrm{Var}(Y) = \mathrm{Var}(ax + b + \varepsilon) = \mathrm{Var}(ax + b) + \mathrm{Var}(\varepsilon) = \mathrm{Var}(\varepsilon) = \sigma^2.$$

With x as a controlled variable, the vertical deviation of the point (x_i, y_i) from the line is

$$e_i = y_i - (ax_i + b).$$

We use the method of least squares for the sum of the squares of the deviations which requires the minimum of

$$f(a, b) = \sum_{i=1}^{n} e_i^2 = \sum_{i=1}^{n} (y_i - ax_i - b)^2$$

(see Section 28.7 for a full derivation of a and b). The minimum occurs where $\partial f/\partial a = \partial f/\partial b = 0$ and, as in (28.10), the best straight line fit is given by the solution of

$$a\sum_{i=1}^{n} x_i^2 + b\sum_{i=1}^{n} x_i = \sum_{i=1}^{n} x_i y_i,$$

$$a\sum_{i=1}^{n} x_i + bn = \sum_{i=1}^{n} y_i.$$

The solutions of these equations are the **least-squares estimates** for a and b, and using the notation for estimators we shall distinguish them by $\hat{a}$ and $\hat{b}$:

Least-squares estimates:

$$\hat{b} = \bar{y} - \hat{a}\bar{x}, \qquad \hat{a} = \frac{\sum_{i=1}^{n} x_i y_i - n\bar{x}\bar{y}}{\sum_{i=1}^{n} x_i^2 - n\bar{x}^2},$$

where

$$\bar{x} = \frac{1}{n}\sum_{i=1}^{n} x_i, \qquad \bar{y} = \frac{1}{n}\sum_{i=1}^{n} y_i.$$

(41.7)

The least-squares regression estimator $\hat{y}$ is given by

$$\hat{y} = \hat{a}x + \hat{b},$$

and this can be used to estimate y for other values of x. It also defines the equation of the **regression line** of y on x though the data. The regression line of x on y, which generally will be a different line, can be found similarly.

The estimates $\hat{a}$ and $\hat{b}$ have been obtained by least squares. Are they *unbiased estimators* of a and b? We can decide the answer to this question by finding their expected values. Thus, noting that Y_i is the random variable with value x_i and that x_i is a controlled variable,

$$E(\hat{a}) = E\left[\frac{\sum_{i=1}^{n} x_i Y_i - n\bar{x}\bar{Y}}{\sum_{i=1}^{n} x_i^2 - n\bar{x}^2}\right] \quad \text{sdfdsf}$$

$$= \frac{E[\sum_{i=1}^{n} x_i(ax_i + b + \varepsilon_i) + \bar{x}\sum_{i=1}^{n}(ax_i + b + \varepsilon_i)]}{\sum_{i=1}^{n} x_i^2 - n\bar{x}}$$

$$= \frac{\sum_{i=1}^{n} x_i(ax_i + b) - \sum_{i=1}^{n} \bar{x}(ax_i + b)}{\sum_{i=1}^{n} x_i^2 - n\bar{x}} \quad \text{(since } E(\varepsilon_i) = 0\text{)}$$

$$= \frac{a\sum_{i=1}^{n} x_i^2 + bn\bar{x} - na\bar{x}^2 - nb\bar{x}}{\sum_{i=1}^{n} x_i^2 - n\bar{x}} = a.$$

Also, by (41.9) and the result $E(\hat{a})$ above

$$E(\hat{b}) = E(\bar{Y} - \hat{a}\bar{x})$$

$$= \frac{1}{n}E\left[\sum_{i=1}^{n}(ax_i + b + \varepsilon_i)\right] - \bar{x}E(\hat{a})$$

$$= \frac{1}{n}\sum_{i=1}^{n}(ax_i + b) - \bar{x}a$$

$$= a\bar{x} + b - \bar{x}a = b.$$

Hence $\hat{a}$ and $\hat{b}$ are unbiased estimators of a and b respectively.

Regression lines are most easily determined and compared with the data by using computer software. Whilst we have only discussed regression *lines*, in many applications regression *curves* are more appropriate, but the important point is that they must be *linear* in the parameters.

Problems

41.1 Find the mean, median, first and third quartiles, and the interquartile range of the following two data sets:
(a) 10, 11, 11, 15, 17, 20, 25, 25, 27, 30, 38, 42, 47;
(b) 5, 12, 15, 16, 20, 29, 29, 32, 39, 44.
Draw box plots for both sets of data.

41.2 In a university degree examination with four papers each taken by 20 candidates the percentage marks are as shown in Table 41.6. Draw comparable box plots for the results.

Table 41.6

Examination marks (0–100)	
Paper 1	24, 27, 27, 30, 40, 42, 48, 55, 58, 60, 61, 63, 64, 66, 66, 68, 69, 72, 78, 85
Paper 2	30, 35, 36, 38, 39, 40, 44, 45, 48, 51, 54, 58, 61, 64, 65, 65, 69, 70, 81, 90
Paper 3	26, 29, 30, 35, 36, 37, 46, 48, 49, 49, 50, 54, 56, 61, 69, 70, 71, 71, 72, 74
Paper 4	10, 20, 22, 34, 41, 44, 45, 45, 45, 50, 55, 55, 55, 56, 64, 65, 66, 70, 85, 91

41.3 Samples of packets of crisps are weighed at the end of a manufacturing process. Packets have to contain a minimum of 25 g. The sample weights are

25.1, 25.3, 25.0, 25.7, 25.3, 25.2, 25.1, 25.5, 25.7, 25.1.

Calculate the sample mean, mode, and standard deviation.

41.4 In a continuous production process a machine cuts pipes into nominal lengths of 10 metres. The actual lengths in a production run are given in Table 41.7. Draw a histogram over (a) 10 intervals of width 0.1 metres, (b) 5 intervals of width 0.2 metres. Add a frequency polygon to both histograms.

Table 41.7

Length interval	Frequency of pipes	Length interval	Frequency of pipes
$9.5 \leqslant x < 9.6$	1	$10.0 \leqslant x < 10.1$	21
$9.6 \leqslant x < 9.7$	4	$10.1 \leqslant x < 10.2$	15
$9.7 \leqslant x < 9.8$	5	$10.2 \leqslant x < 10.3$	11
$9.8 \leqslant x < 9.9$	12	$10.3 \leqslant x < 10.4$	5
$9.9 \leqslant x < 10.0$	20	$10.4 \leqslant x < 10.5$	2

41.5 In an experiment 127 observations are taken which can be assigned to a maximum of 36 intervals. If you wish to display the data in a histogram, what would be a suitable number of intervals to use?

41.6 A random variable X has a uniform distribution (see Problem 41.4) with pdf

$$f(x) = \begin{cases} 1, & 1 \leqslant x \leqslant 2; \\ 0, & \text{otherwise.} \end{cases}$$

A random sample of size 35 is taken. Find the mean and estimate the variance of the sample. What can you say about the distribution of the sample mean?

41.7 A random sample is taken from a population which has mean μ and variance σ^2. The sample values are

9.71, 10.26, 9.80, 9.85, 9.99, 10.10, 9.79.

Estimate the sample mean and the same variance, σ^2.

41.8 A die is thrown 9000 times, and the number of times face 1 appears is recorded. If T is the random variable for the number of 1s in 9000 throws, calculate k_1 and k_2 such that

$$P(1460 < T < 1540) = \frac{1}{\sqrt{2\pi}} \int_{k_1}^{k_2} e^{\frac{1}{2}x^2} \, dx.$$

41.9 Fuel consumption figures for standard urban cycles of a selection of cars together with their weights are given in Table 41.8. Find the least-squares estimator for a regression line of fuel consumption (c) on weight (w).

Table 41.8

Vehicle	Weight, w (kg)	Fuel consumption, c (km l^{-1})
A	2100	4.96
B	1350	9.10
C	1008	12.04
D	1323	7.68
E	710	15.15
F	1215	10.98
G	1436	7.75
H	1561	8.25
I	2120	4.85
J	1975	4.64
K	1535	5.56

An unbiased estimator for the variance in linear regression is given by

$$\sum_{i=1}^{n} \frac{(y_i - \hat{y}_i)^2}{n - 2},$$

where $\hat{y}_i = \hat{a}x_i + \hat{b}$. Estimate the variance of the regression line.

One point is some distance from the regression line (such rogue values are known as **outliers**). If this particular vehicle is excluded from the data how are the regression line and the estimated variance affected?

One test of whether we are looking at an appropriate measure of the probability p is the behaviour of the expected value of $\hat{p}$. Thus

$$E(\hat{p}) = E(X/n) = \frac{1}{n} E(X) = \frac{np}{n} = p, \tag{4.1}$$

since we are assuming a binomial distribution. Hence the expected value of the estimates gives the probability p, the mean of the sampling distribution of $\hat{p}$. Generally, if the expected value of the estimate equals the parameter being estimated, then the estimate is called **unbiased**; if this is not the case then the estimate is called **biased**.

The spread of the estimate can be found by calculating the expected value $E[(\hat{p} - p)^2]$. Then

$$E[(\hat{p} - p)^2] = E[\{\hat{p} - E(\hat{p})\}^2] = \mathrm{Var}[\hat{p}]$$

by (40.5). Hence, the variance of the sampling distribution of $\hat{p}$ is given by

$$\mathrm{Var}[\hat{p}] = \mathrm{Var}\left[\frac{X}{n}\right]$$

$$= \frac{1}{n^2} \mathrm{Var}[X] \quad \text{(by (40.8(ii))}$$

$$= \frac{np(1 - p)}{n^2} = \frac{p(1 - p)}{n}, \tag{4.2}$$

for the binomial distribution. As we might expect, the variance of the sample means decreases with increasing sample size.

Given the *one* sample at the beginning of this section, the estimate for p is $p = 0.56$. The estimated variance of this single sample replacing p by $\hat{p}$ is

$$\mathrm{Var}[\hat{p}] = \frac{\hat{p}(1 - \hat{p})}{n}$$

$$= \frac{0.56 \times 0.44}{500} \approx 0.005.$$

The corresponding **standard error** for $\hat{p}$ is $\sqrt{\mathrm{Var}(\hat{p})} \approx 0.022$.

41.3 Sample mean and variance, and their estimation

The general sampling problem is as follows. Suppose that we take a sample of n observations $x_1, x_2, \ldots, x_n$ of a population where x_i is a value of a random variable X_i. The random variables (or **random sample**) $X_1, X_2, \ldots, X_n$ are assumed to be independent with the same density distributions. We assume that the sampling distribution can be modelled by a known probability distribution. Estimates, preferably unbiased, are required for the mean and variance of the underlying distribution.

An obvious choice for the mean is simply the **sample mean** which is the average value of the sample.

The **sample mean** is a random variable defined by

Sample mean for sample size n

$$\bar{X} = \frac{X_1 + X_2 + \cdots + X_n}{n}.$$

(41.3)

If $x_1, x_2, \ldots, x_n$ are *values* obtained in a particular sample then its mean is

$$\bar{x} = \frac{1}{n} \sum_{i=1}^{n} x_i.$$

What is the relation between the sample mean and the mean of the population? The expected value of $\bar{X}$ is

$$E(\bar{X}) = E\left(\frac{1}{n} \sum_{i=1}^{n} X_i\right) = \frac{1}{n} \sum_{i=1}^{n} E(X_i)$$

$$= \frac{1}{n} n\mu = \mu.$$

As we might expect, the expected value of the sample mean is the same as the mean of the population.

The variance of the sample mean is, by (40.6),

$$\text{Var}(\bar{X}) = \text{Var}\left(\frac{1}{n} \sum_{i=1}^{n} X_i\right) = \frac{1}{n^2} \sum_{i=1}^{n} \text{Var}(X_i)$$

$$= \frac{n\sigma^2}{n^2} = \frac{\sigma^2}{n}$$

(41.4)

where σ^2 is the unknown variance of the population. Its standard deviation $\sigma/\sqrt{n}$ is known as the **standard error** of the sample mean.

We also need an estimate for the variance σ^2 of the population. We might choose

$$T^2 = \sum_{i=1}^{n} \frac{(X_i - \bar{X})^2}{n},$$

which is the variance of the sample, but is it unbiased? In other words does its expected value equal σ^2? The following algebra supplies the answer:

Find and compare
(a) AB and BA;
(b) $A(BC)$ and $(AB)C$;
(c) $(A + B)^T$ and $A^T + B^T$;
(d) $(AB)^T$ and $B^T A^T$.

2. Find the inverse of

$$\begin{bmatrix} 1 & x_1 & x_1^2 \\ 1 & x_2 & x_2^2 \\ 1 & x_3 & x_3^2 \end{bmatrix}$$

(see Problem 7.18). Find the equation of the parabola of the form $y = a + bx + cx^2$ through the points $(-1, -2)$, $(\frac{1}{2}, -1)$, and $(\frac{5}{2}, 2)$.

3. Let

$$A = \begin{bmatrix} \frac{1}{3} & \frac{1}{3} & \frac{1}{6} & \frac{1}{6} \\ \frac{1}{4} & \frac{1}{2} & \frac{1}{8} & \frac{1}{8} \\ \frac{1}{8} & \frac{1}{4} & \frac{3}{8} & \frac{1}{4} \\ \frac{1}{2} & \frac{1}{6} & \frac{1}{6} & \frac{1}{6} \end{bmatrix}.$$

Find A^2, A^4, A^8, A^{16}. How do you expect A^n to behave as $n \to \infty$?

Chapter 8

1. Let

$$A = \begin{bmatrix} 1 & -1 & 2 & 3 \\ 3 & 1 & 0 & -3 \\ 2 & -1 & 3 & -1 \\ 2 & -1 & 2 & 4 \end{bmatrix}, \qquad B = \begin{bmatrix} 2 & 4 & -3 & 1 \\ 0 & -1 & 4 & 3 \\ -2 & -2 & 3 & 1 \\ -2 & 5 & 6 & -5 \end{bmatrix}.$$

Find $\det A$, $\det B$, $\det A^{-1}$, and $\det AB$. Confirm that
$$\det A^{-1} = 1/\det A, \qquad \det A \det B = \det AB.$$

2. Factorize the following determinants:

(a)
$$\begin{vmatrix} 1 & 1 & 1 \\ a & b & c \\ a^2 & b^2 & c^2 \end{vmatrix};$$

(b)
$$\begin{vmatrix} 1 & 1 & 1 & 1 \\ a & b & c & d \\ a^2 & b^2 & c^2 & d^2 \\ a^3 & b^3 & c^3 & d^3 \end{vmatrix};$$

(c)
$$\begin{vmatrix} 1 & 1 & 1 & 1 \\ a & b & c & d \\ a^2 & b^2 & c^2 & d^2 \\ a^4 & b^4 & c^4 & d^4 \end{vmatrix}.$$

3. Find the values of a for which

$$\begin{vmatrix} 5 & a & -1 & 1 \\ 2 & 1 & a & 2 \\ 3 & a & 1 & 4 \\ -1 & 0 & a & 2 \end{vmatrix}$$

is zero.

Chapter 9

1. Plot the curve which has the position vector

$$r = (2 \cos t)\hat{\imath} + (2 \sin t)\hat{\jmath} + 0.3t\hat{k}$$

from $t = 0$ to $t = 20$. What is the curve called? The position vector represents a particle moving along the curve. Find the velocity vector $\dot{r}$ and the acceleration vector $\ddot{r}$ of the particle. Show that $\dot{r} \cdot \ddot{r} = 0$.

2. Plot the *trefoil knot* given parametrically by

$$r = (1 + a \cos 3t)(\cos 2t\,\hat{\imath} + \sin 2t\,\hat{\jmath}) + a \sin 3t\,\hat{k}$$

with $a \sim 0.25$ and $0 < t \leqslant 2\pi$.

Chapter 10

1. Show that

$$\begin{bmatrix} -2^{-3/2} & 2^{-1/2} & 2^{-3/2}3^{1/2} \\ -2^{-3/2} & -2^{-1/2} & 2^{-3/2}3^{1/2} \\ 2^{-1}3^{1/2} & 0 & 2^{-1} \end{bmatrix}$$

defines a rotation of axes. If each row defines the direction of the X, Y, Z axes in the x, y, z frame, find the equation of the plane $x + 2y - 2z = 1$ in the new axes.

Chapter 11

1. The area of a triangle whose vertices are the points with position vectors a, b, and c is given by the formula

$$\tfrac{1}{2}|b \times c + c \times a + a \times b|.$$

Devise a program based on this formula to determine the area for general vertices. What is the area if $a = (1, 0, 1)$, $b = (2, -1, 1)$, and $c = (1, 1, 2)$? Plot a diagram showing the triangle.

2. A tetrahedron has vertices with position vectors

$$a = (1, -1, 2), \quad b = (-1, 2, 3), \quad c = (2, -1, 3), \quad d = (1, 3, -2).$$

Find its surface area. Draw a three-dimensional plot showing the tetrahedron viewed from the point with position vector $(2.1, -2.4, 1.5)$.

PART
VIII

Projects

42 Applications projects using symbolic computing

CONTENTS

42.1 Symbolic computation

There have been a number of significant advances in symbolic computation and computer algebra manipulation in recent years. These are systems which bring together symbolic, numerical, and graphical operations in one software package. The mathematical methods introduced in this book are particularly appropriate contexts in which to have a first look at such systems.

The software Mathematica† has been used extensively in the production of the drawings of curves and surfaces, and in the checking of examples and problems in this text. At an elementary level, Mathematica is particularly helpful, for example, with operations such as differentiation (including partial derivatives), the construction of Taylor series, elementary algebraic operations involving matrices and linear equations, elementary integration (including repeated integrals), and difference equations; but most topics in this book can be approached to some extent using Mathematica. It is also useful in

---

† Mathematica is a registered trade mark of Wolfram Research Inc.

curve sketching in that a quick view of the general feature of a curve can be obtained, which can then be revised and edited to produce detailed graphs as required.

It is not the purpose of this book to provide an introduction to Mathematica. There are a number of texts which do, including the handbook that comes with the system.

A few useful titles are listed below:

Wolfram, S. (1996). *The Mathematica Book* (3rd ed). Wolfram Media/Cambridge University Press.

Abell, M. L., and Braselton, J. P. (1992). *Mathematica by example.* Academic Press, San Diego, California.

Blackman, N. (1992). *Mathematica: a practical approach.* Prentice Hall, Englewood Cliffs, New Jersey.

Skeel, R. D., and Keeper, J. B. (1993). *Elementary numerical computing with Mathematica.* McGraw-Hill, New York.

42.2 Projects

The following projects are listed by chapter. They are selective samples of problems and do not cover every topic in the book. The intention is that they can be approached using mainly built-in Mathematica commands: very few problems require programming in Mathematica. It is generally inadvisable to attempt these problems by hand, since many could involve a great deal of manipulation, although some projects are prompted by examples and problems in the relevant chapters.

It is worth emphasizing that computer algebra systems usually generate outputs or answers without explanation of how the outputs were arrived at, unless the programming within them is investigated. Outputs can go wrong for many mathematical reasons. For example, a curve can oscillate too frequently for the built-in point spacing to detect, which can result in a false graph. This can be corrected by increasing the number of plot points, but the potential difficulty has to be recognized at the formulation stage. Symbolic computation is not a substitute for understanding mathematical techniques.

Mathematica notebooks for each project are available on the web at

http://www.keele.ac.uk/depts/ma/maths

Any comments should be sent to the authors at

Department of Mathematics, Keele University, Keele, Staffordshire ST5 5BG, UK. (Email: p.smith@keele.ac.uk)

Chapter 1

1. Draw the graphs of $y = x^3$, $y = (x-1)^3$, $y - 1 = x^3$, $y - 1 = (x-1)^3$ for $-1.5 \leqslant x \leqslant 2.5$. How do they differ?

2. **(a)** Plot the points $(n, n^2 + 1)$ for $n = 1, 2, 3, 4, 5$.
  (b) Plot the points in (a) but with successive points joined by straight lines.
  (c) Plot $y = x^2$ between $x = 0$ and $x = 5$.
  (d) Show the curves from (b) and (c) on the same graph.

3. Plot curves defined by the following relations between x and y.
  (a) $x^2 + 3y^2 = 4$; $-2 \leqslant x \leqslant 2$;
  (b) $x^2 + 2y^2 - xy + 2y = 4$; $-3 \leqslant x \leqslant 3$;
  (c) $x^4 + 2y^2 - xy - 2x^2y = 4$; $-2 \leqslant x \leqslant 3$.

4. Define the function $f(x) = x(1 - x^2)$. Plot the graphs
  (a) $y = f(x)$;
  (b) $y = f(1 - x)$;
  (c) $y = f(-x)$;
  (d) $y = f(|x|)$; all for $-2 \leqslant x \leqslant 2$.

5. Define the Heaviside function $H(t)$ and the signum function sgn t. Plot graphs of the following functions on $-4 \leqslant t \leqslant 4$:
  (a) $H(t)$;
  (b) sgn t;
  (c) $H(t) + H(-t)$;
  (d) sgn$(\sin t)$.

6. Plot the graphs of the curves defined by the following polar equations:
  (a) $r = \frac{1}{2}(1 - \cos \theta)$ for $0 \leqslant \theta \leqslant \pi$ (cardioid).
  (b) $r = (4 \sin^2\theta - 1) \cos \theta$ for $0 \leqslant \theta < 2\pi$ (folium).

7. Express
$$\frac{1}{(x-1)(x-2)(x-3)(x-4)(x-5)}$$
in partial fractions.

Chapter 2

1. Define the function
$$f(x) = \frac{x \sin x - 1 + \cos x}{\sin 2x + 2 - 2 e^x}.$$
Find $\lim_{x \to 0} f(x)$. Plot the function for $-0.5 \leqslant x \leqslant -0.001$ and for $0.001 \leqslant x \leqslant 0.5$, and check graphically that this agrees with the limit.

2. Find the derivative of
$$f(x) = 7x^2 + 8x^3 + 9x^4 + 10x^5 + 11x^6 + 12x^7$$
and its values $f'(0.2)$ and $f'(0.4)$.

3. Find the derivative of
$$f(x) = x^4 + 2x^3 - 3x^2 - 2x + 4.$$
Find the approximate values of x where $f'(x) = 0$, using a numerical solution routine. Plot graphs of $y = f(x)$ and $y = f'(x)$ on the same axes and compare the zeros of $f'(x)$ with the zero slopes on $y = f(x)$.

4. Find the equation of the tangent to the curve
$$y = x \sin 2x$$
at $x = 0.7$. Plot the graphs of the curve and its tangent.

5. Find the first three derivatives of
$$f(x) = x \sin^2 x + x^2 \sin(x^2),$$
and confirm that the first nonzero higher derivative at $x = 0$ is $f^{(3)}(0) = 6$.

6. Plot the graphs of $y = f(x)$, $y = f'(x)$, and $y = f''(x)$ for
$$f(x) = x^2(x^2 - 3)$$
in the interval $-2 \le x \le 2.5$. (This should confirm the results from Problem 2.19.)

Chapter 3

1. Display rules for the derivatives of the following general forms:
   (a) $f(x)g(x)$;
   (b) $f(x)/g(x)$;
   (c) $f(g(x))$;
   (d) $f(x)g(x)h(x)$;
   (e) $f(x)g(x)/h(x)$;
   (f) $f(h(x))/h(x)$.

2. Find the first derivatives of
$$f(x) = e^{\sin x \cos^2 x} \sin x.$$
The function is periodic. What is its minimum period? Plot its graph and the graph of $f'(x)$ over one cycle. Estimate where $f(x)$ is stationary and then find each of the roots of $f'(x) = 0$ to 5 decimal places using a root-finding routine.

3. If
$$x^2 + 2y^2 - xy - 2yx^2 = 4,$$
find dy/dx as a function of x and y.

Chapter 4

1. Display rules for the first and second derivatives with respect to x of the following general forms:
   (a) $f(x^2)$;
   (b) $f(\sin x)$;
   (c) $f(\sin(x^2))$.

2. Find the first and second derivatives of
   $$f(x) = 0.1x^5 - 0.5x^4 + 0.2x^3 + x^2 - 0.7x + 2.2.$$
   Estimate the roots of $f'(x) = 0$ from a graph of $y = f(x)$. Then find the roots to 5 decimal places by a root-finding routine. Calculate $f''(x)$ at each stationary point, and confirm the second-derivative test for stationary points. Points of inflection are given by $f''(x) = 0$. Find their locations on the original graph of $y = f(x)$.

3. Plot the graph of
   $$y = \frac{x^2 - 1}{2x + 1},$$
   and its asymptotes $y = \frac{1}{2}x - \frac{1}{4}$ and $x = -\frac{1}{2}$ (see Fig. 4.13).

4. Plot the graph of $y = f(x) = x^5 - 2x^3 + x^2 - 3x + 1$ in the interval $-1 \leqslant x \leqslant 3$, and estimate the roots of $f(x) = 0$ in this interval. Set up a Newton routine
   $$x_{n+1} = x_n - \frac{f(x_n)}{f'(x_n)},$$
   for calculating the roots of $f(x) = 0$, and find, starting at $x = 0.5$ and 1.6, the roots to 10 significant figures. What is the smallest number of iterations required in each case to calculate the roots to 10 significant figures?

5. Plot the graph of $y = x + \sin 5x$ in the interval $0 \leqslant x \leqslant 25$ using
   (a) the default plotting routine,
   (b) plotting with 20 plot points,
   (c) plotting with 50 plot points.
   Explain why the graphs are different for this type of function.

Chapter 5

1. Obtain formulae for the Taylor polynomials for the following functions centred at $x = a$ as far as $(x - a)^3$:
   (a) $f(x)$;
   (b) $[f(x)]^2$;
   (c) $f(x)g(x)$;
   (d) $e^{f(x)}$.
   State the coefficient of $(x - a)^2$ in each case.

2. Find Taylor expansions about $x = 0$ up to and including x^5 for each of the following functions:
   (a) e^x;
   (b) $(x + 1) \cos x$;
   (c) $\ln(1 + \sin x)$;
   (d) $\exp(\sin(e^x - 1))$.

3. Find the Taylor polynomials for $(\sin^2 x)/x^2$ up to and including x^N for $N = 2, 4, 6$. Plot the graphs of the function and its Taylor polynomials for $0.001 \leqslant x \leqslant 2$, and compare them. At approximately what values of x do the Taylor polynomials visibly part company from the exact function?

4. Find the Taylor polynomials for $\ln x$ about $x = 1$ for $N = 6$. Construct an error function which is the difference of $\ln x$ and its Taylor polynomial. Show that, at 2.159 approximately, this error starts to exceed 0.2 as x increases. Plot this error function against x for $1 \leqslant x \leqslant 2.2$.

Chapter 6

1. Solve, for the complex number a, the equation $z = 0$ where
$$z = \frac{(2 + 3i)^4}{(1 - 5i)^3} + \frac{(a - 2i)}{(1 + 5i)^4}.$$

2. If $z = x + iy$, find the real and imaginary parts of $z\, e^z \cos z$.

3. Find the 13 roots of $z^{13} = 1 + i$, and plot the roots on the Argand diagram.

4. Let $z_1 = 1 - 2i$, $z_2 = 3 + i$. Plot the following points on the Argand diagram:
$$z_1 + z_2, \quad \bar{z}_1 + \bar{z}_2, \quad z_1 - z_2, \quad \bar{z}_1 + z_2, \quad z_1 z_2, \quad z_1/z_2.$$

5. Find $|z|$ and $\mathrm{Arg}\, z$, where
$$z = \frac{(1 + 2i)^4}{(1 + 3i)} - \frac{2(3 - 4i)^3}{1 + 4i}.$$

Chapter 7

1. Let
$$A = \begin{bmatrix} 1 & 2 & 3 & 4 \\ -2 & 3 & -4 & 1 \\ 3 & 4 & 1 & 2 \\ 4 & -1 & 2 & 3 \end{bmatrix}, \qquad B = \begin{bmatrix} 1 & 0 & -1 & 0 \\ 1 & -2 & 1 & 2 \\ -3 & 1 & -3 & 1 \\ 2 & 1 & 2 & 1 \end{bmatrix},$$

$$C = \begin{bmatrix} 3 & 1 & 2 & 1 \\ p & p & 1 & 2 \\ 1 & -2 & -3 & 2 \\ 2 & 1 & 0 & -1 \end{bmatrix}.$$

Find and compare
(a) AB and BA;
(b) $A(BC)$ and $(AB)C$;
(c) $(A+B)^\mathsf{T}$ and $A^\mathsf{T}+B^\mathsf{T}$;
(d) $(AB)^\mathsf{T}$ and $B^\mathsf{T}A^\mathsf{T}$.

2. Find the inverse of

$$\begin{bmatrix} 1 & x_1 & x_1^2 \\ 1 & x_2 & x_2^2 \\ 1 & x_3 & x_3^2 \end{bmatrix}$$

(see Problem 7.18). Find the equation of the parabola of the form $y = a + bx + cx^2$ through the points $(-1, -2)$, $(\tfrac{1}{2}, -1)$, and $(\tfrac{5}{2}, 2)$.

3. Let

$$A = \begin{bmatrix} \frac{1}{3} & \frac{1}{3} & \frac{1}{6} & \frac{1}{6} \\ \frac{1}{4} & \frac{1}{2} & \frac{1}{8} & \frac{1}{8} \\ \frac{1}{8} & \frac{1}{4} & \frac{3}{8} & \frac{1}{4} \\ \frac{1}{2} & \frac{1}{6} & \frac{1}{6} & \frac{1}{6} \end{bmatrix}.$$

Find A^2, A^4, A^8, A^{16}. How do you expect A^n to behave as $n \to \infty$?

Chapter 8

1. Let

$$A = \begin{bmatrix} 1 & -1 & 2 & 3 \\ 3 & 1 & 0 & -3 \\ 2 & -1 & 3 & -1 \\ 2 & -1 & 2 & 4 \end{bmatrix}, \qquad B = \begin{bmatrix} 2 & 4 & -3 & 1 \\ 0 & -1 & 4 & 3 \\ -2 & -2 & 3 & 1 \\ -2 & 5 & 6 & -5 \end{bmatrix}.$$

Find $\det A$, $\det B$, $\det A^{-1}$, and $\det AB$. Confirm that

$$\det A^{-1} = 1/\det A, \qquad \det A \det B = \det AB.$$

2. Factorize the following determinants:

(a) $\begin{vmatrix} 1 & 1 & 1 \\ a & b & c \\ a^2 & b^2 & c^2 \end{vmatrix}$;

(b) $\begin{vmatrix} 1 & 1 & 1 & 1 \\ a & b & c & d \\ a^2 & b^2 & c^2 & d^2 \\ a^3 & b^3 & c^3 & d^3 \end{vmatrix}$;

(c) $\begin{vmatrix} 1 & 1 & 1 & 1 \\ a & b & c & d \\ a^2 & b^2 & c^2 & d^2 \\ a^4 & b^4 & c^4 & d^4 \end{vmatrix}.$

3. Find the values of a for which

$$\begin{vmatrix} 5 & a & -1 & 1 \\ 2 & 1 & a & 2 \\ 3 & a & 1 & 4 \\ -1 & 0 & a & 2 \end{vmatrix}$$

is zero.

Chapter 9

1. Plot the curve which has the position vector

$$r = (2\cos t)\hat{\imath} + (2\sin t)\hat{\jmath} + 0.3t\hat{k}$$

from $t = 0$ to $t = 20$. What is the curve called? The position vector represents a particle moving along the curve. Find the velocity vector $\dot{r}$ and the acceleration vector $\ddot{r}$ of the particle. Show that $\dot{r} \cdot \ddot{r} = 0$.

2. Plot the *trefoil knot* given parametrically by

$$r = (1 + a\cos 3t)(\cos 2t\,\hat{\imath} + \sin 2t\,\hat{\jmath}) + a\sin 3t\,\hat{k}$$

with $a \sim 0.25$ and $0 < t \le 2\pi$.

Chapter 10

1. Show that

$$\begin{bmatrix} -2^{-3/2} & 2^{-1/2} & 2^{-3/2}3^{1/2} \\ -2^{-3/2} & -2^{-1/2} & 2^{-3/2}3^{1/2} \\ 2^{-1}3^{1/2} & 0 & 2^{-1} \end{bmatrix}$$

defines a rotation of axes. If each row defines the direction of the X, Y, Z axes in the x, y, z frame, find the equation of the plane $x + 2y - 2z = 1$ in the new axes.

Chapter 11

1. The area of a triangle whose vertices are the points with position vectors a, b, and c is given by the formula

$$\tfrac{1}{2}|b \times c + c \times a + a \times b|.$$

Devise a program based on this formula to determine the area for general vertices. What is the area if $a = (1, 0, 1)$, $b = (2, -1, 1)$, and $c = (1, 1, 2)$? Plot a diagram showing the triangle.

2. A tetrahedron has vertices with position vectors

$$a = (1, -1, 2), \quad b = (-1, 2, 3), \quad c = (2, -1, 3), \quad d = (1, 3, -2).$$

Find its surface area. Draw a three-dimensional plot showing the tetrahedron viewed from the point with position vector $(2.1, -2.4, 1.5)$.

Chapter 12

1. Use a row-reduction routine to solve the linear equations

$$x + 2y - 3z = q,$$
$$2x + py + z = -1,$$
$$x - 2y - z = 4,$$

where p and q are two parameters. Determine for what values of p and q the equation have (a) a unique solution, (b) no solution, (c) an infinite set of solutions.

2. Use a row-reduction method to solve the linear equations

$$x + 2y + pz = 5,$$
$$3x + 2y + z = q,$$
$$2x - y + 4z = 7,$$

where p and q are two parameters. Confirm that

$$z = \frac{63 - 5q}{11 + 7p} \quad (p \neq -\tfrac{11}{7}),$$

and discuss the nature of solutions for all values of p and q.

3. Using a row-reduction instruction, show that

$$x_1 + \qquad 3x_3 \qquad = 5,$$
$$-x_1 + x_2 - x_3 + x_4 = -1,$$
$$x_1 + 2x_2 + 11x_3 \qquad = 4,$$
$$-x_1 + 2x_2 + 3x_3 + x_4 = 3$$

is an inconsistent set of equations.

Chapter 13

1. Find the eigenvalues and eigenvectors of

$$A = \begin{bmatrix} -6 & 1 & 2 & 0 \\ 1 & 0 & -3 & -1 \\ 2 & 1 & -6 & 0 \\ -2 & 2 & 0 & -3 \end{bmatrix}.$$

How many linearly independent eigenvectors does A have?
Find the eigenvalues of the following matrices:
(a) A^{-1}; (b) A^2; (c) $A + k\mathrm{I}$.

2. Find the eigenvalues and eigenvectors of

$$A = \begin{pmatrix} 1 & 2 & 1 \\ 2 & 1 & 1 \\ 1 & 1 & 2 \end{pmatrix}.$$

Construct a matrix C of eigenvectors and confirm that
$$A = CDC^{-1},$$
where D is a diagonal matrix of eigenvalues. Obtain the general formula for
$$A^n = CD^nC^{-1}.$$

3. Find the inverse and transpose of
$$A = \tfrac{1}{3}\begin{bmatrix} 1 & 2 & 2 \\ 2 & 1 & -2 \\ 2 & -2 & 1 \end{bmatrix},$$

and verify that A is an orthogonal matrix. Find the eigenvalues of A. What expected property do they have?

4. Find the eigenvalues of
$$A = \begin{bmatrix} 5 & 5 & -6 & 2 \\ -3 & 13 & -6 & 2 \\ -3 & 7 & 0 & 2 \\ 3 & -15 & 12 & 2 \end{bmatrix}.$$

Find the expression $\det(A - \lambda I_4)$, and demonstrate the Cayley–Hamilton theorem of Problem 13.21.

Chapter 14

1. Plot the graphs of the derivative $dy/dx = \sin 2x$ and the equation of the curve through $(\pi, -1)$ of which this is the derivative (see Example 14.7).

2. Plot the graph of
$$\frac{dy}{dx} = x\,e^{-x} + \sin x - x^2 \cos 2x,$$

for $0 \leqslant x \leqslant 10$. Show that an antiderivative which is zero when $x = 0$ is
$$y = 2 + \tfrac{1}{4}[-4(1 + x)\,e^{-x} - 4\cos x - 2x\cos 2x$$
$$+ \sin 2x - 2x^2 \sin 2x].$$

Plot the graph of the signed area between $x = 0$ and $x = 10$.

Chapter 15

1. Set up a program to compute the area under the curve $y = f(x)$ between $x = a$ and $x = b$ using the approximation
$$h \sum_{n=0}^{N-1} f(x_n),$$

where $h = (b - a)/N$ and $x_n = a + nh$. Apply the method to the following functions, limits, and subdivision numbers:

(a) $f(x) = x^2$, $1 \leqslant x \leqslant 3$, $N = 200$;

(b) $f(x) = x\,e^{-x}$, $0 \leqslant x \leqslant 3$, $N = 20$;

(c) $f(x) = x^3 \sin x$, $0 \leqslant x \leqslant \pi$, $N = 30$;

(d) $f(x) = \cos(e^{-x})$, $0 \leqslant x \leqslant 1$, $N = 25$.

In cases (a), (b), and (c), compare the numerical result with the areas obtained by integration. In these cases, how many subdivisions are required to obtain a numerical result correct to 3 decimal places? In (a), show that over 10 000 steps are required. Why is this?

2. Use a symbolic integration program to obtain the following indefinite integrals:

(a) $\displaystyle\int (\ln x)^3 \, dx$;

(b) $\displaystyle\int \sin^5 x \cos^3 x \, dx$;

(c) $\displaystyle\int x^2 \, e^x \sin x \, dx$;

(d) $\displaystyle\int \sqrt{(1 - x^2)} \, dx$;

(e) $\displaystyle\int \frac{dx}{x(x + 1)(x + 2)(x + 3)}$;

(f) $\displaystyle\int \frac{dx}{(1 - x^3)}$.

Check each answer by recovering the integrands by differentiation.

3. Evaluate the following definite integrals:

(a) $\displaystyle\int_1^2 x(\ln x)^3 \, dx$;

(b) $\displaystyle\int_0^1 \frac{x \, dx}{\sqrt{(5 + 4x - 4x^2)}}$;

(c) $\displaystyle\int_0^{\frac{1}{2}} \frac{x^3 \, dx}{(1 - x^2)^{\frac{5}{2}}}$;

(d) $\displaystyle\int_0^1 \sum_{n=0}^{100} \frac{x^n}{n!} \, dx$.

4. Find

$$I(a) = \int_1^a (\ln x)^3 \, dx.$$

Find the limit

$$\lim_{b \to 0} \frac{bI(1/b)}{(-\ln b)^3}.$$

How does $I(a)$ behave as $a \to \infty$? Does

$$\int_1^\infty (\ln x)^3 \, dx$$

exist?

5. A cylindrical hole of circular cross-section and radius b is drilled through a sphere of radius $a > b$, the axis of the hole passing through the centre of the sphere. Find the volume of the remaining object. Display a diagram of the object for some values of a and b.

Chapter 16

1. Plot the graph of the polar equation $r = \sin 5\theta$ for $0 \leqslant \theta \leqslant 2\pi$. Find the area enclosed by the five 'petals' of the curve.

Show that the area of the $2n + 1$ petals of $r = \sin(2n + 1)\theta \, (n \geqslant 1)$ is independent of n.

2. Devise a program to generate the trapezium rule:

$$\int_a^b f(x)\,dx \approx \frac{b - a}{N} [\tfrac{1}{2}f(a) + (f(x_1) + f(x_2) + \cdots + f(x_{N-1})) + \tfrac{1}{2}f(b)].$$

Apply the program to the integral

$$\int_0^2 e^{-2x} \sin^2 x \, dx,$$

and compare the result with the exact value of the integral. Investigate how many steps are required to obtain a result accurate to 3 decimal places.

Apply the program also to Problem 16.20.

3. A thin plane metal plate consists of an isosceles triangle of height h and base length $2a$ with a semicircle of radius a attached symmetrically by its diameter to the base of the triangle. Find the location of its centroid on its axis of symmetry.

4. Set up a program to generate Simpson's rule

$$\int_a^b f(x)\,dx \approx \frac{b - a}{3N}\left(f(a) + f(b) + 4 \sum_{k=1}^{\frac{1}{2}N} f(x_{2k-1}) + 2 \sum_{k=1}^{\frac{1}{2}N-1} f(x_{2k}) \right),$$

where N is an even number. Apply the method to $f(x) = e^{-x^2}$, with $b = 1$, $a = 0$. Compare results with the trapezium rule above.

Chapter 17

1. Illustrate the substitution method in integration by writing a program to integrate

$$\int \frac{x-2}{\sqrt{(5+4x-x^2)}}\,dx,$$

using the substitutions $x = u + 2$, $u = 3\sin t$. Integrate directly and through the substitutions.

2. Integrate the following, and compare your answers with computer-integrated ones:

(a) $\displaystyle\int \frac{x\,dx}{4x^2+1};$

(b) $\displaystyle\int \tan x\,dx;$

(c) $\displaystyle\int \cos^4 x\,dx;$

(d) $\displaystyle\int \frac{x\,dx}{\sqrt{(x-1)}};$

(e) $\displaystyle\int \frac{\sin^3 x}{\cos x}\,dx.$

3. Computer-integrate the infinite integrals

$$I_{10} = \int_0^\infty t^{10}\,e^{-t}\,dt, \quad I_{11} = \int_0^\infty t^{11}\,e^{-t}\,dt,$$

and confirm that $I_{11}/I_{10} = 11$.

4. Computer-integrate the following infinite integrals:

(a) $\displaystyle\int_0^\infty e^{-x}\sin x\,dx;$

(b) $\displaystyle\int_1^\infty \frac{\ln x}{x^{10}}\,dx;$

(c) $\displaystyle\int_1^\infty x^3\,e^{-ax^2}\,dx.$

5. Evaluate the integral

$$f(a) = \int_1^a \frac{(\ln x)^6}{x^2}\,dx$$

for $a > 1$. Find $f(10), f(20)$, and $f(\infty)$. The results indicate that $f(a)$ tends to a limit very slowly as $a \to \infty$. Find where

$$g(x) = \frac{(\ln x)^6}{x^2}$$

has a maximum value, and plot the graph $y = g(x)$ for $1 \leqslant x \leqslant 100$.

Chapter 18

1. Solve the differential equation $\dot{x} + x = 0$, for the initial conditions (a) $x(0) = 0$, (b) $x(0) = 1$, (c) $x(0) = 2$, and plot the solutions on the same axes for $0 \leqslant t \leqslant 2$.

2. Solve the differential equations

     (a) $2\ddot{x} + 3\dot{x} + x = 0$,
     (b) $\ddot{x} + 2\dot{x} + 2x = 0$,
     (c) $\ddot{x} + 2\dot{x} + x = 0$,

each for the six sets of initial conditions:

     (i) $x(0) = 0$, $\dot{x}(0) = 1$;
     (ii) $x(0) = 0$, $\dot{x}(0) = 2$;
     (iii) $x(0) = 0$, $\dot{x}(0) = 3$;
     (iv) $\dot{x}(0) = 0$, $x(0) = 1$;
     (v) $\dot{x}(0) = 0$, $x(0) = 2$;
     (vi) $\dot{x}(0) = 0$, $x(0) = 3$.

Plot all solutions on the same axes for each differential equation, for $0 \leqslant t \leqslant 5$.

Chapter 19

1. Solve the differential equation $2\ddot{x} + 3\dot{x} + x = \cos t$ subject to $\dot{x}(0) = 0$, $x(0) = 1$. Plot the solution for $0 \leqslant t \leqslant 50$.

2. Solve the differential equation $\ddot{x} + x = \cos t$ subject to $x(0) = 0$, $\dot{x}(0) = 0$. Plot the solution for $0 \leqslant t \leqslant 20$.

Chapter 20

1. Solve the differential equation $\ddot{x} + x = 0$ subject to the initial conditions $x(0) = 1$, $\dot{x}(0) = 0$. Also solve $\ddot{x} + \sin x = 0$, by a built-in numerical solution method for $0 \leqslant t \leqslant 10$ subject to the same initial conditions. Plot both solutions for $0 \leqslant t \leqslant 10$. Comparison of the plotted solutions will indicate by how much the period decreases when the linear approximation is used. Rerun the programs for different amplitudes $x(0)$.

Chapter 21

1. Draw the phasor diagram of the sum of the three phasors of
$$u(t) = 2 \cos 10t, \quad v(t) = \cos(10t - \tfrac{1}{2}\pi), \quad w(t) = 3 \cos(10t + \tfrac{1}{4}\pi)$$
(see Example 21.6).

Chapter 22

1. Draw the lineal-element diagram of $dy/dx = xy$, produced by a standard package in the square $\{0 \leqslant x \leqslant 1, 0 \leqslant y \leqslant 1\}$ (see Section 22.1). Compare this with the exact solution (see Section 22.1) drawn through the points $(0, 0.2)$, $(0, 0.4)$, and $(0, 0.6)$.

2. Repeat the above process for the differential equation $dy/dx = x - y$ of Example 22.1.

3. Design a program for Euler's method (Section 22.2) for the initial-value problem
$$\frac{dy}{dx} = xy^2, \qquad y(0) = 1$$
(see Example 22.4) with step length $h = 0.2$ and five steps. Run the program for the cases $h = 0.1$ and $h = 0.01$ and compare the results.

4. Plot numerical solutions for
$$\frac{dy}{dx} = \frac{3y - x}{3x - y}$$
(Example 22.14 and Fig. 22.11) using built-in routines. As with many equations of this type it is often easier to solve the equivalent simultaneous equations
$$\frac{dx}{dt} = 3x - y, \qquad \frac{dy}{dt} = 3y - x,$$
numerically for various initial values of $x(0)$ and $y(0)$.

Chapter 23

1. By splitting the differential equation $\ddot{x} + 2x^3 = 0$ into the system
$$\dot{x} = y, \qquad \dot{y} = -2x^3,$$
and plotting four phase paths respectively through the four points
$$(x(0), y(0)) = (0.3, 0), (0.6, 0), (0.9, 0), (1.2, 0)$$
over the interval $-1.5 \leqslant x \leqslant 1.5$, show that the solutions appear to be periodic.

2. Plot phase paths for the van der Pol equation

$$\ddot{x} + 10(x^2 - 1)\dot{x} + x = 0$$

showing the limit cycle. Also show the corresponding (t, x) graph of the periodic solution (the periodic solution has an initial value close to $x(0) = 2$, $\dot{x}(0) = 0$).

Chapter 24

1. Computer algebra systems are quite efficient at finding Laplace transforms of complicated expressions involving standard functions. Test the system with the following transforms:

(a) $L\{t^8 e^{-t}\}$;

(b) $L\{t^2 e^{-t} \cos t\}$;

(c) $L\left\{\dfrac{d^3x}{dt^3}\right\}$;

(d) $L\{f(t)\}$ where $f(t) = \begin{cases} 1 & \text{if } 0 \leqslant t \leqslant a \\ 0 & \text{if } t > a; \end{cases}$

(e) $L\{e^t/t^{\frac{1}{2}}\}$;

(f) $L\{\cosh at\}$.

2. Solve

$$\dot{x} + 2x = e^{-t}, \qquad x(0) = 3,$$

using a Laplace-transform package, and compare the answer with that of Example 24.12. Plot the input e^{-t} and the output against t for $0 \leqslant t \leqslant 3$.

3. Using a Laplace-transform package, solve the system

$$\ddot{x} + 2\dot{x} + x = a \cos \omega t, \qquad x(0) = 0, \dot{x}(0) = 0.$$

Plot the input and output functions for $a = 1$, $\omega = 1$, and $0 \leqslant t \leqslant 30$. Estimate the eventual amplitude of the periodic output.

4. Find the functions whose Laplace transforms are:

(a) $\dfrac{1}{s(s + 1)(s + 2)(s + 3)}$;

(b) $\dfrac{e^{-s}}{(s^2 + 4)(s + 1)}$.

Plot the functions in each case.

5. Consider the function $f(t) = \ln t$. Show the Laplace-transform package produces the transform

$$\frac{1}{s}(\gamma + \ln s),$$

where γ is *Euler's constant* given by

$$\gamma = \lim_{m \to \infty} \left(\sum_{k=1}^{m} \frac{1}{k} - \ln m \right).$$

Derive a program to calculate Euler's constant. It should give $\gamma = 0.577\,215\ldots$.

Chapter 25

1. Find the Laplace transform of the solution of
   $$\ddot{x} + \omega^2 x = a\,\delta(t-1), \qquad x(0) = \dot{x}(0) = 0,$$
   which has impulse input applied at time $t = 1$. Invert the transform and plot the output for $\omega = 4$, $a = 1$ (see Example 25.3).

2. Following the previous project, solve the more complicated problem with two impulses:
   $$2\ddot{x} + 3\dot{x} + 2x = a\,\delta(t-\pi)\cos t + b\,\delta(t-2\pi),$$
   $$x(0) = \dot{x}(0) = 0.$$
   Plot the output for $a = b = 1$.

3. Let $f(t) = t^3$, $g(t) = \cos t$. Find the convolution
   $$\int_0^t f(t-u)g(u)\,du.$$
   Then verify that
   $$L\{f(t)\}L\{g(t)\} = L\left\{ \int_0^t f(t-u)g(u)\,du \right\}.$$

4. A transfer function with a parameter a is given by (Section 25.10)
   $$G(z) = \frac{4z^3 - 8z^2 - 2z + 4}{6z^4 - 6z^3 - 2a^2 z^2 + 3z^2 + 2a^2 z - 2a^2}.$$
   Find the locations of the poles of $G(z)$. For what values of a do all poles lie within the unit circle (indicating transient stability)? Plot the poles on an Argand diagram for $a = 2$.

Chapter 26

1. Consider the period-2 sawtooth function defined over its fundamental interval $-1 < t \leqslant 1$ by $f(t) = t$. Find its general Fourier coefficient and output its first four terms. Plot and compare the graphs of this truncated series and the sawtooth for $-3 < t \leqslant 3$.

2. Repeat the previous problem but with the function

$$f(t) = \begin{cases} 1 & (0 \leqslant t < 1), \\ -1 & (-1 \leqslant t < 0). \end{cases}$$

Plot the graphs of $f(t)$ and the first 12 terms of its Fourier series. The graph should show the **Gibbs' phenomenon**, in which the Fourier series approximation overshoots the function at discontinuities. You can try it with (say) 20 terms or more, but you should include more interpolating points in these cases.

3. Find the Fourier coefficients of the 2π-periodic function defined by

$$f(x) = x^6 - 5\pi^2 x^4 + 7\pi^4 x^2$$

on the interval $-\pi \leqslant x < \pi$. What is the sum of the series

$$\sum_{n=1}^{\infty} \frac{(-1)^{n+1}}{n^6} \, ?$$

Chapter 27

1. Find the Fourier transforms of the following functions:
  (a) the top-hat function $\Pi(t)$;
  (b) the one-sided exponential $e^{-t} H(t)$;
  (c) $e^{-|t|}$;
  (d) $e^{-|t-1|}$;
  (e) $1/(1 + t^2)$.
  Plot the graph of the transform in (e).

2. Find the functions whose Fourier transforms are
  (a) e^{-f^2};
  (b) $1/(4 + f^2)$;
  (c) 2;
  (d) $2 \cos(f - a)$.

Chapter 28

1. Plot the saddle surface $z = x^2 - y^2$ in the cylinder $x^2 + y^2 \leqslant 1$, using a three-dimensional parametric plot routine with parameters r and u where

$$(x, y, z) = (r \cos u, \quad r \sin u, \quad r^2 \cos 2u).$$

Also draw a contour plot of the surface in the (x, y) plane on the square $-1 \leqslant x \leqslant 1, -1 \leqslant y \leqslant 1$.

2. Plot the surface $z = xy(x^2 - y^2)$ in the cylinder $x^2 + y^2 \leqslant 1$ using the same routine as in Project 28.1 above, but with the parametric equations

$$(x, y, z) = (r \cos u, r \sin u, \tfrac{1}{4} r^3 \sin 4u).$$

How would you describe this saddle? Draw its contour plot in the square $-1 \leqslant x \leqslant 1, -1 \leqslant y \leqslant 1$.

3. For the function

$$f(x, y) = e^{x^2 y} \sin(xy) + x \ln(x^2 + y^3),$$

verify that

$$\frac{\partial^2 f}{\partial y \, \partial x} = \frac{\partial^2 f}{\partial x \, \partial y}.$$

4. Plot the surface given by $z = \cos xy$ over $-\pi \leqslant x \leqslant \pi, -\frac{1}{2}\pi \leqslant y \leqslant \frac{1}{2}\pi$. Find the partial derivatives at $(\frac{1}{4}\pi, 1)$ and construct the equation of the tangent plane there. Finally plot the surface and its tangent plane.

5. Find the stationary points of

$$f(x, y) = 0.3x^3 + 0.2y^2 - x^2y - xy + 2y$$

numerically by solving

$$\frac{\partial f}{\partial x} = \frac{\partial f}{\partial y} = 0.$$

Plot the contours on the (x, y) plane for $-3 \leqslant x \leqslant 3, -9 \leqslant y \leqslant 3$.

Find the values of the second derivatives at each stationary point and check the second derivative tests (28.9) at each point.

6. Find the least-squares straight line fit to the points

$$(0, 1.1), \quad (1, 2), \quad (2, 2.9), \quad (3, 3.9), \quad (4, 4.5), \quad (5, 5.1),$$

in the (x, y) plane. Plot the data and the least-squares straight line fit. If you are using a built-in routine, check your results against that given by (28.10).

Chapter 29

1. Find the family of curves orthogonal to that of

$$\frac{dy}{dx} = y \, e^{-x}.$$

Plot both families of curves for $|x| \leqslant 2, |y| \leqslant 2$.

Chapter 30

1. Find where the function

$$f(x, y) = x^3 - 2xy - x + 3y^2$$

is stationary subject to the condition $x^2 + 2y^2 = 1$. Devise a program which uses the Lagrange-multiplier method (30.4): here is a suggested line of approach. First plot the contours of $z = f(x, y)$ and the curve $x^2 + 2y^2 = 1$. Locate the approximate coordinates of any point of tangency. Then use a built-in root-finding scheme to locate the stationary values. There should be four.

Chapter 31

1. Find the equation of the tangent plane to the surface

$$x^3y + zx + xy^2z = -3$$

at $(1, 2, -1)$.

2. Show graphically the intersection of the cylinder $x^2 + y^2 = 1$ and the plane $x + y + z = 1$ (Example 31.9).

3. Find the envelope of the family of curves

$$y(a^2 - 1 + ax) = x$$

with parameter a. Plot the envelope and a sample of touching curves in $-3 \leqslant x \leqslant 3$.

Chapter 32

1. By repeated integration, evaluate the integral

$$\int_{-1}^{1} \int_{0}^{1} (x + y\, e^{-xy} + xy)\, dx\, dy,$$

using a symbolic routine. Plot the surface

$$z = x + y\, e^{-xy} + xy$$

over $0 \leqslant x \leqslant 1, -1 \leqslant y \leqslant 1$. Interpret the integral as the volume under the surface. Does the integral contain 'negative' volumes under the surface? Plot the positive part of the surface over the same rectangle.

2. Evaluate the repeated integral

$$\int_{0}^{a} \int_{-\sqrt{(a^2-y^2)}/a}^{\sqrt{(a^2-y^2)}/a} x^2y\, dx\, dy.$$

Plot the region of integration in the (x, y) plane, and then check that the integral has the same value with the order of the integration reversed.

Chapter 33

1. Let

$$f(x, y, z) = xy\hat{i} + yz\hat{j} + (z - y)x\hat{k}.$$

Find f as a function of t on the line $x = t, y = t, z = t$. Evaluate the line integral

$$\int f \cdot dr$$

on this line between $(0, 0, 0)$ and $(1, 1, 1)$.

Repeat the process with the curve $x = t^2$, $y = t^3$, $z = t^4$, and the same end-points. Plot both paths of integration.

Chapter 34

1. Plot the surfaces defined parametrically by the following position vectors:
   (a) $r = (3 + \cos v) \cos u\, \hat{\imath} + (3 + \cos v) \sin u\, \hat{\jmath} + \sin v\, \hat{k}$ (see Section 34.3);
   (b) $r = (1 + a\, \sin(bu)) \cos v\, \hat{\imath} + (1 + a\, \sin(bu)) \sin v\, \hat{\jmath} + u\hat{k}$, where $a = 0.3$ and $b = 3.5$ (see Section 34.3).

2. Given that
   $$f(x, y, z) = e^{xyz}\hat{\imath} + z\, \cos(xy)\hat{\jmath} + (x^2 + y^2)\hat{k},$$
   find
   (a) div f;
   (b) curl f;
   (c) div curl f;
   (d) curl curl f at the point $(1, 0, -1)$.

3. Using symbolic computation test the validity of the following identities:
   (a) $(F \cdot \text{grad})F = \frac{1}{2}\text{grad}(F \cdot F) - F \times \text{curl } F$;
   (b) $\text{div}(F \times G) = G \cdot \text{curl } F - F \cdot \text{curl } G$;
   (c) $\text{curl}(F \times G) = (G \cdot \text{grad})F - (F \cdot \text{grad})G - G\,\text{div } F + F\,\text{div } G$;
   (d) $\text{div}(U\,\text{grad } V - V\,\text{grad } U) = U\nabla^2 V - V\nabla^2 U$;
   (e) $\text{curl curl } F = \text{grad div } F - \nabla^2 F$.

Chapter 35

1. A and B are the sets of integers defined by
   $$A = \{2n + 5(-1)^n \,|\, n \in \mathbb{N}^+, 1 \leqslant n \leqslant 100\},$$
   $$B = \{n^2 - n + 1 \,|\, n \in \mathbb{N}^+, 1 \leqslant n \leqslant 10\}.$$
   Produce lists of the elements in $A \cup B$ and $A \cap B$. How many elements do each of these sets have?

2. Let A, B, and C be the following sets:
   $$A = \{n(n-1) \,|\, n \in \mathbb{N}^+, 2 \leqslant n \leqslant 100\},$$
   $$B = \{|n^2 - 100n| \,|\, n \in \mathbb{N}^+, 1 \leqslant n \leqslant 160\},$$
   $$C = \{4n \,|\, n \in \mathbb{N}^+, 1 \leqslant n \leqslant 2200\}.$$
   Verify the first distributive law
   $$A \cap (B \cup C) = (A \cap B) \cup (A \cap C).$$
   How many elements are there in the set $A \cap (B \cup A)$?

Chapter 36

1. Design programs to generate the truth tables for the OR gate, the AND gate, the NOT gate, the NAND gate, and the NOR gate.

2. Design a program to simulate the truth table in Example 36.3 which has the output
$$f = \overline{a * b} \oplus b \oplus c$$
for inputs a, b, and c.

Chapter 37

1. In the cutset method applied to the circuit in Fig. 37.23, the currents i_1, i_2, i_3, i_4, i_5 and the voltages v_a, v_b, v_c, v_d satisfy the nine equations
$$i_1 - i_3 + i_2 = 0, \qquad i_X - i_3 + i_2 = 0,$$
$$-i_Y + i_5 - i_3 + i_2 = 0, \qquad -i_Y + i_4 + i_2 = 0,$$
$$i_1 = (v_a - v_b)/R_1, \qquad i_2 = (v_c - v_b)/R_2, \qquad i_3 = (v_b - v_d)/R_3,$$
$$i_4 = (v_c - v_d)/R_4, \qquad i_5 = v_d/R_5,$$
where $i_X = 2$ A, $i_Y = 2$ A, and $R_1 = \frac{1}{2}\,\Omega$, $R_2 = 3\,\Omega$, $R_3 = 1\,\Omega$, $R_4 = 2\,\Omega$, $R_5 = 2\,\Omega$. Solve this set of linear equations for the currents and voltages.

2. Draw the labelled drawings of the bipartite graphs $K_{5,6}$ and $K_{6,6}$. Answer the following for each graph by the built-in diagnostic test.
(a) How many edges has each graph?
(b) Is the graph eulerian? If it is, list an eulerian walk.
(c) Is it hamiltonian? If it is, list a hamiltonian cycle.

3. Check the complete graphs K_n, $2 \leqslant n \leqslant 7$, and the bipartite graphs $K_{i,j}$ ($2 \leqslant i \leqslant 5$; $i \leqslant j \leqslant 6$) for planarity, using a built-in diagnostic test.

Chapter 38

1. Rework Example 38.2 using a symbolic package for solving difference equations. Solve the mortgage difference equation
$$Q_m - (1 + I)Q_{m-1} = -A,$$
with $I = 0.08$ and $Q_0 = P = 50\,000$ (in £). Given that $Q_{25} = 0$, find A. List the outstanding debt Q_m each year m to the nearest £. Plot (a) the outstanding debt against years and (b) the annual interest repayments $A - IQ_m$ against years.

2. Solve the following homogeneous difference equations:

   (a) $u_{n+2} - u_{n+1} - 12u_n = 0$;

   (b) $u_{n+2} + 2u_{n+1} + 2u_n = 0$;

   (c) $u_{n+2} + 4u_{n+1} + 4u_n = 0$;

   (d) $u_{n+3} + 3u_{n+2} + 3u_{n+1} + u_n = 0$, $u_0 = 0$, $u_1 = 1$, $u_2 = -1$.

3. Solve the following inhomogeneous difference equations:

   (a) $u_{n+2} - u_{n+1} - 12u_n = 2 + n + n^2$;

   (b) $u_{n+2} - u_{n+1} + 4u_n = 2^n$;

   (c) $u_{n+3} + 3u_{n+2} + 3u_{n+1} + u_n = n^2$, $u_0 = 0$, $u_1 = 1$, $u_2 = -1$.

4. Devise a program to generate cobweb plots for the first-order difference equation

   $$u_{n+1} = -ku_n + k$$

   for (a) $k = \frac{1}{2}$, (b) $k = \frac{3}{2}$, (c) $k = 1$, with initial value $u_0 = \frac{3}{4}$ in each case (see Example 38.3).

5. Display cobweb plots for the logistic difference equation

   $$u_{n+1} = \alpha u_n (1 - u_n)$$

   for selected values of α. Some suggested values are:

   (a) $\alpha = 2.8$ to show a stable fixed point;

   (b) $\alpha = 3.4$: find the period-2 solution;

   (c) $\alpha = 3.5$: find the period-4 solution;

   (d) $\alpha = 3.7$: chaotic output;

   (e) $\alpha = 3.83$: should be able to locate a stable period-3 solution.

6. Design a program to generate the period-doubling display shown in Fig. 38.11 for the logistic equation $u_{n+1} = \alpha u_n (1 - u_n)$ for α increasing from $\alpha = 2.8$ to $\alpha = 4$.

Chapter 39

1. (See Example 39.8.) A box contains 40 balls of which 7 are red, 12 are white, and 21 are black. In each of the cases $n = 2, 3, 4, 5, 6, 7$, n balls are drawn at random from the box without replacement. What is the total number of n-ball selections which can be made? What is the probability that there are n ($n = 2, 3, 4, 5, 6, 7$) balls of the same colour? Show the probabilities graphically in a bar chart.

Chapter 40

1. List the probabilities of the binomial distribution for $n = 12$ and $p = 0.7$. Check that their sum is 1. Plot this discrete distribution as a bar chart.

2. Plot graphs of the probability density function (pdf) and the cumulative distribution function (cdf) for the standardized normal distribution $N(0, 1)$.

3. Model a sequence of n Bernoulli trials with success/failure equally likely, in which the number of successes is recorded. You could try $n = 50$ run 500 times and count the number of successes i for $i = 0, 1, 2, \ldots, n$. This should approximate to the binomial distribution $_nC_i p^i q^{n-i}$. Plot this distribution and compare it with the simulation.

Chapter 41

1. Devise a program to draw comparative box plots for the examination data given in Problem 41.2.

2. Produce a histogram and frequency polygon for the pipe length data given in the table accompanying Problem 41.4.

3. Some randomized points (x_i, y_i) are generated by the Mathematica command

```
Table[{x+0.2*Random[],x+2+1.2*Random[]},{x,0,6,0.5}].
```

Find the regression lines of y on x, and of x on y, for the data. Plot the data and both regression lines. Also find the *mass centre* of the data, and add this point to the graph. Where does the mass centre lie in relation to the regression lines?

4. Two dice are rolled and the average scores recorded. Compute the probabilities of the possible average scores, and plot them in a bar chart. Repeat the program for four and six dice. Plot bar charts in each case to illustrate the development normal distribution predicted by the central limit theorem.

ANSWERS TO SELECTED PROBLEMS

Chapter 1

1.2 (a) $y = -2x + 3$; (b) $y = 1$; (c) $y = \frac{2}{3}x - \frac{1}{3}$.
Intersections are $A : (2, 1)$, $B : (\frac{5}{4}, \frac{1}{2})$, $C : (1, 1)$.

$AB = \frac{1}{4}\sqrt{13}$, $\quad AC = 1$, $\quad BC = \frac{1}{4}\sqrt{5}$.

1.3 (b) Slope $= \frac{1}{3}$. Intersection with axes at $(2, 0)$, $(0, -\frac{2}{3})$.

1.4 (b) $(y + 2)/(x + 1) = -2$, so $y = -2x - 4$.
(d) $(y - 2)/(x - 1) = 3$, so $y = 3x - 1$.

1.6 Hint: choose α suitably.

1.7 (b) Centre $(1, 0)$, radius 2.
(d) Centre $(\frac{1}{2}, -\frac{1}{2})$. radius $\frac{1}{2}\sqrt{11}$.

1.9 (b) $x = -\frac{3}{5} \pm \frac{1}{5}\sqrt{14}$, $y = -\frac{1}{5} \pm \frac{1}{5}\sqrt{14}$.

1.14 (b) 1. (d) $-1/\sqrt{2}$. (f) $-\sqrt{3}/2$.

1.16 (b) $\cos x$; (d) $-\cos x$.

1.17 (b) $2 \cos \frac{1}{2}(x + y) \sin \frac{1}{2}(x - y)$.

1.18 In the following, n represents any integer:
(b) $\frac{1}{2}\pi + n\pi$; (d) $\frac{1}{6} + \frac{1}{3}n$; (f) $2n$.

1.19 (b) amp. $= 1.5$; ang. freq. $= 0.2$; period $= 31.41$;
phase $= -0.48$.

1.20 (b) $\frac{1}{2}x - \frac{3}{2}$; (d) $\arcsin \frac{1}{2}x$, $0 \le x \le 2$.
(f) $\arccos(\arcsin x)$, $0 \le x \le \sin 1$.
(h) $-\frac{1}{2} + (1 + 4x)^{\frac{1}{2}}$, $x \ge -\frac{1}{4}$.

1.22 (b) $\frac{1}{3}e^2$; (d) $\frac{1}{3}\ln\frac{1}{3}$, or $-\frac{1}{3}\ln 3$; (f) 2; (h) $\pm\sqrt{2}$;
(l) Hint: write $\sinh 2x = \frac{1}{2}(e^{2x} - e^{-2x})$ and obtain a
quadratic equation for e^{2x}. $x = \frac{1}{2}\ln(2 + \sqrt{17})$.

1.26 Hint: $x = \sinh y = \frac{1}{2}(e^y - e^{-y})$. Form an equation for
e^y and solve it.

1.28 $5 \cos(\omega t - 0.927)$.

1.29 $C = 2$, $\alpha = 1.386$, $f(2) = 1/8$.

1.30 Tidal period $= 12.57$ h. It floats for 9.20 h.
Hint: it floats when $\sin 0.5t \ge -0.666$. Sketch $y = \sin 0.5t$
and $y = 0.666$ and find the intersections.

1.33 The vertex is $(-4, 7)$.

1.36 (b) $2/(x + 2) - 1/(x + 1)$.
(d) $1/2x - 1/(x + 1) + 1/2(x + 2)$.
(f) $1/4x - 1/4(x + 2) - 1/2(x + 2)^2$.
(h) $1/2(x - 3) + 1/2(x + 1)$.

1.37 (b) $1/[2(x - 1)] + 1/[2(x^2 + 1)] - x/[2(x^2 + 1)]$.

1.38 (b) $x - 3 - 1/(x + 1) + 8/(x + 2)$.

1.39 (b) $1 + 1/2 + 1/5 + 1/10 + 1/17$.

1.40 (b) $\sum_{n=2}^{6} (\frac{1}{3})^n = (\frac{1}{3})^2 + (\frac{1}{3})^3 + \cdots + (\frac{1}{3})^6$
$= (\frac{1}{3})^2[1 + \frac{1}{3} + \cdots + (\frac{1}{3})^4]$.
Now (1.31) gives the sum in the brackets. Finally we
obtain 121/729.
(e) $-341/1024$.

1.44 (c) 1/99; (e) 30/11.

1.45 (b) 10/9; (d) 2/3.

1.47 (c) 2ml.

1.49 (c) 256; (d) 20; (f) 30.

1.50 (a) 72; (b) 360.

1.51 (b) 24; (d) 96.

1.54 (a) 2880; (b) 720.

1.55 (a) 120; (b) 720; (c) 220; (d) 1110.

1.56 (a) 600; (b) 168.

Chapter 2

2.1 (b) 0.5; (e) 2; (g) 1.

2.2 (c) 6; (e) -0.25; (g) -4.

2.3 (c) $-1/x^2$; (f) $4x$.

2.4 (c) -8.

2.5 (c) 32, -32.

2.8 (c) $dE/dT = 4kT^3$.

2.9 (b) $7x^6 - 18x^5 + 1$.

2.10 $dy/dx = \tan \alpha$, where α is the inclination angle.

2.11 Use the formula for $\tan(A - B)$ in Appendix B(b).

2.12 (b) $\frac{1}{2}$; (d) 1; (g) 2; (i) $\pi/180 = 0.0175$.

2.15 (a) $2 \cos x + 3 \sin x$.

2.16 (b) $y = 24x - 39$; (d) $y = e^{-1}x$.

2.17 (b) $6x - 2$, 6, 0.

2.20 $y = (-x + x_0 + 2a^2x_0^3)/(2ax_0)$.

Chapter 3

3.1 (b) $x \cos x + \sin x$; (f) $2x \ln x + x$.

3.2 (b) $1/(1+x)^2$; (f) $(x^2 - 2x \sin x \cos x)/x^4 \cos^2 x$.
(m) nx^{n-1}.

3.3 (d) $f \dfrac{dg}{dx} + g \dfrac{df}{dx}$, $\ g \dfrac{d^2 f}{dx^2} + 2 \dfrac{df}{dx}\dfrac{dg}{dx} + f \dfrac{d^2 g}{dx^2}$,

$\quad g \dfrac{d^3 f}{dx^3} + 3 \dfrac{d^2 f}{dx^2}\dfrac{dg}{dx} + 3 \dfrac{df}{dx}\dfrac{d^2 g}{dx^2} + f \dfrac{d^3 g}{dx^3}$.

3.4 (b) $-2 \cos x \sin x$; (e) $2 \sin x/\cos^3 x$; (j) $12x^2(x^3 + 1)^3$;
(n) $-3 \, e^{-3x}$; (s) $a^x = e^{x \ln a}$, so $d(a^x)/dx = (\ln a)a^x$.

3.5 (f) $\frac{1}{2}x^{-\frac{1}{2}}$; (i) $-\frac{1}{2}x^{-\frac{3}{2}}$.

3.6 (f) $e^{-t}(\cos t - \sin t)$; (k) $2 \sin x(\cos x - \sin x)/x^3$.

3.9 (c) $(-2x \sin x^2)/\cos x^2$. The original function only has a meaning when $\cos x^2 > 0$.

3.10 (b) $e^t(\cos t + t \cos t - t \sin t)$.

3.11 (b) $dy/dx = -y^{\frac{1}{2}}/x^{\frac{1}{2}}$. This can be written in other ways; for example, put $y^{\frac{1}{2}} = 1 - x^{\frac{1}{2}}$ from the equation of the curve.

3.15 (b) -5.

3.16 (b) $dy/dx = \pm x/[2\sqrt{(1 - (x/2)^2)}]$.

Chapter 4

4.1 (b) $2t^2$; (c) $4t^3$.

4.2 (c) $x = e^{-1}$ (min); (g) $x = 0$ (min);
(i) $x = -1/\sqrt{3}$ (min), $x = 1/\sqrt{3}$ (max);
(t) Points of inflection at $x = n\pi$; maxima at $x = (2n + \frac{1}{2})\pi$; minima at $(2n - \frac{1}{2})\pi$.

4.5 If base $= x$ and rectangle height $= y$, then
$A = xy + \frac{1}{8}\pi x^2$ (constant), and $P = (1 + \frac{1}{2}\pi)x + 2y$.
Substitute for y from the formula for A to express P in terms of x only. The minimum of P is reached when
$x = [2A/(1 + \frac{1}{4}\pi)]^{\frac{1}{2}}$.

4.10 (b) $\delta y \approx -0.2$ (exact value $-0.227\ldots$).
(d) $\delta y \approx -0.4$ (exact value -0.5).

4.11 (a) $\delta v \approx -0.11$; (d) $\delta A \approx -0.08$.

Chapter 5

5.1 (b) $(1 + x)^{\frac{1}{2}} \approx 1 + \frac{1}{2}x - \frac{1}{8}x^2 + \frac{1}{16}x^3$. For 2 decimal places, we need $|\frac{1}{16}x^3| < 0.005$, or $-0.43 < x < 0.43$.
(d) To four terms,

$\quad \sin 2x \approx 2x - 1.333x^3 + 0.267x^5 + 0.025x^7$,

where (for this context) the coefficients are rounded to 3 decimal places. For two-decimal accuracy, we need $-0.79 < x < 0.79$.

5.3 (b) The terms in the expansion of $\sin x$ are of size $|x|^{2n-1}/(2n - 1)!$ with $n = 1, 2, \ldots$. We need to choose n so that this is less than $0.000\,05$ when $x = \pm 2$. The first value within the limits is $n = 7$. The polynomial is

$\quad x - \frac{1}{3!}x^3 + \frac{1}{5!}x^5 - \frac{1}{7!}x^7 + \frac{1}{9!}x^9 - \frac{1}{11!}x^{11} + \frac{1}{13!}x^{13}$.

5.4 (b) $\frac{1}{2}\pi - x$.

5.5 (b) $\frac{1}{2} + \frac{1}{4}x + \frac{1}{8}x^2 + \cdots$, $-2 < x < 2$.
(h) $1 - \frac{1}{2!}x + \frac{1}{4!}x^2 - \cdots$, valid for all x.

5.6 (b) $1 + \frac{1}{2}x - \frac{1}{8}x^2$.

5.7 (b) $\tan x \approx (x - \frac{1}{6}x^3 + \frac{1}{120}x^5)(1 - \frac{1}{2}x^2 + \frac{1}{24}x^4)^{-1}$
$\quad\quad\quad \approx x + \frac{1}{3}x^3 + \frac{2}{15}x^5$.

5.8 (d) $\ln(1 + x + x^2) = \ln[x^2(1 + 1/x + 1/x^2)]$
$\quad\quad\quad\quad\quad\quad\quad\quad = 2 \ln x + \ln(1 + 1/x + 1/x^2)$.

Then treat $1/x + 1/x^2$ as the small variable.

5.11 (b) Suppose that the first nonzero derivative is the N th : $f^{(N)}(c) \neq 0$. Consider whether N is even or odd, and whether $f^{(N)}(c)$ is positive or negative.

5.17 (c) $\frac{1}{2}(e^x + e^{-x})$.

5.18 (c) $\frac{4}{3}$.

Chapter 6

6.1 (b) $3 \pm i$.

6.3 (b) $3 - 5i$; (d) $9 + 3i$; (f) $1 + 6i$.

6.5 (d) $-\frac{13}{25} - \frac{9}{25}i$.

6.6 (a) $-4i$; (c) $-\frac{1}{5} + \frac{8}{5}i$.

6.7 (a) $1 - i$; (c) $-2i$.

6.8 (b) $16.233 - 0.167i$; (d) 88.669.

6.9 (b) $|z_2| = 8$; $\text{Arg } z_2 = \frac{3}{4}\pi$. (d) $|z_4| = 3$; $\text{Arg } z_4 = \pi$.

6.10 (b) $y = 2$; (d) the parabola, $y^2 = 4x$; (f) $y = x \ (x \geqslant 0)$.

6.11 (a) $\sqrt{2} \, e^{\frac{3}{4}\pi i}$; (d) $14 \, e^{-\frac{1}{3}\pi i}$; (g) $e^2 \, e^i$; (j) $\sqrt{2} \, e^{\frac{3}{4}\pi i}$.

6.16 (a) $2n\pi i$ $(n = 0, \pm 1, \pm 2, \ldots)$; (c) $(2n + 1)\pi i$.

6.18 (a) $\cos(\ln 2) + i \sin(\ln 2)$.

6.23 (a) $x^2 - y^2 + 2xyi$.
(d) $\cos x \cosh y - i \sin x \sinh y$.

6.28 $2 - i, -1 - i, -1 + i$.

6.29 (b) $e^{2\cos \theta} \cos(2 \sin \theta)$.

Chapter 7

7.2 $x = -2, y = 1.$

7.6 $A = \begin{bmatrix} -10 & -5 \\ 20 & 10 \end{bmatrix}.$

7.7 $A^2 + C^2 = \begin{bmatrix} -5 & 6 & 16 \\ -8 & 11 & 2 \\ -6 & -6 & -7 \end{bmatrix}.$

7.11 $A^{2n-1}.$

7.16 $x = -17, y = -2, z = 8.$

Chapter 8

8.1 (c) 1; (e) -1.

8.4 (b) 1728; (d) -8132.

8.6 $(b-c)(c-a)(a-b)(a+b+c).$

8.14 $x = a, b, c, -a-b-c.$

8.16 $\det(AB) = -36, A^{-1} = \frac{1}{2}\begin{bmatrix} 7 & 1 & -5 \\ -2 & 0 & 2 \\ 1 & 1 & -1 \end{bmatrix}.$

Chapter 9

9.1 (a) $\overline{PQ} = (5, -3), \overline{QP} = (-5, 3).$

9.2 (f) Length $= 5, \theta = 126.9°.$

9.3 (b) $\left(\dfrac{3}{2}, \dfrac{3\sqrt{3}}{2} \right).$

9.4 $\overline{BE} = (0, -4); BE = 4;$ bearing south.

9.5 (c) $\sqrt{6}.$

9.7 (b) $2a = (6, 4, 6), 3b = (3, 3, 6), 2a - 3b = (3, 1, 0).$

9.10 (a) $(3, 3, -6).$ (b) $(X + 2)^2 + (Y - 1)^2 + (Z + 3)^2 = 1.$

9.16 Speed $10\sqrt{2}$; direction towards north east. (Hint: use $v_{wc} = v_w - v_e$ in components, with $v_N = (u, v).$)

9.22 (b) $\frac{3}{4}a + \frac{1}{4}b$; (c) $\frac{3}{2}a - \frac{1}{2}b.$

9.23 (a) $(a + \lambda b)/(1 + \lambda).$ (b) $(a - \lambda b)/(1 - \lambda).$
(c) The point is on the extension of AB in the direction of $\overline{AB}.$

9.26 (a) $y + z = 1.$ (b) $3x - 2y - z = 0.$

9.27 $\sqrt{2}.$

9.28 (a) $\pm(3/\sqrt{34}, 4/\sqrt{34}, 3\sqrt{34}).$ (b) $\pm(\frac{2}{7}, \frac{3}{7}, \frac{6}{7}).$

9.29 (a) $-3\hat{\imath} + 2\hat{\jmath} + 4\hat{k}.$ Length $\sqrt{29}.$

9.36 $r = \dfrac{1}{2}\left(\dfrac{a}{|a|} + \dfrac{b}{|b|} \right).$ (Hint: draw a diagram involving $\hat{a}$ and $\hat{b}$.)

9.37 The minimum separation occurs when $t = 12\frac{1}{2}$ s.

Chapter 10

10.1 (a) 10. (e) zero.

10.3 If your diagram is a parallelogram $ABCD$, the theorem obtained is $AC^2 + BD^2 = 2(AB^2 + AD^2).$ If you use the triangle rule the result gives the median of a triangle in terms of the sides.

10.5 (a) 6. (b) -5.

10.6 (a) $35.3°.$

10.8 $54.7°.$

10.9 $-33x^2 - 13y^2 + 95z^2 + 48xy - 144yz + 96zx = 0.$

10.10 $78.9°, 68.6°, 32.5°.$

10.12 $F = a + 2b + 3c.$

10.16 $\alpha = -\frac{2}{5}, \beta = \frac{7}{5}, \gamma = \frac{2}{15}.$

10.17 $x = 0, y = 0, z = 1.$

10.18 (a) $2\sqrt{2}, 0.$ (b) $\left(X - \dfrac{1}{\sqrt{2}} \right)^2 + \left(Y + \dfrac{1}{\sqrt{2}} \right)^2 = 1.$

10.19 (c) $(l, m, n) = (\frac{1}{3}, -\frac{2}{3}, -\frac{2}{3}).$

10.21 (a) $\pm(\frac{3}{13}, \frac{4}{13}, \frac{12}{13}).$

10.26 (a) $19.1°.$

10.29 Hint: translate the axes to the point q and so work from the simpler form, (10.24a).

10.30 (a) P_1 is $-y + z = 4, P_2$ is $2x - 2y + z = 5.$
(b) $45°.$ (c) $2\sqrt{2}.$
(d) and (e). The line L is given by $r = \lambda(1, 4, -4).$ Show that intersection with P_1 and P_2 occurs when $\lambda = -\frac{1}{2}.$

10.34 Begin by finding any two points on the line of intersection. (The resulting form is not unique.)

Chapter 11

11.1 (a) $(4, 7, 5).$ (d) $-9.$ (h) $(-24, 3, 15).$

11.9 Hint: the determinant is equal to $a \cdot (b \times c)$, where $\overline{QA} = a$, etc.

11.12 $X = -\frac{1}{6}, Y = -\frac{2}{3}, Z = -\frac{5}{6}.$

11.13 (c) $\lambda = \mu = -\frac{1}{2}, v = \frac{1}{2}. L_3$ meets L_1 at $(1, 1, \frac{1}{2})$ and L_2 at $(\frac{1}{2}, -\frac{3}{2}, \frac{1}{2})$.

11.15 (a) $(\frac{24}{3}, \frac{16}{3}, -\frac{4}{3})$. (b) $(\frac{16}{3}, \frac{24}{3}, -\frac{16}{3})$. (c) $\frac{16}{3}$i.
(Note: the *unit* vector in the direction of $\hat{\imath} - 2\hat{\jmath} - 2\hat{k}$ is $\frac{1}{3}(\hat{\imath} - 2\hat{\jmath} - 2\hat{k})$.)

11.16 (a) -6. (b) 6. (c) 0. (d) 0. (e) $-2\sqrt{3}$.

Chapter 12

12.1 (c) $x_1 = 1, x_2 = -1, x_3 = -5$.
(e) $x_1 = 2, x_2 = -1, x_3 = 2$.

12.7 $x_1 = 40, x_2 = 88, x_3 = -68, x_4 = -59$.

12.9 (b) $\frac{1}{25}\begin{bmatrix} 5 & 0 & -5 \\ -6 & 10 & 1 \\ 7 & 5 & 3 \end{bmatrix}$.

(e) $\begin{bmatrix} 1 & 0 & 0 & 0 & 0 \\ -1 & 1 & 0 & 0 & 0 \\ 0 & -1 & 1 & 0 & 0 \\ 0 & 0 & -1 & 1 & 0 \\ 0 & 0 & 0 & -1 & 1 \end{bmatrix}$.

12.12 The shadow on the z plane has vertices at the points $(-1, 0, 0)$, $(-1, -2, 0)$, $(1, 0, 0)$.

12.16 Non-trivial solutions if $k = 1, -1, 4$.

12.18 Non-trivial solutions if $k = -6, -1, 3, 4$.

12.22 $x_1 = 1.398, x_2 = 1.090, x_3 = -0.2844, x_4 = -0.3697$.

Chapter 13

13.1 (b) Eigenvalues 4, 9. Eigenvectors $\begin{bmatrix} -3 \\ 2 \end{bmatrix}, \begin{bmatrix} 1 \\ 1 \end{bmatrix}$.

(e) Eigenvalues $3 - 4\sqrt{2}, 3 + 4\sqrt{2}$. Eigenvectors

$\begin{bmatrix} -1 - 2\sqrt{2} \\ 7 \end{bmatrix}, \begin{bmatrix} -1 + 2\sqrt{2} \\ 7 \end{bmatrix}$.

13.4 (c) Eigenvalues $(-2, 2, 3)$. Eigenvectors

$\begin{bmatrix} 0 \\ -1 \\ 2 \end{bmatrix}, \begin{bmatrix} 1 \\ 0 \\ 0 \end{bmatrix}, \begin{bmatrix} 0 \\ 2 \\ 1 \end{bmatrix}$.

13.7 $a = -2$ and $a = -\frac{7}{2}$.

13.14 The matrix C is given by

$C = \begin{bmatrix} 7 & -1 & -1 \\ -1 & -1 & 1 \\ -1 & 0 & 2 \end{bmatrix}$.

13.16 $\lim_{n \to \infty} A^n = \frac{1}{3}\begin{bmatrix} 1 & 1 & 1 \\ 1 & 1 & 1 \\ 1 & 1 & 1 \end{bmatrix}$.

13.22 Eigenvalues are 0, 4, 4, 12.

13.26 $A^{3n} = I_3, A^{3n+1} = A, A^{3n+2} = A^2$.

Chapter 14

14.1 (a) $\frac{1}{6}x^6 + C; \frac{3}{2}x^5 + C; \frac{1}{4}x^4 + C; \frac{1}{9}x^3 + C$;
$3x^2 + C; 3x + C; C$.
(g) $e^x + C; -e^{-x} + C; \frac{5}{2}e^{2x} + C; -2e^{-\frac{1}{2}x} + C; -\frac{3}{2}e^{-2x} + C$.
(k) $x + \ln x + C$ (($(x + 1)/x = 1 + x^{-1}$);
$2x - 2x^{\frac{1}{2}} + C; \ln|x| - 2x^{-1} - \frac{1}{2}x^{-2} + C$.

14.2 (b) $-\frac{1}{5}(1 - x)^5 + C; -\frac{2}{3}(8 - 3x)^{-\frac{1}{2}} + C; \frac{3}{2}(1 - x)^{\frac{4}{3}} + C$.

14.3 (b) $-\ln|1 - x| + C; -\frac{1}{5}\ln|4 - 5x| + C$.

14.4 (c) $\frac{3}{8}x + \frac{1}{4}\sin 2x + \frac{1}{32}\sin 4x + C$.

14.5 $x^2 e^x - 2x e^x + 2 e^x + C$.

14.6 (a) 2; (h) $-\ln 2$.

14.7 (c) $4 - x^2 \geqslant 0$ if $-1 \leqslant x \leqslant 2$, and
$4 - x^2 \leqslant 0$ if $2 \leqslant x \leqslant 3$. The geometrical area is
$[F(x)]^2_{-1} - [F(x)]^3_2,$
where $F(x) = 4x - \frac{1}{3}x^3$.

14.8 (a) $At + B$; (b) $\frac{1}{6}t^3 + At + B$.

Chapter 15

15.1 (b) $\lim_{\delta x \to 0} \sum_{x=-1}^{x=1} x^5 \, \delta x = \int_{-1}^{1} x^5 \, dx = [\frac{1}{6}x^6]^1_{-1} = 0$.

15.2 (b) $\int (x + 1)^{\frac{1}{2}} \, dx = \frac{2}{3}(x + 1)^{\frac{3}{2}} + C$.

15.3 (c) $\int_0^2 dx = [x]^2_0 = 2$; (i) $\frac{2}{3}(2^{\frac{3}{2}} - 1)$.

15.4 (b) $\int_{-1}^{1}(x^2 - 1)dx = [\frac{1}{3}x^3 - x]^1_{-1} = -\frac{4}{3}$.

15.5 (b) $\int_0^\infty e^{-\frac{1}{2}v} \, dv = -2 [e^{-\frac{1}{2}v}]^\infty_0 = -2(0 - 1) = 2$.

15.6 (c) $2/\pi$; (h) $\int_0^T (1 - e^{-t}) \, dt = T + e^{-T} - 1$.

$\frac{1}{T}(T + e^{-T} - 1) = 1 + T^{-1} e^{-T} - T^{-1} \to 1$

as $T \to \infty$.

15.7 The integrands are (a) even; (b) odd; (c) odd; (d) odd.

15.9 (b) The exact result is $\sqrt{\pi}/2$.

15.10 (e) $\frac{1}{2}(x+1)^{-\frac{1}{2}}\sin(x+1) - \frac{1}{2}x^{-\frac{1}{2}}\sin x$.

15.11 (b) $x \leqslant -1 : \frac{1}{2}$ (constant); $-1 \leqslant x \leqslant 1 : \frac{1}{2}x^2$; $x \geqslant 1 : \frac{1}{2}$ (constant)

15.14 6.

Chapter 16

16.1 $5.\dot{3} \times 10^{-3}$.

16.2 $\displaystyle\int_2^4 (20 - 10t)\,dt = -20,\ x(4) = -17.$

16.3 (b) $\frac{1}{2}\pi$; (g) π.

16.5 (a) $\frac{4}{3}\pi ab^2$.

16.6 $v = \displaystyle\int_1^2 \pi x^2\,dy = \int_1^2 \pi(2y)^2\,dy = 28\pi/3.$

16.7 Put $x = 0$ at A; moment $= \displaystyle\int_0^L mx\,dx = \frac{1}{2}mL^2.$

16.8 1.27.

16.9 0.015 g.

16.12 A sketch shows that $x(x-1) \geqslant -x$ if $0 \leqslant x \leqslant 2$.

Therefore the area is

$$\lim_{\delta x \to 0} \sum_{x=0}^{x=2} [x(x-1) - (-x)]\delta x = \int_0^2 x^2\,dx = \frac{8}{3}.$$

16.13 $\frac{3}{2}$.

16.14 (b) π.

16.15 In a plane perpendicular to the end, y is downward and x is horizontal; the origin is at the top. Area elements are horizontal strips of width δy in the end face. Force $= \frac{1}{2}\rho g L H^2$. Moment $= \frac{1}{3}\rho g L H^3$.

16.16 Distance of centre of mass from vertex is $\frac{3}{4}H$.

16.17 $\frac{1}{12}\sigma a^3 b$ (σ = mass per unit area).

16.18 (a) $\frac{1}{4}\sigma BH^3$; (b) $\frac{1}{48}\sigma HB^3$, where σ is mass per unit area.

16.23 $8a$.

Chapter 17

17.1 (c) $-\frac{1}{3}e^{-3x} + C$; (f) $-\frac{1}{12}(3 - 2x)^6 + C$;
(j) $(2x - 3)^{\frac{3}{2}} + C$; (n) $\frac{1}{2}\ln|2x + 3| + C$;
(o) $\ln|1 - x| + 1/(1 - x) + C$.

17.2 (b) $-\frac{2}{3}\cos\frac{1}{2}(3t - 1) + C$; (e) $-\frac{2}{3}(-t)^{\frac{3}{2}} + C$;

17.3 (d) $\frac{1}{2}\sin(x^2 + 3) + C$. (j) $\frac{1}{2}\ln(1 + x^2) + C$.

17.4 (c) $\frac{1}{6}\sin^3 2x + C$.
(g) Put $\cot 2x = \cos 2x/\sin 2x$, then $u = \sin 2x$, giving $\frac{1}{2}\ln|\sin 2x| + C$. (j) $\frac{1}{3}\cos^3 x - \cos x + C$.

17.5 (b) 205/32; (e) $-\ln 2$; (h) $\frac{1}{2}\ln 2$;
(k) zero; (n) $(2/\omega)\cos\phi$.

17.6 (b) $\frac{1}{2}\pi$; (d) $\frac{1}{4}\pi + \frac{1}{2}$; (f) $\frac{3}{8}\pi$.

17.7 (e) $\tan x - x + C$; (f) $-x^{-1} - \arctan(x^{-1}) + C$;
(k) $\frac{1}{2}[\arcsin x + x(1 - x^2)^{\frac{1}{2}}] + C$.

17.8 (b) $\frac{1}{2}\ln|x/(x + 2)| + C$.
(d) $\ln|x + 1| - \frac{1}{2}\ln|2x + 1| + C$.
(f) $\ln|x| - \frac{1}{2}\ln(x^2 + 1) + C$.
(i) $\frac{1}{2}\ln[(1 + \sin x)/(1 - \sin x)] + C$.

17.9 (b) $\frac{1}{2}\ln(x^2 - 2x + 3) + C$. (e) $\ln(e^x + e^{-x}) + C$.
(f) $2\ln(x^{\frac{1}{2}} + 1) + C$.

17.10 (b) $\frac{1}{3}xe^{3x} - \frac{1}{9}e^{3x} + C$. (f) $2x\sin\frac{1}{2}x + 4\cos\frac{1}{2}x + C$.
(i) $\frac{1}{2}x^2\ln x - \frac{1}{4}x^2 + C$. (j) $x^{n+1}[\ln x - 1/(n + 1)]/(n + 1)$.
(k) Hint: bring together the two terms $\int(\ln x/x)\,dx$.

17.11 (a) Hint: there are two stages required; see Example 15.20.

17.12 Hint: the same integral occurs on both sides but with a different factor.

17.13 (b) zero; (d) $\frac{1}{2}$; (h) π.

17.15 $F(0) = \frac{1}{2}\pi$, $F(1) = 1$, $F(4) = \frac{3}{16}\pi$, $F(5) = \frac{8}{15}$.

17.16 (a) $2\ln^3 2 - 6\ln^2 2 + 12\ln 2 - 6$.
(b) $F(0) = 2$, $F(1) = \pi$, $F(4) = \pi^4 + 12\pi^2 + 48$,
$F(5) = \pi^5 + 20\pi^3 + 120\pi$.

17.23 (c) $(a/b)\arctan[(a\tan x)/b] + C$; (d) $\ln(\tan\frac{1}{2}x) + C$;
(g) $\ln|\sec x + \tan x| + C$; (j) $\ln[(1 + \sqrt{5})/2]$; (k) $8(6\sqrt{3} + 1)/15$.

17.25 Coordinates of centroid: $(\frac{3}{5}h, 0)$.

Chapter 18

18.2 (b) $x = A\,e^{\frac{1}{2}t}$; (e) $x = A\,e^{-\frac{1}{3}t}$; (i) $x = A\,e^t$.

18.3 (b) $x = e^{\frac{1}{2}(t-1)}$; (d) $x = 10\,e^{-(t+1)}$.

18.4 $I(t) = I_0\,e^{-Rt/L}$. I reduces to a fraction $1/n$ of itself in any interval of length $L\ln n/R$.

18.5 (a) $A(t) = C\,e^{-kt}$ (C arbitrary).

(b) The half-life $T = \dfrac{1}{k}\ln 2$ years. The information

implies that $e^{-20k} = 1 - 0.175 = 0.825$, so $k = 0.0096$. Therefore $T = 72$ years.

18.6 If $N(t)$ is the number, then $\delta N \approx 20(\frac{1}{2}N)\,\delta t$ so the equation is $dN/dt = 10N$. In the second experiment there is an average death-rate of 1 per rabbit per year, so $dN/dt = 9N$.

18.7 (b) $A\,e^t + B\,e^{-2t}$. (e) $A\,e^{t/2\sqrt{3}} + B\,e^{-t/2\sqrt{3}}$.
(l) $A\,e^{-3t} + Bt\,e^{-3t}$.
(n) $A + Bt$ (this is an exception to (18.10)).

18.9 (b) $\frac{2}{3}(e^t - e^{-2t})$.
(d) The general solution is $A\,e^{-x} + Bx\,e^{-x}$, $y = e(x-1)e^{-x}$.

18.10 (b) $A\cos 3t + B\sin 3t$. (d) $A\cos \omega_0 t + B\sin \omega_0 t$.
(f) $e^t(A\cos t + B\sin t)$. (i) $e^{-\frac{3}{2}t}(A\cos \frac{1}{3}\sqrt{2}t + B\sin \frac{1}{3}\sqrt{2}t)$.

18.11 (c) $a\cos \omega_0 t + (b/\omega_0)\sin \omega_0 t$.

18.12 $\theta = \alpha \cos(g/l)^{\frac{1}{2}}t$.

18.13 The initial angular velocity $d\theta/dt$ is v/l;

$$\theta = \frac{v}{(lg)^{\frac{1}{2}}}\sin\left(\frac{g}{l}\right)^{\frac{1}{2}}t.$$

18.14 $\theta = 0.0719\,e^{-0.033t}\sin 0.696t$.

18.18 $A = (Mg/P)\,e^{-\rho g(y-H)P}$.

Chapter 19

19.1 (b) $-\frac{1}{3}t^3 - \frac{1}{3}t^2 - \frac{2}{9}t - \frac{11}{27}$.
(d) $\frac{3}{5}e^{2t}$. (i) $-\frac{2}{15}\sin 3t$.
(k) $-\frac{3}{25}\cos 2t + \frac{4}{25}\sin 2t$.

19.2 (d) $\frac{1}{5}(-6\cos t - 3\sin t)$.
(f) $-\frac{2}{137}(4\cos 2t + 11\sin 2t)$.
(h) $\frac{3}{65}e^t(4\cos 2t + 7\sin 2t)$.

19.3 (b) $-\frac{3}{4}t\cos 2t$.

19.4 (b) $\frac{1}{2}t^2\,e^t$; (e) $\frac{1}{2}te^t\sin t$.

19.5 (c) $A\,e^{\frac{1}{2}t} + B\,e^{-\frac{1}{2}t} - 1 - \frac{3}{17}\cos 2t$.
(i) $A\cos x + B\sin x + x^2 - 1 + \frac{1}{5}e^{3x}$.

19.6 (c) $-\frac{1}{2} + A\,e^{t^2}$. (g) $(\sin x - \cos x - x\cos x + A)/(x+1)$.
(l) $(x+1)\ln|x+1| + 1 + A(x+1)$.

19.9 $11\frac{1}{2}$ minutes.

Chapter 20

20.1 (b) $3\cos(\omega t + \pi)$. (e) $3\cos(2t + \frac{1}{2}\pi)$.
(h) $5\cos(2t + \phi)$, $\phi = -\arctan\frac{4}{3}$.

20.2 (c) x leads y by π.

20.3 (b) (i) 0.318 cycles/s. (ii) 0.316 cycles/s.
(iii) About 3 cycles.

20.4 (b) $C = \sqrt{(4 - \sqrt{6})}$, $\phi = \arctan(-1/\sqrt{6}-1))$,
$(-\frac{1}{2}\pi < \phi < 0)$.

20.7 The solutions are of exponential type.

20.8 $x = e^{-4t} - 4\,e^{-6t}$.

20.9 $A\,e^{-kt} + Bt\,e^{-kt}$.

20.10 (a) Period = 1.0508.
(b) Amplitude = $10/[(36 - \omega^2)^2 + \omega^2]^{\frac{1}{2}}$,
phase = $-\arctan[\omega/(36 - \omega^2)]$.
(c) Resonance: $\omega = 5.958$.

Chapter 21

21.1 (b) $-2e^{\frac{1}{2}\pi i}$ ($2e^{-\frac{1}{2}\pi i}$ in standard form).

21.2 (d) $2e^{-\frac{3}{4}\pi i}$; $2\cos(\omega t - \frac{3}{4}\pi)$.
(i) $e^{1.97i}$; $\cos(\omega t + 1.97)$.

21.3 (b) $1 - e^{-\frac{1}{2}\pi i} = 1 + i = \sqrt{2}\,e^{\frac{1}{4}\pi i}$.

21.4 (b) $1 - 3e^{-\frac{1}{2}\pi i} + e^{\frac{1}{2}\pi i} = 1 + 4i = \sqrt{17}\,e^{i\phi}$, where
$\phi = \arctan 4 = 1.33$.

21.6 (b) $R + \omega Li$. (d) $R/(1 + \omega RCi)$.
(i) $R + i\omega L/(1 - \omega^2 LC)$. (k) $i\omega RL/[R(1 - \omega^2 LC) + i\omega L]$.

21.7 $V = ZI$ and $V = 2$.
(d) $I = 2(1 + i\omega RC)/R$; $|I| = 2(1 + \omega^2 R^2 C^2)^{\frac{1}{2}}/R$;
$\arg I = \arctan(\omega RC)$.

21.8 (b) $V_1/V_0 = \frac{2}{13}(3 - 2i)$; $V_0/I_1 = \frac{1}{2}(5 - i)$.

Chapter 22

22.4 (b) $2x^2 - y^2 = C$. (g) $y = x/(1 + Cx)$.
(k) $x = \pm 2^{-\frac{1}{2}}(C - t^3)^{-\frac{1}{2}}$ for $t^3 < C$.
(n) $\arctan y + \arctan x = \frac{1}{4}\pi$. Take the tangent of this expression and use the formula for $\tan(A + B)$; we find that $y = (x + 1)/(x - 1)$.

22.6 (b) $y = \frac{1}{16}(x^2 + C)^2$ for $x^2 + C > 0$. $y = 0$ is also a solution. (d) Those parts of the curves $y = \sin(\ln|x| + C)$ for which x and dy/dx have the same sign. Also $y = \pm 1$ are solutions.

22.7 (b) $y^3 - 3xy = C$. (d) $xy - y^2 - x^2 = C$.
(f) $y^3 + y - x^3 = C$. (h) $y + \cos y + \sin x = C$.
(j) $e^{x+y} + y - x = C$.

22.8 (b) $xy + y/x = C$; (d) $x/y + y - x = C$;
(e) $y/x - x/y - 1/x = C$; (f) $x^2/(2y^2) + 1/(xy) = C$.

22.12 (b) $x(1 + 2y^2/x^2)^{\frac{1}{4}} = C$.
(d) $x^2 - 4y^2 = Cy^3$.

Chapter 23

23.2 (b) $y = Cx$ (this is not covered by (23.22)).
(d) $xy = C$ (a saddle).

23.4 (b) Saddle (i.e. unstable). $m = \frac{1}{2}(-3 \pm \sqrt{13})$.
(f) Stable spiral; directions are clockwise round origin.

23.5 (b) Equilibrium points at $(1, 1)$. $(1, 1)$ is a stable spiral, anticlockwise about $(1, 1)$. (d) Equilibrium points at $(-1, 0)$, $(0, 0)$, $(0, 1)$; $(0, 0)$ is a centre and $(-1, 0)$, $(1, 0)$ are saddle points.

Chapter 24

24.1 (b) $4/(s+1)$; (d) $6/s^3 - 1/s$; (g) $(3-s)/(s^2+1)$.

24.2 (b) $1/s - 2/(s+2)$; (e) $(3s-4)/(s^2+4)$;
(g) $\frac{1}{2}[1/s - s/(s^2+4)]$.

24.3 (b) $1/(s+2)^2$; (d) $(s-2)/(s^2-4s+5)$;
(i) $(s^2-9)/(s^2+9)^2$; (l) $24/(s+1)^5$.

24.5 (b) 1; (d) $\frac{1}{8}t^4$; (g) $\frac{3}{2}e^{\frac{1}{2}t}$; (k) $\frac{1}{2}e^t + \frac{1}{2}e^{-t}$;
(o) $2\cos 2t - \frac{1}{2}\sin 2t$; (s) $\frac{1}{2}e^t t^2$; (u) $\frac{1}{3}(\cos t - \cos 2t)$.

24.6 (e) $(2s^2+3s-2)X(s) - 10s - 11$.

24.7 (b) $2e^t + e^{-2t}$; (e) $3e^{-t}\cos 2t$;
(f) $y = \frac{1}{4}e^x + \frac{1}{4}e^{-x} + \frac{1}{2}\cos x$.

24.8 (b) $3 - 3\cos t + \sin t$.
(e) $-\frac{1}{8}e^{-t} + \frac{9}{8}e^t - \frac{1}{4}te^t + \frac{1}{4}t^2 e^t$.
(i) $-\frac{7}{6}e^t + \frac{1}{2}e^{-t} + \frac{3}{4}e^{2t} - \frac{1}{12}e^{-2t}$.

24.9 (b) $x = \frac{3}{8} + \frac{5}{8}e^{4t} + \frac{1}{2}te^{4t}$; $y = -\frac{3}{16} + \frac{3}{16}e^{4t} + \frac{1}{4}te^{4t}$.

24.10 (b) $e^t(\frac{1}{2}A + \frac{1}{2}B + \frac{3}{2}) + e^{-t}(\frac{1}{2}A - \frac{1}{2}B + \frac{3}{2}) - 3$, where A and B are arbitrary. This is the same as $Ce^t + De^{-t} - 3$, where C and D are arbitrary.

24.13 $e^{-2}e^{-2s}[(s+1)^2 - 1]/[(s+1)^2+1]^2$
$= e^{-2}e^{-2s}s(s+2)/(s^2+2s+2)^2$.

24.14 (b) $H(t)\sin t - H(t-1)\cos(t-1)$.

24.15 (b) $(\frac{1}{8}e^{2t} + \frac{1}{8}e^{-2t} - \frac{1}{4})H(t)$,
$-(\frac{1}{8}e^{2(t-1)} + \frac{1}{8}e^{-2(t-1)} - \frac{1}{4})H(t-1)$.
(d) $\frac{1}{2}H(t)t\sin t + \frac{1}{2}H(t-\pi)(t-\pi)\sin(t-\pi)$.

Chapter 25

25.3 Hint for working: $s^2 + 2ks + \omega^2$ has real factors when $k^2 > \omega^2$; so put $s^2 + 2ks + \omega^2 = (s-\alpha)(s-\beta)$, where $\alpha, \beta = -k \pm (k^2 - \omega^2)^{\frac{1}{2}}$. Then $x(t)$ is given by
$$(\alpha-\beta)^{-1}[(\alpha+\kappa)e^{\alpha t} - (\beta+\kappa)e^{\beta t}]H(t)$$
$$+ I(\alpha-\beta)^{-1}[e^{\alpha(t-t_0)} - e^{\beta(t-t_0)}]H(t-t_0),$$
where $\kappa = 1 + 2k$.

25.4 By proceeding as suggested, we obtain
$$u(x) = Ax + \frac{1}{6}Bx^3 + (Mg/6K)(x - \frac{1}{2}l)^3 H(x - \frac{1}{2}l).$$
The conditions at $x = l$ give $A = Mgl^2/16K$, $B = -Mg/2K$. This problem could be solved by integrating the equation

four times, and linking the solutions over $[0, \frac{1}{2}l]$ and $[\frac{1}{2}l, l]$ by the condition that $u(x)$, $u'(x)$, $u''(x)$ are continuous at $x = \frac{1}{2}l$, but this is automatically secured in the Laplace-transform method.

25.5 (b) $2s/(6s^2 + s + 1)$.

25.6 (b) $V_2/V_1 = 3/(20s^2 + 12s + 5)$; $V_2/I = 3/(4s^2 + 6s + 1)$.

25.7 (b) t; (f) $1 - \cos t$; (h) $\frac{1}{2}(-t\cos t + \sin t)$;
(j) $n!m!t^{n+m+1}/(n+m+1)!$.

25.8 (b) $\dfrac{1}{2\omega}\displaystyle\int_0^t f(\tau)(e^{\omega(t-\tau)} - e^{-\omega(t-\tau)})\,d\tau$.

25.9 (b) $\cosh t$.

25.19 (a) $x(t) = \delta(t) + 2\delta(t-T) + \delta(t-2T)$,
$X(s) = 1 + 2e^{-sT} + e^{-2sT}$.

25.20 $Y(s) = \displaystyle\sum_{n=0}^{\infty}[3^{-n}\,\delta(t - (n+1)t) + 2\cdot 3^{-n}\,\delta(t - (n+2)T)]$.

25.21 (a) $z^{-1} + 2z^{-2} - z^{-3}$. (b) $1 - z^{-1} + z^{-2} - \cdots = z/(z+1)$.
(c) $2z/(2z-1)$. (d) $z/(z^2-1)$.

25.22 (a) $Tz/(z-1)^2$.

25.23 (a) $(z-1)/(z+1) \cdot g(t) = \{1, -2, 2, -2, \dots\}$.

25.27 (a) Unstable. Poles at $z = \pm 2$, giving growth $\frac{3}{4}2^n$ and $\frac{1}{4}(-1)^n 2^n$. (c) Stable. Poles at $z = \pm\frac{1}{2}i$, giving decay
$$\frac{1}{4}\frac{1}{2^n}\cos\frac{1}{2}\pi n.$$

Chapter 26

26.1 (b) $a_n = 0$, $b_n = -2(-1)^n/n$.
(e) $a_n = 0$, $b_n = \dfrac{2}{\pi n}[1 + (-1)^n - 2\cos(\frac{1}{2}n\pi)]$.

26.2 (b) $b_n = 0$, $a_0 = \dfrac{2\pi^2}{3}$, $a_n = \dfrac{4}{n^2}(-1)^n (n = 1, 2, \dots)$.
(c) $b_n = 0$, $a_n = -\dfrac{4(-1)^n}{\pi(4n^2-1)}$.

26.3 (a) $a_0 = \frac{1}{2}\pi$, $a_{2n} = 0$, $a_{2n-1} = -\dfrac{2}{\pi n^2}$,
$b_n = -\dfrac{(-1)^n}{n}$ $(n = 1, 2, \dots)$.

26.5 Series sum is $\frac{1}{4}\pi$.

26.8 $F = 2$.

26.10 $a_0 = 0$, $a_n = 0$, $b_n = \dfrac{4\beta}{\pi n^3}[1 - (-1)^n](n = 1, 2, \dots)$.

26.16 (a) $\displaystyle\sum_{n=1}^{\infty}\dfrac{4}{(2n-1)\pi}\sin(2n-1)\pi t$.

26.18 $\dfrac{2}{\pi} - \displaystyle\sum_{n=1}^{\infty} \dfrac{4}{\pi(4n^2 - 1)}\cos 2n\omega t.$

26.23 (b) $R(t) = \dfrac{1}{2} + \dfrac{4}{\pi}\left(\cos t + \dfrac{1}{3^2}\cos 3t + \dfrac{1}{5^2}\cos 5t + \cdots\right).$

26.26 (b) $\displaystyle\sum_{n=0}^{\infty} \dfrac{1}{n^2} = \dfrac{\pi^2}{6}.$

26.30 $\dfrac{1}{2} + \dfrac{i}{2\pi}\displaystyle\sum_{n=-\infty}^{\infty} \dfrac{1}{n}e^{i2\pi nt/T}.$

Chapter 27

27.1 $X_s(f) = 4\pi f/(1 + 4\pi^2 f^2);\ X_c(f) = 2/(1 + 4\pi^2 f^2).$

27.9 $x(t) = 2\displaystyle\int_0^{\infty} X(f)\cos 2\pi f t\, df$ where

$X(f) = 2\displaystyle\int_0^{\infty} x(t)\cos 2\pi f t\, dt.$

27.11 (c) $2c$ sinc $cf \cos 2\pi bcf.$

27.12 (a) $\frac{1}{2}$ sinc$^2\frac{1}{2}f.$ (b) $\frac{1}{2}$ sinc$^2\frac{1}{2}f\, e^{-i6\pi f}.$

27.17 $\{1/[\alpha + i(2\pi f + \beta)] + 1/[\alpha + i(2\pi f - \beta)]\}.$

27.19 (b) $1/(1 + i2\pi f)^2.$

27.21 (b) The Fourier transform is sinc$^2(f)\, e^{-i2\pi(a+b)f} \leftrightarrow \Lambda[t - (a + b)].$

27.28 $h(t) = -t^2 + \frac{3}{4}$ for $|t| \leqslant \frac{1}{2},\ \frac{1}{2}(t - \frac{3}{2})^2$ for $\frac{1}{2} \leqslant t \leqslant \frac{3}{2},$ zero elsewhere.

Chapter 28

28.3 (c) $4x - 2y - 1;\ -6y - 2x - 1.$ (f) $y - 2;\ x - 1.$
(i) $2y/(x + y)^2;\ -2x/(x + y)^2.$ (k) $x(x^2 + y^2)^{-\frac{1}{2}};\ y(x^2 + y^2)^{-\frac{1}{2}}.$

28.4 (c) $\dfrac{\partial V}{\partial x} = g'(r)\cos\theta;\ \dfrac{\partial V}{\partial y} = g'(r)\sin\theta.$

28.8 $\partial^2 f/\partial x^2,\ \partial^2 f/\partial y^2,$ and $\partial^2 f/\partial x\,\partial y = \partial^2 f/\partial y\,\partial x$ are given in order: (b) 2, 4, 3. (d) $2y/x^3, 0, -1/x^2.$
(h) $108(3x - 4y)^2,\ 192(3x - 4y)^2,\ -144(3x - 4y)^2.$
(k) $-r^{-3} + 3x^2 r^{-5},\ -r^{-3} + 3y^2 r^{-5},\ 3xyr^{-5},$ where $r = (x^2 + y^2)^{\frac{1}{2}}.$

28.10 (b) $2x + 2y - z = 4;$ one normal is $(2, 2, -1).$
(d) $3x + 4y + 8z = 29;$ one normal is $(-\frac{3}{2}, -2, -1).$

28.11 $78.9°$ or $101.1°.$

28.12 (b) $(1, -1),$ min; (d) $(n\pi, m\pi);$ min if n and m odd, max if n and m even, otherwise saddle;
(h) $(0, 0)$ saddle; $(1, 1)$ minimum; (k) $(0, 0),$ saddle.

28.14 (a) $a = b = c = 7;$ (b) $a = b = c = 4.$

28.15 The maximum is 9, attained at $(2, \pm1).$

28.16 Minimum distance $= \sqrt{2}.$

28.18 (b) Depth $= 2^{-\frac{2}{3}}V^{\frac{1}{3}};$ square base, side $2^{\frac{1}{3}}V^{\frac{1}{3}}.$

28.23 Lowest point is $z = \frac{3}{4}a$ at $(0, a)$ and $(a, a).$

Chapter 29

29.1 (b) $\delta z = 0.0718\ldots$ (exactly). The incremental approximation gives $\delta z \approx 0.0784.$ Error $= 9.1\%.$

29.3 (b) $((\delta y)^2 - 4\,\delta x\,\delta y - 7\,\delta y)/4(1 + \delta y).$

29.6 $-5.7\%.$

29.7 1.67% reduction, approximately.

29.9 (b) $-2\sqrt{2};$ (d) zero (it is the same in all directions).

29.10 (b) $-\frac{3}{4};$ (e) $-\frac{1}{2};$ (j) 1.

29.12 (b) $x_1 x/a^2 + y_1 y/b^2 = x_1^2/a^2 + y_1^2/b^2.$
(f) $ax_1 x + h(y_1 x + x_1 y) + byy_1 + g(x + x_1) + f(y + y_1) + c = 0.$

29.16 (b) $x^{-1} - y^{-1} = $ constant; (d) $e^x + e^y = $ constant.

29.17 (b) $y^2 - x^2 = b^2 - a^2.$

29.19 (b) $49.8°$ or $130.2°.$
(d) Hint: compare Problem 26.12f.

29.21 (b) $(0, \frac{1}{2});$ (d) $(-\frac{1}{4}, 1).$

29.22 (b) $(2, 1)/\sqrt{5}.$

29.23 (b) $\phi = 0.$

Chapter 30

30.2 (b) $-4\sin t\cos t;$ (d) $2\sin(t^2) + 4t^2\cos(t^2).$

30.3 It is easiest to start by expressing the distance D in terms of polar coordinates $(r, \theta),\ (R, \phi)$ by using the cosine rule (Appendix B(f)). Then

$$\frac{dD}{dt} = -\frac{(Rv - rV)\sin(\phi - \theta)}{[R^2 + r^2 - 2Rr\cos(\phi - \theta)]^{\frac{1}{2}}},$$

where $\theta = vt/r,\ \phi = Vt/R.$

30.4 (b) $x = y = 3.$ (e) The coordinates of the nearest point on the given line are $(\frac{3}{5}, \frac{1}{5}).$ Distance $= 2/\sqrt{5}.$

30.5 (b) $(0, 0),\ (2, 0).$ (A suitable parametrization is $x = 1 + \cos t,\ y = \sin t.$)
(d) $(\pm6/\sqrt{5}, \pm4/\sqrt{5}).$ (A suitable parametrization would be $x = 2/\cos t,\ y = 2\tan t.$)

30.8 (b) $\ddot{x} = -2\dot{r}\dot{\theta}\sin\theta + \ddot{r}\cos\theta - \dot{\theta}^2 r\cos\theta - \ddot{\theta}r\sin\theta,$
$\ddot{y} = 2\dot{r}\dot{\theta}\cos\theta + \ddot{r}\sin\theta - \dot{\theta}^2 r\sin\theta + \ddot{\theta}r\cos\theta.$

30.9 (c) $\partial f/\partial u = -2v^2/u^3$, $\partial f/\partial v = 2v/u$.

30.10 (b) $\partial^2 f/\partial u^2 = 12u^2 - 2v^2$, $\partial^2 f/\partial u\,\partial v = -4uv$, $\partial^2 f/\partial v^2 = -2u^2 + 12v^2$.

30.11 It is easiest to put $x^2 - y^2$ in terms of uv. Finally,
$\partial^2 f/\partial u^2 = 16v^2 g''(4uv)$, $\partial^2 f/\partial v^2 = 16u^2 g''(4uv)$,
$\partial^2 f/\partial u\partial v = 4g'(4uv) + 16uvg''(4uv)$.

Chapter 31

31.1 (b) $\delta f \approx -x(x^2 + y^2)^{-\frac{3}{2}} e^{-t}\, \delta x$
$\qquad\qquad - y(x^2 + y^2)^{-\frac{3}{2}} e^{-t}\, \delta y - (x^2 + y^2)^{-\frac{1}{2}} e^{-t}\, \delta t$.
(e) $\delta f \approx 2(x_1 - x_2)\, \delta x_1 - 2(x_1 - x_2)\, \delta x_2$
$\qquad + 2(y_1 - y_2)\, \delta y_1 - 2(y_1 - y_2)\, \delta y_2$.

31.2 -0.07.

31.3 It is easiest to write $\delta(1/R) \approx -\delta R/R^2$. We obtain $\delta R \approx 0.198\, \delta R_1 + 0.018\, \delta R_2 + 0.334\, \delta R_3$. The required δR_3 is -0.108.

31.4 Put $ax^3 - bx - c = f(a, b, c, x)$ and use (31.1).

31.5 (b) Hint: use logarithmic differentiation: $\delta w \approx -3\, \delta x + 3\, \delta z$ and $\delta w \approx 2(\pm 0.6)$. What is the significance of the absence of a term in δy?

31.6 (b) Maximum $|\delta w| \approx 0.14$; max percentage error 10%.

31.8 (b) $2\delta x + 4\delta y - 6\delta z = 0$. For $\partial z/\partial x$, put $\delta y = 0$: $\partial z/\partial x = \frac{1}{3}$. Similarly $\partial z/\partial y = \frac{2}{3}$.

31.11 (b) $(2, -3, 5)$; (d) $(3x^2, 0, 9z^2)$;
(f) $(-x/r^3, -y/r^3, -z/r^3)$, where $r = (x^2 + y^2 + z^2)^{\frac{1}{2}}$.

31.12 (b) $(0, 2y, 2z)$. Unit vector $= (0, y/(y^2 + z^2)^{\frac{1}{2}}, z/(y^2 + z^2)^{\frac{1}{2}})$.

31.13 (b) $\cos \phi = 11/(3\sqrt{14})$, so $\phi = 11.5°$ (i.e. the angle of intersection of smallest magnitude).

31.15 (b) $\hat{s} \cdot (2x, -2y, -3)$.

31.16 (b) (Check that $\hat{s}$ as given is a unit vector.)
$\dfrac{df}{ds} = 7.51$.

31.17 (b) $-2\hat{\imath} - 2\hat{k}$.

31.18 (b) $(\pm 1, 0, 0)$ and $(\pm 1, \frac{1}{4}, \frac{3}{16})$.
(d) $x = y = z$ is a line of stationary points (excluding the origin). (e) $x = y = z = \pm 1$, $\sqrt{3}$, $\lambda = \pm\frac{1}{2}\sqrt{3}$.

31.19 Stationary at $(1, 0, 0)$, $(-1, 0, 1)$, $(-1, 0, -1)$.

31.21 (b) $(3, 3, 3)$; (e) $(a/\sqrt{3}, b/\sqrt{3}, c/\sqrt{3})$; (g) $(\frac{1}{3}, \frac{7}{3}, 1)$.

31.26 (b) $4xy = 1$; (d) $x^2 + y^2 = 1$.

Chapter 32

32.1 (b) $e - 2$; (d) $(d - c)(b - a)$; (i) $-\frac{1}{3}$; (m) $\frac{1}{2}\ln 2$.

32.2 (b) Zero. Refer to the signed-volume analogy (30.2b). (f) $\ln(27/16)$.

32.4 $\frac{4}{3}$.

32.5 (b) $\displaystyle\int_0^1 \int_0^x f(x, y)\, dy\, dx$. (d) $\displaystyle\int_{-1}^1 \int_0^{\sqrt{(1-x^2)}} f(x, y)\, dy\, dx$.

(g) $\displaystyle\int_{-1}^0 \int_0^{1+y} f(x, y)\, dx\, dy + \int_0^1 \int_0^{1-y} f(x, y)\, dx\, dy$.

32.6 (b) $\frac{2}{3}$; (d) $\frac{3}{2}$; (h) $\frac{1}{12}$.

32.7 (b) 1.

32.8 (b) $\frac{1}{4}\pi$; (d) $\frac{15}{8}$; (f) $\frac{\pi}{16}$.

32.9 $2a^2(4 + \pi)/9$.

32.10 (a) $2(u^2 + v^2)$; (b) 2; (c) 1/5; (d) $-2\cosh v$.

32.11 The value of the integral is $2(257 - 129\sqrt{2})/5$.

32.12 Area $= 41/12$.

32.13 $1/e$.

32.14 Volume $= 20$.

32.15 $1/4$.

32.18 (a) $\sqrt{\pi}(|b| - |a|)$.

Chapter 33

33.1 (b) 1.

33.2 (b) $\frac{4}{3}$; (d) $\frac{2}{3}$; (f) 0.

33.3 (b) π; (d) $\frac{8}{3}\pi^3$.

33.5 (b) 2; (d) 0; (g) 0.

33.6 (b) 1; (d) -3; (f) $\frac{15}{2}$.

33.7 (b) 0.

33.8 (b) $-\frac{3}{2}$.

33.9 Zero.

33.11 Put $x = x(u, v)$ and $y = y(u, v)$, where u and v are the new coordinates. Then put $dx = \dfrac{\partial x}{\partial u}\, du + \dfrac{\partial x}{\partial v}\, dv$ etc.

33.14 $\frac{3}{8}\pi$.

33.16 (b) Non-conservative.

Chapter 34

34.1 $\pi[a^3 - (a - h)^3]/a$.

34.2 (a) 1/84; (b) 1/24; (c) 13/384.

34.5 $2\sqrt{6} + 2 \sinh^{-1}(\sqrt{2})$.

34.6 Scalar potential is $e^{xyz} + \cos xy + zx + C$.

34.7 Scale factors are $h_1 = h_2 = \sqrt{(u^2 + v^2)}$, $h_3 = uv$.

34.11 (b) div $F = 2z$.

34.12 (b) curl $F = 2x\hat{\imath} + (x - 2y)\hat{\jmath} + \hat{k}$.

34.14 (a) $3r^2$; (b) 0; (c) $3rr$; (d) 0; (e) 0; (f) $12r$.

34.18 8/3.

Chapter 35

35.1 (c) $-2, -1, 0, 1, 2, 3, 4$; (f) $1, 4, 9$.

35.3 (c) $A \cup B = \{-4, -3, -2, -1, 1, 2, 3, 4\}$.

35.4 (b) $A \cap B = \{x \mid x \in \mathbb{N}^+ \text{ and } -5 \leqslant x \leqslant 2\}$. (d) $A \cap B = \{1\}$.

35.5 (b) $B \backslash (A \cup C)$; (d) $(B \cap D) \backslash A$.

35.6 (b) $S_1 \backslash (S \cup S_2 \cup \cdots \cup S_r)$.

35.7 (b) $[(A \backslash A_1) \backslash B_1] \cup B_2$.

35.10 $A^2 = \{(1, 1), (1, 2), (2, 1), (2, 2)\}$.

35.12 (b) 66.

Chapter 36

36.6 See table below.

36.10 (b) $(a \cdot b) \cdot (b \oplus c)$; (d) $\overline{(\bar{a} \cdot b \oplus a \cdot \bar{b})(c \cdot d)}$.

36.15 (a) If a_1 represents the state of switch S_1, etc., then the switching function is

$$(a_1 \oplus a_2) \oplus [(a_3 \oplus a_4) \cdot a_5].$$

36.16 See the table below.

Solution for 36.6

a	b	c	$(a \oplus \bar{b}) \cdot (a \oplus \bar{c})$
0	0	0	1
0	0	1	0
0	1	0	0
0	1	1	0
1	0	0	1
1	0	1	1
1	1	0	1
1	1	1	1

Solution for 36.16

a_1	a_2	a_3	f
0	0	0	0
0	0	1	1
0	1	0	1
0	1	1	0
1	0	0	1
1	0	1	0
1	1	0	0
1	1	1	1

Chapter 37

37.3 Twenty are planar.

37.4 Five are connected.

37.8 Six not including reversed order.

37.13 There are three different paths between a and e.

37.14 Five vertices.

37.15 $i_1 = -\frac{4}{21}i_0, i_2 = -\frac{1}{21}i_0, i_3 = -\frac{10}{21}i_0, i_4 = -\frac{11}{21}i_0,$ $i_5 = \frac{1}{21}i_0, i_6 = \frac{4}{21}i_0, i_7 = -\frac{3}{7}i_0.$

37.17 The transfer function is

$$Q = \frac{PG_1G_2G_3}{1 - G_2H_1 + G_1G_2G_3H_2}.$$

37.19 (a) The transfer function is

$$Q = \frac{PG_1G_2G_3}{1 + G_2H_1 - G_1G_2G_3H_2}.$$

37.20 (a) $\dfrac{G_1G_3}{(1 - G_1G_2H_2)(1 + G_3H_1)}$.

(d) $\dfrac{G_1G_2G_3G_4}{1 + G_2G_3H_2} + \dfrac{G_5G_6G_7}{1 - H_1}$.

37.24 *SAFT*, length 12.

37.25 One tie.

37.27 (b) Framework is overbraced.

37.28 Two ties.

37.29 Waiting times are $14T/3$ and $9T/2$.

Chapter 38

38.1 £1790.85, 4.87%.

38.2 (b) 16.9 years.

38.3 (b) 0, $\frac{1}{6}(-1 \pm \sqrt{13})/6$.

38.6 $f(n) = (\ln n)/\ln 2$.

38.8 (b) $u_n = A3^n + B(-3)^n$. (c) $u_n = 3^n(A \cos \frac{1}{2}n\pi + B \sin \frac{1}{2}n\pi)$.

38.11 (a)(ii) $u_n = -\frac{3}{16}n + \frac{1}{8}n^2$. (b)(ii) $u_n = \frac{1}{5}n + \frac{11}{25}$. (c)(iii) $u_n = \frac{1}{2}n^2$. (d)(iii) $u_n = \frac{1}{18}n^2 3^n$.

38.13 $D_n(1) = n + 1$.

38.16 $u_n = \frac{5}{4} - \frac{9}{4}(-\frac{1}{3})^n$.

38.17 $d_k = k(N - k)$.

38.19 $s_n = \frac{1}{4}n^2(1 + n)^2$.

38.22 $0 < \alpha < 1$.

38.23 Oscillates between 0.4953 and 0.8124.

38.24 The periodic values of the 2-cycle are 0.4 and 0.8.

Chapter 39

39.1 (a) 31; (b) 216; (c) 12; (d) 63.

39.2 The probability that the score is 7 or less is 7/12.

39.5 $n(A \cup B) = 10$.

39.6 (b) Ace of clubs or ace of spades drawn;
(d) any ace or any heart or any black card drawn;
(f) any heart except the ace of hearts; (h) ace of hearts
or any black card.

39.7 (b) 1/221; (b) 0.004 166… .

39.9 (b) 5040; (d) 7.

39.11 (a) 27 216; (c) 3360.

39.12 (b) 156 849.

39.15 270 725; 0.010 56… .

39.17 (a) 9/209; (c) 16/665; (d) 683/1463.

39.18 (b) 0.872; (c) 0.602.

39.19 (b) 0.453; (c) 0.547.

39.20 With the same probability of failure 0.98,
probability that circuit fails is 0.963.

39.21 (b) 1/495; (c) 4/99.

39.22 Overall probability is approximately 1/53.7.

39.23 Mean number of plays to the end of the game
is $2^{n-1}/n$.

Chapter 40

40.1 $P(X \geq 1) = 0.841$.

40.2 $P(X \geq 6) = 1/32$.

40.3 Mean = 0.0769; standard deviation = 0.0887.

40.4 Mean = $(a + b)/2$; standard deviation = $(b - a)/(2\sqrt{3})$.

40.6 Mean number of non-faulty components to failure
is 82.33; standard deviation of the number of components
to failure is 82.83.

40.7 $1/2^9$.

40.9 Probability that a bottle fails the test is 0.000 67.

40.10 (a) 0.777; (c) 0.223.

40.11 (b) 0.528.

40.13 (b) $P(Z \leq 0.7) = 0.758$.

40.14 On average 30% of operations take longer than
40 seconds.

40.15 Standard deviation of 1 if $a = \sqrt{5}$ and $A = 3/(20\sqrt{5})$.

40.16 Maximum value of standard deviation is 121.6.

40.17 Probability that just two bulbs will be still working
is 0.242.

Chapter 41

41.1 (b) Mean = 24.1; median = 24.5;
interquartile range = 17.

41.3 Sample mean = 25.3; mode = 25.1; variance = 0.0644.

41.5 About 11 intervals.

41.6 Estimated variance of the sample is 1/12.

41.8 $k_1 = -1.1337$; $k_2 = 1.1337$.

41.9 For full data $\hat{a} = -0.019\ 964$; $\hat{b} = 52.998$.

Appendices

Appendix

A

Some algebraical rules

(a) Index laws for real numbers

(i) $a^0 = 1$.

(ii) $a^p a^q = a^{p+q}$.

(iii) $a^{-p} = 1/a^p$.

(iv) $(a^p)^q$ or $(a^q)^p = a^{pq}$ (so $a^{p/q} = (a^p)^{1/q}$ or $(a^{1/q})^p$).

(v) $a^p b^p = (ab)^p$

Conventionally, $a^{\frac{1}{2}}$ and $\sqrt{a}$ represent the **positive root** when we are talking about real numbers (for complex numbers, see Chapter 6). For the rules to hold in all cases, a must be positive so that $a^{p/q}$ is a always real number. For example, $(-8)^{\frac{1}{2}}$ or $\sqrt{(-8)}$ is not real: there is no real number whose square is equal to -8. But $(-8)^{\frac{1}{3}}$ or $\sqrt[3]{(-8)} = -2$.

(b) Quadratic equations

$ax^2 + bx + c = 0$ has the solutions

$$x_1, x_2 = [-b \pm \sqrt{(b^2 - 4ac)}]/2a.$$

(i) In terms of x_1 and x_2, the factors are

$$ax^2 + bx + c = a(x - x_1)(x - x_2).$$

(ii) Sum and product of solutions:

$$x_1 + x_2 = -b/a, \qquad x_1 x_2 = c/a.$$

(c) Binomial theorem

(i) If n is a positive integer (or whole number)

$$(a + b)^n = a^n + na^{n-1}b + \frac{n(n-1)}{2!} a^{n-2} b^2$$

$$+ \frac{n(n-1)(n-2)}{3!} a^{n-3} b^3 + \cdots + b^n$$

$$= \sum_{r=0}^{n} \binom{n}{r} a^{n-r} b^r$$

where the binomial coefficients are

$$\binom{n}{r} = \frac{n!}{(n-r)!r!}.$$

There are $(n+1)$ terms in this sum, and it is symmetrical in a and b.

An important special case is

$$(1+x)^n = 1 + nx + \frac{n(n-1)}{2!}x^2 + \frac{n(n-1)(n-2)}{3!}x^3 + \cdots + x^n.$$

(ii) *Pascal's triangle.* Each entry (apart from the 1s) is the sum of two previous entries – that above, and that above and to the left – as illustrated by the underlined group:

$$
\begin{array}{llllll}
n=1 & 1 & 1 \\
n=2 & 1 & 2 & 1 \\
n=3 & 1 & \underline{3} & \underline{3} & 1 \\
n=4 & 1 & 4 & \underline{6} & 4 & 1
\end{array}
$$

and so on. Thus

$$(1+x)^4 = 1 + 4x + 6x^2 + 4x^3 + x^4.$$

(iii) *Permutations and combinations*

$$_nP_r = \frac{n!}{(n-r)!}, \qquad _nC_r = \frac{n!}{(n-r)!r!} \text{ (see Section 1.17).}$$

(d) Factorization

$$a^2 - b^2 = (a+b)(a-b),$$
$$a^3 - b^3 = (a-b)(a^2+ab+b^2),$$
$$a^3 + b^3 = (a+b)(a^2-ab+b^2).$$

(e) Constants

$$e = 2.718\,281\,82\ldots, \qquad \pi = 3.141\,592\,65\ldots,$$
$$1 \text{ radian} = 57.295\,78\ldots°, \quad 1° = 0.017\,45\ldots \text{ radians},$$
$$360° = 2\pi \text{ radians}.$$

(f) Sums of powers of integers

$$\sum_{r=1}^{n} r = 1 + 2 + 3 + \cdots + n = \tfrac{1}{2}n(n+1)$$

$$\sum_{r=1}^{n} r^2 = 1^2 + 2^2 + 3^2 + \cdots + n^2 = \tfrac{1}{6}n(n+1)(2n+1)$$

$$\sum_{r=1}^{n} r^3 = 1^3 + 2^3 + 3^3 + \cdots + n^3 = \tfrac{1}{4}n^2(n+1)^2.$$

Appendix

B

Trigonometric formulae

(a) Relation between trigonometric functions

$\sin^2 A + \cos^2 A = 1,$

$\tan A = \sin A/\cos A; \quad \sec A = 1/\cos A; \quad \text{cosec } A = 1/\sin A.$

(b) Addition formulae

$\sin(A \pm B) = \sin A \cos B \pm \cos A \sin B,$

$\cos(A \pm B) = \cos A \cos B \mp \sin A \sin B,$

$\tan(A \pm B) = (\tan A \pm \tan B)/(1 \mp \tan A \tan B).$

(c) Addition formulae: special cases

$\sin 2A = 2 \sin A \cos A,$

$\cos 2A = \cos^2 A - \sin^2 A$

$\qquad = 2 \cos^2 A - 1 = 1 - 2 \sin^2 A,$

$\tan 2A = 2 \tan A/(1 - \tan^2 A),$

$\sin 3A = 3 \sin A - 4 \sin^3 A,$

$\cos 3A = 4 \cos^3 A - 3 \cos A.$

(d) Product formulae

$\sin A \sin B = \frac{1}{2}[\cos(A - B) - \cos(A + B)],$

$\cos A \cos B = \frac{1}{2}[\cos(A - B) + \cos(A + B)],$

$\sin A \cos B = \frac{1}{2}[\sin(A - B) + \sin(A + B)].$

$\sin C + \sin D = 2 \sin \frac{1}{2}(C + D) \cos \frac{1}{2}(C - D),$

$\sin C - \sin D = 2 \sin \frac{1}{2}(C - D) \cos \frac{1}{2}(C + D),$

$\cos C + \cos D = 2 \cos \frac{1}{2}(C + D) \cos \frac{1}{2}(C - D),$

$\cos C - \cos D = -2 \sin \frac{1}{2}(C + D) \sin \frac{1}{2}(C - D).$

(e) Product formulae: special cases

$\sin^2 A = \frac{1}{2}(1 - \cos 2A),$

$\cos^2 A = \frac{1}{2}(1 + \cos 2A),$

$\sin^3 A = \frac{1}{4}(3 \sin A - \sin 3A),$

$\cos^3 A = \frac{1}{4}(3 \cos A + \cos 3A).$

(f) Triangle formulae

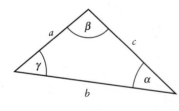

(i) $\alpha + \beta + \gamma = 180°.$

(ii) *Cosine rule*: $a^2 = b^2 + c^2 - 2bc \cos \alpha.$

(iii) *Sine rule*: $\dfrac{\sin \alpha}{a} = \dfrac{\sin \beta}{b} = \dfrac{\sin \gamma}{c}.$

(g) Trigonometric equations

In the following, n represents any integer (i.e. any whole number, positive or negative); x is in radians.

(i) $\sin x = 0$ and $\tan x = 0$ when $x = n\pi$; $\cos x = 0$ when $x = \frac{1}{2}\pi + n\pi$.

(ii) The following formulae show how to obtain all the solutions of certain equations when one solution has been obtained (e.g. a hand calculator or a computer gives only one solution of $\sin x = -\frac{1}{2}$, namely $x = \arcsin(-\frac{1}{2}) = -0.5236...$).

If $\sin \alpha = c$, then all the solutions of $\sin x = c$ are

$$x = n\pi + (-1)^n\alpha.$$

If $\cos \beta = c$, then all the solutions of $\cos x = c$ are

$$x = 2n\pi \pm \beta.$$

If $\tan \gamma = c$, then all the solutions of $\tan x = c$ are

$$x = n\pi + \gamma.$$

(h) Hyperbolic functions

$\cosh x = \frac{1}{2}(e^x + e^{-x})$; $\sinh x = \frac{1}{2}(e^x - e^{-x})$; $\tanh x = \sinh x/\cosh x$;

$\mathrm{sech}\, x = 1/\cosh x$; $\coth x = \cosh x/\sinh x$; $\mathrm{cosech}\, x = 1/\sinh x$,

$\sinh(x \pm y) = \sinh x \cosh y \pm \cosh x \sinh y$,

$\cosh(x \pm y) = \cosh x \cosh y \pm \sinh x \sinh y$,

$\cosh^2 x - \sinh^2 x = 1$,

$\sinh 2x = 2 \sinh x \cosh x$,

$\cosh 2x = \cosh^2 x + \sinh^2 x$,

$\cosh ix = \cos x$; $\sinh ix = i \sin x$;

$\sinh^{-1} x = \ln[x + \sqrt{(x^2 + 1)}]$,

$\cosh^{-1} x = \ln[x + \sqrt{(x^2 - 1)}] \quad (x \geqslant 1)$,

$\tanh^{-1} x = \frac{1}{2}\ln[(1 + x)/(1 - x)] \quad (-1 < x < 1)$.

Appendix

Areas and volumes

(a) The area of a triangle is $\frac{1}{2}bh$, where b is the length of one side and h its height from that side.

(b) The circumference of a circle is $2\pi r$, where r is its radius.

(c) The area of a circle is πr^2, where r is its radius.

(d) The area of a circle sector is $\frac{1}{2}r^2\theta$, wher r is its radius and θ the angle of the sector in radians.

(e) The volume of a sphere is $\frac{4}{3}\pi r^3$, where r is its radius.

(f) The surface area of a sphere is $4\pi r^2$, where r is its radius.

(g) The volume of a cone is $\frac{1}{3}Ah$, where h is its height and A the cross-sectional area of its base.

(h) The area of an ellipse is πab, where a and b are the lengths of its semi-axes.

Appendix
D

A table of derivatives

y	$\dfrac{\mathrm{d}y}{\mathrm{d}x}$
c (constant)	0
x^n (n any constant)	nx^{n-1}
e^{ax}	$a\,\mathrm{e}^{ax}$
k^x ($k > 0$)	$k^x \ln k$
$\ln x$ ($x > 0$)	x^{-1}
$\sin ax$	$a \cos ax$
$\cos ax$	$-a \sin ax$
$\tan ax$	$a/\cos^2 ax$
$\cot ax$	$-a/\sin^2 x$
$\sec ax$	$(a \sin ax)/\cos^2 ax$
$\operatorname{cosec} ax$	$-(a \cos ax)/\sin^2 ax$
$\arcsin ax$	$a/(1 - a^2x^2)^{\frac{1}{2}}$
$\arccos ax$	$-a/(1 - a^2x^2)^{\frac{1}{2}}$
$\arctan ax$	$a/(1 + a^2x^2)$
$\sinh ax$	$a \cosh ax$
$\cosh ax$	$a \sinh ax$
$\tanh ax$	$a/\cosh^2 ax$
$\sinh^{-1} ax$	$a/(1 + a^2x^2)^{\frac{1}{2}}$
$\cosh^{-1} ax$	$a/(a^2x^2 - 1)^{\frac{1}{2}}$
$\tanh^{-1} ax$	$a/(1 - a^2x^2)$
$u(x)v(x)$	$u\dfrac{\mathrm{d}v}{\mathrm{d}x} + v\dfrac{\mathrm{d}u}{\mathrm{d}x}$
$\dfrac{u(x)}{v(x)}$	$\dfrac{1}{v^2}\left(v\dfrac{\mathrm{d}u}{\mathrm{d}x} - u\dfrac{\mathrm{d}v}{\mathrm{d}x}\right)$
$\dfrac{1}{v(x)}$	$-\dfrac{1}{v^2}\dfrac{\mathrm{d}v}{\mathrm{d}x}$
$y(u(x))$	$\dfrac{\mathrm{d}y}{\mathrm{d}u}\dfrac{\mathrm{d}u}{\mathrm{d}x}$
$y(v(u(x)))$	$\dfrac{\mathrm{d}y}{\mathrm{d}v}\dfrac{\mathrm{d}v}{\mathrm{d}u}\dfrac{\mathrm{d}u}{\mathrm{d}x}$

Appendix

E

A table of integrals

$f(x)$	$\int f(x)\,dx$ (C is an arbitrary constant.)
$x^m \ (m \neq -1)$	$\dfrac{1}{m+1}\,x^{m+1} + C$
x^{-1}	$\ln|x| + C$, or $\ln|Cx|$
e^{ax}	$(1/a)\,e^{ax} + C$
$k^x \ (k > 0)$	$k^x/\ln k + C$
$\ln x \ (x > 0)$	$x \ln x - x + C$
$\sin ax$	$-(1/a)\cos ax + C$
$\cos ax$	$(1/a)\sin ax + C$
$\tan ax$	$-(1/a)\ln|\cos ax| + C$ or $-(1/a)\ln|C\cos ax|$
$\cot ax$	$(1/a)\ln|\sin ax| + C$ or $(1/a)\ln|C\sin ax|$
$\sec ax$	$-(1/2a)\ln[(1-\sin ax)/(1+\sin ax)] + C$
$\operatorname{cosec} ax$	$(1/2a)\ln[(1-\cos ax)/(1+\cos ax)] + C$
$\arcsin ax$	$(1/a)(1-a^2x^2)^{\frac{1}{2}} + x\arcsin ax + C$
$\arccos ax$	$-(1/a)(1-a^2x^2)^{\frac{1}{2}} + x\arccos ax + C$
$\arctan ax$	$-(1/a)\ln(1-a^2x^2)^{\frac{1}{2}} + x\arctan ax + C$
$\sinh ax$	$(1/a)\cosh ax + C$
$\cosh ax$	$(1/a)\sinh ax + C$
$\tanh ax$	$(1/a)\ln\{\cosh ax\} + C$
$1/(x^2 + a^2)$	$(1/a)\arctan(x/a) + C$
$1/(x^2 - a^2)$	$(1/2a)\ln|(x-a)/(x+a)| + C$ or $(1/a)\tanh^{-1}(x/a) + C$
$1/(a^2 - x^2)^{\frac{1}{2}}$	$\arcsin(x/a) + C$ (or $-\arccos(x/a) + C$)
$1/(a^2 + x^2)^{\frac{1}{2}}$	$(1/a)\sinh^{-1}(x/a) + C$ or $\ln[x + (x^2 + a^2)^{\frac{1}{2}}] + C$
$1/(x^2 - a^2)^{\frac{1}{2}}$	$\ln[x + (x^2 - a^2)^{\frac{1}{2}}] + C$
$x\,e^{ax}$	$(1/a^2)(ax - 1)\,e^{ax} + C$
$x\cos ax$	$(1/a^2)(\cos ax + ax\sin ax) + C$
$x\sin ax$	$(1/a^2)(\sin ax - ax\cos ax) + C$
$x\ln x$	$\frac{1}{2}x^2\ln x - \frac{1}{4}x^2 + C$
$e^{ax}\cos bx$	$[1/(a^2 + b^2)]\,e^{ax}(a\cos bx + b\sin bx) + C$
$e^{ax}\cos bx$	$[1/(a^2 + b^2)]\,e^{ax}(-b\cos bx + a\sin bx) + C$

Laplace transforms, inverses, and rules

In the following tables, n and m represent a positive integer or zero. The constants k and c are arbitrary unless otherwise indicated.

Transforms		Inverses	
$f(t)$	$\displaystyle F(s) = \int_0^\infty e^{-st} f(t)\,dt$	$F(s)$	$f(t)$
t^n	$\dfrac{n!}{s^{n+1}}$	$\dfrac{1}{s^m}$	$\dfrac{1}{(m-1)!}t^{m-1}$
e^{kt}	$\dfrac{1}{s-k}$	$\dfrac{1}{s-k}$	e^{kt}
$t^n e^{kt}$	$\dfrac{n!}{(s-k)^{n+1}}$	$\dfrac{1}{(s-k)^m}$	$\dfrac{1}{(m-1)!}t^{m-1}e^{kt}$
$\cos kt$	$\dfrac{s}{s^2+k^2}$	$\dfrac{s}{s^2+k^2}$	$\cos kt$
$\sin kt$	$\dfrac{k}{s^2+k^2}$	$\dfrac{1}{s^2+k^2}$	$\dfrac{1}{k}\sin kt$
$t\cos kt$	$\dfrac{s^2-k^2}{(s^2+k^2)^2}$	$\dfrac{s^2-k^2}{(s^2+k^2)^2}$	$t\cos kt$
$t\sin kt$	$\dfrac{2ks}{(s^2+k^2)^2}$	$\dfrac{s}{(s^2+k^2)^2}$	$\dfrac{1}{2k}t\sin kt$
$H(t-c)\ (c>0)$	e^{-cs}/s	$e^{-cs}/s\ (c>0)$	$H(t-c)$
$\delta(t-c)\ (c>0)$	e^{-cs}	$e^{-cs}\ (c>0)$	$\delta(t-c)$

Summary of rules

In the following rules, $F(s) \leftrightarrow f(t)$.

Scale rule (24.5)

$$f(kt) \leftrightarrow \frac{1}{k}F\left(\frac{s}{k}\right) \quad \text{and} \quad F(ks) \leftrightarrow \frac{1}{k}f\left(\frac{t}{k}\right)\ (k>0).$$

Shift rule, or multiplication by e^{kt} (24.7)

If k is any constant,
$$e^{kt}f(t) \leftrightarrow F(s-k).$$

Powers of t (24.8)

If n is a positive integer, then
$$t^n f(t) \leftrightarrow (-1)^n \frac{d^n F(s)}{ds^n}.$$

Derivatives (24.12)

$$\frac{df(t)}{dt} \leftrightarrow sF(s) - f(0), \qquad \frac{d^2 f(t)}{dt^2} \leftrightarrow s^2 F(s) - sf(0) - f'(0).$$

Delay rule (24.15)

If $c>0$, then $e^{-cs}F(s) \leftrightarrow f(t-c)H(t-c)$ (where H is the Heaviside unit function).

$1/s$ as an integration operator (25.1)

If $F(s) \leftrightarrow f(t)$, then
$$\frac{1}{s}F(s) \leftrightarrow \int_0^t f(\tau)\,d\tau.$$

Convolution theorem (25.11)

If $g(t) \leftrightarrow G(s)$ and $f(t) \leftrightarrow F(s)$, then
$$F(s)G(s) \leftrightarrow \int_0^t g(t-\tau)f(\tau)\,d\tau \left(= \int_0^t g(\tau)f(t-\tau)\,d\tau\right).$$

Appendix

Exponential Fourier transforms and rules

General rules

	Signal	Transform
Fourier transform pair	$x(t) = \int_{-\infty}^{\infty} X(f)\, e^{i2\pi ft}\, df$	$X(f) = \int_{-\infty}^{\infty} x(t)\, e^{-i2\pi ft}\, dt$
Linearity	$Ax_1(t) + Bx_2(t)$	$AX_1(f) + BX_2(f)$
Time scaling	$x(At)$	$\lvert A \rvert^{-1} X(A^{-1}f)$
Time reversal	$x(-t)$	$X(-f)$
Time delay	$x(t - B)$	$X(f)\, e^{-i2\pi Bf}$
Frequency scaling	$\lvert C \rvert^{-1} x(C^{-1}t)$	$X(Cf)$
Frequency shift	$x(t)\, e^{i2\pi Dt}$	$X(f - D)$
Modualtion	$x(t)\cos 2\pi Kt$	$\frac{1}{2}[X(f+K) + X(f-K)]$
	$x(t)\sin 2\pi Kt$	$\frac{1}{2}i[X(f+K) - X(f-K)]$
Differentiation	$dx(t)/dt$	$(i2\pi f)X(f)$
	$d^n x(t)/dt^n$	$(i2\pi f)^n X(f)$
Duality	$X(t)$	$x(-f)$
Convolution	$\int_{-\infty}^{\infty} x_1(u)x_2(t - u)\, du = x_1(t) * x_2(t)$ $\;= \int_{-\infty}^{\infty} x_1(t - u)x_2(u)\, du$	$X_1(f)X_2(f)$
Multiplication	$x_1(t)x_2(t)$	$\int_{-\infty}^{\infty} X_1(f - v)X_2(v)\, dv$ $\;= \int_{-\infty}^{\infty} X_2(v)X_2(f - v)\, dv$
Periodic function $x_P(t)$	$x_P(t)$ (period T)	$\sum_{n=-\infty}^{\infty} X_n\, \delta(f - nf_0)$, where $f_0 = 1/T$, $X_n = f_0 \int_{\text{Period}} x_P(t)\, e^{-2\pi i f_0 t}\, dt$

Short table of Fourier transforms

Signal	Transform	Signal	Transform
$\Pi(t) = H(t - \tfrac{1}{2}) - H(t + \tfrac{1}{2})$	$\mathrm{sinc}\, f$	$\dfrac{1}{1 + t^2}$	$\pi\, e^{-2\pi \lvert f \rvert}$
$\mathrm{sinc}\, t$	$\Pi(f)$	$e^{-\pi t^2}$	$e^{-\pi f^2}$
$\Lambda(t) = \begin{cases} 1 + t, & -1 < t < 0 \\ 1 - t, & 0 < t < 1 \\ 0, & \text{elsewhere} \end{cases}$	$\mathrm{sinc}^2 f$	$\delta(t)$	1
		1	$\delta(f)$
$\mathrm{sinc}^2 t$	$\Lambda(f)$	$\cos 2\pi f_0 t$	$\frac{1}{2}[\delta(f + f_0) + \delta(f - f_0)]$
$e^{-t}H(t)$	$1/(1 + i2\pi f)$	$\sin 2\pi f_0 t$	$\frac{1}{2}i[\delta(f + f_0) - \delta(f - f_0)]$
$t\, e^{-t}H(t)$	$1/(1 + i2\pi f)^2$	$\underset{T}{\text{Ш}}(t) = \sum_{n=-\infty}^{\infty} \delta(t - nT) \; (T > 0)$	$f_0 \underset{f_0}{\text{Ш}}(f) \quad (f_0 = 1/T)$
$e^{-\lvert t \rvert}$	$2/(1 + 4\pi^2 f^2)$		

Appendix

Probability distributions and tables

(a) Distributions, means, and variances

(*i*) *Discrete distributions*

Distribution	Probability	Mean (μ)	Variance (σ^2)
Binomial	$\dfrac{n!\,p^r q^{n-r}}{(n-r)!\,r!}$	np	$np(1-p)$
Geometric	$(1-p)^{r-1}p$	$\dfrac{1}{p}$	$\dfrac{1-p}{p^2}$
Poisson	$\dfrac{\lambda^n\,e^{-\lambda}}{n!}$	λ	λ
Pascal	$_{r-1}C_{k-1}\,p^k(1-p)^{r-k}$	$\dfrac{k}{p}$	$\dfrac{k(1-p)}{p^2}$
Hypergeometric	$\dfrac{_{w}C_{r}\,_{b}C_{n-r}}{_{w+b}C_{n}}$	$\dfrac{nb}{w+b}$	$\dfrac{nwb(b+w+n)}{(w+b)^2(w+b-1)}$

(*ii*) *Continuous distributions*

Distribution	Density	Mean (μ)	Variance (σ^2)
Exponential	$\begin{cases}\lambda\,e^{-\lambda x}, & x \geqslant 0 \\ 0, & x < 0\end{cases}$	$\dfrac{1}{\lambda}$	$\dfrac{1}{\lambda^2}$
Uniform	$\begin{cases}1/(b-a), & a < x < b \\ 0, & \text{elsewhere}\end{cases}$	$\tfrac{1}{2}(a+b)^2$	$\tfrac{1}{12}(b-a)^2$
Standardized normal	$\dfrac{1}{\sqrt{2\pi}}\,e^{-\frac{1}{2}x^2}$	0	1

(b) Cumulative normal distribution tables

Standardized cumulative normal distribution giving the values of

$$\Phi(x) = \frac{1}{\sqrt{2\pi}} \int_{-\infty}^{x} e^{-\frac{1}{2}t^2}\,dt$$

for $0 \leqslant x \leqslant 3.0$ at 0.01 intervals. For $x < 0$, $\Phi(x)$ can be calculated from $\Phi(-x) = 1 - \Phi(x)$.

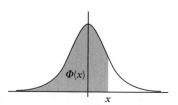

x	0	1	2	3	4	5	6	7	8	9
0.0	0.5000	0.5040	0.5080	0.5120	0.5160	0.5199	0.5239	0.5279	0.5319	0.5359
0.1	0.5398	0.5438	0.5478	0.5517	0.5557	0.5596	0.5636	0.5675	0.5714	0.5753
0.2	0.5793	0.5832	0.5871	0.5910	0.5948	0.5987	0.6026	0.6064	0.6103	0.6141
0.3	0.6179	0.6217	0.6255	0.6293	0.6331	0.6368	0.6406	0.6443	0.6480	0.6517
0.4	0.6554	0.6591	0.6628	0.6664	0.6700	0.6736	0.6772	0.6808	0.6844	0.6879
0.5	0.6915	0.6950	0.6985	0.7019	0.7054	0.7088	0.7123	0.7157	0.7190	0.7224
0.6	0.7257	0.7291	0.7324	0.7357	0.7389	0.7422	0.7454	0.7486	0.7517	0.7549
0.7	0.7580	0.7611	0.7642	0.7673	0.7704	0.7734	0.7764	0.7794	0.7823	0.7852
0.8	0.7881	0.7910	0.7939	0.7967	0.7995	0.8023	0.8051	0.8078	0.8106	0.8133
0.9	0.8159	0.8186	0.8212	0.8238	0.8264	0.8289	0.8315	0.8340	0.8365	0.8389
1.0	0.8413	0.8438	0.8461	0.8485	0.8508	0.8531	0.8554	0.8577	0.8599	0.8621
1.1	0.8643	0.8665	0.8686	0.8708	0.8729	0.8749	0.8770	0.8790	0.8810	0.8830
1.2	0.8849	0.8869	0.8888	0.8907	0.8925	0.8944	0.8962	0.8980	0.8997	0.9015
1.3	0.9032	0.9049	0.9066	0.9082	0.9099	0.9115	0.9131	0.9147	0.9162	0.9177
1.4	0.9192	0.9207	0.9222	0.9236	0.9251	0.9265	0.9279	0.9292	0.9306	0.9319
1.5	0.9332	0.9345	0.9357	0.9370	0.9382	0.9394	0.9406	0.9418	0.9429	0.9441
1.6	0.9452	0.9463	0.9474	0.9484	0.9495	0.9505	0.9515	0.9525	0.9535	0.9545
1.7	0.9554	0.9564	0.9573	0.9582	0.9591	0.9599	0.9608	0.9616	0.9625	0.0633
1.8	0.9641	0.9649	0.9656	0.9664	0.9671	0.9678	0.9686	0.9693	0.9699	0.9706
1.9	0.9137	0.9719	0.9726	0.9732	0.9738	0.9744	0.9750	0.9756	0.9761	0.9767
2.0	0.9772	0.9778	0.9783	0.9788	0.9793	0.9798	0.9803	0.9808	0.9812	0.9817
2.1	0.9821	0.9826	0.9830	0.9834	0.9838	0.9842	0.9846	0.9850	0.9854	0.9857
2.2	0.9861	0.9864	0.9868	0.9871	0.9875	0.9878	0.9881	0.9884	0.9887	0.9890
2.3	0.9893	0.9896	0.9898	0.9901	0.9904	0.9906	0.9909	0.9911	0.9913	0.9916
2.4	0.9918	0.9920	0.9922	0.9925	0.9927	0.9929	0.9931	0.9932	0.9934	0.9936
2.5	0.9938	0.9940	0.9941	0.9943	0.9945	0.9946	0.9948	0.9949	0.9951	0.9952
2.6	0.9953	0.9955	0.9956	0.9957	0.9959	0.9960	0.9961	0.9962	0.9963	0.9964
2.7	0.9965	0.9966	0.9967	0.9968	0.9969	0.9970	0.9971	0.9972	0.9973	0.9974
2.8	0.9974	0.9975	0.9976	0.9977	0.9977	0.9978	0.9979	0.9979	0.9980	0.9981
2.9	0.9981	0.9982	0.9982	0.9983	0.9984	0.9984	0.9985	0.9985	0.9986	0.9986
3.0	0.9987	0.9987	0.9987	0.9988	0.9988	0.9989	0.9989	0.9989	0.9990	0.9990

Table giving x for specified values of $\Phi(x)$ for $0.50 \leq \Phi(x) \leq 0.99$ at 0.01 intervals

$\Phi(x)$	x	$\Phi(x)$	x	$\Phi(x)$	x
0.50	0.0000	0.67	0.4399	0.84	0.9945
0.51	0.0251	0.68	0.4677	0.85	1.0364
0.52	0.0502	0.69	0.4959	0.86	1.0803
0.53	0.0753	0.70	0.5244	0.87	1.1264
0.54	0.1004	0.71	0.5534	0.88	1.1750
0.55	0.1257	0.72	0.5828	0.89	1.2265
0.56	0.1510	0.73	0.6138	0.90	1.2816
0.57	0.1764	0.74	0.6433	0.91	1.3408
0.58	0.2019	0.75	0.6745	0.92	1.4051
0.59	0.2275	0.76	0.7063	0.93	1.4758
0.60	0.2533	0.77	0.7388	0.94	1.5548
0.61	0.2793	0.78	0.7722	0.95	1.6449
0.62	0.3055	0.79	0.8064	0.96	1.7507
0.63	0.3319	0.80	0.8416	0.97	1.8808
0.64	0.3585	0.81	0.8779	0.98	2.0537
0.65	0.3853	0.82	0.9154	0.99	2.3263
0.66	0.4125	0.83	0.9542		

Index

Pages of the main topics are given in heavy type for quick reference.